MAGNETIC RESONANCE IMAGING HANDBOOK

# Imaging of the Pelvis, Musculoskeletal System, and Special Applications to CAD

**Magnetic Resonance Imaging Handbook**

*Image Principles, Neck, and the Brain*
*Imaging of the Cardiovascular System, Thorax, and Abdomen*
*Imaging of the Pelvis, Musculoskeletal System, and Special Applications to CAD*

MAGNETIC RESONANCE IMAGING HANDBOOK

# Imaging of the Pelvis, Musculoskeletal System, and Special Applications to CAD

edited by

Luca Saba

CRC Press is an imprint of the
Taylor & Francis Group, an informa business

CRC Press
Taylor & Francis Group
6000 Broken Sound Parkway NW, Suite 300
Boca Raton, FL 33487-2742

First issued in paperback 2019



ISBN-13: 978-1-4822-1621-9 (hbk)
ISBN-13: 978-0-367-86890-1 (pbk)



**Library of Congress Cataloging-in-Publication Data**

Names: Saba, Luca, editor.
Title: Magnetic resonance imaging handbook / [edited by] Luca Saba.
Description: Boca Raton : Taylor & Francis, 2016. | Includes bibliographical references and index.
Identifiers: LCCN 2015043923| ISBN 9781482216288 (set : hardcover : alk. paper) | ISBN 9781482216134 (v. 1 : hardcover : alk. paper) | ISBN 9781482216264 (v. 2 : hardcover : alk. paper) | ISBN 9781482216219 (v. 3 : hardcover : alk. paper)
Subjects: | MESH: Magnetic Resonance Imaging.
Classification: LCC RC386.6.M34 | NLM WN 185 | DDC 616.07/548--dc23
LC record available at http://lccn.loc.gov/2015043923

**Visit the Taylor & Francis Web site at**
**http://www.taylorandfrancis.com**

**and the CRC Press Web site at**
**http://www.crcpress.com**

*This book is dedicated to Giovanni Saba.*

*Thank you.*

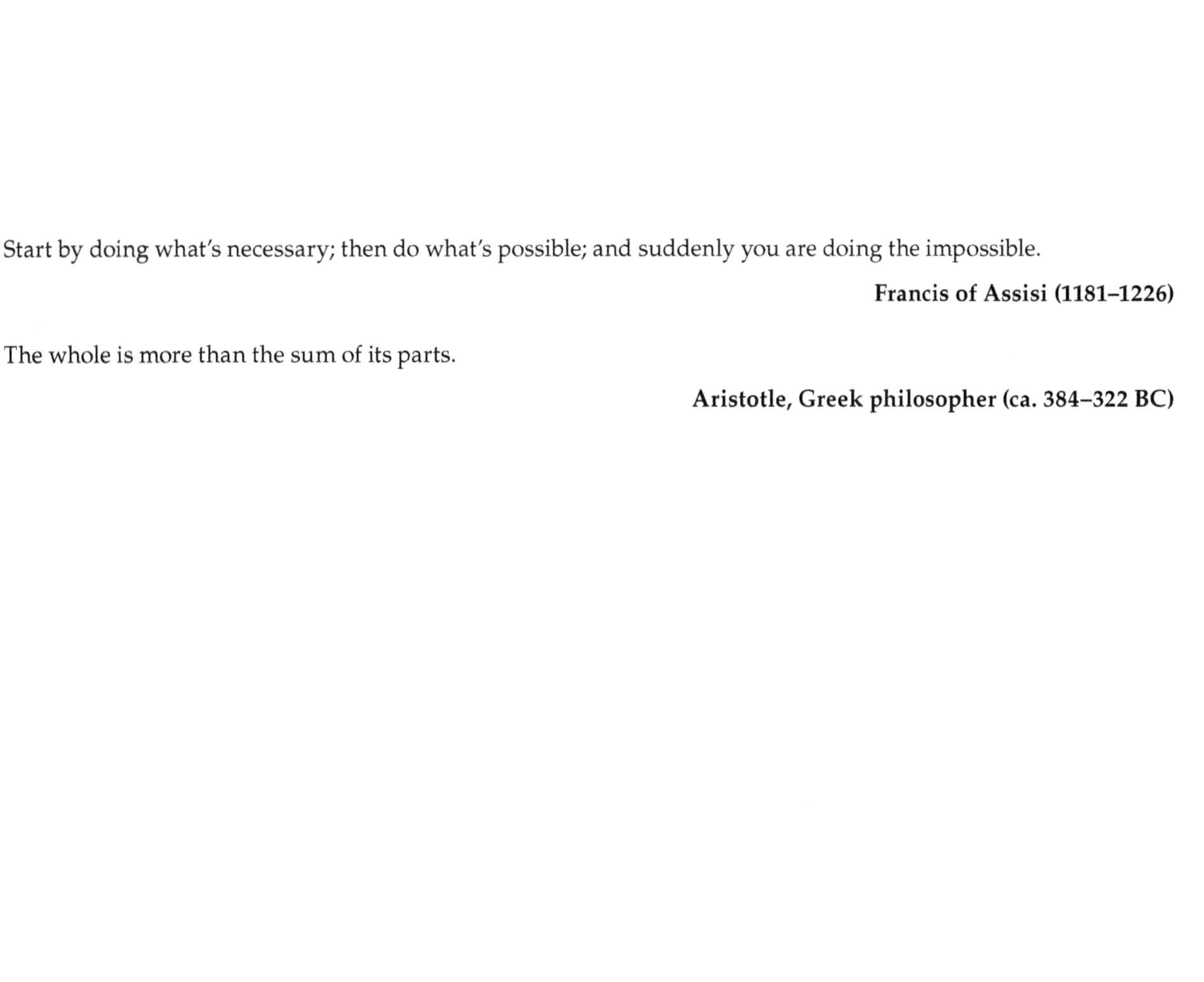

Start by doing what's necessary; then do what's possible; and suddenly you are doing the impossible.

**Francis of Assisi (1181–1226)**

The whole is more than the sum of its parts.

**Aristotle, Greek philosopher (ca. 384–322 BC)**

# Contents

# Preface

Magnetic resonance imaging (MRI) is a medical imaging technique used in radiology to visualize internal structures of the body in detail. The introduction of MRI resulted in a fundamental and far-reaching improvement of the diagnostic process because this technique provides an excellent contrast between the different soft tissues of the body, which makes it especially useful in imaging the brain, muscles, heart, and cancers compared with other medical imaging techniques such as computed tomography or X-rays.

In the past 20 years, MRI technology has further improved with the introduction of systems up to 7 T and with the development of numerous postprocessing algorithms such as diffusion tensor imaging (DTI), functional MRI (fMRI), and spectroscopic imaging. From these developments, the diagnostic potentialities of MRI have impressively improved with exceptional spatial resolution and the possibility of analyzing the morphology and function of several kinds of pathology.

The purpose of this book is to cover engineering and clinical benefits in the diagnosis of human pathologies using MRI. It will cover the protocols and potentialities of advanced MRI scanners with very high-quality MR images. Given these exciting developments in the MRI field, I hope that this book will be a timely and complete addition to the growing body of literature on this topic.

**Luca Saba**
*University of Cagliari, Italy*

# Acknowledgments

It is not possible to overstate my gratitude to the many individuals who helped to produce this book; their enthusiasm and dedication were unbelievable.

I express my appreciation to CRC Press/Taylor & Francis Group for their professionalism in handling this project. Your help was wonderful and made producing this book an enjoyable and worthwhile experience.

Finally, I acknowledge Tiziana.

# Editor

**Professor Luca Saba** earned his MD from the University of Cagliari, Italy, in 2002. Currently, he works at the Azienda Ospedaliero Universitaria of Cagliari. His research is focused on multidetector-row computed tomography, magnetic resonance, ultrasound, neuroradiology, and diagnostics in vascular sciences.

Professor Saba has published more than 180 papers in high-impact factor journals such as the *American Journal of Neuroradiology, Atherosclerosis, European Radiology, European Journal of Radiology, Acta Radiologica, Cardiovascular and Interventional Radiology, Journal of Computer Assisted Tomography, American Journal of Roentgenology, Neuroradiology, Clinical Radiology, Journal of Cardiovascular Surgery, Cerebrovascular Diseases, Brain Pathology, Medical Physics,* and *Atherosclerosis.* He is a well-known speaker and has spoken over 45 times at national and international conferences.

Dr. Saba has won 15 scientific and extracurricular awards during his career, and has presented more than 500 papers and posters at national and international congress events (Radiological Society of North America [RSNA], ESGAR, ECR, ISR, AOCR, AINR, JRS, Italian Society of Radiology [SIRM], and AINR). He has written 21 book chapters and is the editor of 10 books in the fields of computed tomography, cardiovascular surgery, plastic surgery, gynecological imaging, and neurodegenerative imaging.

He is a member of the SIRM, European Society of Radiology, RSNA, American Roentgen Ray Society, and European Society of Neuroradiology, and serves as the reviewer of more than 40 scientific journals.

# Contributors

**Abass Alavi**
Department of Radiology
Hospital of the University of Pennsylvania
Philadelphia, Pennsylvania

**Alberto Alonso-Burgos**
Vascular and Interventional Radiology Unit
Radiology Department
University Hospital Fundación Jiménez Díaz
Madrid, Spain

**Garyfalia Ampanozi**
Institute of Forensic Medicine
University of Zurich
Zurich, Switzerland

**Laura W. Bancroft**
Department of Radiology
Florida Hospital
and
Department of Radiology
University of Central Florida College of Medicine
Orlando, Florida

and

Department of Radiology
Florida State University College of Medicine
Tallahassee, Florida

**Peter Brugge**
Department of Biomedical Imaging and Image-Guided Therapy
Medical University of Vienna
Vienna, Austria

**Kelsey Budd**
Department of Biomedical Imaging and Image-Guided Therapy
Medical University of Vienna
Vienna, Austria

**Valeria Buonocore**
Department of Radiological Sciences
Sapienza University
Rome, Italy

**Gaia Cartocci**
Department of Radiological, Oncological and Anatomo-pathological Sciences
Sapienza University of Rome
Rome, Italy

**Ellen M. Chung**
Department of Radiology and Radiological Sciences
Uniformed Services University of the Health Sciences
Bethesda, Maryland

**Francois Cornelis**
Department of Radiology
Pellegrin Hospital
Bordeaux, France

**Nicole F. Darcy**
Advanced Body and Women's Imaging
Keck School of Medicine
University of Southern California
Los Angeles, California

**Nandita M. deSouza**
MRI Unit
Institute of Cancer Research
and
Royal Marsden Hospital
Surrey, United Kingdom

**Geoff Dougherty**
Department of Radiology
California State University Channel Islands
Camarillo, California

**Alireza Eajazi**
Quantitative Imaging Center
Department of Radiology
Boston University School of Medicine
Boston, Massachusetts

**Lukas Ebner**
Institute of Forensic Medicine
University of Zurich
Zurich, Switzerland

and

Institute of Diagnostic, Interventional and Pediatric Radiology, Inselspital
University Hospital Bern
Bern, Switzerland

**Gary Felsberg**
Department of Radiology
Florida Hospital
and
Department of Radiology
University of Central Florida College of Medicine
Orlando, Florida

**Patricia Mildred Flach**
Institute of Forensic Medicine
University of Zurich
Zurich, Switzerland

**Sabine Franckenberg**
Institute of Forensic Medicine
University of Zurich
Zurich, Switzerland

**Jan Fritz**
Russell H. Morgan Department of Radiology and Radiological Science
Johns Hopkins University School of Medicine
Baltimore, Maryland

**Jurgen J. Fütterer**
Department of Radiology
and Nuclear Medicine
Radboud University Medical Center
Nijmegen, the Netherlands

**Dominic Gascho**
Institute of Forensic Medicine
University of Zurich
Zurich, Switzerland

**Nicolas Grenier**
Department of Radiology
Pellegrin Hospital
Bordeaux, France

**Ali Guermazi**
Quantitative Imaging Center
Boston Medical Center
Boston University School
of Medicine
Boston, Massachusetts

**Mark A. Haidekker**
College of Engineering
University of Georgia
Athens, Georgia

**Brian K. Harshman**
Department of Radiology
Florida Hospital
and
Department of Radiology
University of Central Florida
College of Medicine
Orlando, Florida

**Monica L. Huang**
Department of Radiology
The University of Texas MD
Anderson Cancer Center
Houston, Texas

**Mohamed Jarraya**
Department of Radiology
Boston University School
of Medicine
Boston, Massachusetts

**Yousuke Kakitsubata**
Department of Radiology
Miyazaki Konan Hospital
Miyazaki, Japan

**Nathan J. Kohler**
Department of Radiology
Florida Hospital
and
Department of Radiology
University of Central Florida
College of Medicine
Orlando, Florida

**Stavroula Kyriazi**
Clinical Magnetic Resonance
The Institute of Cancer Research
London, United Kingdom

**Anne-Sophie Lasserre**
Department of Radiology
Pellegrin Hospital
Bordeaux, France

**Ott Le**
Department of Diagnostic
Radiology
Division of Diagnostic Imaging
The University of Texas MD
Anderson Cancer Center
Houston, Texas

**Yann Le Bras**
Department of Radiology
Pellegrin Hospital
Bordeaux, France

**Philip S. Lim**
Department of Radiology
Abington Memorial Hospital
Abington, Pennsylvania

**Ángeles Franco López**
Radiology Department
Cardiac Imaging Unit
University Hospital Fundación
Jiménez Díaz
Madrid, Spain

**Francesca Maccioni**
Department of Radiological
Sciences
Sapienza University
Rome, Italy

**Amy M. Mackey**
Department of Obstetrics
and Gynecology
Abington Memorial Hospital
Abington, Pennsylvania

**Stefano Marcia**
Department of Radiology
Santissima Trinità Hospital
Cagliari, Italy

**Mariangela Marras**
Department of Radiology
Microcitemico Hospital
Cagliari, Italy

**Steven A. Messina**
Department of Radiology
Florida Hospital
and
Department of Radiology
University of Central Florida
College of Medicine
Orlando, Florida

**Mateen C. Moghbel**
Department of Radiology
Hospital of the University
of Pennsylvania
Philadelphia, Pennsylvania

**John M. Morelli**
Department of Radiology
St. John's Medical Center
Tulsa, Oklahoma

**Alessandro Napoli**
Department of Radiological
Sciences
MRgFUS & Cardiovascular
Imaging Unit
School of Medicine
Sapienza University of Rome
Rome, Italy

**M. Cody O'Dell**
Department of Radiology
Florida Hospital
and
Department of Radiology
University of Central Florida
College of Medicine
Orlando, Florida

**Suzanne L. Palmer**
Body Imaging Division
Keck School of Medicine
University of Southern California
Los Angeles, California

**Teresa Pérez de la Fuente**
Plastic and Reconstructive Surgery Department
University Hospital Fundación Jiménez Díaz
Madrid, Spain

**Francois Petitpierre**
Department of Radiology
Pellegrin Hospital
Bordeaux, France

**Michele Porcu**
Department of Radiology
University of Cagliari
Cagliari, Italy

**Daniela Prayer**
Department of Biomedical Imaging and Image-Guided Therapy
Medical University of Vienna
Vienna, Austria

**Armugam Rajesh**
Department of Radiology
Leicester General Hospital
University Hospitals of Leicester
Leicester, United Kingdom

**Frank W. Roemer**
Department of Radiology
University of Erlangen-Nuremberg
Erlangen, Germany

and

Quantitative Imaging Center
Department of Radiology
Boston University School of Medicine
Boston, Massachusetts

**Steffen Günter Ross**
Institute of Forensic Medicine
University of Zurich
Zurich, Switzerland

**Thomas Daniel Ruder**
Institute of Forensic Medicine
University of Zurich
Zurich, Switzerland

**Luca Saba**
Department of Radiology
Azienda Ospedaliero Universitaria of Cagliari
Cagliari, Italy

**Allison D. Salibian**
Department of Radiology
Keck School of Medicine
University of Southern California
Los Angeles, California

**Tsukasa Sano**
Department of Oral Diagnostic Sciences
Showa University School of Dentistry
Tokyo, Japan

**Francisco Sepulveda**
Department of Biomedical Imaging and Image-Guided Therapy
Medical University of Vienna
Vienna, Austria

**Marilyn J. Siegel**
Mallinckrodt Institute of Radiology
Washington University School of Medicine
St. Louis, Missouri

**Claire L. Templeman**
Children's Hospital of Los Angeles
and
Keck School of Medicine
University of Southern California
Los Angeles, California

**Michael Josef Thali**
Institute of Forensic Medicine
University of Zurich
Zurich, Switzerland

**Daphne J. Theodorou**
Department of Radiology
General Hospital of Ioannina & National Healthcare System
Ioannina, Greece

**Stavroula J. Theodorou**
Department of Radiology
University of Ioannina
Ioannina, Greece

**Drew A. Torigian**
Medical Image Processing Group
Department of Radiology
University of Pennsylvania School of Medicine
Philadelphia, Pennsylvania

**Elise Tricaud**
Department of Radiology
Pellegrin Hospital
Bordeaux, France

**Emilio García Tutor**
Plastic and Reconstructive Surgery Department
University Hospital of Guadalajara
Guadalajara, Spain

**Sadhna Verma**
Department of Radiology
University of Cincinnati Medical Center
Cincinnati, Ohio

**Raghu Vikram**
Department of Diagnostic Radiology
Division of Diagnostic Imaging
The University of Texas MD Anderson Cancer Center
Houston, Texas

**Christopher W. Wasyliw**
Department of Radiology
Florida Hospital
and
Department of Radiology
University of Central Florida College of Medicine
Orlando, Florida

# 1

# *Magnetic Resonance Imaging of Kidneys and Ureters*

**Francois Cornelis, Anne-Sophie Lasserre, Elise Tricaud, Francois Petitpierre, Yann Le Bras, and Nicolas Grenier**

**CONTENTS**



## 1.1 Introduction

The role of magnetic resonance imaging (MRI) for the evaluation of renal or ureter abnormalities has progressively increased during this last decade. Following successive technical developments, MRI may now provide specific morphologic and functional information in multiple clinical situations. Advantages of MRI in renal diseases over the other techniques such as ultrasound (US) or computed tomography (CT) may be related to the superior soft tissue contrast observed, to the flexibility of the MR technique allowing imaging of numerous organs and tissues, and to the ability to combine several sequences in order to obtain a multiparametric analysis. Renal MRI provides a powerful tool for the detection and the characterization of abnormalities in particular in case of renal tumor. In addition, MRI may be a useful method for imaging of renal function, including perfusion or glomerular filtration rate, or to evaluate the response after treatment. Compared to CT urography, which is still mainly used for the exploration of the upper urinary tract, MR urography (MRU) can be performed even in case of compromised renal function, severe contrast allergy, or in children or pregnant women when radiation exposure is a problem. All these aspects have been discussed successively in this chapter.

## 1.2 Specific Technical Considerations for MRI of Kidneys and Ureters

### 1.2.1 Standard Morphologic MRI of the Kidneys and Renal Pelvis

To adequately perform an MR exploration of the kidneys and renal pelvis, some specific technical points must be considered. First, the use of phased array body coils is preferable because it improves the signal-to-noise ratio. Furthermore, the patient's arms should be raised above the head or positioned anteriorly to the coronal plane through the kidneys in order to limit the artifacts due to aliasing or movements. As for all moving organs in the abdomen, fast imaging techniques are essential to limit the artifacts due to the respiratory motions observed in the kidneys. Moreover, although the kidney position is more constant in expiration than in inspiration, several techniques may be used to improve the images. Acquisition of the images at the end of expiration facilitates achieving a reproducible position of the diaphragm during image acquisition and thus may decrease misregistration artifact (on subtraction images). Breath holding may be helpful when short, whereas respiratory triggering or navigator pulse must be used when an adequate breath hold cannot be obtained [1].

Although it must be obviously adapted to the purpose of the MR exploration and to the patient's characteristics, the MRI protocol performed to evaluate the kidney, renal pelvis, or upper ureters often includes the following MR sequences (Table 1.1): coronal $T_2$-weighted; axial $T_2$-weighted sequence with or without fat saturation; axial $T_1$-weighted gradient echo (with in-phase and opposed-phase) (Figure 1.1); axial $T_1$-weighted gradient-echo sequence for dynamic imaging using intravenous gadolinium contrast injection, immediately followed by three to five breath-hold periods (at least: precontrast phase, *arterial* corticomedullary phase, and nephrographic phase); and coronal $T_1$-weighted (Figure 1.2). The corticomedullary-phase images are acquired during the peak aortic enhancement and the nephrographic-phase images approximately 1 min after the start of the acquisition of the corticomedullary phase. Alternatively, early and late nephrographic phases can be obtained at 40 s (approximately 1 min after the start of contrast administration) and at 90 s (approximately 2 min after the start of contrast administration), respectively, after the initiation of the corticomedullary phase. Finally, an excretory phase image must be acquired approximately 3 min after the start of the acquisition of the corticomedullary phase. Subtraction images may be therefore obtained and seem particularly useful for the assessment of possible enhancement within renal lesions with increased signal on precontrast $T_1$-weighted images. Diffusion-weighted imaging (DWI) using multiple *b*-values ($b = 0$, 100, 400, 800 s/mm$^2$) may be performed in particular for tissue characterization [2] (Figure 1.2). Adding a delayed $T_1$-weighted sequence (>5 min) may be also helpful in case of heterogeneous tumor [3].

Finally, it has been shown that the development of nephrogenic systemic fibrosis may be related to the administration of gadolinium contrast material in patients with a severe degree of chronic renal insufficiency (estimated glomerular filtration rate [eGFR] below 30 mL/min/1.73 m$^2$) or acute renal insufficiency of any severity [4–6]. In recent years, the incidence of the disease has significantly decreased [7], with now a very low prevalence of nephrogenic systemic fibrosis (0.4%). This may be the consequence of a progressively restricted use of gadolinium-based contrast agents in patients with impaired kidney function and a more widespread application of macrocyclic chelates according to guidelines limiting the use of these contrast agents in at-risk patients [8]. Nevertheless, radiologists must keep abreast of the latest recommendations regarding the use of a gadolinium contrast agents in patients with acute/chronic renal insufficiency. Risks and benefits of administration of a gadolinium contrast agent, as well as risks of alternative diagnostic procedures, should be carefully balanced for individual patients.

**TABLE 1.1**

Example of Magnetic Resonance Imaging Protocol for Renal Tumors Performed at 1.5 T

| | MRI Sequences | | | |
|---|---|---|---|---|
| | $T_1$-Weighted | $T_2$-Weighted | Dynamic | Diffusion |
| **MRI Protocol** | **GRE** | **FastSE** | **GRE** | **EPI** |
| Plane | Axial | Axial | Axial | Axial |
| Fat saturation | No | No | Yes or no | Yes |
| Time to repeat (ms) | 180 | 2100 | 3.9 | 1500 |
| Time to echo (ms) | 4.6–2.3 | 100 | 1.8 | 75 |
| Angulation (°) | 70 | 90 | 10 | 90 |
| Thickness (mm) | 5 | 5 | 4 | 5 |
| FOV (mm) | 375 | 325 | 380 | 400 |
| Matrix (mm × mm) | 220 × 284 | 268 × 344 | 184 × 184 | 180 × 194 |
| Scan time (s) | 20 | 180 | 30 | 180 |
| Delay (s) | | | 0, 40, 120, 250 | |
| *b*-value (s.mm$^{-1}$) | | | | 0, 100, 400, 800 |

*Notes:* MRI, magnetic resonance imaging; FOV, field of view; GRE, gradient echo; SE, spin echo; EPI, echo-planar imaging.

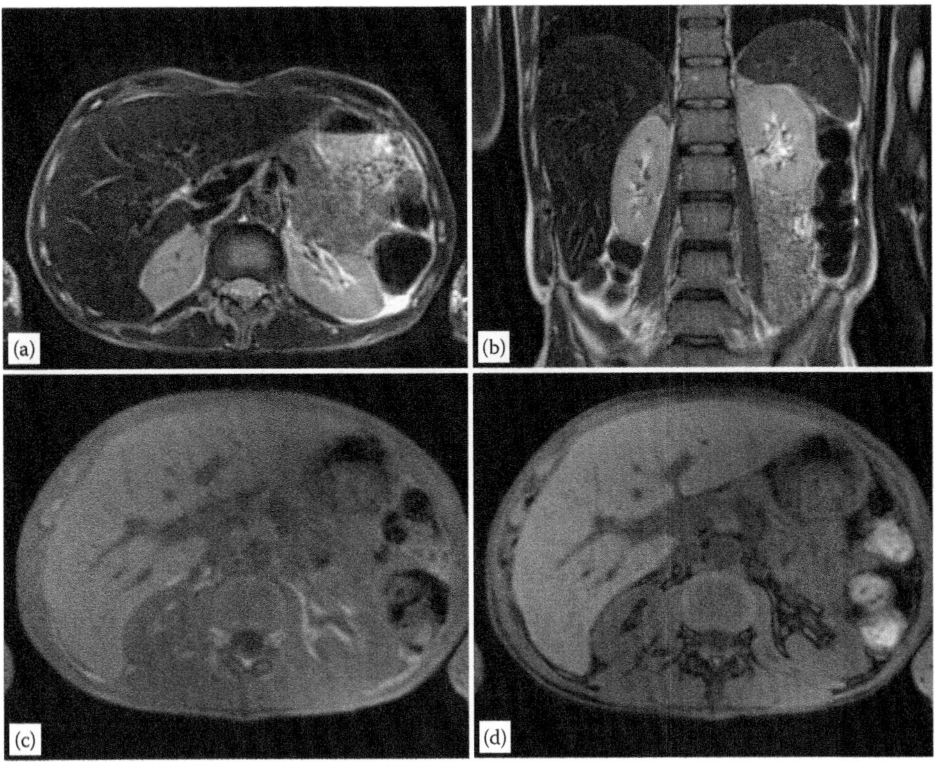

**FIGURE 1.1**
Renal MRI protocol: axial (a) and coronal (b) $T_2$-weighted sequences, and in-phase (c) and out-of-phase (d) dual chemical shift MR images.

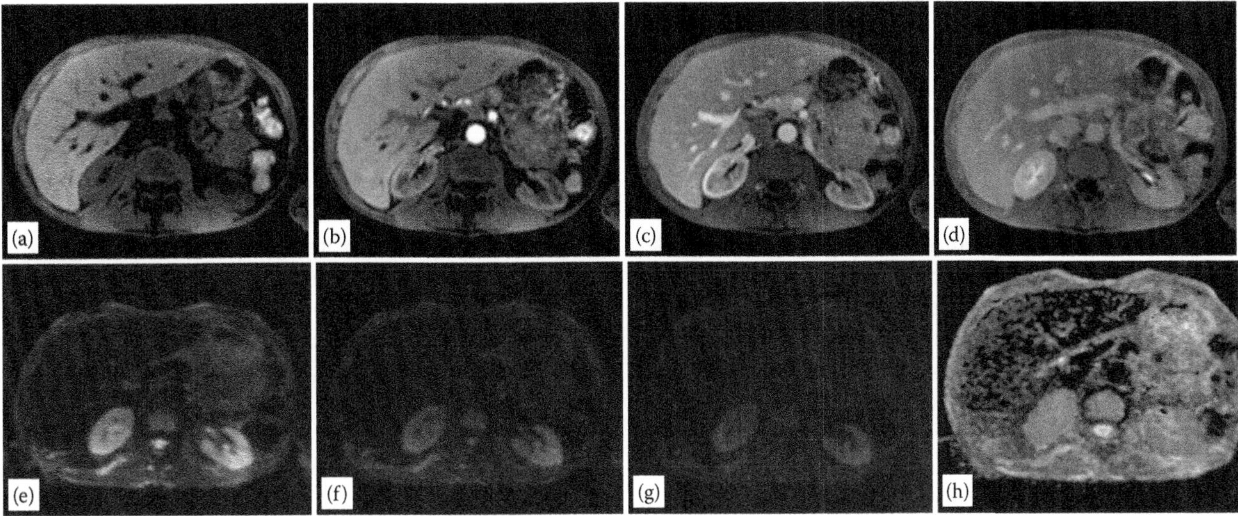

**FIGURE 1.2**
Renal MRI protocol: dynamic contrast-enhanced $T_1$-weighted sequence (a, no contrast; b, *arterial* corticomedullary phase; c, *parenchymal* nephrographic phase; and d, late phase) and diffusion-weighted imaging (e, $b = 100 \times 10^{-3}$ mm$^2$/s; f, $b = 400$; g, $b = 800$; h, corresponding ADC map).

### 1.2.2 MR Urography

To perform an adequate and reliable MRU, some specific sequences must be considered [9]. In general, the renal pelvis and the ureters are well visualized using highly $T_2$-weighted images (with long time to repeat TR) or $T_1$-weighted images with gadolinium contrast injection (Figure 1.3). For static-fluid MRU using highly $T_2$-weighted sequences (often with long TR), the urine in the collecting system is hyperintense due to its long $T_2$ relaxation time, whereas the surrounding tissue is

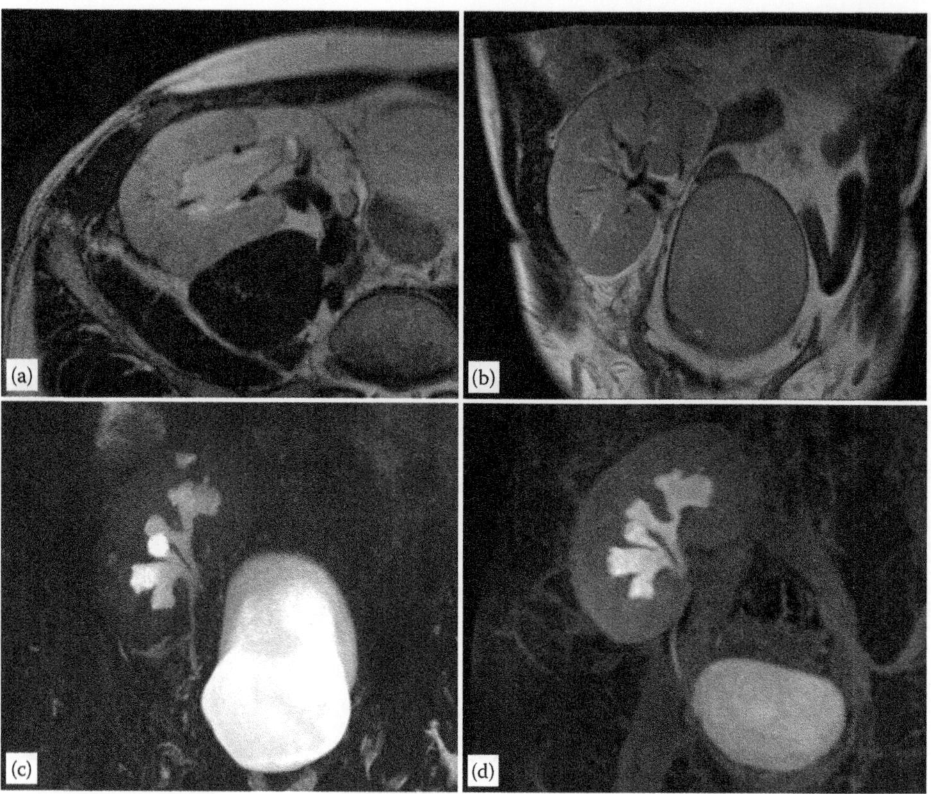


**FIGURE 1.3**
MR urography protocol: axial (a) and coronal (b) $T_2$-weighted sequences, highly $T_2$-weighted urography (c), and postcontrast $T_1$-weighted urography (d).

hypointense [10]. Breath-hold $T_2$-weighted MRU can be obtained with either thick-slab single-shot fast spin-echo techniques or similar thin-section techniques (e.g., half-Fourier rapid acquisition with relaxation enhancement, single-shot fast spin echo, and single-shot turbo spin echo). Obviously, the signal intensity (SI) of the background tissues may be adjusted by modifying the time to echo (TE) or using fat suppression. As performed for the biliary duct exploration, maximum intensity projection may be obtained for an adequate overview of the urinary tract. Reconstruction in two or three dimensions allows an adequate visualization of the cavities. As static-fluid MRU can be obtained in 1–2 s, multiple images can be obtained sequentially in a relatively short period of time and played as a cine loop. Such cine MRU is particularly helpful in confirming the existence of urinary tract stenosis. In case of non-obstructive disease, an injection of diuretic may be useful to allow an adequate $T_2$-weighted MRU because such explorations require that the urinary tract must be sufficiently filled with urine. However, normal and abnormal fluid-filled structures can interfere with static-fluid MRU, since the $T_2$-weighted techniques used to display the urinary tract are not specific for urine.

For contrast-enhanced (or excretory) MRU, intravenous gadolinium can be combined with a $T_1$-weighted 3D gradient-echo sequence, which allows an accurate and reliable visualization of the urinary tract [11]. A coronal through-plane resolution of 2–4 mm must be obtained. A gadolinium-based contrast agent is administered intravenously, and the collecting systems are imaged during the excretory phase. As gadolinium shortens the $T_1$ relaxation time of the urine, the urine appears bright on $T_1$-weighted images. At standard doses of 0.1 mmol/kg, gadolinium-based contrast material is often sufficient. Fat suppression enhances the conspicuity of the ureters and may be recommended: either a 3D soft tissue imaging type sequence such as volumetric interpolated breath-hold examination (VIBE), fast acquisition with multiphase EFGRE 3D (FAME), $T_1$-weighted high-resolution isotropic volume examination (THRIVE), liver acquisition with volume acceleration (LAVA), or MR angiography (MRA) may be used. Motion suppression is critical. If the breath hold cannot be achieved, fast sequences must be used. The images are acquired in the excretory phase, typically 5–8 min after intravenous gadolinium injection. As for CT scan, a dilution of the contrast may be also achieved with an additional injection of a diuretic. It may contribute to limit the $T_2$* effect due to the contrast product, which may cause signal loss [10].

### 1.2.3 Functional MRI of Kidneys

Renal MRI seems to be particularly interesting in order to obtain functional data. Several parameters may be derived from MRI but require specific sequences. At this time, there is still a lack of homogeneity or clinical validation of the technique but these points will be resolved probably in the near future.

Renal blood flow (RBF), or flow rate, may be evaluated relatively easily with MRI using phase-contrast imaging [12–14]. Spins in the renal arteries or veins that are moving in the same direction as a magnetic field gradient develop a phase shift that is proportional to the velocity of the spins. Stationary spins undergo no change in phase after application of bipolar gradients, whereas moving spins will experience a different magnitude of the second gradient compared to the first, because of its different spatial position. RBF is derived from this phase shift between the spins, based on the product of mean velocities in the renal artery and its section area. It refers to the global amount of blood reaching the kidney per unit time and is normally expressed in mL/min. However, motion artifacts in phase-contrast imaging can occur and are caused by the long acquisition time. They can be overcome by using interleaved gradient-echo planar technique, shortening the acquisition time from minutes to about 30 s [15]. To obtain the most accurate measurements in the human arteries, the imaging plane is usually positioned 10–15 mm downstream from the ostium, where respiratory movements are minimal, and perpendicular to the renal artery. The main application of this technique is the kidney transplantation in order to evaluate noninvasively the kidneys in case of healthy donor [16]. However, it is difficult to select the proper velocity-encoding gradient [1] and tortuous venous anatomy. In addition, low flow may limit the use of phase-contrast MRI in potential kidney donors [16]. In a recent study, cine phase-contrast MRI has been used to noninvasively and safely assess renal perfusion during septic acute kidney injury [17]. RBF was consistently reduced as a fraction of cardiac output.

Renal perfusion (or function) may be obtained with several techniques or agents (Figure 1.4). *Renal perfusion* refers to the blood flow that passes through a unit mass of renal tissue (mL/min/g) in order to vascularize it and to

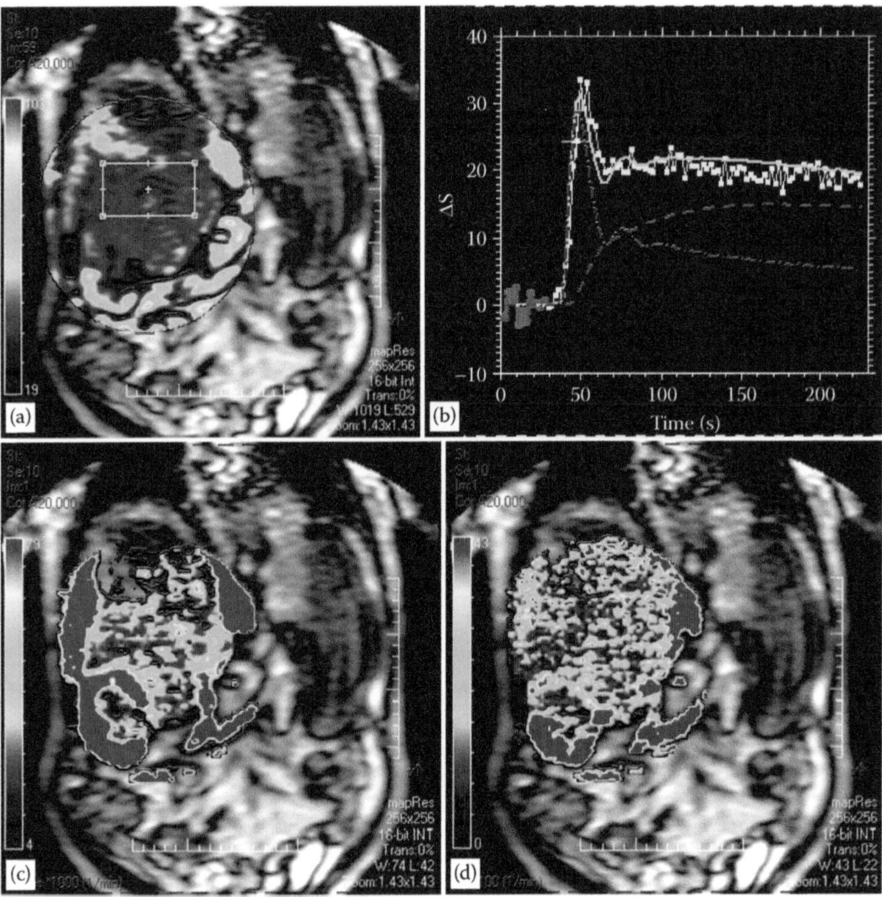


**FIGURE 1.4**
Functional MRI using the St. Lawrence–Lee model: cartography of signal (a), time–intensity curve (b), cartography of permeability (c), and cartography of perfusion (d).

exchange with the extravascular space. The renal perfusion depends on several systemic and local factors such as the arterial flow rate, the regional blood volume, and the vasoreactivity. Renal perfusion parameters, such as renal blood volume (RBV) and RBF, can be subsequently derived and measured after application of mathematical models [18]. However, such evaluations still require technical optimization. At this time, dynamic contrast-enhanced (DCE) $T_1$-weighted and regular low-molecular-weight gadolinium chelates are still the only combination used in routine [19,20] (Figure 1.5). Other types of contrast agents or sequences, such as larger gadolinium chelates or iron oxide particles or spin-labeling, are not fully available or do not have agreement for clinics. With DCE-MRI, the signal–time curve obtained within the renal parenchyma with these diffusible contrast agents is characterized by three discernible phases: a vascular phase with a tight upslope and an early peak, a glomerulotubular phase with a slow uptake (the agent being filtered), and a slowly descending excretory phase. When only perfusion is required, dynamic MR acquisitions can be short in time, compatible with a single breath-hold, because one needs only the first renal pass. However, when complementary parameters are necessary, as permeability or GFR measurements, movement correction must be applied because it requires longer acquisition times. Several methods have been proposed [21] but not always available commercially. Moreover, the role of gadolinium agents in the evaluation of renal perfusion or function still shows some limitations: these agents are freely diffusing into the interstitium, compartment which is usually neglected in most pharmacokinetic models; the relationship between SI and concentration is highly complex, inducing concomitant reduction in $T_1$ and $T_2$ (or $T_2$*), which is not the case for

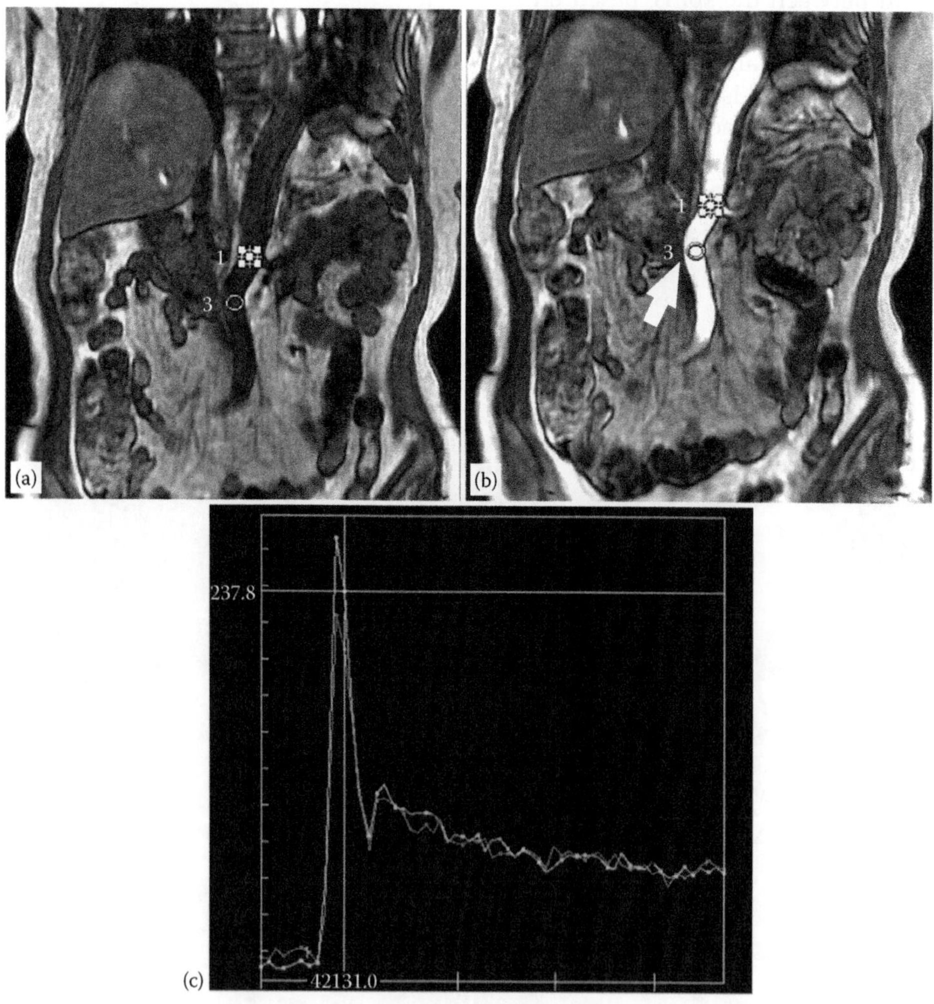


**FIGURE 1.5**
Functional MRI: coronal $T_1$-weighted images before injection (a) and at the arterial phase (b). An adequate arterial input function is obtained (c): 1 and 3 refer to the regions of interest drawn in the aorta and used to establish the input arterial function.

radioactive agents or iodine compounds. Compared with CT where conversion of densities into contrast concentration is straightforward, this conversion is more complex with MRI because it is nonlinear. Measurement of relative values therefore seems to be more easier because it does not require converting changes in SI into changes in $R_1$ ($1/T_1$): only relative values of perfusion (rRBV and rRBF) can be extracted, making only comparative studies possible (one kidney/the other; one area/another area; follow-up of treatments for tumors). In fact, this methodology may be applied in case of tumor but not to explore the renal function (GFR), which render complex such exploration. Pulse sequences used must have a heavy $T_1$-weighting: gradient-echo sequences or sequences with a non-selective magnetization preparation are preferred, combining with very short TR/TE and low flip angles. Concentration of gadolinium within the kidney may be very high due to water reabsorption in the proximal convoluted tubule and in the medulla. Therefore, to avoid $T_2$* contribution to the signal, the injected dose must be lowered between 0.025 and 0.05 mmol/kg, and the patient should be well hydrated [22]. An oblique-coronal plane, passing through the long axis of the kidneys, has to be preferred to an axial plane in order to facilitate the movement correction. To compensate the non-instantaneous bolus of contrast product injected into the blood, the quantification requires an accurate sampling of the vascular phase of the enhancement with a high temporal resolution. It has been therefore recommended to measure the arterial input function (AIF) within the suprarenal abdominal aorta, far from the entry volume to avoid inflow artifacts [20]. However, the AIF can be severely impaired by inflow effects within the aorta. If the AIF is not taken into account, only semi-quantitative parameters as maximal signal change, time to maximal signal change, or wash-in and wash-out slopes can be measured for comparison from right to left kidney, from cortex to medulla, or from one territory to another, or for follow-up of the patients [23–25]. Calculated concentration/time curves must then be processed using specific mathematical models [26]. Dujardin et al. [27,28] generalized the tracer kinetic theory from intravascular to diffusible tracers using deconvolution, which is a model-free approach. From region of interest (ROI) drawn on the aorta and on the renal cortex, tissue concentration/time course has to be deconvolved pixel by pixel with the flow-corrected aortic time course, resulting in an impulse response function (IRF). This method allows getting intrarenal maps of RBF, RBV, and mean transit time. Others used a two-compartment model, giving access to RBV, RBF, and additional parameters, such as GFR, for renal diseases, or tissue permeability, for tumors.

## 1.3 Current Applications for Non-Tumoral Diseases

### 1.3.1 Inflammation and Infection Diseases of the Kidneys

In addition to other imaging techniques, MRI can provide morphological information which could be used as morphologic biomarkers of kidney function [29]. Although nonspecific, the size of the kidneys can be a useful indicator of disease and the residual kidney function. As proposed for other techniques, acquired small kidneys may be the result of vascular disease or glomerular disease, irregular-shaped kidneys may be observed in interstitial kidney disease, and enlarged kidneys may be due to obstruction, diabetes mellitus, infection, amyloidosis, renal vein thrombosis, or polycystic disease. The most widely used morphologic marker is renal length as measured by US but remains imperfect [30]. This limitation may be overcome by the measurement of the longitudinal mid-kidney parenchymal [31], parenchymal thickness [32,33], or using a volumetric technique [34] but still required technical improvement. Computer-assisted detection software may be necessary for this purpose. Interestingly, low signal on $T_2$-weighted may be observed in case of hemolysis due to paroxysmal nocturnal hemoglobinuria, cortical hemosiderin deposition due to mechanical hemolysis, or sickle cell disease; in case of infection as observed in hemorrhagic fever with renal syndrome; or in case of vascular disease such as renal arterial infarction, acute renal vein thrombosis, renal cortical necrosis, transplanted kidney rejection, and acute non-myoglobinuric renal failure [35]. Corticomedullary differentiation (CMD) visible on $T_1$-weighted has been correlated with the degree of renal failure. CMD is supposedly absent when serum creatinine levels reach 3.0 mg/dL [36], whereas it may be preserved in acute renal failure within 2 weeks. However, its degree seems to be independent of serum creatinine level [37].

However, compared to other techniques, MRI may evaluate infection or inflammation diseases more accurately [38]. Moreover, MRI may be able to identify and differentiate acute or chronic diseases. In the particular case of the chronic renal diseases often evolving to end-stage kidney disease, a progressive renal tissue fibrosis at the level of the interstitium or glomeruli may be observed with MRI. Fibrosis results from transformation of the extracellular matrix by cytokines and chemokines released by activated cells in the setting of recurrent episodes of acute inflammation. MRI may therefore image intrarenal inflammation as well as fibrosis and allows the noninvasive evaluation of these processes to improve follow-up and monitoring of drug therapy [39]. The techniques are based on methods of cellular and molecular imaging, and methods of functional, such as DWI or blood oxygen level-dependent (BOLD) MRI, and structural, such as MR elastography, imaging. However, any of these techniques have not been validated clinically.

DWI is the most developed technique but its role must be still clarified. The diagnosis of various renal diseases such as chronic renal failure, renal artery stenosis, and ureteral obstruction can benefit from measuring the diffusion characteristics of the kidney [40]. As the apparent diffusion coefficient (ADC) combines the effects of capillary perfusion and water diffusion in the extracellular extravascular space, ADC mapping shows differences between the lesions and normal tissue and provides information on perfusion and diffusion simultaneously [41].

For acute disease, DW MRI has been applied in patients with pyelonephritis (Figure 1.6) or acute renal failure [42], in particular in case of ischemia suspicion (Figure 1.7). DW images and ADC maps allow to identify lesions, showing very high signal on $b > 400$ s/mm$^2$ images and extremely dark signal on ADC maps. In case of infection, marked restriction of diffusion can

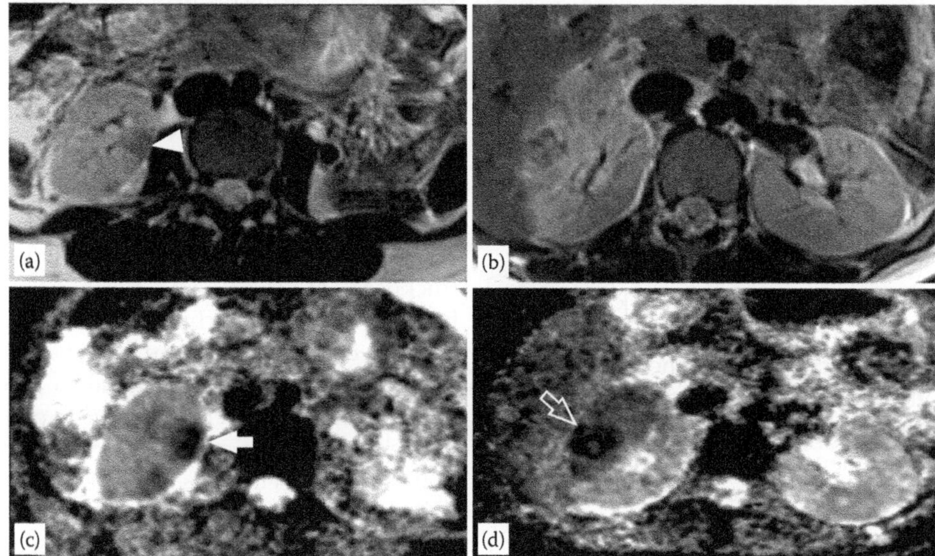


**FIGURE 1.6**
Diffusion-weighted imaging of pyelonephritis: $T_2$-weighted images showed heterogeneous aspects (a and b), whereas DWI clearly demonstrated area of nephritis corresponding to low value of ADC (c and d).

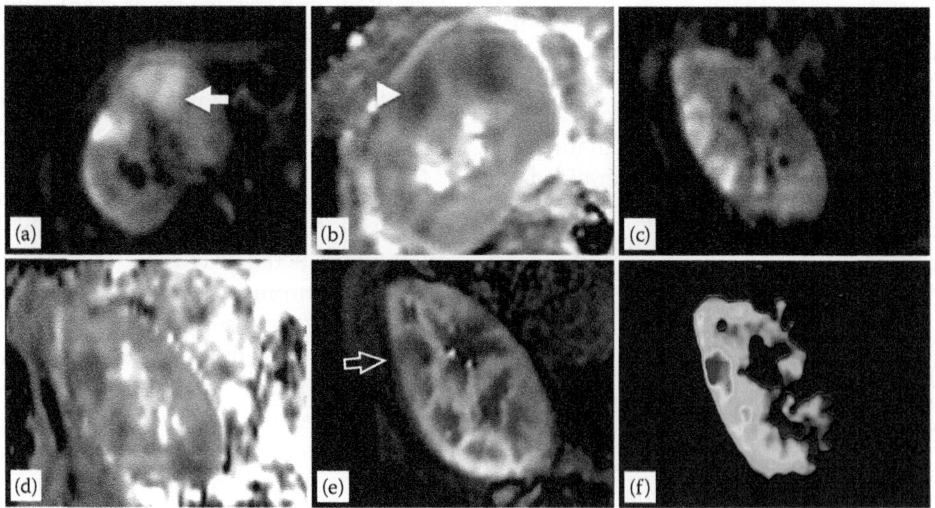


**FIGURE 1.7**
Early control after renal transplantation: diffusion-weighted imaging ($b = 800 \times 10^{-3}$ mm$^2$/s) (a and c) and corresponding ADC map (b and d) showed restriction of diffusion of ischemic area (arrowheads). Postcontrast images showed a cortical defect (e) (arrow) and corresponding image after fusion with DWI (f).

be attributed to the presence of thick viscous pus, consisting of inflammatory cells, bacteria, necrotic tissue, and proteinaceous exudates [43–45]. For acute pyelonephritis, DWI was awarded the highest visibility score compared to DCE-MRI ($p = .05$), while for abscess, DCE-MRI had the highest score ($p = .04$). In a recent study, agreement between DCE-MRI and DWI was 94.3% for the diagnosis of infection (83/88 patients; $p < .05$) [46]. Interestingly, ADC may be lower in the $T_2$ hyperintense non-enhancing fluid intensity regions, which correspond to areas of purulent material [43] (Figure 1.8). Chan et al. found in a limited group of patients that DWI showed a hypointense pyelocalyceal system in hydronephrosis and a hyperintense pyelocalyceal system in pyonephrosis [47] (Figure 1.9). The high SI in pyonephrosis may also be related to the high viscosity of the pus, whereas the free-moving molecules in hydronephrosis cause low SI.

In case of inflammation, which may be more chronic, the ADC values would be decreased due to restriction of the diffusion space in proportion to the development of fibrosis. The clinical value of mono- or biexponential analysis of DWI has been evaluated, and it has been showed that ADC of renal cortex correlates with renal function [48]. In this study, patients with stable function of renal allograft for at least 6 months compared to patients with acute deterioration of allograft function presented significant differences in ADC ($p < .006$), as well as patients who recently underwent transplantation (<14 days) with good renal function compared to others ($p < .04$). The ADC decreased in case of alteration of renal function (mean: $1.961 \pm 0.104$ vs. $1.801 \pm 0.150$, and $2.053 \pm 0.169$ vs. $1.720 \pm 0.191$ mm$^2$/s). This may be particularly useful for renal allograft recipients, when contrast-enhanced MRI may be considered problematic such as in case of allergy or impaired renal function (Figure 1.10). Even being a limitation of the DWI technique [49], anisotropy of kidneys due to the radial orientation of the tubules in the pyramids and the blood vessels in the renal cortex may also be used for DWI. Such characteristic may be used to perform a diffusion tensor imaging and tractography analysis (Figure 1.11). The medullary fractional anisotropy (FA) value seems to be the main parameter for assessing renal damage [50]. Further studies are mandatory to prove the usefulness of these techniques because whereas the FA value in the medulla was significantly lower ($p = .0149$) in patients with renal function impairment compared to patients with normal renal function, a direct correlation between diffusion tensor imaging parameters and the

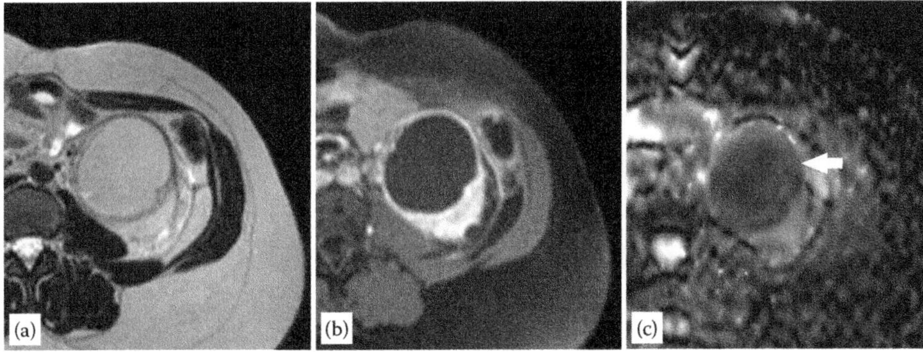


**FIGURE 1.8**
Exploration of a cystic lesion of the left kidney: $T_2$-weighted image showed high signal intensity (a), no specific contrast enhancement may be observed on $T_1$-weighted image (b), but low ADC value of the cyst content (arrow) (c). This lesion corresponded to a renal abscess.

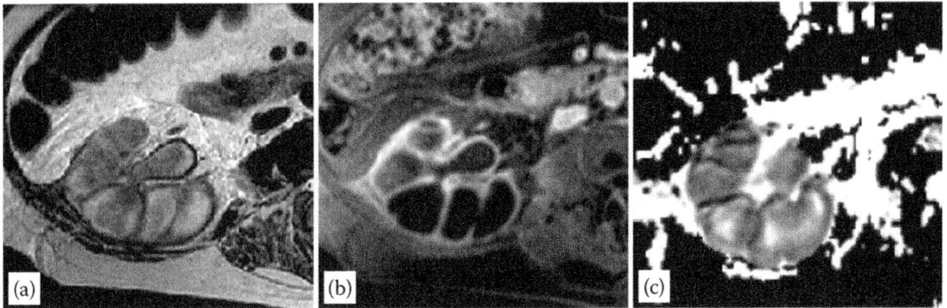


**FIGURE 1.9**
Exploration of a renal obstruction associated with sepsis: $T_2$-weighted image showed dilatation and heterogeneous signal into the renal cavities (a), no enhancement may be observed on postcontrast $T_1$-weighted (b), but low ADC (c) corresponding to a pyonephrosis.

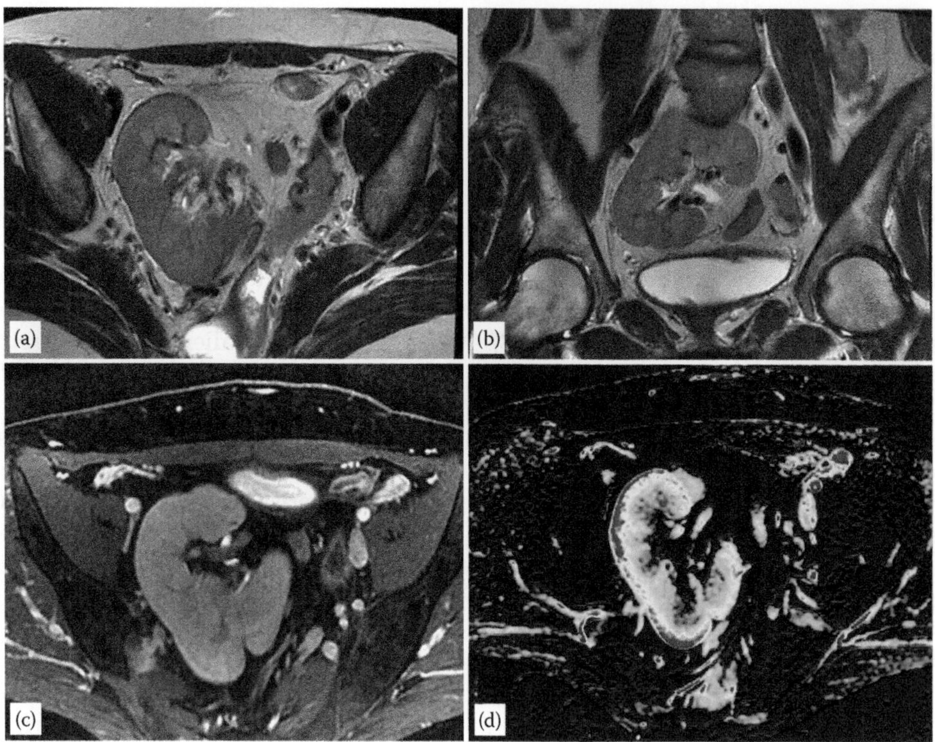


**FIGURE 1.10**
Renal MRI protocol for kidney transplant without DWI: axial (a) and coronal (b) $T_2$-weighted sequences, DCE $T_1$-weighted sequences (c: parenchymal phase), and semi-quantitative analysis of the enhancement (area under the curve) (d).

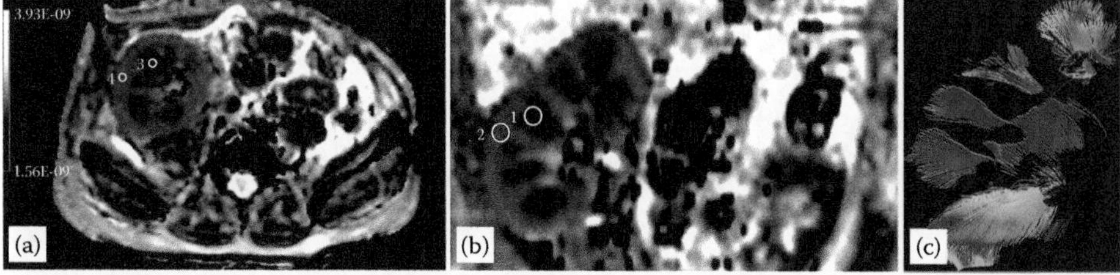


**FIGURE 1.11**
Diffusion tractography of a renal graft: ADC cartography may identify cortex and medulla (a and b). A tractography of the medulla can be obtained (c). The values in the figures correspond to the regions of interest allowing the measurement of the apparent diffusion coefficient in different parts of the kidney.

eGFR was not found. In case of renal transplant, whereas the ADC values in the cortex and in the medulla were lower in transplanted than in healthy kidneys ($p < .01$), differences were more distinct for FA, especially in the renal medulla, with a significant reduction in allografts ($p < .001$) [51]. Furthermore, in transplanted patients a correlation between mean FA in the medulla and estimated GFR has been observed ($r = 0.72, p < .01$).

Other techniques may be used but only preliminary results have been reported. The elastography method requires the application of shear waves by means of a mechanical device to produce tissue displacements on the order of nanometers to micrometers. Propagation of these waves is dependent on viscoelastic properties of the tissue, and its elastic characteristics can be quantified and/or mapped as parametric images [52]. Fibrosis might be therefore evaluated. BOLD MRI has been also proposed [40,53] because the outer medulla seems to be particularly sensitive to hypoxia. The active reabsorption process within the thick ascending loop of Henle requires a high level of oxygen consumption [54]. Therefore, decreased medullary blood flow, as in acute renal failure, or increased tubular reabsorption, as in diabetic nephropathy (at the stage of hyperfiltration), may induce medullary hypoxia and secondary ischemia. In a recent study focus on diabetic nephropathy [55], the mean medullary $R_2$* values were lower in patients with diabetic nephropathy compared to healthy volunteers ($p = .0002$), whereas the cortical $R_2$* values were not significantly different between the two groups. The degree of kidney disease was independently associated with a decrease in medullary $R_2$* values after logistic regression ($p$ = .005). Further, in a study evaluating DWI and BOLD [56], ADC values correlated with eGFRs in the diabetic nephropathy and in patients with chronic kidney disease [$r$(2) = 0.56 and $r$(2) = 0.46, respectively]. Although the $T_2$* values of BOLD MRI and eGFR displayed good correlation in chronic kidney disease [$r$(2) = 0.38], no significant correlations were reported between these values in patients with diabetic nephropathy, suggesting that factors other than tubulointerstitial alteration determine the degree of hypoxia in the renal cortex. Interestingly, neither the $T_2$* nor ADC values correlated with eGFR in patients with acute kidney injury.

### 1.3.2 Vascular Diseases

To obtain an adequate MRA, a 3D $T_1$-weighted gradient-echo sequence with intravenous gadolinium after timing bolus can be used with coronal thin-section reconstructions (Figure 1.12). The accuracy of MRA in the evaluation of the anatomy of renal vessels shows globally similar or even better results compared to digital subtraction angiography and CT angiography as observed in the specific case of vascular exploration before transplantation [57–62]. On the other hand, research in non-contrast MRA was boosted by the advent of nephrogenic systemic fibrosis. This technique is now robust with similar diagnostic quality when compared with contrast-enhanced techniques [63]. However, it may be more time-consuming [64]. For similar reasons, low-dose contrast-enhanced techniques have similarly gained interest.

The 3D dataset can be used for the reconstruction of thin 2D sections for detailed evaluation, as well as for maximum intensity projection reconstructions of the vessels. As for other vascular explorations, relatively large flip angle (up to 40°) can be used to minimize background signal around the high signal of the renal arteries [16]. If a venous angiography sequence is required as in case of preoperative exploration or before transplantation, it must be taken into account that the concentration of gadolinium in the renal veins is lower than in the renal arteries due to the excretion of gadolinium by the kidneys. It causes lower contrast of the veins compared to the background. In this case, a lower flip angle (15°) can be used to compensate for the lower gadolinium concentration, at the expense of more background signal [16]. In an alternate case, the time-of-flight technique can be accurately used to evaluate renal arteries, in particular since this technique has been implemented at 3 T

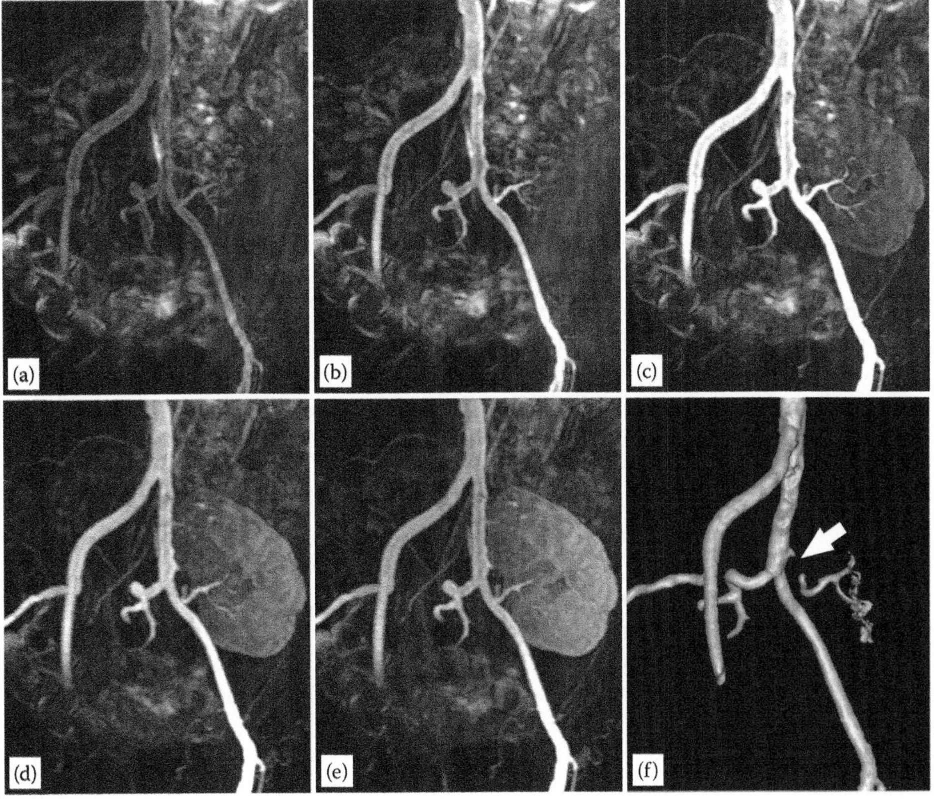


**FIGURE 1.12**
Angiography of the renal artery of a kidney transplant in left iliac fossa: a paraosteal stenosis was observed (arrow). (a–e) Various acquisitions over time and (f) final 3D reconstructions allowing to identify a paraosteal stenosis.

**FIGURE 1.13**
Time-of-flight technique at 3 T can be used to evaluate renal arteries without contrast injection: a stenosis was observed in this renal artery of a kidney transplant (arrow).

(Figure 1.13) [65,66]. It may be used to clarify intraluminal filling defects potentially caused by flow artifacts on gadolinium-enhanced MRA [1]. Recently, 22 hypertensive patients underwent both true fast imaging with steady-state precession (TrueFISP) MRA and contrast-enhanced MRA (CE-MRA) on a 1.5-T MR imager [67]. Volume of main renal arteries, length of maximal visible renal arteries, and number of branches indicated no significant difference between the two techniques ($p > .05$). Stenotic degree of 10 RAS was greater on CE-MRA than on TrueFISP MRA. Qualitative scores from TrueFISP MRA were higher than those from CE-MRA ($p < .05$). A comparative evaluation of kidney transplant unenhanced MRA spatial labeling with multiple inversion pulses (SLEEK) and other techniques has been performed [68]. The authors concluded such evaluation was a reliable diagnostic method for depiction of anatomy and complications of renal vascular transplant because more accessory renal arteries were detected with SLEEK than with Doppler and correlation was excellent between the stenosis degree with SLEEK and subtraction angiography ($r = 0.96$; $p < .05$).

Diagnosis of renal artery stenosis may be obviously achieved with MRA based on morphologic findings. Although the reproducibility of this evaluation is not perfect and as it is not a good predictor of functional improvement after revascularization, such exploration may be useful to identify patient eligible for invasive explorations such as transarterial angiography. In the past, several functional tests have been proposed to characterize *functional stenoses* as captopril-scintigraphy or captopril-MRI but these approaches are not used routinely anymore. However, as MRI allows measuring RBF or renal perfusion, added to the morphologic evaluation in the same examination, it may be possible to establish the hemodynamic significance of stenosis. Schoenberg et al. [69] used first cardiac-gated phase-contrast flow measurement to complete morphological acquisitions. Agreement between the morphological degree of stenosis and changes in the pattern of the flow profile was first documented in animal and then in human studies [13,14]. The loss of the early systolic peak was proposed as a sensitive indicator for the loss of the autoregulatory capacity and the onset of significant mean flow reduction. This flow measurement technique provided a functional grading of the degree of stenosis independent of the morphologic grading. These observations were confirmed by a multicenter trial, showing a significant reduction in interobserver variability and an improvement in overall accuracy compared with digital subtraction angiography, with sensitivities and specificities exceeding 95%. Evaluation of perfusion also provided quantified functional information in these patients [13]. Michaely et al. [23] first calculated semi-quantitative parameters such as mean transit time, maximal upslope of the curve, maximum SI, and time to SI peak after a gamma variate fit of the SI–time curve. Significant differences between patients without stenoses or with low-to-intermediate grade stenoses and patients with high-grade stenoses were found for each of these parameters. Based on a small series of 27 patients, the same group evaluated the diagnostic accuracy of quantified renal perfusion in identifying and differentiating renovascular from renal parenchymal disease [19]. Measurement of MR perfusion yielded a sensitivity of 100% and a specificity of 85% utilizing an optimal plasma flow threshold value of 150 mL/100 mL/min, whereas single MRA achieved a sensitivity of 51.9% and a specificity of 90%. Follow-up of RAS after revascularization is another promising application of these quantitative functional studies, adding objective hemodynamic criteria to the morphologic analysis of the artery [26].

## 1.4 Renal Obstruction

The accuracy of MRU for assessing renal obstruction is similar to CT urography [70,71] (Figure 1.14). $T_2$-weighted MR urograms have proved to be excellent in the visualization of the markedly dilated urinary tract, even if the renal excretory function is quiescent [72]. However, static-fluid MRU is less suitable for imaging of disorders that occur in the non-dilated collecting system or in case of multiple cysts. Gadolinium excretory MRU allows

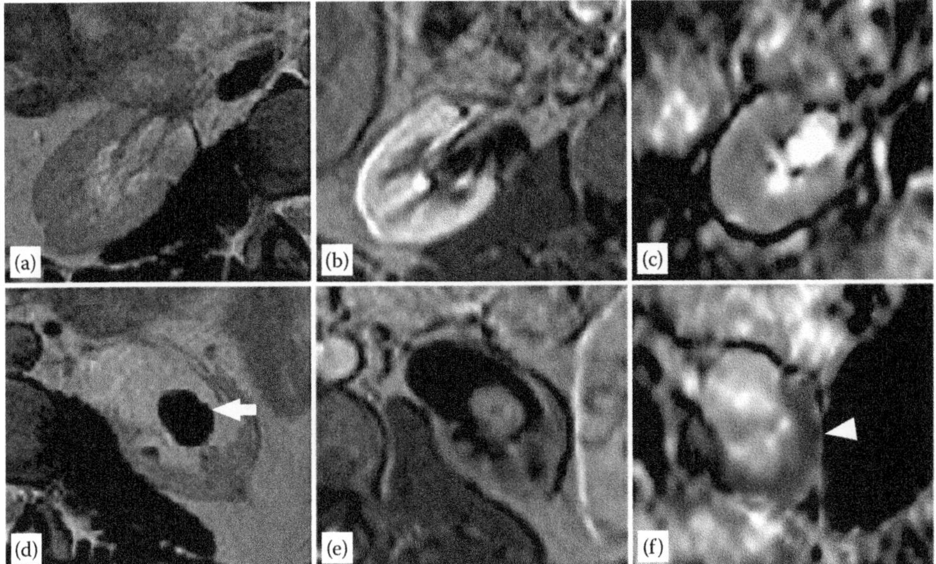


**FIGURE 1.14**
Evaluation of a unilateral chronic obstruction: $T_2$-weighted, postcontrast $T_1$-weighted, ADC for a normal kidney (a–c) and non-functional kidney (d–f) due to chronic lithiasis (arrow). ADC of the non-functional kidney is lower than the healthy kidney (arrowhead).

to obtain high-quality images of both non-dilated and obstructed urinary tracts in patients with a normal or moderately impaired renal function.

In children, the most common indication for MRU is the evaluation of hydronephrosis [73,74], which occurs mainly in the ureteropelvic junction but less frequently elsewhere [75]. MRU allows the differentiation between obstructive and non-obstructive uropathies by the calculation of renal transition time. Moreover, after contrast injection the arterial phase may help to identify crossing vessels [76]. Considering the fact that MRU not only enables the depiction of whole dilated and non-dilated ureters from ureteropelvic junction to ureteric insertion and their analysis from all angles via maximum intensity projection reconstruction, it also enables the examination of the internal outlook of ureters and of external structures, which may lead to its compression; hence, ureteric anatomy and pathology are well demonstrated with MRU [72,77].

On MRU, even if a calculus appears as a signal void, this finding is nonspecific. Blood clots, gas, sloughed papilla, and tumors may also appear as a low signal within the bright signal of urine [10]. Morphologically, MRI has the advantage compared to CT scan to better detect perinephric edema as a secondary sign of obstruction [70]. Gadolinium-enhanced MRU showed renal calculi with considerably higher sensitivity than $T_2$-weighted MRU [71] (Figure 1.15). Sudah et al. [71,78] demonstrated improved sensitivity of ureteral calculi for diuretic-augmented excretory MRU (96.2%–100%) compared with a $T_2$-weighted technique (53.8%–57.7%). If a signal void is not clearly detached from the wall of the pelvis or ureter, additional $T_1$-weighted and contrast-enhanced images are necessary to further characterize the lesion (Figure 1.16). Urinary tract calculi can therefore often be distinguished from blood clots and tumor by analyzing $T_1$-weighted images: the blood clots exhibit high SI elements on unenhanced $T_1$-weighted images, whereas neoplasm enhances after intravenous contrast material administration. Interestingly, MRU seems to be more sensitive and specific for non-calculous urinary tract obstruction than in unenhanced CT [79] due to the higher soft tissue contrast.

The impact of the obstruction may be also achieved with MRI by allowing a stratification of the grade of obstruction. Technically, the acquisition mode for this purpose is the same for perfusion imaging studies but requires injection of furosemide, which is now recommended just before the injection of gadolinium, at a dose of 1 mg/kg in neonates and 0.5 mg/kg in infants and adults. Comparison between both kidneys' function has been performed in a prospective multicentric study in France on 295 patients [80]. This study showed an underestimation of the values for highly dilated kidneys. Grading obstruction by placing an additional ROI on the pyelocaliceal system is more difficult because many factors of variability may occur, such as the level of split function, the volume of cavities, and the degree of bladder filling. Moreover, DWI may also be useful in patients with upper urinary tract obstruction [81]. As hydration is an important factor to increase global ADC values [42], ureteral obstruction (as well as renal artery stenosis) may be evaluated by the decrease in those values of ADC [82,83]. In case of

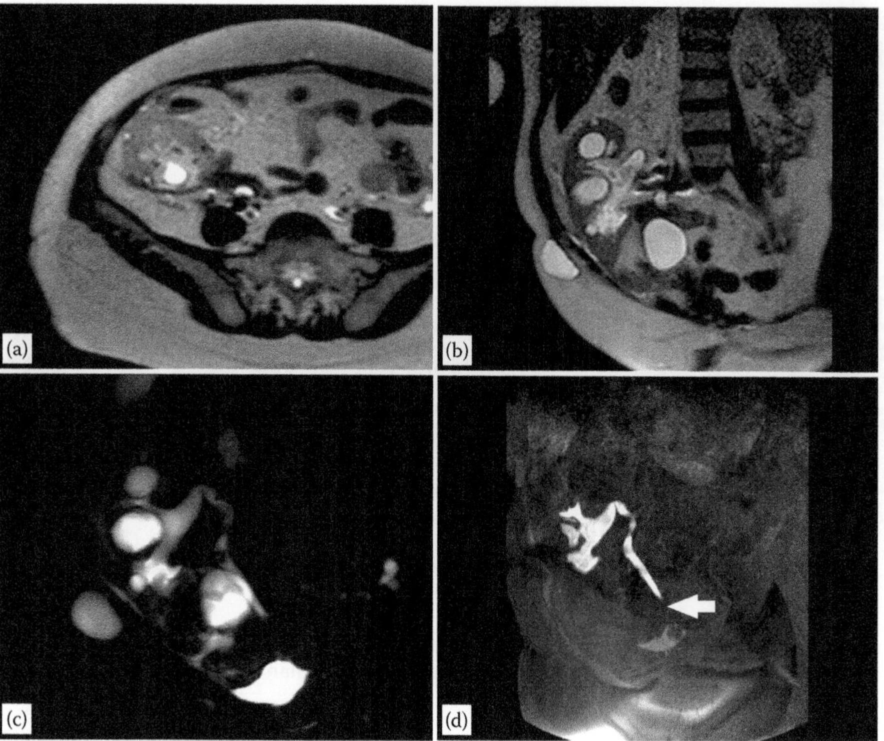


**FIGURE 1.15**
MR exploration performed for the exploration of iterative infection episodes of a renal transplant: axial (a) and coronal (b) $T_2$-weighted sequences, highly $T_2$-weighted urography (c), and postcontrast $T_1$-weighted urography (d). A stricture was observed in the lower part of the reimplanted ureter (arrow).

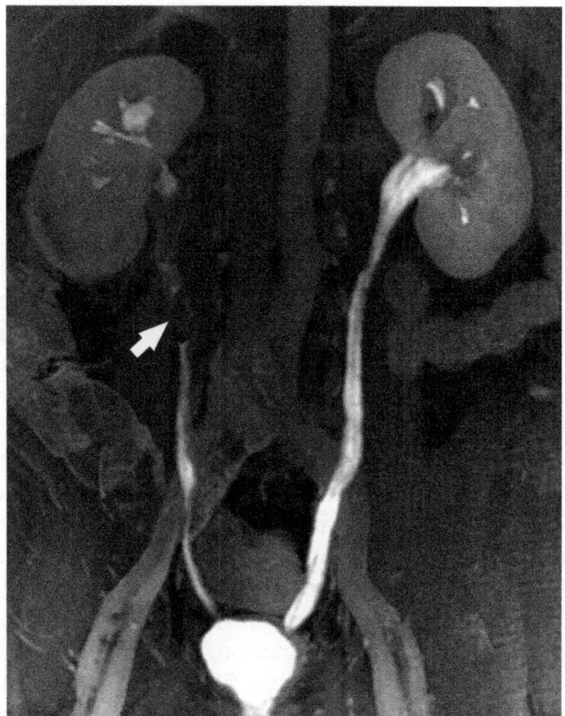

**FIGURE 1.16**
$T_1$-weighted MR urography after contrast injection performed to explore unilateral obstruction. Defect in the right renal pelvis and upper right urinary tract corresponded to an urothelial tumor.

acute or chronic obstructive renal failure, the cortical and medullary ADC values are significantly decreased when compared with normal kidneys and the cortical value decrease seems to be well correlated with serum creatinine levels [83].

In case of pregnant patients, static-fluid MRU must be preferred. Multiple acquisitions (cine MRU) may be necessary to visualize the entire ureters and exclude fixed narrowings or filling defects. Results are excellent in identifying urinary tract dilatation and level of obstruction [84] but it may be challenging to differentiate physiologic hydronephrosis from pathologic obstruction. The analysis of the filling of the lower ureter may be useful.

A recent study [85] reported that BOLD and DW sequences allowed for the monitoring of pathophysiologic changes of obstructed kidneys during obstruction and after its release. During acute obstruction, $R_2$* and perfusion fraction ($F_P$) values were significantly lower in the cortex ($p$ = .020 and $p$ = .031, respectively) and medulla ($p$ = .012 and $p$ = .190, respectively) of the obstructed compared to the contralateral unobstructed kidneys. After release of obstruction, $R_2$* and $F_P$ values increased in both the cortex ($p$ = .016 and $p$ = .004, respectively) and medulla ($p$ = .071 and $p$ = .044, respectively) reaching values similar to those found in the contralateral kidneys.

## 1.5 Applications in Oncology

### 1.5.1 Characterization of Kidney Tumors

Whereas the high prevalence of renal cell carcinoma (RCC) and the lack of reliable imaging criteria for recognition of benign tumors other than typical angiomyolipoma (AML) justify the growing role of pre-therapeutic image-guided biopsy, recent published studies have showed that MRI may help to identify benign renal tumors if considering a multiparametric approach [2,3,86–89] (Figure 1.17). The application of these new imaging criteria, able to accurately differentiate the most common tumor types as well as indolent versus more aggressive malignant histologies, would be useful to reduce the number of unnecessary biopsies, to consider the most appropriate treatment approach for a tumor, and to promote active surveillance with imaging follow-up for eligible patients (Table 1.2).

#### *1.5.1.1 Benign Renal Tumors*

Typical AML diagnosis may be easily assessed with CT and/or MRI. In contrast, the pre-therapeutic characterization of low-fat content AML or renal oncocytoma (RO) remained a challenge [88,89]. Validation of new imaging criteria would be useful to adequately identify low-fat content AML or RO before deciding on any therapeutic option and subsequently avoid resection of these benign tumors.

##### *1.5.1.1.1 Renal AML*

Renal AMLs are benign renal neoplasm composed of vascular, smooth muscle, and fat elements. This benign tumor is encountered in 2%–6% of excised solid masses from surgical series [106,107] but has in fact an incidence of 0.3%–3%. AMLs may be easily identified by unenhanced CT [108], or fat suppression and/or chemical shift gradient-echo MR sequences [109]. However, Dixon-based approaches may provide a more homogeneous fat suppression than spoiled gradient-echo imaging with frequency-selective fat-suppression strategies [110]. The majority of AML is sporadic and is typically identified in adults (mean age of presentation 43 years), with a strong female predilection but may be associated with tuberous sclerosis.

The characteristic appearances of typical AMLs with MRI include variable areas of high SI within the tumor on both $T_1$-weighted and $T_2$-weighted images due to their fat content. Because of this fat content, high SI is visible on non-enhanced $T_1$-weighted images. However, areas of high SI on $T_1$-weighted images are not pathognomonic of fat, and blood and pockets of fluid of high protein content may have a similar appearance. Intratumoral fat is best demonstrated with fat-suppression techniques, and the in-phase and out-of-phase $T_1$-weighted imaging

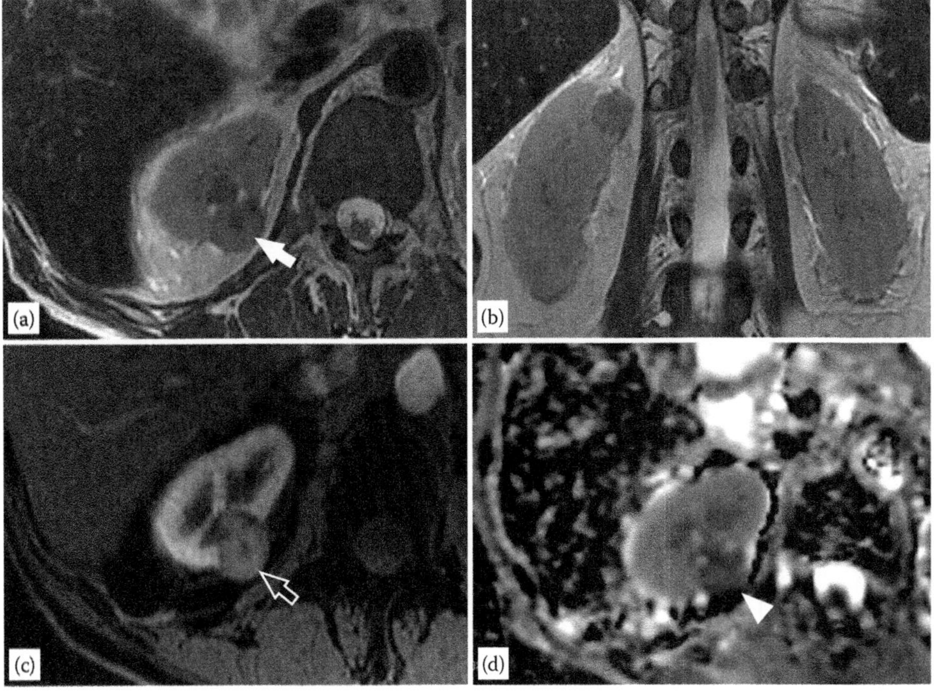


**FIGURE 1.17**
Multiparametric MRI to explore a right renal tumor before biopsy: the tumor presented a low signal on $T_2$-weighted (a and b), a slow enhancement (c, arterial phase), and a restriction on ADC map (d).

**TABLE 1.2**

Evocative MR Features Reported in the Literature of the Most Common Renal Tumors

| Tumor Subtype | % in Surgical Series | $T_2$-Weighted | $T_1$-Weighted | Fat Saturation | Dual Chemical Shift MRI | DCE-MRI | Late Postcontrast $T_1$-Weighted | DWI |
|---|---|---|---|---|---|---|---|---|
| AML | 2–6 | Heterogeneous with high signal intensity | Heterogeneous with high signal intensity | Signal suppression | Interface | Heterogeneous | – | – |
| AML with minimal fat content | | Low signal intensity | – | – | Signal drop | Arterial enhancement | – | Low ADC |
| Renal oncocytoma | 3–7 | Heterogeneous<br>Central area (scar) | – | – | No | Heterogeneous<br>Moderate wash-in and wash-out | Complete late segmental inversion of central scar | – |
| Clear-cell RCC | 75–80 | Heterogeneous<br>Central area (necrosis)<br>Pseudocapsule<br>Vascular invasion<br>Vascular signal void | Heterogeneous<br>High signal intensity of central area | Small area with signal suppression | Signal drop<br>Interface | Heterogeneous<br>High arterial wash-in and quick wash-out | Absence of segmental inversion of necrosis | Heterogeneous |
| Papillary RCC | 10–15 | Homogeneous<br>Low signal intensity<br>Pseudocapsule | – | – | Signal drop | Slow enhancement | – | Low ADC |
| Chromophobe RCC | 5 | Heterogeneous<br>Central area (necrosis) | – | – | No | Moderate wash-in and wash-out | Absence of segmental inversion of necrosis | – |
| References | [90] | [2,3,87,91–95] | [87,96] | [97] | [2,87,97,98] | [2,87,98–100] | [3] | [2,101–105] |

technique is extremely sensitive to small quantities of fat. Most AMLs contain macroscopic fat easily identified with fat-suppression techniques. In opposed phase, an artifact will appear at the interface of AML with kidney.

However, for the unusual AMLs with minimal fat content, no reliable MR criteria have been reported at this time. Specific diagnosis of minimal fat AMLs versus all other tumor types remains uncertain [2]. Even calculation of SI index or tumor-to-spleen SI ratio, based on a measure of SI of renal mass and spleen on opposed-phase and in-phase gradient-echo images [97], is not sufficient to distinguish this type of AML from RCCs [98]. A small lipid content may be found in both types of tumor cells [111] (Figure 1.18). However, as minimal fat AMLs often presented as renal tumors with low $T_2$-weighted SI [95,98,111,112] and as these AMLs may also present a high arterial enhancement [100], Hindman et al. [98] observed that the coexistence of a low SI on $T_2$-weighted images and avid enhancement could help to differentiate minimal fat AML from clear-cell RCC (cRCC).

*1.5.1.1.2 Renal Oncocytoma*

RO is the second most common benign renal neoplasm after typical AML. RO represents around 3%–7% of all renal tumors in surgical series [90]. They typically present sporadically in the sixth to seventh decades with a peak incidence at 55 years of age. There is 2:1 male predilection. Unfortunately, these demographics are similar to RCC. Apart from the exceptional cases of renal carcinomas coexisting as hybrids or separate tumors, especially in Birt–Hogg–Dube syndrome, their prognosis is considered to be excellent [90,113,114]. Based on these considerations, conservative treatment has been recommended [113,115,116]. The importance of preoperative diagnosis is therefore crucial and, although literature on morphologic characteristics of ROs abounds, no one can accurately separate oncocytomas from malignant lesions [114,117,118]. Diagnosis of RO remains based on histopathological examination of surgical or biopsy specimens [88,89] showing large eosinophilic cells with small, round nuclei with large nucleoli, apical Hale coloration, and no expression of cytokeratin 7

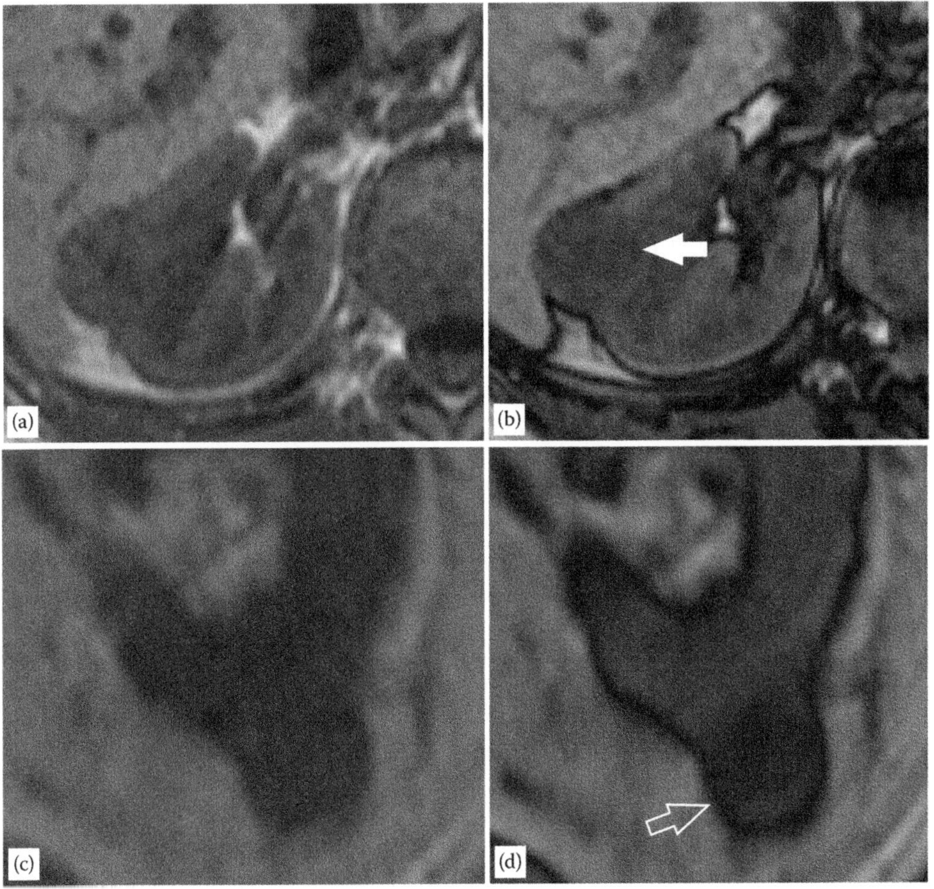


**FIGURE 1.18**
Clear-cell RCC (a and b) and low fat AML (c and d) with drop of signal intensity on out-phase sequence compared to in-phase dual chemical shift MR sequences.

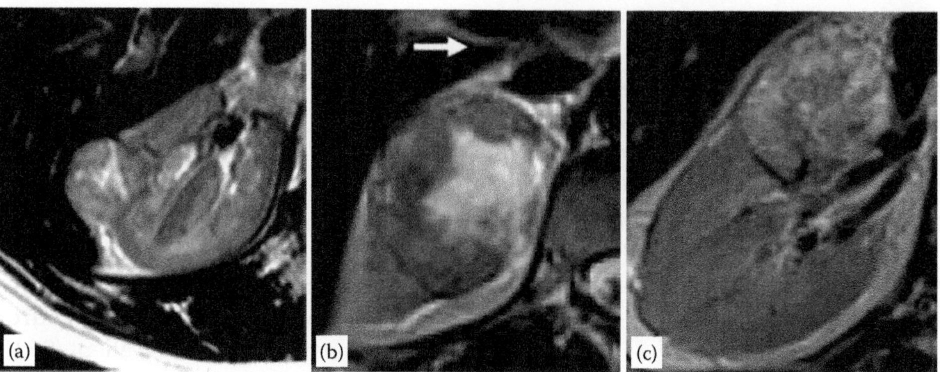

**FIGURE 1.19**
Renal tumor with central area on $T_2$-weighted imaging: small clear-cell RCC with central necrosis (a), large oncocytoma with central scar (b), and large heterogeneous clear-cell RCC (c).

and vimentin [119]. Interestingly, a central area corresponding to a central scar is morphologically found in 54% [117] to 80% [113] of RO. Unfortunately, this central area is classically considered as evocative of this tumor type [91], but is also observed in 30%–40% of chromophobe carcinomas corresponding to central necrosis [96] (Figure 1.19).

MRI of RO may show a low SI on $T_1$-weighted compared to renal cortex and a high SI on $T_2$-weighted compared to cortex renal, and may demonstrate a central scar in low SI and homogeneous enhancement after gadolinium injection [96]. If the maximal diameter is larger than 3 cm, RO often presents a heterogeneous aspect, which could correspond to a central scar. However, when a central area with high SI on $T_2$-weighted images is present within a tumor, the question of central necrosis within classical RCC still arises [120]. Considering the fibrous scars of hepatic focal nodular hyperplasia showing delayed enhancement on gadolinium-enhanced $T_1$-weighted sequences [121], recent studies have attempted to evaluate the central area with contrast-enhanced CT according to the gradual enhancement after contrast injection [122] and the wash-in and wash-out of the contrast agent [99]. Unfortunately, it offered only low sensitivity and specificity. More recently, a segmental enhancement inversion pattern was described within some renal tumors [123,124] as a tumor segment with a lower density at the arterial phase and a higher density at the early excretory phase. This last feature was observed in indeterminate small renal masses without scar and is associated according to Kim et al. [123] with differences in hypocellular hyalinized stroma content within the tumor parenchyma. However, these results were not confirmed by other studies [122,124]. Using MRI, Rosenkrantz et al. [96] observed a comparable contrast inversion in 28.6% of oncocytomas and 13.3% of RCC but without differentiating inversions within the tumor parenchyma, nor with contrast enhancement within the central scar itself. In fact, RO with central areas presented late central SI inversions within the central scar, responsible for a contrast inversion that was either complete or incomplete [3] (Figure 1.20). Adding a delayed (more than 5 min) post-gadolinium $T_1$-weighted allows to better characterizing these lesions. Such differences in the degree of SI inversion of central areas of renal masses were related to heterogeneity in vascularization of the tissue and the presence or absence of residual tumor cells [125]. In addition to a signal loss on opposed-phase images, double-echo chemical shift MR sequences cannot be detected in oncocytomas and chromophobe carcinomas [96], whereas a small amount of fat in clear-cell and papillary carcinomas was observed [111]. The combination of DCE $T_1$-weighted and double-echo gradient-echo MR images in tumors presenting with a central area compatible with central scar or central necrosis can accurately distinguish RO with high $T_2$ SI centrally from RCC [3]. Therefore, absence of any visual SI inversion of this central area and/or presence of decrease in SI on opposed-phase images should rule-out the diagnosis of oncocytoma.

#### 1.5.1.2 Malignant Renal Tumors

RCCs are responsible for 3%–5% of all new cancer diagnoses [126]. Imaging techniques able to differentiate the intrinsic characteristics and grade of RCC would thus be useful before any invasive explorations in order to propose the most appropriate management, particularly when malignant disease is suspected, or to improve the diagnostic accuracy of biopsies. With conventional morphologic MRI, Pedrosa et al. [127] predicted the subtype and grade of a variety of renal malignancies, including low-grade and high-grade cRCCs. The addition of functional parameters to obtain a multiparametric MRI (mpMRI) using chemical shift, diffusion-, and contrast-enhanced sequences may help to improve such differentiation [2,3,87,103]. It could prove clinically useful for

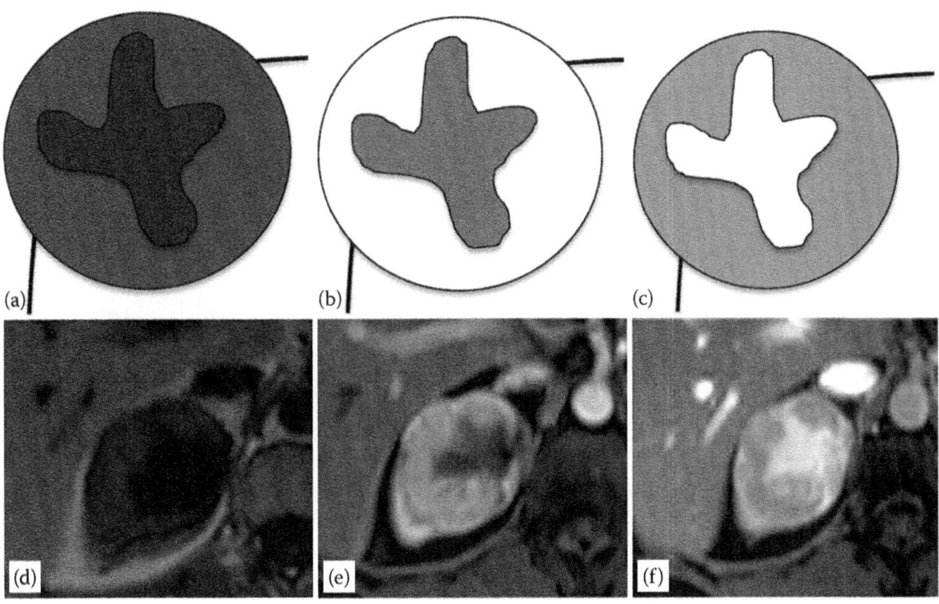


**FIGURE 1.20**
Evolution of the enhancement of the central scar of renal oncocytoma: representation of no contrast (a), arterial-phase (b), and late-phase imaging (c) and corresponding clinical images of a 62-year-old woman with renal tumor corresponding to oncocytoma (d–f). Dynamic axial $T_1$-weighted fat-suppressed images obtained at 40 s after gadolinium injection: peripheral tumor component shows early enhancement, whereas central area shows slow and progressive enhancement. Delayed (15 min) $T_1$-weighted water-select image shows signal intensity inversion between tumor and central area is complete, without discernible washout.

treatment planning in elderly patients with comorbidities in whom biopsy or surgery is high risk, or to promote active surveillance in eligible patients. Although mpMRI is not able to characterize renal masses overall, mpMRI appears useful to develop a diagnostic approach based on these results [2]. cRCCs and papillary RCCs (pRCCs) subtypes have been mainly studied, whereas for other less frequent subtypes additional explorations are mandatory.

#### *1.5.1.2.1 Clear-Cell RCC*

cRCC is encountered in 65%–70% of RCC diagnoses [128]. MRI aspect of cRCC depends on their size, fat content, or necrosis. In general, these tumors are large, with a specific aspect on $T_1$-weigthed sequences but often present a characteristic heterogeneous and high SI on $T_2$-weighted. Roy et al. [95] showed indeed that all cRCC ($n$ = 42) presented high but heterogeneous SI on $T_2$-weighted (Figure 1.19). This particular aspect on $T_2$-weighted allowed the diagnosis of cRCC with a 100% specificity and a 36% sensitivity among a group of 21 pRCCs and 28 cRCCs for Oliva et al. [94]. Intracytoplasmic lipid vacuoles are a typical histological finding in cRCC, which could be well seen on MRI with a loss of SI in solid part of tumors on $T_1$ out-phase sequences compared to on-phase sequences (Figures 1.18 and 1.21). This finding is observed in about 60% of cases. According to Pedrosa et al. [91,92], microscopic lipid content was associated with the cRCC diagnosis ($p$ = .001) in a 79 renal malignant tumor cohort (within 48 cRCC). This drop of SI in out-phase sequences allowed to correctly identify the tumor as an RCC with a sensitivity between 42% and 82% and a specificity between 94% and 100% depending on the study but may confound with minimal fat AML. Necrosis is also a classical finding in cRCC. In MRI, it corresponds to a heterogeneous and hyperintense area on T2 associated with a lack of enhancement (Figure 1.22). For Pedrosa et al. [91], necrosis was directly correlated with the size of lesion and with high nuclear grade cRCC ($p$ = .0001). Indeed, 93% of high-grade cRCC showed necrosis in MRI against 79% of low-grade cRCC, 20% of high-grade pRCC, and 10% of low grade pRCC. No correlation was found between size, nuclear grade, and the diagnosis of cRCC. Pseudocapsule is a hypointense rim on $T_1$-weighted and $T_2$-weighted surrounding a tumor, and its disruption is correlated with a high nuclear grade and with a local advanced pathology. According to Roy et al. [93], pseudocapsule was observed in 40 cRCC among 42 and its absence was associated with pT3 stage.

Epithelial tumor enhancements present similar aspect in MRI than CT scan and give useful information when trying to distinguish subtypes of RCC. The cRCCs usually present an intense and rapid enhancement, during the corticomedullary and nephrographic phases, which help to differentiate it from lower enhancement of pRCC and chromophobe RCC (chRCC) (Figure 1.23). On later phases, cRCC shows rapid washout of contrast, appearing

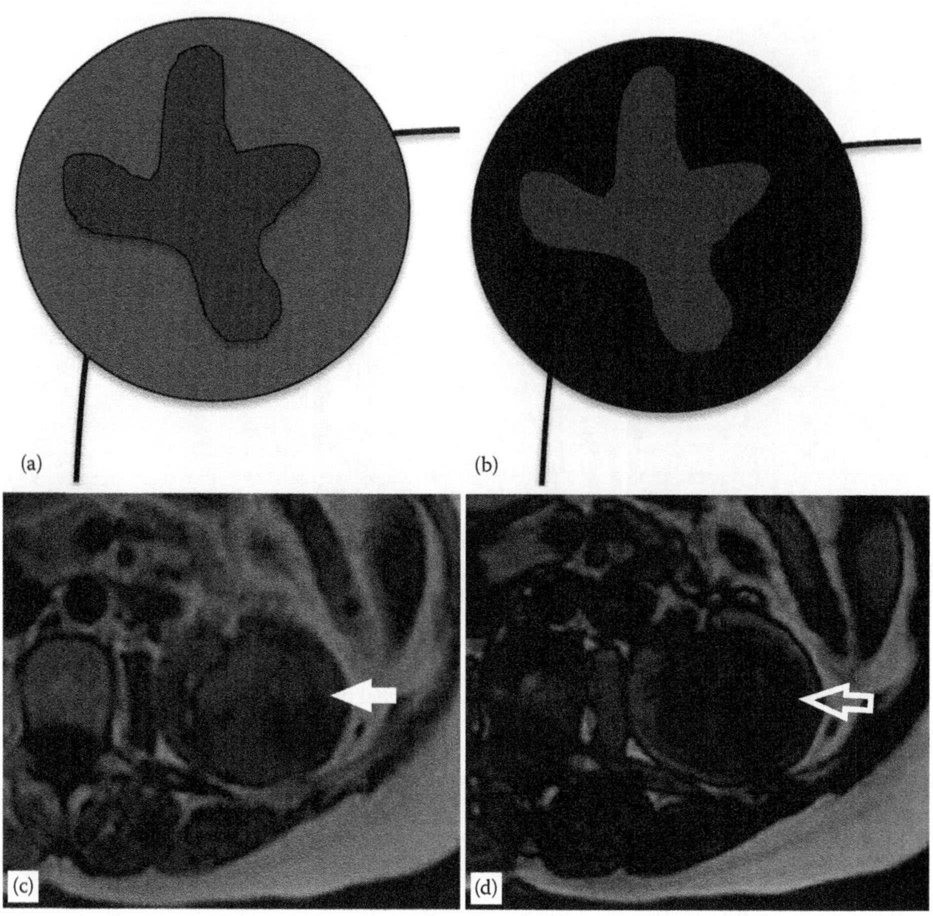


**FIGURE 1.21**
Typical espect of signal drop on dual chemical shift sequences: representation (a and b) and clinical images (c and d). Compared to on-phase sequences (a and c), a loss of signal intensity in solid part of tumors on T1 out-phase sequences is observed (b and d).

hypointense to the renal cortex by the excretory phase. Moreover, because of necrosis, cRCC enhancement is often heterogeneous [93]. DWI showed controversial results in case of cRCC because a wide range of ADC values has been reported [2,103–105]. However, ADC seems to be useful for establishing the tumor grade [129,130]. Other associated loco-regional morphologic criteria seem useful in order to characterize renal carcinomas. Pedrosa et al. [91] showed that thrombosis of renal vein associated with collateral retroperitoneal veins were correlated with high-grade cRCC ($p$ = .0001) (Figure 1.24). Using such features, MRI provides a 92% sensitivity and an 83% specificity for the diagnosis of cRCC.

#### 1.5.1.2.2 Papillary RCC

pRCC accounts for 10%–15% of RCC [128]. Two cellular subtypes of pRCC (types 1 and 2) have been identified but differing in terms of prognosis. No study tried to differentiate such subtypes with imaging. pRCC is most frequently located in the external part of renal parenchyma and is smaller than other carcinomas. Peripheral location was significantly associated with pRCC ($p$ < .05) according to Pedrosa et al. [92]. This tumor type has a lower SI on $T_2$-weighted images than other RCCs [95,111,112]. Roy et al. [95] showed that pRCCs have in general lower SI on $T_2$-weighted and homogeneous or discretely heterogeneous aspect compared to normal renal parenchyma (Figure 1.25). This low SI is probably related to a paramagnetic effect due to intracytoplasmic or interstitial hemosiderin deposit. Among 21 pRCC and 28 cRCC, $T_2$-weighted low SI in a tumor provided specificity from 96% to 100% and sensitivity from 46% to 58% for the diagnosis of pRCC according to Oliva et al. [94]. By using a $T_2$ ratio (IStumoral/IScortical), a value less than or equal to 0.93 (median) provided an 86% specificity and a 96% sensitivity for the diagnosis of pRCC. Similar to cRCC, pRCC presents a pseudocapsule in T1 and T2 as a peripheral rim in low SI. Roy et al. [93] found pseudocapsule in 51 pRCC among 55. The four tumors without this aspect was only of pT1 grade, with a marked low SI on $T_2$ sequence which probably mask the rim.

pRCCs can be distinguished from all other renal tumors, including AMLs, which also show a low

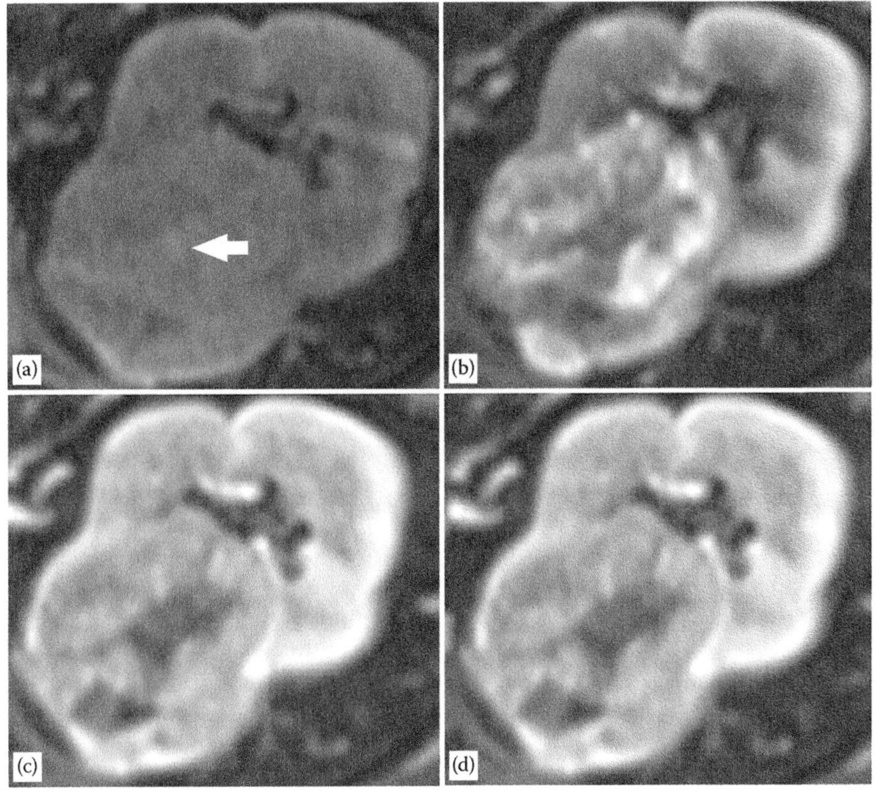


**FIGURE 1.22**
Area of necrosis in a clear-cell RCC before (a) and after enhancement (b–d) on $T_1$ sequences: necrosis can present spontaneously a high signal on $T_1$-weighted images (arrow) whereas no enhancement may be observed thereafter.

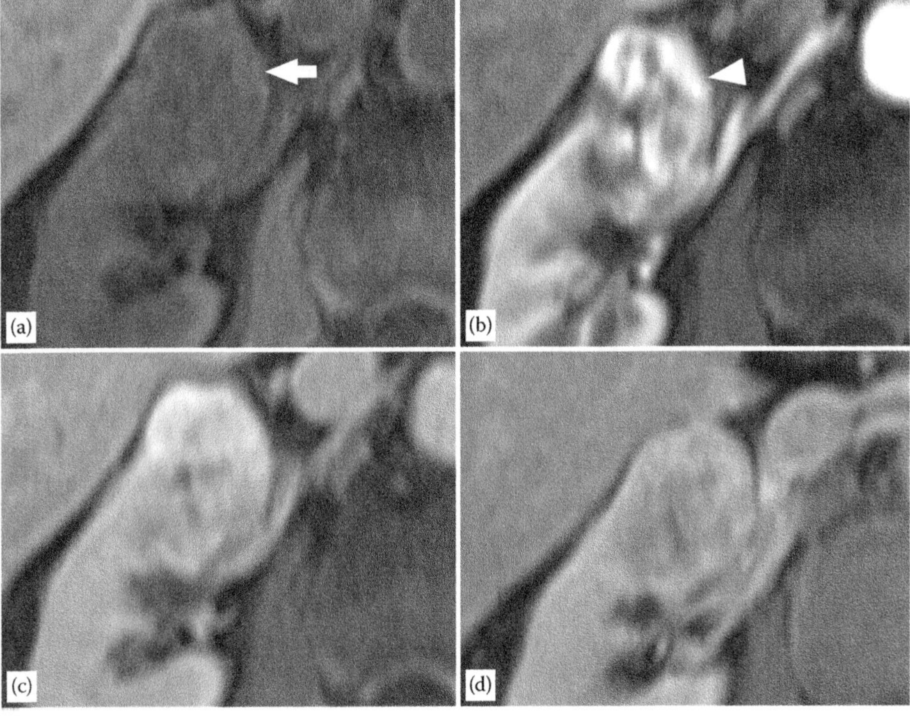


**FIGURE 1.23**
Typical enhancement of clear-cell RCC (a, arrow): enhancement is rapid, intense at the arterial phase but heterogeneous (b). A wash-out may be observed (c and d).

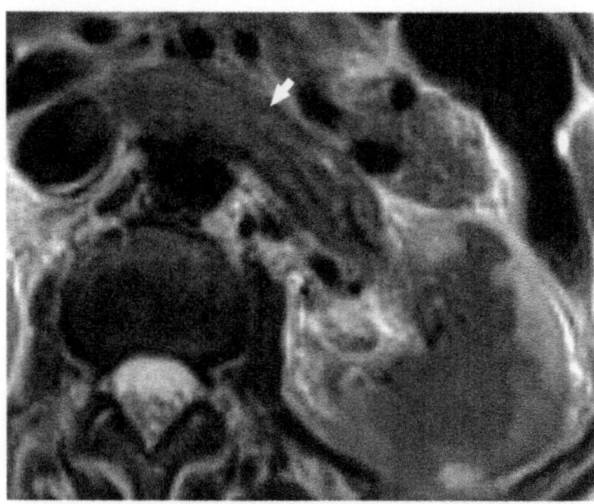

**FIGURE 1.24**
High-grade clear-cell RCC associated with a thrombosis of left renal vein (arrow) extended to inferior vena cava.

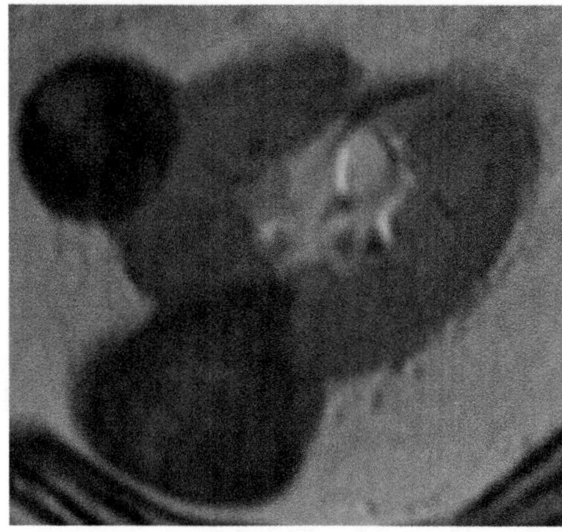

**FIGURE 1.25**
Classical peripheral location and hypointense aspect on $T_2$-weighted images of two synchronous papillary RCC.

$T_2$ SI, based on its evocative low enhancement. DCE sequences appear therefore helpful, showing low and slow enhancement in pRCCs [95] (Figure 1.26). Arterial enhancement of pRCCs was significantly lower than for minimal fat AMLs [2,98]. This is concordant with the already described low enhancement of papillary tumors after injection [131] and high arterial wash-in of AMLs [100]. Roy et al. [95] demonstrated that pRCC enhancement was delayed compared to cRCC, with maximal enhancement peak observed at 250 s (±45) against 56 s (±35) for cRCC. After injection, according to Pedrosa et al. [91], on all phases, pRCC enhanced lower than cRCC when renal cortex was chosen as reference standard ($p = .001$). These results match with those presented by Sun et al. [131] who studied the mean enhancement of cRCC, pRCC, and chRCC at corticomedullary and nephrographic phases after contrast injection in MRI. All differences were significant ($p < .005$) except between chRCC and cRCC at nephrographic phase. ADC of pRCCs may be also discriminant with lower values of ADC than other renal tumors such as oncocytomas or cRCCs [2,105].

### *1.5.1.2.3 Chromophobe RCC*

chRCC is the third most common subtype and accounts for 6%–11% of RCC [128]. No MR features are accurately evocative of chRCC. Morphologically, a central scar is found in 30%–40% of chromophobe carcinomas [96]. A microscopic fat content evaluated with MRI using chemical shift double-echo sequences [109] is exceptional, such content was reported visually in only one case of a chRCC [96]. These results support the findings of Krishnan et al. [132] who noted that only minimal lipid could be detected with electron microscope in chRCC. With DCE $T_1$-weighted images, the diagnosis should be suspected in masses with intermediate enhancement during the corticomedullary phase. A segmental SI inversion of the central area of necrosis may be observed in 13.3%–26.7% of chRCC according to each reader [96] but remains often incomplete [3]. Multiparametric MRI may help to differentiate chRCC from other tumor by studying the wash-in and the wash-out after enhancement [2]. These later parameters seem lower than in case of oncocytoma or cRCC.

### *1.5.1.2.4 Fuhrman Grade*

Although the Fuhrman scoring system is the widely accepted grading system for RCC [133], a wide variability has been observed in the grading accuracy of biopsies due to its subjectivity (46%–76%), but also in fine needle aspiration (28%–72%) [134–137]. Furthermore, the rate of non-diagnostic biopsies is non-negligible, ranging between 5% and 40% [138]. Based on morphologic consideration, size was significantly associated with a higher grade for cRCCs but not for other subtypes [139–142]. In a study of 1867 patients, the distribution of the number of low-grade tumors showed a decreasing trend with the diameter becoming larger in the different tumor subgroups, while the distribution of high-grade tumors showed an opposite trend, and the changes between different subgroups were mostly significant ($p < .05$) [143]. However in the same study, logistic regression analysis predicted that the odds of papillary, chromophobe, and other types versus clear cell decreased as the tumor size increased. In a study of 2675 patients with cRCCs, incidence of high-grade lesions (Fuhrman 3–4) increased from 0% to 59% between tumors of 1 cm and those greater than 7 cm [139]. With conventional morphologic MRI, Pedrosa et al. [127] predicted the subtype and grade of a variety of renal malignancies, including low-grade and high-grade cRCCs. The addition of functional parameters in a multiparametric MRI using chemical shift, diffusion-, and contrast-enhanced sequences may help to improve such differentiation [2,3,87,103]. High tumor grade (grade $\geq 3$) may be associated with larger

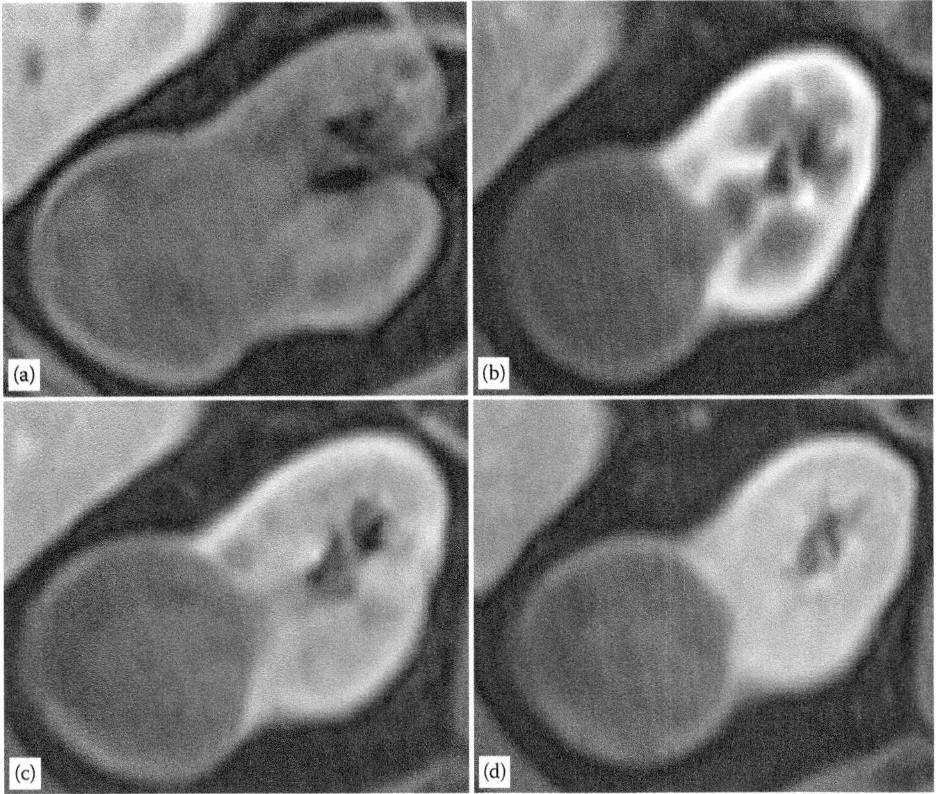


**FIGURE 1.26**
Typical enhancement of a papillary RCC after injection: $T_1$ sequences without injection (a) and then after injection at arterial (b), parenchymal (c), and excretory phases (d).

size, lower parenchymal enhancement, and lower ADC for cRCC. However, for pRCCs or chRCCs, studies failed to establish a preoperative noninvasive assessment of tumor aggressiveness by identifying any significant associations, given that the use of the Fuhrman classification seems better applied to cRCCs than to other subtypes [128,144–146].

Rosenkrantz et al. [103] reported significantly higher ADCs for low-grade tumors (Fuhrman grade ≤ 2) compared to high-grade tumors (Fuhrman grade ≥ 3) ($2.24 \pm 0.50$ vs. $1.59 \pm 0.57$ mm$^2$/s in mean for ADC-400, $p < .001$; $1.85 \pm 0.40$ vs. $1.28 \pm 0.48$ mm$^2$/s for ADC-800; $p < .001$). Goyal et al. [130] concluded that a decreasing trend in ADC values was seen with increasing Fuhrman grade. In this study, the mean ADC of high-grade RCC was significantly lower than low-grade RCC (1.3145 vs. $1.6982 \times 10^{-3}$ mm$^2$/s in mean, $p = .005$). Previously, Sandrasegaran et al. [102] observed no significant differences. These results highlighted the fact that diffusion-weighted MRI may be helpful to establish the aggressiveness of cRCCs [103], as proposed for the Gleason score in prostate carcinoma [147,148]. However in contrast to the Gleason score which is mainly related to cellularity, a correlation with the Fuhrman nuclear grading system requires that ADC must also reflect the tissue properties [103]. As proposed by Rosenkrantz [103], an ADC map may help to adequately target regions within the renal tumor that indicate greater aggressiveness by avoiding regions with low ADC ratios (or ADC) but prospective studies are now mandatory to test this hypothesis.

#### *1.5.1.3 Transitional Cell Carcinomas*

To evaluate transitional cell carcinomas, intravenous urography, CT, and endoscopy are usually performed. However, when CT and endoscopy are not feasible, MRI may be obtained (Figures 1.16 and 1.27). This is particularly the case for patients with poor renal function, in case of ureteral obstruction, or when contrast excretion may be too limited to allow tumor detection. Moreover, the ureter may not be accessible for endoscopy because of fibrosis and stricture of the ureter and ureter ostium or due to previous surgical reconstruction. MRU has several other advantages, including better tissue contrast resolution, greater sensitivity for urothelial enhancement, and the ability to provide additional information regarding tissue properties on the basis of $T_1$ and $T_2$ signal characteristics [149]. However, MRU may have lower sensitivity than CT urography in detecting

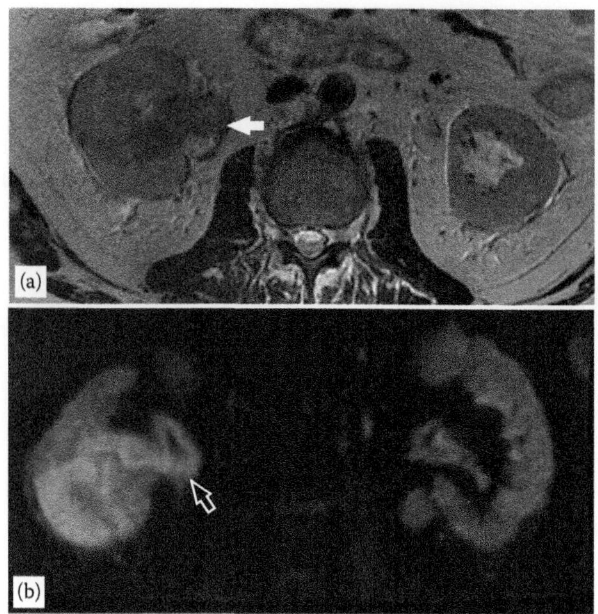


**FIGURE 1.27**
Axial $T_2$-weighted (a) and coronal DWI (b) showed a heterogeneous tumor of the lower part of the right kidney and involving the renal pelvis. It corresponds to an urothelial tumor.

urothelial lesions [150]. The primary reasons for this are the inherent lower spatial resolution of MRI, its greater vulnerability to respiratory motion artifact, as well as the frequent presence of some degree of susceptibility artifact, which could be limited by the use of new contrast agent [150].

Gadolinium-enhanced nephrographic and pyelographic phase images are the most important sequences of MRU for the detection of small urothelial carcinomas [151]. Small urothelial carcinomas appear as focal areas of enhancement or as a segmental area of diffuse thickening and enhancement of the urinary tract wall on nephrographic phase images. On pyelographic phase images, focal filling defects may be observed. Tumors may be visible on only one of the two sequences, particularly in the ureter. Since normal wall of the upper urinary tract only enhances minimally during the nephrographic phase, it is often difficult to follow the entire course of the upper urinary tract with the nephrographic phase images alone. Infiltrative tumor often appears as a large heterogeneously enhancing mass. Enhancing vessels could be mistaken for urinary tract. Therefore, it may be helpful to evaluate both nephrographic phase images and pyelographic phase images simultaneously with the images synchronized on a workstation. In the study of Chahal et al. performed in 23 patients with high clinical suspicion of upper tract transitional carcinoma and hydronephrosis that could not be explained with other imaging modalities, MRU showed 5 renal pelvic transitional cell carcinomas and 8 ureteral transitional cell carcinomas, confirmed by histology. In the remaining patients, no sign of transitional cell carcinoma was observed during 1-year follow-up [152]. When the lesions are obstructive, MRU can detect upper tract urothelial carcinoma with an accuracy of 88% or higher using gadolinium-enhanced pyelographic phase or $T_2$-weighted single-shot fast spin-echo MRU.

Although MRI seems limited for the staging of upper tract urothelial carcinoma, extension of tumor outside the wall of the urinary tract (T3) appears as an irregularity of the tumor border or disruption of the intensely enhancing rim of the ureteral wall on 2–3 min delayed $T_1$-weighted images [153].

According to the appearances of upper tract urothelial carcinoma, differential diagnoses of MRU findings can be categorized into three patterns [151]. First, filling defect may correspond to stone, blood clot or debris, fungus ball, sloughed papilla, and gas. Presence of enhancement is the most important finding to differentiate tumor from these other causes. However, absence of identifiable enhancement does not exclude tumor and often follow-up imaging or further workup is required [154]. In addition, some specific signs must help to characterize the abnormalities. The signal of the filling defect on $T_2$-weighted and $T_1$-weighted pyelographic phase images in a case of tumor tends to be similar to that of soft tissue, whereas signal in a case of stone tends to be signal void. Spontaneous increased signal on unenhanced spoiled $T_1$-weighted is suggestive of a blood clot [77]. Filling defects due to concentrated gadolinium contrast material often appear as multiple filling defects in the central portion of the calyces at early (5 min) pyelographic phase image. A further delayed image should clarify the nature of filling defects [155]. Flow-related artifact is often seen on $T_2$-weighted hydrographic images [9]. Blood vessels may also appear as a filling defect due to signal void. Second, diffuse enhancement or thickening of the urinary tract wall may be related to inflammation in case of recent instrumentation or reflux (such as in case of neobladder and urinary diversion), urinary stasis, ureteral stone, or intravesical chemotherapy for bladder tumor may also cause similar findings. The wall of the urinary tract may falsely appear thickened due to breathing motion or ureteral peristalsis. A small amount of gadolinium contrast material in the collecting system from a test bolus injection may also simulate thickening or enhancement of the urinary tract wall. Finally, the differential diagnoses of infiltrative neoplasm of the upper urinary tract include RCC, and lymphoma or non-neoplastic causes such as retroperitoneal fibrosis, endometrioma, extramedullary hematopoiesis, and inflammatory mass such as xanthogranulomatous pyelonephritis.

### 1.5.2 Therapeutic Evaluation

In recent years, the therapeutic evaluation with imaging has grown significantly due to the evolution of techniques and practices. However, whereas imaging is now fully integrated into the course of patient care and based primarily on the study of morphological criteria, the use of MRI remains low compared to CT scan. As the response criteria based on tumor mensuration (WHO, RECIST) have reported their limits, especially since the introduction of targeted therapy, the use of MRI may increase in the future due to its advantages.

For this purpose, the concept of biomarkers of a disease can be implemented for a number of cases in functional and molecular MRI. But this imagery extends from basic research to clinical applications and only a few techniques have matured to consider its application in daily clinical practice. Among all the biomarkers actually reported (such as the *radio-genomic* for kidney tumors [156], for example), biomarkers promoting angiogenesis, such as vascular endothelial growth factor, appear to be a more specific evaluation of potential target in imaging. These proteins are expressed by tumor cells in response to hypoxia, nutritional stress, or acidosis [157]. The advantage of targeting this pathway is that it has been reported that this biomarker is very common in cancer cell lines as to grow beyond 2 $mm^3$, a tumor must acquire an angiogenic phenotype [158]. Functional MRI seems therefore particularly interesting for the early evaluation of targeted therapies. Indeed, this MRI can be more focused on the result of the mechanism of the drug's action as is the case for the evaluation of tumor perfusion techniques during anti-angiogenic therapy. This is essential to better understand the disease evolution, to limit the duration of treatment, and consequently the side effects and costs. Sahani et al. [159] reported that functional imaging was more sensitive to predict the tumor evolution's simple morphological criteria.

Functional MRI is able to evaluate the blood flow and the microvascular permeability ($K_{trans}$) which may precede the decrease in size of a renal lesion [160] after anti-angiogenic (anti-vascular endothelial growth factor) or anti-tyrosine kinase epidermal growth factor receptor therapies [161] (Figure 1.28). However, at this time there are only few published series and radio-pathological correlations. There is a significant increase in the intensity of the signal T1 renal carcinomas ($p < .0001$) and a significant decrease in enhancement ($p < .0001$) after sorafenib [162], which can be appreciated by Choi criteria. Although the RECIST criteria may significantly decrease after treatment ($p = .005$) for responding patients, stable disease on early evaluation may be observed thereby limiting the impact of these criteria. Using DCE-MRI, Desar et al. [163] also evaluated the early effects of sunitinib in patients and reported significant decrease in relative blood volume of the renal tumors (RBV) and blood flow (RBF) at 3 days ($p = .037$ and $p = .018$) and 10 days ($p = .006$ and $p = .009$). Despite these promising results, Hahn et al. [164] reported after treatment (sorafenib) that the changes of area under the curve (AUC) after injection and $K_{trans}$ were not associated with progression-free survival. Only patients with an initial value elevated $K_{trans}$ had better progression-free survival ($p = .027$). Thus in kidney cancer, permeability ($K_{trans}$) could be used as a prognostic factor [161,164] while its impact in therapeutic evaluation must still be evaluated. Alternatively to DCE-MRI, the development of the technique of arterial spin labeling may be an alternative for such patients not tolerating the contrast (i.e., with renal failure or

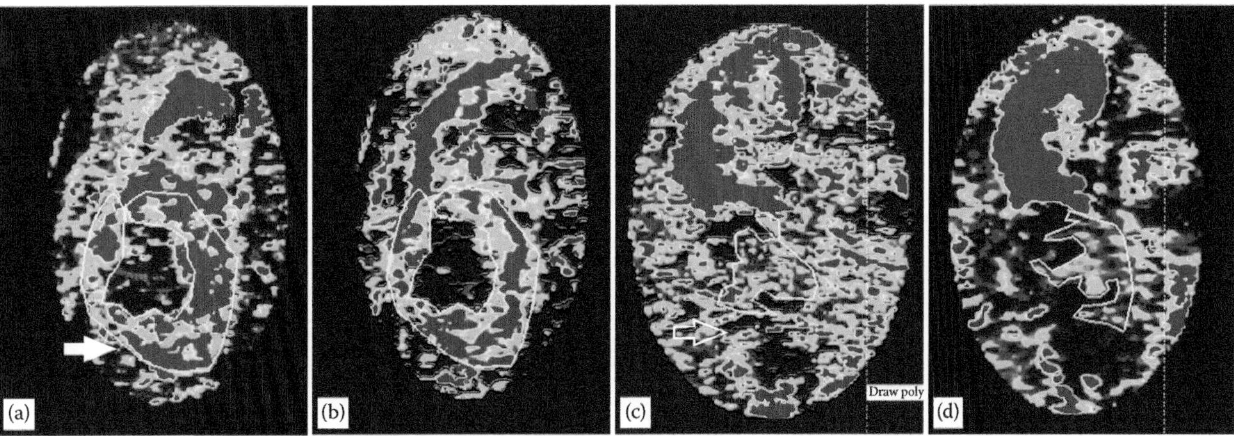


**FIGURE 1.28**
Evaluation after anti-angiogenic drugs of a left renal tumor (arrow): the blood flow of the tumor decreases at 6 months (a and c) as well as the $K_{trans}$ (b and d) (red: high value; blue: low value).

allergy) [165]. Low values of perfusion were correlated with decreased sensitivity to treatment. In a mouse model, it was observed for responders significant variations in blood flow within 30 days after initiation of treatment with sorafenib [166].

Further, diffusion-weighted sequences allow assessing the micro-architecture of the tumors by studying the movement of water molecules they contain over a defined time [167,168]. Desar et al. [163] reported a significant increase in ADC of renal tumors at 3 days ($p$ = .015) followed by a decrease in the baseline at 10 days ($p$ = .001). Moreover, as $K_{trans}$, it has been reported that high-value pre-therapeutic ADC liver metastases of RCC are associated with poor prognosis [169]. MRI with BOLD effect appears to be a sensitive technique for the study of tumor hypoxia [170], but perhaps lacking specificity (Figure 1.29). Several feasibility studies have already proposed this noninvasive technique to map the prostate tumor hypoxia for example [171–174] but need to be evaluated in the kidney.

However, there still remains a lack of standardization of protocols for all these functional techniques. Heye et al. [175] reported by testing the reproducibility of quantitative and semi-quantitative measurement of pharmacokinetic parameters infusion uterine fibroids on different systems, significant differences existed in mean values after normalization. The interobserver agreement was 48.3%–68.8% for $K_{trans}$, 37.2%–60.3% for $k_{ep}$, 27.7%–74.1% for the fifth, and 25.1%–61.2% for the AUC. The intra-class correlation coefficient was low to moderate ($K_{trans}$: 0.33–0.65; $K_{ep}$: 0.02–0.81; $V_e$: −0.03 to 0.72, and AUC: 0.47–0.78). Moreover, post-processing is not homogeneous and often dependent on teams, whether at the level of calculation models (not identical), data collection (ROI in 2D vs. 3D [101]), of performing a normalization. Further evaluations are now mandatory to validate these findings but also to propose a standardization of the techniques.

## 1.6 MR-Guided Procedures on Kidneys

In recent years, *nephron sparing* therapies have become more common for the treatment of small renal tumor although the gold standard is still the surgery [176]. Whereas thermal ablations were initially realized under direct surgical and laparoscopic visualization [177], percutaneous image-guided ablations for RCC are now an important treatment option for many patients [178–180]. The primary ablation techniques are radio-frequency ablation (RFA) and cryoablation. Currently, most centers in the world use CT to guide the ablation of renal tumors [181,182]. In most of the cases, CT is perfectly adapted. Nevertheless in some cases CT combine or not with US is not an ideal imaging guidance modality. For example, the density of renal tumor may be similar to the normal parenchyma on unenhanced CT, and in patients with compromised renal function, iodine contrast may be not permitted. Furthermore when numerous needles are introduced such as in case of cryoablation, CT guiding may expose patients to substantial amounts of ionizing radiation, which could be problematic in few cases. MRI may overcome these limitations because MRI provides an excellent three-dimensional field of view and best natural contrast between tumor and normal parenchyma. Moreover, MRI is temperature sensitive and can be used to monitor thermal ablation [183]. In these situations, MRI guidance can be a very useful alternative.

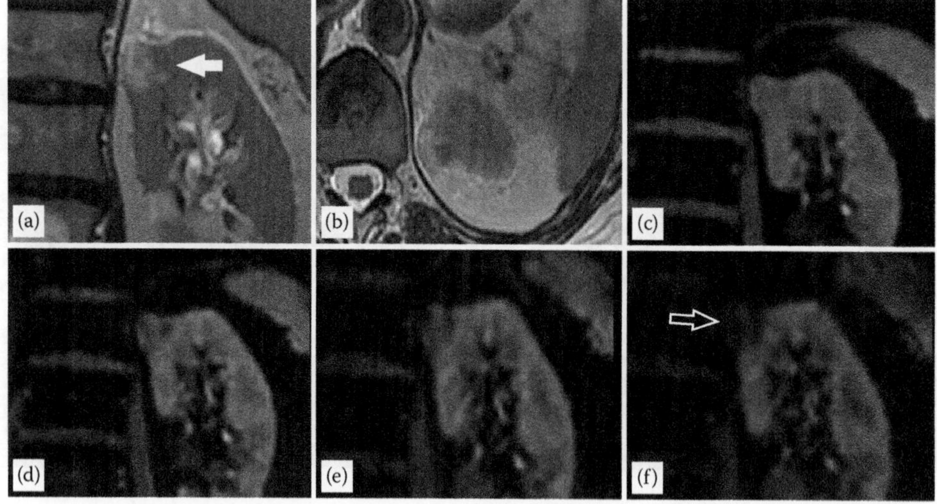


**FIGURE 1.29**
Exploration of a small renal tumor with BOLD imaging (arrow). The tumor is well seen on $T_2$W imaging (a, coronal; b, axial). On BOLD imaging (c–f, various acquisitions over time), the drop of signal is higher in the tumor than in the other parts of the kidney (arrow).

In particular, MRI is the only guidance technique that allows a real-time monitoring ablation [184]. As the interferences between the RFA ablation system and the MRI scanning sequences are consequent (but may overcome by scanning between the ablation cycle [185]), cryoablation is the principal technique used under MR guidance.

In the particular case of renal cryoablation, MRI seems to be the best modality to visualize the frozen tissue. The MR images should be acquired with a scan time that is short enough to monitor the dynamic processes of freezing and thawing. MRI can visualize not only the reversible temperature changes in tissue, but also the tissue damage left behind after treatment. Therefore, two sorts of sequences can be used. The $T_2$-weighted multislice half-Fourier acquired single-shot turbo spin echo (HASTE) or $T_2$-like contrast, steady-state free precession (SSFP) acquisitions can be used for planning and targeting. These sequences are also used during the procedure to verify the position of the applicator [184,186]. For ice ball monitoring, multiplanar images using either the bSSFP or HASTE are collected every 3 min. When two or more cryoprobes are used, HASTE sequence is preferred because of the magnetic susceptibility-induced signal losses, which degraded the quality of the bSSFP acquisition. Real-time imaging using bSSFP acquisitions is used to advance the probes to the desired location. During the procedure, scanning approximately every 2 min is usually adequate to observe changes in the slow iceball growth or the thermal ablation zone after RFA appearing hypointense on $T_2$-weighted. If needed, scanning more frequently could be useful to avoid the damaging of noble surrounding structures. The post-ablative control shows a hypointense T2 ablation area and a hyperintense rim corresponding to the surrounding edema [184]. Local residual tumor contingent can be easily detected on the late control, and complementary treatment can be realized without injection of contrast media.

However, the use of MRI for interventional procedures is still at this time limited by several factors. Performing thermal ablation under MR guidance requires specific material, and interventional MRI suite is not yet a classical component of radiology department. As the space is limited and as the procedures often require to be performed under general anesthesia, the access to the patient may be challenging. Some centers experimented the cryotherapy under local anesthesia with or without local sedation in order to overcome this limitation.

The first experimental trial on MRI-guided cryoablation of kidney tumor has been reported in 2001 with satisfactory results [187,188]. The first studies were performed using open-configuration MRI system with field strengths of only 0.2–0.5 T. In order to have the full benefit of the technique, recent studies report now the feasibility of real-time placement and manipulation of cryoprobes during MRI-guided cryoablation in a closed-bore high magnetic field scanner. In fact, open-configuration MRI allows an easy access to the patient with less good resolution images [188]. The best compromise is probably closed-bore magnet design with wide tunnel [186] or with moving magnet.

As reported by Ogan et al. [189], whatever the technique employed for the ablation of renal tumors under image guidance, margins should be of 3 up to 10 mm to achieve effective ablation [190]. The anatomical location of the kidney should lead to a close inspection of critical structures surrounding the tumor (such as ureters, genitofemoral, or iliofemoral nerves; the psoas; small and large bowel; and the adrenal gland), which could be obtained accurately with MRI [191]. MRI guidance authorizes the same protection procedures than a classical CT guidance. Hydrodissection (5%–30% dextrose ± contrast) or gas dissection (with $CO_2$) can be injected between ablation area and vulnerable structures [192,193]. These techniques permit organ displacement during the renal ablation but explicit attention should be paid to the saline or gas dissipation during the procedure to maintain a security distance [194–196].

## 1.7 MRI Follow-Up after Renal Percutaneous Ablations

After any renal ablations, the aim of the imaging follow-up is to detect a residual tumor or a recurrent malignancy. The control rhythm is not standardized, and there is no clinical trial that has validated a particular imaging scheduled. However, long-term multimodalities surveillance is commonly admitted by most of the investigators. An early control by CT or MRI at 1 or 3 months to be sure that no residual tumor remains is precolonized [179,180]. Follow-up scheduled depends on the clinical conditions of the patient and the comorbid conditions. Usually, the controls' intervals are of 6–12 months during at least 3 years [178,197–199] and yearly thereafter. The relevance of direct post-ablation control within 24 or 48 h is not assessed.

There are two types of imaging findings that are identified after an ablation procedure: those related to zones of decreased perfusion and those in which the SI (at MRI), echogenicity (at US), or attenuation (at CT) is altered. Although there still exists controversy for the terminology of local recurrence versus treatment failure, it is currently admitted that the procedure is considered to be technically successful if the lesion treated did not enhance just after contrast material administration or at least at the first examination followed by the

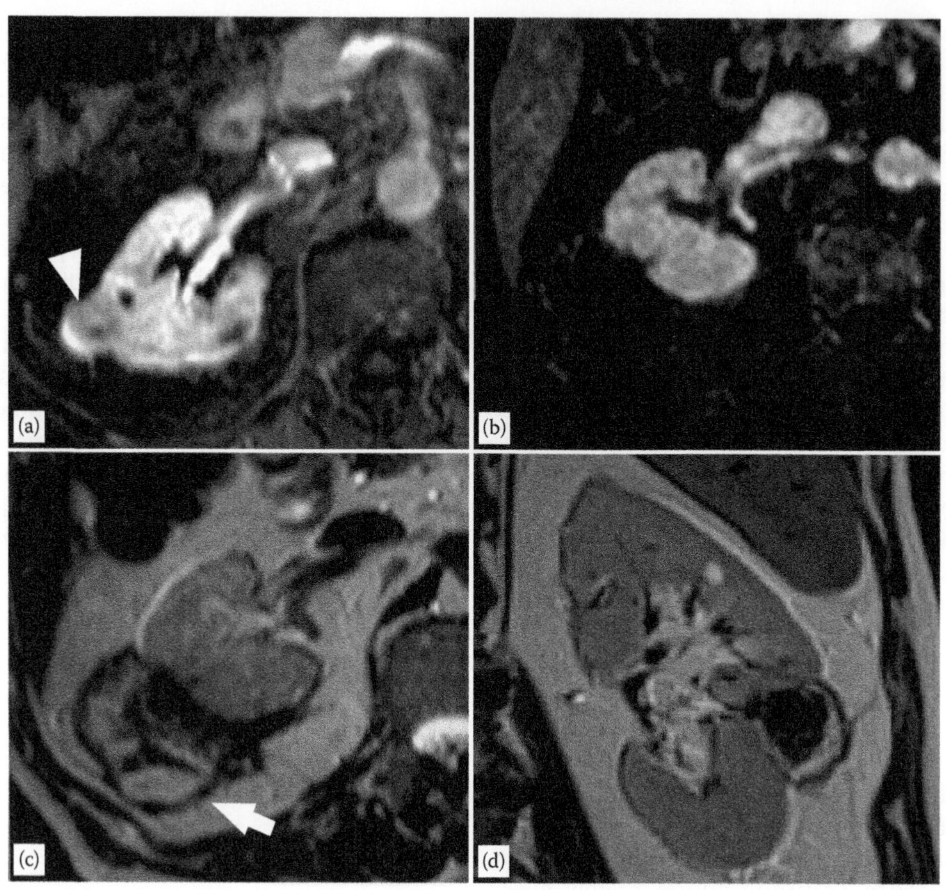


**FIGURE 1.30**
MRI of a clear-cell RCC before (a) and after (b–d) RFA: no residual enhancement is visualized after injection (b), whereas $T_2$-weighted images showed post-ablation aspect with a low-signal intensity ring (arrow) (c, d).

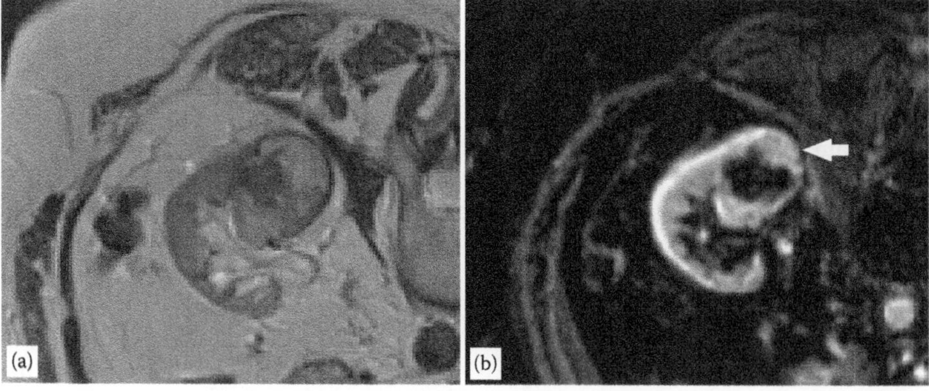


**FIGURE 1.31**
MRI performed 3 months after RFA for clear-cell RCC: $T_2$-weighted (a) and postcontrast $T_1$-weighted (b) images showed residual disease (arrow).

ablation (Figure 1.30). Thereafter, to assess the diagnosis of local recurrence, which corresponds to new nodular enhancement on cross-sectional imaging, examination with contrast injection must be performed (Figure 1.31). Enhancement is considered significant when MRI was greater than 16% on the follow-up examinations [200,201]. For cryoablation, a low-SI ring surrounding the ablative zone in $T_2$-weighted sequence could be observed and seems to be useful for assessing the recurrence.

In addition, it is particularly relevant to have special attention on ablative margins after renal ablation. As for other organs, the ablation of appropriate margins beyond the borders of the renal tumor is necessary to achieve complete tumor destruction. On follow-up, an irregular enhancement may be indicating incomplete local treatment (i.e., residual unablated tumor). While it corresponds to a 0.5–1.0-cm-wide region surrounding the tumor in other organs, this region could be the smallest in the kidney in case of multiple tumor ablations or when nephron-sparing therapies are desired to preserve renal function and avoid dialysis. It is important to note that a transient benign enhancement surrounding the ablative site may also occur and can also mask a residual lesion. This finding can be seen at both pathologic examination and contrast-enhanced imaging and typically suggests a benign physiologic response to thermal injury (initially, reactive hyperemia; subsequently, fibrosis and giant cell reaction). It is most easily appreciated on the arterial phase CT scans, with persistent enhancement that is often seen on delayed MR images but depending on the protocol used for contrast-enhanced imaging. This transient finding can be seen immediately after ablation and can last for up to 6 months after ablation. On the other hand, involution of ablation zone can also occur over time in particular in case of cryoablation. In case of residual tumor or recurrent malignancy, the patient should be referred to the multidisciplinary commission to consider an additional treatment (repeat ablation and surgery).

## 1.8 Conclusion

Compared to other techniques such as US or CT scan, MRI has additional diagnostic value. By allowing concomitantly morphologic and functional evaluation without radiation exposure or iodine contrast injection, the role of MRI for the evaluation of renal, renal pelvis, and ureters may become central. MRI is a suitable tool in the preoperative work-up, to guide therapeutic procedure or to evaluate the therapeutic response. Through further developments, an increasing role for multiparametric MRI in the management of patients with renal or upper urinary tract diseases can be expected.

## References


1. Zhang H, Prince MR. Renal MR angiography. *Magn Reson Imaging Clin N Am*. 2004;12(3):487–503.
2. Cornelis F, Tricaud E, Lasserre AS, Petitpierre F, Bernhard JC, Le Bras Y et al. Routinely performed multiparametric magnetic resonance imaging helps to differentiate common subtypes of renal tumours. *Eur Radiol.* 2014;24(5):1068–80.
3. Cornelis F, Lasserre AS, Tourdias T, Deminiere C, Ferriere JM, Bras YL et al. Combined late gadolinium-enhanced and double-echo chemical-shift MRI help to differentiate renal oncocytomas with high central T2 signal intensity from renal cell carcinomas. *AJR Am J Roentgenol*. 2013;200(4):830–8.
4. Kuo PH, Kanal E, Abu-Alfa AK, Cowper SE. Gadolinium-based MR contrast agents and nephrogenic systemic fibrosis. *Radiology*. 2007;242(3):647–9.
5. Broome DR, Girguis MS, Baron PW, Cottrell AC, Kjellin I, Kirk GA. Gadodiamide-associated nephrogenic systemic fibrosis: Why radiologists should be concerned. *AJR Am J Roentgenol*. 2007;188(2):586–92.
6. Wertman R, Altun E, Martin DR, Mitchell DG, Leyendecker JR, O'Malley RB et al. Risk of nephrogenic systemic fibrosis: Evaluation of gadolinium chelate contrast agents at four American universities. *Radiology*. 2008;248(3):799–806.
7. Becker S, Walter S, Witzke O, Kreuter A, Kribben A, Mitchell A. Application of gadolinium-based contrast agents and prevalence of nephrogenic systemic fibrosis in a cohort of end-stage renal disease patients on hemodialysis. *Nephron Clin Pract*. 2012;121(1–2):c91–4.
8. Daftari Besheli L, Aran S, Shaqdan K, Kay J, Abujudeh H. Current status of nephrogenic systemic fibrosis. *Clin Radiol*. 2014;69(7):661–8.
9. Leyendecker JR, Barnes CE, Zagoria RJ. MR urography: Techniques and clinical applications. *RadioGraphics*. 2008;28(1):23–46; discussion -7.
10. Kawashima A, Glockner JF, King BF. CT urography and MR urography. *Radiol Clin N Am*. 2003;41(5):945–61.
11. Nolte-Ernsting CC, Tacke J, Adam GB, Haage P, Jung P, Jakse G et al. Diuretic-enhanced gadolinium excretory MR urography: Comparison of conventional gradient-echo sequences and echo-planar imaging. *Eur Radiol.* 2001;11(1):18–27.
12. Schoenberg SO, Bock M, Kallinowski F, Just A. Correlation of hemodynamic impact and morphologic degree of renal artery stenosis in a canine model. *J Am Soc Nephrol*. 2000;11(12):2190–8.
13. Schoenberg SO, Rieger JR, Michaely HJ, Rupprecht H, Samtleben W, Reiser MF. Functional magnetic resonance imaging in renal artery stenosis. *Abdom Imaging.* 2006;31(2):200–12.
14. Schoenberg SO, Knopp MV, Londy F, Krishnan S, Zuna I, Lang N et al. Morphologic and functional magnetic resonance imaging of renal artery stenosis: A multireader tricenter study. *J Am Soc Nephrol*. 2002;13(1):158–69.
15. Bock M, Schoenberg SO, Schad LR, Knopp MV, Essig M, van Kaick G. Interleaved gradient echo planar (IGEPI) and phase contrast CINE-PC flow measurements in the renal artery. *J Magn Reson Imaging*. 1998;8(4):889–95.
16. Hussain SM, Kock MCJM, Ijzermans JNM, Pattynama PMT, Hunink MGM, Krestin GP. MR imaging: A "one-stop shop" modality for preoperative evaluation of potential living kidney donors. *RadioGraphics*. 2003;23(2):505–20.
17. Prowle JR, Molan MP, Hornsey E, Bellomo R. Measurement of renal blood flow by phase-contrast magnetic resonance imaging during septic acute kidney injury: A pilot investigation. *Crit Care Med*. 2012;40(6):1768–76.

18. Tofts PS, Cutajar M, Mendichovszky IA, Peters AM, Gordon I. Precise measurement of renal filtration and vascular parameters using a two-compartment model for dynamic contrast-enhanced MRI of the kidney gives realistic normal values. *Eur Radiol.* 2012;22(6):1320–30.
19. Attenberger UI, Sourbron SP, Schoenberg SO, Morelli J, Leiner T, Schoeppler GM et al. Comprehensive MR evaluation of renal disease: Added clinical value of quantified renal perfusion values over single MR angiography. *J Magn Reson Imaging.* 2010;31(1):125–33.
20. Bokacheva L, Rusinek H, Zhang JL, Lee VS. Assessment of renal function with dynamic contrast-enhanced MR imaging. *Magn Reson Imaging Clin N Am.* 2008;16(4):597–611.
21. De Senneville BD, Mendichovszky IA, Roujol S, Gordon I, Moonen C, Grenier N. Improvement of MRI-functional measurement with automatic movement correction in native and transplanted kidneys. *J Magn Reson Imaging.* 2008;28(4):970–8.
22. Rusinek H, Lee VS, Johnson G. Optimal dose of Gd-DTPA in dynamic MR studies. *Magn Reson Med.* 2001;46(2):312–6.
23. Michaely HJ, Schoenberg SO, Oesingmann N, Ittrich C, Buhlig C, Friedrich D et al. Renal artery stenosis: Functional assessment with dynamic MR perfusion measurements—Feasibility study. *Radiology.* 2006;238(2):586–96.
24. Michaely HJ, Kramer H, Oesingmann N, Lodemann K-P, Reiser MF, Schoenberg SO. Semiquantitative assessment of first-pass renal perfusion at 1.5 T: comparison of 2D saturation recovery sequences with and without parallel imaging. *AJR Am J Roentgenol.* 2007;188(4):919–26.
25. Michaely HJ, Kramer H, Oesingmann N, Lodemann K-P, Miserock K, Reiser MF et al. Intraindividual comparison of MR-renal perfusion imaging at 1.5 T and 3.0 T. *Invest Radiol.* 2007;42(6):406–11.
26. Attenberger UI, Morelli JN, Schoenberg SO, Michaely HJ. Assessment of the kidneys: Magnetic resonance angiography, perfusion and diffusion. *J Cardiovasc Magn Reson.* 2011;13:70.
27. Dujardin M, Sourbron S, Luypaert R, Verbeelen D, Stadnik T. Quantification of renal perfusion and function on a voxel-by-voxel basis: A feasibility study. *Magn Reson Med.* 2005;54(4):841–9.
28. Dujardin M, Luypaert R, Vandenbroucke F, Van der Niepen P, Sourbron S, Verbeelen D et al. Combined T1-based perfusion MRI and MR angiography in kidney: First experience in normals and pathology. *Eur J Radiol.* 2009;69(3):542–9.
29. Ribba B, Saut O, Colin T, Bresch D, Grenier E, Boissel JP. A multiscale mathematical model of avascular tumor growth to investigate the therapeutic benefit of anti-invasive agents. *J Theor Biol.* 2006;243(16930628):532–41.
30. Bakker J, Olree M, Kaatee R, de Lange EE, Moons KG, Beutler JJ et al. Renal volume measurements: Accuracy and repeatability of US compared with that of MR imaging. *Radiology.* 1999;211(3):623–8.
31. Cost GA, Merguerian PA, Cheerasarn SP, Shortliffe LM. Sonographic renal parenchymal and pelvicaliceal areas: New quantitative parameters for renal sonographic follow-up. *J Urol.* 1996;156(2 Pt 2):725–9.
32. Roger SD, Beale AM, Cattell WR, Webb JA. What is the value of measuring renal parenchymal thickness before renal biopsy? *Clin Radiol.* 1994;49(1):45–9.
33. Mounier-Vehier C, Lions C, Devos P, Jaboureck O, Willoteaux S, Carre A et al. Cortical thickness: An early morphological marker of atherosclerotic renal disease. *Kidney Int.* 2002;61(2):591–8.
34. Coulam CH, Bouley DM, Sommer FG. Measurement of renal volumes with contrast-enhanced MRI. *J Magn Reson Imaging.* 2002;15(2):174–9.
35. Jeong JY, Kim SH, Lee HJ, Sim JS. Atypical low-signal-intensity renal parenchyma: Causes and patterns. *RadioGraphics.* 2002;22(4):833–46.
36. Semelka RC, Corrigan K, Ascher SM, Brown JJ, Colindres RE. Renal corticomedullary differentiation: Observation in patients with differing serum creatinine levels. *Radiology.* 1994;190(1):149–52.
37. Chung JJ, Semelka RC, Martin DR. Acute renal failure: Common occurrence of preservation of corticomedullary differentiation on MR images. *Magn Reson Imaging.* 2001;19(6):789–93.
38. Laissy JP, Idee JM, Fernandez P, Floquet M, Vrtovsnik F, Schouman-Claeys E. Magnetic resonance imaging in acute and chronic kidney diseases: Present status. *Nephron Clin Pract.* 2006;103(2):c50–7.
39. Grenier N. Imaging and renal failure: From inflammation to fibrosis (Imagerie et insuffisance renale: de l'inflammation a la fibrose). *J Radiol.* 2011;92(4):323–35.
40. Grenier N, Basseau F, Ries M, Tyndal B, Jones R, Moonen C. Functional MRI of the kidney. *Abdom Imaging.* 2003;28(2):164–75.
41. Le Bihan D, Breton E, Lallemand D, Aubin ML, Vignaud J, Laval-Jeantet M. Separation of diffusion and perfusion in intravoxel incoherent motion MR imaging. *Radiology.* 1988;168(2):497–505.
42. Thoeny HC, De Keyzer F, Oyen RH, Peeters RR. Diffusion-weighted MR imaging of kidneys in healthy volunteers and patients with parenchymal diseases: Initial experience. *Radiology.* 2005;235(3):911–7.
43. Goyal A, Sharma R, Bhalla AS, Gamanagatti S, Seth A. Diffusion-weighted MRI in inflammatory renal lesions: All that glitters is not RCC! *Eur Radiol.* 2013;23(1):272–9.
44. Verswijvel G, Vandecaveye V, Gelin G, Vandevenne J, Grieten M, Horvath M et al. Diffusion-weighted MR imaging in the evaluation of renal infection: Preliminary results. *JBR-BTR.* 2002;85(2):100–3.
45. Goyal A, Gadodia A, Sharma R. Xanthogranulomatous pyelonephritis: An uncommon pediatric renal mass. *Pediatr Radiol.* 2010;40(12):1962–3.
46. Faletti R, Cassinis MC, Fonio P, Grasso A, Battisti G, Bergamasco L et al. Diffusion-weighted imaging and apparent diffusion coefficient values versus contrast-enhanced MR imaging in the identification and characterisation of acute pyelonephritis. *Eur Radiol.* 2013;23(12):3501–8.
47. Chan JH, Tsui EY, Luk SH, Fung SL, Cheung YK, Chan MS et al. MR diffusion-weighted imaging of kidney: Differentiation between hydronephrosis and pyonephrosis. *Clin Imaging.* 2001;25(2):110–3.

48. Blondin D, Lanzman RS, Mathys C, Grotemeyer D, Voiculescu A, Sandmann W et al. Functional MRI of transplanted kidneys using diffusion-weighted imaging (Funktionelle MRT der Transplantatnieren: klinische Wertigkeit der Diffusionsbildgebung). *Rofo*. 2009;181(12):1162–7.
49. Fukuda Y, Ohashi I, Hanafusa K, Nakagawa T, Ohtani S, An-naka Y et al. Anisotropic diffusion in kidney: Apparent diffusion coefficient measurements for clinical use. *J Magn Reson Imaging*. 2000;11(2):156–60.
50. Gaudiano C, Clementi V, Busato F, Corcioni B, Orrei MG, Ferramosca E et al. Diffusion tensor imaging and tractography of the kidneys: Assessment of chronic parenchymal diseases. *Eur Radiol*. 2013;23(6):1678–85.
51. Hueper K, Gutberlet M, Rodt T, Gwinner W, Lehner F, Wacker F et al. Diffusion tensor imaging and tractography for assessment of renal allograft dysfunction-initial results. *Eur Radiol*. 2011;21(11):2427–33.
52. Korsmo MJ, Ebrahimi B, Eirin A, Woollard JR, Krier JD, Crane JA et al. Magnetic resonance elastography noninvasively detects in vivo renal medullary fibrosis secondary to swine renal artery stenosis. *Invest Radiol*. 2013;48(2):61–8.
53. Khatir DS, Pedersen M, Jespersen B, Buus NH. Reproducibility of MRI renal artery blood flow and BOLD measurements in patients with chronic kidney disease and healthy controls. *J Magn Reson Imaging*. 2013;40:1091–8.
54. Brezis M, Rosen S. Hypoxia of the renal medulla—Its implications for disease. *N Engl J Med*. 1995;332(10):647–55.
55. Wang ZJ, Kumar R, Banerjee S, Hsu CY. Blood oxygen level-dependent (BOLD) MRI of diabetic nephropathy: Preliminary experience. *J Magn Reson Imaging*. 2011;33(3):655–60.
56. Inoue T, Kozawa E, Okada H, Inukai K, Watanabe S, Kikuta T et al. Noninvasive evaluation of kidney hypoxia and fibrosis using magnetic resonance imaging. *J Am Soc Nephrol*. 2011;22(8):1429–34.
57. Kock MCJM, Ijzermans JNM, Visser K, Hussain SM, Weimar W, Pattynama PMT et al. Contrast-enhanced MR angiography and digital subtraction angiography in living renal donors: Diagnostic agreement, impact on decision making, and costs. *AJR Am J Roentgenol*. 2005;185(2):448–56.
58. Halpern EJ, Mitchell DG, Wechsler RJ, Outwater EK, Moritz MJ, Wilson GA. Preoperative evaluation of living renal donors: Comparison of CT angiography and MR angiography. *Radiology*. 2000;216(2):434–9.
59. Tan SP, Bux SI, Kumar G, Razack AHA, Chua CB, Lee SH et al. Evaluation of live renal donors with three-dimensional contrast-enhanced magnetic resonance angiography in comparison to catheter angiography. *Transplant Proc*. 2004;36(7):1914–6.
60. Rankin SC, Jan W, Koffman CG. Noninvasive imaging of living related kidney donors: evaluation with CT angiography and gadolinium-enhanced MR angiography. *AJR Am J Roentgenol*. 2001;177(2):349–55.
61. Bhatti AA, Chugtai A, Haslam P, Talbot D, Rix DA, Soomro NA. Prospective study comparing three-dimensional computed tomography and magnetic resonance imaging for evaluating the renal vascular anatomy in potential living renal donors. *BJU Int*. 2005;96(7):1105–8.
62. Hodgson DJ, Jan W, Rankin S, Koffman G, Khan MS. Magnetic resonance renal angiography and venography: An analysis of 111 consecutive scans before donor nephrectomy. *BJU Int*. 2006;97(3):584–6.
63. Braidy C, Daou I, Diop AD, Helweh O, Gageanu C, Boyer L et al. Unenhanced MR angiography of renal arteries: 51 patients. *AJR Am J Roentgenol*. 2012;199(5):W629–37.
64. Runge VM. Current technological advances in magnetic resonance with critical impact for clinical diagnosis and therapy. *Invest. Radiol*. 2013;48(12):869–77.
65. Loubeyre P, Trolliet P, Cahen R, Grozel F, Labeeuw M, Minh VA. MR angiography of renal artery stenosis: Value of the combination of three-dimensional time-of-flight and three-dimensional phase-contrast MR angiography sequences. *AJR Am J Roentgenol*. 1996;167(2):489–94.
66. Angeretti MG, Lumia D, Cani A, Barresi M, Nocchi Cardim L, Piacentino F et al. Non-enhanced MR angiography of renal arteries: Comparison with contrast-enhanced MR angiography. *Acta Radiol*. 2013;54(7):749–56.
67. Zhang W, Lin J, Wang S, Lv P, Wang L, Liu H et al. Unenhanced respiratory-gated Magnetic Resonance Angiography (MRA) of renal artery in hypertensive patients using true fast imaging with steady-state precession technique compared with contrast-enhanced MRA. *J Comput Assist Tomogr*. 2014;38:700–4.
68. Tang H, Wang Z, Wang L, Hu X, Wang Q, Li Z et al. Depiction of transplant renal vascular anatomy and complications: Unenhanced MR angiography by using spatial labeling with multiple inversion pulses. *Radiology*. 2014;271:879–87.
69. Schoenberg SO, Knopp MV, Bock M, Kallinowski F, Just A, Essig M et al. Renal artery stenosis: Grading of hemodynamic changes with cine phase-contrast MR blood flow measurements. *Radiology*. 1997;203(1):45–53.
70. Regan F, Kuszyk B, Bohlman ME, Jackman S. Acute ureteric calculus obstruction: Unenhanced spiral CT versus HASTE MR urography and abdominal radiograph. *Br J Radiol*. 2005;78(930):506–11.
71. Sudah M, Vanninen R, Partanen K, Heino A, Vainio P, Ala-Opas M. MR urography in evaluation of acute flank pain: T2-weighted sequences and gadolinium-enhanced three-dimensional FLASH compared with urography. Fast low-angle shot. *AJR Am J Roentgenol*. 2001;176(1):105–12.
72. Vegar-Zubovic S, Kristic S, Lincender L. Magnetic resonance urography in children—When and why? *Radiol Oncol*. 2011;45(3):174–9.
73. Jones RA, Easley K, Little SB, Scherz H, Kirsch AJ, Grattan-Smith JD. Dynamic contrast-enhanced MR urography in the evaluation of pediatric hydronephrosis: Part 1, functional assessment. *AJR Am J Roentgenol*. 2005;185(6):1598–607.
74. McDaniel BB, Jones RA, Scherz H, Kirsch AJ, Little SB, Grattan-Smith JD. Dynamic contrast-enhanced MR urography in the evaluation of pediatric hydronephrosis: Part 2, anatomic and functional assessment of ureteropelvic junction obstruction [corrected]. *AJR Am J Roentgenol*. 2005;185(6):1608–14.
75. Nolte-Ernsting CC, Adam CB, Gunther RW. MR urography: Examination techniques and clinical applications. *Eur Radiol*. 2001;11(3):355–72.

76. Jones RA, Perez-Brayfield MR, Kirsch AJ, Grattan-Smith JD. Renal transit time with MR urography in children. *Radiology*. 2004;233(1):41–50.
77. Blandino A, Gaeta M, Minutoli F, Salamone I, Magno C, Scribano E et al. MR urography of the ureter. *AJR Am J Roentgenol*. 2002;179(5):1307–14.
78. Sudah M, Vanninen RL, Partanen K, Kainulainen S, Malinen A, Heino A et al. Patients with acute flank pain: Comparison of MR urography with unenhanced helical CT. *Radiology*. 2002;223(1):98–105.
79. Shokeir AA, El-Diasty T, Eassa W, Mosbah A, Mohsen T, Mansour O et al. Diagnosis of noncalcareous hydronephrosis: Role of magnetic resonance urography and noncontrast computed tomography. *Urology*. 2004;63(2):225–9.
80. Schuster TG, Ferguson MR, Baker DE, Schaldenbrand JD, Solomon MH. Papillary renal cell carcinoma containing fat without calcification mimicking angiomyolipoma on CT. *AJR Am J Roentgenol*. 2004;183(5):1402–4.
81. Takeuchi M, Matsuzaki K, Kubo H, Nishitani H. Diffusion-weighted magnetic resonance imaging of urinary epithelial cancer with upper urinary tract obstruction: Preliminary results. *Acta Radiol*. 2008;49(10):1195–9.
82. Muller MF, Prasad PV, Bimmler D, Kaiser A, Edelman RR. Functional imaging of the kidney by means of measurement of the apparent diffusion coefficient. *Radiology*. 1994;193(3):711–5.
83. Namimoto T, Yamashita Y, Mitsuzaki K, Nakayama Y, Tang Y, Takahashi M. Measurement of the apparent diffusion coefficient in diffuse renal disease by diffusion-weighted echo-planar MR imaging. *J Magn Reson Imaging*. 1999;9(6):832–7.
84. Roy C, Saussine C, LeBras Y, Delepaul B, Jahn C, Steichen G et al. Assessment of painful ureterohydronephrosis during pregnancy by MR urography. *Eur Radiol*. 1996;6(3):334–8.
85. Giannarini G, Kessler TM, Roth B, Vermathen P, Thoeny HC. Functional multiparametric magnetic resonance imaging of the kidneys using blood oxygen level-dependent and diffusion-weighted sequences: A reliable tool for monitoring acute upper urinary tract obstruction. *J Urol*. 2014;192:434–9.
86. Rosenkrantz AB, Niver BE, Fitzgerald EF, Babb JS, Chandarana H, Melamed J. Utility of the apparent diffusion coefficient for distinguishing clear cell renal cell carcinoma of low and high nuclear grade. *AJR Am J Roentgenol*. 2010;195(5):W344–51.
87. Sasiwimonphan K, Takahashi N, Leibovich BC, Carter RE, Atwell TD, Kawashima A. Small (<4 cm) renal mass: Differentiation of angiomyolipoma without visible fat from renal cell carcinoma utilizing MR imaging. *Radiology*. 2012;263(1):160–8.
88. Lechevallier E, Andre M, Barriol D, Daniel L, Eghazarian C, De Fromont M et al. Fine-needle percutaneous biopsy of renal masses with helical CT guidance. *Radiology*. 2000;216(2):506–10.
89. Eshed I, Elias S, Sidi AA. Diagnostic value of CT-guided biopsy of indeterminate renal masses. *Clin Radiol*. 2004;59(3):262–7.
90. Romis L, Cindolo L, Patard JJ, Messina G, Altieri V, Salomon L et al. Frequency, clinical presentation and evolution of renal oncocytomas: Multicentric experience from a European database. *Eur Urol*. 2004;45(1):53–7; discussion 7.
91. Pedrosa I, Alsop DC, Rofsky NM. Magnetic resonance imaging as a biomarker in renal cell carcinoma. *Cancer*. 2009;115(10 Suppl):2334–45.
92. Pedrosa I, Chou MT, Ngo L, Baroni RH, Genega EM, Galaburda L et al. MR classification of renal masses with pathologic correlation. *Eur Radiol*. 2008;18(2):365–75.
93. Roy C, Sr., El Ghali S, Buy X, Lindner V, Lang H, Saussine C et al. Significance of the pseudocapsule on MRI of renal neoplasms and its potential application for local staging: A retrospective study. *AJR Am J Roentgenol*. 2005;184(1):113–20.
94. Oliva MR, Glickman JN, Zou KH, Teo SY, Mortele KJ, Rocha MS et al. Renal cell carcinoma: T1 and T2 signal intensity characteristics of papillary and clear cell types correlated with pathology. *AJR Am J Roentgenol*. 2009;192:1524–30.
95. Roy C, Sauer B, Lindner V, Lang H, Saussine C, Jacqmin D. MR imaging of papillary renal neoplasms: Potential application for characterization of small renal masses. *Eur Radiol*. 2007;17:193–200.
96. Rosenkrantz AB, Hindman N, Fitzgerald EF, Niver BE, Melamed J, Babb JS. MRI features of renal oncocytoma and chromophobe renal cell carcinoma. *AJR Am J Roentgenol*. 2010;195(6):W421–7.
97. Kim JK, Kim SH, Jang YJ, Ahn H, Kim C-S, Park H et al. Renal angiomyolipoma with minimal fat: Differentiation from other neoplasms at double-echo chemical shift FLASH MR imaging. *Radiology*. 2006;239:174–80.
98. Hindman N, Ngo L, Genega EM, Melamed J, Wei J, Braza JM et al. Angiomyolipoma with minimal fat: Can it be differentiated from clear cell renal cell carcinoma by using standard MR techniques? *Radiology*. 2012;265(2):468–77.
99. Bird VG, Kanagarajah P, Morillo G, Caruso DJ, Ayyathurai R, Leveillee R et al. Differentiation of oncocytoma and renal cell carcinoma in small renal masses (<4 cm): The role of 4-phase computerized tomography. *World J Urol*. 2011;29(6):787–92.
100. Vargas HA, Chaim J, Lefkowitz RA, Lakhman Y, Zheng J, Moskowitz CS et al. Renal cortical tumors: Use of multiphasic contrast-enhanced MR imaging to differentiate benign and malignant histologic subtypes. *Radiology*. 2012;264(3):779–88.
101. Vargas HA, Delaney HG, Delappe EM, Wang Y, Zheng J, Moskowitz CS et al. Multiphasic contrast-enhanced MRI: Single-slice versus volumetric quantification of tumor enhancement for the assessment of renal clear-cell carcinoma Fuhrman grade. *J Magn Reson Imaging*. 2013;37(5):1160–7.
102. Sandrasegaran K, Sundaram CP, Ramaswamy R, Akisik FM, Rydberg MP, Lin C et al. Usefulness of diffusion-weighted imaging in the evaluation of renal masses. *AJR Am J Roentgenol*. 2010;194(2):438–45.

103. Rosenkrantz AB, Niver BE, Fitzgerald EF, Babb JS, Chandarana H, Melamed J. Utility of the apparent diffusion coefficient for distinguishing clear cell renal cell carcinoma of low and high nuclear grade. *AJR Am J Roentgenol.* 2010;195(5):344–51.
104. Rosenkrantz AB, Oei M, Babb JS, Niver BE, Taouli B. Diffusion-weighted imaging of the abdomen at 3.0 Tesla: Image quality and apparent diffusion coefficient reproducibility compared with 1.5 Tesla. *J Magn Reson Imaging.* 2011;33(1):128–35.
105. Wang H, Cheng L, Zhang X, Wang D, Guo A, Gao Y et al. Renal cell carcinoma: Diffusion-weighted MR imaging for subtype differentiation at 3.0 T. *Radiology.* 2010;257(1):135–43.
106. Fujii Y, Komai Y, Saito K, Iimura Y, Yonese J, Kawakami S et al. Incidence of benign pathologic lesions at partial nephrectomy for presumed RCC renal masses: Japanese dual-center experience with 176 consecutive patients. *Urology.* 2008;72:598–602.
107. Milner J, McNeil B, Alioto J, Proud K, Rubinas T, Picken M et al. Fat poor renal angiomyolipoma: Patient, computerized tomography and histological findings. *J Urol.* 2006;176:905– 9.
108. Bosniak MA, Megibow AJ, Hulnick DH, Horii S, Raghavendra BN. CT diagnosis of renal angiomyolipoma: The importance of detecting small amounts of fat. *AJR Am J Roentgenol.* 1988;151:497–501.
109. Kim JK, Kim SH, Jang YJ, Ahn H, Kim CS, Park H et al. Renal angiomyolipoma with minimal fat: Differentiation from other neoplasms at double-echo chemical shift FLASH MR imaging. *Radiology.* 2006;239(1):174–80.
110. Rosenkrantz AB, Raj S, Babb JS, Chandarana H. Comparison of 3D two-point Dixon and standard 2D dual-echo breath-hold sequences for detection and quantification of fat content in renal angiomyolipoma. *Eur J Radiol.* 2012;81(1):47–51.
111. Yoshimitsu K, Irie H, Tajima T, Nishie A, Asayama Y, Hirakawa M et al. MR imaging of renal cell carcinoma: Its role in determining cell type. *Radiat Med.* 2004;22(6):371–6.
112. Pedrosa I, Sun MR, Spencer M, Genega EM, Olumi AF, Dewolf WC et al. MR imaging of renal masses: Correlation with findings at surgery and pathologic analysis. *RadioGraphics.* 2008;28(4):985–1003.
113. Chao DH, Zisman A, Pantuck AJ, Freedland SJ, Said JW, Belldegrun AS. Changing concepts in the management of renal oncocytoma. *Urology.* 2002;59(5):635–42.
114. Newhouse JH, Wagner BJ. Renal oncocytomas. *Abdom Imaging.* 1998;23(3):249–55.
115. De Carli P, Vidiri A, Lamanna L, Cantiani R. Renal oncocytoma: Image diagnostics and therapeutic aspects. *J Exp Clin Cancer Res.* 2000;19(3):287–90.
116. Tan YK, Best SL, Olweny E, Park S, Trimmer C, Cadeddu JA. Radiofrequency ablation of incidental benign small renal mass: Outcomes and follow-up protocol. *Urology.* 2012;79(4):827–30.
117. Marciano S, Petit P, Lechevallier E, De Fromont M, Andre M, Coulange C et al. Renal onocytic adenoma. *J Radiol.* 2001;82(4):455–61.
118. Bandhu S, Mukhopadhyaya S, Aggarwal S. Spoke-wheel pattern in renal oncocytoma seen on double-phase helical CT. *Australas Radiol.* 2003;47(3):298–301.
119. Cochand-Priollet B, Molinie V, Bougaran J, Bouvier R, Dauge-Geffroy MC, Deslignieres S et al. Renal chromophobe cell carcinoma and oncocytoma. A comparative morphologic, histochemical, and immunohistochemical study of 124 cases. *Arch Pathol Lab Med.* 1997;121(10):1081–6.
120. Ball DS, Friedman AC, Hartman DS, Radecki PD, Caroline DF. Scar sign of renal oncocytoma: Magnetic resonance imaging appearance and lack of specificity. *Urol Radiol.* 1986;8(1):46–8.
121. Mortele KJ, Praet M, Van Vlierberghe H, Kunnen M, Ros PR. CT and MR imaging findings in focal nodular hyperplasia of the liver: Radiologic-pathologic correlation. *AJR Am J Roentgenol.* 2000;175(3):687–92.
122. McGahan JP, Lamba R, Fisher J, Starshak P, Ramsamooj R, Fitzgerald E et al. Is segmental enhancement inversion on enhanced biphasic MDCT a reliable sign for the noninvasive diagnosis of renal oncocytomas? *AJR Am J Roentgenol.* 2011;197(4):W674–9.
123. Kim JI, Cho JY, Moon KC, Lee HJ, Kim SH. Segmental enhancement inversion at biphasic multidetector CT: Characteristic finding of small renal oncocytoma. *Radiology.* 2009;252(2):441–8.
124. Millet I, Doyon FC, Hoa D, Thuret R, Merigeaud S, Serre I et al. Characterization of small solid renal lesions: Can benign and malignant tumors be differentiated with CT? *AJR Am J Roentgenol.* 2011;197(4):887–96.
125. Yen TH, Chen Y, Fu JF, Weng CH, Tian YC, Hung CC et al. Proliferation of myofibroblasts in the stroma of renal oncocytoma. *Cell Prolif.* 2010;43(3):287–96.
126. Siegel R, Naishadham D, Jemal A. Cancer statistics, 2013. *CA Cancer J Clin.* 2013;63(1):11–30.
127. Pedrosa I, Chou MT, Ngo L, R HB, Genega EM, Galaburda L et al. MR classification of renal masses with pathologic correlation. *Eur Radiol.* 2008;18(2):365–75.
128. Cheville JC, Lohse CM, Zincke H, Weaver AL, Blute ML. Comparisons of outcome and prognostic features among histologic subtypes of renal cell carcinoma. *Am J Surg Pathol.* 2003;27(5):612–24.
129. Maruyama M, Yoshizako T, Uchida K, Araki H, Tamaki Y, Ishikawa N et al. Comparison of utility of tumor size and apparent diffusion coefficient for differentiation of low- and high-grade clear-cell renal cell carcinoma. *Acta Radiol.* 2014;56:250–6.
130. Goyal A, Sharma R, Bhalla AS, Gamanagatti S, Seth A, Iyer VK et al. Diffusion-weighted MRI in renal cell carcinoma: A surrogate marker for predicting nuclear grade and histological subtype. *Acta Radiol.* 2012;53(3):349–58.
131. Sun MRM, Ngo L, Genega EM, Atkins MB, Finn ME, Rofsky NM et al. Renal cell carcinoma: Dynamic contrast-enhanced MR imaging for differentiation of tumor subtypes—Correlation with pathologic findings. *Radiology.* 2009;250:793–802.
132. Krishnan B, Truong LD. Renal epithelial neoplasms: The diagnostic implications of electron microscopic study in 55 cases. *Human Pathol.* 2002;33(1):68–79.

133. Fuhrman SA, Lasky LC, Limas C. Prognostic significance of morphologic parameters in renal cell carcinoma. *Am J Surg Pathol.* 1982;6:655–63.
134. Gervais L, Lebreton G, Casanova J. The making of a fusion branch in the Drosophila trachea. *Dev Biol.* 2012;362(2):187–93.
135. Volpe A, Panzarella T, Rendon RA, Haider MA, Kondylis FI, Jewett MAS. The natural history of incidentally detected small renal masses. *Cancer.* 2004;100:738–45.
136. Ficarra V, Brunelli M, Novara G, D'Elia C, Segala D, Gardiman M et al. Accuracy of on-bench biopsies in the evaluation of the histological subtype, grade, and necrosis of renal tumours. *Pathology.* 2011;43(2):149–55.
137. Al Nazer M, Mourad WA. Successful grading of renal-cell carcinoma in fine-needle aspirates. *Diagn Cytopathol.* 2000;22(4):223–6.
138. Leveridge MJ, Finelli A, Kachura JR, Evans A, Chung H, Shiff DA et al. Outcomes of small renal mass needle core biopsy, nondiagnostic percutaneous biopsy, and the role of repeat biopsy. *Eur Urol.* 2011;60(3):578–84.
139. Thompson RH, Kurta JM, Kaag M, Tickoo SK, Kundu S, Katz D et al. Tumor size is associated with malignant potential in renal cell carcinoma cases. *J Urol.* 2009;181(5):2033–6.
140. Crispen PL, Wong Y-N, Greenberg RE, Chen DYT, Uzzo RG. Predicting growth of solid renal masses under active surveillance. *Urol Oncol.* 2008;26(5):555–9.
141. Hsu RM, Chan DY, Siegelman SS. Small renal cell carcinomas: Correlation of size with tumor stage, nuclear grade, and histologic subtype. *AJR Am J Roentgenol.* 2004;182:551–7.
142. Delahunt B, Bethwaite PB, Nacey JN. Outcome prediction for renal cell carcinoma: Evaluation of prognostic factors for tumours divided according to histological subtype. *Pathology.* 2007;39(5):459–65.
143. Zhang C, Li X, Hao H, Yu W, He Z, Zhou L. The correlation between size of renal cell carcinoma and its histopathological characteristics: A single center study of 1867 renal cell carcinoma cases. *BJU Int.* 2012;110(11 Pt B):E481–5.
144. Meskawi M, Sun M, Ismail S, Bianchi M, Hansen J, Tian Z et al. Fuhrman grade [corrected] has no added value in prediction of mortality after partial or [corrected] radical nephrectomy for chromophobe renal cell carcinoma patients. *Mod Pathol.* 2013;26(8):1144–9.
145. Goldstein NS. The current state of renal cell carcinoma grading. Union Internationale Contre le Cancer (UICC) and the American Joint Committee on Cancer (AJCC). *Cancer.* 1997;80(5):977–80.
146. Patard JJ, Leray E, Rioux-Leclercq N, Cindolo L, Ficarra V, Zisman A et al. Prognostic value of histologic subtypes in renal cell carcinoma: A multicenter experience. *J Clin Oncol.* 2005;23(12):2763–71.
147. Somford DM, Hambrock T, Hulsbergen-van de Kaa CA, Futterer JJ, van Oort IM, van Basten JP et al. Initial experience with identifying high-grade prostate cancer using diffusion-weighted MR imaging (DWI) in patients with a Gleason score </= 3 + 3 = 6 upon schematic TRUS-guided biopsy: A radical prostatectomy correlated series. Investigative radiology. 2012;47(3):153–8.
148. Oto A, Yang C, Kayhan A, Tretiakova M, Antic T, Schmid-Tannwald C et al. Diffusion-weighted and dynamic contrast-enhanced MRI of prostate cancer: Correlation of quantitative MR parameters with Gleason score and tumor angiogenesis. *AJR Am J Roentgenol.* 2011;197(6):1382–90.
149. Silverman SG, Leyendecker JR, Amis ES, Jr. What is the current role of CT urography and MR urography in the evaluation of the urinary tract? *Radiology.* 2009;250(2):309–23.
150. Dym RJ, Chernyak V, Rozenblit AM. MR imaging of renal collecting system with gadoxetate disodium: Feasibility for MR urography. *J Magn Reson Imaging.* 2013;38(4):816–23.
151. Takahashi N, Kawashima A, Glockner JF, Hartman RP, Kim B, King BF. MR urography for suspected upper tract urothelial carcinoma. *Eur Radiol.* 2009;19(4):912–23.
152. Chahal R, Taylor K, Eardley I, Lloyd SN, Spencer JA. Patients at high risk for upper tract urothelial cancer: Evaluation of hydronephrosis using high resolution magnetic resonance urography. *J Urol.* 2005;174(2):478–82.
153. Obuchi M, Ishigami K, Takahashi K, Honda M, Mitsuya T, Kuehn DM et al. Gadolinium-enhanced fat-suppressed T1-weighted imaging for staging ureteral carcinoma: Correlation with histopathology. *AJR Am J Roentgenol.* 2007;188(3):W256–61.
154. Takahashi N, Kawashima A, Glockner JF, Hartman RP, Leibovich BC, Brau AC et al. Small (<2-cm) upper-tract urothelial carcinoma: Evaluation with gadolinium-enhanced three-dimensional spoiled gradient-recalled echo MR urography. *Radiology.* 2008;247(2):451–7.
155. Ergen FB, Hussain HK, Carlos RC, Johnson TD, Adusumilli S, Weadock WJ et al. 3D excretory MR urography: Improved image quality with intravenous saline and diuretic administration. *J Magn Reson Imaging.* 2007;25(4):783–9.
156. Karlo CA, Di Paolo PL, Chaim J, Hakimi AA, Ostrovnaya I, Russo P et al. Radiogenomics of clear cell renal cell carcinoma: Associations between CT imaging features and mutations. *Radiology.* 2013;270:464–71.
157. Hicklin DJ, Ellis LM. Role of the vascular endothelial growth factor pathway in tumor growth and angiogenesis. *J Clin Oncol Official J Am Soc Clin Oncol.* 2005;23(5):1011–27.
158. Li J, Chen F, Cona MM, Feng Y, Himmelreich U, Oyen R et al. A review on various targeted anticancer therapies. *Target Oncol.* 2012;7(1):69–85.
159. Sahani DV, Jiang T, Hayano K, Duda DG, Catalano OA, Ancukiewicz M et al. Magnetic resonance imaging biomarkers in hepatocellular carcinoma: Association with response and circulating biomarkers after sunitinib therapy. *J Hematol Oncol.* 2013;6:51.
160. Desar IM, van Herpen CM, van Asten JJ, Fiedler W, Marreaud S, Timmer-Bonte JN et al. Factors affecting the unexpected failure of DCE-MRI to determine the optimal biological dose of the vascular targeting agent NGR-hTNF in solid cancer patients. *Eur J Radiol.* 2011;80(3):655–61.

161. Flaherty KT, Rosen MA, Heitjan DF, Gallagher ML, Schwartz B, Schnall MD et al. Pilot study of DCE-MRI to predict progression-free survival with sorafenib therapy in renal cell carcinoma. *Cancer Biol Therapy.* 2008;7(4):496–501.
162. Kang HC, Tan KS, Keefe SM, Heitjan DF, Siegelman ES, Flaherty KT et al. MRI assessment of early tumor response in metastatic renal cell carcinoma patients treated with sorafenib. *AJR Am J Roentgenol.* 2013;200(1):120–6.
163. Desar IM, ter Voert EG, Hambrock T, van Asten JJ, van Spronsen DJ, Mulders PF et al. Functional MRI techniques demonstrate early vascular changes in renal cell cancer patients treated with sunitinib: A pilot study. *Cancer Imaging.* 2011;11:259–65.
164. Hahn OM, Yang C, Medved M, Karczmar G, Kistner E, Karrison T et al. Dynamic contrast-enhanced magnetic resonance imaging pharmacodynamic biomarker study of sorafenib in metastatic renal carcinoma. *J Clin Oncol.* 2008;26(28):4572–8.
165. De Bazelaire C, Rofsky NM, Duhamel G, Michaelson MD, George D, Alsop DC. Arterial spin labeling blood flow magnetic resonance imaging for the characterization of metastatic renal cell carcinoma(1). *Acad Radiol.* 2005;12(3):347–57.
166. Schor-Bardach R, Alsop DC, Pedrosa I, Solazzo SA, Wang X, Marquis RP et al. Does arterial spin-labeling MR imaging-measured tumor perfusion correlate with renal cell cancer response to antiangiogenic therapy in a mouse model? *Radiology.* 2009;251(3):731–42.
167. Le Bihan D, Turner R, Douek P, Patronas N. Diffusion MR imaging: Clinical applications. *AJR Am J Roentgenol.* 1992;159(3):591–9.
168. Le Bihan D, Breton E, Lallemand D, Grenier P, Cabanis E, Laval-Jeantet M. MR imaging of intravoxel incoherent motions: Application to diffusion and perfusion in neurologic disorders. *Radiology.* 1986;161(2):401–7.
169. Cui Y, Zhang X-P, Sun Y-S, Tang L, Shen L. Apparent diffusion coefficient: Potential imaging biomarker for prediction and early detection of response to chemotherapy in hepatic metastases. *Radiology.* 2008;248(3):894–900.
170. Baudelet C, Cron GO, Gallez B. Determination of the maturity and functionality of tumor vasculature by MRI: Correlation between BOLD-MRI and DCE-MRI using P792 in experimental fibrosarcoma tumors. *Magn Reson Med.* 2006;56(5):1041–9.
171. Chopra S, Foltz WD, Milosevic MF, Toi A, Bristow RG, Menard C et al. Comparing oxygen-sensitive MRI (BOLD R2*) with oxygen electrode measurements: A pilot study in men with prostate cancer. *Int J Radiat Biol.* 2009;85(9):805–13.
172. Hoskin PJ, Carnell DM, Taylor NJ, Smith RE, Stirling JJ, Daley FM et al. Hypoxia in prostate cancer: Correlation of BOLD-MRI with pimonidazole immunohistochemistry-initial observations. *Int J Radiat Oncol Biol Phys.* 2007;68(4):1065–71.
173. Diergarten T, Martirosian P, Kottke R, Vogel U, Stenzl A, Claussen CD et al. Functional characterization of prostate cancer by integrated magnetic resonance imaging and oxygenation changes during carbogen breathing. *Invest. Radiol.* 2005;40(2):102–9.
174. Jiang L, Zhao D, Constantinescu A, Mason RP. Comparison of BOLD contrast and Gd-DTPA dynamic contrast-enhanced imaging in rat prostate tumor. *Magn Reson Med.* 2004;51(5):953–60.
175. Heye T, Merkle EM, Reiner CS, Davenport MS, Horvath JJ, Feuerlein S et al. Reproducibility of dynamic contrast-enhanced MR imaging. Part II. Comparison of intra- and interobserver variability with manual region of interest placement versus semiautomatic lesion segmentation and histogram analysis. *Radiology.* 2013;266(3):812–21.
176. Vogelzang NJ, Stadler WM. Kidney cancer. *Lancet.* 1998;352(9141):1691–6.
177. Delworth MG, Pisters LL, Fornage BD, von Eschenbach AC. Cryotherapy for renal cell carcinoma and angiomyolipoma. *J Urol.* 1996;155(1):252–4; discussion 4–5.
178. Gervais DA, Arellano RS, Mueller PR. Percutaneous radiofrequency ablation of renal cell carcinoma. *Eur Radiol.* 2005;15(5):960–7.
179. Zagoria RJ, Pettus JA, Rogers M, Werle DM, Childs D, Leyendecker JR. Long-term outcomes after percutaneous radiofrequency ablation for renal cell carcinoma. *Urology.* 2011;77(6):1393–7.
180. Zagoria RJ, Traver MA, Werle DM, Perini M, Hayasaka S, Clark PE. Oncologic efficacy of CT-guided percutaneous radiofrequency ablation of renal cell carcinomas. *AJR Am J Roentgenol.* 2007;189(2):429–36.
181. Zagoria RJ, Childs DD. Update on thermal ablation of renal cell carcinoma: Oncologic control, technique comparison, renal function preservation, and new modalities. *Curr Urol Rep.* 2012;13(1):63–9.
182. Zagoria RJ, Hawkins AD, Clark PE, Hall MC, Matlaga BR, Dyer RB et al. Percutaneous CT-guided radiofrequency ablation of renal neoplasms: Factors influencing success. *AJR Am J Roentgenol.* 2004;183(1):201–7.
183. Tuncali K, Morrison PR, Winalski CS, Carrino JA, Shankar S, Ready JE et al. MRI-guided percutaneous cryotherapy for soft-tissue and bone metastases: Initial experience. *AJR Am J Roentgenol.* 2007;189(1):232–9.
184. Boss A, Clasen S, Kuczyk M, Schick F, Pereira PL. Image-guided radiofrequency ablation of renal cell carcinoma. *Eur Radiol.* 2007;17:725–33.
185. Lewin JS, Nour SG, Connell CF, Sulman A, Duerk JL, Resnick MI et al. Phase II clinical trial of interactive MR imaging-guided interstitial radiofrequency thermal ablation of primary kidney tumors: Initial experience. *Radiology.* 2004;232(3):835–45.
186. Fischbach F, Fischbach K, Ricke J. Percutaneous interventions in an open MR system: Technical background and clinical indications (Perkutane Interventionen in einem offenen MR-System: Technischer Hintergrund und klinische Indikationen). *Radiologe.* 2013;53(11):993–1000.
187. Shingleton WB, Sewell PE, Jr. Percutaneous renal tumor cryoablation with magnetic resonance imaging guidance. *J Urol.* 2001;165(3):773–6.
188. Harada J, Mogami T. Minimally invasive therapy under image guidance—Emphasizing MRI-guided cryotherapy. *Rinsho Byori.* 2004;52(2):145–51.

189. Ogan K, Jacomides L, Dolmatch BL, Rivera FJ, Dellaria MF, Josephs SC et al. Percutaneous radiofrequency ablation of renal tumors: Technique, limitations, and morbidity. *Urology*. 2002;60(6):954–8.
190. Campbell SC, Krishnamurthi V, Chow G, Hale J, Myles J, Novick AC. Renal cryosurgery: Experimental evaluation of treatment parameters. *Urology*. 1998;52(1):29–33; discussion -4.
191. Hwang JJ, Walther MM, Pautler SE, Coleman JA, Hvizda J, Peterson J et al. Radio frequency ablation of small renal tumors: Intermediate results. *J Urol*. 2004;171(5):1814–8.
192. Yamakado K, Nakatsuka A, Akeboshi M, Takeda K. Percutaneous radiofrequency ablation of liver neoplasms adjacent to the gastrointestinal tract after balloon catheter interposition. *J Vasc Interv Radiol*. 2003;14(9 Pt 1):1183–6.
193. Buy X, Tok CH, Szwarc D, Bierry G, Gangi A. Thermal protection during percutaneous thermal ablation procedures: Interest of carbon dioxide dissection and temperature monitoring. *Cardiovasc Interventional Radiol*. 2009;32(3):529–34.
194. Gervais DA. Cryoablation versus radiofrequency ablation for renal tumor ablation: Time to reassess? *J Vasc Interv Radiol*. 2013;24(8):1135–8.
195. Gervais DA, Arellano RS, McGovern FJ, McDougal WS, Mueller PR. Radiofrequency ablation of renal cell carcinoma: Part 2, Lessons learned with ablation of 100 tumors. *AJR Am J Roentgenol*. 2005;185:72–80.
196. Tateishi R, Shiina S, Teratani T, Obi S, Sato S, Koike Y et al. Percutaneous radiofrequency ablation for hepatocellular carcinoma. An analysis of 1000 cases. *Cancer*. 2005;103(6):1201–9.
197. Atwell TD, Farrell MA, Leibovich BC, Callstrom MR, Chow GK, Blute ML et al. Percutaneous renal cryoablation: experience treating 115 tumors. *J Urol*. 2008;179(6):2136–40; discussion 40–1.
198. Atwell TD, Schmit GD, Boorjian SA, Mandrekar J, Kurup AN, Weisbrod AJ et al. Percutaneous ablation of renal masses measuring 3.0 cm and smaller: Comparative local control and complications after radiofrequency ablation and cryoablation. *AJR Am J Roentgenol*. 2013;200(2):461–6.
199. Atwell TD, Callstrom MR, Farrell MA, Schmit GD, Woodrum DA, Leibovich BC et al. Percutaneous renal cryoablation: local control at mean 26 months of followup. *J Urol*. 2010;184(4):1291–5.
200. Levinson AW, Su L-M, Agarwal D, Sroka M, Jarrett TW, Kavoussi LR et al. Long-term oncological and overall outcomes of percutaneous radio frequency ablation in high risk surgical patients with a solitary small renal mass. *J Urol*. 2008;180:499–504.
201. Davenport MS, Caoili EM, Cohan RH, Ellis JH, Higgins EJ, Willatt J et al. MRI and CT characteristics of successfully ablated renal masses: Imaging surveillance after radiofrequency ablation. *AJR Am J Roentgenol*. 2009;192:1571–8.

# 2

# *Carcinoma of the Bladder and Urethra*

**Ott Le, Sadhna Verma, Armugam Rajesh, and Raghu Vikram**

**CONTENTS**



## 2.1 Bladder Carcinoma

### 2.1.1 Introduction/Epidemiology

Bladder carcinoma is the most common malignancy of the urinary system with an estimated 141,610 new cases expected to be diagnosed in the year 2014 accounting for 5,897 deaths in the United States alone.[1] The peak age of incidence is between 50 and 70 years with a male–female ratio of 3:1. Bladder cancer is the most frequently recurring cancer known to man thereby needing long-term surveillance after initial therapy. Moreover, patients with bladder cancer survive longer than most other cancers making it one of the most expensive cancers to manage.[2]

### 2.1.2 Etiology/Pathology

Transitional cell (urothelial) carcinoma accounts for more than 90% of all bladder cancers. Squamous cell carcinoma accounts for 6%–8% and glandular neoplasms such as adenocarcinoma account for the rest. There are several known and potential risk factors for development of bladder carcinoma. Smoking is implicated in nearly 60% of bladder carcinomas. The risk of bladder cancer in smokers is twofold to sixfold than those in non-smokers. Bladder cancer is also associated with occupational exposures to chemicals such as analine dye and chronic overuse of drugs such as phenacetin and chlornaphazine. Chronic cystitis caused by *Schistosoma haematobium* and chronic inflammation due to any cause including bladder stones are associated with squamous cell carcinoma.[3]

Urothelial carcinomas are either muscle invasive or non-muscle invasive. Squamous cell carcinomas and adenocarcinomas on the other hand are always muscle invasive. A vast majority (80%–90%) of urothelial carcinomas are superficial or non-muscle invasive in nature. These commonly have a papillary growth pattern and are frequently multifocal. However, these tumors are very recurrent and need to enter a rigorous surveillance program. These show a unique molecular and genetic pathway involving the RTK/RAS signaling pathway. On the other hand, invasive variety of urothelial carcinoma is triggered by defects in the p53 and retinoblastoma tumor suppressor pathways.[4,5] Muscle invasive tumors are usually of high grade and show a tendency to metastasize early.

### 2.1.3 Anatomy of Urinary Bladder

Urinary bladder is a muscular reservoir situated in the pelvis posterior to the symphysis pubis. The superior surface of the bladder is covered by peritoneal lining and the remainder is extra peritoneal in location. An important landmark within the bladder is the bladder trigone, which is a triangular area located between the two vesicoureteric junctions and the urethra opening. The wall of the urinary bladder is made up of four predefined layers: (1) the inner urothelium or transitional epithelium, (2) lamina propria or submucosa, (3) muscularis propria, and (4) outer serosa. The epithelium of the urinary bladder and much of the urinary tract are lined by transitional cells, which are composed of multiple layers of cuboidal cells with a domed apex. The function of the transitional cells is to accommodate fluctuation of volume in an organ. The lamina propria is highly vascular and its thickness varies with the degree of distension of the urinary bladder. The muscularis propria is composed of multiple layers of smooth muscle that makes up the detrusor muscle. The outer serosa is a thin layer of loose connective tissue, which allows for the rapid changes in the volume of the organ.

### 2.1.4 Staging

Bladder is staged using the American Joint Committee for Cancer TNM staging system. Papillary noninvasive tumors are staged as Ta (Figures 2.1 through 2.3). Tis is

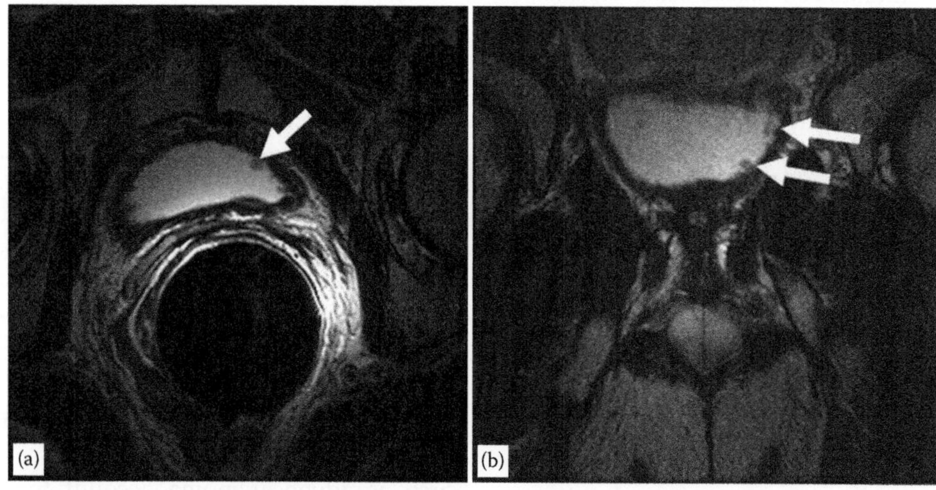


**FIGURE 2.1**
(a) Axial and (b) coronal $T_2$W images of the bladder showing small filling defect with intermediate signal intensity (white arrows) compatible with multifocal papillary bladder carcinoma. Stage Ta.

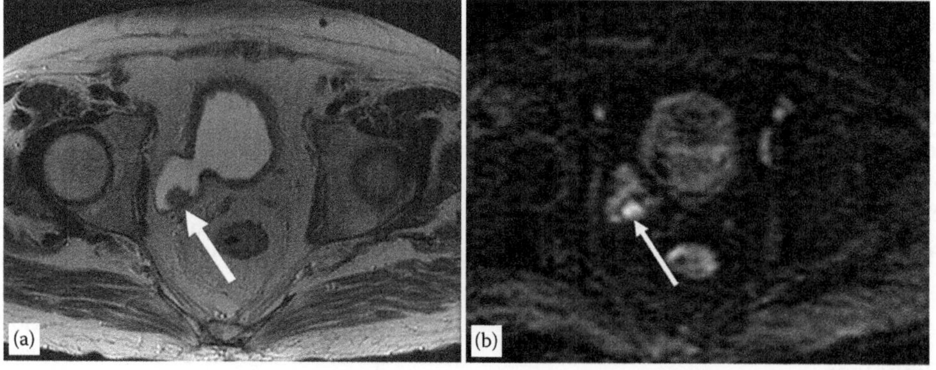


**FIGURE 2.2**
(a) Axial $T_2$W image. Intermediate signal lesion in a bladder diverticulum from a bladder carcinoma. There is no perivesicle invasion. (b) Axial DWI shows a high-signal-intensity lesion. Notice a subtle linear low signal intensity due to the fibrotic stalk giving rise to an *inchworm sign.* Stage Ta.

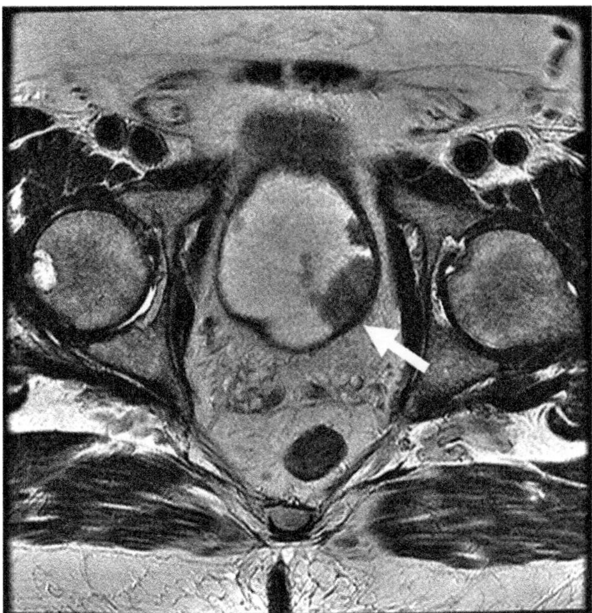

**FIGURE 2.3**
Axial $T_2$W image with large multifocal masses within the bladder. The low-signal layer of the detrusor muscle is uninterrupted. Stage Ta/T1.

the term used to describe carcinoma *in situ* (CIS) or flat tumors. Tumors invading the subepithelial connective tissue are T1 (Figure 2.3). Invasion into the muscularis is considered T2 (Figures 2.4 through 2.7). Extension into the perivesicle fat is staged as T3 (see Figures 2.7, 2.9, and 2.10 later in the chapter), and involvement of adjacent organs such as the prostatic stroma, uterus, vagina, or the pelvic side wall/abdominal wall is staged as T4 disease[6] (see Figure 2.11 later in the chapter). The depth of invasion offers prognostic information. A multi-institutional study reported 5 years survival rates of 81%, 74%, 47%, and 38%, respectively, in patients with node-negative pT1, pT2, pT3, and pT4 disease.[7,8] Further iterations of the survival data have also shown that there is a significant difference in survival between organ-confined (pT1-T2) and non-organ-confined (pT3-T4) disease.[9,10] Involvement of prostate gland through stromal invasion, whether through extension via the prostatic urethra or through the bladder neck, has a better prognosis and comparable survival to the organ-confined disease versus involvement of the prostate gland from the outside which portends a poorer prognosis[10] (see Figure 2.11 later in the chapter). While organ-confined status is an important prognostic indicator, muscle invasion is the most important factor considered in making treatment decisions. Papillary and non-muscle-invasive tumors are treated by transurethral resection, and patients with muscle-invasive tumors are treated with total and occasionally partial cystectomy or adjuvant therapies. In addition to the depth of invasion, nodal involvement is crucial for prognosis. Node-positive patients had a 5-year recurrence-free survival rate of 35% after radical cystectomy and pelvic lymph node section compared to overall survival of the entire cohort of 68% in a study reporting results of 1054 patients with invasive disease.[8] Up to 50% of patients with muscle-invasive disease eventually develop metastasis.[11]

In most centers, bladder carcinoma is staged with a combination of a bimanual examination, cystoscopy and deep biopsy (include all layers of bladder wall), and cross-sectional imaging such as CT. This approach can inaccurately stage bladder carcinoma in a significant proportion of patients. Identification of muscle invasion, extra vesicle spread, and nodal metastases is of paramount importance in this group. Multidetector CT is generally the imaging modality of choice for detecting nodal and distant metastasis but is inferior to magnetic resonance imaging (MRI) for local staging. MRI with its multiplanar capabilities and superior contrast resolution may be exploited to demonstrate the different layers of the bladder wall and enable assessment of the depth of invasion.

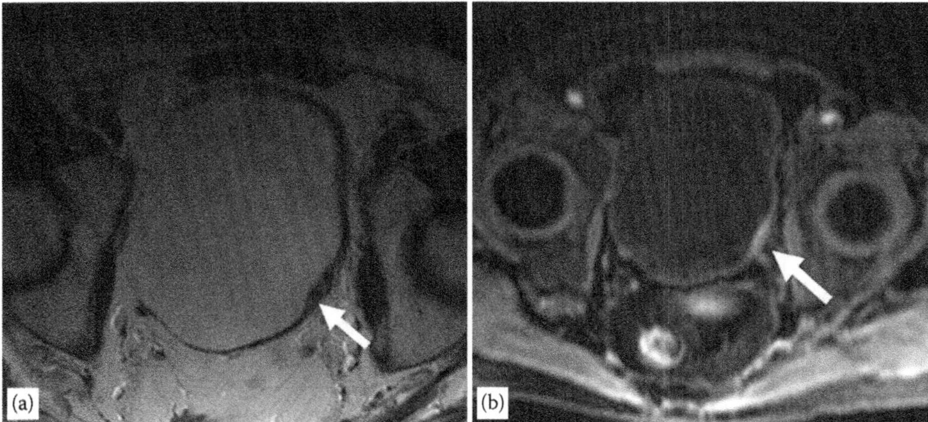


**FIGURE 2.4**
(a and b) Axial $T_2$W image and contrast-enhanced 3D gradient echo image showing a flat plaque like tumor in the left posterior wall of the urinary bladder. The bladder wall shows intermediate signal on $T_2$W image. The CE $T_1$W image shows a subtle non-enhancing wall. Stage T1/T2.

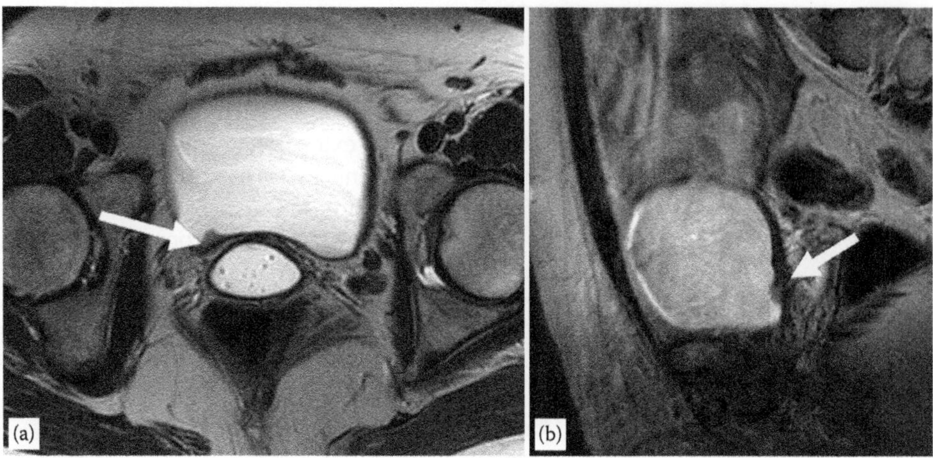


**FIGURE 2.5**
(a) Axial $T_2$W and (b) sagittal $T_2$W showing small lesion with intermediate signal intensity (white arrow) compatible with bladder carcinoma. The coronal image shows partial interruption of the muscle wall indicating a T2 lesion.

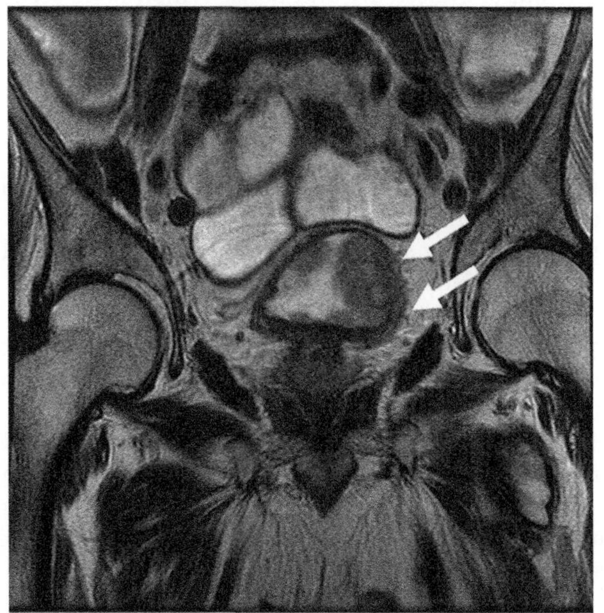

**FIGURE 2.6**
Coronal $T_2$W image with large multifocal masses within the bladder. The low-signal layer of the detrusor muscle in its entire thickness is lost indicating deep muscle invasion. Stage T2.

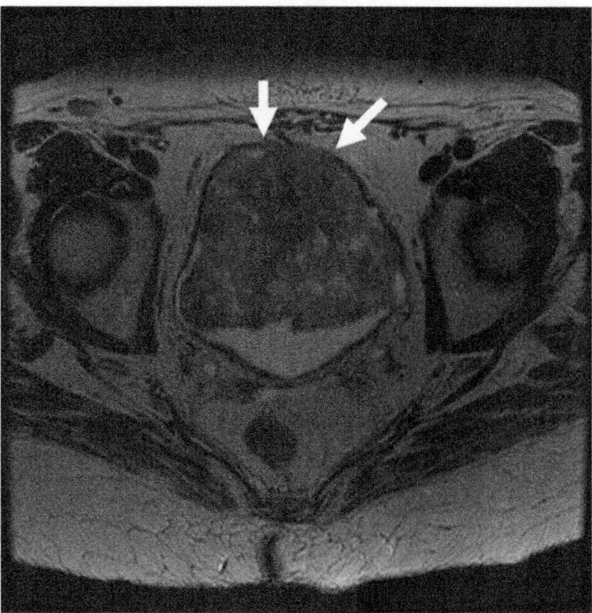

**FIGURE 2.7**
Axial $T_2$W image with large bladder mass. The low-signal layer of the detrusor muscle in its entire thickness is lost in the anterior wall indicating deep muscle invasion. Stage T2/T3.

### 2.1.5 MRI Technique

MRI of the bladder is best performed with use of high spatial resolution achieved typically with the use of a combination of external pelvic-phased array coils such as a cardiac coil, slice thickness of not more than 3 mm with no interslice gap, and large matrix size.[2,12] A three-plane localizer is used to evaluate coil placement and bladder distention. A field of view of 28–32 is used to evaluate the bladder and surrounding soft tissue structures. An optimal echo time (60–100 ms) is used to achieve a high contrast-to-noise ratio. Optimal distension of the bladder is important for accurate diagnosis and can generally be achieved by asking the patient to void 2 h prior to imaging.

$T_1$-weighted ($T_1$W) images are obtained in axial plane to give an overview of the pelvis. They are useful in evaluating the perivesicle fat for involvement and detecting enlarged nodes and any bony lesions in the pelvis. On $T_1$W images, the bladder wall and the tumor have intermediate signal intensity, perivesicle fat has high signal intensity, and urine has low signal intensity.

$T_2$-weighted ($T_2$W) images are important as they demonstrate the detrusor muscle as low-signal linear

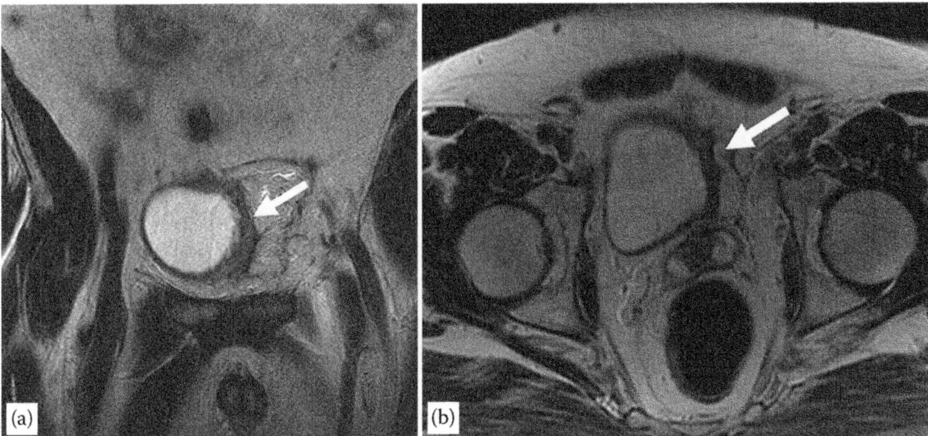


**FIGURE 2.8**
(a) Coronal $T_2$W and (b) axial $T_2$W images showing chronic thickening of the left wall of the bladder with distortion of contours secondary to fibrosis from repeated transurethral resection and BCG instillation.

band making up most of the bladder wall. These are obtained in all three orthogonal planes to be able to assess for muscle invasion. The detrusor muscle has low signal intensity on $T_2$W images and is interrupted in case of muscle-invasive tumors (Figures 2.5 through 2.7, 2.9, and 2.10). Posttreatment changes, for example, inflammation and scarring post-BCG therapy or post-transurethral resection, may result in bladder wall thickening and mimic muscle-invasive disease (Figure 2.8). The overall staging accuracy of $T_2$W imaging is between 40% and 67% with over staging the most common error.[13–17]

This is generally followed by a 3D gradient-echo fat-suppressed dynamic contrast-enhanced sequence. The clinical utility of this is debatable. Transitional cell carcinomas tend to enhance early (20 s after injection) and to a greater extent than the bladder wall. The bladder wall enhances late and can be seen as an interrupted line in muscle-invasive disease (Figure 2.9b). Some authors report the usefulness of submucosal linear enhancement which is seen in the early phase of contrast injection compared to the muscle which does not enhance immediately (Figure 2.4b). Intact submucosal enhancement is considered low-stage or non-muscle-invasive disease.[18] However, up to 60% of the tumors show identical enhancement characteristics to the submucosal line thereby making it difficult to recognize muscle invasion. The overall staging accuracy of dynamic contrast-enhanced MRI is reported to be 52%–85%.[13,19,20]

Diffusion-weighted imaging (DWI) is added as it is valuable in detection of tumors and has been shown to add an incremental value to T staging. One of the most useful reported signs is the *inchworm sign* on DWI, which is typically seen in Ta (papillary) tumors (Figure 2.2b). This is due to the low signal of the stalk surrounded by the high signal intensity of the tumor.[19] DWI is also useful in differentiating thickened submucosa from

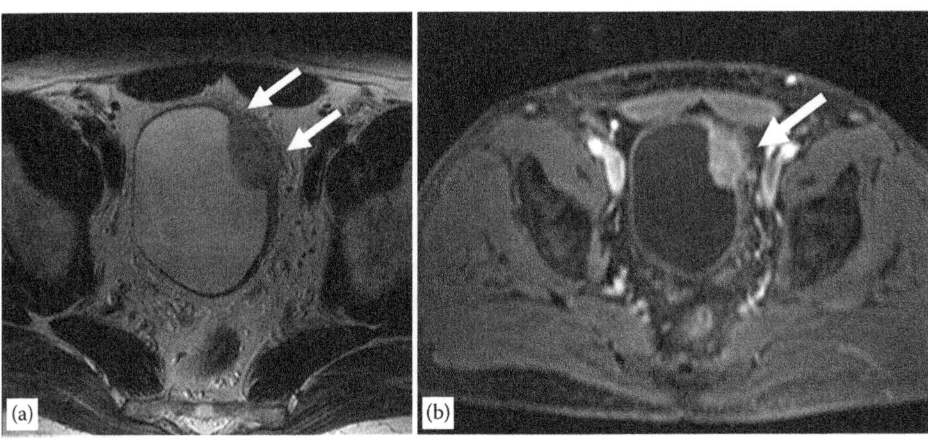


**FIGURE 2.9**
(a and b) Axial $T_2$W image and contrast-enhanced 3D GRE image showing a large mass with full thickness bladder wall invasion and perivesicle extension. Note that the low signal of the detrusor is lost in the $T_2$W image. Stage T3.

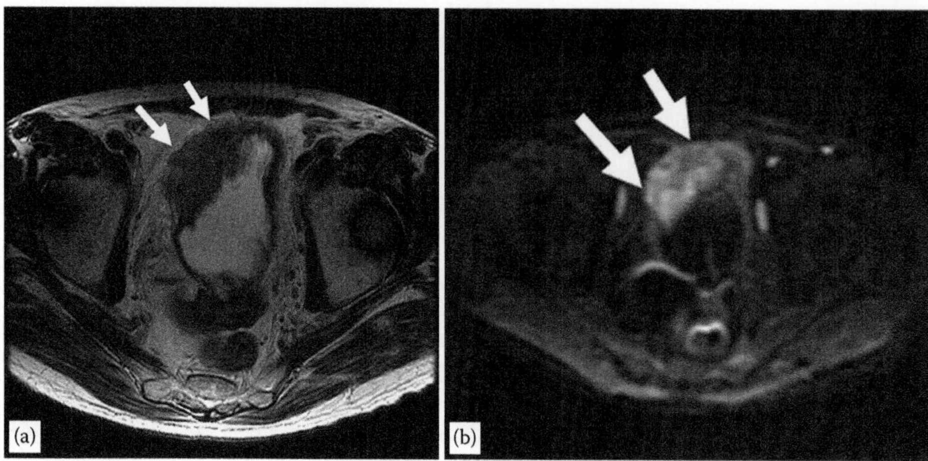


**FIGURE 2.10**
(a) Axial $T_2$W image. Large mass in the right lateral and anterior wall of the bladder with full thickness involvement and perivesicle fat invasion. (b) Axial DWI shows a high-signal-intensity lesion with full thickness involvement and perivesicle fat invasion.

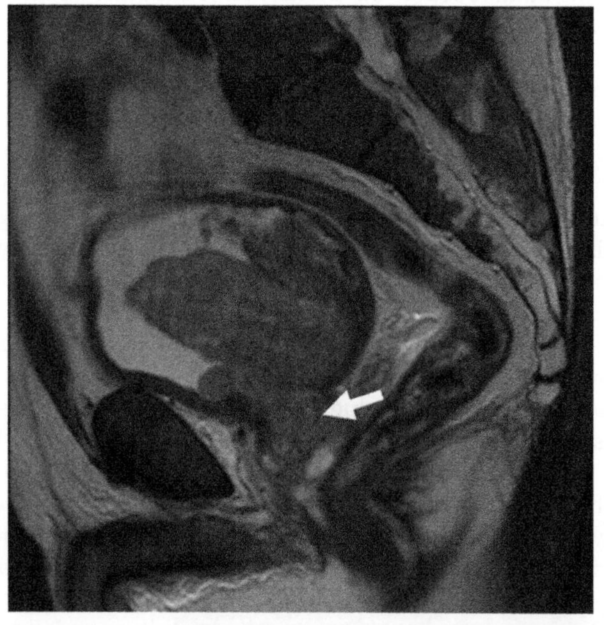

**FIGURE 2.11**
Sagittal $T_2$W image showing a large bladder tumor that involves the prostate by direct contiguous invasion of the prostatic urethra and the bladder base. Stage T4.

inflammation which mimics muscle invasion. Similar advantage is also seen when differentiating perivesicle invasion versus inflammatory or reactive stranding[17] (Figure 2.10b).

Conventional MRI is as sensitive as CT in detecting nodal metastasis when size criterion is used. However, size criterion (>1 cm) is known to be notoriously inaccurate as several metastatic nodes are smaller than 1 cm. Use of ultrasmall superparamagnetic iron oxide particles in combination with $T_2$* imaging improves nodal staging significantly.[21] However, their use is not routine due to lack of commercial availability. DWI improves the detection of lymph nodes but is not accurate in characterization (Figure 2.11).

### 2.1.6 Conclusion

Accurate preoperative staging of bladder carcinoma is the basis for planning therapy in all patients. The conventional clinical method of staging is far from accurate, and there is a need to improve staging. Currently, MRI is the most accurate noninvasive modality for T-staging bladder carcinoma, and its use in addition to conventional staging protocols would improve staging accuracy. MRI however has a tendency to overstage and should be interpreted with caution.

## 2.2 Urethral Carcinoma

### 2.2.1 Introduction/Epidemiology

Epithelial neoplasms of the male and the female urethra are very rare and account to less than 1% of all malignancies involving the genitourinary system.[22] Due to the rarity of these tumors, there are no universally accepted standards of care. Female urethra carcinomas have different clinical and pathologic features when compared to the male counterpart.[23] The differences have been explained by the anatomic and histologic variations between the two sexes. It is rare to find benign epithelial tumors in the urethra of either sex.[23] Primary urethral carcinoma more commonly occurs in men than women, its annual incidence rate increases with age (more often in the seventh decade), and it is two times more common in African Americans than whites.[23–25]

### 2.2.2 Etiology

Risk factors for the development of primary urethra cancer also differ between the sexes. Urethral strictures account for 25%–76%, followed by sexually transmitted disease (24%–50%), human papillomavirus (HPV) (4%), and trauma (7%) in men.[22] In women, it has been found that chronic irritation (also from HPV), diverticula, sexual activity, and child birth are associated with urethral carcinoma.[22] Squamous cell carcinoma of the urethra is the resulting most common in both sexes (80% of cases for men and 60% for women), followed by transitional cell carcinoma according to Bertolotto et al.[24] However, Swartz et al. in their study have challenged that assumption, demonstrating 55% of the cases were urothelial carcinomas, 22% squamous cell carcinomas, 16% adenocarcinomas, and remaining rare (melanoma or unclassified entities).[26]

### 2.2.3 Male Urethra

#### *2.2.3.1 Anatomy*

It is important to understand the anatomy of the male and female urethra. Their difference influences the histology, pathology, spread of disease, and resulting treatment options. The male urethra measures approximately 18–20 cm in length, begins from the bladder neck to the penile meatus, and is anatomically segmented into a prostatic, membranous, bulbar, and pendulous portion. The prostatic and membranous portions are also the posterior urethra, while the bulbar and the pendulous components form the anterior urethra.[27,28] The urethra is lined by mucosa, submucosal stroma, and the surrounding corpus spongiosum.[6] The corpus spongiosum along the penis houses the anterior urethra, while the posterior urethra pierces through the urogenital diaphragm and superficial and deep muscles of the pelvic floor. The same muscle of the pelvic floor and perineum forms the necessary support and accompanying sphincter mechanism.[27] The posterior urethra consists of the prostatic urethra (composed of transitional cell epithelium) and membranous urethra within the urogenital diaphragm (composed of stratified columnar cells). The anterior urethra begins with the bulbar urethra, which begins distal to the membranous urethra and ends at the penile urethra or penile scrotal junction. It makes up most of the anterior urethra and consists of columnar epithelium and pseudostratified cells. Finally, the penile urethra is the most distal segment as it traverses the glans penis and is composed mainly of stratified squamous epithelium.

Several glands and ducts enter the urethra. In the prostate, paired ejaculatory ducts carry secretions and semen into the prostatic urethra at the verumontanum. Cowper's glands are also paired and empty their secretions into the membranous urethra, while the glands of Littre similarly secrete into the pendulous urethra.

Arterial supply for the urethra is maintained by the paired bulbourethral arteries, which arises from the pudendal artery. The deep dorsal veins provide venous drainage. Lymphatic drainage is dependent on the location of the urethra. If anterior, the lymphatic drainage is to the groin and into the hypogastric, internal iliac and remainder of the pelvic lymph nodes if the urethra is posterior.

#### *2.2.3.2 Urogenital Diaphragm*

When evaluating the anatomy and disease process of the urethra, familiarity of the urogenital diaphragm is required. Treatment of urethral malignancies is most often dependent on involvement of this structure. The urogenital diaphragm is a disc-like structure that defines the deep perineal space, is encased in a superficial and deep fascia, consists of the external sphincter of the urethra and the deep transverse perineal muscle, and has also been called the *perineal membrane*, *triangular ligament*, or *deep transverse muscle of the perineum*.[29] It lies just inferior to the pelvic diaphragm and anterior to the anorectum (Figure 2.12).[30] In the male, it is situated between the root of the penis and prostate, and it is penetrated by the urethra and vagina in the female.[30] Within the urogenital diaphragms, paired Cowper's gland, fat, vessels, and lymphatics are present. Surrounding structures that provide support and attachments consist of the vagina, perineal body, external sphincter, and

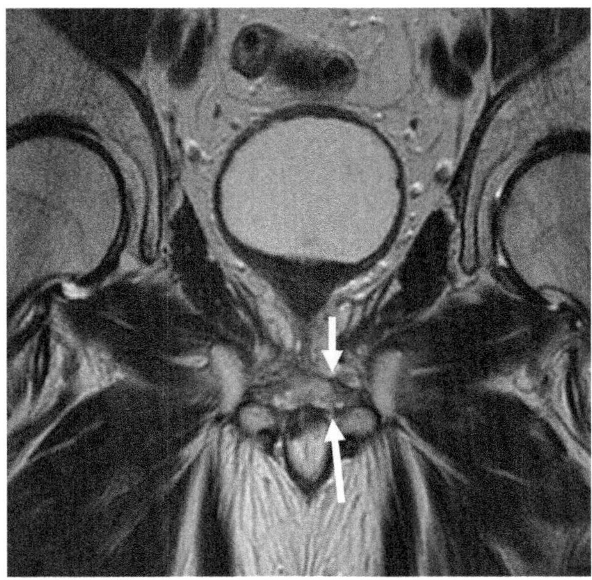

**FIGURE 2.12**
The urogenital diaphragm seen on the coronal view. The urogenital diaphragm is a disc-like structure (in between the arrows) that lies immediately inferior to the prostatic apex.

bulbocavernous muscle.[30] The urogenital diaphragm houses the membranous urethra and can act as a natural barrier against the spread of diseases such as ascending urinary tract infections.[29,31,32]

#### 2.2.3.3 Magnetic Resonance Imaging

Urethral cancer can be typically evaluated with cystourethroscopy with biopsy and retrograde urethrogram (RUG). MRI with its superb soft tissue contrast is especially useful for evaluating urethral carcinoma. High-resolution MRI has become crucial for staging and optimizing treatment planning for urethral as well as penile cancer and is now considered the gold standard.[33] It is able to evaluate the periurethral anatomy, unlike a conventional RUG. MRI allows for assessment of the local extent of disease, such as whether there is invasion of the corpus spongiosum and cavernosum, rectum, prostate, pelvic bones, lymph nodes, and other abnormalities.[26]

Since tumor of the urethra is rare, there is lack of MRI protocol standardization. Typical sequence utilizes T1, T2, and post-gadolinium contrast in mulitplanar orthogonal views utilizing a surface coil. At our institution, MRI of the urethra is performed similarly to MRI of the penis. That is, the patient is placed supine, a folded towel is placed in between the patient's legs inferior to the perineum to elevate the scrotum and penis, and then the penis is dorsiflexed and taped against the anterior abdominal wall.

The urethra on MRI is typically low to intermediate in signal intensity on T2 relative to the corpus spongiosum (Figures 2.13 and 2.14). It can be seen as a tubular structure within the prostate on the axial view and a

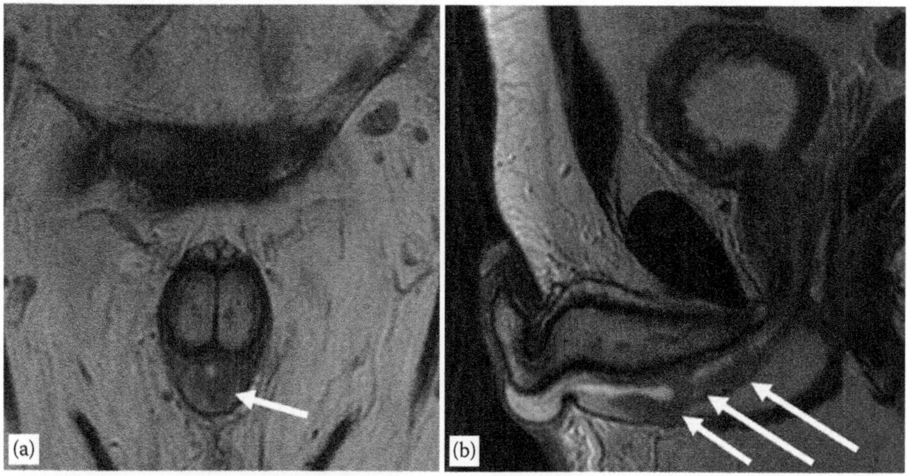


**FIGURE 2.13**
Typical appearance of a male urethra tumor (arrow), seen as low to intermediate signal abnormality on T2 relative to the corpus spongiosum seen on the coronal and sagittal view (a and b). The tumor involves the corpus spongiosum and does not invade the corpora cavernosa.

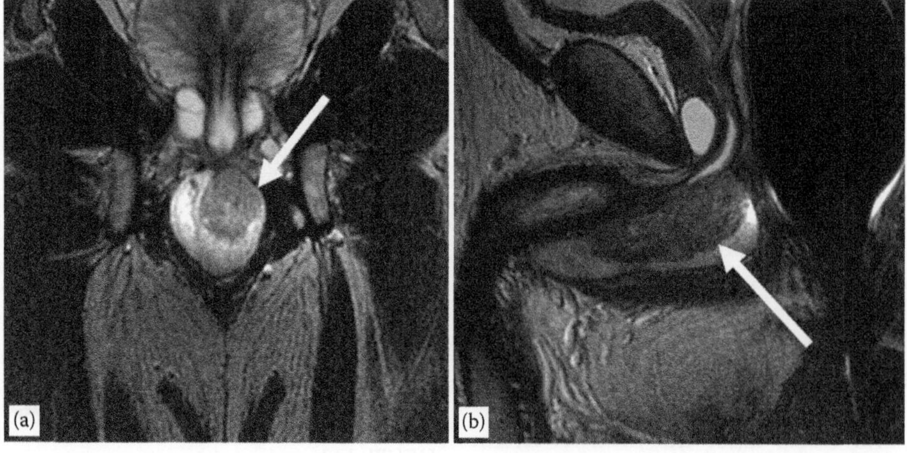


**FIGURE 2.14**
Coronal $T_2$W (a) and sagittal $T_2$W (b) images of a patient with squamous cell carcinoma of the bulbar urethra (arrow) with invasion into the corpus spongiosum.

linear enhancing structure post-gadolinium contrast.[34] Urethral cancer in males manifests as a soft tissue mass with similar or lower signal intensity on $T_1$W images, lower signal intensity on $T_2$W images, and mild gadolinium contrast enhancement compared to the corpus spongiosum.[24]

### 2.2.4 Pathology

Location of the tumor in the urethra is of paramount importance in determining the clinical and biological behavior and prognosis. Macroscopically, on cystourethroscopy, they may be ulcerative, nodular, papillary, cauliflower-like, and ill-defined.[23] The appearance can also be indicative of the histology. For example, grayish–white or pearly with necrosis is associated with squamous cell carcinoma, while mucoid, gelatinous, or cystic changes can be found in adenocarcinomas.[23] Erythematous erosions can be reflective of urothelial CIS.[23]

The bulbomembranous urethra is the most commonly affected by tumor in men (59%), followed by the anterior (33%) and prostatic (7%).[22] The pathology also corresponds with the histological location. For instance, since the penile urethra is composed of stratified squamous epithelium, it is subject to squamous cell carcinoma. The prostatic urethra consists of transitional cell epithelium; therefore, urothelial carcinoma commonly occurs. Approximately 75% of carcinomas are squamous cell carcinoma (involving usually the penile and bulbomembranous urethra) and the rest are urothelial carcinomas (usually prostatic) and adenocarcinomas (bulbomembranous urethra).[23] Note that the histology of a tumor is not specific to any one location of the urethra. Transitional cell carcinomas can be found in the penile urethra.[35,36]

Tumor anatomical location along the urethra and stage are two important prognostic factors, but survival overall remains poor. Distal urethral tumors have better prognosis than proximal tumors, which is due to the ability to achieve local control with treatment strategies.[22] In retrospective studies with a median survival of 48–125 months, the survival has been found to be between 42% and 52%.[22]

A TNM staging system is used for urethral carcinomas, but a separate subset is used specifically for the prostatic urethra.[6] In short, T1 tumor invades the subepithelial connective tissue; T2 tumor invades the corpus spongiosum (Figures 2.13 and 2.14), prostate, or periurethral muscle; T3 invades the corpus cavernosum, beyond the prostatic capsule, anterior vagina, or bladder neck (Figures 2.15 and 2.16); and T4 invades other adjacent organs (Figure 2.17).[6]

Other tumors of the urethra are rare, including epithelial, non-epithelial, and the benign varieties. They can include melanoma, lymphoma, leiomyoma, metastasis, and hemangioma.

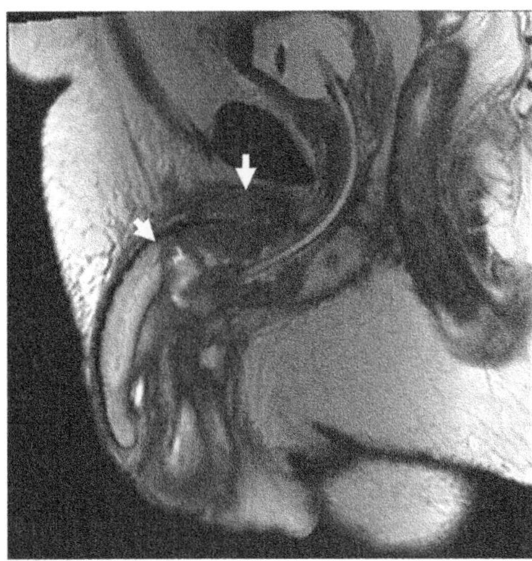

**FIGURE 2.15**
Sagittal $T_2$W image of a patient with squamous cell carcinoma of the urethra. The low-signal-intensity tumor is seen invading the corpus cavernosa (white arrows).

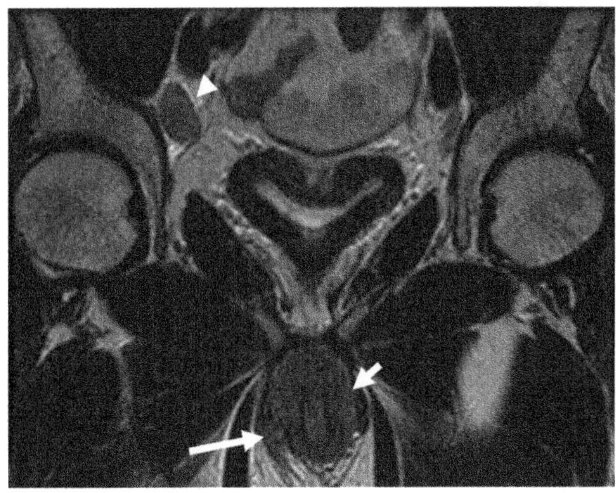

**FIGURE 2.16**
Coronal $T_2$W image of a patient with carcinoma of the penile urethra. Note the tumor invades the corpora cavernosa (white arrows). There is also an enlarged internal iliac lymph node (arrowhead).

Tumors can also arise from the associated ducts, such as from the bulbourethral gland or periurethral ducts of the prostate or glands of Littre.[37] Cowper's gland urethral tumor may be seen as soft tissue mass arising from and distorting the urogenital diaphragm.[24] Primary transitional cell carcinoma of the prostatic periurethral ducts has also been described.[37]

### 2.2.5 Female Urethra

#### 2.2.5.1 Anatomy

The female urethra is shorter than the male urethra. It is a short muscular tube approximately 4 cm long lined by mucous membrane and can be divided into superior

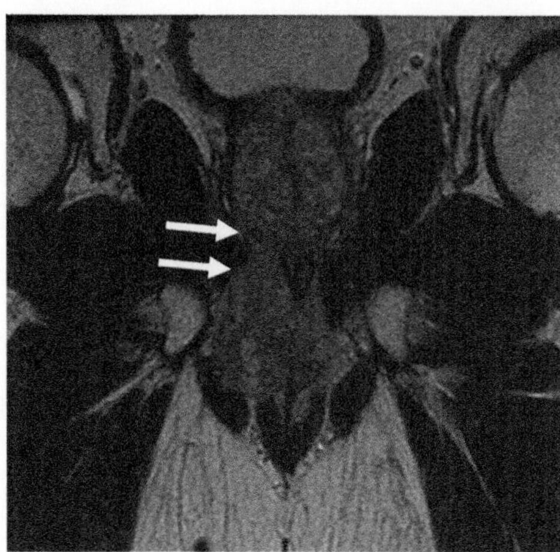

**FIGURE 2.17**
Coronal $T_2$W image of a patient with adenocarcinoma of the membranous urethra. The tumor invades the urogenital diaphragm laterally, the pelvic side walls and the prostate gland superiorly.

and inferior components which are analogous to the anterior and posterior division of the male urethra. It extends from the internal urethral meatus at the bladder neck and traverses through the urogenital diaphragm to the external urethral meatus anterior to the vaginal opening.[28,32] It runs anteroinferiorly from bladder and posteroinferiorly to the pubic symphysis, and then it's orifice exits at the vestibule of the vagina. Inferiorly, the urethra is closely intimate with the vagina, where both passes through the pelvic and urogenital diaphragm and perineal membrane. The sphincter urethrae muscle surrounds the inferior end of the urethra and some of its fibers enclose also the vagina.[32] Multiple tiny urethral glands, called *paraurethral glands of Skene*, which are homologous to the prostate, line the superior urethra on each side and their ducts open near the urethral orifice.[28] Squamous epithelium lines the distal two-third of the urethra and the proximal one-third with urothelium (transitional epithelium).

#### 2.2.5.2 Magnetic Resonance Imaging

Conventional imaging with RUG, voiding cystourethrography (VCUG), double-balloon catheter urethrography, or ultrasonography in combination with examination under anesthesia is used to assess the female urethra. However, they can be invasive and may be limited evaluation of the periurethral tissues. MRI provides excellent soft tissue contrast and multiplanar capability to evaluate both the urethra and periurethral tissues.

T1, T2, and contrast media are typically the sequences acquired for MRI of the female urethra. The urethra is best imaged along orthogonal planes (axial, sagittal, and coronal), and signal to noise ratio can be improved with a surface coil or endovaginal or endorectal coils.[28] In the female urethra, the axial plane perpendicular to the anterior wall of the vagina is particularly important.[38] Although endocavitary probes improve spatial resolution, their small field of view and high signal intensity in the near field may degrade image quality.[28] Artificial distension of the vagina with use of gel improves the detection of vaginal wall involvement. On the axial $T_2$W images or gadolinium contrast-enhanced $T_1$W images, a characteristic target appearance can be seen in a normal female urethra. The target appearance is made of four concentric rings: an outer low-signal-intensity ring, a middle high-signal-intensity ring, an inner low-signal-intensity ring, and a high-signal-intensity center.[28] The outer ring represents striated muscle, middle ring consists of submucosa and smooth muscle, and the central ring contains mucosa and stratified epithelium.[38] The outer ring is thicker along the proximal urethra.[39]

#### 2.2.5.3 Pathology

Primary carcinoma of the female urethra is rare. It is classified as either anterior (vulvourethral or distal one-third) or entire.[28,40] Female urethral cancer is more commonly found in the distal urethra (35%–50%), the whole urethra (43%), and less in the proximal urethra (9%–18%).[22] Presence of urethra diverticulum is associated with an increased risk of carcinoma (Figures 2.18 and 2.19). Squamous cell carcinoma is the predominant histology in females (39%–70%), followed by adenocarcinomas (15%–35%) and urothelial carcinomas (15% of cases).[22] The symptoms of urethral carcinoma are nonspecific and women may present with recurrent urinary tract infections, obstructive and irritative symptoms, spotting, and hematuria.

Urethral carcinoma is staged according to the depth of invasion (Figures 2.20 through 2.22). T1 tumors invade the subepithelial connective tissue. T2 tumors invade the periurethral muscle. T3 tumors involve the bladder neck and the anterior vaginal wall. T4 tumors invade the adjacent organs. Distant metastasis at the time of presentation is rare (up to 6%), and regional lymph nodes are seen in up to a third of patients (anterior tumors are more likely to spread to the inguinal lymph nodes and posterior tumors to the pelvic lymph nodes).[22]

MRI has shown to be useful for female urethra for local staging and treatment planning. Urethral tumors demonstrate low $T_1$W and relatively high $T_2$W signals. The key role of MRI is in the detection of urethral tumor extension into paraurethral tissues and anterior vaginal wall, and involvement of the base of the urinary bladder which renders the tumor a high stage.[41] The reported accuracy of MRI in determining local tumor extension has been found to be 90%.[38] MRI is also very useful in assessing the response to therapy in patients treated

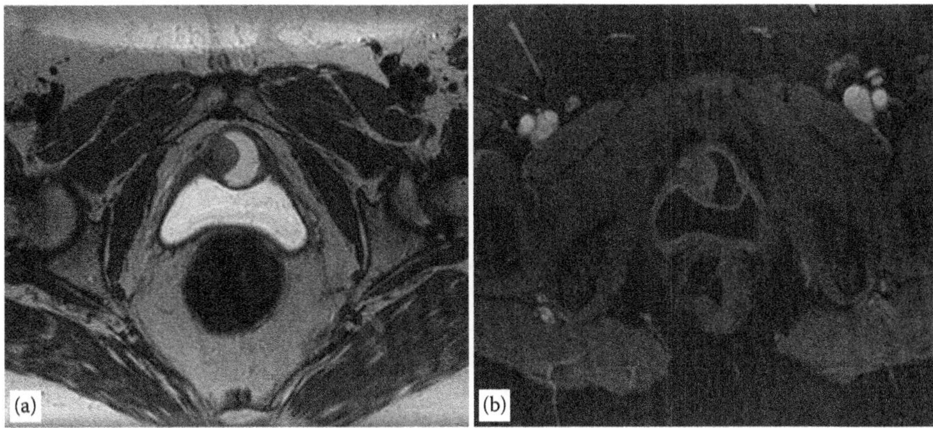

**FIGURE 2.18**
Axial $T_2$W (a) and contrast-enhanced axial $T_1$W (b) images of a patient with a urethral diverticulum. Urethral diverticula are associated with an increased incidence of carcinoma.

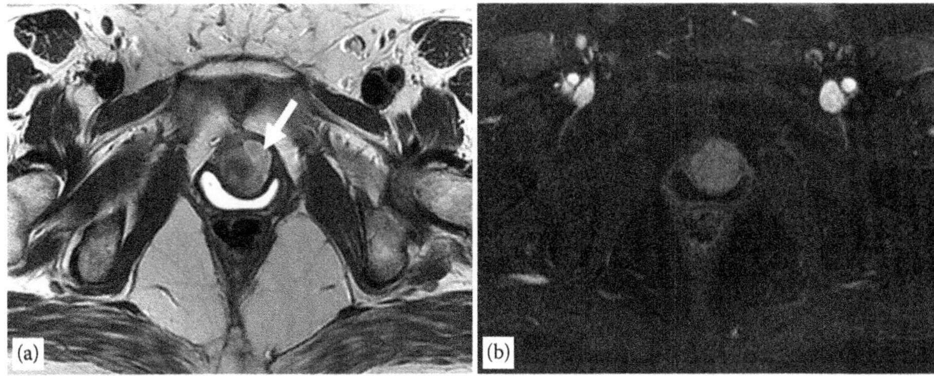

**FIGURE 2.19**
Axial $T_2$W (a) and contrast-enhanced axial $T_1$W (b) images of a patient with clear-cell adenocarcinoma of the urethra in a diverticulum. The $T_2$W image shows intermediate signal tumor in the diverticulum. Postcontrast image shows intensely enhancing tumor filling the diverticular sac.

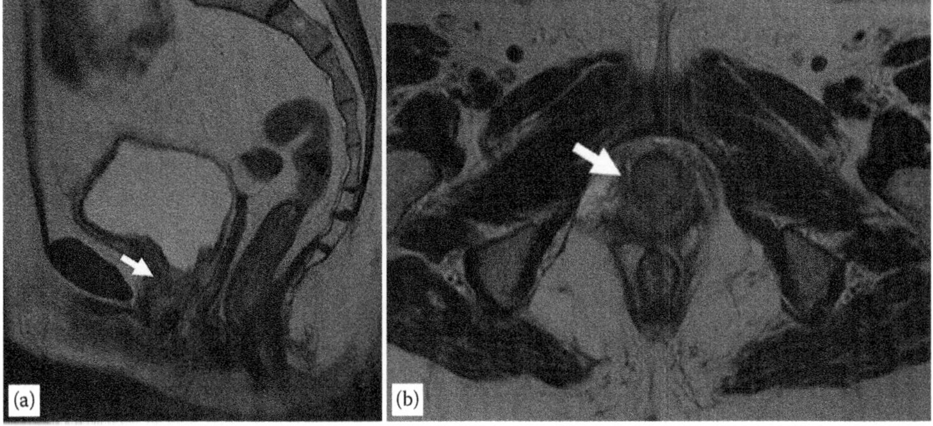

**FIGURE 2.20**
Sagittal $T_2$W (a) and axial $T_2$W (b) images of a patient with urethral carcinoma. The urethra is distended with tumor. Note the fat plane separating the tumor from the anterior wall of the vagina.

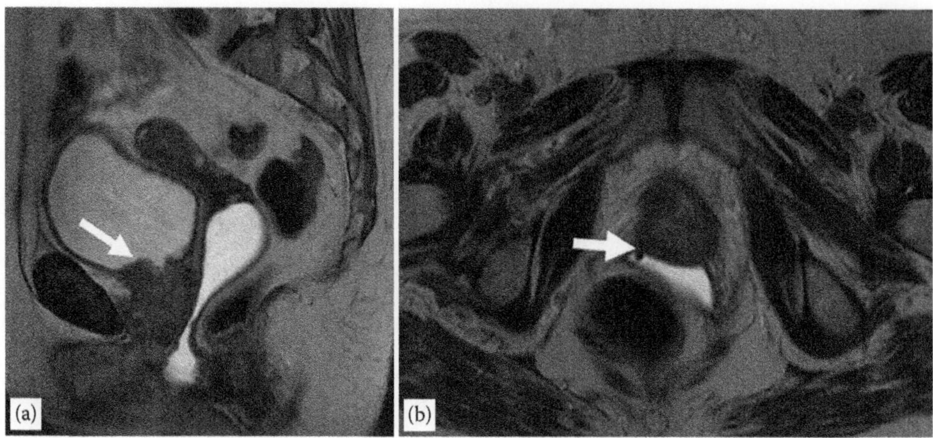

**FIGURE 2.21**
Sagittal $T_2$W image (a) and axial $T_2$W image (b) of a patient with urethra carcinoma extending to the base of the urinary bladder. Use of vaginal gel helps improve stage the vaginal wall involvement which can otherwise be challenging.

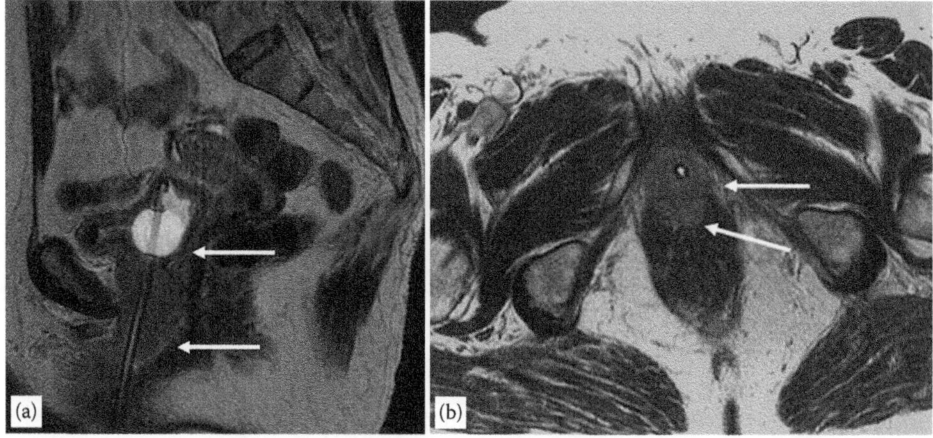

**FIGURE 2.22**
(a) Sagittal $T_2$W image of a patient with carcinoma of the urethra extending from the base of the urinary bladder and involving the entire length of the urethra. (b) Axial $T_2$W image shows periurethral invasion and involvement of the anterior wall of the vagina.

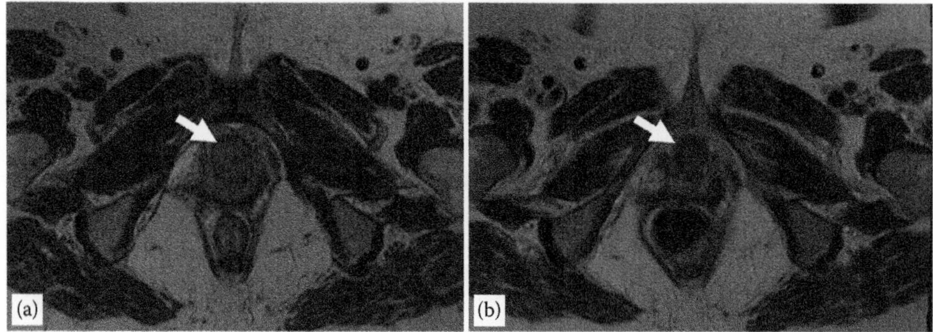

**FIGURE 2.23**
Pretreatment (a) and posttreatment (b) axial $T_2$W images of a patient with urethral carcinoma. MRI has been shown to be an excellent modality to monitor response to therapy.

with adjuvant therapy and evaluating the extent of surgery prior to exenterative surgery (Figure 2.23).[42]

Besides the urethra, tumors can also arise along the accessory glands (Skene gland, which is the female prostate[43]) and preexisting diverticular formations.

MRI is an excellent modality for visualizing and evaluating urethral diverticulum with a reported sensitivity and specificity of 100% in one study.[44] A urethral diverticulum needs to be differentiated from a submucosal cyst or a Skene gland abscess, both of which can present as a periurethral cystic mass and result in different treatment strategies. A communication between the urethra and the cystic mass is specific for a diverticulum, and they are typically found at the mid-urethra posteriorly with a hyperintense signal on the $T_2$W imaging.[45] Infections, stones, and carcinomas can develop within a preexisting diverticulum. The tumor within the diverticulum can be visualized as an enhancing soft tissue mass (Figure 2.21). Similar to the main urethra, transitional cell carcinoma, adenocarcinoma, and squamous cell carcinoma can develop within the diverticulum.[45]

Primary urethral carcinoma is rare, and cancers arising from accessory gland are even rarer, with a few case of adenoid cystic carcinoma reported in the literature.[46]

### 2.2.6 Conclusion

Primary carcinoma of the urethra is again rare. Although related to the lower urinary tract, it is unique and varied in histopathology. The male urethra and female urethra share differences and similarities in regard to their anatomy, function, and histopathology. The anatomic location of the tumor corresponds to its histopathology and prognosis. Squamous cell carcinoma is most commonly found in both sexes, followed usually by transitional cell carcinoma and adenocarcinoma. Tumors in the periurethral and preexisting diverticulum are even rarer. High-resolution MRI plays an important role in the evaluation and local staging of primary carcinoma of the urethra. The normal female urethra is recognized as a target appearance. Urethral tumors typically present as soft tissue mass with predominantly low signal intensity on $T_1$W and $T_2$W images and heterogeneous contrast enhancement.

## References


1. American Cancer Society. Cancer facts & figures 2014. Atlanta, GA: American Cancer Society; 2014.
2. Verma S, Rajesh A, Prasad SR et al. Urinary bladder cancer: Role of MR imaging. *RadioGraphics* 2012;32:371–87.
3. Knowles MA, Sidransky D, Cordon-Cardo C, Jones PA, Cairns P, Simon R, Amin MB, Tyczynski JE. Infiltrating urothelial carcinoma. In: JN Eble, JI Epstein, IA Sesterhenn, eds. *Pathology and Genetics of Tumours of the Urinary System and Male Genital Organs*. Lyon, France: IARC Press; 2004:93–109.
4. Cote RJ, Dunn MD, Chatterjee SJ et al. Elevated and absent pRb expression is associated with bladder cancer progression and has cooperative effects with p53. *Cancer Research* 1998;58:1090–4.
5. Wu XR. Urothelial tumorigenesis: A tale of divergent pathways. *Nature Reviews Cancer* 2005;5:713–25.
6. Edge S. *AJCC Cancer Staging Handbook*. 7th ed. Chicago, IL: American Joint Committee on Cancer; 2010.
7. Takahashi A, Tsukamoto T, Tobisu K et al. Radical cystectomy for invasive bladder cancer: Results of multi-institutional pooled analysis. *Japanese Journal of Clinical Oncology* 2004;34:14–9.
8. Stein JP, Lieskovsky G, Cote R et al. Radical cystectomy in the treatment of invasive bladder cancer: Long-term results in 1,054 patients. *Journal of Clinical Oncology: Official Journal of the American Society of Clinical Oncology* 2001;19:666–75.
9. Cornu JN, Neuzillet Y, Herve JM, Yonneau L, Botto H, Lebret T. Patterns of local recurrence after radical cystectomy in a contemporary series of patients with muscle-invasive bladder cancer. *World Journal of Urology* 2012;30:821–6.
10. Dalbagni G, Genega E, Hashibe M et al. Cystectomy for bladder cancer: A contemporary series. *Journal of Urology* 2001;165:1111–6.
11. Beyersdorff D, Zhang J, Schoder H, Bochner B, Hricak H. Bladder cancer: Can imaging change patient management? *Current Opinion in Urology* 2008;18:98–104.
12. De Haas RJ, Steyvers MJ, Futterer JJ. Multiparametric MRI of the bladder: Ready for clinical routine? *AJR American Journal of Roentgenology* 2014;202:1187–95.
13. Tekes A, Kamel I, Imam K et al. Dynamic MRI of bladder cancer: Evaluation of staging accuracy. *AJR American Journal of Roentgenology* 2005;184:121–7.
14. El-Assmy A, Abou-El-Ghar ME, Mosbah A et al. Bladder tumour staging: Comparison of diffusion- and T2-weighted MR imaging. *European Radiology* 2009;19: 1575–81.
15. Saito W, Amanuma M, Tanaka J, Heshiki A. Histopathological analysis of a bladder cancer stalk observed on MRI. *Magnetic Resonance Imaging* 2000;18:411–5.
16. Watanabe H, Kanematsu M, Kondo H et al. Preoperative T staging of urinary bladder cancer: Does diffusion-weighted MRI have supplementary value? *AJR American Journal of Roentgenology* 2009;192:1361–6.
17. Takeuchi M, Sasaki S, Naiki T et al. MR imaging of urinary bladder cancer for T-staging: A review and a pictorial essay of diffusion-weighted imaging. *Journal of Magnetic Resonance Imaging* 2013;38:1299–309.
18. Hayashi N, Tochigi H, Shiraishi T, Takeda K, Kawamura J. A new staging criterion for bladder carcinoma using gadolinium-enhanced magnetic resonance imaging with an endorectal surface coil: A comparison with ultrasonography. *BJU International* 2000;85:32–6.
19. Takeuchi M, Sasaki S, Ito M et al. Urinary bladder cancer: Diffusion-weighted MR imaging—Accuracy for diagnosing T stage and estimating histologic grade. *Radiology* 2009;251:112–21.
20. Kim B, Semelka RC, Ascher SM, Chalpin DB, Carroll PR, Hricak H. Bladder tumor staging: Comparison of contrast-enhanced CT, T1- and T2-weighted MR imaging, dynamic gadolinium-enhanced imaging, and late gadolinium enhanced imaging. *Radiology* 1994;193:239–45.
21. Deserno WM, Harisinghani MG, Taupitz M et al. Urinary bladder cancer: Preoperative nodal staging

with ferumoxtran-10-enhanced MR imaging. *Radiology* 2004;233:449–56.

22. Dayyani F, Hoffman K, Eifel P et al. Management of advanced primary urethral carcinomas. *BJU International* 2014;114:25–31.
23. Hartmann F. Tumours of the urethra. In: JN Eble, G Sauter, JI Epstein, IA Sesterhenn, eds. *WHO Pathology and Genetics of Tumors of the Urinary System and Male Genital Organs*. Lyon, France: IARC Press; 2004:154–7.
24. Bertolotto M, Valentino M, Barozzi L. Neoplasms of the urethra. In: SD Vikram, GT MacLennan, eds. *Genitourinary Radiology: Kidney, Bladder adn Urethra*. London: Springer; 2013:355–69.
25. Swartz MA, Porter MP, Lin DW, Weiss NS. Incidence of primary urethral carcinoma in the United States. *Urology* 2006;68:1164–8.
26. Grivas PD, Davenport M, Montie JE, Kunju LP, Feng F, Weizer AZ. Urethral cancer. *Hematology Oncology Clinics of North America* 2012;26:1291–314.
27. Carroll PR, Dixon CM. Surgical anatomy of the male and female urethra. *Urologic Clinics of North America* 1992;19:339–46.
28. Ryu JA, Kim B. MR imaging of the male and female urethra. *RadioGraphics* 2001;21:1169–85.
29. Mirilas P, Skandalakis JE. Urogenital diaphragm: An erroneous concept casting its shadow over the sphincter urethrae and deep perineal space. *Journal of American College of Surgeons* 2004;198:279–90.
30. Stoker J, Halligan S, Bartram CI. Pelvic floor imaging. *Radiology* 2001;218:621–41.
31. Kureel SN, Gupta A, Gupta RK. Surgical anatomy of urogenital diaphragm and course of its vessels in exstrophy-epispadias. *Urology* 2011;78:159–63.
32. Moore KL. The pelvis and peritoneum. In: KL Moore, ed. *Clinically Oriented Anatomy*. Baltimore, MD: Williams adn Wilkins; 1992:243–322.
33. Stewart SB, Leder RA, Inman RA. Imaging tumors of the penis and urethra. *Urologic Clinics of North America* 2010;37:353.
34. Kim B, Kawashima A, LeRoy AJ. Imaging of the male urethra. Seminars in Ultrasound CT and MR 2007;28:258–73.
35. Wolski Z, Tyloch J, Warsinski P. Primary cancer of the anterior urethra in a male patient. *Central European Journal of Urology* 2011;64:266–9.
36. Hayashi Y, Komada S, Maruyama Y, Hirao Y, Okajima E. Primary transitional cell carcinoma of the male urethra: Report of a case. *Hinyokika Kiyo* 1987;33:428–32.
37. Sawczuk I, Tannenbaum M, Olsson CA, deVere White R. Primary transitional cell carcinoma of prostatic periurethral ducts. *Urology* 1985;25:339–43.
38. Hricak H, Secaf E, Buckley DW, Brown JJ, Tanagho EA, McAninch JW. Female urethra: MR imaging. *Radiology* 1991;178:527–35.
39. Macura KJ, Genadry R, Borman TL, Mostwin JL, Lardo AC, Bluemke DA. Evaluation of the female urethra with intraurethral magnetic resonance imaging. *Journal of Magnetic Resonance Imaging* 2004;20:153–9.
40. Fisher M, Hricak H, Reinhold C, Proctor E, Williams R. Female urethral carcinoma—MRI staging. *AJR American Journal of Roentgenology* 1985;144: 603–4.
41. Surabhi VR, Menias CO, George V, Siegel CL, Prasad SR. Magnetic resonance imaging of female urethral and periurethral disorders. *Radiologic Clinics of North America* 2013;51:941–53.
42. Gourtsoyianni S, Hudolin T, Sala E, Goldman D, Bochner BH, Hricak H. MRI at the completion of chemoradiotherapy can accurately evaluate the extent of disease in women with advanced urethral carcinoma undergoing anterior pelvic exenteration. *Clinical Radiology* 2011;66:1072–8.
43. Kazakov DV, Stewart CJ, Kacerovska D et al. Prostatic-type tissue in the lower female genital tract: A morphologic spectrum, including vaginal tubulosquamous polyp, adenomyomatous hyperplasia of paraurethral Skene glands (female prostate), and ectopic lesion in the vulva. *The American Journal of Surgical Pathology* 2010;34:950–5.
44. Dwarkasing RS, Dinkelaar W, Hop WCJ, Steensma AB, Dohle GR, Krestin GP. MRI evaluation of urethral diverticula and differential diagnosis in symptomatic women. *AJR American Journal of Roentgenology* 2011;197:676–82.
45. Hahn WY, Israel GM, Lee VS. MRI of female urethral and periurethral disorders. *AJR American Journal of Roentgenology* 2004;182:677–82.
46. Ueda Y, Mandai M, Matsumura N et al. Adenoid cystic carcinoma of skene glands: A rare origin in the female genital tract and the characteristic clinical course. *International Journal of Gynecological Pathology* 2012;31:596–600.

# 3

# *Male Pelvis (Prostate, Seminal Vesicles, and Testes)*

**Jurgen J. Fütterer**

CONTENTS



## 3.1 Background

Urogenital organs include the testes, epididymis, vas deferens, ejaculatory ducts, urethra, penis, prostate, and seminal vesicles. The male reproductive system consists of these organs whose function is to accomplish reproduction. It consists of the testes, which produce spermatoza and hormones, a series of ducts that store and transport the sperm, the seminal vesicles, the prostate, and the penis. Imaging has given radiology a significant role in the characterization, diagnosis, and staging of tumors in these organs.

The seminal vesicles and vas deferens are ancillary but essential urogenital organs [1]. Primary malignant pathologies involving the seminal vesicles are extremely rare with most of the reported cases being carcinomas, whereas secondary neoplasms are much more common [2]. Prostate cancer is a multifactorial disease. Age and a positive family history of prostate cancer are the main risk factors. Other factors are the type of diet, lifestyle-related factors, and certain genetic defects. Prostate cancer is the second most frequently diagnosed malignancy and the sixth leading cause of cancer-related death mortality in males worldwide [3]. From autopsy studies, it is known that prostate cancer can be found in 55% and 64% of men in their fifth and seventh decades, respectively [4,5]. The widespread use of prostate-specific antigen (PSA) as a screening test for prostate cancer resulted in earlier detection of prostate cancer at a curative stage. A major concern related to prostate cancer screening and early detection is overdiagnosis and overtreatment of indolent disease. Strategies to reduce overdiagnosis are necessary, as are strategies to differentiate indolent from aggressive tumors [6,7].

The incidence of testicular cancer has almost doubled in the past 30 years [3]. Bilateral tumors are found in 0.7% of men with germ cell tumors at diagnosis, and 1.5% of patients develop metachronous lesions within 5 years [8]. Once the leading cause of cancer deaths in men between 15 and 35 years of age, it has now proved to be a model of success with a 5-year survival rate exceeding 95% [9]. About 36,000 men are diagnosed with testicular cancer each year. Imaging plays an important role in assessing the primary tumor, accurately staging the disease, and conducting follow-up imaging surveillance [10].

Imaging has become an indispensable tool in cancer research, clinical trials, and medical practice. Improvement of the spatial and temporal resolution to provide anatomic detail and functional imaging techniques greatly enhanced tumor detection and staging. This chapter focuses on magnetic resonance (MR) imaging (MRI) of the prostate, seminal vesicles, and testes.

## 3.2 Prostate and Seminal Vesicles

The current diagnostic pathway in men with elevated serum PSA levels and/or abnormal digital rectal examination consists of a random standardized systematic transrectal ultrasound (TRUS)-guided prostate biopsy. The clinical stage is identified using clinical variables (digital rectal examination, PSA, and Gleason score) and is expressed in the tumor node metastasis (TNM) staging classification. The Gleason score is a histopathologic score that correlates with prostate cancer prognosis. In order to determine the treatment that optimally suits the individual patient, it is necessary to evaluate all patient and clinical characteristics.

### 3.2.1 Prostate Anatomy

The prostate gland is a complex organ composed of four different zones that have different embryological origins and different functional activities. Its walnut-sized gland is situated directly caudal to the bladder. It envelops the prostatic urethra and the ejaculatory ducts. The seminal vesicles are paired grapelike pouches filled with fluid that are located caudolateral to the corresponding deferent duct, between the bladder and the rectum.

The prostate is subdivided into the apex, the mid prostate, and the base (directed upward to the inferior border of the bladder). The neurovascular bundle courses bilaterally along the posterolateral aspect of the prostate and is a preferential pathway of tumor spread. The prostate is anatomically divided on the basis of embryological origins into four zones that are eccentrically located around the urethra: (1) the transition zone, which contains 5% of the glandular tissue; (2) the central zone, which contains 20% of the glandular tissue; (3) the peripheral zone, which contains 70%–80% of the glandular tissue; and (4) the anterior fibromuscular zone, which is nonglandular [11]. The central zone has histologic features more akin to those of the ejaculatory ducts and seminal vesicles, suggesting that it is a Wolffian duct derivate, whereas the remaining prostate (including the transition and peripheral zones) derives from the urogenital sinus [12,13].

Knowledge of the zonal anatomy of the prostate is very useful, considering that many prostatic diseases have a zonal distribution. Up to 70%–80% of prostate cancer is located in the peripheral zone, whereas about 20% emerges in the transition zone and 10% in the central zone [5,14]. Prostate cancers in the transition zone have a relatively lower Gleason score, local stage, and biochemical recurrence rate compared to cancers located in the peripheral zone [15–17]. Radiologically, in older patients, it is difficult to differentiate the transition zone from the central zone due to compression of the central zone by benign prostatic hyperplasia (BPH). Most transition zone tumors have additional tumor foci in the peripheral zone. Another aspect of the prostate anatomy that is relevant to radiologic imaging relates to the prostatic capsule.

Worldwide, 30 million men have symptoms related to BPH. According to the National Institutes of Health, BPH affects more than 50% of men over age 60 and as many as 90% of men over the age of 70. BPH is a disease of the prostate gland and consists of nodular hypertrophy of the fibrous, muscular, and glandular tissue within the periurethral and transition zones. The exact pathophysiology of BPH in the prostate is unknown. It is probably associated with hormonal changes that occur as men age.

There is still a debate about the prostate having a capsule or not. The prostate is contained within a thin, fibrous, adherent capsule that is most apparent posteriorly and posterolaterally to the prostate [18]. The *true* prostatic capsule is a thin (0.5–2 mm) layer of connective tissue located external to the peripheral zone. Around this layer, there is the pelvic fascia, often called the *false* prostatic capsule.

### 3.2.2 Multiparametric MRI

Due to its high soft tissue contrast, high resolution, and ability to simultaneously image functional imaging parameters, MRI provides the best visualization of the prostate compared to other available imaging methods. MRI is currently the most accurate modality to preoperatively stage prostate cancer. It has a higher accuracy in the assessment of intraprostatic disease (T2 disease), extracapsular extension, and seminal vesicle invasion (T3 disease) as well as invasion of periprostatic structures, compared to other current available imaging modalities. Anatomical MRI using $T_2$-weighted imaging alone fares poorly in making the diagnosis of prostate cancer, with a maximum joint sensitivity and specificity rate of 74% [19]. Functional imaging techniques are being applied to overcome the limitations of anatomical imaging.

The following functional MRI techniques are clinically applied in the prostate: (1) diffusion-weighted MRI (DWI), (2) dynamic contrast-enhanced MRI (DCE-MRI), and (3) MR spectroscopy imaging [20–27]. This multiparametric MRI (mpMRI) approach (Figure 3.1) has resulted

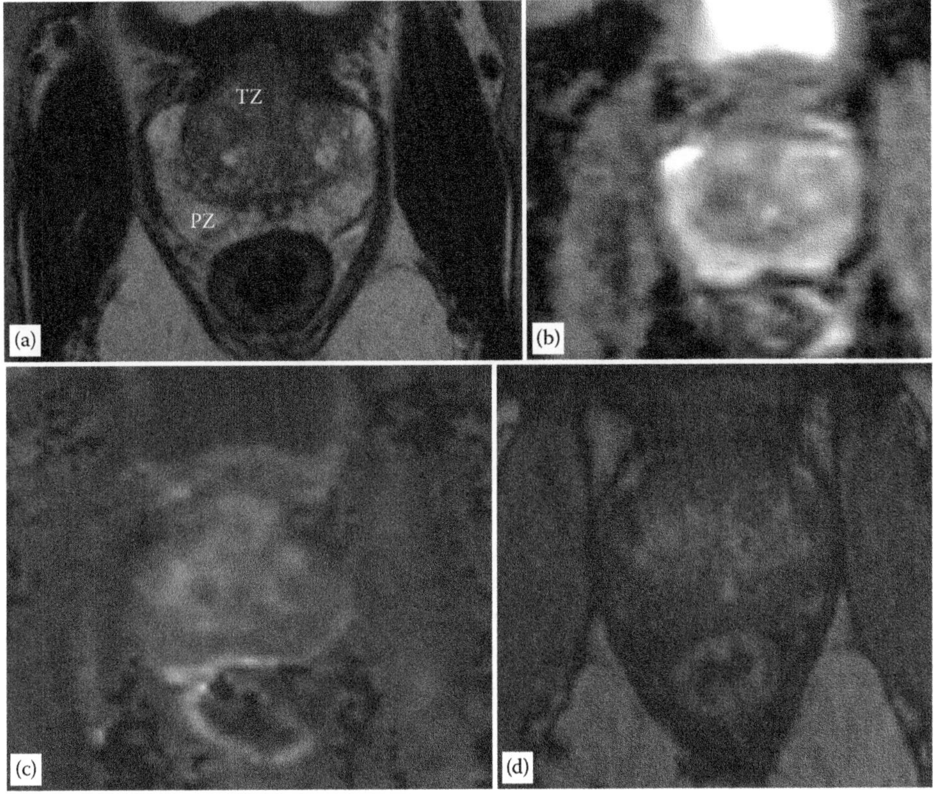


**FIGURE 3.1**
A 67-year-old male with one negative TRUS-guided biopsy session and a PSA of 7.6 ng/mL. The MR examination did not reveal signs of prostate cancer. (a) Axial $T_2$-weighted MR image demonstrates a relatively higher signal intensity in the peripheral zone (PZ) compared to the transition zone (TZ). (b) The ADC map at the same level as (a) demonstrates no restricted diffusion in the peripheral zone. (c) The high $b$-value image ($b = 1400$) does not display high signal in the peripheral or transition zone. (d) The postcontrast $T_1$-weighted MR image demonstrates no early enhancement in the peripheral zone.

in accurate prostate cancer detection and localization [27–29]. Multiparametric prostate MRI plays an important role in the detection, localization, staging, image-guided targeted prostate biopsy, and assessment of posttreatment changes in prostate cancer.

#### 3.2.2.1 $T_1$-Weighted MRI

Morphologic $T_1$-weighted MRI has a limited role for prostate cancer detection. The $T_1$-weighted image contrast in the prostate is low. Therefore, the zonal anatomy cannot be discerned using $T_1$-weighted MRI. The neurovascular bundles can be identified because they are embedded in the periprostatic bed. $T_1$-weighted imaging is mainly used for the detection of TRUS-guided biopsy hemorrhage or artifacts. The latter can be a confounding factor on $T_2$-weighted MR images [30]. The degree of biopsy hemorrhage is significantly less in areas of prostate cancer than in regions where biopsy shows benign changes [31]. This may be explained by the anticoagulant effect of citrate (produced within the prostate gland) [32,33]. The level of citrate is presumably lower in an area of cancer than in the normal peripheral zone, and as a result, hemorrhagic foci in tumorous areas may resolve more rapidly. The latter is coined as the hemorrhage exclusion sign and can be used to potentially take advantage of the distribution of the hemorrhage as a tool to aid tumor localization [29,34].

#### 3.2.2.2 $T_2$-Weighted MRI

$T_2$-weighted MRI of the prostate provides anatomical information and the zonal anatomy is clearly distinguished. This anatomical imaging technique is the cornerstone of mpMRI of the prostate. The signal intensity of healthy peripheral zone tissue appears as an intermediate-to-high signal, whereas the transition zone has lower signal intensity than the peripheral zone. Furthermore, the latter zone has variable amounts of intermediate signal intensity, which is often replaced by well-circumscribed BPH nodules (Figure 3.2). The central zone is symmetric and elicits homogeneous low signal intensity [13]. The degree of signal intensity on

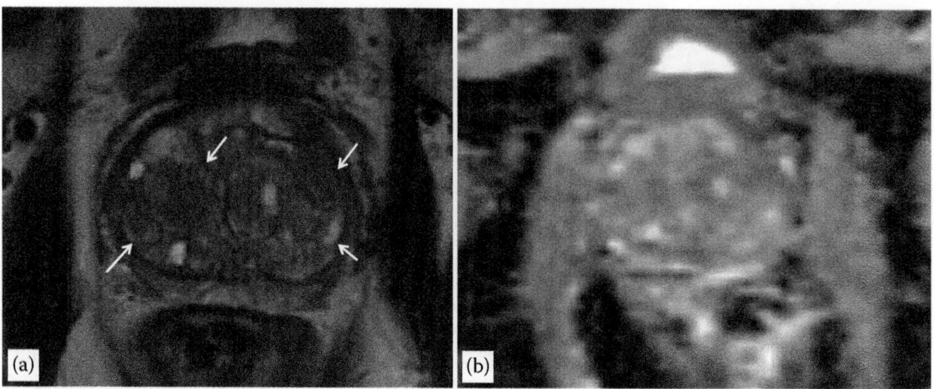

**FIGURE 3.2**
A 76-year-old male patient who complains of lower urinary tract symptoms underwent MRI of the prostate. (a) Well-circumscribed BPH nodules in the transition zone (arrows). (b) The ADC map demonstrate variable amounts of ADC restriction in these BPH nodules.

the $T_2$-weighted MR image decrease may differ with the Gleason score [35]. Higher Gleason grade component 4 or 5 has shown lower signal intensities than does lower Gleason grade component 2 or 3.

High in-plane resolution $T_2$-weighted fast or turbo spin-echo MR sequences, performed with endorectal and/or external pelvic phased array coils, are generally used to depict prostate anatomy and the surrounding tissue. The seminal vesicles demonstrate an intraluminal high signal. The seminal vesicle wall has a well-defined low-signal-intensity rim.

On MRI, prostate cancer is characterized as an area of low signal intensity within the brighter, healthy peripheral zone using a $T_2$-weighted MRI sequence [36] (Figure 3.3). However, some lesions are iso- to hyperintense, which implies a limited sensitivity. In addition to carcinoma, the differential diagnosis of an area of low signal intensity in the peripheral zone includes postbiopsy hemorrhage, prostatitis, BPH, effects of hormone or radiation treatment, scars, calcifications, smooth muscle hyperplasia, and fibromuscular hyperplasia. Consequently, a focus or area of low signal intensity on $T_2$-weighted images has a low specificity. Therefore, functional techniques are added to the anatomical information in order to improve this low specificity.

In the transition zone, prostate cancer is not as clearly discernable because the transition zone generally has lower signal intensity than the peripheral zone and due to its inhomogeneous nature. Transition zone cancers are generally of lower grade and stage. However, zonal location of high Gleason grade prostate cancer did not influence biochemical relapse-free survival [37]. In the elderly males, owing to variable extension of the transition zone due to BPH, the size and signal intensity of the prostate transition zone may vary. Benign nodular hyperplasia of the prostate gives rise to nodular adenomas in the transition zone, and with time these will compress the central zone, which thus forms a pseudocapsule, thereby occupying the complete transition zone. The peripheral zone is almost never affected by BPH and will retain its own histological characteristics. BPH is a round or oval nodule with a well-defined low-signal-intensity rim (capsule-like). The nodule is inhomogeneous with the (variable) intermediate signal intensity and the low-signal-intensity rim that surrounds the expanded transition zone. Important signs

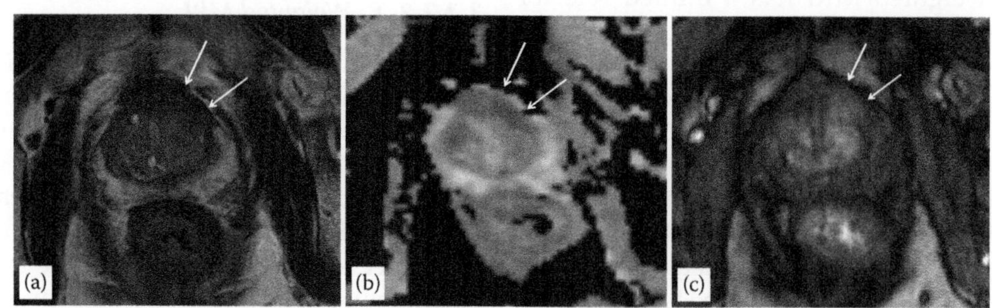

**FIGURE 3.3**
A 67-year-old patient with biopsy-proven prostate cancer (Gleason score of 7 [3 + 4] and PSA of 12 ng/mL). (a) The axial $T_2$-weighted anatomical MR image shows a lesion in the left transition zone (arrows). (b) This is confirmed on the ADC map showing restricted diffusion (arrows). (c) The postcontrast $T_1$-weighted MR image demonstrates an increased asymmetric enhancement compared to the contralateral site (arrows).

of a transition zone cancer include homogeneous $T_2$ low signal intensity, disruption of the low-signal-intensity rim of a BPH nodule, and mass effect [38]. Unless transition zone cancers have a large volume (>4 $cm^3$), a higher Gleason grade (grade 4–5), or both, their $T_2$-weighted MRI detection sensitivity is low (8%–30%) [39].

#### 3.2.2.3 Diffusion-Weighted MRI

Free water molecules exhibit constant random motion in tissue (Brownian effect), which is related to thermal kinetic energy [40]. DWI can quantify this water motion in an indirect manner [41]. The DWI pulse sequence labels hydrogen nuclei in space, of which most will be part of water molecules at any moment, and determines the length of the path that water molecules travel over a short period of time. However, DWI is not able to determine the distance traveled by individual hydrogen nuclei but is capable of estimating the mean distance traveled by all hydrogen nuclei in every voxel of imaged tissue. This quantification of proton diffusion properties in water is used to produce image contrast. Images that reflect proton diffusion are acquired by applying motion-encoding gradients, which cause phase shifts in moving protons, depending on the direction and quantity of their movement [41].

The greater the mean distance, the more self-diffusion of water molecules has taken place in a certain time interval [42]. The attenuation of the MR signal in DWI is expressed with the Stejskal–Tanner equation [41]. The $b$-value and the apparent diffusion coefficient (ADC) are components in this equation. The amount of diffusion in tissue is determined by the diffusion coefficient ($D$). In a volume of pure water, this self-diffusion is equal in all directions, hence isotropic, and not restricted by any barrier. As diffusion in tissue is limited by cellular structures, to establish a reliable estimate of this mean distance traveled by hydrogen nuclei, an ADC is acquired in at least three different orthogonal directions [42,43]. This phenomenon of varying restriction of self-diffusion along different axes is called anisotropy and can also be used for tissue characterization, whereas in linear aligned tissue, this anisotropy will be more pronounced as there will be one direction that contributes most to the ADC. The ADC reflects the movement of the water molecules between the two pulsed field gradients (of duration $\delta$). The ADC quantifies the combined effects of both capillary perfusion and diffusion. DWI has $T_2$-weighted MRI and DWI characteristics. The $b$-value specifies the sensitivity of diffusion [43].

DWI has to be obtained with at least two $b$-values to correct for this $T_2$-weighted effect in every diffusion-weighted sequence, which may be present in lesions with high $T_2$ signal intensity. For differentiation between a lesion with a low ADC value and one that displays high $T_2$-signal intensity, at least two $b$-values have to be used and the images consequently subtracted, typically a low, $b$ = 0–50, and a high diffusion-weighted acquisition. Correctly assigning the $b$-value for DWI is critical important because it can significantly affect ADC estimates [44]. ADC values can produce a similar discriminant performance in distinguishing prostate cancer from normal tissue region of interests and in correlation with the Gleason score, but an appropriate ADC cutoff value needs to be selected specifically for each $b$-value combination. For prostate cancer, $b$-values of 50–100 and 800 s/$mm^2$ are typically used [24]. High $b$-values, that is, ≥1000, may increase the accuracy of prostate cancer detection within the transition zone [45]. However, high $b$-values are affected by lower signal-to-noise ratio [46]. Image quality is dependent on gradient system and field strength. To minimize the influence of bulk motion as a distorting factor, typically a time to echo as short as possible is chosen for DWI sequences.

Healthy prostate tissue in the peripheral zone is rich in tubular structures. This will allow for abundant self-diffusion of water molecules in their contents and thus provides high ADC values. Prostate cancer tissue destroys the normal glandular structure of the prostate and replaces the ducts. It also has a higher cellular density than does healthy prostate peripheral zone tissue. In most cases, the peripheral zone can be easily discriminated from the transition zone on DWI, as it will display relative higher ADC values [47]. On ADC maps, therefore, prostate cancer often shows lower ADCs compared to surrounding healthy peripheral zone prostate tissue [48–50].

BPH is defined by hyperplasia of all cells that constitute the transition zone, with glandular, muscular, and fibrous compartments more or less evenly involved. This nodular hyperplasia gives rise to inhomogeneous diffusion patterns, and as tubular structures often remain in place, the increased cellular density of hyperplasia, which is far less predominant than in prostate carcinoma, might explain the observed reduction in ADC levels of the central gland on DWI, due to decreased ratio of extracellular to intracellular volume. However, as BPH has inhomogeneous diffusion characteristics, an increase in ADC has been observed as well [51]. ADC values have a good discriminatory power to distinguish tumors with Gleason score of 7 and those with Gleason score <7 [52]. DWI-directed MRI-targeted biopsies significantly improve pretreatment risk stratification by obtaining biopsies that are representative of true Gleason grade [53] (Figure 3.4).

Chronic prostatitis is histopathological characterized by extracellular edema surrounding the involved prostatic cells with concomitant aggregation of lymphocytes, plasma cells, macrophages, and neutrophils in the

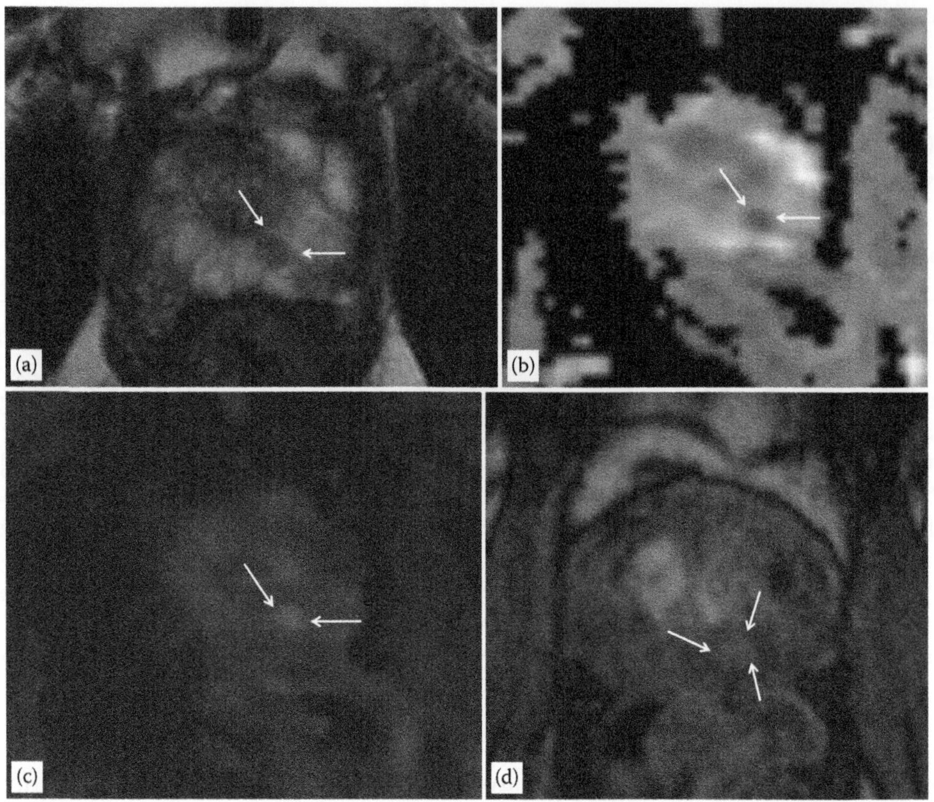

**FIGURE 3.4**
A 73-year-old male with biopsy-proven prostate cancer in the left apical peripheral zone (Gleason score of 7 and PSA of 4.2). (a) The axial $T_2$-weighted anatomical MR image shows a lesion in the left peripheral zone (arrows). (b) This is confirmed on the ADC map showing restricted diffusion (arrows). (c) The high *b*-value MR image demonstrates high signal intensity in the left peripheral zone. (d) The postcontrast $T_1$-weighted MR image demonstrates an increased asymmetric enhancement compared to the contralateral site (arrows).

prostatic stroma. This abundance in cells compared to normal prostatic tissue may lead to a decreased ADC due to decreased extracellular-to-intracellular fluid volume ratio. DWI shows differences between prostatitis and prostate cancer in both the peripheral zone and the central gland, with higher ADC values in prostatitis compared with prostate cancer, although its usability in clinical practice is limited as a result of significant overlap in ADCs (Figure 3.5) [54,55].

### *3.2.2.4 Dynamic Contrast-Enhanced MRI*

The microvascular network formed by the capillaries supplies the tissues and permits their function. It provides a considerable surface area for exchanges between the blood and the tissues [56]. All pathological conditions cause changes in the microcirculation. Angiogenesis (the sprouting of new capillaries from existing blood vessels) and vasculogenesis (the *de novo* generation of new blood vessels) are the two primary methods of vascular expansion by which nutrient supply to tumor tissue is adjusted to match physiologic needs [57]. Within prostate cancer tissue, angiogenesis is induced by secretion of vascular growth factors, in reaction to the presence of local hypoxia or lack of nutrients [58]. Resulting changes in vascular characteristics can be studied well with DCE-MRI. These include (1) spatial heterogeneity and chaotic structure; (2) poorly formed fragile vessels with high permeability to macromolecules because of the presence of large endothelial cell gaps, incomplete basement membrane, and relative lack of pericytes or smooth muscle association with endothelial cells; (3) arteriovenous shunting; (4) intermittent or unstable blood flow; and (5) extreme heterogeneity of vascular density, with areas of low vascular density mixed with regions of high angiogenic activity [59].

There are two different approaches in acquiring functional DCE-MRI data [60]. First, dynamic susceptibility MR sequences ($T_2$*-weighted) are sensitive to the vascular phase of contrast medium delivery, which reflect on tissue perfusion and blood volume. However, limited evidence is gathered for clinical application in the prostate and currently not used in clinical practice. Second, $T_1$-weighted sequences designed to be sensitive

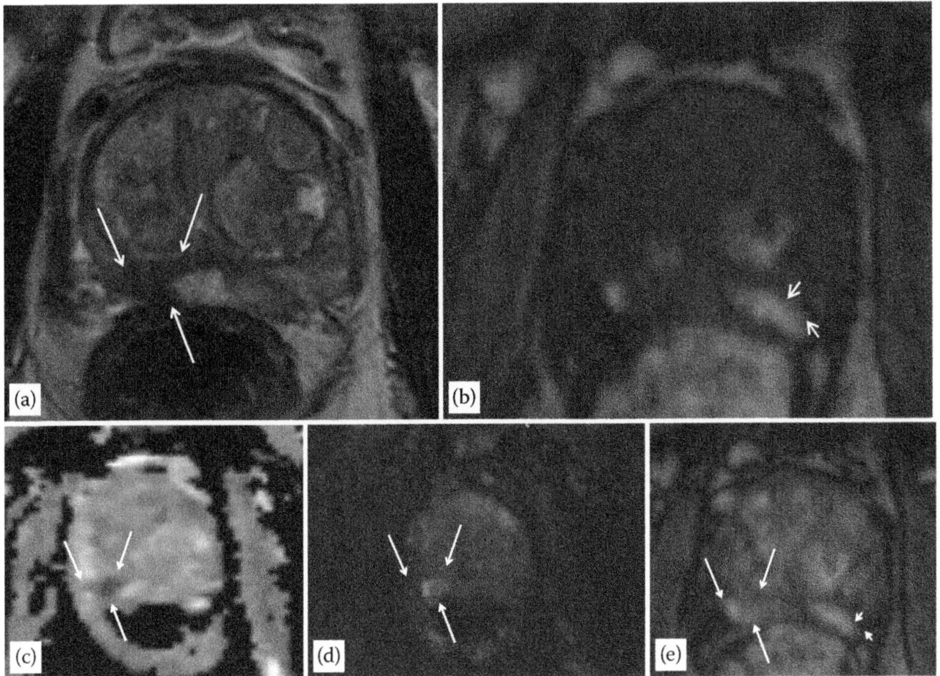


**FIGURE 3.5**
A 72-year-old male with biopsy-proven prostate cancer in the right peripheral zone (Gleason score of 8, 3 out of 6 cores positive, and PSA level of 6.8). Stage T3a prostate cancer. (a) The axial $T_2$-weighted anatomical MR image shows a lesion in the right peripheral zone with extension toward the rectum (arrows), suggesting extracapsular extension. (b) A second lesion (small arrows) is visible in the left peripheral zone; however, the $T_1$-weighted sequence demonstrates high signal intensity in this area confirming biopsy artifact. (c) The lesion in the right peripheral zone is confirmed on the ADC map showing restricted diffusion (arrows). (d) The high *b*-value MR image demonstrates high signal intensity in the right peripheral zone (arrows). (e) The postcontrast $T_1$-weighted MR image demonstrates an early enhancement in the right peripheral zone (arrows) and unchanged signal intensity in the left peripheral zone (small arrows).

to the presence of contrast medium in the extravascular, extracellular space, thus reflecting on microvessel perfusion, permeability, and extracellular leakage space (relaxivity-based methods). The latter is currently the most often used method in prostate mpMRI.

The three most important aspects of DCE-MRI include (1) intravenous contrast agent administration, (2) fast temporal resolution imaging covering the entire prostate, and (3) qualitative or quantitative estimation of signal intensity changes to determine the pharmacokinetic parameters of perfusion [45]. DCE-MRI consists of a series of fast $T_1$-weighted sequences covering the entire prostate before and after rapid injection of a bolus of a low-molecular-weight gadolinium chelate. Because the prostate as a whole is vascularized, a simple comparison of pre- and postgadolinium images is insufficient to discern prostate cancer. Subtraction technique can be applied to demonstrate early enhancement.

Assessment of signal intensity changes on $T_1$-weighted DCE-MR images in order to estimate contrast agent uptake *in vivo* can be performed qualitatively, semiquantitatively, or quantitatively.

The *qualitative analysis* of signal intensity changes can be achieved by assessing the shape of the signal intensity–time curve. This approach is most commonly used in daily practice.

The *semiquantitative* approach describing signal intensity changes by using a number of parameters such as (1) the onset time of the signal–intensity curve ($t_0$ = time from appearance in an artery to the arrival of contrast agent in the tissue of interest); (2) the slope and height of the enhancement curve (time-to-peak); (3) maximum signal intensity (peak enhancement); and (4) persistent delayed phase (type 1 curve), plateau delayed phase (type 2 curve), and washout delayed phase (type 3 curve) [61].

The *quantitative* approach uses pharmokinetic modeling, which is usually applied to changes in the contrast agent concentrations in tissue. Concentration–time curves are mathematically fitted by using one of many described pharmokinetic models, and quantitative kinetic parameters are derived. These include (1) transfer constant of the contrast agent ($KT_{rans}$), (2) rate constant ($K_{ep}$), and (3) interstitial extravascular, extracellular space ($V_e$) [62].

##### *3.2.2.5 Local Staging*

Local staging is accomplished by examining the capsule and seminal vesicles. mpMRI is currently the most accurate imaging modality to preoperatively stage prostate cancer [63]. MRI has a higher accuracy in the assessment of intraprostatic disease, extracapsular extension, and seminal vesicle invasion as well as invasion of periprostatic structures, compared to other imaging techniques. In patients in whom the diagnosis of cancer has been established, reliably determining its local stage, along with the localization of tumor within the gland, is an important element of prostate MRI. In the last decade, the focus of MRI in prostate cancer moved from staging to localization and detection of prostate cancer. Information regarding the location of the tumor, capsule involvement, tumor volume, and neurovascular bundle integrity is becoming more important than *simple* stage information.

On $T_2$-weighted images, extracapsular extension can be detected by visualizing the direct extension of the tumor into the periprostatic fat. Indirect imaging criteria for the detection of extracapsular extension are asymmetry of the neurovascular bundle, obliteration of the rectoprostatic angle, tumor bulge into the periprostatic fat, broad tumor contact with the surface of the capsule, and capsular retraction [63–67]. DCE-MRI and DWI techniques improve local staging [68]. Most likely, the acquired functional information may be used to guide and draw the attention of the less experienced radiologist to a particular area.

## 3.3 Testes

The first step in diagnosing testicular cancer is usually through self-examination. Testicular cancer most commonly presents as a painless scrotal mass. Although representing 1% of all cancers in males, it is the most common neoplasm in boys and young adults from 15 to 34 years old [69]. The number of new cases of testis cancer is 5.6 per 100,000 men per year. The number of deaths is 0.3 per 100,000 men per year. These rates are age adjusted and based on 2008–2012 cases and deaths [70]. The 5-year survival rate is exceeding 95%. The high success rate in treatment is related to improved staging and treatment methods. Imaging plays a central role in assessment of tumor bulk, sites of metastases, monitoring response to therapy, surgical planning, and accurate assessment of disease at relapse [71].

The incidence of testicular cancer has doubled in the past 40 years. Bilateral tumors are found in 0.7% of men with germ cell tumors at diagnosis, and 1.5% of patients develop metachronous lesions within 5 years.

### 3.3.1 Anatomy

The testes and their efferent duct systems, epididymis, and vas deferens develop from the gonadal ridges and mesonephric duct. The adult testis consists of densely packed seminiferous tubules, which are septated by thin fibrous septa. The testes lie within the scrotum, a sac composed of internal cremasteric and external fascial layers, dartos muscle, and skin [72]. They are encased by the tunica albuginea, a fibrous capsule that invaginates into the testis posteriorly to form the mediastinum testis. The tunica albuginea is covered by a flattened layer of mesothelium, the tunica vaginalis. In a normal adult testis, there are 200–300 lobules, each of which contains 400–600 seminiferous tubules.

### 3.3.2 Histopathology

Ninety-five percent of all testicular tumors are germ cell tumors. The remaining are lymphomas (4%) and Leydig or Sertoli cell tumors. Testicular germ cell tumors are derived from spermatogenic cells and may be classified as unipotential or totipotential. Unipotential tumors are seminomas, which comprise 35%–50% of all germ cell tumors. Nonseminomatous germ cell tumors are considered to be totipotential.

Serum tumor markers are especially helpful to differentiate germ cell tumors from each other and from other malignancies. Serum concentrations of alpha-fetoprotein (AFP) and/or beta-human chorionic gonadotropin (beta-hCG) are elevated in 80%–85% of nonseminomas. By contrast, serum beta-hCG is elevated in fewer than 25% of testicular seminomas, and AFP is not elevated in pure seminomas. However, these tumor markers cannot accurately assess disease bulk or locate sites of tumor spread [71].

### 3.3.3 Staging of Testicular Cancer

As soon as the diagnosis of germ cell cancer has been pathologically confirmed, further staging examinations are warranted to examine the extent of disease. The TNM staging is used to stage testicular cancer. Currently, abdominal and chest computed tomography is the standard technique in primary staging.

The most common sites for metastases are via the lymphatic system to the retroperitoneal nodes and via the hematogenous route to the lungs and less common to the liver, brain, and bone. In general, advanced stage disease will be treated primarily with chemotherapy. Nonseminomatous germ cell tumors appear as multiple small peripheral nodules, whereas seminoma metastases tend to be larger masses [71]. Rare sites of hematogenous metastases include the adrenals, kidneys, spleen, pleura, pericardium, and peritoneum [71].

Lymphatic spread occurs via lymphatic channels (from the spermatic cord and testicular vessels to retroperitoneal lymph nodes). Usually, right-sided testicular neoplasms spread to the right side of the retroperitoneum. Lymph node metastases can be identified around the inferior vena cava and between the level of right renal hilum and the aortic bifurcation. However, lymph node metastases of left-sided testicular cancer may be found adjacent to the abdominal aorta and just below the left renal vein. Contralateral involvement is uncommon but may occur with a larger disease burden [73]. Pelvic lymph adenopathy is uncommon in the absence of bulky disease [74].

### 3.3.4 MR Imaging

Ultrasonography is currently the primary imaging modality in the assessment of scrotal disease. However, because of its wide field of view, multiplanar capabilities, and intrinsic high soft tissue contrast, MRI may represent an efficient supplemental technique for scrotal imaging [75]. Tumors are typically isointense with normal testis on $T_1$-weighted MR images, becoming inhomogeneous and slightly darker on proton density-weighted MR images and moderately darker on $T_2$-weighted images.

MRI is performed in supine positioning and surface coils (phased array) are positioned above the testicles. The scrotum is elevated by means of a support between the tights (folded towel is placed between the tights) to ensure that both testes are in the same horizontal plane for proper coronal imaging. The penis is angled to the side and the whole region is draped.

$T_1$- and $T_2$-weighted sequences in at least two planes are acquired (preferably in the coronal and axial planes). The coronal plane is the optimal plane for imaging the scrotum. The coronal orientation covers from the posterior aspect of the scrotum to the anterior aspect of the external inguinal ring. DCE subtraction MRI can be used to differentiate testicular diseases from scrotal disorders [76]. In general, DCE-MRI is reserved for patients with subtle, atypical, or complex findings. For staging purposes, $T_1$-weighted axial images of the abdomen should be obtained to detect lymph node metastases.

The normal testis is a sharply demarcated oval structure of intermediate homogeneous signal intensity to muscle on $T_1$-weighted images and homogeneous higher signal intensity (less than fluid signal intensity) on $T_2$-weighted images. Signal intensity of the epididymis is low on both $T_1$- and $T_2$-weighted images. The epididymis is more clearly differentiated from the testis on $T_2$-weighted images, because it is lower in signal intensity than the adjacent testis. The testis is completely surrounded by the tunica albuginea, a thick layer of dense fibrous tissue. The tunica albuginea and testicular septa appear as low-signal-intensity structures [75]. The mediastinum testis can be identified as a low-signal-intensity band within the posterior testis on $T_2$-weighted images (Figure 3.6).

The MRI appearance of seminoma, like its histology, has been consistent. Seminoma is usually homogeneous in appearance and relatively isointense to the normal testicular parenchyma on $T_1$-weighted images and low signal intensity on $T_2$-weighted images (Figure 3.7). However, MRI cannot reliably predict the histological type [77]. Although, atypical, some seminomatous lesions may bleed internally, resulting in a focus of different signal dependent on the age of the hemorrhage.

Nonseminomatous germ cell tumors are more likely to have cystic areas than seminomas. The former tumors, given their histologic patterns, are markedly inhomogeneous on MR images, which represent their most distinctive feature compared with seminomatous lesions. At MRI, these tumors are usally iso- to hyperintense on $T_1$-weighted MR images and hypointense on $T_2$-weighted MR images. The overall heterogeneous

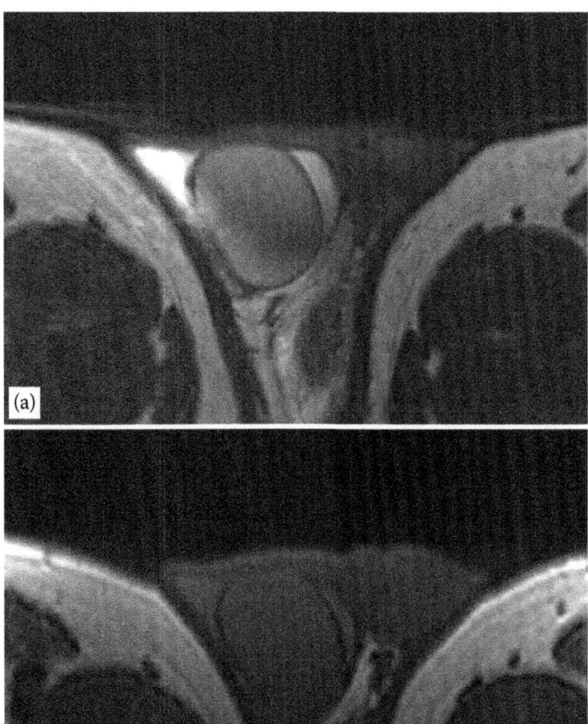


**FIGURE 3.6**
A 40-year-old male with normal testes. (a) The $T_2$-weighted image demonstrates a testis with homogeneous higher signal intensity (less than fluid signal intensity) to muscle. (b) The $T_1$-weighted image demonstrates a sharply demarcated oval structure of intermediate homogeneous signal intensity to muscle.

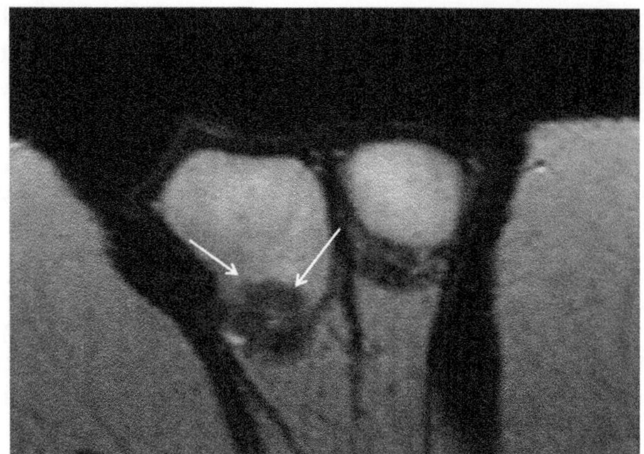

**FIGURE 3.7**
A 26-year-old male with a painless testicular mass. The $T_2$-weighted MR image demonstrates a small lesion in right testis (low signal mass, arrows) representing testicular seminoma.

appearance is mostly due to the presence of mixed cell types, hemorrhage, and necrosis.

On occasion, primary testicular tumors may undergo spontaneous regression. These lesions are usually termed *burned-out* germ cell tumors [78]. At MRI, $T_2$-weighted images demonstrate a focal low-signal-intensity area of distortion of the normal testicular architecture, without a visible mass. The appearance can resemble segmental infarction.

## References


1. Kim B, Kawashima A, Ryu JA et al. Imaging of the seminal vesicle and vas deferens. *RadioGraphics* 2009;29:1105–1121.
2. Amirkhan RH, Molber KH, Huley EL, Nurenberg P, Sagalowsky AI. Primary leiomyosarcoma of the seminal vesicle. *Urology* 1994;44:132–135.
3. Siegel RL, Miller KD, Jemal A. Cancer statistics, 2015. *CA Cancer J Clin* 2015;65(1):5–29.
4. Zlotta AR, Egawa S, Pushkar D et al. Prevalence of prostate cancer on autopsy: Cross-sectional study on unscreened Caucasian and Asian men. *J Natl Cancer Inst* 2013;105(14):1050–1058.
5. McNeal JE. Normal and pathologic anatomy of the prostate. *Urology* 1981;17:11–16.
6. Loeb S, Bjurlin MA, Nicholson J et al. Overdiagnosis and overtreatment of prostate cancer. *Eur Urol* 2014;65(6):1046–1055.
7. Krakowsky Y, Loblaw A, Klotz L. Prostate cancer death of men treated with initial active surveillance: Clinical and biochemical characteristics. *J Urol* 2010;184:131–135.
8. Comiter CV, Benson CJ, Capelouto CC et al. Nonpalpable intratesticular masses detected sonographically. *J Urol* 1995;154:1367–1369.
9. Ries LAG, Melbert D, Krapcho M et al. (eds). *SEER Cancer Statistics Review, 1975–2005.* National Cancer Institute, Bethesda, MD. http://seer.cancer.gov/csr/1975_2005.
10. Moreno CC, Small WC, Camacho JC et al. Testicular tumors: What radiologists need to know-differential diagnosis, staging, and management. *RadioGraphics* 2015;35:400–415.
11. Kundra V, Silverman PM, Matin SF, Choi H. Imaging in oncology from the University of Texas M. D. Anderson Cancer Center: Diagnosis, staging, and surveillance of prostate cancer. *AJR Am J Roentgenol* 2007;189(4):830–844.
12. Quick CM, Gokden N, Sangoi AR, Brooks JD, McKenney JK. The distribution of PAX-2 immunoreactivity in the prostate gland, seminal vesicle, and ejaculatory duct: Comparison with prostatic adenocarcinoma and discussion of prostatic zonal embryogenesis. *Hum Pathol* 2010;41(8):1145–1149.
13. Vargas HA, Akin O, Franiel T et al. Normal central zone of the prostate and central zone involvement by prostate cancer: Clinical and MR imaging implications. *Radiology* 2012;262(3):894–902.
14. McNeal JE, Redwine EA, Freiha FS, Stamey TA. Zonal distribution of prostatic adenocarcinoma: Correlation with histologic pattern and direction of spread. *Am J Surg Pathol* 1988;12:897–906.
15. Augustin H, Erbersdobler A, Hammerer PG, Graefen M, Huland H. Prostate cancers in the transition zone. Part 2: Clinical aspects. *BJU Int* 2004;94:1226–1229.
16. Greene DR, Wheeler TM, Egawa S, Weaver RP, Scardino PT. Relationship between clinical stage and histological zone of origin in early prostate-cancer—Morphometric analysis. *Br J Urol* 1991;68:499–509.
17. King CR, Ferrari M, Brooks JD. Prognostic significance of prostate cancer originating from the transition zone. *Urol Oncol* 2009;27:592–597.
18. Boonsirikamchai P, Choi S, Frank SJ et al. MR Imaging of prostate cancer in radiation oncology: What radiologists need to know. *RadioGraphics* 2013;33:741–761.
19. Sonnad SS, Langlotz CP, Schwartz JS. Accuracy of MR imaging for staging prostate cancer: A meta-analysis to examine the effect of technologic change. *Acad Radiol* 2001;8(2):149–157.
20. Pinto PA, Chung PH, Rastinehad AR et al. Magnetic resonance imaging/ultrasound fusion guided prostate biopsy improves cancer detection following transrectal ultrasound biopsy and correlates with multiparametric magnetic resonance imaging. *J Urol* 2011;186(4):1281–1285.
21. Sciarra A, Panebianco V, Salciccia S et al. Modern role of magnetic resonance and spectroscopy in the imaging of prostate cancer. *Urol Oncol* 2011;29:12–20.
22. Puech P, Sufana Iancu A, Renard B, Villers A, Lemaitre L. Detecting prostate cancer with MRI: Why and how. *Diagn Interv Imaging* 2012;93:268–278.
23. Tamada T, Sone T, Higashi H et al. Prostate cancer detection in patients with total serum prostate-specific antigen levels of 4–10 ng/mL: Diagnostic efficacy of diffusion-weighted imaging, dynamic contrast-enhanced MRI, and T2-weighted imaging. *AJR Am J Roentgenol* 2011;197:664–670.

24. Hoeks CM, Barentsz JO, Hambrock T et al. Prostate cancer: Multiparametric MR imaging for detection, localization, and staging. *Radiology* 2011;261(1):46–66.
25. Turkbey B, Pinto PA, Mani H et al. Prostate cancer: Value of multiparametric MR imaging at 3 T for detection—Histopathologic correlation. *Radiology* 2010;255(1):89–99.
26. Delongchamps NB, Rouanne M, Flam T et al. Multiparametric magnetic resonance imaging for the detection and localization of prostate cancer: Combination of T2-weighted, dynamic contrast-enhanced and diffusion-weighted imaging. *BJU Int* 2011;107(9):1411–1418.
27. Delongchamps NB, Beuvon F, Eiss D et al. Multiparametric MRI is helpful to predict tumor focality, stage, and size in patients diagnosed with unilateral low-risk prostate cancer. *Prostate Cancer Prostatic Dis* 2011;14(3):232–237.
28. Scheenen TW, Futterer J, Weiland E et al. Discriminating cancer from noncancer tissue in the prostate by 3-dimensional proton magnetic resonance spectroscopic imaging: A prospective multicenter validation study. *Invest Radiol* 2011;46(1):25–33.
29. Rosenkrantz AB, Taneja SS. Radiologist, be aware: Ten pitfalls that confound the interpretation of multiparametric prostate MRI. *AJR Am J Roentgenol* 2014;202(1): 109–120.
30. White S, Hricak H, Forstner R et al. Prostate cancer: Effect of postbiopsy hemorrhage on interpretation of MR images. *Radiology* 1995;195:385–390.
31. Tamada T, Sone T, Jo Y et al. Prostate cancer: Relationships between postbiopsy hemorrhage and tumor detectability at MR diagnosis. *Radiology* 2008;248(2):531–539.
32. Janssen MJ, Huijgens PC, Bouman AA, Oe PL, Donker AJ, van der Meulen J. Citrate versus heparin anticoagulation in chronic haemodialysis patients. *Nephrol Dial Transplant* 1993;8(11):1228–1233.
33. Schiebler ML, Schnall MD, Pollack HM et al. Current role of MR imaging in the staging of adenocarcinoma of the prostate. *Radiology* 1993;189:339–352.
34. Barrett T, Vargas HA, Akin O, Goldman DA, Hricak H. Value of the hemorrhage exclusion sign on T1-weighted prostate MR images for the detection of prostate cancer. *Radiology* 2012;263:751–757.
35. Wang L, Mazaheri Y, Zhang J, Ishill NM, Kuroiwa K, Hricak H. Assessment of biologic aggressiveness of prostate cancer: Correlation of MR signal intensity with Gleason grade after radical prostatectomy. *Radiology* 2008;246(1):168–176.
36. Bezzi M, Kressel HY, Allen KS et al. Prostatic carcinoma: Staging with MR imaging at 1.5 T. *Radiology* 1988;169(2):339–346.
37. King CR, Ferrari M, Brooks JD. Prognostic significance of prostate cancer originating from the transition zone. *Urol Oncol* 2009;27(6):592–597.
38. Hricak H, Choyke PL, Eberhardt SC, Leibel SA, Scardino PT. Imaging prostate cancer: A multidisciplinary perspective. *Radiology* 2007;243(1):28–53.
39. Hoeks CM, Hambrock T, Yakar D et al. Transition zone prostate cancer: Detection and localization with 3-T multiparametric MR imaging. *Radiology* 2013;266(1):207–217.
40. Lansberg MG, Norbash AM, Marks MP, Tong DC, Moseley ME, Albers GW. Advantages of adding diffusion-weighted magnetic resonance imaging to conventional magnetic resonance imaging for evaluating acute stroke. *Arch Neurol* 2000;57:1311–1316.
41. Stejskal EO, Tanner JE. Spin diffusion measurements: Spin echoes in the presence of a time-dependent field gradient. *J Chem Phys* 1965;42(1):288–292.
42. Bammer R, Skare S, Newbould R et al. Foundations of advanced magnetic resonance imaging. *NeuroRx* 2005;2:167–195.
43. Basser PJ. Inferring microstructural features and the physiological state of tissues from diffusion-weighted images. *NMR Biomed* 1995;8:333–344.
44. Peng Y, Jiang Y, Antic T et al. Apparent diffusion coefficient for prostate cancer imaging: Impact of B values. *AJR Am J Roentgenol* 2014;202(3):W247–W253.
45. Katahira K, Takahara T, Kwee TC et al. Ultra-high-b-value diffusion-weighted MR imaging for the detection of prostate cancer: Evaluation in 201 cases with histopathological correlation. *Eur Radiol* 2011;21(1):188–196.
46. Maas MC, Fütterer JJ, Scheenen TW. Quantitative evaluation of computed high B value diffusion-weighted magnetic resonance imaging of the prostate. *Invest Radiol* 2013;48(11):779–786.
47. Somford DM, Fütterer JJ, Hambrock T, Barentsz JO. Diffusion and perfusion MR imaging of the prostate. *Magn Reson Imaging Clin N Am* 2008;16(4):685–695.
48. Kumar V, Jagannathan NR, Kumar R et al. Apparent diffusion coefficient of the prostate in men prior to biopsy: Determination of a cut-off value to predict malignancy of the peripheral zone. *NMR Biomed* 2007;20:505–511.
49. Kim CK, Park BK, Lee HM, Kwon GY. Value of diffusion-weighted imaging for the prediction of prostate cancer location at 3T using a phased-array coil: Preliminary results. *Invest Radiol* 2007;42:842–847.
50. Tamada T, Sone T, Toshimutsu S et al. Age-related and zonal anatomical changes of apparent diffusion coefficient values in normal human prostatic tissues. *J Magn Reson Imaging* 2008;27:552–556.
51. Ren J, Huan Y, Wang H et al. Diffusion-weighted imaging in normal prostate and differential diagnosis of prostate diseases. *Abdom Imaging* 2008;33(6):724–728.
52. Nowak J, Malzahn U, Baur AD et al. The value of ADC, T2 signal intensity, and a combination of both parameters to assess Gleason score and primary Gleason grades in patients with known prostate cancer. *Acta Radiol* December 12, 2014. pii: 0284185114561915. [Epub ahead of print].
53. Hambrock T, Hoeks C, Hulsbergen-van de Kaa C et al. Prospective assessment of prostate cancer aggressiveness using 3-T diffusion-weighted magnetic resonance imaging-guided biopsies versus a systematic 10-core transrectal ultrasound prostate biopsy cohort. *Eur Urol* 2012;61(1):177–184.
54. Nagel KN, Schouten MG, Hambrock T et al. Differentiation of prostatitis and prostate cancer by using diffusion-weighted MR imaging and MR-guided biopsy at 3 T. *Radiology* 2013;267(1):164–172.

55. Esen M, Onur MR, Akpolat N, Orhan I, Kocakoc E. Utility of ADC measurement on diffusion-weighted MRI in differentiation of prostate cancer, normal prostate and prostatitis. *Quant Imaging Med Surg* 2013;3(4): 210–216.
56. Cuenod CA, Balvay D. Perfusion and vascular permeability: Basic concepts and measurement in DCE-CT and DCE-MRI. *Diagn Interv Imaging* 2013;94(12):1187–1204.
57. Padhani AR, Harvey CJ, Cosgrove DO. Angiogenesis imaging in the management of prostate cancer. *Nat Clin Pract Urol* 2005;2:596–607.
58. Bonekamp D, Macura KJ. Dynamic contrast-enhanced magnetic resonance imaging in the evaluation of the prostate. *Top Magn Reson Imaging* 2008;19:273–284.
59. Hambrock T, Padhani AR, Tofts PS, Vos P, Huisman HJ, Barentsz JO. Dynamic contrast-enhanced MR imaging in the diagnosis and management of prostate cancer. *RSNA Categorical Course in Diagnostic Radiology: Genitourinary Radiology* 2006;61–77.
60. Beyersdorff D, Lüdemann L, Dietz E et al. Dynamic contrast-enhanced MRI of the prostate: Comparison of two different post-processing algorithms. *Rofo* 2011;183(5):456–461.
61. Alonzi R, Padhani AR, Allen C. Dynamic contrast enhanced MRI in prostate cancer. *Eur J Radiol* 2007;63(3):335–350.
62. Huisman HJ, Engelbrecht MR, Barentsz JO. Accurate estimation of pharmacokinetic contrast-enhanced dynamic MRI parameters of the prostate. *J Magn Reson Imaging* 2001;13(4):607–614.
63. Futterer JJ. MR imaging in local staging of prostate cancer. *Eur J Radiol* 2007;63(3):328–334.
64. Tempany CM, Rahmouni AD, Epstein JI, Walsh PC, Zerhouni EA. Invasion of the neurovascular bundle by prostate cancer: Evaluation with MR imaging. *Radiology* 1991;181(1):107–112.
65. Turkbey B, Albert PS, Kurdziel K, Choyke PL. Imaging localized prostate cancer: Current approaches and new developments. *AJR Am J Roentgenol* 2009;192(6): 1471–1480.
66. Yu KK, Hricak H, Alagappan R, Chernoff DM, Bacchetti P, Zaloudek CJ. Detection of extracapsular extension of prostate carcinoma with endorectal and phased-array coil MR imaging: Multivariate feature analysis. *Radiology* 1997;202(3):697–702.
67. Outwater EK, Petersen RO, Siegelman ES, Gomella LG, Chernesky CE, Mitchell DG. Prostate carcinoma: Assessment of diagnostic criteria for capsular penetration on endorectal coil MR images. *Radiology* 1994;193(2):333–339.
68. Fütterer JJ, Engelbrecht MR, Huisman HJ et al. Staging prostate cancer with dynamic contrast-enhanced endorectal MR imaging prior to radical prostatectomy: Experienced versus less experienced readers. *Radiology* 2005; 237(2):541–549.
69. Garner MJ, Turner MC, Ghadirian P, Krewski D. Epidemiology of testicular cancer: An overview. *Int J Cancer* 2005;116, 331–339.
70. http://seer.cancer.gov/statfacts/html/testis.html.
71. Dalal PU, Sohaib SA, Huddart R. Imaging of testicular germ cell tumours. *Cancer Imaging* 2006;6:124–134.
72. Semelka RC. Kidneys, testes. In: Semelka RC (ed.) *Abdominal-Pelvic MRI*, 1st edn. Wiley-Liss, New York, 2002; pp. 741–1007.
73. Donohue JP, Zachary JM, Maynard BR. Distribution of nodal metastases in nonseminomatous testis cancer. *J Urol* 1982;128:315–320.
74. Mason MD, Featherstone T, Olliff J, Horwich A. Inguinal and iliac lymph node involvement in germ cell tumours of the testis: Implications for radiological investigation and for therapy. *Clin Oncol* 1991;3:147–150.
75. Rholl KS, Lee JKT, Ling D, Heiken JP, Glazer HS. MR imaging of the scrotum with a high resolution surface coil. *Radiology* 1987;163, 99–103.
76. Watanabe Y, Dohke M, Ohkubo K et al. Scrotal disorders: Evaluation of testicular enhancement patterns at dynamic contrast-enhanced subtraction MR imaging. *Radiology* 2002;217:219–227.
77. Tsili AC, Argyropoulou MI, Astrakas LG, Ntoulia EA, Giannakis D, Sofikitis N, Tsampoulas K. Dynamic contrast-enhanced subtraction MRI for characterizing intratesticular mass lesions. *AJR Am J Roentgenol* 2013;200(3):578–585.
78. Patel MD, Patel BM. Sonographic and magnetic resonance imaging appearance of a burned-out testicular germ cell neoplasm. *J Ultrasound Med*. 2007 Jan;26(1):143–6.

# 4

# *Uterus and Vagina*

**Suzanne L. Palmer, Nicole F. Darcy, Claire L. Templeman, and Allison D. Salibian**

**CONTENTS**



## 4.1 Introduction

The presenting symptoms associated with pelvic pathology are typically nonspecific and differentially broad, including gynecologic, obstetrical, and non-gynecologic etiologies. Ultrasound (US) remains the first-line imaging modality for the evaluation of the female patient presenting with most gynecological symptoms, including abnormal vaginal bleeding, pelvic pain, and follow-up of a previously detected abnormality or congenital anomaly [1–4]. US is well suited for the evaluation of the female pelvis due to its availability, high resolution, ability to focus the examination, and relatively lower cost when compared to computed tomography (CT) and magnetic resonance imaging (MRI). Due to excellent soft tissue contrast, MRI is widely accepted as a useful problem-solving tool if pelvic US is incomplete or indeterminate. MRI is able to differentiate uterine zonal anatomy; accurately identify, localize, and characterize fibroids; confirm the diagnosis of adenomyosis; and evaluate complex uterine anomalies. In the presence of newly diagnosed endometrial cancer, MRI is the imaging study of choice if imaging is indicated for treatment planning. MRI can evaluate the extent of pelvic disease and be used to assess the depth of myometrial and endocervical tumor extent, both before and at follow-up [3]. Although not as accurate for evaluating cervical stromal invasion by cervical cancer, MRI is preferred over CT and US for pretreatment planning of invasive cervical cancer [4].

In this chapter, we will

- Present techniques for imaging the uterus and vagina with MRI,
- Review the normal MRI appearance of the uterus and vagina as well as congenital anomalies,
- Review benign and malignant conditions of the uterus and vagina,
- Identify the information the gynecologist needs to know.

## 4.2 Imaging Techniques

Although MRI of the uterus and vagina may be performed on low field magnets [5], imaging is limited by lower signal-to-noise ratio with lower spatial and temporal resolution when compared to high field (1.5–3 T) imaging. The greatest body of published work and experience supports the use of 1.5 T for imaging the female pelvis; however, with the increased signal-to-noise ratio at 3 T, more work is being published comparing imaging of the female pelvis at 1.5 and 3 T magnet strengths [6,7]. Kataoka et al. focused on the uterine body and cervix and found that there was no significant difference in the overall image quality or uterine zonal anatomy contrast; however, the image contrast of the cervix and vagina was significantly higher at 3 T [7] We have noticed similar findings at our institution (Figure 4.1). They also found artifacts such as image inhomogeneity were more prominent at 3 T, but motion artifact was greater at 1.5 T. Contrast-enhanced sequences benefit the most from the increase in signal-to-noise ratio at 3 T; however, the improved spatial resolution afforded by 3 T imaging may be at the expense of an increase in artifacts, including susceptibility, shading, and chemical shift. There is also the increase in specific absorption rate that limits the use of some sequences at 3 T [6,8]. To date, diagnostic imaging at 1.5 and 3 T appears to be comparable; therefore, the choice of magnet strength remains at the discretion of the radiologist and scanner availability.

Patient preparation is not mandatory for routine pelvic MRI; however, imaging is optimally performed after fasting for at least 6 h to reduce bowel peristalsis. Bowel motion can reduce image quality; therefore in select examinations, especially for the evaluation of endometrial or cervical cancer, an antiperistaltic agent may be used. Frequently used agents include glucagon and butylscopolamine, although the latter is not Food and Drug Administration approved for use in the United States. The patient should be asked to void prior to the examination for patient comfort and to limit motion and mass effect on adjacent organs by an over-distended bladder. Rectal and vaginal contrast is not routinely used; however, it may be helpful in the evaluation of vaginal pathology, pelvic floor abnormalities, and pelvic organ prolapse.

When imaging with a standard pelvic protocol and pelvic phase array coils, MRI is usually diagnostic for the evaluation of normal pelvic anatomy and most pelvic pathology. The choice of sequences will vary depending on clinical question posed; however, standard pelvic MRI protocols typically include a combination of $T_2$-weighted ($T_2$W) and $T_1$-weighted ($T_1$W) sequences without and with fat suppression (Table 4.1) (Figures 4.2 and 4.3). $T_2$W sequences typically used include mutiplanar 2D $T_2$W rapid acquisition with relaxation enhancement where high spatial and contrast resolution allows

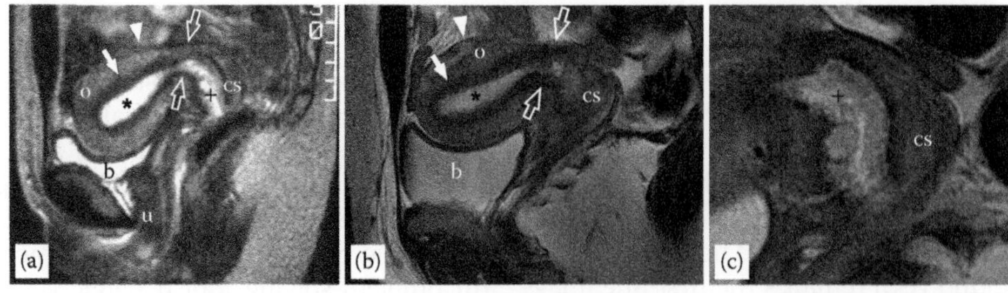


**FIGURE 4.1**
Sagittal T2W images of a normal uterus in a reproductive age female at (a) 1.5 T and (b) 3 T. (c) Magnified view of the cervix at 3 T. The uterine zonal contrast is similar at 1.5 T and 3 T, but resolution of the cervix and vagina is improved at 3 T. Endometrium (*), endocervical canal (+), junctional zone (white arrow), outer myometrium (o), bladder (b), urethra (u), cervical stroma (cs), serosal layer (arrowhead), and internal cervical os (open arrows).

**TABLE 4.1**

University of Southern California Sample 1.5 T Pelvic MRI Sequences

| Sequence | Plane | TR/TE | FOV (cm) | Thickness/Gap (mm) | Flip Angle | Matrix | Echo Train Length |
|---|---|---|---|---|---|---|---|
| Localizer | 3 plane | | 48 | 7/5 | | 128 × 256 | |
| $T_2$W FSE | Sagittal | 3000/110 | 30 | 5/1 | | 256 × 256 | 20 |
| $T_2$W FSE | Coronal | 3000/110 | 20 | 5/1 | | 256 × 256 | 20 |
| $T_2$W FSE + FS | Axial | 3000/110 | 20 | 5/1 | | 256 × 256 | 20 |
| $T_1$W FSE | Axial | 500/20 | 20 | 5/1 | | 512 × 256 | 3 |
| LAVA[a] | Axial | | 40 | 5/0 | 12 | 320 × 192 | |
| LAVA[a] | Coronal | | 40 | 5/0 | 12 | 320/192 | |
| LAVA[a] | Sagittal | | 40 | 5/0 | 12 | 320/192 | |
| $T_2$ ssFSE[b,c] | Uterus long axis | Minimum/120 | 36 | 5/0 | | 256 × 224 | |
| $T_2$ ssFSE[b,c] | Uterus short axis | Minimum/120 | 36 | 5/0 | | 256/224 | |
| DWI $b = 500$ | Axial | 3725/minimum | 38 | 5/0 | | 160 × 160 | |
| DWI $b = 1000$ | Axial | 3950/minimum | 38 | 5/0 | | 160 × 160 | |
| $T_2$W ssFSE[b] | Coronal | 111/3975 | 42 | 6/1 | | 192 × 320 | 20 |

*Note:* 8-channel phase array surface coil used.

[a] Pre- and postcontrast sequences; optional sequences.

[b] Müllerian anomalies.

[c] Endometrial cancer.

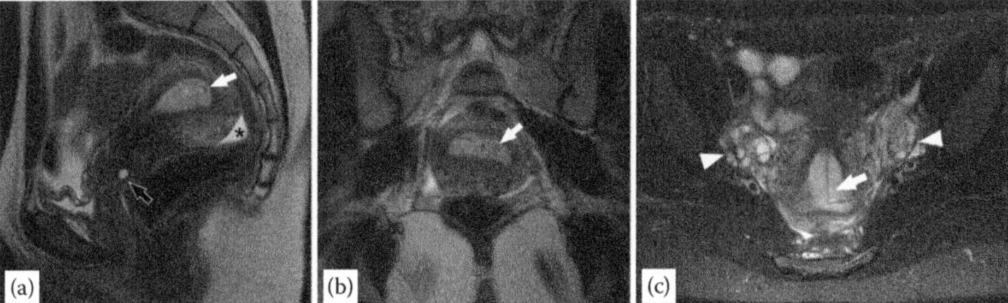

**FIGURE 4.2**

Our standard pelvic protocol includes (a) sagittal, (b) coronal, and (c) axial $T_2$W fast spin-echo sequences for high spatial and contrast resolution. $T_2$W sequences provide excellent anatomical definition. We typically obtain the axial with fat suppression because we feel this allows better visualization of low pelvic (vaginal, ureteral, and anal) structures. When time is a factor, ultrafast single-shot sequences are adequate in most cases except for staging endometrial and cervical cancer. Intrauterine device (white arrow), nabothian cyst (black arrow), free fluid (*), and ovaries (arrowheads).

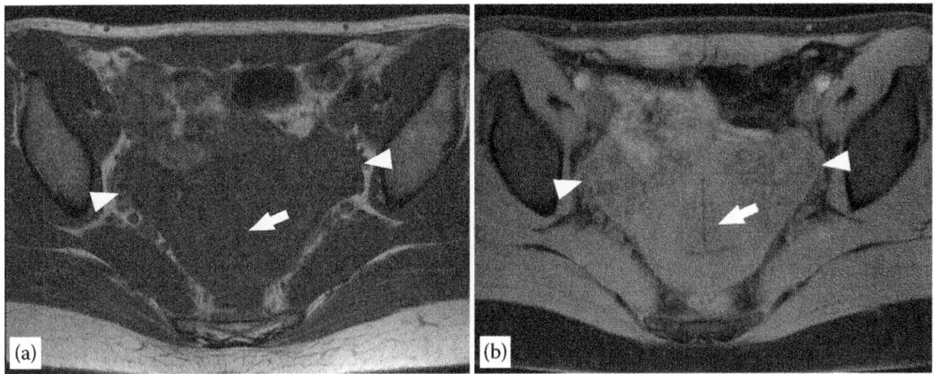

**FIGURE 4.3**

Our standard pelvic protocol includes (a) axial $T_1$W fast spin echo and (b) $T_1$W gradient-echo sequences. If time is a factor, gradient-echo sequences may be used to acquire both the non-fat-suppressed and fat-suppressed images. The uterus is homogeneously intermediate in $T_1$ SI. $T_1$W images are used for tissue characterization; increasing lesion conspicuity, including the presence of fat and blood; and as a mask for subtraction in contrast-enhanced cases where a precontrast structure is T1 bright. Intrauterine device (arrow) and ovaries (arrowheads).

for excellent anatomical definition; these include fast spin-echo (FSE) and turbo spin-echo (TSE) sequences, which have a longer acquisition time and higher resolution than the ultrafast, single-shot sequences. Single-shot sequences include ssFSE, ssTSE, and half-Fourier acquisition single-shot turbo spin echo (HASTE). $T_1$W sequences are typically used for tissue characterization. When high resolution is required, axial 2D $T_1$W FSE/TSE may be used. When more rapid imaging is desired, $T_1$W gradient-echo sequences may be acquired. 2D or 3D, $T_1$W gradient-echo sequences with fat suppression are obtained both before and after intravenous contrast administration. If the clinical question to be answered involves evaluating uterine anatomy or neoplasm, $T_2$W acquisitions oriented parallel to the long and short axis of the uterus should be performed (Figure 4.4). A coronal, large field of view, ultrafast $T_2$W sequence is acquired to evaluate for the presence of normal kidneys in cases of gynecologic anomalies.

We typically administer contrast for our pelvic exams. The choice of non-dynamic versus dynamic-contrast imaging depends on the exam indication; venous-phase acquisition postcontrast imaging (Figure 4.5) may be adequate for imaging leiomyomas, whereas dynamic-contrast enhancement (DCE) is typically obtained for the evaluation of endometrial and cervical cancers. DCE is performed by using a fat-suppressed, three-dimensional gradient-echo $T_1$W sequence after administration of a gadolinium chelate agent (Gd-C). Images oriented to the long and short axis of the uterus are acquired prior to Gd-C injection and then during multiple dynamic phases of enhancement (immediate, 1, 2, and 4 min). Enhancement of normal myometrium peaks at 120 s after the administration of Gd-C and then decreases over time. Injection rate is 0.1–0.2 mmol per kilogram of body weight, at 2 mL/s. The injection should be administered by means of an automatic, MRI-compatible power injector to ensure accuracy of the timed bolus.

The use of contrast and the choice of Gd-C are at the discretion of the individual radiologist or institution; however, the administration of Gd-C should be avoided in patients with acute kidney injury and end-stage or severe chronic kidney disease, as these patients are at

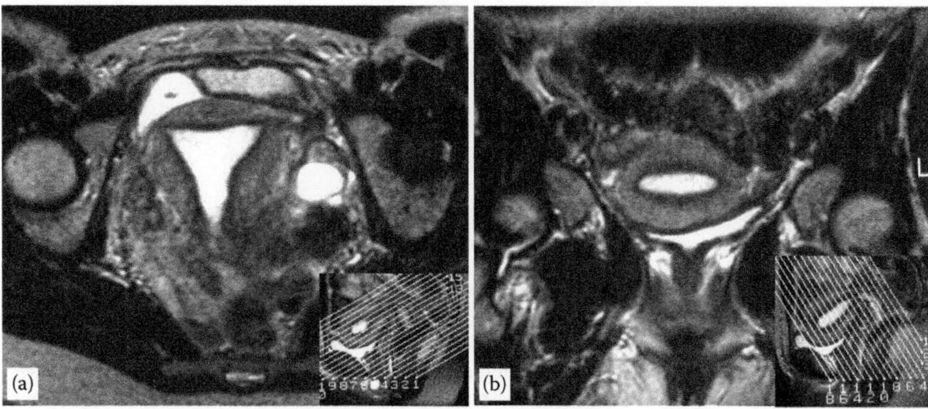

**FIGURE 4.4**
Our standard pelvic protocol includes $T_2$W sequences acquired parallel to the (a) long axis of the uterus to best characterize the external uterine contour and (b) short axis to best evaluate for septa in the case of Müllerian duct anomalies and myometrial invasion in the case of endometrial cancer staging.

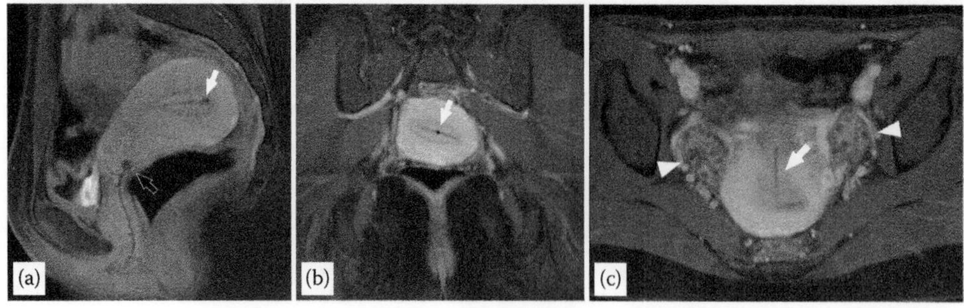

**FIGURE 4.5**
Our standard pelvic protocol includes $T_1$W gradient-echo sequences with fat suppression acquired in the (a) sagittal, (b) coronal, and (c) axial planes after the intravenous administration of a gadolinium chelate agent. The three planes are acquired sequentially after the injection of contrast (non-dynamic) and are adequate for routine pelvic imaging, including evaluation for leiomyoma degeneration. Intrauterine device (white arrow), nabothian cyst (black arrow), and ovaries (arrowheads).

higher risk of nephrogenic systemic fibrosis, a potentially life-threatening condition. There is no absolute estimated glomerular filtration rate cutoff below which Gd-C cannot be administered; however, the use of Gd-C should be avoided when the estimated glomerular filtration rate is <30 ml/min per 1.73 m$^2$ [9,10].

MRI of the female pelvis is generally performed with free breathing, as motion artifacts are thought to be less critical in the pelvis when compared to the abdomen; however, motion artifact from breathing, vessel pulsation, bowel peristalsis, and random patient motion may lead to a marked loss of diagnostic information and a reduction in image quality [11] Techniques that reduce motion artifacts due to periodic motion (respiratory and vessel pulsation) include triggering and gating. Because these techniques lengthen the acquisition time, we favor techniques to reduce the intensity of the artifact or displace the artifact from the region of interest. Increased signal from subcutaneous fat within the anterior abdominal wall accentuates respiratory motion artifact; in field of view saturation bands placed along the anterior body wall fat help reduce the signal intensity (SI) of the subcutaneous fat and thus the artifact [12]. Fat suppression will reduce the SI of all fat, reducing the intensity of respiratory and motion artifacts; however, fat suppression may mask pathology [12].

Random motion artifacts include peristalsis of the bowel, bladder, and uterus. These random movements may cause blurring of the pelvic viscera, leading to degradation in imaging. Bowel peristalsis may be minimized with fasting and antiperistaltic medications and bladder peristalsis by emptying the bladder. Uterine peristalsis, wavelike contractions occurring in the inner myometrium, is different from sporadic myometrial contractions, which can simulate fibroids or adenomyosis (Figure 4.6); however, both may be suppressed by antiperistaltic medications, although not as completely as bowel [11]. Subendometrial myometrial contractility is cyclic, increased in the follicular and periovulatory phases and decreased in the luteal phase [13].

Bulk patient motion may be reduced by utilizing techniques to reduce imaging time, including those that enhance *k*-space filling (parallel, BLADE, and PROPELLER) [14]. The use of single-shot, ultrafast $T_2$W sequences can reduce motion artifact, but also reduces image and contrast resolution. 3D single-shot $T_2$W sequences may provide higher contrast and image resolution when compared to 2D single-shot sequences and are similar to conventional FSE/TSE [15].

There has been work on the benefit of acquiring the $T_2$W imaging in a near isotropic 3D sequence [16]. By acquiring the information as a 3D dataset, only one acquisition is needed as the dataset can be reconstructed into any plane. Theoretically, this should reduce the scan time and ensure appropriate anatomic coverage; however, acquiring near isotropic voxel size datasets is also time consuming. At 3 T, Hecht et al. found that 3D and 2D images were diagnostically comparable with the exception of distinction between fat and fluid, which was thought to be superior with the 2D sequences [16]. Further work is needed in this area.

Diffusion-weighted imaging (DWI) is a part of our routine protocol because of its relatively short acquisition time and its ability to increase conspicuity of certain abnormalities. Correlation of findings on DWI with structural imaging ($T_2$W and $T_1$W sequences) should be performed to avoid misinterpretation of DWI findings [17]. DWI is promising for the evaluation of myometrial invasion of endometrial cancer and cervical stroma invasion of cervical cancer and may be an alternative to DCE for patients in whom contrast is contraindicated. DWI is most commonly acquired in the transaxial plane, the plane associated with the least susceptibility-induced signal loss (Figure 4.7). Theony et al. suggest acquiring thin-section DWI (3 mm or less) in the axial plane to allow for multiplanar reconstructions in any direction [17]. The choice of *b*-values is still somewhat arbitrary, and the values typically range from 400 to 1000 s/mm$^2$. We typically use *b*-values of 0, 500, and 1000 s/mm$^2$. There is overlap in DWI and apparent diffusion coefficient values

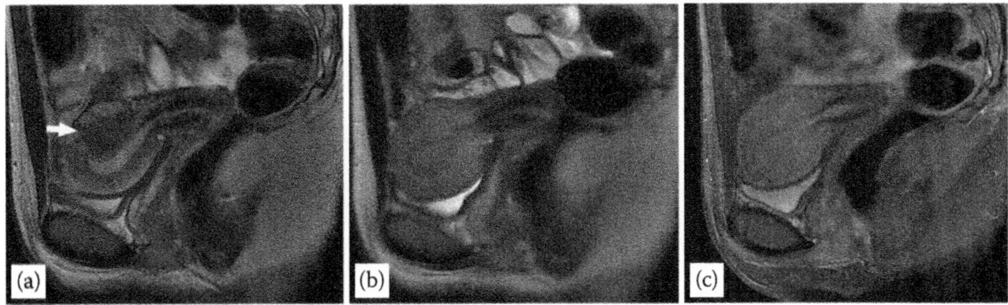


**FIGURE 4.6**
Sporadic myometrial contractions may be differentiated from fibroids and adenomyosis by documenting change in appearance over time. Sagittal $T_2$W images at (a) time = 0, (b) time = 30 min, and (c) time = 45 min demonstrate that the low $T_2$ SI area in the posterior wall (arrow) at time 0 is no longer seen at 30 and 45 min.

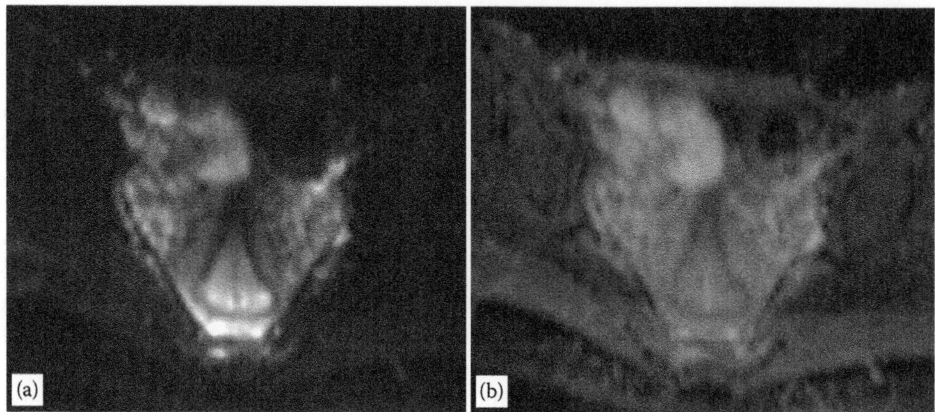


**FIGURE 4.7**
DWI acquired with (a) $b = 0$ and (b) $b = 800$. We typically perform DWI with the same field of view and section thickness as the $T_2$W axial sequence and acquire them before intravenous contrast administration. DWI may be angled parallel to the long or short axis of the uterus for evaluation of myometrial invasion.

of benign and malignant lesions; therefore, further work is needed.

Blood oxygenation level-dependent MRI, perfusion and spectroscopy, and the use of magnetic iron oxide nanoparticles need further study [3]. Quantitative MRI techniques ($T_1$, $T_2$, and perfusion) are being used in gynecological research, including the role of endometrial blood flow dysregulation in association with abnormal uterine bleeding, endometriosis, and infertility. The feasibility of using echo-planar imaging to evaluate perfusion in uterine tissues over the menstrual cycle has been evaluated with the aim of establishing normal ranges for perfusion [18].

**FOCUS POINTS**

- Imaging at 1.5 and 3 T is comparable; therefore, the choice of magnet strength is at the discretion of the radiologist and scanner availability.
- Patient preparation should include fasting and emptying the bladder.
- $T_2$W sequences obtained along the long and short axis of the uterus are useful in the evaluation of Müllerian duct anomalies and endometrial cancer.
- Enhancement of normal myometrium peaks at 120 s after administration of Gd-C and then decreases over time.
- DWI is promising for the evaluation of tumor invasion and may be an alternative to DCE for patients in whom contrast is contraindicated.

## 4.3 Anatomy and MRI Appearance of the Normal Uterus

The uterus, vagina, and fallopian tubes are located in the middle compartment of the pelvis, intimately related to the urethra and bladder anteriorly and the anal canal and lower rectum posteriorly [19–21]. The size and appearance of the uterus vary with patient age, number of gestations, and hormonal status (Figures 4.1, 4.8 through 4.11) [20]. The uterus is comprised of the cervix and corpus, which can be further subdivided into body, cornua, and fundus (Figure 4.12).

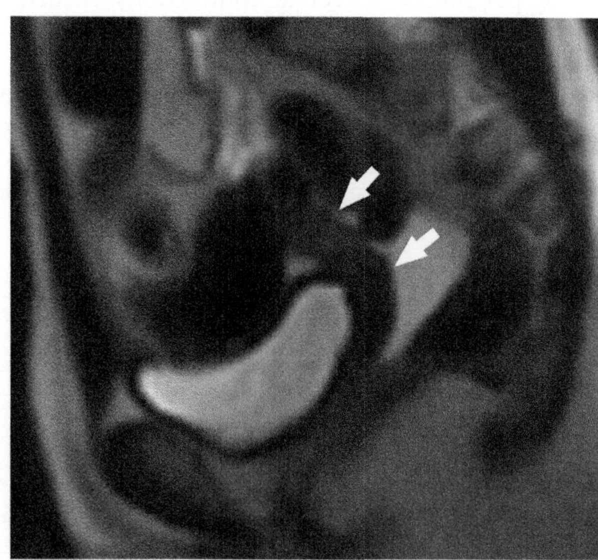

**FIGURE 4.8**
Appearance of the uterus in premenarche—sagittal $T_2$W image of a 6-year-old female. Uterus is intermediate in SI and difficult to separate from adjacent anatomic structures due to its small size (arrows). The cervix is typically larger than the uterine corpus in premenarche.

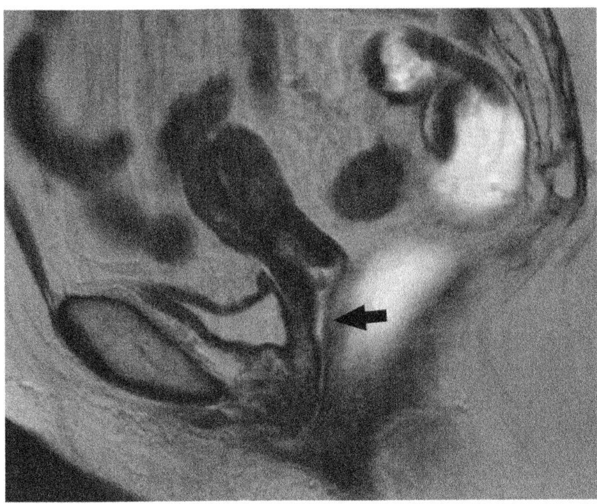

**FIGURE 4.9**
Appearance of the uterus in postmenopause—sagittal $T_2$W image of a 78-year-old female. The uterus atrophies to 3.5–7.5 cm in length. The cervical length typically equals the uterine corpus. The vaginal wall is thin and low in $T_2$ SI (arrow). The high-signal-intensity material in the vaginal canal and rectum is aqueous gel instilled for pelvic floor examination.

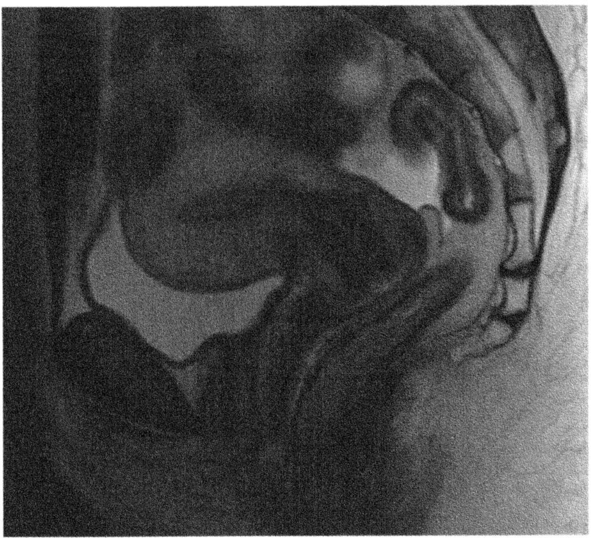

**FIGURE 4.11**
Appearance of the uterus in reproductive age on oral contraceptives. Sagittal $T_2$W image demonstrates hyperintense myometrium and thin endometrium.

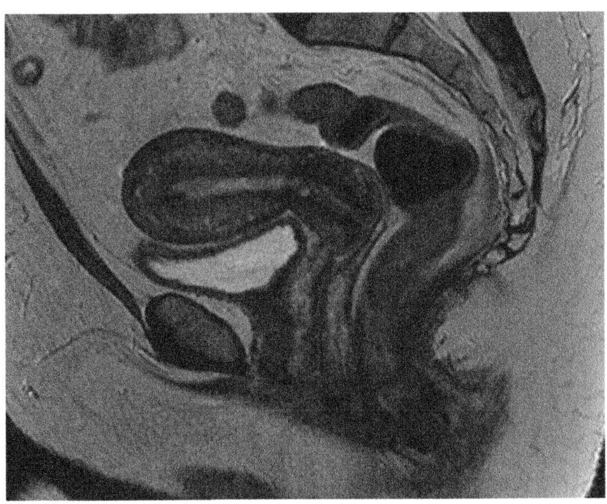

**FIGURE 4.10**
Appearance of the uterus in postmenopause on hormone replacement therapy. Sagittal $T_2$W image demonstrates uterine zonal anatomy and vaginal SI very similar to reproductive age. The relationships to the urethra and bladder within the anterior compartment, and anal canal and lower rectum in the posterior compartment are well demonstrated on midline $T_2$W images.

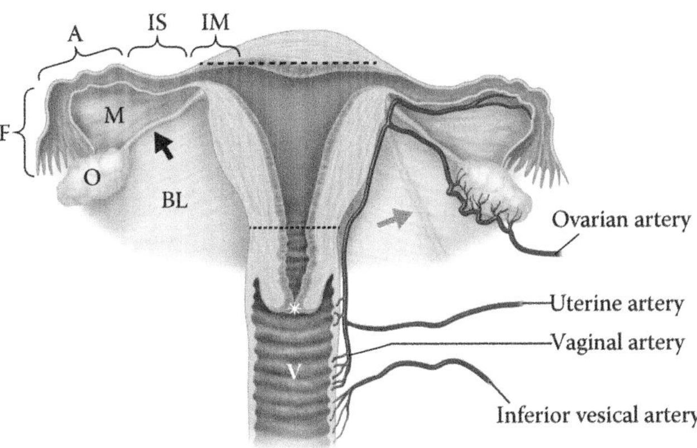


**FIGURE 4.12**
Diagram of uterus in the reproductive age female. The cervix is approximately one-third the length of the entire uterus and extends from the external cervical os (*) to the isthmus (dotted line). Corpus can be subdivided into body (below the dashed line) and fundus (above the dashed line). The fallopian tubes are divided into interstitial (IM), isthmus (IS), ampulla (A), and infundibulum (F). Ovary (O), ovarian ligament (black arrow), round ligament (gray arrow), mesosalpinx (M), broad ligament (BL), and vagina (V).

On histopathology, the corpus wall is comprised of three layers: endometrium, myometrium, and serosa/peritoneum [19,22]. The endometrium is made up of gland-forming columnar epithelium and can be divided into two layers: (1) deep basalis layer, the source of endometrial regeneration; and (2) superficial functionalis layer, the hormonally active layer that changes cyclically [19]. The myometrium can be divided into three layers: (1) subendometrium, which abuts the endometrium and is comprised of densely packed smooth muscle bundles oriented parallel to the endometrium; (2) inner myometrium, which is more vascular and comprised of randomly oriented and loosely organized smooth muscle bundles with traversing arcuate arteries; and (3) subserosa, which is comprised of compact smooth muscle fibers in a circumferential orientation forming a capsular boundary with the serosa/peritoneum [19,22]. Normal myometrium thickness is around 1.5–2.5 cm [19,23].

MRI has been utilized to evaluate the uterus for over 30 years; its capabilities of excellent soft tissue contrast, differentiating uterine wall zonal anatomy, and identifying differences due to hormonal stimuli have been well documented in the radiological and gynecological literature [23–25]. The uterus is of uniform, intermediate SI on $T_1$W sequences, limiting anatomical definition (Figure 4.3). Uterine wall zonal anatomy is best depicted on $T_2$W sequences and is relatively constant throughout the cycle (Figure 4.1). In the reproductive age, normal endometrium is uniformly bright on $T_2$W sequences and should measure less than 1.5 cm in thickness (Figure 4.1). On US, endometrial thickness increases throughout the menstrual cycle [20]. Hoad et al. followed 23 healthy women over a normal menstrual cycle with US and MRI and found that the volume and thickness of the endometrium increased between early follicular and ovulation phases; however, unlike US, the thickness did not increase in the luteal phase on MRI and may have decreased slightly, due to tissue compaction [26]. The discrepancy between US and MRI is thought to be due to differences in conspicuity of the junctional zone (JZ) which is less clear on US during the luteal phase [26].

The myometrium can be divided into two distinct zones on MRI: the inner JZ, which correlates with the anatomic subendometrial myometrium, and the outer myometrium, which correlates with the anatomic inner myometrium (Figure 4.1) [22]. The anatomic outer serosal layer is variably seen. The JZ is typically low on $T_2$W sequences, and the outer myometrium is typically intermediate SI $T_2$W sequences, but may change in SI depending on phase of cycle and hormonal influence. When compared histologically, the JZ has three times the nuclear area compared to the outer myometrium. It has been postulated that the increase in nuclear area is associated with a decrease in the extracelluar matrix per unit volume [27,28]. A decrease in water content within the JZ is thought to be the etiology leading to the low SI of the JZ [27,28]. Hoad et al. found that the volume of the JZ increases throughout the menstrual cycle, but the thickness does not. The JZ may be irregular and eccentric from one area to another; this irregularity is greatest in the luteal phase, the phase with greatest contractions [26]. Demas et al. demonstrated that the JZ was best defined and the myometrial SI greatest in the mid-secretory phase. They also found that the myometrial SI was brighter in women taking oral contraceptives and the endometrial thickness markedly reduced (Figure 4.11) [25]. There is a wide spectrum of normal uterine appearance, however. The anatomic outer serosal layer is variably seen, but if seen, is a thin dark interface between the myometrium and adjacent pelvic structures (Figure 4.1) [22].

The cervical wall is comprised of three histologic layers: mucosa, submucosa, and fibromuscular stroma [19]. The mucosa is comprised of non-keratinized stratified squamous epithelium (characterizing the exocervix) below the squamocolumnar junction and mucus secreting columnar epithelium (characterizing the endocervical canal) above the squamocolumnar junction. The squamocolumnar junction is a dynamic interface (the transformation zone) that is hormonally dependent. The junction is located lower in the cervix in hormonally active women and higher in postmenopausal women. The fibromuscular stroma is comprised of primarily fibrous connective tissue and small amounts (10%–15%) of smooth muscle in a circular arrangement [29].

On MRI, the cervix can be divided visually into endocervix and fibromuscular stroma (Figures 4.1 and 4.13). The endocervix is contiguous with the endometrium and is typically bright on $T_2$W sequences due to the cervical glands (plicae palmatae), cervical mucosa, and secretions. The cervical glands extend with varying depth into the fibromuscular stroma. The fibromuscular stroma can be divided visually into a low $T_2$ SI inner layer, similar to the JZ, and an intermediate $T_2$ SI outer layer. The zonal anatomy of the fibromuscular stroma depicted on MRI does not correspond with the histologic distribution of the smooth muscle and fibrous elements [29].

In the infant, zonal anatomy may be well demonstrated due to residual effect of maternal estrogens. After infancy and before puberty, the uterus is difficult to identify due to its small size: 2–3.5 cm long by 0.5–1 cm width. In premenarche, the uterine body is small; however, the cervix is similar to adult size. Zonal anatomy is indistinct and myometrial SI lower compared to those of reproductive age (Figure 4.8) [25]. Perimenarcheal uterine growth ceases around 7 years after menarche [30]. In the reproductive age, the nulliparous uterus measures approximately 6–9 cm in length and 8–11 cm in length in multiparous females. The cervix measures approximately 2–3 cm of the total uterine length. In the postmenopausal patient, there is involution of the uterus, with indistinct zonal anatomy; the endometrial thickness should be less than 8 mm (Figure 4.9) [20,25]. Uterine size and MRI characteristics may be preserved with hormonal replacement therapy (Figure 4.10).

The vagina is commonly described as a fibromuscular sheath that extends from the vulvar vestibule to the uterus and has a well-developed venous plexus in its wall (Figure 4.13) [19,20,31]. The average vaginal length is from 7 to 9 cm, but can range from 4 to over 12 cm in length [20,31,33]. The vaginal wall is comprised of three histologic layers: (1) mucosa, hormonally sensitive, non-keratinized stratified squamous epithelium

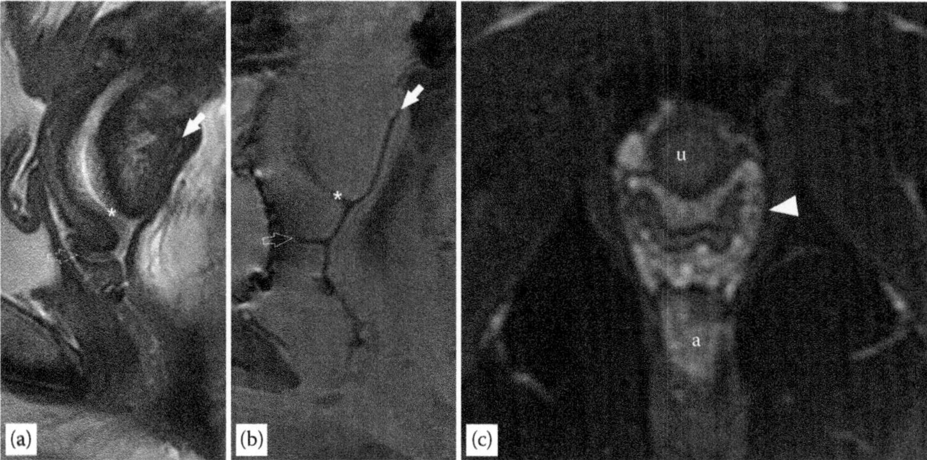


**FIGURE 4.13**
Normal vaginal anatomy in reproductive age. Sagittal (a) $T_2$W and (b) $T_1$W postcontrast, and (c) axial $T_2$W with fat suppression. The vagina is of uniform, intermediate SI on precontrast $T_1$W imaging (not shown), limiting anatomic definition. The length, configuration, and position of the vagina are widely variable, and the shape is determined by the surrounding structures. The posterior wall is longer and ends in the posterior fornix (solid arrow), and the shorter anterior wall ends in the anterior fornix (open arrow). The apposition of the anterior and posterior vaginal walls is due to the lateral connections of the endopelvic fascia with the pelvic wall, creating the slit-like configuration of the vagina. Endocervical canal (*), urethra (u), anus (a), and hyperintense parametrium (arrowhead).

without glands and lubricated by secretions from cervical and Bartholin glands [19]; (2) muscularis, comprised of connective tissues (collagen and elastin) and smooth muscle arranged in inner circular and outer longitudinal layers [19]; and (3) adventitia, fibromuscular elements representing endopelvic fascia that connects the vagina to the surrounding pelvic structures to maintain support [32]. The apposition of the anterior and posterior vaginal walls is due to the lateral connections of the endopelvic fascia with the pelvic wall, creating the slit-like configuration of the vagina [19].

Vaginal wall anatomy is best depicted on $T_2$W sequences (Figure 4.13) [34,35]. The vagina is of uniform, intermediate SI on $T_1$W sequences, limiting anatomical definition. The T2 appearance of the vagina changes with reproductive age and phase of menstrual cycle. The vaginal wall and central mucus are thickest and of the highest $T_2$ SI during the mid-secretory phase of the menstrual cycle. Maximal $T_2$ contrast between the vaginal wall and surrounding pelvic fat occurs during the early proliferative or late secretory phase. In postmenopausal women receiving hormone replacement therapy, the MRI appearance of the vagina is similar to premenopausal women (Figure 4.10). In postmenopausal women, not receiving hormone replacement therapy, the vaginal wall is of diminished $T_2$ SI and the mucous layer is thinned (Figure 4.9) [34,35].

The fallopian tubes are located in the upper margin of the broad ligament and range from 7 to 12 cm in length. The fallopian tubes can be anatomically subdivided into four sections: (1) interstitial, which traverses the myometrium; (2) isthmus, which is a narrow segment closest to the myometrium; (3) ampulla, which is larger in diameter than the isthmus and comprises about one half of the tube; and (4) fimbria or infundibulum, the lateral continuation of the ampulla (Figure 4.12) [19]. Normal fallopian tubes are not well demonstrated on MRI.

## 4.4 Embryology and Anomalies

Sexual differentiation of the reproductive ducts begins in the 7th week of gestation, but gender is not apparent until the 12th week [19,31,37,38]. In the female, the lack of Müllerian-inhibiting substance allows the Müllerian (paramesonephric) ducts to persist and the lack of testosterone allows the Wolffian (mesonephric) ducts to resorb (Figure 4.14). The uterus and upper vagina are formed from the fused caudal ends of the Müllerian ducts and coalescence of surrounding mesenchyme; the fallopian tubes are formed by the unfused cranial ends of the Müllerian ducts; and the distal vagina develops from the bilateral sinovaginal bulbs, which arise from the urogenital sinus, and coalescence of surrounding mesenchyme [19,31,36,37]. Uterovaginal malformations are reported to occur in 0.16% of women and result from incomplete development of one or both Müllerian ducts, failure of distal fusion of the Müllerian duct, reabsorption of the septum between the fused ducts, or absent or incomplete canalization of the vaginal plate [19,31].

Müllerian duct anomalies are estimated to occur in 0.1%–0.5% of women; however, the true incidence may

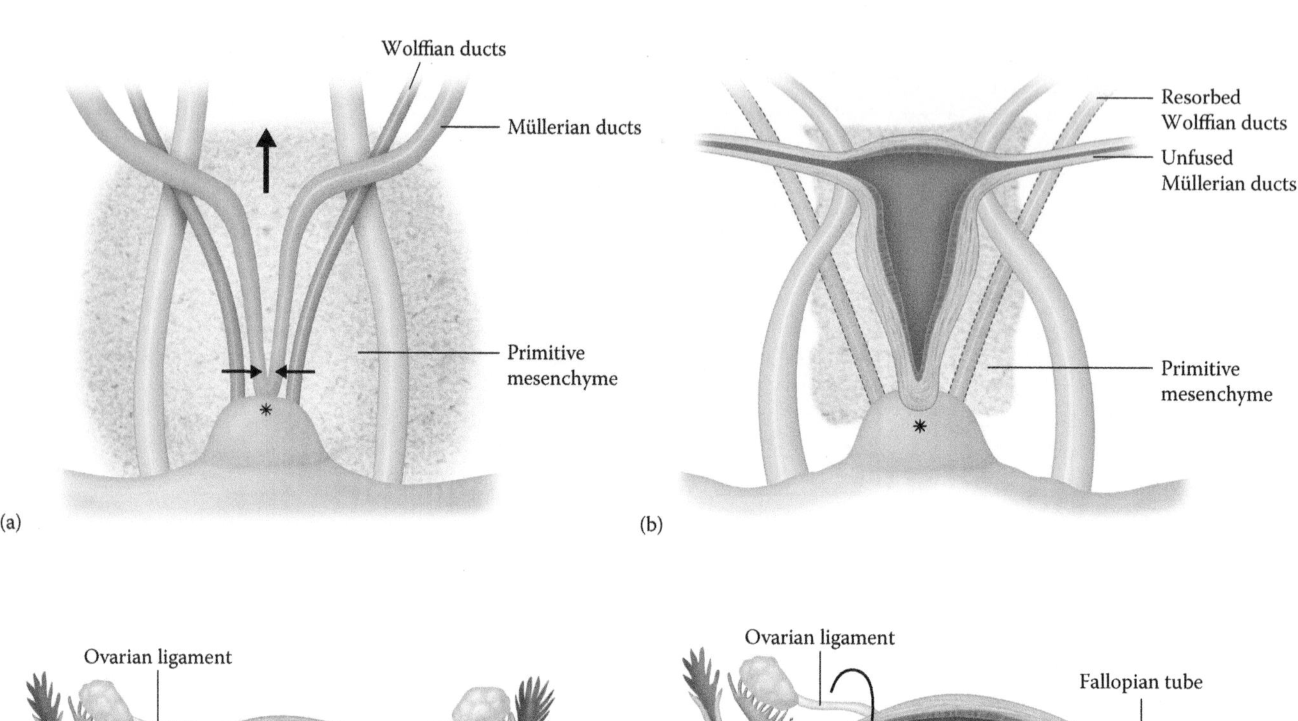


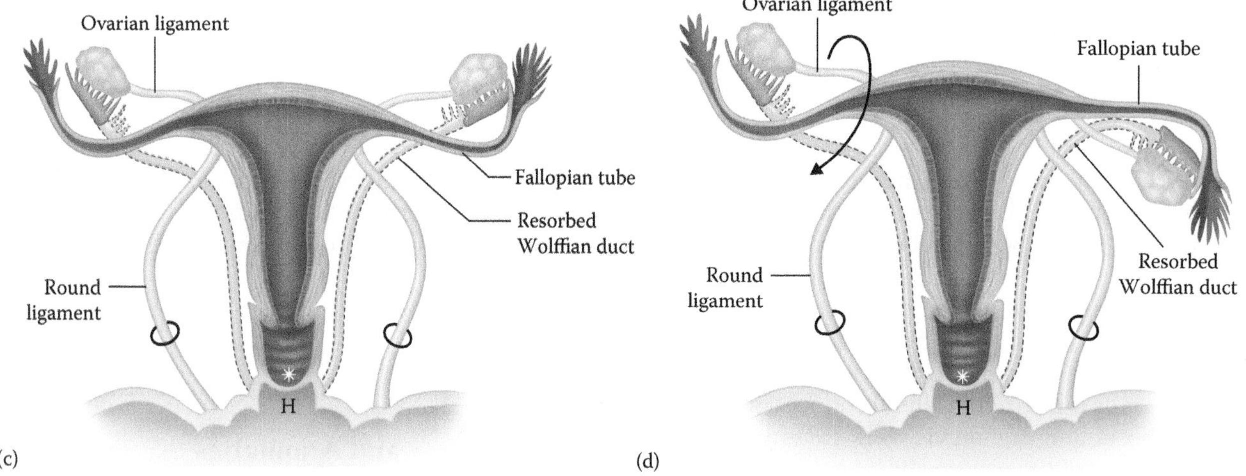


**FIGURE 4.14**
Classic embryology theory. Paramesonephric (Müllerian) ducts form lateral to and grow caudally with the mesonephric (Wolffian) ducts. Subsequently (a) the distal portions of the Müllerian ducts grow medially to fuse in the midline beginning distally and progressing cranially (arrows) to form the uterovaginal canal. The uterovaginal canal inserts into the urogenital sinus (*). (b) The uterovaginal canal and coalescence of surrounding primitive mesenchyme become the lining of the uterus and myometrium/cervical stroma, respectively. (c, d) The distal vagina develops from paired sinovaginal bulbs, which arise from the urogenital sinus, and coalescence of surrounding mesenchyme. The sinovaginal bulbs fuse into a solid mass called the *vaginal plate*, which undergoes canalization in the second trimester, creating the distal vaginal lumen (*). The vagina ends distally at the hymen (H), which becomes perforated in the perinatal period.

be significantly underestimated since most cases are diagnosed during the evaluation for primary amenorrhea, infertility, endometriosis, and obstetrical complications [19,31]. In the absence of symptoms, anomalies may remain undiagnosed. Due to the related development of the genital and urinary systems, the association of Müllerian duct anomalies with renal anomalies occurs in 30%–50% of cases [31,36]. The role of imaging is to help detect, diagnose, and distinguish surgically correctable forms of Müllerian duct anomalies from inoperable forms [38,39]. MRI is an important modality to help define complex uterine and vaginal anomalies, especially when the normal anatomy is distorted in the presence of obstruction and hematometrocolpos. It is also important for defining the presence or absence of a cervix, which is critical to the long-term management of patients with Müllerian anomalies.

The American Fertility Society (AFS) classification system of Müllerian anomalies was created with the goal of developing an easy to use, flexible reporting system [40]. It was published in 1988 and continues to be the most widely used classification system. The AFS classification system is based on the classical theory of Müllerian duct development: caudal to cranial duct fusion and

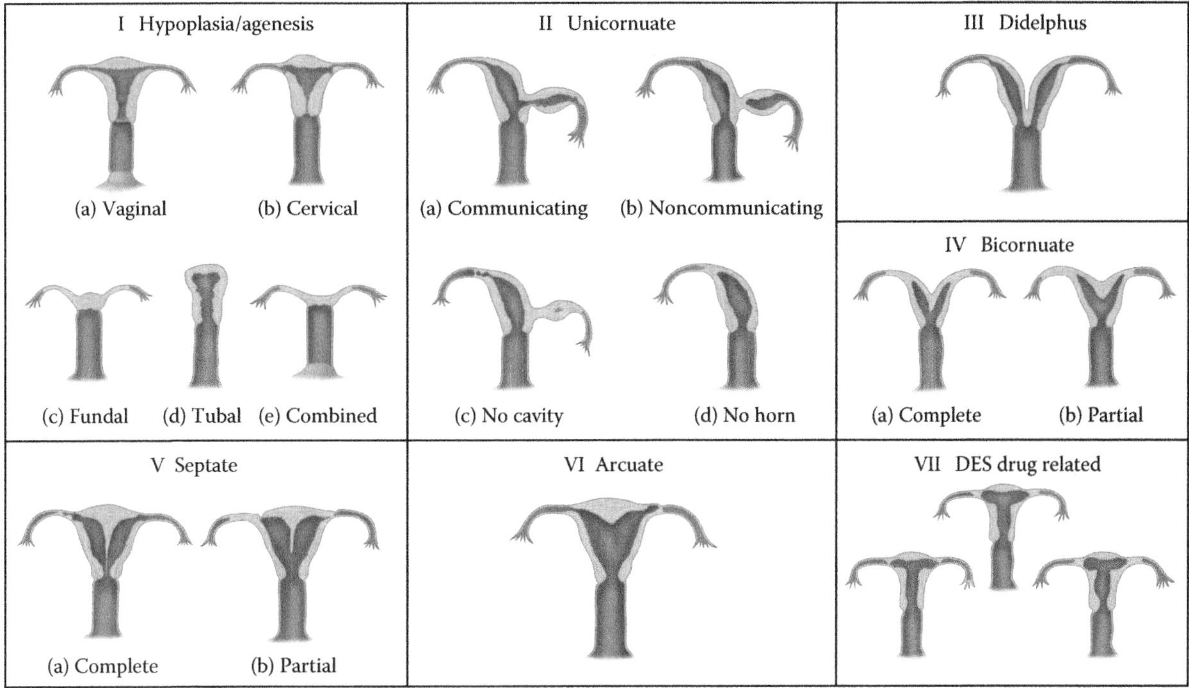


**FIGURE 4.15**
AFS classification system. The system organized anomalies according to the major uterine anatomic types and encouraged the user to indicate the malformation type as well as all associated anomalies involving the vagina, cervix, tubes, ovaries, and urologic system. (Reprinted from *Fertil. Steril.*, 49(6), The American Fertility Society, The American Fertility Society classifications of adnexal adhesions, distal tubal occlusion, tubal occlusion secondary to tubal ligation, tubal pregnancies, Müllerian anomalies and intrauterine adhesions, 944–955, Copyright (1988), with permission of Elsevier.)

resorption (Figure 4.14). At the time, it was accepted that Müllerian duct anomalies were a spectrum of defects and that not all anomalies would fit clearly into one of the AFS classifications. The goal was to develop a registry that could be used to gather enough numbers to be used to further improve the classification system and thereby formulate a more accurate prognosis for each patient. The AFS classification system groups anomalies into the seven classes according to similarities in clinical presentation, reproductive prognosis, and treatment (Figure 4.15).

The AFS classification system is useful as a framework, but incomplete [39,40,41]. The classic theory of Müllerian development does not take into account the presence of complex duplication anomalies. Muller et al. presented an alternative embryologic hypothesis of simultaneous, bidirectional fusion and resorption of the Müllerian ducts, beginning at the isthmus (Figure 4.16) [42]. This was supported by many groups who suggested that a separate classification should be made for anomalies found above and below the uterine isthmus [41,43–45]. The European Society of Human Reproduction and Embryology (ESHRE) and the European Society of Gyneacological Endoscopy (ESGE)

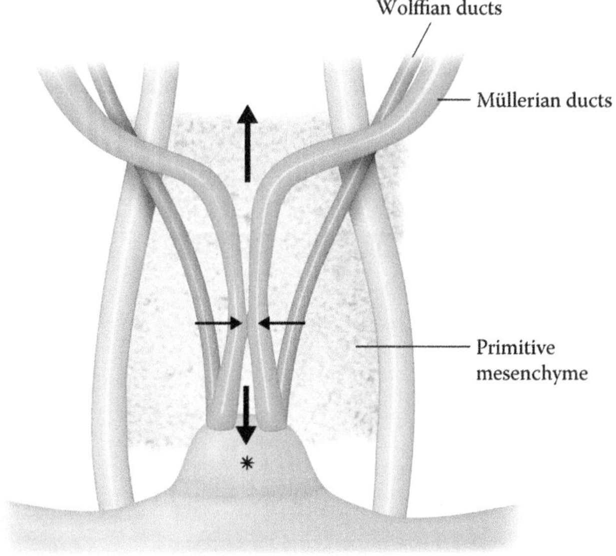


**FIGURE 4.16**
Alternate embryologic hypothesis. Paramesonephric (Müllerian) ducts form lateral to and grow caudally with the mesonephric (Wolffian) ducts. Subsequently the distal portions of the Müllerian ducts grow medially to fuse in the midline beginning at the isthmus and progressing both cranially and caudally (arrows) to form the uterovaginal canal. * urogenital sinus.

established a working group to develop a new, updated classification system presented in 2013 (Figure 4.17) [46]. Cervical and vaginal anomalies are classified independently. ESHRE/ESGE avoids the use of absolute numbers in the definitions of uterine abnormalities and instead compares the abnormality relative to uterine wall thickness [46].

Segmental Müllerian duct hypoplasia or Müllerian agenesis falls under AFS Class 1 and ESHRE/ESGE Class U5. Ninety percent of these anomalies have no identifiable uterus and a blind-ending lower vagina (Mayer–Rokitansky–Kuster–Hauser syndrome) (Figure 4.18). The remaining 10% have variable degrees of utero-vaginal hypoplasia where rudimentary elements of the uterus and/or vagina may be present (Figures 4.19 through 4.21).

Unicornuate uterus (AFS Class 2) and hemi-uterus (ESHRE/ESGE Class U4) incorporate all cases of unilateral Müllerian duct development with either incomplete formation or absence of the contralateral Müllerian duct. This results in a spectrum of findings ranging from uterine horn agenesis to development of a rudimentary horn with functional endometrium (Figure 4.22). Identification of a functional cavity is important clinically since its presence is associated with symptoms and complications including hematometros and ectopic pregnancy; resection is recommended (Figures 4.19 and 4.22) [38,47,48]. Unilateral Müllerian duct development is associated with endometriosis and renal anomalies ipsilateral to the abnormal horn (Figure 4.22d) [49,50].

Uterine didelphys (AFS Class 3), bicornuate uterus (AFS Class 4), and bicorporeal uterus (ESHRE/ESGE Class U3) encompass the Müllerian duct fusion defects (Figures 4.23 through 4.27). All have a fundal defect, completely dividing the uterine corpus in complete non-fusion/didelphys and partially dividing the corpus in incomplete fusion/bicornuate uterus. For ESHRE/ESGE Class U3, the external indentation is defined as exceeding

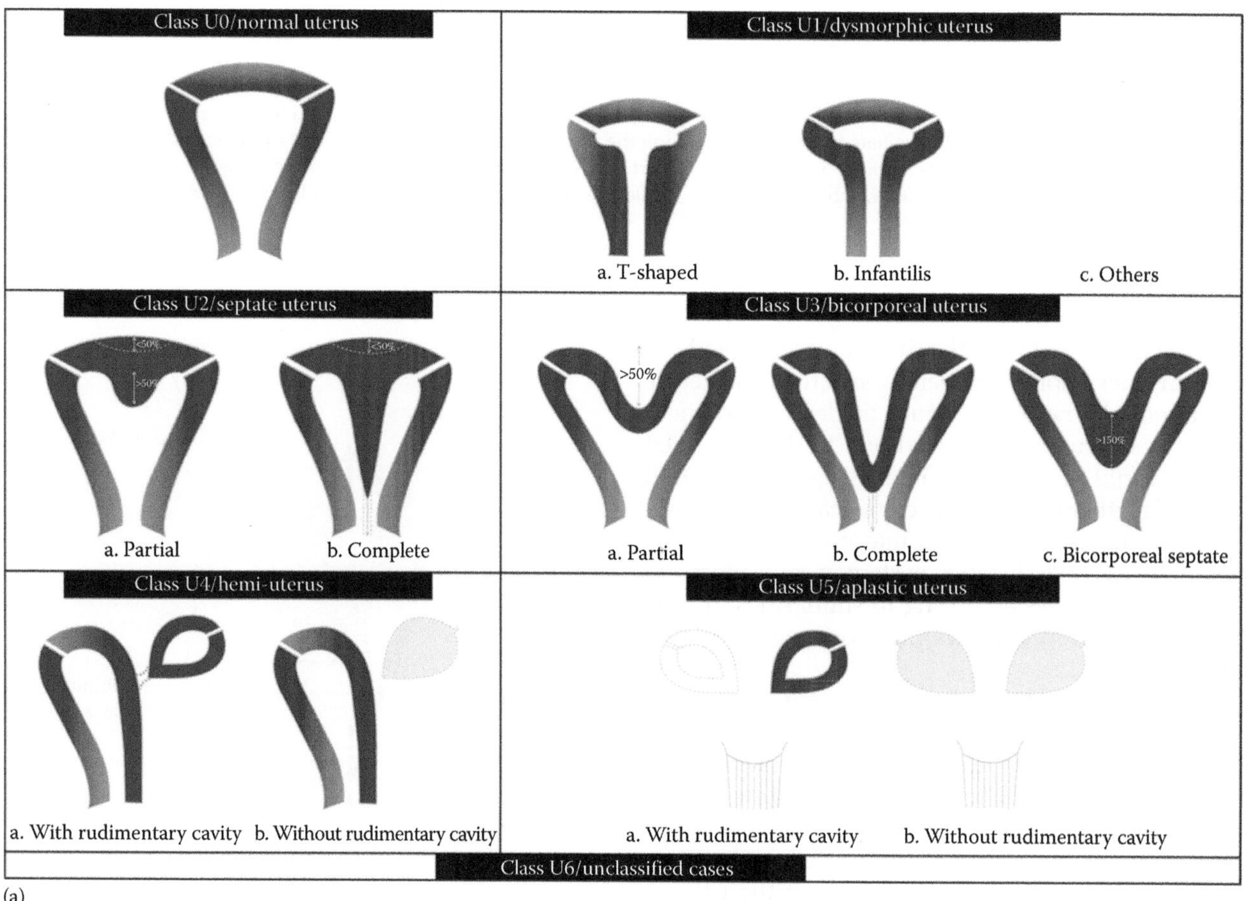


**FIGURE 4.17**
ESHRE/ESGE classification system (a and b). Uterine anomalies are grouped into classes based on deviations derived from the same embryologic origin. Subclasses represent clinically significant uterine deformities within the main class. Extremely detailed subclassification of anomalies is avoided to maintain a simple, easy-to-use system. (Grimbizis GF, Gordts S, Di Spiezio Sardo A et al., The ESHRE/ESGE consensus on the classification of female genital tract congenital anomalies. *Hum Reprod.*, 2013, by permission of Oxford University Press.) *(Continued)*

ESHRE/ESGE classification

Female genital tract anomalies

**Uterine anomaly**

| | Main class | Sub class |
|---|---|---|
| U0 | Normal uterus | |
| U1 | Dysmorphic uterus | a. T-shaped<br>b. Infantilis<br>c. Others |
| U2 | Septate uterus | a. Partial<br>b. Complete |
| U3 | Bicorporeal uterus | a. Partial<br>b. Complete<br>c. Bicorporeal septate |
| U4 | Hemi-uterus | a. With rudimentary cavity (communication or not horn)<br>b. Without rudimentary cavity (horn without cavity/no horn) |
| U5 | Aplastic | a. With rudimentary cavity (bi- or unilateral horn)<br>b. Without rudimentary cavity (bi- or unilateral uterine remnants/aplasia) |
| U6 | Unclassified malformations | |

**Cervical/vaginal anomaly**

| Co existent class | |
|---|---|
| C0 | Normal cervix |
| C1 | Septate cervix |
| C2 | Double normal cervix |
| C3 | Unilateral cervical aplasia |
| C4 | Cervical aplasia |
| | |
| V0 | Normal vagina |
| V1 | Longitudinal non-obstructing vaginal septum |
| V2 | Longitudinal obstructing vaginal septum |
| V3 | Transverse vaginal septum and/or imperforate hymen |
| V4 | Vaginal aplasia |

| U | C | V |
|---|---|---|

Associated anomalies of non-Müllerian origin:

Drawing of the anomaly

(b)

**FIGURE 4.17 (Continued)**
ESHRE/ESGE classification system (a and b).

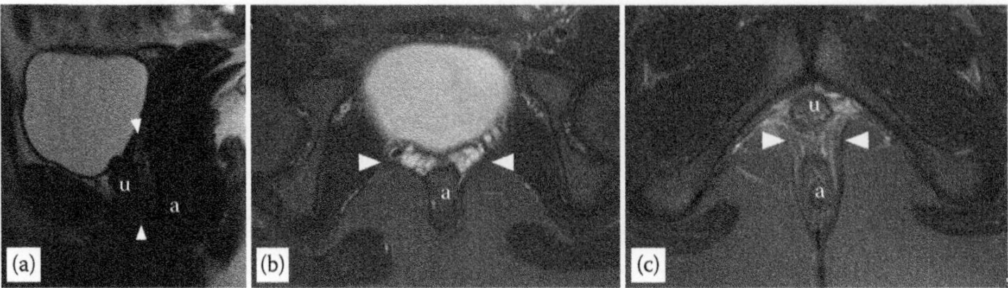


**FIGURE 4.18**
(a) Sagittal $T_2$W image demonstrates no identifiable uterus. (b and c) Transaxial $T_2$W fat-suppressed images demonstrate no vagina (arrowheads). The patient had a blind-ending lower vagina on physical exam. The degree and location of segmental agenesis and hypoplasia are varied and may involve the upper vagina, cervix, corpus, fallopian tubes, and any combination of these areas. Urethra (u) and anus (a).

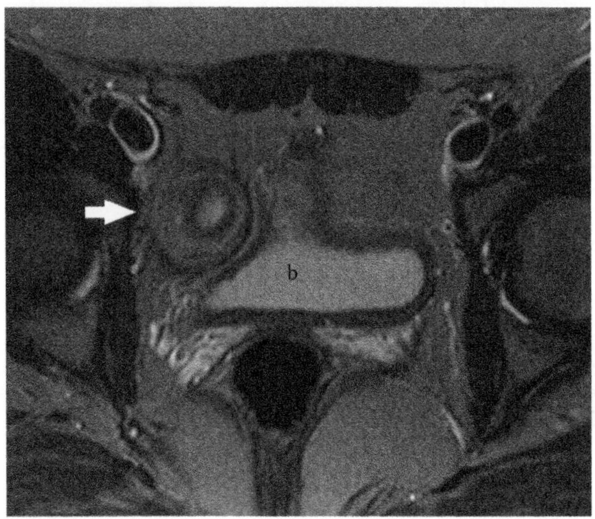


**FIGURE 4.19**
Axial $T_2$W image demonstrates an isolated rudimentary horn (arrow) without a cervix or vagina. There is normal zonal anatomy in the remnant; therefore, resection was performed. Bladder (b).

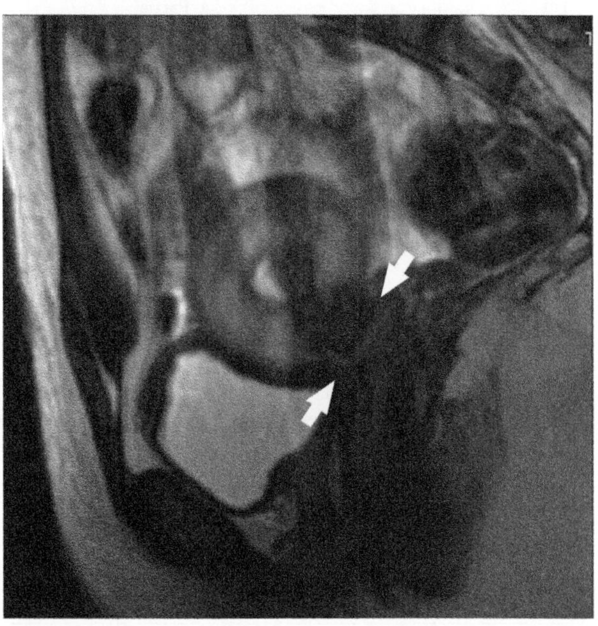

**FIGURE 4.20**
Sagittal $T_2$W image demonstrates complete absence of the cervix (arrows). Image degraded by motion artifact (vertical dark bands through uterine corpus).

50% of the uterine wall thickness [46]. There may be variable fusion at the level of the cervices and coexistent vaginal defects, including a horizontal or longitudinal septum [51]. An upper vaginal septum may be difficult to identify unless there is obstruction of one hemivagina and the presence of hematometrocolpos (Figure 4.25). The increased association between obstruction and endometriosis and pelvic adhesions is thought to be due to retrograde menstrual flow [52,53].

Septate uterus (AFS Class 5 and ESHRE/ESGE Class U2) includes all cases with normal Müllerian duct development and fusion, but abnormal absorption of the midline septum [40,46]. All have a normal convex or flattened uterine fundal contour (no external indentation) and an internal septum of varying length, including into the cervix and vagina (Figures 4.28 through 4.30) [38]. Although the AFS classification does not characterize the external fundal indentation for differentiation between bicornuate and septate uterus, a fundal indentation of 1.0 cm has been found to be a reliable discriminator between the two configurations [38]. ESHRE/ESGE defines the septum as an internal indentation exceeding 50% of the uterine wall at the fundal midline [46]. Measuring angles between uterine horns and intercornual distance is not as predictable since there is overlap in the angles and distortion by fibroids; these are criteria created for hysterosalpingography (HSG) and not useful in differentiating between septate and bicornuate uteri with MRI [38].

Diethylstilbestrol (DES)-related anomalies (AFS Class 7) fall into the ESHRE/ESGE Class U1 (dysmorphic uterus) where there is a normal uterine outline, but the shape of the uterine cavity is abnormal [46]. Dysmorphic uteri are typically smaller than normal and include the T-shaped uterus associated with DES, infantalis configuration (Figure 4.31), and any other abnormal uterine cavity configurations. *In utero* exposure to DES, a synthetic estrogen

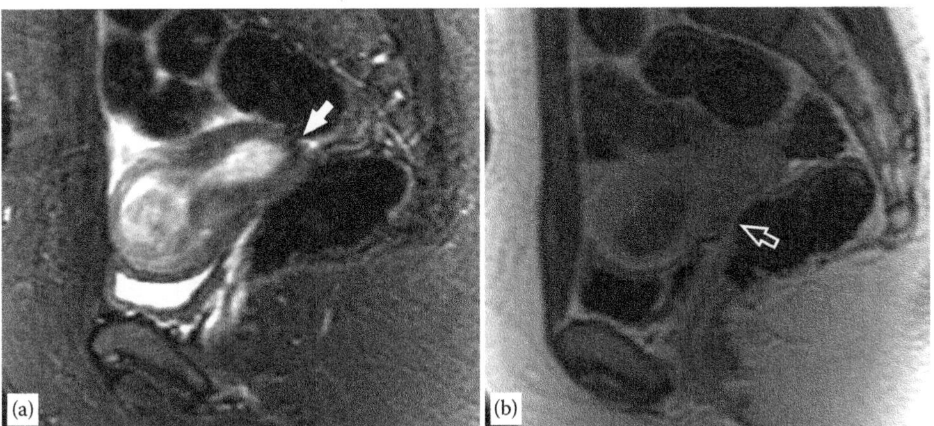

**FIGURE 4.21**
(a) Sagittal $T_2$W image demonstrates agenesis of the lower cervix (white arrow), and (b) sagittal $T_1$W image postcontrast demonstrates agenesis of the proximal vagina (open arrow). Two kidneys were present.

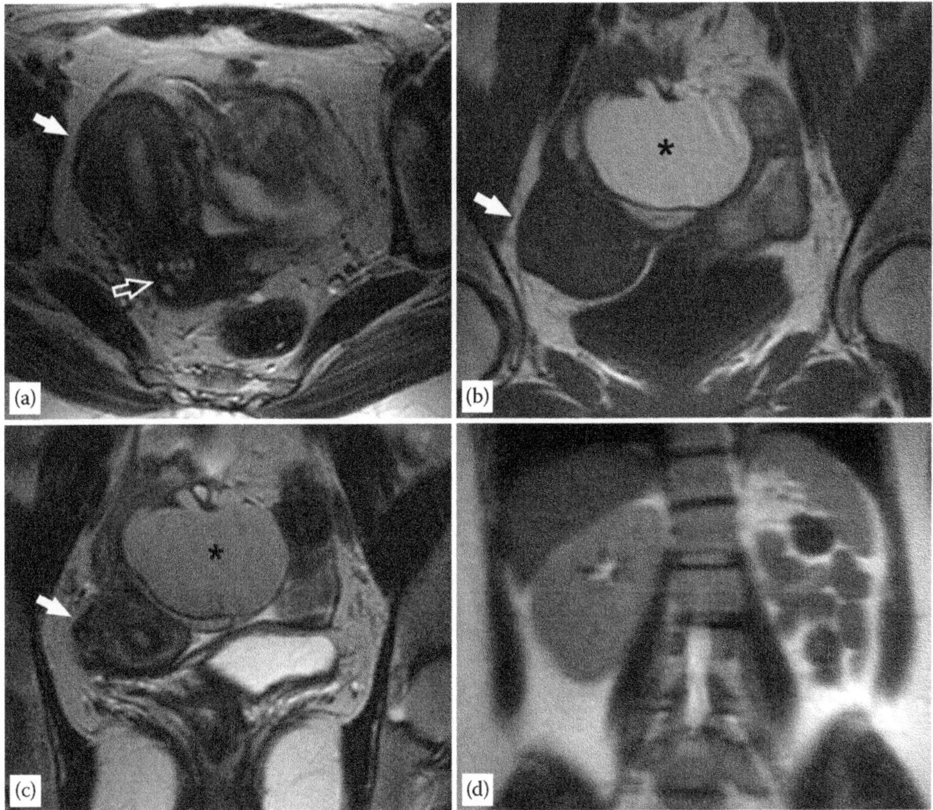

**FIGURE 4.22**
(a) Transaxial $T_2$W image demonstrates elongated right uterine horn with normal zonal anatomy (arrow). Coronal (b) $T_1$W and (c) $T_2$W images demonstrate a rudimentary horn dilated with blood products (*). When interpreting imaging, the radiologist should note the presence of a rudimentary horn, whether it contains endometrium, and whether the cavity communicates with the fully formed unilateral horn as this may change clinical management. (d) Coronal $T_2$W image demonstrates the absence of the left kidney ipsilateral to the rudimentary horn. Nabothian cysts (open arrow).

prescribed to pregnant women from late 1940s through 1970, was linked to an increased incidence of vaginal adenosis and clear cell adenocarcinoma, as well as development of structural anomalies of the uterus and vagina [54–56]. Since 1971 the use of DES has been contraindicated in pregnancy [57].

Arcuate uterus (AFS Class 6) is very close to normal and represents near complete resorption of the septum (Figure 4.32). There is no separate ESHRE/ESGE classification for arcuate configuration. Because anomalies of the cervix and vagina can occur with a normal uterus, ESHRE/ESGE classification added Class U0 for all

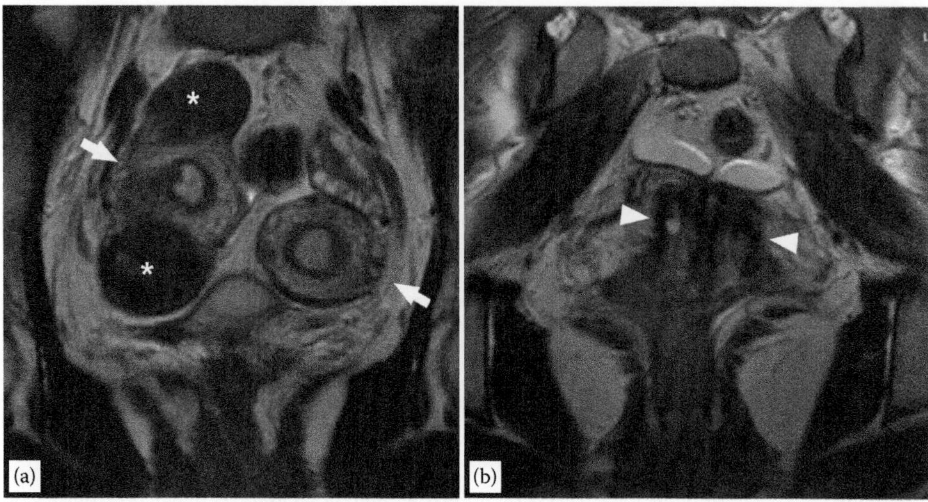


**FIGURE 4.23**
Coronal $T_2$W images demonstrate (a) two widely separated uterine horns (arrows), with normal zonal anatomy and subserosal leiomyomas (*). (b) There is fusion of the cervices (arrowhead) and a single vagina (not shown). Normal kidneys were seen.

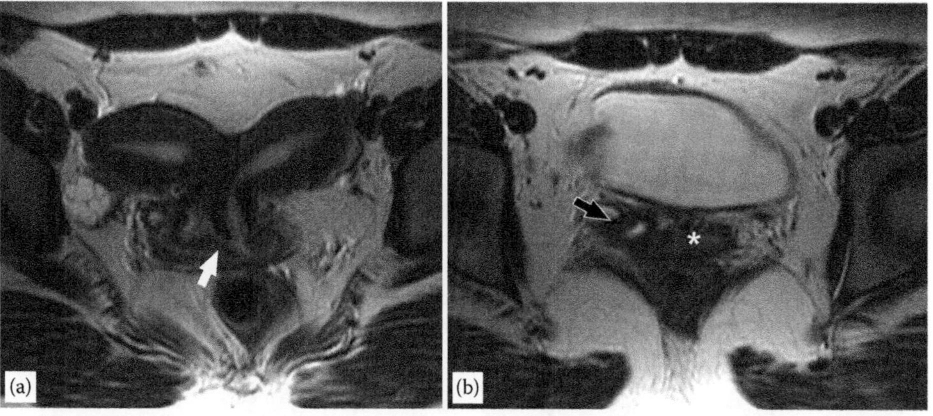


**FIGURE 4.24**
Axial $T_2$W images demonstrate (a) two widely separated uterine horns, with normal zonal anatomy. There is fusion of the cervices and partial fusion of the isthmus. (b) Two upper vaginas are present. The right vagina is blind ending (black arrow); however, there is no evidence of obstruction due to the presence of a small communication immediately below the cervix (white arrow). There is absence of the kidney, ipsilateral to the hypoplastic right vagina (not shown). Left vagina (*).

cases with a normal uterine corpus. Anomalies of the cervix can occur with or without coexisting uterine corpus and vaginal anomalies [46,58]. Cervical anomalies include absorption defects: septate cervix (Figure 4.29); fusion defects: double cervix (Figures 4.23 through 4.26); aplasia and agenesis (Figures 4.20 and 4.21).

Anomalies of the vagina can occur with or without coexisting uterine anomalies (Figure 4.31). A longitudinal vaginal septum results from abnormal lateral fusion or incomplete reabsorption of the paired Müllerian ducts, may be partial or complete, and is typically associated with partial or complete duplication of the cervix (Figures 4.24, 4.25, and 4.33) [19,31]. Longitudinal septa are difficult to identify on MRI unless there is obstruction. A persistent transverse (horizontal) septum results from improper fusion of the Müllerian ducts with the urogenital sinus or incomplete canalization of the vaginal plate; therefore, transverse septa can occur at varying levels in the vagina and be of varying thickness (Figure 4.34) [31,38,59]. Up to 20% of women with longitudinal septa have renal anomalies and almost all patients with obstructed hemivagina have ipsilateral renal agenesis [31,36,38,51,60,61]. Unlike other Müllerian duct anomalies, transverse septa are associated with fewer urologic anomalies.

If the vaginal plate fails to undergo canalization, distal vaginal atresia results. The distal vagina is absent above the hymeneal ring and in its place is 2–3 cm of fibrous tissue. In these women, the upper tract reproductive organs are typically present. MRI is important to evaluate the length of the atresia, the presence of upper tract reproductive organs, and the presence of hematocolpus (Figures 4.35 and 4.36) [31].

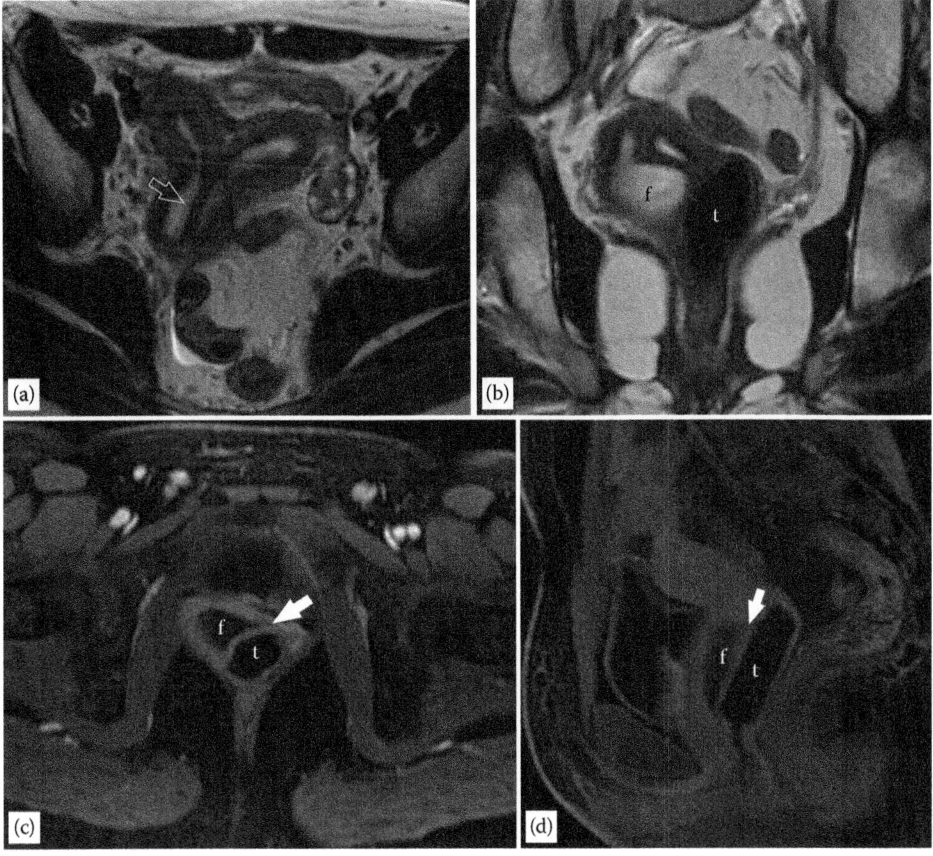


**FIGURE 4.25**
(a) Axial and (b) coronal $T_2$W images demonstrate divergent horns with fusion of the cervices and lower corpus. The myometrium between the fused segments is relatively thin (open arrow). Fundal indentation greater than 50% of the uterine wall thickness is present. (c) Axial and (d) sagittal $T_1$W, postcontrast image demonstrating a longitudinal septum (white arrow) and retention of blood in the right hemivagina (f) likely due to a tampon in the left hemivagina (t).

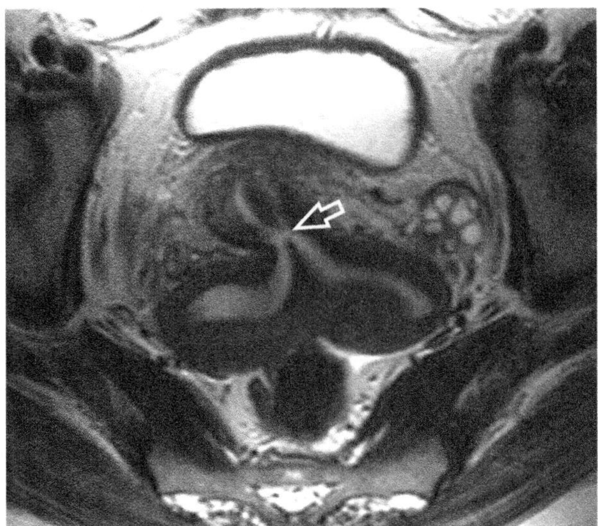

**FIGURE 4.26**
Axial $T_2$W image demonstrates two symmetrical, divergent uterine horns, normal zonal anatomy and cervical duplication. There is communication between the horns at the level of the lower uterine segment (isthmus).

## FOCUS POINTS

- The role of MRI is to detect, diagnose, and distinguish surgically correctable forms of Müllerian duct anomalies from inoperable forms.
- Therapy is based on the component parts of the gynecologic anomalies; therefore, the radiologist should describe fully all corpus, cervical, and vaginal anomalies and not classify them into the *best fit* category.
- Identifying the presence or absence of a cervix is critical in choosing the correct surgical procedure.
- The association of Müllerian duct anomalies with renal anomalies occurs in up to 50% of cases; therefore, a patient with Müllerian anomalies should be evaluated for renal anomalies.

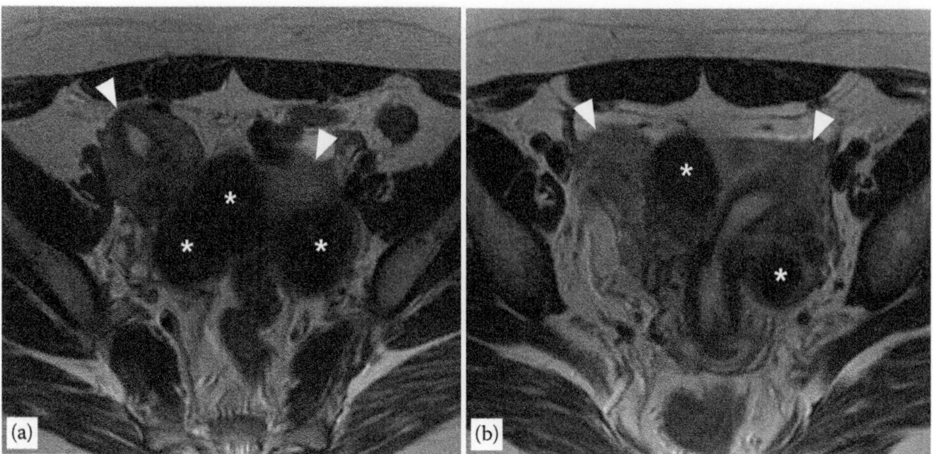


**FIGURE 4.27**
Axial $T_2$W images demonstrate (a) two divergent horns and (b) one cervix. Several leiomyomas (*) distort the uterus making identification of the fundal cleft more challenging.

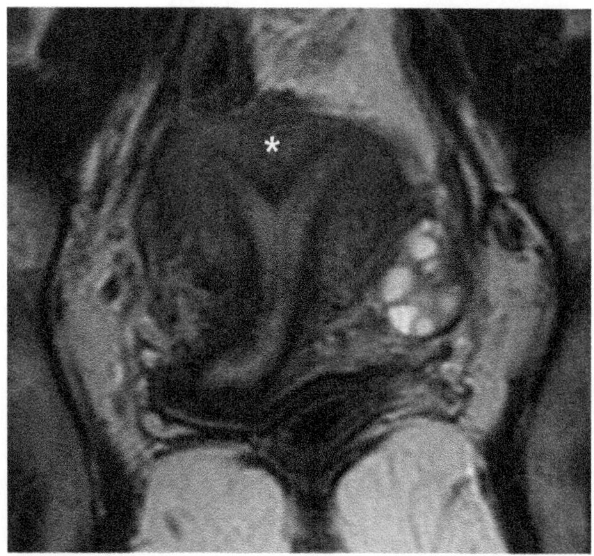

**FIGURE 4.28**
Coronal $T_2$W image demonstrates the absence of a fundal cleft (*), and the uterine horns are not widely separated. The horns are separated by a partial, midline, fibromuscular (intermediate $T_1$ and $T_2$ SI) septum. Description of a septate uterus should include whether the septum is partial or complete.

## 4.5 Benign Diseases of the Myometrium

### 4.5.1 Leiomyomas

Uterine leiomyomas are common [62,63]. The National Institute of Environmental Health Sciences evaluated 1364 randomly selected women living in Washington, DC, for age-specific cumulative incidence of leiomyomas and found the estimated lifetime risk of developing leiomyomas for white women was almost 70% and greater than 80% in black women [64], and black women were significantly more likely to develop leiomyomas at an earlier age and to come to clinical attention [64]. Leiomyomas are of smooth muscle origin, develop primarily in premenopausal women, and are hormonally dependent. The International Federation of Gynecology and Obstetrics (FIGO) classification system subclassifies leiomyomas into eight types (Figures 4.37 through 4.41). (Table 4.2) [65].

In reproductive age women, leiomyomas both grow and shrink at variable rates depending on hormonal stimulation as well as other factors not yet determined [64,66–68]. Typically leiomyomas stabilize or decrease in size after menopause unless the patient is on hormone replacement therapy [66,69]. Most patients are asymptomatic, but when symptoms occur, they are nonspecific and include abnormal bleeding, pain and pressure symptoms, and reproductive dysfunction. Intervention is indicated when the patient is symptomatic or there is concern for leiomyosarcoma. Specific symptoms may be caused by the location of the leiomyoma [69,70]; however, Ruushanen et al. did not find any association between specific MRI characteristics and symptoms of urinary stress incontinence, non-menstrual related lower abdominal pain, and pressure symptoms related to the back. Submucosal leiomyomas with a greater than 50% protrusion into the endometrial cavity contribute to abnormal uterine bleeding [71]; however, there is evidence to suggest that physical compression of the endometrium is not the main reason for abnormal uterine bleeding. Growth factor dysregulation associated with fibroid growth may cause uterine vascular dysfunction

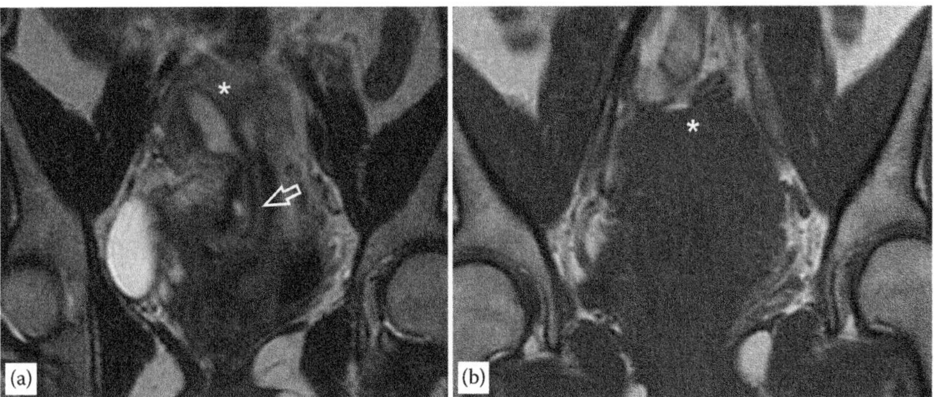


FIGURE 4.29
Coronal (a) $T_2$W and (b) $T_1$W images demonstrate the absence of a fundal cleft (*), and the uterine horns are not widely separated. There is a complete, midline, fibrous (low $T_2$ and $T_1$ SI) septum extending to the external cervical os, but not into the vagina (open arrow). The cephalad portion of the septum is fibromuscular (*). $T_1$W images may be helpful for determining the fundal contour if not well defined on $T_2$W sequences.

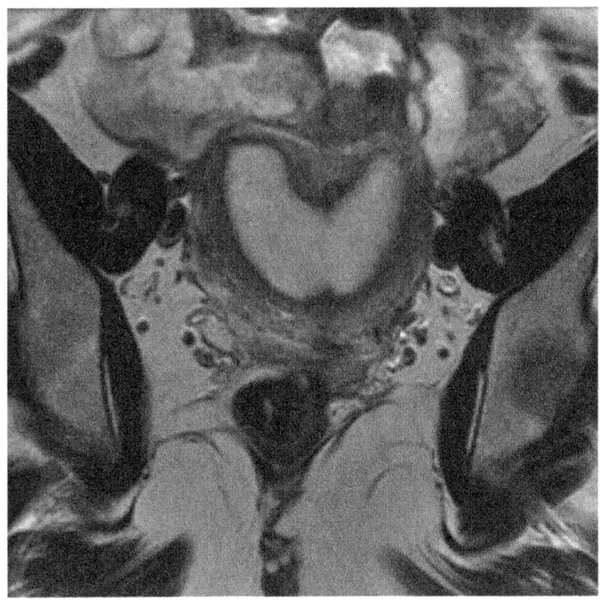

FIGURE 4.30
Appearance after septoplasty. If there is access to preoperative imaging, it is helpful to detail how much, if any residual septum remains. This may determine if a second surgery is required.

leading to increased bleeding, independent of leiomyoma location [70,72].

US is the best first imaging study for differentiating fibroids from other pelvic pathology, and saline infusion ultrasonography can be particularly useful in the assessment of submucosal leiomyomas and the extent to which they involve the myometrium. When the uterus is enlarged, or contains more than several leiomyomas, MRI may be necessary for accurate assessment of the number, size, and location of leiomyomas, especially when planning for and monitoring response to treatment [2,73]. $T_2$W sequences are optimal for evaluating leiomyoma location as contrast between the tumor and the myometrium and/or endometrium is maximized. Gd-C-enhanced MRI is necessary to accurately determine whether there is presence of degeneration and thus can be used to predict and follow up response to therapy [72–77].

In 1986, Hricak et al. first described the typical MRI appearance of leiomyomas as a well-circumscribed mass with sharp margins between it and adjacent, compressed myometrium [73]. Leiomyomas were divided into two groups: (1) those demonstrating homogeneously low $T_2$ SI indicating no degenerative changes (Figures 4.23, 4.27, and 4.38), and (2) those with various degrees of heterogeneity due to increased cellularity (Figure 4.42) and hyaline, myxomatous or lipomatous degeneration [73]. Degenerative changes are considered to be due to inadequate blood supply, and although specific degeneration may have specific MRI characteristics, degenerating leiomyomas are typically very heterogeneous in appearance (Figures 4.43 through 4.46) [74,78]. DWI may be helpful in identifying degenerative changes in cases where Gd-C is contraindicated, but further work is needed [79].

Intravenous leiomyomatosis is an uncommon, benign smooth muscle tumor that invades into uterine and extrauterine veins, with rare extension into the inferior vena cava and right heart [80–82]. Early diagnosis is difficult as extension into small uterine vessels may not be detected by MRI. The best treatment is surgical excision with hysterectomy and bilateral sapingo-oopherectomy (Figure 4.47) [81,82]. Recurrence is rare when complete resection has been achieved [82].

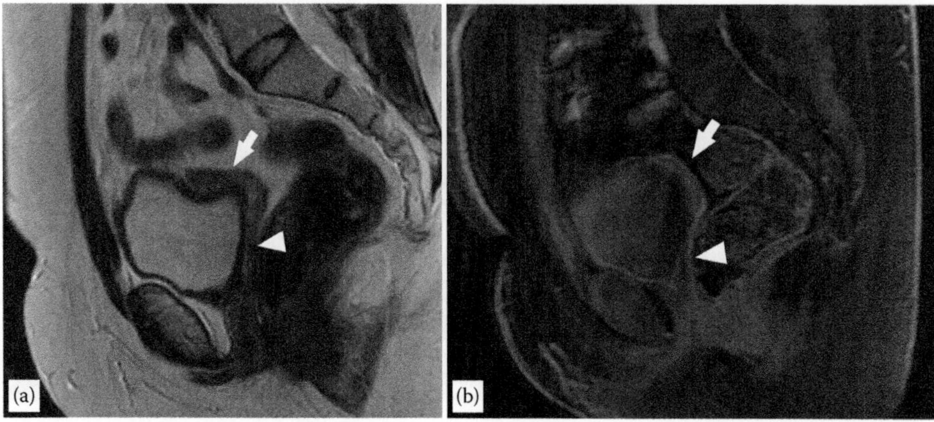


**FIGURE 4.31**
Infantalis uterine configuration in a 40-year-old female with vaginal agenesis and primary amenorrhea (a and b). Sagittal (a) $T_2$W and (b) $T_1$W postcontrast images demonstrate a small uterus (arrow) with no normal zonal anatomy and no normal vagina (arrowhead). Patient ultimately diagnosed with mosaic Turner's syndrome.

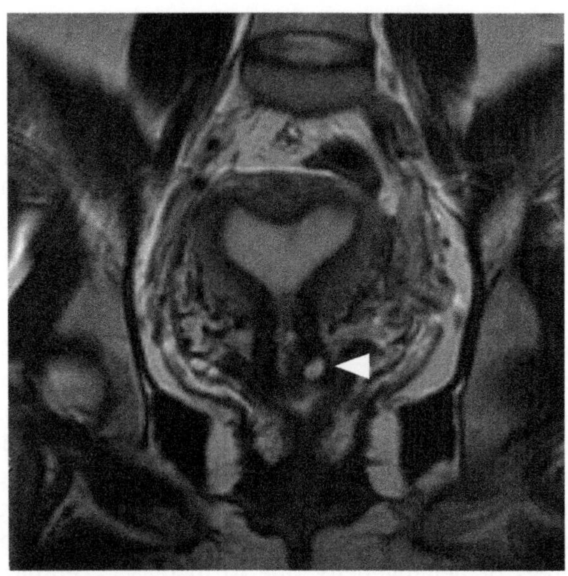

**FIGURE 4.32**
Axial $T_2$W image demonstrates a normal, convex external fundal contour, with mild bulging of the myometrium into the endometrial cavity. This would be classified as arcuate under the AFS system. By ESHRE/ESGE, this would be classified as normal, as the internal indentation is no greater than 50% of the wall thickness. Nabothian cyst (arrowhead).

### 4.5.2 Adenomyosis

Benign mimics of leiomyomas include uterine contractions (Figure 4.6), ovarian fibromas (Figure 4.48) [83], and adenomyosis. Symptoms of adenomyosis are nonspecific and overlap with those of leiomyomas making clinical diagnosis difficult. Adenomyosis is defined as the presence of non-hormonally responsive endometrial glands and stroma within the myometrium, surrounded by smooth muscle hyperplasia [84]. In one study of 1334 patients undergoing hysterectomy, adenomyosis was found in almost 25% of specimens and associated with fibroids in over 23% [84]. The incidence is increased in multiparous women [85,86]. Adenomyosis may be associated with polyps, hyperplasia, endometriosis, and endometrial cancer [87].

The accuracy of MRI in diagnosing adenomyosis has been well established [88–92]. When comparing real-time, transvaginal (TV) US, MRI, and pathology, TVUS and MRI are equally accurate in diagnosing adenomyosis, except when there are coexisting leiomyomas where MRI is more accurate [91,93]. Dueholm et al. found that MRI was more specific than TVUS for the diagnosis of adenomyosis, but both were equally sensitive [92].

Adenomyosis is intermediate in TI SI and diffusely low in $T_2$ SI, with or without scattered foci of increased SI. These $T_2$W bright foci represent endometrial cysts, ectopic endometrial tissue, or hemorrhage and are seen in only 50% of cases (Figures 4.49 and 4.50) [90,91]. Adenomyosis is diagnosed when the JZ appears thickened to greater than 12 mm [91], with a diagnostic accuracy and specificity of 85% and 96%, respectively [93]. Sensitivity is lower at 63% [92]. Normal JZ thickness is 8 mm or less and may be poorly visualized in up to 20% [91]. When the JZ measures between 8 and 12 mm, secondary findings should be used to make the diagnosis of adenomyosis, including asymmetric JZ thickening, poorly defined borders, and foci of high $T_1$ and $T_2$ SI [91]. Adenomyosis enhances earlier relative to myometrium on DCE, best demonstrated on arterial phase [94]. Enhancement of adenomyosis is similar to endometrium on venous phase.

An adenomyoma is a well-circumscribed nodule/mass made up of smooth muscle, endometrial glands, and stroma and may be located either in the myometrium (Figure 4.50b) or in the endometrium (as a polyp) (Figure 4.51) [87]. Mimics for adenomyosis include diffuse muscular hypertrophy of the inner myometrium and uterine peristalsis.

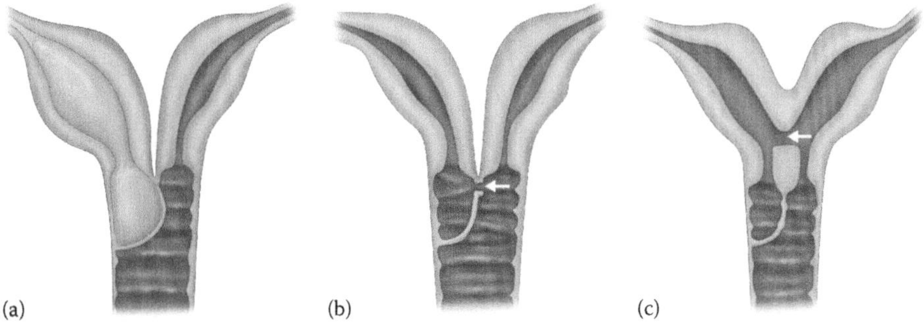


FIGURE 4.33
Horizontal vaginal septum may be partial or complete and may be associated with obstruction (a). There may be fenestrations of the septum (b) or communication between the horns at the level of the isthmus (c) (arrows).

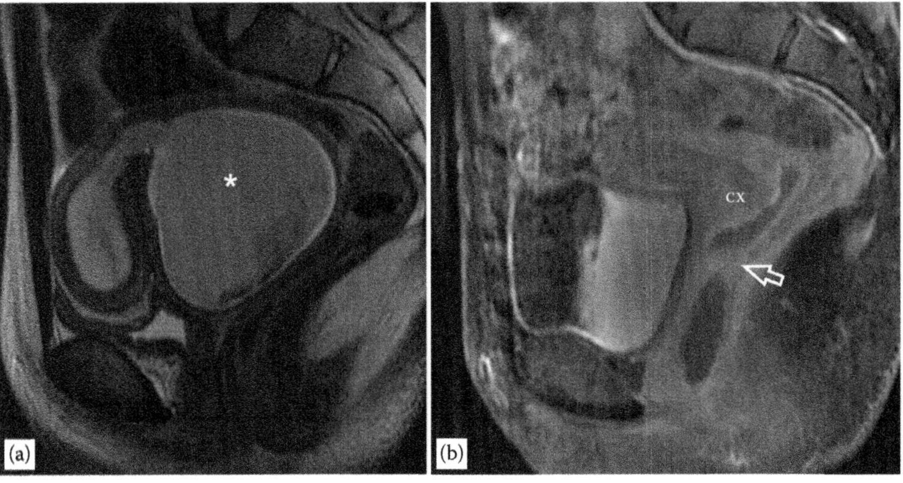


FIGURE 4.34
(a) Sagittal $T_2$W image demonstrates hematometrocolpos (*) to the level of a transverse vaginal septum. The septum is not well characterized until after the fluid is aspirated (transvaginally, under ultrasound guidance). (b) Sagittal $T_1$W postcontrast image allows identification of the location and thickness of the septum (arrow), as well as the presence of a normal cervix (cx). Differentiating a transverse septum from an abnormal cervix and defining the level and thickness of the septum are important for deciding on surgical approach. A high septum is typically thicker than those in the mid- or lower vagina. (From Rock, J.A. et al. *Obstet. Gynecol.*, 59, 448–451, 1982.)

### 4.5.3 Therapy and Post-Therapy Changes

Treatment for leiomyomas is indicated when symptoms interfere with quality of life (abnormal uterine bleeding, urinary symptoms, pelvic pressure, or pain), and when all other causes of recurrent pregnancy loss and infertility have been excluded and a leiomyoma is distorting the endometrial cavity [95]. Treatment options include observation, medical therapy, hysterectomy, myomectomy, uterine artery embolization (UAE), high-energy-focused US, and radio-frequency ablation. Medical therapy may include gonadotropin-releasing hormone agonists, progestins, and oral contraceptives. These agents may be used to temporize symptoms and decrease uterine size prior to surgery [69,72]. In 2007, Viswanathan et al. reviewed the English literature regarding the outcomes of treatment of uterine leiomyomas by various methods and concluded that there are no well-conducted trials directly comparing treatment options and that there is little high-quality evidence supporting the effectiveness of most interventions [96]. Further follow-up is needed.

Hysterectomy is the definitive and most common treatment for leiomyomas (Figure 4.52) [62]. Myomectomy is considered for the symptomatic patient who wants to retain their uterus. A hysteroscopic approach is chosen if the fibroid is submucosal, and laparotomy or laparoscopy is performed for intramural and subserosal leiomyomas. Large leiomyomas are typically morcellated prior to removing them through the laparoscopic port which may lead to peritoneal spread and implantation of leiomyoma remnants. Morcellation has been implicated as a possible cause of peritoneal leiomyomas [62] and peritoneal leiomyosarcoma metastasis [97] found after surgery. Laparoscopic leiomyoma coagulation is a newer surgical technique using radio-frequency thermal ablation [98].

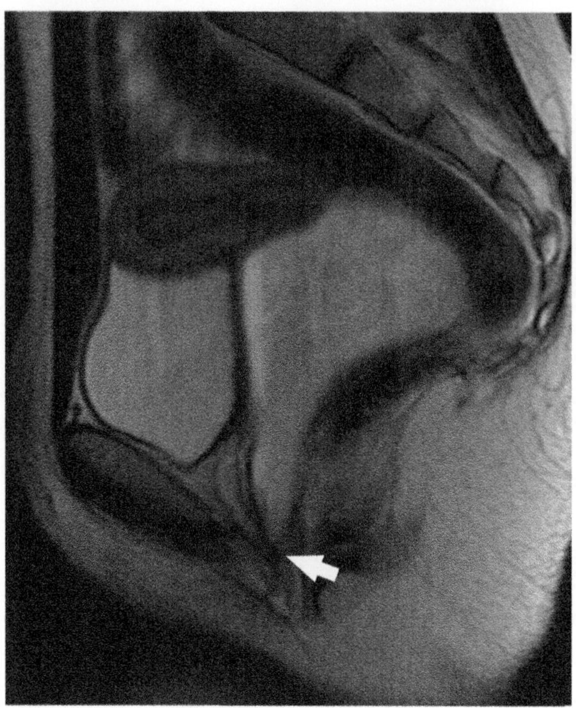

**FIGURE 4.35**
Sagittal $T_2$W image demonstrates distal vaginal agenesis (arrow) with high-signal-intensity fluid distending the vagina. $T_1$W images confirmed blood products (not shown). Uterine anatomy is normal. Determining the length of the atresia, the presence of upper tract reproductive organs and the presence of hematocolpus are important for deciding on surgical approach.

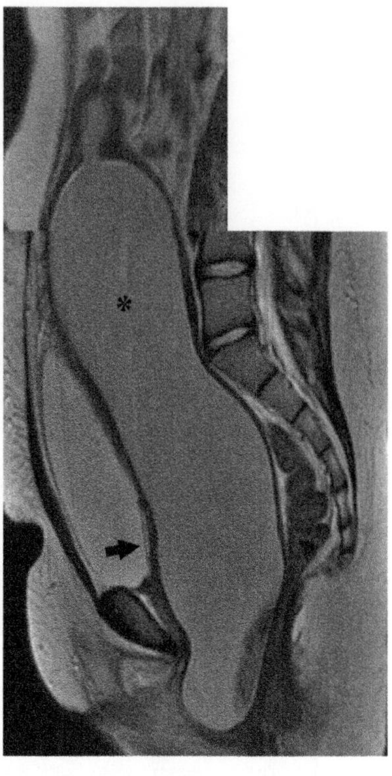

**FIGURE 4.36**
A 14-year-old female with distal vaginal atresia and hematometrocolpos (*). Uterine anatomy was otherwise normal. A low transverse septum may have a similar appearance to distal vaginal agenesis and imperforate hymen. This patient was also found to have an associated duplicated right renal collecting system (not shown) and ureterocele (arrow).

UAE is a uterus-sparing alternative for patients who would otherwise undergo surgical resection. UAE uses embolic particles to occlude end-arterioles of the uterine artery that perfuse the leiomyomas (Figure 4.12). The role of MRI in patients undergoing UAE is to determine which patients will have a successful outcome by evaluating the number, location, and size of the leiomyomas and whether autoinfarction or degeneration is present (Figure 4.53) [99,100]. UAE is not indicated if the dominant leiomyoma is devascularized [79,100]. Embolizing large submucosal, pedunculated intracavitary, and pedunculated subserosal leiomyomas is controversial. Submucosal fibroids may be expelled after UAE and, if large, may cause symptoms (Figure 4.53a). Pedunculated subserosal leiomyomas have a risk of detachment; however, Katsumori et al. found no increase in complications when treating pedunculated leiomyomas with a stalk base measuring 2 cm or larger (Figure 4.53c, f) [101]. Cervical leiomyomas are more resistant to treatment than uterine corpus leiomyomas, likely due to a more complex blood supply from uterine, vaginal, and hemorrhoidal arteries [102]. Post-procedure infarction of leiomyoma tissue (imaged within 72 h of UAE) may be the most important imaging parameter for predicting clinical success, not patient age, uterine, volume, or number, location, and volume of

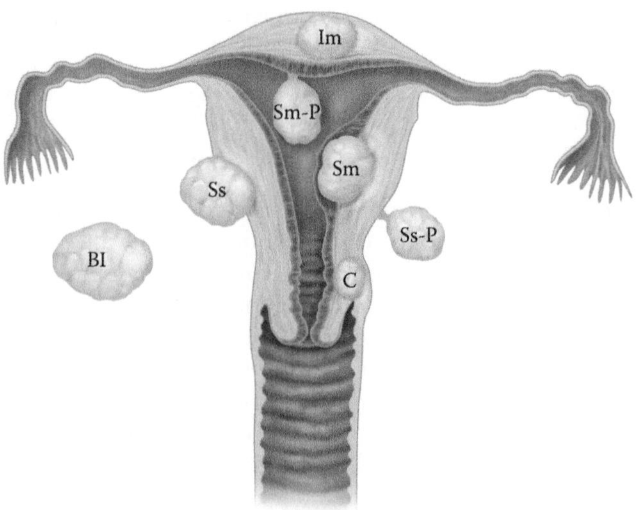


**FIGURE 4.37**
Leiomyomas may be solitary or multiple and may be confined completely within the uterine wall (Im, intramural), protrude into the endometrial cavity (Sm, submucosal), protrude outside the confines of the myometrium (Ss, subserosal), have no involvement with the myometrium (C, cervical; Bl, parasitic), or involve both the endometrial and serosal surfaces (transmural). Pedunculated (-P).

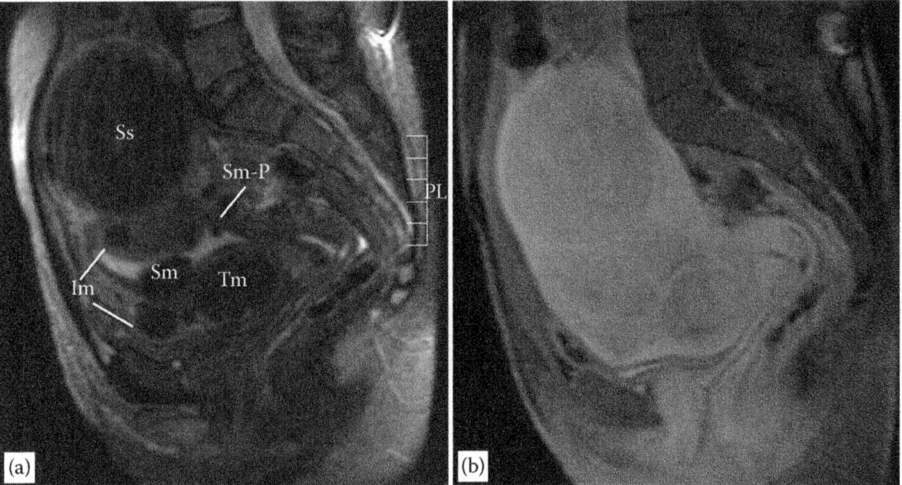


**FIGURE 4.38**
Sagittal (a) $T_2$W and (b) $T_1$W images demonstrate characteristic low $T_2$ SI of leiomyomas. They are well circumscribed, without a capsule and are homogenously hypoenhancing relative to normal myometrium. Submucosal (Sm), intramural (Im), subserosal (Ss), transmural (Tm), and pedunculated (-P). Transmural fibroids should be categorized by their relationship to both the endometrial and the serosal surfaces, with the endometrial relationship noted.

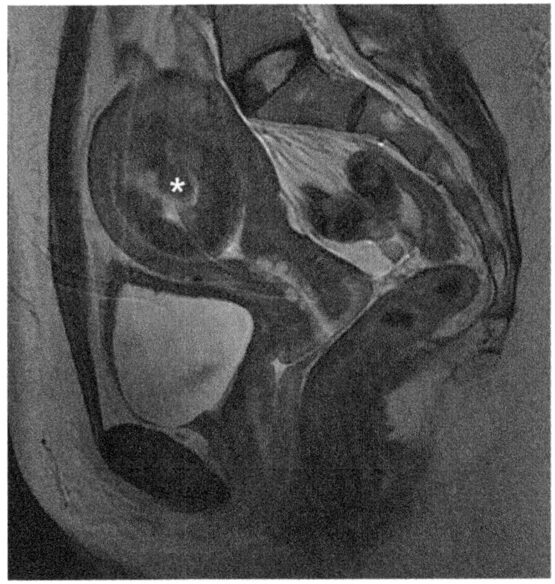

**FIGURE 4.39**
Intracavitary leiomyoma. Sagittal $T_2$W image demonstrates a submucosal leiomyoma (*) projecting into and widening the endometrial canal. The heterogeneous appearance is due to degeneration. Due to symptoms of menorrhagia, anemia, and dysmenorrhea, it was subsequently removed by laparoscopic myomectomy.

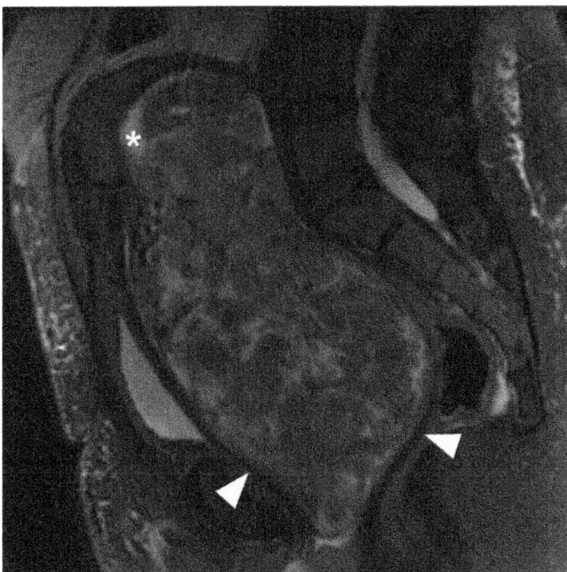

**FIGURE 4.40**
Aborting leiomyoma. Sagittal $T_2$W image demonstrates a large, heterogeneous (degenerating), pedunculated, submucosal leiomyoma projecting into and widening the vaginal canal. The cervix is effaced and difficult to identify. Endometrium (*) and vagina (arrowheads).

uterine leiomyomas [103]. Submucosal leiomyomas have the greatest post-UAE volume reduction, followed by intramural and subserosal leiomyomas [76,104]. UAE has been used as a uterus-conserving procedure for treating adenomyosis; however, the long-term effectiveness needs further study (Figure 4.54) [105].

MRI-guided focused US is a noninvasive, uterus-sparing alternative where high-frequency US is used under MRI guidance to selectively treat targeted tissue by thermal coagulation [106]. Pretreatment selection of leiomyomas may predict the success of the treatment [98,107,108]. Low $T_2$ SI, non-autoinfarcted leiomyomas respond better to therapy [107]; high $T_2$ SI, enhancing leiomyomas required higher US energies to achieve therapeutic temperature increases [108]; and size matters, treating leiomyomas between 5 and 10 cm is optimum.

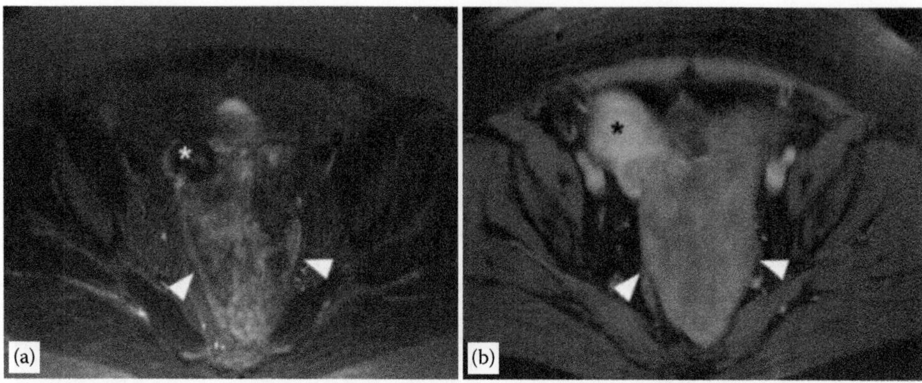


**FIGURE 4.41**
Cervical leiomyoma. Axial (a) $T_2$W with fat suppression and (b) $T_1$W postcontrast images demonstrate a heterogeneous $T_2$ SI mass (arrowheads) with hypoenhancement relative to normal cervical stroma (*).

**TABLE 4.2**
FIGO Classification System

| Localization | Class | Description |
|---|---|---|
| SM-submucosal | 0 | Intracavitary (pedunculated) |
| | 1 | <50% of the fibroid diameter within the myometrium |
| | 2 | ≤50% of the fibroid diameter within the myometrium |
| O-other | 3 | Intramural, but abuts the endometrium |
| | 4 | Completely intramural, no involvement of either the endometrial or serosal surfaces |
| | 5 | Subserosal with ≥50% intramural component |
| | 6 | Subserosal with <50% intramural component |
| | 7 | Subserosal with a stalk (pedunculated) |
| | 8 | No involvement with the myometrium (cervical, broad ligament, and parasitic) |

## 4.6 Malignant Diseases of the Myometrium

### 4.6.1 Uterine Sarcomas

Uterine sarcomas are rare and are associated with a poor prognosis [62,109]. They include leiomyosarcoma, carcinosarcoma (malignant mixed Müllerian tumor), endometrial stromal sarcomas, and adenosarcomas. Leiomyosarcoma accounts for 30% of uterine sarcomas and the risk factors are unknown [110]. Clinical demographics for leiomyosarcoma overlap with benign disease and include older age, postmenopausal status, and presentation with a symptomatic uterine mass. As with the incidence of leiomyomas, black women have a higher incidence of leiomyosarcomas than white women [111]. Most leiomyosarcomas arise *de novo,* via distinct pathogenic pathways and not from sarcomatous degeneration of benign leiomyomas [112,113]. Development of uterine sarcomas has been reported following pelvic radiation and tamoxifen therapy [114] and is associated with a history of childhood retinoblastoma and hereditary leiomyomatosis and renal cell carcinoma syndrome [113].

Differentiating between malignant and benign masses is increasingly important given the increasing utilization of conservative management approaches for patients who want to preserve their uterus. There are numerous less-invasive options for treating benign leiomyomas, many of which do not acquire tissue for pathologic evaluation; laparoscopic myomectomy may include tumor morcellation, which can lead to peritoneal dissemination of tumor cells and may adversely affect survival outcomes [97,115]. There are no findings that reliably predict sarcoma. Most women with rapidly growing masses do not have sarcoma [116]; however, new uterine masses or mass growth after menopause should be evaluated for malignancy [117]. Ultimately, leiomyosarcoma is diagnosed on histologic examination where marginal irregularity, hemorrhage, and/or coagulative tumor cell necrosis are typically seen [110,117]. What makes diagnosis even more difficult is that for certain tumors, histopathological diagnosis may be difficult. These tumors are designated smooth muscle tumors of uncertain malignant potential (STUMP) and may follow MRI signal characteristics of benign leiomyomas [118].

Although non-degenerating leiomyomas have a characteristic appearance, leiomyosarcoma and benign degenerating leiomyoma may have very similar appearances on imaging (Figure 4.55) [73,118–120]. There are no pathognomonic imaging features of leiomyosarcoma; however, there are features that raise the suspicion of malignancy [117]. Goto et al. prospectively evaluated 140 patients and found that the combined use of DCE MRI and serum lactate dehydrogenase (LDH) measurements (frequently elevated in leiomyosarcoma and in some leiomyomas) was useful in differentiating leiomyosarcoma from degenerating leiomyomas prior to treatment. Degenerating leiomyomas demonstrated

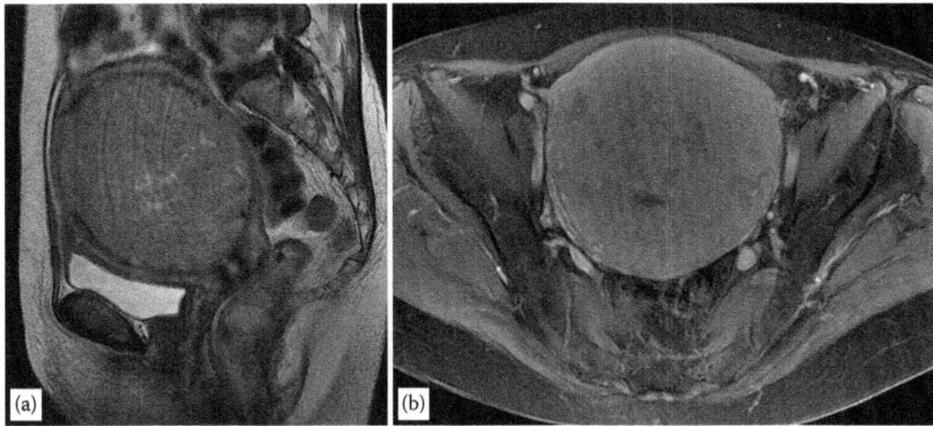

**FIGURE 4.42**
Cellular leiomyoma. (a) Sagittal $T_2$W image demonstrates an intracavitary mass with high $T_2$ and homogenously intermediate $T_1$ (not shown) SI. Cellular leiomyomas are higher in $T_2$ SI due to a decreased amount of collagen within the tumor. (b) The compact smooth muscle strongly enhances on postcontrast images.

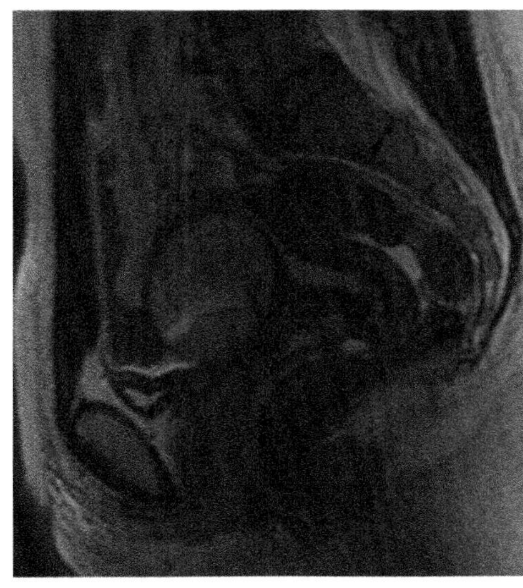

**FIGURE 4.43**
Cystic degeneration. Sagittal $T_2$W image demonstrates a well-circumscribed, transmural mass with heterogeneously increased $T_2$ SI. Central cystic degeneration confirmed on ultrasound (not shown).

no enhancement in the first 60 s after contrast injection, whereas leiomyosarcomas demonstrated early contrast enhancement, within 60 s, in all patients with leiomyosarcoma (total of 10 patients); normal myometrium enhanced later (120–180 s). Although there were 4 false positive cases with DCE (2 cellular and 2 degenerating leiomyomas), their recommendation is to obtain dynamic imaging at 40–80 s after Gd-C administration and combine these findings with LDH levels to help differentiate degenerating fibroids from leiomyosarcoma preoperatively [121]. Tanaka et al. found that the highest accuracy in diagnosing leiomyosarcoma is expected when more than 50% of the tumor is high signal on $T_2$WI, any area of high $T_1$SI is present, and there are areas of non-enhancement [118]. Therefore, in these cases and in cases of early enhancement on DCE, uterus-preserving therapy should be avoided. Further study of the use of MRI for the noninvasive diagnosis of leiomyosarcoma is needed.

### FOCUS POINTS

- TVUS is the best first imaging study for differentiating fibroids from other pathology; however, MRI is recommended when precise evaluation of leiomyoma number, size, position, and degeneration are needed for treatment planning and evaluation of response to therapy.
- MRI is more accurate than real-time TVUS in diagnosing adenomyosis when there are coexisting leiomyomas.
- Removal of leiomyomas is considered in symptomatic patients and in those with submucosal and/or cavity distorting leiomyomas who are attempting pregnancy.
- Rapid uterine growth almost never indicates sarcoma in premenopausal women. After menopause, a new or growing uterine mass warrants further evaluation.
- Preoperative diagnosis of leiomyosarcoma may be possible with a combination of LDH and DCE MRI. Further work is needed, but leiomyosarcomas likely have earlier enhancement compared with degenerating leiomyomas. It may be beneficial to obtain dynamic imaging at 40–80 s after Gd-C administration and combine these findings with LDH.

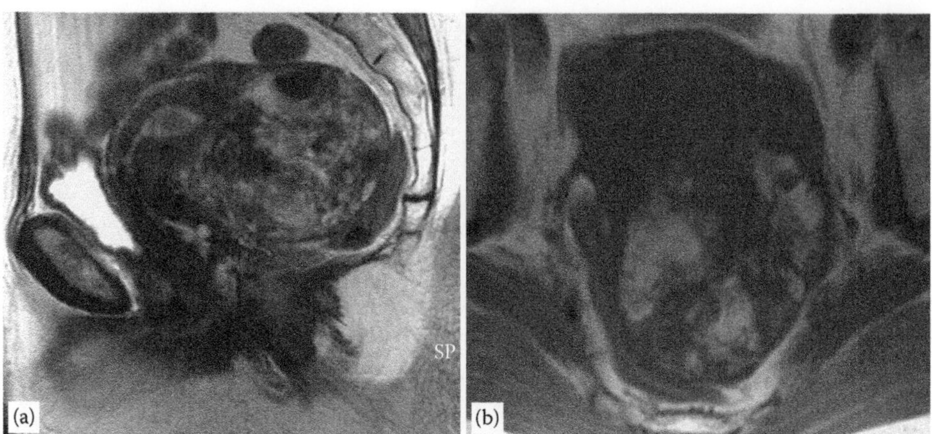


**FIGURE 4.44**
Lipomatous degeneration. (a) Sagittal $T_2$W and (b) axial $T_1$W images demonstrate a well-circumscribed, subserosal mass with less than 50% intramural component. There are areas of increased $T_1$ and $T_2$ SI that drop in SI on fat-suppressed sequences (not shown), consistent with fat.

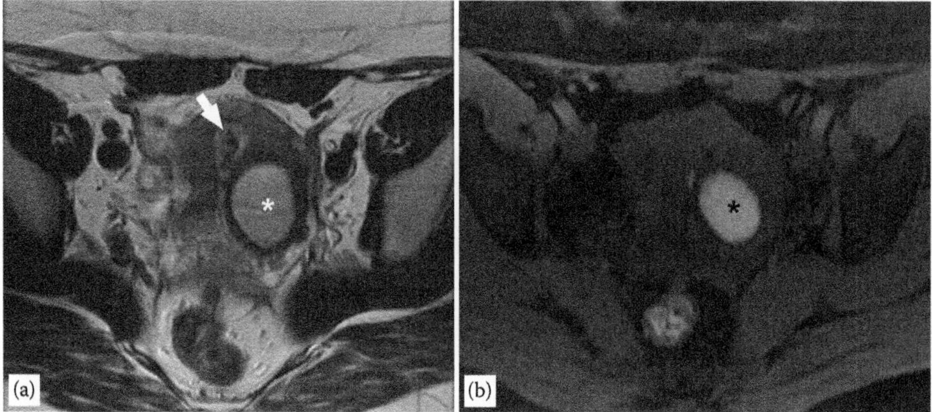


**FIGURE 4.45**
Red degeneration. Axial (a) $T_2$W and (b) $T_1$W with fat-suppression images demonstrate a well-circumscribed, transmural mass of increased $T_1$ and $T_2$ SI, following methemoglobin (*). Hemorrhagic infarction typically occurs during pregnancy. Low attenuation material in cavity is calcifications (arrow).

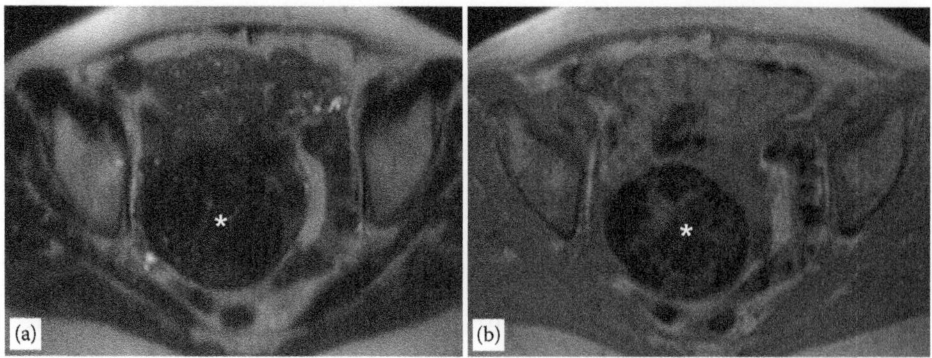


**FIGURE 4.46**
Calcific degeneration. Axial (a) $T_2$W and (b) $T_1$W images demonstrate a well-circumscribed, subserosal mass with diffuse low SI on all sequences (*), consistent with a densely calcified leiomyoma and confirmed on pathology.

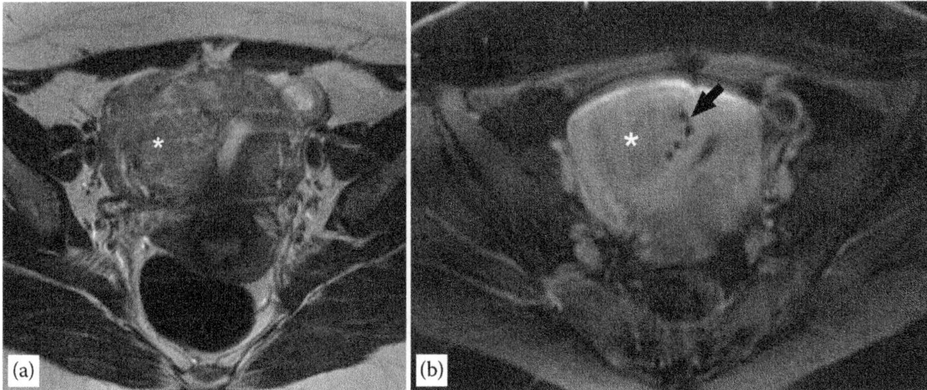


**FIGURE 4.47**
A 37-year-old female one year after myomectomy for uterine sparing removal of a 14-cm mass. Pathologic diagnosis was transmural intravenous leiomyoma. Axial (a) $T_2$W and (b) $T_1$W postcontrast images two years later demonstrate an heterogeneous, poorly defined mass of increased $T_2$ SI and delayed hypoenhancement consistent with residual disease (*). The patient went on to deliver a normal child by caesarian section one year later. Note the susceptibility artifact on the $T_1$W image (arrow) from prior surgical resection.

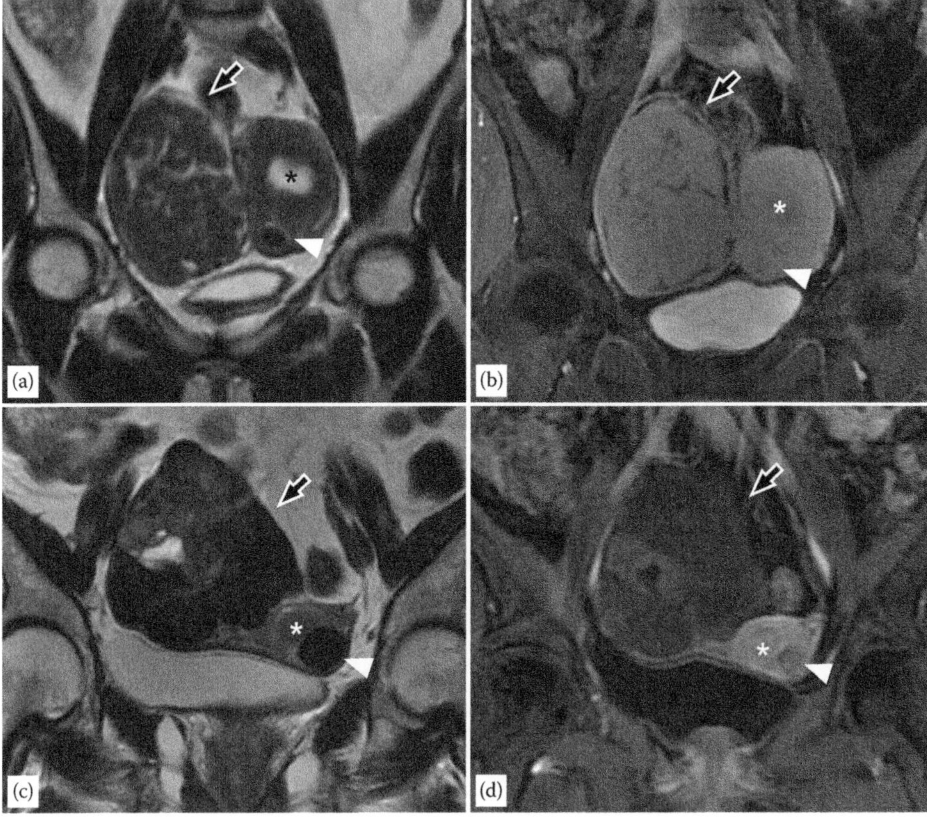


**FIGURE 4.48**
Broad ligament leiomyoma versus ovarian fibroma. Coronal (a) $T_2$W and (b) $T_1$W postcontrast images of a broad ligament leiomyoma (arrow), and coronal (c) $T_2$W and (d) $T_1$W postcontrast images of a right ovarian fibrothecoma (arrow) with dystrophic calcification and hyalinized degeneration. The $T_1$ and $T_2$ SI of both entities are typically similar; however, dynamic contrast enhancement of subserosal uterine leiomyomas has been found to be more intense and earlier than that of ovarian fibromas. Uterus (*); intramural leiomyoma (arrowhead).

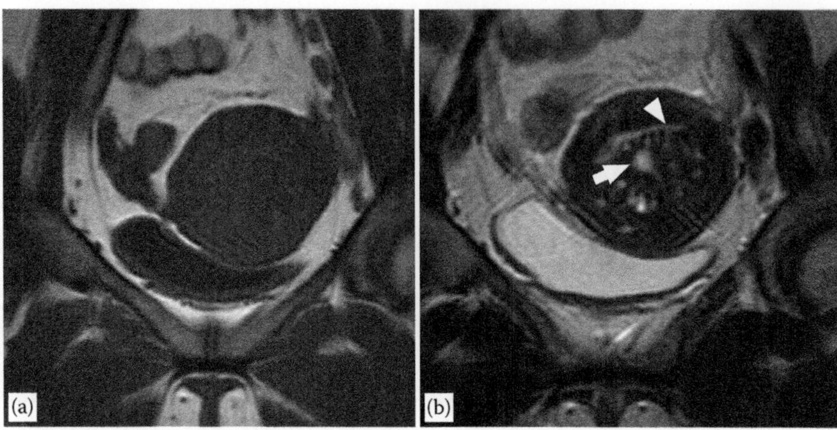

**FIGURE 4.49**
Adenomyosis. Coronal (a) $T_1$W image demonstrates a globular uterus of intermediate SI. (b) $T_2$W image demonstrates thickening of the anterior wall junctional zone with internal foci of increased SI corresponding to islands of heterotopic endometrial tissue with cystic glandular dilatation (arrow). Hyperintense linear striations radiating from the myometrium are also seen (arrowhead). The use of intravenous Gd-C material does not increase accuracy in diagnosing adenomyosis.

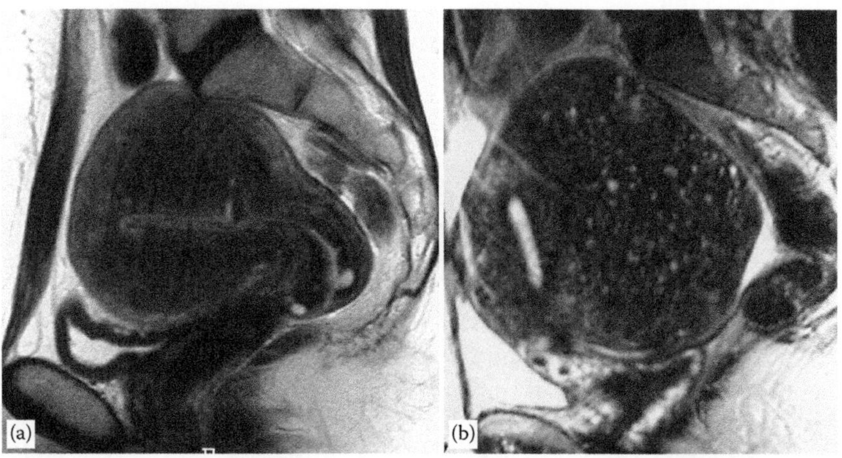

**FIGURE 4.50**
Adenomyosis may be diffused with globular enlargement of the uterus (a), or focal when only part of the junctional zone is involved (b). Focal adenomyosis can simulate a leiomyoma; however, in contrast to leiomyomas, the masslike focal form of adenomyosis (adenomyoma) tends to be oval rather than spherical, poorly marginated, contiguous with junctional zone, and without large peripheral vessels. Adenomyosis does not distort the endometrium, unlike leiomyomas that show more mass effect and distortion. Adenomyosis may still be difficult to distinguish from leiomyomas.

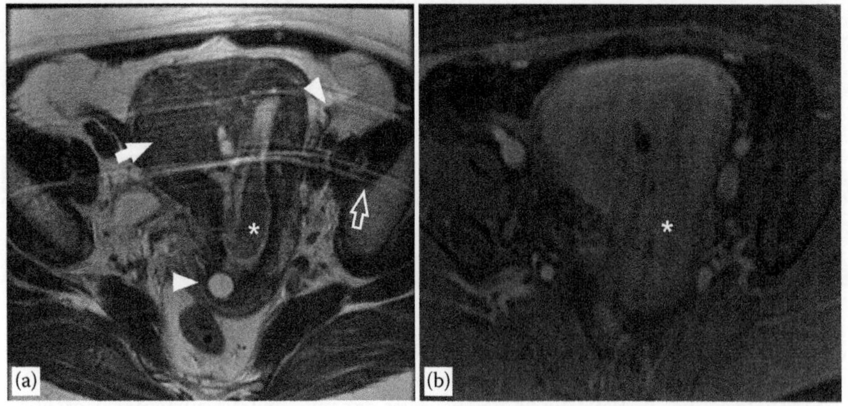

**FIGURE 4.51**
Axial (a) $T_2$W and (b) $T_1$W images demonstrate an intraluminal adenomatous polyp (*) with homogeneous enhancement in a partial septate uterus with adenomyosis involving the right cornua and fundus (arrow). Nabothian cyst (arrowhead); motion artifact (open arrow).

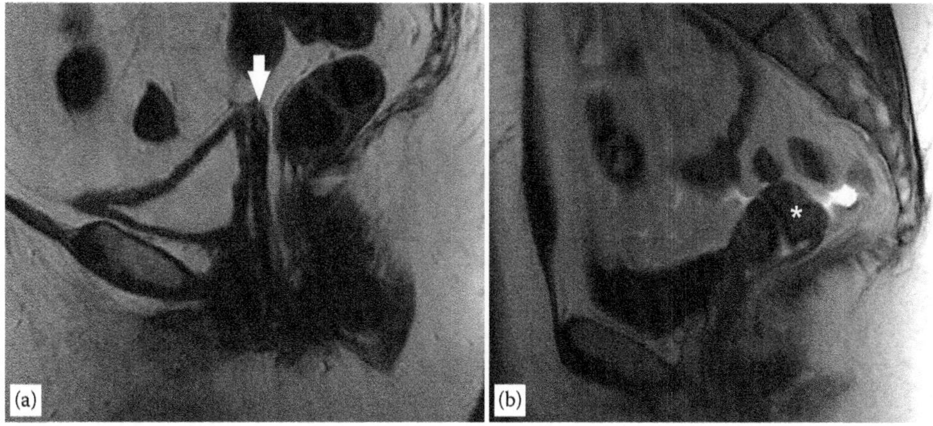


**FIGURE 4.52**
Sagittal $T_2$W images, status post-hysterectomy: (a) total hysterectomy with atrophic vagina (arrow), and (b) supracervical hysterectomy with retention of most of the cervix (*).

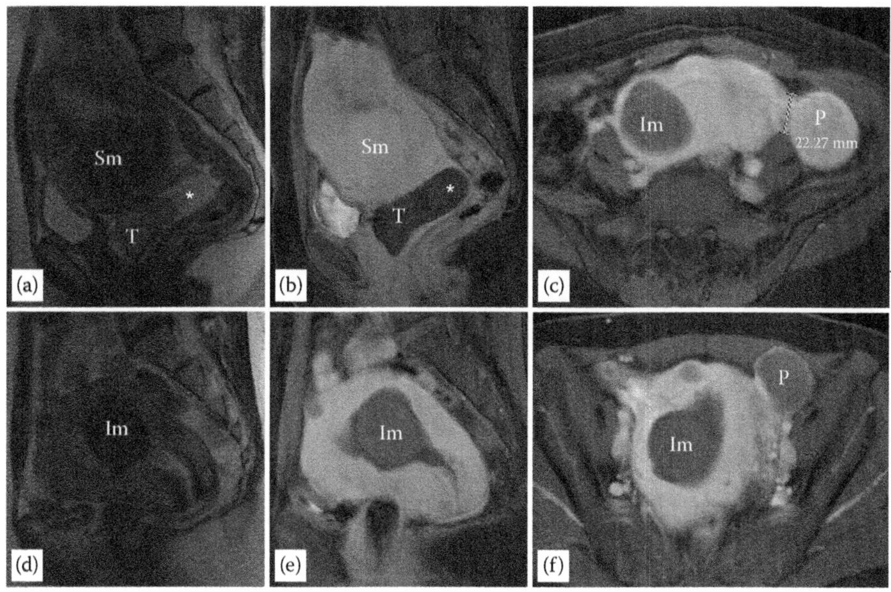


**FIGURE 4.53**
(a) Sagittal $T_2$W, and (b) sagittal and (c) axial $T_1$W postcontrast images demonstrate three fibroids pre-treatment: submucosal with 50% intraluminal component (Sm), pedunculated with 2.2 cm stalk (P), intramural devascularized (Im). Two months after uterine artery embolization, the patient presented with symptoms of pelvic cramping and *passage of solid tissue*. (d) Sagittal $T_2$W, and (e) sagittal and (f) axial $T_1$W postcontrast images demonstrate absence of the large submucosal leiomyoma confirming evacuation of the infarcted tissue. The pedunculated leiomyoma is completely devascularized and the previously devascularized leiomyoma is unchanged in size, but now submucosal in position. Tampon (T); fluid in vagina (*).

## 4.7 Benign Diseases of the Endometrium

### 4.7.1 Hyperplasia and Polyps

Endometrial hyperplasia results from a proliferation of endometrial glands resulting in a greater gland-to-stroma ratio. Hyperplasia is caused by excess estrogen exposure and may progress to or coexist with endometrial carcinoma [122]. Risk factors include obesity, chronic anovulation, unopposed exogenous estrogen therapy without progesterone, and tamoxifen use. Endometrial polyps are benign protrusions of endometrial hyperplasia, with a reported prevalence of approximately 8% in one study, and the prevalence increases with age [123]. Polyps may occur with or without generalized endometrial hyperplasia, and approximately 5% of polyps are malignant [124]. The most common presenting symptom of both hyperplasia and polyps is abnormal uterine bleeding.

The diagnosis of hyperplasia is made histologically. Hyperplasia is classified into four categories: (1) simple

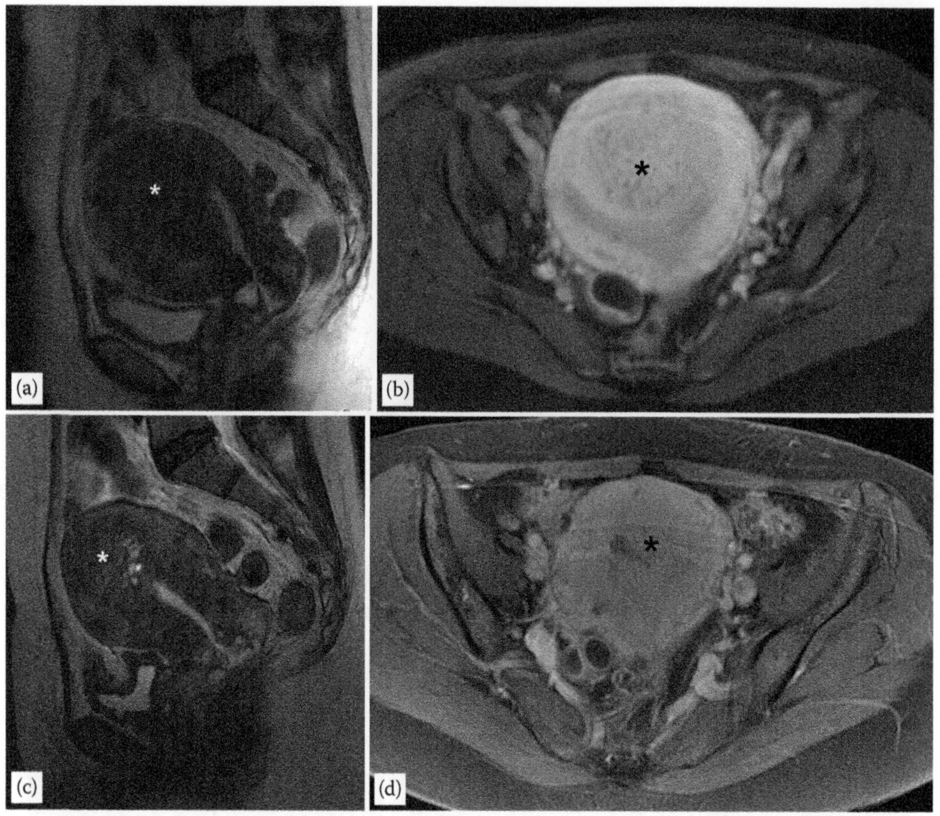


**FIGURE 4.54**
Pre-uterine artery embolization evaluation for leiomyoma size and location. (a) Sagittal $T_2$W and (b) axial $T_1$W images demonstrate focal adenomyosis involving the anterior uterine wall (*). It was decided to go forward with embolization. Four months after uterine artery embolization, (c) sagittal $T_2$W and (d) axial $T_1$W images demonstrate a decrease in anterior wall adenomyosis. Uterine artery embolization has been used to relieve symptoms for some women with adenomyosis. However, success rates vary widely, ranging from 25% to 85%, and approximately 50% of patients still require eventual hysterectomy.

without nuclear atypia, (2) complex without atypia, (3) simple atypical hyperplasia, and (4) complex atypical hyperplasia [125]. A more recent classification called the *endometrial intraepithelial neoplasia system* was proposed to improve the management of endometrial hyperplasia by developing better standardized diagnostic criteria [126]; however, this system has not gained as much favor as the WHO classification system. Identifying the presence of atypia is important because risk of developing endometrial carcinoma is greater in patients with atypia. Lacey et al. found that the cumulative 19-year risk of developing endometrial cancer in patients with endometrial hyperplasia without atypia was less than 5%, whereas the risk rose to 28% in patients with atypia [127].

US is the primary imaging tool for evaluating endometrial pathology; however, MRI may be helpful in patients with unsuccessful or indeterminate endometrial biopsy results. The MRI of endometrial hyperplasia, polyps, and noninvasive endometrial cancer may be nonspecific (Figure 4.56), and all three entities may be present synchronously. DCE may be useful in distinguishing malignant from benign lesions; weak enhancement favors endometrial carcinoma, whereas moderate to strong enhancement favors benign conditions [128]. The most common location of polyps is in the fundus and cornua. Identification of a vascular stalk can help in the diagnosis, but is not typically seen without the presence of fluid within the endometrial canal (Figure 4.57) [129].

### 4.7.2 Tamoxifen-Related Changes

Tamoxifen is an antiestrogen medication used as adjuvant therapy in patients with breast carcinoma. It acts as a weak estrogen agonist on the endometrium and can cause a variety of proliferative changes including hyperplasia, polyps, and carcinoma. Some studies have shown that the degree of proliferative changes corresponds with the duration of tamoxifen therapy [130]. Two MRI appearances of tamoxifen-related changes have been described [131]. In the first, the endometrium is homogeneously hyperintense on $T_2$W images, with enhancement of the endometrial–myometrial

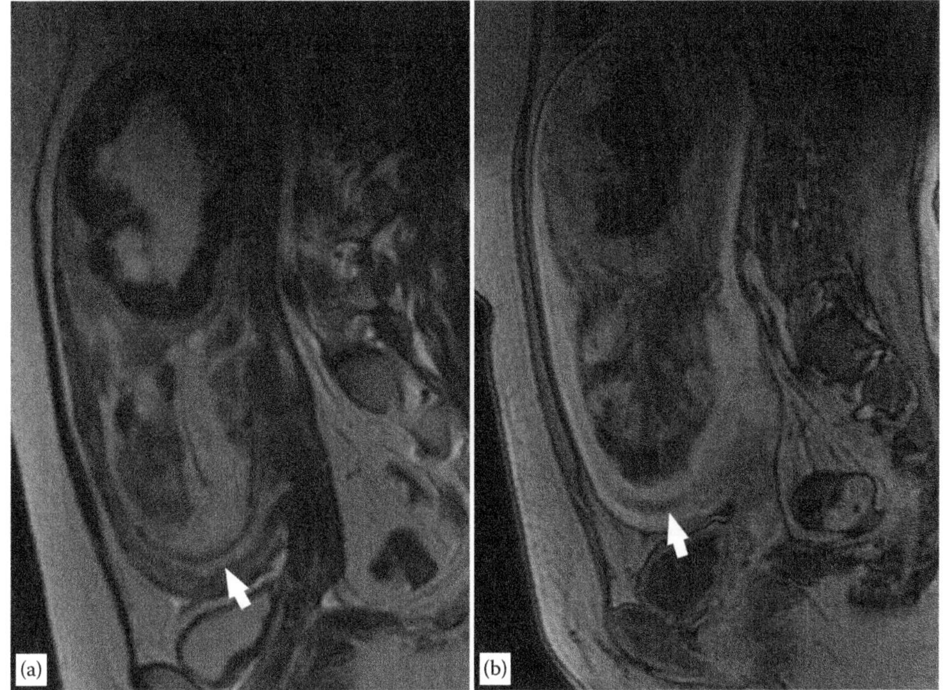


**FIGURE 4.55**
A 28-year-old female presenting with rapidly growing abdominal mass. Sagittal (a) $T_2$W and (b) $T_1$W postcontrast images demonstrate a 22 × 17 × 10 cm heterogeneous, subserosal mass with large areas of necrosis/degeneration. There is no imaging modality that can reliably differentiate between benign and malignant myometrial tumors as heterogeneity in $T_1$ and $T_2$ SI can be seen in both leiomyosarcoma and degenerating leiomyomas. This was diagnosed as a degenerating leiomyoma on pathology. Endometrium (arrow).

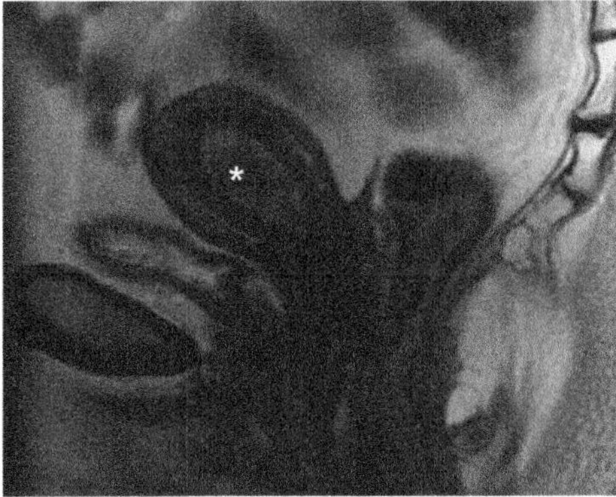

**FIGURE 4.56**
Endometrial hyperplasia. Sagittal $T_2$W image demonstrates diffuse endometrial thickening (*) which is heterogeneous and hypointense relative to normal endometrium. This appearance is nonspecific and the differential includes hyperplasia, polyp, and endometrial cancer.

interface. This pattern is associated with endometrial atrophy or proliferative changes. In the second, the endometrium is thickened and heterogeneous on $T_2$W images, with enhancement of endometrial–myometrial interface and lattice-like enhancement of the endometrium. This pattern is associated with polyps (Figure 4.58) [131].

## 4.8 Malignant Diseases of the Endometrium

### 4.8.1 Endometrial Carcinoma

Endometrial cancer is the most common gynecologic malignancy [132]. Risk factors include personal history of colorectal cancer, endometrial polyps and hyperplasia, and morbid obesity [133]. Endometrial carcinoma is a histologic diagnosis. The most important findings that predict extra-uterine disease and poor outcome include tumor extent, as defined by depth of myometrial invasion, cervical invasion, lymphovascular space involvement, and lymph node metastases, and histologic factors including tumor grade and histologic subtype [134,135]. Although surgery remains the primary modality for the staging of endometrial cancer, accurate pretreatment evaluation with imaging can optimize surgical and nonsurgical treatment. Imaging may help determine whether a patient may undergo fertility preservation options or require surgery, including simple versus radical hysterectomy, lymph node dissection, and the possible need for adjuvant treatment. MRI is most helpful in the evaluation of myometrial invasion, cervical involvement, and detection of lymphadenopathy [3,136].

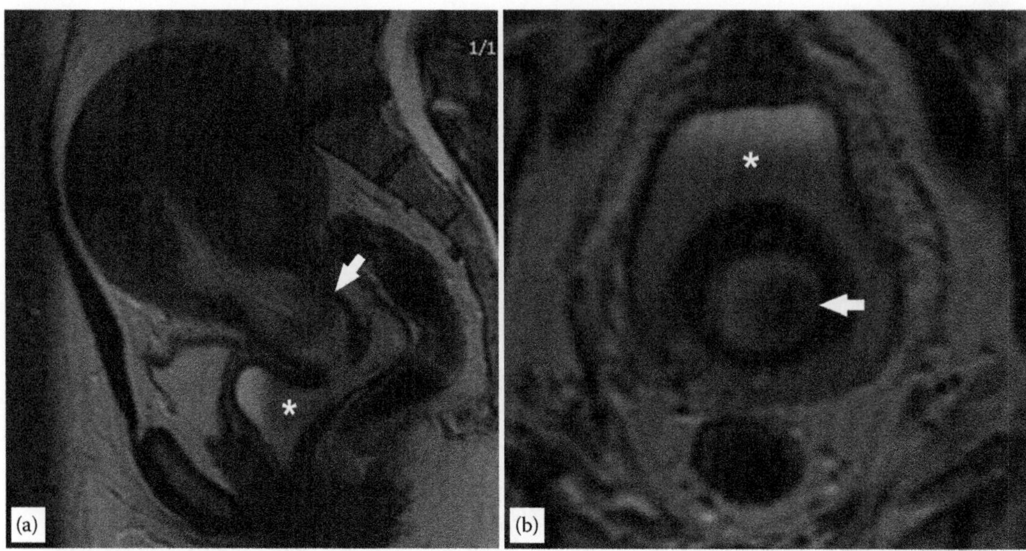


**FIGURE 4.57**
(a) Sagittal and (b) axial $T_2$W images through the cervix show a prolapsing polyp (arrow) with dilatation of the endocervical canal. The corpus is globular due to the presence of diffuse adenomyosis. Vaginal fluid (*).

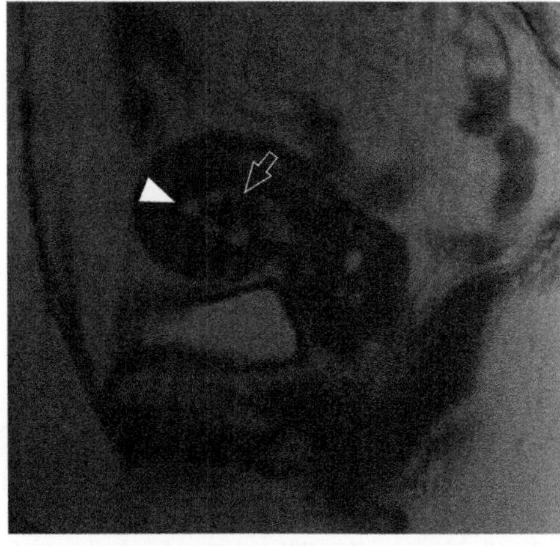

**FIGURE 4.58**
Sagittal $T_2$W image in a patient on tamoxifen demonstrates a polyp consisting of multiple endometrial cysts (arrowhead) and low SI fibrous core (open arrow), two MRI findings which may be helpful in differentiating endometrial polyps from carcinoma.

MR findings of endometrial cancer include an endometrial mass or endometrial thickening. Compared to normal endometrium, endometrial cancer is usually lower in $T_1$ SI and higher in $T_2$ SI (Figure 4.59). DCE MRI further improves staging of endometrial cancer by helping determine the depth of myometrial invasion through differential enhancement of tumor and normal myometrium (Figure 4.60). Normal myometrial enhancement peaks at 120 s and then decreases over time; enhancement of endometrial carcinoma is more gradual and less intense than the myometrium. The maximum contrast between the hyperintense myometrium and hypointense endometrial tumor occurs 50–120 s [137]. Some reported pitfalls in assessing the depth of myometrial invasion include nonvisualization of the mass after curettage, a bulky polypoid tumor, adenomyosis, leiomyoma, a small uterus, and retroversion of the uterus. Caution must be used when the tumor is located in the cornual region of the uterus. The myometrium is thinnest in this location (Figure 4.59b), and therefore, over estimation of myometrial invasion is a risk. Additionally, in postmenopausal women, there is thinning of the myometrium secondary to uterine involution, which can make evaluation of the depth of myometrial invasion more difficult on standard $T_1$ and $T_2$ sequences [138].

The addition of DWI has been shown to increase accuracy in assessing the depth of myometrial invasion compared with DCE [139,140]. Additional benefits of DWI include the lack of need for gadolinium-based contrast agents in patients with impaired renal function. Beddy et al. also found that DWI was not influenced by standard pitfalls (i.e., adenomyosis and leiomyomas) [139].

The FIGO staging system for staging of endometrial carcinoma was recently updated in 2009 [141], with several important changes (Table 4.3). Tumors with superficial myometrial invasion (<50%) are now considered stage IA (Figure 4.59). Stage IB disease is now reserved for tumors with greater than 50% myometrial invasion (Figure 4.60). Determining myometrial invasion

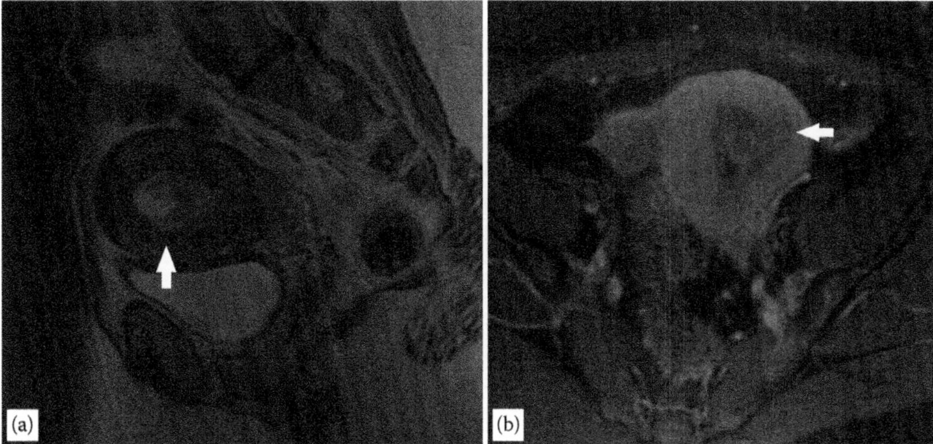

**FIGURE 4.59**
Stage IA endometrial cancer. (a) Sagittal $T_2$W image demonstrates heterogeneous tumor distending the endometrial cavity with superficial myometrial invasion (arrow). Superficial invasion is best seen on dynamic contrast-enhanced sequences. (b) One minute, $T_1$W postcontrast image demonstrates the hypoenhancing tumor invading into the relatively hyperenhancing myometrium (arrow). The depth of myometrial invasion is less than 50%; however, the location near the left cornua may make the invasion appear greater. This is a pitfall to remember.

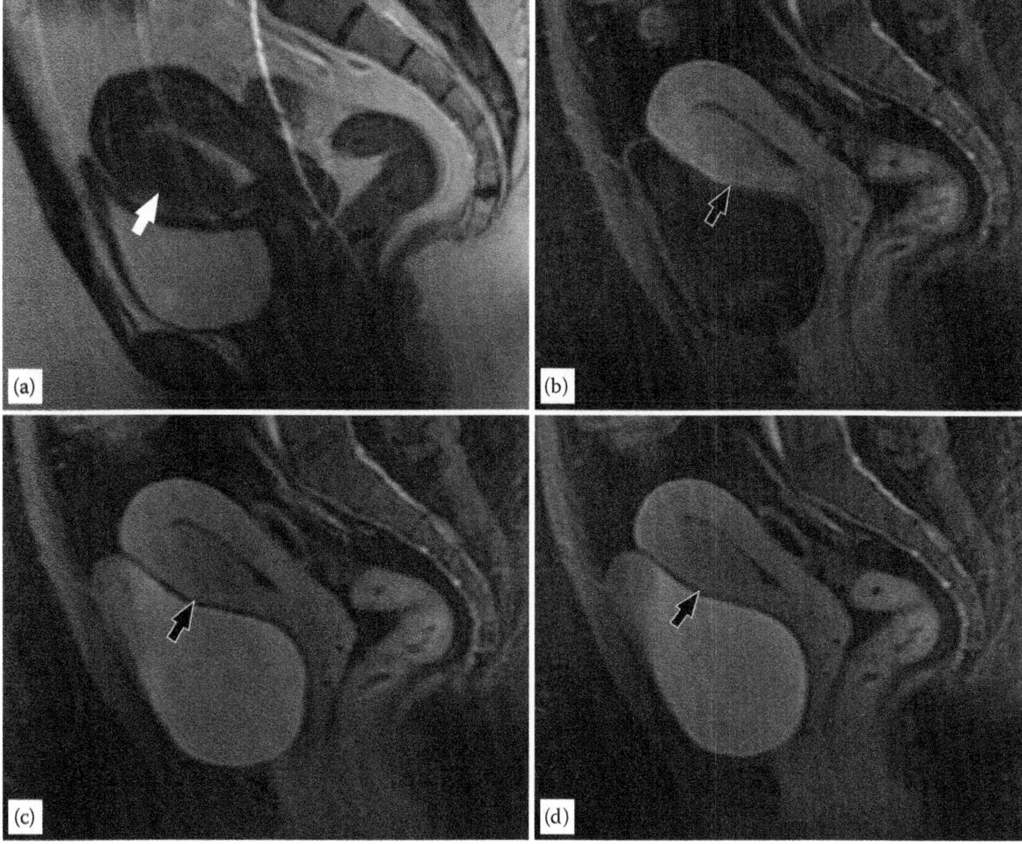

**FIGURE 4.60**
Stage IB endometrial carcinoma. (a) Sagittal $T_2$W image demonstrates a mass isointense to myometrium (white arrow) making evaluation of invasion difficult. Dynamic contrast-enhanced images obtained at (b) 1 min, (c) 3 min, and (d) 5 min better demonstrate the hypoenhancing tumor (black arrow) invading inferiorly into the relatively hyperenhancing myometrium. At surgery, greater than 50% myometrial invasion was seen, compatible with deep invasion.

**TABLE 4.3**

2009 FIGO Staging System Endometrial Carcinoma

| Stage | Description | MR Findings |
|---|---|---|
| IA | Myometrial invasion <50% | Superficial invasion best seen on dynamic contrast-enhanced and diffusion-weighted sequences |
| IB | Myometrial invasion >50% | Deep invasion best seen on dynamic contrast-enhanced and diffusion-weighted sequences |
| II | Cervical stromal invasion | Disruption of low SI inner cervical stroma on $T_2$W images and disruption of cervical epithelium enhancement on contrast-enhanced images |
| IIIA | Serosal or adnexal invasion | Disruption of continuity of outer myometrium and/or presence of nodules on the peritoneal surface or adnexa |
| IIIB | Vaginal or parametrial invasion | Enhancing tumor extending into vagina and parametrium |
| IIIC1 | Pelvic lymph node involvement | Short axis diameter greater than 10 mm, restricted diffusion |
| IIIC2 | Paraaortic lymph node involvement | Short axis diameter greater than 10 mm, restricted diffusion |
| IVA | Bladder or bowel mucosa invasion | Disruption of bladder or rectal wall on $T_2$W images and enhancing tumor on contrast-enhanced images |
| IVB | Distant metastases | Tumor spread beyond pelvis, includes inguinal lymph nodes involvement |

is important because it can predict the likelihood of lymph node metastases. Only one percent of patients with endometrial cancer limited to the endometrium had lymph node metastases, whereas the rate of pelvic node and paraaortic lymph node metastases increased to 25% and 17%, respectively, for deep myometrial invasion [134].

Cervical stromal invasion is defined as stage II disease (Figure 4.61). Preoperative identification of cervical invasion is important because it may change surgical management from a simple hysterectomy to radical hysterectomy with lymphadenectomy. Although diagnosis of cervical invasion is most often performed by endocervical curettage, studies show that MRI has greater specificity and accuracy than endocervical curettage in the diagnosis of cervical stromal invasion [142]. For the evaluation of cervical stromal invasion, delayed-phase images should be obtained 3–4 min after contrast medium administration; during this phase, the presence of normal cervical mucosal enhancement reliably excludes stromal invasion [143].

Lymph node involvement is divided into pelvic and paraaortic involvement (stage III) (Figure 4.62). Stage IV disease indicates bladder or bowel mucosal invasion (Figure 4.63) and distant metastases (Figure 4.64). Detection of lymph nodes metastases with MRI relies on size criteria (short axis diameter of greater than 10 mm) and SI. DWI is particularly useful in increasing the conspicuity of lymph nodes due to their high SI. Some studies suggest that ADC values are able to differentiate metastatic versus benign lymph nodes in patients with endometrial carcinoma [144]. Lin et al. found that combining the size and relative ADC values resulted in better sensitivity (83%) and specificity (99%) [144]. Further work is needed.

Although MRI cannot reliably differentiate between uterine sarcomas and endometrial carcinoma, MRI features that favor the diagnosis of sarcoma over endometrial carcinoma include large tumor size, intermediate mixed $T_2$ SI, hemorrhage and necrosis, areas of delayed enhancement, and invasive or metastatic disease at time of diagnosis [145,146]. DCE may also be useful as sarcoma typically demonstrates earlier and persistent enhancement [147].

**FOCUS POINTS**

- The new FIGO staging system for endometrial carcinoma contains several important changes with regard to MR imaging, including classifying all tumors with <50% myometrial invasion as stage IA, tumors with greater than 50% myometrial invasion as stage IB, and cervical stromal invasion as stage II.
- MRI is most helpful in the evaluation of myometrial invasion and cervical involvement and detection of lymphadenopathy.
- Maximum contrast between the hyperintense myometrium and hypointense endometrial tumor occurs 50–120 s after Gd-C administration.
- MRI with DCE and DWI is preferred for treatment planning of endometrial cancer and is the recommended imaging modality by the American College of Radiology.

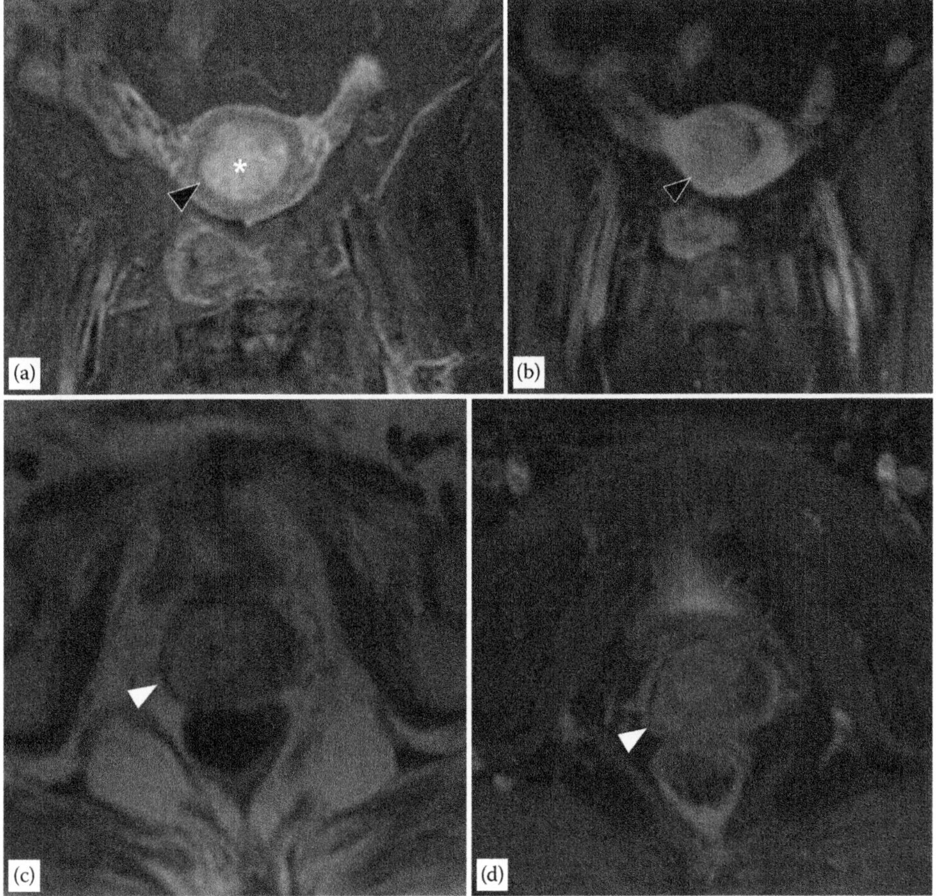

**FIGURE 4.61**
Stage II clear-cell endometrial carcinoma. Coronal (a) $T_2$W fat-suppressed image demonstrates a bright mass distending the endometrial cavity (*). Myometrial invasion is best depicted on (b) coronal $T_1$W postcontrast image. The mass is hypoenhancing with deep (>50%) myometrial invasion (black arrowhead). Axial (c) $T_2$W and (d) $T_1$W postcontrast images at the level of the cervix demonstrate cervical stromal invasion (white arrowhead).

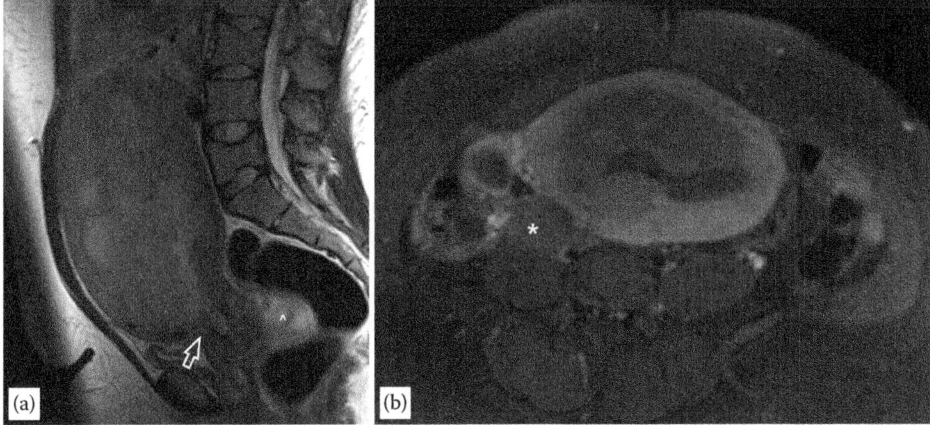

**FIGURE 4.62**
Stage IIIa endometrial adenocarcinoma, endometrioid type. (a) Sagittal $T_2$W and (b) axial $T_1$W, postcontrast images demonstrate a large, $T_2$ heterogeneously hyperintense mass with hypoenhancement relative to normal myometrium. There is an extension into the cervix (arrow), and nodules in the cul-de-sac (^) and right adnexa (*).

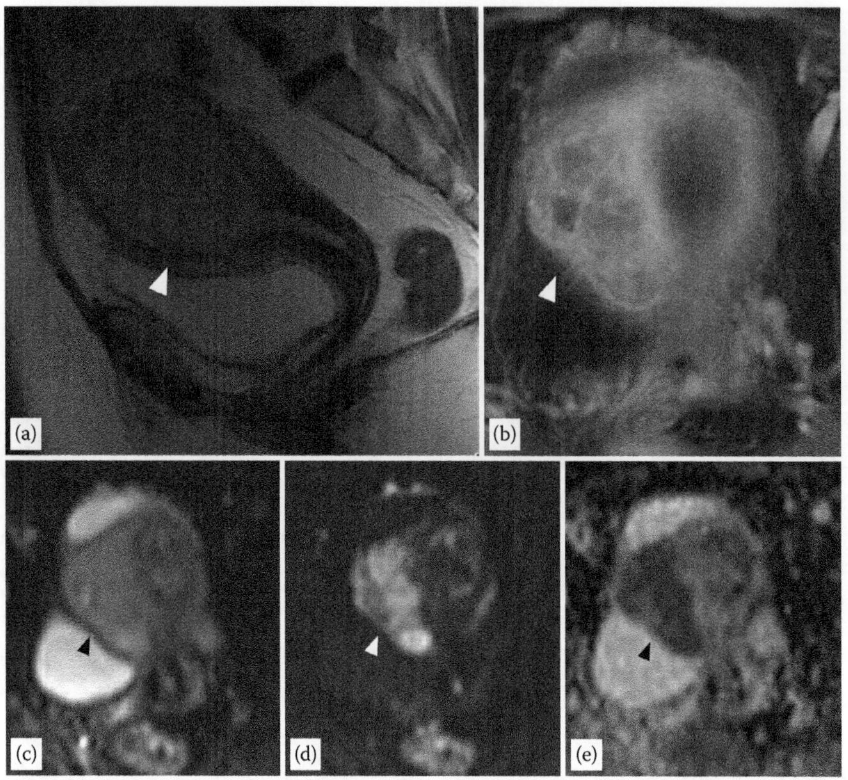


**FIGURE 4.63**
Stage IVa endometrial carcinoma with invasion of the colon and bladder (high-grade neuroendocrine carcinoma). (a) Sagittal $T_2$W and (b) axial $T_1$W postcontrast images demonstrate a large endometrial mass with deep myometrial invasion (arrowhead) and heterogeneous enhancement. Corresponding diffusion-weighted images with (c) *b*-value 0 and (d) *b*-value 800 and (e) ADC map. Endometrial cancer demonstrates restricted diffusion compared with surrounding tissue, with higher SI on diffusion-weighted images and low SI on ADC maps. Bladder and rectal invasion not shown.

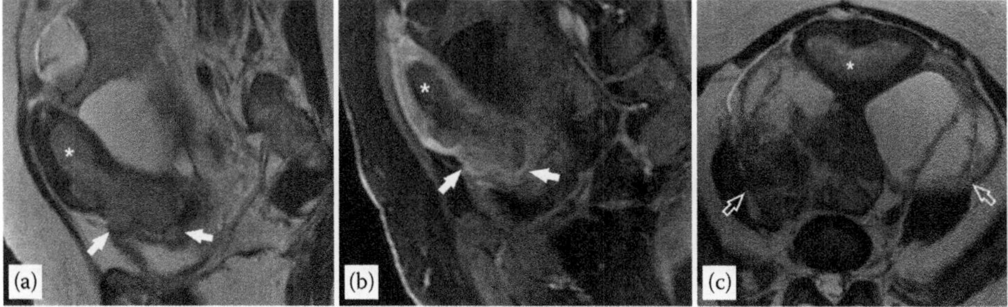


**FIGURE 4.64**
Stage IVb endometrial adenocarcinoma (endometrioid type). Sagittal (a) $T_2$W and (b) $T_1$W postcontrast images demonstrate a mass distending the endometrial cavity (*) and invading through the myometrium and serosa with possible bladder wall invasion (solid arrows). No bladder wall invasion was seen at surgery. (c) Axial $T_2$ image demonstrates large bilateral complex adnexal masses (open arrows), which were found to be ovarian metastases at surgery.

## 4.9 Benign Diseases of the Cervix

Nabothian cysts, also called mucinous retention cysts, are the most common benign lesion of the cervix (Figure 4.65). They are caused by mucus distension of the deep glandular follicles and typically occur in the upper two-thirds of the cervix. Nabothian cysts may also occur following minor trauma or childbirth [148].

## 4.10 Malignant Diseases of the Cervix

### 4.10.1 Cervical Carcinoma

Cervical cancer is the third most common malignancy of the female genital tract [132]. Human papillomavirus infection is detected in virtually all cases and is thought to be the causative factor [149]. The most common

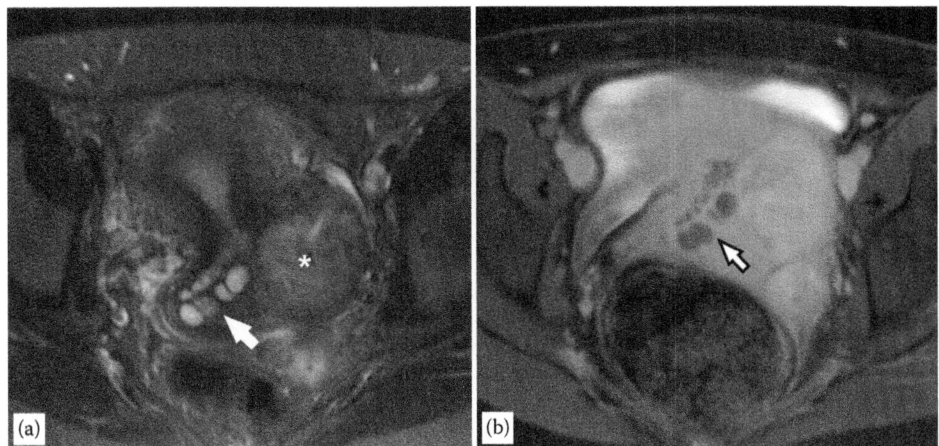

FIGURE 4.65
Nabothian cysts (arrow) in patient with cervical leiomyoma (*). Nabothian cysts have high SI on $T_2$W imaging (a), and variable SI on $T_1$W imaging, due to the varying viscosity of mucus within these cysts. They do not enhance after the administration of gadolinium (b).

histologic types of cervical cancer are squamous cell (85%) and adenocarcinoma (15%). Less common types include neuroendocrine, adenosquamous, and sarcomas [150]. Most are similar in imaging appearance, with the exception of adenoma malignum, a subtype of mucinous adenocarcinoma, which can mimic nabothian cysts on MRI [151,152]. Adenosquamous carcinomas are more aggressive and carry a worse prognosis [153].

Invasive cervical carcinoma is staged clinically according to the FIGO staging system (Table 4.4) [154]. Precise assessment of disease extent is important because it affects choice of treatment: stage IA may be treated with simple hysterectomy or fertility-sparing surgery such as trachelectomy; stage Ib2 and IIa may be treated with radical hysterectomy and pelvic lymph node dissection; and stage Ib2 disease or greater (Figures 4.66 through 4.68) may be treated with radiation therapy. Although clinical staging remains the standard, MRI can play an important role in preoperative assessment, including evaluation for parametrial involvement where MRI has an accuracy of 88%, sensitivity of 100%, and specificity of 80% [155]. Detection of stromal invasion has also been shown to be superior with dynamic MRI [156]. In fact, Hricak et al. found that pretreatment workup with MRI resulted in net cost savings for all patients and significantly fewer procedures and fewer invasive studies, such as cystoscopy, protoscopy, and pelvic exam under anesthesia [157].

$T_2$W sequences are the most important sequence for staging of cervical carcinoma where the tumor SI is greater than the surrounding tissues. The tumor is usually isointense to the surrounding tissues on $T_1$W sequences. The use of contrast does not significantly improve staging accuracy compared to $T_2$W sequences alone, except in the evaluation of stage IB disease.

TABLE 4.4
FIGO Staging System for Cervical Cancer

| Stage | Description | MR Findings |
|---|---|---|
| IA | Confined to cervix, but not clinically visible (microinvasion of tumor) | No MR findings may be present |
| IB | Confined to cervix, clinically visible lesion (clinically invasive) | Partial or complete disruption of the low SI fibrous stroma, but rim of intact cervical tissue surrounding the tumor |
| IIA | Extension beyond the uterus but not the pelvic wall or lower one-third of the vagina, no parametrial invasion | Disruption of the low $T_2$ SI vaginal wall |
| IIB | Parametrial invasion | Irregular cervical margin, parametrial stranding, obliteration of the parametrial fat planes, or disruption of the low SI peripheral stroma |
| IIIA | Involvement of lower one-third of the vagina | Loss of low $T_2$ SI wall of lower vagina |
| IIIB | Extension to the pelvic sidewall, hydronephrosis, or nonfunctioning kidney | Loss of the low SI on $T_2$W images that correspond to the levator ani, pyriformis or obturator internus muscles, and hydronephrosis |
| IVA | Extension beyond the true pelvis or involvement of bladder or rectal mucosa | Disruption of the normal low SI of the bladder or rectal wall on $T_2$W images |
| IVB | Spread to distant organs | Tumor spread beyond pelvis, includes paraaortic and inguinal lymph nodes, lung, liver, and bone |

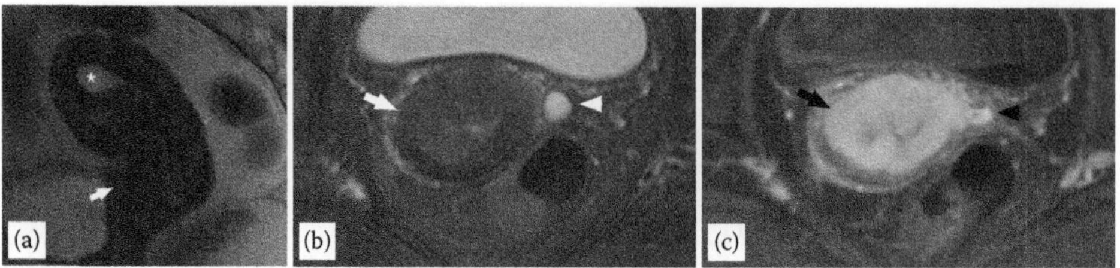

**FIGURE 4.66**
Stage IIb cervical carcinoma. The tumor is difficult to identify on (a) sagittal $T_2$W image; however, it is more conspicuous on (b) axial $T_2$W image with fat suppression (white arrow). (c) Axial $T_1$W postcontrast image shows a lobulated mass hypoenhancing relative to normal cervical stroma (black arrow). Cervical cancer enhances early relative to the cervical stroma. Parametrial invasion is seen on the left (arrowhead). Retained fluid (*) is seen within the endometrial canal secondary to cervical stenosis.

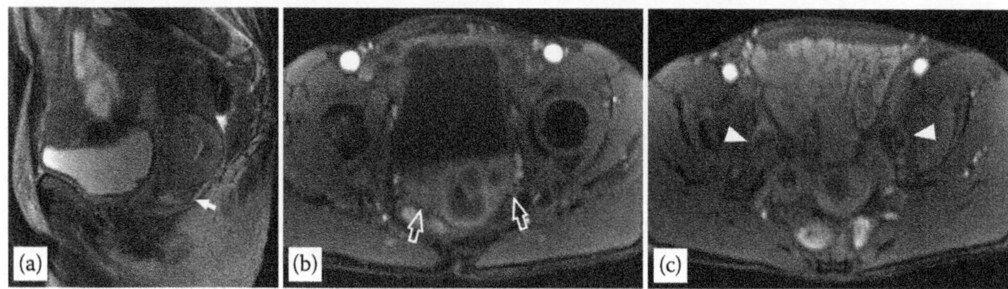

**FIGURE 4.67**
Stage IIIb adenosquamous cervical cancer with extension to the pelvic sidewall. (a) Sagittal $T_2$W and (b, c) axial $T_1$W postcontrast images demonstrate a bulky $T_2$ hyperintense cervical mass with extension into the upper vagina (solid arrow). The tumor is necrotic on postcontrast imaging with invasion of the parametrium (open arrows) and extension to the pelvic sidewall, bilaterally (arrowheads). There was no associated hydronephrosis.

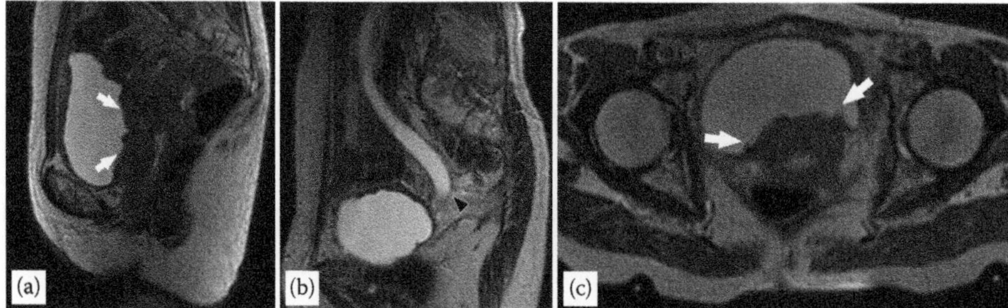

**FIGURE 4.68**
Stage IVa squamous cell carcinoma of the cervix. (a, b) Sagittal and (c) axial $T_2$W images demonstrate an irregular cervical mass with associated bladder wall invasion (arrows). The mass also invades the left ureterovesicular junction with resulting hydroureteronephrosis (arrowhead).

Seki et al. showed that DCE can differentiate deep stromal invasion (>3 mm) from superficial invasion (0–3 mm) [156]. In another study, cervical carcinoma was readily distinguished from the cervical stroma and myometrium in the early dynamic phase (first 30–60 s) with greater tumor to cervix contrast when compared to $T_2$W or non-dynamic contrast-enhanced $T_1$W imaging [158].

DWI can improve lesion detection, especially in early-stage cervical carcinoma [159]. Additionally, some studies have shown that ADC measurements can help in monitoring treatment response in patients treated by chemotherapy and radiation therapy [160]. However, the role of DWI in the evaluation of cervical carcinoma is still uncertain.

Lymph node status is not included in the FIGO classification, but is the most important prognostic factor in cervical carcinoma [161]. In a systematic review comparing the accuracy of sentinel lymph node biopsy, positron emission tomography (PET), CT, and MRI, sentinel node biopsy provided the most accurate assessment of lymph node metastasis, and PET was more accurate than MRI or CT [162]. Hybrid PET/CT is as good as or better than other modalities for the assessment of nodal, extrapelvic, and bone metastases [4].

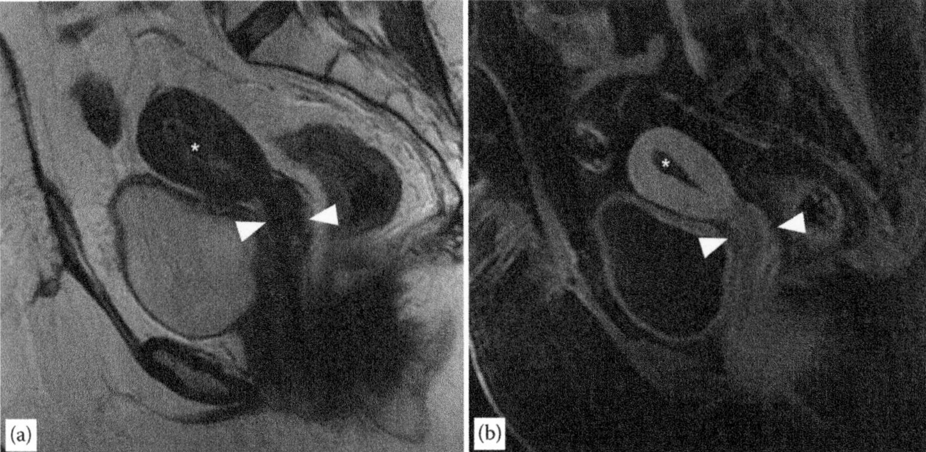

**FIGURE 4.69**
A 44-year-old female with history of cervical carcinoma status post-external beam radiation and chemotherapy. Endometrial fluid was seen on staging CT, and MRI was obtained to evaluate for recurrence. Sagittal (a) $T_2$W and (b) $T_1$W postcontrast images demonstrate reconstitution of zonal anatomy of the cervix and the presence of low SI cervical stroma (white arrowheads). Findings are consistent with post-treatment/radiation changes. No enhancing tumor is seen within the endometrial canal, fluid within the endometrium compatible with degrading blood products (*).

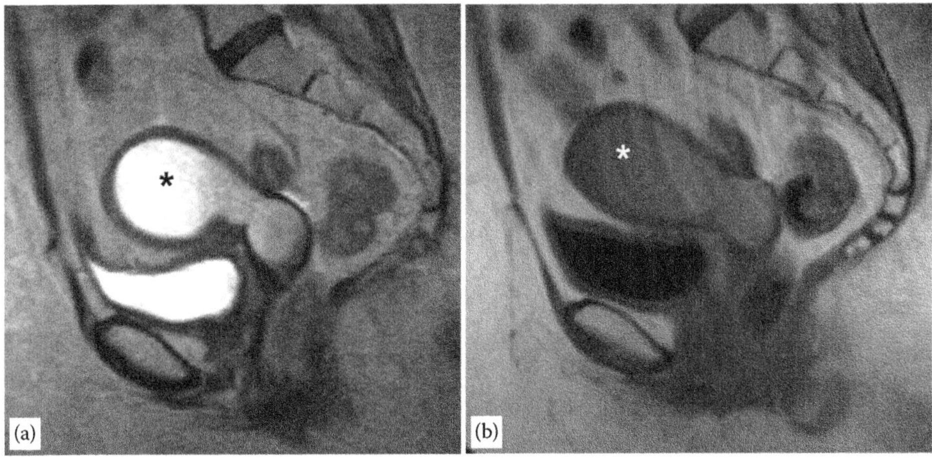

**FIGURE 4.70**
A 62-year-old female with history of stage IB cervical cancer status post-whole pelvic XRT and brachytherapy. Sagittal (a) $T_2$W and (b) $T_1$W images demonstrate a dilated endometrial cavity filled with blood/proteinaceous fluid (*) and diffuse thinning of the uterine wall. There is no evidence of recurrence. This is consistent with post-radiation cervical stenosis.

Chemoradiation therapy is the main modality of treatment in patients with grade Ib2 or greater cervical carcinoma and includes external beam and brachytherapy. The greater the total radiation dose, the greater the risk of complications, both acute and chronic [163]. There are typical imaging characteristics of the cervix that can be seen on MRI post-radiation therapy (Figure 4.69) [164]. The sensitivity and specificity of MRI for the detection of recurrence are high and have been reported as 86% and 94%, respectively [165]. Contrast enhancement has not been shown to improve diagnostic accuracy in determining irradiation changes from tumor recurrence; however, contrast does better demonstrate radiation-induced complications including vesicovaginal and rectovaginal fistulas when compared to non-contrast imaging (Figures 4.70 and 4.71) [164].

**FOCUS POINTS**

- In the pretreatment evaluation of cervical cancer, imaging plays an important role in evaluating tumor size and location, and assessing parametrial, pelvic sidewall, rectal and bladder wall, and lymph node involvement.

- MRI provides the best visualization of the primary tumor and extent of soft tissue disease in the central pelvis.
- FDG PET is as good as or better than other modalities for the assessment of nodal, extrapelvic, and bone metastasis. It is also helpful in predicting patient outcome when maximum standardized uptake values are incorporated into the assessment [4].

## 4.11 Benign Diseases of the Vagina

### 4.11.1 Cysts

Vaginal cysts are common incidental findings found on imaging and are reported in almost 1 in 200 females [166] and can be congenital or acquired. Congenital (Gartner duct) cysts are vestigial remnants of the Wolffian or Müllerian ducts and are typically located in the wall of the upper vagina (Figure 4.72a) and are associated with ipsilateral genitourinary tract anomalies. Vaginal

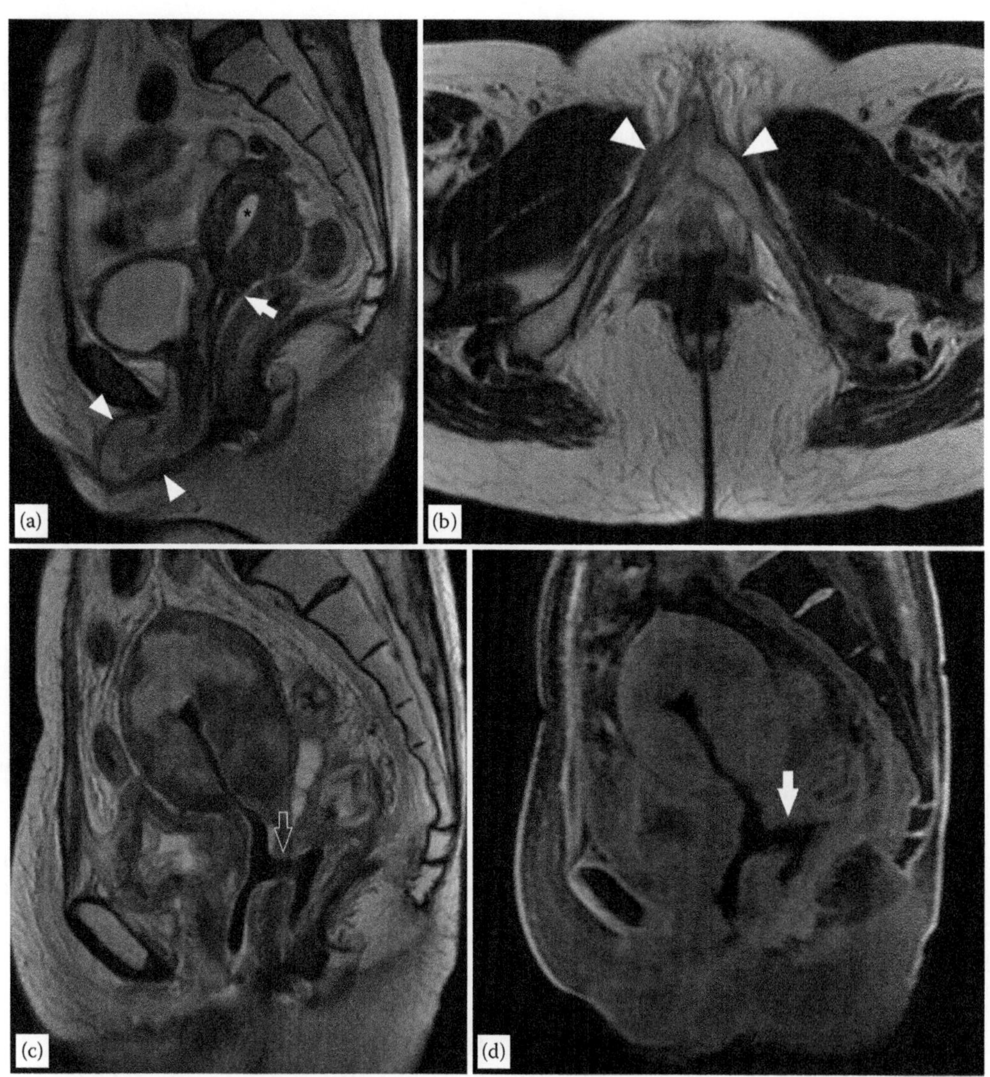

**FIGURE 4.71**
A 53-year-old female with history of stage III cervical cancer status post-chemotherapy and radiation therapy. Six months after radiation, the patient presented clinically with painful clitoral and vulvar swelling. (a) Sagittal and (b) axial $T_2$W images demonstrate high SI enlargement of the clitoris (arrowheads) and labia without discrete mass, this was felt to represent post-radiation changes. The cervix is markedly atrophic (solid arrow) and there is fluid in the endometrial canal (*). Twelve months later, sagittal (c) $T_2$W and (d) $T_1$W postcontrast images show marked progression of disease, including diffuse infiltration of the uterus with tumor, and vaginal, periurethral, perineal, and rectal invasion. A rectovaginal fistula is newly identified (arrow).

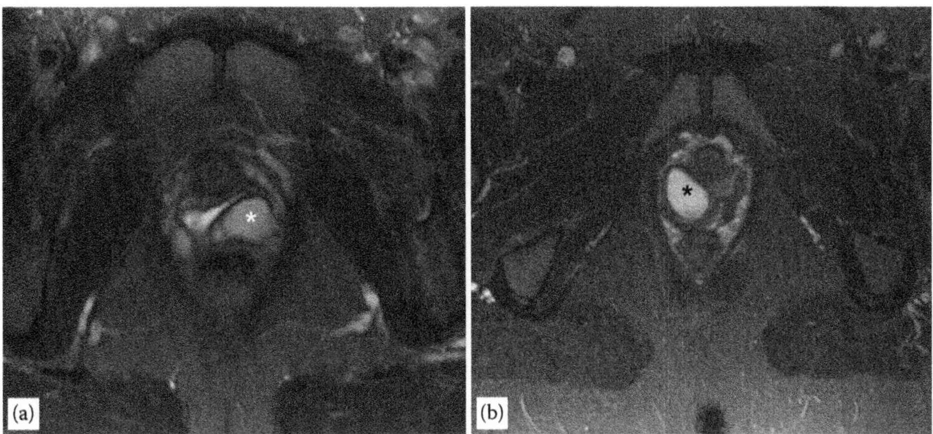


**FIGURE 4.72**
Vaginal cysts. Axial $T_2$W with fat-suppression images demonstrate (a) Gartner's duct cyst and (b) Bartholin cyst. Both typically follow fluid SI; although they contain proteinaceous material, they may appear hyperintense on both $T_1$W and $T_2$W images. The cysts are well-defined, thin-walled, and show no rim enhancement (*). Rim enhancement suggests infection of the cyst.

inclusion cysts (epidermal inclusion cysts) are the most common acquired cystic lesions of the vagina. They are typically found at sites of prior surgery such as episiotomy. These cysts are typically located in the lower posterior or lateral vaginal wall, depending on the location of prior surgery [167]. Bartholin cysts arise from duct obstruction of Bartholin glands resulting in retention of secretions and cyst formation (Figure 4.72b). Etiologies include previous infection, trauma, or chronic inflammation. Both are non-tender, located in the wall of the lower vagina, medial to the labia minora. On MRI, cysts typically follow fluid SI.

### 4.11.2 Benign Vaginal Tumors

Benign tumors of the vagina are rare and most are diagnosed clinically. These include leiomyoma, cavernous hemangioma, fibroepithelial polyp, and rhabdomyoma [168]. Leiomyomas are rare in the vagina and may originate in the smooth muscle of the vagina, local arterial musculature, and the smooth muscle of the bladder or urethra. The anterior vaginal wall is the most common location for vaginal leiomyomas (Figure 4.73) [169]. Due to its increased diagnostic accuracy, MRI is utilized when a vaginal leiomyoma must be distinguished from an aborting uterine fibroid or other atypical vaginal mass (Figure 4.74).

Vaginal foreign bodies are frequent incidental findings on pelvic imaging and include tampons (Figures 4.25 and 4.54), pessaries (Figure 4.75), and hormone rings (Figure 4.76), to name only a few. One should be alert to the presence of foreign bodies in the vagina and uterus when interpreting all studies.

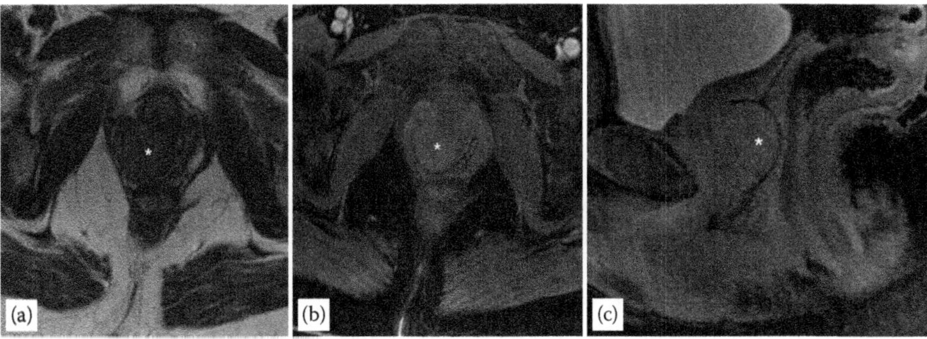


**FIGURE 4.73**
Vaginal leiomyoma. (a) Axial $T_2$W and (b) axial and (c) sagittal $T_1$W postcontrast images demonstrating a well-circumscribed, homogeneously enhancing anterior vaginal wall mass (*).

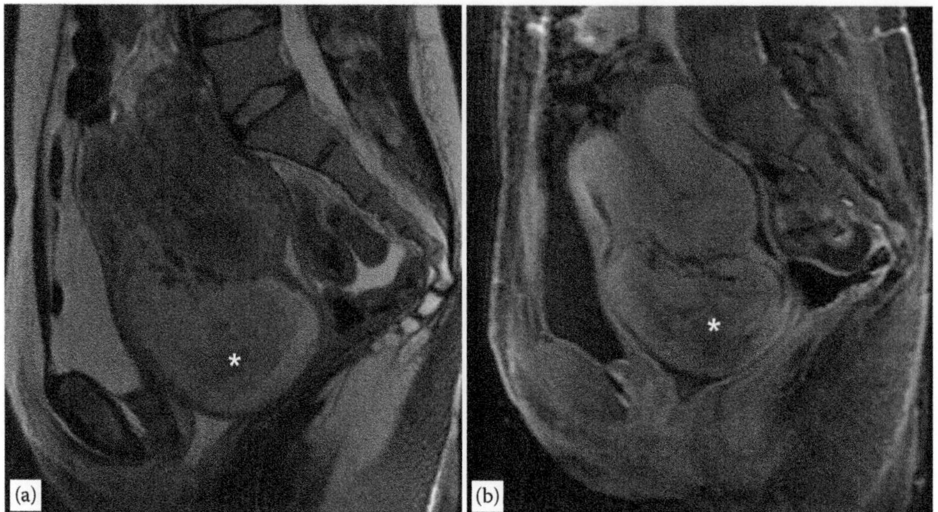


**FIGURE 4.74**
Aborting leiomyoma mimicking a vaginal mass. Sagittal (a) $T_2$W and (b) postcontrast $T_1$W images demonstrate an aborting, pedunculated, submucosal leiomyoma (*) with inversion and prolapse of the corpus.

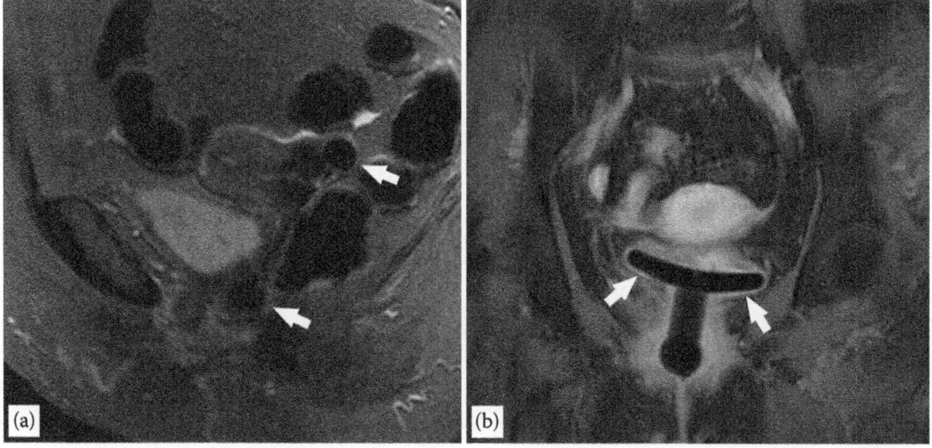


**FIGURE 4.75**
A pessary is inserted into the vagina to help support the pelvic organs in patients with pelvic organ prolapse. They come in various shapes and sizes. (a) Sagittal $T_2$W image with fat suppression and (b) coronal $T_1$W image show two different types (arrows). Pessaries are low in SI on both $T_1$W and $T_2$W sequences.

## 4.12 Malignant Diseases of the Vagina

### 4.12.1 Primary Vaginal Cancer

Primary cancer of the vagina comprises approximately 3% of all malignant neoplasms of the female genital tract [132]. Diagnosis of vaginal cancer is made by biopsy of the suspected lesion, which may appear as a mass, a plaque, or an ulcer. The majority of cases of vaginal cancer are squamous cell carcinomas and, like cervical cancer, are thought to be mediated by human papillomavirus infection. Non-squamous varieties mainly include adenocarcinoma, melanoma, and sarcomas. The majority of vaginal adenocarcinomas are of clear-cell histology and are most commonly associated with women exposed to DES *in utero*. The FIGO system is used for staging of vaginal cancer (Table 4.5).

MR imaging is useful in detection and staging of the vaginal tumor. Although intermediate $T_2$ SI suggests squamous cell carcinoma (Figure 4.77) and high $T_2$ SI on MR is suggestive of a mucin-producing adenocarcinoma (Figure 4.78), neither SI or patterns of contrast enhancement can predict tumor histologic subtype [170].

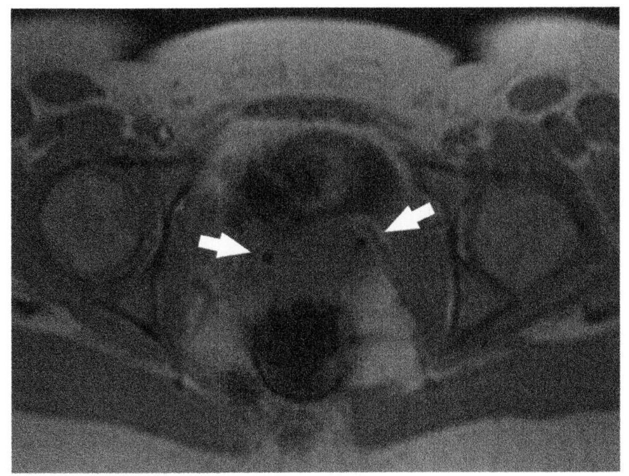

**FIGURE 4.76**
Axial $T_1$W image demonstrates an intravaginal contraceptive ring (arrows). The contraceptive ring is thinner than a pessary.

**TABLE 4.5**
FIGO Staging System for Vaginal Cancer

| Stage | Description | MR Findings |
|---|---|---|
| I | Limited to the vagina | Preservation of normal low $T_2$ SI vaginal wall |
| II | Invasion of the paravaginal tissues, but not to pelvic wall | Loss of the normal bright $T_1$ SI in the paravaginal fat and disruption of the low $T_2$ SI vaginal wall |
| III | Extension to pelvic sidewall | Disruption of the normal low $T_2$ SI of pelvic sidewall muscles and obliteration of the perivaginal fat planes on $T_1$W images |
| IVA | Adjacent organ or direct extension beyond true pelvis | Obliteration of the low $T_2$ SI of rectum, urethra, or bladder |
| IVB | Distant metastases | Tumor extension to distant organs such as liver and lungs |

### 4.12.2 Vaginal Metastases

Vaginal metastases are much more common than primary vaginal tumors [171]. Tumor involvement is most commonly secondary to direct extension into the vagina from adjacent structures (Figure 4.78). However, vaginal metastases may also occur by lymphatic or hematogenous spread. Both primary and metastatic tumors demonstrate similar imaging characteristics.

### 4.12.3 The Fallopian Tubes

Imaging evaluation of the fallopian tubes is performed with TVUS, as normal fallopian tubes are not visualized on routine pelvic MRI. MRI HSG has been evaluated and deemed feasible for the workup of infertility [172,173]; however, X-ray HSG remains the standard of care due to its high resolution and lower cost.

Paratubal cysts are incidental MRI findings found on imaging and typically follow fluid SI. They represent embryonic remnants of the mesonephric (Wolffian) ductal system and are found in approximately 25% of women [36,37]. The role of MRI is to distinguish dilated fallopian tubes from other adnexal masses when US is indeterminate [174]. The SI of the fluid within the tube can be associated with pathologic conditions, for example, bright $T_1$ and $T_2$ SI may be the only finding of endometriosis [174]. Dilated tubes typically have a folded configuration (Figures 4.79 and 4.80). Wall thickening may indicate salpingitis; however, this is not a consistent finding.

Primary fallopian tube carcinoma is thought to be rare [175]; however, preoperative diagnosis is difficult and the incidence may be underestimated due to its similar pathologic appearance to papillary serous ovarian cancer [176]. It is becoming clear that the pathogenesis of ovarian serous cancer is undergoing a paradigm shift where most serous cancers of the ovary are now thought to originate outside the ovary [177]. The hypothesis is that the fallopian tube is the primary site of most high-grade serous carcinomas of the ovary, fallopian tube, and primary peritoneum (Figure 4.81) [177]. With this in mind, any abnormality in the fallopian tubes should be scrutinized (Figure 4.81). Tubal metastases and primaries may be over shadowed by coexisting ovarian disease [178]. Primary tumors that most commonly metastasize to the tubes include ovarian, colorectal, appendiceal, and gastric tumors [179].

**FOCUS POINTS**

- Vaginal and fallopian tube metastases are much more common than primary vaginal and fallopian tube tumors.
- The role of MRI is not as established for pretreatment evaluation of vaginal carcinoma as it is for endometrial and cervical carcinoma. However, MRI may be useful for delineation of tumor size and extent, and evaluation of nodal or distant metastases.
- The fallopian tubes are likely the primary site of most high-grade serous carcinomas of the ovary, fallopian tube, and primary peritoneum.
- Fallopian tube cancers may be misdiagnosed as ovarian cancer on imaging and pathology. Although this does not affect patient outcome, this has implications for the future of ovarian cancer screening.

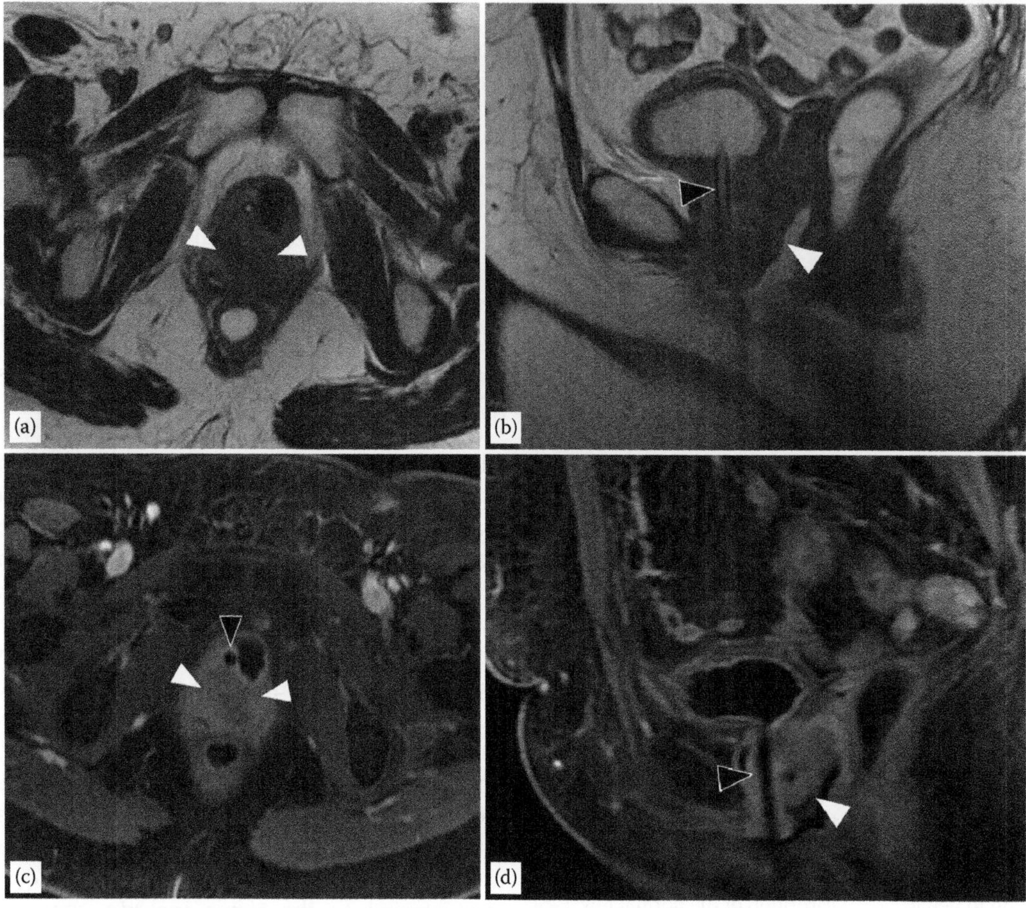

**FIGURE 4.77**
Stage IVa primary vaginal squamous cell carcinoma. (a) Axial and (b) sagittal $T_2$W images demonstrate a round mass (white arrowheads) along the anterior vaginal wall that is isointense relative to the vagina. On (c) axial and (d) sagittal $T_1$W postcontrast images, the tumor (white arrowheads) enhances less than the adjacent paravaginal tissues. There is invasion of the urethra anteriorly. An indwelling Foley catheter (black arrowhead) is present secondary to patient urinary retention. The uterus is not identified and was previously removed secondary to fibroids.

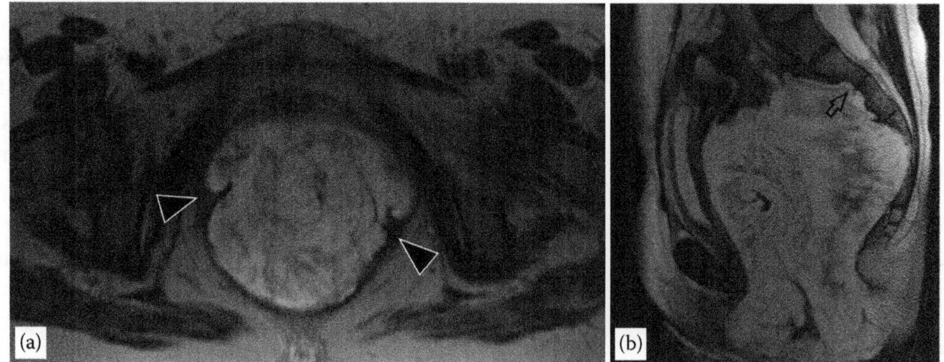

**FIGURE 4.78**
Large enhancing mass in a patient with familial adenomatous polyposis syndrome, status post-subtotal colectomy. (a) Axial and (b) sagittal $T_2$W images demonstrate a mass involving both the vagina and rectum with communication through the posterior vaginal wall (black arrowheads). The mass also erodes the ventral aspect of the sacrum and coccyx (open arrow). When a mass is this large, it is difficult to determine the origin. This is believed to be rectal adenocarcinoma with direct vaginal invasion. Vaginal metastases are much more common than primary vaginal tumors.

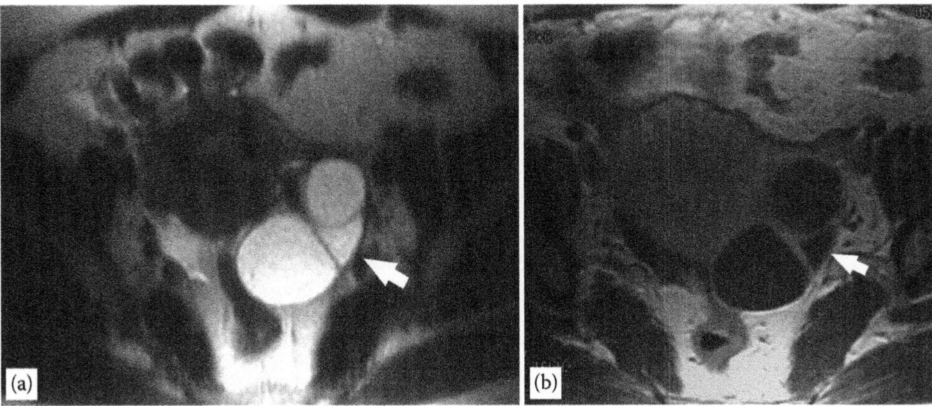

**FIGURE 4.79**
Hydrosalpinx. Axial (a) $T_2$W and (b) $T_1$W images demonstrated a dilated, tubular structure folded upon itself (arrow).

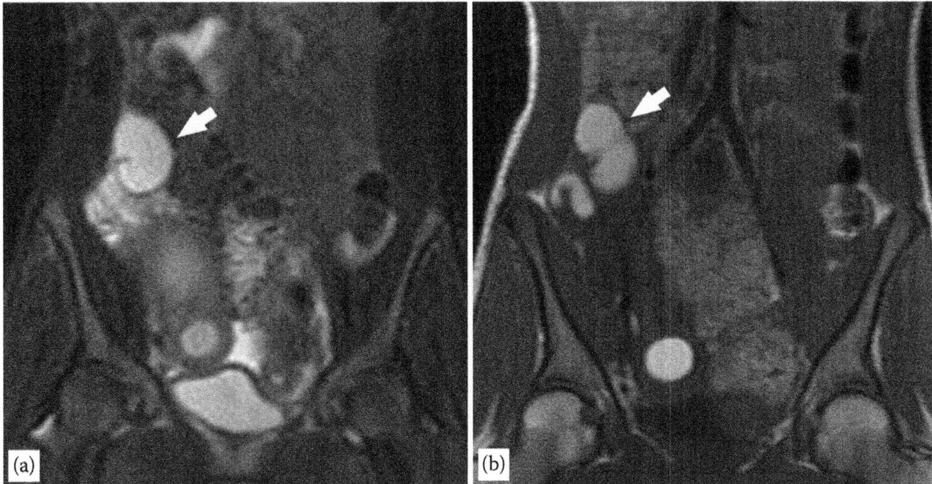

**FIGURE 4.80**
Hematosalpinx. Coronal (a) $T_2$W with fat suppression and (b) $T_1$W images demonstrate a dilated tubular structure containing blood products (arrow). The patient was found to have didelphys uterus with right horn obstruction due to a transverse vaginal septum (not shown), hematosapinx, and absence of right kidney.

## 4.13 Summary

MRI is widely accepted as a useful problem-solving tool, especially if pelvic US is incomplete or indeterminate. MRI may be used to detect, diagnose, and distinguish surgically correctable forms of Müllerian duct anomalies from those that do not require surgical intervention. Because therapy is based on the component parts of the gynecologic anomalies, all findings involving the uterine corpus, cervix, and vagina should be thoroughly described and not classified into the *best fit* category. MRI is recommended when precise evaluation of leiomyoma number, size, position, and degeneration are needed for treatment planning and evaluation of response to therapy. Although there are no pathognomonic findings of leiomyosarcoma, preoperative diagnosis may be possible with a combination of LDH and DCE MRI. MRI has an expanding role for treatment planning of endometrial cancer. DCE can differentiate normal myometrium, where enhancement peaks at 120 after contrast administration, from the more gradual enhancement of endometrial cancer. Maximum contrast between the myometrium and endometrial tumor occurs 50–120 s after contrast administration. DWI is promising for the evaluation of myometrial invasion and may be an alternative to DCE for patients in whom contrast is contraindicated. And, finally, because the fallopian tubes are likely the primary site of origin of serous cancer of the ovary, close attention should be made to any tubal abnormalities found on MRI.

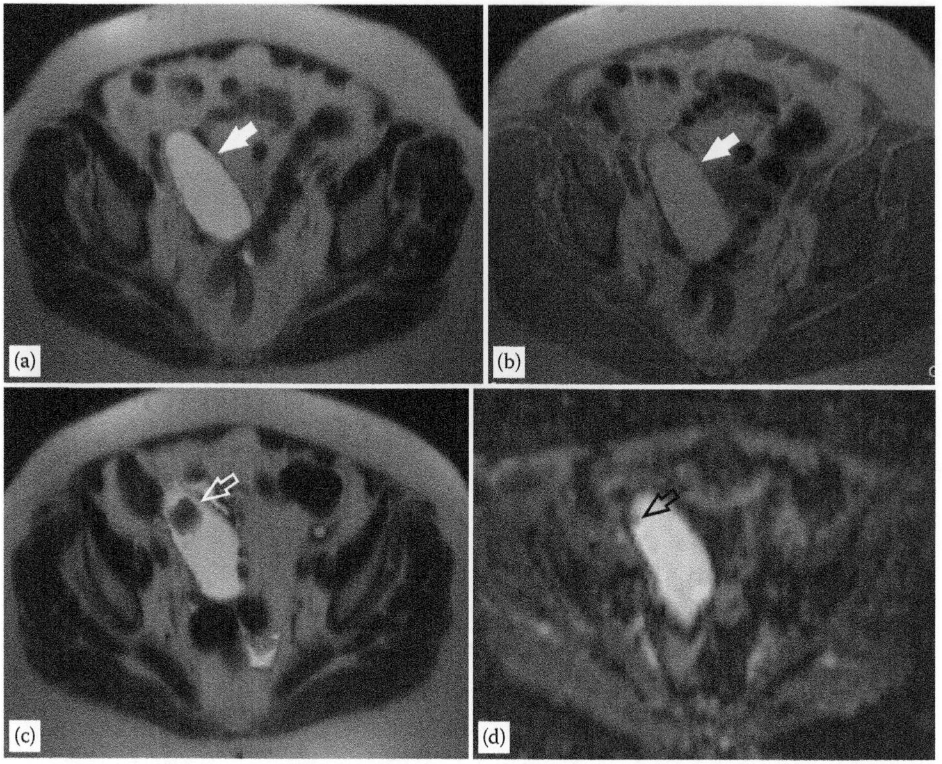

**FIGURE 4.81**
A 71-year-old female with hydrosalpinx incidentally found on ultrasound. Axial (a) $T_2$W and (b) $T_1$W images demonstrate a dilated tubular structure of increased $T_1$ and $T_2$ SI (solid arrow). Three years later, (c) axial $T_2$W image demonstrates interval development of solid components within the dilated tube (open arrow) and a complex right adnexal mass (not shown). In retrospect (d), ADC map demonstrates a small focus of restriction diffusion on the first study (open arrow), indicating that fallopian tube cancer was likely present at the time of the first examination.

## Acknowledgments

We express our gratitude to Yvonne Lin, M.D. and Raymond Azab, M.D. for their assistance in final manuscript review.

## References

1. Bennet GL, Andreotti RF, Lee SI et al. ACR Appropriateness Criteria® abnormal vaginal bleeding. http://www.acr.org/~/media/ACR/Documents/AppCriteria/Diagnostic/AbnormalVaginalBleeding.pdf. American College of Radiology. Last accessed February 21, 2014.
2. Andreotti RF, Lee SI, DeJesus SO et al. ACR Appropriateness Criteria® acute pelvic pain in the reproductive age group. http://www.acr.org/~/media/ACR/Documents/AppCriteria/Diagnostic/AcutePelvicPainReproductiveAgeGroup.pdf. American College of Radiology. Last Accessed February 21, 2014.
3. Lalwani N, Dubinsky T, Javitt MC et al. ACR Appropriateness Criteria® pretreatment evaluation and follow-up of endometrial cancer. http://www.acr.org/~/media/ACR/Documents/AppCriteria/Diagnostic/PretreatmentEvaluationAndFollowUpEndometrialCancer.pdf. American College of Radiology. Last accessed February 21, 2014.
4. Siegel CL, Andreotti RF, Cardenes HR et al. ACR Appropriateness Criteria® pretreatment planning of invasive cancer of the cervix. http://www.acr.org/~/media/ACR/Documents/AppCriteria/Diagnostic/PretreatmentPlanningInvasiveCancerCervix.pdf. American College of Radiology. Last accessed February 21, 2014.
5. Varpula M, Kiilhoma P, Klemi P, Komu M (1994) Magnetic resonance imaging of the uterus in vivo and in vitro at an ultra low magnetic field (0.02): Assessment of its normal structure and of leiomyomas. *Magn Reson Imaging* 12(8):1139–1145.
6. Turnbull L, Booth S (2007) MR imaging of gynecologic dieseases at 3T. *Magn Reson Imaging Clin N Am* 15(3):403–431.
7. Kataoka M, Kido A, Koyama T et al. (2007) MRI of the female pelvis at 3T compared to 1.5T: Evaluation on high-resolution $T_2$-weighted and HASTE images. *J Magn Reson Imaging* 25(3):527–534.

8. Soher BJ, Dale BM, Merkle EM (2007) A review of MR physics: 3T versus 1.5T. *Magn Reson Imaging Clin N Am* 15(3):277–290.
9. US Food and Drug Administration. Information for healthcare professionals: Gadolinium-based contrast agents for magnetic resonance imaging (marketed as Magnevist, MultiHance, Omniscan, OptiMARK, ProHance). http://www.fda.gov/Drugs/DrugSafety/PostmarketDrugSafetyInformationfor Patientsand Providers/ucm142882.htm. Last accessed April 11, 2014.
10. ACR Manual on Contrast Media, Version 9 American College of Radiology Committee on Drugs and Contrast Media. http://www.acr.org/Quality-Safety/Resources/Contrast-Manual. Last accessed April 11, 2014.
11. Nakai A, Togashi K, Kosaka K et al. (2008) Do anticholinergic agents suppress uterine peristalsis and sporadic myometrial contractions at cine MR imaging? *Radiology* 246(2):489–496.
12. Zand KR, Reinhold C, Haider MA et al. (2007) Artifacts and pitfalls in MR imaging of the pelvis. *J Magn Reson Imaging* 26(3):480–497.
13. Lyons EA, Taylor PJ, Zheng XH et al. (1991) Characterization of subendometrial myometrial contractions throughout the menstrual cycle in normal fertile women. *Fertil Steril* 55(4):771–774.
14. Fujimoto K, Koyama T, Tamai K et al. (2011) BLADE acquisition method improves $T_2$-weighted MR images of the female pelvis compared with a standard fast spin-echo sequence. *Eur J Radiol* 80(3):796–801.
15. Sugimura H, Yamaguchi K, Furukoji E et al. (2004) Comparison of conventional fast spin echo, single-shot two-dimensional and three-dimensional half-fourier RARE for $T_2$-weighted female pelvic imaging. *J Magn Reson Imaging* 19(3):349–355.
16. Hecht EM, Yitta S, Lim RP et al. (2011) Preliminary clinical experience at 3T with a 3D $T_2$-weighted sequence compared with multiplanar 2D for evaluation of the female pelvis. *AJR Am J Roentgenol* 197(2):W346–352.
17. Thoeny HC, Forstner R, De Keyzer F (2012) Genitourinary applications of diffusion weighted MR imaging in the pelvis. *Radiology* 263(2):326–342.
18. Hoad CL, Fulford J, Raine-Fenning NJ et al. (2006) In vivo perfusion, $T_1$, and $T_2$ measurements in the female pelvis during the normal menstrual cycle: A feasibility study. *J Magn Reson Imaging* 24(6):1350–1356.
19. Sokol ER, Genadry R, Anderson JR (2012) Anatomy and embryology. In: Berek, JS (ed) *Berek and Novak's Gynecology*, 15th edn. Lippincott Williams & Wilkins, Philadelphia, PA.
20. Levi CS, Lyons EA, Holt SC, Dashefsky SM (2008) Normal anatomy of the female pelvis and transvaginal sonography. In: Callen PW (ed) *Ultrasonography in Obstetrics and Gynecology*, 5th edn. Saunders Elsevier, Philadelphia, PA.
21. Clemente CD (1981) *Anatomy: A Regional Atlas of the Human Body*, 2nd edn. Urban & Schwarzenber, Baltimore, MD.
22. Brown HK, Stoll BS, Nicosia SV et al. (1991) Uterine junctional zone: Correlation between histologic findings and MR imaging. *Radiology* 179(2):409–413.
23. Varpula M, Komu M, Irjala K (1993) Relaxation-time changes of the uterus during the menstrual cycle: Correlation with hormonal status. *Eur J Radiol* 16(2):90–94.
24. Hricak H, Alpers C, Crooks LE, Sheldon PE (1983) Magnetic resonance imaging of the female pelvis: Initial experience. *AJR Am J Roentgenol* 141(6):1119–1128.
25. Demas BE, Hricak H, Jaffe RB (1986) Uterine MR imaging: Effects of hormonal stimulation. *Radiology* 159(1):123–126.
26. Hoad CL, Raine-Fenning NJ, Fulford J et al. (2005) Uterine tissue development in healthy women during the normal menstrual cycle and investigations with magnetic resonance imaging. *Am J Obstet Gynecol* 192(2):648–654.
27. Scoutt LM, Flynn SD, Luthringer DJ, McCauley TR, McCarthy SM (1991) Junctional zone of the uterus: Correlation of MR imaging and histologic examination of hysterectomy specimens. *Radiology* 179(2):403–407.
28. McCarthy S, Scott G, Majumdar S et al. (1989) Uterine junctional zone: MR study of water content and relaxation properties. *Radiology* 171(1):241–243.
29. Scoutt LM, McCauley TR, Flynn SD, Luthringer DJ, McCarthy SM (1993) Zonal anatomy of the cervix: Correlation of MR imaging and histologic examination of hysterectomy specimens. *Radiology* 186(1):159–162.
30. Holm K, Laursen EM, Brocks V et al. (1995) Pubertal maturation of the internal genitalia: An ultrasound evaluation of 166 healthy girls. *Ultrasound Obstet Gynecol* 6(3):175–181.
31. Hoffman BL, Schorge JO, Schaffer JI, Halvorson LM, Bradshaw KD, Cunningham F, Calver LE (eds) (2012) Anatomic disorders (Chapter 18) In: *Williams Gynecology*, 2nd edn. McGraw-Hill, New York.
32. Hoffman BL, Schorge JO, Schaffer JI, Halvorson LM, Bradshaw KD, Cunningham F, Calver LE (eds) (2012) Pelvic organ prolapse (Chapter 24) In: *Williams Gynecology*, 2nd edn. McGraw-Hill, New York.
33. Salem S, Wilson S (2005) Gynecologic ultrasound. In: Rumack CM, Wilson SR, Charboneau JW, Johnson JA, (eds) *Diagnostic Ultrasound*, 3rd edn. Elsevier Mosby, St. Louis, MO.
34. Eisenberg LB, Elias, Jr., J, Qureshi W, Young MK, Semelka RC (2010) Female urethra and vagina. In: Semelka RC (ed) *Abdominal-Pelvic MRI*, 3rd edn. John Wiley and Sons, Hoboken, NJ.
35. Scoutt LM, McCarthy SM (1992) Female pelvis. In: Stark DD, Bradley WG (ed) *Magnetic Resonance Imaging*, 2nd edn. Mosby Year Book, St. Louis, MO.
36. Moore KL, Persaud TVN, Tochia MG (2011) Urogenital system. In: Moore KL (ed) *The Developing Human*, 9th edn. Saunders, Philadelphia, PA.
37. Arey LB (1974) The genital system. In: Arey LB (ed) *Developmental Anatomy*, 7th edn. Saunders, Philadelphia, PA.
38. Troiano RN, McCarthy SM (2004) Mullerian duct anomalies: Imaging and clinical issues. *Radiology* 233(1):19–34.
39. Mueller GC, Hussain HK, Smith YR et al. (2007) Mullerian duct anomalies: Comparison of MRI diagnosis and clinical diagnosis. *AJR Am J Roentgenol* 189(6):1294–1302.

40. The American Fertility Society (1988) The American Fertility Society classifications of adnexal adhesions, distal tubal occlusion, tubal occlusion secondary to tubal ligation, tubal pregnancies, müllerian anomalies and intrauterine adhesions. *Fertil Steril* 49(6):944–955.
41. Acien P, Acien M, Sanchez-Ferrer ML (2009) Müllerian anomalies "without a classification": From the didelphys-unicollis uterus to the bicervical uterus with or without septate vagina. *Fertil Steril* 91(6):2369–2375.
42. Muller P, Musset R, Netter A et al. (1967) State of the upper urinary tract in patients with uterine malformation. Study of 133 cases. *Presse Med* 75(26):1331–1336.
43. Acien P, Acien MI (2011) The history of female genital tract malformation classifications and proposal of an updated system. *Hum Reprod Update* 17(5):693–705.
44. Frontino G, Bianchi S, Ciappina N et al. (2009) The unicornuate uterus with an occult adenomyotic rudimentary horn. *J Minim Invasive Gynecol* 16(5):622–625.
45. Oppelt P, Renner SP, Brucker S et al. (2005) The VCUAM (vagina cervix uterus adnex-associated malformation) classification: A new classification for genital malformations. *Fertil Steril* 84(5):1493–1497.
46. Grimbizis GF, Gordts S, Di Spiezio Sardo A et al. (2013) The ESHRE/ESGE consensus on the classification of female genital tract congenital anomalies. *Hum Reprod* 28(8):2032–2044.
47. Rall K, Barresi G, Wallwiener D, Brucker SY, Staebler A (2013) Uterine rudiments in patients with Mayer-Rokitansky-Kuester-Hauser syndrome consist of typical uterine tissue types with predominantly basalis-like endometrium. *Fertil Steril* 99(5):1392–1399.
48. Fedele L, Bianchi S, Zanconato G, Berlanda N, Bergamini V (2005) Laparoscopic removal of the cavitated noncommunicating rudimentary uterine horn: Surgical aspects in 10 cases. *Fertil Seril* 83(2):432–436.
49. Fedele L, Zamberletti D, Vercellini P, Dorta M, Candiani GB (1987) Reproductive performance of women with unicornuate uterus. *Fertil Steril* 47(3):416–419.
50. Oppelt PG, Lermann J, Strick R et al. (2012) Malformations in a cohort of 284 women with Mayer-Rokitansky-Kuester-Hauser syndrome (MRKH). *Reprod Biol Endocrol* 10:57–64.
51. Fedele L, Motta F, Frontino G, Restelli E, Bianchi S (2013) Double uterus with obstructed hemivagina and ipsilateral renal agenesis: Pelvic anatomic variants in 87 cases. *Hum Reprod* 28(6):1580–1583.
52. Olive DL, Henderson DY (1987) Endometriosis and müllerian anomalies. *Obstet Gynecol* 69(3):412–415.
53. Ugur M, Turan C, Mungan T et al. (1995) Endometriosis in association with mullerian anomalies. *Gynecol Obstet Invest* 40(4):261–264.
54. Herbst AL, Ulfelder H, Poskanzer DC (1971) Adenocarcinoma of the vagina: Association of maternal stilbesterol therapy with tumor appearance in young women. *N Engl J Med* 284(15):878–881.
55. Herbst AL, Kurman RJ, Scully RE (1972) Vaginal and cervical abnormalities after exposure to stilbestrol in utero. *Obstet Gynecol* 40(3):287–298.
56. Van Gils AP, Than RT, Falke TH, Peters AA (1989) Abnormalities of the uterus and cervix after diethylstilbestrol exposure: Correlation of findings on MR and hysterosalpingography. *AJR Am J Roentgenol* 153(6):1235–1238.
57. FDA drug bulletin: Diethylstilbestrol contraindicated in pregnancy, November 1971. http://www.unboundmedicine.com/medline/citation/18730697/Selected_item_from_the_FDA_drug_bulletin_november_1971:_diethylstilbestrol_contraindicated_in_pregnancy_. Last accessed April 11, 2014.
58. Rock JA, Roberts CP, Jones HW Jr (2010) Congenital anomalies of the uterine cervix: Lessons from 30 cases managed clinically by a common protocol. *Fertil Steril* 94(5):1858–1863.
59. Rock JA, Zacur HA, Dlugi AM, Jones HW, TeLinde RW (1982) Pregnancy success following surgical correction of imperforate hymen and complete transverse vaginal septum. *Obstet Gynecol* 59(4):448–451.
60. Junqueira BL, Allen LM, Spitzer RF, Lucco KL, Babyn PS, Doria AS (2009) Mullerian duct anomalies and mimics in children and adolescents: Correlative intraoperative assessment with clinical imaging. *RadioGraphics* 29(4):1085–1103.
61. Oppelt P, von Have M, Paulsen M et al. (2007) Female genital malformations and their associated abnormalities. *Fertil Steril* 87(2):335–342.
62. Hoffman BL, Schorge JO, Schaffer JI, Halvorson LM, Bradshaw KD, Cunningham F, Calver LE (eds) (2012) Pelvic mass (Chapter 9) In: *Williams Gynecology*, 2nd edn. McGraw-Hill, New York.
63. Parker (2012) Uterine fibroids. In: Berek JS (ed) *Berek and Novak's Gynecology*, 15th edn. Lippincott Williams & Wilkins, Philadelphia, PA.
64. Baird DD, Dunson DB, Hill MC, Cousins D, Schectman JM (2003) High cumulative incidence of uterine leiomyoma in black and white women: Ultrasound evidence. *Am J Obstet Gynecol* 188(1):100–107.
65. Munro MG, Critchley HO, Broder MS et al. (2011) FIGO classification system (PALM-COEIN) for causes of abnormal uterine bleeding in nongravid women of reproductive age. *Int J Gynaecol Obstet* 113(1):3–13.
66. Wallach EE, Vlahos NF (2004) Uterine myomas: An overview of development, clinical features, and Management. *Obstet Gynecol* 104(2):393–406.
67. Peddada SD, Laughlin SK, Miner K et al. (2008) Growth of uterine leiomyomata among premenopausal black and white women. *Proc Natl Acad Sci USA* 105(50):19887–19892.
68. Baird DD, Garrett TA, Laughlin SK, Davis B, Semelka RC, Peddada SD (2011) Short-term change in growth of uterine leiomyoma: Tumor growth spurts. *Fertil Steril* 95(1):242–246.
69. Stewart E (2001) Uterine fibroids. *Lancet* 357(9252):293–298.
70. Stovall D (2001) Clinical symptomatology of uterine leiomyomas. *Clin Obstet Gynecol* 44(2):364–371.
71. Ruushanen AJ, Hippelaeinen MI, Sipola P, Manninen HI (2012) Association between magnetic resonance imaging findings of uterine leiomyomas and symptoms demanding treatment. *Eur J Radiol* 81(8):1957–1964.

72. Stewart EA, Nowak RA (1996) Leiomyoma-related bleeding: A classic hypothesis updated for the molecular era. *Hum Reprod Update* 2(4):295–306.
73. Hricak H, Tscholakoff D, Heinrichs L et al. (1986) Uterine leiomyomas: Correlation of MR, histopathologic findings, and symptoms. *Radiology* 158(2):385–391.
74. Okizuka K, Sugimura K, Takemori M, Obayashi C, Kitao M, Ishida T (1993) MR detection of degenerating uterine leiomyomas. *J Comput Assist Tomogr* 17(5):760–766.
75. Yamashita Y, Torashima M, Takahashi M et al. (1993) Hyperintense uterine leiomyoma at $T_2$-weighted MR imaging: Differentiation with dynamic enhanced MR imaging and clinical implications. *Radiology* 189(3):721–725.
76. Jha RC, Ascher SM, Imaoka I et al. (2000) Symptomatic fibroleiomyomata: MR imaging of the uterus before and after uterine artery embolization. *Radiology* 217(1):228–235.
77. Levine DJ, Berman JM, Harris M, Chudnoff SG, Whaley FS, Palmer SL (2013) Sensitivity of myoma imaging using laparoscopic ultrasound compared with magnetic resonance imaging and transvaginal ultrasound. *J Minim Invasive Gynecol* 20(6):770–774.
78. Ueda H, Togashi K, Konishi I et al. (1999) Unusual appearances of uterine leiomyomas: MR imaging findings and their histopathologic backgrounds. *RadioGraphics* 19:S131–S145.
79. Shimada K, Ohashi I, Kasahara I et al. (2004) Differentiation between completely hyalinized uterine leiomyomas and ordinary leiomyomas: Three-phase dynamic magnetic resonance imaging (MRI) vs. diffusion-weighted MRI with very small b-factors. *J Magn Reson Imaging* 20(1):97–104.
80. Worley MJ Jr, Aelion A, Caputo TA et al. (2009) Intravenous leiomyomatosis with intracardiac extension: A single-institution experience. *Am J Obstet Gynecol* 201(6):574.
81. Du J, Zhao X, Guo D, Li H, Sun B (2011) Intravenous leiomyomatosis of the uterus: A clinicopathologic study of 18 cases, with emphasis on early diagnosis and appropriate treatment strategies. *Hum Pathol* 42(9):1240e6.
82. Clay TD, Dimitriou J, McNally OM, Russell PA, Newcomb AE, Wilson AM (2013) Intravenous Leiomyomatosis with intracardiac extension—A review of diagnosis and management with an illustrative case. *Surg Oncol* 22(3):e44–e52.
83. Thomassin-Naggara I, Darai E, Nassar-Slaba J, Cortez A, Marsault C, Bazot M (2007) Value of dynamic enhanced magnetic resonance imaging for distinguishing between ovarian fibroma and suberous uterine leiomyoma. *J Comput Assist Tomogr* 31(2):236–242.
84. Rapkin AJ, Nathan L (2012) Pelvic pain and dysmenorrhea. In: Berek JS (ed) *Berek and Novak's Gynecology*, 15th edn. Lippincott Williams & Wilkins, Philadelphia, PA.
85. Vercellini P, Parazzini F, Oldani S, Panazza S, Bramante T, Crosignani PG (1995) Adenomyosis at hysterectomy: A study on frequency distribution and patient characteristics. *Hum Reprod* 10(5):1160–1162.
86. Templeman C, Marshall SF, Ursin G et al. (2008) Adenomyosis and endometriosis in the California Teachers Study. *Fertil Steril* 90(2):415–424.
87. Bergeron C, Amant F, Ferenczy (2006) Pathology and physiopathology of adenomyosis. *Best Pract Res Clin Obstet Gynaecol* 20(4):511–521.
88. Mark AS, Hricak H, Heinrichs LW et al. (1987) Adenomyosis and leiomyoma: Differential diagnosis with MR imaging. *Radiology* 163(2):527–529.
89. Ascher SM, Arnold LL, Patt RH et al. (1994) Adenomyosis: Prospective comparison of MR imaging and transvaginal sonography. *Radiology* 190(3):803–806.
90. Togashi K, Nishimura K, Itoh K et al. (1988) Adenomyosis: Diagnosis with MR imaging. *Radiology* 166(1):111–114.
91. Reinhold C, McCarthy S, Bret PM et al. (1996) Diffuse adenomyosis: Comparison of endovaginal US and MR imaging with histopathologic correlation. *Radiology* 199(1):151–158.
92. Dueholm M, Lundorf E, Hansen ES, Sorensen JS, Ledertoug S,Olesen F (2001) Magnetic resonance imaging and transvaginal ultrasonography for the diagnosis of adenomyosis. *Fertil Steril* 76(3):588–594.
93. Bazot M, Cortez A, Darai E et al. (2001) Ultrasonography compared with magnetic resonance imaging for the diagnosis of adenomyosis: Correlation with histopathology. *Hum Reprod* 16(11):2427–2433.
94. Outwater Ek, Siegleman ES, Van Deerlin V (1998) Adenomyosis: Current concepts and imaging considerations. *AJR Am J Roentgenol* 170(2):437–441.
95. Falcone T, Parker WH (2013) Surgical management of leiomyomas for fertility or uterine preservation. *Obstet Gynecol* 121(4):856–868.
96. Viswanathan M, Hartmann K, McKoy N et al. (2007) *Management of uterine fibroids: An update of the evidence.* Evidence report/technology assessment no. 154, AHRQ publication no. 07-E011. Agency for Healthcare Research and Quality, Rockville, MD.
97. Kho KA, Nezhat CH (2014) Evaluating the risks of electric uterine morcellation. *JAMA* 311(9):905–906.
98. Shen S-H, Fennessy F, McDannold N, Jolesz F, Tempany C (2009) Image-guided thermal therapy of uterine fibroids. *Semin Ultrasound CT MR* 30(2):91–104.
99. Deshmukh SP, Gonsalves CF, Guglielmo FF, Mitchell DG (2012) Role of MR imaging of uterine leiomyomas before and after embolization. *RadioGraphics* 32(6):E251–E281.
100. Nikolaidis P, Siddiqi AJ, Carr JC et al. (2005) Incidence of nonviable leiomyomas on contrast material-enhanced pelvic MR imaging in patients referred for uterine artery embolization. *J Vasc Interv Radiol* 16(11):1465–1471.
101. Katsumori T, Akazawa K, Mihara T (2005) Uterine artery embolization for pedunculated subserosal fibroids. *AJR Am J Roentgenol* 184(2):399–402.
102. Kim MD, Lee M, Jung DC et al. (2012) Limited efficacy of uterine artery embolization for cervical leiomyomas. *J Vasc Interv Radiol* 23(2):236–240.
103. Kroencke TJ, Scheurig C, Poellinger A, Gronewold M, Hamm B (2010) Uterine artery embolization for leiomyomas: Percentage of infarction predicts clinical outcome. *Radiology* 255(3):834–841.

104. Naguib NNN, Mbalisike E, Nour-Eldin NA et al. (2010) Uterine artery embolization: Correlation with the initial leiomyoma volume and location. *J Vasc Interv Radiol* 21(4):490–495.
105. Kim MD, Kim S, Kim NK et al. (2007) Long-term results of uterine artery embolization for symptomatic adenomyosis. *AJR Am J Roentgenol* 188(1):176–181.
106. Tempany CM, Stewart EA, McDannold N, Quade BJ, Jolesz FA, Hynynen K (2003) MR imaging–guided focused ultrasound surgery of uterine leiomyomas: A feasibility study. *Radiology* 226(3):897–905.
107. Lenard ZM, MacDonnold NJ, Fennessy FM (2008) Uterine leiomyomas: MR imaging-guided focused ultrasound surgery—Imaging predictors of success. *Radiology* 249(1):187–194.
108. Gorny KR, Woodrum DA, Brown DL et al. (2011) Magnetic resonance-guided focused ultrasound of uterine leiomyomas: Review of a 12-month outcome of 130 clinical patients. *J Vasc Interv Radiol* 22(6):857–864.
109. Dowdy, SC, Mariani A, Lurain JR (2012) Uterine cancer. In: Berek JS (ed) *Berek and Novak's Gynecology*, 15th edn. Lippincott Williams & Wilkins, Philadelphia, PA.
110. Moinfar F, Azodi M, Tavassoli FA (2007) Uterine sarcomas. *Pathology* 39(1):55–71.
111. Brooks SE, Zhan M, Cote T, Baquet CR (2004) Surveillance, epidemiology, and end results analysis of 2677 cases of uterine sarcoma 1989–1999. *Gynecol Oncol* 93(1):204–208.
112. D'Angelo E, Prat J (2010). Uterine sarcomas: A review. *Gynecol Oncol* 116(1): 131–139.
113. Stewart EA, Morton CC (2006) The genetics of uterine leiomyomata: What clinicians need to know. *Obstet Gynecol* 107(4):917–921.
114. Wysowski DK, Honig SF, Beitz J (2002) Uterine sarcoma associated with tamoxifen use. *N Engl J Med* 346(23):1832–1833.
115. Parker WH, Fu YS, Berek JS (1994) Uterine sarcoma in patients operated on for presumed leiomyoma and rapidly growing leiomyoma. *Obstet Gynecol* 83(3):414–418.
116. Amant F, Coosemans A, Debiec-Rychter M, Timmerman D, Vergote I (2009) Clinical management of uterine sarcomas. *Lancet Oncol* 10(12):1188–1198.
117. Tanaka YO, Nishida M, Tsunoda H, Okamoto Y, Yoshikawa H (2004) Smooth muscle tumors of uncertain malignant potential and leiomyosarcomas of the uterus: MR findings. *J Magn Reson Imaging* 20(6):998–1007.
118. Schwartz LB, Zawin M, Carcangiu ML, Lange R, McCarthy S (1998) Does pelvic magnetic resonance imaging differentiate among the histologic subtypes of uterine leiomyomata? *Fertil Steril* 70(3):580–587.
119. Cornfield D, Israel G, Martel M, Weinreb J, Schwartz P, McCarthy S (2010) MRI appearance of mesenchymal tumors of the uterus. *Eur J Radiol* 74(1):241–249.
120. Goto A, Takeuchi S, Sugimura K, Maruo T (2002) Usefulness of Gd-DTPA contrast-enhanced dynamic MRI and serum determination of LDH and its isozymes in the differential diagnosis of leiomyosarcoma from degenerated leiomyoma of the uterus. *Int J Gynecol Cancer* 12(4):354–361.
121. Park JY, Park SK, Kim DY et al. (2011) The impact of tumor morcellation during surgery on the prognosis of patients with apparently early uterine leiomyosarcoma. *Gynecol Oncol* 122(2):255–259.
122. Kurman, RJ, Kaminski PF, Norris HJ (1985) The behavior of endometrial hyperplasia: A long-term study of untreated hyperplasia in 170 patients. *Cancer* 56(2):403–412.
123. Dreisler E, Stampe Sorensen S, Ibsen PH et al. (2009) Prevalence of endometrial polyps and abnormal uterine bleeding in a Danish population aged 20-74 years. *Ultrasound Obstet Gynecol* 33(1):102–108.
124. Baiocchi G, Manci N, Pazzaglia M et al. (2009) Malignancy in endometrial polyps: A 12-year experience. *Am J Obstet Gynecol* 201(5):462.
125. Scully RE, Bonfiglio TA, Kurman RJ et al. (1994) Histological typing of female genital tract tumours. In: Scully RE, Poulsen HE, Sobin LH (eds) *World Health Organization International Histological Classification of Tumours*, 2nd edn. Springer-Verlag, Berlin, Germany.
126. Mutter GL (2000) Endometrial intraepithelial neoplasia (EIN): Will it bring order to chaos? The Endometrial Collaborative Group. *Gynecol Oncol* 76(3):287–290.
127. Lacey JV, Sherman ME, Rush BB et al. (2010) Absolute risk of endometrial carcinoma during 20-year follow-up among women with endometrial hyperplasia. *J Clin Oncol* 28(5):788–792.
128. Imaoka I, Sugimura K, Masui T et al. (1999) Abnormal uterine cavity: Differential diagnosis with MR imaging. *Magn Reson Imaging* 17(10):1445–1455.
129. Grasel RP, Outwater EK (2000) Endometrial polyps: MR imaging features and distinction from endometrial carcinoma. *Radiology* 214(1):47–52.
130. Hann LE, Giess CS, Bach AM et al. (1997) Endometrial thickness in tamoxifen-treated patients: Correlation with clinical and pathologic findings. *AJR Am J Roentgenol* 168(3):657–661.
131. Ascher SM, Johnson JC, Barnes WA et al. (1996) MR imaging appearance of the uterus in postmenopausal women receiving tamoxifen therapy for breast cancer: Histopathologic correlation. *Radiology* 200(1):105–110.
132. Siegel R, Ma J, Zou Z, Jemal A (2014) Cancer statistics, 2014. *CA Cancer J Clin* 64(1):9–29.
133. Torres ML, Weaver AL, Kumar S et al. (2012) Risk factors for developing endometrial cancer after benign endometrial sampling. *Obstet Gynecol* 120(5):998–1004.
134. Boronow RC, Morrow CP, Creasman WT et al. (1984) Surgical staging in endometrial cancer: Clinical-pathologic findings of a prospective study. *Obstet gynecol* 63(6):825–832.
135. Briet JM, Hollema H, Reesink N et al. (2005) Lymphovascular space involvement: An independent prognostic factor in endometrial cancer. *Gynecol Oncol* 96(3):799–804.
136. Kinkel K, Kaji Y, Yu KK et al. (1999) Radiologic staging in patients with endometrial cancer: A meta-analysis. *Radiology* 212(3):711–718.
137. Yamashita Y, Harada M, Sawada T et al. (1993) Normal uterus and FIGO stage 1 endometrial carcinoma: Dynamic gadolinium–enhanced MR imaging. *Radiology* 186(2):495–501.

138. Lee EJ, Byun JY, Kim BS et al. (1999) Staging of early endometrial carcinoma: Assessment with $T_2$-weighted and gadolinium-enhanced $T_1$-weighted MR imaging. *RadioGraphics* 19(4):937–945.
139. Beddy P, Moyle P, Kataoka M et al. (2012) Evaluation of depth of myometrial invasion and overall staging in endometrial cancer: Comparison of diffusion-weighted and dynamic contrast-enhanced MR imaging. *Radiology* 262(2):530–537.
140. Rechichi G, Galimberti S, Signorelli M et al. (2010) Myometrial invasion in endometrial cancer: Diagnostic performance of diffusion-weighted MR imaging at 1.5-T. *Eur Radiol* 20(3):754–762.
141. Creaseman W (2009) Revised FIGO staging for carcinoma of the endometrium. *Int J Gynaecol Obstet*. 105(2):109.
142. Haldorsen IS, Berg A, Werner HM et al. (2012) Magnetic resonance imaging performs better than endocervical curettage for preoperative prediction of cervical stromal invasion in endometrial carcinomas. *Gynecol Oncol* 125(3):413–418.
143. Ascher SM, Reinhold C (2002) Imaging of cancer of the endometrium. *Radiol Clin N Am* 40(3):563–576.
144. Lin G, Ho KC, Wang JJ et al. (2008) Detection of lymph node metastasis in cervical and uterine cancers by diffusion-weighted magnetic resonance imaging at 3T. *J Magn Reson Imaging* 28(1):128–135.
145. Sahdev A, Schaib SA, Jacobs I et al. (2001) MR Imaging of uterine sarcomas. *AJR Am J Roentgenol* 177(6):1307–1311.
146. Shapeero LG, Hricak H (1998) Mixed mullerian sarcoma of the uterus: MR imaging findings. *AJR Am J Roentgenol* 153(2):317–319.
147. Ohguri T, Aoki T, Watanabe H et al. (2002) MRI Findings including gadolinium-enhanced dynamic studies of malignant, mixed mesodermal tumors of the uterus: Differentiation from endometrial carcinomas. *Eur Radiol* 12(11):2737–2742.
148. Schnall, MD (1994) Magnetic resonance evaluation of acquired benign uterine disorders. *Semin Ultrasound CT MR* 15(1):18–26.
149. Walboomers JM, Jacobs MV, Manos MM et al. (1999) Human papillomavirus is a necessary cause of invasive cervical cancer worldwide. *J Pathol* 189(1):12–19.
150. Tiltman AJ (2005) The pathology of cervical tumours. *Best Pract Res Clin Obstet Gynaecol* 19(4):485–500.
151. Okamoto Y, Tanaka YO, Nishida M et al. (2003) MR Imaging of the uterine cervix: Imaging-pathologic correlation. *RadioGraphics* 23(2):425–445.
152. Yamashita Y, Takahashi M, Katabuchi H et al. (1994) Adenoma malignum: MR appearances mimicking nabothian cysts. *AJR Am J Roentgenol* 162(3):649–650.
153. Wang SS, Sherman ME, Silverberg SG et al. (2006) Pathological characteristics of cervical adenocarcinoma in a multi-center US-based study. *Gynecol Oncol* 103(2):541–546.
154. Pecorelli S (2009) Revised FIGO staging for carcinoma of the cervix. *Int J Gynecol Obstet*. 105(2):107–108.
155. Sironi S, Belloni C, Taccagni GL et al. (1991) Carcinoma of the cervix: Value of MR in detecting parametrial involvement. *AJR Am J Roentgenol* 156(4):753–756.
156. Seki H, Azumi R, Kimura M et al. (1997) Stromal invasion by carcinoma of the cervix: Assessment with dynamic MR imaging. *AJR Am J Roentgenol* 168(6):1579–1585.
157. Hricak H, Powell CB, Yu KK et al. (1996) Invasive cervical carcinoma: Role of MR imaging in pretreatment workup–cost minimization and diagnostic efficacy analysis. *Radiology* 198(2):403–409.
158. Yamashita Y, Takahashi M, Sawada T et al. (1992) Carcinoma of the cervix: Dynamic MR imaging. *Radiology* 182(3):643–648.
159. Charles-Edwards EM, Messiou C, Morgan V et al. (2008) Diffusion-weighted imaging in cervical cancer with an endovaginal technique: Potential value for improving tumor detection in stage Ia and Ib1 disease. *Radiology* 249(2):541–550.
160. Naganawa S, Sato C, Kumada H et al. (2005) Apparent diffusion coefficient in cervical cancer of the uterus: Comparison with the normal uterine cervix. *Eur Radiol* 15(1):71–78.
161. Singh N, Arif S (2004) Histopathologic parameters of prognosis in cervical cancer—A review. *Int J Gynecol Cancer* 14(5):741–750.
162. Selman TJ, Mann C, Zamora J et al. (2008) Diagnostic accuracy of tests for lymph node status in primary cervical cancer: A systematic review and meta-analysis. *CMAJ* 178(7):855–862.
163. Perez CA, Breaux S, Bedwinek JM et al. (1984) Radiation therapy alone in the treatment of carcinoma of the uterine cervix II analysis of complications. *Cancer* 54(2):235–246.
164. Hricak H, Swift PS, Campos Z et al. (1993) Irradiation of the cervix uteri: Value of unenhanced and contrast-enhanced MR imaging. *Radiology* 189(2):381–388.
165. Weber TM, Sostman HD, Spritzer CE et al. (1995) Cervical carcinoma: Determination of recurrent tumor extent versus radiation changes with MR imaging. *Radiology* 194(1):135–139.
166. Hwang JH, Oh MJ, Lee NW et al. (2009) Multiple vaginal müllerian cysts: A case report and review of literature. *Arch Gynecol Obstet* 280(1):137–139.
167. Eilber KS, Raz S (2003) Benign cystic lesions of the vagina: A literature review. *J Urol* 170(3):717–722.
168. Griffin N, Grant LA, Sala E (2008) Magnetic resonance imaging of vaginal and vulval pathology. *Eur Radiol* 18(6):1269–1280.
169. Shadbolt CL, Coakley FV, Qayyum A (2001) MRI of vaginal leiomyomas. *J Comput Assist Tomogr* 25(3):355–357.
170. Siegelman ES, Outwater EK, Banner MP et al. (1997) High resolution MR imaging of the vagina. *RadioGraphics* 17(5):1183–1203.
171. Chagpar A, Kanthan SC (2001) Vaginal metastasis of colon cancer. *Am Surg* 67(2):171.
172. Wiesner W, Ruehm SG, Bongartz G et al. (2001) Three-dimensional dynamic MR hysterosalpingography. *Eur Radiol* 11:1439–1444.
173. Sadowski EA, Ochsner JE, Riherd JM et al. (2008) MR hystrosalpingography with an angiographic time-resolved 3D pulse sequence: Assessment of tubal patency. *AJR Am J Roentgenol* 191(5):1381–1385.

174. Outwater EK, Siegelman ES, Chiowanich P et al. (1998) Dilated fallopian tubes: MR imaging characteristics. *Radiology* 208(2):463–469.
175. Kalampokas E, Kalampokas T, Tourountous I et al. (2013) Primary fallopian tube carcinoma. *Eur J Obstet Gynecol Reprod Biol* 169(2):155–161.
176. Shaaban AM, Rezvani M (2012) Imaging of primary fallopian tube carcinoma. *Abdom Imaging* 38(3):608–618.
177. Nik NN, Vang R, Shih IM, Kurman RJ (2014) The origin and pathogenesis of pelvic (ovarian, tubal and primary peritoneal) serous carcinoma. *Annu Rev Pathol Mech Dis* 9:27–45.
178. Stewart CJ, Leung YC, Whitehouse A (2012) Fallopian tube metastases of non-gynaecological origin: A series of 20 cases emphasizing patterns of involvement including intra-epithelial spread. *Histopathology* 60(6B):E106–E114.
179. De Waal YR, Thomas CM, Oei AL (2009) Secondary ovarian malignancies: Frequency, origin, and characteristics. *Int J Gynecol Cancer* 19(7):1160–1165.

# 5

# *Benign and Malignant Conditions of the Ovaries and Peritoneum*

**Stavroula Kyriazi and Nandita M. deSouza**

**CONTENTS**

## 5.1 Normal Anatomy

### 5.1.1 Ovaries

The ovaries are paired pelvic organs, each measuring on average 3 × 2 × 2 cm. They are attached to the cornua of the uterus by a condensation of fibrous tissue called the *ligament of the ovary (utero-ovarian ligament).* This is the remnant of the gubernaculum, which plays a role in the descent of the ovary during embryogenesis, and is the homologue of the gubernaculum testis in the male. The *suspensory ligament of the ovary* runs from the pelvic side wall to the ovary. The fallopian tubes extend from the angle of the uterus laterally and end in the expanded fimbriated portion called the *uterine infundibulum* (funnel). They lie in close proximity to the ovaries. The *mesovarium* is a double fold of peritoneum that is attached to the back of the ovary and carries the neurovascular bundle to the ovary across the top of the suspensory ligament. The ureters, gonadal vessels, and bifurcation of the iliac vessels lie deep into the ovaries (Figure 5.1).

The ovarian arteries are branches of the abdominal aorta and arise from just below the level of the renal artery running retroperitoneally deep into the right and left colic vessels to enter the pelvis anterior to the ureter. They enter the ovaries from the lateral aspect of the broad ligament. These arteries also supply the fallopian tubes and anastomose with the uterine artery. The venous drainage of the ovaries via a pair of ovarian veins has different courses on the right and left: the right ovarian vein drains into the inferior vena cava, while the left drains into the left renal vein. They are accompanied by a plexus of lymphatic vessels that drain into the paraaortic nodes alongside the origin of the ovarian arteries at the level of the umbilicus.

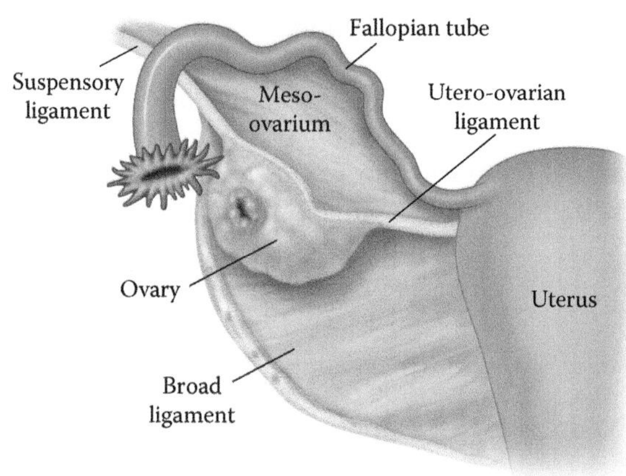


**FIGURE 5.1**
Schematic of adnexal anatomy adapted from www.knowyourbody.net.

### 5.1.2 Microscopy

Histologically, the ovaries consist of an outer cortex and an inner medulla. They are covered by a thin layer of cuboidal epithelium (germinal epithelium). The cortex is made of tightly packed connective tissue and contains the follicles and oocytes at various stages of development and degeneration. At birth, each ovary contains about three million ova derived from endodermal cells from the yolk sac. Each ovum is surrounded by a single layer of follicular or thecal (also known as *granulosa*) cells. The stroma around the follicular epithelium is composed of an inner theca interna and an outer theca externa. A basement membrane separates the follicular cells from the theca interna; no distinct boundary exists between the thecae interna and externa. The theca externa is composed of spindle-shaped cells and is more fibrous. Both thecae are connective tissue derivatives. The spindle-shaped fibroblasts of the cortical connective tissue (unlike other fibroblasts) respond to hormonal stimulation. The medulla consists of loose stromal tissue and is highly vascularized. At the hilum, urothelial-type epithelium may also be present where round or polygonal transitional cells exist in nests (Walthard nests); these may also be seen in the mesovarium and the mesosalpinx.

### 5.1.3 Peritoneal Reflections

The peritoneal reflections of the pelvis are important in understanding patterns of tumor spread from ovarian cancer. The parietal peritoneum lining the internal surface of the anterior abdominal wall is continuous with that lining the viscera—the visceral peritoneum. The ascending and descending colon lie retroperitoneally, being covered only on their anterior aspect by visceral peritoneum. Much of the sigmoid colon lies free within a mesentery attached to the left side of the pelvic brim called the *sigmoid mesocolon*. Laterally, the peritoneum extends to the pelvic wall. Running within the superior free edge of this peritoneal reflection are the fallopian tubes and laterally the ovaries. This reflection is the *broad ligament of the uterus* (Figure 5.1). Inferiorly, the peritoneal reflection covers the top of the rectum and the back of the uterus. The peritoneally lined cul-de-sac between anterior rectum and posterior uterus deep within the pelvis is called the *rectouterine pouch* (or pouch of Douglas). Anterior to this, the visceral peritoneum covers the fundus of the uterus and dome of the urinary bladder. The anterior cul-de-sac between anterior uterus and posterior bladder is shallower than the posterior one and is called the *uterovesical pouch*. These two pouches are the most dependent portions of the peritoneal cavity and therefore are prone to harboring free fluid and peritoneal deposits of tumor.

### 5.1.4 Hormonal Cycling

From menarche to menopause, the normal ovaries (Figure 5.2) are under the influence of pituitary gonadotrophins resulting in the maturation of a dormant follicle within one ovary and the production of one ovum usually every 28 days (Figure 5.3). When the ovum is nearly mature, a surge in pituitary lutenizing hormone weakens the wall of the follicle and facilitates the release of the ovum into the fallopian tube, a process known as *ovulation*. The empty follicle persists as

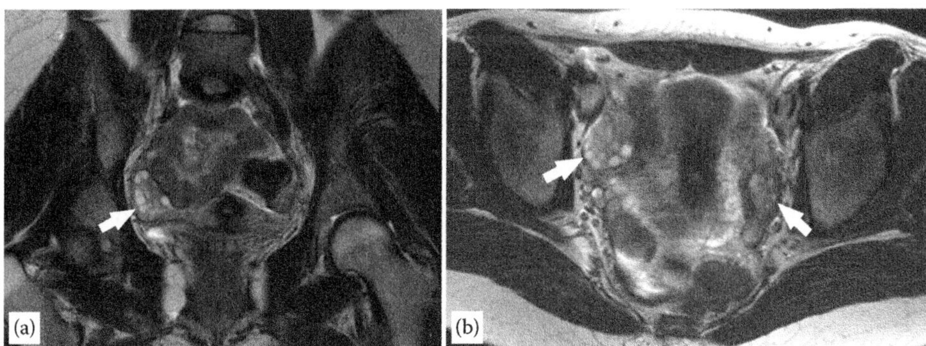


**FIGURE 5.2**
Normal ovaries: $T_2$-weighted coronal (a) and transverse (b) images showing multiple follicles within both ovaries (arrows). Only the right ovary is fully visualized on the slice illustrated in (a).

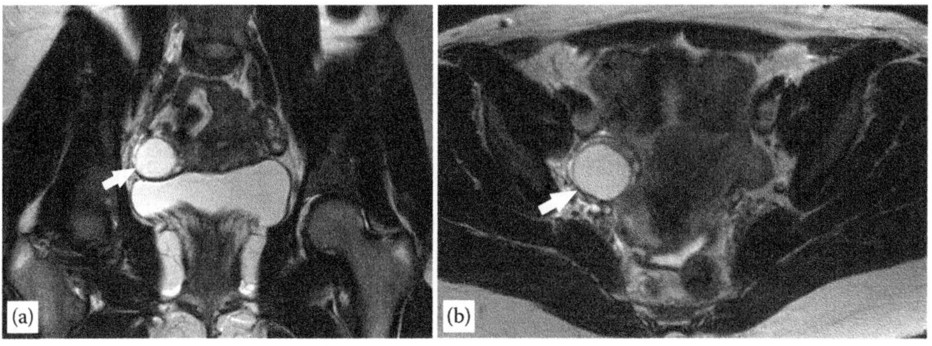


**FIGURE 5.3**
Normal ovaries: $T_2$-weighted coronal (a) and transverse images (b) showing enlarged normal follicle in the right ovary at day 14 of the menstrual cycle prior to ovulation (arrows).

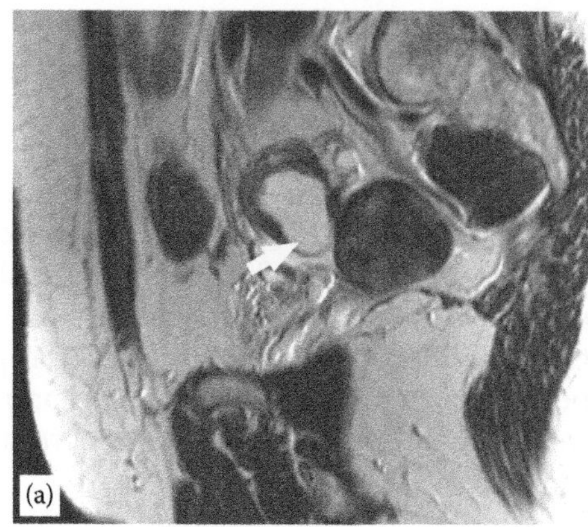


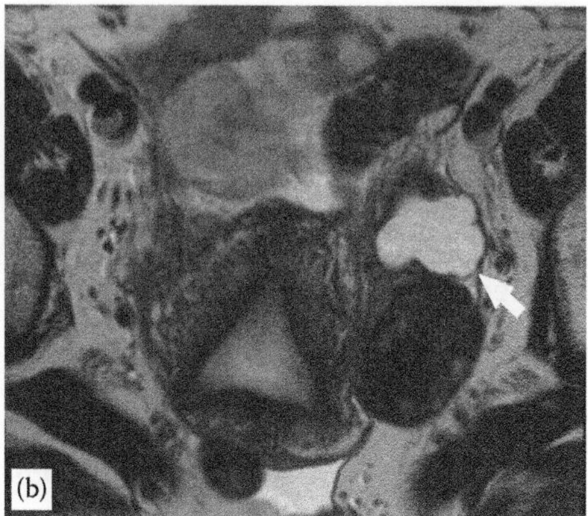


**FIGURE 5.4**
Normal ovaries: $T_2$-weighted sagittal (a) and transverse images (b) showing collapsed corpus luteal cyst in the left ovary prior to menstruation (arrows).

a corpus luteal cyst. Even in the absence of fertilization, these cysts may persist for up to 12 weeks growing to large sizes (up to 10 cm) [1] and give rise to concern on imaging. Longitudinal studies following their course and disappearance usually resolve the issue (Figure 5.4).

### 5.1.5 Polycystic Ovaries

Polycystic ovaries may be recognized in 5%–10% of reproductive age-group females. The diagnosis of the associated syndrome is primarily made on clinical (menstrual irregularity in 80%, hirsutism in 62%, and coexisting obesity in 31% [2]) and biochemical (luteinizing hormone: follicle-stimulating hormone [LH:FSH] ratio exceeding 2:1 and elevated serum androgen level) grounds. In patients (<40 years) diagnosed with endometrial cancer, as many as 25% may have polycystic ovaries [3]. Bilaterally enlarged ovaries (2–5 times normal) with presence of multiple (>10–12), small (<10 mm), peripheral cysts on ultrasound or MRI support the clinical/biochemical diagnosis. Increased stromal echogenicity has also been postulated in two-thirds of cases. The primary use of imaging in suspected polycystic ovary syndrome is to exclude functional ovarian neoplasms such as sex cord-stromal tumors (SCSTs) and Sertoli–Leydig cell tumors (SLCTs), which may be associated with hyperandrogenism [2].

## 5.2 Benign Non-Neoplastic Conditions of the Ovaries

### 5.2.1 Dysfunctional Ovarian Cysts

Follicle cysts of the ovary are the most common cystic structures found in healthy ovaries. These cysts arise from temporary pathologic variations of a normal physiologic process and are not neoplastic. They result from either failure of a dominant mature follicle to rupture or failure of an immature follicle to undergo the normal process of atresia. Many follicle cysts lose the ability to produce estrogen; in other instances, the granulosa cells remain productive, with prolonged secretion of estrogen. Solitary follicle cysts are common and occur during all stages of life, from the fetal stage to the postmenopausal period. Follicle cysts are lined with an inner layer of granulosa cells and an outer layer of theca interna cells. The cysts are thin walled and unilocular, usually ranging from several millimeters to 8 cm in diameter (average, 2 cm). Usually, cysts with dimensions less than 2.5 cm are classified as follicles and therefore are not of clinical significance.

Corpus luteal cysts (Figure 5.3) are less prevalent than follicular cysts. They mainly result from intracystic hemorrhage and may be seen in the second half of the menstrual cycle. They are hormonally inactive but may tend to rupture, when they cause irritation from intraperitoneal blood products. Both follicle and corpus luteal cysts are easily recognized in women of reproductive age on transvaginal sonography as they are usually simple anechoic structures that evolve with time and eventually disappear. They do not usually merit investigation with MRI.

### 5.2.2 Inflammatory Conditions

Inflammatory ovarian pathologies rely on clinical and laboratory as well as imaging findings to achieve correct diagnoses. Acute oophoritis arises from *Neisseria gonorrhoeae* or *Chlamydia trachomatis* ascending infections or may involve multiple organisms. On imaging, the ovaries are enlarged and may show variable enhancement with soft tissue stranding and infiltration of pelvic floor fascial planes, thickening of the uterosacral ligaments, and thickened/dilated fallopian tubes. If prolonged, tubo-ovarian abcesses result, which are recognized as cystic pelvic masses in a febrile, unwell patient. They are primarily treated with antibiotics with or without surgical drainage. Chronic pelvic inflammatory disease, complicated by bacterial infection, may rarely result in xanthogranulomatous oophoritis, and when enlarged, floridly enhancing ovaries are recognized on imaging in a patient with pyrexia and pelvic pain [4]. Rarely, tuberculosis may affect the ovaries and peritoneum and may mimic a disseminated ovarian cancer. Imaging alone is unable to differentiate these entities, and biopsy is required to make the diagnosis. Another rare condition is inflammatory pseudotumor, which clinically manifests with pyrexia, weight loss, and pelvic pain. Multiloculated cystic and solid lesions, which show contrast enhancement, are seen on imaging, which are indistinguishable from ovarian neoplasms. Diagnosis requires histological evaluation. Autoimmune inflammatory conditions of the ovary are also rare but may be suspected when bilateral enlarged multicystic ovaries on imaging are clinically accompanied by ovarian failure and/or autoimmune conditions of other organs (thyroid and adrenals).

## 5.3 Benign Conditions Involving Peritoneum

### 5.3.1 Endometriosis

*Endometriosis,* defined as the presence of endometrial glands and stroma outside the endometrium, is thought to affect ~10% of women of reproductive age with a much higher prevalence in infertile women (25%–40%, [5]). Extrauterine sites are commonly ovarian (endometrioma) or peritoneal/subperitoneal, which may be superficial or deep. However, deposits have also been seen within scar tissue and even in metastatic sites such as the lungs so that suggested theories of the histological origin of endometriosis have included not only retrograde menstruation but also lymphatic or hematogenous spread and metaplasia. The ovarian steroids, estrogen and progesterone, are implicated in the development of endometriosis, as evidenced by the relationship of disease activity to uninterrupted menstrual cycles, onset following menarche and remittance of disease at menopause and the benefit of medical therapy that suppresses ovulation [6].

Pelvic pain and infertility are disabling consequences of endometriosis. Moderate to severe endometriosis can cause tubal damage; lesser degrees of endometriosis, even in the absence of any obvious tubal damage, are also associated with subfertility and increased risk of ectopic pregnancy. More recently, endometriosis has been associated with a significantly increased risk of clear-cell, low-grade serous and endometrioid invasive ovarian cancers [7]. The American Society of Reproductive Medicine classifies endometriosis into four grades of severity based on the presence of lesions and adhesions but it does not take into account the location of lesions nor their depth of infiltration, features that are better identified on imaging. Routine use of imaging to detect the presence of endometriosis and assess its distribution and response to medical therapy is therefore an invaluable clinical tool.

#### 5.3.1.1 Ovarian Endometriosis

Endometriomas (cystic ovarian endometriosis) are recognized as *chocolate cysts.* They are mostly 3–4 cm in diameter but can grow as large as 15 cm and give rise to adhesions with the pelvic side wall and the other pelvic organs. Their primarily cystic nature is evident on imaging, although uncommonly solid elements also may be present within an endometrioma (Figure 5.5). Thus, although mural nodules are considered to be the most important hallmark in the recognition of ovarian cancers that accompany endometriotic cysts, they may be present in the absence of a malignancy [8]. Medical therapy for endometriomas is ineffective, and surgical excision is the preferred management.

#### 5.3.1.2 Peritoneal (Superficial) Endometriosis

Peritoneal endometriotic deposits are often subcentimeter (1–3 mm) and visualized as white, red, or flame-shaped vesicles at laparoscopy. More typically, puckered

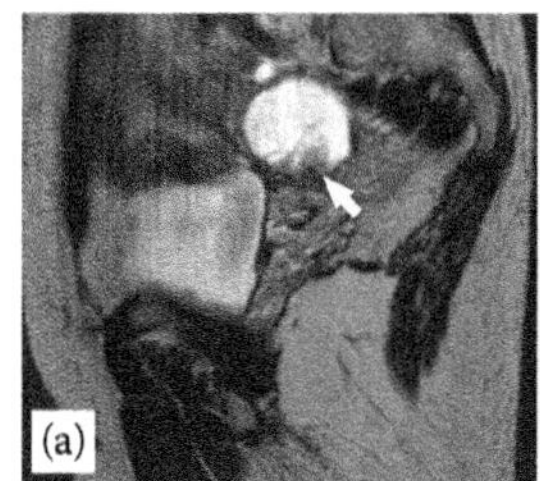

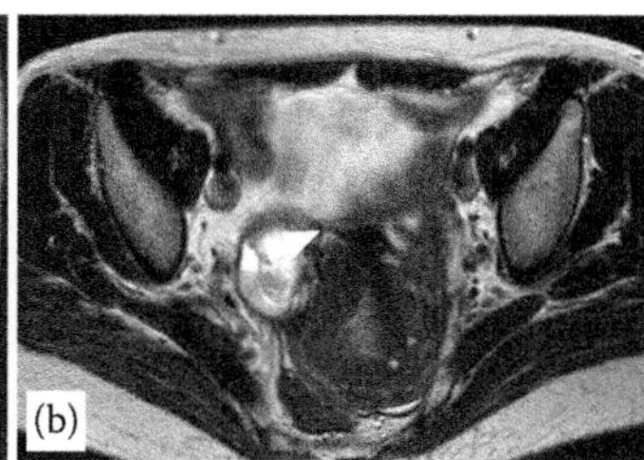

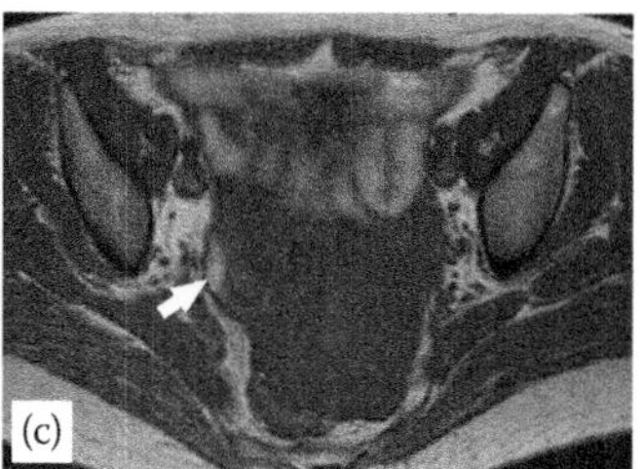


**FIGURE 5.5**
Endometrioma: $T_2$-weighted sagittal (a) and transverse (b) and $T_1$W weighted transverse (c) images showing a complex right adnexal cyst (arrows) in a 30-year-old female. The high signal on $T_1$-weighting indicates the presence of hemorrhage. Laparoscopy confirmed endometriosis.

hemorrhagic lesions generally in a white sclerotic area of scarring may be present. Peritoneal endometriotic deposits are found along any peritoneal surfaces in the abdomen and pelvis from under surface of the diaphragm to pouch of Douglas. In isolation, their clinical relevance is related to the extent of associated scarring and adhesions. Laparoscopy and biopsy remain the gold standard for diagnosis as they are often below the spatial resolution of current imaging techniques. However, they are often associated with deep pelvic endometriosis, where imaging techniques are increasingly invaluable.

#### 5.3.1.3 Subperitoneal and Deep Pelvic Endometriosis

The most common site of subperitoneal endometriosis is in the posterior pelvic compartment (uterosacral ligament or pouch of Douglas), which may be involved in more than a half of cases. The medial portion of the uterosacral ligaments is most frequently involved adjacent to the cervix. From here direct extension to the rectum and the lateral vaginal fornices is common [9,10]. Diagnosis of uterosacral ligament involvement with endometriosis is based on its thickness >9 mm [11] and/or irregularity [12,13]. Pouch of Douglas lesions include the rectovaginal septum in 10%, posterior vaginal wall fornices in 65%, and posterior vaginal wall and anterior rectal wall in 25%. Anterior pelvic compartment endometriosis affects the bladder and ureters and the round ligament [14].

#### 5.3.1.4 Role of Imaging in Endometriosis

The value of MRI in preoperative diagnosis of endometriosis is dependent on lesion size and location. Cystic ovarian endometriomas are hyperintense on both $T_1$W and $T_2$W MRI (Figure 5.5), often with fluid levels resulting from their hemorrhagic components [15]. A cystic corpus luteum can mimic an ovarian endometrioma. Even in the absence of malignancy, solid elements within an endometrioma may show enhancement [8].

Subcentimeter peritoneal lesions require laparoscopy and biopsy for diagnosis: transabdominal ultrasound lacks the spatial resolution, and computerized tomography has insufficient contrast to distinguish these small nodules. $T_2$W MRI is the best imaging option in the pelvis, because of its superior contrast resolution, but fails to detect lesions of <1 cm with any reliability. The inclusion of $T_1$W fat-suppressed sequences is recommended because they facilitate the detection of small endometriomas and aid in their differentiation from mature cystic teratomas [16]. In a prospective study of 152 women, Kruger et al. showed that the highest diagnostic accuracy was found in bladder endometriosis and the lowest in peritoneal endometriosis (sensitivity and specificity for pouch of Douglas: 87.6% and 84.6%; vagina: 81.4% and 81.7%; rectosigmoid colon: 80.2% and 77.5%; USL: 77.5% and 68.2%; ovaries: 86.3% and 73.6%; urinary bladder: 81.0% and 94.7%; peritoneum: 35.3% and 88.1%) [17]. Lesion location also affects interobserver agreement in identifying endometriotic lesions: of 157 lesions in 71 cases, MRI was found to be excellent for the ovaries (kappa = 0.8) and good for the rectovaginal septum (kappa = 0.78) and vaginal fornix (kappa = 0.72), but poor for the uterosacral ligament (kappa = 0.58) [18]. Other studies have confirmed the high accuracy of MRI (83%–100%) in identifying bladder endometriosis [12,18,19].

Opacification of the vagina and rectum has been employed to improve the sensitivity of MRI for the detection of deep pelvic endometriosis by expanding the vagina and rectum and allowing better delineation of the pelvic organs. A significantly improved diagnostic accuracy has been reported for three separate observers in one study, which was especially apparent for lesions localized to the vagina and rectovaginal septum [20]. However, there is no widespread acceptance of this technique as yet. Endometriotic lesions usually enhance after intravenous gadolinium contrast injection, and an improvement in diagnostic accuracy by adding extrinsic contrast agents was initially demonstrated [21]. However, contrast-enhanced imaging has been shown to be unreliable in differentiating infiltrating lesions from other normal fibromuscular pelvic anatomic structures [22] so that it is not routinely used. Techniques that rely on intrinsic contrast properties of tissue such as diffusion-weighted MRI also have been used to characterize endometriosis. An early report indicated that diffusion-weighted MRI showed potential to evaluate cystic contents of ovarian lesions [23], while more recently, a consistently low value for the apparent diffusion coefficient (ADC) of deep pelvic endometriosis has been reported ($0.7–0.79 \times 10^{-3}$ mm$^2$/s) [24]. This has been extended to aid differentiation between infiltrating endometriosis and bowel cancer, although only pilot data indicating its potential are available to date [25].

## 5.4 Neoplastic Conditions

Tumors of the ovary can be placed in three major categories, surface epithelial-stromal, sex cord-stromal, and germ cell, according to the cells from which the tumors originate. Each category contains a mixture of subtypes. Combinations of subtypes, either intimately intermixed or existing side by side within a single tumor, are common. The ovarian surface epithelium is similar to peritoneal mesothelium and is the largest group of ovarian tumors.

**TABLE 5.1**

The World Health Organization Histological Classification

| | | |
|---|---|---|
| Surface epithelial-stromal | Serous | Benign, borderline, and malignant |
| | Mucinous | Benign, borderline, and malignant |
| | Endometroid | Benign, borderline, and malignant |
| | Clear cell | Benign, borderline, and malignant |
| | Transitional cell | Brenner |
| | | Non-Brenner (transitional) |
| | Epithelial-stromal | Adenosarcoma and carcinosarcoma |
| Sex cord-stromal | Granulosa tumors | Fibromas |
| | | Fibrothecomas |
| | | Thecomas |
| | | Malignant granulosa cell |
| | Sertoli | Leydig |
| | Sex cord with annular tubules | |
| | Gynandroblastoma | |
| | Steroid (lipid) cell | |
| Germ cell | Teratoma | Immature |
| | | Mature |
| | Monodermal | |
| | Dysgerminoma | |
| | Yolk sac | |
| | Mixed | |
| Malignant, not otherwise specified | Metastatic cancer from non-ovarian primary | Colon |
| | | Appendix |
| | | Gastric |
| | | Breast |

*Note:* The World Health Organization Histological Classification for ovarian tumors separates ovarian neoplasms by the tissue of origin into surface epithelial (65%), germ cell (15%), sex cord-stromal (10%), metastases (5%), and non-ovarian primary (5%). Surface epithelial tumors are further classified by cell type (serous, mucinous, and endometrioid) and cellular atypia (benign, borderline, and malignant). 90% of malignant tumors are surface epithelial.

The sex cord-stromal group includes tumors of mesenchymal or mesonephric origin, namely, fibromas, thecomas, and granulosa cell tumors, while teratomas both mature and immature and embryonal and yolk-sac carcinomas are of germ cell origin. The WHO classification of ovarian tumors by histological subtype is given in Table 5.1.

### 5.4.1 Benign Cystic Neoplasms (Epithelial-Stromal)

The vast majority of epithelial ovarian tumors are either serous or mucinous and can be either benign (cystadenomas) or malignant (cystadenocarcinomas). They are derived from the layer of cuboidal epithelium covering the ovary, although recently their origin from the epithelial lining of the fimbrial end of the fallopian tube is believed to be more likely. Differentiation between these four subtypes on imaging is not always straightforward but certain characteristics are useful in distinguishing them.

#### *5.4.1.1 Serous Cystadenomas*

Serous cystadenomas tend to be unilocular or multilocular cystic lesions with homogeneous cystic fluid reflected by a homogeneous signal intensity on MRI (Figure 5.6). They may demonstrate a thin regular wall or septum with no soft tissue vegetations.

#### *5.4.1.2 Mucinous Cystadenomas*

*Mucinous cystadenomas* usually present as multiloculated cystic masses and tend to be larger on presentation than serous cystadenomas. Like the former, they have a thin regular wall or a few septae but contain cystic fluid within locules that may differ in signal intensity. Though said to typically contain mucin with its distinctive imaging characteristics, contents of these cysts may include hemorrhage and differing protein/mucin content. As with their serous counterparts, they contain no soft tissue vegetations.

### 5.4.2 Benign Mixed (Cystic-Solid) Neoplasms

These are commonly a mixture of subtypes and are usually malignant. Benign lesions include tumors of transitional cell origin (Brenner's tumors) or a combination of epithelial and sex cord-derived lesions (cystadenofibromas) and mature germ tumors (mature teratoma).

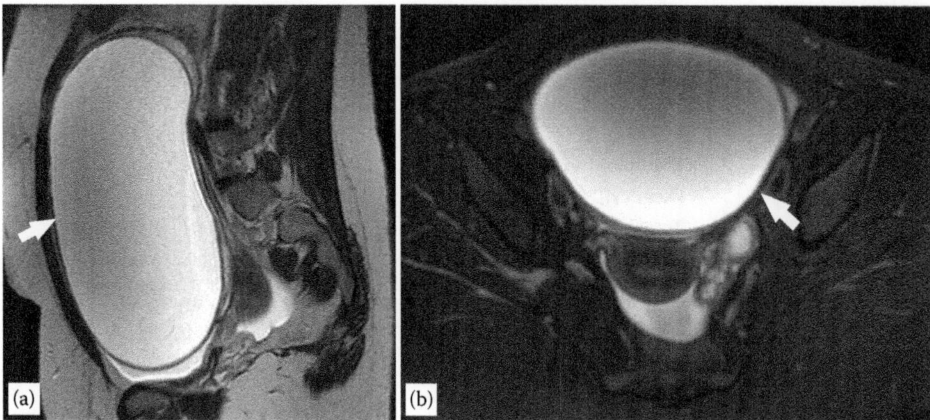


**FIGURE 5.6**
Serous cystadenoma: $T_2$-weighted sagittal (a) and fat-saturated transverse (b) images showing a large cystic mass centrally in the pelvis and lower abdomen (arrow) that appeared to arise from the left ovary. The mass is a simple cyst without any complex features indicating it is likely to be a cystadenoma. This was confirmed at surgical histology.

#### *5.4.2.1 Brenner's Tumors*

Brenner's tumors of the ovary [26] were first described in detail by Fritz Brenner in 1907 [27]. Between 1.1% and 2.5% of all ovarian tumors are thought to be Brenner's tumors; about 99% of them are benign [28–31]. Bilateral lesions are found in 5%–14% of cases [26,30]. Almost all Brenner's tumors are asymptomatic and discovered by chance, for example, at the time of imaging for an unrelated indication or at surgery. In one study of 29 tumors, 71% contained solid components, 50% being purely solid, 17% multilocular-solid, and 4% unilocular-solid. If cystic fluid was present, the echogenicity of the cyst content was most often anechoic or of low echogenicity; papillary projections were found in one benign tumor and irregular internal walls in three, while ascites and fluid in the pouch of Douglas were rare [32]. Brenner's tumors with calcifications have been reported [33] as has prominent vascularization [34], but unfortunately there are no characteristic imaging features (Figure 5.7). Around 1% of Brenner's tumors are malignant. A report of a case with transition of benign to malignant components revealed dense calcifications on computed tomography (CT) in the benign component with very low intensity on $T_2$W images, whereas the malignant component showed high intensity [35].

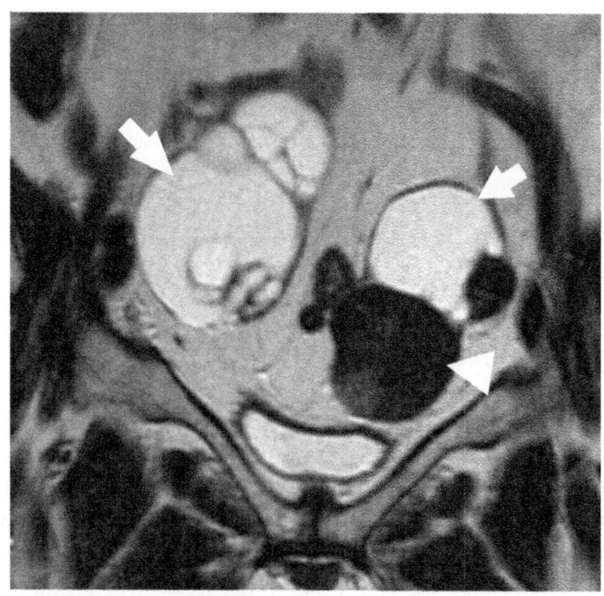

**FIGURE 5.7**
Brenner's tumor with cystadenoma: $T_2$-weighted coronal image showing a complex cystic (arrow) and solid (arrowhead) mass in the left adnexa consistent with a Brenner's tumor. The cystic mass in the right adnexa contained a few thin septations only and no solid components; the left sided mass was confirmed as a Brenner's tumor, the right-sided mass proved to be a cystadenoma.

#### *5.4.2.2 Cystadenofibromas*

These are complex solid masses with cystic, septated components. The largest series characterizing 47 histologically proven lesions on MRI describes a low-signal intensity solid component in 75%, septated cysts in 74%, and 1 lesion as purely cystic. The solid component and the septae show enhancement, but this is at a much lower rate than that of the myometrium [36].

#### *5.4.2.3 Mature Teratoma*

Mature teratomas (dermoids) are the commonest benign ovarian tumor in women under 45 years of age. Overall, they are responsible for up to 25% of all ovarian neoplasms and two-thirds of pediatric ovarian tumors [37]. They are bilateral in up to 25% of cases and present with an abdominal mass in 67% or pelvic pain due to torsion and hemorrhage [38].

Teratomas consist of cells with the potential to differentiate along one or all of the three germ cell lines—that is, endoderm, mesoderm, and ectoderm: mature teratomas, which comprise the vast majority, are derived from these partly or completely differentiated cell lines leading to the production of tissue with a wide phenotype ranging from primitive embryonal cells to mature adult tissue. They produce predominantly cystic lesions with a wide range of appearances on any imaging modality reflecting the potential tissue lines within them. In addition to purely cystic masses, mixed masses with components from any or all of the three germ cell layers or a solid mass partly or wholly composed of fat may be seen (Figure 5.8). In the most common case of a large unilocular cyst, they are filled with sebaceous material and lined by squamous epithelium. This often forms a protuberance into the cyst chamber called the *Rokitansky nodule.* Epithelium-derived structures, such as hair, bone, or even teeth, when present within the cyst, usually arise from or in this nodule [39].

On ultrasound, this spectrum of tissue differentiation gives rise to a typical range of appearances. A cystic lesion appears as a typical anechoic mass often with a densely echogenic focus within its wall (the Rokitansky nodule). It may also appear as a diffusely or partially echogenic mass showing sound attenuation due to the sebaceous content. Occasionally, multiple thin echogenic bands are seen within the cyst cavity due to hair within the cyst. At CT, the finding of fat within an adnexal cyst is diagnostic of a mature cystic teratoma; occasionally, bone or teeth are identified and are particularly well seen on CT acquisition. On MRI, the key imaging findings are the presence of tissue returning signal similar to fat often with fat-fluid levels. Thus, the presence of very high intracystic signal on $T_1$W imaging and intermediate signal on $T_2$W imaging, mirroring the signal characteristics of retroperitoneal fat, is diagnostic. The presence of foci of very low signal on both $T_1$W and $T_2$W imaging reflects the presence of intracystic bone or even teeth. In the case of monodermal teratomas,

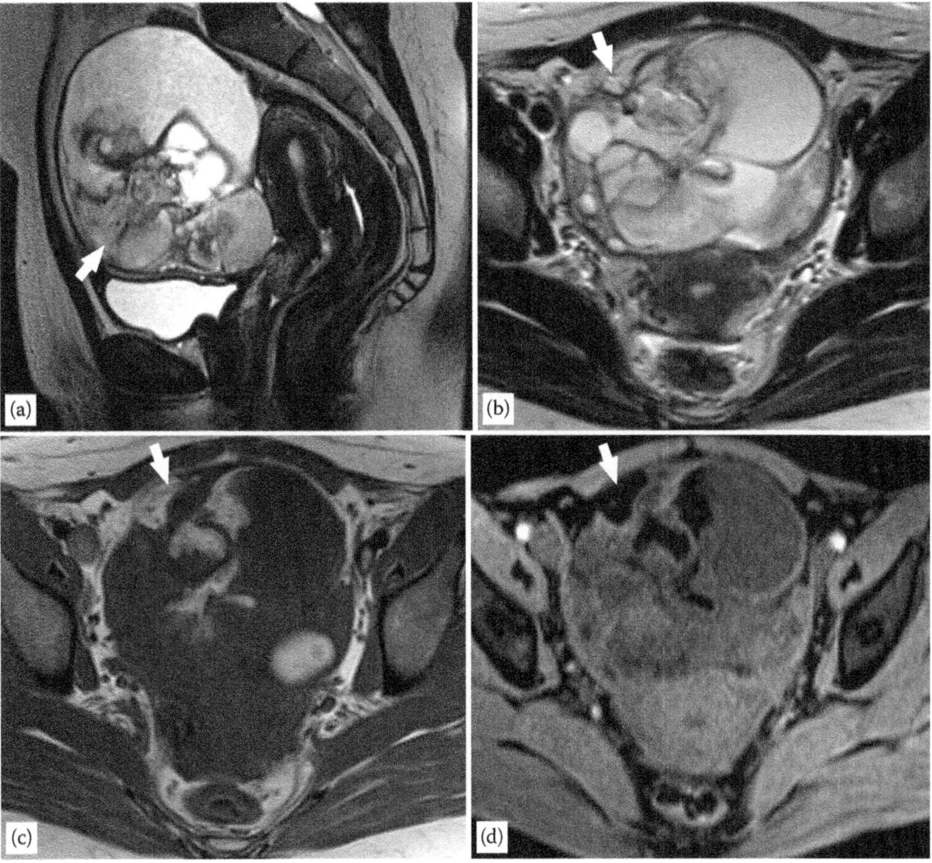


**FIGURE 5.8**
Mature teratoma: $T_2$-weighted sagittal (a) and transverse (b) and $T_1$-weighted transverse without (c) and with fat-saturation (d) images showing a large complex right adnexal mass (arrows). The remnant of the normal right ovary is noted laterally in (b) (short arrow). High signal intensity in (c) (arrow) is saturated in (d) (arrow), indicating the presence of fat which is pathognemonic of a mature teratoma (dermoid cyst).

imaging characteristics depend on the tissue produced within the cyst. These less common benign teratomas often contain mature thyroid tissue (struma ovarii) and as such appear solid or partly solid.

Occasionally, mature cystic teratomas may present with complications such as torsion, hemorrhage, or malignant transformation [40]. If they rupture, spillage of sebaceous material into the peritoneum has been known to cause a chemical granulomatous peritonitis [41].

### 5.4.3 Benign Solid Neoplasms

These lesions are derived from sex cord (stromal) or germ cell elements, the latter being by far the commoner. Sex cord-derived tumors with significant fibrous component are generally of low signal intensity on $T_2$W images and difficult to distinguish from more commonly found leiomyomas, both uterine and broad ligament, especially when they are large because determining their site of origin as ovarian or extraovarian is not always possible on imaging.

#### 5.4.3.1 Ovarian Fibromas

Ovarian fibromas are rare, constituting less than 3% of ovarian tumors. They occur primarily in perimenopausal and postmenopausal patients. They are occasionally bilateral (<10%) and are typically found incidentally [42]. When symptomatic, they most commonly present with abdominal pain, fullness, or discomfort [43]. Fibromas arise from mesenchymal spindle cells from the sex cord, which produce collagen and can be associated with Meig syndrome (ascites, ovarian tumor, and pleural effusion). They are hormonally inactive as they contain no thecal cells. On imaging, they are predominantly solid, but may be partly cystic. MRI features show hypo- or isointensity on $T_1$W, hypointensity on $T_2$W, and contrast enhancement that is variable [44–47]. Homogeneously low signal intensity solid masses on $T_2$W may make them indistinguishable from a leiomyoma. Larger tumors are often heterogeneous with hypo- and hyper-intensity on $T_2$W images because of stromal edema or cystic degeneration [48]. Hemorrhagic infarction also occurs and is recognized as hyper-intensity of the periphery of the mass on $T_1$W with heterogeneous mixed signal intensity centrally on $T_2$W [48,49]. Myxomatous change can also give rise to high signal intensity centrally on $T_2$W [50].

#### 5.4.3.2 Thecomas and Fibrothecomas

Thecomas are rare, solid sex cord-stromal ovarian tumors and account for approximately 0.5%–1% of primary ovarian lesions [51]. Thecomas are often mixed with fibrous components (then called *fibrothecomas*); thecomas and fibrothecomas are now considered to originate from the ovarian medulla, with a different etiology from pure fibromas, which originate from the cortex [52]. These unilateral ovarian tumors are most commonly found in postmenopausal women and are most frequently misdiagnosed as uterine or broad ligament leiomyomas [46]. As with fibromas, they are predominantly solid masses, homogeneously iso-intense on $T_1$W imaging, while appearances on $T_2$W imaging can be more variable, particularly when there is stromal edema or cystic degeneration. If the tumor has a significant fibroma component (i.e., fibrothecoma), the abundant fibrous tissue may produce predominantly low signal intensity on both $T_1$W and $T_2$W sequences (Figure 5.9). Following administration of gadolinium contrast agents, the degree of contrast enhancement varies with the amount of fibrous tissue within the tumor: while thecal cells in the normal ovary are highly vascularized, the fibrous tissue is known for delayed weak enhancement at dynamic contrast-enhanced imaging. MRI therefore can classify them as benign ovarian tumors, but differentiating them from pure fibromas or leiomyomas is unreliable. As with ovarian fibromas, ascites and pleural effusions may accompany these lesions (Meig's syndrome) [53]. Although benign, they can be hormonally

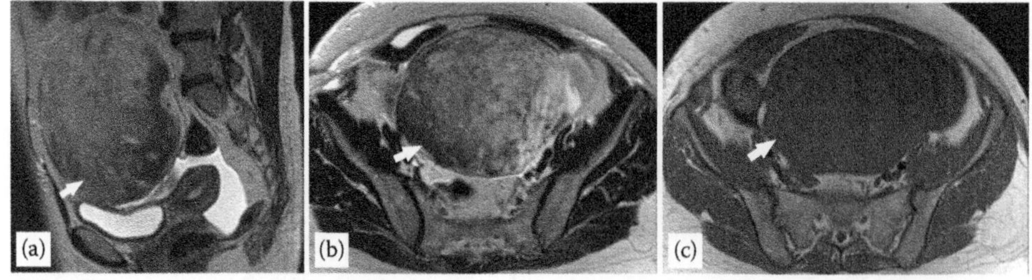


**FIGURE 5.9**
Fibrothecoma: $T_2$-weighted sagittal (a) and transverse (b) and $T_1$-weighted transverse images (c) without fat saturation show a large solid left ovarian mass (arrows) with some ascites in the pouch of Douglas. The mass is heterogeneous in signal intensity and does not contain any fat. The presence of ascites favors a fibroma, as do the very low signal intensity elements on $T_2$-weighting. The intermediate signal intensity components favor thecomatous elements. A fibrothecoma was confirmed at surgical histology.

active, containing thecal cells that produce estrogen. The presence of hyperestrogenic states points to the presence of thecal elements within the tumor and results in secondary features such as endometrial thickening; around 20% are therefore also associated with endometrial carcinoma [53]. A recent study has investigated the use of diffusion-weighted MRI in differentiating these benign ovarian masses; however, while they all showed significantly lower ADC than malignant masses, ADC was unable to distinguish between thecoma and other solid adnexal masses [54].

#### *5.4.3.3 Adnexal Leiomyomas*

Pedunculated uterine and broad ligament leiomyomas often appear as adnexal masses with very low signal intensity on MRI. Differentiating these masses from ovarian-derived masses is often difficult due to their close proximity to the ovary in question. These masses are supplied by uterine vessels, which course through the myometrium to supply them. These vessels are often seen intervening between the leiomyoma and the adjacent uterus. On the other hand, ovarian masses are supplied directly by ovarian vessels or by branches of the uterine artery to the ovary which run along the uterine tubes. This distinction may help in defining the organ of origin and so differentiating ovarian from uterine lesions [35,53].

## 5.5 Borderline Tumors of the Ovary

Borderline ovarian tumors describe proliferative ovarian epithelial tumors that are *intermediate* in their clinicopathologic features between clearly benign cystadenomas and unquestionably malignant cystadenocarcinomas. The absence of obvious stromal invasion is a principal diagnostic criterion. Papillary ovarian cystadenomas with "clinical features that stand on the border of malignancy" were first described by Pfannenstiel in 1898 [55], and the term *semi-malignant* was introduced in 1929 [56]. The term *tumor of low malignant potential* has also been used in a combined classification of the International Society of Gynecologic Pathologists and the WHO [57]. Borderline tumors of every surface epithelial cell type (serous, mucinous, endometrioid, clear cell, transitional cell, and mixed epithelial cell) have been reported, but the serous and mucinous types are the commonest. It is not possible to distinguish between the subtypes of borderline tumors on imaging. However, the key questions on imaging are the differentiation between borderline and frankly malignant tumors. In patients in whom the clinical features are suggestive of a borderline ovarian tumor (young age and normal or minimally elevated CA125), the ability to predict borderline disease based on morphological features observed on MRI is extremely helpful in surgical planning, particularly as it may be possible to offer partial ovarian preservation.

### 5.5.1 Serous Borderline Tumors

SBTs comprise around 10% of all serous neoplasms [58]. The age at presentation is much younger than in frankly malignant disease (mean age around 38 years) [59]. Unlike their malignant counterpart, the majority present as early stage disease (68% are stage I, 11% stage II, 21% stage III, and less than 1% stage IV by the FIGO staging system). Tumors are found in both ovaries in 40% of cases [60]. Papillary excrescences are found in around half of cases on histopathology, which would classify these tumors as stage Ic; however, as they are microscopic, they are not recognized on imaging, and differentiation from a benign cystadenoma may be difficult. The MRI findings in 26 patients with 31 borderline ovarian tumors were described in 4 morphological categories by Bent et al.: group 1, unilocular cysts (19%); group 2, minimally septate cysts with papillary projections (19%) (Figure 5.10); group 3, markedly septate

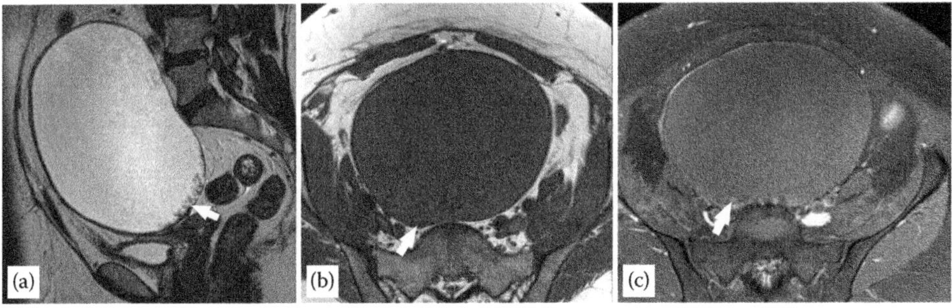


**FIGURE 5.10**
Borderline tumor: $T_2$-weighted sagittal (a), transverse $T_1$-weighted precontrast (b) and postcontrast (c) images showing a large adnexal cyst (arrows) in the center of the pelvis in a 32-year-old female that appeared to arise from the left ovary. Nodules are noted inferiorly in (a) (arrows), which enhance postcontrast in (c). These are relatively small and their limited extent favors a borderline tumor, which was confirmed at surgery.

lesions with plaque-like excrescences (45%); and group 4, predominantly solid with exophytic papillary projections (16%) [61]. A minority of tumors which form a micropapillary variant may be easier to distinguish from benign cystadenomas because of the presence of unusually prominent micropapillae. When compared to typical borderline tumors, micropapillary variants are more often bilateral and have a higher frequency of exophytic surface tumor, and a greater proportion are of advanced stage. Utilizing immunostaining for epithelial markers, microinvasion has been detected in 13% of SBTs [62]. The thickness of septations and the size of solid components were shown to be significantly smaller in borderline compared to serous ovarian cancer, so that these features may be helpful for predicting the likelihood of invasive tumors but neither feature allowed confident differentiation of SBTs from ovarian carcinomas [63]. In a small series of six serous surface borderline epithelial tumors, primarily solid masses were demonstrated with hyperintense papillary architecture and hypointense internal branching on $T_2$W MRI. Five patients had peritoneal implants, and two had lymph node enlargement, and all tumors were accompanied by ascites. In all cases, contralateral ovaries had cystic masses with mural nodules or mixed solid and cystic masses, of which the solid part was similar to the contralateral mass. No evidence of recurrence was noted at a follow-up of >12 months postoperatively [64].

In SBT, peritoneal implants on serosal and omental surfaces are found at the time of initial operation in 20%–46% of patients [59], and paraaortic and pelvic lymph node involvement is found in 7%–23% of cases with node sampling at the time of surgery [65]. It is in these cases that imaging is challenging, not only in accurate detection of these extrapelvic sites, where the volume of disease is small and often beyond the resolution of imaging, but also in differentiating noninvasive from invasive implants, which is currently not possible. Lymph node involvement found during staging procedures generally is of microscopic dimension and prognosis does not seem to be adversely affected. Invasive implants are very uncommon, occurring in only 4%–13% of patients with high-stage disease [59]. If an ovarian exophytic component is present, it is a strong indicator for extraovarian peritoneal disease, which could then be assiduously looked for on imaging. Almost two-thirds of patients whose ovarian tumor has an exophytic surface component have implants, and 94% of patients with implants have an exophytic surface component on their ovarian tumor [60]. Prognosis is excellent for patients with limited extent of tumor and surprisingly good even for those with extensive peritoneal disease. Unresected peritoneal implants often remain dormant, and some have apparently undergone spontaneous regression.

### 5.5.2 Borderline Mucinous Tumors

BMTs are much less common than SBTs and recognized in intestinal (commoner) and Mullerian (less common) types. They occur over a wide age range, but like SBTs occur mainly in the fourth decade of life. Typically, they produce large multicystic masses with smooth outer surfaces that resemble benign mucinous cystadenomas. Over 90% are unilateral. This is a key finding, because bilaterality of a cystic tumor, which shows mucinous histology, should suggest the possibility of a metastatic tumor to the ovaries (e.g., from the appendix or other gastrointestinal sites), rather than a primary ovarian neoplasm, and demands review of the gastrointestinal tract on imaging [66].

The solid component of *mucinous borderline tumors* is in the form of nodules [67], which are predominantly intracystic unlike the frond-like and often exophytic papillary projections of SBTs. Noninvasive mucinous carcinoma occurs in about 15%–55% of otherwise typical borderline mucinous tumors [68]. Almost all borderline mucinous tumors of intestinal type are stage I and have an excellent prognosis following surgical treatment with reported metastatic rates of 0%–3% for those without noninvasive carcinoma and 0%–7% for those with foci of noninvasive carcinoma [69].

## 5.6 Malignant Tumors of the Ovaries and Fallopian Tubes

### 5.6.1 Epidemiology

Ovarian cancer is the second most common gynecological malignancy after endometrial cancer in the industrialized world but accounts for more deaths than the remainder of gynecological cancers added together [70,71]. In the United States, 21,990 new cases and 15,460 deaths are estimated to occur annually [71]. Although a statistically significant rise in 5-year overall survival has been reported in the last decades (37% vs. 45% between 1975 and 2006, respectively), stratified survival according to stage ranges from 28% for overtly metastatic to 73% for regional and 94% for organ-confined disease [71]. Neoplasms of epithelial origin are responsible for 90% of malignant ovarian cancers, whereas other types include sex cord-stromal and germ cell tumors, metastases, and rarely lymphoma [72,73].

The histological subtypes of epithelial ovarian carcinomas are serous (68%–71%), clear cell (12%–13%), endometrioid (9%–11%), mucinous (3%), transitional (1%), undifferentiated (1%), and mixed (5%) [73,74]. Approximately 90% of epithelial ovarian carcinomas are sporadic, while the remainder are associated with high-penetrance (autosomal dominant) cancer susceptibility disorders, most importantly (1) hereditary breast and ovarian cancer syndrome, caused by mutations in *BRCA1/2* tumor suppressor genes; and (2) Lynch syndrome (hereditary non-polyposis colorectal cancer), caused by alterations in mismatch repair genes *MLH1*, *MSH2*, *MSH6*, and *PMS2* [75]. Average cumulative ovarian cancer risks by age 70 years have been reported as 39%–40% and 11%–18% for *BRCA1* and *BRCA2* carriers, respectively, whereas for breast cancer average risks have been reported as 57%–65% and 45%–47%, respectively [76,77].

### 5.6.2 MRI in Lesion Characterization

MRI has been established as a cost-effective modality in the characterization of sonographically indeterminate adnexal lesions due to its detailed depiction of their internal structure and its high specificity in differentiating benign pathology by recognizing the hallmarks of benignancy of fat, hemorrhage, and fibrosis [78,79]. In a meta-analysis, the accuracy of contrast-enhanced MRI ranged from 83% to 89% compared to 63% for grayscale with color Doppler ultrasonography [80]. Another meta-analysis revealed that in sonographically equivocal masses undergoing secondary imaging, the posttest probability for malignancy increased significantly more after contrast-enhanced MRI (premenopausal women, 80%; postmenopausal women, 95%) than after combined grayscale and Doppler ultrasonography (premenopausal women, 30%; postmenopausal women, 69%) or CT (premenopausal women, 38%; postmenopausal women, 76%) [81]. On imaging, serous carcinomas are typically seen as complex cystic masses with solid components in the form of septa, mural nodules, and papillary projections, and occasionally as predominantly solid masses. Papillary vegetations may be endo- or exophytic, are often edematous, and demonstrate high or intermediate $T_2$ signal intensity and strong contrast enhancement [82,83]. Gadolinium enhancement is helpful in distinguishing true papillary projections from non-enhancing debris or clots within a cystic lesion [84]. The principal morphological MRI criteria for diagnosing malignancy are large (>4 cm) lesion size (univariate odds ratio 5), wall or septal thickness >3mm (odds ratio 2–3), septal irregularity and nodularity, presence of vegetations in cystic lesion (odds ratio 30), and necrosis of the solid portion (odds ratio > 100). Ancillary findings associated with malignancy include ascites, peritoneal implants, and lymphadenopathy [85]. In a meta-analysis of 18 primary studies, MRI at 1.5 T achieved a pooled sensitivity of 0.92, specificity of 0.85, and AUROC of 0.95 for the differentiation of borderline and malignant from benign adnexal lesions [86].

Preliminary studies on functional quantitative MRI-based techniques, such as dynamic contrast-enhanced MRI (DCE-MRI) and diffusion-weighted imaging (DWI), have shown promise in incrementing the diagnostic accuracy of conventional MRI. Semiquantitative analysis of enhancement curves on DCE-MRI has been used to characterize ovarian lesions using myometrial enhancement as an internal reference through demonstration of earlier onset, more rapid progression, and increased intensity of contrast uptake in malignancy. Enhancement amplitude, maximal slope, and initial area under the curve have been found significantly higher in malignant lesions with a steep initial rise in the curve (type 3 curve) being specific for invasive tumors [87]. Other parameters, such as maximum enhancement, relative enhancement, and wash-in rate, have been also shown to be significantly higher in borderline/invasive tumors than in their benign counterparts, and use of a threshold significantly improves positive predictive value for malignancy over conventional MRI (86% vs. 62%, respectively) [88]. Preliminary quantitative DCE-MRI data using a two-compartment pharmacokinetic model showed significantly increased tissue blood flow, blood volume fraction, and AUC and decreased interstitial volume fraction in malignant lesions, with the highest discriminatory accuracy exhibited by tissue blood flow (AUROC, 0.86) [89]. However, the enhancement pattern of high-grade serous carcinomas has not been shown as discriminatory from other types of ovarian malignancy [90]. On DWI, the combination of low signal intensity in the high $b$ (1,000 s/mm$^2$) value images with $T_2$ hypointensity within the solid component of an adnexal mass has been found discriminatory of benignity; alternately, an intermediate $T_2$ and high $b_{1,000}$ signal intensity in the solid portion has been shown highly suggestive of malignancy [91]. The incremental value of DCE-MRI and DWI in complex adnexal lesion characterization prompted a correct change of diagnosis in 19%–24% of cases, with no incorrect changes [92]. A recently developed five-point MR scoring system incorporating morphological, perfusion-, and diffusion-based criteria achieved an AUROC of 0.94–0.98 in the differentiation of benign from malignant tumors with excellent interobserver agreement ($\kappa$ = 0.98) [93]. ADC mapping, however, has not been proven valuable as a diagnostic tool due to the broad overlap of mean and lowest ADC values between benign and malignant lesions [91,94].

### 5.6.3 MRI for Staging

Ovarian cancer is routinely staged according to the FIGO surgical–pathological system (Table 5.2) [95]. It spreads by a transcoelomic route (Figure 5.11). CT is the standard modality for preoperative staging and follow-up of ovarian cancer due to its wide availability, with a per-patient

**TABLE 5.2**

Carcinoma of the Ovary

| Stage | |
|---|---|
| I | Growth limited to the ovaries. |
| Ia | Growth limited to one ovary; no ascites present containing malignant cells. No tumor on the external surface; capsule intact. |
| Ib | Growth limited to both ovaries; no ascites present containing malignant cells. No tumor on the external surfaces; capsules intact. |
| Ic | Tumor either stage Ia or Ib, but with tumor on surface of one or both ovaries, or with capsule ruptured, or with ascites present containing malignant cells, or with positive peritoneal washings. |
| II | Growth involving one or both ovaries with pelvic extension. |
| IIa | Extension and/or metastases to the uterus and/or tubes. |
| IIb | Extension to other pelvic tissues. |
| IIc[a] | Tumor either stage IIa or IIb, but with tumor on surface of one or both ovaries, or with capsule(s) ruptured, or with ascites present containing malignant cells, or with positive peritoneal washings. |
| III | Tumor involving one or both ovaries with histologically confirmed peritoneal implants outside the pelvis and/or positive regional lymph nodes. Superficial liver metastases equals stage III. Tumor is limited to the true pelvis, but with histologically proven malignant extension to small bowel or omentum. |
| IIIa | Tumor grossly limited to the true pelvis, with negative nodes, but with histologically confirmed microscopic seeding of abdominal peritoneal surfaces, or histologically proven extension to small bowel or mesentery. |
| IIIb | Tumor of one or both ovaries with histologically confirmed implants, peritoneal metastasis of abdominal peritoneal surfaces, none exceeding 2 cm in diameter; nodes are negative. |
| IIIc | Peritoneal metastasis beyond the pelvis >2 cm in diameter and/or positive regional lymph nodes. |
| IV | Growth involving one or both ovaries with distant metastases. If pleural effusion is present, there must be positive cytology to allot a case to stage IV. Parenchymal liver metastasis equals stage IV. |

*Source:* Benedet, J.L. et al. *Int. J. Gynaecol. Obstetrics*, 70(2), 209–262, 2000.

[a] In order to evaluate the impact on prognosis of the different criteria for allotting cases to stage Ic or IIc, it would be of value to know if rupture of the capsule was spontaneous or caused by the surgeon, and if the source of malignant cells detected was peritoneal washings or ascites.

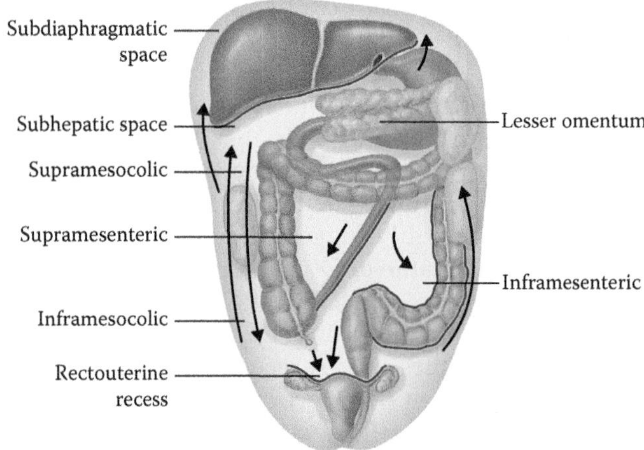


**FIGURE 5.11**
Pathways of spread of ovarian cancer: diagram illustrating pathways of dissemination along the peritoneal space with sites of high deposition of tumor burden shown in red, particularly in dependent regions such as the pouch of Douglas.

sensitivity of 85%–93% [96–99]. However, accuracy in delineation of peritoneal dissemination is highly dependent on implant location and size; in the right subdiaphragmatic space, omentum, root of mesentery, and bowel serosal surface, the reported per-lesion sensitivity of CT is 11%–37% with only moderate to fair inter-rater agreement (κ = 0.35–0.70), whereas sensitivity involving implants smaller than 1 cm declines to 7%–28% [100]. Overall performance of MRI for disease staging has been reported as 95% sensitivity and 82% specificity, with an AUROC of 0.91–0.95, which is not significantly superior to CT (AUROC, 0.85–0.93; $p > .10$) [98,101]. MRI with dual oral and IV contrast material may be advantageous over CT in detecting disease in subphrenic, mesenteric, and serosal sites or peritoneal implants of subcentimeter diameter (sensitivity 40%–77% and 72%–80%, respectively) [102].

The imaging features of peritoneal carcinomatosis range from fat stranding, discrete nodules or plaques, and thickened sheets of soft tissue, progressing to infiltrative masses, stiffening of the mesentery (producing a *pleated* or *stellate* appearance) (Figure 5.12), and diffuse involvement of the greater omentum in the form of *omental cake* [103,104]. Concomitant ascites and abdominopelvic lymphadenopathy are frequently observed. Increased gadolinium enhancement of peritoneal layers compared to liver parenchyma is indicative of infiltration and is optimally assessed 5–10 min after intravenous administration [103].

On source DWIs, peritoneal implants appear as hyperintense foci on a background of suppressed signal from normal peritoneal fat, leading to improved detectability (Figure 5.13). Addition of DWI to conventional MRI has been shown to increase the number of depicted peritoneal lesions by 21%–29% [105]. The combination

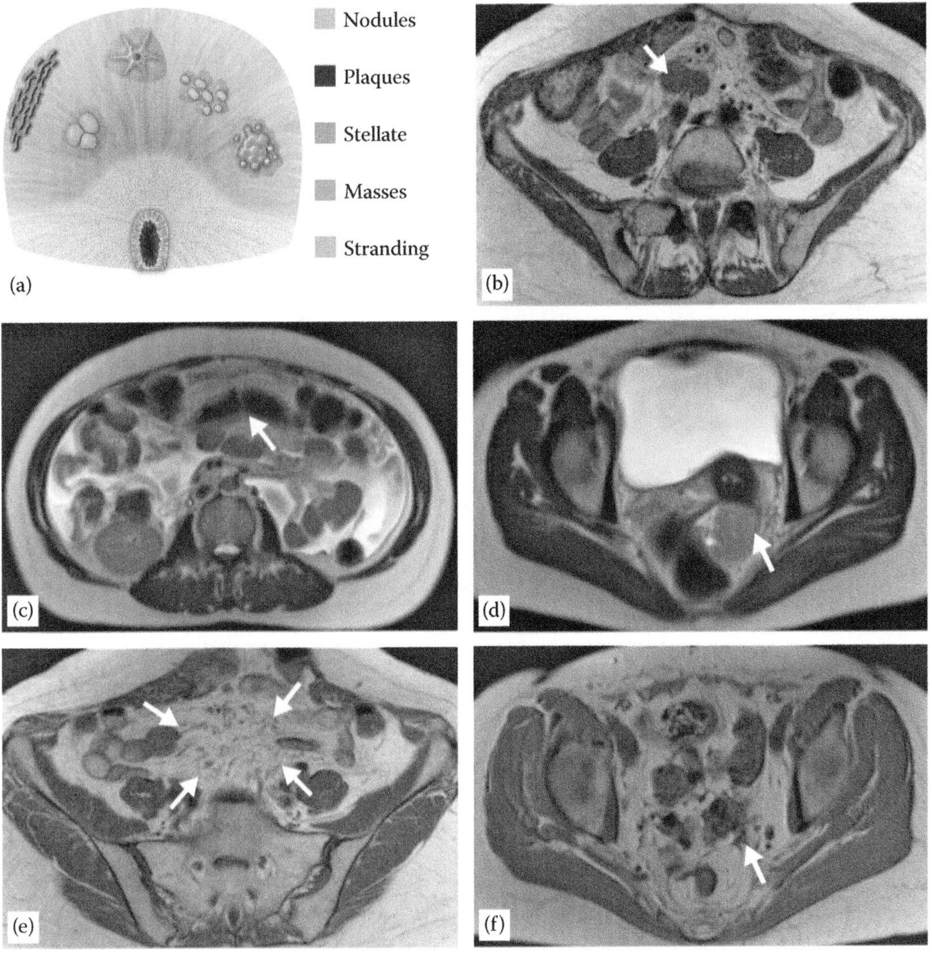


**FIGURE 5.12**
Patterns of tumor deposition of ovarian cancer: Diaphragmatic illustration of morphological patterns of tumor (a) with corresponding $T_2$-weighted transverse images illustrating tumor in nodules (b) plaque-like (c) and mass-like (d) forms and $T_1$-weighted transverse images showing stellate (e) and stranding (f) patterns.

of techniques is superior, achieving an accuracy of 84%–88% compared with 52%–72% for MRI alone and 71%–81% for DWI alone [106]. The overall performance of DWI in detecting peritoneal carcinomatosis has been found comparable to that of positron emission tomography (PET)/CT both on per-patient (sensitivity, 84% vs. 84%; specificity, 82% vs. 73%, respectively) and on per-site analysis (sensitivity, 74% vs. 63%; specificity, 97% vs. 90%, respectively; $p \geq .27$) [107]. Correlation with anatomic imaging is required to avoid false-positive results from inflammatory, hemorrhagic, and densely proteinaceous lesions or tissues with physiologically restricted diffusion (bowel serosa and functional endometrium) and false-negative results in tumors of relatively free diffusion (necrotic and mucinous) [108]. In comparison, nonquantitative DCE-MRI showed 87% sensitivity and 86%–92% specificity, with excellent interobserver agreement ($\kappa = 0.84$) when correlating imaging and surgical findings per peritoneal segment [109].

### 5.6.4 Surface Epithelial Tumors

#### *5.6.4.1 Serous Carcinoma*

Among the different histopathological subtypes of ovarian epithelial cancer, serous tumors have the highest frequency (68%–71%), with 90% of them being high grade [73,74]. The high overall mortality reflects the fact that approximately 60% of all patients and 88% of patients with serous tumors present with extraovarian dissemination (Figure 5.14), either in the form of extrapelvic peritoneal involvement and/or abdominopelvic lymphadenopathy (International Federation of Gynaecology and Obstetrics, FIGO stage III) or parenchymal metastases (FIGO stage IV) [73,74] (Figure 5.15). Recent immunohistochemical research has demonstrated that ovarian serous carcinoma follows a dualistic model of pathogenesis, leading to two distinct types, low grade and high grade, with different underlying molecular processes, biological behavior, and prognosis [110,111]. Low-grade

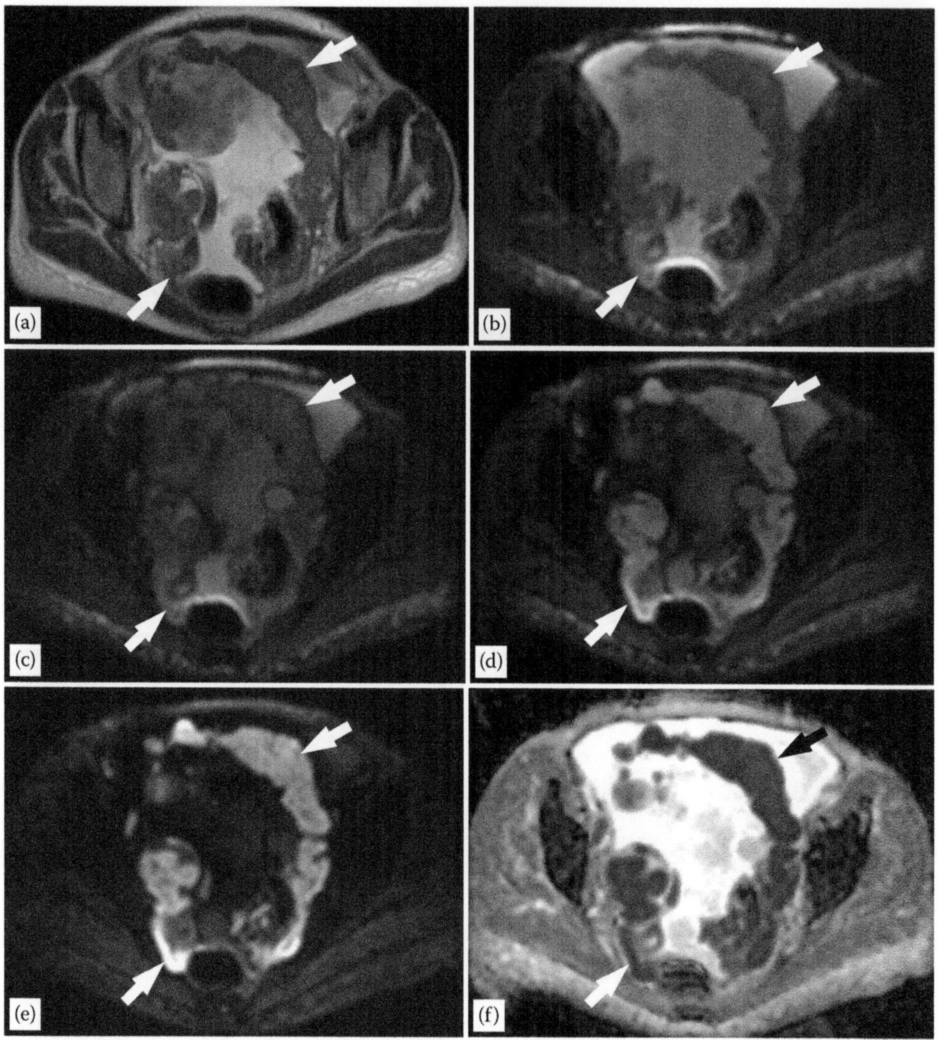

**FIGURE 5.13**
Diffusion-weighted MRI of ovarian metastases: $T_2$-weighted transverse (a) image through the mid-pelvis with corresponding diffusion-weighted images at $b = 0$ (b), $b = 100$ (c), $b = 500$ (d), and $b = 900$ (e) s/mm$^2$. There is increasing loss of signal from fluid and ascites and retention of signal within the metastatic deposits in the peritoneum (arrows). On apparent diffusion coefficient (ADC) map (f), tumor is seen to have a low ADC (arrows).

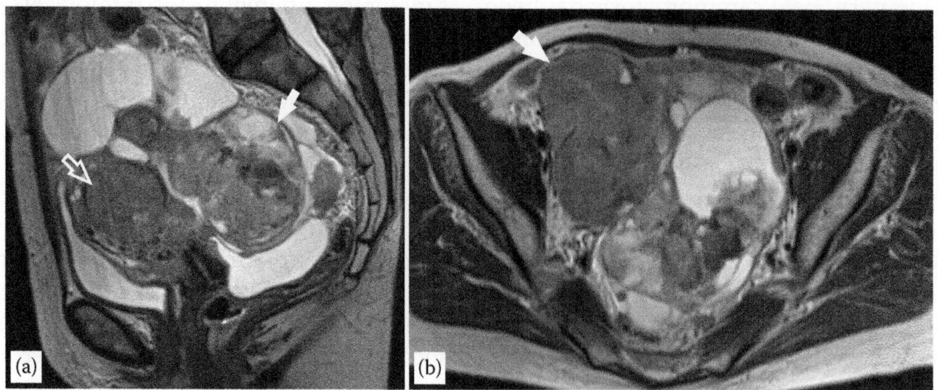

**FIGURE 5.14**
Serous carcinoma: $T_2$-weighted sagittal (a) and transverse (b) images showing bilateral mixed solid and cystic adnexal masses of serous carcinoma (arrows). In (a), involvement of the myometrium with this tumor is appreciated (arrow).

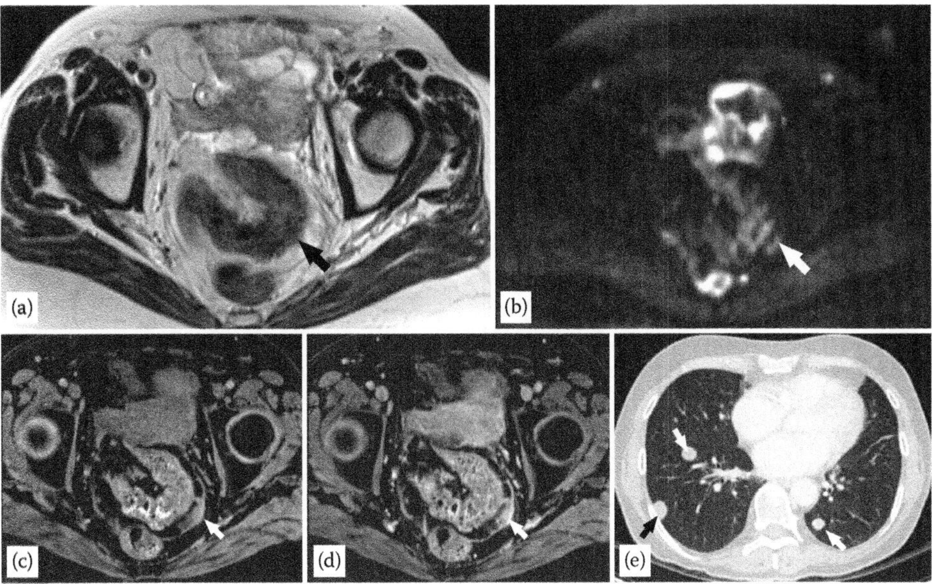


**FIGURE 5.15**
Extensive stage IV epithelial ovarian cancer: Transverse $T_2$-weighted (a), diffusion-weighted ($b$ = 1050 mm$^2$/s, b), $T_1$-weighted fat-saturated images precontrast (c) and postcontrast (d) showing extensive peritoneal involvement around the sigmoid colon and peritoneal reflections (arrows). Disease is seen as restricted diffusion in (b) and enhances after contrast in (d). Lung metastases were demonstrated on CT scan (e, arrows).

serous carcinoma is thought to follow a stepwise continuum of tumor progression from a benign serous cystadenoma through a serous borderline tumor to noninvasive and then invasive micropapillary serous carcinoma [111]. On the contrary, high-grade serous tumors are considered to progress rapidly from the epithelium of the distal fimbria of the fallopian tube and less commonly from ovarian surface epithelium or the epithelium of cortical inclusion cysts without identifiable intermediate forms [110,112]. Low-grade serous carcinomas harbor mutations of *KRAS* or *BRAF* in approximately 20% and 40% of cases, respectively, and are not associated with abnormalities of the tumor suppressor genes *BRCA1/2*, whereas their high-grade counterparts are almost ubiquitously associated with *TP53* mutations and *BRCA* inactivation [113]. A more than threefold variability in the size of tumor cell nuclei serves as the principal diagnostic criterion between low- and high-grade serous carcinomas [114].

#### *5.6.4.2 Mucinous Carcinoma*

Mucinous carcinomas account for 3%–4% of ovarian carcinomas and typically demonstrate gastrointestinal differentiation [113]. They are frequently histologically heterogeneous, with coexisting areas of benign, borderline, noninvasive, and invasive malignant components, suggestive of a stepwise tumorigenic process [113]. An invasive histology has been reported in approximately 3%–12% of primary mucinous tumors [68,115]. Depending on the pattern of stromal invasion, primary mucinous carcinomas can be divided into the *expansile* type (also known as *confluent glandular* or *intraglandular*) and *infiltrative* (or *destructive*) type. The expansile type is defined by the presence of complex malignant glands with minimal or no intervening normal ovarian stroma, exceeding 10 mm$^2$ in area, and is associated with a favorable prognosis [113,115]. The infiltrative type is characterized by the disorderly stromal invasion of glands and cellular nests, accompanied by desmoplastic stromal reaction [113,115]. Generally, invasive mucinous carcinomas carry a better prognosis that their serous counterparts, due to the fact that approximately 80% are diagnosed at stage I [116]. Metastatic behavior is almost invariably associated with the infiltrative invasive type, which is found to be high stage in 14%–25% of cases [115,117].

Typical imaging features of primary mucinous carcinomas include large size (6–40 cm), a multilocular cystic composition, often with honeycomb or *stained glass* appearance, and presence of solid mural nodules with contrast enhancement [83,118,119]. The signal intensity of the cystic portions reflects the concentration of mucin, as loculi with denser mucin demonstrate higher $T_1$- and lower $T_2$-signal intensity.

Primary mucinous cystadenocarcinomas have to be differentiated from ovarian metastases, which account for approximately 80% of mucinous tumors (site of origin in descending order of frequency being the gastrointestinal tract, pancreas, uterine cervix, breast, and endometrium [120]). Bilateral distribution and tumor size <10 cm are strongly associated with metastases (respective frequencies 77%–94% and 87%–95%); indeed, a unilateral lesion larger than 10 cm may predict a primary origin with 84%–90% accuracy [120,121]. Thus, the most pertinent morphological and clinical criteria differentiating primary from metastatic mucinous ovarian lesions include localization of mucin (intra- vs. extracellular), laterality and size of the lesion, presence of peritoneal dissemination, and metastatic growth pattern (defined by ovarian surface involvement, lymphovascular invasion, and desmoplastic response) [122]. In a histopathological study of 43 metastatic and 25 stage I primary mucinous tumors, morphological features indicating a primary versus a metastatic origin were size >10 cm, a smooth surface, expansile invasive pattern, microscopic cysts <2 mm, and benign- or borderline-appearing regions [68]. Findings associated with secondary ovarian deposits were bilaterality, microscopic surface involvement, infiltrative invasive pattern, nodular growth pattern, and hilar involvement [68]. Macroscopic features, which were not found discriminatory, were a predominantly cystic or solid gross appearance and the presence of focal papillary, necrotic, or hemorrhagic areas [68].

Immunochemistry may be helpful in discriminating a primary ovarian mucinous tumor from metastatic colorectal adenocarcinoma. Positive staining for cytokeratins 7 and 20 (CK7+/CK20+) has been reported in 68%–74% of primary ovarian mucinous tumors, whereas the most common immunoprofile in lower intestinal tract tumors is CK7−/CK20+ (69%–79%) [123,124]. In tumors with concomitant expression of CK7 and CK20, the pattern of immunostaining also may be discriminatory; diffuse CK7 positivity, as defined by involvement of >50% of tumor cells, with a focal or patchy (<50% of tumor cells) CK20 distribution is frequently observed in primary ovarian tumors, whereas colorectal and appendiceal tumors typically display patchy CK7 and diffuse CK20 distribution [123,124].

*Pseudomyxoma peritonei* refers to the clinical entity of diffuse presence of mucinous implants on peritoneal surfaces accompanied by gelatinous ascites. Its commonest cause is rupture of a low-grade mucinous neoplasm of the appendix [125]. The historical belief that it represents the typical spread pattern of a ruptured mucinous ovarian tumor, mostly of borderline histology, has been currently refuted; it is now held that ovarian tumors associated with mucinous peritoneal deposits are in the vast majority metastases from appendiceal or colonic carcinomas, after deposition of neoplastic epithelium on the ovarian cortex and dissection of pools of mucin through the ovarian stroma, a condition referred to as *pseudomyxoma ovarii* [115]. A rare ovarian cause of pseudomyxoma peritonei has been attributed to rupture of a mucinous tumor in association with a mature dermoid cyst, the mucinous component being immunohistochemically consistent with an intestinal-type adenoma of teratomatous origin [125].

#### 5.6.4.3 *Endometrioid Carcinoma*

Endometrioid carcinomas represent approximately 7%–11% of ovarian malignant lesions and are mostly low-stage (43% stage I) and low-grade [73,74,126]. Overall 5-year survival is reported between 38% and 70% [126]. They are associated with endometriosis in 15%–20% of cases, particularly when disease is located in the ovary [127]. Synchronous endometrial hyperplasia, either premalignant or overtly malignant, is seen in 15%–35% [128,129]. The imaging features of endometrioid carcinomas are nonspecific, usually presenting as unilateral complex solid/cystic masses (Figure 5.16) but more often predominantly solid compared to other epithelial malignant tumors [119]. The presence of a solid nodule within an endometriotic cyst may be useful in the correct characterization [8]. On MRI, solid components of endometrioid tumors are often of heterogeneous signal intensity, whereas in the setting of endometriosis the solid nodule may display $T_1$-hypointensity and $T_2$-hyperintensity in an otherwise $T_2$-hypointense ovarian lesion. Evaluation of enhancement of the solid components necessitates dynamic subtraction techniques due to the $T_1$-hyperintense background of endometriotic cyst fluid containing products of hemoglobin degradation and to the small size of

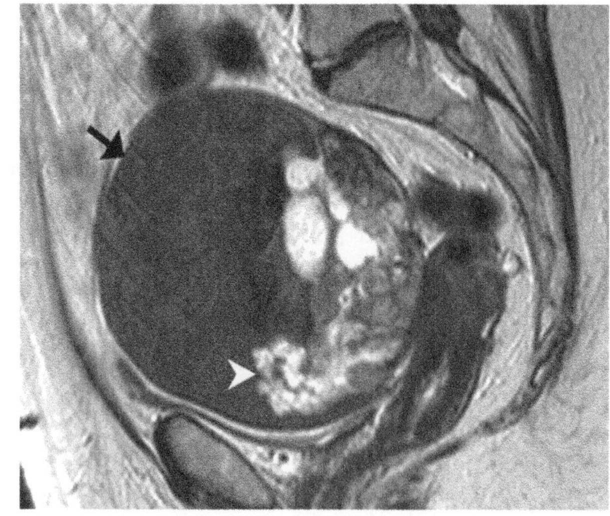

**FIGURE 5.16**
Endometroid carcinoma: $T_2$-weighted sagittal image showing a complex partly cystic, partly solid adnexal mass (arrows). The extent, heterogeneity and complexity of the solid components are indicative of an invasive tumor. This lesion was confined to one ovary and was classified as an endometroid carcinoma at surgical histology.

malignant nodules [8]. Ovarian adenofibroma, a benign epithelial tumor, and decidual change of ectopic endometrial tissue may mimic malignant transformation of an endometrioma [8].

Synchronous endometrial and ovarian carcinomas occur in approximately 9%–12% of patients with ovarian and 5% of patients with endometrial carcinoma. Patients with synchronous primary ovarian and uterine endometrioid carcinomas have distinct demographic characteristics compared to their counterparts with endometrial or adnexal malignancy alone, including younger age, obesity, premenopausal status, and nulliparity [130]. The designation of a dual primary versus metastases may be challenging, since 68%–93% of tumors detected concurrently in the endometrium and ovary are of endometrioid type, with or without squamous differentiation [129,130]. The distinction is critical, since standard treatment of endometrial carcinoma with metastatic adnexal involvement (stage IIIa) comprises surgery and adjuvant chemo- and/or radiotherapy, whereas independent stage I uterine and ovarian endometrioid carcinomas have a better overall prognosis and do not routinely require adjuvant treatment. Independent primaries are most probably represented by low-grade endometrioid adenocarcinomas, notably in the presence of atypical endometrial hyperplasia or ovarian endometriosis [8]. High histological grade of the endometrial tumor and bilateral ovarian malignancy, characteristically with surface involvement and a micronodular pattern, suggests a uterine primary with adnexal metastases.

#### 5.6.4.4 *Clear-Cell Carcinoma*

Clear-cell carcinomas are characterized by growth of cells with clear cytoplasm in glandular, tubular, or solid patterns and presence of *hobnail* cells, which have prominent bulbous nuclei protruding beyond cytoplasmic boundaries [112]. They constitute the histological subtype most frequently associated with ovarian endometriosis (approximately 20%–54%) [7,131]. The highest prevalence is encountered in Japan, where they account for up to 25% of epithelial ovarian tumors [132] in contrast to 5%–12% in Western populations [74,133]. Venous thromboembolic events and hypercalcemia as paraneoplastic phenomena are more often associated with the clear-cell subtype than other epithelial ovarian tumors [134,135]. Clear-cell carcinomas comprise approximately 26% of low-stage (I/II) ovarian tumors as opposed to 5% of their high-stage (III/IV) counterparts [74]. The majority (59%–71%) are detected early (stage I/II) due to their large size and slow growth [136,137]. However, their stage-by-stage prognosis is worse compared to other ovarian malignant tumors due to their relative chemoresistance to conventional platinum-based regimens and high postoperative recurrence rate even of stage I disease (ranging from 18% in stage Ia to 54% in stage Ic) [132,136,137]. On MRI, they often present as unilateral and unilocular large cystic adnexal lesions (mean diameter 13 cm) with variable $T_1$ signal intensity (reflecting hemorrhagic content) and $T_2$ hyperintensity [138]. Solid nodules display intermediate or slightly increased $T_1$ signal intensity, suggestive of hemorrhage, and marked contrast enhancement helps differentiate them from blood clots.

#### 5.6.4.5 *Malignant Brenner's Tumors*

Atypical proliferating (borderline) and malignant Brenner's tumors are found only as unusual pathological variants of their benign counterparts [139]. Due to their small size (<5 cm), they are often asymptomatic, but may also present with abdominal distention, pain, and vaginal hemorrhage secondary to hormonal activity [26]. Borderline and malignant Brenner's tumors present as complicated multicystic masses with papillary projections and solid elements of variable $T_2$ signal intensity and cannot be reliably differentiated from each other or other epithelial tumors on imaging criteria [139–141].

### 5.6.5 Sex Cord-Stromal Tumors

Ovarian SCSTs develop from two groups of cells with distinct embryologic origin: primitive sex cord cells, derived from coelomic epithelium, and stromal cells, derived from the mesenchyma of the genital ridge. Sex cords comprise granulosa cells in the normal ovary, Sertoli cells in the testis and Sertoli cells in ovarian tumors. Stromal cells include fibroblasts, theca cells, and Leydig cells [142]. SCSTs include juvenile- and adult-type granulosa cell tumors, SLCTs, SCLCTs, as well as theca and granulosa-theca tumors, sclerosing stromal tumors, SCSTs with annular tubules, and gynandroblastomas with simultaneous Sertoli and granulosa cell differentiation [143]. They thus present as a heterogeneous predominantly solid masses on MRI. SCSTs account for approximately 8% of ovarian neoplasms; SCSTs affect all age groups, and their vast majority are benign, such as fibrothecomas and sclerosing stromal tumors: the malignant forms are diagnosed early (stage I) in approximately 70% of cases [44]. SCSTs are the most frequent hormonally functioning tumors, presenting with hyperandrogenism or hyperestrogenemia. Under control of LH, thecal cells produce androstenedione and testosterone, whereas in response to follicle-stimulating hormone granulosa cells convert these androgens into estrone and estradiol; therefore, thecomas and granulosa cell tumors may present with symptoms of excessive estrogen production or, less commonly, virilization

[144]. Moreover, Sertoli and steroid cell tumors are notably associated with testosterone secretion [144].

### 5.6.5.1 Granulosa Cell Tumors

Granulosa cell tumors constitute both the commonest malignant SCSTs and hormonally active ovarian neoplasms. The two distinct histological subtypes, adult and juvenile, share indistinguishable radiological features but differ in their age distribution and biological behavior. The adult type accounts for 95% of granulosa cell tumors and affects peri- and postmenopausal women with peak prevalence at 55 years of age [145]. The associated hyperestrogenism may manifest as irregular vaginal bleeding, endometrial hyperplasia (25%–50%), or carcinoma (5%–13%) [146]. The juvenile type is rare in patients older than 30 years, with an average patient age of 13 years, and often presents clinically with isosexual precocious puberty [145,146]. Stage I disease is found in 60%–90% of patients at presentation and is associated with 85%–95% 10-year survival rate [146]. However, granulosa cell tumors have a tendency for delayed recurrence, with a median time to relapse of 4–6 years after initial diagnosis and not infrequently after 10–20 years [146,147]. Histologically, granulosa cell tumors demonstrate multiple patterns, of which most frequent is the microfollicular form with Call-Exner bodies. They present as large (average size 12 cm) unilateral adnexal masses in 95% of cases, typically without extraovarian spread [142]. Imaging characteristics reflect the variety of histological appearances: the macrofollicular pattern manifests as multilocular cystic masses with serous or hemorrhagic content, lacking intracystic papillary projections (compared to ovarian epithelial malignancies); and the trabecular or diffuse pattern presents as homogeneous solid masses with areas of fibrous or hemorrhagic degeneration and infarction recognized as heterogeneously enhancing solid masses [148,149]. On MRI, a typical *sponge-like* appearance with innumerable cystic spaces has been described, probably corresponding with the macrofollicular type [150] (Figure 5.17), and $T_1$-hyperintensity due to intratumoral hemorrhage has been reported in 60%–70% of cases [149,150]. Concomitant uterine changes, such as uterine enlargement, endometrial thickening, adenomyosis, and prominent junctional zone, also have been reported in a large number of patients [149].

### 5.6.5.2 Sertoli-Stromal Cell Tumors

Sertoli-stromal cell tumors (formerly referred to as arrhenoblastomas or androblastomas) encompass tumors with varying components of Sertoli cells, Leydig cells, and fibroblasts. SLCTs are the commonest type in this group and although overall rare (0.5% of ovarian neoplasms) represent the principal ovarian tumor associated with virilization [144]. They involve patients younger than 30 years in 75% of cases and approximately 30% present with clinical hyperandrogenism (amenorrhea and virilized secondary sexual features) [151]. Their size at diagnosis ranges from microscopic (especially in functioning tumors) to 50 cm (average 13 cm), and 97% of tumors are stage I at surgery [151]. Biological behavior depends on degree of differentiation and stage, and approximately 60% of poorly differentiated and 20% of tumors containing heterologous elements (mucinous epithelium, striated muscle, and cartilage) are clinically malignant [151,152]. Overall 10-year survival rate has

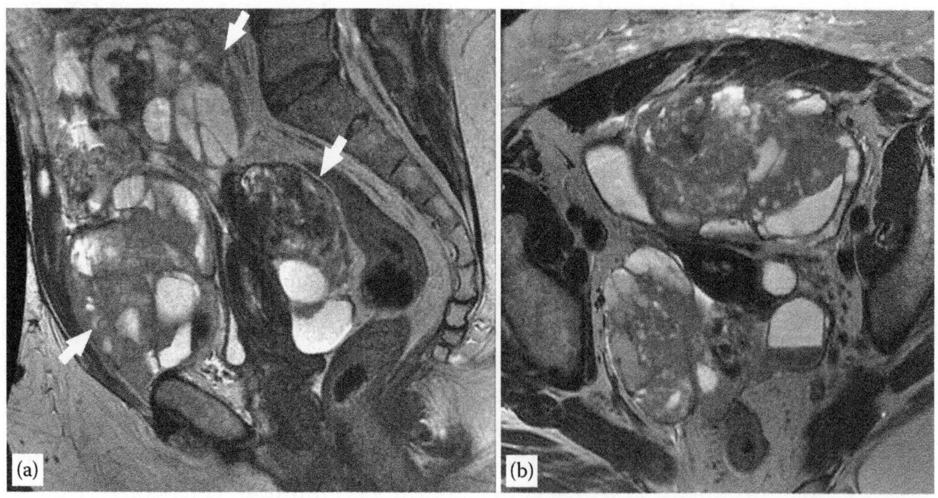


**FIGURE 5.17**
Granulosa cell tumor: $T_2$-weighted sagittal (a) and transverse (b) images showing bilateral solid adnexal masses with cystic elements. The *sponge-like* appearance with the intermediate signal intensity of the solid elements is characteristic of a granulose cell tumor.

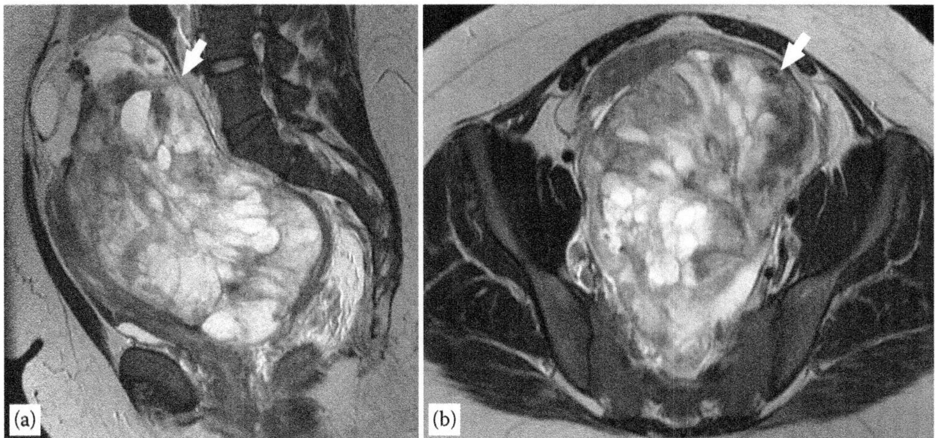

**FIGURE 5.18**
Sertoli–Leydig tumor: $T_2$-weighted sagittal (a), transverse (b), images showing unilateral heterogeneous adnexal mass (arrows) with extensive cystic elements. The high signal intensity throughout the tumor is more indicative of a Sertoli–Leydig histological type.

been reported as 92% [152]. In contrast to granulosa cell tumors, SLCTs tend to recur relatively early after initial diagnosis [44]. Macroscopically, 58% of SLCTs have been reported as mixed solid/cystic (Figure 5.18), 38% as purely solid, and 4% as cystic [151]. On MRI, the solid components are of intermediate or decreased $T_2$ signal intensity correlating with the extent of fibrous stroma, albeit not as low as characteristically seen in fibromas [142]. Heterologous components give rise to multicystic areas, and, in contrast to granulosa cell tumors, intratumoral bleeding is rare [44,142].

#### 5.6.5.3 Steroid Cell Tumors

Steroid cell tumors are composed of cells that resemble typical steroid-secreting cells, such as Leydig cells, stromal lutein cells, and adrenal rest cells, and include stromal luteomas, Leydig cell tumors, and steroid cell tumors, not otherwise specified (NOS). Approximately 75% of steroid cell tumors contain abundant intracellular fat, giving rise to the term *lipoid cell tumors* [142,153]. The hallmark of Leydig cell tumors is the presence of intracytoplasmic crystals of Reinke [142]. Steroid cell tumors, NOS represent tumors not classifiable as either of the former types and account for approximately 60% of steroid cell tumors, and the proportion of malignancy has been reported as 28% [153]. Clinically, approximately 50% of patients present with virilization, although hyperestrogenemia and hypercortisolemia with Cushing's syndrome may occur [153]. On MRI, steroid cell tumors have been described as predominantly unilateral solid masses of heterogeneous intermediate $T_2$ signal intensity, correlating with the amount of fibrous stroma, and intense contrast enhancement, reflecting hypervascularity [154,155]. Increased $T_1$ signal intensity has been attributed to lipid content [44,142].

### 5.6.6 Germ Cell Tumors

Malignant germ cell tumors account for 1%–2% of ovarian malignant neoplasms with a distinct age predilection for patients younger than 21 years; where they make up approximately two-thirds of ovarian carcinomas [156,157]. They can be divided into the following: (1) primitive germ cell tumors, which mainly include dysgerminoma, yolk sac (or endodermal sinus) tumor, embryonal carcinoma, and non-gestational choriocarcinoma; (2) biphasic or triphasic teratomas, including mature (benign) and immature (malignant) teratoma; and (3) monodermal tumors and somatic-type tumors associated with dermoid cysts [58]. Elevated serum alpha-fetoprotein (AFP) and beta-human chorionic gonadotropin (beta-hCG) may help establish the diagnosis [157]. Stage distribution at presentation is distinct from that of epithelial carcinomas, as 60%–70% of malignant germ cell tumors are stage I/II and 30%–40% stage III, with stage IV disease being relatively uncommon [157].

#### 5.6.6.1 Dysgerminoma

Dysgerminoma is considered homologous to testicular seminoma and represents the most common malignant germ cell tumor (0.5%–2% of all malignant ovarian lesions). It differs from other malignant germ cell tumors in its greater likelihood to be diagnosed as stage Ia, its more common bilaterality (5%–15%), its propensity for retroperitoneal lymphadenopathy rather than peritoneal dissemination, and its radiosensitivity [157]. Dysgerminoma does not secrete AFP but may be associated with increased beta-hCG in 5% of cases due to concomitant presence of syncytiotrophoblastic giant cells [158]. Dysgerminoma may develop from a preexisting gonadoblastoma in phenotypic female patients with 46,

XY karyotype and gonadal dysgenesis [159]. On MRI, dysgerminomas are seen as mainly solid masses with avid heterogeneous contrast enhancement, possible areas of hemorrhage or necrosis, and abundant fibrovascular septae [156,160,161].

#### 5.6.6.2 Yolk Sac Tumor

Yolk sac (also known as *endodermal sinus*) tumor accounts for 9%–16% of ovarian neoplasms in children and adolescents, with a peak incidence in the second decade of life [58]. They are considered as the most highly malignant among germ cell tumors and prognosis depends on stage (95% 5-year survival in stage I, 75% in stage II, 30% in stage III, and 25% in stage IV) [162]. It is almost invariably associated with increased serum AFP and normal beta-HCG levels. On MRI, it is usually seen as a large unilateral predominantly solid mass with extensive portions of necrosis and hemorrhage [156,163,164]. Multiple signal void structures and prominent gadolinium enhancement correlate with hypervascularity on pathologic evaluation [164]. Ascites and peritoneal implants may be present, as well as a concurrent dermoid cyst in 14% of cases [156,163].

#### 5.6.6.3 Choriocarcinoma

In patients of reproductive age, an ovarian choriocarcinoma may represent a metastasis from a primary tumor of the uterus or develop from gestational tissue in an ectopic pregnancy (gestational choriocarcinoma) or arise from ovarian germ cells showing trophoblastic differentiation (non-gestational choriocarcinoma). Serum beta-HCG levels are invariably abnormal and determination of origin is not feasible on histological features alone; differentiation between the two types can be reliable only in the prepubertal phase or through identification of paternal DNA contribution [165]. The pure form of non-gestational choriocarcinoma is extremely rare, and most often, mixed forms with other germ cell tumor types are seen. Imaging features are nonspecific and comprise a large unilateral mass with central areas of hemorrhage and necrosis and avid peripheral contrast enhancement [156,166].

#### 5.6.6.4 Immature Teratoma

Immature teratoma is the malignant form of teratoma, containing immature or embryonic tissues derived from all three germ layers with a predominance of neuroectodermal tissue in the form of primitive neuroepithelial rosettes and tubules [58]. Immature teratomas are much rarer than their mature counterparts (<1% of ovarian teratomas), their largest peak is in the first two decades of life, and they are associated with raised serum AFP in 33%–65% of cases [167]. Immature teratomas have been reported to contain macroscopically visible mature cystic teratomas in 26% of cases and to be associated with contralateral dermoid cysts in 10% [168]. The amount of primitive neuroectodermal tissue determines lesion grade. The presence of microscopic foci of yolk sac tumor has been recognized as a major predictor of grade, stage, and overall survival [169]. Immature teratomas usually present as unilateral large (average diameter 14–25 cm) adnexal lesions of predominantly solid (Figure 5.19) or mixed solid/cystic composition. Small foci of fat (fat-suppression techniques useful for distinction from hemorrhage) and scattered calcifications (difficult to detect on MRI) [142,156,170,171] are sometimes seen. They may demonstrate an ill-defined capsule, suggestive of perforation. *Gliomatosis peritonei* refers to the presence of peritoneal implants of mature, low-grade glial tissue, which are indistinguishable from malignant seedings on imaging [172].

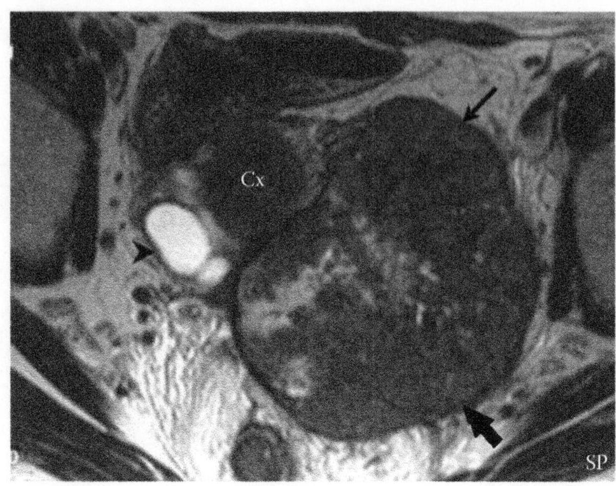


**FIGURE 5.19**
Immature teratoma: $T_2$-weighted transverse image showing a unilateral solid adnexal mass on the left (arrow). It is relatively well defined with some areas of central necrosis. An immature teratoma was confirmed at surgical histology. The right ovary and cervix (Cx) are normal.

### 5.6.7 Primary Peritoneal Serous Carcinoma

Approximately 60% of peritoneal carcinomatosis consists of serous papillary or poorly differentiated adenocarcinomas, and in the female population, 80%–90% of cases represent stage III/IV ovarian carcinoma. In 10%–15% of patients, no malignant process is evidenced in the ovaries, uterus, or fallopian tubes, and the clinical entity is referred to as *primary serous peritoneal carcinoma* (PSPC) [173]. Since PSPC is identical to advanced serous ovarian carcinoma at both histopathological and immunohistochemical level, the following set of criteria has been established to distinguish the two diseases: (1) both ovaries are physiologically normal or enlarged by a benign process; (2) extraovarian involvement is greater than

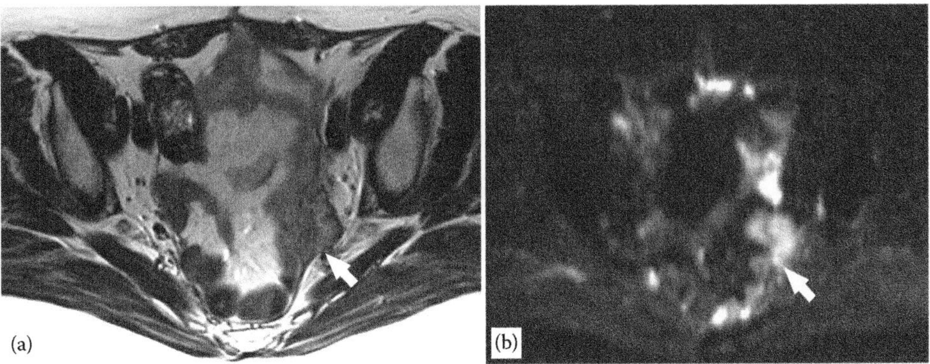


**FIGURE 5.20**
Peritoneal carcinomatosis: Transverse $T_2$-weighted (a) and diffusion-weighted ($b$ = 1050 mm²/s) (b) images showing extensive solid plaques of tumor involving the peritoneal reflection on the left (arrows). The disease shows marked restriction of diffusion in (b).

involvement of the ovarian surface; (3) microscopically the ovarian component is nonexistent or confined to the ovarian surface epithelium with no evidence of cortical invasion or involving the ovarian epithelium/underlying stroma with tumor size less than 5 × 5 mm [173,174]. PSPC affects almost exclusively women, at a slightly older age than ovarian serous carcinoma (median age 55–65), with prevailing clinical findings of ascites and abnormal serum CA125 levels in over 70%–90% of cases [173]. Both somatic and germline mutations of *BRCA* genes occur with the same frequency in patients with PSPC and ovarian serious carcinoma (40%–70% and 5%–10%, respectively) [175]. *BRCA1* mutation carriers who have undergone prophylactic oophorectomy are at higher risk of developing PSPC on follow-up (3.4 cases in 100 women/year) [176,177]. Compared to ovarian carcinoma, PSPC is more often multifocal with a diffuse micronodular spread pattern, leading to a high disease burden that renders optimal surgical debulking difficult [173].

The imaging features of PSPC resemble those of metastatic peritoneal carcinomatosis with ascites, peritoneal thickening or nodules, omental masses, mesenteric changes, and retroperitoneal lymphadenopathy [178,179]. On MRI, peritoneal nodules are $T_1$ hypointense and $T_2$ hyperintense (Figure 5.20), and diffuse contrast enhancement of peritoneal surfaces is seen [179]. Also PSPC has to be differentiated from diffuse peritoneal mesothelioma, lymphomatosis, and tuberculous peritonitis.

### 5.6.8 Fallopian Tube Carcinoma

Primary fallopian tube carcinoma (PFTC) is a rare malignancy accounting for 0.14%–1.8% of female genital tract malignancies and an age-adjusted annual incidence rate of 0.3–0.5 per 100,000 women [180–182]. The histological distribution and biological behavior of PFTC parallel that of primary epithelial ovarian cancer, possibly reflecting the common embryological origin of ovarian surface epithelium and Mullerian ducts, which give rise to the fallopian tubes, endocervix, and endometrium. Approximately 90% of tumors are of serous epithelial type. The distal fallopian tube has been recognized as a potential site of serous carcinogenic pathway [183,184]. Evidence in support of the role of the fimbriae in the pathogenesis of pelvic serous carcinoma includes the finding that approximately half of primary ovarian and peritoneal carcinomas involve the endosalpinx or coexist with an early form of a tubal primary in the form of serous tubal intraepithelial carcinoma [184]. Furthermore, in BRCA positive women undergoing prophylactic bilateral salpingo-oophorectomy, 72%–86% of occult pelvic malignancies were located in the fimbria [86,185,186]. Staging of PFTC follows the FIGO classification for primary ovarian carcinoma. Disease spread is principally via the transcoelomic route with seeding of the peritoneal cavity. A propensity for early lymphatic dissemination due to the rich lymphatic supply of the salpinges has been reported, and pelvic and/or para-aortic lymphadenopathy is encountered.

### 5.6.9 Metastases to the Ovary

Ovarian metastases should be considered in patients with known other malignancy, typically gastrointestinal primary, predominantly gastric (Krukenberg tumors), or from the breast. Lesions are commonly bilateral and present with disorganization of the ovarian architecture (Figure 5.21) or large bilateral masses, often in the context of disseminated malignancy. Diagnosis is therefore by exclusion and depends on the clinical context. If ovarian metastases are suspected, a gastrointestinal workup including endoscopy and biopsy and/or mammography in female patients is warranted.

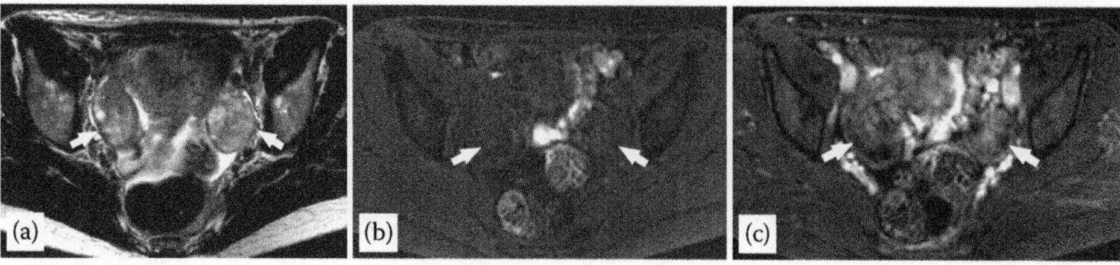


**FIGURE 5.21**
Metastases to the ovaries from breast cancer: $T_2$-weighted transverse (a), $T_1$-weighted transverse precontrast (b) and postcontrast (c) images showing enlargement of both ovaries. Intermediate heterogeneous signal intensity is noted in (a) (arrows) with disruption to the normal follicular architecture. Following contrast, there is bilateral patchy enhancement of this tissue in (c) (arrows).

## 5.7 Summary

This chapter describes the MR imaging appearances of the normal ovaries and adnexa together with the appearances in benign and malignant conditions. It relates imaging appearances to morphological features on histology and links these findings with the clinical context in order to aid diagnosis in individual patients with adnexal or peritoneal pathology. Current knowledge on functional MRI biomarkers as diagnostic tools is included in instances where robust evidence is available. A final differential diagnosis requires a combination of skills in interpreting the imaging information in conjunction with the clinical findings. This is crucial in order to plan an appropriate management strategy. In the majority of cases, the management is primarily surgical, so the role of imaging is in indicating the extent of surgery required or its timing (e.g., whether surgical treatment may be optimized by delaying it to follow neoadjuvant chemotherapy) or even avoiding surgery altogether in appropriate cases.

## References

1. Levine D, Brown DL, Andreotti RF, Benacerraf B, Benson CB, Brewster WR et al. Management of asymptomatic ovarian and other adnexal cysts imaged at US: Society of Radiologists in Ultrasound Consensus Conference Statement. *Radiology*. 2010;256(3):943–54.
2. Shanbhogue AK, Shanbhogue DK, Prasad SR, Surabhi VR, Fasih N, Menias CO. Clinical syndromes associated with ovarian neoplasms: A comprehensive review. *RadioGraphics*. 2010;30(4):903–19.
3. Futterweit W. Polycystic ovary syndrome: Clinical perspectives and management. *Obstetrical & Gynecological Survey*. 1999;54(6):403–13.
4. Lalwani N, Patel S, Ha KY, Shanbhogue AK, Nagar AM, Chintapalli KN et al. Miscellaneous tumour-like lesions of the ovary: Cross-sectional imaging review. *The British Journal of Radiology*. 2012;85(1013):477–86.
5. Ozkan S, Murk W, Arici A. Endometriosis and infertility: Epidemiology and evidence-based treatments. *Annals of the New York Academy of Sciences*. 2008;1127:92–100.
6. Bulun SE. Endometriosis. *The New England Journal of Medicine*. 2009;360(3):268–79.
7. Pearce CL, Templeman C, Rossing MA, Lee A, Near AM, Webb PM et al. Association between endometriosis and risk of histological subtypes of ovarian cancer: A pooled analysis of case-control studies. *The Lancet Oncology*. 2012;13(4):385–94.
8. Tanaka YO, Okada S, Yagi T, Satoh T, Oki A, Tsunoda H et al. MRI of endometriotic cysts in association with ovarian carcinoma. *AJR American Journal of Roentgenology*. 2010;194(2):355–61.
9. Kataoka ML, Togashi K, Yamaoka T, Koyama T, Ueda H, Kobayashi H et al. Posterior cul-de-sac obliteration associated with endometriosis: MR imaging evaluation. *Radiology*. 2005;234(3):815–23.
10. Menada MV, Remorgida V, Abbamonte LH, Fulcheri E, Ragni N, Ferrero S. Transvaginal ultrasonography combined with water-contrast in the rectum in the diagnosis of rectovaginal endometriosis infiltrating the bowel. *Fertility and Sterility*. 2008;89(3):699–700.
11. Tokue H, Tsushima Y, Endo K. Magnetic resonance imaging findings of extrapelvic endometriosis of the round ligament. *Japanese Journal of Radiology*. 2009;27(1):45–7.
12. Bazot M, Darai E, Hourani R, Thomassin I, Cortez A, Uzan S et al. Deep pelvic endometriosis: MR imaging for diagnosis and prediction of extension of disease. *Radiology*. 2004;232(2):379–89.
13. Bazot M, Malzy P, Cortez A, Roseau G, Amouyal P, Darai E. Accuracy of transvaginal sonography and rectal endoscopic sonography in the diagnosis of deep infiltrating endometriosis. *Ultrasound in Obstetrics & Gynecology*. 2007;30(7):994–1001.
14. Novellas S, Chassang M, Bouaziz J, Delotte J, Toullalan O, Chevallier EP. Anterior pelvic endometriosis: MRI features. *Abdominal Imaging*. 2010;35(6):742–9.
15. Zhang H, Zhang GF, He ZY, Li ZY, Zhu M, Zhang GX. Evaluation of primary adnexal masses by 3T MRI: Categorization with conventional MR imaging and diffusion-weighted imaging. *Journal of Ovarian Research*. 2012;5(1):33.
16. Siegelman ES, Oliver ER. MR imaging of endometriosis: Ten imaging pearls. *RadioGraphics*. 2012;32(6):1675–91.

17. Kruger K, Behrendt K, Niedobitek-Kreuter G, Koltermann K, Ebert AD. Location-dependent value of pelvic MRI in the preoperative diagnosis of endometriosis. *European Journal of Obstetrics, Gynecology, and Reproductive Biology*. 2013;169(1):93–8.
18. Saba L, Guerriero S, Sulcis R, Ajossa S, Melis G, Mallarini G. Agreement and reproducibility in identification of endometriosis using magnetic resonance imaging. *Acta Radiologica*. 2010;51(5):573–80.
19. Chamie LP, Blasbalg R, Pereira RM, Warmbrand G, Serafini PC. Findings of pelvic endometriosis at transvaginal US, MR imaging, and laparoscopy. *RadioGraphics*. 2011;31(4):E77–100.
20. Chassang M, Novellas S, Bloch-Marcotte C, Delotte J, Toullalan O, Bongain A et al. Utility of vaginal and rectal contrast medium in MRI for the detection of deep pelvic endometriosis. *European Radiology*. 2010;20(4):1003–10.
21. Ascher SM, Agrawal R, Bis KG, Brown ED, Maximovich A, Markham SM et al. Endometriosis: Appearance and detection with conventional and contrast-enhanced fat-suppressed spin-echo techniques. *Journal of Magnetic Resonance Imaging*. 1995;5(3):251–7.
22. Onbas O, Kantarci M, Alper F, Kumtepe Y, Durur I, Ingec M et al. Nodular endometriosis: Dynamic MR imaging. *Abdominal Imaging*. 2007;32(4):451–6.
23. Moteki T, Ishizaka H. Evaluation of cystic ovarian lesions using apparent diffusion coefficient calculated from reordered turboflash MR images. *Magnetic Resonance Imaging*. 1999;17(7):955–63.
24. Busard MP, Mijatovic V, van Kuijk C, Pieters-van den Bos IC, Hompes PG, van Waesberghe JH. Magnetic resonance imaging in the evaluation of (deep infiltrating) endometriosis: The value of diffusion-weighted imaging. *Journal of Magnetic Resonance Imaging*. 2010;32(4):1003–9.
25. Busard MP, Pieters-van den Bos IC, Mijatovic V, Van Kuijk C, Bleeker MC, van Waesberghe JH. Evaluation of MR diffusion-weighted imaging in differentiating endometriosis infiltrating the bowel from colorectal carcinoma. *European Journal of Radiology*. 2012;81(6):1376–80.
26. Balasa RW, Adcock LL, Prem KA, Dehner LP. The Brenner tumor: A clinicopathologic review. *Obstetrics & Gynecology*. 1977;50(1):120–8.
27. Brenner F. Das Oophoroma folliculare. *Frankf Z Path*. 1907;1:150–71.
28. Ehrlich CE, Roth LM. The Brenner tumor. A clinicopathologic study of 57 cases. *Cancer*. 1971;27(2):332–42.
29. Hermanns B, Faridi A, Rath W, Fuzesi L, Schroder W. Differential diagnosis, prognostic factors, and clinical treatment of proliferative Brenner tumor of the ovary. *Ultrastructural Pathology*. 2000;24(3):191–6.
30. Jorgensen EO, Dockerty MB, Wilson RB, Welch JS. Clinicopathologic study of 53 cases of Brenner's tumors of the ovary. *American Journal of Obstetrics & Gynecology*. 1970;108(1):122–7.
31. Silverberg SG. Brenner tumor of the ovary. A clinicopathologic study of 60 tumors in 54 women. *Cancer*. 1971;28(3):588–96.
32. Dierickx I, Valentin L, Van Holsbeke C, Jacomen G, Lissoni AA, Licameli A et al. Imaging in gynecological disease (7): Clinical and ultrasound features of Brenner tumors of the ovary. *Ultrasound in Obstetrics & Gynecology*. 2012;40(6):706–13.
33. Buy JN, Ghossain MA, Sciot C, Bazot M, Guinet C, Prevot S et al. Epithelial tumors of the ovary: CT findings and correlation with US. *Radiology*. 1991;178(3):811–8. PubMed PMID: 1994423.
34. Alcazar JL, Guerriero S, Pascual MA, Ajossa S, Olartecoechea B, Hereter L. Clinical and sonographic features of uncommon primary ovarian malignancies. *Journal of Clinical Ultrasound*. 2012;40(6):323–9.
35. Takeuchi M, Matsuzaki K, Sano N, Furumoto H, Nishitani H. Malignant Brenner tumor with transition from benign to malignant components: Computed tomographic and magnetic resonance imaging findings with pathological correlation. *Journal of Computer Assisted Tomography*. 2008;32(4):553–4.
36. Tang YZ, Liyanage S, Narayanan P, Sahdev A, Sohaib A, Singh N et al. The MRI features of histologically proven ovarian cystadenofibromas—an assessment of the morphological and enhancement patterns. *European Radiology*. 2013;23(1):48–56.
37. Chu SM, Ming YC, Chao HC, Lai JY, Chen JC, Yung CP et al. Ovarian tumors in the pediatric age group: 37 cases treated over an 8-year period. *Chang Gung Medical Journal*. 2010;33(2):152–6.
38. Park SB, Kim JK, Kim KR, Cho KS. Imaging findings of complications and unusual manifestations of ovarian teratomas. *RadioGraphics*. 2008;28(4):969–83.
39. Ulbright TM. Germ cell tumors of the gonads: A selective review emphasizing problems in differential diagnosis, newly appreciated, and controversial issues. *Modern Pathology*. 2005;18(Suppl 2):S61–79.
40. Hackethal A, Brueggmann D, Bohlmann MK, Franke FE, Tinneberg HR, Munstedt K. Squamous-cell carcinoma in mature cystic teratoma of the ovary: Systematic review and analysis of published data. *The Lancet Oncology*. 2008;9(12):1173–80.
41. Huss M, Lafay-Pillet MC, Lecuru F, Ruscillo MM, Chevalier JM, Vilde F et al. Granulomatous peritonitis after laparoscopic surgery of an ovarian dermoid cyst. Diagnosis, management, prevention, a case report (Peritonite granulomateuse apres traitement coeliochirurgical d'un kyste dermoide de l'ovaire. Diagnostic, prise en charge, prevention, a propos d'un cas). *Journal de gynecologie, obstetrique et biologie de la reproduction*. 1996;25(4):365–72.
42. Sivanesaratnam V, Dutta R, Jayalakshmi P. Ovarian fibroma—Clinical and histopathological characteristics. *International Journal of Gynaecology and Obstetrics*. 1990;33(3):243–7.
43. Leung SW, Yuen PM. Ovarian fibroma: A review on the clinical characteristics, diagnostic difficulties, and management options of 23 cases. *Gynecologic and Obstetric Investigation*. 2006;62(1):1 6.
44. Jung SE, Rha SE, Lee JM, Park SY, Oh SN, Cho KS et al. CT and MRI findings of sex cord-stromal tumor of the ovary. *AJR American Journal of Roentgenology*. 2005;185(1):207–15.

45. Outwater EK, Siegelman ES, Talerman A, Dunton C. Ovarian fibromas and cystadenofibromas: MRI features of the fibrous component. *Journal of Magnetic Resonance Imaging*. 1997;7(3):465–71.
46. Troiano RN, Lazzarini KM, Scoutt LM, Lange RC, Flynn SD, McCarthy S. Fibroma and fibrothecoma of the ovary: MR imaging findings. *Radiology*. 1997;204(3):795–8.
47. Yen P, Khong K, Lamba R, Corwin MT, Gerscovich EO. Ovarian fibromas and fibrothecomas: Sonographic correlation with computed tomography and magnetic resonance imaging: A 5-year single-institution experience. *Journal of Ultrasound in Medicine*. 2013;32(1):13–8.
48. Kitajima K, Kaji Y, Sugimura K. Usual and unusual MRI findings of ovarian fibroma: Correlation with pathologic findings. *Magnetic Resonance in Medical Sciences*. 2008;7(1):43–8.
49. Takehara M, Saito T, Manase K, Suzuki T, Hayashi T, Kudo R. Hemorrhagic infarction of fibroma. MR imaging appearance. *Archives of Gynecology and Obstetrics*. 2002;266(1):48–9.
50. Ohara N, Murao S. Magnetic resonance appearances of ovarian fibroma with myxomatous changes. *Journal of Obstetrics and Gynaecology*. 2002;22(5):569–70.
51. Chen VW, Ruiz B, Killeen JL, Cote TR, Wu XC, Correa CN. Pathology and classification of ovarian tumors. *Cancer*. 2003;97(10 Suppl):2631–42.
52. Nocito AL, Sarancone S, Bacchi C, Tellez T. Ovarian thecoma: Clinicopathological analysis of 50 cases. *Annals of Diagnostic Pathology*. 2008;12(1):12–6.
53. Tanaka YO, Tsunoda H, Kitagawa Y, Ueno T, Yoshikawa H, Saida Y. Functioning ovarian tumors: Direct and indirect findings at MR imaging. *RadioGraphics*. 2004;24(Suppl 1):S147–66.
54. Zhang H, Zhang GF, Wang TP, Zhang H. Value of 3.0 T diffusion-weighted imaging in discriminating thecoma and fibrothecoma from other adnexal solid masses. *Journal of Ovarian Research*. 2013;6(1):58.
55. Pickel H, Tamussino K. History of gynecological pathology: XIV. Hermann Johannes Pfannenstiel. *International Journal of Gynecological Pathology*. 2003;22(3):310–4.
56. Taylor Jr HC. Malignant and semi-malignant tumors of the ovary. *Surgery Gynecology & Obstetrics*. 1929;48:204–30.
57. Scully RE, Sobin LH. *Histologic Typing of Ovarian Tumours*. World Health Organisation International Histological Classification of Tumors. Springer-Verlag, New York; 1999.
58. Scully RE. Common epithelial tumors of borderline malignancy (carcinomas of low malignant potential). *Bulletin du cancer*. 1982;69(3):228–38.
59. Kennedy AW, Hart WR. Ovarian papillary serous tumors of low malignant potential (serous borderline tumors). A long-term follow-up study, including patients with microinvasion, lymph node metastasis, and transformation to invasive serous carcinoma. *Cancer*. 1996;78(2):278–86.
60. Segal GH, Hart WR. Ovarian serous tumors of low malignant potential (serous borderline tumors). The relationship of exophytic surface tumor to peritoneal "implants." *The American Journal of Surgical Pathology*. 1992;16(6):577–83.
61. Bent CL, Sahdev A, Rockall AG, Singh N, Sohaib SA, Reznek RH. MRI appearances of borderline ovarian tumours. *Clinical Radiology*. 2009;64(4):430–8.
62. Hanselaar AG, Vooijs GP, Mayall B, Ras-Zeijlmans GJ, Chadha-Ajwani S. Epithelial markers to detect occult microinvasion in serous ovarian tumors. *International Journal of Gynecological Pathology*. 1993;12(1):20–7.
63. DeSouza NM, O'Neill R, McIndoe GA, Dina R, Soutter WP. Borderline tumors of the ovary: CT and MRI features and tumor markers in differentiation from stage I disease. *AJR American Journal of Roentgenology*. 2005;184(3):999–1003.
64. Tanaka YO, Okada S, Satoh T, Matsumoto K, Oki A, Nishida M et al. Ovarian serous surface papillary borderline tumors form sea anemone-like masses. *Journal of Magnetic Resonance Imaging*. 2011;33(3):633–40.
65. Malpica A, Deavers MT, Gershenson D, Tortolero-Luna G, Silva EG. Serous tumors involving extra-abdominal/extra-pelvic sites after the diagnosis of an ovarian serous neoplasm of low malignant potential. *The American Journal of Surgical Pathology*. 2001;25(8):988–96.
66. Tanaka YO, Okada S, Satoh T, Matsumoto K, Oki A, Saida T et al. Diversity in size and signal intensity in multilocular cystic ovarian masses: New parameters for distinguishing metastatic from primary mucinous ovarian neoplasms. *Journal of Magnetic Resonance Imaging*. 2013;38(4):794–801.
67. Hart WR, Norris HJ. Borderline and malignant mucinous tumors of the ovary. Histologic criteria and clinical behavior. *Cancer*. 1973;31(5):1031–45.
68. Lee KR, Scully RE. Mucinous tumors of the ovary: A clinicopathologic study of 196 borderline tumors (of intestinal type) and carcinomas, including an evaluation of 11 cases with 'pseudomyxoma peritonei'. *The American Journal of Surgical Pathology*. 2000;24(11):1447–64.
69. Rodriguez IM, Prat J. Mucinous tumors of the ovary: A clinicopathologic analysis of 75 borderline tumors (of intestinal type) and carcinomas. *The American Journal of Surgical Pathology*. 2002;26(2):139–52.
70. Colombo N, Peiretti M, Parma G, Lapresa M, Mancari R, Carinelli S et al. Newly diagnosed and relapsed epithelial ovarian carcinoma: ESMO clinical practice guidelines for diagnosis, treatment and follow-up. *Annals of Oncology*. 2010;21(Suppl 5):v23–30.
71. Siegel R, Ward E, Brawley O, Jemal A. Cancer statistics, 2011: The impact of eliminating socioeconomic and racial disparities on premature cancer deaths. *CA: A Cancer Journal for Clinicians*. 2011;61(4):212–36.
72. Jemal A, Siegel R, Xu J, Ward E. Cancer statistics, 2010. *CA: A Cancer Journal for Clinicians*. 2010;60(5):277–300.
73. Seidman JD, Horkayne-Szakaly I, Haiba M, Boice CR, Kurman RJ, Ronnett BM. The histologic type and stage distribution of ovarian carcinomas of surface epithelial origin. *International Journal of Gynecological Pathology*. 2004;23(1):41–4.
74. Kobel M, Kalloger SE, Huntsman DG, Santos JL, Swenerton KD, Seidman JD et al. Differences in tumor

type in low-stage versus high-stage ovarian carcinomas. *International Journal of Gynecological Pathology*. 2010;29(3):203–11.
75. Miesfeldt S, Lamb A, Duarte C. Management of genetic syndromes predisposing to gynecologic cancers. *Current Treatment Options in Oncology*. 2013;14(1):34–50.
76. Antoniou A, Pharoah PD, Narod S, Risch HA, Eyfjord JE, Hopper JL et al. Average risks of breast and ovarian cancer associated with BRCA1 or BRCA2 mutations detected in case Series unselected for family history: A combined analysis of 22 studies. *American Journal of Human Genetics*. 2003;72(5):1117–30.
77. Chen S, Parmigiani G. Meta-analysis of BRCA1 and BRCA2 penetrance. *Journal of Clinical Oncology*. 2007;25(11):1329–33.
78. Sohaib SA, Reznek RH. MR imaging in ovarian cancer. *Cancer Imaging*. 2007;7 Spec No A:S119–29.
79. Spencer JA, Ghattamaneni S. MR imaging of the sonographically indeterminate adnexal mass. *Radiology*. 2010;256(3):677–94.
80. Anthoulakis C, Nikoloudis N. Pelvic MRI as the "gold standard" in the subsequent evaluation of ultrasound-indeterminate adnexal lesions: A systematic review. *Gynecologic Oncology*. 2014;132(3):661–8.
81. Kinkel K, Lu Y, Mehdizade A, Pelte MF, Hricak H. Indeterminate ovarian mass at US: Incremental value of second imaging test for characterization—Meta-analysis and Bayesian analysis. *Radiology*. 2005;236(1):85–94.
82. Outwater EK, Huang AB, Dunton CJ, Talerman A, Capuzzi DM. Papillary projections in ovarian neoplasms: Appearance on MRI. *Journal of Magnetic Resonance Imaging*. 1997;7(4):689–95.
83. Togashi K. Ovarian cancer: The clinical role of US, CT, and MRI. *European Radiology*. 2003;13(Suppl 4):L87–104.
84. Mohaghegh P, Rockall AG. Imaging strategy for early ovarian cancer: Characterization of adnexal masses with conventional and advanced imaging techniques. *RadioGraphics*. 2012;32(6):1751–73.
85. Sohaib SA, Sahdev A, Van Trappen P, Jacobs IJ, Reznek RH. Characterization of adnexal mass lesions on MR imaging. *AJR American Journal of Roentgenology*. 2003;180(5):1297–304.
86. Medeiros LR, Freitas LB, Rosa DD, Silva FR, Silva LS, Birtencourt LT et al. Accuracy of magnetic resonance imaging in ovarian tumor: A systematic quantitative review. *American Journal of Obstetrics & Gynecology*. 2011;204(1):67.e1–10.
87. Thomassin-Naggara I, Bazot M, Darai E, Callard P, Thomassin J, Cuenod CA. Epithelial ovarian tumors: Value of dynamic contrast-enhanced MR imaging and correlation with tumor angiogenesis. *Radiology*. 2008;248(1):148–59.
88. Bernardin L, Dilks P, Liyanage S, Miquel ME, Sahdev A, Rockall A. Effectiveness of semi-quantitative multiphase dynamic contrast-enhanced MRI as a predictor of malignancy in complex adnexal masses: Radiological and pathological correlation. *European Radiology*. 2012;22(4):880 90.
89. Thomassin-Naggara I, Balvay D, Aubert E, Darai E, Rouzier R, Cuenod CA et al. Quantitative dynamic contrast-enhanced MR imaging analysis of complex adnexal masses: A preliminary study. *European Radiology*. 2012;22(4):738–45.
90. Pannu HK, Ma W, Zabor EC, Moskowitz CS, Barakat RR, Hricak H. Enhancement of ovarian malignancy on clinical contrast enhanced MRI studies. *ISRN Obstetrics and Gynecology*. 2013;2013:979345.
91. Thomassin-Naggara I, Darai E, Cuenod CA, Fournier L, Toussaint I, Marsault C et al. Contribution of diffusion-weighted MR imaging for predicting benignity of complex adnexal masses. *European Radiology*. 2009;19(6):1544–52.
92. Thomassin-Naggara I, Toussaint I, Perrot N, Rouzier R, Cuenod CA, Bazot M et al. Characterization of complex adnexal masses: Value of adding perfusion- and diffusion-weighted MR imaging to conventional MR imaging. *Radiology*. 2011;258(3):793–803.
93. Thomassin-Naggara I, Aubert E, Rockall A, Jalaguier-Coudray A, Rouzier R, Darai E et al. Adnexal masses: Development and preliminary validation of an MR imaging scoring system. *Radiology*. 2013;267(2):432–43.
94. Katayama M, Masui T, Kobayashi S, Ito T, Sakahara H, Nozaki A et al. Diffusion-weighted echo planar imaging of ovarian tumors: Is it useful to measure apparent diffusion coefficients? *Journal of Computer Assisted Tomography*. 2002;26(2):250–6.
95. Benedet JL, Bender H, Jones H, 3rd, Ngan HY, Pecorelli S. FIGO staging classifications and clinical practice guidelines in the management of gynecologic cancers. FIGO Committee on Gynecologic Oncology. *International Journal of Gynaecology and Obstetrics*. 2000;70(2):209–62.
96. Coakley FV, Choi PH, Gougoutas CA, Pothuri B, Venkatraman E, Chi D et al. Peritoneal metastases: Detection with spiral CT in patients with ovarian cancer. *Radiology*. 2002;223(2):495–9.
97. Pannu HK, Horton KM, Fishman EK. Thin section dual-phase multidetector-row computed tomography detection of peritoneal metastases in gynecologic cancers. *Journal of Computer Assisted Tomography*. 2003;27(3):333–40.
98. Tempany CM, Zou KH, Silverman SG, Brown DL, Kurtz AB, McNeil BJ. Staging of advanced ovarian cancer: Comparison of imaging modalities—Report from the Radiological Diagnostic Oncology Group. *Radiology*. 2000;215(3):761–7.
99. Woodward PJ, Hosseinzadeh K, Saenger JS. From the archives of the AFIP: Radiologic staging of ovarian carcinoma with pathologic correlation. *RadioGraphics*. 2004;24(1):225–46.
100. Kyriazi S, Kaye SB, deSouza NM. Imaging ovarian cancer and peritoneal metastases—Current and emerging techniques. *Nature Reviews Clinical Oncology*. 2010;7(7):381–93.
101. Kurtz AB, Tsimikas JV, Tempany CM, Hamper UM, Arger PH, Bree RL et al. Diagnosis and staging of ovarian cancer: Comparative values of Doppler and conventional US, CT, and MR imaging correlated with surgery and histopathologic analysis—Report of the Radiology Diagnostic Oncology Group. *Radiology*. 1999;212(1):19–27.
102. Low RN, Barone RM, Lacey C, Sigeti JS, Alzate GD, Sebrechts CP. Peritoneal tumor: MR imaging with dilute oral barium and intravenous gadolinium-containing

contrast agents compared with unenhanced MR imaging and CT. *Radiology*. 1997;204(2):513–20.
103. Levy AD, Shaw JC, Sobin LH. Secondary tumors and tumorlike lesions of the peritoneal cavity: Imaging features with pathologic correlation. *RadioGraphics*. 2009;29(2):347–73.
104. Raptopoulos V, Gourtsoyiannis N. Peritoneal carcinomatosis. *European Radiology*. 2001;11(11):2195–206.
105. Low RN, Gurney J. Diffusion-weighted MRI (DWI) in the oncology patient: Value of breathhold DWI compared to unenhanced and gadolinium-enhanced MRI. *Journal of Magnetic Resonance Imaging*. 2007;25(4):848–58.
106. Low RN, Sebrechts CP, Barone RM, Muller W. Diffusion-weighted MRI of peritoneal tumors: Comparison with conventional MRI and surgical and histopathologic findings—A feasibility study. *AJR American Journal of Roentgenology*. 2009;193(2):461–70.
107. Soussan M, Des Guetz G, Barrau V, Aflalo-Hazan V, Pop G, Mehanna Z et al. Comparison of FDG-PET/CT and MR with diffusion-weighted imaging for assessing peritoneal carcinomatosis from gastrointestinal malignancy. *European Radiology*. 2012;22(7):1479–87.
108. Kyriazi S, Collins DJ, Morgan VA, Giles SL, deSouza NM. Diffusion-weighted imaging of peritoneal disease for noninvasive staging of advanced ovarian cancer. *RadioGraphics*. 2010;30(5):1269–85.
109. Klumpp BD, Aschoff P, Schwenzer N, Fenchel M, Koenigsrainer I, Falch C et al. Peritoneal carcinomatosis: Comparison of dynamic contrast-enhanced magnetic resonance imaging with surgical and histopathologic findings. *Abdominal Imaging*. 2012;37(5):834–42.
110. Singer G, Kurman RJ, Chang HW, Cho SK, Shih Ie M. Diverse tumorigenic pathways in ovarian serous carcinoma. *The American Journal of Pathology*. 2002;160(4):1223–8.
111. Singer G, Stohr R, Cope L, Dehari R, Hartmann A, Cao DF et al. Patterns of p53 mutations separate ovarian serous borderline tumors and low- and high-grade carcinomas and provide support for a new model of ovarian carcinogenesis: A mutational analysis with immunohistochemical correlation. *The American Journal of Surgical Pathology*. 2005;29(2):218–24.
112. McCluggage WG, Wilkinson N. Metastatic neoplasms involving the ovary: A review with an emphasis on morphological and immunohistochemical features. *Histopathology*. 2005;47(3):231–47.
113. Prat J. Ovarian carcinomas: Five distinct diseases with different origins, genetic alterations, and clinicopathological features. *Virchows Archiv: An International Journal of Pathology*. 2012;460(3):237–49.
114. Malpica A, Deavers MT, Lu K, Bodurka DC, Atkinson EN, Gershenson DM et al. Grading ovarian serous carcinoma using a two-tier system. *The American Journal of Surgical Pathology*. 2004;28(4):496–504.
115. Hart WR. Mucinous tumors of the ovary: A review. *International Journal of Gynecological Pathology*. 2005;24(1):4–25.
116. Kikkawa F, Nawa A, Kajiyama H, Shibata K, Ino K, Nomura S. Clinical characteristics and prognosis of mucinous tumors of the ovary. *Gynecologic Oncology*. 2006;103(1):171–5.
117. Hoerl HD, Hart WR. Primary ovarian mucinous cystadenocarcinomas: A clinicopathologic study of 49 cases with long-term follow-up. *The American Journal of Surgical Pathology*. 1998;22(12):1449–62.
118. Jung SE, Lee JM, Rha SE, Byun JY, Jung JI, Hahn ST. CT and MR imaging of ovarian tumors with emphasis on differential diagnosis. *RadioGraphics*. 2002;22(6):1305–25.
119. Wagner BJ, Buck JL, Seidman JD, McCabe KM. From the archives of the AFIP. Ovarian epithelial neoplasms: Radiologic-pathologic correlation. *RadioGraphics*. 1994;14(6):1351–74.
120. Seidman JD, Kurman RJ, Ronnett BM. Primary and metastatic mucinous adenocarcinomas in the ovaries: Incidence in routine practice with a new approach to improve intraoperative diagnosis. *The American Journal of Surgical Pathology*. 2003;27(7):985–93.
121. Khunamornpong S, Suprasert P, Pojchamarnwiputh S, Na Chiangmai W, Settakorn J, Siriaunkgul S. Primary and metastatic mucinous adenocarcinomas of the ovary: Evaluation of the diagnostic approach using tumor size and laterality. *Gynecologic Oncology*. 2006;101(1):152–7.
122. Kelemen LE, Kobel M. Mucinous carcinomas of the ovary and colorectum: Different organ, same dilemma. *The Lancet Oncology*. 2011;12(11):1071–80.
123. Ji H, Isacson C, Seidman JD, Kurman RJ, Ronnett BM. Cytokeratins 7 and 20, Dpc4, and MUC5AC in the distinction of metastatic mucinous carcinomas in the ovary from primary ovarian mucinous tumors: Dpc4 assists in identifying metastatic pancreatic carcinomas. *International Journal of Gynecological Pathology*. 2002;21(4):391–400.
124. Vang R, Gown AM, Barry TS, Wheeler DT, Yemelyanova A, Seidman JD et al. Cytokeratins 7 and 20 in primary and secondary mucinous tumors of the ovary: Analysis of coordinate immunohistochemical expression profiles and staining distribution in 179 cases. *The American Journal of Surgical Pathology*. 2006;30(9):1130–9.
125. Ronnett BM, Seidman JD. Mucinous tumors arising in ovarian mature cystic teratomas: Relationship to the clinical syndrome of pseudomyxoma peritonei. *The American Journal of Surgical Pathology*. 2003;27(5):650–7.
126. Bell KA, Kurman RJ. A clinicopathologic analysis of atypical proliferative (borderline) tumors and well-differentiated endometrioid adenocarcinomas of the ovary. *The American Journal of Surgical Pathology*. 2000;24(11):1465–79.
127. Stern RC, Dash R, Bentley RC, Snyder MJ, Haney AF, Robboy SJ. Malignancy in endometriosis: Frequency and comparison of ovarian and extraovarian types. *International Journal of Gynecological Pathology*. 2001;20(2):133–9.
128. Valenzuela P, Ramos P, Redondo S, Cabrera Y, Alvarez I, Ruiz A. Endometrioid adenocarcinoma of the ovary and endometriosis. *European Journal of Obstetrics, Gynecology, and Reproductive Biology*. 2007;134(1):83–6.
129. Zaino R, Whitney C, Brady MF, DeGeest K, Burger RA, Buller RE. Simultaneously detected endometrial and ovarian carcinomas—A prospective clinicopathologic

study of 74 cases: A gynecologic oncology group study. *Gynecologic Oncology*. 2001;83(2):355–62.
130. Soliman PT, Slomovitz BM, Broaddus RR, Sun CC, Oh JC, Eifel PJ et al. Synchronous primary cancers of the endometrium and ovary: A single institution review of 84 cases. *Gynecologic Oncology*. 2004;94(2):456–62.
131. Somigliana E, Vigano P, Parazzini F, Stoppelli S, Giambattista E, Vercellini P. Association between endometriosis and cancer: A comprehensive review and a critical analysis of clinical and epidemiological evidence. *Gynecologic Oncology*. 2006;101(2):331–41.
132. Sugiyama T, Kamura T, Kigawa J, Terakawa N, Kikuchi Y, Kita T et al. Clinical characteristics of clear cell carcinoma of the ovary: A distinct histologic type with poor prognosis and resistance to platinum-based chemotherapy. *Cancer*. 2000;88(11):2584–9.
133. Chan JK, Teoh D, Hu JM, Shin JY, Osann K, Kapp DS. Do clear cell ovarian carcinomas have poorer prognosis compared to other epithelial cell types? A study of 1411 clear cell ovarian cancers. *Gynecologic Oncology*. 2008;109(3):370–6.
134. Duska LR, Garrett L, Henretta M, Ferriss JS, Lee L, Horowitz N. When 'never-events' occur despite adherence to clinical guidelines: The case of venous thromboembolism in clear cell cancer of the ovary compared with other epithelial histologic subtypes. *Gynecologic Oncology*. 2010;116(3):374–7.
135. Matsuura Y, Robertson G, Marsden DE, Kim SN, Gebski V, Hacker NF. Thromboembolic complications in patients with clear cell carcinoma of the ovary. *Gynecologic Oncology*. 2007;104(2):406–10.
136. Behbakht K, Randall TC, Benjamin I, Morgan MA, King S, Rubin SC. Clinical characteristics of clear cell carcinoma of the ovary. *Gynecologic Oncology*. 1998;70(2):255–8.
137. Takano M, Kikuchi Y, Yaegashi N, Kuzuya K, Ueki M, Tsuda H et al. Clear cell carcinoma of the ovary: A retrospective multicentre experience of 254 patients with complete surgical staging. *British Journal of Cancer*. 2006;94(10):1369–74.
138. Matsuoka Y, Ohtomo K, Araki T, Kojima K, Yoshikawa W, Fuwa S. MR imaging of clear cell carcinoma of the ovary. *European Radiology*. 2001;11(6):946–51.
139. Moon WJ, Koh BH, Kim SK, Kim YS, Rhim HC, Cho OK et al. Brenner tumor of the ovary: CT and MR findings. *Journal of Computer Assisted Tomography*. 2000;24(1):72–6.
140. Outwater EK, Siegelman ES, Kim B, Chiowanich P, Blasbalg R, Kilger A. Ovarian Brenner tumors: MR imaging characteristics. *Magnetic Resonance Imaging*. 1998;16(10):1147–53.
141. Takahama J, Ascher SM, Hirohashi S, Takewa M, Ito T, Iwasaki S et al. Borderline Brenner tumor of the ovary: MRI findings. *Abdominal Imaging*. 2004;29(4):528–30.
142. Outwater EK, Wagner BJ, Mannion C, McLarney JK, Kim B. Sex cord-stromal and steroid cell tumors of the ovary. *RadioGraphics*. 1998;18(6):1523–46.
143. Schultz KA, Schneider DT, Pashankar F, Ross J, Frazier L. Management of ovarian and testicular sex cord-stromal tumors in children and adolescents. *Journal of Pediatric Hematology/Oncology*. 2012;34(Suppl 2):S55–63.
144. Tanaka YO, Saida TS, Minami R, Yagi T, Tsunoda H, Yoshikawa H et al. MR findings of ovarian tumors with hormonal activity, with emphasis on tumors other than sex cord-stromal tumors. *European Journal of Radiology*. 2007;62(3):317–27.
145. Young RH, Dickersin GR, Scully RE. Juvenile granulosa cell tumor of the ovary. A clinicopathological analysis of 125 cases. *The American Journal of Surgical Pathology*. 1984;8(8):575–96.
146. Pectasides D, Pectasides E, Psyrri A. Granulosa cell tumor of the ovary. *Cancer Treatment Reviews*. 2008;34(1):1–12.
147. Crew KD, Cohen MH, Smith DH, Tiersten AD, Feirt NM, Hershman DL. Long natural history of recurrent granulosa cell tumor of the ovary 23 years after initial diagnosis: A case report and review of the literature. *Gynecologic Oncology*. 2005;96(1):235–40.
148. Ko SF, Wan YL, Ng SH, Lee TY, Lin JW, Chen WJ et al. Adult ovarian granulosa cell tumors: Spectrum of sonographic and CT findings with pathologic correlation. *AJR American Journal of Roentgenology*. 1999;172(5):1227–33.
149. Kim SH, Kim SH. Granulosa cell tumor of the ovary: Common findings and unusual appearances on CT and MR. *Journal of Computer Assisted Tomography*. 2002;26(5):756–61.
150. Morikawa K, Hatabu H, Togashi K, Kataoka ML, Mori T, Konishi J. Granulosa cell tumor of the ovary: MR findings. *Journal of Computer Assisted Tomography*. 1997;21(6):1001–4.
151. Young RH, Scully RE. Ovarian Sertoli–Leydig cell tumors. A clinicopathological analysis of 207 cases. *The American Journal of Surgical Pathology*. 1985;9(8):543–69.
152. Zaloudek C, Norris HJ. Sertoli–Leydig tumors of the ovary. A clinicopathologic study of 64 intermediate and poorly differentiated neoplasms. *The American Journal of Surgical Pathology*. 1984;8(6):405–18.
153. Hayes MC, Scully RE. Ovarian steroid cell tumors (not otherwise specified). A clinicopathological analysis of 63 cases. *The American Journal of Surgical Pathology*. 1987;11(11):835–45.
154. Reedy MB, Richards WE, Ueland F, Uy K, Lee EY, Bryant C et al. Ovarian steroid cell tumors, not otherwise specified: A case report and literature review. *Gynecologic Oncology*. 1999;75(2):293–7.
155. Wang PH, Chao HT, Lee RC, Lai CR, Lee WL, Kwok CF et al. Steroid cell tumors of the ovary: Clinical, ultrasonic, and MRI diagnosis—A case report. *European Journal of Radiology*. 1998;26(3):269–73.
156. Brammer HM, 3rd, Buck JL, Hayes WS, Sheth S, Tavassoli FA. From the archives of the AFIP. Malignant germ cell tumors of the ovary: Radiologic-pathologic correlation. *RadioGraphics*. 1990;10(4):715–24.
157. Pectasides D, Pectasides E, Kassanos D. Germ cell tumors of the ovary. *Cancer Treatment Reviews*. 2008;34(5):427–41.
158. Zaloudek CJ, Tavassoli FA, Norris HJ. Dysgerminoma with syncytiotrophoblastic giant cells. A histologically and clinically distinctive subtype of dysgerminoma. *The American Journal of Surgical Pathology*. 1981;5(4):361–7.

159. Kim SK, Sohn IS, Kim JW, Song CH, Park CI, Lee MS et al. Gonadoblastoma and dysgerminoma associated with 46,XY pure gonadal dysgenesis—A case report. *Journal of Korean Medical Science*. 1993;8(5):380–4.
160. Kim SH, Kang SB. Ovarian dysgerminoma: Color Doppler ultrasonographic findings and comparison with CT and MR imaging findings. *Journal of Ultrasound in Medicine*. 1995;14(11):843–8.
161. Tanaka YO, Kurosaki Y, Nishida M, Michishita N, Kuramoto K, Itai Y et al. Ovarian dysgerminoma: MR and CT appearance. *Journal of Computer Assisted Tomography*. 1994;18(3):443–8.
162. Nawa A, Obata N, Kikkawa F, Kawai M, Nagasaka T, Goto S et al. Prognostic factors of patients with yolk sac tumors of the ovary. *American Journal of Obstetrics & Gynecology*. 2001;184(6):1182–8.
163. Levitin A, Haller KD, Cohen HL, Zinn DL, O'Connor MT. Endodermal sinus tumor of the ovary: Imaging evaluation. *AJR American Journal of Roentgenology*. 1996;167(3):791–3.
164. Yamaoka T, Togashi K, Koyama T, Ueda H, Nakai A, Fujii S et al. Yolk sac tumor of the ovary: Radiologic-pathologic correlation in four cases. *Journal of Computer Assisted Tomography*. 2000;24(4):605–9.
165. Koo HL, Choi J, Kim KR, Kim JH. Pure non-gestational choriocarcinoma of the ovary diagnosed by DNA polymorphism analysis. *Pathology International*. 2006;56(10):613–6.
166. Bazot M, Cortez A, Sananes S, Buy JN. Imaging of pure primary ovarian choriocarcinoma. *AJR American Journal of Roentgenology*. 2004;182(6):1603–4.
167. Talerman A. In: Kurman RJ, editor. *Blausteins Pathology of the Female Genital Tract*, 5th Edition. Springer-Verlag, New York; 2002, pp. 994–7.
168. Yanai-Inbar I, Scully RE. Relation of ovarian dermoid cysts and immature teratomas: An analysis of 350 cases of immature teratoma and 10 cases of dermoid cyst with microscopic foci of immature tissue. *International Journal of Gynecological Pathology*. 1987;6(3):203–12.
169. Heifetz SA, Cushing B, Giller R, Shuster JJ, Stolar CJ, Vinocur CD et al. Immature teratomas in children: Pathologic considerations: A report from the combined Pediatric Oncology Group/Children's Cancer Group. *The American Journal of Surgical Pathology*. 1998;22(9):1115–24.
170. Saba L, Guerriero S, Sulcis R, Virgilio B, Melis G, Mallarini G. Mature and immature ovarian teratomas: CT, US and MR imaging characteristics. *European Journal of Radiology*. 2009;72(3):454–63.
171. Yamaoka T, Togashi K, Koyama T, Fujiwara T, Higuchi T, Iwasa Y et al. Immature teratoma of the ovary: Correlation of MR imaging and pathologic findings. *European Radiology*. 2003;13(2):313–9.
172. England RA, deSouza NM, Kaye SB. Gliomatosis peritonei: MRI appearances and its potential role in follow up. *The British Journal of Radiology*. 2007;80(953):e101–4.
173. Pentheroudakis G, Pavlidis N. Serous papillary peritoneal carcinoma: Uprimary tumour, ovarian cancer counterpart or a distinct entity? A systematic review. *Critical Reviews in Oncology/Hematology*. 2010;75(1):27–42.
174. Bloss JD, Liao SY, Buller RE, Manetta A, Berman ML, McMeekin S et al. Extraovarian peritoneal serous papillary carcinoma: A case-control retrospective comparison to papillary adenocarcinoma of the ovary. *Gynecologic Oncology*. 1993;50(3):347–51.
175. Wang PH, Shyong WY, Li YF, Lee HH, Tsai WY, Chao HT et al. BRCA1 mutations in Taiwanese with epithelial ovarian carcinoma and sporadic primary serous peritoneal carcinoma. *Japanese Journal of Clinical Oncology*. 2000;30(8):343–8.
176. Casey MJ, Synder C, Bewtra C, Narod SA, Watson P, Lynch HT. Intra-abdominal carcinomatosis after prophylactic oophorectomy in women of hereditary breast ovarian cancer syndrome kindreds associated with BRCA1 and BRCA2 mutations. *Gynecologic Oncology*. 2005;97(2):457–67.
177. Olivier RI, van Beurden M, Lubsen MA, Rookus MA, Mooij TM, van de Vijver MJ et al. Clinical outcome of prophylactic oophorectomy in BRCA1/BRCA2 mutation carriers and events during follow-up. *British Journal of Cancer*. 2004;90(8):1492–7.
178. Levy AD, Arnaiz J, Shaw JC, Sobin LH. From the archives of the AFIP: Primary peritoneal tumors: Imaging features with pathologic correlation. *RadioGraphics*. 2008;28(2):583–607.
179. Morita H, Aoki J, Taketomi A, Sato N, Endo K. Serous surface papillary carcinoma of the peritoneum: Clinical, radiologic, and pathologic findings in 11 patients. *AJR American Journal of Roentgenology*. 2004;183(4):923–8.
180. Pfeiffer P, Mogensen H, Amtrup F, Honore E. Primary carcinoma of the fallopian tube. A retrospective study of patients reported to the Danish Cancer Registry in a five-year period. *Acta Oncologica*. 1989;28(1):7–11.
181. Riska A, Martinsen JI, Kjaerheim K, Lynge E, Sparen P, Tryggvadottir L et al. Occupation and risk of primary fallopian tube carcinoma in Nordic countries. *International Journal of Cancer*. 2012;131(1):186–92.
182. Stewart SL, Wike JM, Foster SL, Michaud F. The incidence of primary fallopian tube cancer in the United States. *Gynecologic Oncology*. 2007;107(3):392–7.
183. Jarboe E, Folkins A, Nucci MR, Kindelberger D, Drapkin R, Miron A et al. Serous carcinogenesis in the fallopian tube: A descriptive classification. *International Journal of Gynecological Pathology*. 2008;27(1):1–9.
184. Kindelberger DW, Lee Y, Miron A, Hirsch MS, Feltmate C, Medeiros F et al. Intraepithelial carcinoma of the fimbria and pelvic serous carcinoma: Evidence for a causal relationship. *The American Journal of Surgical Pathology*. 2007;31(2):161–9.
185. Finch A, Shaw P, Rosen B, Murphy J, Narod SA, Colgan TJ. Clinical and pathologic findings of prophylactic salpingo-oophorectomies in 159 BRCA1 and BRCA2 carriers. *Gynecologic Oncology*. 2006;100(1):58–64.
186. Powell CB, Swisher EM, Cass I, McLennan J, Norquist B, Garcia RL et al. Long term follow up of BRCA1 and BRCA2 mutation carriers with unsuspected neoplasia identified at risk reducing salpingo-oophorectomy. *Gynecologic Oncology*. 2013;129(2):364–71.

# 6

# *MRI of the Placenta and the Pregnant Patient*

**Philip S. Lim, Amy M. Mackey, and Monica L. Huang**

**CONTENTS**



This chapter will begin with a description of the normal placenta, then focus mostly on MR imaging of placenta accreta, and finally review some non-obstetric etiologies of abdominal pain in pregnancy. Fetal anomalies will not be included in this chapter.

## 6.1 The Normal Placenta: Physiology and Histopathology

The placenta is the site of exchange for nutrients essential for fetal growth and removal of fetal waste. The exchange occurs by several mechanisms. By simple diffusion, oxygen and carbon dioxide cross the syncytiotrophoblast, which creates a continuous uninterrupted layer covering the intervillous space. Oxygen and carbon dioxide are dependent on the concentration gradient to determine the direction of flow of these ions. Facilitated transport relies on membrane proteins to assist in the transport of molecules across the membranes of the placental villi. The transport of glucose molecules across the membrane barrier is an example of this. Its transport is facilitated by the GLUT1 protein. Active transport such as what occurs with the calcium ions requires adenosine triphoshate, an energy source, to move molecules against a concentration gradient. Finally, endocytosis and exocytosis allow larger molecules to move across the maternal–fetal barrier.

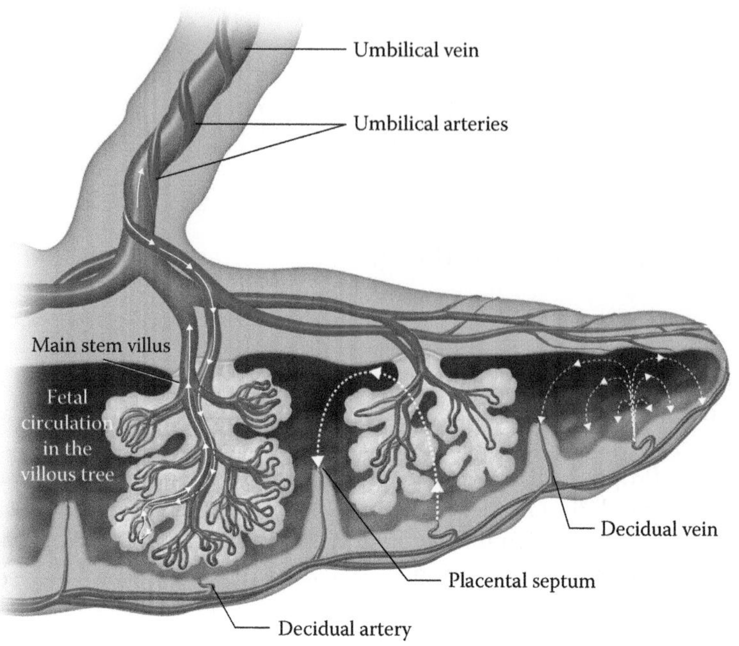


**FIGURE 6.1**
Anatomy of intervillous space. Typical spatial relations between villous trees and the maternal bloodstream. According to the placentone theory, a placentone is one villous tree together with the related part of the intervillous space. In the case of typical placentones that prevail in the periphery of the placenta, the maternal blood (arrows) enters the intervillous space near the center of the villous tree and leaves near the clefts between neighboring villous trees. One or only a few villous trees occupy one placental lobule (cotyledon). In the central parts of the placenta, the villous trees, because of size and nearby location, may partly overlap so that the zonal arrangement of the placentone disappears. (Reprinted with permission from Baergen, R.N., Overview and microscopic survey of the placenta, in: Baergen, R.N., ed, *Manual of Benirschke and Kaufmann's Pathology of the Human Placenta,* 2nd ed, Springer-Verlag, New York, 2011.)

When the placental barrier is intact, the increased excretion of proteins such as alpha fetoprotein can be a marker for abnormal fetal development. Increased maternal serum alpha-fetoprotein values can indicate the presence of an open neural tube defect, twin gestation, fetal demise, or gastroschisis. When placental pathology is present, the balance of molecules and proteins can be disrupted. An increased maternal serum alpha-fetoprotein level that is unexplained by fetal anomalies can indicate disruption of the syncytiotrophoblast and the integrity of the maternal–fetal barrier. Pregnancies with an unexplained elevated maternal serum alpha-fetoprotein have an increase in poor perinatal outcomes such as intrauterine growth restriction and intrauterine fetal demise [1]. Also, in several small studies, an elevated maternal serum alpha-fetoprotein has been found in 45% of patients with a placenta accreta [2,3].

In addition to fetal and maternal exchange, the placenta is an important site of hormone production. Hormones such as progesterone, estrogen, human placental lactogen (hPL), and human chorionic gonadotropin (hCG) are secreted by the syncytiotrophoblast and are important for the ongoing health of the pregnancy.

The intervillous space is the location where maternal blood circulates around the villous trees, which is seen histologically as a system of narrow clefts (Figure 6.1) [4]. Maternal blood is deposited into the intervillous space via the decidual arteries, which are located near the centers of the villous trees. Maternal blood exits through the peripheral maternal decidual veins, which are located in the placental septa. One fetomaternal circulatory unit, called a *placentone,* is composed of one villous tree with the corresponding centrifugally perfused intervillous space, each measuring 1–4 cm in size. There are 40–60 placentones in the placenta, which overlap with each other. Placentones may also have different degrees of maturation.

## 6.2 Developmental Variations in Placental Shape

At term, the normal placenta thins and stretches and has a diameter approximately 18–20 cm by 1.5–2.5 cm thickness [5]. The usual shape of the placenta is round or oval, but other shapes exist [6]. Variations in placental shape are presumed to be determined by location of implantation, atrophy of placenta, and possibly manner of original implantation.

The bilobed placenta, occurring in 2%–8% of placentas, is observed when there are approximately two equal-sized lobes separated by a segment of fetal membranes (Figure 6.2). The umbilical cord inserts usually in a membranous or velamentous fashion. The succenturate lobe, occurring in approximately 5% of placentas, is observed when one lobe is much smaller than the other lobe and the lobes are separated by a fetal membrane. Dynamic placentation is the process where the original implantation is modified so that there is preferential growth in areas of better perfusion and atrophy or infarction in areas of poor perfusion. It is observed that approximately 50% of the succenturiate lobes are associated with infarction or atrophy. Membranous vessels lack the protection from Wharton's jelly and are therefore at risk for compression, rupture, thrombosis, or bleeding. A condition of exposed vessels lying over the cervix is known as *vasa previa*, and this is a particularly dangerous situation during labor and artificial or spontaneous rupture of membranes.

Circumvallate placenta and circummarginate placenta occur when the fetal membranes attach more centrally toward the umbilical cord than the margin of the placenta. Circumvallation occurs from 1% to 6% of placentas and shows a fold or ridge of the fetal membranes containing fibrin, whereas circummargination occurs in up to 25% of placentas and shows flat membranes without a ridge.

Other rare abnormalities of placental shape include extramembranous pregnancy where there is early rupture of the amnion and chorion leading to the fetus in the uterine cavity; placenta membranacea where the thin, disrupted chorionic villi covers nearly all the circumference of the fetal sac; placental fenstrata where a central portion of the placenta is atrophied with only fetal membranes in place; and lastly, annular (zonary) placenta which has a ring shape due to a central focal atrophy of the placenta.

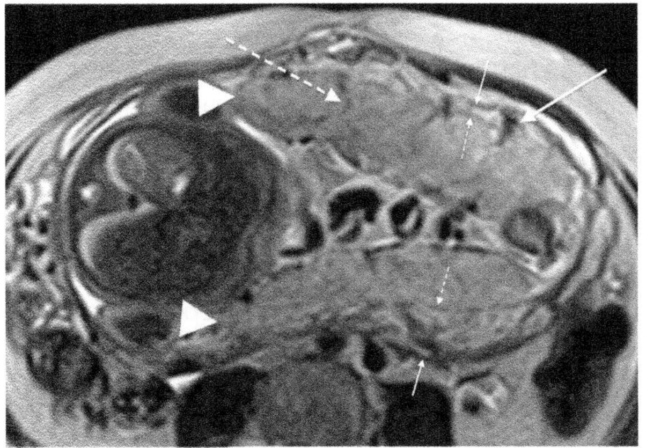

**FIGURE 6.2**
Bilobed placenta in a 29-year-old female at 30 weeks gestation. There are similarly sized separate anterior and posterior lobes of placenta (arrowheads). The normal myometrium is bordered by thin low signal lines (thin dashed and solid arrows). Dark signal vessels are normally coursing through the myometrium (long solid arrow) and thin dark signal vessels (long dashed arrow) are seen normally coursing through the placenta.

## 6.3 Normal Placental Thickness

Placental thickness is usually determined subjectively but is considered normal if it measures between 2 and 4 cm in the second and third trimesters [7]. The true thickness and appearance of the placenta may be altered by the following: uterine contractions, which may artifactually create placental thickening; polyhydramnios, which may compress the placenta making it appear small or thin; or oligohydramnios, which may make the placenta appear too large or thick. Abnormally thickened placentas are greater than 4 cm at its midposition. Differential diagnosis of thickened placentas includes blood group incompatibilities, diabetes mellitus, maternal anemia, thalassemias, triploidy, and fetal neoplasm [8]. Abnormally small placentas have the following etiologies: toxemia, hypertension, chromosomal abnormalities, severe diabetes mellitus, and chronic infection.

## 6.4 Abnormal Placentation: Proposed Pathogenesis and Histological Changes

The term *abnormal placentation* refers to placenta accreta, increta, or percreta. Histologically, *placenta accreta* is defined when anchoring villi implant on uterine smooth muscle without intervening decidua. Clinically, it is defined as adherent placenta that is difficult to separate from the myometrium at delivery. Abundant neovascularization of the placenta and uterus is a prominent clinical feature.

The pathogenesis of placenta accreta is unclear but is strongly associated with previous uterine surgery and uterine scar [9,10]. Several theories of abnormal placentation have been proposed, including abnormal or excessive trophoblast invasion and focal abnormalities in oxygen tension [11]. Oxygen tension regulates the growth and invasion of specialized placental cells called *cytrotrophoblasts*, which in turn influences placental growth and architecture. In normal pregnancies, cytotrophoblasts differentiate into tumor-like cells and invade the uterus and its vasculature to provide blood flow to the placenta. In an *in vitro* hypoxic condition (2% oxygen) that exists at the uterine surface prior to 10 weeks gestation, it has been shown that poorly differentiated cytotrophoblasts continued to proliferate. However, when cultured at 20% oxygen, similar to the area near uterine arterioles, the cytotrophoblasts stopped proliferating and differentiated normally [12].

Another theory is that the normal physiologic conversion of maternal vessels may be focally absent, possibly because of deficiency of decidualization [13]. The lower uterine segment has been observed histologically to have a deficient decidua. Deficiency of formation of placental septa has also been observed, which can create abnormal flow patterns and cause intraplacental thrombosis in the intervillous space which may lead to chronic thrombi or placental infarcts; it is difficult to distinguish chronic thrombi from placental infarcts histologically.

In normal placentas, there is an extensive vascular anastamosis in the subplacental myometrium, which creates arteriovenous shunts in the subplacental myometrium and provides blood to the intervillous space [14]. After a normal delivery, myometrium contractions stop the circulation in this vascular anastamosis. However, abnormal vascular architecture has been observed histologically at the placental–maternal interface in patients with placenta increta where the vessels are sparser and larger, and the vessel distribution (area and distance from the placental–maternal interface) is more heterogeneous compared to normal placentas. Both normal placentas and placenta increta have comparable total areas occupied by vessels [15].

## 6.5 Patients at Risk for Abnormal Placentation

The incidence of placenta accreta has been reported to range from 1 in 533 deliveries in 2005, 1 in 4027 in the 1970s, and 1 in 2510 in 1980s [16–18]. This increase has been attributed to rising cesarean section rates. At our institution, we have an incidence of placenta accreta of 1 in 1000 deliveries.

Patients who are at increased risk of accreta include any patient with a uterine scar. In addition, it has been reported that placenta previa alone carries a 3% risk. In patients with placenta previa, the risk of placenta accreta increases to 11%, 40%, and 61% with one, two, and three previous cesarean sections, respectively [19]. Patients with prior history of accreta found at pathology in a prior pregnancy should also be considered at high risk. Other maternal conditions, which have been reported in the literature to be associated with increased risk of accreta, include the following: prior uterine surgery including curettage, endometrial ablation, myomectomy penetrating the uterine cavity, Asherman's syndrome, uterine irradiation, increasing parity, increasing maternal age, uterine anomalies such as fibroids and septum, smoking, and hypertensive disorders; however, the level of severity of increased risk is considered lower.

## 6.6 Management of Patients with Abnormal Placentation: A Multidisciplinary Approach

When patients with potential placenta accreta are identified, they may be referred to a multidisciplinary team for management because of the complexity of their care and the high morbidity and increased mortality these patients experience at delivery. The multidisciplinary group of healthcare providers in our institution takes care of our most surgically complicated obstetric patients. Representatives from maternal fetal medicine, gynecologic oncologic surgery, anesthesiology, urology, and interventional radiology are critical in planning details of patient preparation and surgical approach such as type of anesthesia, patient positioning, preoperative cystoscopy and ureteral stent placement, skin and uterine incision approach, and possible placement of internal iliac occlusion balloon catheters. Nursing department representatives from obstetrics, neonatal intensive care, and the operating room ensure that proper equipment is made available such as a cell saver in cases of anticipated large volume blood loss and infant warmers. Blood bank representatives provide guidance to massive blood transfusion protocols.

Attempts at removing the placenta when there is abnormal placentation often lead to significant maternal hemorrhage. The average blood loss for patients with the diagnosis of placenta accreta spectrum is between 3000 and 5000 cc [20]. This compares to an average blood loss at vaginal delivery of 500 cc and an average blood loss at cesarean section of 1000 cc. Preoperative knowledge of placenta accreta and its variants allows the surgical team time to plan and prepare for the cesarean section and counsel the patient about her management options [10].

Preoperatively, a multidisciplinary meeting serves as a forum for discussion and planning of patient management during the hospital stay and at the time of delivery. At the meeting, ultrasound and MRI are reviewed, the strength of the diagnosis is discussed, and a management strategy is recommended.

Since placental extraction can lead to significant morbidity and occasionally mortality from blood loss but removal of the uterus will eliminate the patient's future childbearing potential, the correct diagnosis of abnormal placental attachment is critical. Direct comparison of the utility of ultrasound with MRI has been reviewed with mixed results [21–23].

## 6.7 Imaging of Abnormal Placentation: Ultrasound

Ultrasound is the primary screening tool to evaluate the placenta. Most patients with abnormal placentation have multiple findings on ultrasound rather than just one sign. False-positive ultrasound interpretations are more likely if only one finding is seen. Grayscale, color Doppler and power Doppler imaging signs with sensitivity, specificity, positive predictive values (PPVs), and negative predictive values (NPVs) have been reported in two recent prospective studies [24,25]. An understanding of sonographic findings helps one understand the reported MR findings and helps one design an MRI protocol. Sonographic findings may be present as early as the first trimester (see Figure 6.5 later in the chapter) [26].

The following grayscale imaging signs have been reported: complete loss of the retroplacental sonolucent zone; irregular retroplacental sonolucent zone; thinning or disruption of the hyperechoic uterine serosa–bladder interface; the presence of focal exophytic masses invading the urinary bladder; and the presence of abnormal placental lacunae. *Lacunae* are defined as hypoechoic irregular elongated areas in the placenta with turbulent flow. It is the opinion of some authors that lacunae are the most sensitive sign of accreta and that accreta can usually be diagnosed by ultrasound at 18 weeks gestation [27,28]. However, one recent meta-analysis found that the sensitivity of lacunae was 77.43% and specificity was 95.02% [29].

Previously reported color Doppler imaging signs include diffuse or focal lacunar flow pattern; sonolucent vascular lakes with turbulent flow typified by high velocity (peak systolic velocity >15 cm/s) and low resistance waveform; hypervascularity of the uterine serosa–bladder interface with abnormal blood vessels linking the placenta to the bladder; and markedly dilated vessels over the peripheral subplacental region [24].

Three-dimensional (3D) power Doppler imaging signs include: intraplacental hypervascularity, inseparable cotyledonal and intervillous circulations, and tortuous vascularity with chaotic branching defined as vessels having a complex vascular arrangement, varying size caliber of vessels, and tortuous course [24].

One prospective ultrasound study with 6 accretas, 24 incretas, and 9 percretas was compared with 131 placenta previas and found that the sign *numerous coherent vessels* visualized using 3D power Doppler in the basal view was the best single criterion for the diagnosis of abnormal placentation, with a sensitivity of 97% and a specificity of 92% [24]. Considering the presence of at least one sonographic criterion to be diagnostic when using each ultrasound technique, 3D power Doppler had the best PPV (76%), then grayscale (51%), and then color Doppler (47%). In this study, placenta lacunae, which have been commonly cited as the most helpful signs of accreta, had a sensitivity of 54%, specificity of 85%, PPV of 51%, and NPV of 86%. If there is only one ultrasound imaging sign, then the authors recommend weighing the PPV and NPV of the imaging sign to help guide clinical management. A high PPV would favor a diagnosis of placenta accreta and aggressive management, whereas a high NPV would favor placenta previa with attempt at placental removal and uterine-sparing surgery.

Another prospective ultrasound study reviewed the following ultrasound criteria for accreta in 187 patients with placenta previa and history of uterine surgery: Two-dimensional (2D) criteria were the absence or irregularity of the retroplacental clear space, thinning or interruption of the hyperechoic interface between the uterine serosa and bladder wall, and placental lacunae with turbulent high-velocity flow (>15 cm/s); 3D criteria were irregular intraplacental vascularization with tortuous confluent vessels crossing the placental width, and hypervascularity of the uterine serosa–bladder wall interface [25]. They reported that hypervascularity at the uterine serosa–bladder interface on 3D ultrasound was the most accurate criteria for abnormal placentation with sensitivity of 90%, specificity and PPV of 100%, and NPV of 97%. The loss of the echolucent clear space between the uterus and placenta had sensitivity of 90%, specificity of 81%, PPV of 57%, and NPV of 97% on 2D ultrasound. In this study using 3D ultrasound, only placenta percreta had irregular intraplacental vascularization with tortuous confluent vessels crossing the placental width, resembling an aneurysm. This has been proposed as a method to distinguish placenta accreta from percreta, which may alter the timing and management of delivery.

Finally, a recent meta-analysis reviewing 23 studies has reported the use of color Doppler in the prenatal diagnosis of invasive placentation with the best accuracy with a sensitivity of 90.74%, specificity of 87.68%, positive likelihood ratio of 7.77, negative likelihood ratio of 0.17, and diagnostic odds ratio of 69.02 [29]. Overall, the reported ultrasound accuracy for detection of abnormal placental invasion was sensitivity of 90.72%, specificity of 96.94%, positive likelihood ratio of 11.01, negative likelihood ratio of 0.16, and diagnostic odds ratio of 98.59.

## 6.8 Imaging of Abnormal Placentation: MRI

Ultrasound with Doppler examination remains the first-line imaging tool for evaluation of accreta. Patients referred for MRI usually have one of the following: equivocal ultrasound findings, negative ultrasound findings with persistent clinical concern, lateral or posterior location of placenta, and uncertainty of depth of invasion of the placenta into the myometrium in cases with suspicious ultrasound findings.

Prior to ordering or performing MRI, it is important to understand that the interpretation of MRI scans performed for the prenatal diagnosis of accreta is likely dependent on the observer's experience [30]. One study showed higher sensitivity and specificity for placental invasion (90.9% and 75%, respectively) for senior radiologists rather than junior radiologists (81.8% and 61.8%, respectively). Also, senior radiologists had a higher degree of certitude than junior radiologists for placental invasion and depth of invasion ($p$ = .0002 and $p$ = .0282, respectively). This study suggested that MRI can be a reliable, reproducible imaging tool, depending on the interpreter's experience. However, it is challenging for the radiologist to gain experience when the volume of MRI of the placenta and pregnant woman is low at most institutions.

Regardless, MRI is being used more frequently to evaluate the placenta. The following changes in surgical planning after MRI evaluation for abnormal placentation have been reported: modification of surgery date, use of prophylactic ureteral catheterization, use of intraoperative blood salvage, planning for vertical skin incision approach, probability of aortic clamping or segmental myometrial resection, need to investigate subclinical disseminated intravascular coagulation, necessity for posterior pelvic dissection, and possibility of uterine sparing surgery [31,32].

Advantages of using MRI over ultrasound include improved tissue contrast, multiplanar capabilities, lack of ionizing radiation, ability to image the posteriorly implanted placenta, and less operator dependence than sonography. However, disadvantages of MRI include lack of widespread expertise, prolonged imaging time, patient claustrophobia, less established robust imaging to detect abnormal intraplacental vascular flow, unknown risk to fetus by MRI, and increased cost.

## 6.9 MRI Techniques for Placenta Evaluation

The Blue Ribbon Committee from the American College of Radiology has the following recommendations when one considers the use of MRI in pregnant women [33]. First, MR imaging should be considered if the requested information cannot be acquired by ultrasonography. Second, information gained from performing the MRI study may affect the patient or fetus during the pregnancy. Finally, the referring physician does not feel it is prudent to wait until pregnancy is complete. Written informed consent to perform MR imaging in pregnancy is usually obtained before performing the examination.

Patient preparation should include a partially filled urinary bladder so that the dome of the bladder can be assessed relative to the placenta and myometrium. Imaging of the placenta requires utilization of pulse sequences which delineate normal anatomy, are sensitive to pathology, and minimize artifacts. Pulse sequences are chosen to limit fetal exposure to energy deposition.

MR imaging strategy includes the following pulse sequences. At our institution, we use an approximate 38 cm field of view and slice thickness of 5 mm in the sagittal and coronal planes and 5–7 mm in axial planes to cover the uterus, and weighing factors such as signal to noise, artifacts, and patient comfort.

Single-shot fast spin echo (SSFSE) using a time to echo (TE) of 60–90 ms and steady-state free precession (SSFP) sequences in three planes are often the most useful. SSFSE and SSFP sequences are motion insensitive and give good delineation of placental abnormalities. Additionally, SSFP sequences show vessels well as hyperintense signal. Imaging in sagittal, axial, and coronal planes of SSFSE and SSFP should be performed since the placenta usually attaches to uterus in a curved plane. During interpretation of images, it can be useful to analyze the sagittal, axial, and coronal SSFSE and SSFP sequences together in pairs, respectively, so that abnormal intraplacental vasculature can be more easily detected by comparing $T_2$ dark areas on SSFSE with $T_2$ hyperintense areas on SSFP. $T_2$ gradient-echo sequences are useful to delineate low signal areas of hemorrhage which may bloom.

$T_1$-fat-saturated 3D gradient-echo sequence is useful to identify areas of hyperintense hemorrhage within the placenta. Fast spin-echo (FSE) $T_2$ sequence with TE of 100 ms is sometimes useful for looking at adnexal areas but is sensitive to motion artifacts by the fetus and bowel.

Diffusion-weighted imaging (DWI) has been used as an investigative tool to evaluate placenta increta, intrauterine growth retardation, and placental abruption. Contrast between the placenta and myometrium is well seen with DWI [34]. Also, DWI is relatively resistant to motion artifacts. Newer non-contrast MR angiography may also be useful for the detection of intraplacental vascularity.

## 6.10 MRI Appearance of the Placenta in Normal Pregnancies

One study has used MRI to evaluate the appearance of the placenta over a range of gestational ages in normal pregnancies (Figure 6.3) [35]. Between 19 and 23 weeks gestations, there is homogeneous uniform signal of the placenta without lobulation in 85% of patients. Between 24 and 31 weeks gestation, there was slightly lobular appearance of the placenta in 90% of patients. Between 32 and 35 weeks gestation, there was an even more lobular placenta. Pregnancies between 36 and 41 weeks had various appearance of lobulation. Placental infarcts were found in approximately 25% of uncomplicated term pregnancies without any clinical significance.

As the gestation progresses in the third trimester, there is more heterogeneous signal within the placenta as intraplacental calcifications and cotyledons form. On $T_2$-weighted imaging, gray colored linear arcs may be seen in within the placenta in the late third trimester corresponding to cotyledons. Normally, vascular lakes believed to contain maternal blood can form in the placenta; these appear as anechoic areas on ultrasound with vascular flow on Doppler examination.

Regularly spaced septa are seen as $T_2$ low signal thin lines tracking through the placenta. These should be distinguished from the irregular thickened $T_2$ dark bands seen in placenta accreta.

As the placenta develops, fibrin deposits can be seen in the fetal side of the placenta or as an irregular lacelike pattern; this is most frequently seen as a normal finding in late third trimester placentas [36,37]. Pathologists routinely observe placental infarcts occupying less than 5% of the placental surface in normal mature placentas [38]. After delivery of the placenta, whitish nodules can be seen on the fetal side of the placenta on gross examination. These appear as $T_2$ dark nodules and should not be considered $T_2$ dark intraplacental bands found in accreta.

Also, placental infarcts along the margins of the placenta can be a normal finding and will appear as $T_2$ dark areas. Location of these infarcts should be considered when distinguishing them from $T_2$ dark bands of accreta.

## 6.11 Gadolinium Contrast for the Diagnosis of Accreta and Retained Products of Conception

The use of gadolinium contrast in pregnant women must be carefully considered to weigh the risks and benefits [39]. Gadolinium contrast does cross the

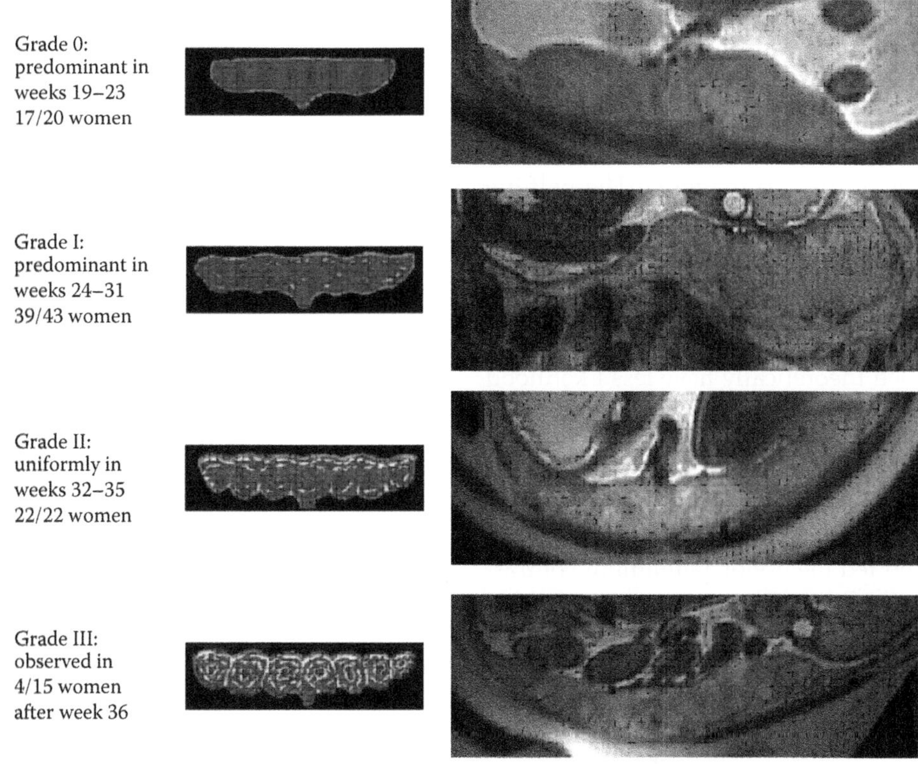


**FIGURE 6.3**
Changes in placental appearance during gestation on MRI. (Reprinted with permission from Blaicher, W. et al., *Eur. J. Radiol.*, 57, 256–260, 2006.)

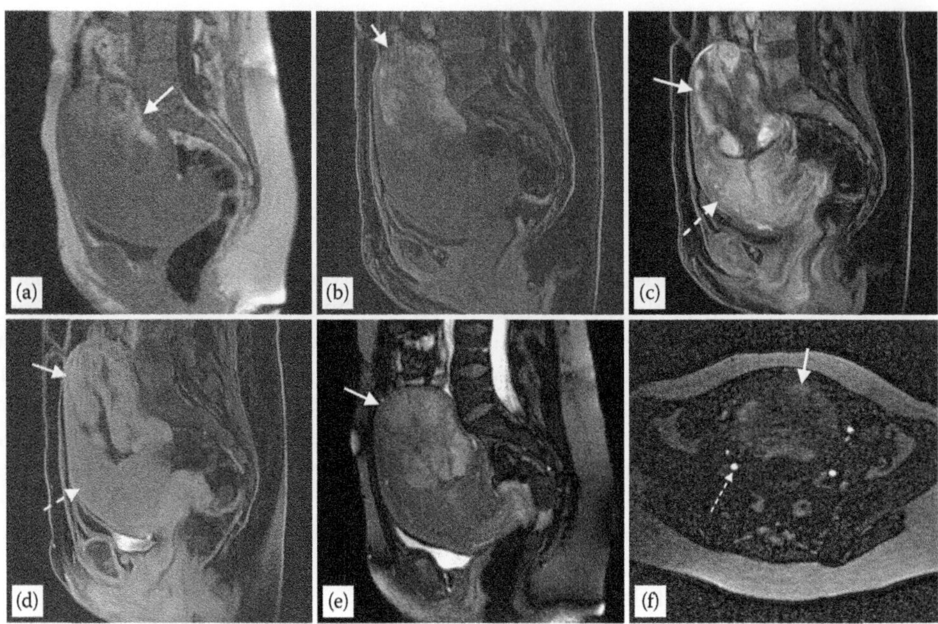


**FIGURE 6.4**
Placenta increta, 2 weeks postpartum. (a) Sagittal $T_1$-weighted in-phase (Sag $T_1$W), (b) Sag $T_1$W precontrast with fat saturation, (c) early arterial phase Sag $T_1$W postcontrast with fat saturation, (d) delayed phase Sag $T_1$W postcontrast with fat saturation, (e) SSFSE $T_2$W, (f) two-dimensional time-of-flight noncontrast MR angiogram. The placenta (arrow) is hyperintense on $T_1$W (a,b) and $T_2$W (e) sequences and enhances more than myometrium (dashed arrow) on arterial phase (c). The placenta (arrow) is nearly isointense to myometrium (dashed arrow) on delayed phase (d) due to contrast wash out. On noncontrast MR angiogram (f), there are no foci of vascular flow (hyperintense signal) in the placenta (arrow) in contrast to the external iliac artery (dashed arrow).

placenta and enters the fetal circulation. Fetal kidneys excrete the contrast into the amniotic fluid for an indefinite amount of time. The potential for dissociation of free gadolinium from its chelate molecule is possible, which in turn may be toxic to the fetus. The half-life of gadolinium in the fetus is unknown. Nephrogenic systemic fibrosis has been strongly associated with prolonged retention of gadolinium contrast in patients with chronic renal failure. Advantages of using macrocyclic gadolinium agents like gadoterate, gadoteridol, or gadobutrol, which theoretically have less likelihood of dissociation than a linear gadolinium agent, are unknown in fetuses [40].

Gadolinium can certainly be administered if fetal delivery is imminent, thereby minimizing long-term fetal exposure to gadolinium contrast. However, one previous study reported the use of gadolinium-contrast MRI in patients typically at 28 weeks gestation to help establish the diagnosis of accreta [21]. Gadolinium contrast was used to help distinguish the placenta which enhances earlier than the uterine myometrium, thereby identifying the border of the placenta from the myometrium. A sensitivity of 88% and a specificity of 100% were reported in this study using contrast MRI in the diagnosis of accreta.

In cases of suspected retained products of conception, the use of dynamic MRI should be strongly considered. In the recent postpartum state, the retained placenta enhances earlier than the myometrium, confirming its presence (Figure 6.4).

## 6.12 MRI Signs of Placenta Accreta

The following MRI findings have been reported to be signs of placenta accreta (Figures 6.5 and 6.6) [41–45].

### 6.12.1 Placenta Previa

The location of the placenta over the internal os can be *central* with implantation centered over the internal cervical os, *complete* with one edge completely covering the internal os, and *partial* where the placental edge lies within 2 cm of the internal os. Incidence of previa at midgestation is 5%, but over 90% of these convert to nonprevia by term through a process of trophotropism where there is marginal atrophy on the one side and

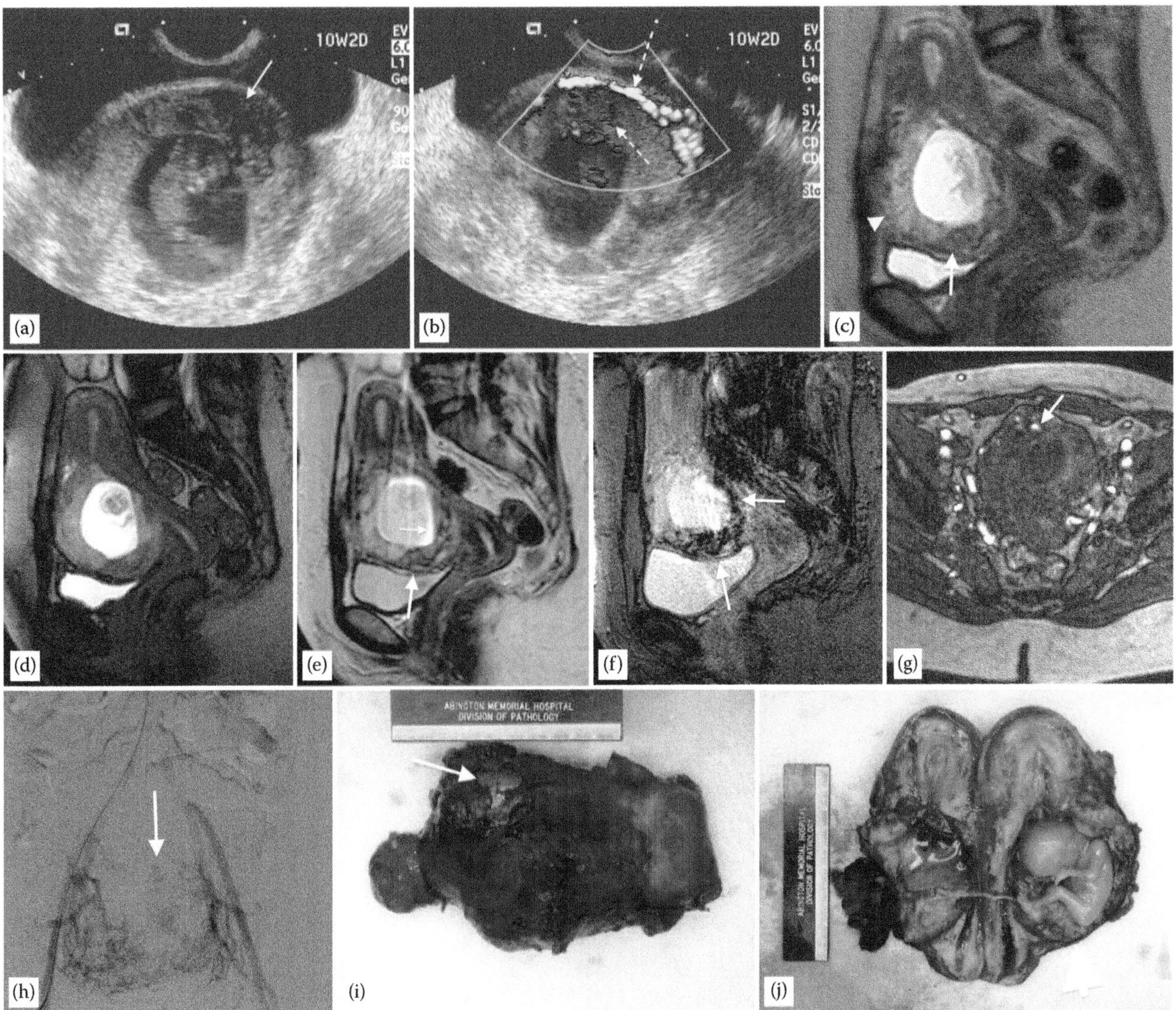


**FIGURE 6.5**
Placenta percreta in a 38-year-old female at first trimester gestation. Grayscale (a) and power Doppler ultrasound (b) of 10 weeks gestation shows lacunae (arrow) with increased vascular flow at the myometrial–bladder interface and some flow in lacunae (dashed arrows) (Ultrasound images courtesy of Stephen Smith, MD, and Frank Craparo, MD). MRI performed with sagittal SSFSE $T_2$ (c), SSFP (d), FSE $T_2$ (e), and gradient-echo $T_2$-weighted (f) at 11 weeks gestation shows dark intraplacental $T_2$ bands (arrows), which bloom on gradient-echo $T_2$ sequence. There is focal bulging of the lower uterine segment with distortion of the normal pear-shaped uterus. There is a low implantation of the gestational sac over the prior cesarean section scar and incomplete distention of the uterine cavity. There is marked thinning of the myometrium with placenta extending to the serosa (arrowhead). Amnionic sac is separate from chorionic sac (small arrow, e). Two-dimensional time-of-flight MR angiogram shows hyperintense vessel (arrow) at the fetal side of the placenta with similar signal intensity to the external iliac arteries (g). Conventional pelvic angiogram prior to uterine artery embolization shows irregular contrast blush (arrow) in the placenta (h) (Angiogram image courtesy of Steven Tey, MD). Embolization was performed to minimize hemorrhage during hysterectomy. Gross specimen shows placenta extending through the serosa (i). Bivalved uterus shows *in situ* fetus with placenta extending to the serosa (arrow) (j).

growth on the other side [7]. With implantation over the cervix and lower uterine segment, there is a deficiency in decidualization since there is little to no normal endometrium and the mucosa does not respond to hormonal signals of decidualization. In normal pregnancies, implantation generally occurs in the uterine fundus due to cellular factors inducing adhesion molecules in the underlying endometrial stroma [46].

### 6.12.2 Uterine Bulging

The normal gravid uterus has a pear shape with a wider superior uterus and thinner inferior uterus. A focal bulge of the placenta into the myometrium, which distorts the normal pear-shaped contour, is another reported sign of accreta [42]. The lower uterine segment may become wider than the superior segment in invasive placentation. This may occur anywhere along the

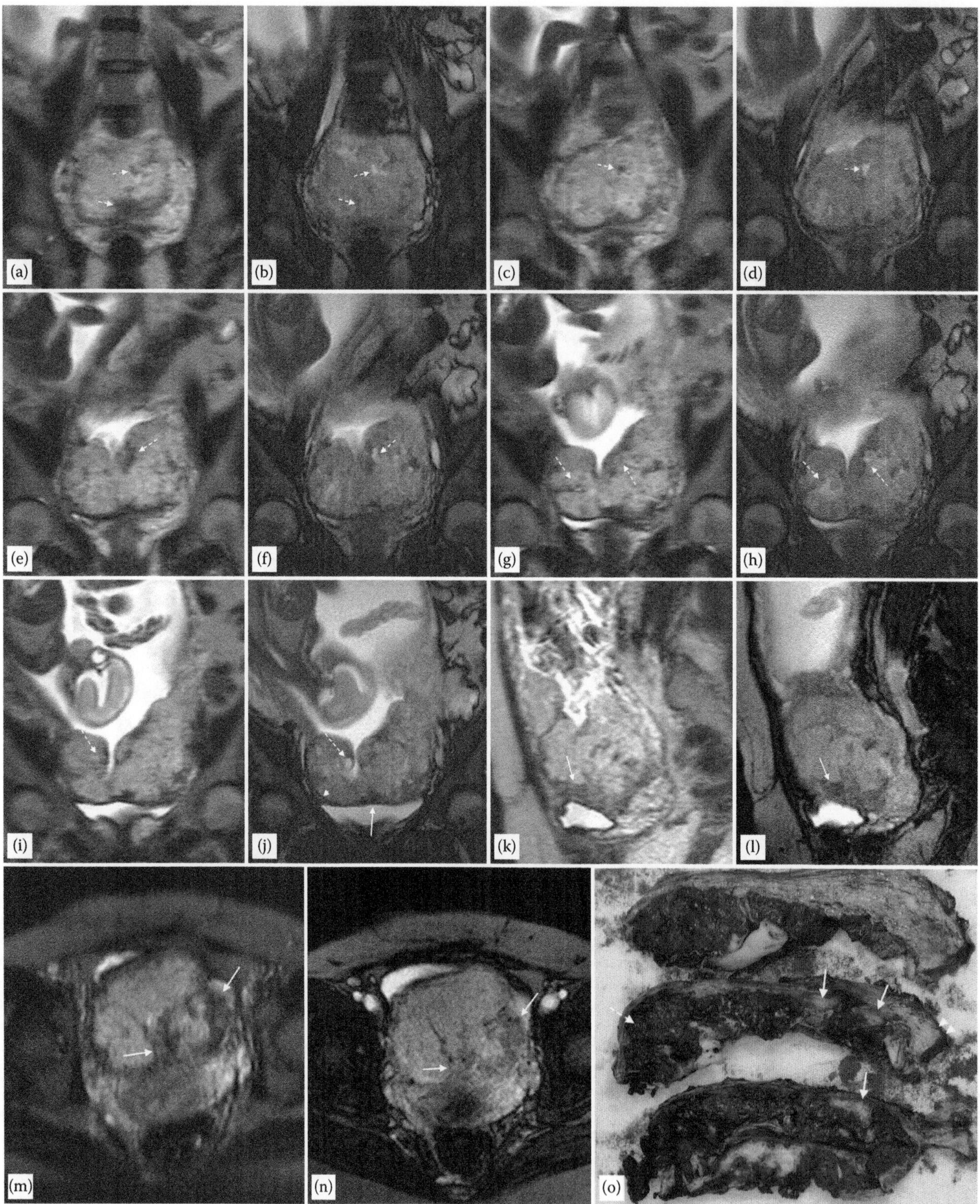


**FIGURE 6.6**
Placenta percreta in a 29-year-old female. Coronal SSFSE (a, c, e, g, i) and SSFP (b, d, f, h, j) show hyperintense intraplacental vessels on SSFP, which correspond to dark intraplacental bands on SSFSE sequences (dashed arrows). (j) There is an intraplacental dark band (solid arrow) just above the urinary bladder with focal loss of myometrium; thin myometrium is seen on the right side (arrowhead). Overall, the placenta is also heterogeneous in signal with areas of gray and dark signal in the placenta. Sagittal SSFSE (k) and corresponding SSFP (l) show intraplacental dark bands (solid arrows) corresponding to placental infarcts at gross pathology. Axial SSFSE (m) and SSFP (n) just above the urinary bladder show two areas of dark bands (solid arrows). Photograph (o) of serial sectioned gross specimen shows dark placenta (*dashed arrow*) extending through full thickness of myometrium (*arrowheads*) and gross infarcts (*solid arrows*) corresponding to low-signal-intensity bands in (k). (Reprinted with permission from Lim, P.S. et al., *AJR Am J Roentgenol*, 197, 1506–1513, 2011.)

cesarean section scar and may be along the spectrum of the placenta growing into a scar line.

### 6.12.3 Heterogeneous Signal Intensity in Placenta

Lack of homogeneous $T_2$ signal throughout the placenta is another reported sign of accreta [42]. This can be difficult to distinguish from the normal heterogeneity of a late third trimester placenta with normal placental aging [23,35]. Previous studies have reported imaging prior to 30 weeks gestation, which may reduce the confounding factor of placental aging. This heterogeneity, especially when it is extensive, corresponds to areas of hemorrhage or lacunae. This sign is also somewhat subjective and may depend on the presence of dark bands.

### 6.12.4 Dark Intraplacental Bands on $T_2$-Weighted Imaging

Intraplacental bands are irregular or polygonal in shape, which differ from the regular thin septa or vessels normally present in the placenta. Often they extend from the maternal side of the placenta toward the fetal side of the placenta. These bands represent areas of hemorrhage, fibrin, or infarcts. Absence of this sign makes placental invasion unlikely [42]. More extensive dark bands may be indicative of more severe invasion such as placenta percreta, and less extensive dark bands may be indicative of less severe invasion like placenta accreta [23]. Dark bands are likely the ultrasound correlate to placental lacunae seen on ultrasound, with more extensive lacunae seen with more severe invasion [47]. In some patients, MRI may be more sensitive than ultrasound in the detection of accreta since MRI is sensitive to detecting hemorrhage [23].

### 6.12.5 Focal Interruption in the Myometrial Wall

The borders of the normal myometrium on MRI have thin $T_2$ linear low signal, but the middle layer has $T_2$ hyperintense signal [41]. The middle layer has prominent vessels. Focal loss of the normal hypointense myometrial wall is a sign of focal placental invasion, but this sign is more difficult to visualize when the normal myometrium becomes thinned at late gestation.

### 6.12.6 Tenting of the Urinary Bladder

The normal smooth contour is disrupted when there is placental invasion with focal angulated margins of the urinary bladder. It is a reported specific sign of percreta [41].

### 6.12.7 Direct Visualization of Invasion of Pelvic Structures by Placental Tissue

This rare, specific sign is only seen with placenta percreta with this severe level of invasion [41].

### 6.12.8 Abnormal Placental Vascularity

The presence of intraplacental focal vessels, which are dark on SSFSE but hyperintense on SSFP, has been reported as another sensitive sign of abnormal placentation in one study [48]. These vessels are hypertrophied, disorganized, and tortuous, similar to sonographic findings on Doppler examination [24]. Non-contrast MR angiography with arterial spin labeling is an area of research and may become a supplemental MRI sequence to help diagnose abnormal placentation.

### 6.12.9 DWI of Accreta

DWI has been used to help define the interface between myometrium and placenta since the placenta itself has high signal intensity at *b*-value of 1000 $s/mm^2$ rather than the myometrium; at a *b*-value of zero, the myometrium is hyperintense relative to the surrounding fat [34]. Image fusion was used to help detect focal myometrial thinning, which can be seen with placenta increta and percreta. However, the authors acknowledge that the myometrium is normally thin in the third trimester and it may be difficult to distinguish normally thin myometrium from placental invasion. Also, since placental attachment to myometrium with absent decidua rather than placental invasion occurs with accreta, DWI may not detect placenta accreta since there is no focal thinning of myometrium.

## 6.13 Authors' Pearls and Pitfalls of Diagnosing Placenta Accreta

Assessment of the patient's risk for placenta accreta begins with the patient's history. If the placenta overlies a known uterine scar from prior instrumentation, this should be considered a strong risk factor for invasive placentation. The presence of placenta previa, increasing number of prior cesarean sections, and prior accreta history will also increase the pretest probability of abnormal placentation. Tobacco smoke or the presence of maternal conditions of hypertension or other maternal–placental insufficiency factors may influence interpretation of imaging findings as these may cause false-positive MR findings (Figure 6.7).

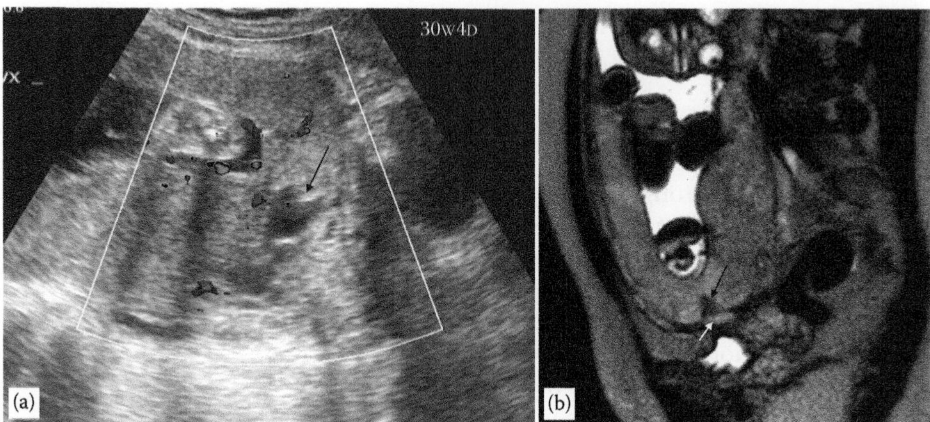


**FIGURE 6.7**
Placental infarct in a mature placenta with complete placenta previa in a 35-year-old female who smoked during pregnancy, and false-positive ultrasound and MRI findings Ultrasound image obtained at 30 weeks gestation shows large hypoechoic intraplacental area interpreted as lacunae (*arrow*) (a). Sagittal SSFSE $T_2$-weighted MR image obtained at 31 weeks shows triangular area of low signal intensity (*black arrow*) identical to ultrasound finding and confirmed on images obtained with other pulse sequences and in other planes which correspond to an infarct at pathology (b). High-signal-intensity area of hemorrhage extends from band toward cervical os (*white arrow*). (Reprinted with permission from Lim, P.S. et al., *AJR Am J Roentgenol*, 197, 1506–1513, 2011.)

MR imaging is best obtained before 30 weeks gestation because the placenta ages and matures as the pregnancy progresses, and the myometrium stretches and thins as the fetus grows, confounding the MR diagnosis of accreta. Placental infarcts, which appear as intraplacental $T_2$ dark bands, especially in the periphery of the placenta or fetal side of the placenta, are present in the placenta of up to 25% of normal pregnancies and more often in late third trimester. Intraplacental $T_2$ dark bands elsewhere, especially on the maternal side, and found at earlier gestational age are more likely pathologic. As the uterus enlarges in the third trimester as the fetus grows, the myometrium becomes thinner, also making it more difficult to visualize focal placental invasion.

Recent MRI literature has reported on the MR imaging signs of intraplacental $T_2$ dark bands and abnormal vessels seen on steady-state free precession sequences [23,48]. MRI interpretation with side-by-side comparison of the uterus in the same plane of orientation using SSFSE and SSFP is helpful. $T_2$ dark bands due to hemorrhage or infarcts will be dark on SSFSE, SSFP, and gradient-echo sequences. Abnormal intraplacental vessels will have $T_2$ intermediate-dark bands on SSFSE but hyperintense on SSFP and not visualized on gradient-echo sequences. As in the ultrasound literature with increasing number of placental lacunae, the presence of larger $T_2$ dark bands will increase likelihood of abnormal placentation and perhaps worsening severity of placental invasion.

Usually, the gestational sac is high in the fundal portion of the endometrium and is completely surrounded by thick myometrium. However, if there is an implantation low in the uterine cavity over a cesarean scar which can be diagnosed with a first trimester ultrasound or if the gestational sac does not fully distend the uterine cavity, then the presence of myometrial thinning may be the solitary finding of placenta accreta. A low implantation of the gestational sac over a scar is covered by myometrium on three sides, unlike a scar pregnancy which is implanted in the scar and not in the uterine cavity. The location of the implantation is much less obvious as the gestational sac fills the uterus.

## 6.14 DWI of Intrauterine Growth Retardation

DWI has also been used in the detection of fetuses at risk for intrauterine growth retardation [49]. A decreased apparent diffusion coefficient and restricted diffusion are seen in placental dysfunction associated with fetal intrauterine growth restriction. It may be used as an early marker of placental dysfunction. Placental diffusion in addition to ultrasound has been reported to improve sensitivity of placental insufficiency from 73% to 100%, increased accuracy from 91% to 99%, and preserved specificity of 99%.

## 6.15 MRI Diagnosis of Placental Abruption

Advantages of MRI over sonography for the diagnosis of placental abruption include better tissue characterization, more accurate assessment of the age of hemorrhage, and improved differentiation of types of fluid collections. One study which included 19 cases of placental abruption showed that sonography was able to detect only 53% of cases, whereas MRI could detect all (100%) cases [50]. The presence of a hematoma with hyperacute or acute MR signal intensity characteristics ($n = 6$) showed correlation in all cases to fetal or maternal distress leading to delivery within 10 days, whereas hematomas with late subacute bleeding ($n = 20$) did not.

Hyperacute hemorrhage has isointense to hypointense $T_1$ signal, hyperintense on $T_2$, and hyperintense to hypointense on diffusion sequence. Acute hemorrhage is isointense to hypoinense on $T_1$, and hypointense on $T_2$ and diffusion-weighted sequence. Early subacute hemorrhage is hyperintense on $T_1$ and hypointense on $T_2$ and diffusion-weighted sequence. Late subacute hemorrhage has hyperintense signal on $T_1$, $T_2$, and diffusion sequence. Chronic hemorrhage has low signal on $T_1$ and $T_2$ and is isointense to hypointense on diffusion sequence. These authors have recommended that MR imaging should be considered after negative ultrasound findings in cases of bleeding in late pregnancy where diagnosis of abruption would change management.

## 6.16 MRI of Non-Accreta Conditions in Pregnant Patients

There is a wide range of pathology that may affect the pregnant patient that is beyond the scope of this chapter. Differential diagnosis is extensive including gastrointestinal etiologies such as acute cholecystitis and bowel obstruction, genitourinary etiologies such as kidney stones and ureteral obstruction, and gynecologic etiologies such as ovarian vein thrombosis and ovarian torsion (Figure 6.8) [51–54]. Evaluation of right lower quadrant pain, usually requested from the Emergency Department for suspected acute appendicitis, is one of the most common MRI requests for pregnant patients in our institution after an equivocal ultrasound.

MRI protocol of right lower quadrant pain is as follows. A torso-phased array coil is used. A field of view of approximately 35 cm is usually adequate. $T_1$ in-phase and opposed-phase gradient-echo sequences will help characterize high signal fat and hemorrhage in collections and lesions. SSFSE $T_2$ with TE of 60–90 ms with slice thickness of 4 mm in three orthogonal planes is motion insensitive and will be helpful in evaluating bowel in patients with right lower quadrant pain. SSFSE $T_2$ with frequency-selective fat saturation is sensitive to edema and inflammation. Short-tau inversion recovery-weighted sequences are also sensitive to edema but are motion sensitive. Some authors advocate $T_2$-weighted time-of-flight 2D sequences from the level of the renal veins to the symphysis pubis to identify venous structures which may be confused with the appendix [52]. Administration of an oral contrast is also somewhat controversial. Imaging without the use of an oral contrast agent keeps the patient in a fasting state and does not prolong the completion of imaging. However, some authors advocate using a negative oral contrast agent mixture of ferumoxsil and barium sulfate 1 h before the study, which then provides dark signal intensity on both $T_1$- and $T_2$-weighted images without significant susceptibility artifacts [51,54]. The presence of a blooming effect within the appendix filled with the negative oral contrast virtually excludes appendicitis and is used as an imaging sign to distinguish it from pelvic varices.

Incidence of acute appendicitis is similar to the general population, but pregnant patients are more likely to present with bowel perforation (43% vs. 4%–19%) due to delay in diagnosis [51,52]. One in 766 pregnancies has been reported as the incidence of acute appendicitis [53]. Fetal mortality is reported to range from 6% to 27% in

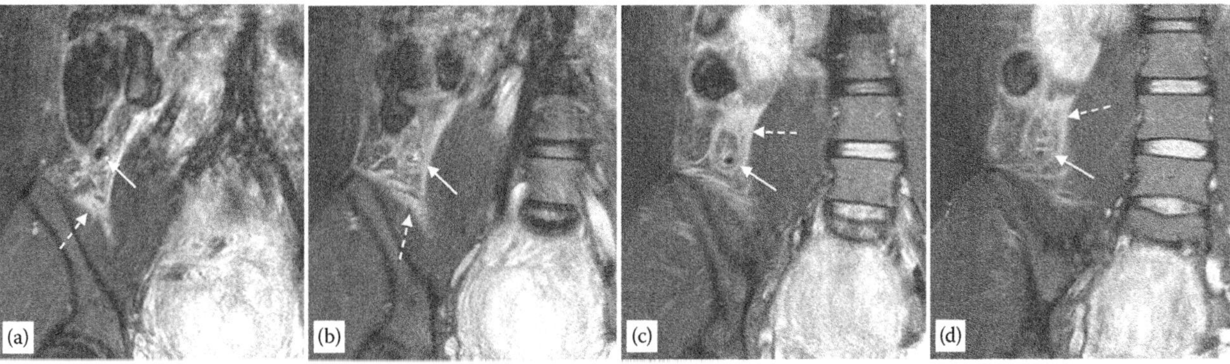


**FIGURE 6.8**
(a–d) Acute appendicitis in a 24-week gestation. Series of sequential coronal short-tau inversion recovery-weighted sequences show infiltration of fat (dashed arrow) around distended appendix (arrow).

cases of ruptured appendicitis. Diagnostic challenges of evaluation of abdominal pain include physiological leukocytosis in normal pregnancy, gastrointestinal symptoms such as nausea and vomiting present in normal pregnancies, and lack of rebound tenderness in pregnant patients with lax abdominal musculature. MRI challenges may include claustrophobia, discomfort from supine positioning, limited breath-hold capabilities, possibly excessive high (>90 decibel) noise during MRI, and displacement of organs by the gravid uterus. MR has been reported to have a high specificity (98%–100%) and high NPV (94%–100%).

There is a lack of consensus on best practices for evaluation of right lower quadrant pain. A recent survey of academic institutions reported that in first trimester pregnancies, MRI was preferred as the next imaging modality after an indeterminate ultrasound result [53]. However, in second and third trimester pregnancies, computed tomography (CT) rather than MRI was the preferred method after an indeterminate ultrasound. It is generally accepted that a normal appendix has a size of 6 mm or smaller without infiltration of surrounding fat. Some studies have used the presence of air or oral contrast in the lumen of the appendix as another sign of a normal appendix [54,55]. Positive MR signs of acute appendicitis generally include an appendix larger than 6 mm in width without or with infiltration of the surrounding fat due to inflammation. It is the opinion of the authors that MRI should be the preferred method for evaluation of abdominal pain in the pregnant patient after an equivocal ultrasound study. CT should be reserved if there are instances where MRI cannot be used such as limited availability and claustrophobia; if CT is needed, imaging protocols should be modified to reduce radiation exposure.

## 6.17 Summary

There is a wide range of uses of MRI in the evaluation of the placenta [56–58]. MRI of the placenta is commonly used to evaluate placenta accreta and complex surgical obstetric cases at our institution, especially if sonography is indeterminate or if clinical suspicion is high. Placental pathology can often be evaluated with three plane SSFSE and SSFP pulse sequences. Gradient echo $T_1$, $T_2$ sequences and diffusion-weighted sequences can be useful to evaluate for hemorrhage. Non-contrast MR angiography may have some potential in accreta evaluation. Contrast MR can be useful to distinguish earlier enhancing placenta from myometrium which enhances later. MR evaluation of abdominal pain in the obstetrics patient rather than CT should be considered after an equivocal ultrasound study.

## References


1. Burton BK. (1988) Outcome of pregnancy in patients with unexplained elevated or low levels of maternal serum alpha-fetoprotein. *Obstet Gynecol* 72:709–713.
2. Kupferminc MJ, Tamura RK, Wigton TR, Glassenberg R, Socol ML. (1993) Placenta accreta is associated with elevated maternal serum alpha-fetoprotein. *Obstet Gynecol* 82:266–269.
3. Zelop C, Nadel A, Frigoletto FD Jr, Pauker S, MacMillan M, Benacerraf BR. (1992) Placenta accreta/percreta/increta: A cause of elevated maternal serum alpha-fetoprotein. *Obstet Gynecol* 80:693–694.
4. Baergen RN (ed). (2011) Overview and microscopic survey of the placenta. In: *Manual of Benirschke and Kaufmann's Pathology of the Human Placenta.* (2nd ed) Springer-Verlag, New York.
5. Kaplan CG. (2007) Basic placental anatomy and development. In: *Color Atlas of Gross Placental Pathology.* (2nd ed) Springer-Verlag, New York.
6. Baergen RN (ed). (2011) Placental shape aberrations. In: *Manual of Benirschke and Kaufmann's Pathology of the Human Placenta.* (2nd ed) Springer-Verlag, New York.
7. Middleton WD, Kurtz AB, Hertzberg BS. Placenta, umbilical cord, and cervix. In: *Ultrasound: The Requisites.* (2nd ed) Mosby, Philadelphia, PA.
8. Spirt BA, Gordon LP. (1998) Placenta and cervix. In: McGahan JP and Goldberg BG (eds) *Diagnostic Ultrasound: A Logical Approach.* Lippincott, Philadelphia, PA.
9. Jauniaux E, Jurkovic D. (2012) Placenta accreta: Pathogenesis of a 20th century iatrogenic uterine disease. *Placenta* 33:244–251.
10. Committee on Obstetric Practice. (2012) Committee opinion no. 529: Placenta accreta. *Obstet Gynecol* 120:207–211.
11. Brosens JJ, Pijnenborg R, Brosens IA. (2002) The myometrial junctional zone spiral arteries in normal and abnormal pregnancies. *Am J Obstet Gynecol* 187:1416–1423.
12. Genbacev O, Zhou Y, Ludlow JW, Fisher SJ. (1997) Regulation of human placental development by oxygen tension. *Science* 277:1669–1672.
13. Baergen RN (ed). (2011) Postpartum hemorrhage, subinvolution of the placental site, and placenta accreta. In: *Manual of Benirschke and Kaufmann's Pathology of the Human Placenta.* (2nd ed) Springer-Verlag, New York.
14. Schaaps JP, Tsatsaris V, Goffin F, Brichant JF, Delbecque K, Tebache M, Collignon L, Retz MC, Foidart JM. (2005) Shunting the intervillous space: New concepts in human uteroplacental vascularization. *Am J Obstet Gynecol* 192:323–332.
15. Chantraine F, Blacher S, Berndt S et al. (2012) Abnormal vascular architecture at the placenta-maternal interface in placenta increta. *Am J Obstet Gynecol* 207:188e1–188e9.
16. Wu S, Kocherginsky M, Hibbard JU. (2005) Abnormal placentation: Twenty-year analysis. *Am J Obstet Gynecol* 192:1458–61.
17. Read JA, Cotton DB, Miller FC. (1980) Placenta accreta: Changing clinical aspects and outcome. *Obstet Gynecol* 56:31–34.

18. Miller DA, Chollet JA, Goodwin TM. (1997) Clinical risk factors for placenta previa-placenta accreta. *Am J Obstet Gynecol* 177:210–214.
19. Silver RM, Landon MB, Rouse DJ et al. National Institute of Child Health and Human Development Maternal-Fetal Medicine Units Network. (2006) Maternal morbidity associated with multiple repeat cesarean deliveries. *Obstet Gynecol* 107:1226–1232.
20. Hudon L, Belfort MA, Broome DR. (1998) Diagnosis and management of placenta percreta: A review. *Obstet Gynecol Surv* 53:509–517.
21. Warshak CR, Eskander R, Hull AD et al. (2006) Accuracy of ultrasonography and magnetic resonance imaging in the diagnosis of placenta accreta. *Obstet Gynecol* 108:573–581.
22. Dwyer BK, Belogolovkin V, Tran L et al. (2008) Prenatal diagnosis of placenta accreta: Sonography or magnetic resonance imaging? *J Ultrasound Med* 27:1275–1281.
23. Lim PS, Greenberg M, Edelson MI, Bell KA, Edmonds PR, Mackey AM. (2011) Utility of ultrasound and MRI in prenatal diagnosis of placenta accreta: A pilot study. *AJR Am J Roentgenol* 197:1506–1513.
24. Shih JC, Palacios-Jaraquemada JM, Su YN et al. (2009) Role of three-dimensional power Doppler in the antenatal diagnosis of placenta accreta: Comparison with gray-scale and color Doppler techniques. *Ultrasound Obstet Gynecol* 33:193–203.
25. Calì G, Giambanco L, Puccio G, Forlani F. (2013) Morbidly adherent placenta: Evaluation of ultrasound diagnostic criteria and differentiation of placenta accreta from percreta. *Ultrasound Obstet Gynecol* 41:406–412. doi:10.1002/uog.12385.
26. Stirnemann JJ, Mousty E, Chalouhi G, Salomon LJ, Bernard JP, Ville Y. (2011) Screening for placenta accreta at 11–14 weeks of gestation. *Am J Obstet Gynecol* 205:547.e1–547.e6.
27. Comstock CH. (2005) Antenatal diagnosis of placenta accreta: A review. *Ultrasound Obstet Gynecol* 26:89–96.
28. Comstock CH. (2013) Re: Morbidly adherent placenta: Evaluation of ultrasound diagnostic criteria and differentiation of placenta accreta from percreta. G. Calì, L. Giambanco, G. Puccio and F. Forlani. *Ultrasound Obstet Gynecol* 41:406–412. doi:10.1002/uog.12453.
29. D'Antonio F, Iacovella C, Bhide A. (2013) Prenatal identification of invasive placentation using ultrasound: Systematic review and meta-analysis. *Ultrasound Obstet Gynecol* doi:10.1002/uog.13194.
30. Alamo L, Anaye A, Rey J, Denys A, Bongartz G, Terraz S, Artemisia S, Meuli R, Schmidt S. (2012) Detection of suspected placental invasion by MRI: Do the results depend on observer' experience? *Eur J Radiol* doi:org/10.1016/j.ejrad.2012.08.022.
31. Palacios Jaraquemada JM, Bruno CH. (2005) Magnetic resonance imaging in 300 cases of placenta accreta: Surgical correlation of new findings. *Acta Obstet Gynecol Scand* 84:716–724.
32. Palacios Jaraquemada JM, Bruno CH, Martín E. (2013) MRI in the diagnosis and surgical management of abnormal placentation. *Acta Obstet Gynecol Scand* 92:392–397.
33. Kanal E, Barkovich AJ, Bell C et al., ACR Blue Ribbon Panel on MR Safety. (2007) ACR guidance document for safe MR practices:2007. *AJR Am J Roentgenol* 188:1447–1474.
34. Morita S, Ueno E, Fujimura M, Muraoka M, Takagi K, Fujibayashi M. (2009) Feasibility of diffusion-weighted MRI for defining placental invasion. *J Magn Reson Imaging* 30:666–671.
35. Blaicher W, Brugger PC, Mittermayer C et al. (2006) Magnetic resonance imaging of the normal placenta. *Eur J Radiol* 57:256–260.
36. Baergen RN (ed). (2011) Miscellaneous placental lesions. In: *Manual of Benirschke and Kaufmann's Pathology of the Human Placenta.* (2nd ed) Springer-Verlag, New York.
37. Kaplan CG. (2007) Lesions of the villous tree. In: *Color Atlas of Gross Placental Pathology.* (2nd ed) Springer-Verlag, New York.
38. Kaplan CG. (1996) Postpartum examination of the placenta. *Clin Obstet Gynecol* 39:535–548.
39. Webb JA, Thomsen HS, Morcos SK. (2005) The use of iodinated and gadolinium contrast media during pregnancy and lactation. Members of the contrast media safety committee of European Society of Urogenital Radiology (ESUR). *Eur Radiol* 15:1234–1240.
40. Kidney Disease: Improving Global Outcomes (KDIGO) CKD Work Group. (2013) KDIGO 2012 clinical practice guideline for the evaluation and management of chronic kidney disease. *Kidney Int Suppl* 3:1–150.
41. Baugman WC, Corteville JE, Shah RR. (2008) Placenta accreta: Spectrum of US and MR imaging findings. *RadioGraphics* 28:1905–1916.
42. Lax A, Prince MR, Mennitt KW, Schwebach JR, Budorick NE. (2007) The value of specific MRI features in the evaluation of suspected placental invasion. *Magn Reson Imaging* 25:87–93.
43. Kim JA, Narra VR. (2004) Magnetic resonance imaging with true fast imaging with steady-state precession and half-Fourier acquisition single-shot turbo spin echo sequences in cases of suspected placenta accreta. *Acta Radiol* 45:692–698.
44. Levine D, Hulka CA, Ludmir J, Li W, Edelman RR. (1997) Placenta accreta: Evaluation with color Doppler US, power Doppler US, and MR imaging. *Radiology* 205:773–776.
45. Teo TH, Law YM, Tay KH, Tan BS, Cheah FK. (2009) Use of magnetic resonance imaging in evaluation of placental invasion. *Clin Radiol* 64:511–516.
46. Kraus FT, Redline RW, Gersell DJ, Nelson DM, Dicke JM. (2004) *Atlas of Nontumor Pathology: Placental Pathology.* American Registry of Pathology, Washington, DC, 49–53.
47. Finberg HJ, Williams JW. (1992) Placenta accreta: Prospective sonographic diagnosis in patients with placenta previa and prior cesarean section. *J Ultrasound Med.* 11:333–343.
48. Derman AY, Nikac V, Haberman S, Zelenko N, Opsha O, Flyer M. (2011) MRI of placenta accreta: A new imaging perspective. *AJR Am J Roentgenol* 197:1514–1521.
49. Bonel HM, Stolz B, Diedrichsen L et al. (2010) Diffusion-weighted MR imaging of the placenta in fetuses with placental insufficiency. *Radiology* 257:810–819. Errata.

Diffusion-weighted MR imaging of the placenta in fetuses with placental insufficiency. *Radiology* 2010;257(3): 810–819.
50. Masselli G, Brunelli R, Di Tola M, Anceschi M, Gualdi G. (2011) MR imaging in the evaluation of placental abruption: Correlation with sonographic findings. *Radiology* 259(1):222–230.
51. Eyvazzadeh AD, Pedrosa I, Rofsky NM, Siewert B, Farrar N, Abbott J, Levine D. (2004) MRI of right-sided abdominal pain in pregnancy. *AJR Am J Roentgenol* 183:907–914.
52. Pedrosa I, Levine D, Eyvazzadeh AD, Siewert B, Ngo L, Rofsky NM. (2006) MR imaging evaluation of acute appendicitis in pregnancy. *Radiology* 238:891–899.
53. Long SS, Long C, Lai H, Macura KJ. (2011) Imaging strategies for right lower quadrant pain in pregnancy. *AJR Am J Roentgenol* 196:4–12.
54. Pedrosa I, Zeikus EA, Levine D, Rofsky NM. (2007) MR imaging of acute right lower quadrant pain in pregnant and nonpregnant patients. *RadioGraphics* 27:721–753.
55. Pedrosa I, Lafornara M, Pandharipande PV, Goldsmith JD, Rofsky NM. (2009) Pregnant patients suspected of having acute appendicitis: Effect of MR imaging on negative laparotomy rate and appendiceal perforation rate. *Radiology* 250:749–757.
56. Masselli G, Gualdi G. (2013) MR imaging of the placenta: What a radiologist should know. *Abdom Imaging* 38:573–587.
57. Nguyen D, Nguyen C, Yacobozzi M, Bsat F, Rakita D. (2012) Imaging of the placenta with pathologic correlation. *Semin Ultrasound CT MRI* 33:65–77.
58. Dekan S, Linduska N, Kasprian G, Prayer D. (2012) MRI of the placenta—A short review. *Wien Med Wochenschr* 162:225–228.

# 7

# *MRI of Pelvic Floor Disorders*

**Francesca Maccioni and Valeria Buonocore**

**CONTENTS**



## 7.1 Introduction

When the pelvic floor is damaged in any of its fascial, muscular, or neural components, several pelvic floor dysfunctions (PFDs) may arise. PFDs are characterized by a variable association of pelvic organ prolapse (POP) and functional disturbances, involving bladder (urinary incontinence [UI] and voiding dysfunction), vagina and/or uterus (sexual dysfunctions and prolapse), and the rectum (obstructed defecation syndrome [ODS]). Thus, any of the three pelvic floor compartments, the anterior or urinary, the middle or genital, and the posterior or anorectal, may be variably and concurrently involved, resulting in a variable association of POP and functional disturbances.

The ODS represents one of the main dysfunctions of the posterior pelvic floor, consisting of constipation due to an inadequate evacuation of fecal contents from the rectum [1–3]. It may be determined by different disorders, including rectocele, rectal invagination, and rectal prolapse. ODS, POP, and UI are the most common PFDs, which develop in nearly 50% of multiparous women over 50 years of age with significant impact on their quality of life. Due to their high incidence and frequent need of surgical repair, PFDs represent a relevant healthcare problem in the Western countries [4,5].

In order to plan an effective surgical or conservative treatment and to avoid frequent post-surgical relapses, a comprehensive and accurate assessment of PFDs is essential. Due to the wide spectrum of symptoms and associated diseases, clinical evaluation alone is not usually sufficient to reach a complete diagnosis. A comprehensive diagnostic approach usually requires the association of several physiological and diagnostic examinations [5]. Recently, MRI's role in the evaluation of PFDs has been extensively investigated because of its specific capability to assess the three pelvic compartments contemporarily with similar accuracy and without invasiveness or discomfort. The aim of this chapter is to review the diagnostic value of MRI in the evaluation of PFDs, particularly those involving the anorectal compartment, commonly defined as *posterior pelvic floor disorders* (PPFDs). The main MRI features of PFDs will be illustrated as well as the functional anatomy of the pelvic floor, the clinical and diagnostic problems, technical aspects, MRI reference parameters, and grading systems.

## 7.2 Functional Anatomy of the Pelvic Floor

A comprehensive knowledge of the pelvic floor's complex anatomy is necessary to understand the specific lesions underlying PFDs and to explain the frequent association of disorders in the different pelvic compartments.

The pelvic floor is a complex integrated system, made up of skeletal and striated muscles, suspensory ligaments, fascial coverings, and an intricate neural network. It does not only provide support for the pelvic viscera (bladder, bowel, and uterus) but also maintains the functioning of these organs, thanks to two major structures: the endopelvic fascia (EPF) and the levator ani muscle (LAM) [1,2,5–8].

The LAM has two main components: the *iliococcygeus* and *pubococcygeus muscles*. Various muscle subdivisions have been assigned to the medial portions of the pubococcygeus, reflecting the attachments of the muscle to the urethra, vagina, anus, and rectum, respectively, named pubourethralis, pubovaginalis, puboanalis, and puborectalis—or collectively pubovisceralis [1,2]. The LAM is made of slow-twitch muscular fibers that continuously contract, thus providing tone to the pelvic floor against gravity and intra-abdominal pressure. Furthermore, contraction of the LAM closes the urogenital hiatus and compresses the urethra, vagina, and anorectal junction (AJR) in the direction of the pubic bone, thus actively acting on the function of these pelvic organs. Particularly, the puborectalis component of the pubococcygeous, namely, the puborectal muscle, acts as an intrinsic part of the anal sphincter complex, playing a primary role in the functioning of posterior pelvic floor compartment. It attaches to the pubic bone and forms a sling around the rectum, aligning with the external anal sphincter. Relaxation of the puborectalis sling opens the anorectal angle (ARA), while its contraction closes this angle, thus impeding defecation (Figure 7.1).

The EPF is a layer of connective tissue anchoring uterus and vagina to the pelvic sidewalls [1–2]. A series of fascial condensations (ligaments) and elastic condensations of the EPF support the uterus and vagina, thus preventing genital organ prolapse. These include uterosacral ligaments, parametrium, and paracolpium in the middle compartment [1,2,9]. In the anterior compartment, the anterior portion of the EPF, named pubocervical fascia, supports the bladder, by attaching to the pubic bones inferiorly, the obturator internus muscles laterally, and the cervix and uterus superiorly [1,2,9]. Its damage may determine urethral hypermobility and UI. The posterior portion of the EPF forms the rectovaginal fascia, also called the *denonvilliers aponeurosis*, made of a thin connective tissue placed within the rectovaginal septum extending from the posterior aspect of the cervix and posterior vaginal wall toward the sacrum, with the cardinal and uterosacral ligaments [7–9]. The damage or weakness of this fascia represents a major pathogenetic cause of rectoceles.

Finally, the central tendon of the perineum, also called the *perineal body*, represents the fibrous site of attachment for the perineal membrane, the LAMs, the external anal sphincter, and the rectovaginal component of the EPF. In men, it is located posterior to the spongious and cavernous bodies; in women, it lies within the anovaginal septum, between the introitus of the vagina and the anal canal [7–9].

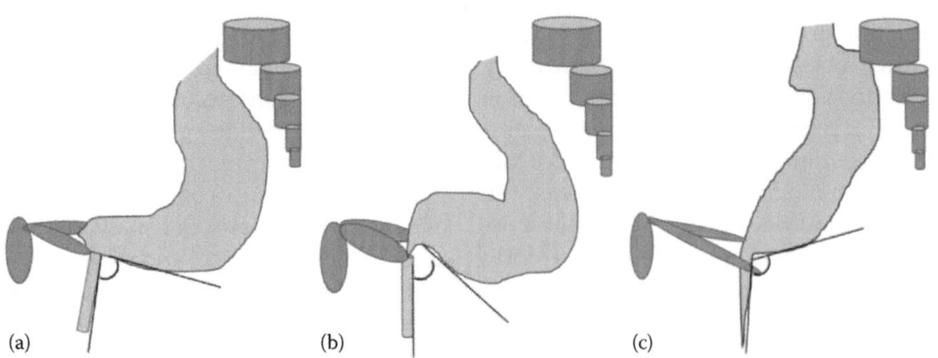


**FIGURE 7.1**
Normal variations of the ARA in correlation with PRM activity: (a) ARA at rest, (b) during squeezing, and (c) straining. The PRM forms a sling around the rectum at rest (a), aligning with the external anal sphincter. At rest, tone helps in maintaining the fecal continence. Contraction of the PRM closes the ARA, thus impeding defecation (b). Relaxation of the puborectalis sling opens the ARA allowing defecation (c).

## 7.3 Pathogenesis of Pelvic Floor Disorders

Although the etiology of pelvic floor failure is multifactorial, obstetric lesions are considered primary causes of pelvic floor damage, due to vaginal birth traumas (e.g., prolonged second stage of labor, forceps delivery, and multiple deliveries) or operative vaginal delivery [9–11]. Lesions of the iliococcygeus muscles are more frequent in the first phase of delivery, while the pubococcygeus muscles may be damaged in the second phase; midline episiotomy or forceps delivery is associated with anal sphincter rupture [10,11].

During delivery, the damage of the anterior portion of EPF (pubocervical fascia) may determine urethral hypermobility and/or cystocele, while a lesion of its posterior portion (rectovaginal fascia) may result in an anterior rectocele or enterocele [9–11].

Moreover, neuromuscular damages, particularly pudendal nerve impairment, which primarily occurs during vaginal delivery for both ischemic and mechanical factors, may diminish the LAM's capability of providing adequate pelvic support.

Aging, increased intra-abdominal pressure due to obesity, chronic conditions, as well as hysterectomy, radical pelvic surgery, and the patient's genetic predisposition may determine pelvic floor weakness and PFDs [3,5,11].

Obstetric traumas may also produce a permanent damage of the anal sphincters causing fecal incontinence, which is frequently observed in association with pelvic floor disorders (PFDs) [10].

## 7.4 PFDs: Classification and Clinical Features

The female pelvic floor is divided into three main functional and anatomic compartments: the anterior, supporting bladder and urethra; the middle, supporting vagina and uterus; and the posterior or anorectal compartment [5–9]. The spectrum of dysfunctions of the pelvic floor is therefore complex and characterized by the association of different symptoms, according to the compartments involved.

When the anterior compartment is predominantly damaged, the main symptoms include dysuria, pollachiuria, cystocele, and/or UI [9]. The weakness of the middle compartment produces vaginal vault or uterine prolapse [9]. Dysfunction or damage of the posterior compartment produces anal or pelvic pain, constipation, rectal prolapse, or fecal incontinence, variably associated [5,6–8]. The term *obstructed defection syndrome* or *outlet obstruction* is commonly used to describe one of the main dysfunctions of the posterior pelvic floor, consisting in severe constipation due to an inadequate evacuation of fecal contents from the rectum, frequently determined by the association of rectocele, rectal invagination, and prolapse.

The compartmentalization of the pelvic floor in anterior or urinary, middle or genital, and posterior or anorectal compartment has resulted in the clinical partitioning of patients into urology, colorectal surgery, or gynecology, respectively, depending on the main presenting symptoms. Such clinical partitioning may lead to an underestimation of the clinical problem, which should be rather faced by multidisciplinary clinical approach.

In most of the cases, in fact, all the three compartments are variably damaged and relative symptoms variably associated. Pelvic floor muscles and fasciae, in fact, tend to act as a unique functional entity, and hence, their dysfunction leads to dysfunction of more than one organ system, although one may predominate over the others.

Maglinte et al. reported that at dynamic colpocystoproctography, 71% of patients with PPFD showed associated cystoceles, 65% hypermobile bladder neck, and 35% severe vaginal vault prolapse [12]. Hence, a comprehensive functional pelvic floor examination is mandatory in patients with defecatory or other pelvic floor disorders.

## 7.5 Diagnosis and Classification of Anterior PFDs

The UI is the main functional disorder of the urinary compartment in women, with prevalence rates of 20%–50% [13,14]. According to the International Continence Society, UI of all types is defined as involuntary loss of urine [13,14]. Common subtypes of UI include stress UI (SUI), urge UI (UUI), and mixed UI (MUI). SUI is characterized by involuntary leakage on effort, whereas UUI by involuntary leakage accompanied by or preceded by urgency. MUI is a combination of SUI and UUI [15]. SUI is the most common type of incontinence in women, with 86% of incontinent women presenting with the symptoms of pure (50%) or mixed (36%) forms of SUI [13].

The precise anatomical causes of SUI remain somewhat unclear, being attributed to urethral hypermobility, to unequal movement of the urethral walls, to defects in the urethral supporting structures, or to intrinsic sphincter deficiency, caused by a poorly functioning urethral sphincter muscle [9,16].

Urodynamic studies (especially medium-fill cystometry) play a major role in the diagnosis and differentiation of

SUI from UUI. However, the correlation between urodynamic findings and UI symptoms is generally poor, particularly in patients with symptoms of MUI. MRI can provide objective documentation of anatomical and structural abnormalities.

## 7.6 Diagnosis and Classification of PPFDs: The ODS

The defecation is a complex process that requires normal colonic transit, anorectal sensation, expulsion force, and coordinated function of the pelvic floor for successful evacuation [17]. Disorders of this process at any level may cause constipation.

Chronic constipation is very common in Western world, affecting nearly 15%–20% of the population [3,4,17,18]. A minority of these patients, mostly women, complains of severe symptoms, which are refractory to common treatments. In these patients, the severe constipation can be determined either by difficult defecation with prolonged and unsuccessful straining, the so-called obstructed defecation, or by slow colonic transit, the so-called idiopathic slow-transit constipation, or by a combination of both.

The ODS can be found in approximately half of severely constipated patients, and it is defined as an "inadequate evacuation of faecal contents from the rectum, related to different anorectal dysfunctions, requiring prolonged repeated straining and digital manouvres to evacuate" [12,15,17,18]. The prevalence of ODS in the Western countries is about 7%, and it is observed predominantly in women. ODS is also the most common reason to request a dynamic evaluation of the pelvic floor with imaging, although pelvic pain, rectal prolapse, and anal incontinence may be other indications.

The so-called idiopathic slow-transit constipation, characterized by a slow colonic transit time, is another recognized cause of chronic constipation, which should always be distinguished from ODS, although the two causes may be associated [19].

The main PFDs underlying an ODS include rectoceles, rectal descents, rectal invaginations and external prolapses, enteroceles, peritoneoceles, anismus, or spastic-diskinetic puborectal muscle syndrome.

Once an ODS is clinically suspected, the patient should be extensively investigated in order to diagnose the underlying pelvic disorder and the associated dysfunctions. Identification of the specific anorectal dysfunction leading to ODS is fundamental for planning an effective treatment, addressing the patient toward surgical or conservative treatments. ODS may be sustained either by mechanical causes (e.g., rectal prolapse, rectal descent, rectal invagination, rectocele, and enterocele) or by a functional disorder (puborectalis syndrome) [5–8]. The distinction between mechanical or functional causes is crucial, since mechanical obstructions are usually addressed toward surgical treatments (e.g., rectopexis or trans-anal rectal resection), while functional ones toward conservative therapy (biofeedback). Furthermore, the choice of the surgical procedure is driven not only by the type and severity of underlying rectal disorder, but also by the presence of associated PFDs, usually requiring a more extensive and combined surgical approach [5–8]. Surgical techniques for PPFD are continuously evolving in order to provide a more affective and comprehensive repair of the pelvic floor and to improve long-term results.

Hence, although functional gastroenterologists and colorectal surgeons are primarily involved in the management of ODS, it is recommended to follow these patients within a multidisciplinary team, including urogynecologists, physicians, radiologists, physiotherapists, and specialized nurses, what it is modernly defined as *pelvic floor unit*.

Therefore, the diagnosis of PPFD is demanding and achieved by associating clinic history, physical examination, physiological testing, and diagnostic imaging. The clinical examination alone either underestimates or results in misdiagnosis of the site of prolapse in over 50%–90% of patients, being not reliable for assessing evacuation abnormalities. Physiological tests more widely used include anorectal manometry, electromyography, and rectal balloon expulsion test. Manometry is particularly useful for differentiating functional disorders, particularly to detect an impaired relaxation of the puborectalis and anal sphincter [18–20]. Endoanal ultrasound is widely used in patients with fecal incontinence to detect tears of the anal sphincters, frequently associated with PFD as results of obstetric traumas [18–20].

The two main radiologic tools for the diagnosis of PPFD include conventional defecography (or dynamic proctography) and dynamic MRI of the pelvic floor (DPF-MRI) or MR-defecography (Figure 7.2).

Conventional defecography has always played a central role in the assessment of the ODS. Being able to document the evacuating process in real time, it may evaluate both functional and anatomical anorectal disorders with high accuracy [12,21–24]. In the past years, however, this technique failed to recognize the frequently associated abnormalities of the anterior and middle compartments.

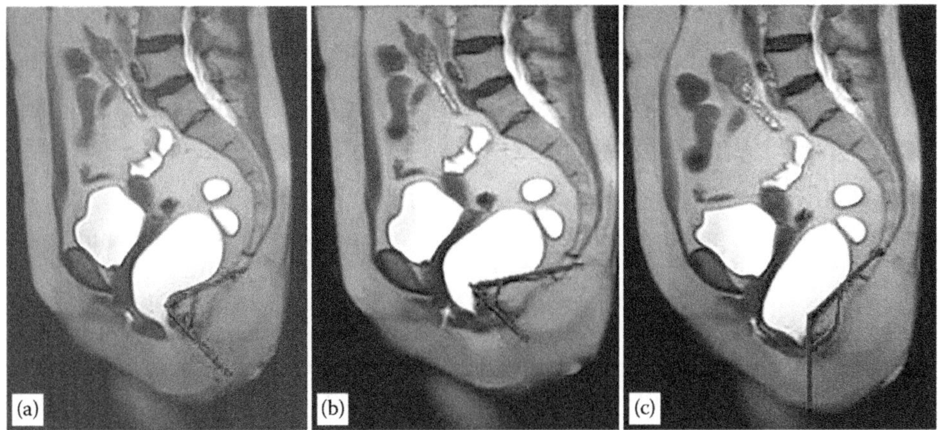


FIGURE 7.2
Normal variations of the ARA at MRI after rectal filling with gel: (a) sagittal-balanced image at rest, (b) sagittal-balanced image during squeezing, and (c) sagittal-balanced image during straining. The ARA measures approximately 90° at rest (a); it reduces during contraction (85°) and opens during straining (>120°) allowing defecation (c). Please note a mild descent of the posterior compartment with respect to the anterior and middle one, in all functional phases.

More recently, its diagnostic value has been significantly improved by the simultaneous opacification of bladder, vagina, and small bowel (colpocystodefecography) [22–25]. In this way, however, the examination has become longer, uncomfortable, and more expensive. Limits of defecography include the limited projectional planes, the inherent radiation risk, and its inability to depict perirectal soft tissues [25,26].

MRI has emerged as a valuable alternative technique for assessing PFDs and staging POP. Thanks to its multiplanar capability and high soft tissue contrast, MRI allows a comprehensive morphologic and functional evaluation of all three compartments at the same time, without ionizing radiations [6–8,27–29] (Figure 7.2).

Nowadays MRI enables also a real-time assessment of functional diseases with dynamic acquisitions, similar to conventional defecography [6–9,27–31]. The main limit of MRI is the obliged supine position, which is necessary if a closed 1.5 T magnet is used. The supine position of MR-defecography has been widely criticized, since the defecation is not evaluated under physiological conditions [27]. MR-defecography can been performed with open magnets too, which allow a physiological sitting position [27,32]. Published comparative studies between open and closed MR-defecography, however, have established a good concordance between findings obtained in sitting and supine position, thus validating the use of closed MRI in the assessment of defecation [27]. So far, open magnets are poorly available, while conventional closed MRI units are widely diffuse and more and more accessible to body imaging.

## 7.7 Functional MRI Evaluation of PFD: Technical and Diagnostic Issues

A standardized technique for the MRI evaluation of posterior PFDs is not available yet, because of the heterogeneity of MRI equipments, of different sequences available on different equipments, and of various rectal contrast agents [6–9,27–34].

Several authors [27–29,31–33] suggest to perform this examination by filling the rectum with various contrast media (ultrasound gel, mashed potatoes mixed with gadolinium, etc.), while others without rectal filling [35].

Some authors consider the complete rectal evacuation as the main part of the exam (MR-defecography), while others base the examination mostly on functional maneuvers (DPF-MRI). Again, others prefer to study the patient in the left lateral rather than in the supine position [36].

As a matter of fact, there is still an open debate on the choice of the technique for evaluating PFDs, each one having specific advantages and limitations.

To date, in fact, no rectal contrast agent is able to reproduce the consistency of normal or hard stools, neither gel or mashed potatoes, that rather mimic diarrheal feces [6,25]. Undoubtedly, to perform a complete evacuation inside the gantry of a closed MRI unit is not a physiological process, neither may be comfortable for the patient. Several authors, however, believe that without evacuation dynamic pelvic floor MRI should be considered incomplete, whereas others are in favor of dynamic examinations without rectal filling and evacuation phase, by affirming that all major pelvic dysfunctions can be diagnosed at maximal straining without the need of a *real* evacuation. Besides, choosing air to distend the rectum, or no contrast at all, maximal straining can be repeated several times to obtain the best evidence of the pathologic process, while the evacuation of the gel or any other contrast agent is a *one-shot* maneuver. Furthermore, the rectal filling with gel could overestimate rectal invagination and prevent, at the same time, the evidence of cystoceles or colpoceles due to the marked filling of the rectum in the posterior compartment. To date, in spite of numerous published papers, there is no evidence of the superiority of one technique on the other.

What is mostly important is to remind that ARA values and main reference parameters adopted for supine-MRI are slightly different from those used in conventional defecography or sitting-MRI. In fact, in the supine position the gravity force acts on the defecation process differently than in the physiological sitting position, thus requiring specific MRI grading systems for rectoceles and other findings, as reported in previous studies [6–8,18,20,27,28,32,34,37].

Independently from the choice of one or another technique, to obtain an effective dynamic MRI examination of the pelvic floor in supine position, several technical aspects are crucial. (1) Before the examination, the patient should be always informed and trained, in order to perform at the best the different functional maneuvers. Patient's cooperation is mandatory to obtain satisfactory results. (2) Static and dynamic acquisitions at rest and during functional maneuvers (contraction and straining) are both needed to perform a exhaustive examination, independently from the rectal contrast used [6–8,27–37] (Figure 7.2). (3) The urinary bladder should not be excessively full but neither empty (containing 50 to 100 cc, approximately), in order to properly assess dysfunctions of the anterior compartment. (4) Although the rectum may be filled in different ways by different contrast agents, the choice of the rectal contrast agent determines the choice of the dynamic fast sequences to acquire. For example, mashed potatoes and gadolinium require fast $T_1$-weighted sequences, while rectal gel filling is better depicted both with fast $T_2$-weighted and balanced steady-state sequences; rectal air distention, instead, is displayed at the best with fast $T_2$-weighted sequences. (5) The evacuation phase should be always performed, or at least substituted by repeated maximal straining phases [6–8,27–37].

## 7.8 MR Imaging Protocols

In our experience [6] according to the patient's compliance and scheduled examination time, we alternatively use two different functional MRI techniques, which are performed in supine position using a closed standard 1.5 T MRI unit (Figure 7.3). In all cases, for hygienic reasons, we ask the patient to preliminary clean the rectum by using two micro-clysters, approximately 12 and 2 h before the examination.

One MRI technique for dynamic pelvic floor imaging may be called *air-balloon technique* (Figure 7.4), being

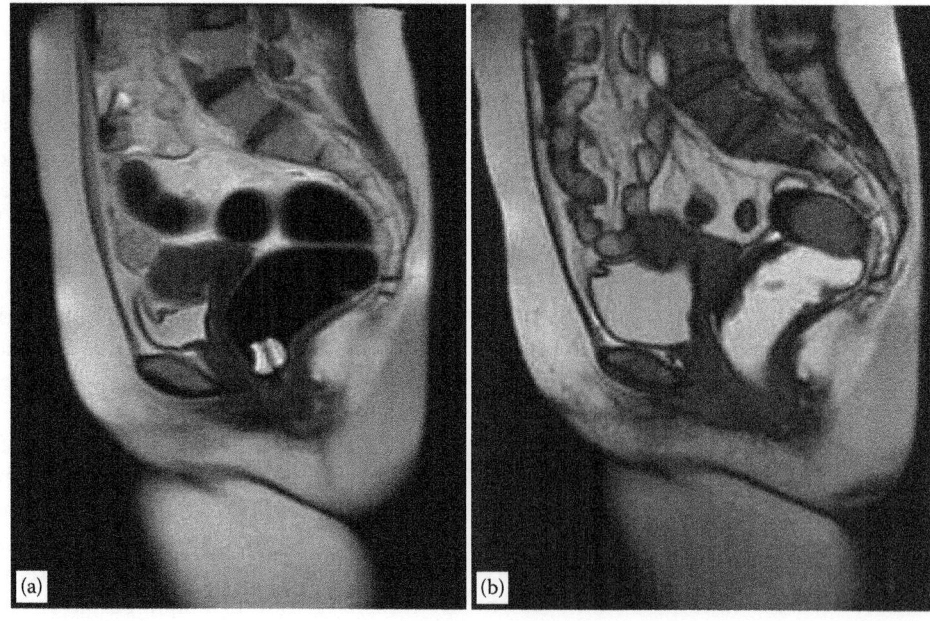

**FIGURE 7.3**
Same patient at rest: comparison between air-balloon technique and gel-filling technique. (a) Rectal filling with air, after placement of a Foley catheter, then at the end balloon filled with saline (HASTE $T_2$-weighted sagittal image). (b) Rectal filling with gel (TrueFISP sagittal image).

based on the use of a Foley catheter to distend the rectum. First, a 16 or 18 French soft Foley catheter is placed in the rectum, then at the end-balloon is distended by 15–20 cc of saline solution and approximately 300 cc of room air is inflated in the rectum. The balloon mimics solid fecal material, while the air distends the rectum during functional maneuvers. Distension of the rectum with air may sometimes be suboptimal but it does not prevent the contemporary evaluation of dysfunctions or prolapses of the anterior and middle compartment, while rectal filling with gel could reduce the evidence of urogenital prolapses. The pelvic floor is examined on static and dynamic images at rest, during contraction and during straining. The *static morphological and functional phase* lasts approximately 10 min, and it includes three consecutive $T_2$-weighted HASTE sequences (half Fourier acquisition single-shot turbo spin echo: matrix 160 × 256, 6 mm thickness, 0.15 mm distance factor, 25 parallel slices, 25 s acquisition time), acquired on axial, sagittal, and coronal planes at rest.

Then two additional sequences are acquired during functional maneuvers on adjacent sagittal planes, asking the patient to hold the pelvic maximum contraction for 25 s (static contraction phase) and then the strain for additional 25 s (a—static straining phase) (Figure 7.4a).

In the second examination phase (*functional dynamic phase*), a $T_2$-weighted HASTE dynamic cine sequence is modified in order to acquire a slice of approximately 6–8 mm of thickness per second on the midsagittal plane, repeated for approximately 50 s, starting with the rest position, then asking the patient to contract pelvic floor muscles progressively from minimum to a maximum, then to strain progressively in the last 30–40 s and finally to relax (Figure 7.4b). The dynamic phase of maximal straining is repeated several times (routinely 2–3 times) to obtain better results. The final evacuation phase may be obtained by asking the patient to expel the balloon although this happens rarely; this technique may resemble the balloon expulsion test, which is a physiological test used to assess the rectal capability of defecating. The overall examination time is approximately 15–20 min, including patients positioning.

The other MRI technique that we use is widely known as *MR-defecography* or *gel-filling technique* (Figures 7.3 and

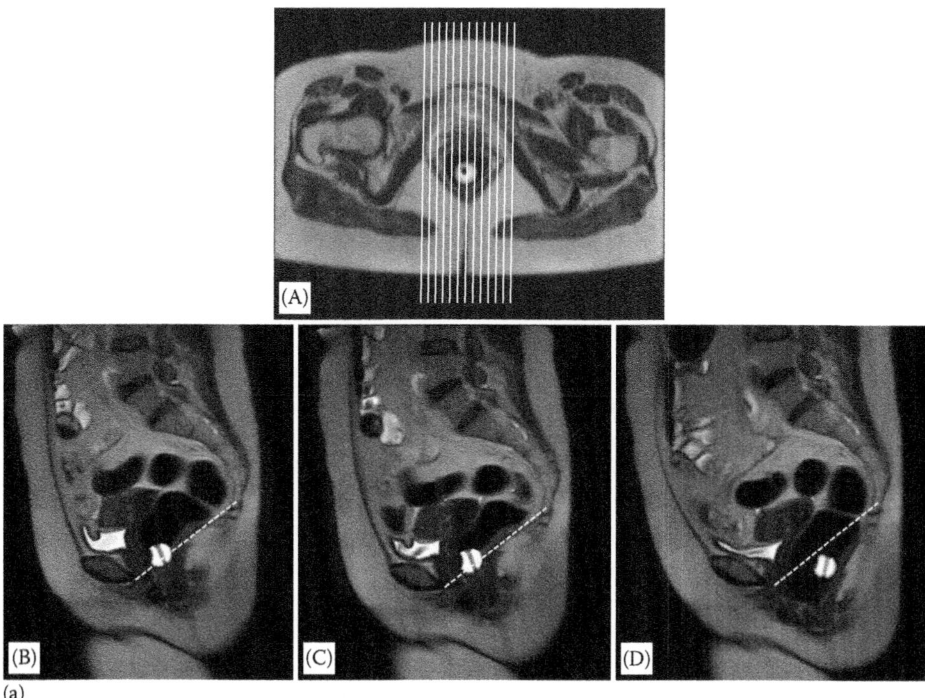


**FIGURE 7.4**
(a) Static morphological and functional phase: It includes evaluation of the pelvic floor at rest, during squeezing (or contraction), and during maximal training, both on static and dynamic images. This figure explains the first phase, the static functional phase. Axial coronal and sagittal images are first acquired at rest, after placement of the balloon catheter and rectum distension with air. An axial HASTE $T_2$-weighted image acquired at rest at the level of the anorectal junction (A) is used to place a slab of parallel contiguous sagittal slices, first acquired at rest (B), then asking the patient to hold at the maximum pelvic floor contraction for approximately 25 s (squeezing) (C), and then to strain maximally for additional 25 s (straining) (D). The contiguous slices acquired in axial, coronal, and sagittal planes at rest, and then on the sagittal plane at contraction and straining, offer a valuable assessment of the entire pelvis in different functional phases. This phase includes not only evaluation of the bladder, uterus, and rectum, but identification of the ovaries and of some portions of the large and small bowel placed in the pelvis as well. Ovarian cysts or diverticula of the sigmoid colon may be detected in this phase. *(Continued)*

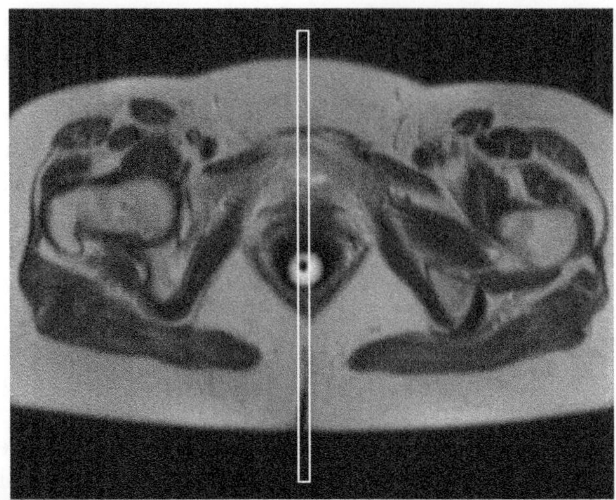

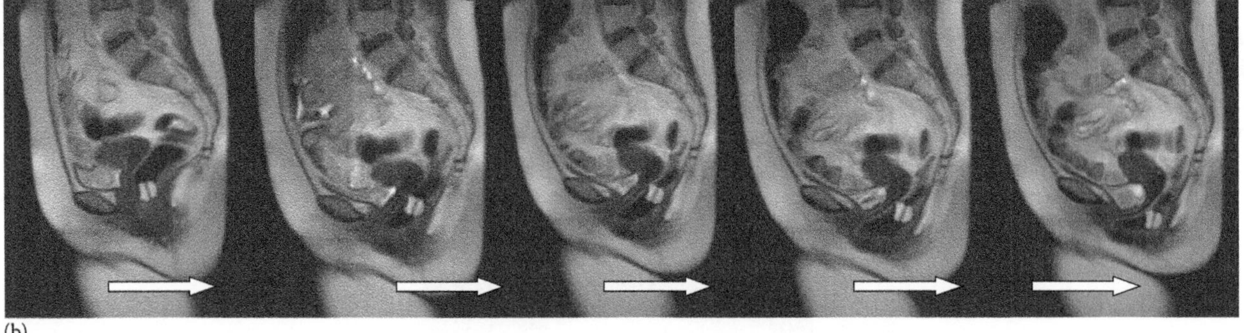

(b)

**FIGURE 7.4 (Continued)**
(b) This figure illustrates the second phase: the dynamic (or cine) functional phase. An axial HASTE $T_2$-weighted image is used to place a single thick (8 mm) sagittal slice, centered on the anal canal (top). This midsagittal slice is therefore repeated for 45 to 60 s (approximately 1 image/s,) by asking the patient to strain progressively to a maximum, finally trying to expel the balloon. Please note the mild bladder descent at the end of the maximal straining (bottom). The dynamic phase is usually repeated two or three times, asking the patient to progressively increase the straining, and it is usually reviewed as a movie in cine-loop. The sagittal images represent only a few of the many (over 45) consecutive acquisitions obtained in a dynamic phase.

7.5 through 7.9). The rectum is first filled with gel (approximately 150–180 cc), then a static morphological–functional phase and a dynamic phase are acquired, similar to the previous technique, with an additional final evacuation phase. The *static morphological and functional phase* lasts approximately 10 min, and it includes balanced or $T_2$-weighted HASTE sequences (matrix 160 × 256, 4 or 5 mm thickness, 0.15 mm distance factor, 25 parallel slices, 25 s acquisition time) acquired on axial, sagittal, and coronal planes at rest for a morphological evaluation of the pelvis, then two sagittal sequences on adjacent planes acquired during functional phases, by asking the patients to hold for 25 s the maximum contraction and then strain for additional 25 s.

In the second phase (dynamic functional phase), a 6–8-mm thickness slice is repeated in the midsagittal plane every second for 50–60 s, starting with the rest position, then asking the patient to contract pelvic floor muscles progressively to a maximum, then straining progressively in the last 40 s, trying to evacuate the all the gel. If the gel is not evacuated at the first attempt, the dynamic sequence is repeated two or three more times. The overall examination time is approximately 20–25 min, including patients positioning and the rectal filling.

We suggest to use one or the other technique, according to the radiologist experience and the procotologist or uro-gynecologist preferences.

The gel-filling technique could be considered as a first- or second-line examination, to use as an alternative to the air-balloon technique whenever findings are doubtful. Definitely, the gel allows a better distension of the rectum and a clearer evaluation of the evacuation phase, being more effective in the diagnosis of rectal invagination. However, it is a one-shot procedure, since it cannot be repeated after a complete evacuation. Hence, it may underestimate some pathologic events such as rectoceles or enteroceles. Once the gel has been eliminated, if the rectal dysfunction has not emerged, it may be completely missed, while the balloon air-filling technique allows repetition of the maximal straining phase for several times, thus increasing the possibility to show the pelvic dysfunction. Moreover, the rectal

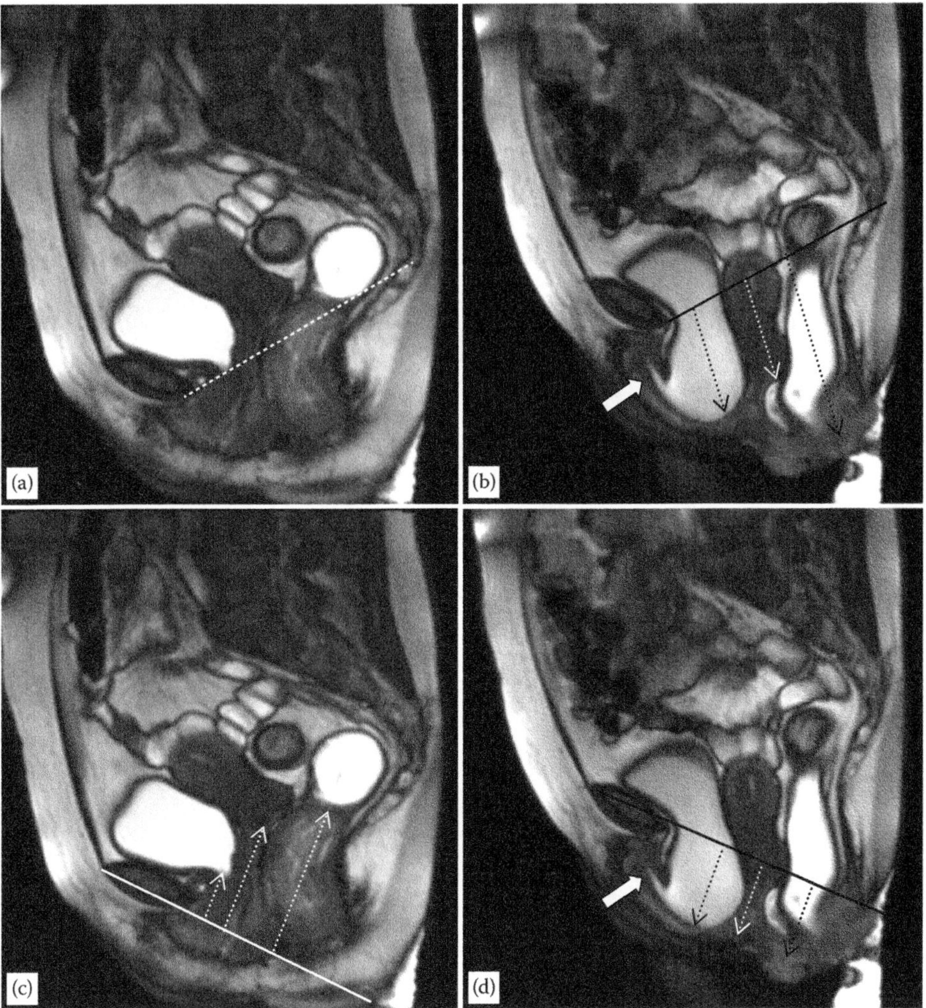


**FIGURE 7.5**
The two more common reference lines used in dynamic MRI of the pelvic floor/MR-defecography: The pubococcygeal line (a) and the mid-pubic line (b). (a, b) The pubococcygeal line has been traced between the inferior aspect of the pubic bone to the last coccygeal joint, at rest and during maximal straining. (c, d) the mid-pubic line is traced extending caudally along the mid-axis of the pubic symphysis. With both systems, the staging of POP is obtained by measuring the perpendicular distance from the anatomical landmark of each compartment (bladder base, uterine cervix, and anorectal junction) to the reference line. In this patient affected by ODS and urinary incontinence, the rectum has been preliminarily filled with gel. The image obtained during straining in the dynamic phase shows rectal descent, severe cystocele associated with urethral hypermobility (large arrow) and colpocele. The three-compartment pelvic floor descent is clearly depicted.

filling with gel may prevent the descent of the bladder and vaginal vault in the presence of cystocele or uterine prolapse. Furthermore, by using the gel, the entire examination becomes slightly longer and less comfortable and hygienic for the patient. According to our experience, both techniques are valuable, although showing favorable and unfavorable aspects; if needed, they may be associated in the same imaging session to reach the maximal accuracy. The consistency of the gel, being similar to diarrhea rather than to normal stools, may determine an overestimation of rectal invagination that, on the contrary, may be underestimated with balloon air-filling. In our experience, a preliminary comparison between the two techniques did not show relevant differences, demonstrating a good agreement with clinical findings and defecography in both cases. The air-balloon MRI showed a tendency to underestimate rectal invagination or small rectoceles, while gel MRI could occasionally miss completely the evidence of rectoceles, cystoceles, and enteroceles.

## 7.9 Reference Lines and Grading Systems for PPFD

To date, several landmarks and different reference lines for measuring and staging POP on MRI have been proposed. The two most commonly used lines are the

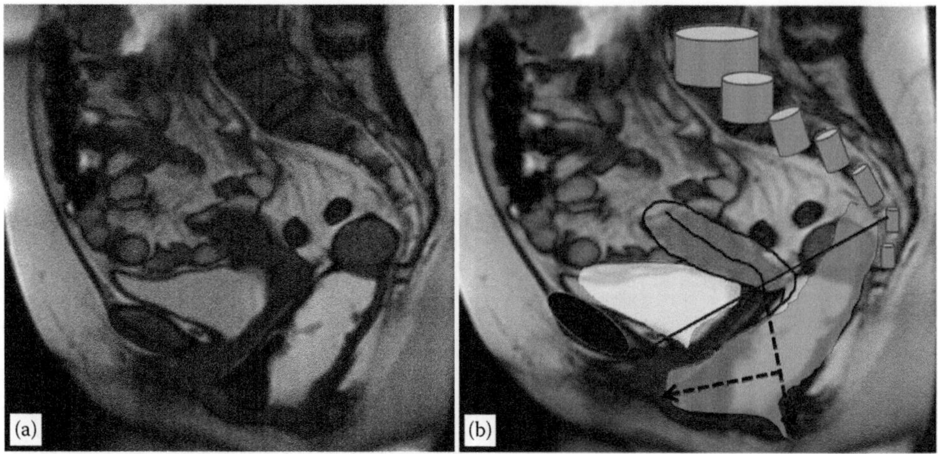


**FIGURE 7.6**
Patient with ODS. (a) Dynamic phase obtained with repeated midsagittal TrueFISP images during straining. An anterior rectocele is clearly depicted. There are no abnormalities in the anterior and middle compartment. (b) Same image with overlapped drawings. The depth of the rectocele (blue) is measured as the distance between the mid-anal axis and anterior rectal wall, approximately 4 cm in this case (discontinuous arrow). Similarly, the degree of posterior compartment descent is measured as the distance between the PCL (pubococcygeal line) and AJR (anorectal junction) at maximal straining (discontinuous arrow). The bladder base (yellow) and vagina vault (pink) are in place.

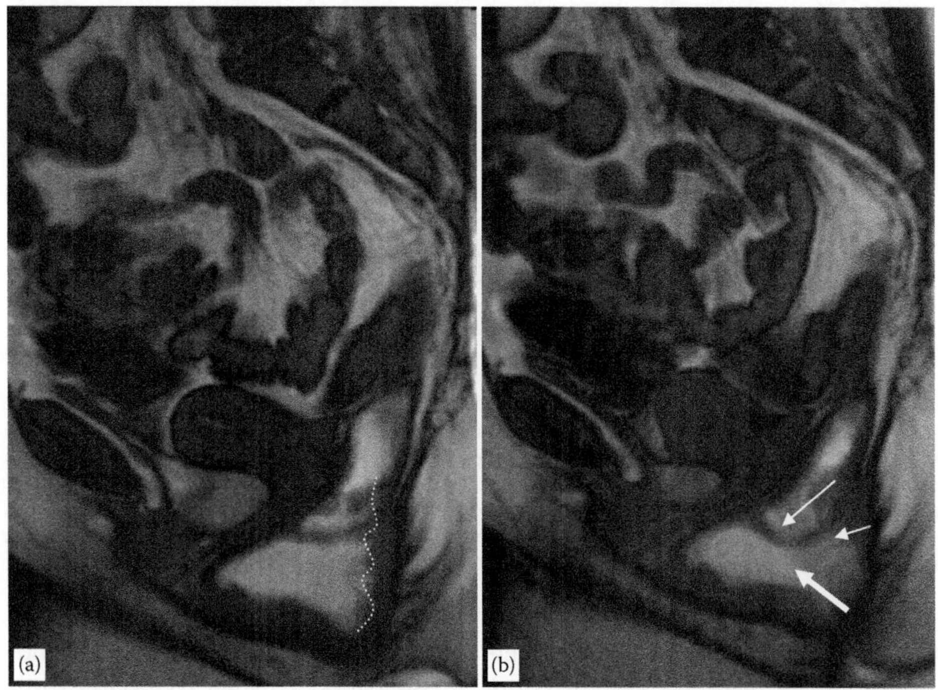


**FIGURE 7.7**
Female patient with severe ODS and rectocele with rectal invagination at MRI. (a, b) The dynamic phase performed after gel filling with a TrueFISP dynamic sequence shows an evident intra-rectal invagination and clear signs of intra-rectal mucosal invagination (b, arrows) as well as mucosal enfolding at the level of the posterior rectal wall (a, curve line), with gel entrapment in the rectocele pouch (b, large arrow).

*pubococcygeal line (PCL),* which connects the inferior aspect of the pubic symphysis to the last coccygeal joint, and the *mid-pubic line (MPL),* extending caudally along the long the mid-axis of the pubic symphysis [6–9,27–29,34,36,37] (Figure 7.5).

To date, neither the PCL nor the MPL has shown to have better agreement with clinical staging. Hence, the choice of the reference lines for MRI interpretation may depend on radiologist experience and referring physician's preference [40].

Once the MRI reference line is chosen, the staging of pelvic dysfunctions and POP is performed by measuring the perpendicular distance from the anatomical landmark of each compartment to the reference line. In the

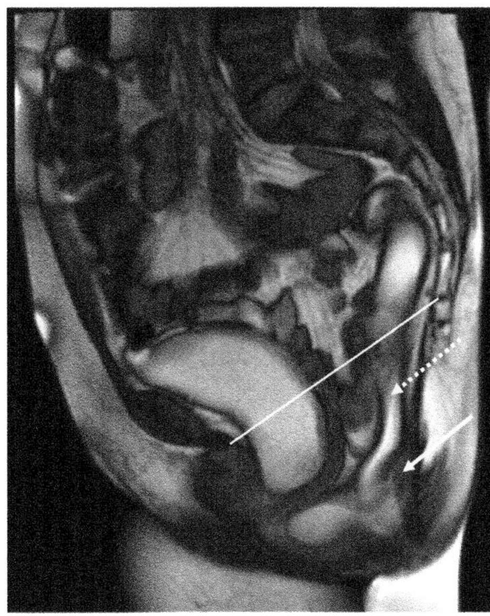

**FIGURE 7.8**
Female patient with ODS. An enterocele is clearly observed at dynamic MRI after gel distention. The small bowel loops deeply prolapse (discontinuous white arrow) in the peritoneal pouch at maximal straining. Please note the concurrent cystocele and the rectocele, with evident rectoanal invagination (white arrow).

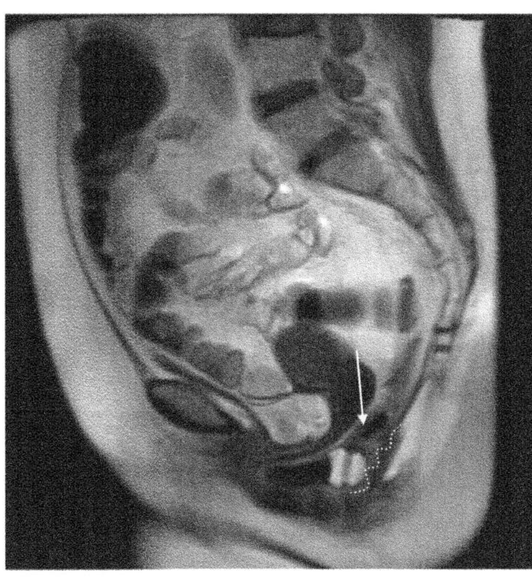

**FIGURE 7.9**
The dynamic phase performed after air filling with a $T_2$-weighted HASTE dynamic sequence shows an evident intra-rectal invagination (white line) and clear signs of intra-rectal mucosal invagination (curve arrow) as well as mucosal enfolding at the level of the posterior rectal wall (a, curve line), with balloon entrapment in the rectocele pouch.

anterior compartment, the landmark is the most inferior aspect of the bladder base. In the middle compartment, the landmark is the uterine cervix, or the vaginal apex in case of previous hysterectomy. In the posterior compartment, reference point is the *AJR* [6,34,37,38,40].

Using the PCL (PCL-POP staging system) at the level of any pelvic compartment, a prolapse is considered *small* if the reference point is placed 1–3 cm below PCL, *moderate* between 3 and 6 cm below the PCL, and *large* over 6 cm below the PCL. The *rule of three* is useful in grading POP with PCL system: a prolapse of an organ below the PCL by ≤3 cm is considered mild, by 3–6 cm moderate, and by >6 cm severe [6,34,37,38].

Instead, using the *MPL* (MPL-POP staging system), any POP is classified as stage 0 when the reference point is 3 cm above the MPL, as stage 1 if more than 1 cm above the MPL, as stage 2 if equal or less than 1 cm below MPL, as stage 3 if over 1 cm below the MPL; stage 4 represents the complete organ eversion [6,40].

Other measurements in the sagittal plane during maximum straining include the H-line, which extends from the inferior aspect of the pubic symphysis to the AJR, representing pelvis iatus widening; and the M-line, a perpendicular line traced from the PCL to the posterior aspect of the H-line [40] representing the iatal descent. To grade PFD, HMO (H line, M line, organ prolapse) system is applied to a midsagittal image obtained during maximal patient straining [31,37,38,40].

Independently from the reference system adopted, a complete three-compartment evaluation is always recommended in the assessment of a patient presenting with PPFD.

The use of the PCL as main anatomical landmark is usually easier for the radiologists.

In healthy women, the PCL represents the site of the pelvic floor, running at rest parallel to the levator plate. At rest, in a normal subject, the base of the bladder, the upper third of the vagina, and the peritoneal cavity should project superior to the PCL (Figures 7.1 and 7.2). The distance from the PCL to the bladder base, uterine cervix, and AJR, measured on images obtained when the patient is at rest and at maximal pelvic strain or evacuation, expresses the severity of the prolapse.

Choosing the PCL reference line to quantify PFD, the same reference parameters used in conventional defecography may be adopted for DPF-MRI. Anatomical landmarks are even better detectable on MRI than in radiologic images so that any measure is easier and more reproducible.

The *ARA* is the angle formed by a line along the posterior border of the rectum and a line along the central axis of the anal canal. Its changes express the functioning of the puborectalis muscle; when the pelvic floor and puborectalis contrast, the angle closes (squeezing), whereas during straining and evacuation the angle opens.

In our experience, according to other authors, in the supine position in healthy subjects, at rest the ARA is comprised between 85° and 95° [6]. During squeezing (maximal pelvic floor contraction), organs elevate in relationship to the PCL, thus sharpening the ARA by 10°–15° (due to contraction of the puborectal muscle): a change in ARA of 5° or less may be considered abnormal. During straining and defecation, the pelvic floor muscles relaxes and the ARA becomes obtuse, typically 15°–25° wider than when measured at rest [28,36,41]. Otto et al. found a rectoanal angle of 100° at rest, 87° at squeezing and 130° at defecation [36].

In healthy subjects when the puborectalis muscle and external sphincter relax, the anal canal opens thus allowing evacuation of at least 2/3 of the contrast material within 30 s, as observed at conventional defecography and defeco-MRI in sitting position [8,18,27]. At dynamic MRI in sitting position, however, the evacuation time maybe more variable, mostly due to the reduced gravity force and uncomfortable position for evacuating.

## 7.10 PFDs: Pathologic, Clinical, and MRI Features

### 7.10.1 Rectocele

It is defined as an outpouching or bulging of the anterior rectal wall during defecation. A small to mild rectocele can be observed in nearly 78%–99% of parous women [42–44], while rarely in men. The underlying etiology is weakening of the support structures of EPF and, particularly, thinning or tears of the rectovaginal fascia (Kenton, felt). Factors that may increase the risk of developing a rectocele include vaginal birth traumas (multiple, difficult, or prolonged deliveries; forceps delivery; perineal tears), constipation with chronically increased intra-abdominal pressure, hysterectomy, aging, and congenital or inherited weaknesses of the pelvic floor support system (Figures 7.6 through 7.8) [44].

In most of the cases, rectoceles develop anteriorly, due to weakness of the structures sustaining the anterior rectal wall. The presence of bony and fibrous structures (sacrocccygeal vertebras) likely reduces the capability of a posterior bulging of the rectal wall. A rectocele rarely develops postero-laterally and, in this case, the lesion occurs laterally, through a puborectalis muscle defect rather than in the midline. Postero-lateral rectoceles are officially named as *posterior perineal hernia* [15,23].

Rectal invagination, either intra-rectal or rectoanal, is frequently associated with rectoceles [23,42,44]; in particular, anterior rectoceles are frequently associated with rectal wall invagination and infolding of the posterior rectal wall. A plausible explanation to this finding could be related to the inability of the posterior rectal wall and mucosa to fully distend, differently from the anterior wall. With the increase in the rectal pressure, the posterior rectal wall tends to fold and invaginate (Figure 7.7) The rectoceles are frequently associated with enteroceles and anismus. Interestingly, enteroceles or peritoneoceles frequently act as compensatory mechanisms, since they may improve the evacuation by reducing fecal trapping in the rectocele (Figure 7.8).

Symptoms related to rectocele may be primarily vaginal or rectal. Vaginal symptoms include vaginal bulging, dyspareunia, and the sensation of a mass in the vagina. Rectal symptoms include defecatory dysfunction, constipation, and sensation of incomplete evacuation. The fecal trapping in the rectocele leads many patients to emptying it by digitating and pressing on the posterior wall of the vagina or perineum.

The sensibility of the clinical examination in the diagnosis of rectocele varies from 31% to 80% [2,4,7,17,45]. The physical examination alone generally does not distinguish an enterocele from a high rectocele. For these reasons, dynamic MRI during maximal straining and/or evacuation is valuable for the diagnosis and grading of rectoceles.

At MRI, a rectocele is measured during the phase of maximal straining and evacuation, as the distance between mid-anal line and the anterior rectal wall, or as the depth of wall protrusion beyond expected anterior rectal wall (Figures 7.6 through 7.8). At MRI performed in supine position, rectoceles are graded small if <2 cm, mild between 2 and 4 cm, and large if >4 cm [27]. This grading system is slightly different from the one adopted in conventional defecography, in order to correct the bias of the supine position. By using this staging system, the sensitivity of MRI in detecting large rectoceles, over 3.5 cm, ranges between 87% and 100% [27]. Large symptomatic rectoceles (more than 3 cm in size) are easily detected at supine MRI with the air-balloon technique as well as with gel filling.

Conversely, smaller rectoceles (less than 2 cm) considered as normal features in most of parous women are hardly detectable at MRI in supine position, whereas usually identified at conventional defecography [42].

MR-defecography does not only provide objective information about the size of the rectocele, but also displays the dynamics of its emptying and detects concurrent rectal invagination or enterocele, which are relevant findings usually missed at physical examination [20,30,33,36,41] (Figures 7.6 through 7.8).

Entrapment of the rectal gel or balloon inside the rectocele pouch is frequently observed at dynamic MRI (Figure 7.7).

The surgical repair of rectocele can be performed with different techniques and approaches, therefore requiring an accurate preparative planning [46]. A rectocele may be treated through a transanal or transvaginal approach, using or not a stapled rectal resection, with or without concurrent open or laparoscopic surgical repair of other pelvic disorders. Moreover, most of these surgical techniques (i.e., stapled trans-anal rectal resection or STARR) are continuously evolving, and there still is a major debate on their appropriateness and long-term efficacy [46].

### 7.10.2 Rectal Intussusception

Rectal intussusception, also named *incomplete* rectal prolapse, is an invagination of the rectal wall [20], located either anteriorly, posteriorly, or circumferentially, which may involve the full thickness of the rectal wall or the mucosa only [7,17,44] (Figures 7.7 and 7.8).

Rectal intussusception is thus classified as intrarectal (if it remains within the rectum), intra-anal (if it extends inside the anal canal), or extra-anal (if it passes the anal sphincter); the latter is also called *complete* or full rectal prolapsed [44].

Small invaginations of the rectal wall are considered as normal findings during defecation, observed in nearly 80% of healthy subjects [42]. An intrarectal intussusception (or incomplete rectal prolapse) is unlikely to obstruct defecation, since it occurs as the rectum collapses, rather producing a sensation of incomplete emptying. Conversely, when the invagination progresses becoming intra-anal, patients most likely experience a sensation of incomplete or obstructed defecation due to outlet obstruction [42,44,47]. High-grade intussusceptions are frequently associated with rectocele and, occasionally, with the solitary rectal ulcer syndrome [47,48]. Whenever the invagination is associated with a rectocele, it may cause sequestration of the rectocoele content and return of the feces back into the rectum during relaxation, thus resulting in incomplete evacuation and ODS as well [17].

According to Bertschinger et al. [27], MRI sensitivity in detecting rectal invagination is lower if the examination is performed in supine position; in this study, 100% of rectal intussusceptions were missed at supine imaging with respect to the sitting position, although this series included only a limited number of patients. More recently, Dvorkin et al. reported an overall MRI sensitivity for intussusceptions of nearly 70% with respect to evacuation proctography [48]. Other authors argue that conventional defecography may rather overestimate rectal invaginations, due to low consistency of the barium paste. In our experience, MR-defecography with gel filling shows a higher sensitivity for low-grade and mild invaginations, particularly during the evacuation phase, with respect to air-balloon technique, due to a better depiction of the rectal mucosa. High-grade invaginations, however, are easily detected with both techniques (Figures 7.8 and 7.9).

### 7.10.3 Rectal Prolapse

Rectal prolapse is defined as an extra-rectal or complete rectal intussusception, when either the mucosal or full-thickness layer of the rectal wall extends through the anal orifice [44]. Typically, rectal prolapse begins with an intra-rectal intussusception and progresses toward a full prolapse. The incidence of rectal prolapse is estimated at approximately four cases per 1000 people; in the adult population, the female to male ratio is 6:1 [44]. Common symptoms include constipation, sensation of incomplete evacuation, fecal incontinence, and rectal ulceration with bleeding [44]. Although rarely, untreated rectal prolapses can lead to incarceration and strangulation.

### 7.10.4 Enterocele

The herniation of the pelvic peritoneal sac into the rectogenital or Douglas pouch may contain fat, the so-called peritoneocele, small bowel loops, properly defined enterocele, or sigmoid colon, defined as sigmoidocele. Occasionally, it may also contain the cecum (Figure 7.8). In patients with PFDs, the incidence of enteroceles ranges between 17% and 37%; they are more common in women and frequently associated with rectoceles. [7,17,20,21]. Uterus removal increases the risk of enterocele, causing a separation of the anterior (pubocervical) from posterior (rectovaginal) wall fascia [7].

Enteroceles may be symptomatic, causing a sense of fullness and incomplete evacuation and occasionally lower abdominal pain. Usually however, an enterocele does not impair evacuation, but it may rather improve the obstructive symptoms associated with a large rectocele, by compressing the rectocele pouch thus reducing the fecal entrapment (Figure 7.8).

According to several authors, MRI seems to be superior to dynamic cystocolpoproctography, which failed to identify up to 20% of enteroceles [20,49] Other authors, however, have reported discordant results, by demonstrating a higher accuracy of conventional proctography [50]. The surgical repair of enteroceles usually consists in the obliteration of the cul-de-sac [46].

## 7.11 Rectal and Perineum Descent

Rectal descent is defined as the descent (over 3 cm) of the AJR below the PCL, usually combined with the abnormal descent of the middle and anterior pelvic floor compartments as well, at variable extents. [6–8,51,52]. Such generalized pelvic floor weakness is also defined as *descending perineal syndrome* (Figure 7.5). Typically, the syndrome is described as ballooning of the perineum several centimeters below the PCL during straining, although a descent in severe cases may also be observed at rest [17].

The major pathogenetic mechanism of the descending perineal syndrome seems to be an excessive and repetitive straining. Chronic straining determines a progressive protrusion of the anterior rectal into the anal canal, determining a sensation of incomplete defecation and weakness of the pelvic floor musculature, that causes more straining, hence establishing a vicious cycle [51,52]. Other causes are weakness of the muscles of the pelvic floor caused by pudendal nerve impairment due to childbirth trauma or neuropathy [52].

At MR-defecography, an abnormal rectal descent is defined as an inferior displacement of the AJR below the PCL, which is considered *mild* between 3 and 5 cm and *severe* when greater than 5 cm. The landmark for the posterior compartment is, in fact, the position of the AJR with respect to the PCL.

Abnormal herniation of the bladder below the PCL, termed cystocele, represents the abnormal anterior compartment descent. Hysteroptosis or descent of the vaginal vault, which is the landmark of the middle compartment, represents descent of the middle compartment [6,7,28,37].

Pelvic floor descent or prolapse may thus be defined as three-compartment descent when the pelvic floor is fully involved, or a bicompartment descent when two out of the three pelvic compartments are involved.

## 7.12 Puborectal Syndrome or Anismus: Pathologic and Clinical Features

Anismus or spastic pelvic floor syndrome (or diskinetic puborectal syndrome or pelvic floor dyssynergy) is a PFD characterized by lack or insufficient relaxation of the puborectal muscle and external anal sphincter during defecation, which causes an ODS [8,17,18,45].

In normal subjects, the puborectal muscle at rest sustains the inferior border of the rectum maintaining the continence (Figures 7.1 and 7.2). Anismus determines constipation and incomplete defecation (ODS) due to a paradoxal contraction of the puborectal muscle during straining and defecation, without significant variation of the ARA in the different functional phases (rest, squeezing, and straining). Associated features may be the lack of normal descent of the pelvic floor during straining, hypertrophy of the puborectal muscle, and occasionally an associated anterior rectocele (Figure 7.10).

At evacuation proctography, a delayed evacuation time (more than 30 s to evacuate 2/3 of rectal content) has 90% of positive predictive value (PPV) for anismus [18,45]. This finding, however, has not been fully investigated with MR-defecography in supine position. At MRI, the paradoxal contraction of the puborectal muscle during straining and defecation causes minimal or absent variation of the ARA in all functional phases (at rest, during contraction straining, and evacuation). At maximal straining, inspite of the increasing intra-abdominal pressure, the ARA does not open, and hence, defecation and evacuation cannot progress (Figure 7.10).

Differently from other PFDs, anismus usually requires conservative treatments, usually a bio-feedback therapy. Not uncommonly, however, it may be associated with rectocele or other mechanical disorders. In these cases, it is very important to identify both disorders; the conservative treatment always precedes the surgical one.

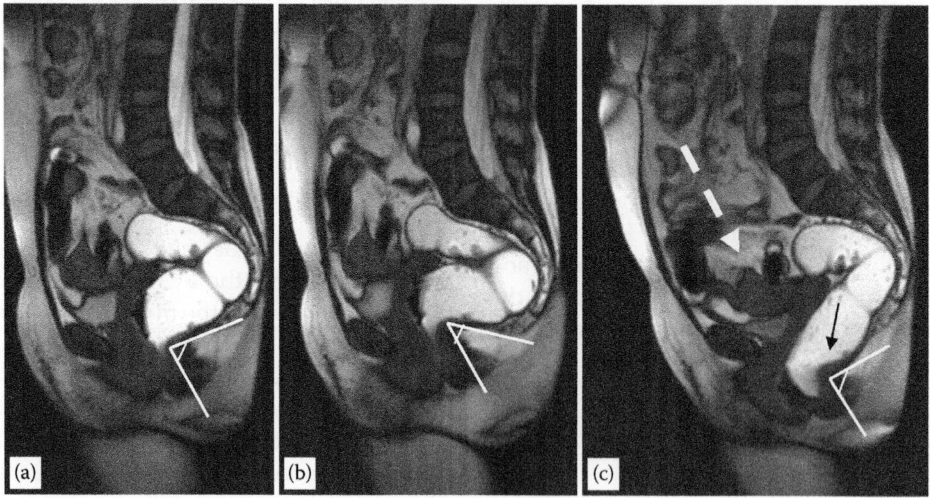


**FIGURE 7.10**
Anismus (or spastic pelvic floor syndrome or dyskinetic puborectal syndrome or pelvic floor dyssynergy). Dynamic MRI with rectal gel distension (a) at rest, (b) during squeezing, and (c) maximal straining. The ARA does not open during maximal straining, due to a lack or insufficient relaxation of the puborectal muscle. At straining (c) in spite of an increased intra-abdominal pressure (discontinuous white large arrow), the ARA does not change in size or rather closes instead of increasing, thus impeding the evacuation. In (c) also a marked impression of the puborectalis muscle in the posterior rectal wall (black arrow) is evident.

## 7.13 Anterior or Urinary PFD

A cystocele is defined as protrusion or bulging of the bladder into the anterior wall of the vagina [6–9,13,34,35] (Figure 7.11). Postmenopausal women are more susceptible to develop a cystocele because estrogens help to keep the supporting muscles and ligaments of the vagina and bladder in good tone. Once estrogen levels drop, these muscles/ligaments become thinner and weaker, thus allowing the bladder to bulge into the vagina. A cystocele can occur as an isolated finding, or it may be associated with other pelvic floor abnormalities (descending perineal syndrome). Cystoceles can be classified by grade according to the degree of bladder descent or by anatomical defect (central, lateral, or combination).

Most grade 1 and 2 cystoceles are asymptomatic but can also be associated with SUI [14,16]. Marked cystocele is commonly symptomatic and can be associated with vaginal bulging, dyspareunia, recurrent urinary tract infections, obstructive voiding symptoms, and urinary retention [6–9,13]. Urethral hypermobility, frequently associated with cystocele, is defined as an excessive change of the urethral axis during urination, exceeding 45°, frequently observed in patients with UI (Figure 7.5).

At MRI, cystocele and urethral hypermobility are easily detected during the maximal straining phase and the evacuation phase.

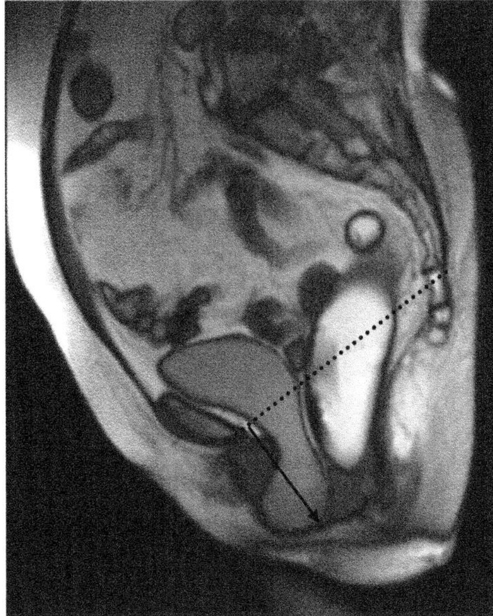

FIGURE 7.11
Severe anterior compartment prolapse in female patient who underwent hysterectomy, complaining recurrent urinary infections and urinary incontinence. The PCL has been traced on an image acquired at maximum straining. A severe cystocele exceeding 5 cm is clearly shown, whereas the posterior compartment is moderately descended.

## 7.14 Uterovaginal Prolapse

A uterine prolapse is the descent of the uterus into the vagina and often beyond the introitus primarily due to damage of the uterosacral ligaments, supporting the upper 20% of the vagina (apex) and the uterus [6–9,34,35,37]. When the uterosacral ligaments break, the uterus begins to descend into the vagina; further uterine descent pulls the rest of the vagina down resulting in apical tears of the pubocervical fascia and rectovaginal fascia from its points of lateral attachment. Continued uterine and vaginal prolapse can result in a complete uterine and vaginal prolapse, such that the uterus falls outside the vaginal opening (Figure 7.5).

Vaginal vault prolapse, which refers to an apical vaginal relaxation post-hysterectomy, involves descent of the apex of the vagina toward, through, or beyond the vaginal introitus after a previously performed total hysterectomy [7,9,34].

Vaginal vault prolapse is almost always associated with the prolapse of other pelvic organs, the most common of which is an enterocele. This reflects a loss of apical level support due to damage of the uterosacral–cardinal complex [9]. Mild uterine prolapse is usually asymptomatic, but higher grades can present as a vaginal mass, with dyspareunia, and low back pain due to stretching of the uterosacral ligaments, occasionally, urinary retention and obstructive uropathy due to ureteral obstruction and/or difficulty in defecating [9].

## 7.15 Conclusions

Dynamic pelvic floor MRI, by providing both morphological and functional information on the structures of the pelvic floor, is able to accurately diagnose and grade most of the urogenital and anorectal PFDs, without radiation exposure and independently from the technique used. The terms PPFD and ODS include many different anorectal dysfunctions, either mechanical or functional, determining similar obstructive symptoms but requiring a specific diagnosis and different surgical or conservative treatments [6,46]. Thanks to the comprehensive diagnostic approach offered by dynamic pelvic floor MRI, the management of all PFDs, and particularly of ODS, is rapidly evolving and improving. Undoubtedly, a tight cooperation between radiologists and clinicians is essential for an effective therapeutic planning of PPFD. The relevant information on PFDs offered by MRI should place the radiologists at the center of the modern pelvic floor units, as the ring of junction between the many different specialists involved in the management of PFDs.

## References

1. DeLancey JOL (1993) Anatomy and biomechanics of genital prolapse. *Clin Obstet Gynecol* 36:897–909.
2. DeLancey JO (1994) The anatomy of pelvic floor. *Curr Opin Obstet Gynecol* 6:313–316.
3. Olsen AL, Smith VJ, bergstrom JO et al. (1997) Epidemiology of surgically managed pelvic organ prolapsed and urinary incontinence. *Obstet Gynecol* 89:501–506.
4. Subak LL, Waetjen LE, van den Eeden S et al. (2001) Cost of pelvic organ prolapse surgery in the United States. *Obstet Gynecol* 98:646–651.
5. Elneil S (2009) Complex pelvic floor failure and associated problems. *Best Pract Res Clin Gastroenterol* 23:555–573.
6. Maccioni F (2013) Functional disorders of the ano-rectal compartment of the pelvic floor: Clinical and diagnostic value of dynamic MRI. *Abdom Imaging* 38:930–951.
7. Mortele KJ, Fairhurst J (2007) Dynamic MR defecography of the posterior compartment: Indications, technique and MRI features. *Eur J Radiol* 61(3):462–472.
8. Fielding JR (2002) Practical MR imaging of female pelvic floor weakness. *RadioGraphics* 22:295–304.
9. Farouk El Sayed R (2013) The urogynecological side of pelvic floor MRI: The clinician's needs and the radiologist's role. *Abdom Imaging* 38:912–929.
10. Fitzgerald MP, Weber AM, Howden N et al. (2007) Risk factors for anal sphincter tear during vaginal delivery. *Obstet Gynecol* 109:29–34.
11. Karasik S, Spettel CM (1997) The role of parity and hysterectomy on the development of pelvic floor abnormalities revealed by defecography. *AJR Am J Roentgenol* 169:1555–1558.
12. Maglinte DDT, Kelvin FM, Fitzgerald K et al. (1999) Association of compartment defects in pelvic floor dysfunction. *AJR Am J Roentgenol* 172:439–444.
13. Abrams P, Cardozo L, Fall M et al. (2002) The standardization of terminology of lower urinary tract function: Report from the standardization subcommittee of the international continence society. *Neurourol Urodyn* 21:1067–1178.
14. Fultz NH, Burgio K, Diokno AC et al. (2003) Burden of stress urinary incontinence for community-dwelling women. *Am J Obstet Gynecol* 189:1275–1282.
15. Kelvin FM, Maglinte DD (2003) Dynamic evaluation of female pelvic organ prolapse by extended proctography. *Radiol Clin N Am* 41(2):395–340.
16. Koelbl H, Mowstin J, Boiteux JP (2002) Pathophysiology. In: Abrams P, Cardozo L, Koury S, Wein A (eds), *Incontinence*, 2nd Edn. Plymouth: Health Publications, pp. 165–201.
17. Ganeshan A, Anderson EM, Upponi S et al. (2008) Imaging of obstructed defecation. *Clin Radiol* 63:18–26.
18. Stoker J, Halligan S, Bartram C (2001) Pelvic floor imaging. *Radiology* 218:621–641.
19. Bharucha AE. (2006) Update of tests of colon and rectal structure function. *J Clin Gastroenterol* 40:96–103.
20. Elshazly WG, ElNekady Hassan H (2010) Role of dynamic magnetic resonance imaging in management of obstructed defecation case series. *Intern J Surg* 8:274–282.
21. Kelvin FM, Maglinte DD, Hornback JA et al. (1992) Pelvic prolapse: Assessment with evacuation proctography (defecography). *Radiology* 184:547–551.
22. Kelvin FM, Maglinte DDT, Hale DS, Benson JT (2000) Female pelvic organ prolapse: A comparison of triphasic dynamic MR imaging and triphasic fluoroscopic cystocoloproctography. *AJR Am J Roentgenol* 174:81–88.
23. Maglinte DDT, Bartram C (2007) Dynamic imaging of posterior compartment pelvic floor dysfunction by evacuation proctography: Techniques, indications, results and limitations. *Eur J Radiol* 61:454–461.
24. Maglinte DDT, Bartram CI et al. (2011) Functional imaging of the pelvic floor. *Radiology* 258:23–29.
25. Maglinte DDT, Hale DS, Sandrasegaran K (2013) Comparison between dynamic cystocolpoproctography and dynamic pelvic floor MRI: Pros and cons: Which the functional examination for anorectal and pelvic floor dysfunction? *Abdom Imaging* 38:952–973.
26. Goei R, Kemerink G (1990) Radiation dose in defecography. *Radiology* 176:137.
27. Bertschinger KM, Hetzer FH, Roos JE et al. (2001) Dynamic MR imaging of the pelvic floor performed with patient sitting in an open-magnet unit versus with patient supine in a closed-magnet unit. *Radiology* 223(2):501–508.
28. Roos JE, Weishaupt D, Wildermuth S et al. (2002) Experience of 4 years with open MR defecography: Pictorial review of anorectal anatomy and disease. *RadioGraphics* 22:817–832.
29. Pannu HK, Kaufman HS, Cundiff GW et al. (2000) Dynamic MR imaging of pelvic organ prolapse: Spectrum of abnormalities. *RadioGraphics* 20(6):1567–1582.
30. Flusberg M, Sahni VA, Erturk SM et al. (2011) Dynamic MR defecography: Assessment of the usefulness of the defecation phase. *AJR Am J Roentgenol* 196(4):W394–W399.
31. Law JM, Fielding JR (2008) MRI of pelvic floor dysfunctions: Review. *AJR Am J Roentgenol* 191:S45–S53.
32. Lamb GM, de Jode MG, Gould SW et al. (2000) Upright dynamic MR defaecating proctography in an open configuration MR system. *Br J Radiol* 73(866):152–155.
33. Solopova AE, Hetzer FH, Marincek B et al. (2008) MR defecography: Prospective comparison of two rectal enema compositions. *AJR Am J Roentgenol* 190:118–124.
34. El Sayed RF, El Mashed S, Farag A (2008) Pelvic floor dysfunction: Assessment with combined analysis of static and dynamic MR imaging findings. *Radiology* 248:518–539.
35. Vanbeckevoort D, Van Hoe L, Oyen R et al. (1999) Pelvic floor descent in females: Comparative study of colpocystodefecography and dynamic fast MR imaging. *J Magn Reson Imaging* 9:373–377.
36. Otto SD, Oesterheld A, Ritz JP et al. (2011) Rectal anatomy after rectopexy: Cinedefecography versus MR-defecography. *J Surg Res* 165:52–58.
37. Reiner CS, Weishaupt D (2013) Dynamic pelvic floor imaging: MRI techniques and imaging parameters. *Abdom Imging* 38:903–911.
38. Betschart C, Chen L, Ashton-Miller JA et al. (2013) On pelvic reference lines and the MR evaluation of genital prolapse: A proposal for standardization using the pelvic inclination correction system. *Int Urogynecol J* 24:1421–1428.

39. Colaiacomo MC, Masselli G, Polettini E et al. (2009) Dynamic MR imaging of the pelvic floor: A pictorial review. *RadioGraphics* 29(3):e35 (Review).
40. Woodfield CA, Krishnamoorthy S, Hampton BS (2010) Imaging pelvic floor disorders: Trend toward comprehensive MRI. *AJR Am J Roentgenol* 194:1640–1649.
41. Lienemann A, Anthuber C, Baron A et al. (1997) Dynamic MR colpo- cystorectography assessing pelvic-floor descent. *Eur Radiol* 7:1309.
42. Shorvon PJ, Marshall MM (2005) Evacuation proctography. In: Wexner SD, Zbar AP, Pescatori M (eds) *Complex Anorectal Disorder: Investigation and Management*. New York: Springer.
43. Kenton K, Shott S, Brubaker L (1999) The anatomic and functional variability of rectoceles in women. *Int Urogynecol J Pelvic Floor Dysfunct* 10(2):96–99.
44. Felt-Bersma RJ, Cuesta MA (2001) Rectal prolapse, rectal intus-susception, rectocele, and solitary rectal ulcer syndrome. *Gastroenterol Clin N Am* 30:199–222.
45. Halligan S, Bartram CI, Park HJ et al. (1995) Proctographic features of anismus. *Radiology* 197:679–668.
46. Zbar AP. (2013) Imaging and surgical decision-making in obstructed defecation. *Abdom Imging* 38:894–902.
47. Dvorkin LS, Gladman MA, Scott MS et al. (2005) Rectal intussusception: A study of rectal biomechanics and visceroperception. *Am J Gastroenterol* 2005(100):1578–1585.
48. Dvorkin LS, Hetzer F, Scott SM et al. (2004) Open-magnet MR defaecography compared with evacuation proctography in the diagnosis and management of patients with rectal intussusception. *Colorectal Dis* 6(1):45–53.
49. Boyadzhyan L, Raman SS, Raz S (2008) Role of static and dynamicMR imaging in surgical pelvic floor dysfunction. *RadioGraphics* 28:949–967.
50. Cappabianca S, Reginelli A, Iacobellis F et al. (2011) Dynamic MRI defecography vs. entero-colpo-cysto-defecography in the evaluation of midline pelvic floor hernias in female pelvic floor disorders. *Int J Colorectal Dis* 26:1191–1196.
51. Parks AG, Porter NH, Hardcastle J (1966) The syndrome of thedescending perineum. *Proc R Soc Med* 59:477–482.
52. Broekhuis SR, Hendrik JCM, Jurgen JF (2010) Perineal descent and patients' symptoms of anorectal dysfunction, pelvic organ prolapse, and urinary incontinence. *Int Urogynecol J* 21:721–729.

# 8

# *Degenerative Disease of the Spine and Other Spondyloarthropathies*

**M. Cody O'Dell, Nathan J. Kohler, Brian K. Harshman, Steven A. Messina, Christopher W. Wasyliw, Gary Felsberg, and Laura W. Bancroft**

**CONTENTS**



## 8.1 Technical Considerations

Technically, imaging of the spine can be complex. The spinal canal is a deep, long cylindrical structure, which requires fine detail for adequate diagnostic evaluation. Specialized coils are required to allow for adequate signal-to-noise ratio for imaging of each spinal segment. Field of view is typically smaller, ranging from 16–24 cm, in order to obtain adequate resolution.

Other considerations, including patient-related factors, such as body size, spinal curvature, and movement are key. Patients being evaluated for low back pain typically have higher levels of anxiety, often requiring sedation. In addition, patients being evaluated for back or neck pain may have more difficulty lying still for the entirety of the examination, necessitating pain medication. Large body habitus can make it difficult or impossible to perform the examination given weight limitations of the table, or inability to fit within the bore of the magnet.

Physiologic motion, such as the cardiac pulsations, diaphragmatic excursion, and swallowing, can also degrade the image and create artifacts. Cerebrospinal fluid (CSF) flow within the thecal sac can create pulsation artifacts as well. These artifacts are most prominent on $T_2$-weighted and gradient-echo sequences. This may lead to areas of decreased signal in the CSF and corresponding areas of increased signal in the spinal cord parenchyma. Such artifacts are less apparent on $T_1$-weighted sequences. Some vendors include software to compensate for these artifacts [1].

Spinal hardware and postoperative changes can pose another set of complications. It should be noted first that when surgical hardware is present, comparison with baseline radiographs is imperative. Traditionally, magnetic resonance imaging (MRI) has not been the modality of choice for imaging postoperative spinal complications because of artifact. There are numerous artifacts that occur with MR spine imaging, the most common of which being susceptibility artifact when spinal hardware is present. Several techniques for decreasing this artifact include more recent use of titanium implants, increasing frequency-encoding gradient strength, orienting the implant along the magnetic field, and the use of fast spin-echo sequences to increase signal near the implant. Other methods for decreasing this artifact include decreasing voxel size, increasing bandwidth, or lowering field strength [2].

## 8.2 Examination Technique

Often the imaging techniques used are as essential as the image interpretation. Patients are placed in the magnet in the supine position. Images are obtained in the axial and sagittal planes. Coronal planes can be helpful in the setting of spinal curvature. According to one study, the prevalence of adult lumbar scoliosis on MRI was 19.9%. It is important, therefore, to include coronal scout images or dedicated coronal sequences to assess spinal curvature [3]. Proper MRI protocol includes thin section axial images from the midbody of L3 to the midbody of S1. It is imperative that contiguous "stacked" images are obtained without skip areas. Obtaining intermittent images at the level of the intervertebral disks may result in missed free fragments, a common etiology for failed back surgery. It is generally not necessary to angle the gantry to be parallel to the end plates. Surface coils are used to increase signal-to-noise ratio and resolution. A special neck coil is helpful with studies of the cervical vertebrae. For studies of the thoracic and lumbosacral spine, a body coil will suffice. Spinal imaging is most often performed on a 1.5 T system. 3.0 T systems are becoming more widely available and are typically more often used in brain imaging due to the increased number of artifacts [1].

Regarding sequences, at the very least $T_1$WI and $T_2$WI should be obtained in sagittal and axial planes. There are many other sequences available with varying degrees of clinical utility. For example, short tau inversion recovery (STIR) is helpful in the setting of trauma and in increasing the conspicuity of spinal cord lesions [4]. Intravenous contrast is of limited use and is discussed in the next section [1].

## 8.3 MR Contrast Agents

There are a limited number of contrast agents used in spinal MRI, all of which are derivatives of gadolinium, a paramagnetic agent [1]. Gadolinium shortens the $T_1$, and to some of degree the $T_2$, effect leading to enhancement in areas of blood brain barrier breakdown or increased vascularity [5]. There are few indications for the use of contrast in spinal imaging. For example, gadolinium should always be used in the setting of postoperative spine imaging, spinal cord imaging, or suspected infection. Likewise, gadolinium is given when an abnormality is identified in the epidural space. Studies have concluded that disk height collapse, Modic changes, and a high level of degenerative changes can lead to contrast enhancement in the intervertebral disk [6]. A separate study showed that there is greater contrast between disk fragments and scar tissue with ionic contrast, due to the less rapid rate of diffusion when compared to non-iodinated contrast [7]. Unfortunately, these findings have not had any clinical utility, and there is currently little or no use for intravenous contrast in the setting of degenerative disk disease.

Gadolinium is excreted in the urine, the majority of which is excreted in the first 6 hours following injection. Calculation of the patient's glomerular filtration rate (GFR) prior to administration of intravenous gadolinium is recommended to prevent nephrogenic systemic fibrosis [5].

## 8.4 Normal Anatomy

An understanding of normal spinal anatomy is critical to comprehend the various forms of degenerative disk disease and the current nomenclature for describing them. The osseous spine is generally divided into five segments. The cervical spine consists of 7 vertebrae. The thoracic spine consists of 12 vertebrae. The lumbar spine typically consists of 5 vertebrae. There is great variability in the number of sacrococcygeal vertebrae, usually with 5 sacral and 3–5 coccygeal vertebral bodies present.

The vertebra consists of the vertebral body and the posterior elements. The vertebral body is generally a rounded structure with internal cancellous bone, marrow, and fat covered by a thin outer layer of compact bone. The posterior elements consist of all the bony structures posterior to the vertebral body including the pedicles. It is important to understand the joints formed between the vertebrae at each level, as this is crucial to understanding degenerative disease.

Unlike other vertebral bodies, the cervical vertebrae have transverse foramina through which the vertebral arteries course (Figure 8.1a). Atlas, the first cervical vertebra, articulates with the occipital bone. Axis, the second cervical vertebra, is unique in that it has a process extending superiorly into the spinal canal, which is termed *the odontoid process* or *dens* (Figure 8.1b). The anatomy of the atlantoaxial junction is most significant in the setting of rheumatoid arthritis (RA) or trauma. The C3–C7 vertebrae have five joints connecting each of them: the intervertebral disk, two uncovertebral joints, and two facet joints. The uncovertebral joints are unique to the cervical vertebrae and can be a significant source of degenerative bone hypertrophy.

Unique to the thoracic vertebrae, are the articulation of the transverse processes with the ribs (Figure 8.2). There are typically 12 thoracic vertebrae articulating to 12 pairs of ribs. The lumbar vertebrae have two articular facet joints and an intervertebral disk joint. The lumbosacral spine serves as a fulcrum of support between the upper and lower body. Therefore, the majority of degenerative disk disease occurs in the lower lumbar spine.

Each vertebral level of the spine has an intervening intervertebral disk complex (Figure 8.3). This complex is composed of a cartilaginous end plate, annulus fibrosis, and nucleus pulposus. The nucleus pulposus is located in the center of the vertebral disk complex, eccentrically posterior. This is circumscribed peripherally by the annulus fibrosis, a fibrous structure that contains the less viscous nucleus. The annular fibers attach superiorly and inferiorly to the vertebral end plates and consist of flat bony disks with elevated edges (ring apophyses).

The intervertebral disk undergoes more changes with age than any other tissue and is the largest tissue in the body lacking a blood supply. The cartilaginous end plate is a hyaline structure that exists between the annulus and the vertebral body end plate. It is rarely discussed in anatomical MRI literature because the imaging appearance has been unknown until recently. Studies using cadaver specimens subjected to ultrashort echo time and FLASH (fast low-angle shot) MR sequences show promise in better understanding the role of the cartilaginous end plate in degenerative spine [8,9].

With some degree of variability, a normal $T_1$-weighted MR scan of the adult spine will show increased signal

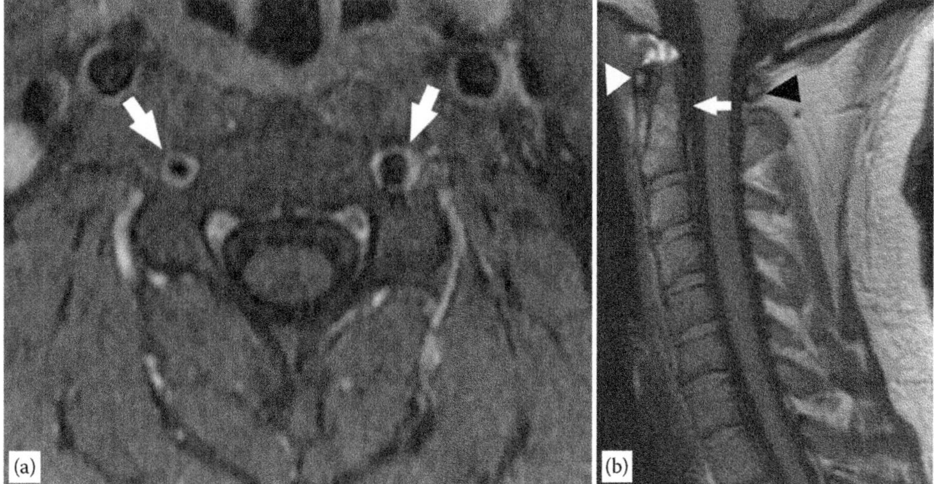


**FIGURE 8.1**
Normal cervical spine MRI in a 32-year-old man. (a) Axial enhanced $T_1$-weighted FS image shows the bilateral vertebral arteries coursing through the arcuate foramina (arrows), which is unique to the cervical spine. The left vertebral artery is larger and dominant. (b) Sagittal $T_1$-weighted image shows the unique ring configuration of C1 (arrowheads) and C2's superiorly protruding dens (arrow).

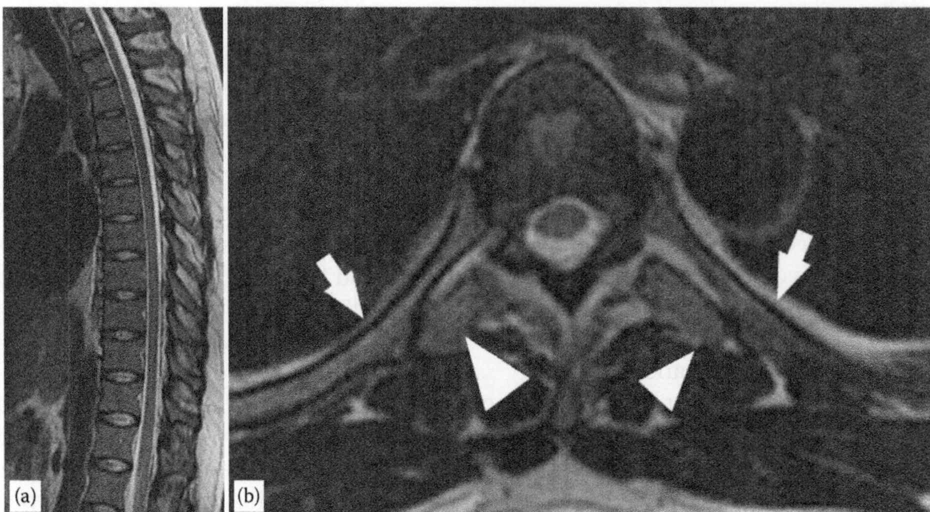


**FIGURE 8.2**
Normal thoracic spine MRI in a 33-year-old man. Sagittal (a) and axial (b) FSE $T_2$-weighted images through the thoracic spine show the unique feature of the thoracic spine transverse processes (arrowheads) articulating with the ribs (arrows).

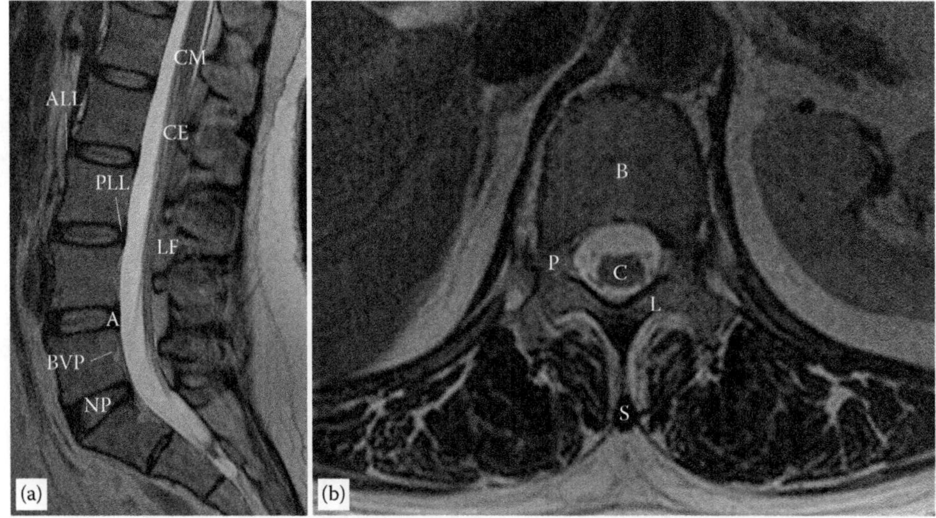


**FIGURE 8.3**
Normal lumbar spine MRI in a 41-year-old woman. Sagittal (a) and axial (b) FSE $T_2$-weighted images. (ALL, anterior longitudinal ligament; PLL, posterior longitudinal ligament; CM, conus medullaris; CE, cauda equinus; LF, ligamentum flavum; A, annulus fibrosis; BVP, basivertebral plexus; NP, nucleus pulposus; B, vertebral body; P, pedicle; C, spinal cord; L, lamina; S, spinous process).

in the bone marrow with intermediate signal in the intervertebral disk and low signal in the CSF [10]). The $T_2$-weighted image will demonstrate low signal intensity in the bone marrow with intermediate to high signal in the intervertebral disk and high signal intensity in the CSF. Adults can occasionally have focal regions of high intensity in the bone marrow on $T_1$WI, which usually represents focal fatty replacement. Normal vertebral body marrow has relatively high intensity on $T_2$-weighted fast spin-echo images without fat suppression, which can make lesion detection difficult. In this setting, STIR sequences will make vertebral body lesions more conspicuous. In children, bone marrow signal is less intense on $T_1$ images secondary to hematopoiesis and frequently enhances.

A normal spine MR will demonstrate a $T_1$ hypointense basivertebral plexus along the posterior aspect of the vertebral body. If the etiology is in question, the basivertebral plexus will usually be $T_2$ hyperintense and enhance with contrast. Also occasionally there is $T_1$ hyperintense epidural fat along the dorsal surface of the thecal sac [5].

There are five important sets of ligaments in the spinal column, which can all be associated with degenerative pathology (Figure 8.3). Anterior to the vertebral column, the anterior longitudinal ligament runs craniocaudad supporting the spinal column anteriorly. The posterior longitudinal ligament (PLL) runs craniocaudad along the backside of the vertebral column anterior to the thecal sac. The intraspinous ligaments link the spinous processes, and the lamina is connected craniocaudally at each level via the ligamentum flavum. Additionally, there are small ligaments within the neural foramina, which are generally not visualized on imaging [5].

## 8.5 Transitional Anatomy

As mentioned previously, there are typically 37–39 vertebral bodies in an individual. The number of vertebral bodies can vary significantly with variability increasing caudally. The most common transitional abnormalities include lumbarization of S1, sacralization of L5, rudimentary L1 ribs, and hypoplastic T12 ribs [4]. Transitional vertebral anomalies are considered a normal variant and have a prevalence in the general population of 4%–30%. Transitional anatomy has a loose association with accelerated degenerative changes of the spine.

Clinically, transitional anatomy is most relevant when surgery is considered, given the risk of wrong site procedure. In fact, one survey revealed that up to 50% of spine surgeons had performed wrong site surgery over the course of their career. Radiologists may not have been helping their cause, as a recent retrospective study showed that transitional anatomy was demonstrated in 10% of a cohort of adolescent patients with idiopathic scoliosis, but the radiology report only reflected that in 0.5% [11]. Lumbosacral transitional vertebrae are best imaged with computed tomography (CT) or anteroposterior (AP) radiographs angled cranially at 30° (i.e., Ferguson radiographs) [12]. Medical errors can be avoided if the radiologist identifies and carefully describes the anatomical variant in a detailed report [4].

## 8.6 Congenital Posterior Element Variability

Facet joint orientation refers to the angle of the facet joint in the transverse plane relative to the sagittal plane. When asymmetry of the facet joint angles occurs, it is referred to as facet tropism. More specifically, the posterior tip of one facet is positioned more medial than its anterior tip. It is believed that when present, vertebral rotation and subluxation can occur. This is an entity whose clinical significance has not yet been established. It is not clear whether facet tropism is a cause or effect of degenerative changes in the facet joint. It has been shown that there is a weak relationship to degenerative spondylolisthesis [5,13].

## 8.7 Degenerative Change of the Spine

### 8.7.1 Degenerative Disk Disease

Degenerative disk disease may be asymptomatic or associated with back pain or radiculopathy. The prevalence of degenerative disk disease among adults is as high as 30% and is the most common cause of disability in patients under 45 years. The etiology is multifactorial including occupational injury, obesity, genetics, and psychosocial factors [4]. It is important to make the discrimination between degenerative disk disease and disk herniation, as these are two separate entities and are managed differently [4]. In addition, some studies have shown a low correlation between degenerative disk disease and low back pain, when controlling for disk herniation and end plate changes [14].

Also of paramount importance is the fact that not all patients with degenerative disk imaging findings have symptoms. The majority of people will experience back pain at some time in their life (approximately two-thirds), and the majority of those will find no definitive reason for it [15]. One study showed no correlation between low back pain and severe degenerative disk disease, when controlling for end plate changes and disk contour [14]. Multiple studies of MR findings in young healthy, asymptomatic adults have demonstrated that at least half of study participants had degenerative changes of at least one disk [16,17]. Given the global economic cost of low back pain, and the fact that degenerative disk is one of the leading causes of functional incapacity, it is essential that we understand the significance of the images. For these reasons, an imaging report should not speculate as to the etiology, nor should it underestimate or overestimate the disease burden [18]. When symptomatic, patients will often present for imaging workup, which typically reveals disk space height loss with degenerative end plate changes, as well as vacuum phenomenon and loss of signal within the intervertebral disk. The intervertebral disk may also show linear enhancement.

### 8.7.2 Nomenclature of Degenerative Disk Disease

In the past, the terminology used to describe degenerative changes in the spine was broad, ambiguous and oftentimes confusing to both neuroradiologists and clinicians. In 2001, the North American Spine Society set out to standardize the language used to describe disk pathology. A multidisciplinary group held a combined task force to accomplish this with the goal of consensus and conciseness in order to improve the care of the patients with degenerative spinal pathology [19]. In this section, we will discuss the results of that combined task force. It should be noted, however, that there still are controversial topics in the field of spinal radiology that will continue to be elucidated each generation. Furthermore, the definitions described in this document refer to imaging findings and do not define clinical syndromes. As with all subspecialty areas, it is important to remain up to date with the most current nomenclature.

In general, diagnostic categories of disk pathology can be grouped according to Fardon and colleagues [19]. We will primarily focus on the degenerative/traumatic lesions and the specific terminology used to describe them.

Annular fissures are defined as separations between annular fibers. These may be avulsions of fibers from the vertebral body insertions or breaks through fibers that extend transversely, radially, or concentrically, possibly involving many layers of the annular lamellae. The term *annular tear* is often used interchangeably; however, some clinicians feel *tear* suggests traumatic etiology. According to the more recent consensus, tear or fissure is appropriate given the common synonymous usage of these terms.

Degeneration describes a myriad of findings including narrowing of the disk space, desiccation of the intervertebral disk, and fibrosis among others. Degenerative changes can be divided into one of two categories: (1) spondylosis deformans or (2) intervertebral osteochondrosis. Spondylosis deformans specifically refers to changes in the disk that can be associated with normal aging, whereas intervertebral osteochondrosis usually suggests a more clear pathologic process. Spondylosis deformans may refer to senescent changes, while an intervertebral osteochondrosis refers to a more specific pathology such as ochronosis or ankylosing spondylitis, for example.

### 8.7.3 Annular Tear

An annular tear is a simple entity that is complicated only by its many synonymous terms—annular fissure, high intensity zone (HIZ), and annular defect, among others. It refers to a defect within the concentric fibers of the annulus fibrosis due to trauma or degenerative change. The result is a HIZ on $T_2$-weighted images or a focally enhancing linear nidus, usually within the posterior annulus (Figure 8.4). The contour of the disk margin has also been shown to correlate with internal disk derangement [20].

Most annular tears are asymptomatic, incidental findings. However, if there is clinical suspicion that the fissure may be a source of the patient's pain, a provocative discography exam may be helpful to illicit symptoms. Surgical intervention is rarely indicated for treatment of a symptomatic fissure [4]. Studies designed to determine the clinical relevance of this entity are still underway, although most agree that a HIZ in the posterior annulus of a patient with discogenic back pain is a reliable marker of symptomatic annular tear [21].

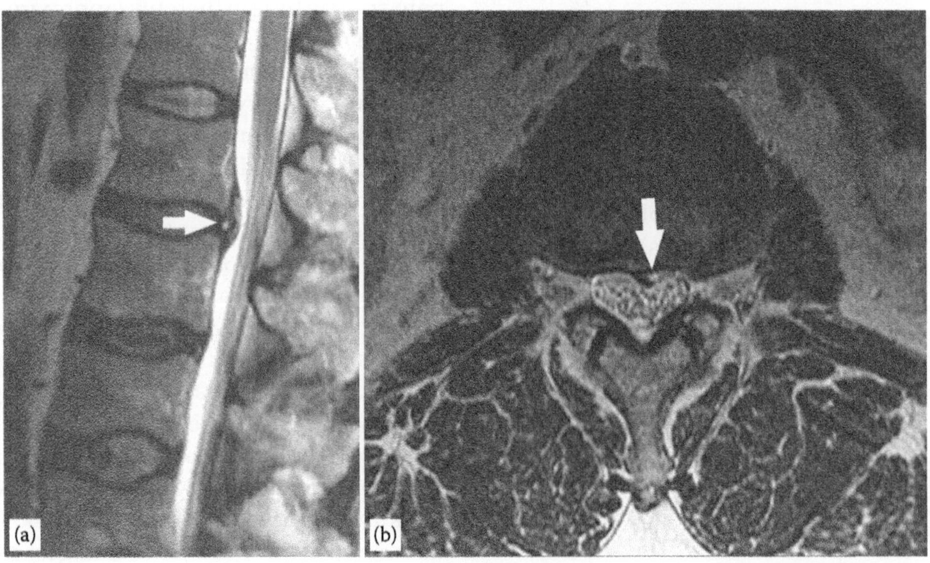


**FIGURE 8.4**
Annular tear in a 68-year-old man with back and left leg pain. Sagittal (a) and axial (b) FSE $T_2$-weighted images show an L2–3 central disk protrusion with hyperintense annular tear (arrows).

### 8.7.4 Disk Bulge

Disk bulge refers to at least 50% circumferential disk expansion beyond the vertebral ring apophyses and is not considered as a form of disk herniation (Figure 8.5) [19]. They most commonly occur at C5–C6 and C6–C7 in the cervical spine, or L4–L5 and L5–S1 in the lumbar spine. Although a disk bulge may cause indentation on the anterior thecal sac, central canal and lateral recess compromise is uncommon. Bulging disks are found in up to 39% of asymptomatic adults. They rarely progress and respond well to conservative therapy in the vast majority of symptomatic cases [4].

### 8.7.5 Disk Herniation

Herniation is a localized displacement of any disk material peripherally beyond the outer edges of the vertebral ring apophyses, excluding osteophytes. Note that herniations can also occur in an axial plane if disk material extends craniad or caudad beyond the end plate (e.g., Schmorl's nodes). This would be more generally termed an *intravertebral herniation*. In order to be classified as a herniation, it must be a localized process involving less than 50% of the disk circumference. A focal herniation involves less than 25% of the disk circumference. A broad-based herniation involves between 25% and 50% of the disk circumference. Any displacement of disk material which occupies greater than 50% of the apophyses is then more appropriately defined as a bulge. It should be noted that a disk bulge is not a subtype of herniation.

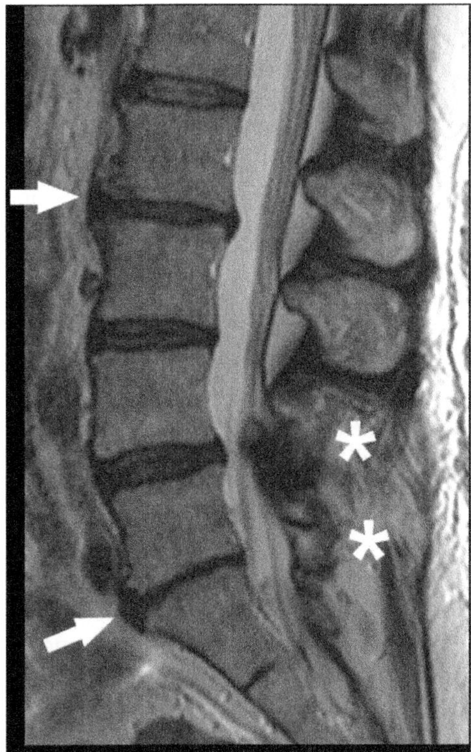

**FIGURE 8.5**
Variable degrees of degenerative disk disease. Sagittal FSE $T_2$-weighted image shows loss of the normal trilaminar appearance and height of the L2–3 and L5–S1 intervertebral disks (arrows), L2–3 disk bulge and L4–5 laminectomies (asterisks).

A disk herniation can be further classified as a protrusion versus extrusion and contained versus uncontained. Protrusion and extrusion are terms that specifically refer to the shape of the displaced material (Figures 8.6 and 8.7). Basically, if the base of the herniation is larger in all dimensions than the rest of the herniation, it is a protrusion. If the base of the herniation is smaller in any dimension (i.e., has a pedunculated appearance), it is an extrusion. Sequestration is present if the extruded material is no longer in communication with the parent disk. Sequestration can be used synonymously with free fragment. Migration is said to have occurred if the extruded material is displaced away from the origin. These characteristics are often only appreciated on sagittal imaging. A herniation is contained if the displaced portion is covered by annulus and is uncontained and if the annular covering is disrupted.

The degree of spinal canal and neural foraminal compromise is graded as mild, moderate or severe, depending on the amount of cross-sectional area displaced by the herniation on axial images. If the herniated disk compromises less than one-third of the spinal canal or neural foramina, it is considered mild. If the degree of compromise is between one-third and two-thirds, it is considered moderate. If the degree of compromise is greater than two-thirds, it is considered severe.

The location of herniated disk material is defined using *zones* and *levels*. On the axial plane, zones are defined as central, subarticular, foraminal, or extraforaminal based on their relationship with the pedicle (Figure 8.8). The central canal is self-explanatory. Moving laterally, the subarticular zone is between the central zone and the pedicle. The foraminal zone encompasses the area immediately above or below the pedicle. Anything lateral to the pedicle is considered extraforaminal. There can be some overlap in these zones, especially between the central and subarticular zones. On the sagittal plane, levels are defined as suprapedicle level, pedicle, infrapedicle, or disk level. These levels are also determined by their location relative to the pedicle.

### 8.7.6 Far Lateral Disk Extrusion and *Far-Out Syndrome*

It is easy to fall into the trap of focusing one's attention on the spinal canal. However, there are several clinical entities that exist in the far lateral space. Several terms are used to describe this area lateral to the facet, including extraforaminal, far lateral, and extreme

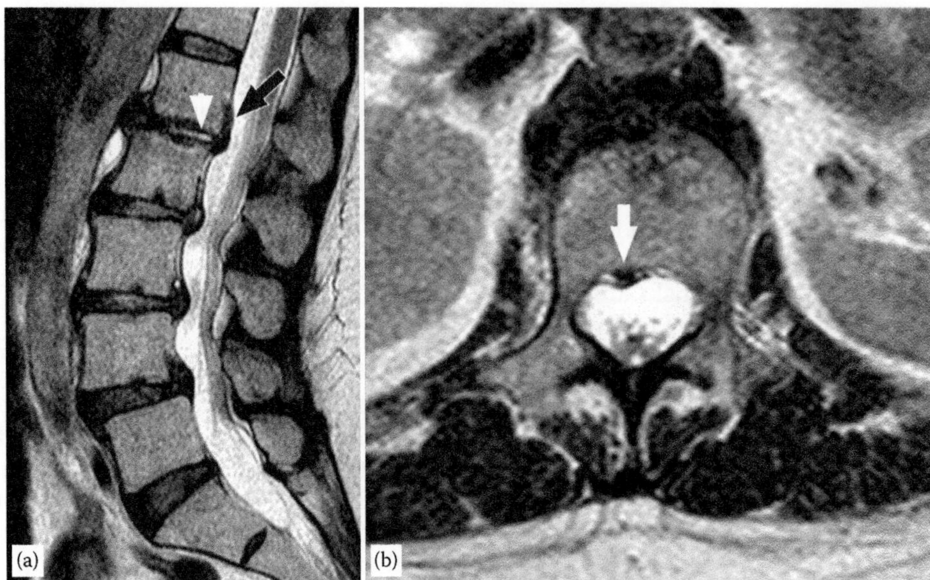


**FIGURE 8.6**
Disk bulge, protrusion, paracentral extrusion, and annular tears in 69-year-old woman with low back pain. Sagittal (a) and axial (b) FSE $T_2$-weighted images demonstrate diffuse degenerative disk disease with L1–2 disk extruded posterosuperior to the L1 vertebral body (arrow). Note the empty posterior half of the L1–2 disk space *(arrowhead).* In addition, there are disk protrusions and annular tears at L2–3 and L3–4, L4–5 disk bulge and grade 1 anterolisthesis.

lateral. Herniated disks and extruded free fragments are not uncommonly found in this space with a reported incidence that varies from 0.7% to 11.7%. The most common locations for this are the lower cervical spine and lower lumbar spine. Careful evaluation on sagittal images can be very helpful. Signal characteristics include isointensity to the parent disk on $T_1$-weighted images, with variable $T_2$ signal characteristics and peripheral enhancement. Given the almost universal need for surgical intervention, it is important to thoroughly evaluate the far lateral space [22]. *Far-out syndrome,* first described in 1984 by Wiltse and colleagues, refers to a more specific clinical syndrome in which there is impingement of the L5 nerve between the L5 transverse process and the ala of the sacrum [23]. Various diagnostic and surgical approaches have since been described, although much research remains on this entity [24,25].

It is considered appropriate for the interpretation to state the degree of confidence of diagnosis. Appropriate terms could include *definite* if there is no doubt, *probable* if there is greater than 50% likelihood but some doubt, and *possible* if there is less than 50% likelihood. It should be stressed that the above recommendations were established primarily to standardize the language used to describe disk pathology specific to the lumbar spine. These terms are generalizable when discussing the findings of the cervical and thoracic spine as well; however, to date there have been no consensus statements specific to the cervical or thoracic spine [19].

### 8.7.7 Schmorl's Nodes

Schmorl's nodes, also termed *intravertebral disk herniation,* result from disk extension into a weakened vertebral body end plate. Their etiology is uncertain, although some have proposed that they are related to axial load trauma, infection, or malignancy. One large study followed edematous Schmorl's nodes over time to monitor evolution. It was concluded that Schmorl's nodes that showed rapid evolution were almost always associated with fracture, infection, or malignancy. However, the majority of edematous Schmorl's nodes showed almost no change over 6 years [26]. In a large retrospective study of Caucasian adults, it was determined that the prevalence in the general population was 3.8% with no correlation to patient gender, age, body mass index (BMI), height, or occupational exposure to heavy lifting [27]. However, classic literature has projected a staggeringly male predilection with prevalences approaching 75% [4].

MR findings include a focal defect in the end plate filled by intervertebral disk on $T_1$-weighted images (Figure 8.9). $T_2$-weighted and post-gadolinium images may demonstrate marginal high signal and enhancement in the acute or subacute stages (Figure 8.9b). STIR sequences will accentuate the edema, if acute. Schmorl's nodes are always contiguous with the parent disk. Follow-up MR may be necessary in cases of unexplained vertebral body edema with localized pain. Schmorl's nodes most commonly occur between T8 and L1 and vary in size from several millimeters to several centimeters. Treatment is almost exclusively conservative [4],

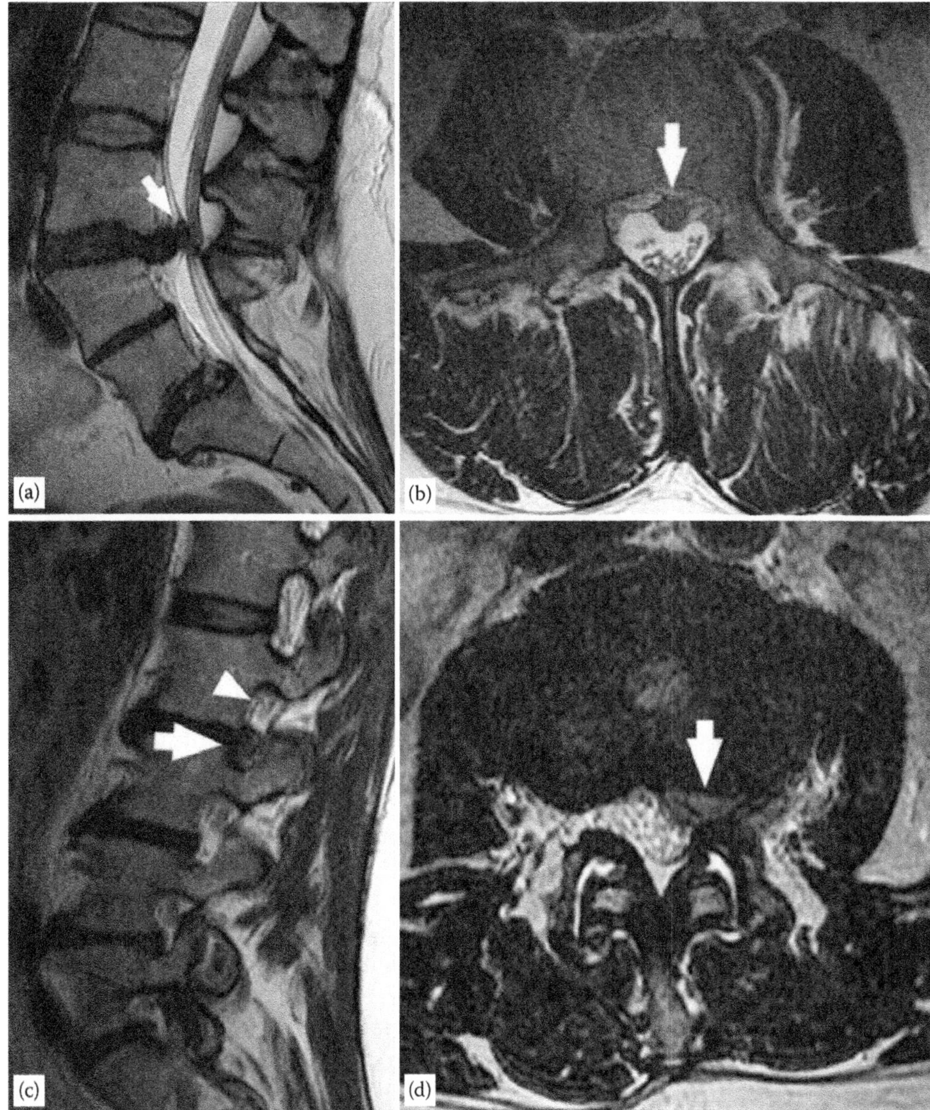


**FIGURE 8.7**
Disk herniation. Sagittal (a) and axial (b) FSE $T_2$-weighted images show a focal central disk extrusion at L3–4 (arrows) in a 43-year-old woman, which abuts several nerve roots. Multifocal degenerative disk disease is also present. (c and d) Left L2–3 foraminal disk protrusion (arrows) in a 49-year-old woman extends inferior to the exiting L3 nerve (arrowhead).

although, in a subset of patients with chronic back pain, there have been several studies reporting significant improvement in symptoms following treatment for presumed symptomatic Schmorl's nodes. Such interventions usually involve fusion or vertebroplasty [28].

## 8.8 Osseous Degenerative Changes

### 8.8.1 Spinal Stenosis

Congenital spinal stenosis is an anatomical abnormality characterized by short pedicles (Figure 8.10). A cervical spinal canal AP diameter of less than 14 mm and a lumbar spine AP diameter of less than 15 mm would suggest the diagnosis. Similarly, Torg ratio (AP canal diameter/AP vertebral body diameter) is typically less than 0.8. In a large cadaveric study, Bajwa and colleagues determined that cervical congenital spinal stenosis was present when the AP diameter was less than 13 mm and the interpedicular distance was less than 23 mm. In the thoracic spine, they determined that canal stenosis was present when the AP diameter was less than 15 mm and the interpedicular distance was less than 18.5 mm [29,30].

Spinal canal stenosis may be inherited in certain clinical conditions such as achondroplasia and mucopolysaccharidosis. Clinically, patients typically present with symptoms similar to those of acquired spinal stenosis.

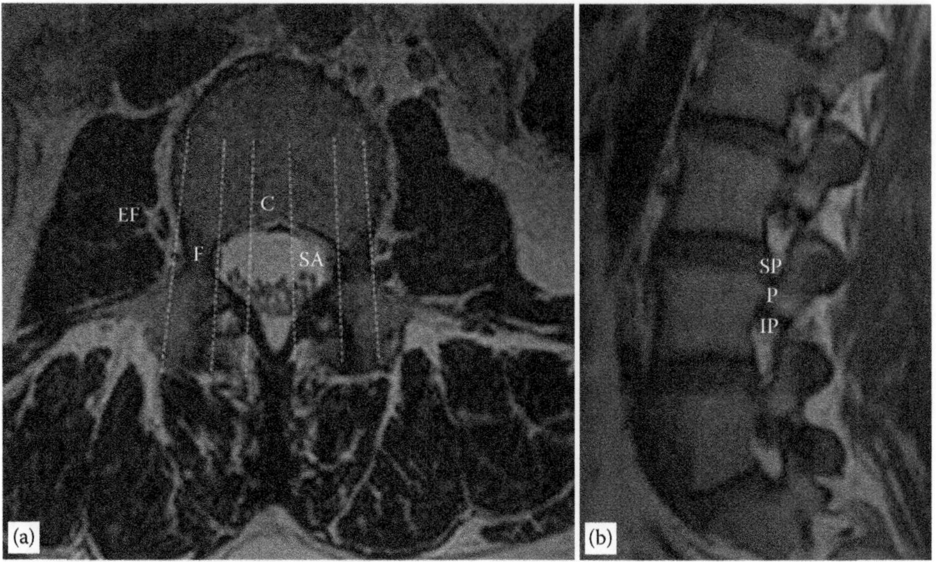


**FIGURE 8.8**
Nomenclature for describing the location of disk herniation. Axial (a) and sagittal (b) images delineate disk herniation locations in the axial (a) and sagittal (b) planes. (EF, extraforaminal; F, foraminal; SA, subarticular; C, central; SP, suprapedicular; P, pedicular; IP, infrapedicular).

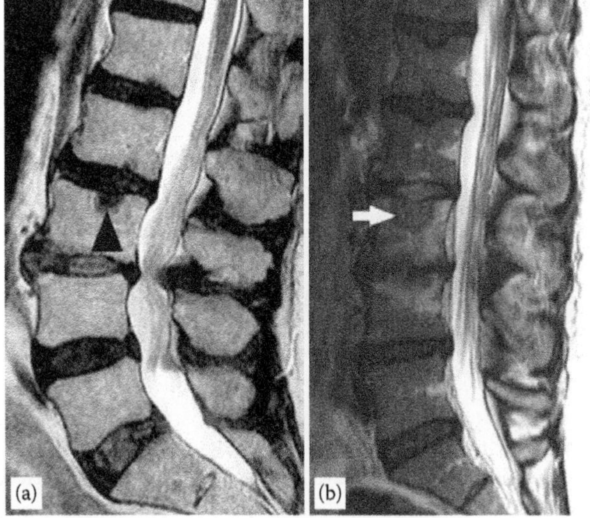


**FIGURE 8.9**
Schmorl's nodes. (a) Sagittal FSE $T_2$-weighted image in a 75-year-old man shows L1–3 Schmorl's nodes, most marked involving the L3 superior end plate (arrowhead). Multilevel degenerative disease is most prominent at L3–4, with disk bulge, ligamentum flavum hypertrophy, and moderate central canal narrowing. (b) Sagittal FSE $T_2$-weighted image demonstrates multiple end plate Schmorl's nodes in a 47-year-old woman with low back pain and sciatica with prominent edema about the L3 superior end plate node (arrow). Multifocal degenerative disk disease and type 1 Modic changes at L3–4 are also noted.

A classic clinical presentation in young athletes is a temporary neurological deficit following a traumatic insult that subsequently resolves. Surgical management typically involves some form of posterior decompression (Figure 8.10c) [4]).

Acquired lumbar central stenosis refers to spinal canal narrowing secondary to multifactorial degenerative changes (Figure 8.11). Patients usually have a combination of disk bulge, facet hypertrophy, and ligamentum flavum hypertrophy, leading to a trefoil appearance of the spinal canal on axial images. Congenitally short pedicles are a common contributory factor. Typically, an AP diameter of the spinal canal less than 1.2 cm would suggest the diagnosis. On sagittal $T_2$-weighted images, the thecal sac has an hourglass appearance with evidence of redundant low signal nerve roots. Clinically, patients may present with chronic low back pain, neurogenic claudication, and rarely bladder or sexual dysfunction. Operative management usually includes posterior decompression, which typically yields complete relief of preoperative symptoms in greater than 70% of patients [4].

### 8.8.2 Spondylosis

Spondylosis generally refers to neural foraminal and spinal canal narrowing secondary to degenerative changes. Imaging findings include multilevel disk desiccation, uncovertebral and facet hypertrophy, disk herniations and bulges as well as annular fissures (Figure 8.12). Effacement of the subarachnoid space may lead to a *washboard spine* appearance on sagittal images. Clinically, patients present with an insidious onset of chronic neck pain and radiculopathy with steady progression of disability. Patients may also present with more severe myelopathic symptoms including Lhermitte, Brown-Séquard, or brachialgia cord syndrome [4].

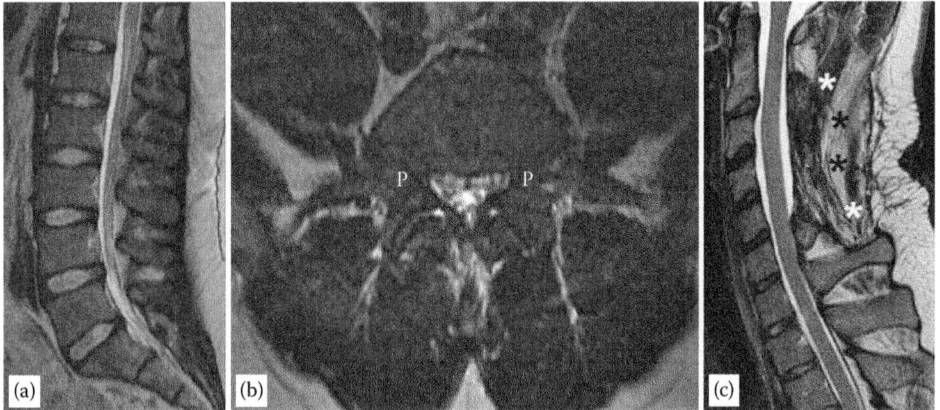


**FIGURE 8.10**
Congenital stenosis. Sagittal (a) and axial (b) FSE $T_2$-weighted images in a 26-year-old man with chronic back pain display diffuse narrowing of the central canal and shortened pedicles (P), resulting in congenital spinal stenosis. (c) Sagittal FSE $T_2$-weighted image after C2–6 decompressive laminectomy in a 17-year-old boy with congenital stenosis shows absent posterior elements (asterisks) and normal caliber central canal.

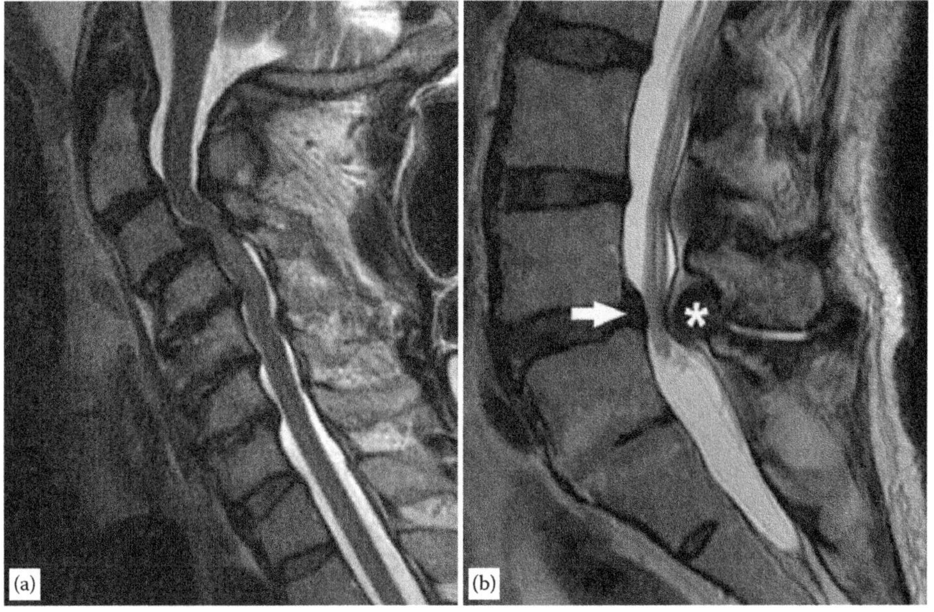


**FIGURE 8.11**
Acquired spinal stenosis. (a) Sagittal FSE $T_2$-weighted image in a 52-year-old man with arm numbness, inability to walk, and generalized weakness shows severe central canal stenosis with cord compression, grade 2 anterolisthesis of C3 upon C4, and diffuse degenerative disk disease. (b) L5–S1 multifactorial canal stenosis in a 68-year-old woman with back pain radiating into her left leg. Sagittal FSE $T_2$-weighted image demonstrates L4–5 disk bulge (arrow), grade 1 anterolisthesis, and ligamentum flavum hypertrophy (asterisk) resulting in severe multifactorial canal stenosis. Partial fusion of L5–S1 and degenerative change of the posterior elements was also present, consistent with Bastruup disease. Patient underwent subsequent posterior decompression and fusion.

Cervical spondylosis has also been associated with severe complications such as vertebrobasilar insufficiency and cervical cord compression. If vertebrobasilar insufficiency is suspected, computed tomography angiography and magnetic resonance angiography are the imaging modalities of choice to determine the degree of vessel compression [31]. If there is suspected impingement on the spinal cord, STIR and fast spin-echo $T_2$-weighted images can be very helpful. More recent research has suggested that MR spectroscopy and diffusion-weighted imaging can provide additional benefit in determining the clinical significance of spinal cord abnormalities seen on MR that are produced by advanced cervical spondylosis [32,33].

### 8.8.3 Spondylolysis and Spondylolisthesis

*Spondylolisthesis* refers to anterior or posterior displacement of the vertebral body relative to the one below. This occurs secondary to degeneration of supporting

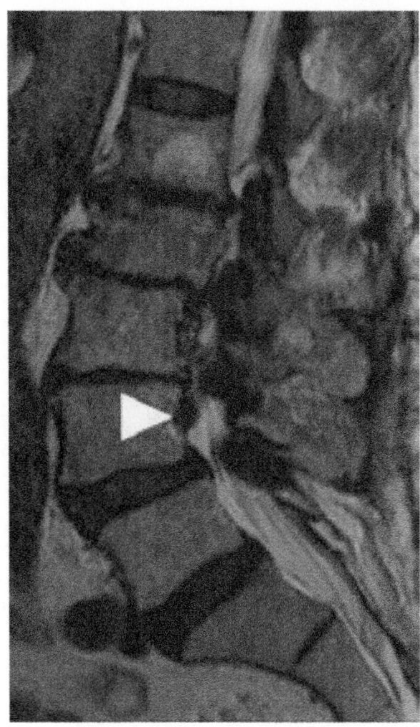

**FIGURE 8.12**
Spondylosis in a 90-year-old man with back pain. Sagittal FSE $T_2$-weighted image shows extensive multilevel lumbar spinal spondylosis, with marked central canal and foraminal stenosis. L3–4 subarticular disk extrusion (arrowhead) was also present.

structures and frequently the intervertebral disk itself [34,35]. Spondylolisthesis may be asymptomatic in many individuals. Instability is uncommon, with 1–3 mm of translation apparent in most normal individuals on flexion and extension views. Some have advocated dynamic images with axial loading to determine the degree of instability [36]. Spondylolisthesis is graded on a scale of 1–4, based on the amount of displacement of the upper vertebral body in relation to the lower (Figure 8.13). If the body is displaced anteriorly <25% relative to the lower vertebra, it is a grade I spondylolisthesis. If between 25% and 50%, it is a grade II spondylolisthesis. If >50%, but less than 75%, it is a grade III spondylolisthesis. And, if greater than 75%, it is a grade IV spondylolisthesis. Complete inferior and anterior displacement of the upper vertebral body in relation to the lower vertebral body is termed *spondyloptosis.*

Spondylolysis is a defect of the pars interarticularis and most often occurs at L5. The etiology is debatable, although the most accepted theory is repetitive microtrauma. Spondylolysis is somewhat common with a prevalence of 4.4% by age 6. It is included in this section due to its association with spondylolisthesis, although the etiology is likely not degenerative. It is notoriously difficult to detect on MR. $T_1$ and $T_2$ sequences show focally decreased signal within the pars interarticularis

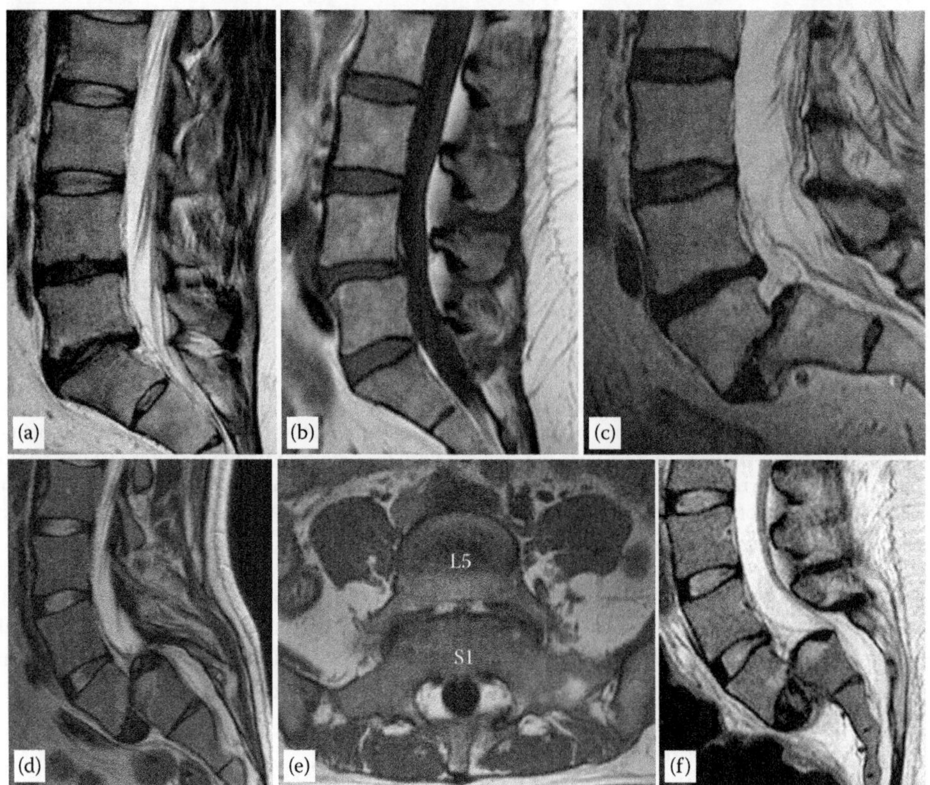


**FIGURE 8.13**
Grading of spondylolisthesis. (a) Grade I retrolisthesis of L5 upon S1. (b) Grade I anterolisthesis of L4 upon L5. (c) Grade II anterolisthesis of L5 upon S1. (d and e) Grade III anterolisthesis of L5 upon S1. (f) Grade IV anterolisthesis of L5 upon S1.

on sagittal and axial images. On $T_2$ images, there may be hyperintensity within the pars defect if acute. MR is significantly less sensitive than CT for the detection of spondylolysis and is only helpful in identifying nerve root compression or acute edema [37].

### 8.8.4 Modic End Plate Changes

Degenerative end plate changes, originally classified by Modic in 1987 as type I–III changes based on histological/MR correlation, refer to replacement of the cortical end plates with fibrovascular marrow, fatty marrow, or bony sclerosis secondary to disk degeneration [18,37]. Modic classification is based on acuity, with type I changes (Figure 8.14) reflecting acute disk degeneration and type III changes reflecting chronic bony sclerosis that would be conspicuous on radiography or CT. Normal end plates are termed *Modic type 0*. Several studies have documented instability in significantly more patients with Modic I changes when compared to II and III (Figures 8.15 and 8.16) [37,38]. However, there is no significant correlation between Modic types and surgical outcomes [39]. Modic changes are dynamic and typically evolve over time with relative predictability [40]. It should be noted that there are several mimickers of Modic end plate changes including disk space infection, hemodialysis spondyloarthropathy, seronegative spondyloarthropathy, metastatic disease, and gout. Additionally, mixed lesions are possible [4].

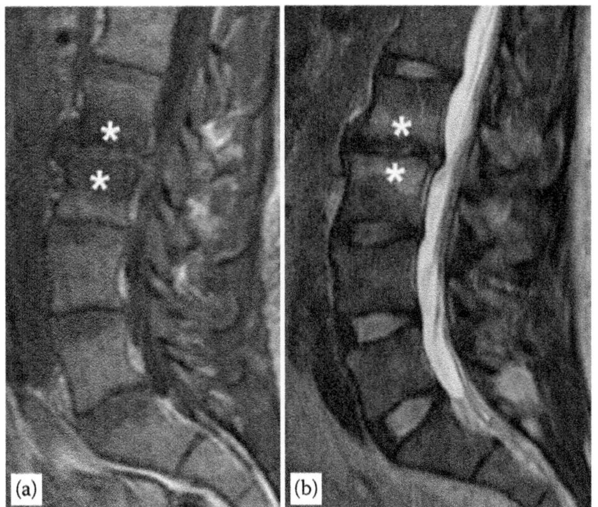


**FIGURE 8.14**
Modic 1 end plate changes in 35-year-old man with back pain and normal white blood cell count. Sagittal $T_1$ (a) and FSE $T_2$-weighted (b) FS images demonstrate bands of hypointense $T_1$-weighted and hyperintense $T_2$-weighted end plate signal change at L2–3 (asterisks), mild end plate irregularity, degenerative disk changes, and bulge. Of note, the L2–3 disk is $T_2$-hypointense and there were no clinical findings for disciitis.

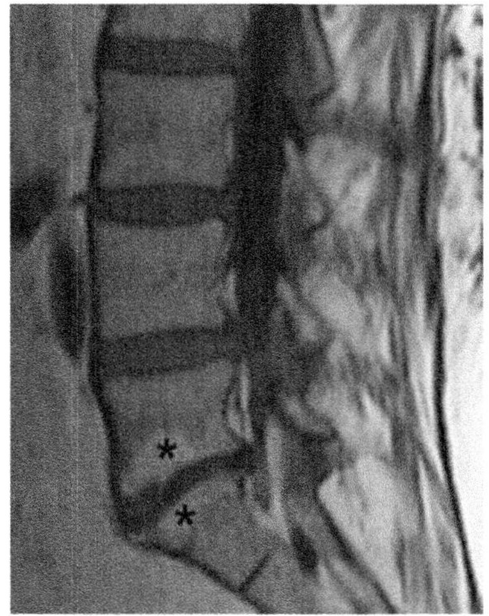

**FIGURE 8.15**
Type 2 Modic changes in a 70-year-old woman with back pain. Hyperintense $T_1$-weighted signal (asterisks) corresponds to fatty end plate changes.

### 8.8.5 Facet Arthropathy

*Facet hypertrophy* refers to osseous facet overgrowth with joint space narrowing and sclerosis. Vacuum phenomenon and enhancing inflammatory soft tissue changes surrounding the facet joint are common. The clinical significance of facet hypertrophy is that neuroforaminal compromise can be seen with extensive facet hypertrophy. Facet hypertrophy typically begins in the first two decades of life and is practically universal after age 60. These lesions are best imaged with thin section CT. Post-gadolinium MR images may demonstrate enhancement of the surrounding soft tissues, and $T_2$-weighted images may show facet effusions as a linear hyperintensity. It should be noted that $T_2^*$ images, while showing osseous changes well, may overemphasize the degree of foraminal narrowing. Given the lack of consensus regarding grading of the degree of facet hypertrophy, it is acceptable to describe facet hypertrophy as mild, moderate, or severe depending on the extent of joint space narrowing and osteophytic spurring [4].

*Facet joint synovial cyst* refers to extrusion of the synovium lining the facet joint. This may occur along the anterior surface of the facet joint, producing a synovial cyst that can then abut or compress upon the thecal sac. Symptoms can range from mild radiculopathy to acute sciatica to cauda equina syndrome [41–43]. If the synovial cyst occurs along the posterior aspect of the facet joint, it is typically asymptomatic. Synovial facet cysts range in incidence from 0.8% to 2.0% on imaging studies. Up to 90% of facet synovial cysts occur in the lumbar spine, most commonly at L4–L5 [44]. It is important to note communication with the synovium of the facet joint, as extruded disk fragments, meningoceles, and Tarlov cysts may have a similar appearance.

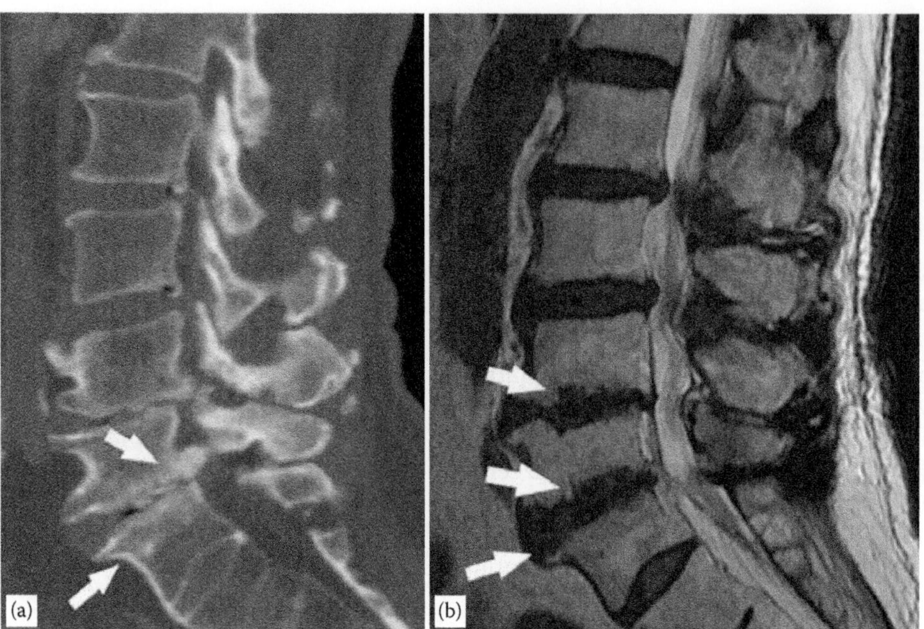

**FIGURE 8.16**
Type 3 Modic changes in a 79-year-old man with severe back pain. (a) Sagittal 2D reformatted CT shows sclerotic, degenerative L4–5 end plates (arrows), disk space narrowing, and vacuum phenomenon. Note is also made of canal stenosis and degenerative change of the spinous processes. (b) Sagittal FSE $T_2$-weighed MRI confirms CT findings and demonstrates low $T_2$ signal in the L3-5 end plates (arrows), also present on $T_1$-weighted image (not shown).

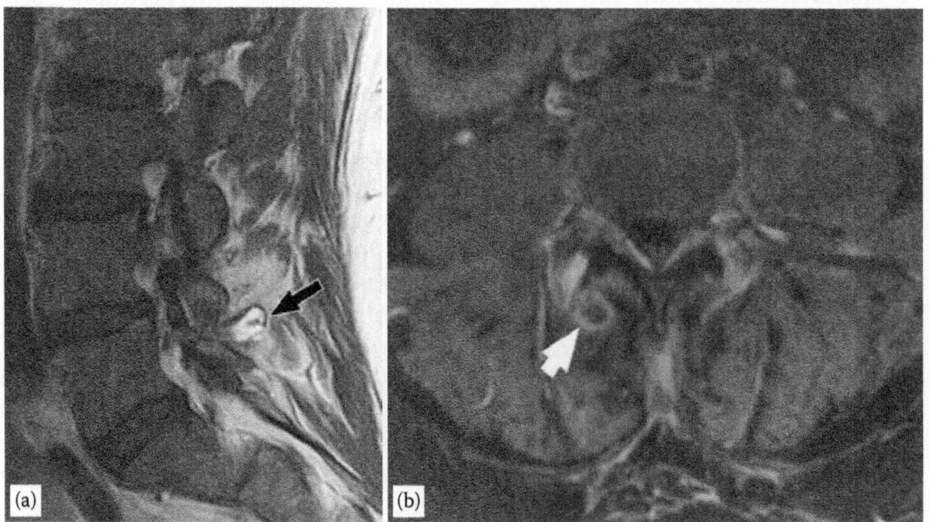

**FIGURE 8.17**
Facet joint synovial cyst. (a) Parasagittal FSE $T_2$-weighted image in a 59-year-old man complaining of back pain shows a minimally complex cyst (arrow) extending posterior to the right L4–5 facet joint. (b) Axial-enhanced $T_1$-weighted FS image through L4–5 in a 57-year-old woman with long-standing back pain demonstrates a peripherally enhancing cyst (arrow) extending posterior to the right facet joint.

MR is the modality of choice for imaging facet synovial cysts and will have a typical appearance of a $T_1$ hypointense/$T_2$ hyperintense cyst that communicates directly with the facet synovium (Figure 8.17). Contrast administration may demonstrate an enhancing wall (Figure 8.17b). Occasionally, synovial cysts will be hyperintense on $T_1$ images in the setting of hemorrhagic or proteinaceous content. Etiologies include stress loading, joint fluid accumulation, and synovial proliferation. This entity has an increased association with RA and calcium pyrophosphate deposition disease (CPPD) arthropathy and is more common in females.

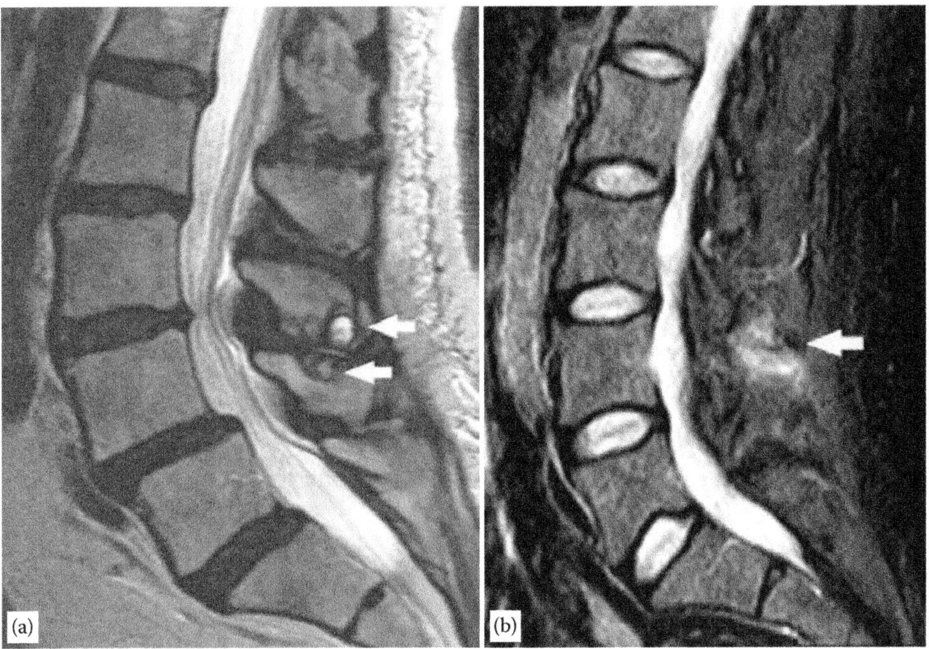


**FIGURE 8.18**
Baastrup disease. (a) Sagittal FSE $T_2$-weighted image delineates interspinous joint space narrowing and subcortical cystic changes (arrows) in a 73-year-old woman. (b) Sagittal FSE $T_2$-weighted FS image shows L3–4 interspinous edema-like signal changes (arrow) in a 55-year-old woman.

Inflammation may have a role in their etiology as many have resolved with the use of oral anti-inflammatories and local steroid injection [45]. In the event that the cyst needs to be drained, which is the majority of cases, this can be done percutaneously or surgically. There is a higher likelihood of percutaneous success when the cyst is $T_2$ hyperintense, owing to the increased fluid content [44]. Minimally invasive resections have been described with good success rates [46,47].

### 8.8.6 Baastrup Disease

Baastrup disease, also called *kissing spinous process disease* or *spinous process impingement syndrome*, is a clinical entity characterized by approximation and contact of adjacent spinous processes. There is sclerosis, enlargement, and flattening of the opposing interspinous surfaces. There may also be neoarthrosis and formation of adventitious bursa. Clinically, patients present with back pain exacerbated by extension and relieved with flexion. Baastrup disease is more common with increasing age, central canal stenosis, and anterolisthesis. Treatment options include steroid injection into the bursa, placement of an interspinous spacing device, or resection of the ascending spinous process. MRI may demonstrate height loss of the interspinous ligament and ligamentum flavum hypertrophy. $T_2$ sequences may demonstrate hyperintensity of the interspinous ligament, also called *Baastrup's sign*, or effusions within the facet joints (Figure 8.18). There can be nonspecific enhancement of the interspinous ligament with contrast [4]. Although there have been reported cases in young athletes, Baastrup disease occurs with high frequency in the elderly and is considered by many to be a normal part of aging [48]. There are many case reports in the literature of patients that respond well to local injection with corticosteroids [49].

### 8.8.7 Bertolotti's Syndrome

Bertolotti's syndrome is characterized by back pain related to unilateral or bilateral enlargement of the transverse process of the most caudal lumbar vertebrae. This enlarged transverse process may or may not fuse with the sacrum or ilium. This abnormal L5 vertebra is considered a transitional vertebra because it has characteristics of the vertebra directly above and below it [50]. It is hypothesized that the mechanism of the back pain is due to reduced and asymmetrical motion at the lumbosacral junction. Causal relationship of this clinical entity is controversial given the fact that of patients with transitional lumbar anatomy, 7% have low back pain and 4%–6% are asymptomatic [4].

MR findings include transitional caudal segment anatomy. There is also a reported increased incidence of disk herniations above the transitional segment. Nuclear medicine bone scans often demonstrate increased uptake near the anomalous articulation. Treatment is usually conservative with nonsteroidal anti-inflammatory drugs (NSAIDs) and physical therapy; however, more invasive techniques have been utilized, including minimally invasive resection of the anomalous transverse process and its articulation [4,51–53]. Some pain physicians advocate for a combination of interventions, given the multiple etiologies of pain that arise from this disease [54].

## 8.9 Other Spondyloarthropathies

### 8.9.1 Diffuse Idiopathic Skeletal Hyperostosis

Diffuse idiopathic skeletal hyperostosis (DISH), also known as *Forestier's disease,* is characterized by flowing anterior vertebral ossification of the entheses (the bony attachments of tendons, ligaments, and joint capsules) with minimal spondylotic changes. Diagnostic criteria include anterior ossification of at least four contiguous vertebral bodies with no apophyseal or sacroiliac (SI) joint ankylosis and minimal degenerative changes (Figure 8.19). Diagnosis can be made with radiography or CT, with MR being reserved to evaluate the uncommon occurrence of spondylosis related or traumatic cord compression. When present, DISH almost always involves the thoracic spine and is usually asymmetrically more prominent on the right, perhaps due to aortic pulsations on the left. DISH is typically an asymptomatic incidental finding on imaging. Fractures are rare in patients with DISH, but are severe when they occur. It should be noted that bulky anterior osteophytes might compress the esophagus leading to dysphagia or rarely compression of the trachea leading to stridor [55–58].

### 8.9.2 Ossification of the PLL

Ossification of the posterior longitudinal ligament (OPLL), as the name implies, is characterized by flowing multilevel ossification of the PLL. Etiology is uncertain; however, there is a high association in patients with DISH, ankylosing spondylitis, and ossification of the ligamentum flavum. OPLL is more common in people of Asian descent and is twice as common in men. Patients typically present with cervical myelopathy symptoms in their 40s or 50s [59,60].

Diagnostic criteria include absence of facet ankylosis and minimal degenerative disk disease (Figure 8.20). The most significant complication of OPLL is cord compression, typically in the cervical spine. On MR, the ossification usually has low signal on all sequences and does not enhance. Axial images may demonstrate a characteristic *bowtie* configuration. $T_2$ sequences may show cord hyperintensity secondary to myelomalacia or cord edema at the level of maximum compression. This can be helpful for surgical decompression planning as it denotes the level of compression. Increased $T_2$ signal also indicates an increased risk for poor postoperative outcome [59–62].

Ossification of the ligamentum flavum is a similar pathology that is characterized by multilevel low intensity in the ligamentum flavum on $T_2$ sequences, or ossification of the ligamentum flavum on CT. CT is the most sensitive imaging modality due to increased lesion conspicuity [59].

### 8.9.3 Scheuermann Disease

Scheuermann disease, also called *juvenile kyphosis,* is an osteochondrosis characterized by multilevel

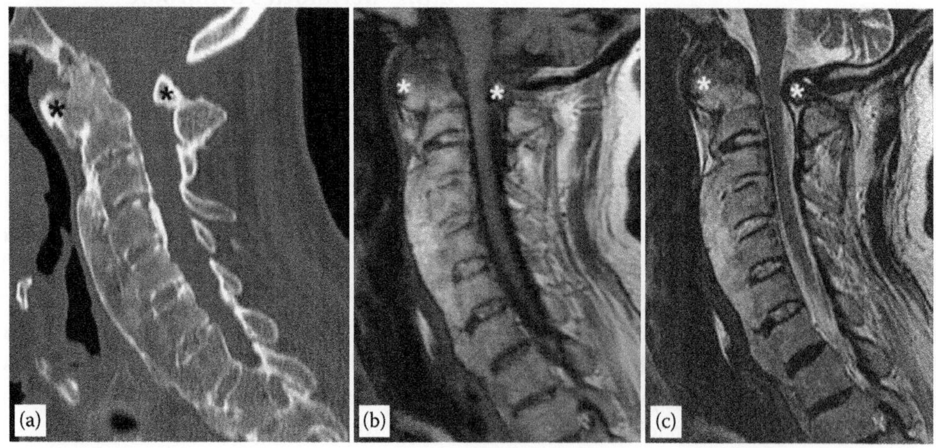


**FIGURE 8.19**
Diffuse idiopathic skeletal hyperostosis in an 80-year-old man. (a) Sagittal 2D reformatted CT shows extensive flowing anterior cervical spine osteophytes, with ankylosis of the vertebrae. Notice the ankylosed, anterior displaced C1 (asterisks), resulting in narrowing of the central canal. Sagittal $T_1$ (b) and FSE $T_2$-weighted (c) images show contiguous marrow along the hyperostotic spine and narrowing of the central canal due to fused and displaced C1 (asterisks).

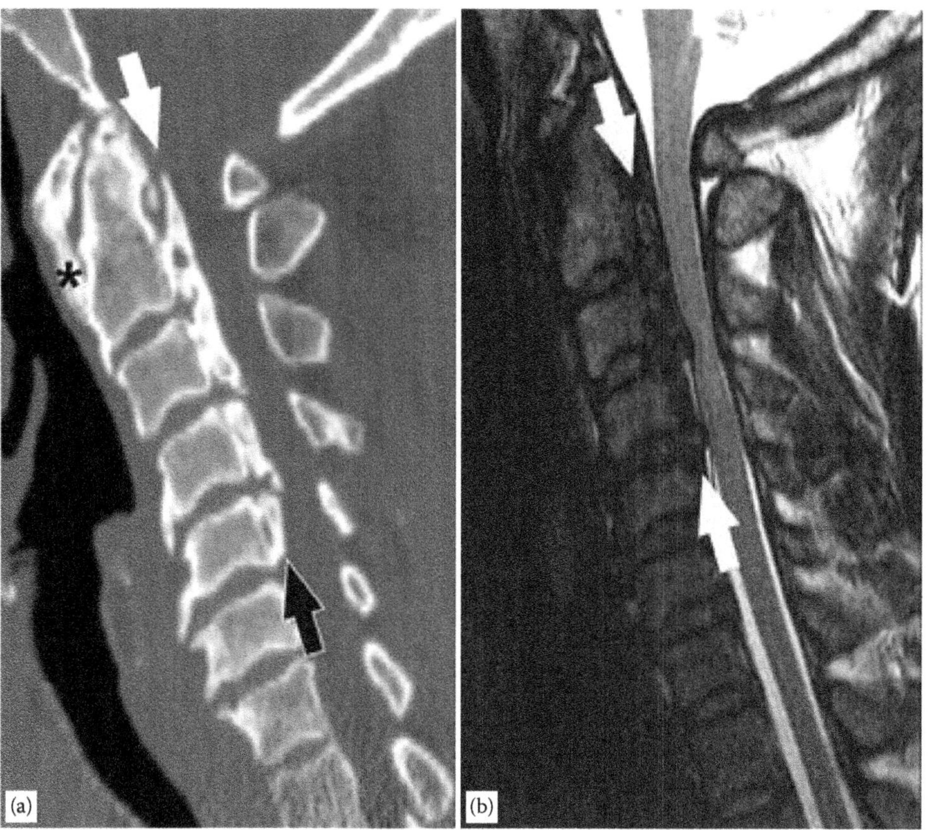


**FIGURE 8.20**
Ossification of the posterior longitudinal ligament. (a) Sagittal 2D reformatted CT shows extensive ossification of the posterior longitudinal ligament (arrows), with narrowing of the central canal. Note the less extensive ossification along the anterior longitudinal ligament (asterisk). (b) Sagittal FSE $T_2$-weighted FS image shows corresponding low signal intensity calcification, with diffuse upper and mid-cervical central canal stenosis (arrows).

wedge-shaped thoracic vertebrae with anterior disk narrowing and multiple well-defined Schmorl's nodes (Figure 8.21). Seventy-five percent of vertebrae affected are thoracic with the remainder being in the thoracolumbar region. Peak incidence of the disease is during adolescence and usually presents as thoracic spine pain worsened by activity. There is increased incidence in weight lifters and gymnasts, and there is a familial tendency. The age of incidence is between 13 and 17 years with a prevalence of approximately 0.4%–0.8% [4].

MR findings include disk degeneration, disk herniations, and Schmorl's nodes with either low or high $T_2$ signal. Diagnostic criteria require greater than three contiguous vertebrae, with at least 5° of wedging at each level. Schmorl's nodes do not have to be present to make the diagnosis [63]. Surgical treatment is warranted when there is greater than 75° of kyphosis in a skeletally immature patient [4]. Surgical techniques include posterior and anterior fusion, while more minimalistic thorascopic anterior release procedures have been described [64].

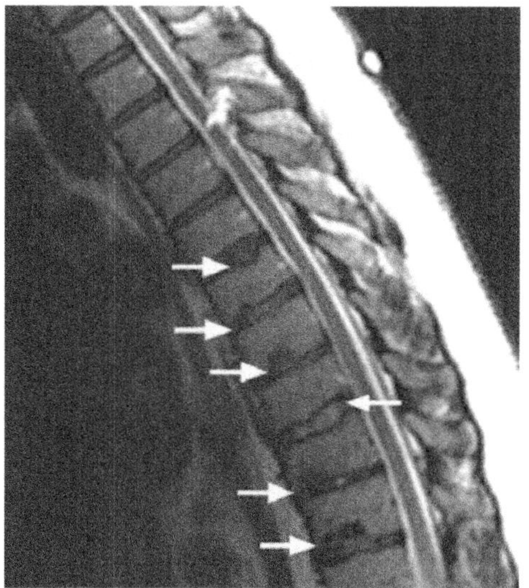

**FIGURE 8.21**
Scheuermann disease in a 32-year-old man with chronic back pain. Sagittal FSE $T_2$-weighted image demonstrates multiple Schmorl's nodes (arrows), mild anterior vertebral body wedging, and mild kyphosis.

### 8.9.4 Adult Rheumatoid Arthritis

Adult RA is an inflammatory arthropathy of unknown etiology that affects the spine in 60% of RA patients. MRI is useful in detecting treatable, reversible changes in patients with RA prior to the detection on radiographs and is also helpful in monitoring response to treatment. Cervical spine pathology includes erosions of the dens, uncovertebral joints, and facet joints. In addition, synovial proliferation can lead to pannus formation with erosion of the transverse ligament. On MRI, $T_1$ and $T_2$ sequences will show low intensity within the pannus with diffuse heterogeneous enhancement (Figure 8.22). If calcifications are present, or there is lack of enhancement within the pannus, an alternative diagnoses such as CPPD should be considered. Erosion of the transverse ligament can result in atlantoaxial subluxation, which may be seen in up to 5% of patients with cervical RA. Greater than 9 mm of separation between the inferior margin of C1 and the dens has a strong correlation with neurological symptoms.

### 8.9.5 Juvenile Idiopathic Arthritis

Juvenile idiopathic arthritis represents a spectrum of inflammatory arthropathies that occur in pediatric patients. Patients typically present with pauciarticular pain, limited neck movement, and/or peripheral tenosynovitis. Girls are typically more commonly involved, except in juvenile onset ankylosing spondylitis. There are five categories of juvenile idiopathic arthritis: (1) oligoarticular *juvenile RA* (40% of cases, involves one to several large joints); (2) still disease (20% of cases, under 5 years old, involves fever, anemia, and hepatosplenomegaly); (3) seronegative polyarticular *juvenile* RA (25% of cases, seronegative); (4) juvenile onset adult (5% of cases, rheumatoid factor positive, more common involvement of cervical spine); and (5) juvenile onset ankylosing spondylitis (10% of cases, HLA-B27 positive, primarily involves lumbar spine and lower extremity joints, enthesopathy is prominent feature).

MR is the most sensitive diagnostic examination for early diagnosis. In the cervical spine, pannus formation has a similar appearance to RA. There will be greater than 5 mm of dynamic widening between the posterior edge of the anterior arch of C1 and the anterior surface of the dens. Compression fractures can occur and are usually in the thoracolumbar spine in patients with long history of steroid use. Additional findings include effusions in the facet and SI joints, facet joint ankylosis, and periostitis (Figure 8.23). There may also be effusions within tendon sheaths as well as synovial thickening. Imaging may also demonstrate periarticular marrow edema and enhancement within the joint space and surrounding bone [4,65].

### 8.9.6 Axial Spondyloarthritis

*Axial spondyloarthritis* refers to an inflammatory arthritis and enthesopathy affecting the spine and SI joints in a patient with a negative rheumatoid factor.

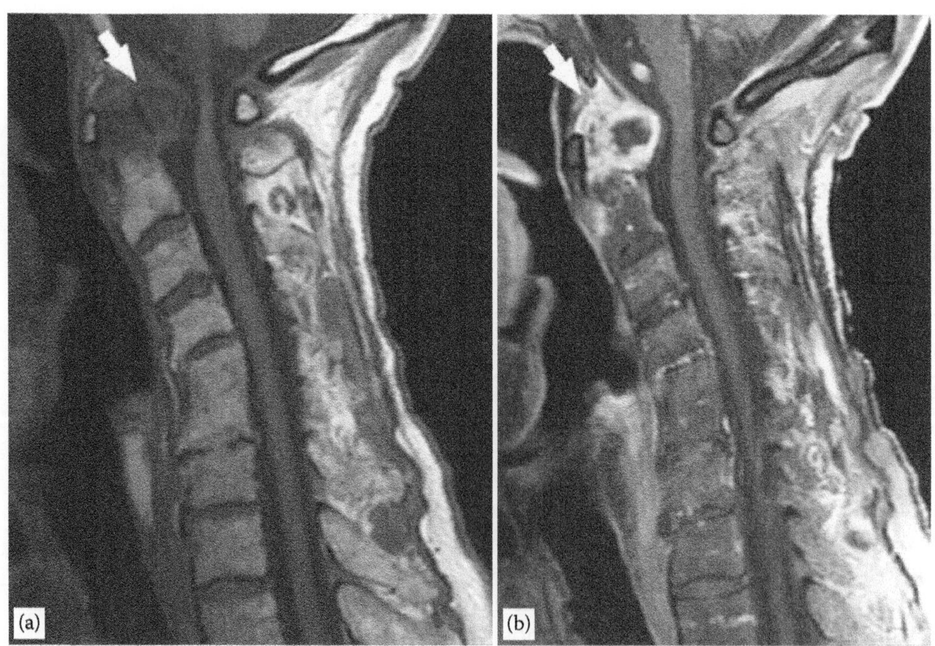


**FIGURE 8.22**
Pannus about the dens in woman with rheumatoid arthritis. Sagittal $T_1$ (a) and enhanced $T_1$-weighted (b) images show a large, enhancing pannus (arrows) surrounding the dens, resulting in central canal narrowing. C5–6 and partial C6–7 vertebral body fusion and mild kyphosis is also present.

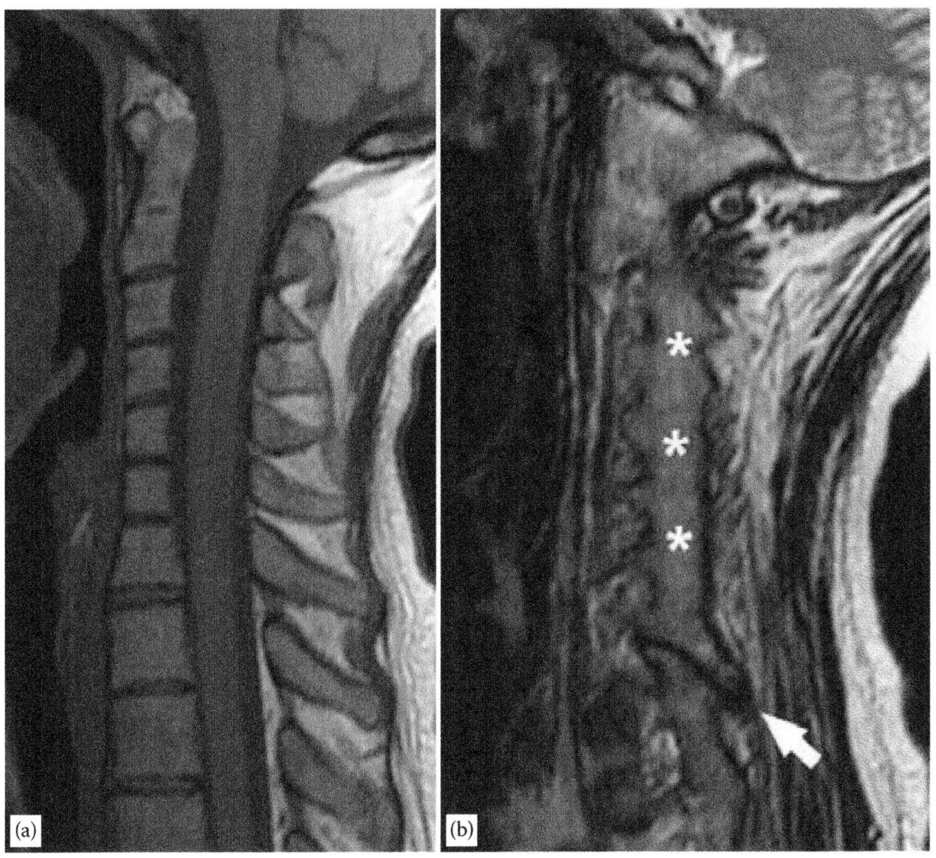


**FIGURE 8.23**
Juvenile idiopathic arthritis. Sagittal $T_1$ (a) and parasagittal FSE $T_2$-weighted (b) images in a patient with childhood history of juvenile idiopathic arthritis show diminutive cervical vertebrae due to stunted growth from early-onset ankylosed facet joints (asterisks). Notice the advanced facet arthropathy at the unfused cervicothoracic junction (arrow).

It is subdivided into five types including intestinal bowel disease-associated spondyloarthropathy, psoriatic spondyloarthropathy, reactive spondyloarthropathy, ankylosing spondylitis, and undifferentiated spondyloarthropathy. Each type occurs in early to mid-adulthood and is more common in men. There is a strong association with HLA-B27 haplotype [66,67].

Imaging findings include ossification of the paraspinous ligaments and annulus fibrosis, corner erosions of the vertebral bodies, and erosive arthritis of the SI joints (Figures 8.24 through 8.27). On MRI, early vertebral body corner and SI joint erosions will appear as high $T_2$ and low $T_1$ signal. There may also be increased $T_2$ signal in the adjacent bone marrow with enhancement noted in active disease. MR is indicated to evaluate cord status in the setting of trauma. Treatments include NSAIDs, sulfasalazine, and methotrexate and regimens are largely dependent on type and symptomatology [66,67].

Ankylosing spondylitis is the prototypical disease in the spectrum of axial spondyloarthritis. HLA-B27 is positive in approximately 90% of patients. The earliest manifestation of this disease is sacroiliitis and then spinal involvement leading eventually to flowing

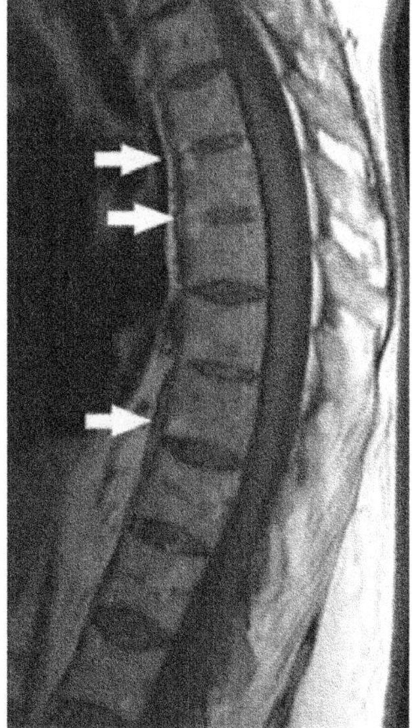

**FIGURE 8.24**
Axial spondylitis with Romanus lesions. Sagittal $T_1$-weighted image shows multiple foci of hyperintense signal intensity at the anterior corners of several thoracic vertebrae (arrows). These are the MRI equivalents of radiographic shiny corners caused by enthesitis.

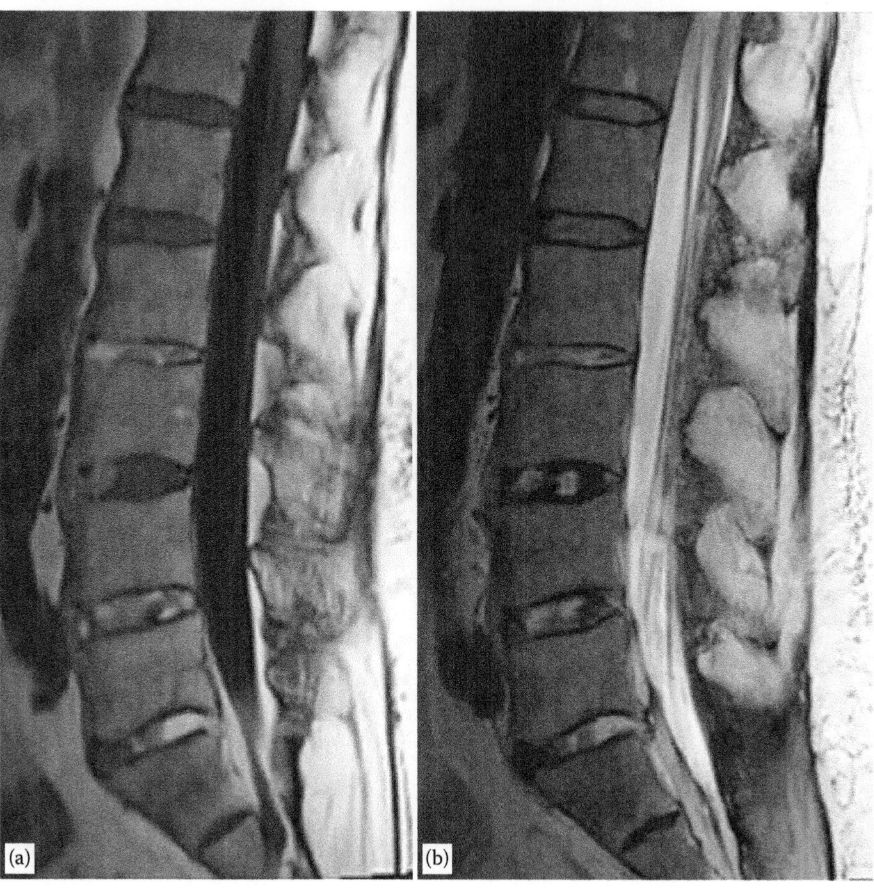


**FIGURE 8.25**
Axial spondylitis in 64-year-old man. Sagittal $T_1$ (a) and FSE $T_2$-weighted (b) images demonstrate bridging syndesmophytes, fusion of the spinous processes, partial fusion of multiple facet joints, and marrow within the disk spaces.

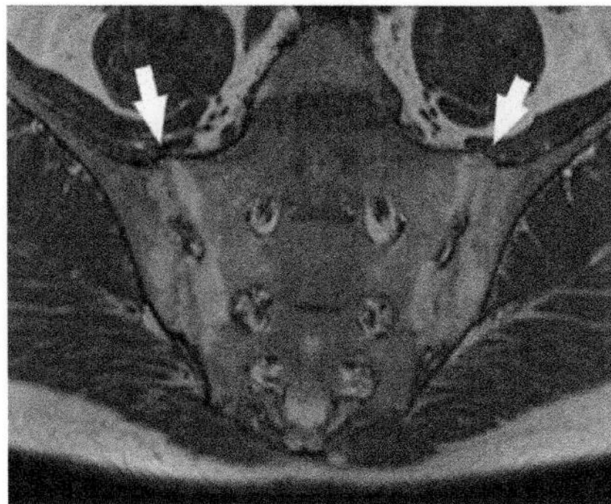

**FIGURE 8.26**
Sacroiliac joint ankylosis in axial spondylitis. Coronal oblique $T_1$-weighted image through the sacrum and sacroiliac joints demonstrates solid ankylosis across both joints (arrows), with adjacent chronic fatty marrow changes.

syndesmophyte formation. This is typically present with joint pain and restricted spinal movement. The most serious complication is spinal fracture, since the fractures usually involve three columns and are therefore unstable. MR is most helpful in this setting given that most of these fractures are initially occult.

### 8.9.7 Crystal Deposition Disease and Periodontoid Pseudotumor

There are three crystals that lead to crystal deposition diseases: hydroxyapatite deposition disease (HADD), CPPD, and monosodium urate deposition disease (gout). *Longus colli calcific tendinosis* refers to focal calcification and soft tissue inflammation of the longus coli muscle secondary to deposition of hydroxyapatite crystals (Figure 8.28). Clinically, patients present with neck pain and stiffness, and less commonly mild leukocytosis with fever. The longus colli tendon traverses craniocaudally, immediately anterior to the anterior longitudinal ligament. With calcific tendinosis of this structure, MRI demonstrates low intensity calcific foci on all sequences [4]. $T_2$-weighted images show increased intensity within soft tissue extending from the skull base to around the level of C5. The adjacent soft tissue may show varying degrees of enhancement. This is a self-limiting malady that usually resolves within 1–2 weeks with NSAIDs. Important differential considerations include retropharyngeal abscess and cervical osteomyelitis [68].

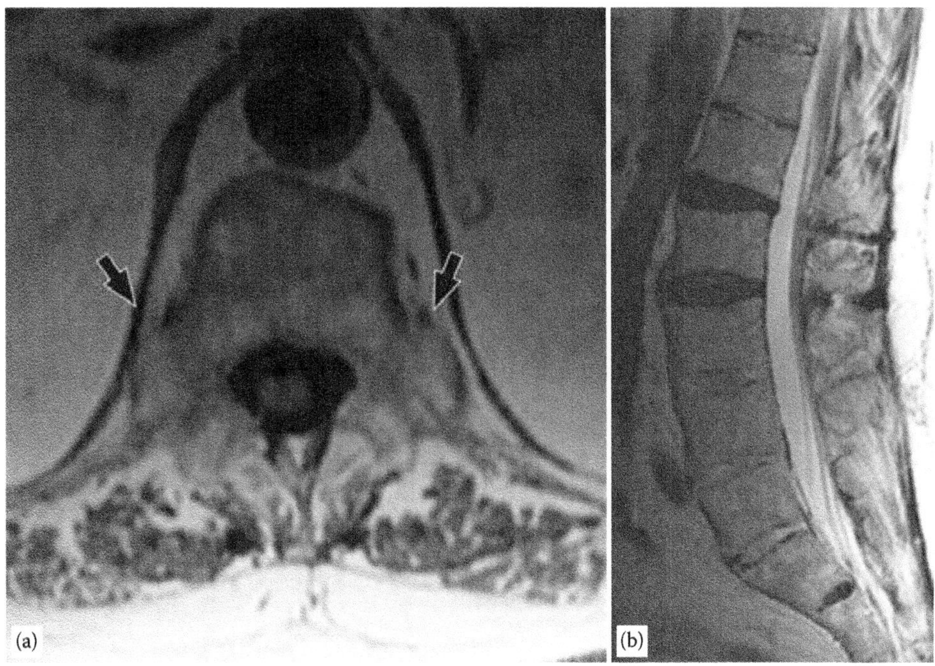

**FIGURE 8.27**
Advanced changes of axial spondylitis in 72-year-old man. Axial $T_1$ (a) and sagittal FSE $T_2$-weighted (b) images show ankylosis across the costovertebral joints (arrows) and multiple disk spaces.

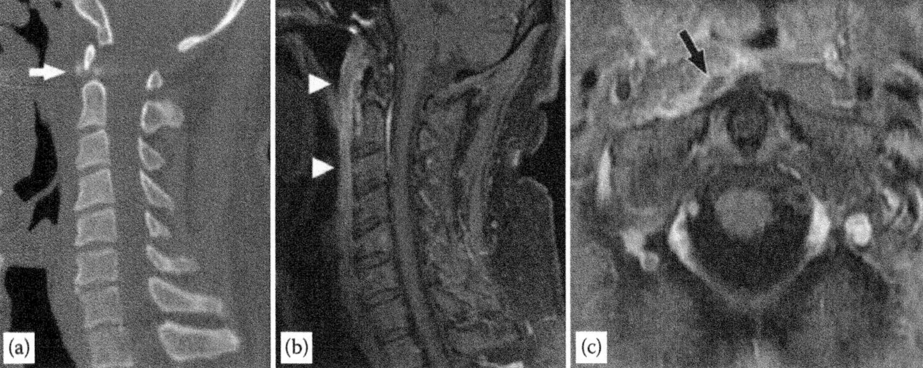

**FIGURE 8.28**
Longus colli calcific tendinosis in a 45-year-old man with neck pain and sensation of retained foreign body when swallowing. (a) Sagittal 2D reformatted CT show focal calcification (arrow) within and posterior to the longus colli muscle. Sagittal-enhanced (b) and axial-enhanced (c) $T_1$-weighted images show diffuse cervical prevertebral soft tissue swelling (arrowheads) and low signal intensity focus (arrow), corresponding to calcification. Symptoms resolved with narcotics and conservative treatment.

CPPD is one that is common in patients over the age of 50. Sometimes referred to as *pseudogout*, this entity may result in chronic asymptomatic degenerative disk disease, or can result in acutely painful episodes, similar to the presentation of gout. CPPD usually manifests on MR studies as soft tissue calcium that is typically low signal intensity on all sequences. In the spine, the intervertebral disk will be narrowed with low signal intensity, indistinguishable from degenerative disk disease. It is much more common in the spine than once thought, often being present in the spine solely. Vertebral end plate erosions are common. There may be ligament thickening of the PLL or ligamentum flavum. Calcified soft tissue masses may be noted adjacent to the vertebrae with low signal on all sequences and minimal enhancement. There can be facet arthropathy with joint effusions [69].

Periodontoid pseudotumor, also known as *crowned dens* syndrome, refers to calcification of the transverse ligament. Patients may present with fever, elevated C-reactive protein, leukocytosis, and cervico-occipital junction pain. MRI of the cervical spine typically demonstrates a low-signal intensity mass projecting posterior to the dens, representative of amorphous calcification and

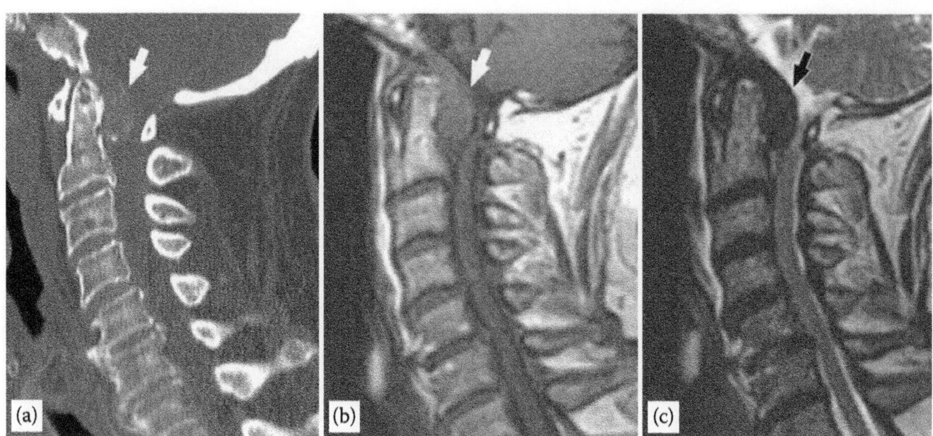


**FIGURE 8.29**
Periodontoid pseudotumor in an 87-year-old woman with gout and chronic renal failure. (a) Sagittal 2D reformatted CT image shows partially mineralized pannus around the dens (arrows), with subcortical cystic change of the dens. Sagittal $T_1$ (b) and FSE $T_2$ (c) images delineate the marked central canal stenosis caused by the mass (arrows).

inflammation of the transverse ligament (Figure 8.29). Heterogeneous peripheral enhancement may be present. Homogeneous enhancement should favor another etiology. Bulky soft tissue calcifications and thickening of the PLL or the ligamentum flavum may lead to myelopathy and can be differentiated from RA by demonstrating calcifications on CT or radiographs.

Gout is an inflammatory arthropathy caused by urate crystal deposition. Urate crystal deposition may occur secondary to an inborn error of metabolism such as in Lesch–Nyhan syndrome, or more commonly, secondary to chronic disease such as renal failure or myeloproliferative disorder. This inflammatory process typically involves peripheral joints, but rarely can involve the spine. When intraspinal, it is generally of the chronic tophaceous subtype, involves only one or two levels, and is more common in the lumbar spine. MR findings include *punched out* lesions of the vertebral bodies and facets that are of low intensity and all sequences with variable enhancement [4,70,71]. Spinal tophi can present with back pain and are easily mistaken for infection or metastatic disease. A distinguishing feature is that gout usually lacks inflammatory change in the adjacent intervertebral disk and vertebra [71].

### 8.9.8 Neuropathic Spondyloarthropathy

Neuropathic change of the spine is an uncommon destructive arthropathy that occurs when pain and proprioception are impaired. The most common cause is diabetes mellitus, followed by traumatic paraplegia. Charcot originally described this entity in association with neurosyphilis. In the spine, it almost always occurs in the lumbar region and can result in significant destructive changes in a short period of time. There is destruction of the end plates and facets, and there can be numerous vertebral subluxations and often an associated heterogeneously enhancing soft tissue mass. There are only a few disease processes that can cause such significant distraction in such a short time frame including infection and metastatic disease. The classic *6 Ds* are helpful descriptors of neuropathic arthropathy and include distention, density, debris, disorganization, dislocation, and destruction. These can be helpful when trying to distinguish neuropathic joint from disk infection. More specific imaging findings for Charcot spine proposed by Wagner et al. include vacuum disk, facet involvement, vertebral body spondylolisthesis, joint disorganization and debris, and gadolinium enhancement along the periphery of the intervertebral disk and diffusely throughout the vertebral body (Figure 8.30) [72].

### 8.9.9 Hemodialysis Spondyloarthropathy

Long-term hemodialysis can lead to destructive spondyloarthropathy in a subset of chronic renal failure patients. The exact etiology is unknown, although amyloidosis, inflammation, and hydroxyapatite deposition have all been implicated. The lower cervical spine is most commonly affected. MRI demonstrates findings similar to those seen in discitis, though there is usually less $T_2$ signal than would be expected with infection. There is typically end plate destruction and amorphous material in the disk, prevertebral soft tissues, and spinal canal. These findings in a patient with a long history of renal failure are virtually diagnostic [4,73].

In conclusion, MRI is an advanced imaging modality that is useful in the evaluation of the soft tissue and osseous components of degenerative disk disease and other spondyloarthropathies. Intervertebral disk pathology is comprised of annular tear, disk bulge, disk herniation, and Schmorl's nodes. MRI is also useful in the

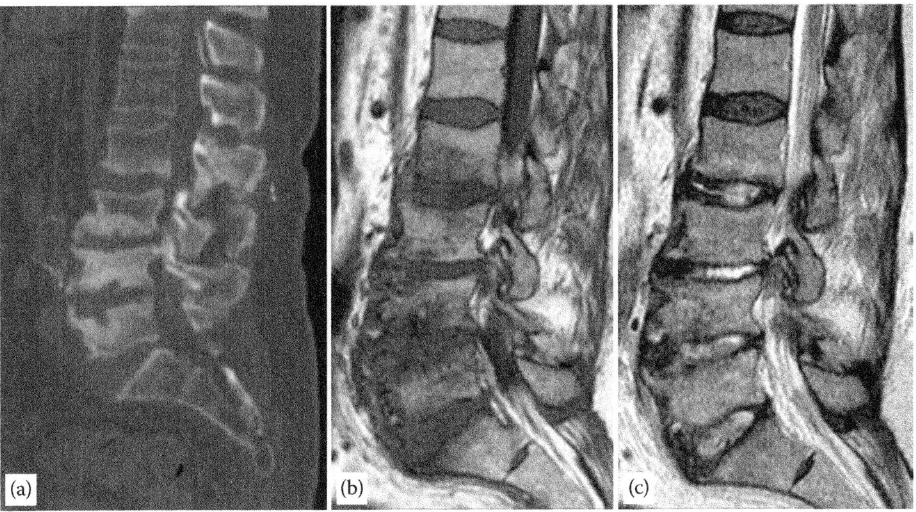


**FIGURE 8.30**
Neuropathic spine in 58-year-old paraplegic man. (a) Sagittal 2D reformatted CT shows extensive sclerosis and end plate irregularity of the lower lumbar spine. Sagittal $T_1$-weighted (b) and FSE $T_2$-weighted (c) images show corresponding low $T_1$ end plate signal changes and hyperintense $T_2$-weighted signal changes throughout multiple lumbar intervertebral disks.

preoperative evaluation of symptomatic patients with congenital and acquired spinal stenosis, extreme cases of diffuse idiopathic skeletal hyperostosis, and ossification of the PLL. Furthermore, MRI is useful in detecting treatable, reversible changes in patients with RA prior to the detection on radiographs and is also helpful in monitoring response to treatment.

## References

1. Berquist TH. *MRI of the Musculoskeletal System.* 6th ed. Philadelphia, PA: Lippincott Williams & Wilkins; 2013.
2. Berquist TH. Imaging of the postoperative spine. *Radiologic Clin N Am.* 2006; 44:407–418.
3. Anwar Z, Zan E, Gujar SK et al. Adult lumbar scoliosis: Underreported on lumbar MR scans. *Am J Neuroradiol.* 2010; 31:832–837.
4. Ross JS, Moore KR, Shah LM, Borg B, Crim J, eds. *Diagnostic Imaging: Spine.* 2nd ed. Manitoba, Canada: Amirsys; 2010.
5. Yousem DM, Grossman RI. *Neuroradiology: The Requisites.* 3rd ed. Philadelphia, PA: Mosby Elsevier; 2010.
6. Tibiletti M, Galbusera F, Ciavarro C, Brayda-Bruno M. Is the transport of a gadolinium-based contrast agent decreased in a degenerated or aged disc? A post contrast MRI study. *PLoS One.* 2013; 8:1–9.
7. Haughton V, Schreibman K, De Smet A. Contrast between scar and recurrent herniated disk on contrast-enhanced MR images. *Am J Neuroradiol.* 2002; 23:1652–1656.
8. Moon SM, Yoder JH, Wright AC et al. Evaluation of intervertebral disc cartilaginous endplate structure using magnetic resonance imaging. *Eur Spine J.* 2013; 22:1820–1828.
9. Bae WC, Statum S, Zhang Z et al. Morphology of the cartilaginous endplates in human intervertebral disks with ultrashort echo time MR imaging. *Radiology.* 2013; 266:564–574.
10. Helms CA, Major NM, Anderson MW et al. *Musculoskeletal MRI.* 2nd ed. Philadelphia, PA: Saunders Elsevier; 2009.
11. Ibrahim DA, Myung KS, Skaggs DL. Ten percent of patients with adolescent idiopathic scoliosis have variations in number of thoracic or lumbar vertebrae. *J Bone Joint Surg Am.* 2013; 95:828–833.
12. Konin GP, Walz DM. Lumbosacral transitional vertebrae: Classification, imaging findings, and clinical relevance. *Am J Neuroradiol.* 2010; 31:1778–1786.
13. Kalichman L, Suri P, Guermazi A et al. Facet orientation and tropism: Associations with facet joint osteoarthritis and degenerative spondylolisthesis. *Spine.* 2009; 34:579–585.
14. Kovacs FM, Arana E, Royuela A et al. Disc degeneration and chronic low back pain: An association which becomes nonsignificant when endplate changes and disc contour are taken into account. *Neuroradiology.* 2014; 56:25–33.
15. Emch TM, Modic MT. Imaging of lumbar degenerative disk disease: History and current state. *Skeletal Radiol.* 2011; 40:1175–1189.
16. Takatalo J, Karppinen J, Niinimaki J et al. Prevalence of degenerative imaging findings in lumbar magnetic resonance imaging among young adults. *Spine.* 2009; 34: 1716–1721.
17. Kanayama M, Togawa D, Takahashi C et al. Cross-sectional magnetic resonance imaging study of lumbar disc degeneration in 200 health individuals. *J Neurosurg Spine.* 2009; 11:501–507.
18. Modic MT, Ross JS. Lumbar degenerative disk disease. *Radiology.* 2007; 245:43–60.
19. Fardon DF, Milette PC. Nomenclature and classification of lumbar disc pathology. Recommendations of the

Combined Task Forces of the North America Spine Society, American Society of Spine Radiology, and American Society of Neuroradiology. *Spine*. 2001; 26:93–113.
20. Bartynski WS, Rothfus WE. Peripheral disc margin shape and internal disc derangement: Imaging correlation in significantly painful discs identified at provocation lumbar discography. *Interventional Neuroradiol.* 2012; 18:227–241.
21. Peng B, Hou S, Wu W et al. The pathogenesis and clinical significance of a high-intensity zone (HIZ of lumbar intervertebral disc on MR imaging in the patient with discogenic low back pain. *Eur Spine J*. 2006; 15:583–587.
22. Kim DG, Eun JP, Park JS. New diagnostic tool for far lateral lumbar disc herniation: The clinical usefulness of 3-Tesla magnetic resonance myelography comparing with the discography CT. *J Korean Neurosurg Soc.* 2012; 52:103–106.
23. Wiltse LL, Guyer RD, Spencer CW, Glenn WV, Porter IS. Alar transverse process impingement of the L5 spinal nerve: the far-out syndrome. *Spine* (Philadelphia, PA 1976). 1984; 9(1):31–41.
24. Kikuchi K, Abe E, Miyakoshi N et al. Anterior decompression for far-out syndrome below a transitional vertebra: A report of two cases. *Spine*. 2013; 13:21–25.
25. Kitamura M, Eguchi Y, Inoue G et al. A case of symptomatic extra-foraminal lumbosacral stenosis ("far-out syndrome") diagnosed by diffusion tensor imaging. *Spine*. 2012; 37:854–857.
26. Wu HH, Morrison WB, Schweitzer ME. Edematous Schmorl's nodes on thoracolumbar MR imaging: Characteristic patterns and changes over time. *Skeletal Radiol.* 2006; 35:212–219.
27. Sonne-Holm S, Jacobsen S, Rovsing H, Monrad H. The epidemiology of Schmorl's nodes and their correlation to radiographic degeneration in 4,151 subjects. *Eur Spine J*. 2013; 22:1906–1902.
28. Mattei TA, Rehman AA. Schmorl's nodes: Current pathophysiological, diagnostic, and therapeutic paradigms. *Neurosurg Rev.* 2014; 37:39–46.
29. Bajwa NS, Toy JO, Young EY, Ahn, NU. Establishment of parameters for congenital stenosis of the cervical spine: An anatomic descriptive analysis of 1066 cadaveric specimens. *Eur Spine J*. 2012; 21:2467–2474.
30. Bajwa NS, Toy JO, Ahn, NU. Establishment of parameters for congenital thoracic stenosis. A study of 700 postmortem specimens. *Clin Orthop Relat Res.* 2012; 470:3195–3201.
31. Denis DJ, Shedid D, Shehadeh M et al. Cervical spondylosis: A rare and curable cause of vertebrobasilar insufficiency. *Eur Spine J*. 2013; 23:206–213.
32. Salamon N, Ellingson BM, Nagarajan R et al. Proton magnetic resonance spectroscopy of human cervical spondylosis at 3T. *Spinal Cord.* 2013; 51:558–563.
33. Ellingson BM, Salamon N, Holly LT. Advances in MR imaging for cervical spondylotic myelopathy. *Eur Spine J*. 2013; 24:197–208.
34. Perie D, Curnier D. Effect of pathology type and severity on the distribution of MRI signal intensities within the degenerated nucleus pulposus: Application to idiopathic scoliosis and spondylolisthesis. *BMC Musculoskel Disord.* 2010; 11:1–7.
35. Gervais J, Perie D, Parent S et al. MRI signal distribution within the intervertebral disc as a biomarker of adolescent idiopathic scoliosis and spondylolisthesis. *BMC Musculoskelet Disorder.* 2012; 13:1–10.
36. Ozawa H, Kanno, H, Koizumi Y et al. Dynamic changes in the dural sac cross-sectional area on axial loaded MR imaging: Is there a difference between degenerative spondylolisthesis and spinal stenosis? *Am J Neuroradiol.* 2012; 33:1191–97.
37. Leone A, Cianfoni A, Cerase A et al. Lumbar spondylolysis: A review. *Skeletal Radiol.* 2011; 40:683–700.
38. Lee JM, Nam KH, Lee IS. Modic degenerative marrow changes in the thoracic spine: A single center experience. *J Korean Neurosurg Soc.* 2013; 54:34–37.
39. Yu LP, Qian WW, Yin GY, Ren YX, Hu ZY. MRI assessment of lumbar intervertebral disc degeneration with lumbar degenerative disease using the Pfirrmann grading systems. *PLoS One.* 2012; 7:1–7.
40. Mann E, Peterson CK, Hodler J et al. The evolution of degenerative marrow (Modic) changes in the cervical spine in neck pain patients. *Eur Spine J*. 2014; 23:584–589.
41. Arthur B, Lewkonia P, Quon JA et al. Acute sciatica and progressive neurological deficit secondary to faet synovial cysts: A report of two cases. *J Can Chiropr Assoc.* 2012; 56:173–178.
42. Muir JJ, Pingree MJ, Moeschler SM. Acute cauda equine syndrome secondary to lumbar synovial cyst. *Pain Physician.* 2012; 15:435–440.
43. Mun JH, Lee RS, Lim BC et al. Intraspinal ganglion cyst. *Chonnam Med J*. 2012; 48:183–184.
44. Cambron SC, McIntyre JJ, Guerin SJ, Pastel DA. Lumbar facet joint synovial cysts: Does $T_2$ signal intensity predict outcomes after percutaneous rupture? *Am J Neuroradiol.* 2013; 34:1661–1664.
45. Mattei TA, Goulart CR, McCall TD. Pathophysiology of regression of synovial cysts of the lumbar spine: The 'anti-inflammatory hypothesis.' *Med Hypothesis.* 2012; 79:813–818.
46. Rhee J, Anaizi AN, Sandhu FA et al. Minimally invasive resection of lumbar synovial cysts from a contralateral approach. *J Neurosurg Spine.* 2012; 17:453–458.
47. Ganau M, Ennas F, Ambu R et al. Excision of synovial cysts: Pathology matters. *J Neurosurg Spine.* 2013; 19:266–267.
48. Kwong Y, Rao N, Latief K. MDCT findings in Baastrup disease: Disease or normal feature of the aging spine? *AJR Am J Roentgenol.* 2011; 196:1156–1159.
49. Lamer TJ, Tiede JM, Fenton DS. Fluoroscopically-guided injections to treat "kissing spine" disease. *Pain Physician.* 2008; 11:549–554.
50. Almeida DB, Mattei TA, Soria MG et al. Transitional lumbosacral vertebrae and low back pain. *Arq Neuropsiquiatr.* 2009; 67:268–272.
51. Quinlan JF, Duke D, Eustace S. Bertolotti's syndrome. A cause of back pain in young people. *J Bone Joint Surg.* 2006; 88:1183–1186.
52. Paraskevas G, Tsaveas A, Koutras G, Natsis K. Lumbosacral transitional vertebra causing Bertolotti's

syndrome a case report and review of the literature. *Cases J.* 2009; doi:10.4076/1757-1626-2-8320.

53. Ugokwe KT, Chen TL, Klineberg E et al. Minimally invasive surgical treatment of Bertolotti's syndrome: Case report. *Neurosurgery.* 2008; 62:454–455.
54. Jain A, Agarwal A, Jain S, Shamshery C. Bertolotti syndrome: A diagnostic and management dilemma for pain physicians. *Korean J Pain.* 2013; 26:368–373.
55. Mazieres B. Diffuse idiopathic skeletal hyperostosis (Forestier-Rotes-Querol disease): What's new? *Joint Bone Spine.* 2013; 80: 466–470.
56. Olivieri I, D'Angelo S, Palazzi C, Padula A. Spondyloarthritis and diffuse idiopathic skeletal hyperostosis: Two different diseases that continue to intersect. *J Rheumatol.* 2013; 40:8:1251–1253.
57. Guo Q, Ni B, Yang J et al. Simultaneous ossification of the posterior longitudinal ligament and ossification of the ligamentum flavum caising upper thoracic myelopathy in DISH: Case report and literature review. *Eur Spine J.* 2011; 20:195–201.
58. Taljanovic MS, Hunter TB, Wisneski RJ et al. Imaging characteristics of diffuse idiopathic skeletal hyperostosis with an emphasis on acute spinal fractures: Review. *AJR Am J Roentgenol.* 2009; 193:S10–S19.
59. Smith ZA, Colin CB, Raphael D et al. Ossification of the posterior longitudinal ligament: Pathogenesis, management, and current surgical approaches. *Neurosurg Focus.* 2011; 30:1–10.
60. Saetia K, Cho D, Lee S et al. Ossification of the posterior longitudinal ligament: a review. *Neurosurg Focus.* 2011; 30:1–16.
61. He S, Hussain N, Li S, Hou T. Clinical and prognostic analysis of ossified ligamentum flavum in a Chinese population. *J Neurosurg Spine.* 2005; 3:348–354.
62. Wang LF, Liu FZ, Zhang YZ et al. Clinical results and intramedullary signal changes of posterior decompression with transforaminal interbody fusion for thoracic myelopathy caused by combined ossification of the posterior longitudinal ligament and ligamentum flavum. *Chin Med J.* 2013; 126:3822–3827.
63. Makurthou AA, Oei L, Saddy SE et al. Scheuermann disease. *Spine.* 2013; 38:1690–1694.
64. Pompeo E. Minimalistic thoracoscopic anterior spinal release in Scheuermann kyphosis. *J Thorac Cardiovasc Surg.* 2013; 146:490–491.
65. Johnson K. Imaging of juvenile idiopathic arthritis. *Pediatr Radiol.* 2006; 36:743–758.
66. Canella C, Schau B, Ribeiro E et al. MRI in seronegative spondyloarthritis: Imaging features and differential diagnosis in spine and sacroiliac joints. *AJR Am J Roentgenol.* 2013; 200:149–157.
67. Carmona R, Harish S, Linda D et al. MR imaging of the spine and sacroiliac joints for spondyloarthritis. *Radiology.* 2013; 269:208–215.
68. Maeseneer MD, Vreugde S, Laureys S et al. Calcific tendinitis of the longus colli muscle: Case report. *Head and Neck.* 1997; 19:545–548.
69. Feydy A, Liote F, Carlier R et al. Cervical spine and crystal-associated diseases: Imaging findings. *Eur Radiol.* 2006; 16:459–468.
70. Beier CP, Hartmann A, Woertgen C et al. A large, erosive intraspinal and paravertebral gout tophus. *J Neurosurg Spine.* 2005; 3:485–487.
71. Khoo JN, Tan SC. MR imaging of tophaceous gout revisited. *Singapore Med J.* 2011; 52:840–847.
72. Wagner SC, Schweitzer ME, Morrison WB et al. Can imaging findings help differentiate spinal neuropathic arthropathy from disk space infection? Initial experience. *Radiology.* 2000; 214:693–699.
73. Theodorou DJ, Theodorou SJ, Resnick D. Imaging in the dialysis patient: Imaging in dialysis spondyloarthropathy. *Semin Dial.* 2002; 15:290–296.

# 9

# *Spine Infections*

**Mariangela Marras, Luca Saba, and Stefano Marcia**

**CONTENTS**



## 9.1 Introduction

Skeletal localization of an infectious process depends on the age of the patient and the type of pathogenic microorganism.[1,2] The involvement of the appendicular skeleton is more frequent in childhood and adolescence, whereas the involvement of the axial skeleton is more frequent in adults.[3,4]

Magnetic resonance imaging (MRI) is the method of choice in the diagnosis of infection of the spine. The infectious processes can involve all components of the spine. The infectious spondylitis, which constitutes about 5% of pyogenic osteomyelitis, is distinct in two relation causative agents: to the route of spread and to the types of spine infection.[5,6]

## 9.2 Types of Spine Infections

Spine infections are classified as follows:

- Disc space infection and vertebral osteomyelitis with extension into epidural space, paravertebral region, and psoas muscles
- Primary discitis (childhood and adolescence)
- Isolated epidural abscess without disc space infection from hematogenous spread or direct extension
- Subdural empyema
- Meningitis
- Intramedullary spinal cord abscess
- Septic arthritis/facet joint

## 9.3 Etiopathogenesis

### 9.3.1 Pyogenic Forms

The most frequent form of osteomyelitis is supported by gram-positive bacteria (*Staphylococcus aureus, Streptococcus pyogenes, Streptococcus pneumoniae*), through hematogenous spread from septic foci as respiratory system, skin, and genitourinary (GU) tract.[7,8]

Gram-negative bacteria forms are less frequent, in part supported by gram-negative cocci and Enterobacteriaceae (*Escherichia coli, Proteus mirabilis, Pseudomonas aeruginosa, Klebsiella, Salmonella*) and are more commonly found in immunocompromised patients.[9,10]

Spinal trauma, spinal and abdominal surgery, drug use, diabetes conditions of immunosuppression, and treatment with corticosteroid drugs are predisposing factors.[11,12]

Pyogenic infection pathoanatomy differs for adults and children: in adults, the vertebral endplates are infected first, with subsequent spread to adjacent disc space, with diffusion to adjacent vertebral body, paravertebral tissues, and epidural space, whereas in children, the vascular channels cross growth plate allowing primary infection of intervertebral disc, with secondary infection of vertebral body.

### 9.3.2 Non-Pyogenic Forms (Granulomatous Forms)

These forms of infectious spondylitis are incurred by microorganisms such as *Mycobacterium tuberculosis, Cryptococcus neoformans, Brucella,* and *Aspergillus fumigatus* that induce a granulomatous inflammation on the target tissue.

Other rare etiologic agents are *Treponema pallidum* and fungi.[13,14]

### 9.3.3 Pathways of Spread

Pathways of spread of infectious spondylitis can be hematogenous or secondary to contiguous infection[15,16]:

- Arterial and venous spread
- Contiguous infection either without or with vascular insufficiency
- Direct injection (penetrating injuries, spine surgery, catheter placement)

## 9.4 Pyogenic Osteomyelitis

It is a suppurative infection involving disc, vertebrae, vertebral marrow, and adjacent soft tissue with endplate erosion. Intravenous drug use, immunocompromised state, chronic medical illnesses, cirrhosis, cancer, and diabetes are the main predisposing factors.

The etiological agents most commonly involved are *S. aureus, E. coli,* and *Salmonella*; the last is more common in patients with sickle cell disease.

The hematogenous spread is among most common routes of infection, secondary to infections of GU or gastrointestinal tract, lungs, and cardiac, mucous, and cutaneous membranes; also secondary infection of the intervertebral disc and adjacent vertebra is a frequent condition.

Primitive disc infection is a rare disease, typical of childhood; the disc infection is the first site of "primary discitis" due to presence yet of vascularity, a condition present up to 20 years of age.

Direct inoculation from a penetrating trauma, surgical intervention, or diagnostic procedures and epidural injection/catheter is less frequent. All spinal segments, such as lumbar spine (48%), thoracic spine (35%), and cervical spine (6.5%), are involved (Figure 9.1). Spinal meningitis and myelitis may be associated. X-ray findings may be negative up to 2–8 weeks after onset of symptoms. Afterward, initial endplate and vertebral osteolysis followed by increased bone density are the first signs, whereas fusion across disc space arises late in course of disease.

Computed tomography (CT) direct study highlights endplate osteolytic and osteosclerotic changes, bony sequestra, spinal deformity associated with iso- to hypodense paraspinal soft tissue enlargement, and

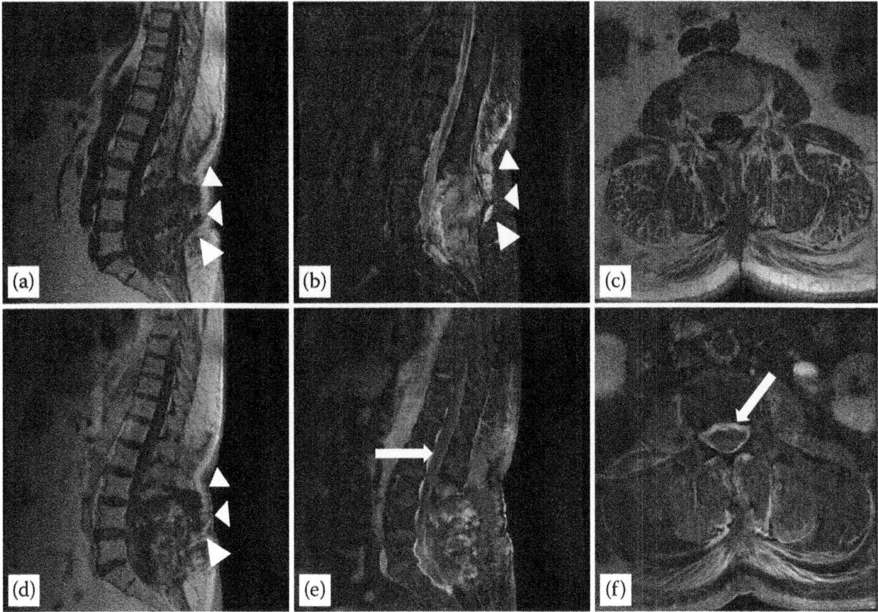


**FIGURE 9.1**
A patient with recent laminectomy at L3, L4, L5, and S1 levels, complicated by an infective lumbar arachnoiditis. The outcomes of the surgical intervention (arrowheads) are quite evident on $T_1$W precontrast sequences on the sagittal (a) and axial (c) planes, and on $T_2$W sagittal sequences (b). On $T_1$W postcontrast images (d), an anomalous enhancement in the region of the cauda equina (arrow) is evident, and on $T_1$W fat-saturated sequences on the sagittal (e) and axial (f) views, even along the epidural space (arrow).

the possible presence of soft tissue gas. CT exam with contrast agent shows enhancing disc, marrow, and paravertebral soft tissue.

MRI is highly sensitive (96%) and specific (92%); magnetic resonance (MR)-specific features are disc space narrowing, hypointense on $T_1$-weighted imaging ($T_1$WI) and variable, typically hyperintense on $T_2$-weighted imaging ($T_2$WI), and hyperintense on fat-saturated $T_2$WI or short time inversion recovery (STIR).[17–19]

Enhancement of post-gadolinium is diffuse; vertebral marrow and disc signal are abnormal. After gadolinium administration, the enhancement is prominent from paraspinal and epidural phlegmon or abscess with rim enhancement. MRI also allows to highlight the possible involvement and compression of the spinal cord.[20–24]

Differential diagnosis is with tuberculous vertebral osteomyelitis, degenerative endplate changes, and chronic hemodialysis spondyloarthropathy.

## 9.5 Tuberculous Spondylitis

It is a granulomatous infection of the spine and adjacent soft tissue typically caused by tuberculosis (TB), most frequently located at mid-thoracic or thoracolumbar tract than lumbar and cervical tracts.

Vertebral collapse and gibbus deformity are observed with relatively intact intervertebral discs and large paraspinal dissemination of disease with the formation of abscesses.

TB is more aggressive in children, where kyphosis and cord compression are more common, whereas intramedullary abscess and arachnoiditis are rare complications.

X-ray findings may not be present until weeks after onset of infection. Initial endplate irregularity and osteolysis with vertebral diffuse sclerosis are afterward observed. CT direct study allows detection of calcifications of chronic paravertebral abscesses, osseous destruction, and bony sequestra.

The contrast agent highlights diffuse or peripherally enhancing epidural and paraspinal soft tissue. MR-specific features are hypointense marrow in adjacent vertebrae.[25–28] On $T_1$WI, intraosseous, extradural, and paraspinal abscesses are hypointense; on $T_2$WI, marrow, disc, soft tissue infections appear hyperintense; on STIR, marrow, disc, phlegmon, and abscess are hyperintense.[29–31] The acquisition of postcontrast agent highlights marrow, subligamentous, discal, and dural contrast enhancement; the contrast uptake may be diffused with involvement of the soft tissue.[32–34]

Moreover, cord displacement or compression from epidural abscess can be observed. The main differential diagnosis is with pyogenic spondylitis, fungal spondylitis, and spinal metastases.

## 9.6 Facet Joint Septic Arthritis

The pyogenic facet joint infection is a suppurative bacterial infection with involvement of adjacent soft tissue, most frequently caused by *S. aureus*. The lumbar spine is most frequently affected (97%), generally one-sided and single level. It is clinically indistinguishable from spondylodiscitis.

X-ray findings are poor in the early stage and may be negative up to 2–6 weeks after outset of infection. In the later stage, the most X-ray findings are osteolytic and sclerotic facet joint with reduced transparency of surrounding soft tissue. CT examination best characterizes the bone osteolytic/sclerotic changes, spondylolisthesis, and juxta articular, epidural, and paraspinal spreads.

The contrast agent better highlights extraarticular extension, paraspinal phlegmon or abscess, and epidural involvement. MR findings show hyperintensity within facet joint with periarticular soft tissue edema on $T_2$WI, hypointensity within facet joint and ill-defined hypointensity in the facet joint on $T_1$WI, hyperintensity in the facet bone marrow edema and epidural involvement on STIR; after gadolinium administration, on $T_1$WI, fat-saturated images are evident diffuse or rim enhancement within joint, contiguous soft tissue enhancement, epidural or paraspinal extension, and walls of abscess uptake. The contrast agent also allows to evaluate engagement of central canal and neural foramina.

The differential diagnosis is with rheumatoid arthritis, which generally affects the cervical spine with atlantoaxial subluxation and dent erosion associated with synovial hypertrophy and facet joint osteoarthritis, which is characterized by the presence of vacuum phenomenon, osteophytes, symmetric bilaterally, without surrounding soft tissue edema or abscess, and metastasis which generally are multifocal and ill-defined lesion.[35–37]

## 9.7 Dural Abscess: Epidural and Subdural

Spinal epidural abscess is an extradural infection with abscess formation (spinal dural empyema). It is generally associated with spondylodiscitis for suppurative processes of epidural space, most frequently localized in the epidural space (80%), although at the sacral level, it can have a circumferential distribution.

X-ray findings are nonspecific and indirect signs such as endplate erosion, vertebral height loss, and scalloping of the posterior somatic wall. CT with contrast agent evaluation points out enhancing epidural mass with central canal narrowing, vertebral collapse, and cortical destruction. At MRI findings, the areas of abnormal signal are iso- to hypointense to cord on $T_1$WI and hyperintense on $T_2$WI and STIR; gadolinium highlights homogeneous enhancing phlegmon and peripherally enhancing necrotic abscess; signal cord alteration can be observed in case of cord compression, direct ischemia, and direct infection.[38–41]

Spinal subdural abscess is a collection of pus in space between the dura and the arachnoid, which frequently involves the thoracolumbar region with multilevel extension. The X-ray study is not sensible and specific. CT highlights increased subdural density and the contrast agent highlights homogeneous subdural enhancing or peripherally enhancing fluid collection. MR examination shows signal alteration and intermediate signal intensity on $T_1$WI, and hyperintense signal on $T_2$WI and STIR with cord displacement; evaluation with gadolinium highlights heterogeneous and diffuse subdural enhancement, rim enhancement of fluid collection, and epidural fat enhancement.

## 9.8 Paraspinal Abscess

Paraspinal abscess is the infection of soft tissues surrounding the spine; the involvement of the paraspinal plans is generally due to suppuration from direct extension or by hematogenous dissemination of the infectious process. The paravertebral space and paravertebral soft tissues (psoas, iliacus, and posterior paraspinous muscle) are involved with subligamentous or muscle plane extension. X-ray common findings are paraspinal soft tissue reduced diaphany, enlarged psoas shadow, endplate osteolysis, and vertebral collapse.

CT highlights hypodensity of paraspinal soft tissue with gas into the abscess and calcification in case of tuberculous paraspinal abscess, spinal deformity, and endplate destruction; contrast agents highlight diffuse or peripheral enhancement and disc space enhancement. The MR examination shows focal hypointense signal on $T_1$WI and hyperintense signal on $T_2$WI and STIR, with diffuse uptake of gadolinium or peripheral enhancement. MRI allows to detect intraspinal extension with cord compression.

Differential diagnosis is with retroperitoneal hematoma, which shows fluid–fluid level, hyperdensity on CT, and hematic signal on MRI with mild contrast enhancement.[42–45]

## 9.9 Myelitis

The spinal cord is characterized by a relatively poor blood supply, so rarely it is involved in infectious processes from other locations. Myelitis is usually caused by viral agents, in particular herpes simplex and herpes zoster and human immunodeficiency virus (HIV).

Diagnosis of nature can be made by isolating the agent from the cerebrospinal liquor. The distribution of the lesions depends on neurotropism elective of each etiological agent: in the case of herpes zoster, alterations are charged to the posterior horn of the gray matter and white matter adjacent to the central level of the column back and sides; in the case of herpes simplex, there is bone marrow involvement with diffuse hemorrhagic–necrotic components.

### 9.9.1 Viral Myelitis

Viral myelitis is an acute inflammatory insult of the spinal cord due to direct viral infection or postviral immunologic attack. It is characterized by thickening and edema of the spinal cord, with involvement of more contiguous tracts. Cervical and thoracic spinal tracts are the most affected.

CT study is not sensitive and specific. MR highlights an expanded cord that fills canal on $T_1$WI, diffuse increase in signal intensity through involved segment on $T_2$WI, and variable non-focal enhancement of involved cord segment on contrast-enhanced $T_1$WI.

Differential diagnosis is with idiopathic transverse myelitis, with acute disseminated encephalomyelitis (ADEM), multiple sclerosis, and optic neuromyelitis.

### 9.9.2 HIV Myelopathy

The acquired immunodeficiency syndrome patients may develop a myelopathy with a prevalence that varies from 20% to 50% of cases depending on the series; the onset occurs late in relation to other complications and may be secondary to metabolic disorders (deficit vitamin $B_{12}$) and especially to opportunistic infections with bone marrow involvement.

There is also a direct HIV lesion with demyelination of the lateral cordons, with axonal degeneration. HIV myelopathy results from primary HIV infection, localized to the thoracic spinal tract more frequently than cervical tract.

MR findings are spinal cord focal atrophy or normal signal on $T_1$WI, hyperintense either diffusely or involving white matter tracts laterally and symmetrically on $T_2$WI, and foci of hyperintensity on STIR. In severe cases, the signs of vacuolar myelopathy in the posterior and lateral cordons are present.

Differential diagnosis is with varicella zoster virus, cytomegalovirus myelitis, transverse myelitis, and lymphoma.

## 9.10 Myelitis Spinal Cord Abscess

Myelitis cord abscess is a rare condition, characterized by an infectious process of spinal cord with necrosis and intramedullary abscess; pyogenic infection is the most common form. The abscesses are formed in response to bacterial embolism in areas of venous infarction, which is followed by a process of colonization.

Majority of cases in adults are idiopathic and cryptogenic, but in children they may be secondary to open dysraphism. Abscess can be of variable size, usually less than 2 cm, and oval in shape with associated edema.

At MR evaluation, the lesion shows ill-defined hypointense signal from the expanded cord, whereas abscess cord may show focal low signal on $T_1$WI; $T_2$WI demonstrates increased signal of the abscess core and surrounding edema with cord expansion, whereas diffusion-weighted imaging (DWI) may show positive diffusion (reduced apparent diffusion coefficient). The evaluation with gadolinium highlights irregular ring enhancement of lesion with cord enlargement.[46–49]

Differential diagnosis is with acute transverse myelitis, acute viral myelitis, and multiple sclerosis.

## 9.11 Spinal Meningitis

The infection of spinal cord leptomeninges and subarachnoid space is almost always bacterial in acute form, whereas in chronic forms, it is generally due to tuberculous and fungal pathogens.

CT examination documents the increase in the density of the cerebrospinal fluid (CSF) and meningeal enhancement.[50–53] MRI evaluation is more sensitive and specific; it points out increased CSF intensity on $T_1$WI, but on $T_2$WI hyperintense cord signal, nodular defects in subarachnoid space, and focal or diffuse cord swelling; the contrast agent documents nodular meningeal enhancement, homogeneously CSF enhancement, smooth nerve root enhancement, and segmental versus focal intramedullary enhancement.

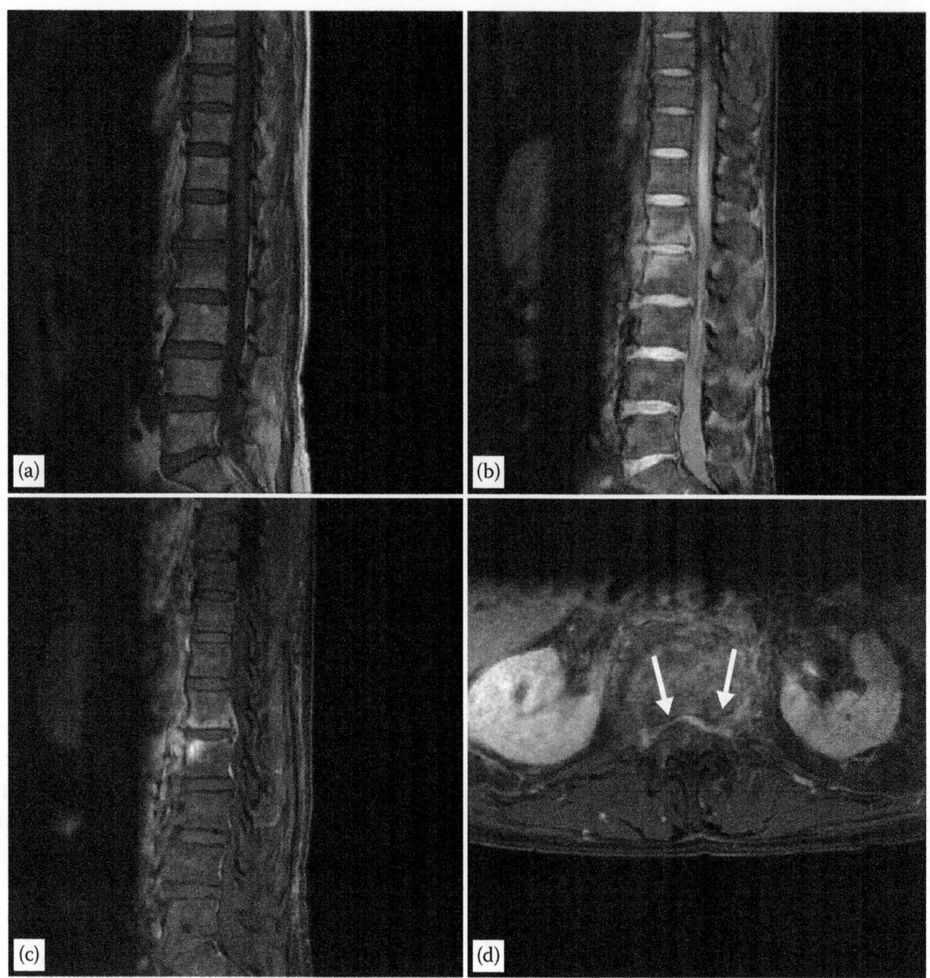


**FIGURE 9.2**
A patient with infective spondylodiscitis, complicated by epidural space involvement and osteomyelitis (inside of the circle). At the L1–L2 level, the lower part of the body of L1 and the upper part of the body of L2 appear hypointense on $T_1$W precontrast sagittal sequences (a), $T_2$W sagittal sequences (b), and $T_1$W fat-saturated postcontrast sagittal sequences (c) as for osteomyelitis. The involvement of the anterior epidural space is more evident on $T_1$W fat-saturated axial sequences after intravenous gadolinium-based contrast medium administration (d, see the arrows).

The associated anomalies are spondylodiscitis (Figures 9.2 and 9.3), spinal epidural abscess, myelitis, and syringomyelia.

Differential diagnosis is with carcinomatous meningitis, sarcoidosis, lumbar arachnoiditis, and Guillain–Barré syndrome.

## 9.12 Spinal Parasitosis

The main forms of parasitic infections are supported by tapeworm of genus *Echinococcus*, trematodes of genus *Schistosoma*, and *Taenia solium*.

### 9.12.1 Echinococcosis

Bone involvement is rare in course of parasitic infections; the vertebrae are the most frequent bone localization, especially in the thoracic segments. Process disease can be extended to the ribs and paraspinal tissues; sometimes it may be associated with the compression of cord.

X-ray findings are poorly specific as multiloculated osteolytic lesions with a "bunch-of-grapes" morphology. CT better defines multiloculated osteolytic lesions in vertebral body and spinal mass associated with minimal enhancement of vertebral body cysts and paraspinal tissues. MRI findings are hyperintense multiseptated/multicystic lesions on $T_1$WI and hyperintense on $T_2$WI and STIR with minimal gadolinium uptake of large cysts.

Differential diagnosis is with primary bone neoplasm, metastatic disease, and TB.

### 9.12.2 Schistosomiasis

The central nervous system (CNS) involvement is rare and may occur at any stage of disease; the damaged area is the CNS, with necrotic thoracic myelopathy.

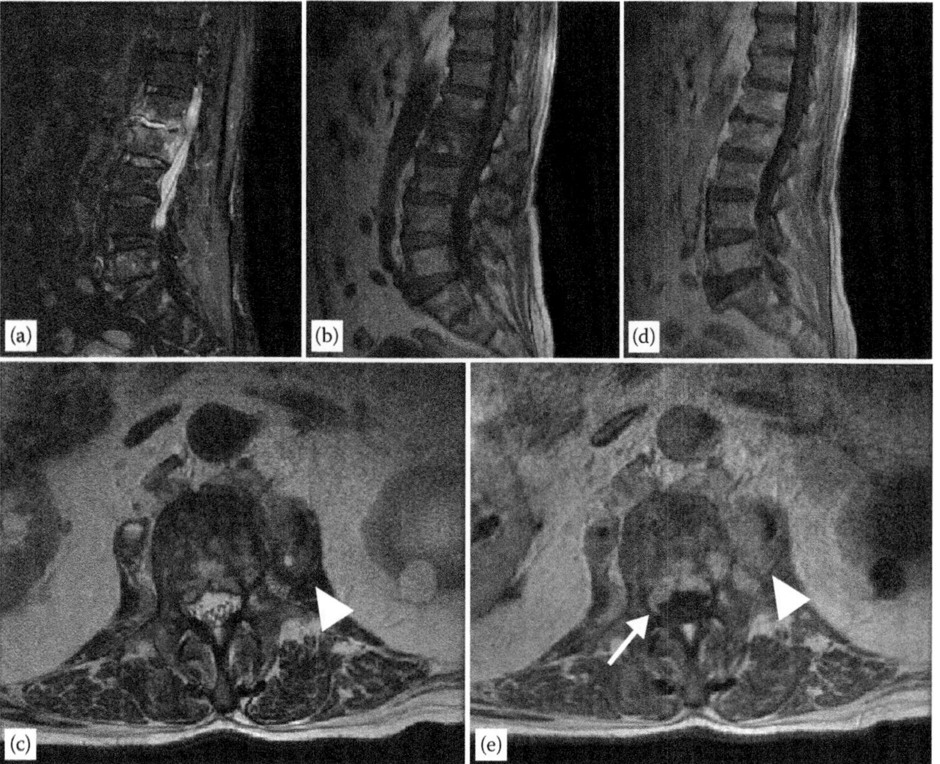

FIGURE 9.3
A patient with infective spondylodiscitis, complicated by epidural involvement, osteomyelitis, and psoas abscess. At the L1–L2 level, the condition of spondylodiscitis characterized by the high signal of the intervertebral disk and of the body of L1 and L2 is quite evident (inside of the circle) on $T_2$W fat-saturated sagittal sequences (a); the vertebral bodies enhance on $T_1$W sequences after the intravenous administration of gadolinium-based contrast medium (d) compared with the $T_1$W precontrast sequences (b). On $T_1$W postcontrast axial sequences (e), the involvement of the anterior epidural space (arrow) is more evident, and the presence of a homogeneous rounded area (arrowhead) located in the left psoas muscle, inhomogeneous, is evident on $T_2$W axial sequences (c), to be referred to an abscess.

MR is the most sensitive examination, but the MRI findings are nonspecific; the enlarged cord and enlarged conus medullaris are the main findings on $T_1$WI, diffuse increased signal within cord over several segments is evident on $T_2$WI, single or multifocal and heterogeneous cord enhancement is evident on postcontrast $T_1$WI.

Differential diagnosis is with multiple sclerosis, intramedullary neoplasm, pyogenic abscess, viral or idiopathic transverse myelitis, ADEM, and TB.

### 9.12.3 Cysticercosis

Neurocysticercosis is intramedullary spinal cord cysticercosis, caused by *Taenia solium*. Cysticercosis spinal involvement is uncommon; extraspinal (involvement of vertebral body) and intraspinal (involvement of extradural, subarachnoid, and intramedullary) are the localizations. Spinal injuries most frequently found are the subarachnoid cysts, foci of arachnoiditis, and less frequently intramedullary cysts.

The MRI findings are nonspecific. MRI shows subarachnoid signal cyst lesions with variable mass effect on the adjacent cord, focal intramedullary cystic lesions, or syrinx cavitation; hyperintense on $T_1$WI; hyperintense on $T_2$WI; peripheral cyst; and subarachnoid space enhancement. MR aspects of arachnoiditis are indistinguishable from those of arachnoiditis to other etiology.

Differential diagnosis is with pyogenic abscess, granulomatous disease, arachnoid cyst, syrinx, and schwannoma.

## 9.13 Inflammatory and Autoimmune Spine Diseases

### 9.13.1 Multiple Sclerosis

It is an autoimmune, cell-mediated inflammatory process focused on CNS myelin. Spinal cord multiple sclerosis is a primary demyelinating disease of the CNS characterized by multiple lesions disseminated over time and space. Spinal cord localizations are associated with intracranial lesions in periventricular, subcallosal, brainstem or cerebellar white matter.

Cervical tract is the most commonly affect spinal cord segment. In 10%–20% of cases, the disease is isolated to the spinal cord; the demyelinating plaques, with common involvement of dorsal horns, do not respect gray–white boundary and show a wedge-shaped region on axial MRI, with size generally less than half cross-sectional area of the spinal cord and less than two vertebral segments in length. The distribution of plaques does not correspond to vascular territories.

MRI (Figure 9.4) shows iso- to hypointense lesions on $T_1$WI; unlike the brain, the lesions in the cord are rarely visible as hypointense on $T_1$WI. The lesions are well-circumscribed hyperintense areas on $T_2$WI, when demyelination is complete, whereas they are ill-defined mildly hyperintense areas when demyelination is partial. Chronic lesions show gliosis and cavitation.

STIR sequence improves lesion detection; DWI sequence shows the increase of mean diffusivity. During acute or subacute phase after gadolinium administration, solitary or multifocal lesions show homogeneous, nodular, or ring enhancement, which lasts 1–2 months, for alterations of the blood–brain barrier. Contrast enhancement is not present during chronic phase. Other findings are mild focal cord expansion in the case of cord edema (pseudotumor form) and cord atrophy in late stage.

Intramedullary neoplasm, idiopathic transverse myelitis, and syringohydromyelia are some diseases that come into differential diagnosis, especially when the spinal cord injury is unique; in this case, the study of the brain is recommended.

### 9.13.2 Acute Disseminated Encephalomyelitis Spinal Cord

ADEM is an immune-mediated disorder pre–post infections of the white matter, localized anywhere in the spinal cord with brain involvement. It is characterized by multifocal white matter lesions with relatively little mass effect or vasogenic edema; the lesions size is punctate to segmental distribution.

MRI findings are focal low signal and slight cord swelling on $T_1$WI; multifocal flame-shaped white matter lesions are appreciable with slight cord swelling on $T_2$WI and may show gray matter involvement. With gadolinium, the enhancement is variable, depending on the stage of disease; can be punctate, ring-shaped, or fluffy enhancement; and may show nerve enhancement. Radiological findings can mimic cord neoplasm.[54–61]

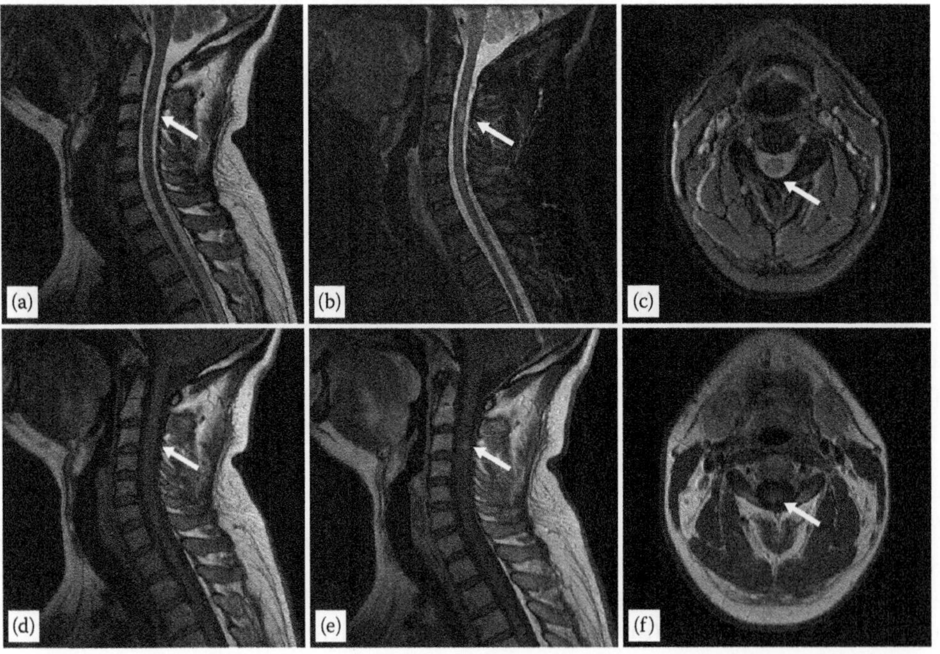


**FIGURE 9.4**
A patient with single multiple sclerosis plaque located in the spinal cord at the level of the body of the third cervical vertebrae. On $T_2$W images on the sagittal plane (a), the presence of a hyperintense oval-shaped swelling lesion, with vanish margins, inside of the spinal cord, is evident. The lesion is more evident on $T_2$W fat-saturated images on the sagittal (b) and axial (c) views, and not well evident on $T_1$W sagittal sequences (d). After intravenous gadolinium-based contrast medium administration on sagittal (e) and axial (f) sequences, the lesion slightly enhances.

Differential diagnosis is with multiple sclerosis, immune-mediated vasculitis, viral or idiopathic myelitis, and cord infarct.

### 9.13.3 Idiopathic Acute Transverse Myelitis

Idiopathic acute transverse myelitis is an inflammatory disorder, characterized by perivascular inflammation and demyelination. Association with autoimmune phenomenon and previous viral infection or vaccination is possible in some cases; it is associated with demyelinating process.

The disease involves both halves of the spinal cord resulting in bilateral motor, sensory, and autonomic dysfunction, characterized by central cord lesion extended more than two vertebral segments in length. The size of the lesions is more than two-thirds of the cross-sectional area of the cord on axial imaging, with more than two vertebral segments in length, commonly three to four segments. The thoracic portion is more commonly involved.

On MRI, smooth cord expansion and iso- to hypointense solitary or multifocal lesions are evident on $T_1$WI, which are less extensive than $T_2$ signal abnormality. On $T_2$WI, hyperintense focal central area surrounded by edema is evident. Post-gadolinium enhancement is variable and more frequent in subacute than in acute or chronic stage. Meningeal uptakes of gadolinium can be associated. Lesion's enhancement is not predictive of clinical course.

Spinal cord neoplasm, multiple sclerosis, syringohydromyelia, and cord infarct are the main differential diagnoses.

## 9.14 Sarcoidosis

Neurosarcoidosis is a noncaseating granulomatous disease of the spine and spinal cord, in course of systemic disease.

CT evaluation highlights multiple lytic lesions with sclerotic margins in the spine; sclerotic or mixed lytic and sclerotic lesions at the same time are also possible. MR shows fusiform cord enlargement, iso- to hypointense lesions, and cord atrophy in late stages on $T_1$WI, and focal or diffuse spinal cord hyperintensity on $T_2$WI and on STIR; after administration of the contrast agent, smooth or nodular leptomeningeal enhancement and intramedullary mass-like enhancement (focal or multifocal) are appreciated.

Radiological differential diagnosis is with intramedullary neoplasm, multiple sclerosis, spinal cord ischemia and infarct, and idiopathic transverse myelitis.

## 9.15 Chronic Inflammatory Demyelinating Polyneuropathy

It is a condition characterized by chronic acquired, immune-mediated demyelinating neuropathy, with relapsing or progressive muscle weakness and sensory loss. This neuropathy affects primarily spinal nerve and proximal nerve trunks. It involves both cellular and humoral immune factors. Peripheral nerves and spinal nerve roots may be affected simultaneously; lumbar tract is most frequently affected with thickening of the nerve roots.

MR characteristic findings are enlargement and abnormal hyperintensity of cauda equina and proximal peripheral nerves on $T_2$WI and isointense enlargement of cauda equina and proximal peripheral nerves on $T_1$WI, with mild or moderate nerve enhancement.

Differential diagnosis is with Guillain–Barré, neurofibromatosis type 1, and inherited demyelinating neuropathy.

### 9.15.1 Polyradiculoneuropathy

#### *9.15.1.1 Dejerine–Sottas Syndrome*

It is a hereditary radiculoneuropathy that manifests in childhood or during adolescence associated with hypertrophy of the popliteal nerves; ulnar, median, and radial nerves; and back of the neck, which is well appreciable on palpation. For the demonstration of the MR imaging of polyradiculoneuropathy, the administration of contrast medium for a vein is required. On $T_1$WI, intradural root impregnation of variable degree, associated with swelling and dorsal root ganglia, is evident.

#### *9.15.1.2 Guillain–Barré Syndrome*

It is a disorder characterized by acute inflammatory demyelinating polyradiculoneuropathy (cranial nerves, peripheral nerves, and nerve roots) with immune activation after infection or vaccination and autoimmune or viral etiology. Both cell-mediated and humoral mechanisms are involved in pathogenesis.

The main MR findings, highlighted after administration of gadolinium, are avid enhancement of cauda equina, slightly thickened roots, with fusiform nerve enlargement, and pial surface enhanced on $T_1$WI post-gadolinium; normal conus should be seen and slight prominence of root thickness may be seen on $T_2$WI.

Radiological differential diagnosis is with physiological nerve root enhancement, vasculitic neuropathy, acute transverse myelitis, and chronic polyneuropathies.

## 9.16 Chronic Adhesive Arachnoiditis

It is a rare pathological condition characterized by postinflammatory adhesion and agglomeration of nerve roots, in the lumbar spine and cauda equina. The etiology is typically characterized by diffuse meningeal inflammation or post-lumbar surgery with "fibrin exude, fibrous septa" and adhesion of the nerve roots in the dural sac.

Intradural calcifications may be associated, especially in posttraumatic forms, in case of subarachnoid hemorrhage. The nerve roots are plastered against periphery of thecal sac, with multiple adhesions compartmentalizing of the arachnoid space.

CT shows calcific densities within clumped nerve roots; the calcifications may surround conus medullaris and cauda equina. MRI on $T_2$WI highlights discrete reduction of nerve roots, clumped nerve roots forming cords: on type 1 centrally and type 2 peripherally, and thickened thecal sac wall. When nerve roots adhere to the walls of thecal sac, there is "empty thecal sac sign" and the CSF appears without nerve roots.

The thickening of the thecal sac wall causes reduction of the sac diameter, union of nerve roots, and obliteration of the subarachnoid space. Nerve root and dural enhancement is minimal to mild, with linear, nodular, and intradural mass-like enhancement. In the cervical and thoracic tracts, the diagnosis of arachnoiditis is more difficult for the smaller number and length of the nerve roots.

Differential diagnosis is with spinal stenosis, cauda equina neoplasms, carcinomatous meningitis, intradural metastases, and post-actinic myelitis.

### 9.16.1 Spinal Cord Bechet's Disease

The alterations in the CNS include the areas of demyelination, glial proliferation, and Wallerian degeneration . The commitment of the CNS is rare at onset and appears in 10%–50% of patients in the course of disease. The alterations that can be highlighted with MRI are nonspecific. Variable single and multiple areas of hypersignal are seen on $T_2$W images; not corresponding to vascular territories, injuries are poorly detectable on $T_1$WI.[62–64] The enhancement after gadolinium is variable in relation to the phase of the disease in the treatment with corticosteroids.

### 9.16.2 Spinal Cord: Systemic Lupus Erythematosus

Neurological complications from this systemic pathology are possible in about 20%–50% of patients; the lupus myelitis is a rare occurrence and usually appears after several years of diagnosis of the disease. In the course of systemic lupus erythematosus, the areas of medullary infarction and paintings of compressive myelopathy from subdural hematoma secondary to coagulopathy lupus can also be highlighted. The exhibits that can be highlighted with MRI are nonspecific. Variable single and multiple areas of hypersignal are seen on $T_2$W images; not corresponding to vascular territories, injuries are poorly detectable on $T_1$WI. The enhancement after gadolinium is variable in relation to the phase of the disease in the treatment with corticosteroids.

### 9.16.3 Devic's Syndrome or Devic's Disease

Devic's syndrome is a condition characterized by the presence of demyelinating lesions in the spinal cord cervical, brain, and optic nerve. Typically, it occurs in the course of multiple sclerosis or systemic lupus erythematosus. Devic's disease is a rare progressive disease, of unknown etiology, which presents itself as a kind of progressive necrotizing myelitis, associated with optic neuritis, without brain involvement.

On MRI, the alterations of the spinal cord are the increase in volume of the spinal cord, extended for long stretches, especially in the cervical and upper thoracic spine. Cavitated lesions are isointense to CSF on $T_1$WI and $T_2$WI; after gadolinium, lesions show variable enhancement, no enhancement, or ring enhancement. For diagnosis, it is useful to extend the examination to the brain and orbits.

## References

1. Coqueugniot H, Dutailly B, Desbarats P, Boulestin B, Pap I, Szikossy I, Baker O et al. Three-dimensional imaging of past skeletal TB: From lesion to process. *Tuberculosis (Edinb).* 2015 Jun;95 Suppl 1:S73–9.
2. Hamilton SM, Bayer CR, Stevens DL, Lieber RL, Bryant AE. Muscle injury, vimentin expression, and nonsteroidal anti-inflammatory drugs predispose to cryptic group A streptococcal necrotizing infection. *J Infect Dis.* 2008 Dec;198(11):1692–8.
3. Xie S, Sun T, Tian R, Xu T, Jia Y, Shen Q. Analysis of risk factors of axial symptoms after single door laminoplasty for cervical myelopathy. *Zhongguo Xiu Fu Chong Jian Wai Ke Za Zhi.* 2014 May;28(5):620–4.
4. Nardone R, Tezzon F, Lochner P, Trinka E, Brigo F. Atlanto-axial erosion as the presenting manifestation of systemic tuberculosis: A tricky diagnosis in western countries. *Acta Neurol Belg.* 2014 Dec;114(4):321–3.
5. Jeong SJ, Choi SW, Youm JY, Kim HW, Ha HG, Yi JS. Microbiology and epidemiology of infectious spinal disease. *J Korean Neurosurg Soc.* 2014 Jul;56(1):21–7.

6. Fantoni M, Trecarichi EM, Rossi B, Mazzotta V, Di Giacomo G, Nasto LA, Di Meco E, Pola E. Epidemiological and clinical features of pyogenic spondylodiscitis. *Eur Rev Med Pharmacol Sci.* 2012 Apr;16 Suppl 2:2–7.
7. Andrews JA, Rizzato Lede D, Senderovsky M, Finn BC, Emery N, Bottaro F, Bruetman JE, Young P. Septic arthritis of the pubic symphysis in two athletes. *Medicina (B Aires).* 2012;72(3):247–50.
8. Hagen R. Osteomyelitis after operative fracture treatment. A report of 62 cases treated with radical surgery and lincomycin (Lincocin). *Acta Orthop Scand.* 1978 Dec;49(6):542–8.
9. Kang SJ, Jang HC, Jung SI, Choe PG, Park WB, Kim CJ, Song KH. Clinical characteristics and risk factors of pyogenic spondylitis caused by gram-negative bacteria. *PLoS ONE.* 2015 May;10(5):e0127126.
10. Yang SC, Fu TS, Chen HS, Kao YH, Yu SW, Tu YK. Minimally invasive endoscopic treatment for lumbar infectious spondylitis: A retrospective study in a tertiary referral center. *BMC Musculoskelet Disord.* 2014 Mar;15:105. doi: 10.1186/1471-2474-15-105.
11. Sugrue PA, O'Shaughnessy BA, Nasr F, Koski TR, Ondra SL. Abdominal complications following kyphosis correction in ankylosing spondylitis. *J Neurosurg Spine.* 2009 Feb;10(2):154–9.
12. Mückley T, Schütz T, Kirschner M, Potulski M, Hofmann G, Bühren V. Psoas abscess: The spine as a primary source of infection. *Spine (Phila Pa 1976).* 2003 Mar;28(6):E106–13.
13. Mavrogenis AF, Igoumenou V, Tsiavos K, Megaloikonomos P, Panagopoulos GN, Vottis C, Giannitsioti E, Papadopoulos A, Soultanis KC. When and how to operate on spondylodiscitis: A report of 13 patients. *Eur J Orthop Surg Traumatol.* 2015 Jul 20. [Epub ahead of print].
14. Esendagli-Yilmaz G, Uluoglu O. Pathologic basis of pyogenic, nonpyogenic, and other spondylitis and discitis. *Neuroimaging Clin N Am.* 2015 May;25(2):159–61.
15. Skoura E, Zumla A, Bomanji J. Imaging in tuberculosis. *Int J Infect Dis.* 2015 Mar;32:87–93.
16. Wu SY, Wei TS, Chen YC, Huang SW. Vertebral osteomyelitis complicated by iliopsoas muscle abscess in an immunocompetent adolescent: Successful conservative treatment. *Orthopedics.* 2012 Oct;35(10):e1576–80.
17. Lee KS, Kong S, Kim J, Kim T, Choi CB, Kim YS, Lee KH. Osteomyelitis of bilateral femoral heads after childbirth: A case report. *Ann Rehabil Med.* 2015 Jun;39(3):498–503. doi:10.5535/arm.2015.39.3.498.
18. Alexiou E, Georgoulias P, Valotassiou V, Georgiou E, Fezoulidis I, Vlychou M. Multifocal septic osteomyelitis mimicking skeletal metastatic disease in a patient with prostate cancer. *Hell J Nucl Med.* 2015 Jan–Apr;18(1):77–8. doi:10.1967/s002449910168.
19. Leone A, Dell'Atti C, Magarelli N, Colelli P, Balanika A, Casale R, Bonomo L. Imaging of spondylodiscitis. *Eur Rev Med Pharmacol Sci.* 2012 Apr;16 Suppl 2:8–19.
20. Kowalski TJ, Layton KF, Berbari EF, Steckelberg JM, Huddleston PM, Wald JT, Osmon DR. Follow-up MR imaging in patients with pyogenic spine infections. lack of correlation with clinical features. *AJNR Am J Neuroradiol.* 2007 Apr;28(4):693–9.
21. Abe E, Yan K, Okada K. Pyogenic vertebral osteomyelitis presenting as single spinal compression fracture: A case report and review of the literature. *Spinal Cord.* 2000 Oct;38(10):639–44.
22. Shih TT, Huang KM, Hou SM. Early diagnosis of single segment vertebral osteomyelitis—MR pattern and its characteristics. *Clin Imaging.* 1999 May–Jun;23(3):159–67.
23. Friedmand DP, Hills JR. Cervical epidural spinal infection: MR imaging characteristics. *AJR Am J Roentgenol.* 1994 Sep;163(3):699–704.
24. Smith AS, Weinstein MA, Mizushima A, Coughlin B, Hayden SP, Lakin MM, Lanzieri CF. MR imaging characteristics of tuberculous spondylitis vs vertebral osteomyelitis. *AJR Am J Roentgenol.* 1989 Aug;153(2):399–405.
25. Kilborn T, Janse van Rensburg P, Candy S. Pediatric and adult spinal tuberculosis: Imaging and pathophysiology. *Neuroimaging Clin N Am.* 2015 May;25(2):209–31. doi:10.1016/j.nic.2015.01.002.
26. Tali ET, Oner AY, Koc AM. Pyogenic spinal infections. *Neuroimaging Clin N Am.* 2015 May;25(2):193–208. doi:10.1016/j.nic.2015.01.003.
27. Lang N, Su MY, Yu HJ, Yuan H. Differentiation of tuberculosis and metastatic cancer in the spine using dynamic contrast-enhanced MRI. *Eur Spine J.* 2015 Aug;24(8): 1729–37. doi:10.1007/s00586-015-3851-z.
28. Thammaroj J, Kitkhuandee A, Sawanyawisuth K, Chowchuan P, Promon K. MR findings in spinal tuberculosis in an endemic country. *J Med Imaging Radiat Oncol.* 2014;58(3):267–76.
29. Park JH, Shin HS, Park JT, Kim TY, Eom KS. Differentiation between tuberculous spondylitis and pyogenic spondylitis on MR imaging. *Korean J Spine.* 2011 Dec;8(4):283–7. doi:10.14245/kjs.2011.8.4.283.
30. Jain AK, Sreenivasan R, Saini NS, Kumar S, Jain S, Dhammi IK. Magnetic resonance evaluation of tubercular lesion in spine. *Int Orthop.* 2012 Feb;36(2):261–9. doi:10.1007/s00264-011-1380-x.
31. Zaidi H, Akram MH, Wala MS. Frequency and magnetic resonance imaging patterns of tuberculous spondylitis lesions in adults. *J Coll Physicians Surg Pak.* 2010 May;20(5):303–6. doi:05.2010/JCPSP.303306.
32. Pieri S, Agresti P, Altieri AM, Ialongo P, Cortese A, Alma MG, de' Medici L. Percutaneous management of complications of tuberculous spondylodiscitis: Short- to medium-term results. *Radiol Med.* 2009 Sep;114(6):984–95. doi:10.1007/s11547-009-0425-3.
33. Kumar R, Das RK, Mahapatra AK. Role of interferon gamma release assay in the diagnosis of Pott disease. *J Neurosurg Spine.* 2010 May;12(5):462–6. doi:10.3171/2009.10.SPINE093.
34. Anik Y, Ciftçi E, Sarisoy HT, Akansel G, Demirci A, Anik I, Buluç L, Ilgazli A. MR spectroscopy findings in tuberculous spondylitis; comparison with Modic type-I end-plate changes and metastatic vertebral disease. *Eur J Radiol.* 2009 Aug;71(2):324–32. doi:10.1016/j.ejrad.2008.05.002.
35. Mas-Atance J, Gil-García MI, Jover-Sáenz A, Curià-Jové E, Jové-Talavera R, Charlez-Marco A, Ibars-Valverde Z, Fernández-Martínez JJ. Septic arthritis of a posterior

lumbar facet joint in an infant: A case report. *Spine (Phila Pa 1976)*. 2009 Jun 1;34(13):E465–8. doi:10.1097/BRS.0b013e3181a4e64b.
36. Ram S, Osman A, Cassar-Pullicino VN, Short DJ, Masry WE. Spinal cord infarction secondary to intervertebral foraminal disease. *Spinal Cord*. 2004 Aug;42(8):481–4.
37. Fujiwara A, Tamai K, Yamato M, Yoshida H, Saotome K. Septic arthritis of a lumbar facet joint: Report of a case with early MRI findings. *J Spinal Disord*. 1998 Oct;11(5):452–3.
38. Shen WC, Lee SK. Chronic osteomyelitis with epidural abscess: CT and MR findings. *J Comput Assist Tomogr*. 1991 Sep–Oct;15(5):839–41.
39. Numaguchi Y, Rigamonti D, Rothman MI, Sato S, Mihara F, Sadato N. Spinal epidural abscess: Evaluation with gadolinium-enhanced MR imaging. *RadioGraphics*. 1993 May;13(3):545–59; discussion 559-60.
40. Jacques C, Boukobza M, Polivka M, Ferrario A, George B, Merland JJ. Cranial epidural tuberculoma. A case report. *Acta Radiol*. 2000 Jul;41(4):367–70.
41. Gabelmann A, Klein S, Kern W, Krüger S, Brambs HJ, Rieber-Brambs A, Pauls S. Relevant imaging findings of cerebral aspergillosis on MRI: A retrospective case-based study in immunocompromised patients. *Eur J Neurol*. 2007 May;14(5):548–55.
42. Holloway A, Dennis R, McConnell F, Herrtage M. Magnetic resonance imaging features of paraspinal infection in the dog and cat. *Vet Radiol Ultrasound*. 2009 May–Jun;50(3):285–91.
43. Harada Y, Tokuda O, Matsunaga N. Magnetic resonance imaging characteristics of tuberculous spondylitis vs. pyogenic spondylitis. *Clin Imaging*. 2008 Jul–Aug;32(4):303–9. doi:10.1016/j.clinimag.2007.03.015.
44. Ng AW, Chu WC, Ng BK, Li AM. Extensive paraspinal abscess complicating tuberculous spondylitis in an adolescent with Pott kyphosis. *Clin Imaging*. 2005 Sep–Oct;29(5):359–61.
45. Kowalski TJ, Layton KF, Berbari EF, Steckelberg JM, Huddleston PM, Wald JT, Osmon DR. Follow-up MR imaging in patients with pyogenic spine infections: Lack of correlation with clinical features. *AJNR Am J Neuroradiol*. 2007 Apr;28(4):693–9.
46. DeSanto J, Ross JS. Spine infection/inflammation. *Radiol Clin N Am*. 2011 Jan;49(1):105–27. doi:10.1016/ j.rcl.2010.07.018.
47. Murphy KJ, Brunberg JA, Quint DJ, Kazanjian PH. Spinal cord infection: Myelitis and abscess formation. *AJNR Am J Neuroradiol*. 1998 Feb;19(2):341–8.
48. Friess HM, Wasenko JJ. MR of staphylococcal myelitis of the cervical spinal cord. *AJNR Am J Neuroradiol*. 1997 Mar;18(3):455–8.
49. Smith AS, Blaser SI. MR of infectious and inflammatory diseases of the spine. *Crit Rev Diagn Imaging*. 1991;32(3):165–89.
50. Tali ET, Oner AY, Koc AM. Pyogenic spinal infections. *Neuroimaging Clin N Am*. 2015 May;25(2):193–208. doi:10.1016/j.nic.2015.01.003.
51. Lummel N, Koch M, Klein M, Pfister HW, Brückmann H, Linn J. Spectrum and prevalence of pathological intracranial magnetic resonance imaging findings in acute bacterial meningitis. *Clin Neuroradiol*. 2014 Sep.
52. Ishizaka S, Hayashi K, Otsuka M, Fukuda S, Tsunoda K, Ushijima R, Kitagawa N, Suyama K, Nagata I. Syringomyelia and arachnoid cysts associated with spinal arachnoiditis following subarachnoid hemorrhage. *Neurol Med Chir (Tokyo)*. 2012;52(9):686–90.
53. Ranasinghe MG, Zalatimo O, Rizk E, Specht CS, Reiter GT, Harbaugh RE, Sheehan J. Idiopathic hypertrophic spinal pachymeningitis. *J Neurosurg Spine*. 2011 Aug;15(2):195–201. doi:10.3171/2011.4.SPINE1037.
54. Lim CC. Neuroimaging in postinfectious demyelination and nutritional disorders of the central nervous system. *Neuroimaging Clin N Am*. 2011 Nov;21(4):843–58, viii. doi:10.1016/j.nic.2011.08.001.
55. Cañellas AR, Gols AR, Izquierdo JR, Subirana MT, Gairin XM. Idiopathic inflammatory-demyelinating diseases of the central nervous system. *Neuroradiology*. 2007 May;49(5):393–409.
56. Singh S, Prabhakar S, Korah IP, Warade SS, Alexander M. Acute disseminated encephalomyelitis and multiple sclerosis: Magnetic resonance imaging differentiation. *Australas Radiol*. 2000 Nov;44(4):404–11.
57. Singh S, Alexander M, Korah IP. Acute disseminated encephalomyelitis: MR imaging features. *AJR Am J Roentgenol*. 1999 Oct;173(4):1101–7.
58. Feydy A, Carlier R, Mompoint D, Clair B, Chillet P, Vallee C. Brain and spinal cord MR imaging in a case of acute disseminated encephalomyelitis. *Eur Radiol*. 1997;7(3):415–17.
59. Tanaka Y, Matsuo M. Serial magnetic resonance imaging of acute disseminated encephalomyelitis, including evaluation of the contrast-enhancing effect of lesions by Gd-DTPA. *Nihon Igaku Hoshasen Gakkai Zasshi*. 1996 Jan;56(1):25–31.
60. Hasegawa H, Bitoh S, Koshino K, Obashi J, Iwaisako K, Fukushima Y. Acute relapsing disseminated encephalomyelitis (ARDEM) mimicking a temporal lobe tumor. *No Shinkei Geka*. 1994 Feb;22(2):185–8.
61. Araki Y, Kohmura E, Nakamura H, Tsukaguchi I. MR imaging of acute disseminating encephalomyelitis. *Radiat Med*. 1993 Nov–Dec;11(6):263–6.
62. Horger M, Maksimovic O, Kötter I, Ernemann U. Neuro-Behçet's disease: MR-imaging findings. *Rofo*. 2008 Aug;180(8):691–7. doi:10.1055/s-0028-1082142.
63. Cakirer S. Isolated spinal neurobehçet disease. MR imaging findings. *Acta Radiol*. 2003 Sep;44(5):558–60.
64. Vuolo L, Bonzano L, Roccatagliata C, Parodi RC, Roccatagliata L. Reversibility of brain lesions in a case of Neuro-Behçet's disease studied by MR diffusion. *Neurol Sci*. 2010 Apr;31(2):213–15. doi:10.1007/s10072-009-0205-9.

# 10

## *Traumatic Disease of the Spine*

**Nathan J. Kohler, M. Cody O'Dell, Steven A. Messina, Brian K. Harshman, Christopher W. Wasyliw, Gary Felsberg, and Laura W. Bancroft**

**CONTENTS**



### 10.1 Introduction

The spine is a segmental structure comprised of vertebral bones, which are held in alignment through soft tissues comprised of intervertebral disks, joints, ligaments, and musculature. While the spinal vertebral bodies are largely responsible for maintaining height in the presence of axial load, the buttressing soft tissues of the neck, thorax, abdomen, and pelvis are largely responsible for maintaining alignment in the coronal and sagittal planes and preventing pathologic rotatory changes. The segmental anatomy of the spine together with its supporting structures allows for spinal motion in the confines of physiological spinal stability.

Spinal instability was defined by Panjabi and White in their authoritative text as "the loss of the ability of the spine under physiologic loads to maintain its pattern of displacement so that there is no initial or additional neurological deficit, no major deformity, and no incapacitating pain" [1]. Prior to the mainstream use of magnetic resonance imaging (MRI) and the advent of multi-detector computed tomography (CT) for evaluation of spine trauma, radiography was utilized for the assessment of spinal instability in the presence of trauma. They wrote: "Radiographic examination is the most often used objective means of determining the relative positions of the vertebrae in a potentially unstable spine. Therefore, it is important to give some consideration to the accurate interpretation of linear radiographic measurements."

The first experiments to assess the damage necessary to cause spinal instability were conducted during the 1970s with *ex vivo* experimentation on a series of cadaver spines. Serial controlled defects in the soft tissues of cadaver cervical spines were created and tested under various loading conditions to assess for failure [2,3]. Anterior elements were defined as the posterior longitudinal ligament and all tissues anterior to it. Posterior elements were defined as all structures posterior to the posterior longitudinal ligament. Their studies had a number of groundbreaking findings: (1) Small changes were observed by the sectioning of elements, followed by a sudden complete disruption of the spine. (2) Removal of the facet joint reduces angular displacement and allows more horizontal displacement. (3) Spinal instability occurs when all of the anterior or posterior elements are disrupted or unable to function, in the setting of more than 3.5 mm of horizontal displacement of one vertebral body with respect to an adjacent vertebral body on lateral radiograph, or more than 11° of angular rotation between two adjacent vertebral bodies is seen on neutral view or flexion-extension radiographs.

CT in the assessment of the trauma patient has revolutionized the rapid evaluation of soft tissue and bony abnormalities. CT allows visualization of bony abnormalities and the rapid assessment of gross soft tissue changes including associated hemorrhage. In the setting of acute spinal trauma, CT remains the best modality for detecting bony abnormalities within the cervical, thoracic, and lumbar spine. Furthermore, CT allows a surveyed assessment for secondary findings in the setting of trauma within the head, thorax, abdomen, and pelvis. CT is largely a static imaging modality, and a limited number of studies have demonstrated the utilization of dynamic CT assessment of spinal stability [4]. However, several factors limit the use of dynamic CT imaging including patient condition and positioning. Therefore, it has not gained widespread acceptance for the evaluation of spinal stability in the setting of trauma.

MRI is the only technique that can reliably directly identify intrinsic injury to the spinal cord, ligaments, and soft tissues in the setting of trauma. Histopathological samples have been correlated with MRI signal features, validating imaging findings in the setting of spinal cord, ligamentous, and bony injury. These correlations lead to the acceptance of MRI-based diagnosis in the setting of spinal trauma [5–9]. MRI was validated in the detection of occult cervical spine injury in a study of 174 posttraumatic patients who had a clinical potential for spinal injury in the presence of minimally abnormal or normal radiography [10]. Of the 174 patients evaluated in this study, 62 had soft tissue abnormalities identified by MRI, including 27 with disk interspace disruption (ventral and dorsal ligamentous injury in 4 patients, ventral ligamentous injury in 3 patients, dorsal ligamentous injury alone in 18 patients, and 2 without dorsal or ventralinjury). In addition, there was isolated ligamentous injury observed in 35 patients (8 with ventral and dorsal ligamentous injury, 5 with ventral ligamentous injury alone, and 22 with dorsal ligamentous injury alone). These findings changed management in all 62 patients. It was determined that $T_2$-weighted sagittal images were most useful in the setting of acute soft tissue injury. Further, a negative MRI may be considered as confirmation of a negative or cleared subaxial cervical spine. It has been recommended, in the setting of occult ligamentous or soft tissue injury, that this MRI be performed within 48 h [11].

## 10.2 Spinal Cord Injury

MRI assessment of the neuroaxis in the setting of spinal trauma requires an extended period of time, which can impede the care of the critically injured patient. Therefore, an understanding of patient evaluation and treatment in the setting of trauma is critical. Targeted imaging of the neuroaxis can be based on physical examination findings, which can potentially diagnose and localize spinal cord injury. The presence or absence of spinal cord injury is based on clinical examination findings as outlined by the American Spinal Injury Association (ASIA) and International Standards for Neurological Classification of Spinal Cord Injury [12]. The evaluation is based on motor and sensory criteria, which are used to identify a neurological level and a potential zone of partial preservation. The ASIA impairment scale grades the injury from A (complete) to E (normal). A patient meeting criteria for A has a complete spinal cord injury having no motor or sensory function preserved at S4–S5. B rating indicates an incomplete spinal cord injury where sensory function is maintained, but no motor function is preserved below the level of injury, including S4–S5. C and D ratings are based on the patient's strength below the level of injury. Both patients maintain strength below the level of injury. However, a patient classified as C has motor strength less than 3 below the level of injury. ASIA E indicates that all motor and sensory function is normal and is used primarily in follow up to indicate neurological function has normalized.

The presenting symptoms of patients with spinal cord injury are diverse. A basic understanding of symptomology is helpful in the localization of a suspected traumatic lesion. There are several spinal cord syndromes that have a classic constellation of examination findings [13]. Knowledge of common spinal cord syndromes is useful in communication with the referring clinician. Bulbar-cervical dissociation syndrome occurs when the neuroaxis injury is at or above C3. This produces immediate pulmonary and often cardiac arrest. Patients usually present with quadriplegia and are ventilator dependent secondary to phrenic nerve/diaphragmatic dysfunction. Central cord syndrome is similar to the presentation of syringomyelia. These patients present motor weakness of the upper extremities, greater than the lower extremities, and have varying sensory disturbance below the level of the lesion. Myelopathy, including anal sphincter dysfunction can also be associated with central cord syndrome. Older patients often have pathology related to central canal narrowing while younger patients may have disk protrusions, subluxations, and fractures.

Anterior cord syndrome classically occurs in the presence of anterior spinal artery infarction. Patients presenting with anterior cord syndrome experience loss of pain and temperature sensation with the preservation of fine touch and proprioception. This may occur secondary to anterior cord compression, dislocated bony fragment, or herniated disk. Brown–Sequard syndrome occurs in the presence of spinal cord hemisection. Classically, these patients present with motor paralysis on the ipsilateral side with dissociated sensory loss on the contralateral side. This is usually due to penetrating trauma such as knife or gunshot wounds. Finally, patients presenting with posterior cord syndrome have pain and paresthesias in the neck, upper arms, and torso.

## 10.3 Indications

Once the trauma patient has been assessed, often with the aid of CT, MRI is used as an adjunctive means of clinical/pathological correlation in the assessment of possible treatment. MRI has some established guidelines for both the referring clinicians and the interpreting radiologists. An understanding of both criteria is helpful in facilitating triage, supportive care, and treatment. In North America, "Guidelines for the Management of Acute Cervical Spine and Spinal Cord Injuries" were created by the American Association of Neurological Surgeons and the Congress of Neurological Surgeons, section on disorders of the spine and peripheral nerves, which outlined appropriateness criteria for imaging of the cervical spine in the setting of trauma [14]. The document is an evidence-based approach of the existing literature.

Radiographic assessment in the setting of spine trauma is reserved for symptomatic trauma patients. Traditionally, these patients were evaluated with cervical spine radiographs. However, CT evaluation has become the initial diagnostic modality of choice in high-risk patients. A negative radiographic evaluation allows the discontinuation of cervical spine immobilization in patients without neck pain.

MRI should be performed in patients with fracture/dislocation injuries who cannot be examined during attempted closed reduction or prior to open posterior reduction. Soft tissue abnormalities including the presence of disk herniation in association with fracture/dislocations may be radiographically occult and, if not identified, can cause additional trauma during treatment. In addition, MRI is recommended for those patients who fail multiple attempts at closed reduction.

Particular attention is placed on MR evaluation in the setting of occipitocervical or atlantoaxial dislocation injuries [15–18]. When high cervical spine injuries are suspected, radiography and CT may be insufficient for providing the diagnosis. The presence of upper cervical prevertebral soft tissue swelling is a useful indicator for further evaluation. MRI may provide useful information on the presence or absence of soft tissue injury [19].

It is not uncommon for patients to present with a spinal cord injury in the presence of an otherwise normal cervical spine series or CT [20–22]. These two imaging modalities do not exclude soft tissue injuries, which may be radiographically occult. Such patients are classified in the literature as having spinal cord injury without radiographic abnormality. A targeted MRI of the region of suspected neurologic injury may provide useful diagnostic information and demonstrate soft tissue changes such as edema or hemorrhage.

The American College of Radiology has also established appropriateness criteria for the evaluation of patients with suspected spine trauma [23–25]. This work has been based on the National Emergency X-Radiography Utilization Study and the Canadian C-spine Rule [26,27]. In patients with suspected acute cervical spine trauma in the presence of myelopathy, MRI evaluation is recommended and found to be complementary to CT evaluation. Similar to the clinically based guidelines, in patients with suspected acute spine trauma, MRI used in conjunction with CT may be helpful for treatment planning of unstable spines to identify possible associated soft tissue injuries. Patients with suspected acute cervical spine trauma may undergo delayed MRI if they require cardiopulmonary support in conjunction with medical paralysis. MRI is recommended for the evaluation of ligamentous injury, cord pathology, and edema. In addition to bony and ligamentous injury, thrombosis of the carotid and vertebral arteries may occur due to possible dissection or transection, particularly in the setting of cervical spine fracture. Thrombosis of these vessels may lead to cerebral ischemia and may require anticoagulation and/or intervention. Patients with suspected arterial injury following cervical spine trauma may be evaluated by either MR angiography or CT angiography [28–33].

## 10.4 MRI Sequences in the Setting of Trauma

As the field of MRI changes, there are a multitude of new sequences to define and diagnose spine trauma. Characterization of the spine in the setting of trauma may provide useful information to help determine the cause and extent of the neurologic deficit, probable mechanism of injury, and the presence of spinal instability [34]. However, in the setting of trauma, such patients may be too unstable to endure long imaging evaluation often due to the need for cardiopulmonary support. Therefore, a standard base protocol to assess soft tissue injury in the setting of trauma is critical.

Bozzo et al. conducted a literature review to define the recommended protocol for MRI in the setting of acute spinal cord injury, to assess whether or not MRI was necessary in the initial management and whether MRI was helpful in predicting the patient's long-term neurologic outcome [35]. Based on their extensive review, sagittal $T_2$ sequences were found to be of particular value in the diagnosis of soft tissue injury with typical TR/TE values of 2000 and 80 ms, respectively. The typical $T_1$ sequences used in the setting of trauma had a TR value of 600 ms and a TE value of 15–20 ms. Gradient recalled echo sequences were used in a limited number of articles and were found to be particularly sensitive to breathing. Fluid sensitive short tau inversion recovery (STIR) or gradient recalled acquisition in the steady state imaging was also evaluated in the presence of spine trauma to assess the integrity of soft tissue injury. A study by Lee and colleagues found a high correlation with increased STIR signal and intraoperative findings of posterior ligamentous injury [36]. Based on their literature review, sagittal $T_2$ imaging was recommended for the assessment of possible cord compression, edema, and hemorrhage; sagittal gradient recalled echo imaging was recommended for possible cord hemorrhage; and axial $T_2$ imaging was recommended for possible disk herniation and cord compression. Newer MRI modalities including diffusion tensor imaging of white matter tracts and susceptibility-weighted imaging continue to be promising on an investigational basis.

## 10.5 MRI Safety in the Trauma Patient

The trauma patient requires a heightened attentiveness to safety. In addition to traditional contraindications to MRI such as pacemakers or aneurysm clips, additional factors must be addressed in the setting of trauma [13]. As stated above, patients presenting with spinal cord injury often have symptoms of spinal shock, which necessitates additional cardiopulmonary monitoring and support during the course of imaging evaluation. Symptoms may include hypotension (systolic blood pressure [SBP] $< 80$), loss of vascular tone and muscular tone resulting in hypovolemia, loss of parasympathetics resulting in bradycardia, and hypoxia due to phrenic nerve dysfunction. Assessment of life support needs should be taken into account as additional personnel may be required to monitor the patient during the examination.

Special attention must be paid to spinal alignment. Movement of the trauma patient should be limited to experienced personnel. Every effort should be made to limit the possibility of continued trauma to the spinal cord due to the possibility of spinal instability. Precautions should be undertaken for patient movement and transfer such as the logroll technique. Prior to traumatic injury, certain types of patients may have higher risk factors for spinal cord injury, such as those with axialspondylitis (AS). The loss of spinal segmentation secondary to anterior longitudinal ligament and disk space calcification leads to a high risk of traumatic spinal cord injury secondary to the increased moment of inertia of the fused spine [37]. Patients with AS have an increased risk of spinal cord injury in the presence of intubation due to neck flexion or extension. In addition, patients with predisposing factors including AS may be deceptively stable with little symptoms other than pain. Their pain and muscle spasm may act to keep the spine in alignment by increased paraspinal muscular tone. If such a patient should lose muscular tone, such as in the setting of anesthesia or muscular block, the alignment of the neuroaxis may be lost resulting in catastrophic spinal cord injury.

An extensive medical history is normally not available in the presentation of trauma. A detailed physical examination and radiographic analysis prior to MRI may reveal metal objects, which are incompatible with MRI equipment. In addition to standard contraindications to MRI, bullet fragments and retained metallic objects may be a significant diagnostic conundrum for the referring clinician and radiologist. A retrospective review of 19 patients presenting with bullet fragments in the region of the spine was conducted by Finitsis et al. [38]. The reviewers determined that these patients did not suffer injury secondary to the retained fragments and the information gained from the imaging investigation was necessary in their care.

### 10.5.1 Occipito-Atlantal Injuries

Craniocervical junction injuries resulting from ligamentous are frequently fatal at the time of the traumatic insult and may result from traumatic hyperflexion or hyperextension with associated rotation [18]. Often grossly unstable, they result in neurologic and/or vascular compromise. This injury occurs more frequently in pediatric patients, possibly secondary to differences in condyle curvature, increased ligamentous laxity, and increased cranial to body weight ratio. Patients with occipito-atlantal injuries present with diverse symptoms including bulbar-cervical dissociation injury, lower cranial nerve palsies, or alternatively may present neurologically intact.

MRI evaluation in the presence of occipito-atlantal dislocation injury is largely focused on the presence of soft tissue changes between the occipito-atlantal interfaces. Occipito-atlantal distraction injuries may consist of ligamentous tear or avulsion, displaced occipital condyle fractures (Figure 10.1), and/or tectorial membrane disruption. MR evaluation may demonstrate prevertebral soft tissue swelling or fluid within the articular capsules, nuchal ligament, and interspinous ligaments. Atlantoaxial distraction injuries result in disruption of the articular capsules, hilar ligaments, transverse ligament, and tectorial membrane. MR imaging may demonstrate prevertebral fluid, interspinous, or nuchal ligament edema, and facet widening.

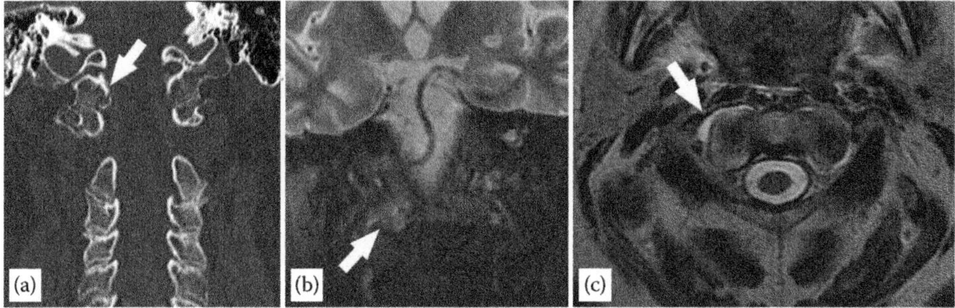


**FIGURE 10.1**
Right occipital condyle fracture in a 61-year-old man after fall. (a) Coronal 2D reformatted image from CT demonstrates a non-displaced fracture through the right occipital condyle (arrow). (b) Coronal FSE $T_2$-weighted image through the craniocervical junction shows edema-like signal changes in the right occipital condyle (arrow). (c) Axial FSE $T_2$-weighted image demonstrates small amount of fluid (arrow) extending about the fracture site.

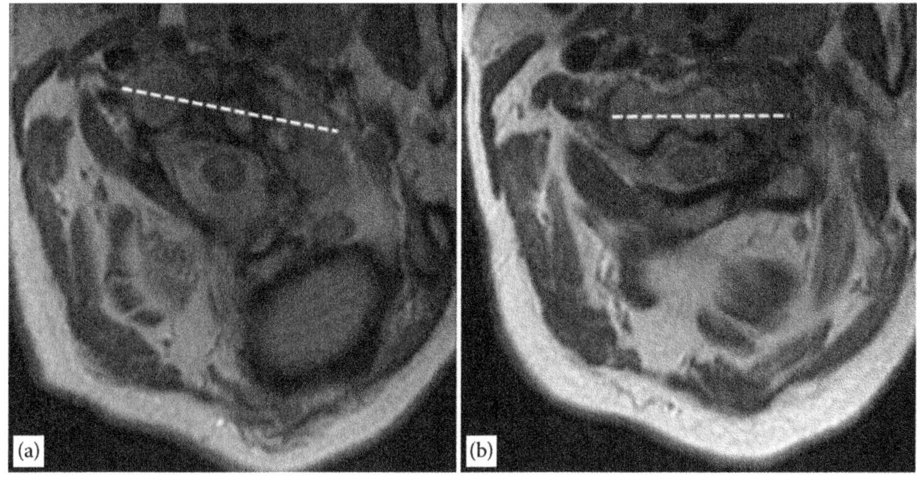


**FIGURE 10.2**
Rotatory subluxation in a 65-year-old woman with 2 months of neck pain. Adjacent axial proton density images through C1 (a) and C2 (b) show 20° rotation of C1 relative to C2 and the remaining cervical spine (dotted lines = axes of C1 and C2).

## 10.6 Cervical Spine Trauma

### 10.6.1 Atlantoaxial Subluxation/Dislocation

There are three types of atlantoaxial subluxation (Figures 10.2 through 10.4)—rotatory, anterior, and posterior. Rotatory subluxation is commonly seen in children and is associated with trauma, rheumatoid arthritis, and respiratory infections. Rotatory deformity at the atlantoaxial junction commonly is of short duration and is easily corrected. The deformity may occur at the occipito-atlantal or the atlantoaxial articulations. The assessment of the transverse atlantal ligament is particularly important in MRI evaluation. With an intact transverse atlantal ligament, rotation may occur without displacement. If the transverse and lateral ligaments are incompetent, there may be anterior displacement or posterior displacement. Anterior atlantoaxial subluxation may take on one of two forms including disruption of the transverse atlantal ligament in which the atlantodental interval will be

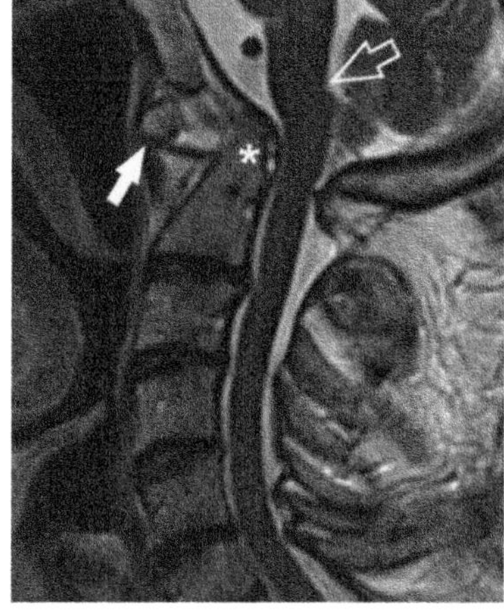


**FIGURE 10.3**
Atlantoaxial widening, transverse ligament rupture, and basilar invagination in a 67-year-old man. Sagittal FSE $T_2$-weighted image shows posterior displacement of the dens (asterisk) relative to the anterior arch of C1 (closed arrow), narrowing of the adjacent central spinal canal, basilar invagination, and mild deformity of the medulla (open arrow) between the tip of the dens and cerebellar tonsil.

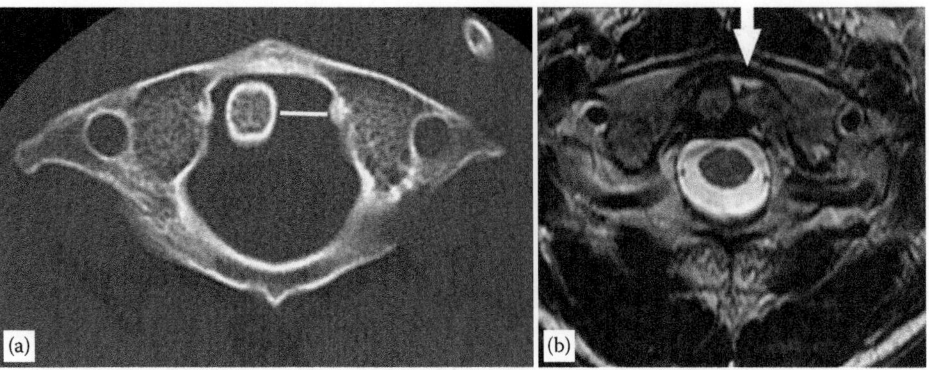


**FIGURE 10.4**
C1–2 instability and left alar ligament tear in a 17-year-old boy who fell off a skateboard while holding onto a moving vehicle. (a) Axial CT demonstrates displacement of the dens to the right, with asymmetric widening between the left lateral mass of C2 and the dens (line). (b) Axial $T_2$-weighted image shows fluid (arrow) within the torn left alar ligament, with corresponding displacement of the dens to the right.

increased (Figure 10.4), or incompetence of the odontoid process, which may occur in odontoid fractures or congenital hypoplasia. MR angiography may be used to image the transverse atlantal ligament directly. Findings may include disruption of the ligament on axial series, which may show high signal intensity on gradient-echo MRI, loss of continuity, or blood at the insertion site.

## 10.7 Axial Loading Injuries

Axial loading injuries are caused by direct forces onto the head, either from objects dropped from above or from driving forces of the head into a fixed object. Resultant injuries are typically burst fractures, most notably the Jefferson fracture, in which the occiput and lateral masses of C2 compress and fracture the ring of C1 (Figure 10.5). Peripheral displacement of fracture fragments typically results in a stable injury, although MRI may be helpful to assess for any mass effect from concomitant hemorrhage.

## 10.8 Extension Injuries

In the setting of trauma, extension injuries may be occult on radiographs [39]. These patients may present with neck pain and have cervical instability without findings of fracture. Furthermore, they may present with realignment following the removal of the impacting force. Soft tissue widening is the most reliable indicator of injury (Figure 10.6). CT may underrepresent the extent of injury due to the realignment of the cervical spine following the injury. MR findings may demonstrate increased $T_2$-weighted signal within the anterior longitudinal ligament and/or intervertebral disk, rupture of the anterior longitudinal ligament, periosteal stripping or rupture of the posterior longitudinal ligament, and injury of the posterior ligamentous complex (Figure 10.7). Hyperintense $T_1$ signal in the prevertebral soft tissues may represent acute hemorrhage in the setting of trauma.

Hangman's fractures are bilateral C2 avulsion fractures through the lamina and/or pedicles, with occasional extension through the posterior vertebral body and transverse foramina (Figure 10.8). The large diameter of the spinal canal typically precludes neurologic injury with these fractures. Hyperextension teardrop fractures are serious injuries that result from avulsion of the intact fibers of the anterior longitudinal ligament from the inferior end plate of the vertebral body. MR may be helpful in assessing any associated soft tissue injuries.

Dens fractures may result from multiple mechanisms of force, most notably hyperextension. Type 1 fractures are rare and caused by avulsion of the tip of the dens from its attachment site of the alar ligaments. Type 2 is the most common dens fracture, with transverse fracture extension through the base of the dens (Figure 10.9). Displacement more than 50% correlates with higher rate of non-union (Figure 10.10). Type 3 fractures horizontally extend through the superior body of C2.

## 10.9 Flexion Injuries

Similar to extension injuries, hyperflexion injuries may also be inconspicuous on radiographs and CT if completely reduced. Pronounced findings of hyperflexion injuries on radiographic imaging and CT include widening (*fanning*) of the spinous processes with or without facet fracture or dislocations, widening of the posterior disk space, and abrupt angulation of more than 11 degrees at a single interspace. MRI may detect traumatic disk protrusions/extrusions (Figure 10.11), increased

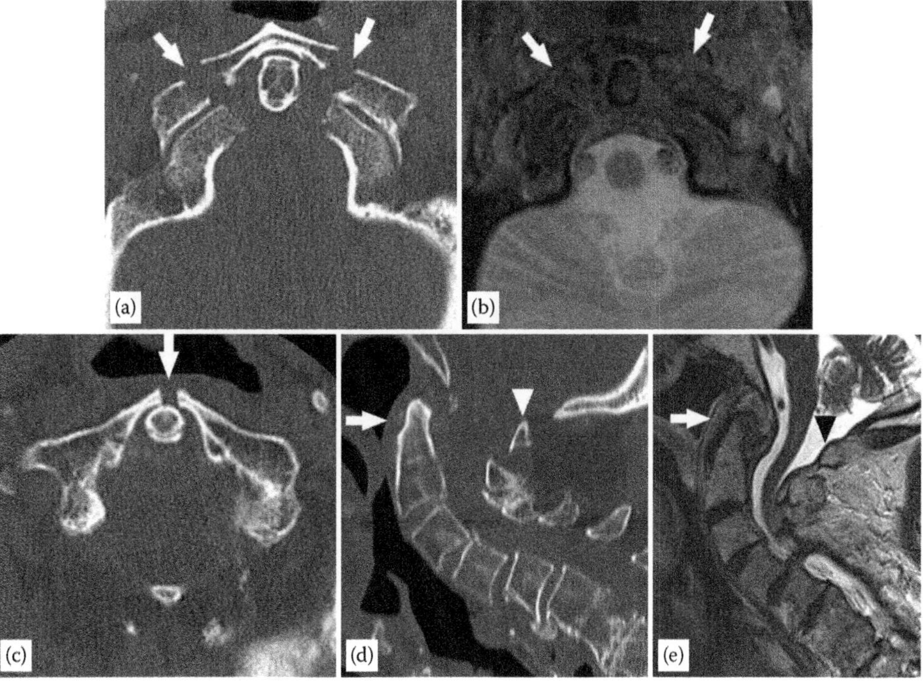


**FIGURE 10.5**
Jefferson fracture. Axial CT (a) and FSE $T_2$-weighted MRI (b) images show a comminuted fracture (arrows) through C1 in a 90-year-old woman who was accidentally thrown from her wheelchair headfirst onto the pavement. (c) Minimally displaced Jefferson fracture (arrow) on axial CT in a 94-year-old man who slipped and fell. Corresponding sagittal 2D reformatted CT (d) and sagittal FSE $T_2$-weighted MRI (e) images show a void through the fracture cleavage plane (arrows) and posterior displacement of the posterior arch of C1 (arrowheads) relative to the remaining spinolaminar line.

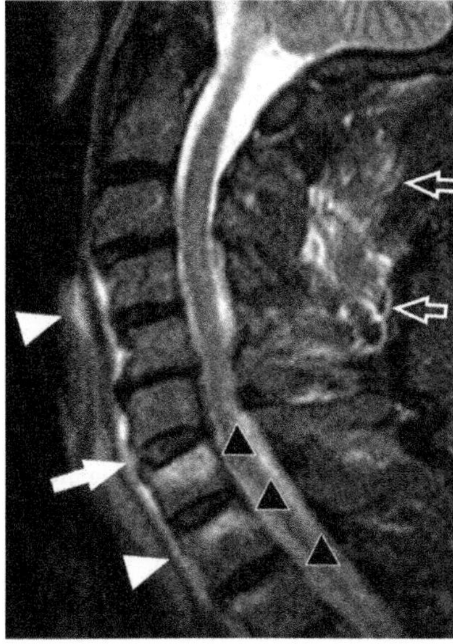

**FIGURE 10.6**
Hyperextension and hyperflexion injuries in a 69-year-old man status post fall. Sagittal FSE $T_2$-weighted image shows disruption of the anterior longitudinal ligament at the level of C6–7 (arrow), prevertebral edema (white arrowheads), mild C7–$T_2$ compression fractures (black arrowheads), and interspinous ligament sprain (open arrows).

signal intensity within the posterior elements on the basis of ligamentous injury, facet widening or increased $T_2$ signal within the joint space on the basis of edema, or flexion deformity of the spinal column (Figure 10.12). Hyperflexion wedge fractures are fractures of the superior end plate. These fractures are deemed simple if unaccompanied by posterior ligamentous disruption. In the presence of posterior ligamentous injury, they are classified as unstable.

Bilateral facet dislocation (*jumped facets*) is an unstable hyperflexion injury resulting in disruption of all ligamentous structures, anterior displacement of the superior vertebra at least 50% of its width, and will invariably lead to cord compression and/or signal abnormalities. Hyperflexion forces with some component of rotation may result in unilateral facet dislocations, with less than 50% anterior displacement of the superior vertebra and asymmetric ligamentous injuries (Figure 10.13). Facet subluxation or *perched facets* may be unilateral or bilateral, with the articular masses of the involved vertebra perched on the superior articular process of the subjacent vertebra. These injuries are deemed unstable, but may not present with neurologic compromise.

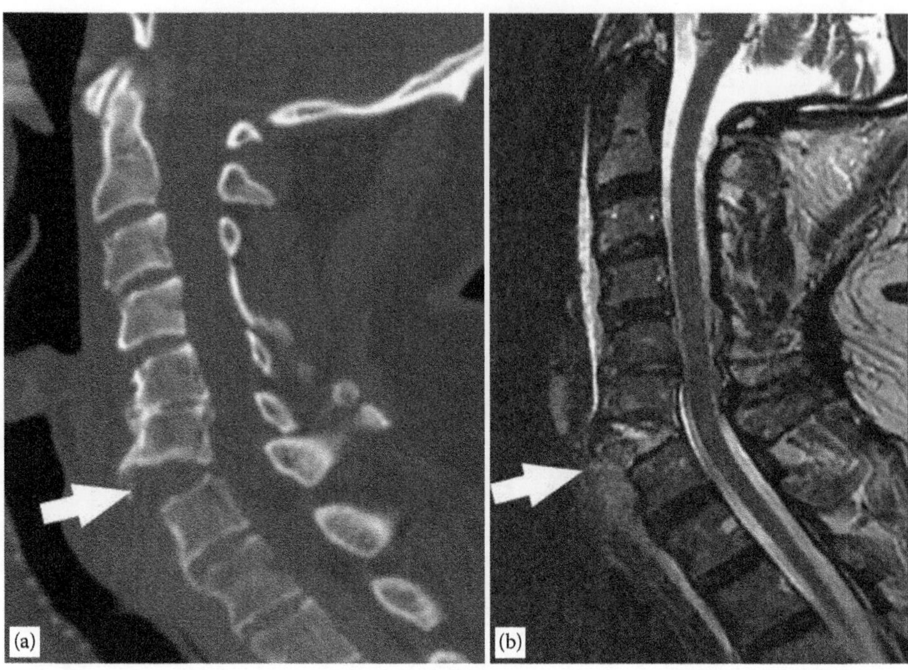

**FIGURE 10.7**
Hyperextension injury sustained in an 81-year-old man who was rear-ended in a motor vehicle accident. (a) Sagittal 2D reformatted image from CT shows widening of the C6–7 anterior disc space and focal spinal angulation at this level. Fractures of the right C6 pedicle and pars and left C6 inferior articular process were also present (not shown). (b) Sagittal FSE $T_2$-weighted image demonstrates disruption of the anterior longitudinal ligament at C6–7 (arrow), with associated hemorrhage in the C6–7 disk and prevertebral soft tissue swelling.

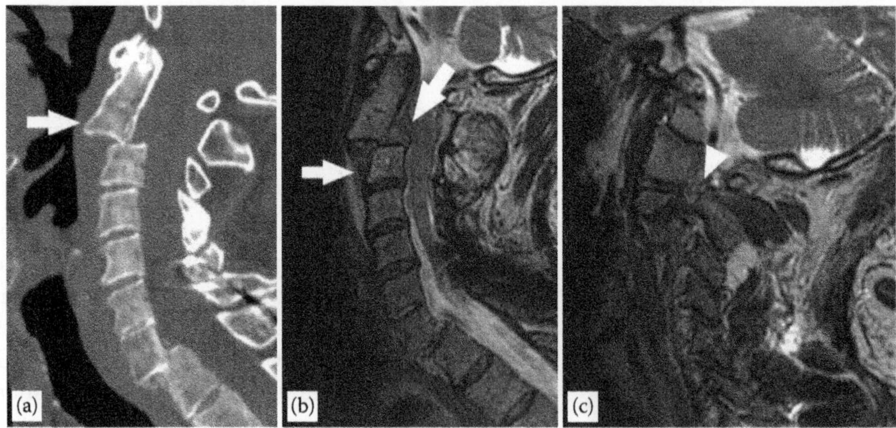

**FIGURE 10.8**
Hangman's fracture (type IIA) in a 71-year-old man who was a backseat passenger in a motor vehicle accident. (a) Sagittal 2D reformatted image from noncontrast CT displays grade 3 anterolisthesis and rotation of the C2 vertebral body (arrow) relative to C3. The posterior spinolaminar line is disrupted due to posterior displacement of the C2 posterior elements. Also shown are prior C6–7 fusion and old displaced C7 fracture from motor vehicle accident 5 years prior, with fusion of the anteriorly displaced C7 vertebra to $T_1$. (b) Sagittal FSE $T_2$ FS image shows stripping of the anterior and posterior longitudinal ligaments (arrows) at the level of the C2–3, with inferiorly dissecting prevertebral hemorrhage. (c) Parasagittal image demonstrates the distracted fracture through the left pedicle (arrowhead). Right-sided fracture is not shown. Patient underwent reduction in the C2–3 subluxation and posterior arthrodesis extending from the occiput to C5.

## 10.10 Thoracic and Lumbar Spine Trauma

Traditional assessment of the thoracolumbar spine was based on the three-column model by Dennis [40]. This model attempted to identify factors of instability of the thoracolumbar spine. The three-column model defines the thoracolumbar spine into the anterior column (anterior half of the disk and vertebral body, anterior annulus fibrosus, and anterior longitudinal ligament),the middle column (posterior half of the disk and vertebral body, posterior wall of the vertebral body, posterior annulus

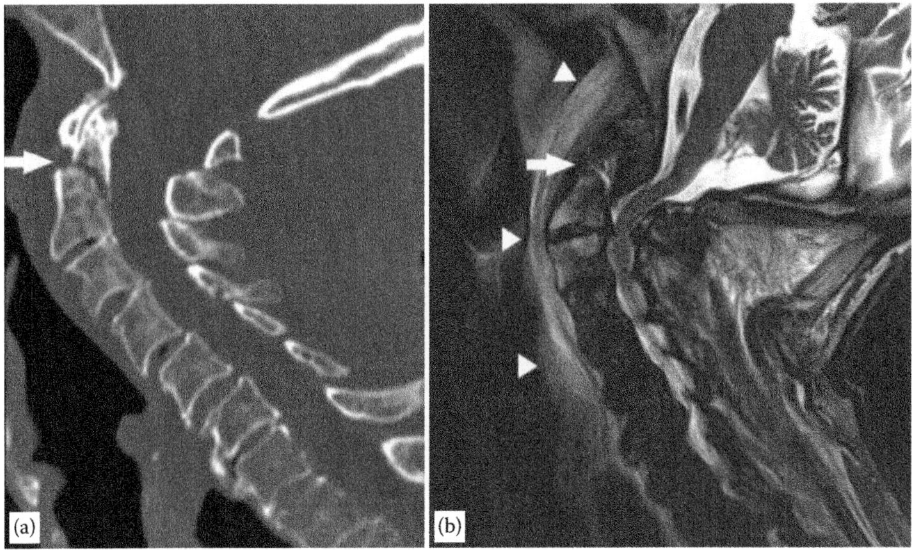


**FIGURE 10.9**
Type 2 dens fracture in an 88-year-old man with neck pain who sustained a fall 1 week prior. (a) Sagittal 2D reformatted CT image shows an oblique fracture (arrow) through the base of the dens, with 3 mm posterior displacement of the dens and C1. (b) Sagittal FSE $T_2$-weighted FS image shows fluid in the dens fracture (arrow) and dissecting prevertebral hemorrhage (arrowheads). Note the diffuse degenerative disk disease and central canal narrowing.

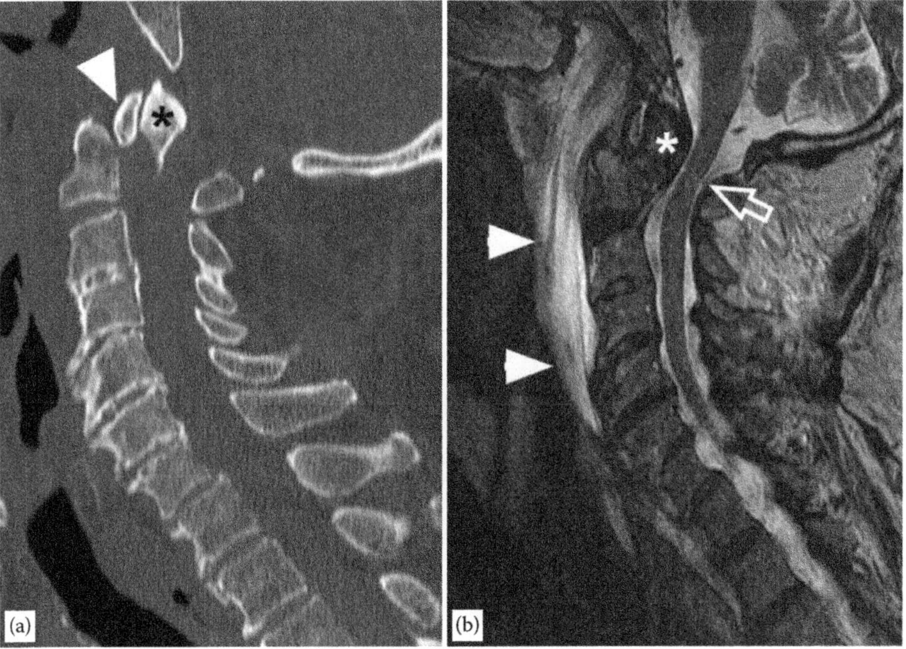


**FIGURE 10.10**
Displaced type 2 dens fractures and cord contusion in a 91-year-old man with progressive neck pain after fall the prior day. (a) Sagittal 2D reformatted CT image demonstrates a type 2 dens fracture, with 1.2 cm posterior displacement of the dens (asterisk) and C1 (arrowhead). (b) Sagittal FSE $T_2$-weighted image shows (arrowheads = prevertebral hemorrhage) Patient underwent subsequent L1-2 arthrodesis with lateral mass and pars screw fixation.

fibrosus, posterior longitudinal ligament, and pedicles), and the posterior column (posterior bone complex with interposed posterior ligamentous complex, supraspinous and interspinous ligaments, facet joints and capsule, and ligamentum flavum). Dennis categorized the instability as first degree (mechanical instability), second degree, (neurological instability), and third degree (mechanical and neurological instability).

More recently, attempts have been made to classify the injuries of the thoracolumbar spine into major or minor

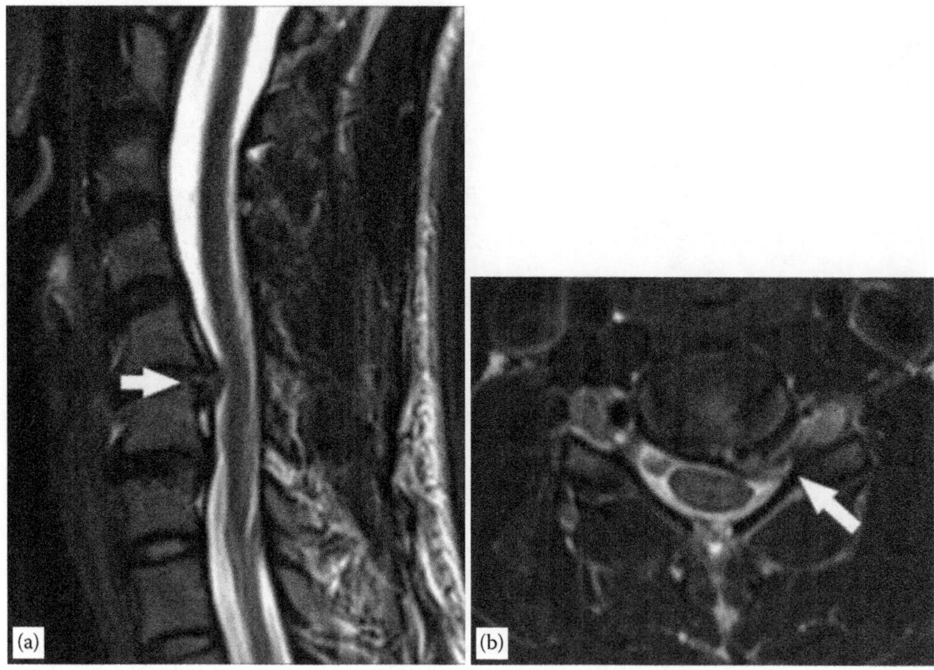


**FIGURE 10.11**
Traumatic disk extrusion in a 38-year-old stunt woman with progressive left arm numbness and pain. Sagittal (a) and axial (b) FSE $T_2$-weighted images demonstrate a left subarticular disk extrusion with effacement of the left C5–6 neural foramen (arrows).

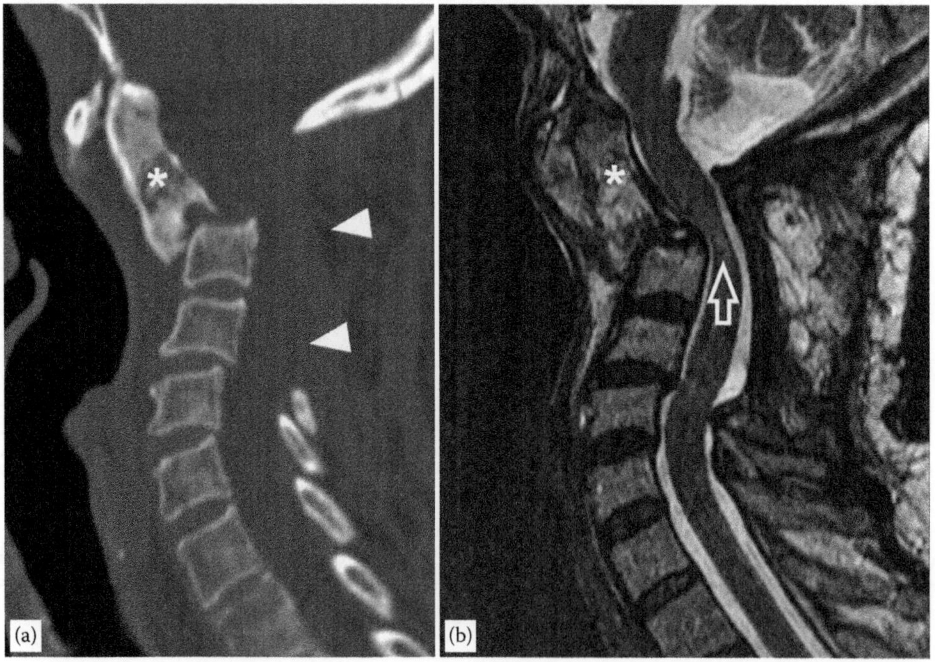


**FIGURE 10.12**
Traumatic C2–3 anterolisthesis after fall in a 50-year-old woman with bilateral arm and leg weakness and numbness. (a) Sagittal 2D reformatted CT image shows grade 3 anterolisthesis and flexion deformity of C2 (asterisk) upon C3 from gross ligamentous instability (arrowheads), and posterior C1–4 decompression. (b) Sagittal FSE $T_2$-weighted image shows hyperintense central cord signal (open arrow) at the level of C3, consistent with contusion, and stable displacement of C2 (asterisk).

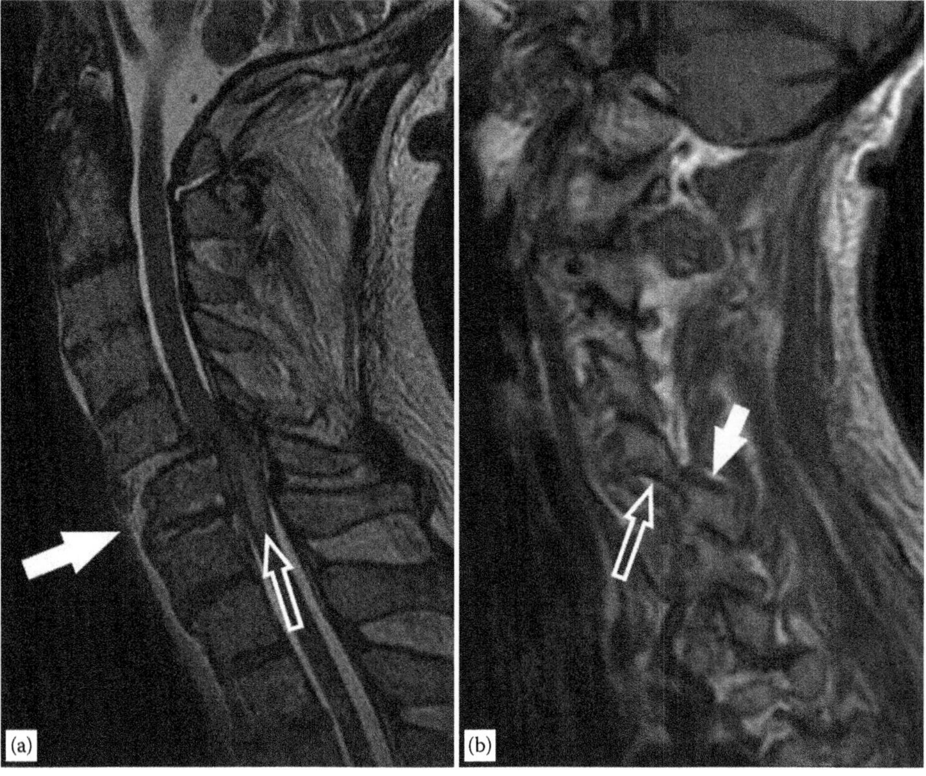


**FIGURE 10.13**
Back pain in unilateral interfacetal dislocation in an 82-year-old man who fell. (a) Sagittal FSE $T_2$-weighted image demonstrates less than 50% anterior displacement of C5 upon C6, rupture of the anterior longitudinal ligament (arrow), traumatic disk protrusion, and cord contusion (open arrow). (b) Parasagittal image through the left facet joints shows dislocation of the inferior articular process C5 (open arrow) relative to the superior articular process of C6 (arrow).

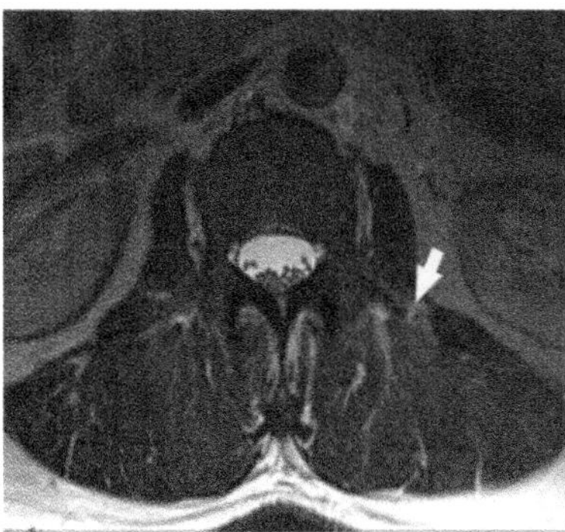

**FIGURE 10.14**
Left L3 transverse process fracture in a 49-year-old woman who fell 5 feet from a ladder. Axial FSE $T_2$-weighted image shows a nondisplaced fracture (arrow) through the left L3 transverse process. The fracture was deemed stable and patient was treated with a brace.

categories. Minor injuries include the following: (1) fracture of a transverse process (Figure 10.14), (2) fracture of an articular process or pars interarticularis (Figure 10.15), (3) isolated fractures of the spinous process, and (4) isolated laminar fractures. Major injuries include the following: (1) compression fracture of the anterior column (the middle column remains intact) (Figures 10.16 through 10.18); (2) burst fracture, which occurs in pure axial load, representing a failure of the anterior and middle columns (Figure 10.19); (3) seat belt fracture (compression of anterior column and distraction fracture of both middle and posterior columns) (Figure 10.20); and (4) fracture-dislocation, which represents a failure of all three columns due to compression, tension, rotation, or shear leading to subluxation or dislocation (Figure 10.21).

The thoracolumbar injury classification and severity score have been proposed taking into account both radiographic and neurological findings [41,42]. The radiographic findings include fracture type and/or dislocation along with the integrity of the posterior elements. Similar to major and minor grading criteria, anterior column injuries are usually stable, burst fractures are considered injuries of the anterior and middle columns, seat belt fractures consist of three column injuries, and fracture-dislocations are three-column injuries with associated dislocation.

The major advantage of MRI in the assessment of the thoracolumbar spinal fractures is in the characterization of posterior ligamentous injury. Using the new thoracolumbar injury classification and severity score, recent literature has compared MRI to CT in the assessment of vertebral fractures in trauma [36,43] MRI offered

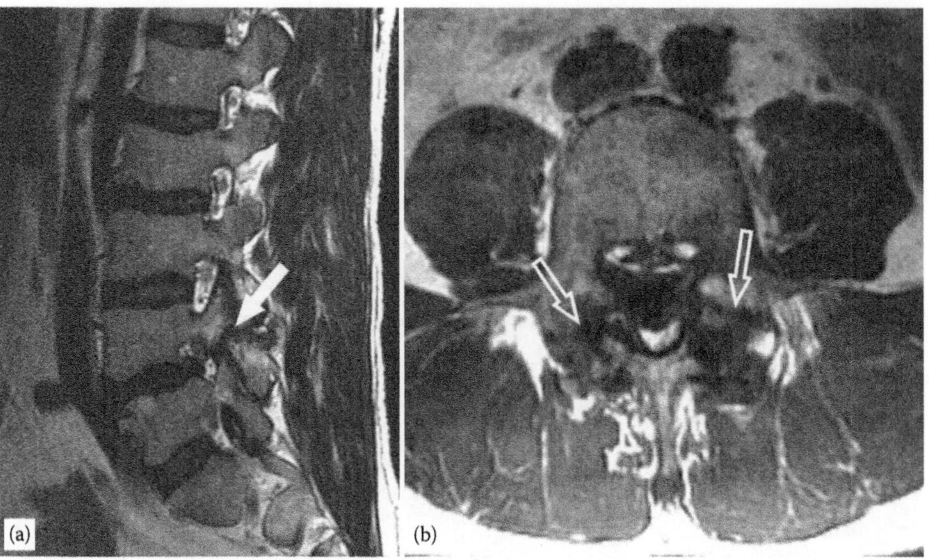


**FIGURE 10.15**
L4 pars intraarticularis defects in a 38-year-old man. Sagittal (a) and axial (b) $T_1$-weighted images show non-displaced bilateral L4 pars intraarticularis defects (arrows), with adjacent hyperintense fatty marrow indicating chronic changes.

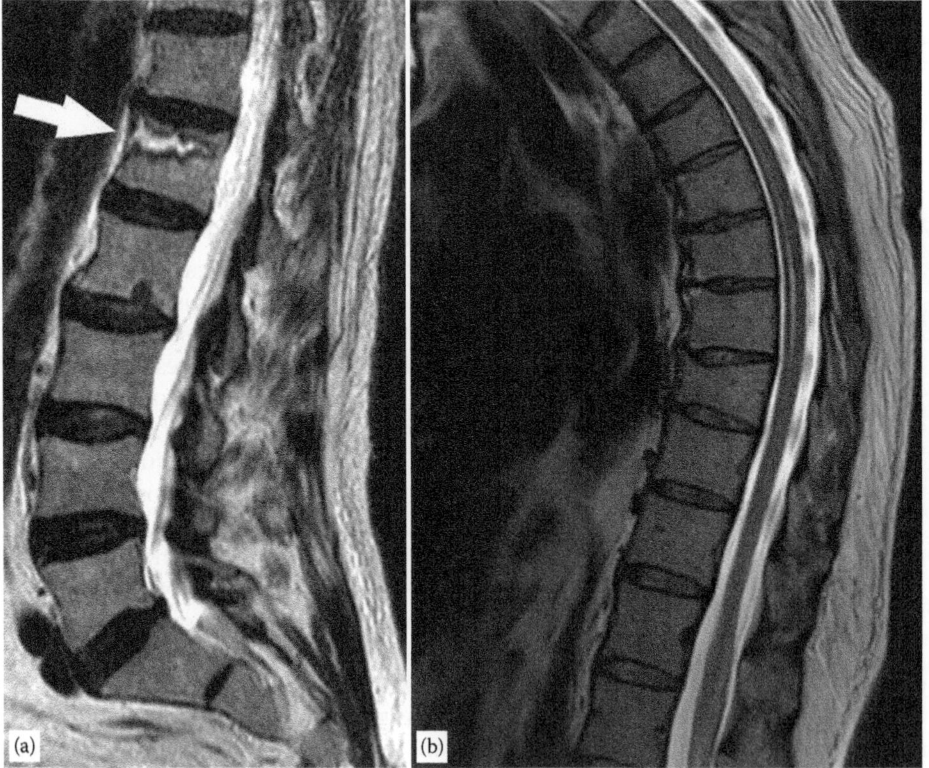


**FIGURE 10.16**
Compression fractures. (a) Acute mild L1 compression fracture in an 86-year-old man who fell the day prior at a wedding. Sagittal FSE $T_2$-weighted image shows fluid within the transverse fracture (arrow) through the superior L1 vertebral body. (b) Multiple chronic mild mid- and lower thoracic compression fractures in a 56-year-old man on alendronate therapy with back pain. Sagittal FSE $T_2$-weighted image demonstrates multiple mild thoracic compression fractures with normal marrow signal intensity, consistent with remote fractures. Degenerative disk disease at the level of the pain marker likely accounts for patient symptoms.

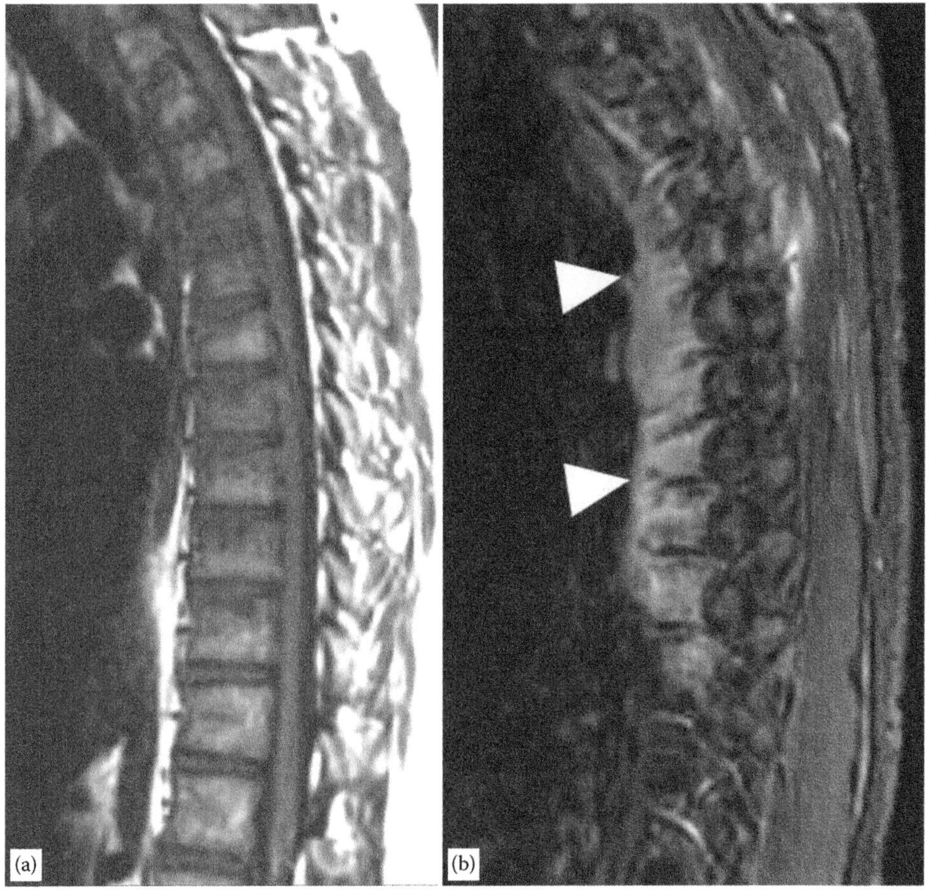

**FIGURE 10.17**
Multiple mild compression fractures in 19-year-old man who was a pedestrian hit by a car. (a) Sagittal $T_1$-weighted image shows linear hypointense signal foci in multiple anterior superior thoracic vertebral bodies, consistent with non-displaced fractures from hyperflexion injury. (b) Parasagittal FSE $T_2$-weighted FS image shows extensive paravertebral hematoma (arrowheads).

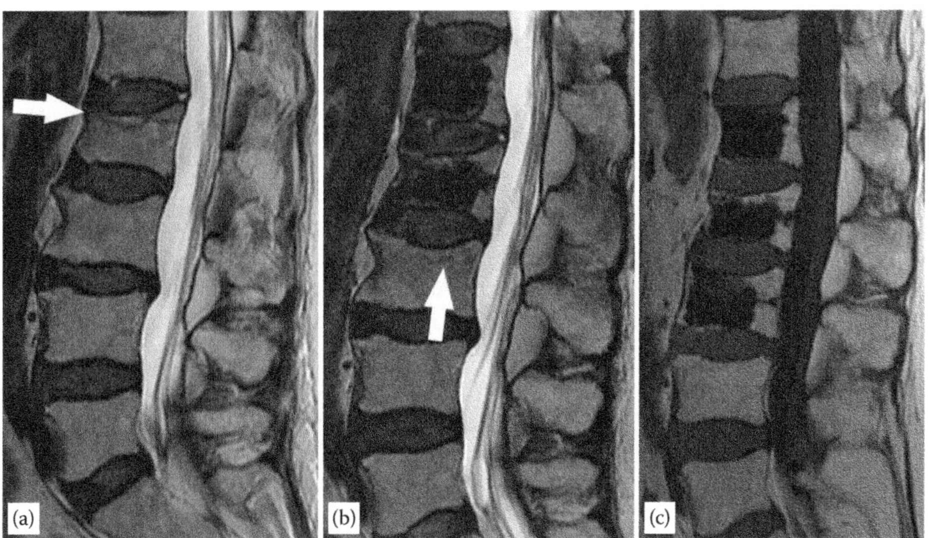

**FIGURE 10.18**
Sequential vertebroplasties for osteopenic compression fractures in a 71-year-old woman. (a) Sagittal FSE $T_2$-weighted image shows a mild L2 compression fracture (arrow) paralleling the superior endplate. (b) Repeat MRI performed for recurrent back pain after interval L1 and L2 vertebroplasties demonstrates a new mild L3 superior compression fracture adjacent to the L2 cement (arrow). (c) MRI after L3 vertebroplasty shows healing of all treated fractures.

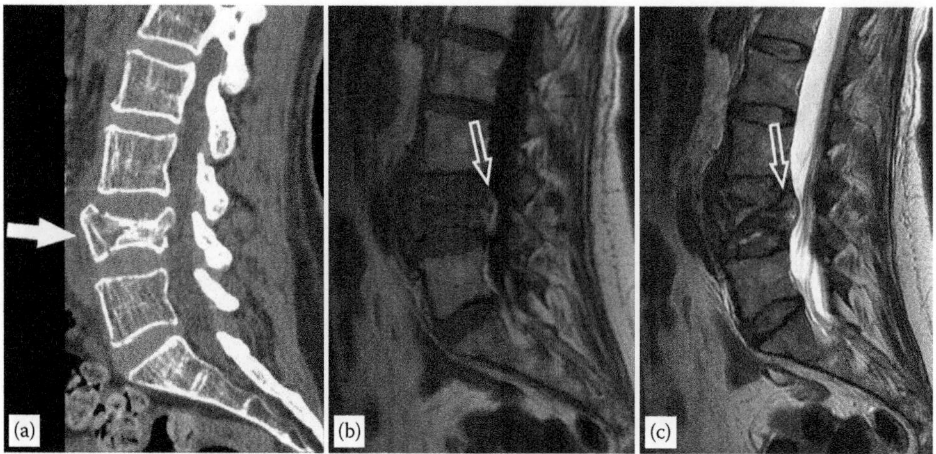


**FIGURE 10.19**
Burst fracture of the L4 vertebral body in a 73-year-old woman with severe back pain status post fall. Sagittal 2D reformatted CT (a) and sagittal $T_1$-weighted (b) and FSE $T_2$-weighted (c) images show a severe L4 burst fracture (arrow) with retropulsion of bony fragments (arrows), resulting in moderate central canal stenosis.

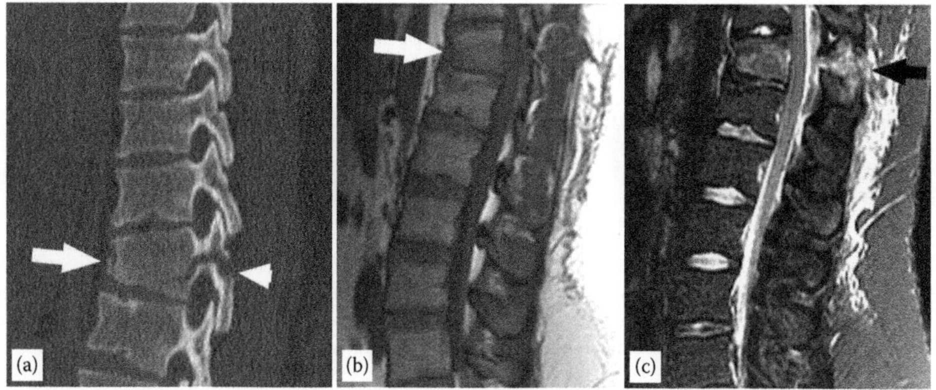


**FIGURE 10.20**
Chance fracture through the T12 vertebra in a 21-year-old restrained passenger involved in a motor vehicle accident. (a) Parasagittal 2D reformatted CT demonstrates mild T12 compression fracture, distracted fracture through the right pedicle, and mild kyphotic deformity. Sagittal $T_1$-weighted (b) and FSE $T_2$-weighted (c) images demonstrate the compressive injury through the vertebral body (white arrow) and distractive injury through the disrupted interspinous ligaments (black arrow). The man underwent subsequent T11-L1 pedicle screw fixation.

moderate reproducibility in the assessment of vertebral fractures pursuant to the AO (*Arbeitsgemeinschaft für Osteosynthesefragen*) classification, slightly better than the CT characterization.

A prospective cohort study was conducted evaluating MRI accuracy in diagnosing posterior ligamentous damage using a fast spin-echo $T_2$ STIR sequencing [44]. The cohort group consists of 58 patients with vertebral fractures. MRI sensitivity for injury diagnosis of the posterior ligamentous complex ranged between 92.3% (interspinous ligament) and 100% (ligamentum flavum).

MRI is commonly utilized for the assessment of compression fractures. One common indication for MRI is the diagnosis of benign versus malignant fractures. A recent meta-analysis of 31 studies identified several imaging features of malignant compression fractures. These features include signal intensity ratio on opposed phase imaging of 0.8 or more, arterial proton density (APD) on echo planer DWI of $1.5 \times 10^{-3}$ mm$^2$/s or less with a *b*-value of 500 s/mm$^2$, presence of non-characteristic vertebral lesions or paraspinal mass, involvement of the posterior elements, replacement of the marrow, presence of epidural mass, and convexity of the posterior vertebral border [45]. Further, MRI has been used to assess for the extent of healing [46]. Increased $T_2$ signal within the vertebral body has been correlated with

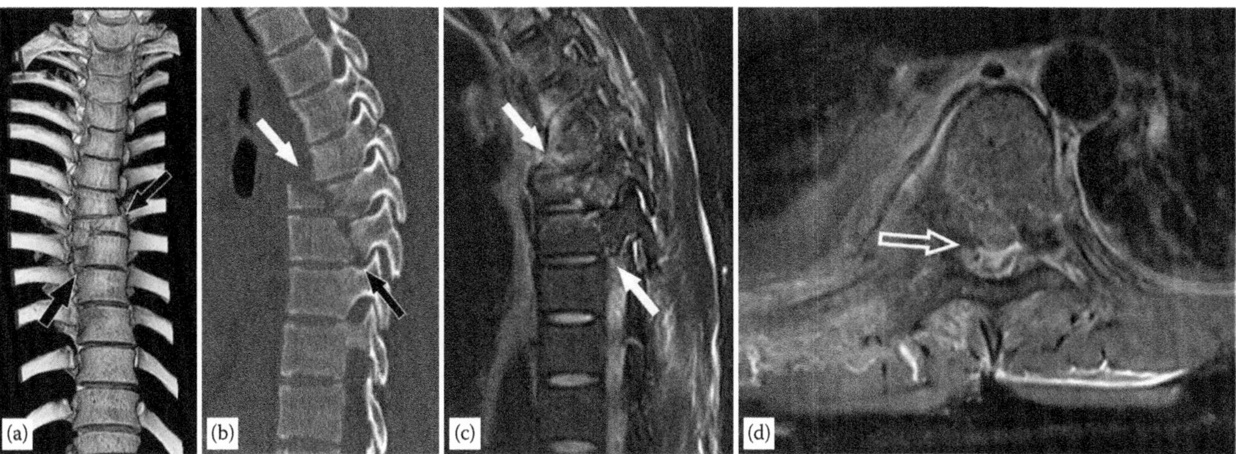

**FIGURE 10.21**
Thoracic fracture-dislocation in a 25-year-old man with severe back pain after a ladder landed on his back. 3D surface rendered (a) and sagittal 2D reformatted (b) images demonstrate oblique, displaced fractures through T7 and T8 vertebral bodies (arrows), with retrolisthesis of T7 upon T8. Sagittal (c) and axial (d) FSE $T_2$-weighted images redemonstrate the unstable fracture dislocation (white arrows), retropulsion of fragments (open arrow), and narrowing of the central canal. Right rib fractures and small right hemothorax were also evident. Patient underwent T4–11 spinal instrumentation, T5–9 posterior decompression and repair of right T8 nerve root avulsion without residual neurologic deficit.

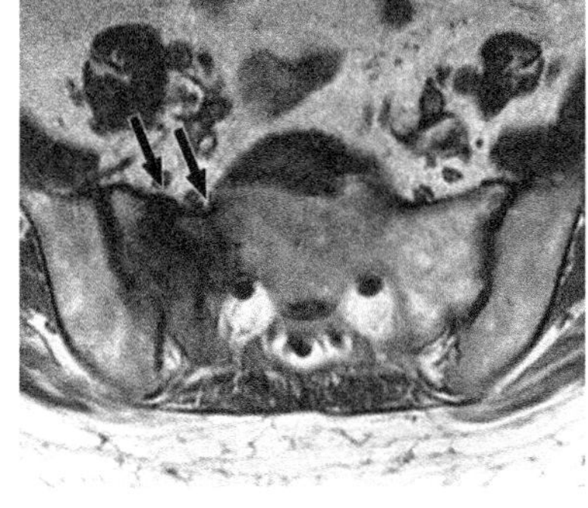

**FIGURE 10.22**
Right sacral wing fracture in an 81-year-old woman. Axial $T_1$-weighted MRI displays non-displaced fractures (arrows) extending through the right sacral wing.

increased edema and non-healed trauma. These findings have provided the basis for treatment of compression fractures with vertebroplasty and kyphoplasty cement.

### 10.10.1 Sacral Fractures

Along with CT and skeletal scintigraphy, MRI has been utilized in the diagnosis of sacral fractures (Figure 10.22) [47]. A retrospective review of MRI in the assessment of sacral fractures was conducted which demonstrated hypointense signal of edema along the fracture line on $T_1$-weighted imaging and hyperintense edema on $T_2$-weighted imaging. Occasionally, MRI will demonstrate the characteristic H-shaped fracture pattern, similar to the classically described scintigraphic *Honda sign*.

## 10.11 Assessment of Intradural Compartment in the Setting of Spinal Cord Injury

MRI has been found to be critical in the assessment of acute spinal cord injury due to its ability to clearly depict lesion location, extent, and severity [39]. In the setting of trauma, sagittal gradient-echo imaging is particularly useful in the presence of subdural or intramedullary blood products. While spinal cord edema and contusions may be inconspicuous on CT, $T_2$-weighted imaging demonstrates increased signal within the spinal cord. Further, $T_2$ hypointensity may be seen in the setting of acute hemorrhage. However, as the pathology evolves, the $T_2$ signal may change to hyperintensity, which may correlate with areas of edema formation, white matter myelin degeneration, and surrounding hemorrhage with necrotic and inflammatory change. MRI has been utilized to predict outcomes in the management of chronic spinal cord injury. While not widely adopted, diffusion tensor imaging can demonstrate the extent of damage to white matter tracts, which correlates with a worse outcome. In the acute setting, hemorrhage in the presence of trauma portends a worse long-term outcome.

In conclusion, the segmental anatomy of the spine together with its supporting structures allows for spinal motion in the confines of physiological spinal stability. Although radiographs and CT are the most appropriate first imaging studies in the setting of acute trauma, MRI is invaluable for the assessing the spinal ligaments, nerve roots, spinal cord, and intervertebral discs in patients who have sustained stable and unstable spinal injuries inflicted by axial loading, hyperextension, and hyperflexion mechanisms.

## References

1. White AA, Panjabi MM. Clinical biomechanics of the spine. 2nd ed. Philadelphia, PA: Lippincott, 1990.
2. Panjabi MM, White AA, 3rd, Johnson RM. Cervical spine mechanics as a function of transection of components. *Journal of Biomechanics*. 1975;8(5):327–36.
3. Schlicke LH, White AA, 3rd, Panjabi MM, Pratt A, Kier L. A quantitative study of vertebral displacement and angulation in the normal cervical spine under axial load. *Clinical Orthopaedics and Related Research*. 1979;34(140):47–9.
4. Spiteri V, Kotnis R, Singh P et al. Cervical dynamic screening in spinal clearance: Now redundant. *The Journal of Trauma*. 2006;61(5):1171–7; discussion 7.
5. Dekutoski MB, Hayes ML, Utter AP et al. Pathologic correlation of posterior ligamentous injury with MRI. *Orthopedics*. 2010;33(1):53.
6. Dlouhy BJ, Dahdaleh NS, Howard MA, 3rd. Radiographic and intraoperative imaging of a hemisection of the spinal cord resulting in a pure Brown-Sequard syndrome: Case report and review of the literature. *Journal of Neurosurgical Sciences*. 2013;57(1):81–6.
7. Martin D, Schoenen J, Lenelle J, Reznik M, Moonen G. MRI-pathological correlations in acute traumatic central cord syndrome: case report. *Neuroradiology*. 1992;34(4):262–6.
8. Rauschning W, McAfee PC, Jonsson H, Jr. Pathoanatomical and surgical findings in cervical spinal injuries. *Journal of Spinal Disorders*. 1989;2(4):213–22.
9. Weirich SD, Cotler HB, Narayana PA et al. Histopathologic correlation of magnetic resonance imaging signal patterns in a spinal cord injury model. *Spine*. 1990;15(7):630–8.
10. Benzel EC, Hart BL, Ball PA, Baldwin NG, Orrison WW, Espinosa MC. Magnetic resonance imaging for the evaluation of patients with occult cervical spine injury. *Journal of Neurosurgery*. 1996;85(5):824–9.
11. Lorenz F, Kespert FW. Radiographic assessment of the cervical spine in symptomatic trauma patients. *Neurosurgery*. 2002;50 (3 Suppl):S36–43.
12. Krassioukov A, Biering-Sorensen F, Donovan W et al. International standards to document remaining autonomic function after spinal cord injury. *The Journal of Spinal Cord Medicine*. 2012;35(4):201–10.
13. Greenberg MS, Greenberg MS. MRI of the bones. *Handbook of Neurosurgery*. Greenberg MS, ed. 7th ed. Tampa, FL: Greenberg Graphics, 2010.
14. Apuzzo MLJ. Guidelines for management of acute cervical spinal injuries. Introduction. *Neurosurgery*. 2002;50(3 Suppl):S1.
15. Dullerud R, Gjertsen O, Server A. Magnetic resonance imaging of ligaments and membranes in the craniocervical junction in whiplash-associated injury and in healthy control subjects. *Acta Radiologica*. 2010;51(2):207–12.
16. Dundamadappa SK, Cauley KA. MR imaging of acute cervical spinal ligamentous and soft tissue trauma. *Emergency Radiology*. 2012;19(4):277–86.
17. Dvorak J, Schneider E, Saldinger P, Rahn B. Biomechanics of the craniocervical region: The alar and transverse ligaments. *Journal of Orthopaedic Research: Official Publication of the Orthopaedic Research Society*. 1988;6(3):452–61.
18. Deliganis AV, Baxter AB, Hanson JA et al. Radiologic spectrum of craniocervical distraction injuries. *RadioGraphics*, A review publication of the Radiological Society of North America, Inc. 2000;20 Spec No:S237–50.
19. Anagnostara A, Athanassopoulou A, Kailidou E, Markatos A, Eystathidis A, Papageorgiou S. Traumatic retropharyngeal hematoma and prevertebral edema induced by whiplash injury. *Emergency Radiology*. 2005;11(3):145–9.
20. Spinal cord injury without radiographic abnormality. *Neurosurgery*. 2002;50(3 Suppl):S100–4.
21. Lamothe G, Muller F, Vital JM, Goossens D, Barat M. Evolution of spinal cord injuries due to cervical canal stenosis without radiographic evidence of trauma (SCIWORET): A prospective study. *Annals of Physical and Rehabilitation Medicine*. 2011;54(4):213–24.
22. Kasimatis GB, Panagiotopoulos E, Megas P et al. The adult spinal cord injury without radiographic abnormalities syndrome: Magnetic resonance imaging and clinical findings in adults with spinal cord injuries having normal radiographs and computed tomography studies. *The Journal of Trauma*. 2008;65(1):86–93.
23. Anderson RE, Drayer BP, Braffman B et al. Spine trauma. American College of Radiology. ACR Appropriateness Criteria. *Radiology*. 2000;215 Suppl:589–95.
24. Daffner RH, Hackney DB. ACR appropriateness criteria on suspected spine trauma. *Journal of the American College of Radiology*. 2007;4(11):762–75.
25. Keats TE, Dalinka MK, Alazraki N et al. Cervical spine trauma. American College of Radiology. ACR Appropriateness Criteria. *Radiology*. 2000;215 Suppl:243–6.
26. Stiell IG, Wells GA, Vandemheen KL et al. The Canadian C-spine rule for radiography in alert and stable trauma patients. *JAMA*. 2001;286(15):1841–8.
27. Hendey GW, Wolfson AB, Mower WR, Hoffman JR, National Emergency XRUSG. Spinal cord injury without radiographic abnormality: Results of the National Emergency X-Radiography Utilization Study in blunt cervical trauma. *The Journal of Trauma*. 2002;53(1):1–4.
28. Friedman D, Flanders A, Thomas C, Millar W. Vertebral artery injury after acute cervical spine trauma: Rate of occurrence as detected by MR angiography and assessment of clinical consequences. *American Journal of Roentgenology*. 1995;164(2):443–7; discussion 8–9.
29. Giacobetti FB, Vaccaro AR, Bos-Giacobetti MA et al. Vertebral artery occlusion associated with cervical spine trauma. A prospective analysis. *Spine*. 1997;22(2):188–92.
30. Parbhoo AH, Govender S, Corr P. Vertebral artery injury in cervical spine trauma. *Injury*. 2001;32(7):565–8.

31. Torina PJ, Flanders AE, Carrino JA et al. Incidence of vertebral artery thrombosis in cervical spine trauma: Correlation with severity of spinal cord injury. *American Journal of Neuroradiology*. 2005;26(10):2645–51.
32. Veras LM, Pedraza-Gutierrez S, Castellanos J, Capellades J, Casamitjana J, Rovira-Canellas A. Vertebral artery occlusion after acute cervical spine trauma. *Spine*. 2000;25(9):1171–7.
33. Weller SJ, Rossitch E, Jr., Malek AM. Detection of vertebral artery injury after cervical spine trauma using magnetic resonance angiography. *The Journal of Trauma*. 1999;46(4):660–6.
34. Provenzale J. MR imaging of spinal trauma. *Emergency Radiology*. 2007;13(6):289–97.
35. Bozzo A, Marcoux J, Radhakrishna M, Pelletier J, Goulet B. The role of magnetic resonance imaging in the management of acute spinal cord injury. *Journal of Neurotrauma*. 2011;28(8):1401–11.
36. Lee HM, Kim HS, Kim DJ, Suk KS, Park JO, Kim NH. Reliability of magnetic resonance imaging in detecting posterior ligament complex injury in thoracolumbar spinal fractures. *Spine*. 2000;25(16):2079–84.
37. Chaudhary SB, Hullinger H, Vives MJ. Management of acute spinal fractures in ankylosing spondylitis. *ISRN Rheumatology*. 2011;2011:150484.
38. Finitsis SN, Falcone S, Green BA. MR of the spine in the presence of metallic bullet fragments: Is the benefit worth the risk? *American Journal of Neuroradiology*. 1999;20(2):354–6.
39. Rao SK, Wasyliw C, Nunez DB, Jr. Spectrum of imaging findings in hyperextension injuries of the neck. *RadioGraphics*, A review publication of the Radiological Society of North America, Inc. 2005;25(5):1239–54.
40. Denis F. The three column spine and its significance in the classification of acute thoracolumbar spinal injuries. *Spine*. 1983;8(8):817–31.
41. Rihn JA, Anderson DT, Sasso RC et al. Emergency evaluation, imaging, and classification of thoracolumbar injuries. *Instructional Course Lectures*. 2009;58:619–28.
42. Vaccaro AR, Oner C, Kepler CK et al. AOSpine thoracolumbar spine injury classification system: Fracture description, neurological status, and key modifiers. *Spine*. 2013;38(23):2028–37.
43. Pizones J, Izquierdo E, Alvarez P et al. Impact of magnetic resonance imaging on decision making for thoracolumbar traumatic fracture diagnosis and treatment. *European Spine Journal*, Official publication of the European Spine Society, the European Spinal Deformity Society, and the European Section of the Cervical Spine Research Society. 2011;20(Suppl 3):390–6.
44. Pizones J, Sanchez-Mariscal F, Zuniga L, Alvarez P, Izquierdo E. Prospective analysis of magnetic resonance imaging accuracy in diagnosing traumatic injuries of the posterior ligamentous complex of the thoracolumbar spine. *Spine*. 2013;38(9):745–51.
45. Thawait SK, Marcus MA, Morrison WB, Klufas RA, Eng J, Carrino JA. Research synthesis: What is the diagnostic performance of magnetic resonance imaging to discriminate benign from malignant vertebral compression fractures? Systematic review and meta-analysis. *Spine*. 2012;37(12):E736–44.
46. Brown DB, Glaiberman CB, Gilula LA, Shimony JS. Correlation between preprocedural MRI findings and clinical outcomes in the treatment of chronic symptomatic vertebral compression fractures with percutaneous vertebroplasty. *American Journal of Roentgenology*. 2005;184(6):1951–5.
47. Lyders EM, Whitlow CT, Baker MD, Morris PP. Imaging and treatment of sacral insufficiency fractures. *American Journal of Neuroradiology*. 2010;31(2):201–10.

# 11

# *Neoplastic Disease of the Spine*

**Michele Porcu**

**CONTENTS**



## 11.1 Introduction

As the other district of the body, a neoplastic process can involve the spine, the spinal cord (SC), and all the tissues that constitute the spine; it could be benignant or malignant, primitive or secondary.

These conditions could be silent for the whole life, founded occasionally during the autopsy, or generate different symptoms, according to their localization and the structures involved. When signs and symptoms appear, often they are nonspecific of a neoplastic involvement, and moreover the physical and instrumental evaluation, imaging investigations are always required in order to reach the correct diagnosis.

*Conventional radiographs*, for example, are useful in the evaluation of bone structures, including any changes in the normal bone structures (differences in the physiologic curvature of the spine, changes in the shape or in the density and quality of the bone).

*Computed tomography (CT)* is the best diagnostic investigation in order to have a complete and multiplanar evaluation of the bone structures of the spine. The use of intravenous iodinate contrast medium is useful to better characterize normal and pathologic findings of the soft tissues (in particular their vascularization) and the relations with neighboring structures.

*Magnetic resonance (MRI)* is the method of choice for the evaluation of the SC and of the soft tissues inside the spinal canal. It is moreover a complementary examination for the study of the neoplastic diseases of the bones.

In this chapter, it will be exposed a synthetic analysis of the most important vertebral structures necessary for the correct interpretation of the MRI examinations; it will be analyzed then the main MRI findings of the principal neoplastic diseases that can affect the spine and the SC, exposing the main characteristics helpful in the differential diagnosis not only between neoplastic and non-neoplastic diseases, but even between different tumors.

## 11.2 Normal Anatomy

The spine consists of two main structures: the SC and vertebral column.

The *SC* is the structure responsible for the transmission of the nervous pulses from the periphery to the central nervous system (CNS) and vice versa. It has a cylindrical shape; it is contiguous cranially to the brain stem and terms generally at the level of the 12° dorsal vertebra or 1° lumbar vertebra; the terminal portion at this level is tapered, with a conical shape, and it is called *conus medullaris;* it continues with a long fascicle of nerves called *cauda equina* up to the sacral channel. The conus medullaris is linked to the internal coccygeal surface with a fibrous ligament called *filum terminale,* which consists of the three meningeal layers strictly adhered one to each other (see above). From the SC originates the anterior (motor) and posterior (sensitive) roots of the spinal nerves, which at the level of the vertebral foramen join together and give birth to the spinal nerves.

At the center of the SC, there is a long and narrow channel called *ependymal channel,* directly in connection with the caudal end of the fourth cerebral ventricle, and responsible for the production and transit of the CSF: this channel is entirely covered by ependymal cells.

In the SC, on an axial, we can recognize, from the center to the outside, the ependymal channel in the center, and around it the *gray matter* with the classical *butterfly* or *H* shape, which constitutes moreover by neurons and glial cells; externally to it, there is the white matter, which constitutes moreover by nerve fibers and glial cells. The posterior half of the SC is mostly important for the tactile, proprioceptive, pain and thermic sensibility, whereas the role of the anterior half (in particular in the para-median region) is the transmission of the motor commands from the periphery by the reflexes or the motor areas.

Externally, the SC is completely covered by the *meninges,* structures made of connective tissues, important for their role of support inside the vertebral channel, in continuity with the intracranial meninges. The meninges consist of three thin layers, which are from the inside to the outside: the pia mater, the arachnoid mater, and the dura mater.

The pia mater is strictly adhered to the SC and the nerve roots.

The space between the pia mater and the arachnoid mater is filled by the CSF, which is an important support mechanism for the SC during the movement of the body.

The dura mater is strictly adhered to the arachnoid mater and is the external sac that contains the SC. Between them there is a virtual space called *subdural space.* The dura mater is linked to the foramen magnum of the occipital bone and to the periostium of the vertebral column at the level of the 2° and 3° cervical vertebra (CV) and caudally in the lumbar and coccygeal region. The space between the dura mater and the osseous vertebral channel is called *epidural space,* and normally, it contains adipose tissues, veins, and lymphatic vessels; this space end laterally at the level of the origin of spinal nerves; in fact, the three layers end together at the origin of the spinal nerves in the neural foramen, merging together with the epineurium of the spinal nerves.

From the SC goes out 8 pairs of cervical nerves, 12 thoracic, 5 lumbar, 5 sacral, and 3 coccygeal, with a metameric distribution (i.e., for every vertebra, there is a correspondent pair of nerves). The first cervical spinal nerve (C1N) goes out from the incisures between the occipital bone and the atlas (C1); the cervical nerves C2N–C7N go out from the neural foramens between the corresponding and the overlying vertebra (e.g., C3N between CV2 and CV3), and C8N pairs go out from the neural foramens between CV7 and TV1. The other nerves go out from the neural foramens delimited by the corresponding and the underlying vertebral vertebra (e.g., L4N pairs between L4V and L5V).

The *arterial vascularization* of the SC is furnished by the three spinal arteries:

- *One anterior artery*: Originates by the vertebral arteries at the level of the brain stem and course in the anterior median fissure of the SC. This artery is well anastomosed at the thoracic level with the spinal branch of the dorsal branch of the intercostal arteries; in particular, the largest of them is the Adamkiewicz artery, located generally at the level of the 7° TV.
- *Two posterior arteries*: Originate directly by the vertebral arteries, narrower than the anterior; they descend parallel along the posterior surface of the spine up to the conus medullaris; at this level, they are linked with the anterior medullary artery.

The *venous vascularization* of the SC consists of the two medullary veins (one anterior and one posterior) that drain into the correspondent *venous epidural plexus*.

The vertebral column is the rigid part of the spine. It is composed of 7 cervical vertebrae, 12 thoracic, 5 lumbar, the sacrum, and the coccygeal bone.

Every vertebra consists of a *body*, located anteriorly, and an *arch*, located posteriorly; they both delimit a space called *vertebral foramen*, in which is located the SC. The arch consists of two pedicles and two laminae, and on this originates two transverse processes, pair and symmetric, the article facets (two superior and two inferior), and one spinous process (located posteriorly). On the inferior surface of the pedicles, there is an incisure called *transverse incisure*, which delimits the neural foramens with the transverse process of the underlying vertebra. The spinal nerves go out from the spinal canal via the neural foramens.

The first cervical better is called *atlas*, and it is articulated cranially with the occipital bone and caudally with the second cervical vertebra, called *epistropheus*. Every vertebra is linked with the underlying vertebra (and the fifth lumbar vertebra with the sacrum bone) anteriorly with the intersomatic disk (a fibrocartilagineus structure) and posteriorly with the facet joint.

The main structures that gave stability to the spine are as follows:

- The anterior and posterior longitudinal ligaments, which course along the anterior and posterior (intracanalar) surface of the vertebral bones
- The flavum ligaments, placed on the anterior surface of the vertebral arch, inside the spinal canal
- The interspinous and sovraspinous ligaments, between and posteriorly to the spinous processes
- The extrinsic musculature of the spine

The arterial vascularization of the spine is furnished directly by the branches of the posterior branchv of the intercostal and lumbar arteries; the venous blood drains anteriorly via the basivertebral vein into the anterior external venous plexus and the anterior epidural plexus, and posteriorly into the posterior epidural plexus and the posterior external plexus.

## 11.3 General Classification and MRI Examination Protocol

In general, the CNS expansive lesions can be classified according to their location into two different groups [1]: intra-axial lesions and extra-axial lesions.

Sometimes, especially when the lesion is very small, it is difficult to distinguish between intra- and extra-axial lesions.

When the radiologist has to approach with a neoplastic lesion of the spine, this classification can be implemented. It is thus possible following this scheme [2,3]:

- Intradural tumors
  - Intramedullary
  - Extramedullary
- Extradural tumors

Intradural tumors are not common neoplasms; they represent 30% of all tumors of the spine [2], whereas extradural tumors account for 60% of the total amount [2]; the rest of the tumors of the spine can involve both the intramedullary and extramedullary spaces (10% of the total amount) [2,4,5].

The most frequent extradural tumors (>90%) are metastases located in the vertebral bodies [3]. Intradural intramedullary tumors are characterized primarily by a SC expansion (symmetric or asymmetric); on MRI, it is possible to identify the SC expansion with a complete subversion of the normal gray–white matter distribution. In intradural extramedullary tumors, the differentiation between white and gray matter is still conserved, even if distorted, the dura mater is still recognizable. In extradural tumors, the dura mater is abnormally displaced from its original position.

The classification mostly used in the clinical practice is the WHO 2007 Classification of Tumours of the Central Nervous System [6]; even in this chapter, the discussion of the single histological entities will refer to this classification. In particular, it is important remembering the grading system of this classification (Grade I–IV) indispensable for a correct prognostic evaluation (see Table 11.1) [6].

**TABLE 11.1**

WHO 2007 Classification of Tumors of the Central Nervous System

| Grading | Characteristics |
|---|---|
| I | Lesions with low-level proliferative potential; rarely recurrence after complete surgical eradication. |
| II | Infiltrative lesions with a low-level proliferative rate; often recurrence, even after complete surgical eradication. Some of them tend to progress to higher grade. |
| III | Infiltrative lesions with histological findings of malignancy, including nuclear atypia and high mitotic activity; different cases require adjuvant radiation or chemotherapy, and recurrence is highly frequent. |
| IV | Diffusively infiltrative lesions with high mitotic rate, marked cellular atypia, prone to necrosis. Rapid unfavorable pre- and postoperative evolution. Some of these lesions are associated with dissemination at distance. |

*Source:* Louis, D.N. et al., *Acta Neuropathol.*, 114, 97–109, 2007.

A primary evaluation of the spine can be obtained with plain radiograph or non-contrast CT in order to evaluate the presence of skeletal deformities and bone expansive or lytic lesions that could even involve the vertebral canal. The CT could be performed with or without iodinate contrast medium intravenous infusion (i.v.); in rare cases in which the MRI could not be performed, a CT myelography with the intrathecal injection of iodinate-contrast medium could be helpful for the diagnosis of gross intradural lesions. MRI is considered as the diagnostic tool of choice for the evaluation of this kind of lesions [7], because of its high contrast resolution.

A correct MRI study approach should be performed with the use of sagittal $T_1$-weighted ($T_1$W) and $T_2$-weighted ($T_2$W) sequences, and axial and sagittal $T_1$ sequences after the intravenous administration (i.v.) of paramagnetic gadolinium-based contrast medium (GCM), with a slice thickness of 3–4 mm and a gap of 0.5–1 mm [5,7]. The use of diffusion-weighted imaging (DWI) sequences could be useful as a complement of the examination; the use of short-tau inversion recovery (STIR) sequences can be useful in the evaluation of bone marrow, SC, and edema of the soft tissues [7]. The gradient-echo sequences were used for the evaluation of deposition of hemosiderin. GCM i.v. is indispensable for the individuation of the solid and more vascularized part of the tumor (although if many types could not show intense contrast enhancement), the individuation of neoplastic and non-neoplastic cysts inside and outside the lesion, and a possible spread of the disease in the rest of the spine. If a spinal neoplasia is detected, a brain MRI study should be used as a diagnostic completion.

## 11.4 Intradural Tumors

The tumoral lesions of the intradural compartment can be divided into two main groups:

- *Intramedullary tumors*: Represent 20% of all intradural tumors in adults (30%–35% in the pediatric population) [7]; 90% of them are represented by glial tumors [7].
- *Extramedullary tumors*: Represent 80% of all intradural tumors in adults (60%–75% in the pediatric population) [7].

From a clinical point of view these lesions have a nonspecific pattern of signs and symptoms and, moreover, the first symptom could appear in an advanced stage of the disease [7].

The *symptomatology* could vary according to the position and level of the lesion, the growth rate, and the presence of other associated pathological conditions. The most common signs are represented by the appearance of progressive back pain, weakness, numbness of the distal limb or arm (monolateral or bilateral), paresthesia, gait problems, bladder and bowel dysfunction, impotence, and incontinence [7].

The *signs* are represented by deficit in the thermal, pain, tactile or proprioceptive sensitivity, hyperreflexia, clonus, or the Babinski sign [7].

In the pediatric population, the intramedullary tumors could present with symptoms like back pain, motor regression, and frequent falling [8], the extramedullary tumors could present with progressive myelopathy, weakness, and diffuse back pain [9], whereas in both cases signs of skeletal deformity could be evident [7,8].

Now it will be analyzed in detail the principle characteristics of the most frequent histologic types of the intramedullary and extramedullary tumors.

### 11.4.1 Intramedullary Tumors

#### *11.4.1.1 General Characteristics and Differential Diagnosis with Non-Tumoral Conditions*

Intramedullary tumors can be divided according to their histological origin in glial and non-glial cells. The presence of an intramedullary tumor has to be suspected in those cases in which it is detected a lesion inside the SC with the following three characteristics in an MRI examination [5,7,8,10]:

1. All of them present with focal enlargement of the SC.
2. Commonly neoplastic lesions are associated with the presence of cystic areas. These areas

could be neoplastic or non-neoplastic, located inside the tumor or adjacent or near the lesion, along the SC. Intense contrast enhancement (c.e.) of the walls is typical of the neoplastic cystic lesions, whereas non-neoplastic cyst-like lesions, in particular those located at the poles of the lesion, could be reactive enlargements of the ependymal canal. Cystic areas inside the tumor could be necrotic areas or contain mucinous material, and could present themselves parietal c.e. [10].

3. Almost always the tumoral lesions present some areas of enhancement after GCM i.v. administration.

The absence of these MRI findings could let the radiologist to consider other lesions that could affect the SC, even if some tumors could not even show one of these characteristics. In order to reach the diagnosis, it is indispensable every time to evaluate the imaging findings with the clinical and laboratory data of the patient.

An isolated cystic lesion without parietal c.e. should suggest us the diagnostic hypothesis of *syringohydromielia*, or *ventriculus terminalis* if it is placed in the conus medullaris [10].

If after a spinal trauma is seen in the SC in one or multiple areas of high signal intensity on $T_2$W images, with associated swelling of the SC and hemorrhagic areas (in the acute phase, hyperintense on $T_1$W and $T_2$W images), especially if epidural hematoma and/or vertebral fractures and/or abnormal signal of paravertebral muscles are present, an *SC contusion* has to be considered [10,11].

In the case of a sudden onset of para/tetraparesis (especially in association with aneurism or dissection of the aorta or of the vertebral arteries) or other neurological symptoms with no previous history of trauma, in relation with hyperintense $T_2$W signal of the gray matter or of the entire cross section of the SC (often of the lower thoracic tract), accompanied sometimes by slight regular enlargement of SC and signs of restrictions in DWI, should suggest us an *SC arterial infarction* [10,12]; the so-called owl's eyes sign is a typical finding of infarction of the anterior half of the SC that sometimes is evident in the axial views, and it is characterized by a hyperintense signal of the anterior horns of the gray matter in the $T_2$W sequences [12]. An *SC venous infarction* could occur sometimes as a complication of a spinal dural arteriovenous fistula, and in this case, the SC appears inhomogeneously hyperintense on $T_2$W sequences, with irregular aspect of the surface of the cord, and the $T_1$W sequences after GCM i.v. administration show enhancement of the vessels and delayed enhancement of the rest of the SC [12].

*Transverse myelitis* is quite similar to the SC infarction; the MRI findings comprehend an area hypointense on $T_1$W sequences and hyperintense on $T_2$W sequences, with a variable enhancement, often patchy, located generally at the center of the SC, which extends for over 3 vertebral segments and most commonly involves more than 50% of the cross section of the SC, accompanied by slight swelling of the SC [10,12,13]; moreover, the onset of neurologic symptoms (both motor and sensory, with transverse segmental diffusion) differs because it is progressive rather than sudden and dramatically responds to the steroid therapy [10].

*SC bacterial abscess* has to be suspected in particular in immunocompromised, septic, or drug-user patients and in those who suffer from meningitis or other related conditions; the symptomatology is characterized by the appearance of neurologic deficit in immunocompromised, drug-user, or septic patients (attention in particular in those patients who suffer from meningitis) associated with fever, leukocytosis, elevated C-reactive protein (CRP) values, and neurologic deficit [10]; the principal MRI findings are the presence of an intramedullary expansive lesions (generally lesser than 2 cm), dishomogeneously hypointense on $T_1$W sequences, hyperintense on $T_2$W sequences with associated edema, restriction in DWI sequences if the content is pus, and above all a well-defined ring-enhancement in $T_1$W sequences after i.v. GCM [10,12].

*Multiple sclerosis* is a demyelinating disease that has to be considered in young or middle-aged patients with a history of different neurological symptoms with sudden appearance that can even spontaneously regress in part or totally in a variable range of time; this disease can have a relapse/remitting or a progressive course [10]. Multiple sclerosis plaques in the SC are present in 74%–85% of the cases [14]; they are very difficult to detect on $T_1$W sequences, but appear as focal ovular-shaped lesions developed along the venular axis of the SC, hyperintense on $T_2$W sequences, more frequently located in the dorsolateral tracts of the cervical spine with no respect for the boundaries between white and gray matter; active lesions could show c.e. after GCM i.v. administration, cord swelling, and perilesional edema [10,12]. The finding of a lesion in the brain can help in the diagnosis, together with the instrumental and laboratory data [12].

The SC lesions of the *acute disseminated encephalomyelitis*, a rare and dangerous disease that affects children and young adults, are multiple, flame-shaped lesions, hyperintense on $T_2$W sequences, with patchy peripheral enhancement in the $T_1$W sequences after GCM i.v. administration; they are accompanied by cerebral irregular lesions hyperintense on $T_2$-sequences, spared, located in the white–gray matter junction, and can involve thalamus and basal ganglia [12].

The *B12 vitamin deficiency* manifests with demyelinating lesions hyperintense on $T_2$W sequences longitudinally oriented in the white matter of the lateral and posterior regions of the SC that sometimes could resemble an inverted V shape, which do not enhance in $T_1$W sequences after GMC i.v. administration [12,15].

*Spinal neurosarcoidosis* is another rare granulomatous disease of unknown origin that has to be considered in the differential diagnosis with the intradural intramedullary neoplasms, in particular in patient with a known diagnosis of sarcoidosis or with abnormal findings on chest X-ray or total body CT examination, increased blood or CSF levels of angiotensin-converting enzyme (ACE), and CD4:CD8 ratio [12,16]; the MRI shows a fusiform expansion of the SC, with diffuse or focal hyperintense signal on $T_2$W sequence, with diffuse or patchy enhancement after GCM i.v. infusion in $T_1$W sequences, accompanied by leptomeningeal enhancement [10,12,16]. The MRI findings (in particular the leptomeningeal enhancement) are very similar to those of the *Lyme disease* [12].

In the end, in some endemic regions, parasitic disease can affect the SC, even if it is a very rare event. *Spinal hydatid disease* can localize in different structures of the rachis: these cystic lesions when localized in the SC have a typical sausage-like shape, appear hyperintense on $T_2$W sequences with hypointense thin walls because of the fibrotic reaction, and in $T_1$W sequences, the walls appear slightly more hypointense in comparison with the hypointense $T_1$W signal of the cystic liquid; the walls moreover enhance after GCM i.v. administration [17]. *Spinal schistosomiasis* is evidenced as lesions that determine a cord expansion with hypointense appearance on $T_1$W sequences and patchy hyperintense appearance on $T_2$W images [18].

For a more detailed discussion of these single pathological entities, the readers are suggested to read the other chapters of this book.

#### 11.4.1.2 General Characteristics of the Most Common Histotypes

Intradural intramedullary tumors can originate from different cellular types. Primitive tumors are absolutely the most represented, whereas metastatic tumors account only for 1%–3% [2]. Among all histotypes, in 80% of the cases these lesions originate from glial cells (ependymoma and astrocytoma for example) [2]. The two most common glial cell tumors are the *astrocytoma* and the *ependymoma*: the first one is more common in the pediatric population (80%–90% of all intradural intramedullary tumors of glial origin in the childhood and up to 60% in the adolescence), and the second one is typical of the adult population.

Of course, the histotype of the lesion can be suggested on MRI, but the correct diagnosis can only be performed by the pathologist, in concordance with the clinical and imaging data.

##### *11.4.1.2.1 Astrocytoma*

Astrocytoma is a glial tumor that is slightly more common in male than in female [5]. According to the WHO 2007 classification of CNS tumors [6], tumors are classified into four grades of aggressiveness according to the histotype:

1. The *low-grade group* (I and II) is represented by pylocitic (I) and fibrillary (II) astrocytomas.
2. The *high-grade group* (III and IV) is represented by anaplastic astrocytoma (III) and glioblastoma multiforme (IV). This group accounts for just 10% of all astrocytomas of the SC [2].

Generally, they are located more in the thoracic SC (65%–70% of the cases) than in the cervical tract, rare in the filum terminale, involve less than 4 SC segments even if sometimes can present with a multisegmental or holocord diffusion, especially the pylocitic histotype [7] and rarely are exophytic [5,7]. On MRI, generally astrocytomas appear as a fusiform enlargement of the SC, usually eccentric, iso-, or hypointense on $T_1$W sequences and hyperintense on $T_2$W sequences and with variable and irregular enhancement after GCM i.v. administration [2,5,7]. The rostral and caudal cysts that could be present associated with the tumor and are not included inside it tend to be benign, with proteinaceous or hemorrhagic content [10].

Low-grade astrocytomas show little or no peripheral edema or hemorrhage [2]; in $T_1$W sequences after GCM i.v. administration, they generally show inhomogeneous enhancement, but sometimes there could be no enhancement at all [2,5,7]. Pylocitic type could show well-defined margins and tend to displace than infiltrate the SC [19], whereas the fibrillary type generally has ill-defined margins and can show no enhancement at all [2] (Figure 11.1).

The high-grade astrocytomas show often edema, intratumoral cysts, and necrosis [2]. In particular, glioblastoma multiforme in 60% of the cases is associated with leptomeningeal spread of the disease [5] (Figure 11.2).

In conclusion, we must remember that the astrocytoma generally differs from the ependymoma (see below) because astrocytoma tends to be eccentric, often located in the thoracic tract, with ill-defined margins, rarely shows hemorrhage, and shows patchy and irregular enhancement after GCM i.v. administration [2,4,5,19].

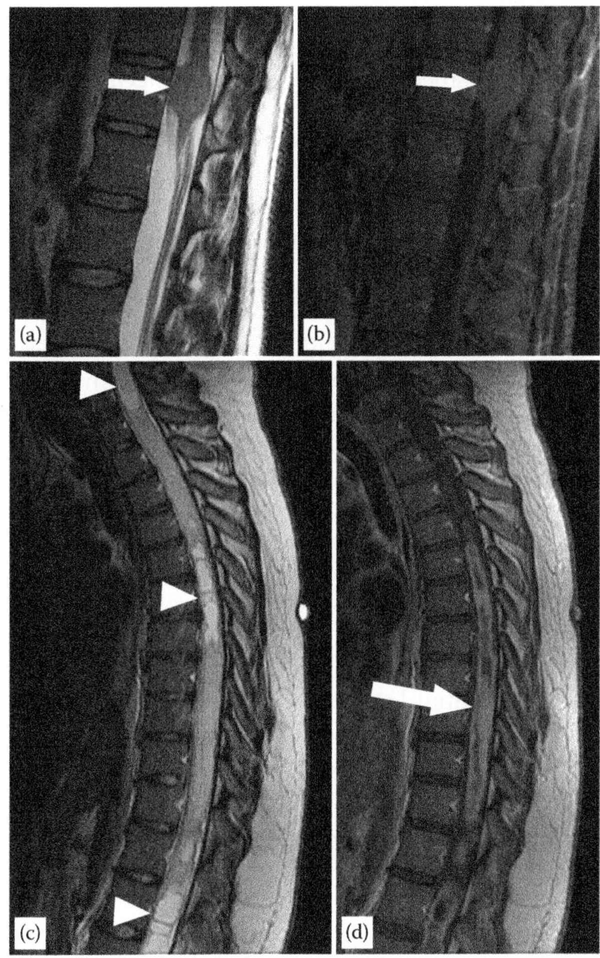


**FIGURE 11.1**
Pilocytic astrocytoma in two different patients. Sagittal FSE $T_2$-weighted (a) and enhanced $T_1$-weighted FS (b) images in a 39-year-old man presenting with left foot drop demonstrate minimal peripheral enhancement of the expansile, solid mass in the conus meduallris. Mass proved to be an infiltrating grade I pilocytic astrocytoma of the conus medullaris at surgical debulking. Pilocytic astrocytoma and syringomyelia in a 17-year-old girl presenting with back pain, right foot numbness, right upper extremity weakness and decreased proprioception. Sagittal FSE $T_2$-weighted (c) and enhanced $T_1$-weighted FS (d) images show extensive area of syringomyelia throughout the entire spinal cord (arrowheads) and heterogeneous contrast enhancement, most marked in the lower thoracic spine (arrow). Grade 1 pilocytic astrocytoma was found at surgical biopsy, and patient was subsequently referred for chemotherapy and radiation therapy.

*11.4.1.2.2 Ependymoma*

Ependymoma is a glial tumor that has a peak of incidence in the fourth and fifth decades and represents 60% of all intadural intramedullary tumors of the adult population [2]. In the pediatric population, they could be associated with neurofibromatosis type 2 [19].

According to the WHO 2007 classification of CNS tumors [6], almost all the ependymomas are low-grade tumors (I–II) [5]. From a histological point of view, the most common variants are the cellular and the myxopapillary [2,5]; the cellular variant is more frequently present in the upper SC, whereas the myxopapillary variant is almost always located in the filum terminale [5,7]. Generally, they involve 3–4 SC segments, even if in the literature are reported cases which have involved up to 15 segments [5].

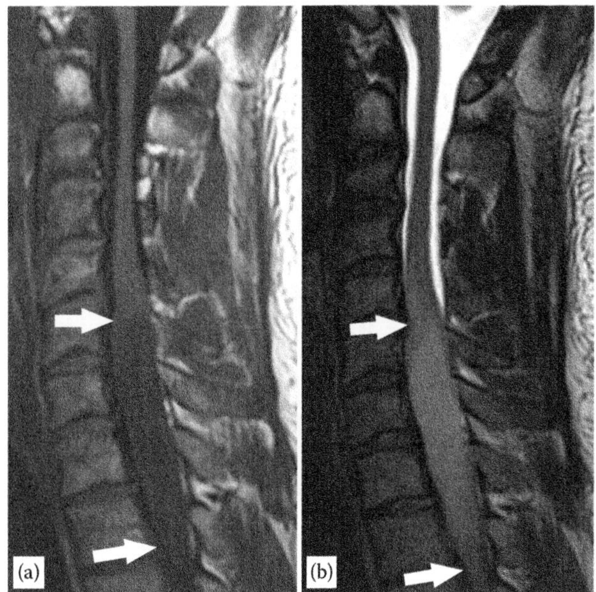


**FIGURE 11.2**
Astrocytoma of the cervical cord in a 31-year-old man with progressive quadriparesis. Sagittal $T_1$-weighted (a) and FSE $T_2$-weighted (b) images delineate an infiltrating, expansile intramedullary mass involving the lower cervical and upper thoracic spinal cord. Pathology from debulking revealed grade II diffuse infiltrating astrocytoma.

On MRI, they show iso-hypointense signal on $T_1$W sequences and inhomogeneously hyperintense signal on $T_2$W sequences; the myxopapillary type could be sometimes hyperintense on both $T_1$W and $T_2$W sequences, because of the mucinous or hemorrhagic content [5]. Since generally ependymomas tend to compress than infiltrate the neighboring tissues [5,19], they appear well circumscribed by a pseudocapsule and show intense homogeneous or sometimes heterogeneous enhancement after GCM i.v. administration [19].

Frequently, they are associated with hemorrhages [5]. A typical sign of the ependymoma is the so-called cap sign [5,19] which when present is (20% of the cases) highly suggestive for ependymoma [19], and it is a rim of low-signal on $T_2$W sequences in the border of the neoplasm because of the deposition of hemosiderin for a previous hemorrhage [5]; this sign could be present even in other highly vascular tumors such as hemangioblastoma (Figure 11.3) [5].

SC edema is present approximately in 60% of the cases, and often, polar cysts and syringohydromyelia are present [5]; intratumoral cysts are less common [19].

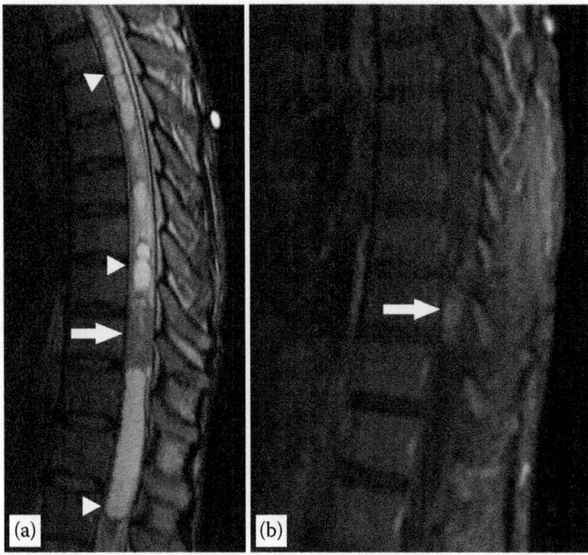


**FIGURE 11.3**
Ependymoma in a 39-year-old man with progressive lower back pain, bilateral leg numbness and weakness, and saddle-type paresthesia. Sagittal FSE $T_2$-weighted (a) and enhanced $T_1$-weighted FS (b) images show a heterogeneously enhancing intramedullary mass (arrows) in the lower thoracic spine, resulting in cord expansion and adjacent syringomyelia (arrowheads). Surgical resection revealed grade II ependymoma.

The 5-year survival rate for patient with low-grade ependymoma is among 83% and 100% [2]; in the pediatric population, the complete removal of the tumor is possible in 50% of the cases, and the 5-year survival rate is 85%, whereas in the case of subtotal removal, this rate decreases up to 57% [19].

In conclusion, we must remember that the ependymoma generally differs from astrocytoma because it is located centrally and symmetrically in the SC, especially in the conus and in the filum terminale, it has well-defined margins, more often tends to be hemorrhagic (in 20% of the cases, it is possible to recognize the *cap sign*), it is often associated with syringohydromyelia [20], and it shows intense enhancement after GCM i.v. administration [2,4,5,19].

### 11.4.1.2.3 Other Glial Neoplasms

*Ganglioglioma* represents about 1% of all SC tumors [2,5], but up to 15% in the pediatric group [19]. Generally, they are low-grade tumors that, from a histological point of view, consist of two different cellular population, one neuronal and one glial. They tend to be very similar and not at all easily distinguishable from astrocytomas, although Patel et al. [21] have suggested the mixed signal in $T_1$W sequences as a peculiar characteristic, very uncommon for the other histotypes. Other characteristics are the possible presence of calcification inside the lesion and the absence of edema; after GCM iv. Administration, they show patchy enhancement in the greatest part of the cases, accompanied or not by leptomeningeal enhancement [5,19,21].

*Subependymoma* represents very rare tumors (less than 50 cases in literature up to 2007 [2]) that are very similar to the ependymomas on MRI, but differs from them because it has been reported that generally it is eccentric [5].

### 11.4.1.2.4 Hemangioblastoma

*Hemangioblastoma* is a non-glial low-grade capillary-rich tumor (WHO grade I) that represents less than the 10% of all the SC tumors, which occur generally in patient <40 years old, and that it is very rare in the pediatric population; in approximately one-third of the cases, they are associated with the Von Hippel–Lindau (VHL) syndrome [19], in particular if they are multiple [5]; if hemangioblastoma is discovered in a pediatric patient, it is fundamental looking for mutation of the VHL gene located in the chromosome 3 [19]. In 75% of the cases, it just has an intramedullary development [5].

The preferential location of the tumor is in the thoracic cord (in about 50% of the cases), less common in the cervical tract, and rare in the caudal tract [4,22]. On MRI, it manifests with a cord swelling with the peculiar characteristic that the SC enlargement extends over the enhancing tumor and it is distinct from the associated syrinx (when the last one is present) [22]. Hemangioblastoma shows a variable signal on $T_1$W sequences, hyperintense on $T_2$W images, with characteristic intermixed focal flow-voids because of its important vascularization, and it enhances homogeneously on $T_1$W sequences after GCM i.v. administration [4,5,19]. Often, this tumor is associated with feeding artery and draining veins, and adjacent cysts, syringohydromyelia (40% of the cases, with a cyst-like or regular shape [22]), and sometimes edema could be present [2,4,5,19,22]. Intraparenchymal or subarachnoid hemorrhage is rare [22]. Some of them could grow up exophytically or be extramedullary [22]. Some cases could present with the so-called aspect *cystic mass with an enhanced mural nodule* typical of the cerebellar hemangioblastoma (Figure 11.4) [4].

They are generally removed with surgical intervention, but if it is not possible radiosurgery and gamma knife are the alternatives [19]. The differential diagnosis must consider arteriovenous malformations, cavernous malformations, and hypervascular cord neoplasms [19].

### 11.4.1.2.5 Angiomatous Lesions

*Angiomatous lesions*, often called *hemangiomas*, generally are considered more vascular malformations than true neoplasms and quite rarely affect pediatric and

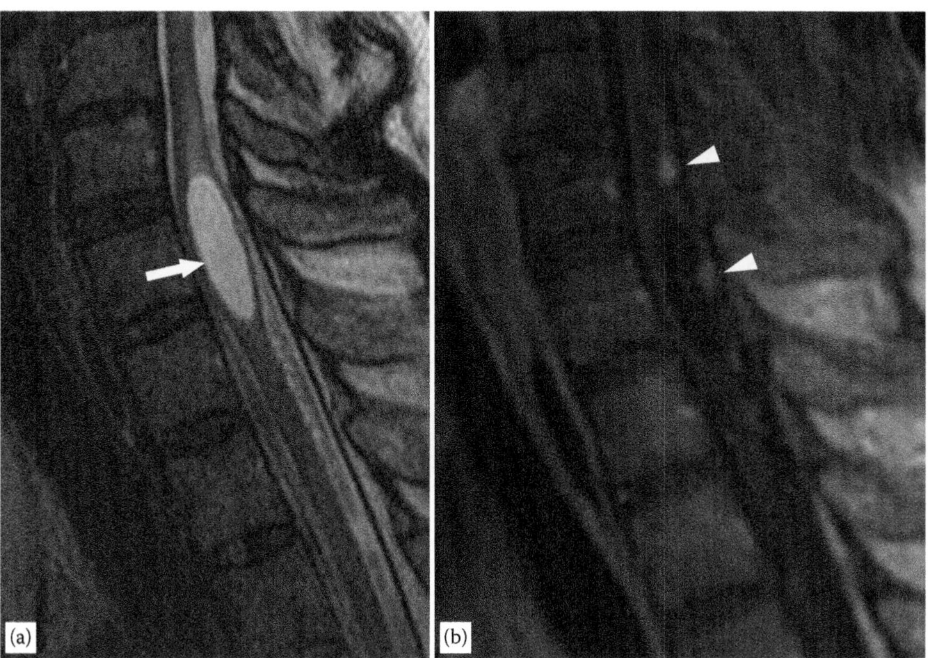


**FIGURE 11.4**
Hemangioblastoma of the thoracic spine in a 54-year-old woman with Von Hippel–Lindau disease and previously resected cerebellar hemangioblastoma. Sagittal FSE $T_2$-weighted (a) and enhanced $T_1$-weighted (b) images show a cystic, expansile lesion (arrow) in the central cord centered at T9, with small nodular enhancing foci (arrowheads). Patient underwent resection of the grade I hemangioblastoma, with T8–10 laminectomies and nerve root microdissection.

adult population [2]. From a histological point of view, they consist of non-organized vessels with no interposition of brain parenchyma. According to the size of the dominant vessels of which are they made of, they could be divided into capillary and cavernous hemangioma [2].

On MRI, if they are not complicated and just incidental findings, generally they manifest as focal lesion without SC swelling or edema [2]. If a complication occurs (such as venous congestion or thrombosis of the vessels of the lesion), they could be responsible for the onset of the symptoms and SC swelling and surrounding edema could appear [2].

*Cavernous hemangioma* appears on MRI with the typical *pop corn* appearance, that is, a focal mass inhomogeneously hyperintense on $T_1$W and $T_2$W sequences (due to subsequent events of micro-thrombosis and micro-hemorrhages inside of the lesion), surrounded by a hemosiderin hypointense ring, more visible on gradient-echo sequences [23]. Usually, they do not enhance after GCM i.v. administration [2]. If an SC cavernous hemangioma is discovered, it is important to look for other lesions of this type in the rest of the neuraxis because often they are multiple and could be associated with *familiar cavernomatosis* [23]. Generally, the total surgical removal is possible with excellent outcomes, and if not operated, the rate of spontaneous hemorrhages is 1%–4% [2].

### *11.4.1.2.6 Other Non-Glial Neoplasms*

*Lipoma* represents about 1%–11% of all the spinal tumors, located preferentially in the cervical and dorsal tract of the SC [2]. On MRI, they are hyperintense on $T_1$W images and hypointense on $T_2$W and STIR sequences.

Gangliogliomas account for just 1% of the total amount of SC intramedullary tumors [2], even if in the pediatric population the incidence grows up to about 15% [19]. They are composed of two cellular populations, one neuronal and one glial [2,19]. They typically are WHO low-grade tumors (I) even if they have a high recurrence after surgical removal [19]. On MRI, they are typically eccentrically located, with mixed signal on $T_1$W sequences, cord medullary surface, and patchy tumor enhancement on $T_1$W sequences after GCM i.v. administration, absence of edema, and accompanied by tumoral cysts and syringohydromyelia [2,19].

*Teratoma* is a very unusual finding in the SC, and generally, it is a benign lesion located in the lumbar SC [2,4]. They are composed of two different parts, one fatty and one more dense or liquid, in different proportions [4]. On MRI, the lipidic part appears hyperintense on $T_1$ sequences and hypointense on fat-suppression $T_1$W sequences; on $T_2$W sequences, teratomas could appear hyper- or hypointense according to the amount of liquid and lipid inside [4]. Sometimes, they can even present as intradural extramedullar lesions (Figure 11.5).

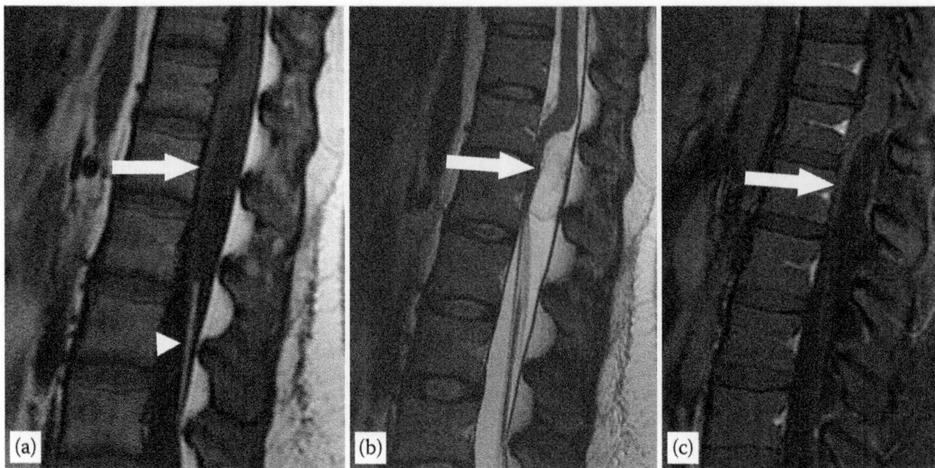


**FIGURE 11.5**
Epidermoid tumor in the intradural thacolumbar spine in a 33-year-old woman presenting with low back pain. (a–c) Sagittal $T_1$-weighted (a), FSE $T_2$-weighted (b), and enhanced $T_1$-weighted FS (c) images show a complex mass (arrow) that is primarily cystic and contains a linear fatty component (arrowhead). The distal spinal cord is anteriorly deviated and compressed by the mass.

Primitive neuroectodermal tumors are a group of tumors with different grades of differentiations, which range from the undifferentiated neuroblastoma to the mixed ganglioneuroblastoma and the more differentiated mature ganglioneuroma [4]. They are commonly located at the cervical spine [2] and are rare but very aggressive tumors, which rarely origin as primitive neoplasm of the SC [19]. Because of the extreme rarity of this type of tumors, consistent and typical MRI findings have not been described in literature and show inhomogeneous intense enhancement after GCM i.v. administration and often leptomeningeal enhancement because of the spread via the CSF [4,19].

*Primary lymphoma* of the SC is extremely rare (1% of all the CNS lymphomas) [4]; the most represented is the non-Hodgkin type, and the dominant cellular population is represented by B-lymphocytes [4,7]. Generally, it is located in the cervical SC, with plurime localizations, and involves even the epidural space and the vertebral bodies [4,7]. On MRI, they are characteristically isointense on $T_1$W sequences with intense but variable contrast enhancement after GCM i.v. administration and hyperintense on $T_2$W sequences (in the brain, they are hyperintense on $T_2$W sequences) [4]; sometimes, it could show a cystic appearance [24].

*Intramedullary metastases* of the SC are rare findings, less than 5% of all spinal metastases [25], generally not isolated and associated with the presence of metastases in other districts of the body [2], and are less common than the leptomeningeal metastasis [7]. Generally, the primitive tumor is located in the lung (the small cell type is the most represented among the lung-cancer histotypes); less commonly SC metastases derive from breast cancer, melanoma, and renal cancer [2]. They are mostly located in the cervical tract of the SC [2]. Their appearance on MRI is heterogeneous, not specific, and indistinguishable often from that of other SC intramedullary lesions [4]; anyway, the sudden onset of neurologic symptoms in a patient with a note primitive metastatic tumor, and an anomalous finding of an SC lesion associated with enhancement, edema, and SC swelling, should suggest us this diagnostic hypothesis.

### 11.4.2 Extramedullary Tumors

Intradural extramedullary tumors represent about 70% of all intradural tumors [2] and are located in the subarachnoid and subdural space. Because of their location, they tend to displace the SC and are not associated with an SC swelling, even if changes in the normal liquoral dynamic and signs of medullary sufferance could be associated if the tumor is big enough.

Among them, nerve sheath tumors and meningiomas are the most common and account for 90% of all the intradural extramedullary tumors [26]. The meninges could be even the location of metastatic diffusion of several neoplastic processes. Now they will be exposed as the principal tumors.

#### *11.4.2.1 Meningioma*

*Meningiomas* are tumors that do not arise from meningeal cells, but from persistent arachnoid cells located along the meninges [26]; sometimes, they could be extradural, but commonly an intradural development is seen [2].

They are benign, well-defined, and slow-growth tumors that affect patients of fifth or sixth decade (female sex in 70% of the cases) and are located in the posterolateral region of the thoracic portion of SC (80% of the cases), attached to the meninges along the SC or to those which encase the spinal nerve roots [26]. Usually, they present as single lesions, but multiple locations could be even associated with the neurofibromatosis type 2 [2]. From a histological point of view, meningothelial and psammomatous (WHO grade I) are the most common histotypes, whereas atypical (WHO grade II) and anaplastic (WHO grade III) variants, together with hemangiopericytoma (WHO grade II), are definitely less frequent [2,6].

On CT, frequently they show calcifications [26] and also hyperostosis, even if less frequently compared with the intracranial meningiomas [9]. On MRI, they show an iso-hypointense signal on $T_1$W sequences, hyperintense on $T_2$W images, and homogeneous enhancement after GCM i.v. infusion, except the calcified areas that are shown as areas with low signal both on $T_1$W and $T_2$W sequences [2,9,26]. Because of their characteristics, these tumors tend to compress than to infiltrate the neighboring tissues, and on MRI, the finding of a hyperintense signal of the SC on $T_2$W sequences could represent a sign of SC sufferance (Figure 11.6).

The total surgical resection of the tumor is possible in 90% of the patients, with a recurrence rate between 3% and 7% [2]. In the case of subtotal resection or recurrence after surgical intervention, radiotherapy should be considered as an alternative and complementary treatment [2].

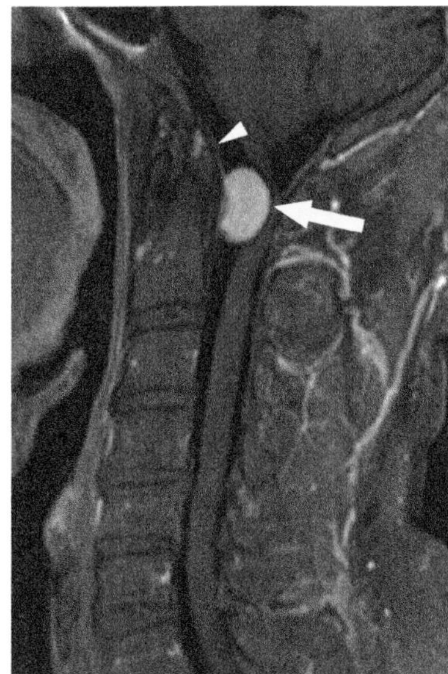

**FIGURE 11.6**
Meningioma in a 54-year-old man with left arm numbness. Sagittal-enhanced $T_1$-weighted FS image shows a homogeneously enhancing dural-based nodule (arrow) at the level of the dens with an enhancing *tail* (arrowhead) extending along the craniocervical junction and mass effect upon the spinal cord. Resected specimen proved to be a grade II meningioma.

#### *11.4.2.2 Nerve Sheath Tumors*

This kind of tumor accounts for about 30% of all intraspinal mass lesions, even if in the pediatric population are less frequent, and the peak of incidence is the fourth decade [2,26]. The greatest part of them is benign tumors, usually *neurofibromas* or *schwannomas* (both WHO grade I) [6,9]; schwannomas are more frequent than neurofibromas [2]. Sometimes, they could present as malignant tumors, and are called *malignant peripheral nerve sheath tumors* (MPNSTs, WHO grade II–IV), and are generally associated with neurofibromatosis type 1, even if only few patients affected by this genetic disease present MPNSTs [2,6].

The main difference between these two different histotypes is due to the fact that neurofibromas in comparison with schwannomas tend more to infiltrate than to compress the nerve, and because of this characteristic it is difficulty dissociable from the nerve during the surgical intervention, and the sacrifice of the nerve is often required [9]. When multiple, neurofibromas could be associated with neurofibromatosis type 1 [9]. As we have seen before, schwannomas are encapsulated and tend more to compress than infiltrate and usually are easily dissociable from the nerve root [9]. When multiple, schwannomas could be associated with neurofibromatosis type 2 [9].

Generally, they affect the dorsolumbar spinal nerves [2]. In more than 50% of the cases, they have an intradural extramedullary development; in 25% of the cases, they have a completely extradural development; and because of the fact that these tumors grow up along the nerve, in 15% they have both intra- and extradural development; rarely, they are completely intramedullary [2,26].

On CT, both the tumors could erode the bone structures of the spine and enlarge the neural foramen [9]. On MRI, schwannomas and neurofibromas are very difficult to distinguish one from each other: both the histotypes are isointense on $T_1$W sequences and hyperintense on $T_2$W sequences, even if on $T_2$W sequences schwannomas may have mix signal whereas neurofibromas could appear hypointense in the center, and this sign is called *target sign* [9]. When these tumors are located intramedullary, the SC appears enlarged by the presence of a usually well-defined iso-hypointense on $T_1$W sequences, hyperintense on $T_2$W sequences, with variable enhancement after GCM i.v. infusion, with minimal perilesional edema, and generally, syringomyelia is not associated [27]. MPNSTs generally have an irregular shape, with ill-defined borders undistinguishable from the surrounding tissues, often located in the neural foramen with associated bone destruction and infiltration, and sometimes show central hyperintensity on $T_2$W sequences because of the presence of cystic components (Figures 11.7 through 11.9) [28].

Surgical management is the best therapeutic option, with a recurrence of 5% of cases, whereas

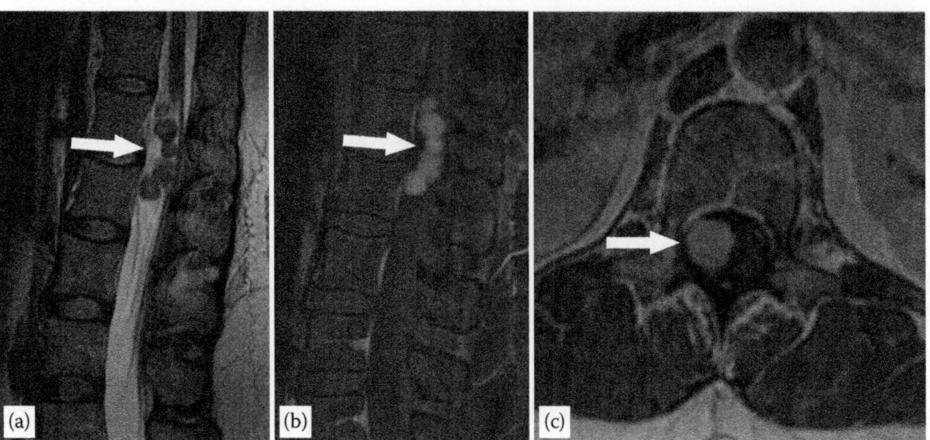


**FIGURE 11.7**
Neurofibroma of the filum terminale in a 38-year-old woman with a history of progressive lower thoracic and upper lumbar pain worse at night. (a) Sagittal FSE $T_2$-weighted image shows a multilobulated mass along the conus tip and proximal cauda equina, with central areas of higher signal intensity. Sagittal-enhanced (b) and axial-enhanced (c) $T_1$-weighted images demonstrate fairly homogeneous contrast enhancement of the multilobulated mass. Surgical resection revealed a neurofibroma; this proved to be a solitary tumor upon further investigation.

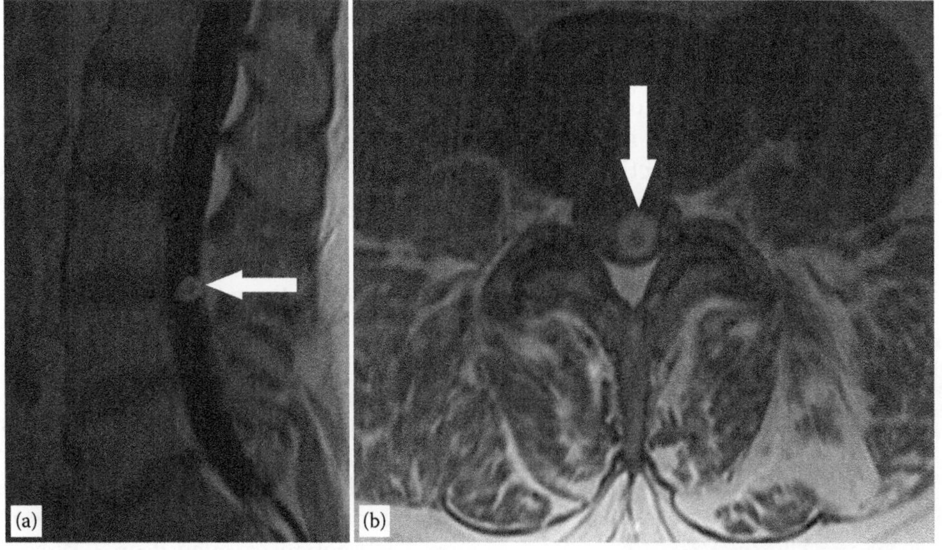


**FIGURE 11.8**
Schwannoma involving a nerve root at the level of L3–4 in a 66-year-old man. Sagittal-enhanced (a) and axial-enhanced (b) $T_1$-weighted images demonstrate a subcentimeter, peripherally enhancing nodule along the lumbar nerve roots at the level of L3–4. The nodule was resected and proved to be a schwannoma. Of note, two intracranial meningiomas were discovered on brain MRI, raising the question of neurofibromatosis.

chemiotherapy and radiotherapy are indicated for tumors with malignant histological characteristics [2].

#### *11.4.2.3 Myxopapillary Ependymoma*

*Myxopapillary ependymoma* is a low-grade tumor (WHO grade I), which originates from the filum terminale or the conus medullaris and occasionally spreads via the CSF along the subarachnoid space [2,9].

These are typical benign, well-defined tumors, distinguished from the ependymomas because they consist of tumoral cells that undergo mucinous degeneration [2]. Because of these characteristics, on MRI they appear isointense on $T_1$W sequences, even if the myxoid mucinous contents could generate an hyperintense appearance on $T_1$W sequences [9]; they appear hyperintense on $T_2$ images, even if the borders could appear hypointense because of hemorrhage (see above *cap sign* of ependymomas), and usually show an homogeneous contrast enhancement [9]. According to the dimension of the tumor, it could determine even scalloping of the spinal canal and could extend through the neural foramens [9].

Surgical intervention is the treatment of choice, and recurrence is rare [2,9].

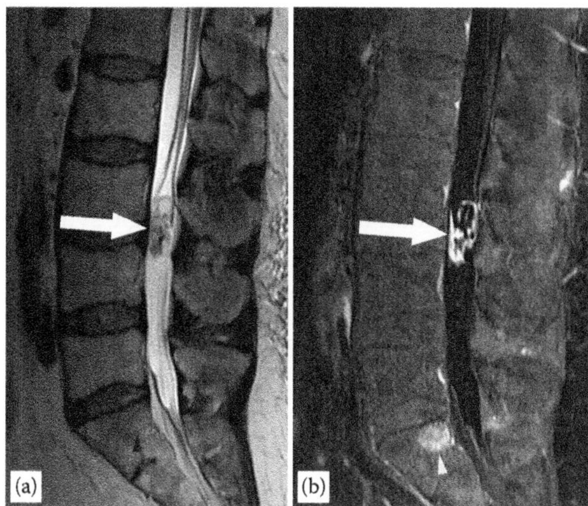

**FIGURE 11.9**
Schwannoma with atypical MRI appearance in a 68-year-old man with refractory back pain and history of melanoma and prostatic carcinoma. Sagittal FSE $T_2$-weighted (a) and enhanced $T_1$-weighted FS (b) images show a complex, heterogeneously enhancing intradural nodule located between several nerve roots at the level of L3–4 (arrows). The nodule was resected because it did not have the classic MRI appearance for a nerve sheath tumor or neurofibroma, and the patient had a history of melanoma and prostatic carcinoma. The nodule proved to be a grade I schwannoma (arrowhead = hemangioma).

### 11.4.2.4 Paraganglioma

*Paraganglioma* is a neuroendocrine tumor, generally benign and located in the adrenal glands, in the jugular foramen, in the carotid bogy, or in the proximity of the nervus vagus [2,5]. When it is located in the SC, it appears on MRI as a well-circumscribed mass, generally isointense on $T_1$W sequences (sometimes with a typical *salt-and-pepper* appearance after GCM i.v. administration) and iso-/hyperintense on $T_2$W sequences [2,5]; sometimes, the *cap sign* could be visible [5] (see ependymoma). Serpentine flow void inside and on the surface of the lesion could be present [5]. This kind of tumor could be well pointed out in nuclear medicine after the administration of metaiodobenzylguanidine (mIBG) [2].

### 11.4.2.5 Other Intradural Extramedullary Tumors

*Lipoma* is an uncommon intradural extramedullary tumor, and in one-third of the cases, it is associated with spinal dysraphism [2]. Lipomas are generally located in the lower thoracic and lumbar tract of the SC, and on MRI, they appear hyperintense on $T_1$W images and hypointense on $T_2$W and fat-suppressed sequences (Figure 11.10) [2].

*Intradural extradural metastases* are found in 5% of autopsies of patients died of cancer, and they originate from the hematic dissemination of a neoplastic process, which originate in particular from lung, breast, or prostate, but even melanoma and lymphoma could involve the intradural compartment (Figure 11.11) [2,7].

Metastasis can even spread via the CSF through the subarachnoid space: in this case, the tumor of origin is represented by primitive tumor of the CNS, as for example medulloblastoma or high-grade gliomas, in particular in the pediatric population [2,7,9].

They are most commonly located in the thoracolumbar tract, generally present as multiple lesions [2,9,29]. On MRI, they can show different characteristics on $T_1$W and $T_2$W sequences, even if generally they present with an irregular thickened focal aspect of the meninges, with variable length and dimensions, and enhance after GCM i.v. administration (Figure 11.12) [29].

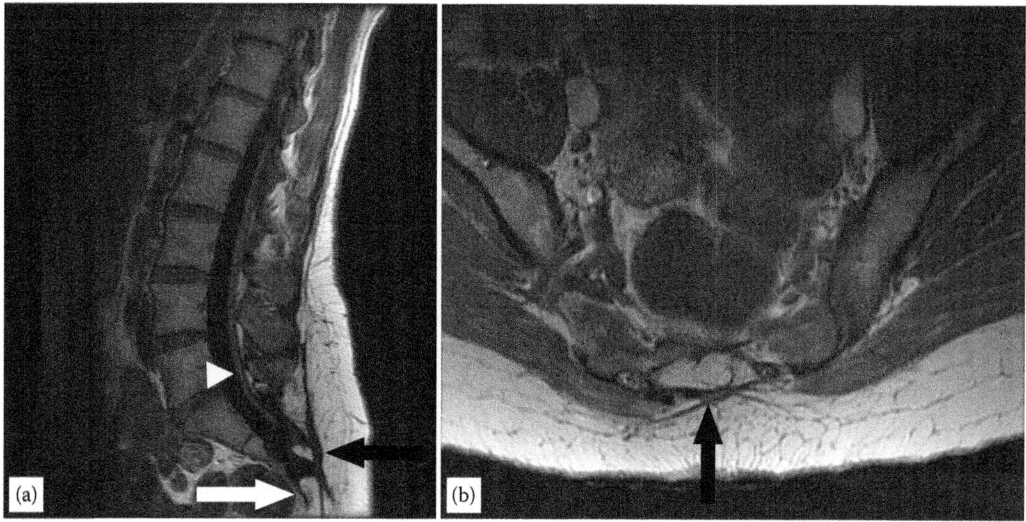

**FIGURE 11.10**
Lipoma and tethered cord in a 42-year-old woman. Sagittal (a) and axial (b) $T_1$-weighted images of the lumbosacral spine demonstrate intradural lipomas (arrows) and spinal dysraphism, with dysmorphic and enlarged thecal sac and tethered cord (arrowhead).

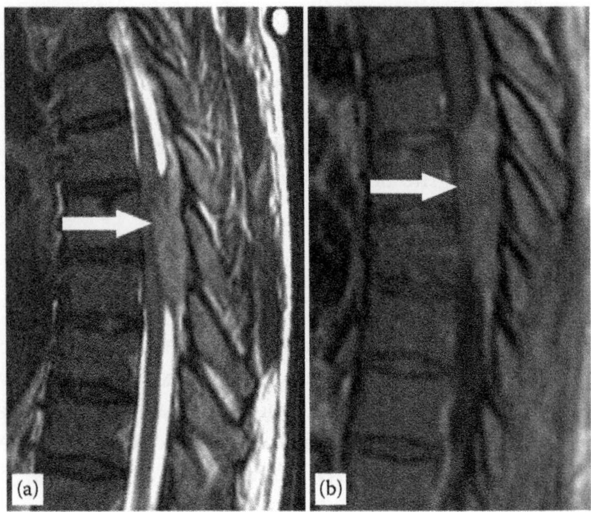

**FIGURE 11.11**
Systemic B-cell lymphoma with extradural component resulting in cord compression and new-onset lower extremity paralysis in a 55-year-old man. Sagittal FSE $T_2$-weighted FS (a) and enhanced $T_1$-weighted (b) images demonstrate two-level thoracic cord compression by extensive, enhancing extradural soft tissue (arrows), which proved to be B-cell lymphoma on surgical decompression.

If they are accidentally founded during an MRI examination of the SC, it is necessary to perform an MRI of the brain in order to detect any other possible localizations of the disease.

## 11.5 Extradural Tumors

Among the tumors that affect the spine, the extradural ones are the most frequent [3,26,30]. As the name said, this kind of lesion develops in the extradural space, and the most frequent location is the vertebral body [26]. In literature, these lesions are variously classified, but the most practical classification used in this chapter divides them into two types: benign tumor and malignant tumor.

From a clinical point of view, the benign lesions are present asymptomatically, and in the greatest part of the cases, they are incidental findings discovered during radiologic examinations performed on the basis of different clinical suspicions [26]. Localized or diffused back pain (in particular at rest or which appear during the night), associated or not with neurologic symptoms (specially in pediatric population), is generally the most common presentation of malignant lesions [3,26].

Of course, the age of the patient plays a pivotal role: in fact, among patients under 30 years, extradural malignant tumors are extremely rare (except osteosarcoma and Ewing sarcoma), whereas among patients over 30 years, most of the tumors are malignant, with the exception of hemangiomas and bone islands [30].

From a radiological point of view, this kind of tumors can be well evaluated on CT and MRI, whereas plain radiographs are usually relegated as first approach.

CT generally is executed with a large field of view (if the location of the suspicious lesion is unknown), and usually, iodinate contrast medium i.v. administration is not required [30]. CT examinations are performed in order to evaluate the matrix of the lesion (included the presence or absence of calcifications) and the involvement of the osseous structures, in particular the presence of bone erosions, sclerosis, and/or remodeling [26,30].

MRI examinations are generally performed on axial and sagittal plane in order to evaluate as best as possible the proper spine structures, whereas the coronal plane could be useful for the evaluation of the paravertebral tissues [30]. The following sequences are those considered the most used and useful for the characterization of the lesions [26,30,31,32]:

- $T_1W$ *spin-echo sequences* are the sequences of choice for the evaluation of the structure of the bone marrow and the evaluation of a sub-acute hemorrhage and even the presence of fat; usually, the focal lesions of the bone marrow hyperintense on $T_1W$ sequences are benign [32].

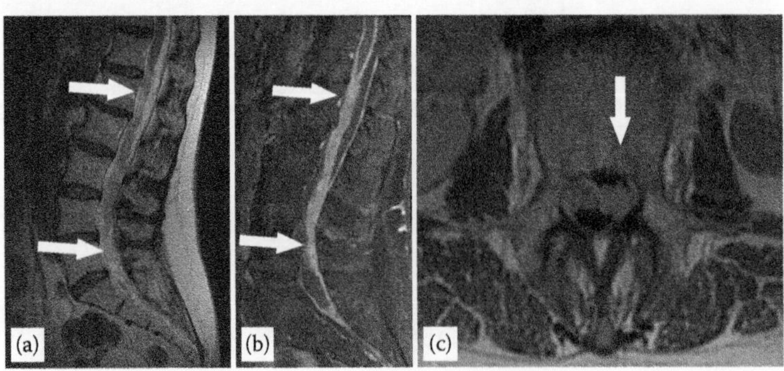

**FIGURE 11.12**
(a–c) Extensive enhancing soft tissue deposits (arrows) along with cauda equina and distal cord in a 69-year-old woman with lung carcinoma, presumed leptomeningeal metastases.

- $T_2W$ *fast spin-echo (FSE) sequences* are the sequences of choice for the evaluation of the bone marrow, of the acute vertebral fractures, of the spinal canal space, and of the possible involvement of the SC.
- $T_1W$ *sequences after GCM i.v. administration*: These sequences are important because the contrast enhancement is directly proportional to the amount of the soft tissue vascularity, allowing, for example, the differentiation between cystic and non-cystic lesions, or a detailed evaluation of the epidural extension of the lesion [30].
- *Fat saturation sequences (STIR,* $T_2W$ *FatSat, or* $T_1W$ *FatSat after GCM i.v. administration)*: These sequences contribute to better characterize the lesions thanks to their high contrast resolution and their intrinsic quality of suppression of the signal of the fat tissues.

As for all the lesions that could affect the human body, when the radiologist has to describe an SC extradural lesion, must describe the *number*, the *topographic location*, the *dimensions*, the *margins*, the *locoregional extension*, and the *matrix*, including other features such as the presence or absence of cysts or necrotic areas [30].

Now there will be discussed the principal benign and malignant extradural tumors, but we must point out that among the benign lesions will be considered even the local aggressive lesions.

### 11.5.1 Benign Tumors

#### *11.5.1.1 Hemangioma*

*Hemangiomas* are vascular tumors, constituted by endothelial-lined vascular structures with interposed fat [33]. They are the most common extradural benign tumors of the spine [26,33], frequently discovered in the population who undergo MRI of the spine (27% of all the patients who perform an MRI), with a slight higher prevalence in the female population [33].

Usually, they are located in the thoracic tract of the lumbar spine (more than 50% of all the cases), and in one-third of the cases, they have multiple locations [26]. Generally, they are confined to the vertebral body, but in 10% of the cases they extend to the posterior portions of the vertebra [26]. Hemangioma can be classified as *asymptomatic* and *aggressive*; in the second case, they show compression of the SC [26].

On diagnostic imaging, hemangiomas usually don't represent a difficult challenge for the radiologist. On CT, they typically show a *jail bar* appearance on sagittal and coronal view and a *polka dot* aspect on axial plane: this fact is due to the replacement of the normal bone by vascular disorganized structures that appear hypodense, whereas the remaining trabeculae appear thickened (hyperdense dots/column) [26,33]. On MRI, usually they appear hyperintense both on $T_1W$ and $T_2W$ sequences, but sometimes, especially in the case of aggressive hemangiomas, they could appear hypointense on $T_1W$ sequences [26,33]; after GCM i.v. administration, they generally enhance, in particular if they are aggressive (Figure 11.13) [26].

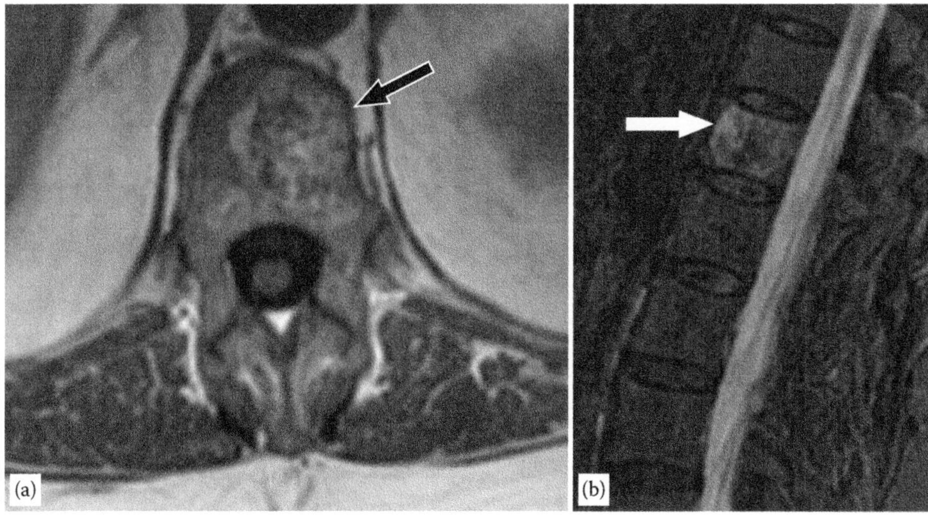


**FIGURE 11.13**
Hemangioma in the T12 vertebral body in a 73-year-old woman. (a) Axial $T_1$-weighted image displays a hetereogeneous mass (arrow) in the vertebral body which is primarily comprised of fatty signal with rounded and ovoid hypointense foci corresponding to corrugated appearance of the hemangioma in cross section. (b) Sagittal FSE $T_2$-weighted image through the mass (arrow) shows fat-suppressed fatty signal interspersed with heterogeneously hyperintense and isointense signal. Lesion had characteristic imaging findings for hemangioma; it was stable for over a year on follow-up imaging and was not biopsied.

#### 11.5.1.2 Osteoid Osteoma

The *osteoid osteoma* is a benign tumor, histologically characterized by the presence of a central *nidus*, which consists of osteoid tissue or mineralized immature bone [26,30,33].

From a clinical point of view, osteoid osteomas manifest with back pain more pronounced during the night at rest and disappear with the use of non-steroid anti-inflammatory drugs (NSAD) [26,30,33]. They more frequently affect patients between the second and third decades, and more in the male than in the female population (M:F = 2–3:1) [26,30].

The majority of them are localized in the lumbar spine, in particular in the neural arch of the vertebrae [26,30,33]. The dimension of the nidus must be smaller than 2 cm; lesions larger than this measure are classified as osteoblastoma (see below) [30,33].

The best imaging modality in order to evaluate osteoid osteoma is CT: generally, they have a round or oval shape, and the nidus appears as an osteolytic lesion with a sclerotic hyperdense center surrounded by periosteal thickening [26,30,33].

MRI in the majority of the cases is not useful in the characterization of the lesion and often may be misleading because of the impossibility to correctly recognize the nidus [33]. On MRI, the calcifications of the nidus appear as spots of low-intensity signal on both $T_1$W and $T_2$W sequences, whereas on $T_2$W the fibrotic matrix generally appears hyperintense as also the bone marrow, which surrounds the lesion [26,30]. The nidus can strongly enhance on $T_1$W sequences after GCM i.v. administration [26,30].

The management of the patient with this type of lesion could be variable: osteoid osteoma often tends to spontaneous regression, and a conservative management with NSAD could be sufficient [30,33]; in other cases in which the symptomatology is stronger and limits the quality of life of the patient, the treatment options are surgical removal or image-guided percutaneous radiofrequency ablation [33].

#### 11.5.1.3 Osteoblastoma

From a histological point of view, osteoblastoma is quite similar to osteoid osteoma and substantially differs from it because of the dimensions, localization, and clinical progression [26,30,33].

The nidus of the osteoblastoma is bigger than 2 cm [30,33]. They are equally distributed along the cervical, thoracic and lumbar spine, moreover in the posterior elements, and could extend in the spinal canal [26,30,33]. This tumor tends to be more aggressive than osteoid osteoma and surgical treatment is often required [26,30]. The findings on CT and MRI are substantially the same of the osteoid osteoma (see above) (Figure 11.14) [26,30,33].

#### 11.5.1.4 Aneurysmal Bone Cysts

Aneurysmal bone cysts are tumor-like benign lesions with a multiloculated, blood-filled cystic appearance [3,26,30,33]. They affect patients usually in the first two decades, with a slight prevalence in female population [33]. Usually, they are located in the posterior elements of the thoracic and lumbar spine [3,30,33]. These lesions tend to grow up and expand into the adjacent structures, causing different neurological problems [3,26,30,33].

CT and MRI are the diagnostic imaging modality of choice. On CT, these lesions generally present as expansile, well-defined lytic lesions, with a typical eggshell layer of cortical bone that surrounds the cyst, and sometimes, it would be visible even in internal septa and fluid–fluid levels [3,26,30]. On MRI, they have an heterogeneous appearance on $T_1$W and $T_2$W sequences,

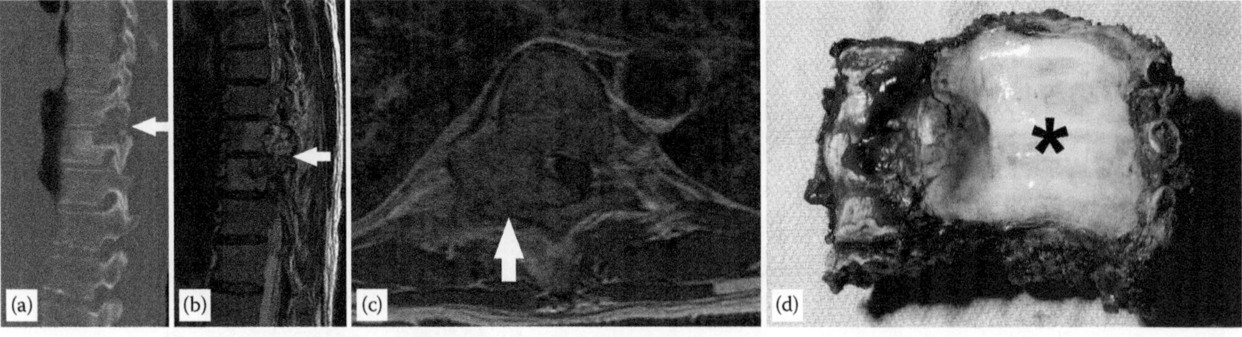


**FIGURE 11.14**
Osteoblastoma of T9–10 in a 51-year-old man with progressive right-sided thoracic back pain. (a) Sagittal 2D reconstruction from non-contrast CT shows a large lytic lesion (arrow) with subtle osteoid matrix involving both the T9–10 posterior elements and posterior vertebral bodies. Sagittal FSE $T_2$-weighted (b) and axial-enhanced $T_1$-weighted (c) images show a corresponding heterogeneous, enhancing mass (arrows) resulting in central canal narrowing and deviation of the spinal cord to the left. Patient underwent subsequent T8–11 decompressive laminectomies, intralesional excision of confirmed osteoblastoma, T9–10 hemivertebrectomies and anterior interbody fusion. (d) Resected specimen shows the completely resected osteoblastoma (asterisk).

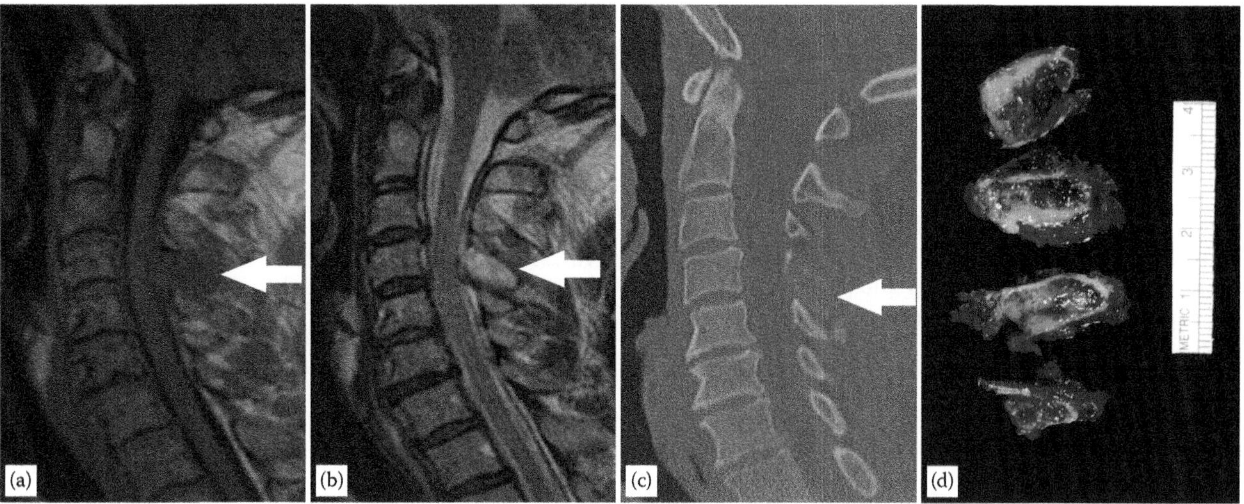

FIGURE 11.15
Aneurysmal bone cyst in a 51-year-old man who developed severe neck pain after falling on the ice several months prior. Sagittal $T_1$-weighted (a) and FSE $T_2$-weighted (b) images demonstrate an expansile mass in the C4 spinous process (arrows) which is isointense to skeletal muscle on $T_1$-weighted imaging and heterogeneous hyperintense on $T_2$-weighted imaging. (c) Sagittal 2D reconstruction from non-contrast CT displays the lytic lesion (arrow) with marked thinning of the cortex by the expansile remodeling. (d) Gross specimen after intralesional curettage shows hemorrhagic cystic foci throughout the spinous process.

and loculations with fluid–fluid levels are more easily identifiable [3,26,30,33]. The septa show contrast enhancement on $T_1$W after GCM i.v. administration (Figure 11.15) [30]. The treatment f choice still remains the total surgical resection [3,26,30].

#### 11.5.1.5 Osteochondroma

This lesion could be considered more a developmental lesion than a true bone neoplasm, because it originates from the separation of a fragment of the cartilaginous growth plate [30]; because of the presence of a cartilaginous cap that acts as the same way as the physeal plate, the lesion tends to grow up exophytically because of an endochondral ossification process up to the skeletal maturity [30].

This type of lesion can grow up from any part of the spine, and generally, they develop from the tip of the spinous or transverse process, in particular in the cervical tract of the spine [2,30].

On CT, it is possible to demonstrate the continuity of this lesion with the vertebra, whereas on MRI this lesion shows a hypointense rim due to the presence of the cortical rim and centrally a fat signal intensity that corresponds to the presence of normal yellow bone marrow [30]. The presence of a thin cartilaginous cap is normal in childhood and adolescence, whereas in adults, especially if its thickness is more than 1.5 cm, it is not normal and exostotic chondrosarcoma must be taken into consideration [30]. In this last case, if the diagnosis of chondrosarcoma has been confirmed, surgical resection is considered mandatory [2,30].

#### 11.5.1.6 Chondroblastoma

*Chondroblastoma* is a benign tumor that originates from cartilaginous cells, typical of the growing skeleton [3,30,33]. Rarely, this tumor affects the spine (about 1.8% of all the cases of chondroblastoma) and generally occurs during the third decade of life with a male predominance (M:F = 2–3:1) [30,33]. From a clinical point of view, it generally manifests with localized back pain, even if other neurological symptoms could be present according to the dimensions and the localization of the lesion [3,30,33].

This tumor can involve both the anterior and posterior elements of the vertebrae, and the involvement of the spinal canal is not unusual [3,30,33].

On CT, it appears as an osteolytic lesion with a geographic shape, sclerotic margins, and typical chondroid matrix characterized by the presence of some mineralized areas with a *rings and archs* or *flocculent* aspect because of the tendency of the lesion to develop in a lobular pattern [3,30,33].

On MRI, the greatest part of the lesions shows low-to-intermediate signal on $T_1$W sequences; on $T_2$W sequences, they appear homogeneously hyperintense on $T_2$W because of the presence of an important hydric component in the hyaline cartilage, whereas the hypointense areas correspond to calcifications and hemosiderinic deposits inside of the lesion [30,33].

The differential diagnosis with chondrosarcoma is almost always impossible to reach just on the imaging appearance because of its local aggressiveness [33]. The surgical option is the gold-standard treatment [3,30,33].

#### 11.5.1.7 Giant Cell Tumor

**Giant cell tumors** are considered as benign tumors, even if they could be locally very aggressive [3,26,30,33]. From a histological point of view, they consist of interposed layers of stromal ovoid cells and giant osteoblatic cells.

They usually affect patients between the second and the fourth decades, with a net prevalence in female population [3,26,30,33]; characteristically they grow dramatically during the pregnancy [33]. As other benign but local aggressive tumors, they present usually with local back pain and sometimes with other associated neurological symptoms [3,30].

The most frequent location is the sacrum (90% of the cases) [30]. When they affect the other tract of the spine, they develop in the vertebral body and then extend in the posterior elements, often invading the spinal canal [3,30,33].

On CT, they appear as expansive osteolytic lesions in the absence of a sclerotic rim and lack of mineralization of the matrix [26,30,33].

On MRI, they appear as multiloculated lesions with low-to-intermediate signal on both $T_1$W and $T_2$W sequences; often, internal cysts, fluid–fluid level, and hyperintensities on $T_1$W sequences related to recent hemorrhagies are visible inside the lesion [30]. The surgical treatment, when possible, is the treatment of choice [3,26,30,33], often accompanied by radiotherapy [3].

#### 11.5.1.8 Eosinophilic Granuloma

The *eosinophilic granuloma* is one of the manifestation of Langerhans cell histiocytosis, known previously as *histiocytosis X* [26,30,33]. It is considered as a *benign non-tumor* of unknown origin, which contains in particular eosinophil cells and lymphocytes [26,33].

It affects in particular patients of the first decade, more frequent in men than in women [26]. From a clinical point of view, it could present in various ways, but generally it presents with localized back pain when patient lies down at rest that regresses when the patient stands; neurological complications are rare and these symptoms are sensible to the therapy with NSAD [26,30].

The disease could present as single or multiple lesions, but generally a single lesion affects the cervical tract of the lumbar spine, in particular the second vertebral body [26].

On CT, it appears as a lytic lesion with sharp borders, which is generally located in the vertebral body and could determine the collapse of the vertebra (a condition noted as *vertebra plana*) but the posterior elements generally are not involved [26,30]; the disc space is preserved as the same as the paravertebral tissues, and the vertebra assumes an hyperdense aspect [30].

MRI has a limited role in the evaluation of this lesion [33]. MRI findings show a variable aspect on $T_1$W sequences, hyperintense on $T_2$W sequences, and strongly enhance after GCM i.v. administration [26].

### 11.5.2 Malignant Tumors

#### 11.5.2.1 Metastases

Metastases are the most frequent tumors that affect the spine [3,26,30,33]. They can be divided into *osteoblastic* (typically hyperdense on CT) and *osteolytic* metastases (typically hypodense on CT) [26,30,33]. Osteoblastic metastases are typical of prostate carcinoma in the male population and of breast cancer in female population (even if this one could even generate osteolytic lesions) and less commonly with hepatic carcinoma, other tumors of the gastrointestinal tract, and bladder carcinoma [30,33]. Osteolytic lesions are commonly associated with lung adenocarcinoma, thyroid, and kidney cancer [30,33].

From a clinical point of view, they usually are detected after the diagnosis of the primary tumor during the imaging examinations routinely performed to stage the disease, and they are not frequently found during imaging diagnostic investigations performed to evaluate a localized low back pain and/or other neurological symptoms in absence of the diagnosis of a primary tumor.

Metastases typically present as multiple lesions diffused along the spine, more frequently evidentiated in the thoracic spine (70% of the cases), followed by the lumbar (20%) and the cervical tract (10%) [26]. Their appearance on CT is absolutely variable, but the distinction in osteoblastic and osteolytic type is generally the most practical.

On MRI, their appearance is variable, but generally they show hypointense signal on $T_1$W sequences, hyperintense on $T_2$W images (usually hypointense if sclerotic) and with a strong and intense contrast enhancement on $T_1$W sequences after GCM i.v. administration [26]. Sometimes, they show restriction on DWI sequences [26]. MRI is useful in particular to evaluate the extension of the metastasis inside the epidural space (and even the possible involvement of the SC) and the presence of associated compression fracture [26,30,34]. MRI is also useful to distinguish an osteoporotic compression fracture from a metastatic compression fracture: in fact, the latter one shows complete substitution of the yellow bone marrow with soft tissues (which could extend over the boundaries of the same vertebra), whereas the osteoporotic type usually shows only a band of edema inside the vertebral body, visible on $T_2$W and fat-saturation sequences (Figures 11.16 through 11.18) [33].

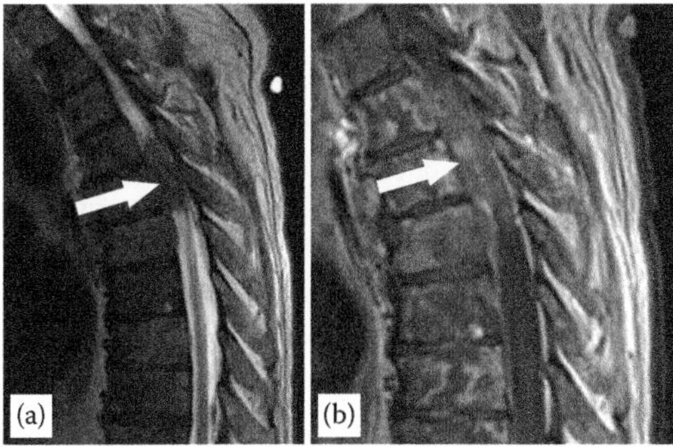

**FIGURE 11.16**
Sagittal FSE $T_2$ (a) and enhanced $T_1$-weighted (b) images in a 78-year-old man with prostatic carcinoma demonstrate multifocal osseous marrow replacing lesions with multilevel epidural and single-level intradural (arrows) extension of tumor.

The most sensible imaging modality in the diagnosis and localization of metastatic disease of the spine still remains the technetium 99$^m$ nuclear bone scan [33].

### 11.5.2.2 Plasmocytoma and Multiple Myeloma

*Plasmocytoma* (or plasmacytoma) and *multiple myeloma* are commonly considered two different stages of the same disease [3,26,30,33], even if it has been demonstrated that plasmocytoma could be present many years before the laboratory evidence of multiple myeloma [26]. Plasmocytoma is a tumor originating by the bone marrow and is considered as the solitary presentation of multiple myeloma, a malignancy that originates from B-lymphocytes [3,26,30,33].

*Solitary plasmocytoma* is a rare tumor that affects more than 70% of the cases over 60 years old [3,30,33]. It generally manifests with localized back pain and/or other neurologic symptoms according to the presence or less of the involvement of the neuroradicular spaces or of the spinal canal, and/or the presence of an associated vertebral fracture; not usually it can be completely asymptomatic and founded as an incidental finding [3,26,30,33]. The most common site of localization is the vertebral body, but generally it involves the posterior elements [26,30]. Usually, it presents with a single fractured vertebra [30]. As the tumor tends to replace the trabecular structure of the bone more than the cortical elements, on CT usually it appears as an expansile osteolytic lesion, characterized by the thickening of the cortical bone and of some remaining trabeculae, giving it the so-called mini-brain aspect on an axial plane [26,30], even if in about 30% of the cases it appears less specific, with a *soap bubble* multicystic appearance, and generally does not involve the disc space metastases [30]. On MRI, as other tumor, it appears hypointense on $T_1$W sequences, hyperintense on $T_2$W sequences, with a strong and homogeneous enhancement in $T_1$W sequences after GCM i.v. administration [30].

As we have seen before, *multiple myeloma* is considered as the evolution of plasmocytoma, and it is characterized by the presence of multiple osteolytic lesions along the spine, often accompanied by multilevel vertebral fractures [3,26,30,33]. About 20% of the patients with a note laboratoristic diagnosis of multiple myeloma shows no abnormalities of the bone marrow on MRI (Figure 11.19) [30].

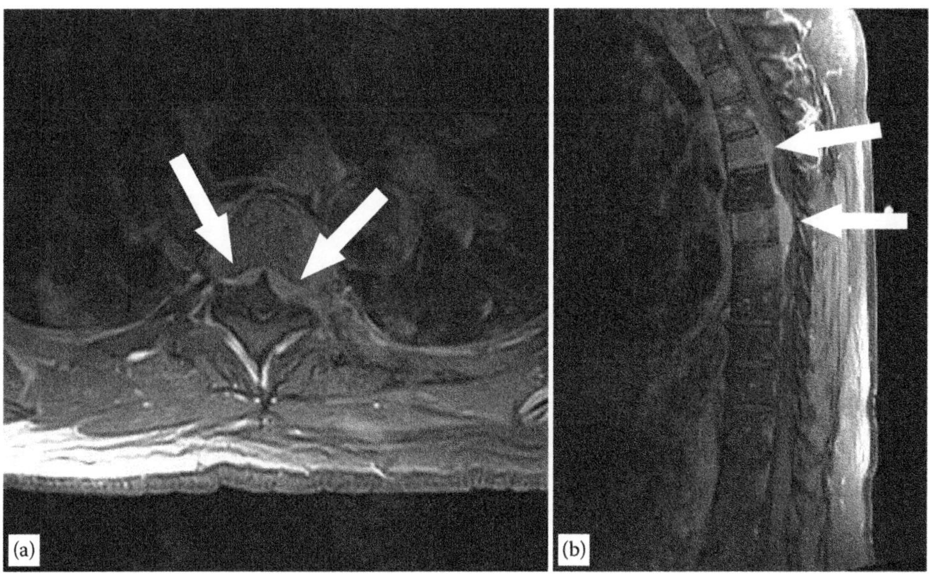

**FIGURE 11.17**
Thoracic spine metastases with epidural tumor extension in a 61-year-old man with prostatic carcinoma. Sagittal FSE $T_2$-weighted (a) and enhanced $T_1$-weighted (b) images show multifocal osseous marrow-replacing lesions with two-level epidural extension of tumor (arrows) and narrowing of the central canal.

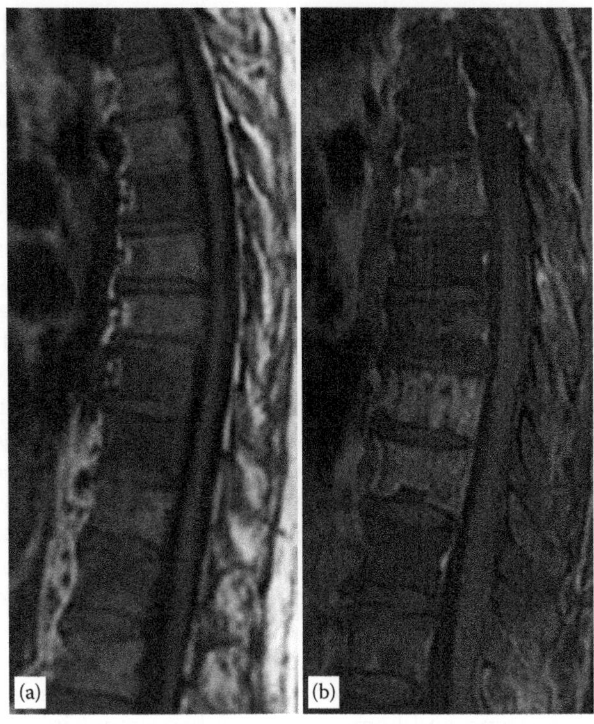


**FIGURE 11.18**
Spinal metastases in a 70-year-old man with non-small cell lung adenocarcinoma. Sagittal $T_1$-weighted images before (a) and after (b) contrast administration show multifocal marrow replacing lesions isolated to the bones.

### 11.5.2.3 Lymphoma

Primary lymphoma of the bone is a very rare tumor, accounting for about 1%–3% of all the primitive lymphoma [30,33]. From a histological point of view, they are more frequently represented by B-cell non-Hodgkin lymphomas [30].

There is a predilection for the male sex (M:F = 8:1) and the maximum incidence is between the fifth and the seventh decades [30,33]. The appearance on CT or MRI is not specific [30,33], even if on CT the sclerotic and mixed pattern is more specific for Hodgkin lymphoma (Figure 11.20) [30].

### 11.5.2.4 Chordoma

*Chordoma* is the most frequent primitive tumor of the spine (even if it is very rare), which originates from remnants of the notochord [3,26,30,33]. It is a low-grade locally aggressive tumor, with a slow growing, that macroscopically presents as a large soft lobulated mass covered by a thin fibrous layer [3,30]. They generally manifest between the fifth and sixth decade, with a male predominance (M:F = 2:1) [3,26,30].

As different other extradural tumors, it can manifest with local back pain, accompanied or less by other neurological symptoms, according to the local extension of

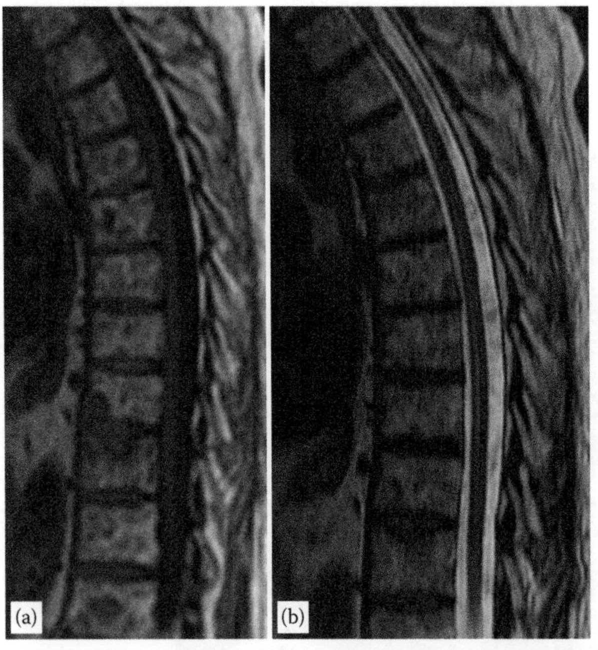


**FIGURE 11.19**
Myelomatous infiltration of the spine in a 75-year-old man. Sagittal $T_1$-weighted (a) and FSE $T_2$-weighted (b) images demonstrate multifocal marrow replacing lesions throughout the imaged spine. Peripheral blood analysis and T10 vertebral body biopsy performed 1 year later prior to vertebroplasty for compression fracture both confirmed myeloma.

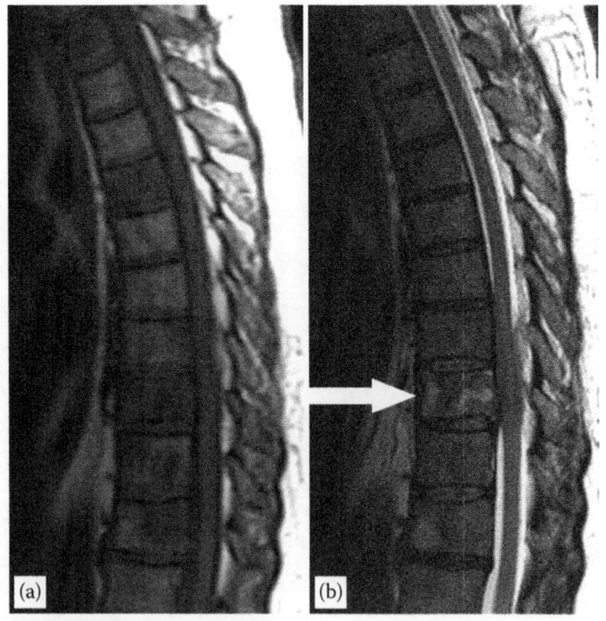


**FIGURE 11.20**
B-cell lymphoma involving the lumbar spine in a 57-year-old man. (a) Sagittal $T_1$-weighted image demonstrates multifocal rounded areas of marrow replacement as well as complete replacement of the vertebral bodies. (b) The lesions are less conspicuous on sagittal FSE $T_2$-weighted image, except for a lower thoracic vertebral body (arrow) due to mild pathologic compression fracture with mild bulging of the posterior cortex.

it in the spinal canal and in the other spinal and paraspinal structures [3,26,30,33].

Most commonly, it is located in the sacrum (50% of the cases); other locations are the clivus of the occipital bone (35%) and the vertebral bodies (15%) [30].

On CT, it usually appears as a destructive mass with a *mushroom* appearance and a *dumbbell* shape, which extends along several metamers involving the disks, with some calcifications in the context (35% of the tumors located in the sacral region, 90% of those located in the other portions of the spine) [30].

On MRI, it shows low-to-intermediate signal intensity on $T_1$W sequences, even if some of them could show hyperintense signal due to the mucinous and myxoid content [30]; on $T_2$W sequences, it appears strongly hyperintense, with subtle hypointense areas because of the presence of fibrous septa and/or hemosiderin deposits [30]. On $T_1$W sequences after GCM i.v. administration, this tumor enhances heterogeneously, but has been described even other enhancement patterns such as *ring and arc* or peripheral (Figure 11.21) [30].

This lesion must be differentiated from *giant nothocordal rest*, which on CT presents surrounded by normal or sclerotic bone, whereas on MRI shows low intensity signal on $T_1$W sequences and high signal on $T_2$W sequences, without a soft tissues involvement [30]. When a lesion with the characteristics of a giant notochordal rest is found, radiological surveillance has to be performed in order to detect signs of bone destruction, which are typical of the chordoma [30].

#### 11.5.2.5 Osteosarcoma

Osteosarcoma is an osteoblastic high-grade lesion, characterized by variable osteoid, fibrous, and cartilaginous tissue production [30]. It generally affects patients of the fourth decade (*primary osteosarcoma*), or patients older than 60 years who in their past have suffered of Paget's disease or have undergone radiation therapy (*secondary osteosarcoma*) [3,30,33]. As for other benign and malignant tumors, from a clinical point of view, it manifests with local back pain and different neurological symptoms according to the location and the extension [3,30,33].

The thoracolumbar tract is the most common localization in the spine, and it is located in the posterior elements in around 80% of the cases, often involving the spinal canal and the neighboring soft tissues, and two vertebrae are involved in about 17% of the cases [30,33].

The tumors of this type generally show on CT ill-defined borders, typically (80% of cases) with an osteoid matrix, whereas rarely the vertebrae involved by the tumoral lesion appear with the so-called ivory aspect, typical of the sclerosing osteoblastic type [30,33]. On MRI, the findings are not specific, but MRI is useful in the evaluation of the SC involvement [3,30,33]. The treatment of choice is the surgical resection, followed by radiotherapy in selected cases [30].

#### 11.5.2.6 Chondrosarcoma

*Chondrosarcoma* is a malignant tumor that derives from the cartilaginous cells, the second more frequent primitive tumor of the spine after chordoma [30,33]. It is more frequent in the male population, with a peak incidence in the fifth decade [30,33].

The clinical symptom is variable according to the structure involved, but generally the patients suffer

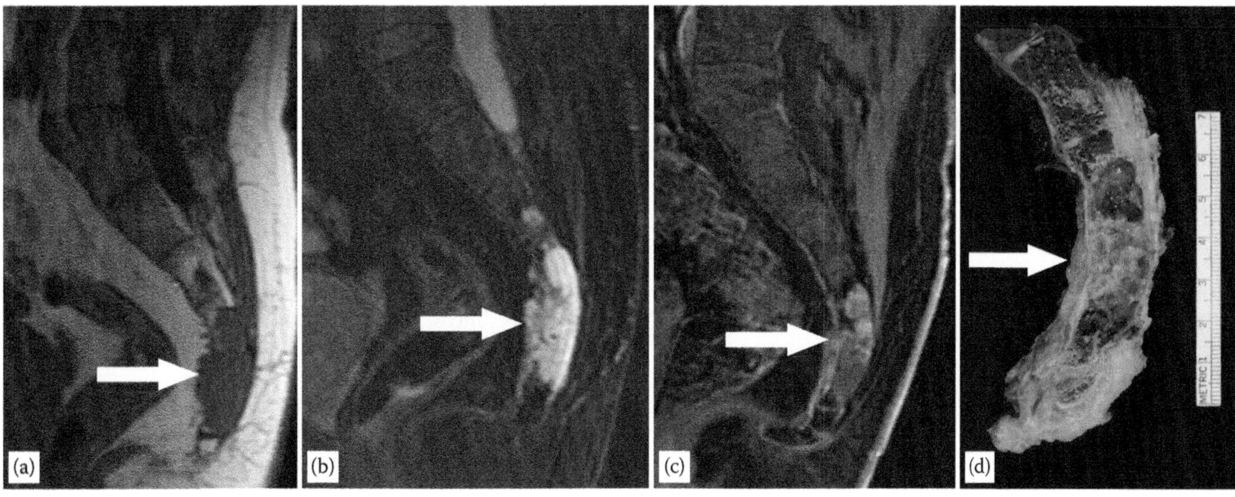


**FIGURE 11.21**
Sacral chordoma in a 64-year-old man presenting with coccygeal pain. Sagittal $T_1$-weighted (a), FSE $T_2$-weighted (b), and enhanced $T_1$-weighted FS (c) images delineate a heterogeneously enhancing midline mass (arrows) involving the S3–5 bodies with extraosseous soft tissue extension into the presacral space. (d) Resected sacral specimen delineates the extent of the chordoma (arrow).

from local back pain with variable neurological symptoms associated [33].

The most common locations are the thoracic and lumbar spine, more frequently in both the body and posterior elements of the vertebra (45%) compared to the posterior elements (40%) or vertebral body alone (15%) [30]. In 35% of the cases, the involvement of the disc structures has been described [30].

On CT, they manifest as large calcified mass with a chondroid matrix and plurime calcifications in the context [30]. On MRI, they appear with low-to-intermediate signal on $T_1$W sequences and high signal on $T_2$W sequences, and on $T_1$W sequences after GCM i.v. administration, they enhance with an *arc and rings* pattern [30]. The complete surgical resection, when possible, is the treatment of choice [30].

#### *11.5.2.7 Ewing Sarcoma*

Ewing sarcoma is an undifferentiated high-grade tumor that consists of round-blue cell of unknown origin [3,30,33], even if it shows similarities with other neuroectodermal tumors [3]. The spine is a very rare site of primitive Ewing sarcoma, and it is seen in particular in the second decade of life with a slight predilection for male population [3,30,33]. Generally, the patient resents with a long-standing story of low back pain and various neurological defects [3].

The lumbosacral region is the most common location, with a predilection for the ala in the sacrum (69% of the cases) and the posterior elements (60%) when it affects the vertebral bodies [30,33].

On CT, the lesion could appear lytic, sclerotic, or with a mixed aspect, even if almost all of them (93%) show an aggressive morphological pattern [30]. An osteoid matrix is usually recognizable [33]. Sometimes, they could present with an *ivory vertebra* aspect [30,33]. MRI findings are not specific but are indispensable in the evaluation of the involvement of the spinal canal and of the paraspinal tissues [3,30,33]. The treatment of choice generally is based on the combination of radio- and chemotherapy [30].

## References


1. Brant, William E. et al. *Fundamentals of Diagnostic Radiology*, 4th Edition. Lippincott Williams & Wilkins, Philadelphia, PA, 2012.
2. Traul, David E. et al. Part I: Spinal-cord neoplasms—Intradural neoplasms. *Lancet Oncol* 2007; 8: 35–45.
3. Sansur, Charles A. et al. Part II: Spinal-cord neoplasms—Primary tumours of the bony spine and adjacent soft tissues. *Lancet Oncol* 2007; 8: 137–147.
4. Do-Dai, Manuel D. et al. Magnetic resonance imaging of intramedullary spinal cord lesions: A pictorial review. *Curr Probl Diagn Radiol* 2010; 39: 160–185.
5. Koeller, Kelly K. et al. Neoplasms of the spinal cord and filum terminale: Radiologic-pathologic correlation. *RadioGraphics* 2000; 20: 1721–1749.
6. Louis, David N. et al. The 2007 WHO classification of tumours of the central nervous system. *Acta Neuropathol* 2007; 114: 97–109.
7. Abul-Kasim, K. et al. Intradural spinal tumors: Current classification and MRI features. *Neuroradiology* 2008; 50: 301–314.
8. Smith, Alice B. et al. Radiologic-pathologic correlation of pediatric and adolescent spinal neoplasms: Part 1, Intramedullary spinal neoplasms. *AJR Am J Roentgenol* 2012; 198: 34–43.
9. Soderlund, Karl A. et al. Radiologic-pathologic correlation of pediatric and adolescent spinal neoplasms: Part 2, Intradural extramedullary spinal neoplasms. *AJR Am J Roentgenol* 2012; 198: 44–51.
10. Do-Dai, Daniel D. et al. Magnetic resonance imaging of intramedullary spinal cord lesions: A pictorial review. *Curr Probl Diagn Radiol* 2010; 39: 160–185.
11. Demaerel, P. et al. Magnetic resonance imaging of spinal cord trauma: A pictorial essay. *Neuroradiology* 2006; 48: 223–232.
12. Sheerin, F. et al. Magnetic resonance imaging of acute intramedullary myelopathy: Radiological differential diagnosis for the on-call radiologist. *Clin Radiol* 2009; 64: 84–94.
13. Borchers, Andrea T. et al. Transverse myelitis. *Autoimmunity Rev* 2012; 11: 231–248.
14. Lycklama, G. et al. Spinal-cord MRI in multiple sclerosis. *Lancet Neurol* 2003; 2: 555–562.
15. Jain, Krishan K. et al. Prevalence of MR imaging abnormalities in vitamin B12 deficiency patients presenting with clinical features of subacute combined degeneration of the spinal cord. *J Neurol Sci* 2014; 342: 162–166.
16. Koyama, T. et al. Radiologic manifestations of sarcoidosis in various organs. *RadioGraphics* 2004; 24: 87–104.
17. Pamir, MN. et al. Spinal hydatid disease. *Spinal Cord* 2002; 40: 153–160.
18. Saleem, S. et al. Spinal cord schistosomiasis: MR imaging appearance with surgical and pathologic correlation. *Am J Neuroradiol* 2005; 26: 1646–1654.
19. Smith, Alice B. et al. Radiologic-pathologic correlation of pediatric and adolescent spinal neoplasms: Part 1, Intramedullary spinal neoplasms. *AJR Am J Roentgenol* 2012; 198: 34–43.
20. Kim, DH. et al. Differentiation between intramedullary spinal ependymoma and astrocytoma: Comparative MRI analysis. *Clin Radiol* 2014; 69: 29–35.
21. Patel, U. et al. MR of spinal cord ganglioglioma. *Am J Neuroradiol* 1998; 19: 879–887.
22. Baker, Kim B. et al. MR imaging of spinal hemangioblastoma. *AJR Am J Roentgenol* 2000; 174: 377–382.

23. Hegde AN. CNS cavernous haemangioma: "Popcorn" in the brain and spinal cord. *Clin Radiol* 2012; 67: 380–388.
24. Thomas, Adam G. et al. Extranodal lymphoma from head to toe: Part 1, The head and spine. *AJR Am J Roentgenol* 2011; 197: 350–356.
25. Sung, Wen-Shan et al. Intramedullary spinal cord metastases: A 20-year institutional experience with a comprehensive literature review. *World Neurosurg* 2013; 79(3/4): 576–584.
26. Van Goethem, JWM. et al. Spinal tumors. *Eur J Radiol* 2004; 50: 159–176.
27. Ho, T. et al. Intramedullary spinal schwannoma: Case report and review of preoperative magnetic resonance imaging features. *Asian J Surg* 2006; 29(4): 306–308.
28. Lang, N. et al. Malignant peripheral nerve sheath tumor in spine: Imaging manifestations. *Clin Imaging* 2012; 36: 209–215.
29. Mut, M. et al. Metastasis to nervous system: Spinal epidural and intramedullary metastases. *J Neuro Oncol* 2005; 75: 43–56.
30. Rodallec, MH. et al. Diagnostic imaging of solitary tumors of the spine: What to do and say. *RadioGraphics* 2008; 28:1019–1041.
31. Shah, LM. et al. MRI of spinal bone marrow: Part 1, Techniques and normal age-related appearances. *AJR Am J Roentgenol* 2011; 197: 1298–1308.
32. Hanrahan, CJ. et al. MRI of spinal bone marrow: Part 2, $T_1$-weighted imaging-based differential diagnosis. *AJR Am J Roentgenol* 2011; 197: 1309–1321.
33. Motamedi, K. et al. Imaging of the lumbar spine neoplasms. *Semin Ultrasound CT MRI* 2004; 25: 474–489, Elsevier.
34. Prasad, D. et al. Malignant spinal-cord compression. *Lancet Oncol* 2005; 6: 15–24.

# 12

## *MR Pathology of Sacrum and Ilium*

**Daphne J. Theodorou, Stavroula J. Theodorou, and Yousuke Kakitsubata**

**CONTENTS**

## 12.1 Introduction

In a busy radiology practice, imaging of the bony pelvis and lumbosacral spine is commonly requested to investigate the source of lower back and/or hip pain. Radiographs are the initial performed studies. The complex anatomy of the pelvic skeleton, with a number of overlapping structures, makes the imaging assessment and interpretation of many disorders difficult and fraught with the potential to err. Often, pelvic radiographs are suboptimal because of poor positioning of the patient or severe pain. Sacral lesions are frequently overlooked because of the nonspecificity of associated symptoms and because the posterior curvature of the sacrum and the presence of overlying bowel contents obscure anatomic details on plain films. If radiographs are unrevealing, computed tomography (CT) is requested for imaging the sacrum, the presacral space, and the iliac bones. When specific questions arise in cases of suspected trauma or tumor, MR imaging is performed adding multiplanar capability and improved soft tissue contrast.

Space-occupying lesions in the pelvic region comprise a large spectrum of disorders that may be neoplastic or non-neoplastic in nature. Neoplasms can be benign or malignant and may be classified into subtypes according to their predominant tissue production (i.e., osteoid-producing and cartilage-producing) and so on. In any suspected neoplastic lesion, a pattern-based diagnostic approach may generate essential diagnostic information for accurate evaluation [1]. Characterization of the morphology of the process, including the pattern of bone destruction, the tumor's size, shape, and margin, the presence and nature of visible tumor matrix, any internal or external trabeculation, the presence of cortical erosion, penetration, and expansion, the type of periosteal response, and the presence of a soft tissue mass provide significant diagnostic information regarding the aggressive or nonaggressive behavior of the lesion [2–4]. Further, tumor mimics can be classified as normal variants, metabolic, posttraumatic, and inflammatory conditions. The addition of clinical information, including the age of the patient, comparison of data with those from prior imaging examinations, information derived from complementary imaging studies, and clinical chemistry, and, in some cases, histologic data also are essential in the analysis of tumors and tumor-like lesions.

Acute injury to the pelvis encompasses a great variety of polymorphous lesions, differing from each other by their specific biomechanics, anatomic details, imaging appearances, and treatment options. Pelvic fractures often occur in violent trauma and are associated with severe visceral lesions (i.e., gastrointestinal and genitourinary) and complications, putting vital prognosis at stake. Stress fractures are very prevalent in the osseous pelvis and may be clinically and radiologically overlooked [5]. Again, detection of a given lesion certainly depends on level of suspicion, location of the disorder, and imaging characteristics.

The variable imaging presentations of a plethora of disease processes in the osseous pelvis definitely challenge diagnosis and require a systematic and multidisciplinary approach. Clearly, an overall awareness of the clinicopathologic and MR imaging findings of common and uncommon abnormalities affecting the pelvic bones is important for effective diagnosis.

## 12.2 Neoplastic Lesions

The osseous pelvis can be the site of origin of a great number of benign or malignant tumors and tumor-like lesions. Imaging is fundamental in the assessment of tumors, particularly for determining anatomic origin and extent, response to therapy, and recurrence. The imaging diagnosis of primary pelvic tumors is often challenging, owing to the varied appearance of neoplasms, and the presence of nonspecific and even overlapping imaging features. Analysis of location of lesion and clinical history may provide additional diagnostic information. Radiography, despite its inherent limitations (i.e., bowel gas obscuring lesions and suboptimal technical projections), is the initial screening modality that plays an important role in the imaging workup of pelvic masses. CT enables more accurate evaluation of tumor location and anatomic extent, margins, visualization of osseous or chondroid calcification, cortical involvement, and associated periosteal reaction. MR imaging has proven superior for assessing the morphology, composition, and extent of tumor and its relationships to the adjacent structures (i.e., soft tissue involvement and intra-articular extension), and for detecting hemorrhage and intratumoral necrosis. When used after the intravenous administration of contrast material, both CT and MR imaging may provide an indication of the vascularity of a tumor. However, these features are not specific for benign or malignant tissue tumors. Thus, biopsy remains the definitive diagnostic procedure.

### 12.2.1 Benign Bone Lesions and Tumors

Benign pelvic bone lesions and tumors are a diverse group of abnormalities with a great variety of histologic lineages. The imaging evaluation usually includes the use of plain radiographs to detect and localize the lesion and of cross-sectional imaging with CT or MR imaging to further characterize the abnormality and define its extent. In some cases, the combined clinical and imaging findings may suggest the diagnosis.

#### 12.2.1.1 Giant Cell Tumor

Giant cell tumor is the most common benign sacral neoplasm [6,7], representing almost 70% of benign sacral tumors, and the second most common primary sacral tumor after chordoma. Women, usually between the ages of 15 and 40 years, are affected almost twice as often as men. Patients present with pain (often with radicular distribution), weakness, and sensory neurologic deficits. Sacral giant cell tumors are usually large, eccentric lesions that may cause destruction of the sacral foramen, and abut or extend across the sacroiliac (SI) joint to the ilium [1,8]. Giant cell tumors are purely lytic, destructive, expansile lesions with a nonsclerotic margin composed of osteoclastic, multinuclear giant cells within a spindle cell stroma [9]. Although usually benign, giant cell tumors may be malignant in 5%–10% of the cases and can metastasize [6,10].

The tumor appears as a well-defined lesion with soft tissue attenuation on CT images and low-to-intermediate signal intensity on $T_1$- and $T_2$-weighted MR images [3,9] (Figure 12.1). In this instance, hypointense signal on $T_2$-weighted images, presumably owing to the deposition of collagen or hemosiderin may help narrow the differential diagnosis because most other tumors usually show high $T_2$-weighted signal intensity [8]. Giant cell tumors are often heterogeneous owing to the presence of cystic areas and intratumoral hemorrhage, which may produce fluid–fluid levels [8]. Marginal sclerosis, tumor matrix calcifications, or septations are usually absent. After the intravenous administration of contrast material, enhancement of the lesion is prominent.

#### 12.2.1.2 Intraosseous Cyst (Pneumatocyst)

Although relatively rare, benign juxta-articular bone cysts may be seen in the pelvic bones, and usually the iliac bones. Recognition of this lesion is important, however, as the differential diagnosis can be complicated by serious other abnormalities as infection or neoplasm.

On radiographs and CT scans, subchondral bone cysts appear as lucent intraosseous gas collections that are surrounded by a discernible sclerotic rim [11] (Figure 12.2). With CT, density measurements easily confirm the presence of gas.

#### 12.2.1.3 Aneurysmal Bone Cyst

Fifty percent of aneurysmal bone cysts involving flat bones are located in the pelvis [12]. Sacral aneurysmal bone cyst accounts for less than 20% of all spinal aneurysmal bone cysts. Most aneurysmal bone cysts occur in patients under the age of 30 years, with a slight female predominance. Patients present with localized pain and neurologic symptoms [13].

Radiographs demonstrate a well-defined, radiolucent and occasionally trabeculated, expansile lesion surrounded by a thin sclerotic margin. Soft tissue extension or a compression fracture may be present. CT portrays the osteolytic and expansile nature of lesion and defines any soft tissue extension. On MR images, aneurysmal bone cyst appears as a lobulated or septated mass that is surrounded by a thin, well-defined rim of low signal intensity on both $T_1$- and $T_2$-weighted images (Figure 12.3). Signal intensity of the lesion is usually increased on $T_2$-weighted images. Multiple fluid–fluid levels indicating intralesional hemorrhage with sedimentation can be seen [13]. After the intravenous administration of contrast material, the septations and rim generally enhance.

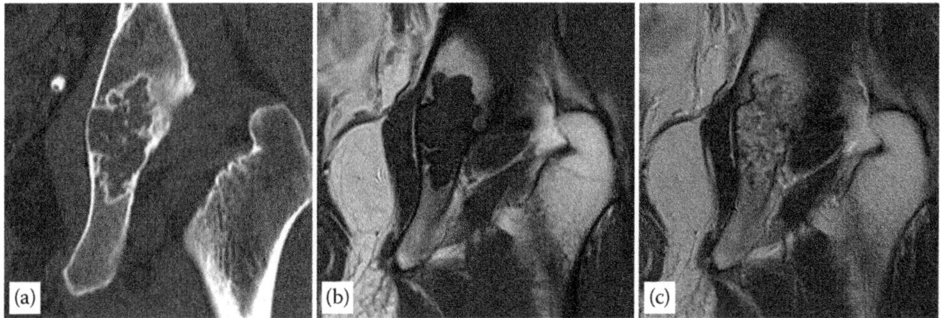

**FIGURE 12.1**
Giant cell tumor in a 40-year-old man with pain localized in the left buttock. No mass could be palpated. (a) On the coronal reconstructed CT image a large, well-defined osteolytic lesion of the left iliac bone is seen. The cortex is thinned, but there is no soft tissue involvement. (Coronal $T_1$-weighted (b) and $T_2$-weighted (c) MR images show that the tumor has low signal intensity in (b) and inhomogeneous intermediate to high signal intensity in (c).

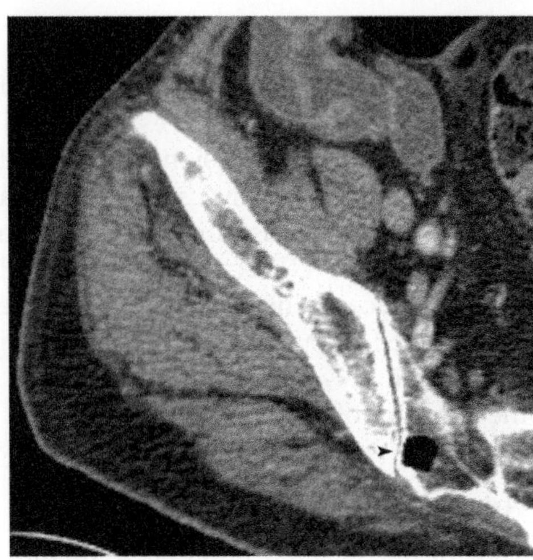

**FIGURE 12.2**
This 84-year-old man presented with low back pain localized to the right sacroiliac joint. Axial CT scan shows a localized, lucent juxta-articular gas collection (arrowhead) with an attenuation coefficient of −900 HU involving the subchondral sacrum.

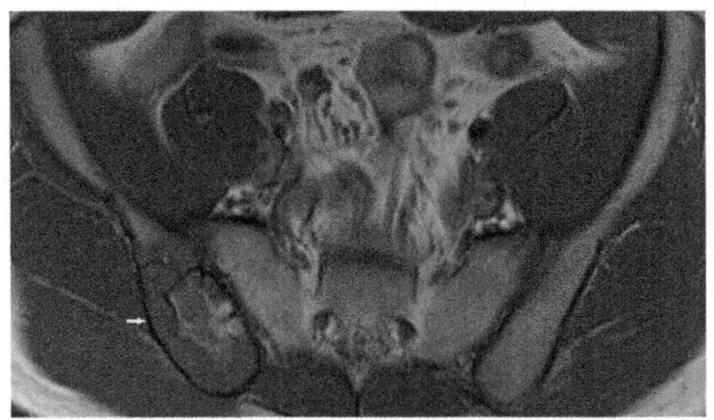

**FIGURE 12.3**
Aneurysmal bone cyst in a 18-year-old female. Axial $T_1$-weighted MR image reveals an expansile, lobulated lesion of low signal intensity in the right iliac bone (arrow) without extraosseous extension.

#### *12.2.1.4 Enostosis (Bone Island)*

Enostoses, or bone islands, are common, most likely developmental lesions that involve predominantly the flat and irregular bones [7]. Enostoses may occur in all age groups, in both men and women, and usually are incidental findings on radiographs obtained for unrelated reasons.

On radiographs, enostoses appear as single or multiple, ovoid, or round intraosseous sclerotic areas with discrete margins and thorny, radiating spicules that may simulate skeletal metastasis. Of diagnostic help in such cases, bone scintigraphy almost uniformly is negative in instances of enostoses, whereas in patients with osteoblastic metastasis, accumulation of the radiopharmaceutical agent is expected.

Giant bone islands simulating osteoblastoma or low-grade osteosarcoma on radiographic examination may involve the pelvis and the femur. CT and MR imaging can be used to characterize the imaging features of enostoses. On MR images, lesions are of low signal intensity on all pulse sequences. They may be located within the medullary cavity of the bone or abut the endosteal surface of the cortex. In the latter situation, the term *endosteoma* also may be applied to the lesion.

#### *12.2.1.5 Osteoid Osteoma*

Sacral osteoid osteomas constitute 2% of all spinal osteoid osteomas [14]. The tumors usually present in the first to second decades, with a male preponderance (2–3:1). Osteoid osteoma typically arises from the articular process of S1. Patients experience pain, which may be radicular in type, and is often worse at night and during spinal motion. Pain usually is relieved by salicylates.

On radiographs, osteoid osteoma appears as an osteolytic lesion less than 2 cm in diameter (representing the nidus) that is surrounded by osteosclerosis. Central calcification within the nidus may be present. Surrounding sclerosis, soft tissue edema, or bone marrow edema, however, may obscure the nidus. CT is valuable for detecting the nidus, which appears as a low attenuation focus surrounded by osteosclerosis. MR imaging is generally considered less useful than CT scanning in the detection of the nidus. On MR images, osteoid osteoma appears as a lesion of low-to-intermediate signal intensity on $T_1$-weighted images and intermediate-to-high signal intensity on $T_2$-weighted images. An associated soft tissue mass may be noted. Enhancement of the nidus and any associated soft tissue mass is seen after the intravenous administration of a gadolinium compound. Juxtanidal edema in marrow and in soft tissue may present as regions of high signal intensity on $T_2$-weighted MR images, with enhancement of signal intensity following the intravenous administration of a gadolinium compound.

#### *12.2.1.6 Osteochondroma*

Osteochondroma is a common benign lesion of bone, comprising a cartilage-covered bony excrescence (exostosis) that usually points away from the nearby joint. Lesions usually occur in patients who are younger than 20 years of age. The innominate bone is involved in 5% of cases [12,15]. Osteochondroma typically presents as a slowly growing, nontender, painless mass. Pelvic osteochondromas may increase in size causing displacement of adjacent soft tissue. The lesion can be symptomatic in the presence of compression of adjacent structures (i.e., nerves or vessels), fracture of the exostosis, or inflammatory changes of the adventitious bursa overlying the cartilaginous cap. The onset of new pain in a previously asymptomatic lesion, however, may indicate malignant transformation.

On radiographs and CT scans, the outgrowths may be pedunculated (with a narrow stalk and bulbous tip) or sessile (with a broad, flat base). A distinctive feature of osteochondroma is continuity between the cortex and spongiosa of the excrescence and the underlying parent bone. Histologically, osteochondroma is characterized by a benign, hyaline cartilage cap, and the presence of fat or hematopoietic elements within its deeper portions. Because intraosseous calcification, indicative of a cartilaginous tumor, may be better delineated by CT than by conventional radiography, CT has been used to define the thickness of a cartilaginous cap of an osteochondroma, a finding of great clinical importance, owing to the thicker cartilage that characterizes a chondrosarcoma. A small, well-defined cap with regular, stippled calcification is most compatible with a benign outgrowth, whereas a large, poorly defined cap, containing irregular calcification is suspicious for malignancy. Any increase in the thickness of the cartilage cap after puberty or a cartilage cap more than 2.0 cm thick is suspicious for malignant transformation [12]. Osteochondromas arising in the innominate bone frequently are large, forming a soft tissue mass with displacement of adjacent structures, and irregular calcification. In this instance, differentiation of a benign osteochondroma from one that has undergone malignant transformation is not without difficulty.

On MR images, the detection of continuity of the cortical and medullary bone in the outgrowth with that of the parent bone is virtually diagnostic of an osteochondroma (Figure 12.4). The cartilaginous tissue in the cap of an osteochondroma is of high signal intensity on $T_2$-weighted MR images and this tissue is covered with perichondrium that appears to be of low signal intensity on these images. After the intravenous administration of contrast material, osteochondroma shows enhancement that usually is limited to the fibrovascular tissue that covers the nonenhancing cartilage cap.

#### 12.2.1.7 Chondromyxoid Fibroma

Chondromyxoid fibroma is an extremely rare benign cartilaginous tumor, representing less than 1% of all primary bone tumors. The tumor offends usually the long bones, with only five cases reported in the sacrum. With a 1.5:1 predilection for males, chondromyxoid fibromas are found in the second or third decade of life. The tumor can be asymptomatic, while large lesions may be associated with pain, swelling, or distortion of bone [16].

The radiographic and CT appearance of chondromyxoid fibroma is that of an expansile, osteolytic, lobulated lesion with a sclerotic rim. Occasionally, septations may be seen. On MR images, chondromyxoid fibromas display low signal intensity on $T_1$-weighted images and high signal intensity on $T_2$-weighted images. After the intravenous administration of gadolinium-based contrast material, heterogeneous enhancement of signal intensity is noted.

#### 12.2.1.8 Osteoblastoma

Sacral osteoblastomas represent 17% of spinal osteoblastomas. Men are affected more frequent (2:1) than women, and patients are affected between the ages of 20 and 30 years [7]. Focal dull pain, scoliosis, and neurologic symptoms are common manifestations of disease [17]. Because osteoblastomas share similar clinical and pathologic characteristics with osteoid osteomas, they are considered variants of the same process.

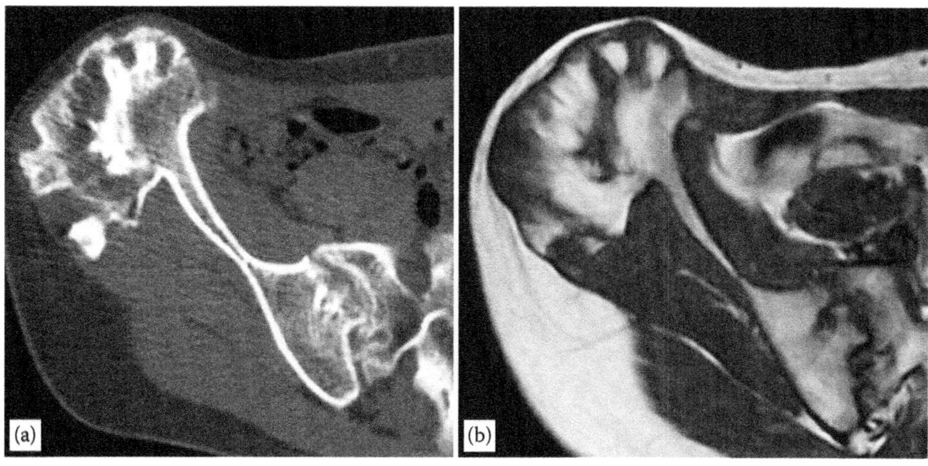


**FIGURE 12.4**
Osteochondroma of right iliac bone in a 15-year-old girl with a palpable and tender mass. (a) Axial CT scan shows anterior superior iliac spine osteochondroma with pathognomonic marrow and cortical continuity. (b) Axial $T_1$-weighted MR image reveals the signal intensity properties of yellow marrow within the osteochondroma.

Osteoblastoma presents as an expansile, purely osteolytic lesion larger than 1.5 cm in diameter surrounded by a peripheral sclerotic rim. Multifocal calcifications may be present. Radiographs and CT images may reveal a purely osteolytic, mixed, or osteosclerotic lesion that may contain calcifications. An aggressive pattern of tumor involvement with cortical destruction and extension into adjacent soft tissue occasionally may be noted, in some patients. MR imaging better delineates the extent of tumor in bone and soft tissue. Osteoblastomas exhibit variable low-to-intermediate signal intensity on $T_1$-weighted images and intermediate-to-high signal intensity on $T_2$-weighted images [14] (Figure 12.5). Peritumoral edema in bone marrow and soft tissues (flare phenomenon), reflecting inflammatory reaction, is a characteristic feature. Osteoblastomas reveal enhancement of signal intensity following the intravenous administration of contrast material [18].

#### 12.2.1.9 Fibrous Dysplasia

Fibrous dysplasia is a skeletal developmental anomaly of unknown cause, in which osteoblasts fail to undergo normal morphologic differentiation and maturation. As a result, there is deposition of abnormal, dysplastic bone and fibrous tissue, both of which compromise skeletal strength. Gross skeletal deformities are seen including bowing, angular and curvilinear distortion of bones, fusiform expansion, and linear growth discrepancies [19]. The disorder may be monostotic (accounting for 75%–80% of the cases) or polyostotic [19]. The ages of patients with monostotic disease have ranged from 10 to 70 years, although recognition is most frequent in the second and third decades of life. The age distribution is considerably younger in polyostotic fibrous dysplasia, with two-thirds of patients being symptomatic before the age of 10 years. Whereas monostotic disease may run asymptomatic, polyostotic involvement is usually correlated with clinical symptoms as pain and pathologic fracture. Fibrous dysplasia may be associated with abnormalities of the skin, skeleton, central nervous system, and endocrine glands. For example, polyostotic fibrous dysplasia may be seen in association with endocrine dysfunction—typically manifested by precocious female sexual development and cutaneous pigmentation—known as the *McCune–Albright syndrome*. Polyostotic fibrous dysplasia may involve the pelvis and spine. With regard to skeletal distribution of abnormalities, the innominate bones are affected most commonly, whereas the sacrum is occasionally involved. Solitary involvement of the ilium, however, is infrequent and in most cases occurs concurrently with involvement of the femur [19]. Severe changes in pelvic bones with accompanying acetabular deformity, including protrusion, may result in telescoping of the bones about the hip.

CT scans efficiently delineate the extent of skeletal involvement in fibrous dysplasia affecting the complex bony pelvis. On CT images, expansile, well-defined osseous lesions containing hazy, amorphous contents and subtle mineral deposits are seen. On MR imaging, the signal intensity characteristics of fibrous dysplasia are variable. Abnormal areas show low signal intensity on $T_1$-weighted images and low, intermediate, or high signal intensity on $T_2$-weighted images (Figure 12.6). After the intravenous administration of gadolinium-based contrast material, the degree and pattern of enhancement of signal intensity also is variable.

#### 12.2.1.10 Hemangioma

Sacral hemangiomas are rare. Women, usually between the ages of 40 and 60 years, are affected two times more frequently than men. Although most patients

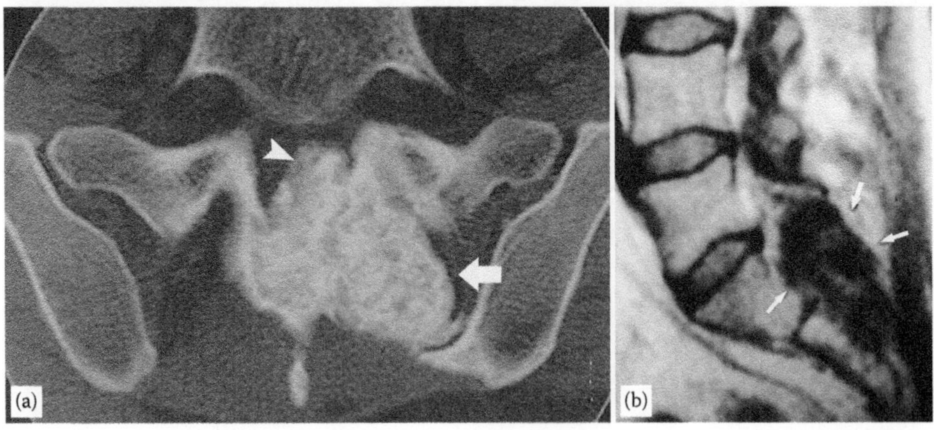


**FIGURE 12.5**
Osteoblastoma of sacrum in a 37-year-old man with low back pain and sciatica. (a) Axial CT scan shows markedly expansile osteosclerotic lesion (arrow) extending into the spinal canal (arrowhead). (b) Sagittal $T_2$-weighted MR image shows lobulated mass (arrows) involving the sacrum. Because of its osteosclerotic nature, the mass is of predominant low signal intensity on $T_2$-weighted images.

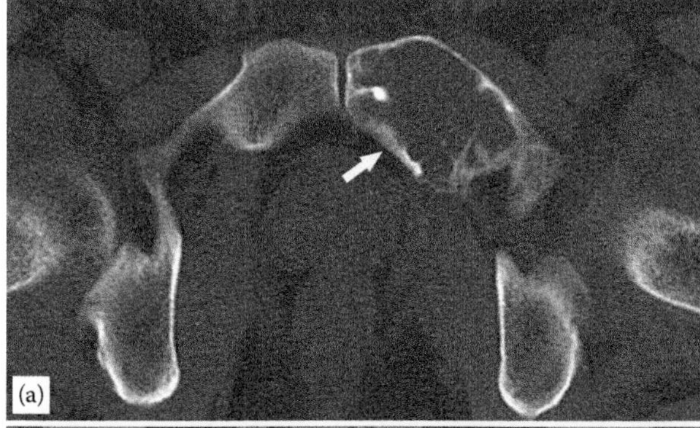


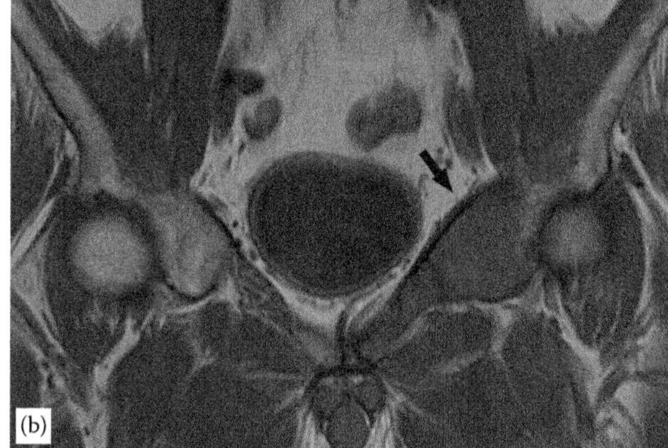


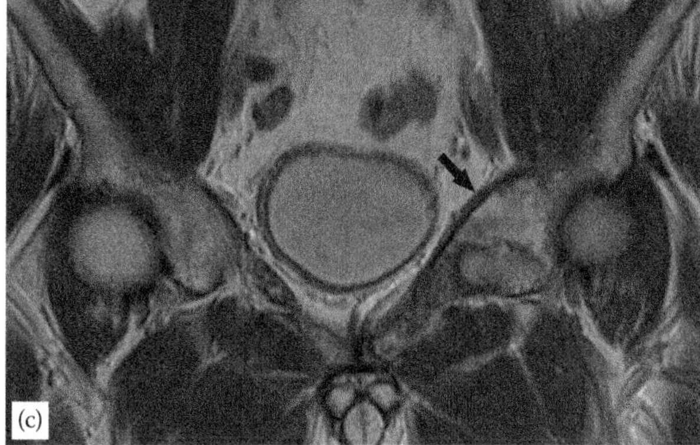


**FIGURE 12.6**
A 61-year-old man with fibrous dysplasia. (a) Axial CT scan shows expansile, osteolytic lesion (arrow) in the left pubic bone. Coronal $T_1$-weighted (b) and $T_2$-weighted (c) MR images reveal the expansile lesion (arrow). It is of low signal intensity in (b) and high signal intensity in (c).

are usually asymptomatic, they may experience pain owing to sacral canal hemorrhage, nerve root compression, or pathologic fracture. Extension of the lesion into surrounding soft tissue may be seen. Radiographs show coarse vertical trabeculation. At CT, thickened trabeculae appear as radiodense foci. Hemangiomas display variable intermediate-to-high signal intensity on $T_1$-weighted images and high signal intensity on $T_2$-weighted images, depending on the fatty and vascular components of the lesion [20,21] (Figure 12.7). MR images obtained after intravenous administration of contrast material show avid homogeneous enhancement of the hemangioma. Atypical hemangiomas such as epithelioid hemangioma and epithelioid hemangio endothelioma may show heterogeneous signal intensity corresponding to the absence of fat and predominant inflammatory changes [22].

*Angiomatosis* is defined as diffuse infiltration of bone or soft tissue by hemangiomatous or lymphangiomatous lesions [23]. Patients are usually affected in the first three decades of life. In patients with diffuse skeletal angiomatosis, multiple osteolytic lesions of variable size may be visualized in the pelvis [20] (Figure 12.8). The CT and MR imaging features of these lesions, however, are identical to those of solitary osseous hemangiomas.

#### *12.2.1.11 Nerve Sheath Tumor*

Benign peripheral nerve sheath tumors have been classified into neurilemoma (schwannoma) and neurofibroma [24]. Peripheral nerve sheath tumors may arise from sacral nerve roots. Nerve sheath tumors usually affect patients between 20 and 30 years old, with no sex predilection. Presenting symptoms may include pain and neurologic deficits.

The radiographic findings are those of a small, well-defined soft tissue mass, which may cause widening of the sacral foramen, with or without sacral bone destruction [25]. CT frequently reveals a mass of low attenuation along the course of the affected sacral nerve root. On MR imaging, nerve sheath tumors usually display low signal intensity on $T_1$-weighted images and high signal intensity on $T_2$-weighted images (Figure 12.9). Marked and uniform enhancement of nerve sheath tumors is seen after intravenous administration of a contrast agent.

### 12.2.2 Malignant Bone Tumors

Malignant tumors of pelvic bones may be primary or secondary. Primary malignant tumors of the sacrum are uncommon, whereas metastatic disease is quite frequent. Diagnostic difficulty arises in the presence of an isolated sacral bone tumor unrelated to known malignancy elsewhere in the body. Malignant tumors usually manifest as painful, rapidly growing, infiltrative masses. Although radiography, the technique most often used for initial evaluation of a lesion, is helpful for detecting cortical bone destruction, CT proves more sensitive for detecting cortical involvement and calcified tumor matrix. MR imaging enables more accurate localization of the tumor and is helpful for determining the extent of tumor invasion and for detailed tissue characterization. Even when the findings are not specific, however, imaging is important for tumor staging, therapeutic planning, and follow-up of patients.

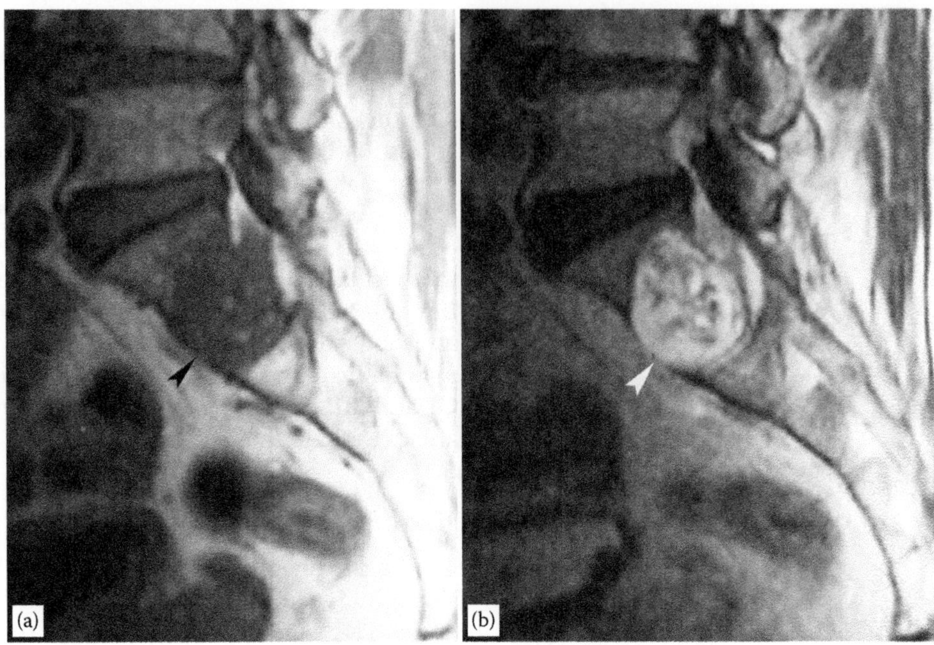


**FIGURE 12.7**
Sacral hemangioma in a 76-year-old man with back pain. Sagittal $T_1$-weighted (a) and $T_2$-weighted (b) MR images show lesion (arrowhead) of intermediate signal intensity in (a) and high signal intensity in (b). Lesion is more conspicuous in (b).

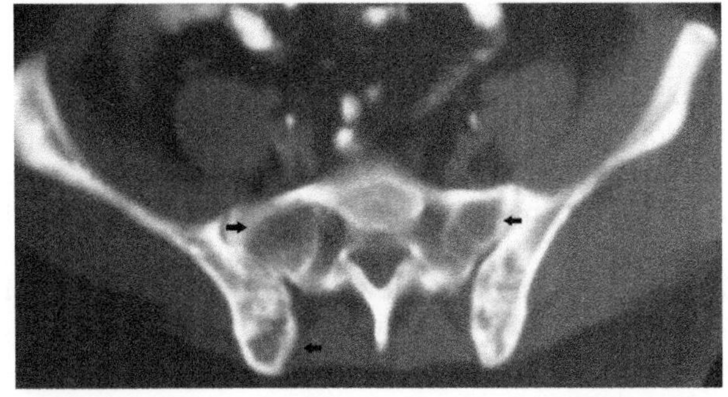

**FIGURE 12.8**
Cystic angiomatosis in a 30-year-old man. Axial CT image reveals multiple osteolytic lesions of variable size, symmetrically distributed in sacrum and iliac bones (arrows). Associated bone sclerosis in the iliac wings is seen.

#### *12.2.2.1 Chordoma*

Although chordoma is an uncommon tumor, accounting for 2%–4% of all primary malignant bone tumors, it represents the most common primary malignant sacral neoplasm [1,26]. Almost 50% of all chordomas originate in the sacrococcygeal region and particularly the fourth and fifth sacral segments [26]. This tumor arises from remnants of the notochord, which is the earliest fetal axial skeleton. Chordoma occurs in middle-aged patients (aged 30–60 years), and men are affected two to three times more often than women. Chordomas are slow-growing tumors, which may reach a huge size when arising from the sacrum. Because tumors spread anteriorly into the spacious pelvic cavity, symptoms usually occur late when chordoma has already reached massive proportions [27]. Presenting symptoms include pain, numbness, motor weakness, and incontinence or constipation in sacral tumors [7]. Tumors may recur locally, and sacral chordomas may metastasize more often than skull base chordomas [28]. Pathologically, typical chordomas contain clear cells with intracytoplasmic vacuoles (physaliphorous cells) and abundant intracellular and extracellular mucin. In dedifferentiated chordomas, the mucinous matrix is replaced by sarcomatous fibrous, chondroid, or osteoid elements. Chordomas are often contained within a fibrous pseudocapsule.

On radiographs, chordoma appears as a large destructive, osteolytic and often expansile lesion centered in the midline, with an associated soft tissue mass [29]. Extraosseous extension of tumor may be present in the presacral region or the sacral canal. Sacrococcygeal chordomas may extend across the SI joint. Additional findings may include amorphous intratumoral calcification (50%–70% of patients) and osteosclerosis. CT shows bone destruction and surrounding soft tissue involvement. Coronal oblique CT images of sacrococcygeal lesions allow better delineation of sacral neural foraminal and SI joint involvement [7]. Chordoma exhibits low-to-intermediate signal intensity on $T_1$-weighted MR images and intense high signal on $T_2$-weighted MR images, corresponding to

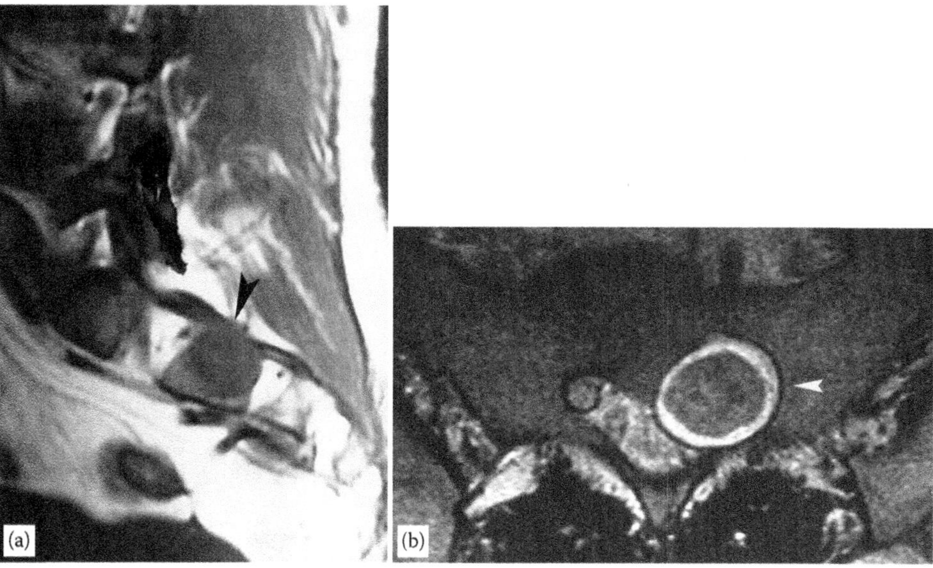

**FIGURE 12.9**
Neurofibroma in a 34-year-old woman, who presented with intractable low back pain and a limp. (a) Sagittal $T_1$-weighted MR image shows lesion of intermediate signal intensity (arrowhead) in the sacrum. (b) Axial $T_2$-weighted MR image reveals large mass (arrowhead) of intermediate signal intensity surrounded by rim of high signal, causing widening of left sacral foramen.

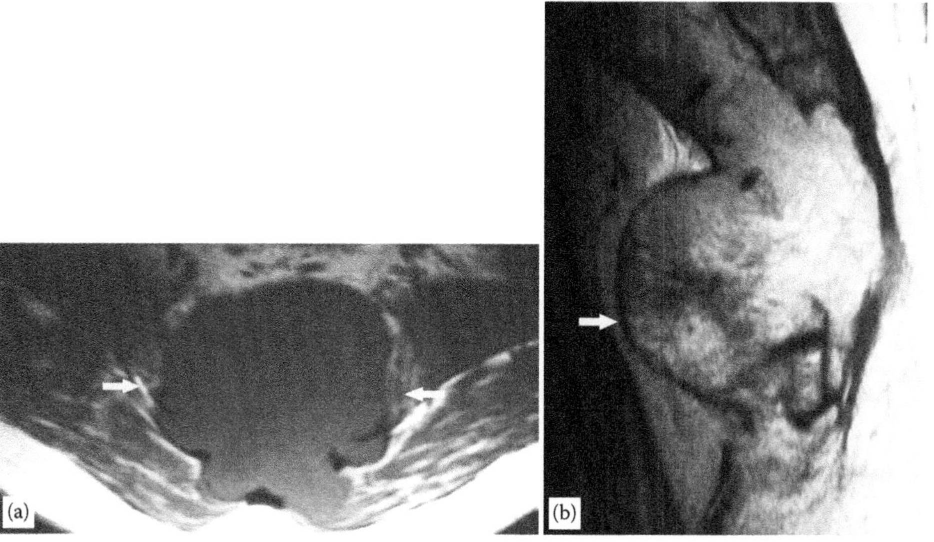

**FIGURE 12.10**
Chordoma of the sacrum in a 54-year-old man with pain and rectal discomfort. (a) Axial $T_1$-weighted MR image shows a large, expansile sacral lesion (arrows) which has intermediate signal intensity. The tumor extends to gluteal muscles. (b) Sagittal $T_2$-weighted MR image shows mass causing gross destruction of the distal sacrum and coccyx (arrow). The mass extends to the presacral space and also invades posterior soft tissues and musculature.

the intratumoral accumulation of mucin (Figure 12.10). Internal septations dividing the gelatinous components of tumor display low signal intensity on $T_2$-weighted images. Hemosiderin deposition related to recurrent intratumoral hemorrhage produces regions of low signal intensity. After the intravenous administration of contrast material, variable moderate to prominent enhancement is usually seen.

#### *12.2.2.2 Chondrosarcoma*

Chondrosarcoma is the second most common nonlymphoproliferative primary malignant tumor of the spine following chordoma in adults, accounting for 7%–12% of these lesions, albeit sacral involvement is rare [7]. The mean age of patients with chondrosarcoma is 45 years. Males are affected two to four times more than females [28]. Pathologically, conventional spinal chondrosarcomas are relatively low-grade tumors composed of lobules of hyaline cartilage that are separated by fibrovascular septations. There are primary (arising *de novo*) and secondary forms of chondrosarcomas; the latter arising from malignant transformation of enchondromas (central) or osteochondromas (peripheral). Most chondrosarcomas are primary and central in origin. Chondrosarcoma arising from osteochondroma is seen as thickening at the cartilaginous cap associated with a large mass at this site [7].

Radiographs and CT images reveal large destructive lesions with characteristic chondroid matrix mineralization, in the form of rings and arcs. Tumors have a lobulated contour, which produces endosteal scalloping. Cortical destruction and extension into the surrounding soft tissues is almost always present. On CT images, chondrosarcomas are hypodense relative to skeletal muscle because of nonmineralized hyaline cartilage. With MR imaging, chondrosarcomas show low-to-intermediate signal intensity on $T_1$-weighted images and high signal intensity on $T_2$-weighted images (Figure 12.11). Areas of mineralization/calcification are manifested as signal void on MR images. Enhanced MR images show mild nodular, peripheral, diffuse, or septal enhancement, the latter in the form of rings and arcs that correspond to vascular septations between the cartilaginous lobules [28,30].

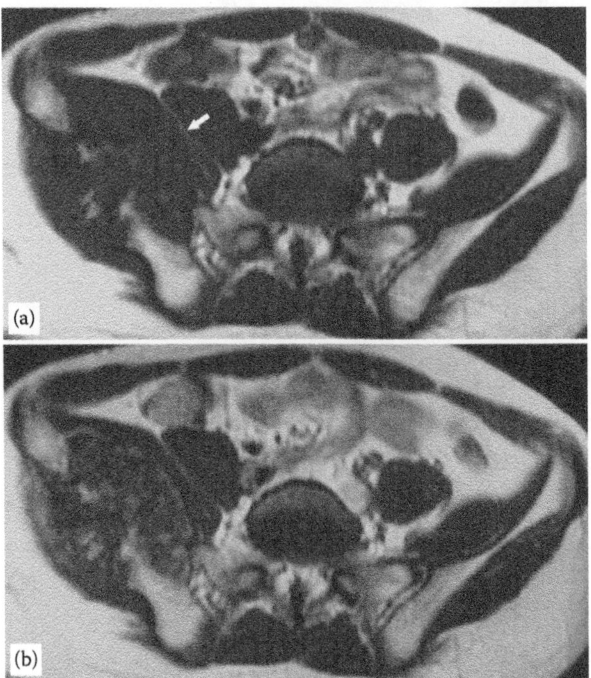


**FIGURE 12.11**
Chondrosacroma of right iliac wing in a 17-year-old boy with severe bone pain. (a) Axial $T_1$-weighted MR image reveals prominent destruction of iliac wing and associated large soft tissue mass (arrow). (b) Axial $T_1$-weighted MR image after gadolinium administration shows inhomogeneous enhancement of osteolytic lesion containing areas of matrix calcification.

#### 12.2.2.3 Ewing's Sarcoma and Primary (Primitive) Neuroectodermal Tumor

Almost 3%–10% of all primary Ewing's sarcomas and primitive neuroectodermal tumors (PNET) are located in the spine and particularly in the lumbosacral region and the sacral ala [28,31]. Ewing's sarcoma, however, is the most common nonlymphoproliferative primary malignant tumor of the spine and pelvic bones in children [32]. In Ewing's sarcoma, there is proliferation of undifferentiated small, round cells. Patients are affected between the ages of 5 and 14 years, with a male predominance [32,33]. The similarities between PNET and Ewing's sarcoma are great, and evidence exists that Ewing's sarcoma may indeed constitute a more undifferentiated form of PNET of bone [31]. Patients present with pain and neurologic symptoms that may be associated with fever and leukocytosis.

Radiographs and CT images may reveal an expansile, osteolytic, mixed, or osteosclerotic lesion residing in the involved bone. Extraosseous extension of the tumor is common. CT and MR imaging are helpful for the depiction of the extent of osseous involvement, associated paraspinal soft tissue mass and invasion of the spinal canal and neural foramina. The MR imaging features of spinal Ewing's sarcoma are nonspecific, with lesions usually showing intermediate signal intensity on $T_1$-weighted images and intermediate-to-high signal intensity on $T_2$-weighted images (Figure 12.12). The tumor can be inhomogeneous due to hemorrhage, calcification, or necrosis. Marked enhancement of signal intensity is seen following the intravenous administration of contrast material.

#### 12.2.2.4 Osteosarcoma

In the spine, osteosarcomas account for 5% of all primary malignant spine tumors and for less than 3% of all osteosarcomas [7,30]. Spinal osteosarcomas may be associated with Paget's disease or previous irradiation [17]. Compared with appendicular osteosarcomas, spinal osteosarcomas present at an older age with peak incidence in the fourth decade of life. Men are affected more frequently than women. The lumbosacral spine is the site of involvement in 60%–70% of the cases where lesions typically arise from the vertebral body. Loss of vertebral body height with sparing of the adjacent disk is common. Tumor manifests with pain, neurologic symptoms, and a growing mass.

Radiographs and CT images reveal a predominantly osteosclerotic, mixed, or purely osteolytic lesion [34]. CT

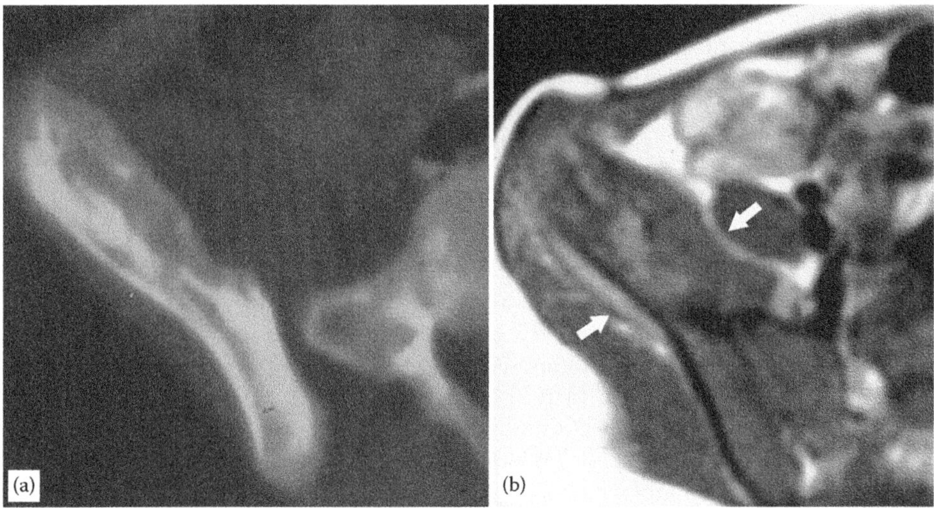

FIGURE 12.12
Ewing sarcoma in a 7-year-old boy with pain and a limp. (a) Axial CT image shows large osteolytic lesion of right iliac wing. (b) Axial $T_1$-weighted MR image shows large lesion (arrows) of intermediate signal intensity and associated pelvic soft tissue mass.

delineates both the matrix mineralization and soft tissue mass within the epidural space and paraspinal region. The MR imaging appearance of spinal osteosarcoma is nonspecific, with lesions showing low-to-intermediate signal intensity on $T_1$-weighted images and high signal intensity on $T_2$-weighted images. Dense matrix mineralization exhibits signal void with all sequences. Heterogeneous enhancement is observed after intravenous administration of contrast material.

#### 12.2.2.5 Paget's Sarcoma

Sarcomatous degeneration is a rare complication of Paget's disease that has been reported to develop in approximately 1% of patients. In patients with polyostotic involvement, sarcomatous transformation may occur in as many as 5%–10% of cases [2]. Patients with sarcomatous degeneration in Paget's disease usually are between 55 and 80 years old; men are affected slightly more frequently than women and at a younger age. Osteosarcoma (50%–60%), fibrosarcoma (20%–25%), and chondrosarcoma (10%) are the neoplasms most commonly arising in sarcomatous transformation of pagetic bone. The pelvis is frequently involved. Pagetic sarcomas may occur close to sites of previous, healed fractures. The diagnosis of neoplasm complicating Paget's disease usually is not difficult when increased pain and a soft tissue mass are observed. Osteolysis rather than osteosclerosis is the major radiographic feature of the sarcoma, although osteosclerosis is occasionally seen. Other findings indicative of a superimposed neoplasm include cortical disruption, bony spiculation, soft tissue mass, and a pathologic nonhealing fracture [2]. CT, in common with MR imaging, provides useful information about the extent and neurovascular compromise of pagetic sarcomas. MR imaging will show the variable signal characteristics of pagetic bone and a soft tissue mass of intermediate signal intensity on $T_1$-weighted images and areas of increased signal intensity on $T_2$-weighted images.

#### 12.2.2.6 Ependymomas (Myxopapillary Ependymoma)

Ependymomas are the most common, albeit quite rare, intramedullary spinal cord tumors in adults. Derived from the ependymal glia, ependymomas are further divided into cellular and myxopapillary subtypes. Cellular ependymomas usually occur in the cervical cord, whereas myxopapillary ependymomas occur almost uniformly in the conus medullaris and filum terminale [35]. Although intradural ependymomas are usually benign, extradural ependymomas may metastasize in up to 27% of reported cases [25,35]. Small tumors may displace the lumbosacral nerve roots, whereas large tumors expand the spinal canal causing bone destruction. The mean age of patients with myxopapillary ependymomas at presentation is 35 years, with a slight male preponderance. Low back pain, sciatica, neurologic deficits and GI or GU dysfunction are among the presenting symptoms.

Radiographs and CT images of myxopapillary ependymomas may show widening of spinal canal, enlargement of neural foramina, bone destruction, and a soft tissue mass. On MR images, tumors are isointense relative to the spinal cord on $T_1$-weighted images and hyperintense on $T_2$-weighted images. Occasionally, myxopapillary ependymomas may be hyperintense relative to the spinal cord on $T_1$-weighted images, owing to the presence of hemorrhagic or mucinous material [36]. Intense homogeneous enhancement of myxopapillary lesions is noted after the intravenous administration of contrast material.

#### 12.2.2.7 Metastatic Disease

The bones of the pelvis are among the usual locations for skeletal metastasis [37]. Metastases from lung, breast, gastrointestinal, prostate, kidney, and skin cancers are common in the sacrum [17]. The routes allowing spread of tumor to the pelvic bones are hematogenous dissemination and direct extension from the primary neoplastic site or via lymphatic spread [38]. Patterns of involvement include osteolytic, osteosclerotic, or mixed osteolytic-osteosclerotic lesions. Although a solitary bone metastasis may occasionally simulate a primary tumor of bone, multiple lesions of varying sizes are highly suggestive of metastatic disease.

Radiographs often are suboptimal to delineate underlying pathology, owing to overlapping osseous and soft tissue structures. On $T_1$-weighted MR images, metastatic deposits usually are of low signal intensity and show variable signal intensity (usually high) on $T_2$-weighted images (Figure 12.13). Surrounding edema on $T_2$-weighted and short-tau inversion recovery (STIR) images may produce a *target-like* appearance of the lesions [1].

### 12.2.3 Myeloproliferative Disorders

Myeloproliferative disorders include a distinct group of diseases caused by the abnormal growth of hematopoietic marrow elements, such as plasma cells in multiple myeloma, white blood cells in leukemia, red blood cells in polycythemia vera, and fibrous tissue in myelofibrosis.

#### 12.2.3.1 Multiple Myeloma and Plasmacytoma

Multiple myeloma is a malignant disease involving excessive monoclonal proliferation of abnormal plasma cells in the bone marrow and other tissues [39]. Multiple myeloma is a common disease, representing 45% of vertebral tumors. The average age of involved patients is 60–70 years, with a slight male predominance. Plasmacytoma is the solitary form of multiple myeloma whose frequent conversion to disseminated disease suggests a definite relationship between the two processes. It affects younger patients at a mean age of about 50 years. Plasmacytoma is most frequent in the spine (50% of plasmacytomas are localized in the spine) and pelvis.

Patients with spinal plasmacytomas or multiple myeloma can be asymptomatic or may experience pain (sciatica) related to nerve root pressure and paraplegia.

Radiographic and CT changes of solitary or multiple myelomatous involvement include osteolytic lesions with discrete margins, which may cause osseous expansion. Sclerotic lesions also may be present. Calcification, vertebral fracture, and paraspinal and extradural extension of tumor are quite characteristic of myelomatous involvement. Myelomatous lesions produce a variety of MR imaging findings including focal, diffuse homogeneous, and stippled or multifocal infiltration of the bone marrow [1]. On MR images, plasmacytomas and multiple myeloma lesions show decreased signal intensity

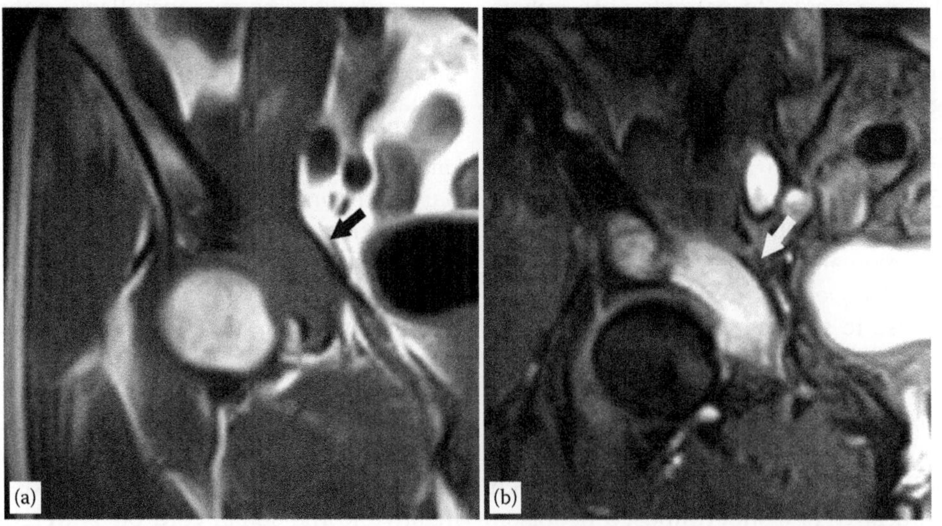


**FIGURE 12.13**
A 65-year-old man with lung cancer and metastasis to right acetabulum. Coronal $T_1$-weighted (a) and STIR (b) MR images reveal the metastatic lesion (arrow). It is of low signal intensity in (a) and high signal intensity in (b).

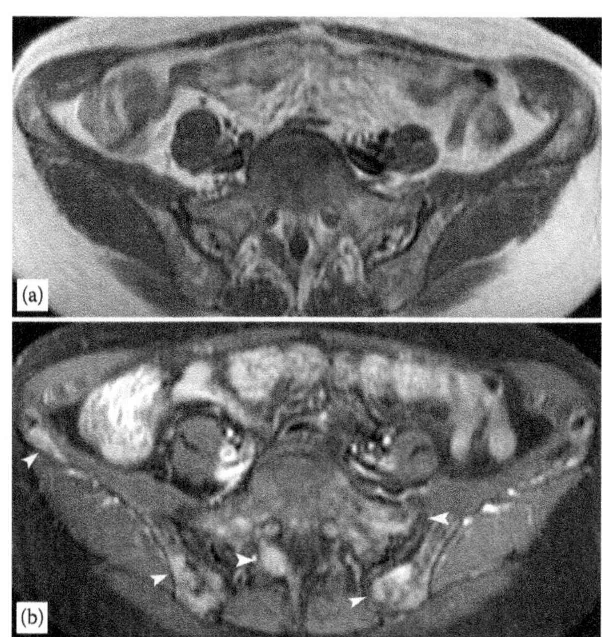


**FIGURE 12.14**
A 66-year-old woman with multiple myeloma and diffuse bone pain. (a) Coronal $T_1$-weighted MR image shows diffusely decreased signal intensity in pelvic bones. (b) Corresponding $T_1$-weighted MR image with fat saturation after gadolinium administration shows enhancement of myelomatous lesions (arrowheads).

relative to normal marrow on $T_1$-weighted images and increased signal intensity on $T_2$-weighted images (Figure 12.14). Prominent enhancement of myelomatous lesions is seen after the intravenous administration of gadolinium-containing contrast agent.

#### 12.2.3.2 Leukemias

Leukemia represents abnormal, diffuse infiltration of the bone marrow by leukemic cells, with definitive skeletal manifestations. The more significant osseous alterations are associated with the more aggressive varieties of leukemias and with the younger patients. Excessive proliferation of white cells in the confined space of the marrow cavity is associated with bone pain and fractures. In the bony pelvis, leukemic infiltration of the bone marrow is manifest as diffuse osteopenia, and multiple (or solitary) discrete osteolytic lesions. Periosteal bone formation may be an associated finding. Widespread or multifocal bone sclerosis is rare, presumably related to diffuse marrow fibrosis.

Despite the nonspecific MR imaging findings in leukemia (as well as other disorders of the bone marrow), MR imaging permits assessment of the extent of involvement of the bone marrow prior to the initiation of therapy, evaluation of the response of the disease to therapy, and detection of skeletal complications of the disease or therapeutic regimen [39]. With MR imaging, leukemic marrow depicts abnormal decreased signal intensity on $T_1$-weighted images and variable increased signal intensity on $T_2$-weighted and STIR images. Increased signal intensity is observed in the bone marrow on $T_1$-weighted spin-echo images after intravenous injection of gadolinium-based contrast agent.

#### 12.2.3.3 Lymphomas

Lymphoreticular neoplasms (lymphomas) may arise in lymphocytic cells, reticulum cells, or primitive precursor cells. Neoplasms may arise in extraskeletal sites and can eventually lodge in the bone marrow, and disseminate throughout the skeleton. Alternatively, lymphoreticular neoplasms may arise as a primary process of bone [39]. Constitutional symptoms include fevers, night sweats, weight loss, and skeletal manifestations as pain, a palpable mass, or swelling.

In widespread non-Hodgkin's lymphoma, skeletal alterations occur in 10%–20% in adults and 20%–30% in children with lesions predominating in the spine, pelvis, and skull. Radiographs and CT scans may reveal multiple osteolytic lesions, with moth eaten or permeative bone destruction, endosteal scalloping, cortical disruption and spread to adjacent soft tissues. In Hodgkin's disease, the most common sites of involvement are the spine, pelvis, ribs, and femora. Radiographic and CT findings include osteosclerosis, osteolysis, or osteosclerosis combined with osteolysis.

On MR images, lymphomatous infiltration of the marrow creates focal or diffuse regions of low signal intensity on $T_1$-weighted images with corresponding inhomogeneous intermediate-to-high signal intensity on $T_2$-weighted MR images. The presence of fibrous tissue in some of the lymphomatous tumors may explain variability in signal intensity characteristics.

#### 12.2.3.4 Primary Bone Lymphoma

Primary lymphoma of bone is a relatively rare neoplasm accounting for less than 5% of all malignant bone tumors and almost 7% of all extranodal lymphomas [40]. Primary non-Hodgkin's lymphoma of bone is a malignant hematologic tumor arising in the medullary cavity of a single bone without concurrent regional lymph node or visceral involvement. Primary non-Hodgkin's lymphoma may affect any bone and predominantly involves those of the lower extremities [41]. Primary lymphoma of the sacrum, however, predominantly affects men in the second and third decades of life. Males are affected two times more than females [28]. Spinal cord or radicular involvement reportedly occurs in 0.1%–5.8% of patients with non-Hodgkin's lymphoma [42]. On histopathologic examination, the most prominent feature of the tumorous lesion is replacement of the normal marrow constituents by neoplastic lymphoid elements.

The radiographic appearance of primary non-Hodgkin's lymphoma of bone characteristically varies widely; osteolysis with a permeative of *moth-eaten* pattern of bone destruction is most common, although sclerotic and mixed osteolytic-osteosclerotic radiographic types may also occur [41,43]. Associated cortical destruction with soft tissue extension of the tumor is frequently present [40,44]. With MR imaging, lymphoma of bone may exhibit abnormal low-to-intermediate signal intensity on $T_1$-weighted spin-echo images and variable intermediate-to-high signal intensity on $T_2$-weighted spin-echo images [45,46] (Figure 12.15). On the STIR MR images, however, focal areas of abnormal intermediate-to-high signal intensity may be seen within the marrow [47,48], which enhance after contrast administration.

#### 12.2.3.5 Polycythemia Vera

Primary polycythemia (polycythemia vera) represents hyperplasia of all of the cellular elements in the bone marrow, and primarily erythrocytes. The disease usually occurs in middle-aged or elderly patients, predominantly men. The musculoskeletal manifestations of this disease are few and may include osteonecrosis, particularly of the femoral head, patchy radiolucent lesions throughout affected bones, and generalized increased radiodensity of the skeleton.

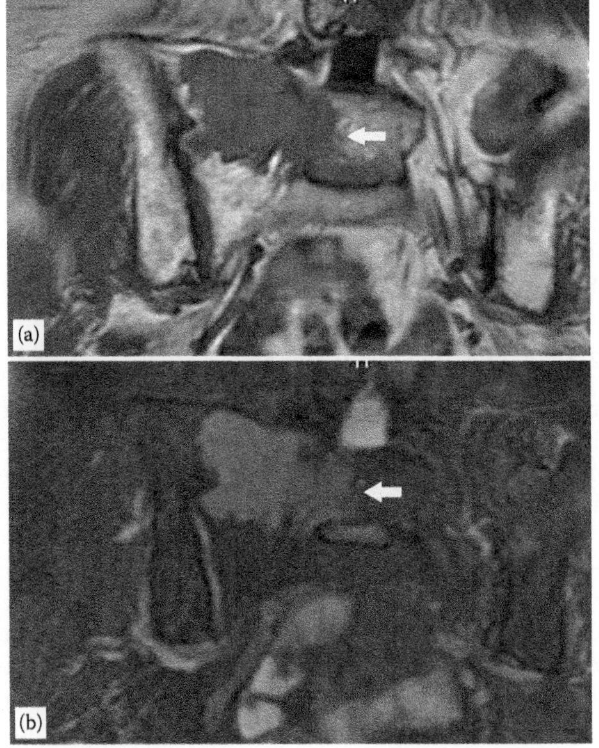


**FIGURE 12.15**
A 77-year-old woman with sacral lymphoma. Coronal $T_1$-weighted (a) and fat-suppressed $T_2$-weighted (b) MR images display the lesion (arrow) in right upper hemisacrum. It is of intermediate signal intensity in a and high signal intensity in b. The sacroiliac joint is spared.

#### 12.2.3.6 Myelofibrosis

Myelofibrosis is an infrequent disease associated with fibrotic or sclerotic bone marrow and extramedullary hematopoiesis [39]. In either its primary (idiopathic) or secondary forms, myelofibrosis encompasses fibrosis of the bone marrow, which, in some instances, may replace almost the entire marrow tissue. As a result, proliferation of potential bone marrow elements occurs in the spleen, the liver, the lymph nodes, and the bones. Myelofibrosis generally is a disease of middle-aged and elderly men and women. Skeletal alterations occur initially in the normal sites of active hematopoiesis in the adult (the vertebrae, the pelvis, and the ribs) and account for clinically evident bone pain and tenderness.

On radiographs and CT scans, there is detectable bone sclerosis. In some patients, normal or osteopenic bone and osteolytic lesions are observed as well. Periostitis generally is mild. On MR images, fibrotic replacement of the marrow, focal or diffuse, is seen as decreased signal intensity on both $T_1$-weighted and $T_2$-weighted images.

## 12.3 Traumatic Injuries

Pelvic trauma encompasses a great range of very polymorphous lesions, differing mostly by anatomical location and severity. Fractures of the pelvic ring and acetabulum are a common consequence of high-speed collisions, usually related to motor vehicle accidents (i.e., car accidents, car-pedestrian crash, motorcyclists, cyclists). Occupational accidents (i.e., falls from height and crushing injuries) and minor falls are additional causes of injury. In general, pelvic fractures imply a violent trauma mechanism with significant morbidity and mortality derived from associated injuries (i.e., gastrointestinal, genitourinary, respiratory, and vascular) and complications [49,50]. Late complications of pelvic fractures, however, can be debilitating and include nonunion, malunion, leg length discrepancy, and low back pain [50].

Because of the complex anatomy of the pelvic region with a number of overlapping structures, and the usually critical condition of patients suffering severe trauma, radiographic assessment of pelvic injury is difficult [51]. Pelvic CT provides detailed information for delineation of osseous anatomy and definition of the pattern of pelvic injury useful to orthopedic surgeons for making the correct therapeutic decisions.

### 12.3.1 Pelvic Ring Fractures

Pelvic ring fractures account for 1.5% of all joint fractures [52,53]. Simple fractures without marked displacement usually occur in elderly women with low-energy trauma. Severe displaced lesions affect younger patients in the context of high-energy trauma, resulting in multiple trauma (polytrauma) in 75% of cases [54,55]. Many systems for classification of pelvic ring fractures have been used in past years according to the principal elementary lesion, degree of instability, mechanism of injury, and therapeutic consequences. The Tile/AO classification suggests three subdivided main categories [56,57]:

1. Type-A fracture, an incomplete fracture with no disruption of the pelvic ring, or the posterior arch
2. Type-B fracture, related to disruption (fracture) of the symphysis pubis and anterior (unilateral or bilateral) SI joint disruption
3. Type-C fracture, related to complete disruption of the posterior–SI complex (Figure 12.16)

Type-A lesions represent up to 52% of all cases, type-B up to 27%, and type-C up to 21% of cases [58].

### 12.3.2 Acetabular Fractures

Classification of acetabular fractures is based on studies of anatomy and acetabular trauma performed by Judet and Letournel [59,60]. This major classification system is divided into ten patterns, consisting of five elementary and five associated patterns. Elementary fractures, running in a single plane, are further divided into anterior wall fractures, anterior column fractures, posterior wall fractures, posterior column fractures, and transverse fractures. Associated fractures are combinations of elementary patterns and are divided into T-shaped fractures, complete two-column fractures, transverse and posterior wall fractures, posterior column and posterior wall fractures, and anterior column posterior hemi-transverse fractures.

Among these ten fracture types, five of them—namely, the complete two-column, transverse, T-shaped, transverse with posterior wall, and posterior wall fractures are very common, constituting almost 90% of acetabular fractures [61,62].

### 12.3.3 Stress Fractures

Stress fractures of the pelvic bones are common in the elderly and represent the response of normal or abnormal bone to repetitive cyclic loading, with the load being less than that which causes acute fracture. In particular, stress fractures are classified into two distinctive types: fatigue fractures, which result from abnormal stress placed on bone of normal elastic resistance; and insufficiency fractures, which occur when a bone of deficient elastic resistance is subjected to normal stress [5].

#### *12.3.3.1 Fatigue Stress Fractures*

Fatigue stress fractures of the pelvic bones are very uncommon and have been documented in military recruits or long-distance runners (Figure 12.17). Sacral fatigue stress fractures are reported in female athletes with secondary amenorrhea, supporting a clear relationship between loss of ovarian function, osteopenia/osteoporosis, and fractures. Clinical findings are nonspecific and patients present with low back or buttock pain that is relieved by rest. Radiographic examination usually reveals no bony changes. MR imaging is diagnostic due to its high sensitivity and specificity to detect abnormal marrow signal. On MR images, stress fractures appear as regions of decreased signal intensity on $T_1$-weighted

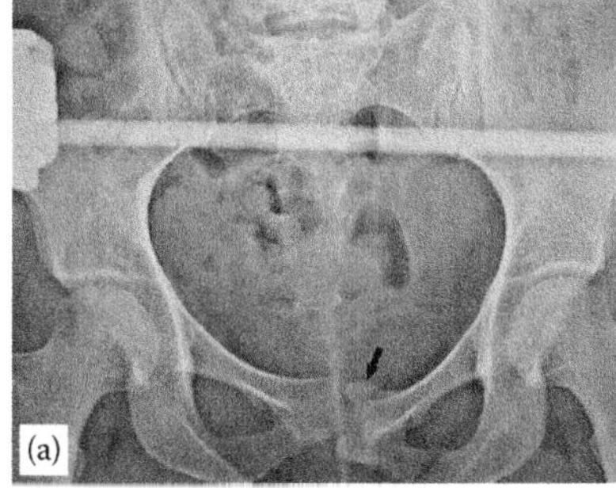


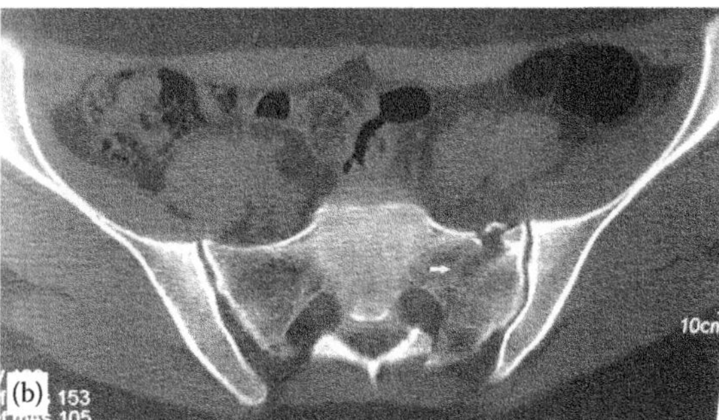


**FIGURE 12.16**
A 29-year-old woman with pelvic ring fracture caused by high-speed motorvehicle accident. (a) Frontal radiograph of pelvis shows left upper pubic ramus fracture (arrow) and diastasis of symphysis pubis. (b) Axial CT image depicts fracture (arrow) of left superior sacral ala.

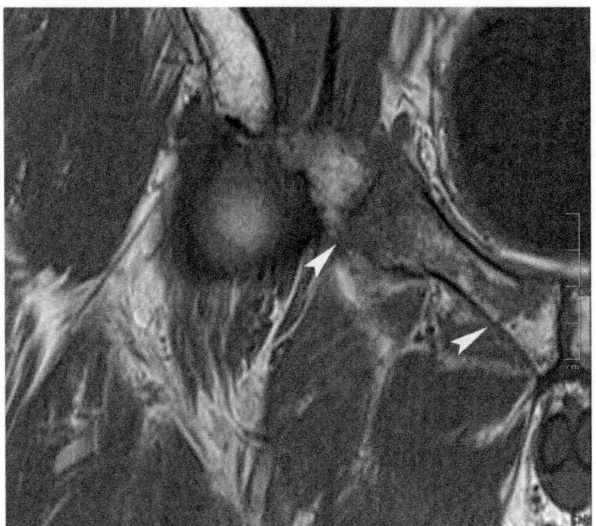

**FIGURE 12.17**
Coronal $T_1$-weighted MR image in a 56-year-old marathon runner shows fractures (arrowheads) of right upper pubic ramus.

images and increased signal intensity on $T_2$-weighted images, with or without a discrete fracture line.

#### *12.3.3.2 Insufficiency Stress Fractures*

Pelvic insufficiency fractures have been described in the sacrum, ilium, and pubic bones, as well as the para-acetabular region [5,63,64]. Various underlying conditions associated with weakened bones and pelvic insufficiency fractures have been reported including postmenopausal or senile osteoporosis, rheumatoid arthritis, radiation therapy, and corticosteroid medications [5,65,66]. Nonspecific low back, buttock, or groin pain that is usually not associated with trauma, or occurs after minor trauma challenges diagnosis [67–69]. In elderly women who have undergone previous radiotherapy for pelvic malignancy, however, clinical manifestations of intractable bone pain may simulate those of metastatic bone disease or local recurrence of tumor [5,70]. Pelvic insufficiency fractures are difficult to diagnose, if not suspected, as they are often radiographically occult or reveal only minimal changes.

Occasionally, sacral stress fractures may be seen on radiographs or CT images as unilateral or bilateral, vertical sclerotic bands or frank fracture lines that parallel to the SI joints and correspond with the classic H-shaped area of increased uptake on bone scans. Associated pubic fractures are found in almost one-third of the patients with sacral insufficiency fractures [67]. Pubic rami fractures may involve both the iliac and ischial pubic rami and can be bilateral, especially when they involve the parasymphyseal pubic joint [71]. Para-acetabular insufficiency stress fractures run parallel to the superior acetabulum in a curvilinear fashion or are obliquely oriented [5]. MR imaging is an important imaging method in the diagnosis of occult insufficiency fractures in the pelvis, because of the contrast it typically provides between the abnormal and normal bone marrow. With MR imaging, bone stress injury exhibits decreased signal intensity on $T_1$-weighted images and increased signal intensity on $T_2$-weighted images consistent with marrow edema, with or without a discrete fracture line (Figure 12.18). This abnormal signal intensity in the marrow is most evident on the STIR and fat-suppressed fast spin-echo $T_2$-weighted sequences, which in addition, may improve detection of the fracture line. Typically, the fracture line is visualized as a band of low signal intensity on both the $T_1$- and $T_2$-weighted images. After the intravenous administration of contrast material, fractures demonstrate marked enhancement of adjacent bone indicating increased vascularity or inflammation, or both.

### 12.3.4 Avulsion Injuries

Apophyseal avulsion injuries of the pelvis and hip may occur in physically active individuals. These injuries are classified as apophyseal avulsion fractures and apophysitis. Apophyseal avulsion fractures of the

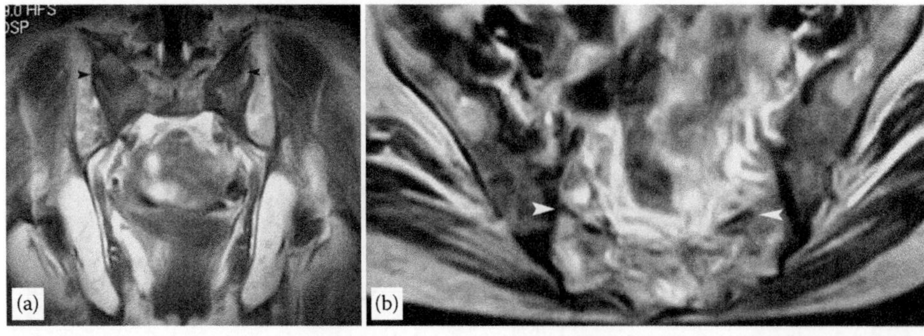


**FIGURE 12.18**
An 85-year-old woman with sacral insufficiency fractures. (a) Coronal $T_1$-weighted MR image shows abnormal decreased signal intensity in bone marrow of the sacrum. Fracture lines appear as discrete low signal intensity bands (arrowheads). (b) Axial $T_2$-weighted MR image displays diffuse high signal intensity in sacrum, and fracture lines of low signal intensity (arrowheads).

pelvis are uncommon injuries seen in individuals who are involved in competitive or recreational sports activities. Sprinters, gymnasts, football, baseball, and track athletes are prone to apophyseal avulsion fractures [72]. In about two-thirds of the cases, avulsion fractures occur in male adolescent athletes [73]. In the skeletally immature patient, avulsion fractures encompass separation of the ossification center, whereas in adults, pelvic avulsion fractures involve fragmentation of the apophysis. The mechanism of injury in athletes is the result of eccentric strenuous contraction or passive lengthening of the affected muscle(s). The most common sites of pelvic avulsion injury are the iliac crest (origin of the abdominal muscles), anterior–superior iliac spine (sartorius muscle and tensor fascia latae), anterior–inferior iliac spine (rectus femoris muscle), ischial tuberosity (hamstring muscles), and the body of pubis and inferior pubic ramus (origin of the gracilis and adductor muscles). Patients relate a history of acute injury occurring during a certain activity. Radiographs reveal an associated fracture through the offended apophysis, with a bony fragment that is situated adjacent to, or displaced from the parent bone (Figure 12.19). Additional findings in acute avulsion fractures may include the formation of hematoma in surrounding soft tissue and callus formation in chronic avulsion fractures. Avulsion fractures occurring in the absence of a specific traumatic event, however, must raise the suspicion of an underlying malignancy or infection [74].

Apophysitis is an inflammation of the apophysis resulting from repetitive pull of muscles. This is a painful chronic stress injury unrelated to previous, major skeletal trauma. On MR images, apophysitis is seen as bone marrow edema at the insertion site of the involved muscle, with enlargement and edema of the tendinous end.

### 12.3.5 Infectious Disorders

#### *12.3.5.1 Osteomyelitis*

Infection of pelvic bones (osteomyelitis) and joints (septic arthritis) usually relates to spread from a contiguous source of infection. For example, pelvic abscesses and soft tissue infections, as those associated with decubitus ulcers, can extend into the neighboring osseous and articular structures and subsequently involve the ilium or sacrum, or the SI joint. Other routes of contamination involve hematogenous spread via the Batson plexus from infection of the gastrointestinal or genitourinary tract. Direct implantation of infection may occur in penetrating traumatic (i.e., vehicular trauma and war wounds) or iatrogenic injuries (i.e., gluteal and SI joint injections, sacral biopsy, and postoperative infection). Nevertheless, in any particular case, more than one potential mechanism by which infection can reach the osseous and articular structures may apply. Infective organisms usually include *Staphylococcus aureus*, *Proteus mirabilis*, *Escherichia coli*, *Bacteroides* species, and group A and other streptococci. Attention should be paid to the atypical mycobacterial infections, especially in the elderly and immunocompromised patients [75].

On radiographs and CT images, osteomyelitis appears as a destructive, poorly defined lytic lesion associated with diffuse soft tissue inflammation, or the formation of bone and soft tissue abscess. In chronic infection,

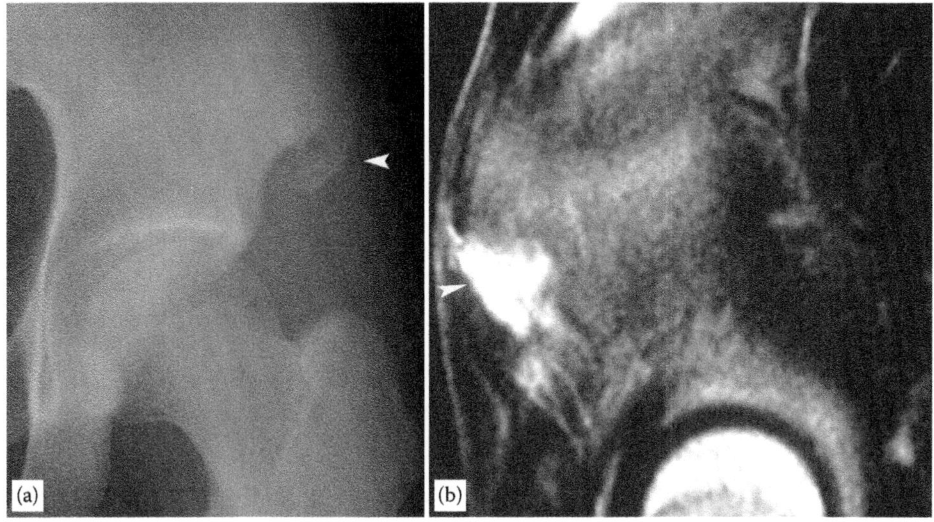


**FIGURE 12.19**
Avulsion injury of left anterior superior iliac spine in a 14-year-old boy runner. (a) Frontal radiograph reveals bone irregularity (arrowhead) as the result of stress at the origin of the tensor fasciae femoris and the sartorius muscle. (b) Sagittal $T_2$-weighted MR image shows high signal intensity at the site of avulsion injury (arrowhead).

osseous sclerosis or sequestra may be present. On MR images, acute osteomyelitis of any pelvic bone (or site within that bone) typically displays low signal intensity on $T_1$-weighted MR images and high signal intensity on $T_2$-weighted and STIR MR images, with subacute and chronic osteomyelitis having a more variable MR imaging appearance (Figure 12.20). Additional MR imaging alterations in either acute or chronic osteomyelitis include cortical erosion or perforation, periosteal bone formation, and soft tissue involvement and, in chronic osteomyelitis, abscesses, bone sequestration, and sinus tracts. After intravenous administration of contrast medium, areas of vascularized inflammatory tissue reveal enhancement of signal intensity, but abscess collections show either no enhancement or enhancement at the margin of the lesion. Brodie's abscesses appear as well-defined intraosseous regions of low signal intensity on $T_1$-weighted MR images and of high signal intensity on $T_2$-weighted MR images.

#### 12.3.5.2 SI Joint Infection

Pyogenic infection of the SI joint(s) may occur in cases of pelvic osteomyelitis. The SI joint can be contaminated by the hematogenous route, by contamination from a contiguous source of infection, and by direct implantation of infectious material into the joint, or after surgery. Not infrequently, the exact route of contamination at this site remains unclear. Because the subchondral circulation of the ilium is slow, resembling that in the metaphysis of long bones in children, hematogenous implantation of infection at this site is not difficult. Indeed, the ilium is the most frequently infected flat bone of the body [76]. From this location, infection can spread into the SI joint. In addition to hematogenous contamination of the SI joint from infection in a distant site or in intravenous drug abuse, various suppurative conditions of the pelvis or previous pelvic surgical procedures (i.e., septic abortions) may spread hematogenously to the SI joint via the paravertebral venous system of Batson. Osseous destruction occurring in both the iliac and the sacral sides of the joint in the early phases of the infectious process, however, has important implications for a direct hematogenous intra-articular contamination.

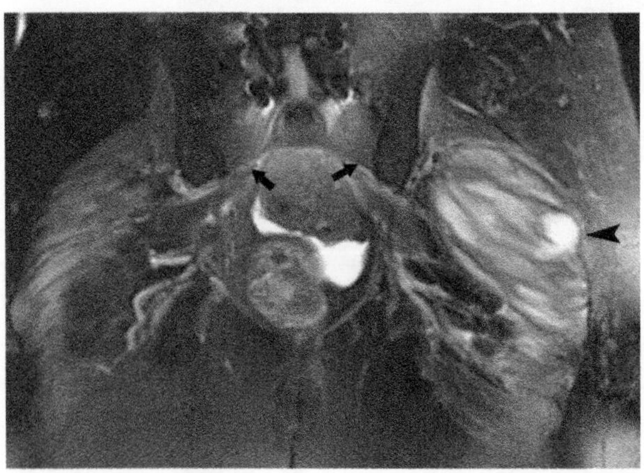

**FIGURE 12.20**
In a 78-year-old woman with pyogenic myositis (pyomyositis) in gluteal muscles, coronal STIR MR image reveals abnormal high signal intensity in the sacrum (arrows), consistent with spread of infection to bone. A soft tissue abscess is present (arrowhead).

Adjacent infection can spread to the SI joint and neighboring bone. Pelvic abscesses associated with local muscle trauma or various infectious processes of the gastrointestinal and genitourinary system can disrupt the anterior articular capsule or the periosteum and cortex of the ilium or sacrum, causing osteomyelitis and septic sacroiliitis by contiguous contamination [76,77]. Pressure sores, or bedsores, which are frequent in the sacral region of immobilized patients, can lead to subsequent osseous and SI joint infection. Septic intragluteal injections of medications can also lead to osteomyelitis and SI joint pyogenic arthritis [78]. Finally, suppurative spinal infection can spread beneath the spinal ligaments into the pelvis and SI joints. An additional route of contamination of the SI joints is that of direct implantation of pathogens following diagnostic or surgical procedures. Examples of iatrogenic spread of infection include needle aspiration of the SI articulation or closed or open biopsy procedures of the adjacent bone. Variable causative microorganisms may be identified in cases of SI joint septic arthritis including staphylococci, streptococci, pneumococci, Proteus, Klebsiella, Pseudomonas, Brucella, mycobacteria, and fungi. Gram-negative pathogens are common in SI joint infectious arthritis, in intravenous drug abusers [76].

The imaging features of SI joint infection usually are noted about the inferoanterior aspect of the joint. SI joint infection typically is unilateral in distribution, a feature that helps differentiation from other articular processes. In septic sacroiliitis, MR imaging may reveal marrow edema in the sacrum and ilium, erosive alterations of subchondral bone on either side of the joint space, narrowing or widening of the interosseous space, joint fluid, and muscle edema, with or without the formation of abscesses, sinus tracts, and fistulae (Figure 12.21). Intravenous administration of gadolinium-containing contrast material may reveal to better extent the MR imaging abnormalities of pyogenic arthritis and can outline infection of adjacent soft tissue.

### 12.3.6 Noninfectious Inflammatory Diseases

In the pelvis, rheumatoid arthritis and the seronegative spondyloarthropathies (ankylosing spondylitis, psoriasis, and Reiter's syndrome) may involve synovial and cartilaginous joints, bursae, tendon sheaths, entheses,

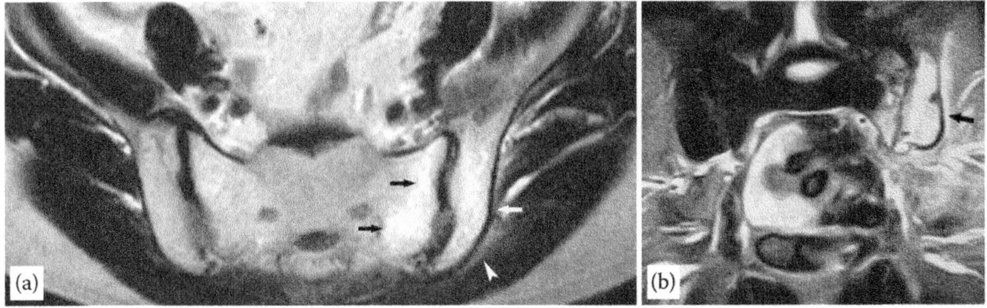

**FIGURE 12.21**
In a 50-year-old woman, axial $T_2$-weighted MR image (a) shows the infectious process as areas of high signal intensity (arrows) on both sides of the left sacroiliac joint. High signal intensity is present in adjacent soft tissue (arrowhead). (b) Coronal STIR MR image better delineates the bone and soft tissue abnormalities, which are of high signal intensity (arrow).

tendons, ligaments, soft tissues, and bones, similar to any other anatomic site.

In rheumatoid arthritis, bilateral and symmetric articular involvement of the hips is typical, whereas involvement of the SI joints is occasionally seen. Cartilaginous joints such as the symphysis pubis may also be involved, as well as tendinous and ligamentous attachments to bone although the severity of alterations at these sites apparently is less striking in rheumatoid arthritis than in the seronegative spondyloarthropathies [64]. Abnormalities in the rheumatoid hip include early and progressive loss of joint space, axial migration of the femoral head, marginal and central osseous defects, mild osteophytosis, and bone thickening along the femoral neck (buttressing). Abnormalities in the SI joints can be bilateral or unilateral in distribution. SI joint space narrowing in rheumatoid arthritis ranges from mild to complete obliteration of the articulation, and subchondral osseous erosions have predilection for the iliac aspect of the joint. Changes in the symphysis pubis of rheumatoid arthritis patients include subchondral erosion, mild eburnation, and narrowing of the interosseous space. Superficial erosive changes in the ischial tuberosities and the iliac crests may be present. Associated findings may include parasymphyseal insufficiency fractures of the os pubis.

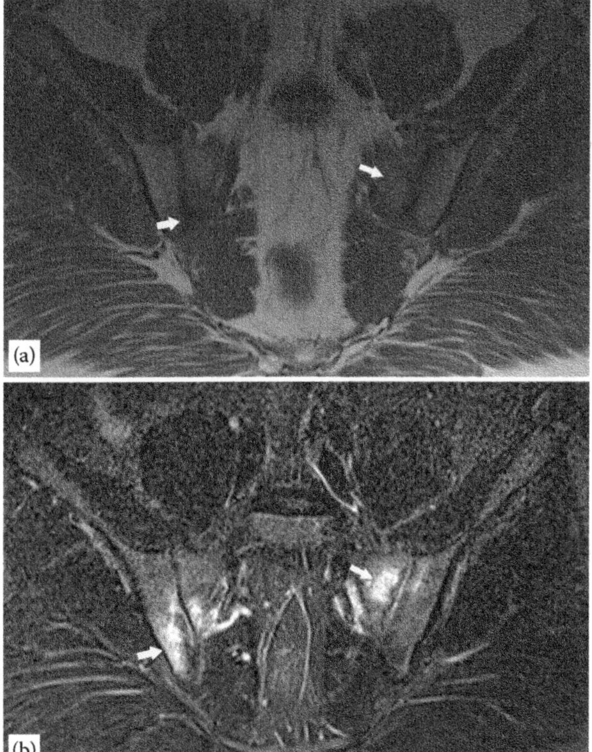

**FIGURE 12.22**
In this 27-year-old man with ankylosing spondylitis, axial $T_1$-weighted MR image (a) shows regions of low signal intensity about both sacroiliac joints (arrows). In an axial STIR MR image (b) the signal intensity in the periarticular bone marrow is increased (arrows).

Ankylosing spondylitis is a disease with variable musculoskeletal manifestations. The hallmark of the disorder is sacroiliitis, which typically is bilateral and symmetric in distribution (Figure 12.22). In fact, it is this symmetric pattern of involvement in ankylosing spondylitis that may facilitate its differentiation from the other seronegative spondyloarthropathies, such as psoriasis and Reiter's syndrome that affect the SI joints, where alterations are usually asymmetric or, rarely, unilateral. Similar to the pattern of sacroiliitis in classic ankylosing spondylitis is the bilateral and symmetric sacroiliitis of inflammatory bowel disease (ulcerative colitis, Crohn's disease, and Whipple's disease), however. In both ankylosing spondylitis and inflammatory bowel disease, associated sacroiliitis appears as poorly defined erosive changes with adjacent sclerosis, particularly in the ilium, joint space narrowing, progressive fusion (ankylosis) of the articulation, and ligamentous ossification [79]. In psoriasis and Reiter's syndrome, extensive bony eburnation usually is not associated with intra-articular osseous fusion.

Bilateral and symmetric changes of the SI articulations are noted in osteitis condensans ilii, gouty arthritis, and

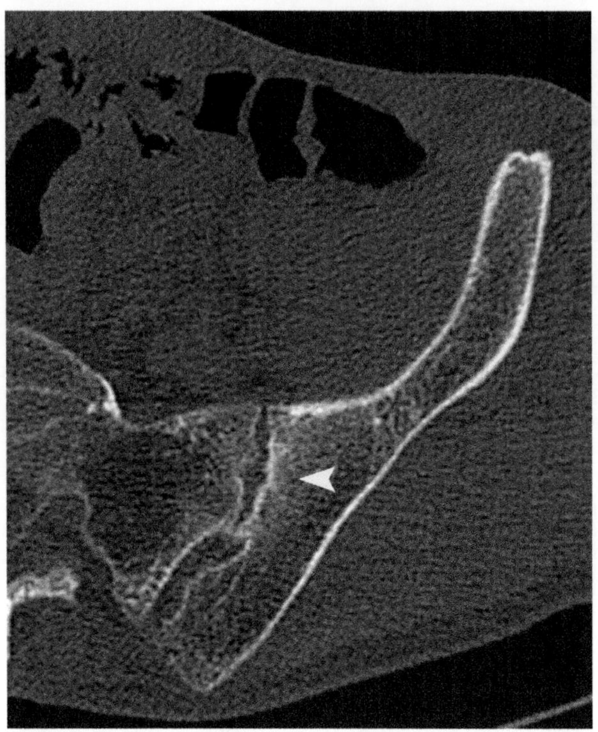

**FIGURE 12.23**
In this 65-year-old man with low back pain, degenerative changes in the sacroiliac joint are seen as bony erosion and eburnation, predominantly in the ilium (arrowhead) and focal narrowing of the articular space.

degenerative joint disease, although in the last two processes, an asymmetric or unilateral distribution can be present (Figure 12.23). Osteitis condensans ilii is seen most commonly in multiparous women and is believed to result from instability of the pubic symphysis. In osteitis condensans ilii, SI joint involvement is seen as a triangular area of subchondral sclerosis on the iliac side of the joint, without erosions or narrowing of the articular space. In tophaceous gouty arthritis offending the SI joint, large, bony defects with surrounding sclerosis are observed. In SI joint degenerative arthritis, abnormalities include joint space narrowing, subchondral sclerosis, and the formation of anterior osteophytes.

Osteitis pubis is a painful condition of the symphysis pubis, which may become manifest after delivery or pelvic operations. Clinical findings include local pain and tenderness, and unstable gait. On radiographs, osteitis pubis appears as irregularity of subchondral bone of the symphysis pubis with resorption or sclerosis (Figure 12.24). Changes usually involve both pubic bones, in a symmetric fashion.

Paget's disease features an inflammatory component at the initial lytic phase similar to that seen in the noninfectious arthropathies [80]. Paget's disease may as well involve the iliac bones and sacrum, with imaging findings in pagetic bone reflecting the stage (lytic, mixed lytic-sclerotic, and sclerotic) of disease activity. As such, the MR imaging findings can be quite variable: in the lytic phase, pagetic bone is hypointense on $T_1$-weighted images and hyperintense on $T_2$-weighted images; and in the sclerotic phase, affected bone is hypointense on both $T_1$- and $T_2$-weighted images (Figure 12.25). Radiographs and CT images supplement imaging investigation of the disease.

### 12.3.7 Developmental Abnormalities

Spinal dysraphism is a developmental anomaly encompassing various combinations of spina bifida, meningocele, myelomeningocele, lipomyelomeningocele, intradural lipoma, intradural dermoid cyst, and tethering of the spinal cord [81]. Both CT and MR imaging are invaluable in delineating these abnormalities [25]. MR imaging also helps in the detailed characterization of congenital lesions and allows definition of their exact relationship to pelvic structures. Meningoceles result from herniation of the meninges through an osseous defect of neural foramen. Anterior sacral meningoceles result from herniation of the caudal meninges through an anterior sacral defect. These lesions may occasionally be symptomatic by compressing pelvic viscera (bladder and rectum).

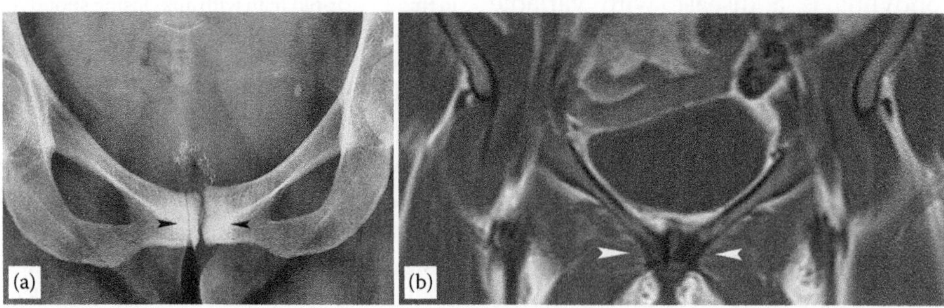


**FIGURE 12.24**
In this 28-year-old woman, local pain and tenderness about the symphysis pubis developed three months after gynaecological surgery. (a) Frontal radiograph depicts marked bone sclerosis on both sides of the symphysis pubis (arrowheads). (b) Coronal $T_1$-weighted MR image shows low signal intensity in the involved bones (arrowheads).

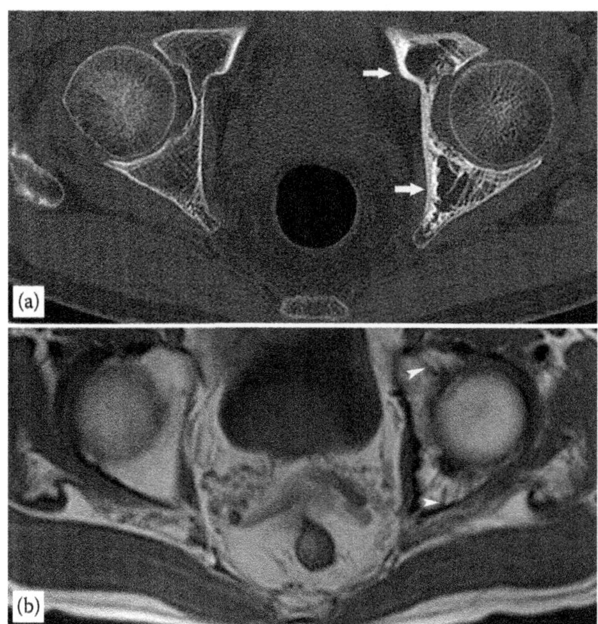

**FIGURE 12.25**
An 80-year-old man with Paget's disease (mixed phase). (a) Axial CT image of pelvis through the hips reveals areas of osteolysis, coarsened trabeculae, and cortical thickening in superior ramus of pubis and ischium (arrows). (b) Axial $T_1$-weighted MR image shows thickened trabeculae (arrowheads) within maintained yellow marrow.

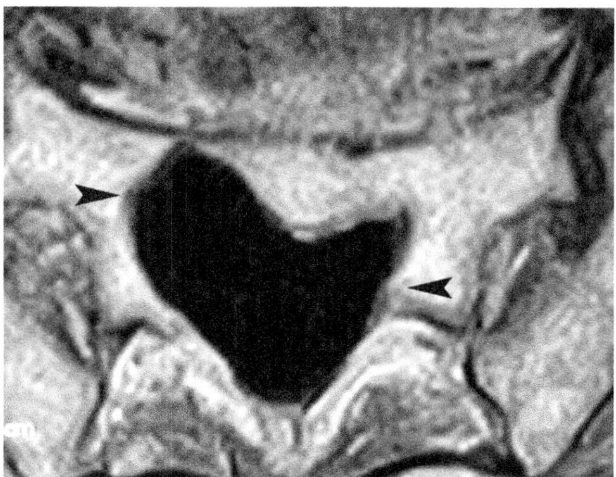

**FIGURE 12.26**
An 88-year-old woman with lower back pain. Axial $T_1$-weighted MR image shows a cystic mass within the sacral canal (arrowheads), consistent with sacral meningeal cyst (Tarlov cyst).

Posterior meningoceles are more common than their anterior counterparts and may be classified as myeloceles, myelomeningoceles, or lipomyelomeningoceles depending upon the contents of the herniated lesion. Posterior meningoceles are frequently associated with a tethered spinal cord. MR imaging is of great value in depicting the connection between the lesion and the thecal sac.

Sacral meningeal cysts are common and often are detected as incidental findings on MR images of the lumbar spine [17]. These developmental lesions are also known as *perineural cysts*, *Tarlov cysts*, or *sacral arachnoid cysts*. Sacral meningeal cysts are abnormal dilatations of the meninges within the sacral canal or neural foramina [82]. Growing of cysts may be associated with erosion and enlargement of osseous structures in the sacrum, while in the presence of large cysts patients may experience neurologic symptoms and signs. The MR imaging findings are quite characteristic and exhibit a cystic lesion containing cerebrospinal fluid (Figure 12.26).

## 12.4 Conclusion

Having completed this long-term journey in the plethora of disease processes affecting the osseous pelvis, the authors of this chapter can tell with certainty that pathology of pelvic structures is broad and multi-faceted. As a result in any given clinical case, the investigating clinician and the radiologist alike should realize that the differential considerations are broad and may relate to abnormalities offending not only bone, but also the other structural (i.e., cartilage, meninges, and neural elements) and surrounding tissues of the pelvis. Careful consideration of the many disease processes described in this work and their manifestations would help clinicians and radiologists arrive at a certain diagnosis, or at least a short list of pertinent differential diagnostic possibilities. Eventual establishment of the correct diagnosis will guide appropriate management of these disorders.

## References


1. Long S, Yablon C, Eisenberg R (2010) Bone marrow signal alteration in the spine and sacrum. *AJR Am J Roentgenol* 195:W178–W200.
2. Resnick D (1995) Tumors and tumor-like lesions of bone: Radiographic principles. In: Resnick D (ed) *Diagnosis of Bone and Joint Disorders*, 3rd edn. WB Saunders, Philadelphia, PA.
3. Theodorou DJ, Theodorou SJ, Sartoris D (2008) An imaging overview of primary tumors of the spine. Part 1. Benign tumors. *Clin Imaging* 32:196–203.
4. Theodorou DJ, Theodorou SJ, Sartoris D (2008) An imaging overview of primary tumors of the spine. Part 2. Malignant tumors. *Clin Imaging* 32:204–211.
5. Theodorou SJ, Theodorou DJ, Schweitzer M, Kakitsubata Y, Resnick D (2006) Magnetic resonance imaging of para-acetabular insufficiency fractures in patients with malignancy. *Clin Radiol* 61:181–190.

6. Nguyen T, Burk L (1995) Musculoskeletal case of the day. Giant cell tumor of the sacrum. *AJR Am J Roentgenol* 165:201–202.
7. Murphey M, Andrews C, Flemming D et al. (1996) Primary tumors of the spine: Radiologic-pathologic correlation. *RadioGraphics* 16:1131–1158.
8. Kwon J, Chung H, Cho E et al. (2007) MRI findings of giant cell tumors of the spine. *AJR Am J Roentgenol* 189:246–250.
9. Theodorou DJ, Theodorou SJ, Sartoris D (2009) Primary benign tumors of the spine: An imaging synopsis of disease findings. *Contemp Diagn Radiol* 32:1–6.
10. Murphey M, Nomikos G, Flemming D et al. (2001) Imaging of giant cell tumor and giant cell reparative granuloma of bone: Radiologic-pathologic correlation. *RadioGraphics* 21:1283–1309.
11. Ramirez H, Blatt E, Cable H et al. (1984) Intraosseous pneumatocysts of the ilium. *Radiology* 150:503–505.
12. Girish G, Finlay K, Morag Y (2012) Imaging review of skeletal tumors of the pelvis—Part I: Benign tumors of the pelvis. *Sci World J* doi:10.1100 / 2012 / 290930 / 1 - 10.
13. Capanna R, Van Horn J, Biagini R et al. (1989) Aneurysmal bone cyst of the sacrum. *Skeletal Radiol* 18:109–113.
14. Greenspan A (1993) Benign bone-forming lesions. *Skeletal Radiol* 22:485–500.
15. Resnick D, Kyriakos M, Greenway G (1995) Tumors and tumor-like lesions of bone: Imaging and pathology of specific lesions. In: Resnick D (ed) *Diagnosis of Bone and Joint Disorders*, 3rd edn. WB Saunders, Philadelphia, PA.
16. Mehta S, Szklaruk J, Faria S et al. (2006) Chondromyxoid fibroma of the sacrum and left iliac bone. *AJR Am J Roentgenol* 186:467–469.
17. Motamedi K, Ilashan H, Seeger L (2004) Imaging of the lumbar spine neoplasms. *Semin Ultrasound CT MRI* 25:474–489.
18. Shaikh M, Saiffudin A, Pringle J et al. (1999) Spinal osteoblastoma: CT and MR imaging with pathological correlation. *Skeletal Radiol* 28:33–40.
19. Feldman F (1995) Tuberous sclerosis, neurofibromatosis, and fibrous dysplasia. In: Resnick D (ed) *Diagnosis of Bone and Joint Disorders*, 3rd edn. WB Saunders, Philadelphia, PA.
20. Wenger D, Wold L (2000) Benign vascular lesions of bone: Radiologic and pathologic features. *Skeletal Radiol* 29:63–74.
21. Baudrez V, Galant C, Vande Berg B (2001) Benign vertebral hemangioma: MR-histological correlation. *Skeletal Radiol* 30:442–446.
22. Vilanova J, Barcelo J, Smirniotopoulos J et al. (2004) Hemangioma from head to toe: MR imaging with pathologic correlation. *RadioGraphics* 24:367–385.
23. Murphey M, Fairbairn J, Parman L et al. (1995) Musculoskeletal angiomatous lesions. Radiologic-pathologic correlation. *RadioGraphics* 15:893–917.
24. Murphey M, Smith S, Smith SE et al. (1999) Imaging of musculoskeletal neurogenic tumors: Radiologic-pathologic correlation. *RadioGraphics* 19:1253–1280.
25. Batnizky S, Soye I, Levine E et al. (1982) Computed tomography in the evaluation of lesions arising in and around the sacrum. *RadioGraphics* 2:500–528.
26. Farsad K, Kattapuram S, Sacknoff S et al. (2009) Best cases of the AFIP. Sacral chordoma. *RadioGraphics* 29:1525–1530.
27. Smith J, Ludwig R, Marcove R (1987) Sacrococcygeal chordoma. A clinicoradiological study of 60 patients. *Skeletal Radiol* 16:37–44.
28. Thornton E, Krajewski K, O'regan K et al. (2012) Imaging features of primary and secondary malignant tumours of the sacrum. *Br J Radiol* 85:279–286.
29. Venkatanarasimha N, Brown S, Suresh P (2010) Coccydynia with midline sacral mass. *AJR Am J Roentgenol* 195:S29–S31.
30. Lauger J, Palmer J, Amores S et al. (2000) Primary tumors of the sacrum: Diagnostic imaging. *AJR Am J Roentgenol* 174:417–424.
31. Ilashan H, Sundaram M, Unni K et al. (2004) Priamary Ewing's sarcoma of the vertebral column. *Skeletal Radiol* 33:506–513.
32. O'Connor J, Martin L, Chen H et al. (1991) Pediatric case of the day. Ewing sarcoma of pubic bone. *AJR Am J Roentgenol* 156:1314–1320.
33. Edeiken J, Dalinka M, Karasick D (1990) Bone tumors and tumorlike conditions. In: Edeiken J (ed) *Roentgen Diagnosis of Diseases of Bone*, 4th edn. Williams and Wilkins, Baltimore, MD.
34. Patel D, Hammer R, Levin B et al. (1984) Primary osteogenic sarcoma of the spine. *Skeletal Radiol* 12:276–279.
35. Shors S, Jones T, Jhaveri M et al. (2006) Myxopapillary ependymoma of the sacrum. *RadioGraphics* 26: S111–S116.
36. Faingold R, Saigal G, Azouz E et al. (2004) Imaging of low back pain in children and adolescents. *Semin Ultrasound CT MR* 25:490–505.
37. Disler D, Miklic D (1999) Imaging findings in tumors of the sacrum. *AJR Am J Roentgenol* 173:1699–1706.
38. Szklaruk J, Tamm E, Choi H et al. (2003) MR imaging of common and uncommon large pelvic masses. *RadioGraphics* 23:403–424.
39. Resnick D, Haghighi P (1995) Myeloproliferative disorders. In: Resnick D (ed) *Diagnosis of Bone and Joint Disorders*, 3rd edn. WB Saunders, Philadelphia, PA.
40. Krishnan A, Shirkhoda A, Tehranzadeh J et al. (2003) Primary bone lymphoma: Radiographic-MR imaging correlation. *RadioGraphics* 23:1371–1387.
41. Resnick D (1995) Lymphomas. In: Resnick D (ed) *Diagnosis of Bone and Joint Disorders*, 3rd edn. WB Saunders, Philadelphia, PA.
42. Epelbaum R, Haim N, Ben-Shahar M et al. (1986) Non-Hodgkin's lymphoma presenting with spinal epidural involvement. *Cancer* 58:2120–2124.
43. Edeiken-Monroe B, Edeiken J, Kim E (1990) Radiologic concepts of lymphoma of bone. *Radiol Clin N Am* 28:841–864.
44. Dorfman H, Czerniak B (1998) Non-Hodgkin's lymphoma. In: Dorfman H, Czerniak B, (eds) *Bone Tumors*, 1st edn. Mosby, St. Louis, MO.
45. Theodorou DJ, Theodorou SJ, Sartoris DJ et al. (2000) Delayed diagnosis of primary non-Hodgkin's lymphoma of the sacrum. *J Clin Imaging* 24: 169–173.

46. Hermann G, Klein M, Fikry-Abdelwahab I et al. (1997) MRI apperance of primary non-Hodgkin's lymphoma of bone. *Skeletal Radiol* 26:629–632.
47. Daffner R, Lupetin A, Dash N et al. (1986) MRI in the detection of malignant infiltration of bone marrow. *AJR Am J Roentgenol* 146:353–358.
48. Ostrowski M, Unni K, Banks P et al. (1986) Malignant lymphoma of bone. *Cancer* 58:2646–2655.
49. Llopis E, Higueras V, Aparisi P et al. (2008) Acute osseous injury to the pelvis and acetabulum. In: Pope T (ed) *Imaging of the Musculoskeletal System*, 1st edn. WB Saunders, Philadelphia, PA.
50. Leone A, Cerase A, Priolo F et al. (1997) Lumbosacral junction injury associated with unstable pelvic fracture: Classification and diagnosis. *Radiology* 205:253–259.
51. Kirby M, Spritzer C (2010) Radiographic detection of hip and pelvic fractures in the emergency department. *AJR Am J Roentgenol* 194:1054–1060.
52. Melton L, Sampson J, Morrey F et al. (1981) Epidemiologic features of pelvic fractures. *Clin Orthop* 155:43–47.
53. Mucha P, Farnell M (1984) Analysis of pelvic fracture management. *J Trauma* 24:379–385.
54. Eastridge B, Burgess A (1997) Pedestrian pelvic fractures: 5-year experience of a major urban trauma center. *J Trauma* 42:695–700.
55. Gansslen A, Pohlemann T, Paul C et al. (1996) Epidemiology of pelvic ring injuries. *Injury* 27Suppl:13–20.
56. Tile M (1995) *Fractures of the Pelvis and Acetabulum*. Williams and Wilkins, Baltimore, MD.
57. Muller M, Allgower M, Schneider R et al. (1990) *Manual of Internal Fixation*, 3rd edn. Springer, New York.
58. Pohlemann T, Richter M, Otte D et al. (2000) Mechanism of pelvic girdle injuries in street traffic. Medical-technical accident analysis. *Unfallchirurg* 103:267–274.
59. Letournel E, Judet R (1993) *Fractures of the Acetabulum*, 2nd edn. Springer, New York.
60. Judet R, Judet J, Letournel E (1964) Fractures of the acetabulum: Classification and surgical approaches for open reduction. *J Bone Joint Surg* 46:1615–1646.
61. Brandser E, El-Khoury G, Marsh J (1995) Acetabular fractures: A systematic approach to reclassification. *Emerg Radiol* 2:18–28.
62. Young J, Resnik C (1990) Fracture of the pelvis: Current concepts of classification. *AJR Am J Roentgenol* 155:1169–1175.
63. Cabarrus M, Ambekar A, Lu Y et al. (2008) MRI and CT of insufficiency fractures of the pelvis and the proximal femur. *AJR Am J Roentgenol* 191:995–1001.
64. Gibbon W, Hession P (1997) Diseases of the pubis and pubic symphysis: MR imaging appearances. *AJR Am J Roentgenol* 169:849–853.
65. Cooper K, Beabout J, Swee R (1985) Insufficiency fractures of the sacrum. *Radiology* 156:15–20.
66. Kwon J, Huh S, Yoon Y et al. (2008) Pelvic bone complications after radiation therapy of uterine cervical cancer: Evaluation with MRI. *AJR Am J Roentgenol* 191:987–994.
67. De Smet A, Neff J (1985) Pubic and sacral insufficiency fractures: Clinical course and radiologic findings. *AJR Am J Roentgenol* 145:601–606.
68. Peh W, Khong PL, Yin Y et al. (1996) Imaging of pelvic insufficiency fractures. *RadioGraphics* 16:335–348.
69. Lyders E, Whitlow C, Baker M et al. (2010) Imaging and treatment of sacral insufficiency fractures. *Am J Neuroradiol* 31:201–210.
70. Blomlie V, Lien H, Iversen T et al. (1993) Radiation-induced insufficiency fractures of the sacrum: Evaluation with MR imaging. *Radiology* 188:241–244.
71. Casey D, Mirra J, Staple T (1984) Parasymphyseal insufficiency fractures of the os pubis. *AJR Am J Roentgenol* 142:581–586.
72. Hebert K, Laor T, Divine J et al. (2008) MRI appearance of chronic stress injury of the iliac crest apophysis in adolescent athletes. *AJR Am J Roentgenol* 190:1487–1491.
73. Fernbach S, Wilkinson R (1981) Avulsion injuries of the pelvis and proximal femur. *AJR Am J Roentgenol* 137:581–584.
74. Bui-Mansfield L, Chew F, Lenchik L et al. (2002) Nontraumatic avulsions of the pelvis. *AJR Am J Roentgenol* 178:423–427.
75. Theodorou DJ, Theodorou SJ, Kakitsubata Y et al. (2001) Imaging characteristics and epidemiologic features of atypical mycobacterial infections involving the musculoskeletal system. *AJR Am J Roentgenol* 176:341–349.
76. Resnick D, Niwayama G (1995) Osteomyelitis, septic arthritis, and soft tissue infection: Axial skeleton. In: Resnick D (ed) Diagnosis of bone and joint disorders, 3rd edn. WB Saunders, Philadelphia, PA.
77. Theodorou SJ, Theodorou DJ, Resnick D (2007) MR imaging findings of pyogenic bacterial myositis (pyomyositis) in patients with local muscle trauma: Illustrative cases. *Emerg Radiol* 14:89–96.
78. Theodorou SJ, Theodorou DJ, Resnick D (2008) Imaging findings of complications affecting the upper extremity in intravenous drug users: Featured cases. *Emerg Radiol* 15:227–239.
79. Resnick D, Niwayama G, Goergen T (1977) Comparison of radiographic abnormalities of the sacroiliac joint in degenerative disease and ankylosing spondylitis. *AJR Am J Roentgenol* 128:189–196.
80. Theodorou DJ, Theodorou SJ, Kakitsubata Y (2011) Imaging of Paget disease of bone and its musculoskeletal complications: Review. *AJR Am J Roentgenol* 196:S64–S75.
81. Wetzel L, Levine E (1990) MR imaging of sacral and presacral lesions. *AJR Am J Roentgenol* 154:771–775.
82. Diel J, Ortiz O, Losada R et al. (2001) The sacrum: Pathologic spectrum, multimodality imaging, and subspecialty approach. *RadioGraphics* 21:83–104.

# 13

# *Magnetic Resonance Imaging of Soft Tissues*

**Alireza Eajazi, Mohamed Jarraya, Ali Guermazi, and Frank W. Roemer**

**CONTENTS**

## 13.1 Introduction

Soft tissue lesions are frequently encountered by radiologists in everyday clinical practice either for further diagnostic assessment of a known but nonspecific lesion, for pre-therapeutic evaluation, especially whenever surgery is considered, or as incidental findings. Characterization of soft tissue lesions remains problematic as many of these lesions have a nonspecific appearance and common radiologic methods such as radiography and computed tomography (CT) are characterized by poor soft tissue contrast. The application of magnetic resonance imaging (MRI) in the assessment of soft tissue lesions has markedly altered the treatment algorithms for a number of lesions [1,2]. In contrast to conventional radiography and CT, MRI provides excellent intrinsic soft tissue contrast, which is the main advantage over other tomographic techniques like CT. MRI uses a three-dimensional (3D) tomographic approach and is able to differentiate different tissue components. Contrast-enhanced imaging adds information on vascularity and dynamic contrast-enhanced imaging gives additional insight into vitality of a lesion and helps evaluating nonsurgical treatment approaches of soft tissue tumors [3]. MRI has become an important tool in pre-operative planning, especially to define vital neurovascular structures. Although soft tissue lesion characterization and consequent definite diagnosis is not always possible due to the inherent nonspecificity of the modality, MRI is crucial in describing extent of lesion, involvement of neighboring tissues and vascularity [4]. MRI findings should be interpreted in conjunction with clinical history, physical examination, laboratory results and potentially all other available imaging modalities to avoid misinterpretations [5,6].

In this chapter, first, we will briefly review some technical considerations relevant specifically for the assessment of soft tissue pathologies. The main focus of this chapter will then be on common soft tissue tumors and their MRI characteristics on non-neoplastic soft tissue lesions including cystic, traumatic, infectious and inflammatory pathologies and their specific MR imaging characteristics. Congenital abnormalities will not be discussed in this chapter.

## 13.2 Technical Considerations

Due to its unequaled soft tissue contrast and multiplanar imaging capability, MRI is the modality of choice to image soft tissue tumors. An impressive arsenal of sequence types has become available, especially with regard to fast MRI, fat-suppression techniques and contrast-enhanced studies. Careful clinical assessment of the region of clinical concern should precede any imaging to ensure its complete coverage with the most appropriate coil and to avoid waste of time due to repositioning of the patient after the first imaging sequence.

Tissue characterization is based on several imaging parameters [7]. Some of them are signal-intensity related: signal homogeneity, changing pattern of homogeneity, and the presence of hemorrhage and peritumoral edema. This information is obtained by comparison of the signal characteristics on $T_1$- and $T_2$-weighted sequences. Extra information can be derived from the addition of fat-suppression to a $T_1$- as well as a $T_2$-weighted sequence. Fat-suppressed $T_1$-weighted imaging is well known for its capacity to differentiate between fatty tissue and melanin or methemoglobin, which all show high signal intensity on conventional $T_1$-weighted images. A chemical shift-based, fat-suppressed $T_1$-weighted sequence decreases the signal intensity of fatty tissue, while melanin and methemoglobin remain hyperintense on this sequence [8,9]. Fat-suppressed $T_2$-weighted imaging is used to increase not only the conspicuity of a lesion and its surrounding edema and reactive zone but also of non-lipomatous components in lipomatous tumors, the latter helping in distinguishing lipoma from well-differentiated liposarcoma [10].

The use of intravenous contrast for lesion evaluation may be helpful for evaluation of vascularity of a given lesion [11,12]. Enhancement characteristics may aid in differentiating benign from malignant soft tissue lesions [11,13,14]. Benign lesions usually reveal less enhancement overall or a delayed rate of enhancement [3], while malignant lesions show commonly avid enhancement at least in portions of the lesion. [15].

### 13.2.1 Static-Enhanced MRI

Compared with unenhanced $T_1$-weighted imaging, contrast-enhanced $T_1$-weighted imaging improves the delineation of a lesion in terms of tumor-to-muscle contrast but not superiorly compared with $T_2$-weighted imaging [14,16]. On the other hand, non-enhanced $T_1$-weighted imaging decreases or even obscures the tumor-to-fat contrast, which can be counteracted by the use of fat-suppression. Since fat-suppressed $T_1$-weighted images after i.v. administration of gadolinium show areas of contrast enhancement with a greater conspicuity than $T_1$-weighted images without fat suppression, subsequently resulting in images that are easier to interpret, the use of this sequence has become very popular [14,16]. However, one should be aware of the risk of misinterpretation of fat-suppressed, enhanced $T_1$-weighted images, since a high signal intensity of a lesion can be the consequence of two variables: real Gd enhancement or apparent Gd uptake due to the scaling effect caused by the fat-suppression technique [17]. Suppression of the high signal intensity of fat induces a rescaling, a redistribution of gray levels, so that minor differences in signal intensity between tissues on non-fat-suppressed $T_1$-weighted images are magnified. The same rescaling effect of fat suppression is responsible for an apparently obvious Gd enhancement of tissue that only shows minimal enhancement on non-fat-suppressed $T_1$-weighted images [17,18]. The use of fat-suppressed, enhanced $T_1$-weighted imaging certainly is not a routine requirement [17].

### 13.2.2 Dynamic Contrast-Enhanced MRI

Dynamic contrast-enhanced MRI provides additional physiological information about tissue perfusion and vascularization, capillary permeability, and the volume of the interstitial space, which is not available on static contrast-enhanced MRI [19,20]. The analysis of the enhancement pattern improves differentiation between highly vascularized, less well vascularized and necrotic tumor areas. The most important advantages of this technique are its abilities to monitor response to chemotherapy, and to help assessing tumor recurrence after therapy [21,22]. It further narrows down the differential diagnosis and helps to increase the suspicion of malignant lesions in addition to unenhanced and static-enhanced MRI [3]. Dynamic MRI should be considered when conventional MRI results in indeterminate findings [21].

Attempts have been made to use the slope of time-intensity curves (TIC) as a differential diagnostic criterion to differentiate benign (low-slope) from malignant (high-slope) lesions [16,19]. Although there is a highly statistically significant difference in slope values of benign and malignant lesions, there is some overlap in some highly vascularized or perfused benign lesions which have slope values in the same range as malignant tumors. Due to this overlap between benign and malignant lesions, TICs and slope values should only be used in conjunction with conventional spin-echo images and other radiological, anatomical, and clinical data to narrow down the differential diagnostic possibilities, rather than to predict the benignity or malignancy of a lesion. The slope of the TIC, the time of onset of enhancement (relative to the onset of enhancement in a local artery) and the type of curve are not helpful to differentiate benign from malignant lesions [23,24].

Some centers have investigated the possible role of diffusion-weighted MRI in tumor characterization. Apparent diffusion coefficient values of benign soft tissue tumors and sarcomas overlap and cannot be used to differentiate between the bulk of benign and malignant tumors [25,26]. True diffusion coefficients are reported to be significantly lower in malignant than in benign soft tissue masses; however, there is still a substantial overlap between both [26].

Gradient-echo sequences are very helpful in characterizing hemorrhage due to their sensitivity in regard to magnetic susceptibility, which makes them an important tool for hemosiderin detection, which may present as a so-called blooming effect [27]. Short-tau inversion recovery (STIR) is highly sensitive in depicting edema and fluid.

Most soft tissue MR imaging is performed using two-dimensional (2D) fast spin-echo sequences. However, 3D sequences have also been used and have the advantage of acquiring thin continuous slices which reduces the effects of partial volume averaging. With recent advances in MR technology, 3D sequences with isotropic resolution have been developed. These sequences allow high quality multi-planar reformations to be obtained following a single acquisition, thereby eliminating the need to repeat sequences with identical tissue contrast in different planes. The use of 3D isotropic resolution sequences in clinical practice could significantly decrease MR examination times which would improve patient comfort, reduce motion artifact, and increase the clinical efficiency of the MR scanner [28].

## 13.3 Neoplastic Soft Tissue Lesions

Although soft tissues, and in particular skeletal muscles, represent about 40% of the total body weight, they are infrequently affected by primary tumors and even more rarely by metastatic lesions.

### 13.3.1 Lipomatous Tumors

#### *13.3.1.1 Lipoma*

A lipoma is a benign neoplasm composed of mature adipose tissue. It is the most common soft tissue neoplasm and represents about 50% of all soft tissue tumors. The incidence is approximately 2.1 per 100 people [29,30]. There is no sex predilection for these lesions [31–33]. Lesions are typically small, with 80% measuring less than 5 cm [31]. Superficial lipomas are most commonly located in the trunk, shoulders, upper arm, and neck and are unusual in the hand or foot [31]. Deep lipomas (intramuscular and intermuscular) commonly affect the large muscles of the lower extremity (45%), trunk (17%), shoulder (12%), and upper extremity (10%) [34]. Radiologic evaluation by CT and MRI is diagnostic in up to 71% of cases [34]. On MRI, lipomas present as homogeneous, well-circumscribed and encapsulated masses of fatty nature, without enhancement after intravenous contrast administration. The differentiation between the superficial lipoma and the surrounding fat may be difficult due to their similarity in signal intensity. Regular, thin septations can be seen on MRI [35,36]. The septae appear as low-intensity strands on $T_1$-weighted MRI, which may appear hyperintense on $T_2$-weighted MRI. The width of these septae can be immeasurable, with slight enhancement after intravenous contrast administration, which is more clearly demonstrated on fat-suppressed $T_1$-weighted MR images [37] (Figure 13.1).

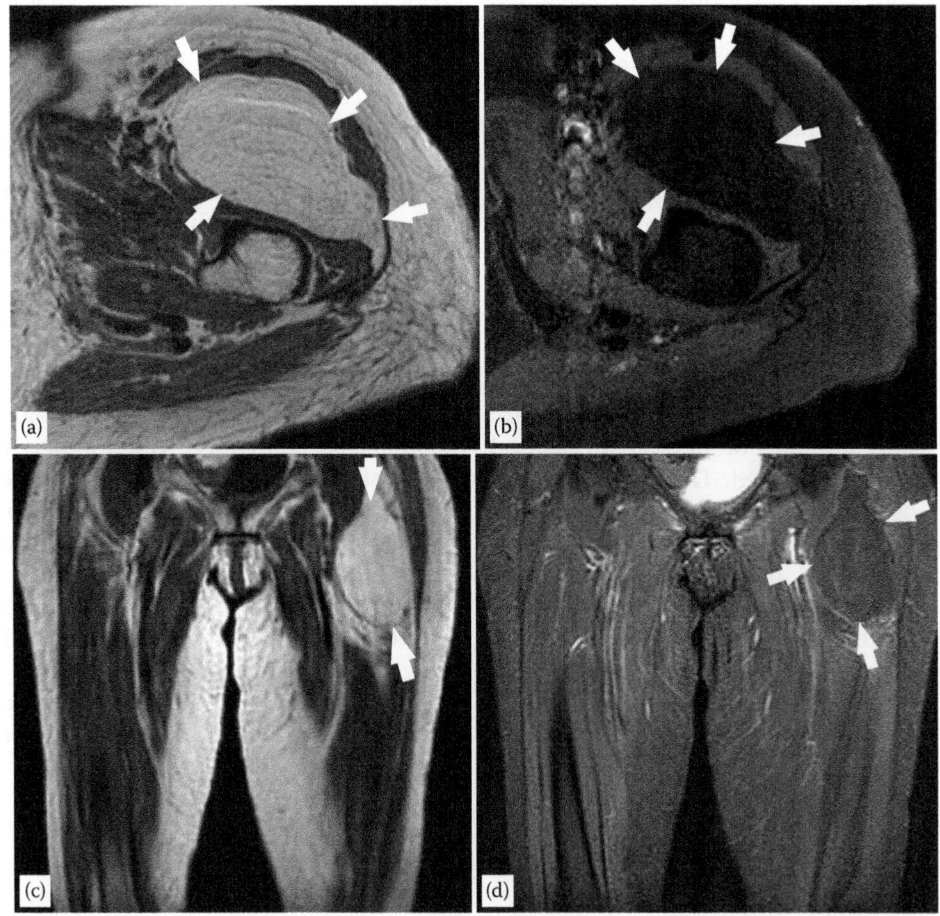

**FIGURE 13.1**
Intermuscular lipoma in a 45-year-old woman with left thigh swelling. (a) Axial $T_1$-weighted MR image shows a homogeneous hyperintense lobulated lesion of intermuscular location within the anterior compartment of the left thigh (arrows). (b) Axial $T_2$-weighted fat-suppressed MR image shows lesion with fat-equivalent signal intensity (arrows). (c) Coronal $T_1$-weighted MR image of the lesion shows the proximal-distal extension of the lesion (arrows). (d) Contrast-enhanced fat-suppressed $T_1$-weighted image of the lesion (arrows) depicts absence of intralesional contrast uptake. Histology post-resection confirmed a diagnosis of lipoma without signs of malignancy.

#### 13.3.1.2 Liposarcomas

Liposarcomas [32,38] may contain mature adipose tissue, but are considerably more heterogeneous than their benign counterparts. Liposarcomas are among the most common soft tissue sarcomas in adults, and typically present in the same age group as benign lipomas (fifth and sixth decades). In general, the malignancy of liposarcomas varies inversely with the degree of differentiation (i.e., the amount of mature intralesional fat). Types of myxoid, liposarcoma include well-differentiated, pleomorphic, or round cell types (Figure 13.2).

*Myxoid variant* is the most common (50%) and typically arises in the lower limbs. This type demonstrates very little fat on MRI and is usually slightly heterogeneous with intermediate signal intensity and a swirled pattern on $T_1$-weighted images. On $T_2$-weighted images myxoid variant is extremely hyperintense, occasionally mimicking cysts. Gd-enhanced $T_1$-weighted image shows a definite and very heterogeneous enhancement. Enhanced $T_1$-weighted images not only improve the delineation but also the evaluation of the internal structure of a tumor. It helps to differentiate well-perfused, viable tumor from tumor necrosis, cysts or cystic parts of a tumor from myxoid ones, and intratumoral hemorrhage from hematoma [39].

*Well-differentiated liposarcomas*, on the other hand, contain predominantly fat, and can in rare cases mimic benign lipomas. These have been termed *atypical lipomas* in the extremities, and exhibit a propensity to recur locally, but do not metastasize (see Table 13.1). The same lesions in the retroperitoneum are simply called well differentiated liposarcomas. $T_1$-weighted MRI shows a large signal abnormality with a loculated pattern and the $T_2$-weighted image shows the high signal features of a fatty tumor [40]. Myxoid and well-differentiated liposarcomas carry significantly better prognoses than pleomorphic and dedifferentiated liposarcomas [41].

*Dedifferentiated liposarcomas*: Although dedifferentiated liposarcoma has no specific presentation that allows distinguishing it from well differentiated liposarcoma, this subtype may be suspected when a non-lipomatous component appears, in a previously known, well-differentiated liposarcoma [42].

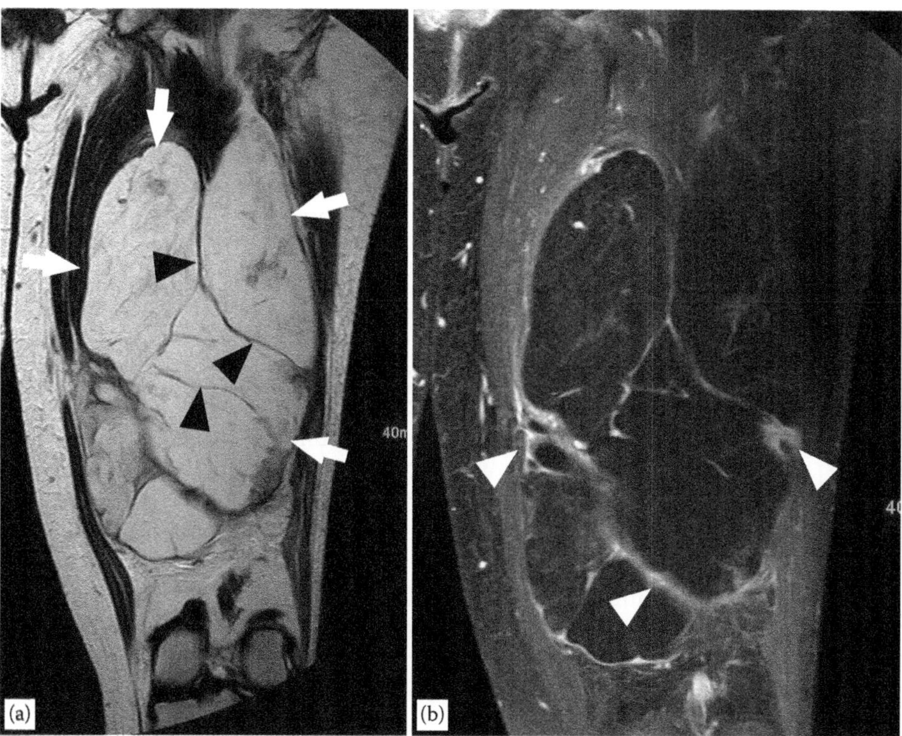


**FIGURE 13.2**
Low-grade liposarcoma in a 34-year-old female who presented with a soft tissue mass of the posterior thigh. (a) Coronal $T_1$-weighted image shows a large well-circumscribed, multilobular lesion of the posterior thigh (arrows) exhibiting intralesional septations (black arrowheads). (b) Contrast-enhanced fat-suppressed $T_1$-weighted image shows a predominantly fatty signal of the lesion with components of nodular and septal enhancement (arrowheads). Pathology confirmed low-grade liposarcoma.

**TABLE 13.1**

Differential Diagnostic Criteria between Lipoma and Liposarcoma

| Characteristics | Lipoma | Intramuscular Lipoma | Intermuscular Lipoma | Well-Differentiated Liposarcoma |
|---|---|---|---|---|
| Location | Subcutaneous fat (back, shoulder, neck, arm, thigh, abdomen) | Within muscle (thigh, shoulder, arm) | Between muscle fasciae (thigh, shoulder, arm) | Within muscle or interfascial (thigh, retroperitoneum) |
| Size | <5 cm (80%)<br><10 cm (95%) | Frequently >5 cm | | Exceptionally <5 cm |
| Margins | Well defined | Infiltrating/well defined | Well defined | Usually well defined |
| Shape | Uninodular | Uninodular (>85%) | Lobulated | Multinodular |
| Septa or nodules | Absent or <2 mm | Discontinuous streaks of variable thickness | Thin, regular, and continuous streaks | Thick (>2 mm), linear or nodular streaks |
| Enhancement | No | No/faint | No/faint | Strong |

*Sources:* Cotten, A., *JBR-BTR*, 85(1), 14–19, 2002; Nishida, J. et al., *J. Orthopaedic Sci.*, 12(6), 533–541, 2007.

The well-differentiated and dedifferentiated liposarcomas most commonly arise in the retroperitoneum, while myxoid liposarcoma is generally a tumor of the extremities.

### 13.3.2 Tumors of Connective Tissue

Fibroblastic/myofibroblastic tumors represent a very large subset of mesenchymal tumors. Many lesions in this category contain cells with both fibroblastic and myofibroblastic features, which may in fact represent functional variants of a single cell type. The relative proportions of these cell types vary not only between individual cases but also within a single lesion over time (often in proportion to cellularity).

#### *13.3.2.1 Benign Proliferative Lesions*

*13.3.2.1.1 Nodular Fasciitis*

Nodular fasciitis is a benign fibroblastic proliferation characterized by rapid growth and high mitotic activity, often simulating an aggressive lesion such as sarcoma. It is the most common tumor or tumor-like condition of fibrous tissue [45]. Nodular fasciitis typically affects patients aged between 20 and 40 years, with no gender predilection [45–47]. Lesions typically present as a rapidly growing painless mass that may cause mild pain or tenderness in approximately 50% of cases [45]. The upper extremity is involved in 46% of cases, particularly the volar forearm. Other common locations include the head/neck (20%), the trunk (18%) and the lower extremity (16%) [36]. The size of this lesion can vary from 0.5 to 10 cm, but most (71%) are 2 cm or smaller [48]. Nodular fasciitis has three common locations: subcutaneous, fascial, and intramuscular. Subcutaneous lesions are 3 to 10 times more common than other sites, followed by fascial and intramuscular locations [49]. The deeper intramuscular form is usually larger and is the most likely to be mistaken for sarcoma [49–51]. Recurrence of nodular fasciitis is rare even after partial resection [48].

Calcification or ossification is rarely seen on radiographs [52]. On $T_1$-weighted MR images, nodular fasciitis has signal intensity similar to or slightly higher than skeletal muscle [49,53]. On $T_2$-weighted sequences, the lesion most often has a high signal intensity (>subcutaneous fat) but may demonstrate intermediate signal intensity [51]. Lesions are frequently homogeneous on $T_1$-weighted sequences and heterogeneous on longer repetition time (TR) acquisitions [53] and show homogeneous enhancement after administration of a gadolinium contrast agent [49,51]. Linear extension along the fascia (fascial tail sign) may suggest the diagnosis, and mild surrounding edema may also be present [36]. The differential diagnosis on MR imaging includes benign fibrous histiocytoma, extraabdominal desmoids tumor, neurofibroma, and malignant fibrous histiocytoma or fibrosarcoma [27]. Treatment is conservative and spontaneous regression of the lesion within few weeks is usually observed. Recurrence is exceedingly rare [49].

*13.3.2.1.2 Fibroma of the Tendon Sheath*

Fibroma of the tendon sheath is a rare, benign, slow-growing tumor of the extremities. Some controversy exists regarding whether it represents a reactive versus a neoplastic process [54]. The tumor affects and adheres to the tendons, tendon sheaths, and sometimes the neurovascular bundles at the affected site [55]. The hand, especially the thumb and index and middle fingers, as well as the wrist, are the most common sites affected (80%) [55]. Males are more frequently affected than females and present most commonly in the fourth decade [55]. It is uncommon in children but affects patients aged 19 years or less in 15%–20% of cases, even patients as young as 5 months [54,55].

At MR imaging, the diagnosis of fibroma of the tendon sheath may be suggested if there is a well-defined, lobulated, homogeneous nodular mass with low signal intensity on all sequences and minimal or no enhancement in

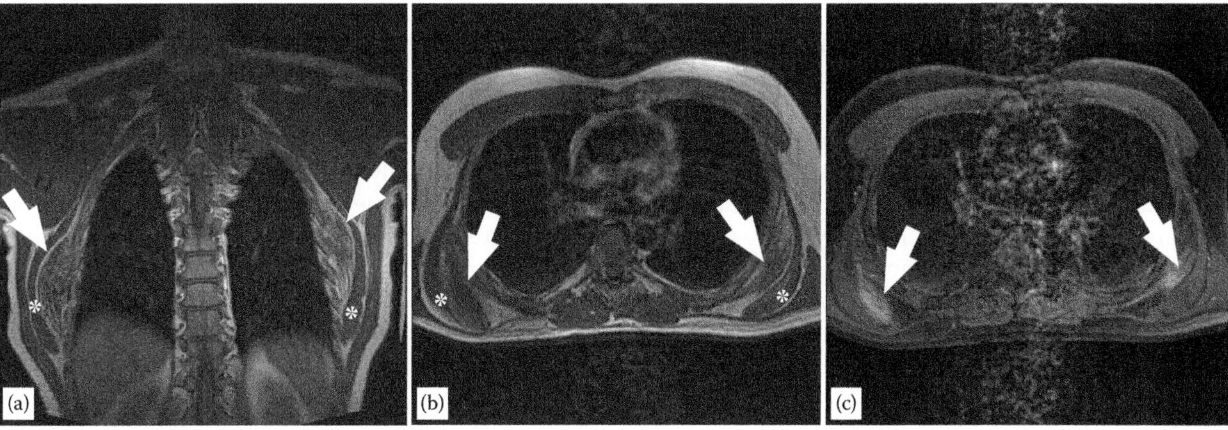


**FIGURE 13.3**
Bilateral elastofibroma in a 58-year-old woman. (a) Coronal $T_1$-weighted image shows bilateral infrascapular soft tissue masses (arrows) lying deep into the serratus anterior muscle (asterisks). (b) Axial $T_1$-weighted image shows the lesions (arrows) lying deep to the latissimus dorsi (asterisks) muscles and adjacent to the chest wall. (c) Axial contrast-enhanced fat-suppressed $T_1$-weighted image shows moderate intralesional enhancement (arrows). The masses present alternating bands of low (fibrous tissue) and high (fatty tissue) signal on $T_1$-weighted images and show mild homogeneous enhancement after contrast administration. The infrascapular location is characteristic of elastofibroma.

the vicinity of tendon sheath [56]. However, the signal intensity of the mass can be quite variable on $T_2$-weighted images. Heterogeneous signal intensity and enhancement patterns can also be seen, reflecting different cellular and stromal elements within the tumor [56]. The tumor may sometimes be difficult to differentiate from other soft tissue tumors commonly found in the extremities, such as giant-cell tumor of the tendon sheath and pigmented villonodular synovitis [54–57].

*13.3.2.1.3 Elastofibroma*

Elastofibroma is a fibrous pseudotumor that is considered to result from chronic mechanical irritation [45]. A genetic predisposition also has been described and the identification of clonal chromosomal changes has led to a suggestion that the lesion may represent a fibroblastic neoplasm [58]. Elastofibroma is a slow-growing lesion that is most frequently encountered in the connective tissue between the posterior chest wall and the inferomedial border of the scapula, usually in patients older than 55 years (mean age, 70 years) [59]. CT of the chest has demonstrated elastofibromas in 2% of patients older than 60 years [60]. Subscapular elastofibromas are bilateral in about 25% of cases [32,61]. Extrascapular sites, which are much less common, include regions of the greater trochanter at the hip and the olecranon at the elbow [59]. Symptomatic lesions generally have a diameter greater than 5 cm. The most commonly reported clinical symptoms are stiffness (approximately 25% of patients) and pain (10% of patients) [59]. Elastofibromas consist of accumulations of collagen and abnormal elastic fibers, with interspersed fat cells and spindle-shaped fibroblasts and myofibroblasts [45,62]. The differential diagnosis of lesions that are relatively hypocellular and contain abundant collagen includes extraabdominal desmoid tumor, as well as neurofibroma and malignant fibrous histiocytoma.

On MR images (Figure 13.3), elastofibromas are typically well-defined, heterogeneous soft tissue masses with signal intensity similar to that of skeletal muscle and frequently with intermixed linear or curvilinear streaks of fat signal intensity, findings representative of the histologic constituents of the lesion [63,64]. Contrast enhancement is typically heterogeneous. Given the characteristic location, patient age and signal characteristics of elastofibroma, a prospective MR imaging diagnosis often can be made with a high degree of confidence, particularly when bilateral lesions are identified [65,66]. Local excision is the treatment of choice for symptomatic lesions. Local recurrence is rare and likely is due to incomplete excision [59]. To our knowledge, no reports of malignant transformation exist in the literature [45].

#### *13.3.2.2 Fibromatoses*

*13.3.2.2.1 Palmar Fibromatosis (Dupuytren's Disease)*

Palmar fibromatosis (Dupuytren's disease) is the most common of the superficial fibromatoses, affecting 1% to 2% of the population [36]. These lesions occur 3 to 4 times more commonly in men and most frequently in patients older than 65 years (up to 20%) [45,67]. Bilateral lesions are present in 40% to 60% of cases [36]. The lesions are painless slow-growing palmar nodules, which may cause a flexion contracture most commonly affecting the flexor tendons of the fourth finger [68]. Patients with palmar fibromatosis commonly have other types of fibromatoses, including plantar fibromatosis (5%–20%), Peyronie disease—an acquired inflammatory condition of the penis (2%–4%), and knuckle pad fibromatosis [36,45].

MR imaging typically shows multiple nodular or cordlike superficial soft tissue masses, which involve the aponeurosis of the volar aspect of the hand, extending superficially in parallel to the flexor tendons. Nodules may progress slowly (months to years) into fibrous cords, which attach to and cause traction on the underlying flexor tendons, resulting in flexion contractures of the digits (Dupuytren contractures) [69]. Lesion signal intensity on $T_1$-weighted and $T_2$-weighted images is low (similar to tendon), reflecting hypocellularity and dense collagen. MR imaging can be helpful for surgical planning because relatively immature lesions demonstrate intermediate to higher signal on $T_1$-weighted and $T_2$-weighted images, reflecting the high cellularity, and have a higher local recurrence rate after local resection. Mature lesions with low $T_1$-weighted and $T_2$-weighted signal intensity are less likely to locally recur [68,70,71]. Lesions show diffuse enhancement, which is more prominent in lesions with higher cellularity.

##### *13.3.2.2.2 Plantar Fibromatosis (Ledderhose's Disease)*

Plantar fibromatosis (Ledderhose's disease) occurs less frequently than Dupuytren's disease, with an unknown incidence [72]. Similar to palmar fibromatosis, incidence increases with advancing age, but 44% of patients were younger than 30 years in a large Armed Forces Institute of Pathology study (501 patients) [45,73]. Men are affected twice as often as women, and lesions are bilateral in 20% to 50% of cases [74,75]. Patients present with one or more subcutaneous nodules, which most frequently affect the medial aspect of the plantar arch (78%) [76] and can extend to the skin or deep structures of the foot. Nodules may be multiple in 33% of cases [76]. The lesions are typically painless, but patients may have pain with prolonged standing or walking. With MR imaging, well- or ill-defined superficial lesions along the deep plantar aponeurosis typically blend with the adjacent plantar musculature. Lesions typically show heterogeneous signal (92%), which is isointense to hypointense to skeletal muscle on $T_1$-weighted (100%) and $T_2$-weighted (78%) sequences. The degree of enhancement has been reported as marked in approximately 60% and mild in 33% of cases [76]. Linear tails of extension (fascial tail sign) along the aponeurosis are frequent and best seen after intravenous contrast administration [45,69].

##### *13.3.2.2.3 Desmoid Fibromatosis*

This rare fibroproliferative disorder has benign histologic features, but is locally invasive, aggressive, and prone to recurrence following excision [77]. It has a peak incidence in the third decade of life, but up to one-third of the cases are diagnosed in patients less than 20 years of age, with a mean age of 13 years [78]. The lesion arises in any musculoaponeurotic structure of the head and neck, trunk, or extremities. Microscopically, the tumor is well differentiated with uniform mature fibrocytes in an intercellular matrix [79]. The etiology is unknown, but has been postulated to reflect a deregulation of connective tissue growth [77], possibly associated with trauma [80]. Fibromas associated with Gardner's syndrome, an autosomal dominant form of colonic polyposis together with tumors outside the colon, are usually seen in childhood and early adolescence. Although some lesions do regress spontaneously, most progress and require treatment. Multicentric disease does occur, but hematogenous or lymphatic metastases are not reported. Treatment goals include local control, acceptable cosmesis and function [77]. MR imaging of fibromatosis in older children shows variable characteristics. Most lesions are of heterogeneous intermediate or low signal intensity on $T_1$-weighted images, and range from very hypointense to hyperintense signal intensity on $T_2$-weighted images [81]. Margins may be sharp or ill-defined. Margins adjacent to fat are often sharper than margins adjacent to muscle [77]. Enhancement pattern after intravenous contrast injection is variable. Lesions that are of increased signal intensity on $T_2$-weighted images tend to enhance more diffusely. The signal characteristics reflect the relative collagenous and cellular components [82]. MR imaging is used to delineate the extent of the lesion and to monitor for residual or recurrent disease after therapy. Surgery with a wide margin of resection remains the primary mode of therapy. The role of chemotherapy and radiation therapy is evolving [77].

#### ***13.3.2.3 Fibrosarcoma: (Infantile and Adult Types)***

Infantile and adult fibrosarcomas share histologic similarities but are markedly different in terms of prognosis. The infantile form, also known as *congenital fibrosarcoma*, is a malignant proliferation of fibroblasts often seen in the first 5 years of life, with about one-third present at birth [83]. They are most often seen in the extremities (74%), followed by the head and neck (15%) [84]. Initially, they grow rapidly and may become quite large in proportion to the size of the child. Metastases, mainly pulmonary, are seen in 5%–10% of cases, with local recurrence reported to be between 17% and 43% [84]. However, overall prognosis is better than in adults; with a greater than 80% 5-year survival rate [84].

MR imaging appearances of infantile fibrosarcomas are nonspecific, as they present as well-demarcated lesions that are isointense to muscle on $T_1$-weighted images and heterogeneously hyperintense on $T_2$-weighted images [85,86]. A heterogeneous enhancement pattern is seen after injection of a gadolinium contrast agent both centrally and in the periphery of the lesions [87]. MRI has a role in follow-up and early detection of local tumor recurrence. Use of Gd contrast agents may aid in differentiating between fibrosis and tumor recurrence [88].

Adult-type fibrosarcoma is histologically similar to the infantile form but shows different cytogenetics [84], is more frequently seen in the proximal regions of the extremities and trunk with a 5-year survival rate less than 60% [89].

Sparse information is available in the literature about the imaging appearances of adult-type fibrosarcoma. Lesions may be homogeneous or heterogeneous. On $T_1$-weighted images, the lesions are usually isointense to muscle, whereas on $T_2$-weighted images they may show areas of low signal intensity on a background of moderate to high signal intensity [57] (Figures 13.4 and 13.5). Dermatofibrosarcoma protuberans is a rare indolent cutaneous tumor that arises in the dermis of the trunk, proximal extremities, and in the head and neck area of adults. Dermatofibrosarcoma protuberans rarely metastasizes through the bloodstream or, less often, to locoregional lymph nodes after multiple local recurrences. Fibrosarcomatous areas within the tumor result in a more aggressive course (Figure 13.6) [90,91].

Following administration of a gadolinium contrast agent, most lesions show an intense peripheral enhancement with an occasional spoked-wheel appearance [92]. Although surgery may be curative, it may be potentially mutilating. Preoperative chemotherapy is recommended [93,94].

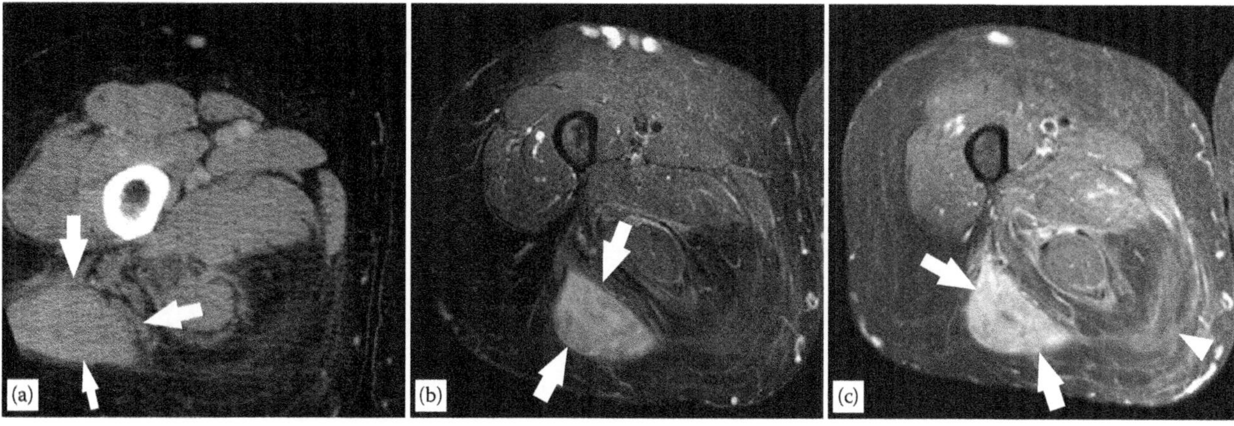

**FIGURE 13.4**
Fibrosarcoma in a 63-year-old woman who presented with a palpable right gluteal mass. (a) Axial contrast-enhanced CT image demonstrates an enhancing ovoid lesion within the right gluteus maximus (arrows). (b) Axial STIR MR image shows lesion with heterogeneous hyperintense signal (arrows). (c) Contrast-enhanced fat-suppressed $T_1$-weighed MR image exhibits intense contrast enhancement (arrows). Also note enhancement within the subcutaneous fat of the posterior compartment, suggesting diffuse infiltration (arrowhead). Histology revealed a cellular and vascular infiltrative tumor with nuclear pleomorphism, cellular atypia and high mitotic count consistent with high-grade fibrosarcoma.

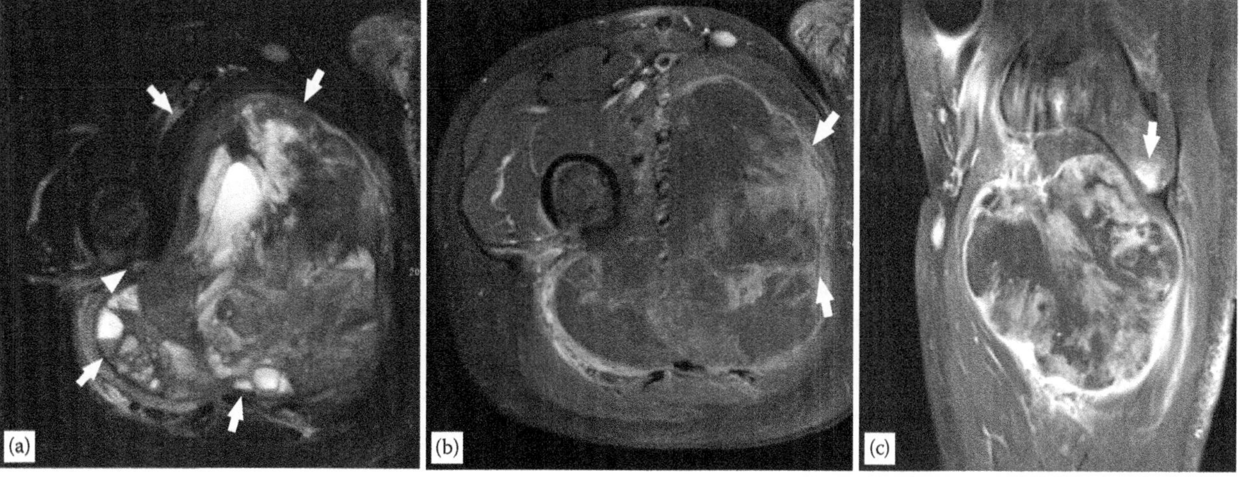

**FIGURE 13.5**
A 32-year-old man with increasing swelling of his right medial thigh for several months. Increasing pain for the last 3 weeks. (a) Axial $T_2$-weighted fat-saturated MRI shows a large inhomogeneous mass infiltrating the adductor muscles (arrows) and femoral cortex (arrowhead). Lesion contains solid and cystic parts including sedimentation due to hemorrhage. (b) Axial contrast-enhanced $T_1$-weighted fat-saturated image shows large areas of necrosis with no enhancement and peripheral nodular solid areas exhibiting avid enhancement (arrows). (c) Sagittal $T_1$-weighted contrast-enhanced image shows hyperintensity in the ischiac bone in addition (arrow). MRI is not able to differentiate infiltrative spread from reactive edema in this case. Histologic diagnosis proved high-grade fibroblastic fibrosarcoma.

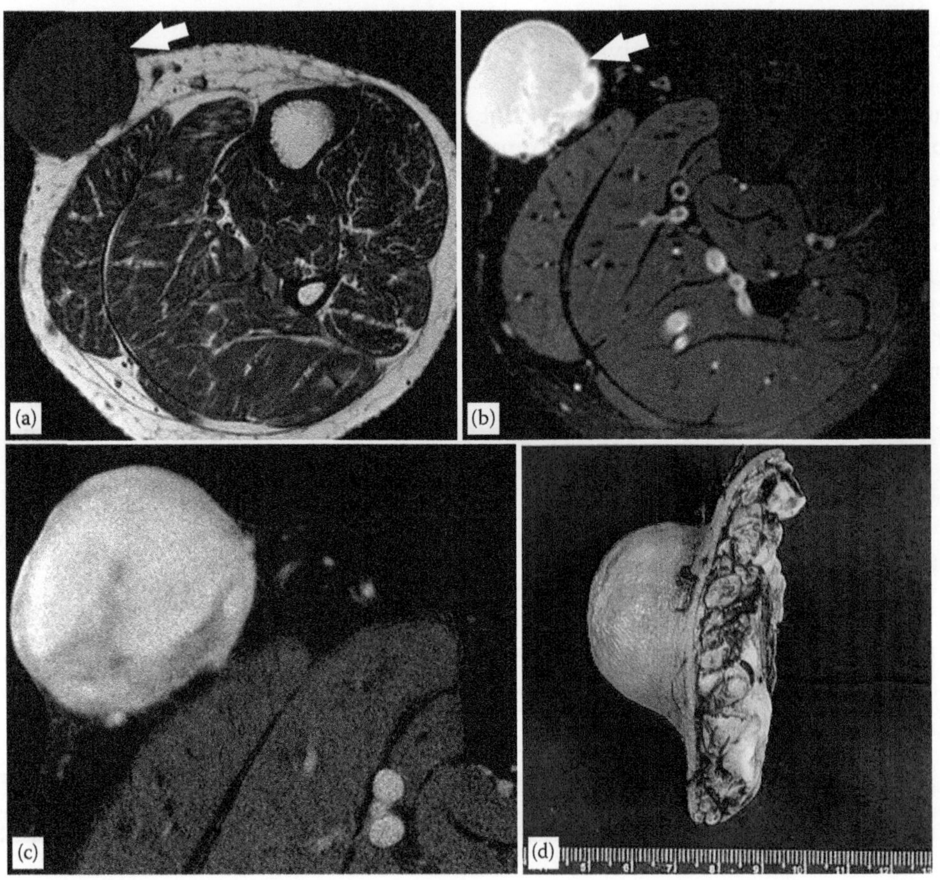


**FIGURE 13.6**
A 55-year-old woman with painless focal nodular swelling at the posterior medial calf. (a) Axial $T_1$-weighted image shows homogeneous hypointense lesion confined to the cutis and subcutis without muscular involvement (arrow). (b) Corresponding $T_2$-weighted fat-suppressed image shows relatively homogeneous almost fluid-equivalent hyperintensity suggesting a subcutaneous cystic or myxomatous lesion (arrow). (c) $T_1$-weighted image after i.v. contrast administration shows diffuse enhancement. (d) The lesion was excised with wide margins and histologic diagnosis confirmed dermatofibrosarcoma protuberans.

### 13.3.3 Tumors of Vascular Origin

Vascular anomalies comprise a broad spectrum of lesions involving all regions of the body. They constitute the most common cause of pediatric soft tissue masses [95]. Their classification and imaging evaluation have markedly changed over the past years, and optimal treatment requires proper lesion classification. MRI has emerged as the preeminent imaging modality for the assessment of vascular malformations and tumors. MR angiography (MRA), more specifically dynamic time-resolved contrast-enhanced MRA provides valuable information about the vascular supply and hemodynamic of these lesions, comparable to conventional digital subtraction angiography.

The diagnosis of a soft tissue vascular anomaly is primarily based on the clinical examination. Imaging is usually reserved for therapeutic planning and for assessment of lesions with unclear diagnosis or deep-tissue involvement. Currently, ultrasound and MRI are the two noninvasive imaging techniques of choice. Grayscale ultrasound with color Doppler assessment is a good imaging modality for the initial assessment and characterization of such lesions [96] because it enables differentiation of high- from low-flow lesions [97]. Its main limitations are insufficient FOV and tissue penetration as well as operator dependency [98]. MRI helps overcome all these limitations and provides superior tissue characterization. It is unsurpassed in showing lesion extent and involvement of different tissue planes and joints [96]. The anatomic extent of the lesion is best seen on fluid-sensitive sequences, such as fat-suppressed fast spin-echo (FSE) $T_2$-weighted or STIR imaging. Many vascular malformations grow in an infiltrative fashion and can involve multiple tissue planes.

#### *13.3.3.1 Hemangioma*

Soft tissue hemangiomas [32,99] can be cutaneous, subcutaneous, intramuscular or synovial. Superficial hemangiomas are often palpable or visible as a bluish discoloration of the skin, and as such do not pose a significant diagnostic challenge. However, deep hemangiomas, such as those in muscle, are difficult to diagnose clinically. Intramuscular hemangiomas are relatively rare (0.8% of all hemangiomas), and tend to arise in childhood or adolescence. Intramuscular hemangiomas also typically contain fat, sometimes as to be virtually indistinguishable from lipomas. This is particularly true of cavernous (large vessels) hemangiomas, which tend to contain greater amounts of nonvascular elements than do capillary (small vessels) hemangiomas. Intramuscular hemangiomas commonly present as a poorly marginated intramuscular mass containing a dense collection of serpiginous and branching vascular channels lined by parallel arrays of fat. Intralesional vessels generally contain slow-flowing blood and exhibit high signal intensity on $T_2$-weighted images and contrast enhancement following intravenous Gd-DTPA injection. Occasionally, the vascular channels are extremely large and sinusoidal. Phleboliths are frequently visible on plain radiography or CT (Figure 13.7). Hemangiomas go through three stages of development and decay: In the proliferation stage, a hemangioma grows very quickly. This stage can last up to 12 months. In the rest stage, there is very little change in a hemangioma's appearance. This usually lasts until the infant is one to two years old. In the involution phase, a hemangioma finally begins to diminish in size. 50% of lesions will have disappeared by 5 years of age, and the vast majority will have gone by 10.

### 13.3.3.2 Glomus Tumor

Glomus tumor is a neoplasm that develops from the neuromyoarterial glomus body [36,100]. The estimated incidence is 1.6% of soft tissue tumors. There is no gender predilection overall, but there is a 3:1 female predominance for subungual lesions [101]. Multiple glomus tumors (nearly 10% of patients) may be present in neurofibromatosis type 1 [102,103]. The lesion is most frequently diagnosed between 20 and 40 years of age. The most common site is the subungual location (65%) [104] in the finger, but other locations include the palm, wrist, forearm, and foot [45]. The average lesion size is approximately 7 mm in the upper extremity and 13 mm in the lower extremity in a recent series [104]. The most frequent clinical presentation is a small red-blue nodule causing paroxysms of pain radiating away from the lesion, which is often elicited by cold or pressure. Imaging reveals a small mass related to the nail bed, with erosion of bone in 22% to 82% of cases [36]. MR imaging reveals a small mass with homogeneous high signal on $T_2$-weighted and intermediate to low signal intensity noted on $T_1$-weighted images (Figure 13.8). Rarely, glomus tumors may show cystic change [105]. Enhancement is typically prominent and diffuse. A high-resolution surface coil has proven useful to demonstrate cortical bone erosion [106]. Therapy aims at complete excision, still leaving a recurrence rate of 10%.

### 13.3.3.3 Angiosarcoma

Angiosarcomas usually occur in the liver, spleen, heart, thyroid gland, breast, bone, and oral cavity as well as in soft tissue, including skin. Angiosarcomas of the skin, including the scalp and face, are more common in men older than 60 years of age, and multiple angiosarcomas sometimes occur [107–110]. Lesions are bluish purple and tend to bleed. They easily spread through the surrounding skin surface and into the deep structures. They can also cause neck lymphadenopathy and lung metastases. The average life expectancy after diagnosis is approximately 30 months [109]. MRI findings in angiosarcoma are only sparsely reported and are

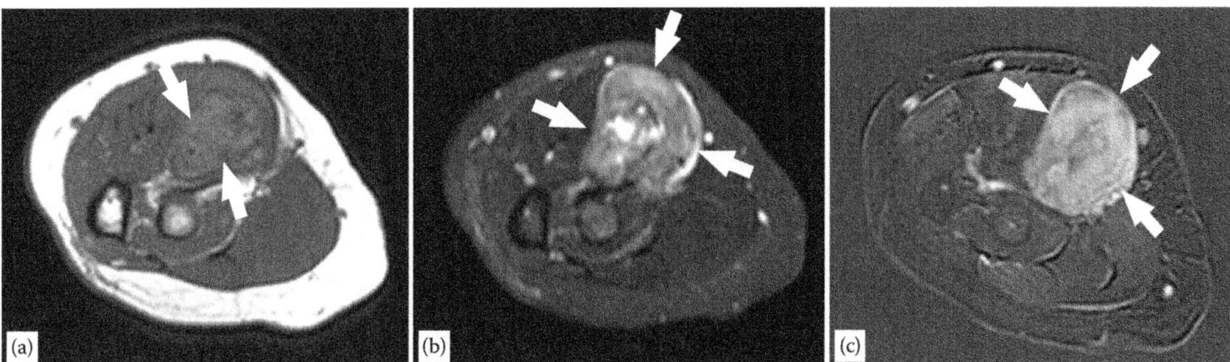


**FIGURE 13.7**
Intramuscular hemangioma in a 51-year-old female with a one year history of right forearm pain and swelling. (a) Axial $T_1$-weighted image shows an intramuscular lesion within the anterior compartment of the right forearm heterogeneous hyperintense signal consistent with the presence of fat and hemoglobin. (b) Axial STIR image. The muscular lesion shows a heterogeneous hyperintensity (arrows). (c) Contrast-enhanced subtracted $T_1$-weighted image of the lesion shows diffuse and intense enhancement (arrows). The lesion was resected and pathology demonstrated epithelioid hemangioma.

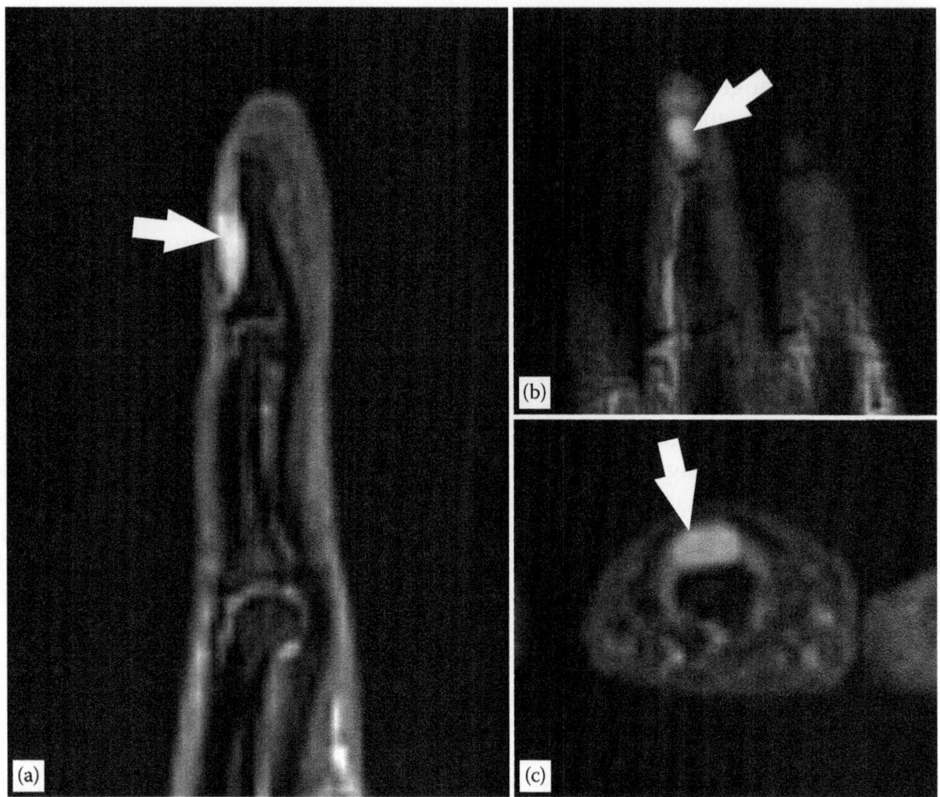


**FIGURE 13.8**
Glomus tumor of the right index finger in a 30-year-old female with history of paroxysmal pain of the tip of her right index finger. (a) Sagittal, (b) coronal, and (c) axial fat-suppressed $T_2$-weighted MR images show a subungual mass adjacent to the distal phalangeal bone, which is homogeneously hyperintense, consistent with the typical appearance of a glomus tumor (arrows).

nonspecific. MRI shows brightly enhancing soft tissue mass, often hyperintense on $T_2$-weighted imaging, with prominent flow voids [111]. $T_1$-weighted SE sequences before and after contrast administration show variable enhancement characteristics depending on tumor grade. The more differentiated the tumor, the more intensely it enhances. Some authors mention the presence of prominent serpentine vessels suggestive for the diagnosis [112].

#### *13.3.3.4 Hemangiopericytomas*

Musculoskeletal hemangiopericytomas are location-specific subtype of hemangiopericytomas. These tumors are most frequently seen in middle-aged adults (~fourth decade). They, along with hemangioendotheliomas and angiosarcomas are tumors that arise from vascular structures. In the case of hemangiopericytomas, they arise from the cells of Zimmerman, which surround vessels. Both benign and malignant forms are encountered. Typically they have large vessels especially located at its periphery, and commonly involve the lower limbs (35% of cases), especially the thigh, pelvis and retroperitoneum (25%). Primary osseous lesions are rare. Radiologically they appear similar to angiosarcomas and hemangioendotheliomas. MRI shows brightly enhancing soft tissue mass, often hyperintense on $T_2$-weighted imaging, with prominent flow voids [113].

### 13.3.4 Tumors of Muscular Origin

#### *13.3.4.1 Intramuscular Myxoma*

Intramuscular myxoma is a benign neoplasm, which occurs almost exclusively in individuals between the fifth and seventh decades [114,115]. It shows fairly homogeneous fluid-equivalent hyperintensity on fluid-sensitive sequences. The contrast enhancement pattern of a myxoma is usually relatively mild and has been described as lacy or patchy. Myxoma may show a rim of fat, especially at the superior and inferior poles of the lesion, as well as leakage of contents, with a brush like $T_2$-hyperintense appearance within the muscle proximal or distal to the lesion. Combinations of these features may yield a nearly pathognomonic appearance, but reasonable caution still requires biopsy of these lesions (Figure 13.9). Following surgery, recurrence of intramuscular myxoma is rare.

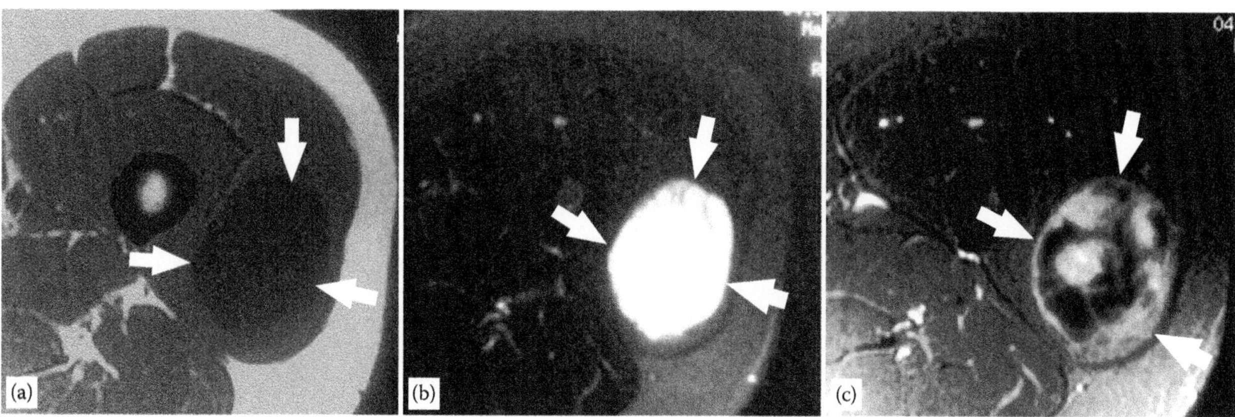

**FIGURE 13.9**
Intramuscular myxoma in a 50-year-old female with lower extremity swelling. (a) Axial $T_1$-weighted MR image displays a markedly and homogeneously hypointense well-circumscribed intramuscular mass within the vastus lateralis muscle. (b) Axial STIR image of the lesion shows a homogeneously hyperintense intramuscular mass. (c) Contrast-enhanced fat-suppressed $T_1$-weighted MR image displays heterogeneous enhancement of the lesion (arrows). Pathology established the diagnosis of intramuscular myxoma.

#### 13.3.4.2 Leiomyoma

Leiomyoma is a benign tumor of smooth muscle. Leiomyomas of the deep soft tissues are extremely rare and can be divided into two groups, one occurring primarily in the pelvic retroperitoneum in women and the other in the deep somatic tissue in patients of either sex. Leiomyomas in this latter group often arise in skeletal muscle or in the deep subcutaneous soft tissues, most frequently in the lower extremities. They often affect middle-aged or young adults and are extremely rare in children [116]. Little information regarding the MR imaging appearance of deep somatic soft tissue leiomyomas is available [116,117]. In the few cases described in the literature, leiomyoma appears as a well-circumscribed mass slightly hyperintense to muscle on $T_1$-weighted images, with heterogeneous high signal intensity on $T_2$-weighted images, and with enhancement after i.v. contrast administration. The tumor may contain dystrophic calcifications that can be visible at radiography and MR imaging as foci of low signal intensity.

### 13.3.5 Skeletal Muscle Lymphoma

Lymphoma is a heterogeneous disease representing approximately 4% of new cases of cancer in the United States; almost all extranodal types of tissue may be involved [118]. Involvement of skeletal muscle, however, is rare. Unlike other malignant soft tissue tumors, skeletal muscle lymphoma responds well to chemotherapy with or without radiation therapy [119,120]. Differentiation of skeletal muscle lymphoma from various neoplastic and inflammatory diseases, however, often is difficult on the basis of clinical and imaging findings alone [121,122]. Therefore, specific MRI findings of skeletal lymphoma would be of value in differential diagnosis. On $T_1$-weighted images, the tumors have equal or slightly increased signal intensity compared with normal muscle. On $T_2$-weighted images, all lesions have intermediate signal intensity. On fat-suppressed $T_2$-weighted images more lesions (62%) have intermediate signal intensity, and less have low signal intensity. On contrast-enhanced MRI, skeletal muscle lymphoma shows homogeneous diffuse enhancement in most cases (Figure 13.10). At MRI the involvement of skeletal muscle by lymphoma can be suspected in cases of diffuse muscle enlargement with long segmental or multi-compartmental involvement, deep fascial involvement, and the presence of traversing vessels in the lesion, subcutaneous stranding, and skin thickening. Skeletal muscle lymphoma can exhibit peripheral thick band like enhancement or marginal septal enhancement in addition to the well-known finding of diffuse homogeneous contrast enhancement [123].

### 13.3.6 Synovial Tumors

#### 13.3.6.1 Giant Cell Tumors

Giant cell tumor of tendon sheath (GCTTS) is clinically a slow growing soft tissue mass that develops over a period of months to years [124,125]. Clinically these masses typically present in the hand with localized swelling with or without pain. They are slow growing and typically, they present in third to fifth decades and have a slight female predilection. In addition to having a fibrous stroma, this tumor contains a benign proliferation of hemosiderin-laden histiocytes and giant cells. These tumors are closely related to pigmented villonodular synovitis, except that instead of involving the synovium of large joints, such as the knee, GCTTS arise from the synovial lining of tendon sheaths, most commonly the fingers and toes. Clinically, GCTTS presents as a slow growing non-calcified mass

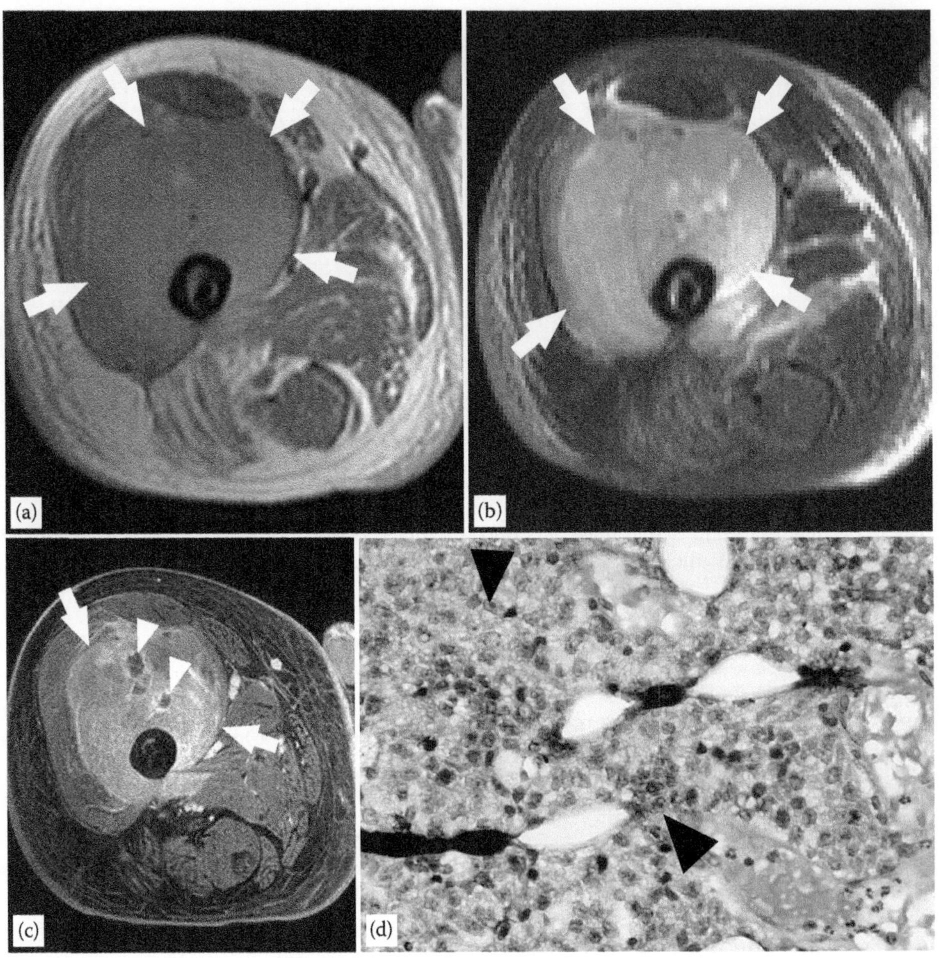


**FIGURE 13.10**
A 58-year-old man with right thigh swelling, 12 months after right hip arthroplasty. Computed tomography showed a soft tissue mass within the anterior compartment of the right thigh that was initially thought to represent hematoma (not shown). Axial (a) $T_1$-weighted and (b) $T_2$-STIR MR images show a large mass located within the vastus intermedius muscle. The lesion shows discrete hyperintensity on the $T_1$-weighted and marked high signal on the $T_2$-weighted fat-suppressed sequence. Axial fat-saturated $T_1$-weighted images after gadolinium administration (c) show heterogeneous contrast enhancement (arrows) with central necrotic areas (arrowheads). (d) Core biopsy revealed cells with pleomorphic, atypical nuclei (arrowheads) and immunohistochemistry confirmed lymphoid lineage consistent with large B-cell lymphoma.

in the adult. On MRI, the lesions are well marginated, homogeneous and show low signal intensity on both $T_1$-weighted and $T_2$-weighted images owing to abundant collagen and hemosiderin. On gradient-echo sequences, the paramagnetic effect of hemosiderin is further exaggerated, resulting in areas of very low signal due to the blooming artifact [126]. An adjacent tendon is invariably seen. Pressure erosion of the adjacent bone often occurs. The appearance is distinctively nonaggressive in contrast to that of desmoids tumors. Local surgical excision usually suffices, with local recurrence (seen in 10%–20% of cases) requiring more extensive surgery with or without radiotherapy being uncommon. There is a marked disparity in prognosis and outcome among patients with giant cell tumors of the soft tissues. When treated adequately by complete surgical excision, a benign course is expected in numerous cases. If not resected completely recurrences may occur and metastases may appear within a relatively short interval after local treatment [127] (Figure 13.11).

### 13.3.6.2 Synovial Sarcoma

Synovial sarcoma is the most common soft tissue non-rhabdomyosarcoma in children [128]. It can be found anywhere in the body, but most commonly in the lower extremities [129]. This lesion usually presents as a painless, enlarging mass. Two histologic subtypes include a biphasic type containing epithelial and spindle cell elements, and a monophasic type composed of only spindle cells [130]. Approximately 30% of patients with synovial sarcoma are under 20 years of age [128]. Metastases are usually to lung and severely limit survival [131]. Radiographs show calcification in the soft tissues in approximately one-third of synovial cell sarcomas [130].

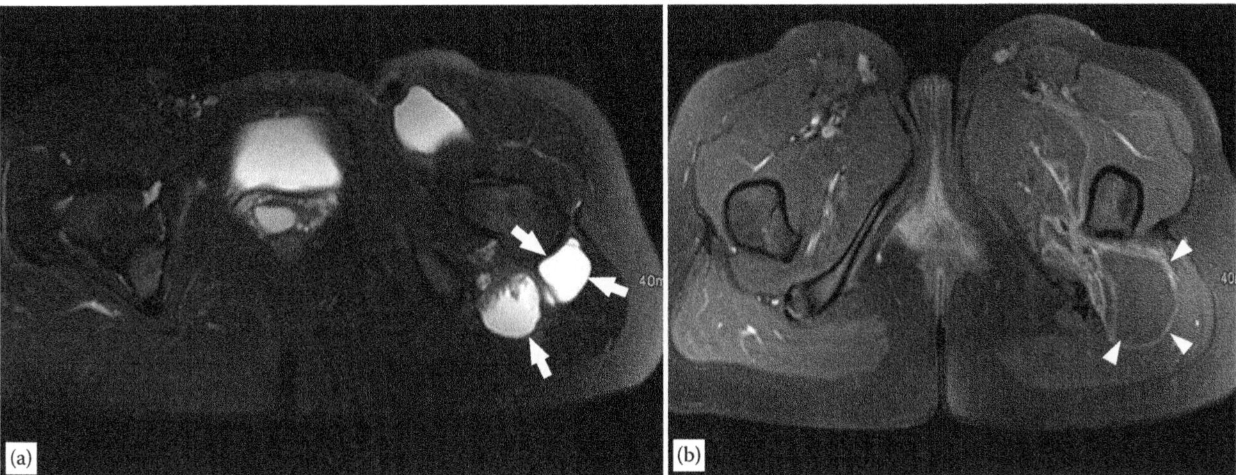


**FIGURE 13.11**
A 32-year-old woman with pain in her left buttock for 2 months. (a) Axial $T_2$-weighted MR image with fat suppression shows a multilobulated mass of the posterior left thigh in contact with the hamstring tendon origin (arrows). The lesion signal is equivalent to fluid. (b) Axial contrast-enhanced fat-saturated $T_1$-weighted MR image displays peripheral enhancement of the lesion (arrowheads). The lesion seems to have a non-infiltrative mass effect on surrounding structures. The lesion was excised and histology proved a giant cell tumor of the tendon sheath.

MRI is the modality of choice for staging of the tumor. The mass is usually large and variably well-defined (smaller lesions tend to be better circumscribed). On $T_1$-weighted sequences it is iso- or slightly hyperintense to muscle or it can be heterogeneous. On $T_2$-weighted images it is mostly hyperintense, very heterogeneous with areas of necrosis and bands of fibrosis and fluid–fluid levels are seen in up to 10%–25% of cases [132]. On contrast-enhanced MRI with Gd, enhancement is usually prominent and can be diffuse (40%) heterogeneous (40%) or peripheral (20%). MR imaging often shows an ill-defined mass in deep dermis and subcutaneous tissues with intermediate signals on both $T_1$ and $T_2$ images. The lesion is often heterogeneous showing occasionally fluid–fluid levels and can be mistaken for a benign cystic mass [130]. Hypervascularity may be quantified by a dynamic contrast-enhanced MR examination. Enhancement of tumor within 7 s after arterial enhancement is a reliable sign, occurring consistently in synovial sarcoma [133]. Other previously described so-called malignant dynamic contrast-enhanced MRI features, such as early plateau or washout phase and peripheral enhancement are not always found in synovial sarcoma [133]. Well defined or reasonably well-defined margins are seen in some cases of synovial sarcoma. This finding has been described as a probable sign of benignity [134]. This may be one of the reasons why synovial sarcoma is the malignant tumor most frequently misdiagnosed as benign. Therapy is based on excision with tumor-free margins of the primary tumor at time of diagnosis, and adjuvant radiation therapy to treat microscopic residual disease [129]. Chemotherapy does not impact survival, although some studies suggest that chemotherapy may be useful for extremity lesions [131]. Lesions that are greater than 5 cm have a high rate of local recurrence following surgery [128].

### 13.3.7 Extraskeletal Cartilaginous and Osseous Tumors

#### *13.3.7.1 Extraskeletal Chondroma*

Extraskeletal chondroma is a relatively rare, benign, slow-growing soft tissue tumor that usually occurs in the soft tissues about the joints of the hands and feet [135] and usually measure less than 3 cm in diameter. This tumor is thought to arise from the fibrous stroma of soft tissues, rather than originating from mature cartilaginous or osseous tissue. Extraskeletal chondroma typically affects adults, usually between the ages of 30 and 60 years [136]. MRI delineates the lesion location and margins, but the appearance is nonspecific. Extraskeletal chondromas have been described as showing low-to-intermediate signal intensity on $T_1$-weighted images and heterogeneous intermediate to-hyperintense signal from the cartilages on $T_2$-weighted images [135–137].

On MRI, soft tissue chondromas appear on $T_1$-weighted images as well-circumscribed masses that are isointense relative to skeletal muscle. On $T_2$-weighted images the hyaline cartilage typically presents with very high signal intensity, greater than that of fat. Mineralized areas cause foci of signal void on all sequences. Diagnosis of chondromas without foci of calcification is more difficult, and these lesions have to be differentiated from other soft tissue masses, especially those of synovial origin. After intravenous injection of gadolinium the majority of chondromas exhibit marked and mostly peripheral contrast enhancement [138].

#### *13.3.7.2 Extraskeletal Mesenchymal Chondrosarcoma*

Chondrosarcoma ranks in third position with regard to incidence among malignant bone tumors. They are graded on a scale from 1 to 3, based upon nuclear size, staining pattern, mitotic activity, and degree of cellularity. Grade 1 chondrosarcomas were reclassified as *atypical cartilaginous tumors* [139]. Grade 2 chondrosarcomas are more cellular with less chondroid matrix than grade 1 tumors. Grade 3 chondrosarcomas are highly cellular, with nuclear pleomorphism and easily detected mitoses [140,141]. Extraskeletal chondrosarcoma is rare, making up only 1% of reported chondrosarcomas [142]. The two most important subtypes of extraskeletal chondrosarcoma include mesenchymal and myxoid chondrosarcoma [143]. Extraskeletal mesenchymal chondrosarcoma (EMC) has some of the characteristics of benign tumors, while having a similar imaging presentation compared to other malignant soft tissue tumors. Thus, a higher rate of misdiagnosis is usually obtained. Mesenchymal chondrosarcoma is a type of highly malignant cartilaginous tumor derived from bone or soft tissue, and lesions in soft tissue account for approximately 30%–75% of all mesenchymal chondrosarcoma lesions [143]. Unlike other types of chondrosarcoma to which males are more susceptible, EMC occurs preferentially in women. EMC usually occurs in the lower limbs especially the thigh and head–neck regions. A peripheral distribution of EMC is found in patients aged 50 years, while a central distribution is found at age 30 years [142–144]. EMC should be distinguished from extraskeletal osteosarcoma, malignant fibrous histiocytoma, synovial sarcoma, and myositis ossificans. EMC or peripheral EMC can be diagnosed as follows: (1) calcification is dense, especially the ring-and-arc calcification; (2) On $T_1$-weighted images, there is equal or low signal intensity; (3) on $T_2$-weighted images, there is high signal intensity surrounding low signal intensity, or the *black pepper* sign is exhibited; (4) diffuse heterogeneous enhancement or nodular enhancement is visible, and calcified areas are also strengthened [142]. The occurrence of a combination of these characteristics can greatly enhance the accuracy of diagnosis.

### 13.3.8 Primitive Neuroectodermal Tumors

#### *13.3.8.1 Extraskeletal Ewing's Sarcoma*

Ewing's sarcoma is a malignant neoplasm of uncharacterized mesenchymal cell origin arising in the bones, rarely in soft tissues [145]. Extraskeletal Ewing's sarcoma (EES), the extra-osseous form of Ewing's sarcoma, was first described by Tefft et al. in 1969 [146]. Most EES patients are between 10 and 30 years of age, with a peak incidence at ~20 years of age. The most common sites are the chest wall, extremities, buttocks, and retroperitoneal space. However, a number of cases have been reported to arise in various locations, such as the small intestine, vagina, kidney, skin, larynx, esophagus, and paravertebral region [45]. The mean age at presentation was 19 years (range: 4–38 years), with 73% of the patients being between 10 and 20 years old. Male patients predominated (66%). The male-female ratio of 1.5:1 in spinal epidural EES resembles that in Ewing's sarcoma of bone.

MRI is the most useful tool for early diagnosis to determine the extent of the tumor, but it is nonspecific. MR imaging demonstrates a soft tissue mass with heterogeneous iso signal intensity on $T_1$-weighted images and intermediate to high signal intensity on $T_2$-weighted images [147]. High signal intensity on long TR images predominates in most cases [147], which are likely due to a high degree of cellularity, as in osseous lesions. Areas of hemorrhage appear as high signal intensity on all pulse sequences and are not uncommon; Focal areas of necrosis with low signal intensity on $T_1$-weighted images and high signal intensity on $T_2$-weighted images are also frequent [147]. Fluid–fluid levels may also be present and can be related to either hemorrhage, or necrosis or both. As in other soft tissue masses, MR imaging is also useful for tumor staging and to evaluate the extent of involvement of surrounding structures [148]. An additional imaging feature in MR imaging of extraskeletal Ewing sarcoma is the presence of serpentine high-flow vascular channels, which have low signal intensity with all pulse sequences [147]. This finding is not unique to extraskeletal Ewing sarcoma and can be seen in higher-grade vascular lesions (hemangioendothelioma, hemangiopericytoma, and angiosarcoma), rhabdomyosarcoma, synovial sarcoma, and alveolar soft tissue sarcoma. However, the presence of this feature in a young person with a large intramuscular mass should raise the possibility of extraskeletal Ewing sarcoma [147]. As with other soft tissue masses, both benign and malignant, a pseudocapsule with relatively well-defined margins may be seen at MR imaging of extraskeletal Ewing sarcoma [147]. Direct invasion of bone is more common in the terminal stage of the disease [148,149]. Prominent contrast enhancement is seen at both CT and MR imaging [147].

### 13.3.9 Peripheral Nerve Sheat Tumors

#### *13.3.9.1 Peripheral Nerve Sheath Tumors*

Peripheral nerve sheath tumors (PNSTs) are classified separately as neurogenic tumors by the WHO and comprise benign and malignant PNSTs [36,150]. Benign PNSTs include both schwannomas (neurilemomas) and neurofibromas, which together account for 10% of benign soft tissue tumors [151]. PNSTs can manifest with both motor and sensory nerve disturbances [152]. Schwannomas and neurofibromas can be difficult to distinguish from each other at imaging [153]. Either tumor can appear as a well-defined smooth-bordered fusiform mass that is aligned along the nerve. Occasionally on MR images (Figure 13.12), a schwannoma can be distinguished from a neurofibroma by its location relative to the nerve: The schwannoma can be eccentric to and separable from the nerve, whereas the neurofibroma is intrinsic to it [154]. The *split fat sign* can be associated with PNSTs: As the tumor enlarges, a surrounding rim of normal fat is maintained [153]. Benign PNSTs are typically isointense to muscle on $T_1$-weighted MR images and slightly hyperintense to fat on $T_2$-weighted MR images [153,155] but are nonspecific in terms of their SI. Nevertheless, on cross-sectional MR images, a *target sign* appearance may be seen on $T_2$-weighted images in some benign PNSTs, more commonly in neurofibromas than schwannomas [154,156]. The central area of low $T_2$ SI histologically corresponds to fibrocollagenous tissue, whereas the outer area of high $T_2$ SI corresponds to myxomatous tissue [155]. Contrast enhancement in benign PNSTs is variable. Malignant PNSTs account for 6% of soft tissue sarcomas and are associated with type 1 neurofibromatosis in 50% of cases [154]. Malignant PNSTs can be difficult to differentiate from benign PNSTs; however, malignant PNSTs are typically larger and have ill-defined margins, rapid growth, and central necrosis [153,154,157].

### 13.3.10 Soft Tissue Metatases

#### 13.3.10.1 Soft Tissue Metastasis

According to Liotta and Stetler-Stevenson, 30% of patients with newly diagnosed solid tumors (except skin neoplasms other than melanoma) have detectable secondary lesions, and 60% have microscopic or macroscopic metastases

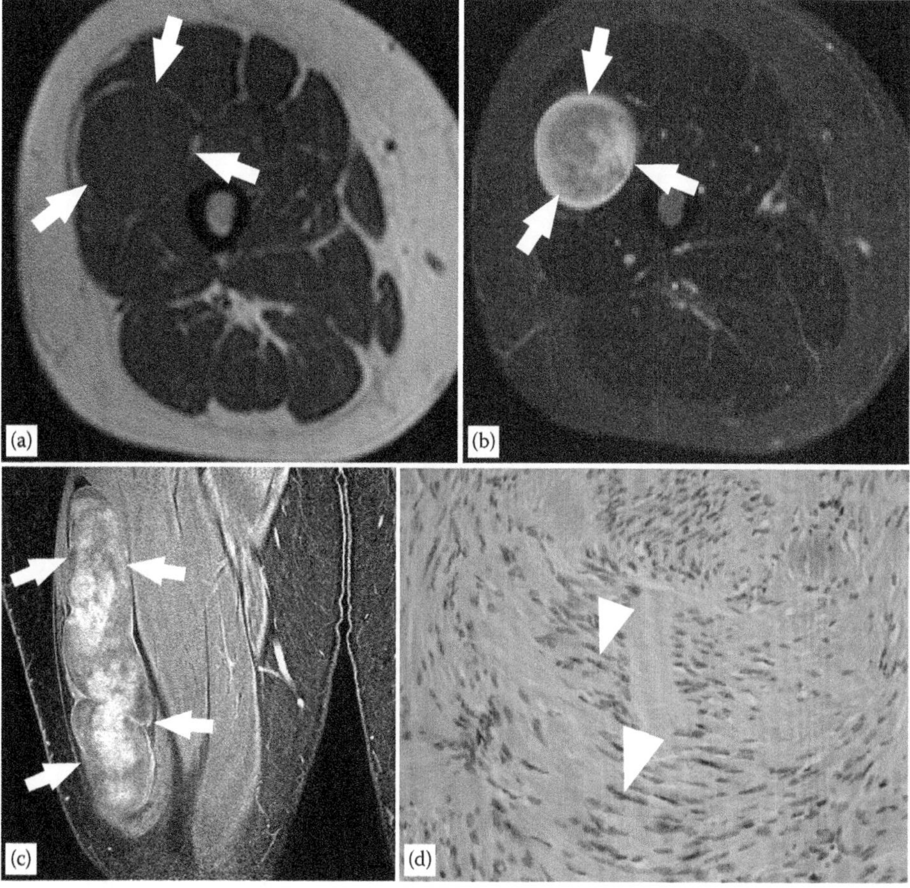


**FIGURE 13.12**
A 33-year-old woman with 5 month history of right thigh swelling. (a) Axial $T_1$-weighted MR image demonstrates an ovoid lesion extending along anterior compartment fascial planes, which is isointense to the adjacent muscle (arrows). (b) Axial fat-suppressed $T_2$-weighted MR image. The lesion appears hyperintense with central hypointensities consistent with the *target sign* (arrows). Coronal $T_1$-weighted MR image after contrast administration (c) shows a lobulated mass with heterogeneous enhancement, predominantly central (arrows). Resection of the lesion revealed an encapsulated mass. (d) Frozen section with H&E stain showed elongated cells arranged in fascicles in histology (arrowhead), consistent with a schwannoma.

as early as the time of primary tumor treatment [158]. Although soft tissues, and in particular skeletal muscles, represent about 40% of the total body weight, they are infrequently affected by primary tumors and even more rarely by metastatic lesions. Rates of 1% of all neoplasms, but 6% in patients under the age of 25, are reported [159–161]. In comparison to primary malignant tumors, metastases to the soft tissues are sparsely reported. In various series of autopsies the proportion of metastases to soft tissues varies considerably, ranging from 0.8% in the study of Willis to 16% in that by Pearson and 52% in the Buerger study [162,163]. Overall, the most common cause of soft tissue metastases are lung, renal and colon primaries [164,165]. A significant number of cases are due to unknown primary carcinoma and the histopathological features are of undifferentiated tumor. Histological diagnosis most commonly seen is adenocarcinoma predominantly from lung and gastrointestinal tract [166]. Squamous cell carcinomas and renal clear-cell carcinomas are other frequent cause of metastases [167]. Sarcomas are uncommon cause of metastases to the soft tissues. MRI offers much better soft tissue contrast resolution than all other techniques previously reported, and it has multiplanar imaging capability useful in tumor assessment and preoperative planning [168,169]. MRI of soft tissue metastases is described in only a small number of cases (Figure 13.13). In these cases the lesions had lower signal intensity than fat and were isointense to muscle on $T_1$-weighted images; they were hyper- or isointense relative to fat, and always presented with a higher signal intensity than muscle on $T_2$-weighted images [170–172]. These features are generally considered nonspecific. In certain cases, for example in metastases of mucinous adenocarcinoma of the colon or pancreas, marked hypointensity is seen on $T_2$-weighted

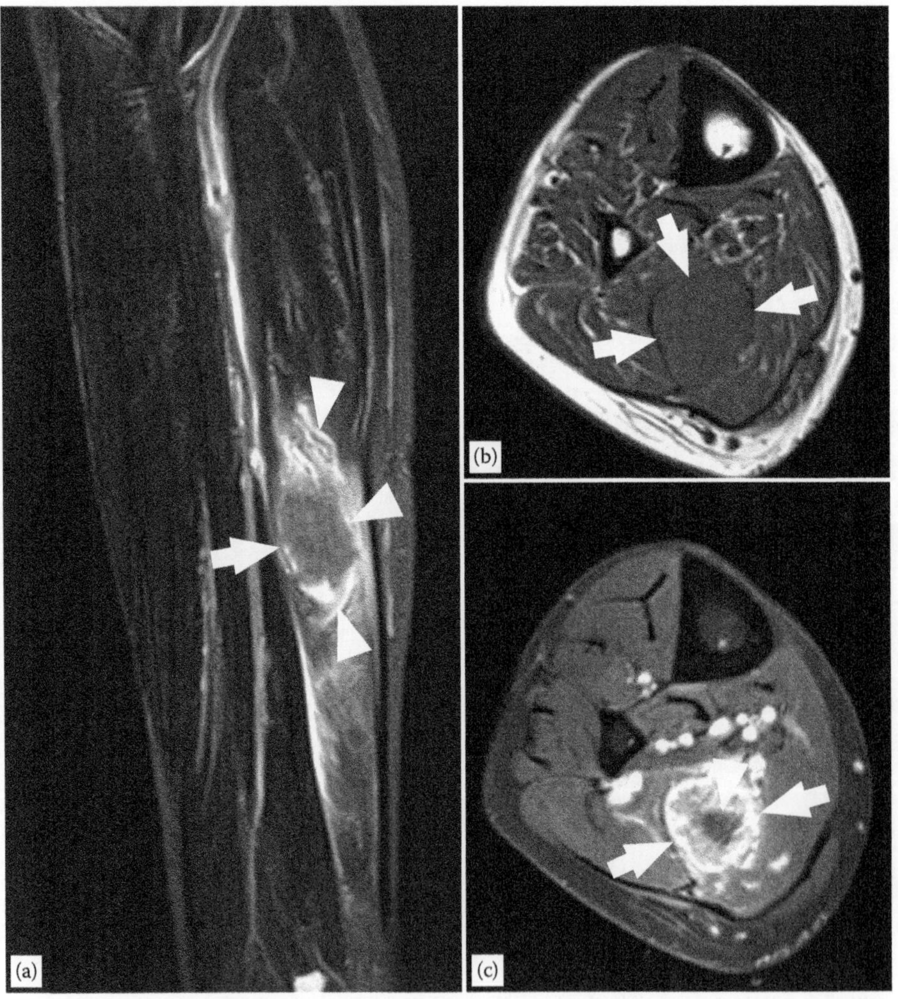


**FIGURE 13.13**
Painful calf after minor trauma in a 73-year-old man. Persistent pain for 4 weeks. (a) Sagittal $T_2$-weighted fat-saturated image shows hypointense lesion (arrow) with an irregular hyperintense rim (arrowheads). (b) $T_1$-weighted axial image shows homogeneous hypointensity within lesion (arrows). (c) Axial contrast-enhanced fat-suppressed $T_1$-weighted image shows thick peripheral enhancement and (arrows) central necrosis (white arrowhead). Biopsy revealed metastasis from prostate cancer.

images as a consequence of the high ratio between nuclei and cytoplasm and the limited extracellular space. Metastases from melanoma show high signal intensity on $T_1$-weighted images; and intermediate signal intensity on $T_2$-weighted images. These signal intensity characteristics are the consequence of the paramagnetic stable free radicals in melanin, although some authors suggest an influence of iron associated with hemorrhage or chelated metal ions [173,174]. Comparable MRI features have also been described in cases of pleomorphic liposarcoma and intratumoral hemorrhage [175]. Secondary tumoral lesions may present either as a regular or an irregular, well-marginated mass or as an ill-defined mass [176]. It may show homogeneous or heterogeneous signal intensity on both $T_1$- and $T_2$-weighted images and a moderate to strong enhancement after intravenous injection of paramagnetic contrast. Small masses tend to be homogeneous and large ones more heterogeneous. As shown by Kransdorf, the probability of malignancy of a mass increases with its size and heterogeneity. This is a consequence of the outgrowth of their vascular supply, which can provoke subsequent infarction and necrosis. Intratumoral bleeding can also be seen in metastases [177]. Although many authors state that MRI cannot reliably distinguish between benign and malignant soft tissue tumors, the use of criteria such as signal intensity, homogeneity, margins, neurovascular invasion, growth rate, septation in the tumor, extension beyond one compartment, bone destruction, and signal changes in adjacent tissues, together with clinical findings often affords better distinction between benign and malignant lesions [168,178]. The use of gadolinium chelates may facilitate distinction between tumor, muscle, and edematous tissue and provides information on tumor vascularity and necrosis [179]. The use of color Doppler imaging, blood pool scintigraphy, and dynamic MRI contrast studies may be useful in monitoring results of chemotherapy [12,180]. A decrease in vascularity and perfusion is correlated with good response and vice versa. Extraskeletal or periosteal osteosarcoma may mimic metastatic disease and needs to be incorporated in the differential diagnosis (Figure 13.14).

## 13.4 Non-Neoplastic Soft Tissue Lesions

### 13.4.1 Cystic Lesions

Cystic-appearing lesions are commonly seen in clinical practice at imaging of the extremities. However, only some of these lesions are truly cystic lesions (e.g., ganglia or synovial cysts, bursa) and may be managed conservatively. A wide variety of cystic lesions may be encountered in the soft tissues and bones during routine MR imaging of the knee. These lesions represent encapsulated fluid collections and exhibit low signal intensity on $T_1$-weighted images and high signal intensity on $T_2$-weighted images because of their high content of free water.

#### *13.4.1.1 Meniscal Cyst*

Meniscal cyst is seen in association with meniscal tear, usually one with a horizontal component (>90% of cases) [181]. Synovial fluid accumulates in the degenerated tissue and may be contained within the torn meniscus (intrameniscal cyst) or, more commonly, displaced through the tear, expanding the meniscocapsular margin and displacing the capsule outward into the adjacent tissues (parameniscal cyst) [181]. At MR imaging, meniscal cysts are well circumscribed, often septated fluid collections in continuity with a horizontal cleavage or complex meniscal tear. On $T_1$-weighted images, meniscal cysts may appear isointense relative to muscle due to hemorrhage or high proteinaceous content.

#### *13.4.1.2 Ganglion Cyst*

Ganglion cysts that involve the dorsal or volar aspect of the wrist are one of the few types of soft tissue lesions for which a physical examination is characteristic enough to label the lesion as determinate. These lesions need no imaging other than conventional radiography and can be treated as the patient and physician prefer. Ganglion cysts also commonly occur in the foot and ankle, although they can be associated with any tendon sheath, labrum, or capsule. Unlike a Baker cyst, a ganglion cyst usually does not communicate with the joint. When such a lesion is found in a location other than the wrist, additional imaging should be performed. On MRI, the lesion appears smooth and round or ovular, and it may have septations (Figure 13.15). The wall of the cyst is well-circumscribed and forms a distinct border against the adjacent tissues. Although the septations may enhance with gadolinium administration, the lesion itself should not. The lesion appears with a signal intensity similar to that of water, bright on $T_2$-weighted imaging or STIR [2,182,183] and dark on $T_1$-weighted imaging (Figure 13.16).

#### *13.4.1.3 Baker Cyst*

Originally described by Baker [184], the synovial cyst (now known as the Baker cyst) forms when synovial fluid insinuates from the knee joint into communicating bursa or by causing herniation of the synovial membrane itself. This lesion occurs most commonly under the medial head of the gastrocnemius muscle. The lesion is often asymptomatic, but it can cause pain. In addition, the cyst may rupture or leak, causing resultant swelling and/or pain in the affected limb; thrombophlebitis,

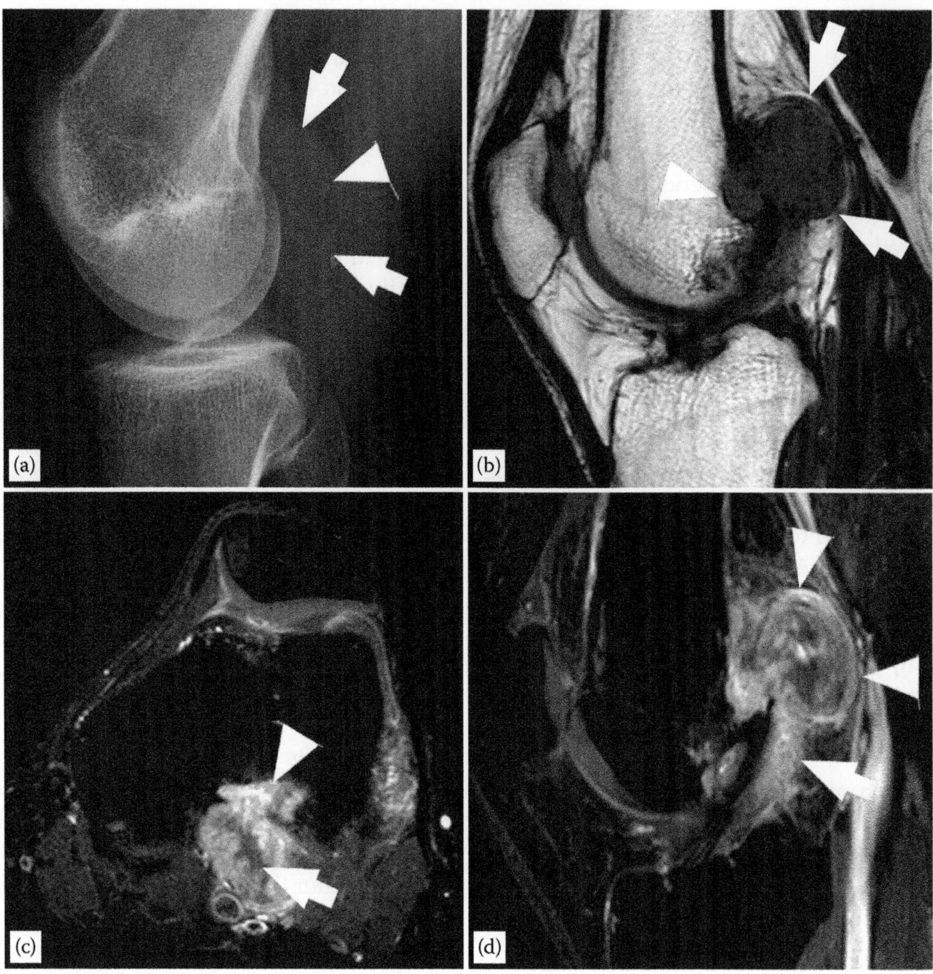


**FIGURE 13.14**
Differential diagnosis of metastatic disease: periosteal sarcoma. A 37-year-old patient with incidental finding of a painless palpable mass in the popliteal fossa. (a) Lateral radiograph shows soft tissue attenuation in the popliteal fossa adjacent to posterior femoral cortex (arrows). No apparent osseous involvement is seen. A small calcification seems to delineate the lesion posteriorly (arrow). (b) Sagittal $T_1$-weighted non-enhanced image depicts lesion as relatively homogeneous hypointense mass (arrows) clearly showing bony involvement (arrowhead). Adjacency to the popliteal vessels suggests extra-articular involvement. (c) $T_1$-weighted axial MR image after i.v. contrast administration shows avid enhancement of lesion and discrete infiltration of tumor into the posterior femur (arrowhead). Note central necrosis reflected as non-enhancing center of lesion (arrow). (d) Corresponding sagittal $T_1$-weighted fat-suppressed contrast-enhanced image shows posterior delineation of lesion (arrowheads) and also involvement and infiltration of dorsal parts of proximal anterior cruciate ligament (arrow). Tumor was treated with local total resection.

compartment syndrome, and claudication can occur [185,186]. The cause of the lesion is intra-articular and can include abnormalities ranging from anterior cruciate ligament damage to osteoarthritis, although meniscal tears are the most common cause [187]. Ultrasound may be useful for diagnosis, but MRI is the study of choice because it can delineate underlying abnormalities such as meniscal or anterior cruciate ligament tears. MRI findings of a Baker cyst (Figure 13.17) are similar to those of a ganglion cyst. The Baker cyst is an ovular, well-defined mass found posterior to the knee joint. The fluid appears dark on $T_1$-weighted images and bright on $T_2$-weighted and STIR imaging. The signal intensity of the lesion matches that of joint fluid on all pulse sequences [188].

### 13.4.2 Traumatic Pathology

#### *13.4.2.1 Hematomas*

It is important to keep hematoma in the differential diagnosis of soft tissue lesions. The patient often relates a history of trauma, and residual ecchymosis can sometimes be seen during the physical examination. When a history of trauma is not present, it is important to ask the patient about systemic anticoagulation or clotting deficiencies because these conditions can be associated with chronic expanding hematomas [189]. Most such lesions will resolve on their own, but the natural history of the lesion may also follow two other courses: the hematoma may begin to calcify on the periphery

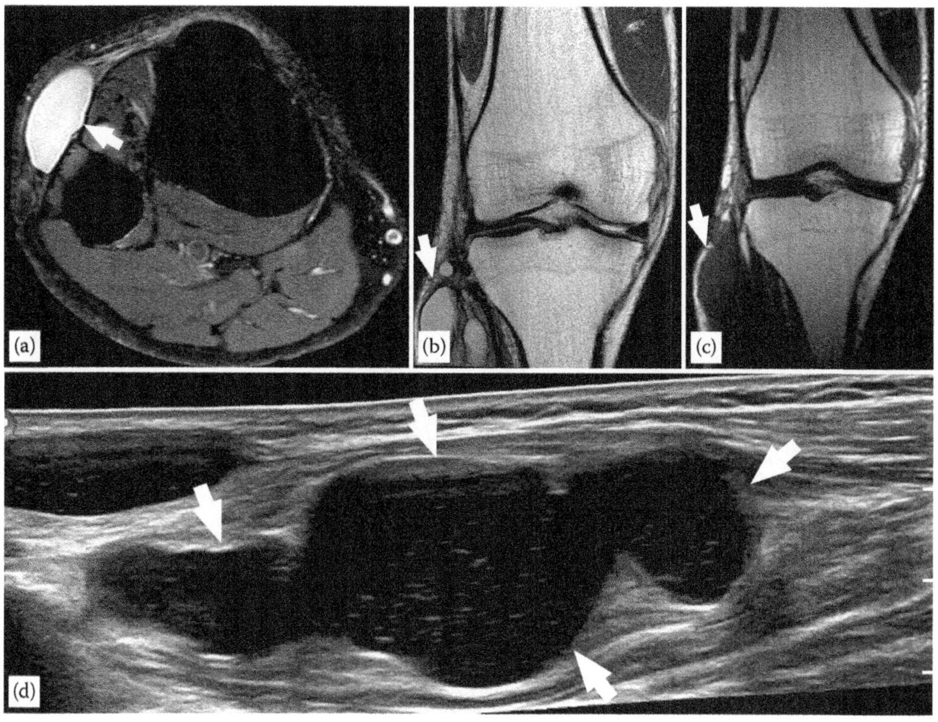


**FIGURE 13.15**
A 34-year-old woman with lateral knee swelling and feeling of local pressure. (a) Axial $T_2$-weighted fat-suppressed MRI shows well-delineated fluid equivalent hyperintense lesion in the subcutis of the proximal calf (arrow). (b) Coronal $T_2$-weighted image shows multi-lobular lesion in close proximity to the tibiofibular joint (arrow). (c) Corresponding $T_1$-weighted image depicts lesions as homogeneous hypointensity (arrow). (d) Ultrasound of lesion shows fluid equivalent echogeneity and multi-lobular appearance of this tibiofibular joint ganglion cyst (arrows).

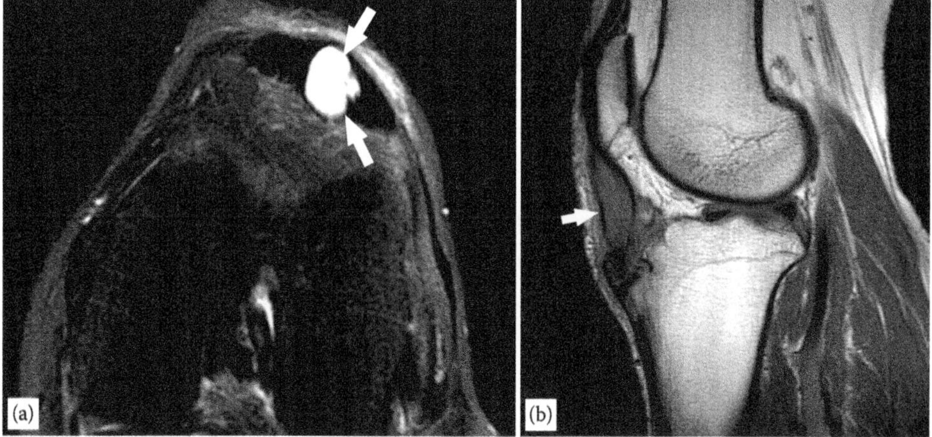


**FIGURE 13.16**
A 57-year-old man with chronic anterior knee pain. (a) Axial $T_2$-weighted MR image shows a well-delineated fluid-containing lesion within the patellar tendon representing an intratendinous ganglion cyst (arrows). (b) Sagittal $T_1$-weighted image shows intermediate signal intensity within cyst (arrow) suggesting proteinaceous content. Longitudinal extent is only fully appreciated on the sagittal image.

(eventually becoming myositis ossificans) or it may continue to expand, which may be related to continual irritation caused by hemosiderin breakdown products. In such instances, the lesion does not heal, and it chronically expands secondary to continued capillary bleeding [190,191]. Continued expansion of the lesion may cause compression of neurovascular structures or pressure erosion of nearby osseous structures [190]. $T_1$-weighted imaging shows a heterogeneous pattern. Areas of high signal intensity represent areas of continuing hemorrhage and granulation tissue. Overall, the lesion is well defined and does not invade other

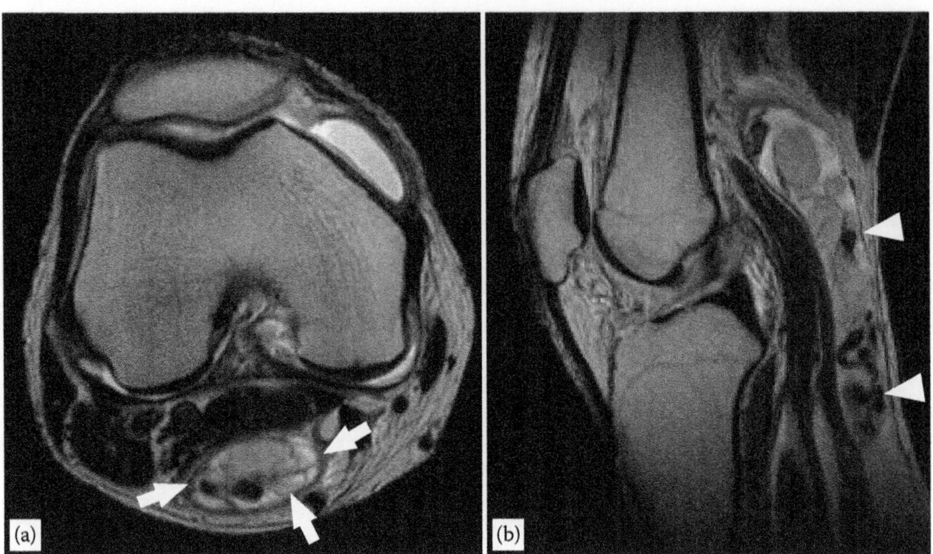


**FIGURE 13.17**
A 48-year-old patient with posterior knee pain. (a) Axial $T_2$-weighted image exhibits a popliteal cyst in the typical location extending through the gastrocnemius and semimembranosus tendons posteriorly (arrows). (b) Sagittal $T_2$-weighted image shows longitudinal extent of cyst and also shows solid components representing synovial debris of intermediate signal and hypointense calcifications/loose bodies within the cyst (arrowheads).

muscle compartments. $T_2$-weighted images show heterogeneity (Figures 13.18 and 13.19). Areas of low signal intensity, that is, signal dropout, occur in areas of hemosiderin deposition. A low-signal pseudo-capsule often is seen. Fluid-fluid levels have also been described [192]. Gradient-echo imaging can be used to investigate further the areas of hemosiderin deposition [189,191] and will show the profound signal dropout (dark areas) that is characteristically seen on gradient-echo imaging in the presence of hemosiderin or other ferromagnetic materials such as metallic implants. Gadolinium-enhanced images can be used to confirm the identity of the lesion. When there is no enhancement with gadolinium contrast, the lesion can be considered determinate.

#### 13.4.2.2 Muscle Contusion

This type of injury is caused by a direct blow to the muscle, usually by a blunt object, resulting in injury to the deep layers of the muscle belly as the muscle is compressed against the underlying bone [193]. MR imaging shows focal high signal intensity on fat-suppressed $T_2$-weighted and STIR sequences that often has an indistinct, feathery appearance [194,195]. With sufficient force, focal edema is often accompanied by the development of an intramuscular hematoma, with myonecrosis and myositis ossificans being potential complications [193]. The MR imaging appearance of an intramuscular hematoma depends on the age of the lesion, with acute hematoma isointense to muscle on $T_1$- and hypointense on $T_2$-weighted sequences. As the hematoma becomes subacute, the proportion of extracellular methemoglobin increases, which results in high signal on $T_1$- and progressively increased signal on $T_2$-weighted sequences. Chronic hematoma is characterized by hemosiderin deposition with consequent low signal on $T_1$- and $T_2$-weighted sequences and blooming artifact on gradient-echo imaging [196]. There is often enlargement of the involved muscle but usually no evidence of muscle fiber discontinuity or laxity [197].

#### 13.4.2.3 Muscle Tear (Strain)

Muscle tears are among the most common complaints treated by physicians and account for the majority of sport-related injuries. These lesions are common among athletes and male sex predominance [198,199]. Muscle tear can be clinically classified as grade 1 (Figures 13.20 and 13.21), grade 2 (Figure 13.22), and grade 3 based on absent, mild, or complete loss of muscle function, respectively [200]. MRI is the best imaging modality for making the diagnosis of a muscle tear. $T_1$-weighted imaging shows the ruptured muscle or the formation of scar, or both. The time relative to the injury dictates the muscle's appearance on $T_2$-weighted imaging. Acute injuries show hyperintense lesions with surrounding edema. Chronic injuries show low to intermediate signal intensity. In grade 1 strains, MRI shows intramuscular *feathery* hyperintensity on fluid-sensitive sequences without muscle fiber disruption. In grade 2, MRI shows hyperintensity (edema and hemorrhage) intramuscularly or at the musculotendinous junction (MTJ), irregularity and

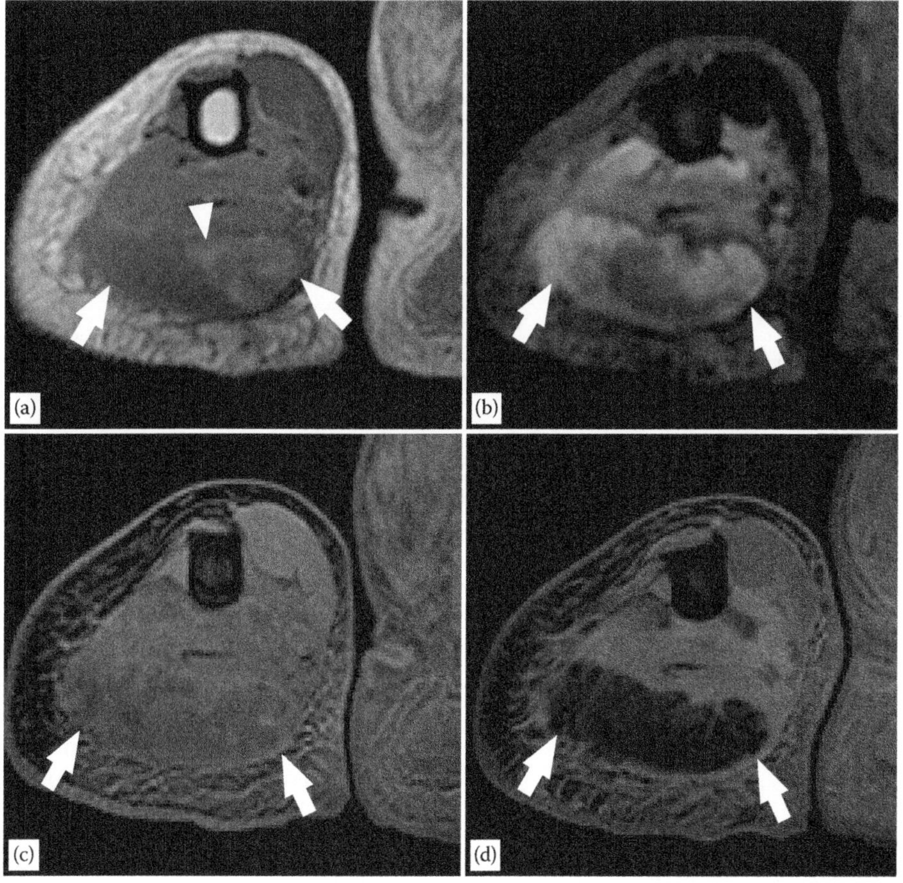

**FIGURE 13.18**
A 62-year-old woman with recent history of fall and upper extremity swelling. (a) Axial $T_1$-weighted MR image demonstrates an ill-defined lesion within the triceps brachii muscle (arrows), which is heterogeneous and exhibits hyperintense components (arrowhead). (b) Axial STIR image demonstrates heterogeneous hyperintensity of the lesion (arrows). Axial fat-suppressed $T_1$-weighted MR images pre- (c) and post-contrast administration (d) show absent enhancement of the lesion. History of a fall, imaging features, and subsequent resolution were all consistent with hematoma.

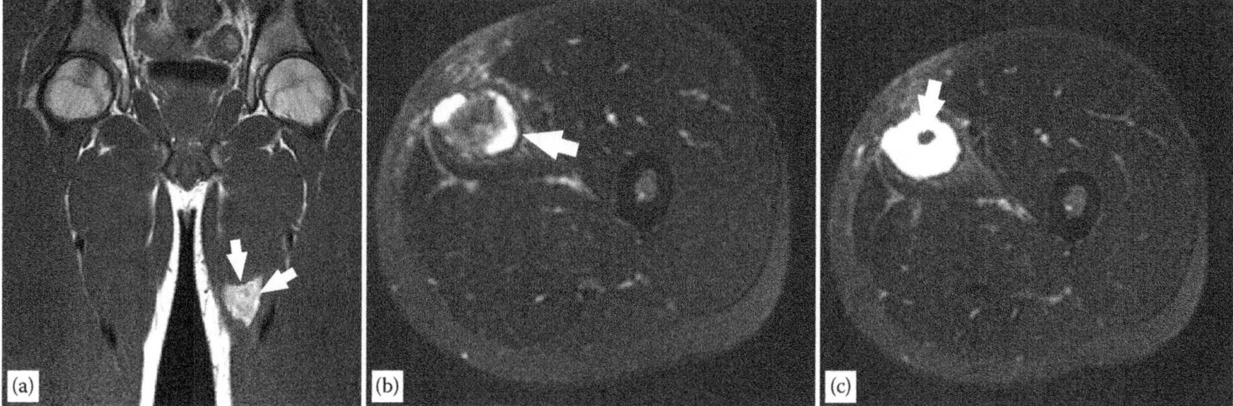

**FIGURE 13.19**
A 30-year-old tennis player who suffered sudden onset of pain during competition. (a) Coronal $T_1$-weighted MRI shows well-defined hyperintense lesion in the adductor longus muscle (arrows). (b) Corresponding axial $T_2$-weighted MRI depicts well circumscribed lesion of partial fluid-equivalent hyperintensity and partial hypointensity consistent with hematoma. Note that medial muscle bundles are still preserved. (c) Image obtained slightly below the image shown in (b) exhibits fluid equivalent alterations with central hypointensity due to clot formation (arrow).

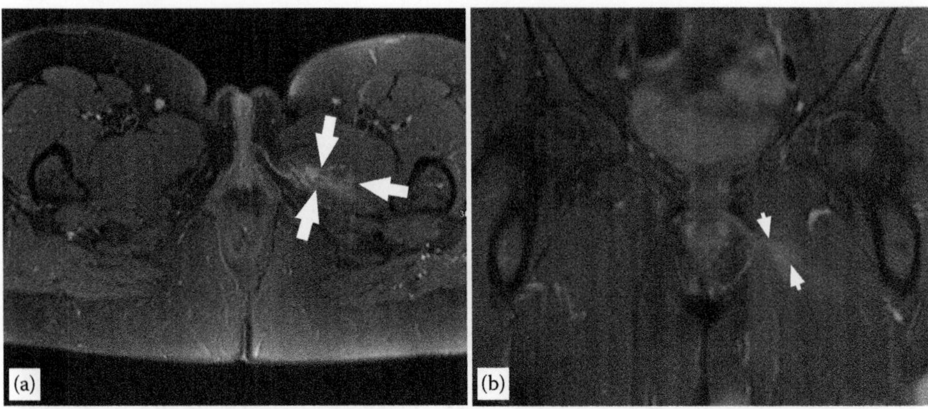


**FIGURE 13.20**
Adductor muscle strain. A 45-year old woman who suffered an indirect trauma due to a mis-step while climbing stairs. (a) Axial fat-saturated intermediate-weighted image shows circumscribed area of diffuse hyperintensity in the left adductor brevis muscle (arrows). (b) Coronal MRI shows longitudinal extent of lesion (arrowheads). Finding is consistent with a grade 1 muscle strain in the adductor brevis muscle.

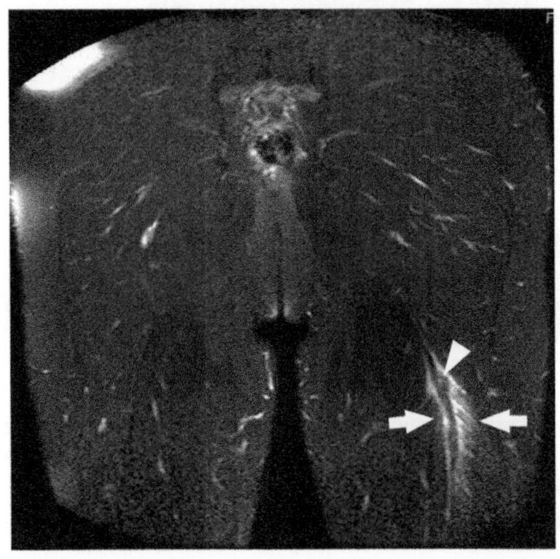

**FIGURE 13.21**
Coronal intermediate-weighted fat-saturated MR image of both posterior thighs in a 25-year-old professional soccer player shows typical feathery appearance of a grade 1 muscle strain (arrows). Note central tendon involvement including hyperintensity and thickening of the tendon of the long head of the biceps (arrowhead).

mild laxity of tendon fibers and hematoma at the MTJ which is pathognomonic. In grade 3, complete discontinuity of muscle fibers associated with extensive edema and hematoma, and possible retraction of tendon can be seen in MR imaging [201].

### 13.4.3 Inflammatory Pathologies

#### *13.4.3.1 Myositis Ossificans*

Myositis ossificans is a localized, self-limiting, reparative lesion of muscle, which at pathology has three components: a central zone of proliferating fibroblasts, a middle zone containing osteoblasts and foci of immature bone, and a peripheral layer with mature bone trabeculae [202]. The cause of myositis ossificans is soft-tissue injury, in most cases secondary to trauma, which is well documented in 60%–75% of patients [203]. In the remaining cases, minor repetitive trauma, ischemia, and inflammation are thought to be the causative agents. The lesion may develop anywhere in the body, but it occurs more frequently in those areas more exposed to trauma, especially the anterior compartments of the thigh and arm. In children, it is usually seen in those older than 10 years [202].

Several subtypes of myositis ossificans exist: posttraumatic myositis ossificans, myositis ossificans progressive (MOP), and neurogenic myositis ossificans.

##### *13.4.3.1.1 Posttraumatic Myositis Ossificans*

Posttraumatic myositis ossificans (Figure 13.23) also called Myositis ossificans circumscripta (MOC), localized myositis ossificans, and myositis ossificans traumatica is the most common form of myositis ossificans, comprising up to 75% of all cases of myositis ossificans. It is a benign, usually solitary reactive form of heterotopic ossification typically occurring within skeletal muscle. A history of prior trauma is usually given, but this may be absent in up to 40% of cases [203]. The likelihood of developing MOC increases with the severity of preceding trauma. MOC may occur at any age, but is usually seen in the second and third decades, with few reported cases involving children <10 years of age [204,205]. There is a slight male predominance, but this is most likely related to differences in levels of physical activity between males and females. MOC most commonly affects the thigh (vastus lateralis) and arm (brachialis), with anterior muscle groups affected more frequently than the posterior groups [206]. Proximal regions of an extremity are more frequently affected than distal parts. Other commonly involved sites include the intercostal spaces, erector spinae, pectoralis, gastrocnemius, and gluteal muscles.

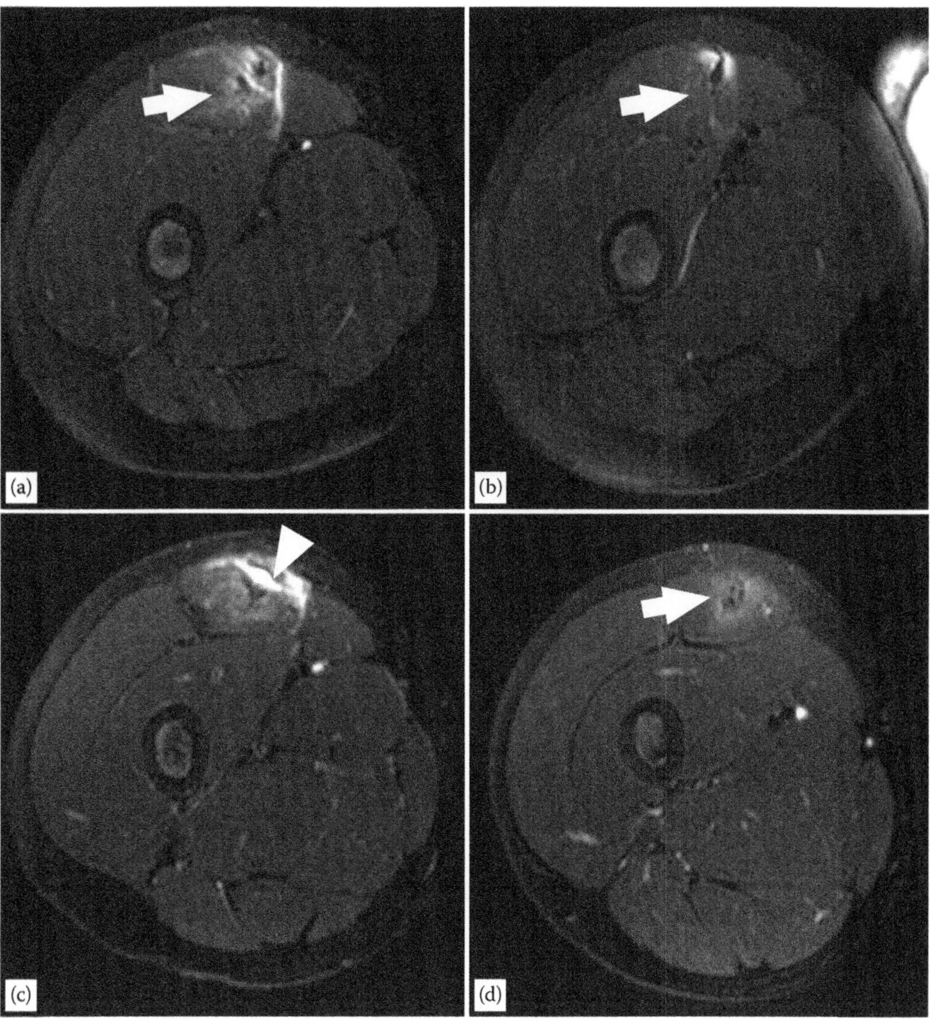


**FIGURE 13.22**
A 24-year-old soccer player with acute pain while sprinting. No loss of function but increase in pain on knee extension was noted clinically. (a) Axial $T_2$-weighted fat-suppressed MRI shows diffuse edema in the rectus femoris muscle (arrow). (b) 3 cm further distal of image depicted in (a) and it is shown that the edema is surrounding the central tendon (arrow). (c) Another 3–4 cm further distally there is a fluid equivalent component with the rectus femoris defining this lesion as a grade 2 muscle injury on MRI (arrowhead). (d) At distal margin of lesion, only subtle extension of edema is seen (arrow).

Established cases may have associated joint contractures. Patients may be asymptomatic but usually present with a painful, tender focal swelling, often with associated cutaneous erythema. Fever may or may not be present, and the erythrocyte sedimentation rate may be raised in the active phase of ossification. Although MOC is a benign self-limiting condition, imaging is an important tool to exclude infection or malignancy. Imaging findings typically change with lesion maturity.

MRI demonstrates acute lesions preceding the appearance of calcification [207]. Early lesions appear as a swelling or nodule isointense to the surrounding musculature on $T_1$-weighted scans, with diffuse or peripheral enhancement [204,208]. Lesions may be difficult to differentiate from adjacent muscle due to marked surrounding edema in the early stages [203]. Intermediate to high signal (hyperintense to fat) is seen centrally on $T_2$-weighted images with or without a low signal continuous or noncontinuous rim [202]. Small foci of $T_2$ hypointensity may be present within the lesion, corresponding to small areas of calcification. The surrounding muscles fibers are usually markedly edematous in the early stages, a feature not frequently seen in sarcomas and an important diagnostic finding. Although the underlying bones usually show no associated abnormality, periostitis, bone marrow edema, and reactive joint effusions may be present in the acute setting. However, the underlying cortex remains intact. In the intermediate stage, lesions demonstrate a variable appearance. The center is isointense or hyperintense to normal muscle on $T_1$-weighted images. On $T_2$-weighted/STIR sequences, lesions tend to be inhomogeneous, with

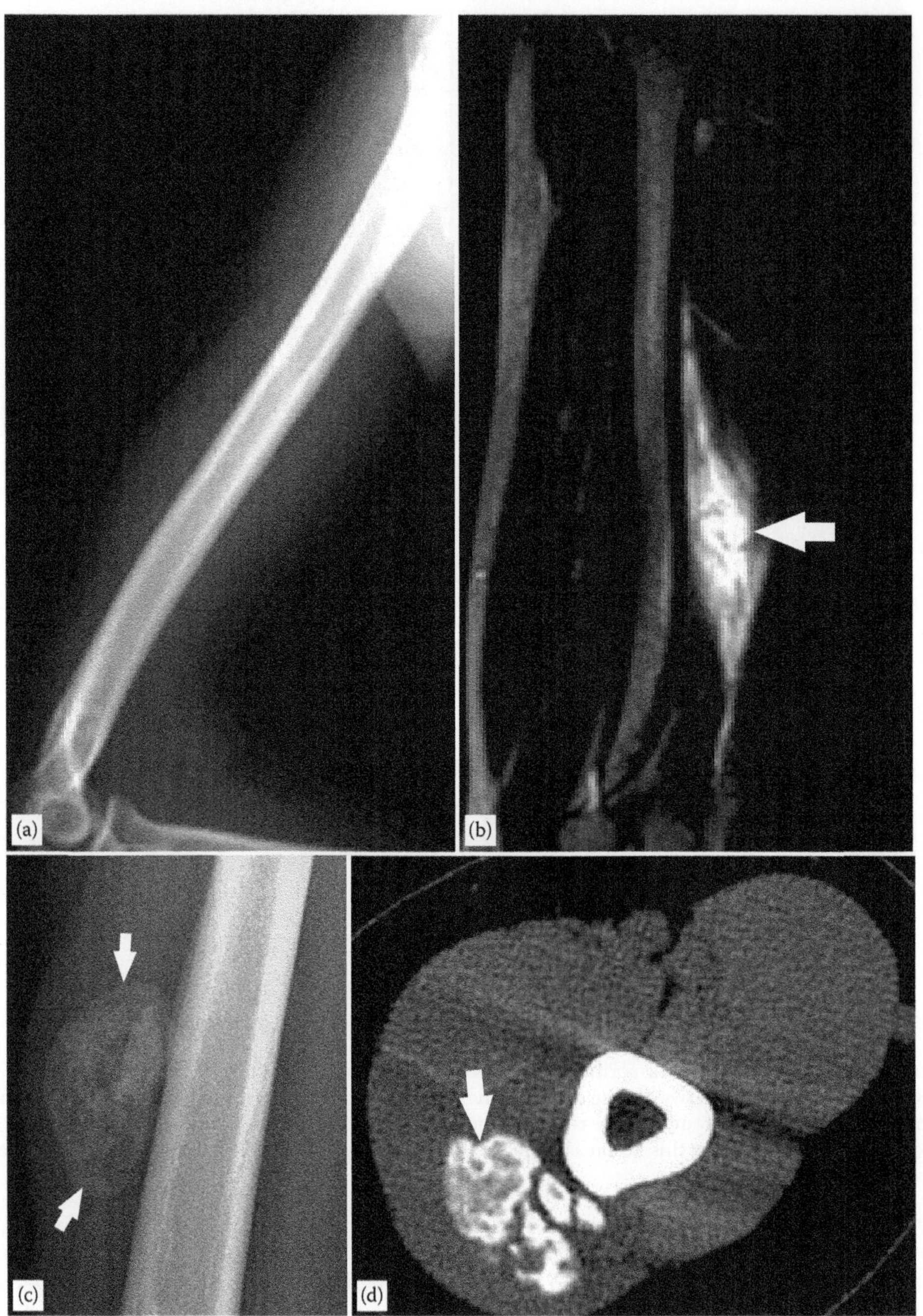


**FIGURE 13.23**
A 29-year-old woman with recent right arm trauma. (a) Initial lateral radiograph demonstrates no focal abnormality within the soft tissue. (b) Coronal $T_2$-weighted MR image demonstrates a nonspecific, ill-defined hyperintensity lesion within the triceps (arrow). (c) Radiograph and (d) axial CT scan performed 3 months later demonstrate ossification within the soft tissue (arrows), consistent with myositis ossificans.

a variable, but predominantly high signal center, and irregular focal areas of intralesional decreased signal intensity [208]. Perilesional edema decreases and a nonspecific pattern of enhancement is seen. A variably thick rim of low signal on all pulse sequences corresponds to the calcified peripheral zone—the classical imaging characteristic of MOC, which is best appreciated on $T_2$-weighted and STIR sequences [207].

*13.4.3.1.2 Myositis Ossificans Progressiva*

Myositis ossificans progressiva (MOP) (synonym: fibrodysplasia ossificans progressiva) is an extremely rare and disabling genetic disorder with a mean age of onset of 3 years. Most affected patients only survive to the fifth or sixth decades. The incidence is 1 in 2 million with >700 cases reported in the literature. [209,210] MOP is characterized by heterotopic soft tissue ossification and skeletal deformities. Patients usually present in childhood with large erythematous and painful tumor like swellings of fibroproliferative tissue involving ligaments, tendons, and skeletal muscle. Initial lesions frequently occur in the neck, and torticollis resulting from a mass in the sternocleidomastoid is the most common presenting complaint. Although some lesions may regress completely, most progress over a period of 2 to 8 months, with a reduction in size, pain, and inflammation of the lesion and the development of a hard mass. These repeated inflammatory flare-ups progressively transform the soft tissues into sheets of heterotopic bone. The calcification progresses cranial to caudal and is first seen in the dorsal, axial, and cranial regions, later progressing to involve the appendicular and distal regions. Erratic remissions and exacerbations are typical, with the most rapid progression seen in the second decade of life [211]. The rate of progression is variable, but most patients are significantly immobile due to loss of joint movement by the third decade. MR findings include well-defined hyperintense soft tissue swellings with surrounding edema are seen acutely on $T_2$-weighted images displacing fascial planes [211]. This $T_2$ hyperintensity decreases with lesion maturation, and areas of low signal appear as the mass calcifies. The spread of lesions along fascial planes is characteristic of MOP and provides a helpful diagnostic feature.

##### 13.4.3.1.3 Neurogenic Myositis Ossificans

First described by Dejereine and Ceillier in paraplegic patients [212], neurogenic myositis ossificans usually occurs in bed-ridden and comatose patients with varied underlying neurological disorders such as spinal cord or head injury, and other protracted diseases of the central nervous system. It is a disabling condition affecting large joints, and characterized by two complications: limitation of joint mobility and neurovascular compromise. The location and incidence depend upon the underlying disease. Shoulders and elbows are more often affected after head injury, whereas hip and knee are more frequently affected in patients with spinal injury [213,214]. After head injury, symptoms usually develop within 2 to 3 months [215]. Although the physiopathology is unclear, many hypotheses have been suggested. One of them suggests that the induction of enchondral ossification may result from repeated tendinous or muscular microtrauma caused by passive mobilization of the paralyzed limb during rehabilitation [216]. Surgical removal of mature heterotopic bone is often required in order to regain functional range of movement and release neurovascular structures. However, surgical treatment is risky because of potential complications including injury to demineralized bones or to compressed or entrapped neurovascular structures within the lesion [217–219]. MRI shows early edema and contrast enhancement in acute or intermediate stage lesions. Findings are similar to those of MOC.

#### 13.4.3.2 Autoimmune Myositis

Inflammatory myopathies can be classified as idiopathic or secondary. Idiopathic inflammatory myopathies encompass a group of heterogeneous muscle diseases that share the clinical features of slowly progressive weakness of skeletal muscles and muscle fatigue. Polymyositis is a rare autoimmune and sometimes paraneoplastic inflammatory myositis. The diagnosis is based on a typical clinical presentation, elevated serum skeletal muscle enzymes, and findings on electromyography and muscle biopsy. MRI accurately documents the extent and intensity of the muscle abnormalities. The inflammation is usually symmetric and classically involves the proximal muscle groups in both polymyositis and dermatomyositis, but muscle involvement can also be patchy and asymmetric (Figure 13.24). High signal intensity is seen in the active phase on STIR and fat-saturated gadolinium-enhanced $T_1$-weighted images. Sometimes inflammation may extend only along or around individual muscles and muscle groups (myofascial distribution) [220]. In the chronic phase, fatty atrophy of the musculature is seen on $T_1$-weighted images. In dermatomyositis, the subcutaneous connective tissue septa and sometimes even the muscle fasciae are also involved. Juvenile dermatomyositis generally takes a more severe clinical course, which is reflected by the extent and intensity of cutaneous, subcutaneous, and muscular signal abnormalities on MRI (Figure 13.25). Inclusion body myositis presents with specific inclusions of amyloid-β protein and is refractory to treatment. Focally increased signal is seen on STIR and fat-saturated gadolinium-enhanced $T_1$-weighted images, predominantly in the anterior thigh compartment [221].

#### 13.4.3.3 Foreign Body Reactions

Foreign body granulomas can appear very nonspecific On MRI. They are typically heterogeneous in composition. On $T_1$-weighted imaging, foreign body granulomas appear hypointense. They may be hypointense to hyperintense on $T_2$-weighted imaging depending on the amount of scarring and fibrosis within the granuloma. A distinct low signal area within a surrounding hyperintense area on the $T_2$-weighted image may represent the foreign body [222].

#### 13.4.3.4 Myonecrosis

Myonecrosis may present as a soft tissue lesion. This condition is associated most commonly with diabetes, but

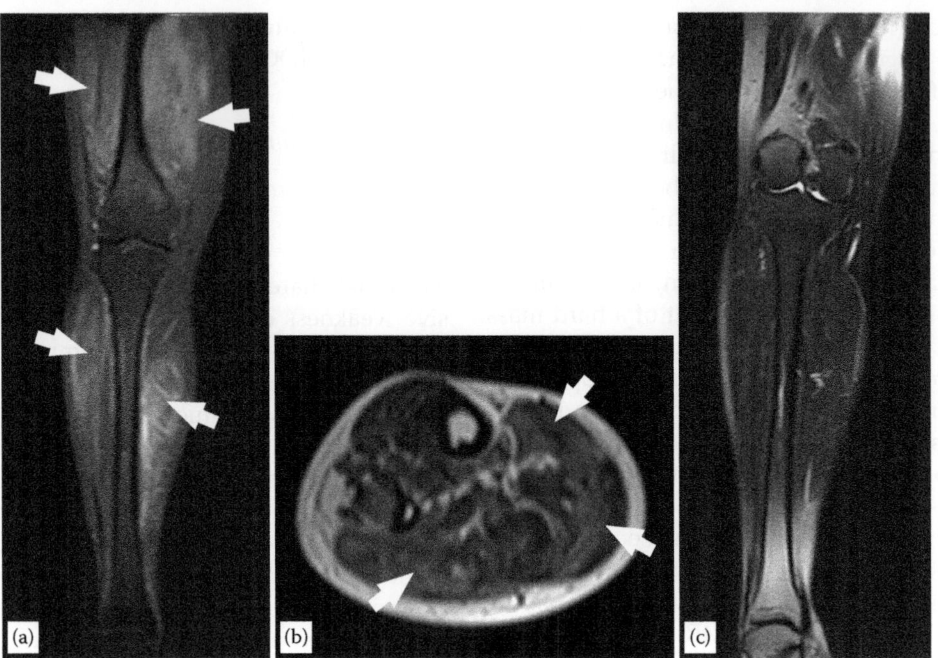

**FIGURE 13.24**
Acute bilateral onset of pain in both lower extremities in a 27-year-old female. (a) Baseline coronal $T_2$-weighted MRI reveals extensive areas of hyperintensity in the thigh and calf regions (arrows). (b) Axial $T_2$-weighted MRI confirms intensive edematous changes in the calf area (arrows) suggesting a diagnosis of nonspecific myositis. (c) Complete resolution after 3 weeks established a diagnosis of acute transient myositis, a rare entity usually observed in children.

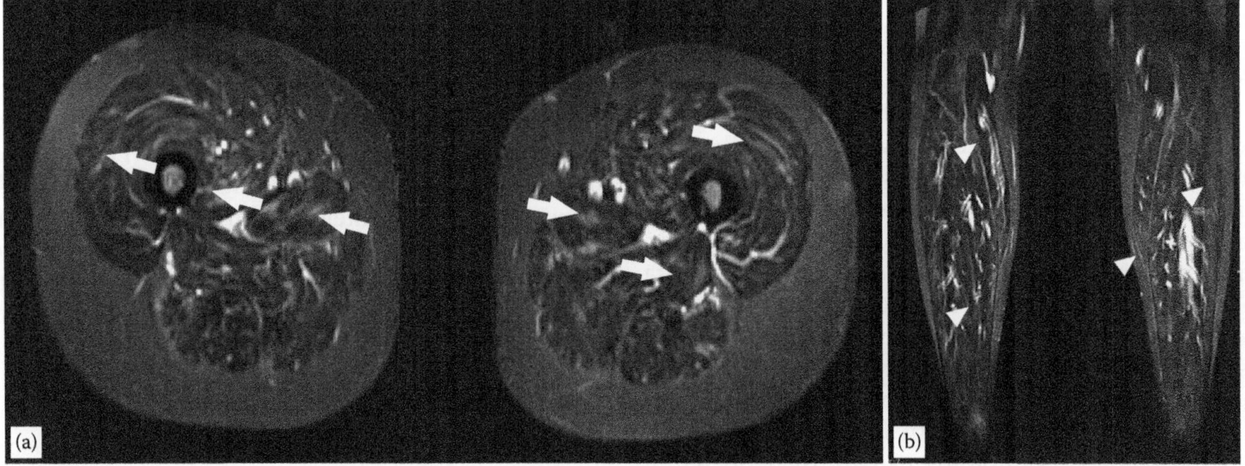

**FIGURE 13.25**
Diffuse muscle pain for 3 weeks in a 52-year-old woman. No known history of a rheumatologic disease at presentation. (a) Axial intermediate-weighted fat-saturated image shows multifocal hyperintensity of speckled appearance in all muscle compartments of both thighs suggesting diffuse myositis (arrows). (b) Coronal intermediate-weighted MR image of the calves showed similar diffuse muscle involvement (arrowheads). The clinical course including typical dermatologic signs and muscle biopsy helped establishing a diagnosis of dermatomyositis.

other predisposing factors include alcohol consumption. Diabetic myonecrosis can even occur as the initial event in a patient's history, although most patients have severe neuropathy or other more severe sequel in the disease process [223,224]. This painful lesion develops quickly with or without a history of trauma. It is thought that the lesion develops secondary to endothelial damage from diabetic microangiopathy, combined with activated coagulation factors and the continued Presence of fibrin degradation products [225]. MRI is the diagnostic modality of choice. Swelling of the muscle occurs and can be visualized best on $T_1$-weighted imaging, although fascial layers and muscle fiber patterns are maintained. $T_2$-weighted imaging shows the lesion more clearly, with diffuse hyperintensity. The overall picture is mixed, with areas of necrosis and muscle regeneration. The lack of invasion of surrounding structures differentiates this lesion from sarcomas. Gadolinium contrast may be used, with enhancing areas corresponding to healing or viable tissue, whereas non-enhancing areas correspond to loci of tissue necrosis [224,226].

### 13.4.4 Infectious Pathologies

Prompt and appropriate imaging work-up of the various musculoskeletal soft tissue infections aids early diagnosis and treatment and decreases the risk of complications resulting from misdiagnosis or delayed diagnosis. The signs and symptoms of musculoskeletal soft tissue infections can be nonspecific, making it clinically difficult to distinguish between disease processes and the extent of disease. MRI is the imaging modality of choice in the evaluation of soft tissue infections [5,227,228]. CT, ultrasound, radiography and nuclear medicine studies are considered ancillary [5].

#### 13.4.4.1 Cellulitis

Cellulitis is a superficial soft tissue infection which involves the skin and subcutaneous tissues. It is clinically characterized by soft tissue swelling, erythema, and warmth, usually resulting from infection with gram positive cocci [229,230]. Diagnosis is usually made clinically, but MRI may be ordered to exclude the presence of complications, including more complex or deep infection [231]. MRI of cellulitis shows thickening of the subcutaneous tissues, diffuse linear or ill-defined increased signal intensity of superficial soft tissue on fluid-sensitive sequences, and corresponding low signal intensity on $T_1$-weighted sequences consistent with edema. Imaging after intravenous administration of contrast agent may show diffuse enhancement of the same areas, consistent with inflammation, although the extent and degree of enhancement is variable [232–234]. No focal fluid collections are present in uncomplicated cellulitis, and no abnormal signal intensity should be seen in the underlying muscle [229].

#### 13.4.4.2 Abscess

When a localized infection progresses, an abscess may form in the subcutaneous or deeper soft tissues [235]. A well-demarcated fluid collection differentiates this process from third-spacing of fluid, fasciitis, or focal cellulitis, but clinically their appearance may be similar. An abscess, however, requires drainage in addition to appropriate antibiotic treatment. A soft tissue abscess is characterized by a localized fluid collection with intermediate to low signal intensity on $T_1$-weighted images and bright signal intensity on fluid sensitive sequences. The abscess wall is often irregular and thick and enhances peripherally after intravenous administration of gadolinium contrast material and centrally, it lacks enhancement [236]. The presence of gas within the cystic mass is a feature that is nearly diagnostic of an abscess (Figure 13.26).

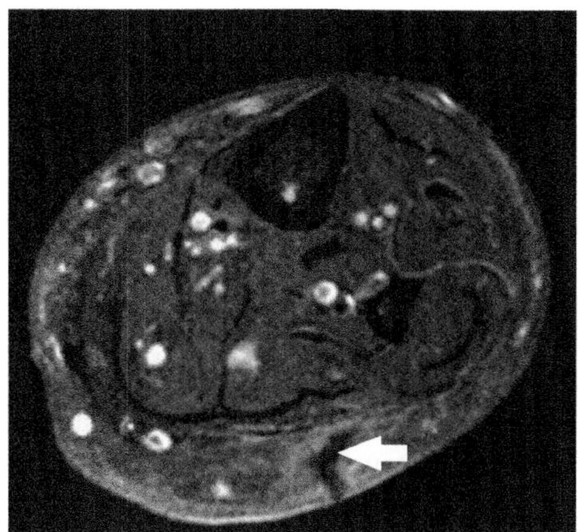

**FIGURE 13.26**
A 74-year-old woman with calf swelling and skin ulceration. Axial fat-suppressed contrast-enhanced $T_1$-weighted MR image shows marked subcutaneous enhancement with central hypointense lesion that extends to skin surface consistent with small abscess and sinus tract (arrow).

#### 13.4.4.3 Necrotizing Fasciitis

Necrotizing fasciitis is a life-threatening soft tissue infection of bacterial origin, which involves mainly the deep fascia. Early recognition of this condition may be hampered by the uncommon nature of the disease and nonspecificity of initial clinical signs and symptoms in less fulminant cases, making the role of imaging important. MRI is the most useful imaging modality in the diagnosis of necrotizing fasciitis. Presence of thick (>3 mm) hyperintense signal in the deep fascia (particularly intermuscular fascia) on fat-suppressed $T_2$-weighted or STIR images is an important marker for necrotizing fasciitis. Contrast enhancement of the thickened necrotic fascia can be variable, with a mixed-pattern of enhancement being more commonly encountered. Involvement of multiple musculofascial compartments increases the likelihood of necrotizing fasciitis. It is important to remember that $T_2$-hyperintense signal in the deep fascia is not specific to necrotizing fasciitis and can also be seen in cases such as noninfective inflammatory fasciitis or muscle tear [237].

#### 13.4.4.4 Infectious Bursitis

Bursitis usually comprises a sterile inflammation of the bursa. When an infectious agent is involved it is usually *S. aureus* [238,239], but other organisms, fungi, tuberculosis or atypical mycobacteria may be encountered. Any bursa can be involved, but the olecranon and prepatellar bursa are most commonly affected, possibly due to their superficial location and susceptibility to direct stress and trauma. The subacromial-subdeltoid, iliopsoas, infrapatellar, and gastrocnemius-semimembranosus bursa may also become involved. There may be communication with the adjacent joint, resulting in or arising from septic arthritis [233]. MRI is very sensitive for detection of bursal fluid and variable amounts of peribursal edema and fluid which have hyperintense signal on fluid-sensitive sequences with corresponding $T_1$ hypointensity. Air bubbles can be seen as punctate signal voids. There may be thickening of the overlying skin. There is usually enhancement of the bursal wall and surrounding tissues following administration of gadolinium-based intravenously administered contrast medium. The infected bursal contents typically do not enhance [239,240].

#### 13.4.4.5 Pyomyositis

A pyogenic infection of the muscle is termed pyomyositis. Over three-fourths of cases are caused by the organism *S. aureus* [241]. Today, the increased prevalence of HIV infection accounts for a large number of cases of bacterial myositis [242]. Other risk factors include rhabdomyolysis and muscle trauma with a hematoma serving as the initial source of infection, diabetes, connective tissue disorders, and conditions that produce an immune deficiency. Clinically, a patient with pyomyositis may depict three stages [241]. Initially, edema leads to localized pain and leukocytosis predominates although there may be erythema as well. As the disease progresses, the patient develops a fever as the myositis evolves to a suppurative stage with abscess formation. If not treated, sepsis and life-threatening toxicity can lead to death. Generally, one muscle is involved but multiple sites can be affected in up to 40% of patients [241]. On MR imaging, the affected muscle may appear enlarged with loss of the architectural definition with heterogeneous areas of low signal intensity on $T_1$-weighted images [243]. In the absence of an abscess, edema may be the only finding manifested by corresponding high signal intensity on fluid-sensitive sequences. On $T_2$-weighted images, disorganized phlegmonous collections herald the development of abscess formation. The abnormal signal intensity within the muscle may be poorly marginated and heterogeneous enhancement may be seen as suppuration develops. On $T_1$-weighted images, an intramuscular abscess may show variable signal intensity depending on the proteinaceous content of the fluid and low signal intensity periphery on both $T_1$- and $T_2$-weighed images [244]. After intravenous administration of a gadolinium contrast agent (Figure 13.27), one should see characteristic rim enhancement and central liquefaction, although when chronic, it may contain areas of intermediate signal intensity surrounded by a low signal intensity rim of fibrous tissue.

## 13.5 Conclusion

MRI is the study of choice for evaluation of soft tissue tumors and tumor-like conditions because of its superior soft tissue contrast, multiplanar imaging capability, and lack of radiation exposure. MRI is valuable for lesion detection, diagnosis, and staging. The signal characteristics and the location of the soft tissue mass

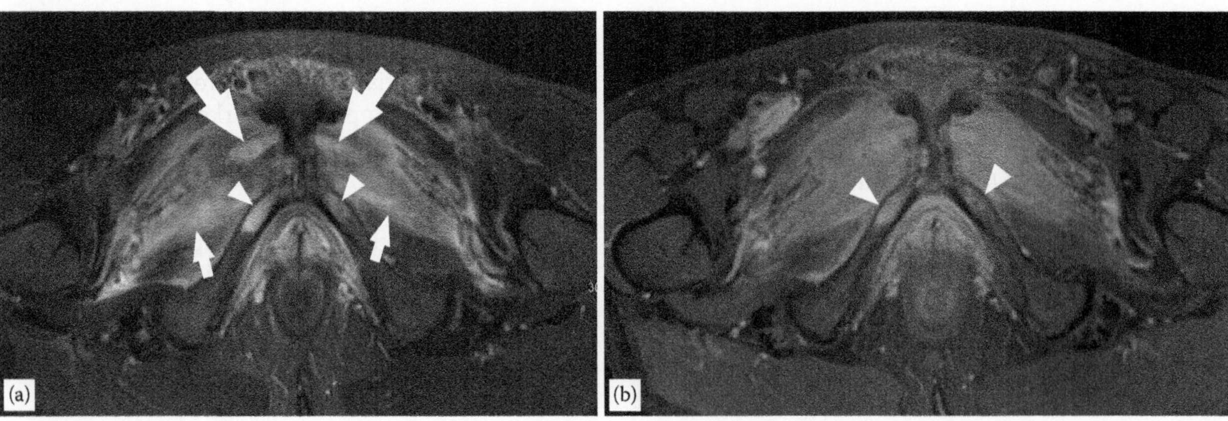


**FIGURE 13.27**
Septic myositis in a 68-year-old man. (a) Axial fat-suppressed $T_2$-weighted MR image shows diffuse hyperintensity of the adductor muscles on both sides (small arrows). In addition, there are hyperintensity alterations within the pubic bones (arrowheads). Note intramuscular perisymphyseal fluid equivalent signal bilaterally consistent with abscesses (large arrows). (b) Axial fat-suppressed contrast-enhanced $T_1$-weighted MR image depicts diffuse avid enhancement of the adductor muscles and adjacent pubic bone marrow (arrowheads).

will help potentially identify the mass. Still at best, MRI offers only suggestive diagnoses due to the nonspecific appearance of a majority of soft tissue lesions. MRI helps in defining the tumor margins, location and involvement of surrounding structures and is best for assisting in developing a surgical plan for excision or biopsy of a soft tissue mass. Gadolinium enhancement may also improve the ability of an MRI in differentiating a malignant lesion from other benign cystic structures.

## References

1. Hanna SL, Fletcher BD. MR imaging of malignant soft-tissue tumors. *Magn Reson Imaging Clin N Am.* 1995;3(4):629–50.
2. Woertler K. Soft tissue masses in the foot and ankle: Characteristics on MR imaging. *Semin Musculoskelet Radiol.* 2005;9(3):227–42, Thieme.
3. Van Rijswijk CS, Geirnaerdt MJ, Hogendoorn PC, Taminiau AH, van Coevorden F, Zwinderman AH et al. Soft-tissue tumors: Value of static and dynamic gadopentetate Dimeglumine-enhanced MR imaging in prediction of Malignancy. *Radiology.* 2004;233(2):493–02.
4. Wu JS, Hochman MG. Soft-tissue tumors and tumor-like lesions: A systematic imaging approach. *Radiology.* 2009;253(2):297–16.
5. Turecki MB, Taljanovic MS, Stubbs AY, Graham AR, Holden DA, Hunter TB et al. Imaging of musculoskeletal soft tissue infections. *Skeletal Radiol.* 2010;39(10):957–71.
6. Chihara S, Segreti J. Osteomyelitis. *Dis Month.* 2010;56(1):6–31.
7. De Schepper AM. Grading and characterization of soft tissue tumors. In: De Schepper AM, Parizel PM, De Beukeleer L, Vanhoenacker F (eds) *Imaging of Soft Tissue Tumors*, 2nd edn. Springer, New York; 2001: pp. 123–41.
8. Mirowitz SA. Fast scanning and fat-suppression MR imaging of musculoskeletal disorders. *AJR Am J Roentgenol.* 1993;161(6):1147–57.
9. Soulié D, Boyer B, Lescop J, Pujol A, Le Friant G, Cordoliani Y. Liposarcome myxoïde: Aspects en IRM. *J Radiol.* 1995;76(1):29–36.
10. Galant J, Marti-Bonmati L, Saez F, Soler R, Alcala-Santaella R, Navarro M. The value of fat-suppressed $T_2$ or STIR sequences in distinguishing lipoma from well-differentiated liposarcoma. *Eur Radiol.* 2003;13(2):337–43.
11. Beltran J, Chandnani V, McGhee R, Kursunoglu-Brahme S. Gadopentetate dimeglumine-enhanced MR imaging of the musculoskeletal system. *AJR Am J Roentgenol.* 1991;156(3):457–66.
12. Verstraete KL, De Deene Y, Roels H, Dierick A, Uyttendaele D, Kunnen M. Benign and malignant musculoskeletal lesions. Dynamic contrast-enhanced MR imaging—Parametric first-pass images depict tissue vascularization and perfusion. *Radiology.* 1994;192(3):835–43.
13. Benedikt RA, Jelinek JS, Kransdorf MJ, Moser RP, Berrey BH. MR imaging of soft-tissue masses: Role of gadopentetate dimeglumine. *J Magn Reson Imaging.* 1994;4(3):485–90.
14. Erlemann R, Reiser M, Peters P, Vasallo P, Nommensen B, Kusnierz-Glaz C et al. Musculoskeletal neoplasms: Static and dynamic gd-DTPA—Enhanced MR imaging. *Radiology.* 1989;171(3):767–73.
15. Mirowitz SA, Totty WG, Lee JK. Characterization of musculoskeletal masses using dynamic gd-DTPA enhanced spin-echo MRI. *J Comput Assist Tomogr.* 1992;16(1):120–5.
16. Verstraete K, Vanzieleghem B, De Deene Y, Palmans H, De Greef D, Kristoffersen D et al. Static, dynamic and first-pass MR imaging of musculoskeletal lesions using gadodiamide injection. *Acta Radiol.* 1995;36(1):27–36.
17. Gielen J, De Schepper A, Parizel P, Wang X, Vanhoenacker F. Additional value of magnetic resonance with spin echo $T_1$-weighted imaging with fat suppression in characterization of soft tissue tumors. *J Comput Assist Tomogr.* 2003;27(3):434–41.
18. Helms CA. The use of fat suppression in gadolinium-enhanced MR imaging of the musculoskeletal system: A potential source of error. *AJR Am J Roentgenol.* 1999;173(1):234–6.
19. Verstraete K, Dierick A, De Deene Y, Uyttendaele D, Vandamme F, Roels H et al. First-pass images of musculoskeletal lesions: A new and useful diagnostic application of dynamic contrast-enhanced MRI. *Magn Reson Imaging.* 1994;12(5):687–702.
20. Verstraete KL, Woude HVd, Hogendoorn PC, De Deene Y, Kunnen M, Bloem JL. Dynamic contrast-enhanced MR imaging of musculoskeletal tumors: Basic principles and clinical applications. *J Magn Reson Imaging.* 1996;6(2):311–21.
21. Shapeero LG, Vanel D, Verstraete KL, Bloem JL. Fast magnetic resonance imaging with contrast for soft tissue sarcoma viability. *Clin Orthop.* 2002;397:212–27.
22. Verstraete K, Lang P. Bone and soft tissue tumors: The role of contrast agents for MR imaging. *Eur J Radiol.* 2000;34(3):229–46.
23. Verstraete KL et al. Dynamic contrast enhanced MRI of musculoskeletal neoplasms: Different types and slopes of TICs (abstract). *Proceedings of Society of Magnetic Resonance in Medicine.* Berkeley, CA; 1992: p. 2609.
24. Van Der Woude H, Verstraete K, Taminiau A, Hogendoorn P, Vanzieleghem B, Bloem J. Double slice dynamic contrast-enhanced subtraction MR imaging in 60 patients with musculoskeletal tumors or tumor-like lesions. *Eur Radiol.* 1995;5:181.
25. Einarsdóttir H, Karlsson M, Wejde J, Bauer HC. Diffusion-weighted MRI of soft tissue tumours. *Eur Radiol.* 2004;14(6):959–63.
26. Van Rijswijk CS, Kunz P, Hogendoorn PC, Taminiau AH, Doornbos J, Bloem JL. Diffusion-weighted MRI in the characterization of soft-tissue tumors. *J Magn Reson Imaging.* 2002;15(3):302 7.
27. Walker EA, Fenton ME, Salesky JS, Murphey MD. Magnetic resonance imaging of benign soft tissue neoplasms in adults. *Radiol Clin N Am.* 2011;49(6):1197–217.

28. Kijowski R, Gold GE. Routine 3D magnetic resonance imaging of joints. *J Magn Reson Imaging.* 2011;33(4):758–71.
29. Myhre-Jensen O. A consecutive 7-year series of 1331 benign soft tissue tumours: Clinicopathologic data. Comparison with sarcomas. *Acta Orthopaedica.* 1981;52(3):287–93.
30. Ronan SJ, Broderick T. Minimally invasive approach to familial multiple lipomatosis. *Plast Reconstr Surg.* 2000;106(4):878–80.
31. Rydholm A, Berg NO. Size, site and clinical incidence of lipoma: Factors in the differential diagnosis of lipoma and sarcoma. *Acta Orthopaedica.* 1983;54(6):929–34.
32. Kransdorf M, Moser R, Meis J, Meyer C. Fat-containing soft-tissue masses of the extremities. *RadioGraphics.* 1991;11(1):81–106.
33. Leffert RD. Lipomas of the upper extremity. *J Bone & Joint Surg.* 1972;54(6):1262–6.
34. Murphey MD, Carroll JF, Flemming DJ, Pope TL, Gannon FH, Kransdorf MJ. From the archives of the AFIP benign musculoskeletal lipomatous Lesions1. *RadioGraphics.* 2004;24(5):1433–66.
35. Gaskin CM, Helms CA. Lipomas, lipoma variants, and well-differentiated liposarcomas (atypical lipomas): Results of MRI evaluations of 126 consecutive fatty masses. *AJR Am J Roentgenol.* 2004;182(3):733–9.
36. Kransdorf MJ, Murphey MD. *Imaging of Soft Tissue Tumors.* Saunders, Philadelphia, PA; 1997: pp. 3–36, 57–102.
37. Hosono M, Kobayashi H, Fujimoto R, Kotoura Y, Tsuboyama T, Matsusue Y et al. Septum-like structures in lipoma and liposarcoma: MR imaging and pathologic correlation. *Skeletal Radiol.* 1997;26(3):150–4.
38. London J, Kim EE, Wallace S, Shirkhoda A, Coan J, Evans H. MR imaging of liposarcomas: Correlation of MR features and histology. *J Comput Assist Tomogr.* 1989;13(5):832–5.
39. Sung MS, Kang HS, Suh JS, Lee JH, Park JM, Kim JY et al. Myxoid liposarcoma: Appearance at MR imaging with histologic correlation. *RadioGraphics.* 2000;20(4):1007–19.
40. Kransdorf MJ, Bancroft LW, Peterson JJ, Murphey MD, Foster WC, Temple HT. Imaging of fatty tumors: Distinction of lipoma and well-differentiated liposarcoma 1. *Radiology.* 2002;224(1):99–104.
41. Tateishi U, Hasegawa T, Beppu Y, Kawai A, Moriyama N. Prognostic significance of grading (MIB-1 system) in patients with myxoid liposarcoma. *J Clin Pathol.* 2003;56(8):579–82.
42. Vanhoenacker F, Marques M, Garcia H. Lipomatous tumors. In: De Schepper AM, Vanhoenacker FM, Gielen J, Parizel PM (eds) *Imaging of Soft Tissue Tumors.* Springer, New York; 2006: p. 227–61.
43. Cotten A. Imaging of lipoma and liposarcoma. *JBR-BTR.* 2002;85(1):14–9.
44. Nishida J, Morita T, Ogose A, Okada K, Kakizaki H, Tajino T et al. Imaging characteristics of deep-seated lipomatous tumors: Intramuscular lipoma, intermuscular lipoma, and lipoma-like liposarcoma. *J Orthopaedic Sci.* 2007;12(6):533–41.
45. Weiss SW, Goldblum JR, Enzinger FM. Benign fibroblastic/myofibroblastic proliferations. In: Weiss SW, Goldblum JR (eds) *Enzinger and Weiss' Soft Tissue Tumors*, 5th edn. Mosby Elsevier, Philadelphia, PA; 2008: p. 175–225.
46. Meister P, Bückmann F, Konrad E. Nodular fasciitis (analysis of 100 cases and review of the literature). *Pathol Res Pract.* 1978;162(2):133–65.
47. Dinauer PA, Brixey CJ, Moncur JT, Fanburg-Smith JC, Murphey MD. Pathologic and MR imaging features of benign fibrous soft-tissue tumors in Adults. *RadioGraphics.* 2007;27(1):173–87.
48. Bernstein KE, Lattes R. Nodular (pseudosarcomatous) fasciitis, a nonrecurrent lesion: Clinicopathologic study of 134 cases. *Cancer.* 1982;49(8):1668–78.
49. Leung L, Shu S, Chan A, Chan M, Chan C. Nodular fasciitis: MRI appearance and literature review. *Skeletal Radiol.* 2002;31(1):9–13.
50. Shimizu S, Hashimoto H, Enjoji M. Nodular fasciitis: An analysis of 250 patients. *Pathology.* 1984;16(2):161–6.
51. Wang X, De Schepper A, Vanhoenacker F, De Raeve H, Gielen J, Aparisi F et al. Nodular fasciitis: Correlation of MRI findings and histopathology. *Skeletal Radiol.* 2002;31(3):155–61.
52. Broder MS, Leonidas JC, Mitty HA. Pseudosarcomatous fasciitis: An unusual cause of soft-tissue calcification. *Radiology.* 1973;107(1):173–4.
53. Meyer CA, Kransdorf MJ, Jelinek JS, Moser Jr RP. MR and CT appearance of nodular fasciitis. *J Comput Assist Tomogr.* 1991;15(2):276–9.
54. Pulitzer DR, Martin PC, Reed RJ. Fibroma of tendon sheath: A clinicopathologic study of 32 cases. *Am J Surg Pathol.* 1989;13(6):472–9.
55. Chung E, Enzinger FM. Fibroma of tendon sheath. *Cancer.* 1979;44(5):1945–54.
56. Fox MG, Kransdorf MJ, Bancroft LW, Peterson JJ, Flemming DJ. MR imaging of fibroma of the tendon sheath. *AJR Am J Roentgenol.* 2003;180(5):1449–53.
57. Laffan EE, Ngan B, Navarro OM. Pediatric soft-tissue tumors and pseudotumors: MR imaging features with pathologic correlation part 2. Tumors of fibroblastic/myofibroblastic, so-called fibrohistiocytic, muscular, lymphomatous, neurogenic, hair matrix, and uncertain origin. *RadioGraphics.* 2009;29(4):e36.
58. Hisaoka M, Hashimoto H. Elastofibroma: Clonal fibrous proliferation with predominant CD34-positive cells. *Virchows Archiv.* 2006;448(2):195–9.
59. Nagamine N, Nohara Y, Ito E. Elastofibroma in Okinawa. A clinicopathologic study of 170 cases. *Cancer.* 1982;50(9):1794–805.
60. Brandser EA, Goree JC, El-Khoury GY. Elastofibroma dorsi: Prevalence in an elderly patient population as revealed by CT. *AJR Am J Roentgenol.* 1998;171(4):977–80.
61. Marin ML, Perzin KH, Markowitz AM. Elastofibroma dorsi: Benign chest wall tumor. *J Thorac Cardiovasc Surg.* 1989;98(2):234–8.
62. Järvi OH, Länsimies PH. Subclinical elastofibromas in the scapular region in an autopsy series. *Acta Pathol Microbiol Scandinavica Sec A Pathol.* 1975;83(1):87–108.
63. Kransdorf MJ, Meis JM, Montgomery E. Elastofibroma: MR and CT appearance with radiologic-pathologic correlation. *AJR Am J Roentgenol.* 1992;159(3):575–9.
64. Massengill AD, Sundaram M, Kathol MH, El-Khoury GY, Buckwalter JH, Wade TP. Elastofibroma dorsi: A radiological diagnosis. *Skeletal Radiol.* 1993;22(2):121–3.

65. Naylor MF, Nascimento AG, Sherrick AD, McLeod RA. Elastofibroma dorsi: Radiologic findings in 12 patients. *AJR Am J Roentgenol*. 1996;167(3):683–7.
66. Joseph SY, Weis LD, Vaughan LM, Resnick D. MRI of elastofibroma dorsi. *J Comput Assist Tomogr*. 1995;19(4):601–3.
67. Mikkelsen OA. Dupuytren's disease—Initial symptoms, age of onset and spontaneous course. *Hand*. 1977;9(1):11–5.
68. Yacoe ME, Bergman AG, Ladd AL, Hellman BH. Dupuytren's contracture: MR imaging findings and correlation between MR signal intensity and cellularity of lesions. *AJR Am J Roentgenol*. 1993;160(4):813–7.
69. Murphey MD, Ruble CM, Tyszko SM, Zbojniewicz AM, Potter BK, Miettinen M. Musculoskeletal fibromatoses: Radiologic-pathologic Correlation. *RadioGraphics*. 2009;29(7):2143–83.
70. Robbin MR, Murphey MD, Temple HT, Kransdorf MJ, Choi JJ. Imaging of musculoskeletal Fibromatosis. *RadioGraphics*. 2001;21(3):585–600.
71. Rombouts J, Noël H, Legrain Y, Munting E. Prediction of recurrence in the treatment of Dupuytren's disease: Evaluation of a histologic classification. *J Hand Surg*. 1989;14(4):644–52.
72. De Bree E, Zoetmulder FA, Keus RB, Peterse HL, van Coevorden F. Incidence and treatment of recurrent plantar fibromatosis by surgery and postoperative radiotherapy. *Am J Surg*. 2004;187(1):33–8.
73. Fetsch JF, Laskin WB, Miettinen M. Palmar-plantar fibromatosis in children and preadolescents: A clinicopathologic study of 56 cases with newly recognized demographics and extended follow-up information. *Am J Surg Pathol*. 2005;29(8):1095–105.
74. Lee T, Wapner K, Hecht P. Current concepts review: Plantar fibromatosis. *J Bone Joint Surg*. American volume. 1993;75(7):1080–4.
75. Aviles E, Arlen M, Miller T. Plantar fibromatosis. *Plast Reconstr Surg*. 1971;48(3):295.
76. Morrison WB, Schweitzer ME, Wapner KL, Lackman RD. Plantar fibromatosis: A benign aggressive neoplasm with a characteristic appearance on MR images. *Radiology*. 1994;193(3):841–5.
77. Spiegel DA, Dormans JP, Meyer JS, Himelstein B, Mathur S, Asada N et al. Aggressive fibromatosis from infancy to adolescence. *J Pediatr Orthopaedics*. 1999;19(6):776.
78. Coffin CM, Dehner LP. Fibroblastic-myofibroblastic tumors in children and adolescents: A clinicopathologic study of 108 examples in 103 patients. *Fetal & Pediatr Pathol*. 1991;11(4):569–88.
79. Faulkner LB, Hajdu SI, Kher U, La Quaglia M, Exelby PR, Heller G et al. Pediatric desmoid tumor: Retrospective analysis of 63 cases. *J Clin Oncol*. 1995;13(11):2813–8.
80. Pignatti G, Barbanti-Brodano G, Ferrari D, Gherlinzoni F, Bertoni F, Bacchini P et al. Extraabdominal desmoid tumor: A study of 83 cases. *Clin Orthop*. 2000;375:207–13.
81. Liu Q, Chen J, Liang B, Li H, Gao M, Lin X. Imaging manifestations and pathologic features of soft tissue desmoid-type fibromatosis. *Chin J Cancer*. 2008;27:535–40.
82. Liu P, Thorner P. MRI of fibromatosis: With pathologic correlation. *Pediatr Radiol*. 1992;22(8):587–9.
83. Pousti TJ, Upton J, Loh M, Grier H. Congenital fibrosarcoma of the upper extremity. *Plast Reconstr Surg*. 1998;102(4):1158–62.
84. Muzaffar AR, Friedrich JB, Lu KK, Hanel DP. Infantile fibrosarcoma of the hand associated with coagulopathy. *Plast Reconstr Surg*. 2006;117(5):81e–6e.
85. Eich G, Hoeffel J, Tschäppeler H, Gassner I, Willi UV. Fibrous tumours in children: Imaging features of a heterogeneous group of disorders. *Pediatr Radiol*. 1998;28(7):500–9.
86. Lee MJ, Cairns RA, Munk PL, Poon PY. Congenital-infantile fibrosarcoma: Magnetic resonance imaging findings. *Can Assoc Radiol J* (Journal l'Association canadienne des radiologistes). 1996;47(2):121–5.
87. Boon LM, Fishman SJ, Lund DP, Mulliken JB. Congenital fibrosarcoma masquerading as congenital hemangioma: Report of two cases. *J Pediatr Surg*. 1995;30(9):1378–81.
88. Vinnicombe S, Hall C. Infantile fibrosarcoma: Radiological and clinical features. *Skeletal Radiol*. 1994;23(5):337–41.
89. Cecchetto G, Carli M, Alaggio R, Dall'Igna P, Bisogno G, Scarzello G et al. Fibrosarcoma in pediatric patients: Results of the Italian cooperative group studies (1979–1995). *J Surg Oncol*. 2001;78(4):225–31.
90. Mendenhall WM, Zlotecki RA, Scarborough MT. Dermatofibrosarcoma protuberans. *Cancer*. 2004;101(11):2503–8.
91. Riggs K, McGUIGAN KL, Morrison WB, Samie FH, Humphreys T. Role of magnetic resonance imaging in perioperative assessment of dermatofibrosarcoma protuberans. *Dermatol Surg*. 2009;35(12):2036–41.
92. De Schepper A, De Beuckeleer L, Vandevenne J, Somville J. Magnetic resonance imaging of soft tissue tumors. *Eur Radiol*. 2000;10(2):213–23.
93. Bravo SM, Winalski CS, Weissman BN. Pigmented villonodular synovitis. *Radiol Clin N Am*. 1996;34(2):311–26, x–xi.
94. Llauger J, Palmer J, Roson N, Cremades R, Bague S. Pigmented villonodular synovitis and giant cell tumors of the tendon sheath: Radiologic and pathologic features. *AJR Am J Roentgenol*. 1999;172(4):1087–91.
95. Navarro OM, Laffan EE, Ngan B. Pediatric soft-tissue tumors and pseudo-tumors: MR imaging features with pathologic correlation part 1. Imaging approach, pseudotumors, vascular lesions, and adipocytic tumors. *RadioGraphics*. 2009;29(3):887–906.
96. Dubois J, Alison M. Vascular anomalies: What a radiologist needs to know. *Pediatr Radiol*. 2010;40(6):895–905.
97. El-Merhi F, Garg D, Cura M, Ghaith O. Peripheral vascular tumors and vascular malformations: Imaging (magnetic resonance imaging and conventional angiography), pathologic correlation and treatment options. *Int J Cardiovasc Imaging*. 2013;29(2):379–93.
98. Moukaddam H, Pollak J, Haims AH. MRI characteristics and classification of peripheral vascular malformations and tumors. *Skeletal Radiol*. 2009;38(6):535–47.
99. Greenspan A, McGahan JP, Vogelsang P, Szabo RM. Imaging strategies in the evaluation of soft-tissue hemangiomas of the extremities: Correlation of the findings of plain radiography, angiography, CT, MRI, and ultrasonography in 12 histologically proven cases. *Skeletal Radiol*. 1992;21(1):11–8.

100. Baek HJ, Lee SJ, Cho KH, Choo HJ, Lee SM, Lee YH et al. Subungual tumors: Clinicopathologic correlation with US and MR imaging Findings. *RadioGraphics.* 2010;30(6):1621–36.
101. Shugart R, Soule E, Johnson Jr E. Glomus tumor. *Surg Gynecol Obstet.* 1963;117:334.
102. Sawada S, Honda M, Kamide R, Niimura M. Three cases of subungual glomus tumors with von recklinghausen neurofibromatosis. *J Am Acad Dermatol.* 1995;32(2):277–8.
103. Okada O, Demitsu T, Manabe M, Yoneda K. A case of multiple subungual glomus tumors associated with neurofibromatosis type 1. *J Dermatol.* 1999;26(8):535.
104. Park E, Hong SH, Choi J, Lee MW, Kang H. Glomangiomatosis: Magnetic resonance imaging findings in three cases. *Skeletal Radiol.* 2005;34(2):108–11.
105. Tachibana R, Hatori M, Hosaka M, Yamada N, Watanabe M, Moriya T et al. Glomus tumors with cystic changes around the ankle. *Arch Orthop Trauma Surg.* 2001;121(9):540–3.
106. Drape J, Idy-Peretti I, Goettmann S, Wolfram-Gabel R, Dion E, Grossin M et al. Subungual glomus tumors: Evaluation with MR imaging. *Radiology.* 1995;195(2):507–15.
107. Park SI, Choi E, Lee HB, Rhee YK, Chung MJ, Lee YC. Spontaneous pneumomediastinum and hemopneumothoraces secondary to cystic lung metastasis. *Respiration.* 2003;70(2):211–3.
108. Rosai J, Sumner HW, Kostianovsky M, Perez-Mesa C. Angiosarcoma of the skin: A clinicopathologic and fine structural study. *Hum Pathol.* 1976;7(1):83–109.
109. Mark RJ, Tran LM, Sercarz J, Fu YS, Calcaterra TC, Juillard GF. Angiosarcoma of the head and neck: The UCLA experience 1955 through 1990. *Arch Otolaryngol Head & Neck Surg.* 1993;119(9):973.
110. Holden CA, Spittle MF, Jones EW. Angiosarcoma of the face and scalp, prognosis and treatment. *Cancer.* 1987;59(5):1046–57.
111. Choi JJ, Murphey MD. Angiomatous skeletal lesions. *Semin Musculoskeletal Radiol.* 2000;4(1):103–12, Thieme.
112. Jie X, Ruo-Fan M, Deng L, Liang-Ping L, Zhi-Qing C, Wen-Wu D et al. Epithelioid angiosarcoma of bone: A neoplasm with potential pitfalls in diagnosis. *Open J Orthopedics.* 2012;2:80–4.
113. Murphey MD, Fairbairn KJ, Parman LM, Baxter KG, Parsa MB, Smith WS. From the archives of the AFIP. Musculoskeletal angiomatous lesions: Radiologic-pathologic correlation. *RadioGraphics.* 1995;15(4):893–917.
114. Brady P, Spence L. Chronic lower extremity deep vein thrombosis associated with femoral vein compression by a lipoma. *AJR Am J Roentgenol.* 1999;172(6):1697–8.
115. Dei Tos AP, Mentzel T, Newman PL, Fletcher CD. Spindle cell liposarcoma, a hitherto unrecognized variant of liposarcoma analysis of six cases. *Am J Surg Pathol.* 1994;18(9):913–21.
116. Yamato M, Nishimura G, Koguchi Y, Saotome K. Calcified leiomyoma of deep soft tissue in a child. *Pediatr Radiol.* 1999;29(2):135–7.
117. Seynaeve PC, De Visschere PJL, Mortelmans LL, De Schepper AM. Tumors of muscular origin. In: De Schepper AM, Vanhoenacker F, Gielen J, Parizel PM (eds) *Imaging of Soft Tissue Tumors,* 3rd edn. Springer-Verlag, Berlin, Germany; 2006: pp. 293–310.
118. Rademaker J. Hodgkin's and non-hodgkin's lymphomas. *Radiol Clin N Am.* 2007;45(1):69–83.
119. Hampson F, Shaw A. Response assessment in lymphoma. *Clin Radiol.* 2008;63(2):125–35.
120. Suresh S, Saifuddin A, O'Donnell P. Lymphoma presenting as a musculoskeletal soft tissue mass: MRI findings in 24 cases. *Eur Radiol.* 2008;18(11):2628–34.
121. Beggs I. Primary muscle lymphoma. *Clin Radiol.* 1997;52(3):203–12.
122. Hwang S. Imaging of lymphoma of the musculoskeletal system. *Radiol Clin N Am.* 2008;46(2):379–96.
123. Guermazi A, Ooi CG. Extranodal hodgkin disease. In: *Radiological Imaging in Hematological Malignancies.* Springer, New York; 2004: p. 48–71.
124. Karasick D, Karasick S. Giant cell tumor of tendon sheath: Spectrum of radiologic findings. *Skeletal Radiol.* 1992;21(4):219–24.
125. Sherry CS, Harms SE. MR evaluation of giant cell tumors of the tendon sheath. *Magn Reson Imaging.* 1989;7(2):195–201.
126. Wan J, Magarelli N, Peh W, Guglielmi G, Shek T. Imaging of giant cell tumour of the tendon sheath. *Radiol Med.* 2010;115(1):141–51.
127. Reilly KE, Stern PJ, Dale J. Recurrent giant cell tumors of the tendon sheath. *J Hand Surg.* 1999; 24(6):1298–302.
128. Andrassy RJ, Okcu MF, Despa S, Raney RB. Synovial sarcoma in children: Surgical lessons from a single institution and review of the literature. *J Am Coll Surg.* 2001;192(3):305–13.
129. Fisher C. Synovial sarcoma. *Ann Diagn Pathol.* 1998;2(6):401–21.
130. McCarville MB, Spunt SL, Skapek SX, Pappo AS. Synovial sarcoma in pediatric patients. *AJR Am J Roentgenol.* 2002;179(3):797–801.
131. Mullen JR, Zagars GK. Synovial sarcoma outcome following conservation surgery and radiotherapy. *Radiother Oncol.* 1994;33(1):23–30.
132. Valenzuela RF, Kim EE, Seo J, Patel S, Yasko AW. A revisit of MRI analysis for synovial sarcoma. *Clin Imaging.* 2000;24(4):231–5.
133. Van Rijswijk C, Hogendoorn P, Taminiau A, Bloem J. Synovial sarcoma: Dynamic contrast-enhanced MR imaging features. *Skeletal Radiol.* 2001;30(1):25–30.
134. Jones BC, Sundaram M, Kransdorf MJ. Synovial sarcoma: MR imaging findings in 34 patients. *AJR Am J Roentgenol.* 1993;161(4):827–30.
135. Kudawara I, Ueda T, Araki N. Extraskeletal chondroma around the knee. *Clin Radiol.* 2001;56(9):779–82.
136. Papagelopoulos PJ, Savvidou OD, Mavrogenis AF, Chloros GD, Papaparaskeva KT, Soucacos PN. Extraskeletal chondroma of the foot. *Joint Bone Spine.* 2007;74(3):285–8.
137. De Riu G, Meloni SM, Gobbi R, Contini M, Tullio A. Soft-tissue chondroma of the masticatory space. *Int J Oral Maxillofac Surg.* 2007;36(2):174–6.
138. Kransdorf MJ, Meis JM. From the archives of the AFIP. Extraskeletal osseous and cartilaginous tumors of the extremities. *RadioGraphics.* 1993;13(4):853–84.
139. Hogendoorn P, Bovee J, Nielsen G. Chondrosarcoma (grades I-III), including primary and secondary variants and periosteal chondrosarcoma. In: Fletcher CD, Bridge

JA, Hogendoorn PC (eds) *World Health Organization Classification of Tumours of Soft Tissue and Bone*. IARC Press, Lyon, France; 2013: pp. 264–8.
140. Angelini A, Guerra G, Mavrogenis AF, Pala E, Picci P, Ruggieri P. Clinical outcome of central conventional chondrosarcoma. *J Surg Oncol*. 2012;106(8):929–37.
141. Evans HL, Ayala AG, Romsdahl MM. Prognostic factors in chondrosarcoma of bone. A clinicopathologic analysis with emphasis on histologic grading. *Cancer*. 1977;40(2):818–31.
142. Shapeero L, Vanel D, Couanet D, Contesso G, Ackerman L. Extraskeletal mesenchymal chondrosarcoma. *Radiology*. 1993;186(3):819–26.
143. Murphey MD, Walker EA, Wilson AJ, Kransdorf MJ, Temple HT, Gannon FH. From the archives of the AFIP imaging of primary chondrosarcoma: Radiologic-pathologic Correlation. *RadioGraphics*. 2003;23(5):1245–78.
144. Okamoto Y, Minami M, Ueda T, Inadome Y, Tatsumura M, Sakane M. Extraskeletal mesenchymal chondrosarcoma of the cervical meninx. *Radiat Med*. 2007;25(7):355–8.
145. Ushigome S, Machinami R, Sorensen P. Ewing sarcoma/primitive neuroectodermal tumour (PNET). In: Fletcher CD, Unni KK, Mertens F (eds) *World Health Organization Classification of Tumours: Pathology and Genetics, Tumours of Soft Tissue and Bone*, IARC Press, Lyon, France; 2002: pp. 298–300.
146. Tefft M, Vawter G, Mitus A. Paravertebral "round cell" tumors in children. *Radiology*. 1969;92(7):1501–9.
147. Grier HE. The ewing family of tumors: Ewing's sarcoma and primitive neuroectodermal tumors. *Pediatr Clin N Am*. 1997;44(4):991–1004.
148. Kennedy JG, Eustace S, Caulfield R, Fennelly DJ, Hurson B, O'Rourke KS. Extraskeletal Ewing's sarcoma: A case report and review of the literature. *Spine*. 2000;25(15):1996–9.
149. Angervall L, Enzinger F. Extraskeletal neoplasm resembling ewing's sarcoma. *Cancer*. 1975;36(1):240–51.
150. Kleihues P, Cavenee WK. *Pathology and Genetics of Tumours of the Nervous System*. International Agency for Research on Cancer, Lyon, France; 2000.
151. Kransdorf MJ. Malignant soft-tissue tumors in a large referral population: Distribution of diagnoses by age, sex, and location. *AJR Am J Roentgenol*. 1995;164(1):129–34.
152. Beggs I. Pictorial review: Imaging of peripheral nerve tumours. *Clin Radiol*. 1997;52(1):8–17.
153. Vilanova JC, Woertler K, Narváez JA, Barceló J, Martínez SJ, Villalón M et al. Soft-tissue tumors update: MR imaging features according to the WHO classification. *Eur Radiol*. 2007;17(1):125–38.
154. Banks KP. The target sign: Extremity. *Radiology*. 2005;234(3):899–900.
155. Murphey MD, Smith WS, Smith SE, Kransdorf MJ, Temple HT. From the archives of the AFIP imaging of musculoskeletal neurogenic tumors: Radiologic-pathologic Correlation. *RadioGraphics*. 1999;19(5):1253–80.
156. Suh J, Abenoza P, Galloway H, Everson L, Griffiths H. Peripheral (extracranial) nerve tumors: Correlation of MR imaging and histologic findings. *Radiology*. 1992;183(2):341–6.
157. Frassica FJ, Khanna JA, McCarthy EF. The role of MR imaging in soft tissue tumor evaluation: Perspective of the orthopedic oncologist and musculoskeletal pathologist. *Magn Reson Imaging Clin N Am*. 2000;8(4):915–27.
158. Liotta L, Stetler-Stevenson W. Principles of molecular cell biology of cancer: Cancer metastasis. In: *Cancer: Principles and Practice of Oncology*, vol 1. Lippincott, Philadelphia, PA; 1993: pp. 134–49.
159. Bongartz G, Vestring T. Soft tissue tumours. MR State of the art ECR. 1991;91:193–8.
160. Boothroyd A, Carty H. The painless soft tissue mass in childhood—Tumour or not? *Postgrad Med J*. 1995;71(831):10–6.
161. Vezeridis MP, Moore R, Karakousis CP. Metastatic patterns in soft-tissue sarcomas. *Arch Surg*. 1983;118(8):915.
162. Buerger LF, Monteleone PN. Leukemic–lymphomatous infiltration of skeletal muscle: Systematic study of 82 autopsy cases. *Cancer*. 1966;19(10):1416–22.
163. Willis RA. *The Spread of Tumours in the Human Body*. Butterworths, London; 1952: pp. 284–285.
164. Araki K, Kobayashi M, Ogata T, Takuma K. Colorectal carcinoma metastatic to skeletal muscle. *Hepatogastroenterology*. 1994;41(5):405–8.
165. Avery G. Case report: Metastatic adenocarcinoma masquerading as a psoas abscess. *Clin Radiol*. 1988;39(3):319–20.
166. Tuoheti Y, Okada K, Osanai T, Nishida J, Ehara S, Hashimoto M et al. Skeletal muscle metastases of carcinoma: A clinicopathological study of 12 cases. *Jpn J Clin Oncol*. 2004;34(4):210–4.
167. Cohen HJ, Laszlo J. Influence of trauma on the unusual distribution of metastases from carcinoma of the larynx. *Cancer*. 1972;29(2):466–71.
168. Chang AE, Matory YL, Dwyer AJ, Hill SC, Girton ME, Steinberg SM et al. Magnetic resonance imaging versus computed tomography in the evaluation of soft tissue tumors of the extremities. *Ann Surg*. 1987;205(4):340.
169. Demas BE, Heelan RT, Lane J, Marcove R, Hajdu S, Brennan MF. Soft-tissue sarcomas of the extremities: Comparison of MR and CT in determining the extent of disease. *AJR Am J Roentgenol*. 1988;150(3):615–20.
170. Herring Jr CL, Harrelson JM, Scully SP. Metastatic carcinoma to skeletal muscle: A report of 15 patients. *Clin Orthop*. 1998;355:272–81.
171. Perrin A, Goichot B, Greget M, Lioure B, Dufour P, Marcellin L et al. Métastases musculaires révélatrices d'un adénocarcinome. *La Revue de médecine interne*. 1997;18(4):328–31.
172. Sudo A, OgiharaY, Shiokawa Y, Fujinami S, Sekiguchi S. Intramuscular metastasis of carcinoma. *Clin Orthop*. 1993;296:213–7.
173. Atlas S, Braffman B, LoBrutto R, Elder D, Herlyn D. Human malignant melanomas with varying degrees of melanin content in nude mice: MR imaging, histopathology, and electron paramagnetic resonance. *J Comput Assist Tomogr*. 1990;14(4):547–54.
174. Woodruff W, Djang W, McLendon R, Heinz E, Voorhees D. Intracerebral malignant melanoma: High-field-strength MR imaging. *Radiology*. 1987;165(1):209–13.

175. McGuire MH, Herbold DR, Beshany SE, Fletcher JW. High signal intensity soft tissue masses on $T_1$ weighted pulsing sequences. *Skeletal Radiol.* 1987;16(1):30–6.
176. Patten RM, Shuman WP, Teefey S. Subcutaneous metastases from malignant melanoma: Prevalence and findings on CT. *AJR Am J Roentgenol.* 1989;152(5):1009–12.
177. Peh W, Shek TW, Wang S, Wong JW, Chien EP. Osteogenic sarcoma with skeletal muscle metastases. *Skeletal Radiol.* 1999;28(5):298–304.
178. Moulton JS, Blebea JS, Dunco DM, Braley SE, Bisset GS,3rd, Emery KH. MR imaging of soft-tissue masses: Diagnostic efficacy and value of distinguishing between benign and malignant lesions. *AJR Am J Roentgenol.* 1995;164(5):1191–9.
179. Schoenberg NY, Beltran J. Contrast enhancement in musculoskeletal imaging. Current status. *Radiol Clin N Am.* 1994;32(2):337–52.
180. Verstraete KL. Categorical course: Introductory and advanced MRI: Techniques with clinical applications. *Presented at the Joint Meeting of the Society of MR and the Society for MR in Medicine and Biology.* Nice, France; August 1995.
181. McCarthy CL, McNally EG. The MRI appearance of cystic lesions around the knee. *Skeletal Radiol.* 2004;33(4):187–209.
182. Frassica FJ, Thompson Jr RC. Instructional course lectures, the american academy of orthopaedic surgeons-evaluation, diagnosis, and classification of benign soft-tissue tumors. *J Bone & Joint Surg.* 1996;78(1):126–40.
183. Yilmaz E, Karakurt L, Özercan İ, Özdemir H. A ganglion cyst that developed from the infrapatellar fat pad of the knee. *Arthroscopy J Arthroscopic Related Surg.* 2004;20(7):e65–8.
184. Baker WM. On the formation of synovial cysts in the leg in connection with disease of the knee-joint. *Clin Orthop.* 1994;299:2–10.
185. Fritschy D, Fasel J, Imbert J, Bianchi S, Verdonk R, Wirth CJ. The popliteal cyst. *Knee Surg Sports Traumatol Arthroscopy.* 2006;14(7):623–8.
186. Zhang WW, Lukan JK, Dryjski ML. Nonoperative management of lower extremity claudication caused by a baker's cyst: Case report and review of the literature. *Vascular.* 2005;13(4):244–7.
187. Miller TT, Staron RB, Koenigsberg T, Levin TL, Feldman F. MR imaging of baker cysts: Association with internal derangement, effusion, and degenerative arthropathy. *Radiology.* 1996;201(1):247–50.
188. Papp DF, Khanna AJ, McCarthy EF, Carrino JA, Farber AJ, Frassica FJ. Magnetic resonance imaging of soft-tissue tumors: Determinate and indeterminate lesions. *J Bone Joint Surg.* 2007;89(suppl_3):103–15.
189. Liu PT, Leslie KO, Beauchamp CP, Cherian SF. Chronic expanding hematoma of the thigh simulating neoplasm on gadolinium-enhanced MRI. *Skeletal Radiol.* 2006;35(4):254–7.
190. Aoki T, Nakata H, Watanabe H, Maeda H, Toyonaga T, Hashimoto H et al. The radiological findings in chronic expanding hematoma. *Skeletal Radiol.* 1999;28(7):396–401.
191. Reid JD, Kommareddi S, Lankerani M, Park MC. Chronic expanding hematomas. *JAMA.* 1980;244(21):2441–2.
192. Keenan S, Bui-Mansfield LT. Musculoskeletal lesions with fluid-fluid level: A pictorial essay. *J Comput Assist Tomogr.* 2006;30(3):517–24.
193. Nelson EN, Kassarjian A, Palmer WE. MR imaging of sports-related groin pain. *Magn Reson Imaging Clin N Am.* 2005;13(4):727–42.
194. Elsayes KM, Lammle M, Shariff A, Totty WG, Habib IF, Rubin DA. Value of magnetic resonance imaging in muscle trauma. *Curr Probl Diagn Radiol.* 2006;35(5):206–12.
195. May DA, Disler DG, Jones EA, Balkissoon AA, Manaster B. Abnormal signal intensity in skeletal muscle at MR imaging: Patterns, pearls, and pitfalls. *RadioGraphics.* 2000;20(suppl 1):S295–315.
196. Bencardino JT, Kassarjian A, Palmer WE. Magnetic resonance imaging of the hip: Sports-related injuries. *Top Magn Reson Imaging.* 2003;14(2):145–60.
197. Napier N, Shortt C, Eustace S. Muscle edema: Classification, mechanisms, and interpretation. *Semin Musculoskeletal Radiol.* 2006;10(4):258.
198. Peterson L, Stener B. Old total rupture of the adductor longus muscle: A report of seven cases. *Acta Orthopaedica.* 1976;47(6):653–7.
199. Rubin SJ, Feldman F, Staron RB, Zwass A, Totterman S, Meyers SP. Magnetic resonance imaging of muscle injury. *Clin Imaging.* 1995;19(4):263–9.
200. Bencardino JT, Rosenberg ZS, Brown RR, Hassankhani A, Lustrin ES, Beltran J. Traumatic musculotendinous injuries of the knee: Diagnosis with MR imaging. *RadioGraphics.* 2000;20 Spec No:S103–20.
201. Hayashi D, Hamilton B, Guermazi A, de Villiers R, Crema MD, Roemer FW. Traumatic injuries of thigh and calf muscles in athletes: Role and clinical relevance of MR imaging and ultrasound. *Insights Imaging.* 2012;3(6):591–601.
202. Kransdorf MJ, Meis JM, Jelinek JS. Myositis ossificans: MR appearance with radiologic-pathologic correlation. *AJR Am J Roentgenol.* 1991;157(6):1243–8.
203. Parikh J, Hyare H, Saifuddin A. The imaging features of post-traumatic myositis ossificans, with emphasis on MRI. *Clin Radiol.* 2002;57(12):1058–66.
204. Hanquinet S, Ngo L, Anooshiravani M, Garcia J, Bugmann P. Magnetic resonance imaging helps in the early diagnosis of myositis ossificans in children. *Pediatr Surg Int.* 1999;15(3–4):287–9.
205. Micheli A, Trapani S, Brizzi I, Campanacci D, Resti M, de Martino M. Myositis ossificans circumscripta: A paediatric case and review of the literature. *Eur J Pediatr.* 2009;168(5):523–9.
206. Hait G, Boswick Jr JA, Stone NH. Heterotopic bone formation secondary to trauma (myositis ossificans traumatica): An unusual case and a review of current concepts. *J Trauma Acute Care Surg.* 1970;10(5):405–11.
207. Lacout A, Jarraya M, Marcy PY, Thariat J, Carlier RY. Myositis ossificans imaging: Keys to successful diagnosis. *Indian J Radiol Imaging.* 2012;22(1):35–9.
208. Wang X, Malghem J, Parizel PM, Gielen J, Vanhoenacker F, De Schepper A. Pictorial essay. myositis ossificans circumscripta. *JBR-BTR: organe de la Societe royale belge de radiologie (SRBR)= orgaan van de Koninklijke Belgische Vereniging voor Radiologie (KBVR).* 2002;86(5):278–85.

209. Mahboubi S, Glaser DL, Shore EM, Kaplan FS. Fibrodysplasia ossificans progressiva. *Pediatr Radiol.* 2001;31(5):307–14.
210. Reinig J, Hill S, Fang M, Marini J, Zasloff M. Fibrodysplasia ossificans progressiva: CT appearance. *Radiology.* 1986;159(1):153–7.
211. Resnick D. *Diagnosis of Bone and Joint Disorders*, 4th edn. WB Saunders, Philadelphia, PA; 2002: pp. 4658–4665.
212. Dejerine M, Ceillier A. Trois cas d'ostéomes: Ossifications périostées juxta-musculaires et interfasciculaires chez les: Paraplégiques par lésion traumatique de la moelle epiniére. *Rev Neurol.* 1918;25:159–72.
213. Bravo-Payno P, Esclarin A, Arzoz T, Arroyo O, Labarta C. Incidence and risk factors in the appearance of heterotopic ossification in spinal cord injury. *Spinal Cord.* 1992;30(10):740–5.
214. Mielants H, Vanhove E, Neels Jd, Veys E. Clinical survey of and pathogenic approach to para-articular ossifications in long-term coma. *Acta Orthopaedica.* 1975; 46(2):190–8.
215. Carlier R, Safa D, Parva P, Mompoint D, Judet T, Denormandie P et al. Ankylosing neurogenic myositis ossificans of the hip an enhanced volumetric CT study. *J Bone Joint Surg*, British Volume. 2005;87(3):301–5.
216. Izumi K. Study of ectopic bone formation in experimental spinal cord injured rabbits. *Spinal Cord.* 1983;21(6):351–63.
217. Della Santa DR, Reust P. Heterotopic ossification and ulnar nerve compression syndrome of the elbow. A report of two cases. *Ann Chir Main Memb Super.* 1990;9(1):38–41.
218. Brooke M, Heard DL, De Lateur B, Moeller DA, Alquist AD. Heterotopic ossification and peripheral nerve entrapment: Early diagnosis and excision. *Arch Phys Med Rehabil.* 1991;72(6):425–9.
219. Gallien P, Nicolas B, Le Bot MP, Robineau S, Rivier I, Sarkis S et al. Heterotopic ossification and vascular compression. *Rev Rhum Ed Fr.* 1994;61(11):823–8.
220. Chen Y, Lin Y, Wang S, Lin S, Shung KK, Wu C. Monitoring tissue inflammation and responses to drug treatments in early stages of mice bone fracture using 50MHz ultrasound. *Ultrasonics.* 2014;54(1):177–86.
221. Schulze M, Kötter I, Ernemann U, Fenchel M, Tzaribatchev N, Claussen CD et al. MRI findings in inflammatory muscle diseases and their noninflammatory mimics. *AJR Am J Roentgenol.* 2009;192(6):1708–16.
222. Monu JU, McManus CM, Ward WG, Haygood TM, Pope TL, Jr, Bohrer SP. Soft-tissue masses caused by long-standing foreign bodies in the extremities: MR imaging findings. *AJR Am J Roentgenol.* 1995;165(2):395–7.
223. Bunch M, Jared T, Birskovich M, Lorraine M, Eiken M, Patrick W. Diabetic myonecrosis in a previously healthy woman and review of a 25-year mayo clinic experience. *Endocrine Pract.* 2002;8(5):343–6.
224. Kattapuram TM, Suri R, Rosol MS, Rosenberg AE, Kattapuram SV. Idiopathic and diabetic skeletal muscle necrosis: Evaluation by magnetic resonance imaging. *Skeletal Radiol.* 2005;34(4):203–9.
225. Bjornskov EK, Carry MR, Katz FH, Lefkowitz J, Ringel SP. Diabetic muscle infarction: A new perspective on pathogenesis and management. *Neuromuscular Disorders.* 1995;5(1):39–45.
226. Khoury NJ, El-Khoury GY, Kathol MH. MRI diagnosis of diabetic muscle infarction: Report of two cases. *Skeletal Radiol.* 1997;26(2):122–7.
227. Beltran J. MR imaging of soft-tissue infection. *Magn Reson Imaging Clin N Am.* 1995;3(4):743–51.
228. Beltran J, Noto AM, McGhee RB, Freedy RM, McCalla MS. Infections of the musculoskeletal system: High-field-strength MR imaging. *Radiology.* 1987;164(2):449–54.
229. Ma LD, Frassica FJ, Bluemke DA, Fishman EK. CT and MRI evaluation of musculoskeletal infection. *Crit Rev Diagn Imaging.* 1997;38(6):535–68.
230. Kothari NA, Pelchovitz DJ, Meyer JS. Imaging of musculoskeletal infections. *Radiol Clin N Am.* 2001;39(4):653–71.
231. Swartz MN. Cellulitis. *N Engl J Med.* 2004;350(9):904–12.
232. Struk DW, Munk PL, Lee MJ, Ho SG, Worsley DF. Imaging of soft tissue infections. *Radiol Clin N Am.* 2001;39(2):277–303.
233. Gylys-Morin VM. MR imaging of pediatric musculoskeletal inflammatory and infectious disorders. *Magn Reson Imaging Clin N Am.* 1998;6(3):537–59.
234. Towers JD. The use of intravenous contrast in MRI of extremity infection. *Semin Ultrasound CT MR.* 1997;18:269–75.
235. Bureau NJ, Chhem RK, Cardinal É. Musculoskeletal infections: US Manifestations. *RadioGraphics.* 1999;19(6):1585–92.
236. Ma LD, McCarthy EF, Bluemke DA, Frassica FJ. Differentiation of benign from malignant musculoskeletal lesions using MR imaging: Pitfalls in MR evaluation of lesions with a cystic appearance. *AJR Am J Roentgenol.* 1998;170(5):1251–8.
237. Ali SZ, Srinivasan S, Peh WC. MRI in necrotizing fasciitis of the extremities. *Br J Radiol.* 2014;87(1033): 20130560.
238. Chau C, Griffith J. Musculoskeletal infections: Ultrasound appearances. *Clin Radiol.* 2005;60(2):149–59.
239. Small LN, Ross JJ. Suppurative tenosynovitis and septic bursitis. *Infect Dis Clin N Am.* 2005;19(4):991–1005.
240. Zimmermann B 3rd, Mikolich DJ, Ho G Jr. Septic bursitis. *Semin Arthritis Rheum.* 1995;24:391–410.
241. Bickels J, Ben-Sira L, Kessler A, Wientroub S. Primary pyomyositis. *J Bone Joint Surg.* 2002;84(12):2277–86.
242. Restrepo CS, Lemos DF, Gordillo H, Odero R, Varghese T, Tiemann W et al. Imaging findings in musculoskeletal complications of AIDS1. *RadioGraphics.* 2004;24(4):1029–49.
243. Soler R, Rodríguez E, Aguilera C, Fernández R. Magnetic resonance imaging of pyomyositis in 43 cases. *Eur J Radiol.* 2000;35(1):59–64.
244. Gordon BA, Martinez S, Collins AJ. Pyomyositis: Characteristics at CT and MR imaging. *Radiology.* 1995;197(1):279–86.

# 14

# *Temporomandibular Joints*

**Tsukasa Sano**

**CONTENTS**



Temporomandibular joint (TMJ) imaging studies can provide us with an understanding of both the normal and pathologic anatomy of the joint and surrounding structures.

Since its clinical introduction in the mid-1980s, the magnetic resonance imaging (MRI) has evolved as the prime diagnostic method for soft tissue abnormalities in the TMJ. It is noninvasive and more accurate than arthrography; it requires less operator skills and is well tolerated by the patient. MRI can also identify other entities that occur in the TMJ.

## 14.1 Disk Displacement

Disk displacement categories include anterior, anterolateral, anteromedial, lateral, medial, and posterior displacements.[1] The combination of anterior and lateral or medial displacement is called *rotational displacement*, whereas pure lateral or pure medial displacement is called *sideways displacement*.[2] Pure lateral and pure medial sideways displacements of the disk also occur, but not as commonly as in combination with an anterior displacement.[2–4]

The most common findings are the anterior Figures 14.1 through 14.3 and the anterolateral (Figure 14.4) displacements when interpreting MR images of patients with clinical signs and symptoms of internalderangement.[1,5]

The functional categories of disk displacement include displacement with or without reduction. In disk displacement with reduction (Figure 14.2), the anteriorly displaced disk reverts to a normal superior position during opening. When a displaced disk reduces to a normal position, a click is usually heard. In disk displacement without reduction (Figure 14.3), the disk lies anterior to the condyle during all mandibular movements, and the normal condyle-disk relationship is not reestablished. In the initial stages of this condition, jaw opening is typically limited and the jaw deviates to the side of the affected joint. However, this clinical characteristic is typical only during the initial (early) phase; with time the opening capacity of the TMJ increases and the jaw no longer deviates to the affected side. This is the result of stretching or progressive elongation of the posterior disk attachment and to a lesser extent deformation of the disk itself. In the early stage, disk displacement without reduction is usually not associated with joint sounds.

Disk displacement could be complete or partial (Figure 14.5).[6] In complete disk displacement, the entire mediolateral dimension of the disk is displaced anterior to the condyle. In partial disk displacement, only the medial or lateral part of the disk is displaced anterior to the condyle. Most frequently, the lateral part is anteriorly displaced and the more medial part of the disk is still in a normal superior position. Partial disk displacement is frequently seen in joints with disk displacement with reduction; the disk is anteriorly displaced in the lateral part of the joint and is in a normal position in the medial part of the joint (Figure 14.5).

Disk displacement is found in approximately 80% of patients with symptoms of temporomandibular disorder (TMD) referred for MRI.[1] Several forms of disk displacement are also seen in up to one-third of asymptomatic volunteers.[1,7,8] This may initially appear confusing but

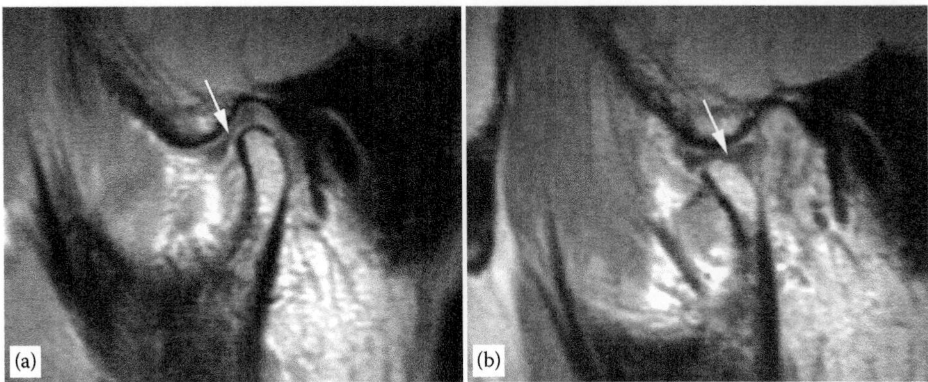

**FIGURE 14.1**
Normal TMJ on proton density MR images. Sagittal image with the jaw closed (a). The biconcave, low-signal intensity disk (arrow) is situated in the proper relationship between the anterior convexity of the condyle inferiorly and the superior convexity of the temporal bone. Sagittal image with the jaw maximally opened (b) shows the biconcave lens-like configuration of the disk (arrows) in a correct anatomic portion to the convex surface of the condyle inferiorly and the convex surface tubercle of the temporal bone superiorly. (From Sano, T. et al., *Neuroimag. Clin. N. Am.*, 13, 573–595, 2003.)

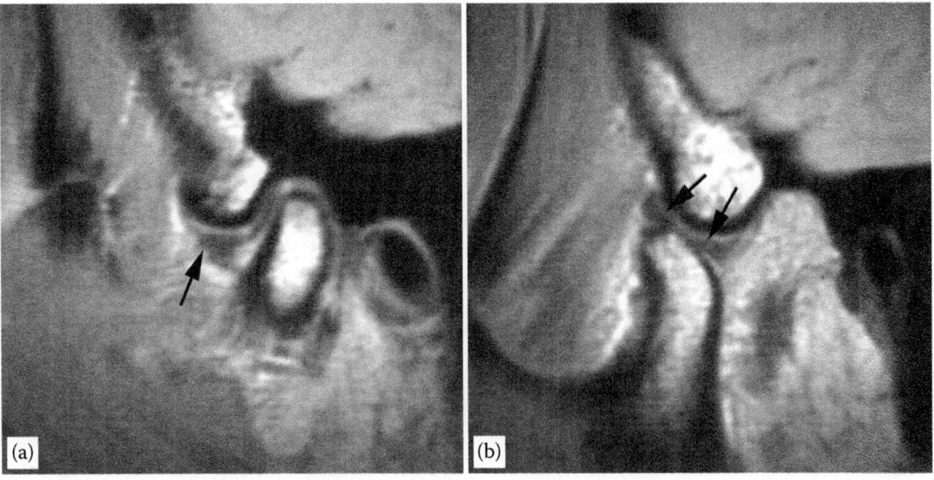

**FIGURE 14.2**
Anterior disk displacement with reduction. Proton density sagittal images in closed-mouth (a) and open-mouth position (b), showing disk displacement with reduction. In the closed-mouth image (a), the disk (arrow) is located anterior to the condyle. At the maximal mouth-open image (b), the disk (arrow) is located in normal relationship to the condyle. The disk is normally biconcave. (From Sano, T. et al., *Neuroimag. Clin. N. Am.*, 13, 573–595, 2003.)

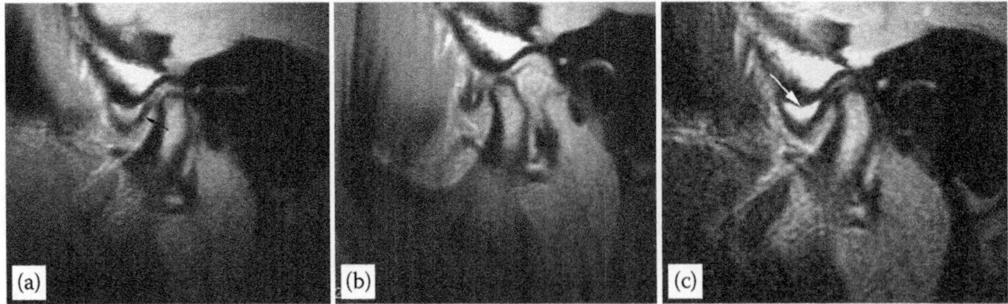

**FIGURE 14.3**
Anterior disk displacement without reduction and joint effusion. The disk (arrow) is anterior to the condyle in the closed-mouth position (a). On mouth opening (b), the disk remains anterior to the condyle, indicating anterior disk displacement without reduction. On the $T_2$-weighted image (c), there is a large effusion in the upper joint space (arrow). (From Sano, T. et al., *Neuroimag. Clin. N. Am.*, 13, 573–595, 2003.)

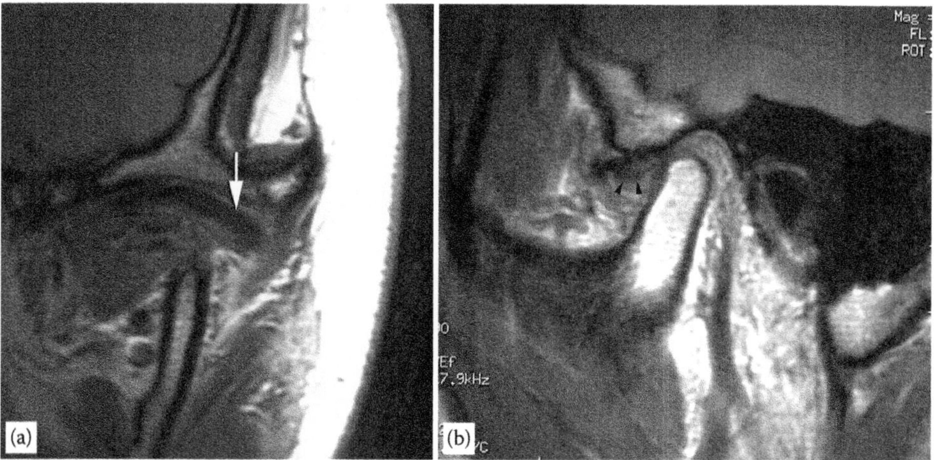


**FIGURE 14.4**
Anterolateral disk displacement of the TMJ. The sagittal proton density image (b) shows the anterior displaced disk to the condyle. On the coronal image of the same joint (a), the slight lateral displacement of the disk (arrowheads) is demonstrated. (From Sano, T. et al., *Neuroimag. Clin. N. Am.*, 13, 573–595, 2003.)

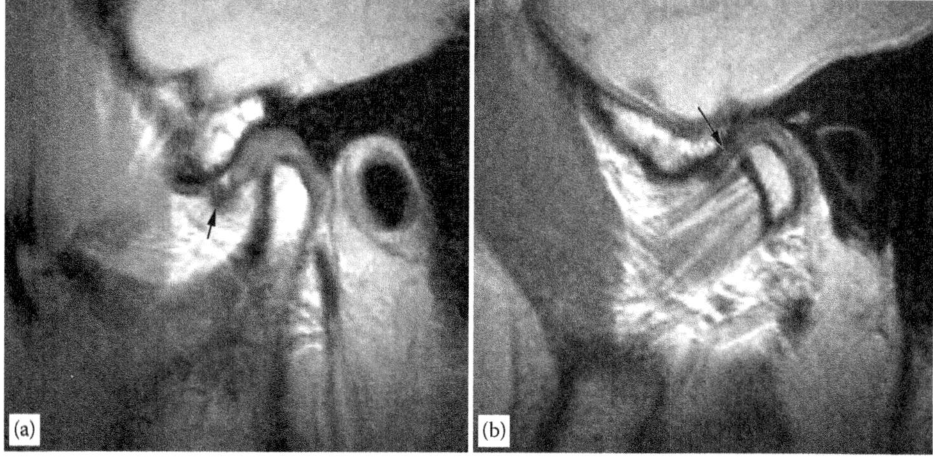


**FIGURE 14.5**
Partial disk displacement. On the closed-mouth sagittal proton density image, the disk (arrow) is anteriorly displaced in the lateral part of the joint (a). In the medial part (b), the same disk (arrow) is located in a normal superior position. This indicates a partial anterior disk displacement. (From Sano, T. et al., *Neuroimag. Clin. N. Am.*, 13, 573–595, 2003.)

a closer analysis of the type of disk displacement in patients and asymptomatic volunteers demonstrates that most, if not all, the volunteers have early stage disk displacement includes partial displacement whereas patients are likely to have later stage displacement, such as complete disk displacement.[6]

## 14.2 Osteoarthritis

*Osteoarthritis* is characterized radiographically by flattening of and irregularities in articular surfaces, osteophytosis, and erosion (Figures 14.6 and 14.7). However, it is often difficult to differentiate radiographically between advanced remodeling and degenerative joint disease. One study has suggested that, in joints with persistent non-reducing disk displacement, flattening and deformation of the articular eminence and regression of condylar size were likely to occur, even after symptoms and signs of TMJ disorders had resolved or reduced.[9] Generally, in Japan, we adopt erosion, osteophyte, and deformity as inclusion criteria for osteoarthritis (Figures 14.8 through 14.10).

Osseous changes involving the condyle and temporal bone often occur as sequelae of disk displacement.[10–15] Osteoarthritis is frequently seen in joints with long-standing disk displacement without reduction (Figure 14.11).[13,14] Disk displacement seems to be a

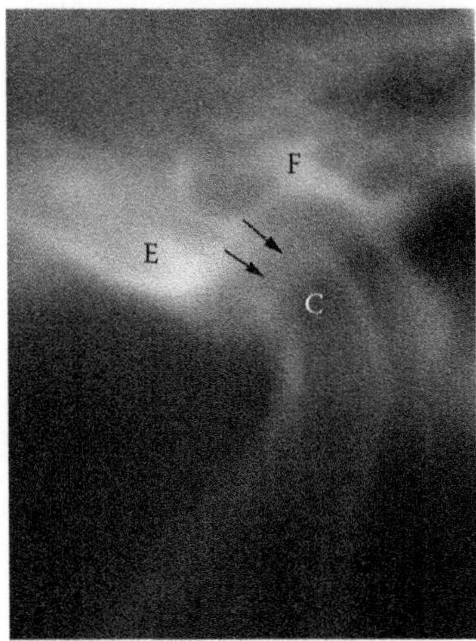


**FIGURE 14.6**
Sagittal tomogram of TMJ. Erosion (arrows) is seen on the upper portion of the condyle (C) indicating osteoarthritis. F: glenoid fossa, E: articular eminence. (From Sano, T. et al., *Neuroimag. Clin. N. Am.*, 13, 573–595, 2003.)

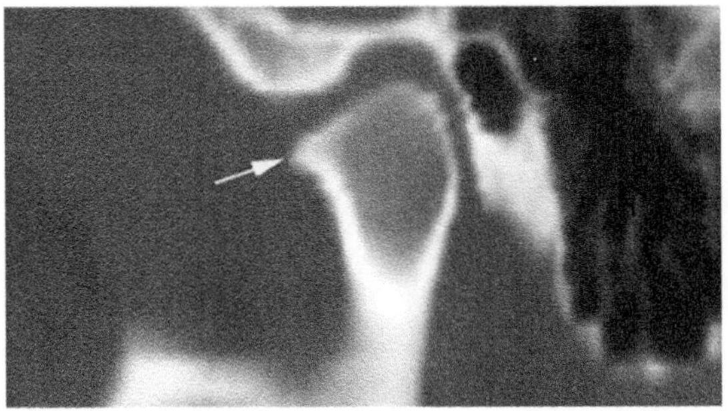

**FIGURE 14.7**
Reformatted sagittal CT images of TMJ. The osteophyte (arrow) on the anterior portion of the condyle is revealed, indicating osteoarthritis. (From Sano, T. et al., *Neuroimag. Clin. N. Am.*, 13, 573–595, 2003.)

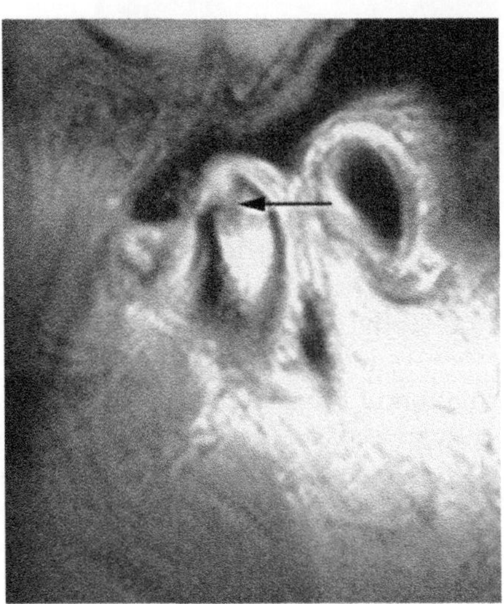

**FIGURE 14.8**
Parasagittal proton density-weighted image in the closed-mouth position shows a discontinuity of the cortical bone (arrow) on the surface of the condyle, called *erosion*. (From Otonari-Yamamoto, M. et al., *Oral Radiol.*, 31, 41–48, 2014.)

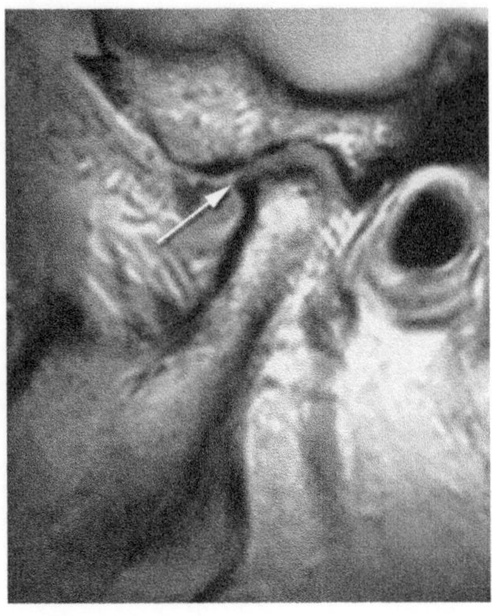

**FIGURE 14.9**
Parasagittal proton density-weighted image in the closed-mouth position shows a thickening of the cortical bone (arrow) on the anterior portion of the condyle, called *osteophyte*. (From Otonari-Yamamoto, M. et al., *Oral Radiol.*, 31, 41–48, 2014.)

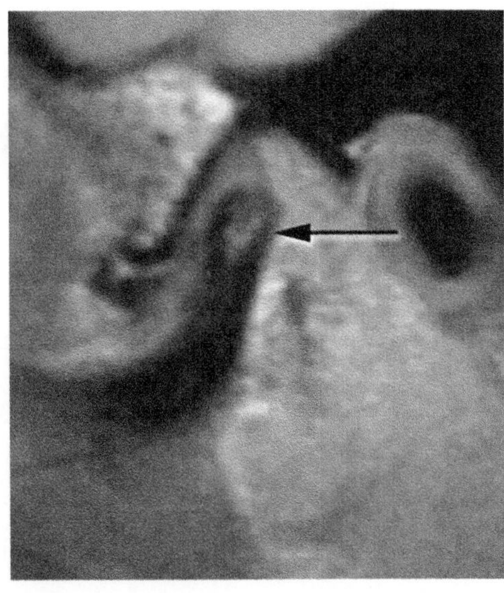

**FIGURE 14.10**
Parasagittal proton density-weighted image in the closed-mouth position shows a condyle reduced in size (arrow), called *deformity*. (From Otonari-Yamamoto, M. et al., *Oral Radiol.*, 31, 41–48, 2014.)

precursor of osteoarthritis, which is rare in joints with a normal superior disk position, although it is seen occasionally in disk displacement with reduction, and more frequently when disk displacement without reduction has been present for some time. Imaging evidence of osteoarthritis can be seen in young persons (teenagers) with disk displacement and no reduction. Disk displacement and internal derangement is, however, only one cause of osteoarthritis, which results from a multitude of primary joint lesions.

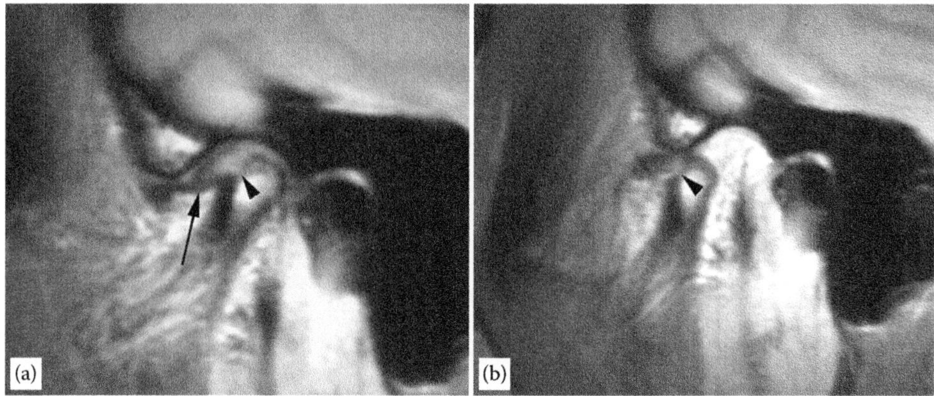


**FIGURE 14.11**
Parasagittal proton density-weighted closed-mouth image (a) shows erosion of the condyle (arrowhead). The disk (arrow) is anterior to the condyle. On mouth opening (b), the disk remains anterior. (From Sano, T. et al., *Neuroimag. Clin. N. Am.*, 13, 573–595, 2003.)

Osteoarthritis has been suggested as a source of TMJ pain,[16] although one study found that osteoarthritis was not a significant factor in TMJ pain,[17] according to regression analysis. Osteoarthritis is also present in a large proportion of older persons and is usually completely asymptomatic. Symptoms related to temporomandibular dysfunction decrease with age and are often remitting and self-limiting.[18–21] The discrepancy that can occur between findings obtained by imaging and patient symptomatology highlights the need for effective clinical examination in determining which findings are significant.

## 14.3 Bone Marrow Abnormalities in Mandibular Condyle

Several MRI studies have described abnormalities in the mandibular condyle with a similar appearance to osteonecrosis of the femoral head.[22–26] This has led to the assumption that osteonecrosis can also affect the mandibular condyle.[22–26] This issue was being discussed long before MRI became available,[27] but it has remained controversial, as no histologic correlations have been available. Larheim and Westesson et al. analyzed core biopsies from fifty mandibular condyles and correlated histologic findings with MR signal patterns categorized as normal, edema, or osteonecrosis (Figures 14.12 through 14.14).[28] They found that edema and osteonecrosis occurred in the marrow of the condyle.[28] Histologic evidence of bone marrow edema was also found where there was no evidence of osteonecrosis, suggesting that edema is a precursor in osteonecrotic development, as in other joints.[28,29] They also concluded that osteoarthritis developed secondary to osteonecrosis in the TMJ, as four out of nine joints with a histological diagnosis of marrow edema showed normal cortical bone on MRI.[28] This contradicts suggestions in earlier publications that bone marrow abnormalities were secondary to osteoarthritis.[30,31] However, their findings were in accordance with those of an analysis by Sano et al.[32] which revealed that nearly half of all joints with bone marrow abnormalities imaged with MR showed normal contours of the mandibular condyle and temporal joint component, which was then interpreted as no evidence of osteoarthritis.[28] This report also suggested that conditions in the TMJ are similar to those in other joints: namely, that abnormalities suggestive of edema or osteonecrosis in the bone marrow are initially separate entities from osteoarthritis.[32] The study also demonstrated that the majority of joints with bone marrow abnormalities had disk displacements with no reduction and that this was more prevalent in joints with osteonecrosis than in those without.[32]

Disk displacement may be a factor in bone marrow change.[24,25,27,28,32,33] However, it should be noted that a relatively small proportion of joints with internal derangement have demonstrated bone marrow change.[32] Abnormal bone marrow in the mandibular condyle is seen in less than approximately 10% of joints with TMD.[30,32]

In an earlier study, we found that joints with bone marrow abnormalities in the mandibular condyle were markedly more painful than those without.[34] The increased intra-articular pressure in conditions such as synovitis and hemophilia has also been suggested as an etiology for osteonecrosis.[35] A correlation between osteonecrosis and increased joint fluid has also been reported.[26,28] In another study, we suggested that pain is more severe in TMJs with marrow edema of the mandibular condyle than in those with osteonecrosis.[29] However, another

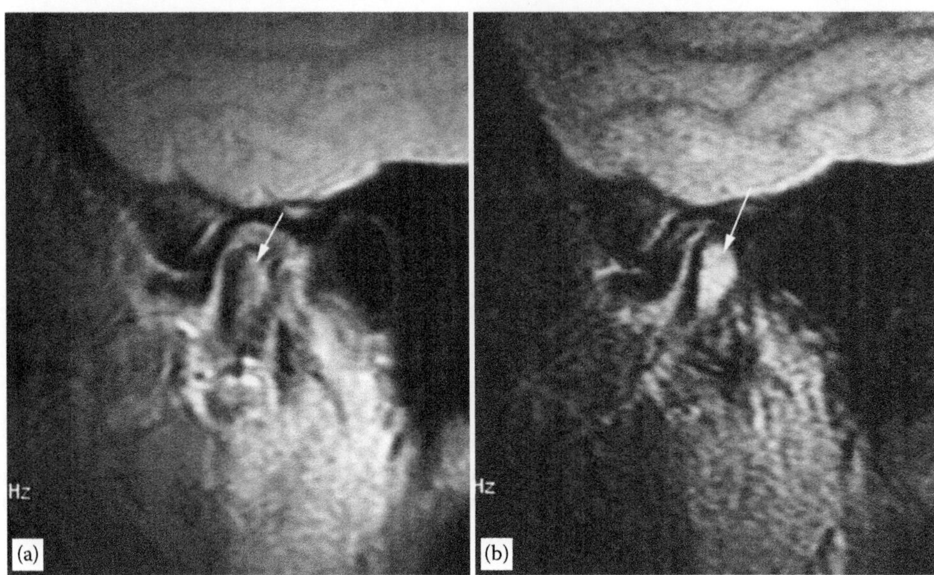


**FIGURE 14.12**
Marrow abnormalities. Parasagittal proton density-weighted image (a) shows intermediate signal from the bone marrow of the condyle (arrow). The disk is anteriorly displaced and deformed. Parasagittal $T_2$-weighted MR images (b) shows increased signal from the bone marrow of the condyle (arrow) suggesting bone marrow edema. There are also small joint effusions in the upper and lower joint spaces. (From Sano, T. and Westesson, P.L., *Oral Surg. Oral. Med. Oral Pathol. Oral Radiol. Endod.*, 79, 511–516, 1995.)

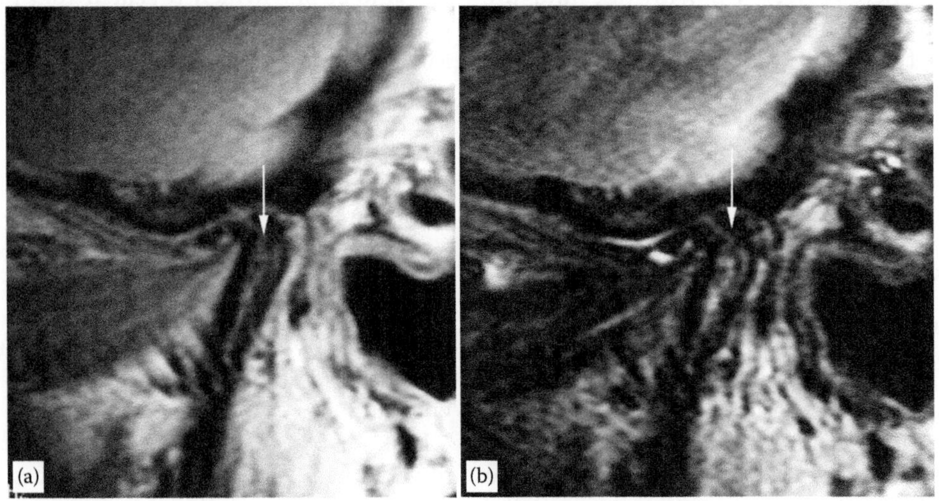


**FIGURE 14.13**
Marrow abnormalities. Parasagittal proton density-weighted image (a) shows decreased signal from the bone marrow of the condyle (arrow). There is also irregularity of the upper surface of the condyle and the temporal joint component suggestive of osteoarthritis. The disk is anteriorly displaced and deformed. Parasagittal $T_2$-weighted image (b) shows decreased signal from the bone marrow condyle (arrow) suggestive of osteonecrosis, and there is a moderate effusion in both upper and lower joint spaces. (From Sano, T. and Westesson, P.L., *Oral Surg. Oral Med. Oral Pathol. Oral Radiol. Endod.*, 79, 511–516, 1995.)

study has suggested that bone marrow edema pattern in the mandibular condyle does not always contribute to the occurrence of joint pain in patients with TMJ disorders.[36]

In a study, we found that symptomatic osteoarthritic TMJ could accompany bone marrow change in the upper portion of the condyle, adjacent to osseous changes, showing increased signals on proton density images.[37]

This result showing increased proton density-weighted signal in TMJ with symptomatic osteoarthritis may reflect early stage edema.

In view of all these findings, osteoarthritis and related entities may be reflected in clinical symptoms

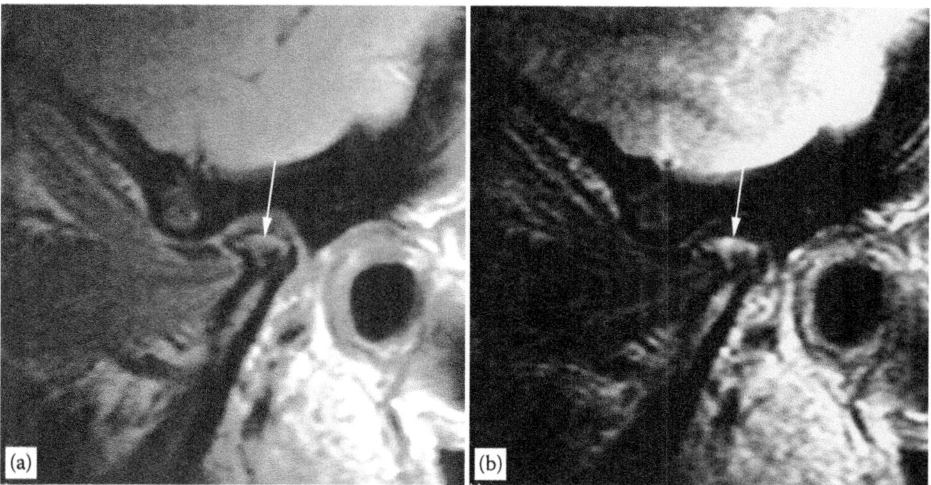


**FIGURE 14.14**
Marrow abnormalities. Parasagittal proton density-weighted MR image (a) shows an area of low signal in the inferior portion of the condyle and an area of higher signal in the more superior portion of the condyle (arrow). There is also displacement and deformation of the disk and an anterior osteophyte of the condyle indicating osteoarthritis. Parasagittal $T_2$-weighted MR image (b) shows an area of increased signal in the upper portion of the condyle (arrow) and decreased signal in the lower portion of the condyle. This is consistent with the combined edema and sclerotic pattern suggestive of osteonecrosis. (From Sano, T. and Westesson, P.L., *Oral Surg. Oral Med. Oral Pathol. Oral Radiol. Endod.*, 79, 511–516, 1995.)

of problems with the TMJ, but there is still sometimes a discrepancy between findings obtained by imaging and patient symptomatology.[38] This highlights the need for effective clinical examination in determining which findings are significant.

## 14.4 Joint Effusion

Joint effusion is significantly more prevalent in painful joints than in nonpainful joints.[39] Joint effusion is a pathological collection of joint fluid in the joint spaces associated with inflammatory condition and usually defined as a large amount of joint fluid. The fluid is best seen on $T_2$-weighted image (Figure 14.3). The large accumulations of joint fluid (Figure 14.3) are seen only in symptomatic patients; however, small amounts of fluid are seen in asymptomatic individuals.

Not all individuals with pain have joint effusion, but a significant accumulation of joint fluid is likely associated with a pain and the status of the disk.[17,40,41] Joint effusion may also have some diagnostic value as the fluid outlines the morphology of the disk and may outline a perforation in the disk and retrodiskal tissue, which is called *arthrographic effect*.[37]

Although much research has been carried out on joint effusion, its detection by MRI remains problematic, and identification by $T_2$-weighted images alone is difficult. Some studies have investigated joints with joint effusion to determine the potential of fluid-attenuated inversion recovery (FLAIR) sequence image. FLAIR sequences revealed that joint effusion was not just water content, but a fluid accumulation containing elements such as protein.[42] However, another study has also had a similar result from the joints with minimal fluid seen in the normal joints, and it is concluded that minimal fluid may contain elements such as protein that are capable of inducing a shortened $T_1$ relaxation time on MR images.[43]

## 14.5 Osteochondritis Dissecans

A loose body seldom found in the joint space is associated with a defect in the condyle of the same size. This condition can be characterized as osteochondritis dissecans (Figure 14.15).[25] This condition has been associated with osteonecrosis, although the relationship between the two conditions is not fully understood.[25]

## 14.6 Tumors and Tumor-Like Conditions

Tumors and tumor-like conditions include synovial chondromatosis (Figure 14.16),[44–47] pigmented villonodular synovitis,[48–54] osteochondroma, and calcium pyrophosphate dehydrate deposition disease (pseudogout).[55–60]

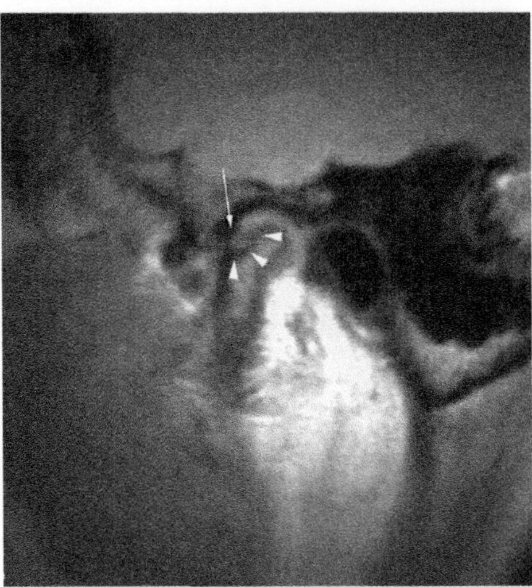

**FIGURE 14.15**
Osteochondritis dissecans. On the proton density-weighted closed-mouth image, a fragment (arrow) from the condyle is dislocated into the joint space, with a corresponding defect (arrowheads) in the condylar head. This observation is consistent with osteochondritis dissecans. The disk is anteriorly displaced and deformed. (From Sano, T. et al., *Neuroimag. Clin. N. Am.*, 13, 573–595, 2003.)

The condyle and the temporal bone are rarely affected by tumors and tumor-like conditions. MRI and computed tomography (CT) are routine procedures in patients with tumors and tumor-like conditions of the TMJ. The most common neoplastic lesion is synovial chondromatosis (Figure 14.16).[46,47] This tumor can be locally aggressive, and cases with intracranial extension have been described.[46,47] Pigmented villonodular synovitis is a benign proliferative disease and rarely involves the TMJ.[48–54] Osteochondroma is probably the second most common lesion affecting the TMJ. Calcium pyrophosphate dehydrate deposition disease, or pseudogout, rarely affects the TMJ.[55–60] More extensive cases present with enlarging masses and show significant erosion of the contiguous skull base, as well as the condyle.[58–60]

## 14.7 Acute Trauma

Acute trauma to the mandible can be evaluated with plain films, panoramic examinations, or CT (Figure 14.17). MRI may occasionally show fractures and effusions not seen on such imaging modalities. Both CT and MRI can be helpful in cases with intracapsular fractures (Figure 14.18).

## 14.8 Arthritides

Arthritides include rheumatoid arthritis, ankylosing spondylitis, and psoriatic arthritis.[61,62] The TMJ is involved in approximately 50% of patients with such rheumatoid diseases. However, MRI is not routinely used for such patients. In patients with rheumatoid arthritis and other arthritides, TMJ involvement may mimic the more common TMDs. Using MRI, it is possible to distinguish these patients from those with no synovial proliferation in most cases.[63]

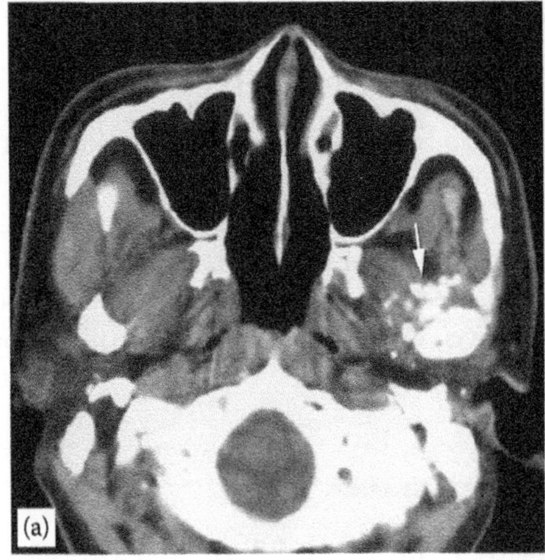


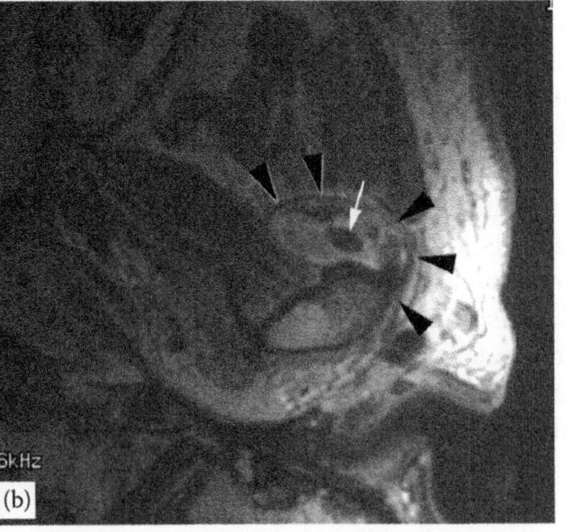


**FIGURE 14.16**
Synovial chondromatosis. The axial CT image (a) shows multiple calcifications (white arrow) anterior to the condyle. On the axial $T_1$-weighted MR image (b), there is expansion of the anterolateral capsule wall (arrowheads). This indicates a neoplastic intra-articular process. There are multiple areas (white arrow) of low signal within the expanded joint capsule, corresponding to the calcifications on the CT image. These findings are consistent with synovial chondromatosis. (From Sano, T. et al., *Neuroimag. Clin. N. Am.*, 13, 573–595, 2003.)

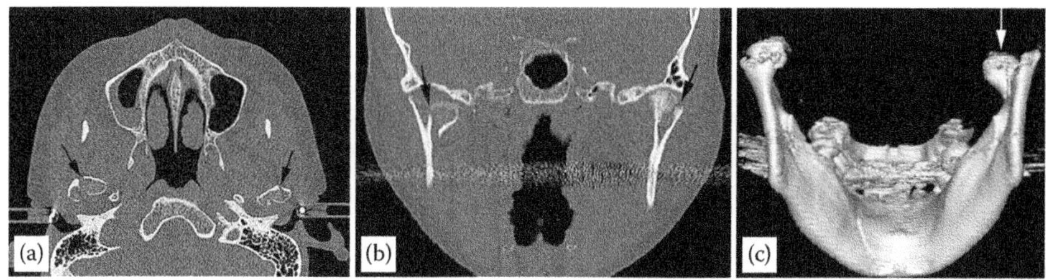

FIGURE 14.17
Acute trauma. Axial (a) and coronal (b) bone window CT images show the fracture lines of the condyle on both sides (arrow). On 3D CT image (c), the small fragment is displaced medially and inferiorly to the original position on the right condyle.

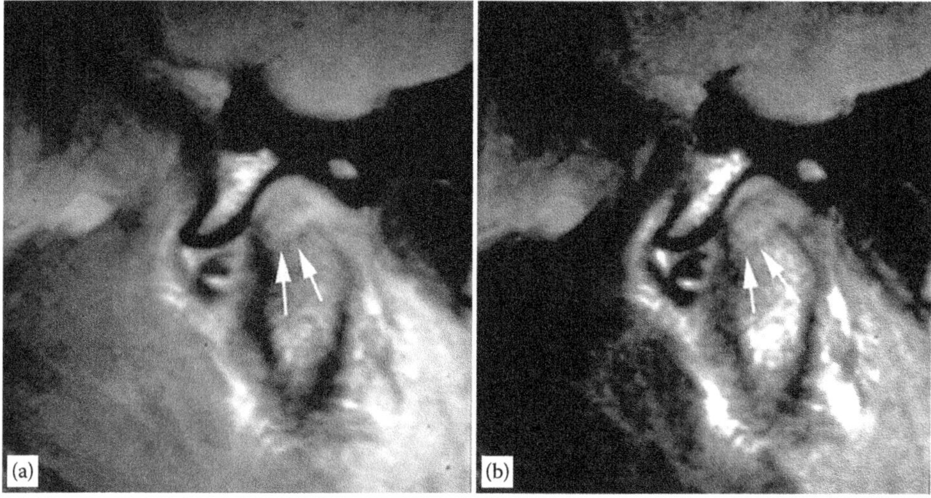

FIGURE 14.18
Intracapsular fracture of the mandibular condyle. Parasagittal proton density-weighted (a) and $T_2$-weighted (b) MR images. On top of the mandibular condyle, consecutive delineation of cortical bone was not preserved (arrows), suggestive of fracture. Instead, there was a lesion with an intermediate signal intensity on $T_2$-weighted image (b) as well as slightly higher signal on proton density-weighted image (a), suggesting relatively acute hematoma. (From Yamamoto, M. and Sano, T. et al. *Dental Radiol.*, 38, 42–43, 1998.)

## References


1. Tasaki MM, Westesson PL, Isberg AM, Tallents RH. Classification and prevalence of temporomandibular joint disk displacement in patients and symptom-free volunteers. *Am J Orthod Dentofacial Orthop.* 1996; 109(3): 249–262.
2. Katzberg RW, Westesson PL, Tallents RH, Anderson R, Kurita K, Manzione JV Jr et al. Temporomandibular joint: MR assessment of rotational and sideways disc displacements. *Radiology.* 1988; 169(3): 741–748.
3. Liedberg J, Westesson PL. Sideways position of the temporomandibular joint disk: Coronal cryosectioning of fresh autopsy specimens. *Oral Surg Oral Med Oral Pathol.* 1988; 66(6): 644–649.
4. Liedberg J, Westesson PL, Kurita K. Sideways and rotational displacement of the temporomandibular joint disk: Diagnosis by arthrography and correlation to cryosectional morphology. *Oral Surg Oral Med Oral Pathol.* 1990; 69(6): 757–763.
5. Paesani D, Westesson PL, Hatala M, Tallents RH, Kurita K. Prevalence of temporomandibular joint internal derangement in patients with craniomandibular disorders. *Am J Orthod Dentofacial Orthop.* 1992; 101(1): 41–47.
6. Larheim TA, Westesson P, Sano T. Temporomandibular joint disk displacement: Comparison in asymptomatic volunteers and patients. *Radiology.* 2001; 218(2): 428–432.
7. Katzberg RW, Westesson PL, Tallents RH, Drake CM. Orthodontics and temporomandibular joint internal derangement. *Am J Orthod Dentofacial Orthop.* 1996; 109(5): 515–520.
8. Katzberg RW, Westesson PL, Tallents RH, Drake CM. Anatomic disorders of the temporomandibular joint disc in asymptomatic subjects. *J Oral Maxillofac Surg.* 1996; 54(2): 147–153.
9. Kurita H, Uehara S, Yokochi M, Nakatsuka A, Kobayashi H, Kurashina K. A long-term follow-up study of radiographically evident degenerative changes in the temporomandibular joint with different conditions of disk displacement. *Int J Oral Maxillofac Surg.* 2006; 35(1): 49–54.

10. Wilkes CH. Arthrography of the temporomandibular joint in patients with the TMJ pain-dysfunction syndrome. *Minn Med.* 1978; 61: 645–652.
11. Wilkes CH. Structural and functional alterations of the temporomandibular joint. *Northwest Dent.* 1978; 57: 287–294.
12. Wilkes CH. Internal derangements of the temporomandibular joint. *Arch Otolaryngol.* 1989; 115: 469–477.
13. Westesson P-L, Rohlin M. Internal derangement related to osteoarthritis in temporomandibular joint autopsy specimens. *Oral Surg Oral Med Oral Pathol.* 1984; 57(1): 17–22.
14. Westesson P-L. Structural hard-tissue changes in temporomandibular joints with internal derangement. *Oral Surg Oral Med Oral Pathol.* 1985; 59(2): 220–224.
15. Dimitroulis G. The prevalence of osteoarthrosis in cases of advanced internal derangement of the temporomandibular joint: A clinical, surgical and histological study. *Int J Oral Maxillofac Surg.* 2005; 34(4): 345–349.
16. Emshoff R, Brandimaier I, Bertram S, Rudisch A. Magnetic resonance imaging findings of osteoarthrosis and effusion in patients with unilateral temporomandibular joint pain. *Int J Oral Maxillofac Surg.* 2002; 31(6): 598–602.
17. Larheim TA, Westesson PL, Sano T. MR grading of temporomandibular joint fluid: Association with disk displacement categories, condyle marrow abnormalities and pain. *Int J Oral Maxillofac Surg.* 2001; 30(2): 104–112.
18. Pereira FJ Jr, Lundh H, Westesson PL, Carlsson LE. Clinical findings related to morphologic changes in TMJ autopsy specimens. *Oral Surg Oral Med Oral Pathol.* 1994; 78(3): 288–295.
19. Lundh H, Westesson PL, Kopp S. A three-year follow-up of patients with reciprocal temporomandibular joint clicking. *Oral Surg Oral Med Oral Pathol.* 1987; 63(5): 530–533.
20. Rasmussen OC. Description of population and progress of symptoms in a longitudinal study of temporomandibular arthropathy. *Scand J Dent Res.* 1981; 89(2): 196–203.
21. Randolph CS, Greene CS, Moretti R, Forbes D, Perry HT. Conservative management of temporomandibular disorders: A posttreatment comparison between patients from a university clinic and from private practice. *Am J Orthod Dentofacial Orthop.* 1990; 98(1): 77–82.
22. Schellhas KP, Wilkes CH, Omlie MR, Peterson CM, Johnson SD, Keck RJ, Block JC, Fritts HM, Heithoff KB. The diagnosis of temporomandibular joint disease: Two-compartment arthrography and MR. *AJR Am J Roentgenol.* 1988; 151: 341–350.
23. Schellhas KP. Internal derangement of the temporomandibular joint: Radiologic staging with clinical, surgical, and pathologic correlation. *Magn Reson Imaging.* 1989; 7: 495–515.
24. Schellhas KP, Wilkes CH. Temporomandibular joint inflammation: Comparison of MR fast scanning with $T_1$- and $T_2$-weighted imaging techniques. *AJR Am J Roentgenol.* 1989; 153: 93–98.
25. Schellhas KP, Wilkes CH, Fritts HM, Omlie MR, Lagrotteria LB. MR of osteochondritis dissecans and avascular necrosis of the mandibular condyle. *AJR Am J Roentgenol.* 1989; 152: 551–560.
26. Schellhas KP. Temporomandibular joint injuries. *Radiology.* 1989; 173: 211–216.
27. Reiskin AB. Aseptic necrosis of the mandibular condyle: A common problem? *Quintessence Int.* 1979; 2: 85–89.
28. Larheim TA, Westesson P-L, Hicks DG, Eriksson L, Brown D. Osteonecrosis of the temporomandibular joint: Correlation of magnetic resonance imaging and histology. *J Oral Maxillofac Surg.* 1999; 57: 888–898.
29. Sano T, Westesson PL, Yamamoto M, Okano T. Differences in temporomandibular joint pain and age distribution between marrow edema and osteonecrosis in the mandibular condyle. *Cranio.* 2004; 22(4): 283–288.
30. Lieberman JM, Gardner CL, Motta AO, Schwartz RD. Prevalence of bone marrow signal abnormalities observed in the temporomandibular joint using magnetic resonance imaging. *J Oral Maxillofac Surg.* 1996; 54: 434–439.
31. Katzberg RW. Discussion. Prevalence of bone marrow signal abnormalities observed in the temporomandibular joint using magnetic resonance imaging. *J Oral Maxillofac Surg.* 1996; 54: 439–440.
32. Sano T, Westesson P-L, Larheim TA, Rubin SJ, Tallents RH. Osteoarthritis and abnormal bone marrow of the mandibular condyle. *Oral Surg Oral Med Oral Pathol Oral Radiol Endod.* 1999; 87: 243–252.
33. Emshoff R, Gerhard S, Ennemoser T, Rudisch A. Magnetic resonance imaging findings of internal derangement, osteoarthrosis, effusion, and bone marrow edema before and after performance of arthrocentesis and hydraulic distension of the temporomandibular joint. *Oral Surg Oral Med Oral Pathol Oral Radiol Endod.* 2006; 101(6): 784–790.
34. Sano T, Westesson PL, Larheim TA, Takagi R. The association of temporomandibular joint pain with abnormal bone marrow of the mandibular condyle. *J Oral Maxillofac Surg.* 2000; 58(3): 254–257.
35. Resnick D, Sweet DE, Madewell JE. Osteonecrosis and osteochondrosis. In Resnick D, ed. *Bone and Joint Imaging.* 2nd Ed. WB Saunders, Philadelphia, PA, 1996: 941–959.
36. Chiba M, Kumagai M, Fukui N, Echigo S. The relationship of bone marrow edema pattern in the mandibular condyle with joint pain in patients with temporomandibular joint disorders: Longitudinal study with MR imaging. *Int J Oral Maxillofac Surg.* 2006; 35(1): 55–59.
37. Yajima A, Sano T, Otonari-Yamamoto M, Otonari T, Ohkubo M, Harada T, Wakoh M.MR evidence of characteristics in symptomatic osteoarthritis of the temporomandibular joint: Increased signal intensity ratio on proton density-weighted images of bone marrow in the mandibular condyle. *Cranio.* 2007; 25(4): 250–256.
38. Sano T, Otonari-Yamamoto M, Otonari T, Yajima A. Osseous abnormalities related to the temporomandibular joint. *Semin Ultrasound CT MR.* 2007; 28(3): 213–221.
39. Westesson PL, Brooks SL. Temporomandibular joint: Relationship between MR evidence of effusion and the presence of pain and disk displacement. *AJR Am J Roentgenol.* 1992; 159(3): 559–563.

40. Larheim TA, Katzberg RW, Westesson PL, Tallents RH, Moss ME. MR evidence of temporomandibular joint fluid and condyle marrow alterations: Occurrence in asymptomatic volunteers and symptomatic patients. *Int J Oral Maxillofac Surg*. 2001; 30(2): 113–117.
41. Rudisch A, Innerhofer K, Bertram S, Emshoff R. Magnetic resonance imaging findings of internal derangement and effusion in patients with unilateral temporomandibular joint pain. *Oral Surg Oral Med Oral Pathol Oral radiol Endod*. 2001; 92(5): 566–571.
42. Imoto K, Otonari-Yamamoto M, Nishikawa K et al. Potential of fluid-attenuated inversion recovery (FLAIR) in identification of temporomandibular joint effusion compared with $T_2$-weighted images. *Oral Surg Oral Med Oral Pathol Oral Radiol Endod*. 2011; 112: 243–248.
43. Hanyuda H, Otonari-Yamamoto M, Imoto K et al. Analysis of elements in a minimal amount of temporomandibular joint fluid on fluid-attenuated inversion recovery magnetic resonance images. *Oral Surg Oral Med Oral Pathol Oral Radiol*. 2013; 115: 114–120.
44. Heffez LB. Imaging of internal derangements and synovial chondromatosis of the temporomandibular joint. *Radiol Clin N Am*. 1993; 31(1): 149–162.
45. Nomoto M, Nagao K, Numata T. Synovial osteochondromatosis of the temporo-mandibular joint. *J Laryngol Otol*. 1993; 107(8): 742–745.
46. Quinn PD, Stanton DC, Foote JW. Synovial chondromatosis with cranial extension. *Oral Surg Oral Med Oral Pathol*. 1992; 73(4): 398–402.
47. Sun S, Helmy E, Bays R. Synovial chondromatosis with intracranial extension. A case report. *Oral Surg Oral Med Oral Pathol*. 1990; 70(1): 5–9.
48. Rickert RR, Shapiro MJ. Pigmented villonodular synovitis of the temporomandibular joint. *Otolaryngol Head Neck Surg*. 1982; 90(5): 668–670.
49. Lapayowker MS, Miller WT, Levy WM, Harwick RD. Pigmented villonodular synovitis of the temperomandibular joint. *Radiology*. 1973; 108(2): 313–316.
50. Curtin HD, Williams R, Gallia L, Meyers EN. Pigmented villonodular synovitis of the temporomandibular joint. *Comput Radiol*. 1983; 7(4): 257–260.
51. O"Sullivan TJ, Alport EC, Whiston HG. Pigmented villonodular synovitis of the temporomandibular joint. *J Otolaryngol*. 1984; 13(2): 123–126.
52. Song MY, Heo MS, Lee SS, Choi SC, Park TW, Lim CY. Diagnostic imaging of pigmented villonodular synovitis of the temporomandibular joint associated with condylar expansion. *Dentomaxillofac Radiol*. 1999; 28(6): 386–390.
53. Bemporad JA, Chaloupka JC, Putman CM, Roth TC, Tarro J, Mitra S. Pigmented villonodular synovitis of the temporomandibular joint: Diagnostic imaging and endovascular therapeutic embolization of a rare head and neck tumor. *Am J Neuroradiol*. 1999; 20(1): 159–162.
54. Klenoff JR, Lowlicht RA, Lesnik T, Sasaki CT. Mandibular and temporomandibular joint arthropathy in the differential diagnosis of the parotid mass. *Laryngoscope*. 2001; 111(12): 2162–2165.
55. Dijkgraaf LC, Liem RS, de Bont LG. Temporomandibular joint osteoarthritis and crystal deposition diseases: A study of crystals in synovial fluid lavages in osteoarthritic temporomandibular joints. *Int J Oral Maxillofac Surg*. 1998; 27(4): 268–273.
56. Goudot P, Jaquinet A, Gilles R, Richter M. A destructive calcium pyrophosphate dihydrate deposition disease of the temporomandibular joint. *J Craniofac Surg*. 1999; 10(5): 385–388.
57. Jibiki M, Shimoda S, Nakagawa Y, Kawasaki K, Asada K, Ishibashi K. Calcifications of the disc of the temporomandibular joint. *J Oral Pathol Med*. 1999; 28(9): 413–419.
58. Olin HB, Pedersen K, Francis D, Hansen H, Poulsen FW. A very rare benign tumour in the parotid region: Calcium pyrophosphate dihydrate crystal deposition disease. *J Laryngol Otol*. 2001; 115(6): 504–506.
59. Jordan JA, Roland P, Lindberg G, Mendelsohn D. Calcium pyrophosphate deposition disease of the temporal bone. *Ann Otol Rhinol Laryngol*. 1998; 107(11): 912–916.
60. Vargas A, Teruel J, Trull J, Lopez E, Pont J, Velayos A. Calcium pyrophosphate dihydrate crystal deposition disease presenting as a pseudotumor of the temporomandibular joint. *Eur Radiol*. 1997; 7(9): 1452–1453.
61. Larheim TA. *Imaging of the Temporomandibular Joint in Rheumatic Disease*. Baltimore, MD, Williams and Wilkins, 1991; 1(1): 133–153.
62. Smith HJ, Larheim TA, Aspestrand F. Rheumatic and nonrheumatic disease in the temporomandibular joint: Gadolinium-enhanced MR imaging. *Radiology*. 1992; 185(1): 229–234.
63. Larheim TA. Role of magnetic resonance imaging in the clinical diagnosis of the temporomandibular joint. *Cells Tissues Organs*. 2005; 180(1): 6–21.
64. Otonari-Yamamoto M, Sano T, Okano T, Wakoh M. Association between osseous changes of the condyle and temporomandibular joint (TMJ) fluid in osteoarthritis. *Oral Radiol*. 2014; 31: 41–48.

# 15

## *MR-Guided Interventional Radiology of the Musculoskeletal System*

**John M. Morelli and Jan Fritz**

**CONTENTS**



### 15.1 Introduction

While the primary use of magnetic resonance imaging (MRI) in clinical practice remains in the diagnostic realm, the modality presents several opportunities for imaging guidance of percutaneous procedures not available with computed tomography (CT), ultrasound, or fluoroscopy. Procedures performed with CT and fluoroscopy expose both the patient and operator to potentially harmful ionizing radiation. This is of particular concern in pediatric patients. The unparalleled soft tissue contrast of MRI also enables procedural targeting of structures not visible with other imaging modalities. Needle placement itself is facilitated by the ability to evaluate the needle in relationship to the imaging target in any arbitrary imaging plane. Finally, functional techniques, enabling visualization such as flow, perfusion, and diffusion may be utilized to identify appealing biopsy targets in the setting of oncologic imaging and intervention. The following chapter will focus on current and future applications of interventional MRI with respect to the musculoskeletal system.

## 15.2 Interventional MR Imaging: Technical Factors

### 15.2.1 Field Strength

Interventional MRI consists of utilization of MR imaging for anatomic target visualization, for accurate needle placement, injectant monitoring, and visualization of injection distribution.[1,2] With interventional MRI, the unparalleled soft tissue contrast of MRI is combined with multiplanar imaging capabilities as well as the possibility for real-time image acquisition, selective tissue and contrast weighting including fat suppression, and the use of non-ionizing radiation. Analogous to advantages of MR imaging of the joints and spine over other modalities, such as radiography and CT, MR imaging guidance likewise provides excellent visualization of joints and neural tissues for targeting during injection procedures.

MR-guided interventions have long been performed with open low-field MR systems[3,4]; however, the widespread clinical adoption of 1.5 T MR systems with a wide bore magnet enabled new opportunities for MR-guided procedures.[3] Specifically, the advent of MR-guided procedures on clinical wide-bore 1.5 T MR systems help transition interventional MR for a technique requiring a dedicated interventional MR system with limited availability to a technique that can be performed on any clinical, wide-bore MR system. The increasing number of wide bore systems adopted worldwide, used predominantly in the setting of imaging large and obese patients, has thus theoretically greatly increased the number of sites with interventional MR capabilities. The bore of so-called open systems measures 70 cm in diameter, providing ample space between the patient and bore for easy access to the site of intervention and adequate space for needle manipulation with the patient inside the MR system.[1,5] The spatial considerations with such open bore scanners are thus similar to those of dual-source CT systems.

Open bore 1.5 T scanners also benefit from higher signal-to-noise ratios than their low field (typically 0.2–0.5 T static magnetic field strength) counterparts traditionally used for spine interventions.[2,6,7] Benefits of greater signal-to-noise include improved depiction of injection targets such as the nerves, joints, and epidural space. Higher bandwidths which would lead to a prohibitive loss of signal-to-noise on lower field systems can be implemented enabling improved, more reliable needle visualization with less associated artifact.[8] Because chemical shift between fat and water bound protons increases proportional to field strength, spectral fat-suppression techniques are more robust. Thus, visualization of the injected material is improved on post-procedural fat-suppressed images. Signal-to-noise is also sufficient to enable use of short-tau inversion recovery (STIR) sequences for injectate visualization during the procedure, minimizing loss of fat saturation due to B0 field heterogeneity induced by the needle. The increased signal-to-noise also allows for implementation of parallel imaging. The number of scan averages can also be kept relatively low due to the higher intrinsic signal-to-noise, reducing the duration of each scan or alternatively improving image resolution.

### 15.2.2 Safety

The same safety factors in play for diagnostic MR imaging are relevant for the use of MRI for imaging guidance. Similar to diagnostic MR imaging, an effective screening procedure for patients and other individuals, including operators, entering the MR environment should be implemented. This is required to ensure patient safety and avoid accidents.[9] General contraindications to MR imaging include brain aneurysm clips, implanted neural stimulators, pacemaker/defibrillators, cochlear implants, metallic ocular foreign bodies, and insulin pumps. Pregnancy is an additional consideration. While no adverse effects are thought to occur with fetal development, MR imaging is often avoided in the first trimester. Claustrophobic patients prove difficult for both diagnostic and therapeutic MR applications; although, repeated movement of the patient out of the bore for MR-guided interventions may help decrease anxiety of some claustrophobic patients. Obesity is of particular concern for MR interventions, as there must be ample space within the bore to maintain a sterile environment and allow for needle placement. If a gadolinium chelate contrast agent is to be utilized, the patient must be specifically consented for this purpose and screened for renal disease, pregnancy, and hemoglobinopathy. Patients are generally monitored through an in-room camera, especially at field strengths greater than 2.5 T. Hearing protection is recommended for patients and operators during MR-guided procedures. Although noise levels within MR systems do not typically exceed 120 dB, reversible hearing damage can occur at 90 dB. This can hinder communication between the operator and technologists as well as between the operator and patient. In addition to static field effects, gradient fields produce associated, perpendicular electric effects which may lead to nerve stimulation, flashes of light, vertigo, nausea and sensations of metallic taste. These are more likely to occur with high field MR systems (3 T and higher). Reassurance usually proves effective for management. The MR operator can also rarely experience such effects during real-time image guidance, particularly nerve stimulation.

The radiofrequency fields utilized for MR imaging and guidance deposit energy into the body, which increases body temperature through microwave heating. The specific absorption ratio (SAR, W/kg body weight) expresses the degree of energy deposition through this mechanism and should not exceed 1 W/kg. With 1 W/kg, the whole body temperature increases by 0.5°C. Most MR systems have safeguards in place to prevent excessive radiofrequency energy deposition. SAR is greatest for brain, liver, and CSF.

### 15.2.3 Needle Selection and Artifacts

While ferromagnetic devices such as stainless steel needles can be utilized for fluoroscopy, CT, and ultrasound-guided procedures, traction forces acting upon these types of hardware during MR-guided procedures can result in serious complications.[10] The heating of ferromagnetic materials by radiofrequency fields can likewise result in burn injuries to the patient.[11,12] The extent of artifact produced by these needles as well as by needles utilized for skin and subcutaneous numbing is also unacceptable. A variety of FDA approved MR-compatible needles are currently commercially available from several vendors for safe use up to 1.5 T.

In addition to safety, needles utilized for MR-guided procedures must be consistently visualized. Mechanisms ensuring accurate visualization are more complex for MR-guided procedures[8,12,13] relative to CT or fluoroscopy where differences in X-ray attenuation provide the primary mechanism for visualization. Needle composition, geometry, and the utilized field strength influence the appearance of associated artifacts. MR compatible needles consist of non-ferromagnetic materials which have lower magnetic susceptibility compared to stainless steel, thus minimizing spatial misregistration of artifact and resulting image distortions. While the artifact should be sufficient to enable needle detection, it should not be so large so as to impair evaluation of adjacent structures or overestimate needle length. Passive visualization of the needle tip is safe and effective, requiring no recalculation or MR image post-processing.[14] The artifact is created by local spin dephasing created by regional field differences caused by magnetic susceptibility of the needle and local gradient disturbances.[15] When pulse sequences are optimized for needle visualization, the passive needle artifact displays actual tip location to within 1 mm at 1.5 T.[13] Artifacts can be more pronounced at higher field strengths.

Multiple other factors influence the appearance of the needle on the MR images utilized for guidance. These factors include pulse sequence type and parameters (voxel size, echo time, readout and radiofrequency bandwidth) as well as the angle of the needle to the static magnetic field.[1,13] In terms of pulse sequences, the lack of 180° pulses refocusing transverse magnetization results in a build-up of field inhomogeneity with gradient-echo sequences, increasing the extent of needle artifact present. This effect is more pronounced at higher field strengths.[1,13] For gradient-echo sequences, minimization of echo times yields less time for dephasing and subsequently less artifact. Although dephasing is less with turbo spin-echo sequences, short echo times and echo spacing may decrease the extent of artifact with these sequences as well. Maximized bandwidth and increasing readout gradient strength may similarly reduce needle artifacts through a reduction in frequency shifts. A wide receiver bandwidth also reduces echo spacing with the detrimental effect of increased image noise. Related to frequency shift, placing the direction of frequency encoding perpendicular to the needle long axis may reduce tip error.[1,13] Improved spatial resolution through decreased voxel sizes improves delineation of the needle tissue interface. Decreasing voxel sizes results in an apparent improvement of the needle tissue interface through improved spatial resolution. Thin slice thickness is likewise desirable but is associated with increased image noise.

Needle artifacts are most pronounced when the needle is oriented perpendicular to the static magnetic field (B0) and smallest when the orientation is parallel.[1,13] However, oftentimes this parameter is not readily modifiable with a horizontal bore system as B0 is oriented along the long axis of the patient's body, whereas the needle will often be placed perpendicular to this since most injections and other MR procedures are performed in the axial plane. Thus artifact is maximized. In-plane medial and lateral angulation of the needle does not affect artifact.

### 15.2.4 Image Acquisition for Guidance

If short, open bore systems are utilized, MR fluoroscopy is the preferred technique. MR fluoroscopy refers to the continuous acquisition and display of MR images on an MR-compatible in-room display[16] This is similar to CT fluoroscopy; however, the operator's hand can be placed within the magnet bore allowing for real-time manipulation without considerations of ionizing radiation exposure. Operation of the MR system can be performed by the technologist in the room with the patient, or remotely if a means of communication (i.e., a headset for the operator) is available. MR fluoroscopy is used for determination of the skin entry site, interactive needle placement, and injection monitoring.[5,17] The implementation of interventional MRI on higher field strength systems enables real-time imaging with higher spatiotemporal resolution.[18] Spoiled and balanced gradient-echo sequences can achieve frame rates of less than one second with sufficient spatial and contrast resolution.[1,5,19]

With the needle in place on MR fluoroscopy images, higher resolution static images are often performed. If the imaging target or pathology is readily visible, then proton-density weighted images are preferred due to the higher signal to noise and resulting advantages in spatial resolution. $T_1$ or $T_2$-weighted or fat-suppressed images can alternatively be acquired if the structure of interest is only visualized on those images. If MR fluoroscopy is not available or real-time needle guidance cannot be performed (i.e., such as in long-bore systems), then the patient can be repetitively scanned with higher resolution static sequences (usually between 20 and 40 s) and moved repetitively in and out of the bore for needle adjustments. Utilizing fat-suppressed imaging techniques for needle guidance can be problematic due to limitations with spectral fat suppression due to the presence of the needle. Thus, STIR (short tau inversion recovery) images are commonly utilized for this purpose. For biopsies, once the needle is in place, position is confirmed periodically until a satisfactory number of samples have been obtained. For pain injections, a test injection is usually given as described further below. After the full injection has been delivered, spectral fat-suppressed $T_2$-weighted images are obtained to further delineate the extent of the injection.

### 15.2.5 Injectant

While confirmatory injections are not typically performed prior to soft tissue or bone biopsies, in the setting of perineural blocks or MR-guided spine injections, the reliable visualization of injectant is essential in order to assess technical adequacy and, in turn, the validity of the pain response.[1,13] Injectants can be visualized based on their $T_1$ and $T_2$ properties. Adding a small amount of gadolinium chelate contrast agent to normal saline for an injection aids in visualization of the injectant through $T_1$ shortening effects. Non-fat saturated or spectral fat saturated $T_1$-weighted images can then be acquired to enable selective injectant visualization. Spectral fat saturated techniques are limited by field inhomogeneities induced by the needle if performed intra-operatively. Gadolinium chelate contrast agents (i.e., gadolinium-DTPA) in a ratio of 1:300–600 yields high contrast to noise between the injected material and surrounding tissue.[1,5,19] Injectants may also be visualized with $T_2$-weighted sequences and optional spectral fat saturation, based on the intrinsically long $T_2$ constant of water-based injectants.[1,5,19–21] Both techniques can be equally effective. Gadolinium-based injectants are particularly useful for MR fluoroscopy monitored injections.[22,23] Perineural gadolinium chelate injections appear to be safe based on the available literature.[24] However, saline injectants with subsequent $T_2$-weighted image confirmation are typically preferred for visualization as this obviates the need for administration of a gadolinium chelate contrast agent.

### 15.2.6 Navigation Systems for MR-Guided Procedures

Free-hand, real-time, and intermittent MR imaging guidance are generally reliable techniques for accurately guiding percutaneous procedures through needle manipulation inside the bore. However, spatial boundaries of the bore may prevent use of longer needles, and the distance to isocenter may impede positioning accuracy. Additionally longer bore systems may prohibit the operator from physically reaching within the bore to manipulate the needle. One alternative to these limitations is the performance of the puncture for a percutaneous procedure outside the bore. This is typically accomplished by an indirect navigation system utilizing an in-room monitor with guidance via intermittent MR image acquisition.[1,25] Thus, needle placement can be performed in multiple consecutive steps with the needle being positioned and advanced outside the bore and the patient being moved inside the magnet for image acquisition and assessment of the needle tip location and distance relative to the targeted structure. Due to the visual and spatial disconnect between the information on the in-room display and needle within the patient, accuracy may be reduced resulting in potentially greater tissue damage and time required to perform a procedure. Navigation systems attempt to ameliorate these problems with conventional outside the bore needle placement approaches, providing guidance information regarding the appropriate skin entry site and trajectory.[25] An added benefit of navigation systems is the potential for expedited interventional MRI procedures on more widely available, closed-bore systems.

Both virtual and augmented reality navigation systems have been proposed.[25,26] Augmented reality consists of the operator being able to simultaneously visualize MR imaging data and the patient.[26] In distinction, virtual reality consists of an entirely virtual environment, whereby a three-dimensional view of the patient and imaging targets are digitally co-registered and fused with the MR dataset.[27] The virtual reality approach requires powerful data processing capabilities, enabling real-time updates of the virtual environment. Augmented reality systems are somewhat less complex and rely on simultaneous patient and MR data visualization via co-registration techniques.[26] From a technical standpoint, this can be accomplished via a semi-transparent mirror installed above the patient outside the bore. This mirror then projects MR imaging data into a line of sight of the operator through a projection from a monitor installed above the system.[28] Available software also enables display of isotropic MR imaging data for target determination and planning of the ideal needle path. Image overlay technology thus allows simultaneous visualization of MR images and the patient.[28] Thus the need to mentally transfer MR imaging data onto the patient is obviated, leading to a more intuitive approach.

Augmented reality systems relying on image overlay navigation is advantageous relative to virtual reality systems in that they do not require additional visualization devices such as head-mounted displays.[26] An augmented reality overlay prototype can be constructed for a total cost of $4,000 to $13,000 based on whether a more expensive shielded monitor is utilized. A cheaper regular monitor can be used if the MR system configuration allows location of the system permanently outside of the 5 Gauss zone.[28] Clinically, the augmented reality overlay technology has been utilized for MR-guided spinal and pelvic bone biopsy as well as for shoulder and hip injections.[29] The latter enabled performance of MR-guided injection and diagnostic MR arthrography in a single setting.[23]

### 15.2.7 Considerations at 3 Tesla

Much in the way transition from low-field to 1.5 T MR systems has yielded benefits for diagnostic MRI, the transition from interventional 1.5 to 3 T MR may provide similar benefits. Minimally invasive neurosurgery procedures have been performed at 3 T[30,31] and the feasibility of other key percutaneous procedures demonstrated. The wide bore design of newer 3 T clinical MR systems (i.e., 70 cm diameter, similar to the 1.5 T systems) open new possibility for musculoskeletal MR-guided procedures in the future.

The underlying benefit of 3 T MR imaging is the linear relationship between the strength of the static magnetic field and spin polarization. At 3 T there is a 10 per million difference in parallel to antiparallel hydrogen spin populations at body temperature.[32] Thus, a larger total number of protons contribute to the MR signal without a commiserate increase in noise. Hence signal to noise is theoretically doubled at 3 T relative to 1.5 T.

Unfortunately the theoretical doubling in signal to noise from 1.5 to 3 T is mitigated by tissue relaxation changes and differing radiofrequency conditions at 3 T.[33] Other factors further deteriorate the achievable signal to noise. For example, chemical shift artifacts are also more pronounced at 3 T, thus necessitating use of wider readout frequency bandwidths, which increase image noise. Other negative factors on signal to noise include susceptibility effects, challenges in coil design, radiofrequency homogeneity, and limitations in the development of radiofrequency energy. With all of these factors considered, the total gain in signal to noise at 3 T relative to 1.5 T is between 1.7 and 1.8 times.[33]

Signal to noise functions as a currency of sorts in MRI, and the higher levels obtainable at 3 T enable image acquisition with higher spatial resolutions at comparable scan time, potentially allowing improved visualization of small injection targets. Alternatively, the added signal to noise can be utilized to provide higher temporal resolution for MR fluoroscopic guidance. 3 T MRI also generates improved contrast to noise in the musculoskeletal system,[34] increasing the conspicuity of pathologic changes, improving the depiction of normal anatomy, and aiding in the visualization of small volume injectants. Reducing the number of scan averages from two to one at 3 T will decrease imaging times by half, while still maintaining an approximate 40% increase of signal to noise relative to 1.5 T.

Both the MAGNETOM Verio and Skyra 3 T systems (Siemens Healthcare, Erlangen, Germany) have an open bore configuration of 70 cm that enables 36% greater patient volume. This is similar to the MAGNETOM Espree (Siemens Healthcare, Erlangen, Germany) system which is widely utilized for interventions at 1.5 T. One important difference between these systems is the longer bore with the Verio and Skyra (176 cm) relative to the Espree (125 cm). In the authors' experience, this results in the bore isocenter being inaccessible during many interventions, eliminating the possibility of real-time MR interventional guidance. Intermittent MR imaging guidance helps overcome this obstacle, but is generally less time-efficient than real-time MR fluoroscopy-guided procedures. The use of navigation systems therefore may be particularly helpful at 3 T.

To provide safe, accurate needle placements under 3 T MR guidance, several factors much be addressed. Magnetic field distortions related to para and diamagnetic tissue properties increase proportional to field strength. Thus, artifact associated with needles are theoretically twice as large at 3 T compared to 1.5 T. Passive needle visualization is the technique utilized almost exclusively for guidance at 1.5 T and has been shown to have a relatively low error rate (margin of 1 mm). At 3 T, susceptibility artifacts from the needle and local field distortions may be more severe, detrimentally affecting spatial registration. This is especially problematic given the typical near 90° orientation of the needle to the main magnetic field which maximizes susceptibility artifacts.[1,13] Non-metallic needles made of carbon and ceramic[35] may thus prove more useful for 3 T interventions. Carbon in particular has little difference in susceptibility relative to human tissue, thus resulting in little if any susceptibility-related artifacts.

Unfortunately initial attempts to visualize carbon fiber needle tips via MRI have proven problematic.[35]

The precession frequency of protons at 3 T is increased relative to 1.5 T. Thus excitation pulses with shorter wavelengths (and thus higher energy) are required. The increase in power deposition is proportional to the square of the Larmor frequency if all other characteristics of the radiofrequency pulse are held equal.[32] This increased energy deposition leads to limitations with respect to specific absorption limits (SAR). MR studies utilizing large transmitter coils are operated in normal mode with a power deposition up to 2 W/kg and in the first level mode with a deposition between 2 and 4 W/kg. A risk-benefit assessment must be made before proceeding in first-level mode. The patient must be closely supervised and may experience a warming sensation. Because power deposition also increases in proportion to the square of the flip angle, conventional fast spin echo and steady-state free precession sequences may be problematic. Potential solutions include utilization of different types of radiofrequency pulses, shorter pulse durations, decreased flip angles, and fewer slices. Smaller transmit coils can also be utilized. Gradient-echo sequences exhibit more favorable SAR properties due to the lack of 180° refocusing pulses and 90° flip angles; however, gradient-echo sequences are prone to susceptibility related artifacts, rendering their use for needle localization at 3 T somewhat limited.[1]

Heating of the metallic needles may prove problematic as well; although, our experience with smaller needles utilized for perineural injections indicates this to not be an issue. Heating of metallic devices in MR occurs as a result of radiofrequency energy deposition or alternatively via electromagnetic resonance effects. A needle may become prone to the latter when subject to the 3 T static field strength, and the extent of heating via this mechanism is dependent upon the injection needle length as well as the wavelength of the radiofrequency pulse. When the length of the needle is half the length of the radiofrequency pulse, the needle essentially serves as the antenna for the electromagnetic field of the pulse, inducing a current.[36] A needle of 13 cm in length placed in water containing tissue might thus act as an antenna for a 126 MHz radiofrequency pulse.[32] At lower field strengths, the critical lengths for needles are substantially longer (i.e., 23 cm at 1.5 T). Thorough testing must be pursued before MR-guided interventions at 3 T are routinely performed.

### 15.2.8 Cost Considerations

While the convenience of performing interventional MRI procedures has been greatly aided by the advent and widespread adoption of wide, short bore MR systems, cost considerations remain, focused on the cost for use of imaging systems, use of devices such as MR compatible needles, drills, and probes, and staff costs. The cost for MR imaging systems is typically based on procedure length; although, MR scanner time is typically more extensive than CT time due to higher purchase and maintenance costs. The length of time for MR-guided spine injection procedures at higher field strengths approximate those of CT procedures.[37] Cost differences in CT and MRI technologists are usually not substantive.

The device costs for MR compatible equipment is currently the rate limiting factor for MR to become cost competitive with CT and other modalities. MR equipment tends to be more expensive due to the use of more expensive materials such as titanium (rather than steel) and smaller quantities of needles, drills, and probes sold. A recent cost comparison study from Germany assessed costs of MR versus CT in nerve root spinal injections, finding that MR-guided procedures (177 Euros) cost approximately two times CT-guided procedures (88 Euros).[37] The increased time of the MR procedure (93 min) relative to the CT procedure (29 min) was the major factor accounting for this difference in cost. The cost differences in staff (43 versus 35 Euros for MR and CT, respectively) and injection needles (22.6 vs. 2.85 for MR and CT, respectively) also played a role. Similar results were found in the evaluation of low-field MR-guided versus CT-guided bone biopsies in Finland.[38] In that study, MR guidance cost 1205 Euros versus 472 Euros for CT. In this case, the higher cost was primarily related to more expensive MR compatible devices (i.e., drill system, 5.57 fold) in addition to higher staff costs (2.73 fold). Cost data with respect to other MR-guided interventions such as osteoid osteoma ablations have been mixed.

## 15.3 Back Pain Management

Musculoskeletal pain is one of the leading reasons for which individuals seek medical care.[39,40] An accurate physical examination in combination with laboratory data and diagnostic imaging are critical for identifying underlying causes of pain.[41] Back pain is frequently due to degenerative disk disease and osteoarthrosis involving the lumbar spine in older individuals. In younger individuals, back pain can be a symptom of inflammatory arthropathies such as juvenile idiopathic arthritis and seronegative spondyloarthropathies.

The initial management of non-emergent low back pain tends to be conservative.[42,43] Upon failure of conservative treatment over a period of 7–10 weeks, low back pain is determined to be chronic. Chronic low back pain has a prevalence of 15% in adults and 44% at the age of 70.[44,45] The source of such pain must be determined so as to establish appropriate treatment. Due to the large number of structures which can cause back pain, identification of the pain generator may prove difficult.[46] This is confounded by nonspecific or inconclusive imaging of spinal structures.[42] One report found that spinal pain generators can only be reliably identified by imaging in patients with back pain in about 15% of patients.[47,48] Paraspinal and spine injections thus provide a powerful tool for hypothesis testing in determining which structures are responsible for patient pain.[48]

Structures responsible for chronic back pain must meet several criteria to be considered as etiologic: the structure should be innervated, able to produce pain (ideally shown to produce pain in provocation tests), susceptible to painful pathologic changes, and shown to cause back pain in patients. Studies have identified facet joints, intervertebral discs, sacroiliac joints and spinal nerves as fulfilling the above criteria, representing common pain generators.[48] Diagnostic injections of such structures can be attempted for multisegmental abnormalities and other pre-surgical assessments. Other indications include situations where symptom location conflicts with imaging abnormalities, no imaging correlates to pain are identified, and when adjacent segment disease is interrogated after spinal surgery.[5] In addition to aiding with diagnosis, injections are also therapeutic and can be attempted for treatment within the context of conservative therapy, symptom relief in non-surgical cases, post-surgical pain, pain resulting from an adverse drug reaction, or adjacent segment pain following spinal fusion.[5]

Studies have likewise identified the sacroiliac joints as one substantial source of low back pain.[49,50] Major etiologies of chronic sacroiliac pain are overuse, osteoarthrosis or inflammatory arthropathy. Mechanical pain originating from the sacroiliac joints is difficult to diagnose and evaluate clinically because of other potentially painful spinal structures.[51] Inflammatory arthropathy of the sacroiliac joints (sacroiliitis) is a hallmark of seronegative spondyloarthropathy, which presents with a relatively specific, subchondral, inflammatory bone marrow edema pattern on $T_2$- and STIR-weighted MR images in the acute phase. In patients with suspected mechanical back pain originating from the sacroiliac joints, no specific imaging findings exist and protocols comprised of a series of intra-articular sacroiliac joint blocks are required to determine if the sacroiliac joints are a source of pain.

### 15.3.1 Alternative Modalities

Accuracy of intra-articular injections is optimized through imaging guidance. Potential targets include small and complex entities such as facet joints, spinal nerves, the epidural space, and the sacroiliac joints.[52] The available modalities are X-ray fluoroscopy, CT, ultrasonography, and interventional MR imaging. X-ray fluoroscopy and CT guidance are widely used and preferred techniques,[53] whereas ultrasound guidance is limited for osseous, anatomically complex, and deep targets.[54,55]

Fluoroscopic guidance is most commonly used for such guidance, is widely available, and is relatively cost-effective. Fluoroscopy provides guidance in variable projections and real-time imaging guidance. However, limitations include a lack of soft tissue visualization, operator dependence, and the projectional images that superimpose anatomic structures. Fluoroscopically guided injections may prove more difficult for complex targets such as the sacroiliac joints and facet joints.

CT guidance enables cross-sectional imaging guidance without projectional superimposition of anatomic structures. Spatial resolution is high with good osseous detail. Real-time or near real-time fluoroscopic modes are also available. CT guidance is thus preferred for injections of complex joints such as the sacroiliac and facet joints. However, several limitations do exist such as low contrast resolution of soft tissues and image acquisition limited to the axial plane.

Perhaps the most serious limitation of X-ray fluoroscopy and CT is the reliance on ionizing radiation and resulting patient and operator exposure during the exam. Ionizing radiation is associated with an increased lifetime cancer risk,[56] and thus imaging guidance utilizing ionizing radiation requires consideration of alternative modalities and techniques. This is part of the ALARA (as low as reasonably achievable) principle. There is a clear link between early age of radiation exposure and an increased life-time risk of cancer.[56] Children and young adults are more susceptible to ionizing radiation, and thus ionizing radiation dependent imaging guidance should be utilized with particular caution in such patients.

Tissues susceptible to ionizing radiation include gonadal, gastric, colonic, pulmonary, breast, and thyroid tissue. These, along with red bone marrow, are all located within the potential radiation field of spinal injection procedures. The current theories of radiation exposure dictate that below deterministic threshold doses, exposure to ionizing radiation exhibits a stochastic cancer risk. That is, the effects of radiation are statistical. Radiation protection principles are currently based on the somewhat controversial linear no threshold principle.[57,58] This presupposes that no threshold exists for stochastic effects, and that they can and do occur at any dose level, the risk of such effects increasing linearly with dose. Of note, no dose-severity effect exists, meaning that stochastic risk and the dose defines if the irradiated person will develop a fatal cancer (rather than cancer incidence) during life or not.

Patient age is the other major risk factor in addition to stochastic risk and dose. Adolescents and young adults are more susceptible to ionizing radiation, and therefore have a higher associated life-time risk of cancer.[56] A small but definite risk associated with radiation exposure has been shown.[56,59]

The expected radiation dose for a spine injection tends to be lower than those associated with body CT examinations with a mean effective dose of 1.34 mSv (CI 95%, 1.30–1.38) for CT-guided lumbar epidural injections versus 1.38 mSv (CI 95%, 1.32–1.44) for CT-guided periradicular injections.[60] The cutoff between acceptable and unacceptable radiation risks is an important consideration for the provider and patient alike.

Interventional MR imaging complies with the ALARA concept because it utilizes no ionizing radiation. It represents a valuable option for image-guided injection of complex joints and spinal targets and is of particular value for procedures performed in children, adolescents, young adults, during pregnancy, and for protocols employing a series of injections in order to avoid cumulative radiation doses.[56,59]

### 15.3.2 Technique

The lumbar facet joints (zygapophysial joints) are a significant source of lower back pain[61] and are often failed to be diagnosed as a pain nidus by noninvasive tests. Facet injections either involving the medial branch of the dorsal rami or injecting the facet joint itself are utilized to attempt to prove the facet joints as a pain generator.[62] The presence of pain relief following facet joint injections helps predict the outcomes of spinal fusion and neurotomy.[63] Medial branch blocks are performed by injecting 0.5–1 mL of local anesthetic to both supplying medial branches. These are targeted at the ipsilateral transverse process both at the level of the lumbar facet and at the level below. The L5 dorsal ramus can be targeted at the notch of the sacral ala. An alternative approach is the injection of local anesthetic within the actual joint of the facet in question. However, medial branch blocks may be more reliable due to delivery of anesthetic directly to the nerve in question.[64] The long-term effectiveness of therapeutic facet joint injections is unclear.

MR-guided facet injections have been described for low, mid-, and high field MR systems. Success rates of attempted intra-articular facet joint injections range from 75% to 89%[63,65,66] secondary to osteophytes which may potentially interfere with needle access to the joint. In distinction to CT, osseous findings and these small osteophytes in particular may be poorly visualized with low field MRI. Utilization of 1.5 or 3 T MR systems may aid in this regard. In cases where the facet joint cannot be accessed percutaneously, an MR guided median branch block may be attempted. Procedural times range from 30 to 40 min.

Radicular pain often relates to irritation of axons and the dorsal root ganglia. While MRI is sensitive for the detection of abnormalities affecting spinal nerves, such as disc herniation and stenosis, correlation between the presence and extent of abnormalities on MRI and clinical symptoms is lacking.[67] Additionally, chemically mediated radicular pain has no imaging correlation.[68] Anesthetic injections targeted to the spinal nerve roots (0.5–1 mL of local anesthetic) are utilized to determine whether spinal nerve roots, the dorsal root ganglion, or the neurodiscal interface are contributing to patient symptoms. This information can be important for predicting response to surgical intervention.[69]

Epidural injections have long been utilized as a minimally-invasive technique to provide pain relief in the setting of disc herniations and radiculopathy allowing time for the disc to resorb. Two approaches to epidural injections have been utilized including a neuroforaminal approach or direct interlaminar approach from which the epidural space is accessed via the ligamentum flavum.[5,70] For diagnostic injections, distribution of anesthetic across multiple lumbar spinal levels should be limited to avoid false-positive results. In distinction, larger volumes (i.e., 3–5 mL of steroid and anesthetic) can be utilized for therapeutic injections enabling spread along multiple spinal levels.[70]

Interventional MR imaging is particularly well-suited for spinal and perineural injections as the spinal nerves may be directly visualized under MR. The ability to directly visualize spinal nerves under MR guidance ensures accurate targeting and positioning of the needle adjacent to the nerve. Perineural spread of the injectant can be directly confirmed, including spread into the anterior and posterior epidural space as well as into the dorsal root ganglion. Injection of gadolinium chelate-enhanced injectant with subsequent fat-suppressed $T_1$-weighted images is particularly useful in this regard, enabling exclusive visualization of the high-signal-intensity injected material relative to the low signal intensity epidural fat and cerebrospinal fluid. $T_2$-weighted images can also be utilized with a simple saline injection with a similar rate of success and safety profile.

One important consideration with epidural injections and indeed all spinal injections is intravascular uptake. This occurs in anywhere from 2% to 22% of lumbar spine injections.[71–73] The significance of intravascular uptake in isolation is unknown; however, it has been rarely linked to spinal cord infarction following both transforaminal and interlaminar injections.[74,75] Intravascular uptake of particulate steroids (i.e., triamcinolone, betamethasone acetate, and methylprednisolone) with subsequent embolization has been implicated as the cause of this condition. It is possible that the ability to perform non-contrast MR angiography images prior to spinal procedures could allow identification of variant vascular anatomy (i.e., low origin of a dominant artery of Adamkiewicz) predisposing to cord infarction prior to injection. Real-time MR monitoring of epidural injections may likewise be utilized to visualize uptake of contrast within the vasculature.[66] Alternatively, a time-resolved MR angiography sequence could be utilized to detect vascular uptake via monitoring of the test gadolinium chelate-enhanced saline injectate. Gadolinium chelate contrast agents used in conjunction with steroids has been previously demonstrated to be safe in lumbar spinal injections.

Discography is a functional test enabling the assessment of discogenic lower back pain. Discography may provide important pre-surgical information prior to surgical or disc intervention, and primary discogenic pain is reported in up to a quarter of patients when no other cause is suspected.[76] With discography contrast agent is injected into the nucleus pulposus. The theory behind provocative pain with this maneuver is that the outer third of the annulus fibrosus is innervated by autonomic nerves, the sinuvertebral nerve, and gray rami communicantes of the spinal nerves.[77] In distinction, the nucleus pulposus is believed to lack innervation. Thus, annulus fibrosus tears lead to leakage of the contents of the nucleus pulposus outward resulting in discogenic pain. Such tears correspond to the MR finding of annular fissures, which are manifest as high-signal-intensity zones within the annulus on $T_2$-weighted images.[78]

Discography can be performed with X-ray fluoroscopy or CT in addition to MRI. With the latter MR is utilized to target the nucleus pulposus, followed by injection of 1–2 mL of a water soluble contrast material into the nucleus. The patient's back pain may be reproduced via the increase in intradiscal pressure. After the provocation test, a small amount of local anesthetic is typically injected for pain relief. Multiple levels may be injected in a single setting, usually with the intention of blinding the patient to the site of the injection, so that the true spinal level corresponding to painful injection can be reliably localized.

With discography there is a low risk (0.17%) of discitis, a risk that can be reduced by a two-needle co-axial technique and prophylactic antibiotics.[79] Double needle techniques with concomitant administration of intravenous antibiotics may also be useful in reducing discitis risk. With MRI guidance, gadolinium chelate-based injectates are better suited for provocation discography, whereas normal saline injectants are not feasible.[6,17,80] CT examinations are more sensitive for evaluation of annular tears following discography than is MRI.

Sacroiliac joint pain may prove a difficult diagnosis to make clinically. For diagnostic blocks, 1–1.5 mL of local anesthetic are typically injected directly into the synovial compartment of the sacroiliac joint ensuring intra-articular distribution without extra-capsular spread to adjacent structures. Upon failure of conservative treatment, larger volumes of steroid and anesthetic (3–5 mL) can be injected. The effectiveness of therapeutic injections has been shown in sacroiliitis due to seronegative spondyloarthropathy; serial treatments may be required.[81]

1.5 T MRI enables a combination of diagnostic MR imaging of the sacroiliac joints combined with MR-guided sacroiliac injections in the same setting, potentially resulting in an efficacious efficient way to treat sacroiliac joint pain.[82] This one-stop-shop technique has the added benefit of decreasing patient trips to the hospital or clinic and leading to fewer days spent away from work. Many patients with inflammatory sacroiliitis require multiple injections.[81] In this setting, MR-guided injections are particularly valuable due to the fact that repeated radiation to the pelvis can be avoided. Similar considerations are relevant in the treatment of children with juvenile idiopathic arthritis and inflammatory sacroiliitis. MR-guided steroid injections decrease inflammation associated with juvenile idiopathic arthritis with the benefit of no exposure to ionizing radiation.[82]

Image-guided selective drug delivery to lumbar sympathetic nerves is occasionally requested for conditions such as complex regional pain or phantom limb syndrome of the lower extremity, circulatory insufficiency, lower extremity frostbite and hyperhidrosis. Anesthetic can be injected via MR guidance. The lumbar sympathetic plexus is targeted for this procedure within the retroperitoneum between the anterolateral L3 vertebrae and medial psoas. Due to variable innervation schema, injection of various levels may be useful. Important structures to avoid during this injection, such as the aorta, inferior vena cava, and ureters are readily visualized on non-contrast-enhanced MRI. The visualization of the later can be improved by intravenous administration of furosemide. Errant needle placement is rare but can result in ureteral necrosis. The effectiveness of this procedure has been shown via quantification of lower extremity blood flow following sympathetic nerve blocks.[83]

Targeted anesthetic injections of the preaortic and celiac plexus can also be performed for palliative purposes, such as in the setting of pancreatic or other abdominal malignancies. Neurolytic drug injections, for example the injection of alcohol, have also been utilized.[84] MR imaging guidance may be useful in this regard due to excellent soft tissue differentiation enabling visualization of the celiac plexus along with the distribution of injected material on $T_2$-weighted MR images. For neurolysis 40–50 mL of alcohol and lidocaine is administered. Key structures to avoid include regional vascular structures such as the celiac trunk and superior mesenteric arteries along with other regional vascular structures.

The most inferior sympathetic ganglion is the solitary ganglion impar. This represents the convergence of the sympathetic chain anteriorly along the coccyx at the level of the sacrococcygeal joint at midline. The ganglion appears to play a role in coccydynia or coccyx pain syndrome. Block of this ganglion may relieve such pain. Several techniques for ganglion impar blocks have been described including a transsacrococcygeal approach, requiring forceful penetration of the sacrococcygeal disk. Fluoroscopy can be utilized for this approach; although, the ganglion impar is only able to be directly visualized by MR guidance. Injection techniques are similar to those utilized for other perineural and ganglion injections.

## 15.4 Temporomandibular Joint

The temporomandibular joint is a synovial joint with upper and lower compartments separated by a fibrocartilagenous disc located between the mandibular fossa and condyle. The anatomic complexity of the joint poses a particular challenge for intra-articular injection. While CT and X-ray fluoroscopic guidance can be utilized, the proximity of organs susceptible to the effects of ionizing radiation, such as the salivary glands and thyroid, bone marrow, and eyes, renders this technique problematic. In pediatric patients, the temporomandibular joints are frequently involved in juvenile idiopathic arthritis. Intra-articular injections may be useful in the treatment of temporomandibular joint arthropathy in children and in augmenting the diagnostic workup of temporomandibular joint disorders in adults. For purposes of diagnosis, intra-articular injection of a gadolinium chelate contrast agent enables detection of disc perforations and adhesions. On a therapeutic basis, anti-inflammatory agents, analgesics, autologous blood products, and hyaluronic acid may be injected. Injections can be performed into either of the two synovial compartments. Because of potentially severe complications resulting from incorrect needle placement (i.e., neurovascular injury or entry into the middle cranial fossa), imaging guidance is essential for temporomandibular injections.

MR guidance is a highly accurate technique for image-guided temporomandibular joint injections, avoiding potentially deleterious side effects of ionizing radiation. One study examining MR-guided temporomandibular joint injections demonstrated apparent intra-articular needle position on imaging in joints in which the contrast injection was not intra-articular, thus suggesting the necessity of real-time injection monitoring.[85]

## 15.5 Joint Injections for Arthrography

Similar to the spine procedures discussed above, MR-guided injections for direct arthrography also possess several advantages relative to conventional X-ray fluoroscopic and CT guidance. In addition to the lack of ionizing radiation required for needle guidance, MR-guided arthrography allows the potential for a *one-stop shop* approach for direct MR arthrography. That is, the same MR scanner can be utilized for both needle guidance for the injection and for subsequent arthrographic imaging without requiring an additional co-located modality. In distinction, conventional approaches to direct arthrography require separate appointments for the fluoroscopic/CT guidance portion of the exam as well as the MR imaging portion of the examination.

Both free-hand and augmented reality-guided approaches for arthrogram injections of the hip and shoulder have been described.[23] Guidance is achieved as discussed above. Typically a 22 gauge MR compatible needle is utilized for the injection. The injectate is typically a 10 mL gadolinium-enhanced saline solution. Real time monitoring of MR imaging is ideal utilizing a $T_1$-weighted fast low-angle shot two-dimensional MR imaging sequence. The length of time for this approach (approximately 14 min) is competitive with times required to perform joint injections under other forms of guidance. Following the MR arthrography joint injection, MR imaging can be pursued with the standard protocol.

## 15.6 MR-Guided Pelvic and Peripheral Nerve Injections

There is a high prevalence of pelvic pain within the general population, particularly women, with a large resulting in significant societal impact.[86] The diagnosis of causes of pelvic pain is difficult due to the multidisciplinary approach required including urologic, gynecologic, gastrointestinal, and neurologic causes in addition to pelvic floor dysfunction. Peripheral neuropathies and pelvic neuropathies are an important group of disorders resulting from nerve compression/entrapment, micro or direct trauma, and post-operative processes.

Nerve blocks are often useful in the diagnosis of such conditions, and furthermore therapeutic perineural injections may be utilize as adjuncts to conservative treatment, to expedite recovery following surgery, or in inoperable neuropathic conditions. While ultrasound guidance is advocated for superficial targets, ultrasound is limited in patients with a large body habitus and deep pelvic nerves are frequently not able to be visualized due to inadequate penetration depth or lack or acoustic window. X-ray fluoroscopic guidance can guide injections utilizing osseous landmarks; however, nerves cannot be directly visualized. CT-guided injections of peripheral nerves other than the sciatic are likewise difficult due to the poor soft tissue resolution. CT and fluoroscopy also rely on the use of ionizing radiation.

MR neurography is a particularly powerful tool for pre-procedural planning of MR-guided pelvic and peripheral nerve injections due to its ability to visualize and determine the course of pelvic nerve targets.[87] A complete high resolution MR neurography study is often performed prior to the MR-guided injection. For the injection itself, a limited MR neurography examination is first performed in order to map the course of the relevant nerve or muscle and for planning of the needle trajectory. MR fluoroscopy is then performed to identify a skin entry point either utilizing a syringe tip or the operator's finger. Needle placement is then performed under fluoroscopy followed by confirmation of adequate needle placement with fast proton-density weighted MR images. An initial test injection is performed or alternatively the entire injectate can be administered under MR fluoroscopy guidance. Finally, post-procedural MR imaging is performed to visualize the injectant and its relationship to the targeted and surrounding structures. The authors utilize a combination 0.5% ropivacaine, and 40 mg/mL of triamcinolone acetonide for the injectate. Pain responses should be assessed one-half to a full hour following the procedure. A positive block is defined as over 50% of pain relief following a technically successful nerve block with recurrence of patient's pain following the expected anesthetic half-life. Reliability of pain response can be confirmed with a second or third block either utilizing the same anesthetic (confirmatory), a placebo (placebo-controlled) or a longer acting anesthetic (comparative). Target-controlled blocks can also be performed whereby the same anesthetic is injected three times but utilizing differing target nerves.

### 15.6.1 Obturator Nerve

The obturator nerve arises from the L2-L4 posterior divisions descending within the psoas. It emerges medially at the sacroiliac joint level, following the lateral pelvic wall to the obturator canal and then making its exit from the pelvis via the obturator foramen.[88] Subsequent to its exit, the nerve divides into anterior, posterior, and cutaneous braches. The anterior branches innervate to the hip joints and hip adductors. The posterior branches innervate the deeper hip adductors and the posterior knee. Cutaneous branches innervate the skin and superficial tissues of the medial distal two thirds of the thigh.

Obturator neuropathy is a clinical syndrome presenting with groin, hip and adductor pain associated with exercise.[89,90] Surgical neurolysis can provide definitive treatment once pathology is isolated to this nerve. A positive response to an obturator nerve injection predicts a positive response to surgical neurolysis. Precise targeting of the obturator nerve typically requires imaging guidance.[91] The obturator nerve is targeted within the obturator canal near the superior obturator foramen superior and superficial to the obturator internus.

### 15.6.2 Pudendal Nerve Entrapment

The pudendal nerve descends between the coccygeal and piriformis muscles, exiting the pelvis through the inferior portion of the greater sciatic foramen.[87] The nerve then travels near the medial ischial spine between the sacrotuberous and sacrospinous ligament extending anteriorly into the pudendal (Alcock's) canal. The pudendal nerve terminates as the perineal and dorsal nerves of the penis or clitoris. The perineal branches overlap with the innervation of the perineal branch of the posterior femoral cutaneous nerve.

Pudendal neuralgia consists of pain worsened by sitting, not awakening the patient and night without objective sensory loss associated. Diagnostic blocks also play an important role in the diagnosis of pudendal neuralgia.[92] Pudendal nerve entrapment can occur at several different locations[93]: the greater sciatic foramen, ischial spine, and Alcock's canal. Some of these locations are more conducive to localized injections than others. Selective injections at the greater sciatic foramen and ischial spine are more difficult to achieve, the latter due to the proximity to the posterior femoral cutaneous nerve. Blockade of the perineal branches of the posterior femoral cutaneous nerve may lead to a false positive test. The optimal location for a pudendal nerve injection is within Alcock's canal due to the fact that the canal boundary provides a barrier to other neural structures. If too much injectate is administered, flow of injectate can extend retrograde to the greater sciatic foramen.

### 15.6.3 Posterior Femoral Cutaneous Nerve

The posterior femoral cutaneous nerve has only sensory components which descend through the greater sciatic foramen along with the sciatic and pudendal nerves.[87] The posterior femoral cutaneous nerve travels anterior to the piriformis and exits the pelvis inferior to the piriformis. The posterior femoral cutaneous nerve then descends laterally with the sciatic lateral to the gluteus maximus, whereas the pudendal nerve curves medially into the pudendal canal. At the inferior most aspect of the gluteus maximus muscle, the inferior cluneal branches arise from the posterior femoral cutaneous nerve. The perineal branches arise just inferior to the sacrotuberous ligament termination. These perineal branches innervate the posterolateral perineum, proximal medial thigh, and portions of the scrotum and labia. Thus, neuropathy of the posterior femoral cutaneous nerve involving the perineal branches can mimic pudendal neuralgia. The posterior femoral cutaneous nerve eventually separates from the sciatic traveling posteriorly along the long head of the biceps.

Posterior femoral cutaneous neuropathy can be related to compression, entrapment, trauma from cycling, direct trauma, or tendinosis at the hamstring origin. Symptoms of posterior femoral cutaneous neuropathy are variable but often include posterior thigh pain or paresthesias in addition to pain in the perineal or inferior lateral buttock (i.e., cluneal nerve distribution) pain.[94] Selective diagnostic blocks of the posterior femoral cutaneous nerve can be useful for determining if it is the pain generating nerve. The nerve is best accessed percutaneously within the posterior thigh. More proximal injections may be confounded by anesthetic effects on the sciatic nerve.

### 15.6.4 Sciatic Nerve Injections

The sciatic is the largest nerve in the human body and has both cutaneous and motor branches to the back of the lower extremity. The sciatic nerve, like the pudendal and posterior femoral cutaneous nerves, exits via the greater sciatic foramen and runs posterior to the adductor magnus and anterior to the gluteus maximus.[87] Multiple sources of sciatic neuropathy exist including diabetes, trauma/iatrogenic injuries, radiation therapy, endometriosis, and piriformis syndrome. The sciatic nerve can be accessed reliably via ultrasound just distal to the piriformis.[95] Deeper sciatic injections at the level of the sciatic foramen are better performed with MR guidance.

The piriformis muscle arises from S2–4, the superior greater sciatic foramen, and the sacrotuberous ligament coursing through the greater sciatic foramen and inserting upon the greater trochanter of the femur. The piriformis divides the greater sciatic foramen into a superior portion including the superior gluteal nerve and vessels and an inferior portion containing the inferior gluteal, internal pudendal vessels, as well as the inferior gluteal, pudendal, sciatic, and posterior femoral cutaneous nerves.

Piriformis syndrome is a cause of chronic pelvic pain syndrome which can affect any nerve coursing through the inferior aspect of the greater sciatic foramen.[96] The proposed mechanism is inflammation and spasticity of the piriformis muscle with compression of the nerve against the pelvis. Asymmetry of the piriformis muscles as well as anatomic variation of the piriformis muscles may relate to entrapment syndromes; however, these may also be seen in asymptomatic patients. Abnormal signal intensity within the sciatic nerve in addition to an abnormal morphology of the sciatic nerve help confirm the diagnosis of piriformis syndrome.

Perineural sciatic nerve injections of anesthetic and steroid can provide pain relief and are often combined with intramuscular injection of steroids, anesthetics, and/or botulism toxin into the piriformis muscle. Particular care must be taken in performing perineural anesthetic injections of the sciatic nerve due to its motor components. Specifically, patients undergoing such injections should be forewarned that temporary lower extremity weakness is a potential effect of such injections. Appropriate arrangements for safe transport of patients undergoing sciatic nerve blocks should be arranged until the anesthetic effect subsides. Botulism neurotoxin A inhibits acetylcholine release at the neuromuscular junction, producing muscle paralysis and eventually muscle atrophy. Thus, such injections are thought to potentially relieve mass effect of the piriformis muscle on the sciatic nerve. 100 to 200 units of botulism toxin are typically utilized.

### 15.6.5 Lateral Femoral Cutaneous Perineural Injections

The lateral femoral cutaneous nerve is a sensory nerve arising from the L2–3 nerve roots providing anterior lateral thigh sensation superior to the level of the knee.[87] The lateral femoral cutaneous nerve penetrates the psoas major and runs obliquely anterior to the iliacus. The nerve exits the pelvis inferior to the inguinal ligament just medial to the anterior superior iliac spine. In a minority of patients the nerve courses lateral to the anterior superior iliac spine or pierces the inguinal ligament.[97]

Meralgia paresthetic refers to neuropathy of the lateral femoral cutaneous nerve secondary to either entrapment or injury.[98] The clinical presentation of this syndrome is characterized by proximal anterior and/or lateral thigh pain. Microtrauma, fascial entrapment, or scar encasement can cause these symptoms. While subject to entrapment throughout its course, the lateral femoral cutaneous nerve is most commonly compressed near the inguinal ligament where it exits the pelvis.

Nerve blocks, either diagnostic or therapeutic, of the lateral femoral cutaneous nerve can be performed without imaging guidance, but because the target is typically near the inguinal ligament and the nerve's course is variable in this area, there is a relatively low success rate. Ultrasound may be helpful to improve guidance in this region. However, intrapelvic blocks are best done under MR guidance. Therapeutic injections are associated with a high level of success and may serve as an adequate bridge to surgery.

## 15.7 Musculoskeletal Biopsy and Aspiration

Image-guided percutaneous biopsies, whether they be done via ultrasound, CT, fluoroscopy, or MRI represent the typical first invasive step in the diagnostic workup of musculoskeletal lesions. CT and fluoroscopy are limited by the use of ionizing radiation, soft tissue contrast are limited with both, and fluoroscopy is further problematic due to superimposition of overlying tissues. Ultrasound is limited by poor tissue penetration in addition to limited evaluation in the setting of soft tissue gas. Benefits of MR imaging guidance include a lack of ionizing radiation, as well as unparalleled soft tissue contrast and the ability to perform functional imaging. Thus, MRI-guided biopsy proves useful in biopsies of lesions that are not visible with other imaging modalities and also when a particular area of tumor is best visualized under MR guidance.[99] In particular many malignant bone and cartilaginous lesions are poorly visualized without intravenous contrast, and the use of MR allows targeting of areas of contrast-enhancement. There are some disadvantages to MR guidance relative to CT due to the fact that spatial resolution of CT is overall greater, thus making it the guidance modality of choice within the cervical spine or in sclerotic lesions. MR compatible bone biopsy needles are also composed of relatively soft titanium alloys, utilized in lieu of steel. Such thin needles are suboptimal for cortical penetration. MR-guided biopsy is slower and more expensive than CT-guided procedures, but less invasive and cheaper than open surgery.

The previously discussed guidance techniques for MR-guided perineural and spine procedures are readily adopted for imaging of bone and soft tissue biopsy targets. Unlike most perineural interventions, both local anesthesia and moderate sedation are utilized with general anesthesia reserved for pediatric patients or adults with potentially challenging airways. Fine needle aspiration is also routinely performed with both bone and soft tissue biopsies and has been shown to identify malignancies missed on histology in a small number of cases.

In terms of biopsy technique itself, for a bone biopsy autoclavable tools are utilized. The skin entry site is identified via a saline syringe or the operator's finger. After sterilization of the field, local anesthetic is applied and intravenous moderate sedation given. With a spinal needle further anesthetic is applied to the periosteum. A small skin incision is made and a trochar is advanced with a stylet under real-time fluoroscopic guidance or alternatively intermittent MR guidance. The stylet is utilized to penetrate thick cortical bone and once medullary access is achieved fine needle aspiration specimens are taken and the drill is utilized to take bone biopsy sample. Lytic lesions or lesions with prominent soft tissue components can also be biopsied utilizing automatic or semi-automatic soft tissue biopsy guns.

Soft tissue biopsies likewise are performed via a co-axial technique. Once local anesthesia and moderate sedation are achieved, a co-axial puncture needle is advanced into the outer lesion border. Subsequently an automatic or semi-automatic biopsy tool is utilized to obtain samples from the lesion. Fine needle aspiration can be performed through the co-axial puncture needle either before or after core biopsies are taken.

In addition to soft tissue and bone biopsies performed under MR guidance, various percutaneous aspirations can be performed throughout the musculoskeletal system. The main indication for aspiration of a joint is typically infection, and given their urgent nature, these aspirations are most frequently performed under readily available modes of imaging guidance, such as ultrasound or fluoroscopy. Aspiration of juxta-articular collections is less well-established via fluoroscopic approaches and CT, ultrasound, and even MRI guidance can be attempted. These include aspiration and steroid injection into inflamed bursae both deep and superficial including the infrapatellar, iliopsoas, trochanteric, and olecranon bursae. Other juxta-articular cysts and ganglia are also commonly aspirated including parameniscal cysts of the knee and paralabral cysts of the shoulder and hip. Following aspiration, steroids are typically injected under imaging guidance. Lymphoceles and hematomas can also be treated with percutaneous drainage. These may involve instillation of appropriate pharmaceutical agents to speed collection dissolution or leaving a drain in place following initial aspiration. Ganglia aspirations must be performed with caution as any fluid collection within an extremity could potentially represent a malignant sarcoma or myxoid tumor.

One advantage of MRI for soft tissue and bone biopsy relative to other modalities is the ability to utilize functional imaging in the context of biopsy planning. For example, in the setting of transrectal prostate biopsy, targeting of samples to areas of low apparent diffusion coefficient (ADC) value aids in biopsying the highest grade areas of the prostate malignancy.[100] The theory behind diffusion weighted imaging is that the motion of water molecules is more restricted in tissues with high cellularity, intact cell membranes, and reduced extracellular space (stroma).[101] Thus, when faced with a tumor, such as a heterogeneous sarcoma or bone tumor, the highest yield biopsy specimens may come from the areas of the greatest cellularity rather than from areas of necrosis, where the necrotic cells available offer little histologic information as to the tumor source, or areas with lower cellularity, which may result in a false diagnosis of a lower grade tumor than is actually present. These areas of increased cellularity result in restricted diffusion and thus decreased ADC values.[102] Therefore, heterogeneous sarcomas or bone tumors may be amenable to ADC-guided targeting under MR guidance; although, the efficacy of this approach has not yet been proven. Other specific areas can also be targeted based on functional imaging, such as the area of greatest contrast enhancement or greatest contrast enhancement in the arterial phase. Such targeting may increase the yield of biopsy. In the future, it may be possible to utilize MR spectroscopy to guide needle placement to the most cellular regions within a given heterogeneous tumor.

## 15.8 Vertebroplasty

X-ray fluoroscopy and CT guidance are most frequently utilized to guide vertebroplasty; however, the associated ionizing radiation exposure and low contrast resolution of both techniques renders some neoplastic lesions of the spine difficult to visualize. The high contrast resolution of MRI and lack of ionizing radiation are thus advantageous. Tumor lesions not identifiable fluoroscopically or by CT can be precisely targeted and treated under MR guidance. Particularly in the setting of vertebroplasty following a minimally invasive therapy such as cryoablation or radiofrequency ablation, MR may prove useful. Critical adjacent structures such as spinal nerves, the spinal cord, and nearby vessels can be identify and safely avoided utilizing MRI for guidance.

Both conventional real-time fluoroscopic guidance as well as image overlay guidance can be used for MR-guided vertebroplasty. Both transpedicular and para pedicular approaches can be utilized. Currently the cement utilized is conventional polymethylmethacrylate (PMMA). This produces low signal intensity on MR images performed for vertebroplasty guidance due to a lack of mobile protons. Some authors have suggested cement containing both saline, PMMA, and gadolinium chelate contrast agents or alternatively hydroxyapatite, PMMA, and a gadolinium chelate with the goal of producing high signal intensity on $T_1$-weighted images.[103] The saline or hydroxyapatite produces mobile protons and the gadolinium chelate results in $T_1$ relaxation. While addition of these components to PMMA does increase its signal intensity on $T_1$-weighted images, a side effect is that the cement mixture loses mechanical strength. Even with conventional PMMA cement, extra-vertebral extravasation of contrast as well as intravasation of contrast into the basivertebral plexus can be readily visualized.

## 15.9 Bone Tumor Therapeutic Applications

Metastatic disease is the most commonly encountered lesion of bone overall. Patients with bone metastases are rarely candidates for surgical resection due to the advanced stage of disease implied by the presence of metastases and poor expected prognosis overall. Diffuse metastatic disease is best treated systemically; however, palliative therapies are often pursued with these patients, most commonly radiation therapy. Several percutaneous therapies are also available including thermal methods such as laser, focused ultrasound, radiofrequency, and cryoablation. These are particularly well-suited for MR monitoring due to the fact that MRI can assess tissue temperature directly through MR thermography. In distinction, utilization of ethanol to ablate tumors is less well-established due to the liquid substance being applied. Bone lesions with some degree of osteolytic or soft tissue component are the lesions on which ablation procedures are most easily performed. Osteoblastic lesions are not as easily accessed and dedicated vertebroplasty or bone biopsy needles may be required for this purpose.

The entry procedures utilized for MR-guided interventions in bone are similar to those for bone biopsy. However, the use of MR-guidance for percutaneously delivered therapy into osseous structures is less well-established. Bone tumor management by this means can be palliative or curative. Tumor ablation consists of the application of chemical, thermal, or physical energy to a tumor in order to destroy tumor cells. Thermal methods include both cold techniques, such as cryoablation, and heat-based techniques such as microwave, radiofrequency, laser, or focused ultrasound.[104] Focused ultrasound, cryoablation, and laser ablation are best suited for MR guidance.

Focused ultrasound utilizes an ultrasound probe placed outside the body to destroy tissue, and represents a true noninvasive technology. Because bone has a high rate of acoustic absorption, pronounced heating of the bone and periosteum occurs upon application of focused ultrasound. Pain relief with focused ultrasound of bone lesions is believed to relate to destruction of periosteal nerves through this heating process. Preliminary results with this technique have been promising for the treatment of pain for metastatic disease. Complications include skin burns and other soft tissue injuries related to heat deposition.

Prior to the procedure, thorough planning of the therapeutic percutaneous intervention is necessary including the assessment of critical surrounding nerves, vascular structures, and, if applicable, bowel and urogenital structures. For thermal procedures, installation of saline or gas can be performed near the catheter to provide thermal protection. In the setting of palliative ablation for bone metastasis, the approach is typically to treat the tumor-bone interface. In the rare case where local disease control is the goal, the treatment must include ablation not only of the tumor, but also beyond the margin of the tumor, avoiding critical structures nearby.

As opposed to metastatic lesions, most primary bone tumors are treated via surgery. However, several focal benign lesions are commonly treated by ablative methods including osteoid osteoma and osteoblastoma. Unicameral bone cysts can also be treated via curettage, steroid injection, and bone marrow injections—therapies which can theoretically be delivered percutaneously as well. Treatment results of osteoid osteoma via MR-guidance are similar to those with other modalities.[105,106]

Cryoablation is a minimally invasive technique whereby cryoprobes are placed into the tumor under imaging guidance. The Joule–Thompson effect achieves tissue freezing with argon exiting and expanding the probe tip and helium thawing the tissue.[107] Several freeze-thaw cycles, usually at least 2, are employed for local therapy. While the cryoablation ice ball may be isointense to ablated tissue with CT-guidance, on MR the ice ball is well-demarcated demonstrated low-signal intensity as a result of a lack of mobile protons within ice. Multiple cryoablation probes can be utilized to *sculpt* the ice ball, and use of several probes may be required for larger lesions. Pain palliation for soft tissue and bone tumors treated by this method has been impressive and associated complication rates low.

Laser ablation utilizes infrared or near infrared optical fibers to induce heat and resulting necrosis within tissue via minimally-invasive percutaneous access. The optimal fibers utilized for this technique are MR compatible, and fibers are small enough to enable insertion through an 18 gauge needle. For osseous ablations, the probe can be exposed to bone after access is obtained with an introducer needle. Cooled applicator systems introduced co-axially can increase the size of the ablation volume. With MRI, real-time guidance of laser induced thermal effects is possible. Laser ablations can be used for treatment of osteoid osteoma and other bone lesions.

Treatment of other musculoskeletal lesions is likewise possible under MR guidance. Specifically, cystic bone lesions such as unicameral, aneurysmal, and posttraumatic bone cysts can be treated, usually simply through biopsy; although, mechanical curettage is possible as well. If liquid sclerosant is utilized for such purposes, both MR and fluoroscopy should be used if a hybrid system is available—MRI to target the initial lesion and fluoroscopy to monitor therapy and ensure that the injection is not intra-vascular. The application of sclerosants can also be helpful for the treatment of aneurysmal bone cysts. Agents utilized for this purpose include ethibloc and polidocanol.

Osteochondral defects in adults, if unstable, are treated surgically. In children, conservative therapy is the usual initial treatment; however, oftentimes surgery is pursued if this fails. Under MR guidance if the overlying cartilage is intact, percutaneous retrograde drilling of this entity can be performed under MRI guidance. Similarly avascular necrosis is often treated with core decompression. When small Ficat stage 1 or 2 lesions are present, MRI can be utilized to find the best target for therapy (i.e., usually with the most edema), and to guide drilling of small channels into the necrotic lesion. Typically 3 mm drills are utilized.

Other potential applications for interventional MRI in the musculoskeletal system are promising. Intraoperative MRI guidance is usually reserved for the realm of neurosurgery, but could play a role in musculoskeletal imaging and intervention as well. In particular, bone lesions poorly visualized optically could be marked under MRI guidance, similar to a needle localization procedure in breast imaging. A path of a surgical approach could be similarly determined or delineated based on MR guidance either pre-procedure or with an operative grade MR system. For example, an intraoperative MR-guided pediatric physeal bone bridge removal was recently performed under MR guidance.[108]

## 15.10 Summary

MR-guided procedures are still in their early stages of clinical acceptance and adoption in the field of musculoskeletal radiology. The primary advantages of utilizing MR imaging for procedural guidance are the lack of ionizing radiation associated, the unparalleled soft tissue contrast, and the ability to assess functional parameters such as temperature in the course of the MR evaluation. In the future, there is great potential for MR-guided procedures within the musculoskeletal system. MRI can detect early cartilage changes, and these findings could be used to guide early intervention, such as targeted injection of regenerative agents. Some chondral restoration procedures and microfracture repairs currently done arthroscopically or as an open surgical procedure could also theoretically be performed under MR guidance. Continued multidisciplinary collaboration among musculoskeletal radiologists, interventional radiologists, orthopedic surgeons, and rheumatologists will continue to guide the future of this exciting field.

## References


1. Fritz J, Pereira PL. MR-Guided pain therapy: Principles and clinical applications. *Rofo* 2007;179:914–24.
2. Carrino JA, Blanco R. Magnetic resonance—Guided musculoskeletal interventional radiology. *Semin Musculoskelet Radiol* 2006;10:159–74.
3. Pereira PL, Gunaydin I, Trubenbach J et al. Interventional MR imaging for injection of sacroiliac joints in patients with sacroiliitis. *AJR Am J Roentgenol* 2000;175:265–6.
4. Lewin JS, Petersilge CA, Hatem SF et al. Interactive MR imaging-guided biopsy and aspiration with a modified clinical C-arm system. *AJR Am J Roentgenol* 1998;170:1593–601.
5. Fritz J, Niemeyer T, Clasen S et al. Management of chronic low back pain: Rationales, principles, and targets of imaging-guided spinal injections. *RadioGraphics* 2007;27:1751–71.
6. Sequeiros RB, Klemola R, Ojala R, Jyrkinen L, Vaara T, Tervonen O. Percutaneous MR-guided discography in a low-field system using optical instrument tracking: A feasibility study. *J Magn Reson Imaging* 2003;17:214–9.
7. Smith KA, Carrino J. MRI-guided interventions of the musculoskeletal system. *J Magn Reson Imaging* 2008;27:339–46.
8. Graf H, Lauer UA, Klemm T, Schnieder L, Schick F. Artifacts in MRT caused by instruments and implants. *Z Med Phys* 2003;13:165–70.
9. Kanal E, Borgstede JP, Barkovich AJ et al. American College of Radiology White Paper on MR Safety: 2004 update and revisions. *AJR Am J Roentgenol* 2004;182:1111–4.
10. Fritz J, Konig CW, Gunaydin I et al. Magnetic resonance imaging—Guided corticosteroid-infiltration of the sacroiliac joints: Pain therapy of sacroiliitis in patients with ankylosing spondylitis. *Rofo* 2005;177:555–63.
11. Dempsey MF, Condon B, Hadley DM. Investigation of the factors responsible for burns during MRI. *J Magn Reson Imaging* 2001;13:627–31.
12. Shellock FG, Crues JV. MR procedures: Biologic effects, safety, and patient care. *Radiology* 2004;232:635–52.
13. Lewin JS, Duerk JL, Jain VR, Petersilge CA, Chao CP, Haaga JR. Needle localization in MR-guided biopsy and aspiration: Effects of field strength, sequence design, and magnetic field orientation. *AJR Am J Roentgenol* 1996;166:1337–45.
14. Lufkin R, Teresi L, Chiu L, Hanafee W. A technique for MR-guided needle placement. *AJR Am J Roentgenol* 1988;151:193–6.
15. Ludeke KM, Roschmann P, Tischler R. Susceptibility artefacts in NMR imaging. *Magn Reson Imaging* 1985;3:329–43.
16. Artner J, Cakir B, Reichel H, Lattig F. Radiation dose reduction in CT-guided sacroiliac joint injections to levels of pulsed fluoroscopy: A comparative study with technical considerations. *J Pain Res* 2012;5:265–9.
17. Streitparth F, Hartwig T, Schnackenburg B et al. MR-guided discography using an open 1 Tesla MRI system. *Eur Radiol* 2011;21:1043–9.
18. Tsao J, Kozerke S. MRI temporal acceleration techniques. *J Magn Reson Imaging* 2012;36:543–60.
19. Streitparth F, Walter T, Wonneberger U et al. Image-guided spinal injection procedures in open high-field MRI with vertical field orientation: Feasibility and technical features. *Eur Radiol* 2010;20:395–403.
20. Ojala R, Vahala E, Karppinen J et al. Nerve root infiltration of the first sacral root with MRI guidance. *J Magn Reson Imaging* 2000;12:556–61.
21. Sequeiros RB, Ojala RO, Klemola R, Vaara TJ, Jyrkinen L, Tervonen OA. MRI-guided periradicular nerve root infiltration therapy in low-field (0.23-T) MRI system using optical instrument tracking. *Eur Radiol* 2002;12:1331–7.
22. Fritz J, Bizzell C, Kathuria S et al. High-resolution magnetic resonance-guided posterior femoral cutaneous nerve blocks. *Skeletal Radiol* 2013;42:579–86.
23. Fritz J, U-Thainual P, Ungi T et al. Augmented reality visualization with use of image overlay technology for MR imaging-guided interventions: Assessment of performance in cadaveric shoulder and hip arthrography at 1.5 T. *Radiology* 2012;265:254–9.
24. Shetty SK, Nelson EN, Lawrimore TM, Palmer WE. Use of gadolinium chelate to confirm epidural needle placement in patients with an iodinated contrast reaction. *Skeletal Radiol* 2007;36:301–7.
25. Moche M, Trampel R, Kahn T, Busse H. Navigation concepts for MR image-guided interventions. *J Magn Reson Imaging* 2008;27:276–91.
26. Fichtinger G, Deguet A, Fischer G et al. Image overlay for CT-guided needle insertions. *Comput Aided Surg* 2005;10:241–55.
27. Jaramaz B, Eckman K. Virtual reality simulation of fluoroscopic navigation. *Clin Orthop Relat Res* 2006;442:30–4.
28. Fritz J, U-Thainual, Ungi T et al. Augmented reality visualization with image overlay for MRI-guided intervention: Accuracy for lumbar spinal procedures with a 1.5-T MRI system. *AJR Am J Roentgenol* 2012;198:W266–73.
29. Fritz J, U-Thainual, Ungi T et al. Augmented reality visualization using image overlay technology for MR-guided interventions: Cadaveric bone biopsy at 1.5 T. *Invest Radiol* 2013;48:464–70.
30. Hall WA, Galicich W, Bergman T, Truwit CL. 3-Tesla intraoperative MR imaging for neurosurgery. *J Neurooncol* 2006;77:297–303.

31. Truwit CL, Hall WA. Intraoperative magnetic resonance imaging-guided neurosurgery at 3-T. *Neurosurgery* 2006;58:ONS-338–45; discussion ONS-45–6.
32. Springer F, Martirosian P, Boss A, Claussen CD, Schick F. Current problems and future opportunities of abdominal magnetic resonance imaging at higher field strengths. *Top Magn Reson Imaging* 2010;21:141–8.
33. Machann J, Schlemmer HP, Schick F. Technical challenges and opportunities of whole-body magnetic resonance imaging at 3T. *Phys Med* 2008;24:63–70.
34. Kornaat PR, Reeder SB, Koo S et al. MR imaging of articular cartilage at 1.5T and 3.0T: Comparison of SPGR and SSFP sequences. *Osteoarthr Cartilage* 2005;13:338–44.
35. Thomas C, Wojtczyk H, Rempp H et al. Carbon fibre and nitinol needles for MRI-guided interventions: First in vitro and in vivo application. *Eur J Radiol* 2011;79:353–8.
36. Schaefers G, Melzer A. Testing methods for MR safety and compatibility of medical devices. *Minim Invasive Ther Allied Technol* 2006;15:71–5.
37. Maurer MH, Schreiter N, de Bucourt M et al. Cost comparison of nerve root infiltration of the lumbar spine under MRI and CT guidance. *Eur Radiol* 2013;23:1487–94.
38. Alanen J, Keski-Nisula L, Blanco-Sequeiros R, Tervonen O. Cost comparison analysis of low-field (0.23 T) MRI- and CT-guided bone biopsies. *Eur Radiol* 2004;14:123–8.
39. Frymoyer JW, Cats-Baril WL. An overview of the incidences and costs of low back pain. *Orthop Clin N Am* 1991;22:263–71.
40. Guo HR, Tanaka S, Halperin WE, Cameron LL. Back pain prevalence in US industry and estimates of lost workdays. *Am J Public Health* 1999;89:1029–35.
41. Brant-Zawadzki MN, Dennis SC, Gade GF, Weinstein MP. Low back pain. *Radiology* 2000;217:321–30.
42. Van den Hoogen HM, Koes BW, van Eijk JT, Bouter LM. On the accuracy of history, physical examination, and erythrocyte sedimentation rate in diagnosing low back pain in general practice. A criteria-based review of the literature. *Spine* (Phila Pa 1976) 1995;20:318–27.
43. Van den Hoogen HJ, Koes BW, Deville W, van Eijk JT, Bouter LM. The prognosis of low back pain in general practice. *Spine* (Phila Pa 1976) 1997;22:1515–21.
44. Andersson GB. Epidemiological features of chronic low-back pain. *Lancet* 1999;354:581–5.
45. Manchikanti L, Singh V, Pampati V, Beyer CD, Damron KS. Evaluation of the prevalence of facet joint pain in chronic thoracic pain. *Pain Physician* 2002;5:354–9.
46. Jarvik JG, Deyo RA. Diagnostic evaluation of low back pain with emphasis on imaging. *Ann Intern Med* 2002;137:586–97.
47. Frymoyer JW. Back pain and sciatica. *N Engl J Med* 1988;318:291–300.
48. Bogduk N. International Spinal Injection Society guidelines for the performance of spinal injection procedures. Part 1: Zygapophysial joint blocks. *Clin J Pain* 1997;13:285–302.
49. Fortin JD, Aprill CN, Ponthieux B, Pier J. Sacroiliac joint: Pain referral maps upon applying a new injection/arthrography technique. Part II: Clinical evaluation. *Spine* (Phila Pa 1976) 1994;19:1483–9.
50. Fortin JD, Dwyer AP, West S, Pier J. Sacroiliac joint: Pain referral maps upon applying a new injection/arthrography technique. Part I: Asymptomatic volunteers. *Spine* (Phila Pa 1976) 1994;19:1475–82.
51. Slipman CW, Sterenfeld EB, Chou LH, Herzog R, Vresilovic E. The value of radionuclide imaging in the diagnosis of sacroiliac joint syndrome. *Spine* (Phila Pa 1976) 1996;21:2251–4.
52. Gangi A, Dietemann JL, Mortazavi R, Pfleger D, Kauff C, Roy C. CT-guided interventional procedures for pain management in the lumbosacral spine. *RadioGraphics* 1998;18:621–33.
53. Gilula LA, Lander P. Management of spinal pain with imaging-guided injection. *RadioGraphics* 2003;23:189–90; author reply 90–1.
54. Parra DA, Chan M, Krishnamurthy G et al. Use and accuracy of US guidance for image-guided injections of the temporomandibular joints in children with arthritis. *Pediatr Radiol* 2010;40:1498–504.
55. Pekkafahli MZ, Kiralp MZ, Basekim CC et al. Sacroiliac joint injections performed with sonographic guidance. *J Ultrasound Med* 2003;22:553–9.
56. Pearce MS, Salotti JA, Little MP et al. Radiation exposure from CT scans in childhood and subsequent risk of leukaemia and brain tumours: A retrospective cohort study. *Lancet* 2012;380:499–505.
57. Nussbaum RH. The linear no-threshold dose-effect relation: Is it relevant to radiation protection regulation? *Med Phys* 1998;25:291–9; discussion 300.
58. Health Physics Society. The linear no-threshold model: Is it still valid for the prediction of dose-effects and risks from low level radiation exposure? Proceedings of a conference to honor Victor Bond in his 75th year. November 1994. *Health Phys* 1996;70:775–882.
59. Brenner DJ, Hall EJ. Cancer risks from CT scans: Now we have data, what next? *Radiology* 2012;265:330–1.
60. Artner J, Lattig F, Reichel H, Cakir B. Effective dose of CT-guided epidural and periradicular jnjections of the lumbar spine: A retrospective study. *Open Orthop J* 2012;6:357–61.
61. Hirsch C, Ingelmark BE, Miller M. The anatomical basis for low back pain. Studies on the presence of sensory nerve endings in ligamentous, capsular and intervertebral disc structures in the human lumbar spine. Acta Orthop Scand 1963;33:1–17.
62. Bogduk N. Evidence-informed management of chronic low back pain with facet injections and radiofrequency neurotomy. *Spine J* 2008;8:56–64.
63. Cohen SP, Hurley RW. The ability of diagnostic spinal injections to predict surgical outcomes. *Anesth Analg* 2007;105:1756–75, table of contents.
64. Marks RC, Houston T, Thulbourne T. Facet joint injection and facet nerve block: A randomised comparison in 86 patients with chronic low back pain. *Pain* 1992;49:325–8.

65. Fritz J, Clasen S, Boss A et al. Real-time MR fluoroscopy-navigated lumbar facet joint injections: Feasibility and technical properties. *Eur Radiol* 2008;18:1513–8.
66. Fritz J, Thomas C, Clasen S, Claussen CD, Lewin JS, Pereira PL. Freehand real-time MRI-guided lumbar spinal injection procedures at 1.5 T: Feasibility, accuracy, and safety. *AJR Am J Roentgenol* 2009;192:W161–7.
67. Boos N, Rieder R, Schade V, Spratt KF, Semmer N, Aebi M. 1995 Volvo Award in clinical sciences. The diagnostic accuracy of magnetic resonance imaging, work perception, and psychosocial factors in identifying symptomatic disc herniations. *Spine* (Phila Pa 1976) 1995;20:2613–25.
68. Mulleman D, Mammou S, Griffoul I, Watier H, Goupille P. Pathophysiology of disk-related low back pain and sciatica. II. Evidence supporting treatment with TNF-alpha antagonists. *Joint Bone Spine* 2006;73:270–7.
69. North RB, Kidd DH, Zahurak M, Piantadosi S. Specificity of diagnostic nerve blocks: A prospective, randomized study of sciatica due to lumbosacral spine disease. *Pain* 1996;65:77–85.
70. Vad VB, Bhat AL, Lutz GE, Cammisa F. Transforaminal epidural steroid injections in lumbosacral radiculopathy: A prospective randomized study. *Spine* (Phila Pa 1976) 2002;27:11–6.
71. Furman MB, O'Brien EM, Zgleszewski TM. Incidence of intravascular penetration in transforaminal lumbosacral epidural steroid injections. *Spine* (Phila Pa 1976) 2000;25:2628–32.
72. Lee CJ, Kim YC, Shin JH et al. Intravascular injection in lumbar medial branch block: A prospective evaluation of 1433 injections. *Anesth Analg* 2008;106:1274–8, table of contents.
73. Goodman BS, Lincoln CE, Deshpande KK, Poczatek RB, Lander PH, Devivo MJ. Incidence of intravascular uptake during fluoroscopically guided lumbar disc injections: A prospective observational study. *Pain Physician* 2005;8:263–6.
74. Jones RL, Landers MH. Has rare case of paraplegia complicating a lumbar epidural infiltration been reported? *Ann Phys Rehabil Med* 2011;54:270.
75. Thefenne L, Dubecq C, Zing E et al. A rare case of paraplegia complicating a lumbar epidural infiltration. *Ann Phys Rehabil Med* 2010;53:575–83.
76. Manchikanti L, Singh V, Pampati V et al. Evaluation of the relative contributions of various structures in chronic low back pain. *Pain Physician* 2001;4:308–16.
77. Hurri H, Karppinen J. Discogenic pain. *Pain* 2004;112:225–8.
78. Aprill C, Bogduk N. High-intensity zone: A diagnostic sign of painful lumbar disc on magnetic resonance imaging. *Br J Radiol* 1992;65:361–9.
79. Sharma SK, Jones JO, Zeballos PP, Irwin SA, Martin TW. The prevention of discitis during discography. *Spine J* 2009;9:936–43.
80. Sequeiros RB, Niinimaki J, Ojala R et al. Magnetic resonance imaging-guided diskography and diagnostic lumbar 0.23T MRI: An assessment study. *Acta Radiol* 2006;47:272–80.
81. Gunaydin I, Pereira PL, Fritz J, Konig C, Kotter I. Magnetic resonance imaging guided corticosteroid injection of sacroiliac joints in patients with spondylarthropathy. Are multiple injections more beneficial? *Rheumatol Int* 2006;26:396–400.
82. Fritz J, Tzaribachev N, Thomas C et al. Evaluation of MR imaging guided steroid injection of the sacroiliac joints for the treatment of children with refractory enthesitis-related arthritis. *Eur Radiol* 2011;21:1050–7.
83. Sze DY, Mackey SC. MR guidance of sympathetic nerve blockade: Measurement of vasomotor response initial experience in seven patients. *Radiology* 2002;223:574–80.
84. Eisenberg E, Carr DB, Chalmers TC. Neurolytic celiac plexus block for treatment of cancer pain: A meta-analysis. *Anesth Analg* 1995;80:290–5.
85. Fritz J, Pereira PL, Lewin JS. Temporomandibular joint injections: Interventional MR imaging demonstrates anatomical landmark approach to be inaccurate when compared to direct visualization of the injectant. *Pediatr Radiol* 2010;40:1964–5; author reply 6–7.
86. Fritz J, Chhabra A, Wang KC, Carrino JA. Magnetic resonance neurography-guided nerve blocks for the diagnosis and treatment of chronic pelvic pain syndrome. *Neuroimaging Clin N Am* 2014;24:211–34.
87. Chhabra A, Lee PP, Bizzell C, Soldatos T. 3 Tesla MR neurography—Technique, interpretation, and pitfalls. *Skeletal Radiol* 2011;40:1249–60.
88. Soldatos T, Andreisek G, Thawait GK et al. High-resolution 3-T MR neurography of the lumbosacral plexus. *RadioGraphics* 2013;33:967–87.
89. Sorenson EJ, Chen JJ, Daube JR. Obturator neuropathy: Causes and outcome. *Muscle Nerve* 2002;25:605–7.
90. Brukner P, Bradshaw C, McCrory P. Obturator neuropathy: A cause of exercise-related groin pain. *Phys Sportsmed* 1999;27:62–73.
91. Wassef MR. Interadductor approach to obturator nerve blockade for spastic conditions of adductor thigh muscles. *Reg Anesth* 1993;18:13–7.
92. Labat JJ, Riant T, Robert R, Amarenco G, Lefaucheur JP, Rigaud J. Diagnostic criteria for pudendal neuralgia by pudendal nerve entrapment (Nantes criteria). *Neurourol Urodyn* 2008;27:306–10.
93. Stav K, Dwyer PL, Roberts L. Pudendal neuralgia. Fact or fiction? *Obstet Gynecol Surv* 2009;64:190–9.
94. Darnis B, Robert R, Labat JJ et al. Perineal pain and inferior cluneal nerves: Anatomy and surgery. *Surg Radiol Anat* 2008;30:177–83.
95. Gelfand HJ, Ouanes JP, Lesley MR et al. Analgesic efficacy of ultrasound-guided regional anesthesia: A meta-analysis. *J Clin Anesth* 2011;23:90–6.
96. Douglas S. Sciatic pain and piriformis syndrome. *Nurse Pract* 1997;22:166–8, 70, 72 passim.

97. Ray B, D'Souza AS, Kumar B et al. Variations in the course and microanatomical study of the lateral femoral cutaneous nerve and its clinical importance. *Clin Anat* 2010;23:978–84.
98. Stevens H. Meralgia paresthetica. *AMA Arch Neurol Psychiatry* 1957;77:557–74.
99. Kerimaa P, Marttila A, Hyvonen P et al. MRI-guided biopsy and fine needle aspiration biopsy (FNAB) in the diagnosis of musculoskeletal lesions. *Eur J Radiol* 2013;82:2328–33.
100. Chen YJ, Pu YS, Chueh SC, Shun CT, Chu WC, Tseng WY. Diffusion MRI predicts transrectal ultrasound biopsy results in prostate cancer detection. *J Magn Reson Imaging* 2011;33:356–63.
101. Malayeri AA, El Khouli RH, Zaheer A et al. Principles and applications of diffusion-weighted imaging in cancer detection, staging, and treatment follow-up. *RadioGraphics* 2011;31:1773–91.
102. Subhawong TK, Durand DJ, Thawait GK, Jacobs MA, Fayad LM. Characterization of soft tissue masses: Can quantitative diffusion weighted imaging reliably distinguish cysts from solid masses? *Skeletal Radiol* 2013;42:1583–92.
103. Bail HJ, Sattig C, Tsitsilonis S, Papanikolaou I, Teichgraber UK, Wichlas F. Signal-inducing bone cements for MRI-guided spinal cementoplasty: Evaluation of contrast-agent-based polymethylmethacrylate cements. *Skeletal Radiol* 2012;41:651–7.
104. Goetz MP, Callstrom MR, Charboneau JW et al. Percutaneous image-guided radiofrequency ablation of painful metastases involving bone: A multicenter study. *J Clin Oncol* 2004;22:300–6.
105. Fuchs S, Gebauer B, Stelter L et al. Postinterventional MRI findings following MRI-guided laser ablation of osteoid osteoma. *Eur J Radiol* 2014;83:696–702.
106. Ronkainen J, Blanco Sequeiros R, Tervonen O. Cost comparison of low-field (0.23 T) MRI-guided laser ablation and surgery in the treatment of osteoid osteoma. *Eur Radiol* 2006;16:2858–65.
107. Bickels J, Kollender Y, Merimsky O, Isaakov J, Petyan-Brand R, Meller I. Closed argon-based cryoablation of bone tumours. *J Bone Joint Surg Br* 2004;86:714–8.
108. Blanco Sequeiros R, Vahasarja V, Ojala R. Magnetic resonance-guided growth plate bone bridge resection at 0.23 Tesla: Report of a novel technique. *Acta Radiol* 2008;49:668–72.

# 16

# *MR of the Lymphatics*

**Alberto Alonso-Burgos, Emilio García Tutor, Teresa Pérez de la Fuente, and Ángeles Franco López**

**CONTENTS**



The lymphatic vasculature constitutes a highly specialized part of the vascular system that is essential for the maintenance of interstitial fluid balance, uptake of dietary fat, and immune response.

Lymphedema describes a progressive pathologic condition of the lymphatic system in which there is interstitial accumulation of protein-rich fluid and subsequent inflammation, adipose tissue hypertrophy, and fibrosis. It can be a difficult condition to treat and one that causes significant morbidity, both physical and psychological, for patients. In addition, it is frequently underdiagnosed and undertreated, which can add to patients' frustration at their chronic and debilitating disease. However, new advances in the field, including expanded diagnostic and management options, are making strides toward improved care for affected patients.

A conservative approach is the mainstay of treatment for patients suffering from lymphedema, whereby most patients are managed appropriately without surgical intervention. For this reason, surgery is performed only in selected patients when physiotherapy has definitively proven unsuccessful. The development of microsurgical techniques of the lymphatic vessels has resulted in the need for an uncomplicated and safe imaging tool to assess the lymphatic system preoperatively and to evaluate the postoperative outcome. Magnetic resonance lymphangiography (MRL) is a safe and accurate diagnostic imaging method for this purpose.

## 16.1 Lymphatic Vascular System

Once blood constituents leave the microvascular exchange blood vessels, it becomes interstitial fluid. *Lymph* is the fluid that is formed when interstitial fluid enters the initial lymphatic vessels of the lymphatic system (Figure 16.1). This lymphatic vascular system serves key physiological functions:

- Maintains fluid homeostasis by absorbing water and macromolecules from the interstitium
- Enables uptake of dietary lipids and vitamins in the intestine and serves as a trafficking route for immune cells

The lymphatic system can be broadly divided into the conducting system and the lymphoid tissue.

- The *conducting system* carries the lymph and consists of tubular vessels that include the lymph capillaries, the lymph vessels, and the right and left thoracic ducts. *The study of this conducting system is the purpose and objective of this chapter.*
- The *lymphoid tissue* is primarily involved in immune responses and consists of lymphocytes and other white blood cells enmeshed in connective tissue through which the lymph passes.

The organization of lymph nodes and drainage follows the organization of the body into external and internal regions; therefore, the lymphatic drainage of the head, limbs, and body cavity walls follows an external route, and the lymphatic drainage of the thorax, abdomen, and pelvic cavities follows an internal route. Eventually, the lymph vessels empty into the lymphatic ducts, which drain into one of the two subclavian veins (near the junctions of the subclavian veins with the internal jugular veins) (Figure 16.2).

### 16.1.1 Mechanisms of Lymph Transport

A deeper-valved subfascial system of lymphatics is responsible for the drainage of lymph from the fascia, muscles, joints, ligaments, periosteum, and bone. This subfascial system parallels the deep venous system of the extremity (Figure 16.3). The epifascial and subfascial systems normally function independently, although valved connections do exist in the popliteal, inguinal, antecubital, and axillary regions where lymph nodes form interconnected chains. These connections probably do not function under normal conditions; however, in lymphedema, some reversed flow through perforators from the epifascial to the subfascial system may occur as a mechanism of decompression of the epifascial system.

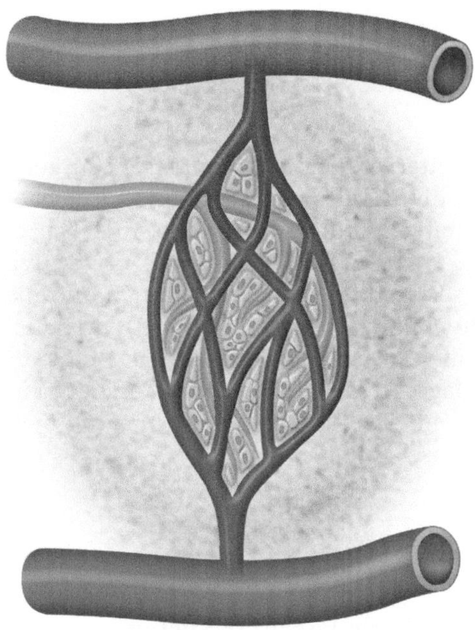

**FIGURE 16.1**
Schematic drawing of the third space. Once blood leaves the microvascular exchange blood vessels (red and blues), lymph is the fluid that is formed when interstitial fluid enters the initial lymphatic vessels (green) of the lymphatic system.

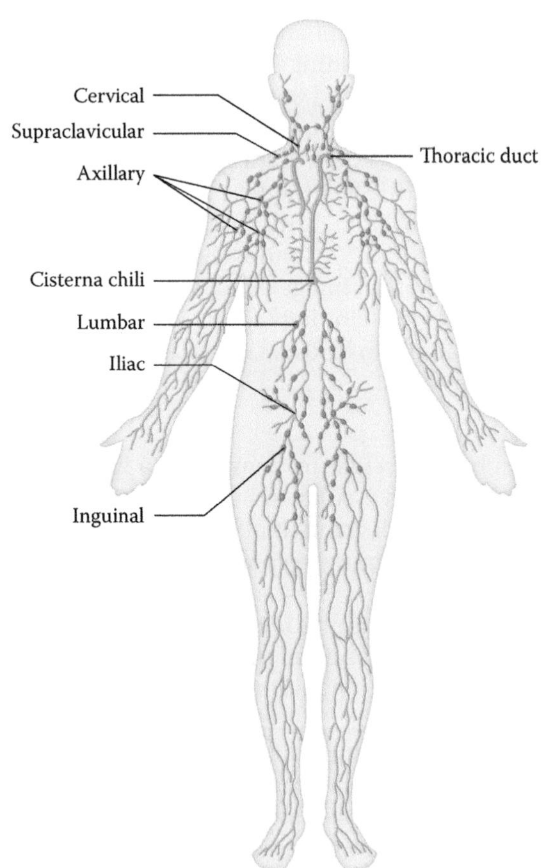


**FIGURE 16.2**
Schematical drawing. The lymphatic system is a linear system in which the lymphatic capillaries at the peripheral tissues drain the fluid (lymph) and transport it back to the blood vascular system through the lymphatic–blood junction at the end of the thoracic duct or the right lymphatic trunk, which drains lymph into the subclavian veins.

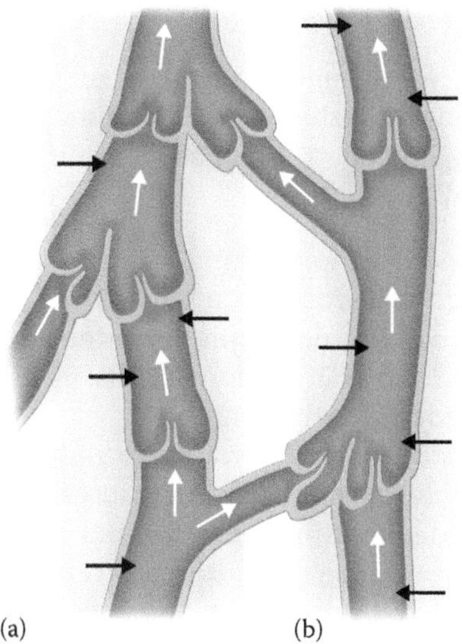


**FIGURE 16.3**
Schematical drawing where basic understanding of the lymphatic system can be achieved. Epifascial (a) and subfascial (b) lymphatic system can be observed as well as valves and connections between both systems. Once the insterstitial fluid comes into the lymphatics as lymph (arrows) its flows as shows (open arrows).

Lymphatic valves contain two semilunar leaflets (Figures 16.3 and 16.4). High lymph pressure upstream of a valve opens the valve and enables lymph flow, whereas reverse flow pushes the leaflets against each other and closes the valve. Therefore, opening and closing of the valve depend on periodic changes in fluid pressure within collecting vessels. Cyclical compression and expansion of lymphatic vessels by surrounding tissues and intrinsic pump forces generated by the spontaneous phasic contraction of smooth muscle cells regulate lymph propulsion.

Both the lymph ducts return the lymph to the blood stream by emptying into the subclavian veins (Figure 16.2). However, the derangement in lymphedema is almost always exclusive to the epifascial lymphatic system, with the subfascial system being uninvolved. Thus, the surgical approaches to lymphedema focus on the epifascial system.

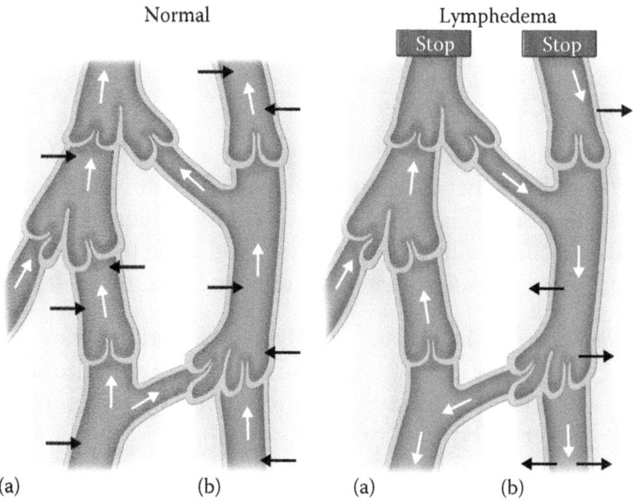


**FIGURE 16.4**
Normal lymphatic flow in (a) subfascial and (b) epifascial as well as interconnections. Whether the cause is acquired blockade of the lymph nodes or disruption of the local lymphatic channels, the result is a failure to drain protein-rich lymphatic fluid from the tissue, causing interstitial edema with swelling of the affected site.

## 16.2 Lymphedema

Lymphedema is caused by a compromised lymphatic system that impedes and diminishes lymphatic return (Figure 16.4). Although etiology determines the classification of lymphedema as either primary or secondary, it rarely impacts the choice of treatment.

### 16.2.1 Classification

#### *16.2.1.1 Primary Lymphedema*

Some defects of vascular morphogenesis in human develop in lymphedema syndromes and some rare but debilitating diseases in which lymphatic vasculature are suggested to play a central role (summarized in Tables 16.1 and 16.2). These conditions involve the lower extremities almost exclusively. All are caused by a congenital abnormality

**TABLE 16.1**

Syndromes with Lymphedema as a Primary Manifestation

| | Inheritance | Main Manifestations |
|---|---|---|
| Hereditary lymphedema IA (Milroy disease) | AD with reduced penetrance | Congenital lymphedema, chylous ascites caused by hypoplasia of lymphatic vessels |
| Hereditary lymphedema IB | AD with reduced penetrance | Lymphedema of lower limbs, nature of lymphatic vascular defects is unknown |
| Hereditary lymphedema IC | AD | Lymphedema of limbs, age of onset 1–15 years, nature of lymphatic vascular defects is unknown |
| Hereditary lymphedema II (Meige disease) | Unknown | Puberty onset lymphedema, nature of lymphatic vascular defects is unknown |

*Note:* AD, Autosomal dominant.

**TABLE 16.2**

Syndromes with Lymphedema as a Consistent Feature

| | Inheritance | Main Manifestations |
|---|---|---|
| Anhidrotic ectodermal dysplasia with immunodeficiency, osteopetrosis and lymphedema | X-linked recessive | Severe infections, osteopetrosis, nature of lymphatic vascular defects is unknown |
| Cholestasis-lymphedema syndrome (Aagenaes syndrome) | AR | Severe neonatal cholestasis, neonatal or childhood onset lymphedema caused by hypoplasia of lymphatic vessels |
| Hennekam lymphangiectasia-lymphedema syndrome | AR | Lymphedema of limbs, intestinal lymphangiectasia, mental retardation, facial anomalies |
| HLT syndrome | AD | Alopecia, ectatic blood vessels, lymphedema, nature of lymphatic vascular defects is unknown |
| Lymphedema, microcephaly, chorioretinopathy síndrome | AD | Congenital microcephaly and lymphedema, nature of lymphatic vascular defects is unknown |
| Lymphedema-choanal atresia síndrome | AR | Blockage of nasal passage (choana), lymphedema of lower legs at 4–5 years, nature of lymphatic vascular defects is unknown |
| Lymphedema–distichiasis syndrome, yellow nail syndrome | AD | Late onset leg lymphedema and metaplasia of Meibomian glands (distichiasis), impaired lymphatic drainage caused by incompetent lymphatic valves |
| Persistence of Mullerian derivatives with lymphangiectasia and postaxial polydactyly (Urioste syndrome) | AR? | Intestinal and pulmonary lymphangiectasia, protein-losing enteropathy, polydactyly, and Mullerian duct remnants |
| Pulmonary congenital lymphangiectasia | Unknown | Congenital pulmonary lymphangiectasia, subcutaneous edema, nonimmune hydrops, chylothorax |

*Notes:* AD, Autosomal dominant; AR, Autosomal recessive.

in the lymphatic system, although these defects may not always be clinically evident until later in life, when a triggering event or worsening of the condition causes the lymphatic transport capacity to exceed the volume of interstitial fluid formation; in such cases, the patient is unable to maintain normal lymphatic flow.

#### *16.2.1.2 Secondary Lymphedema*

Secondary lymphedema is caused by an acquired defect in the lymphatic system and is commonly associated with obesity, infection, neoplasm, trauma, and therapeutic modalities.

- *Filariasis*: The most common cause of secondary lymphedema worldwide is *filariasis*. Commonly occurring in developing countries around the world, this infection results in permanent lymphedema of the limb.
- *Malignancy and cancer treatment*: In the industrialized world, the most common causes of secondary lymphedema are malignancy and its treatment. This means that the disease can arise from obstruction from metastatic cancer or primary lymphoma or can be secondary to radical lymph node dissection and excision. Although lymphatics are thought to regenerate after transection via surgery, when combined with radiotherapy to the area, the risk of lymphedema increases because of scarring and fibrosis of the tissue. The most commonly affected area is the axillary region after mastectomy and radical dissection for breast cancer although lymphedema can also be seen after regional dissection of pelvic, para-aortic, and neck lymph nodes. Other associated neoplastic diseases are Hodgkin lymphoma, metastatic prostate cancer, cervical cancer, breast cancer, and melanoma.
- *Other causes*: Lymphedema is also associated with the following etiologies such as trauma, varicose vein surgery, congestive heart failure, portal hypertension, peripheral vascular surgery, lipectomy, burns, burn scar excision, insect bites, and extrinsic pressure.

### 16.2.2 Clinical Presentation

Clinical manifestations of lymphedema occur secondarily to the subcutaneous accumulation of edematous fluid and adipose tissue (Figure 16.5). Clinical classification of lymphedematous swelling has been defined by the International *Society of Lymphology* using the following parameters:

- *Stage 0*: Latent or subclinical condition where swelling is not evident despite impaired lymph transport. It may exist months or years before overt edema occurs (stages I–III).
- *Stage I*: Early accumulation of fluid relatively high in protein content (e.g., in comparison with *venous* edema) that subsides with limb elevation. Pitting may occur.
- *Stage II*: Pitting may or may not occur as tissue fibrosis develops. Limb elevation alone rarely reduces tissue swelling.

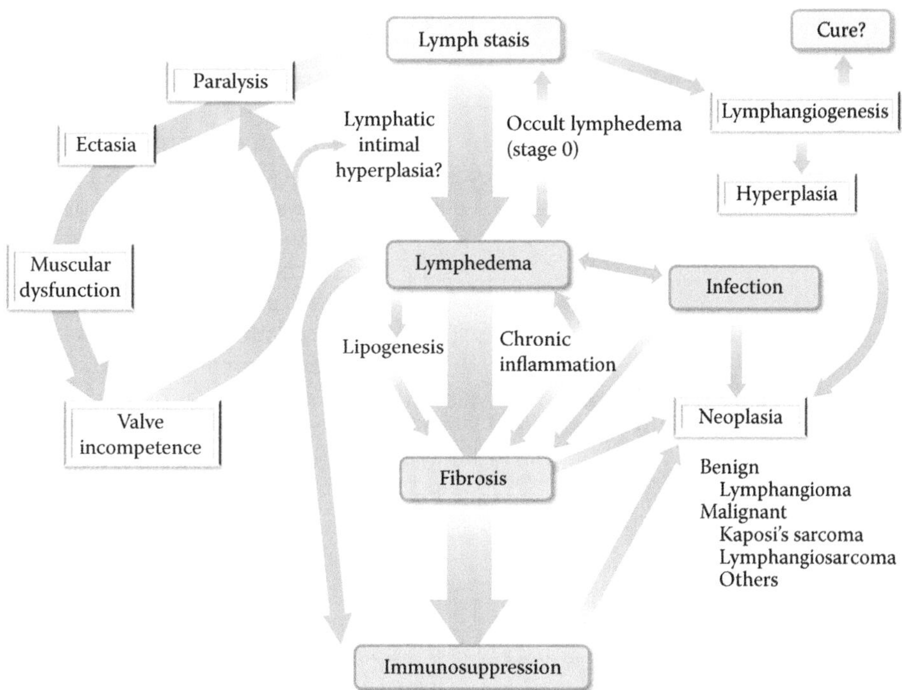


**FIGURE 16.5**
Schematic diagram of the pathophysiology of lymphedema.

- *Stage III*: Lymphostatic elephantiasis where pitting is absent. Trophic skin changes, such as acanthosis, fat deposits, and warty overgrowths, often develop.

### 16.2.3 Treatment

#### 16.2.3.1 Conservative Therapy

For patients suffering from lymphedema, a conservative approach is the mainstay of treatment, keeping in mind that most patients will be well managed without surgical intervention: more than 80% of patients with primary or secondary lymphedema of the lower extremities are treated successfully with complex decongestive physiotherapy.

The mainstay of conservative therapy relies on the finding that reduction of pitting edema can be obtained by compression, which is often achieved using multilayer inelastic lymphedema bandaging or controlled compression therapy. These therapies will only alleviate their symptoms and are not curative of their underlying lymphatic dysfunction.

#### 16.2.3.2 Surgical Therapy

None of the available surgical techniques completely address patients underlying lymphatic dysfunction, although patients can achieve resolution of their symptoms and extremity swelling with sustained adherence to compression garment regimens postoperatively. In selected patients for whom an adequate trial of conservative therapy has not proven effective, likely due to the presence of adipose tissue hypertrophy, surgical evaluation should be considered.

Surgical treatment can be broadly divided into three main approaches: resection procedures, microsurgical interventions, and the use of suction-assisted lipectomy (liposuction).

- *Resection approach* or *debulking*: This approach involves the surgical excision of subcutaneous tissue, which may or may not include excision of the overlying skin. Debulking procedures are not designed to directly address lymphatic vessel dysfunction but instead serve to provide improved comfort by removing redundant skin and subcutaneous tissues. Therefore, as with all currently available therapies, the underlying pathology remains and limb edema may return.
- *Microsurgical techniques*: These techniques attempt to directly correct underlying lymphatic pathology. Approaches include the creation of anastomoses between lymphatic vessels and veins, between lymph nodes and veins, and between distal and proximal lymphatics (Figure 16.6). They have shown good long-term results for peripheral lymphedema, with a great volume reduction as well as reduction in the incidence of cellulitis.
- *Removal of subcutaneous fatty tissue*: This is the most recently developed technique for the

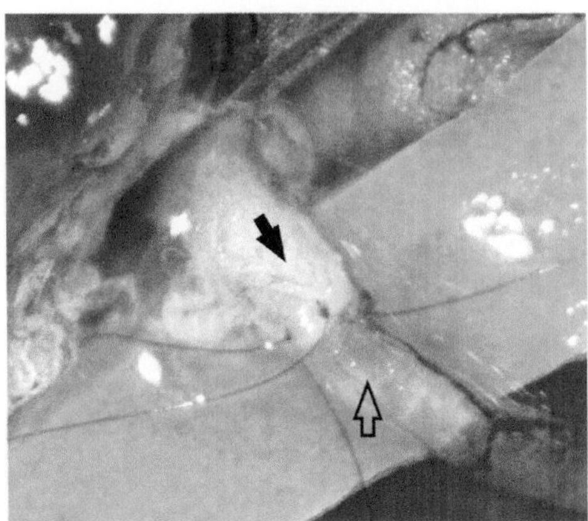

**FIGURE 16.6**
Intraoperative image taken from a microsurgical creation of anastomoses between lymphatic vessel (open arrow) and vein (arrow) as treatment of lymphedema.

surgical treatment of lymphedema which has been done through circumferential liposuction of the affected limb. Results from the largest published case series show significant improvement in appearance and symptoms. However, combination therapy with compression garment use may prove the optimal approach.

## 16.3 MR for the Lymphatics: How and When?

The development of microsurgical techniques of the lymphatic vessels has resulted in the need for an uncomplicated and safe imaging tool to assess the lymphatic system preoperatively and to evaluate the postoperative outcome.

For decades, conventional lymphography has been considered the standard of reference. This technique, using an iodine oil agent that is capable of visualizing the lymphatics, is no longer routinely performed because it can lead to life-threatening complications and is difficult to perform (Figure 16.7a through c).

Lymphoscintigraphy with colloid-bound technetium-99 m is nowadays the primary imaging investigation for peripheral lymphedema. However, spatial and temporal resolution of this technique is low resulting in an insufficient resolution to accurately outline the internal anatomy of lymph nodes and lymphatic vessels (Figure 16.8). The measurement of tracer clearance from the injection point and accumulation of tracer in the inguinal lymph node may be valuable in functional analysis, as well as ionization exposure is required.

MRI has a number of potential advantages compared with lymphoscintigraphy, including higher spatial resolution enabling depiction of lymphatic channels, higher temporal resolution, capability of three-dimensional (3D) rendered images, and the absence of exposure to ionizing radiation. From the previously commented physiological bases of interstitial fluid and lymph, any gadolinium-based contrast media in the interstitial space might be collected by the lymphatic system. Thus, this contrast will be circulating in the lymphatic system and, under an appropriate MR acquisition; these lymphatics should be observed using MR technique (Figure 16.9).

MRL is then a new diagnostic imaging method for the lymphatics with a high resolution. The technique has proved to be safe and technically feasible in patients with primary and secondary lymphedema. Going forward of the diagnosis of primary or secondary lymphedema, the focus of the MRL and the goal of this technique are the assessment and planning of the surgical treatment of the lymphedema with a lymphovenous anastomosis.

### 16.3.1 Establishing a Reference System and How to Locate a Lymphatic

One of the most important steps when performing an MRL is to establish an arbitrary reference system known

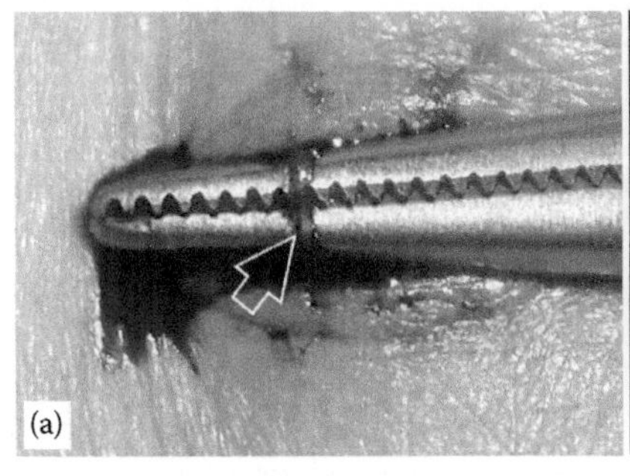


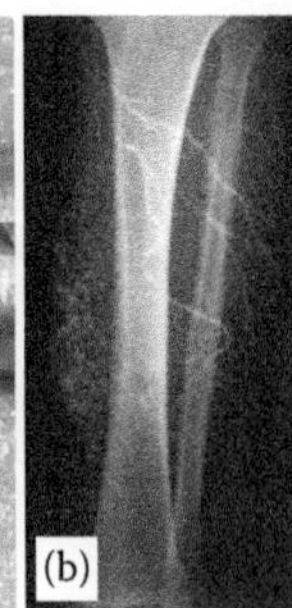


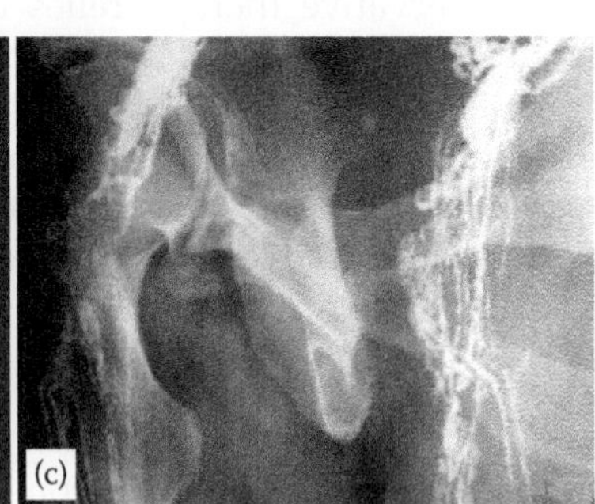


**FIGURE 16.7**
(a) Conventional direct lymphography: After intradermal injection of methylene blue, a periferal subcutaneous lymphatic vessel (open arrow) is dissected and canulated in order to inject iodine oil agent. (b) Conventional direct lymphography. Lower limb image taken after contrast injection. Image courtesy of Prof. Dr. B. Perez-Villacastín. U.H. Fundación Jiménez Díaz (Madrid-Spain). (c) Conventional direct lymphography. Pelvis image (oblique view) taken after contrast injection. (Courtesy of B. Perez-Villacastín, U.H. Fundación Jiménez Díaz, Madrid, Spain.)

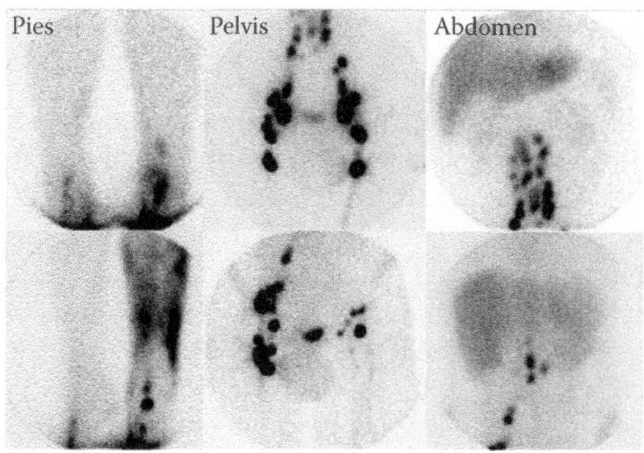


**FIGURE 16.8**
Lymphoscintigraphy with colloid-bound technetium-99 m with pathological uptake in a patient with lymphoma.

**FIGURE 16.9**
The basis of MR study of lymphatic system underline in that any gadolinium-based contrast media in the interstitial space might be collected by the lymphatic system. Then, this contrast will be circulating in the lymphatic system.

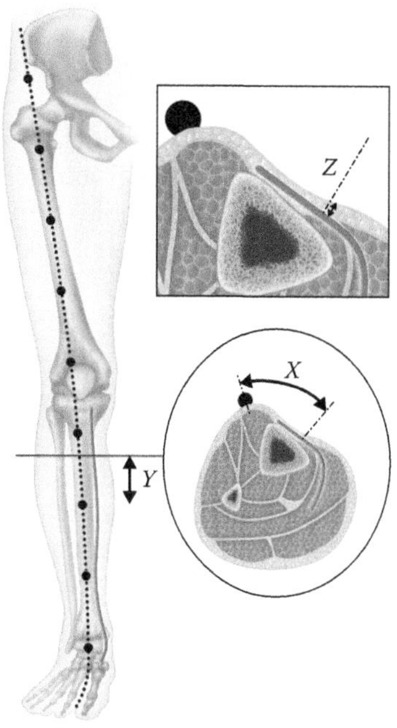


**FIGURE 16.10**
The first step to perform a MRL study is to establish references for a Cartesian coordinate system for the location of the lymphatics. For lower limbs we draw a line in surgical position between antero-superior iliac spine and first metatarsophalangeal joint and place MR-compatible skin marks each 10 cm in caudocraneal sense. We numbered these marks and after select a lymphatic and the point for the lymphaticovenous anastomosis the *X* in the Cartesian system depends on internal or external location referred to the line between marks, the *Y* depends on a craneo or caudal location referred to a mark and *Z* how deep the limphatic is.

both for surgeon and radiologist. Based on it, a Cartesian coordinate system can be applied (*X*, *Y*, and *Z*) and any lymphatic at any level can be assessed and located. Considered lower or upper limb as a territory where natural landmarks are absent, we need to recreate them.

### *16.3.1.1 Lower Limbs Studies*

First, we consider a line running through the anterior aspect of the lower limb that connects the anterosuperior iliac spine and the first metatarsal-phalangeal joint (Figure 16.10). Then, we dispose MR-compatible skin marks each 10 cm, easily seen in volume rendering (VR) and maximum-intensity projection (MIP) images (Figure 16.11), in caudocraneal sense. These marks can be numbered for an easy reporting.

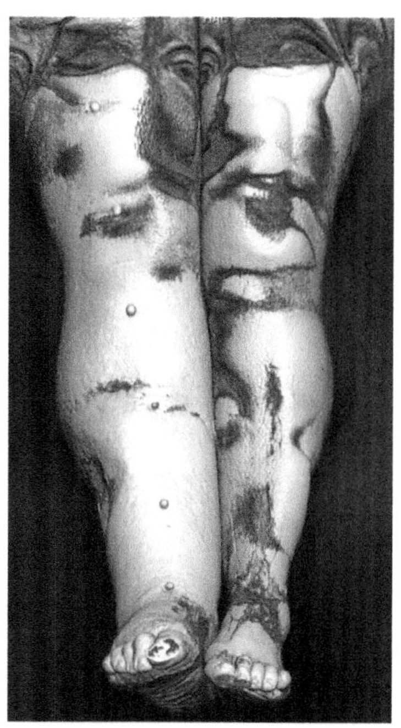

**FIGURE 16.11**
MRL VR image where MR-compatible skin marks can be easily observed in right lower limb.

#### *16.3.1.2 Upper Limbs Studies*

In these studies, we apply the same philosophy and we trace a line running through the anterior aspect of the upper limb that connects the acromioclavicular joint and the second metacarpal-phalangeal joint (Figure 16.12). Then, we dispose MR-compatible skin marks each 10 cm in caudocraneal sense. These marks can be numbered for an easy reporting.

#### *16.3.1.3 Locating a Lymphatic*

As in any Cartesian system, we need an *X*, *Y*, and *Z*. Once a lymphatic has been selected, we establish its location (location in the subcutaneous tissue) according to one of these numbered skins marks. We use to choose the nearest: *X* will be determined, in millimeters, according to internal or external location (positive or negative values of *X*) and *Y* will be given by cranial or caudal disponse (positive or negative values of *Y*). At last, *Z* will be how deep the lymphatic is (Figures 16.10, 16.13, and 16.14).

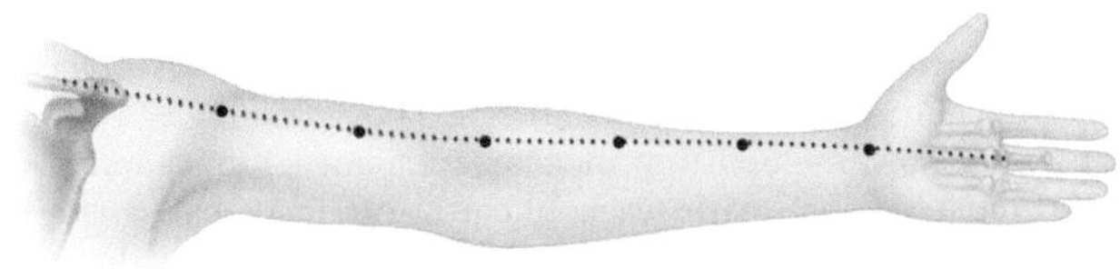

**FIGURE 16.12**
Same philosophy for the lower limb is applied for upper limb, with a line between acromioclavicular and third metacarpophalangeal joint.

### 16.3.2 MRL Technique

Technical parameters in MRL are similar to MR angiography for which dedicated coil designs and parallel imaging techniques have proven especially beneficial to enhance image quality and spatial and temporal resolution. Artifacts related to 3.0 T have to be expected. The introduction of higher field strengths at 3.0 T and multichannel coils provides a higher signal gain, which can be invested in high spatial resolution and isotropy to generate 3D datasets with options for 3D postprocessing. Furthermore, parallel imaging techniques reduce acquisition time with negligible signal penalty. This high-resolution depiction of lymph vessels may help to enhance the understanding of anatomical and morphological details. For this reason, 3 T equipments are mandatory for MRL.

#### *16.3.2.1 MRL Technical Parameters*

The entire extremity can be examined in several steps with fat-saturation techniques rendering image subtraction allowed for time-efficient acquisition.

In our institution, all examinations were performed on a 3.0 T MR system (Magnetom VERIO, Siemens Medical Solutions) mainly according to the Barcelona (Alomar et al.) Munich and Freiburg (Notohamiprodjo et al. and Lohrmann et al., respectively) working groups. For signal reception, we used a dedicated lower extremity 36-element coil. MRL was performed using a

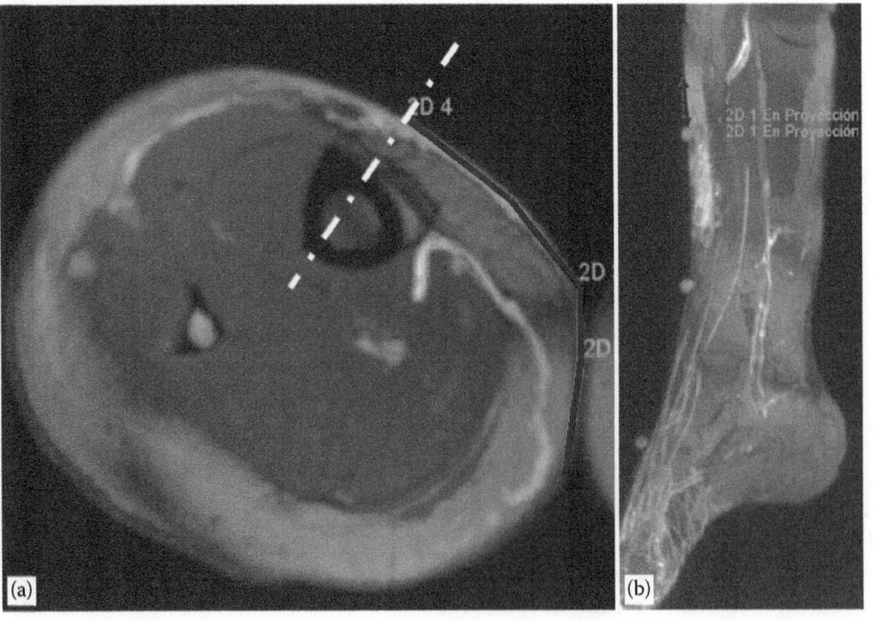


**FIGURE 16.13**
Lower limb MRL reformatted into axial (a) and saggital (b) 5 mm MIP images. For the location of a selected lymphatic vessel in a Cartesian coordinate system, axial images are mainly used for gaining *X* and *Z* values and saggital views are routinely performed for the assesment of Y value.

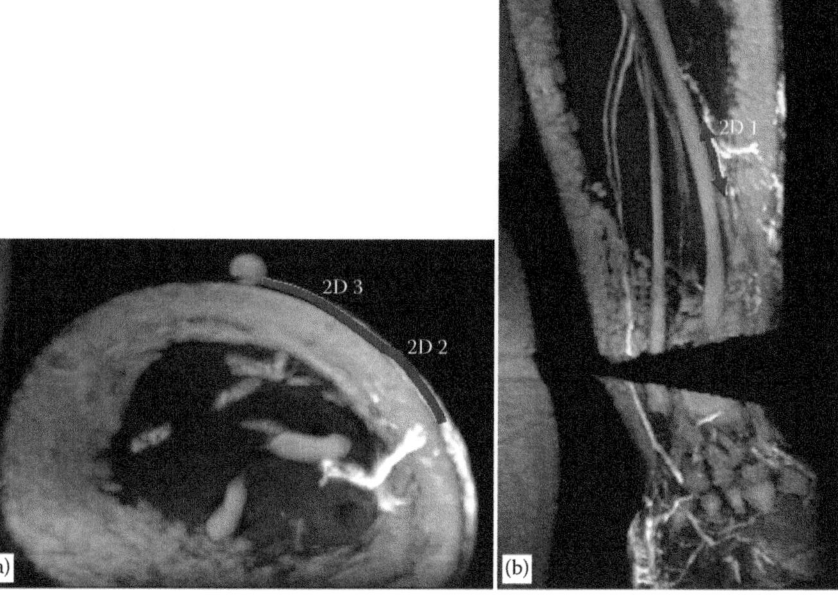


**FIGURE 16.14**
Upper limb MRL reformatted into axial (a) and saggital (b) 5 mm MIP images. Applying the same philosophy used in lower limbs, axial images are mainly used for gaining $X$ and $Z$ values and saggital views are routinely performed for the assesment of $Y$ value in a Cartesian coordinate system.

coronal $T_1$-weighted 3D GRE sequence (fast low angle shot magnetic [FLASH]) with spectral fat-saturation and sequence parameters as summarized in Table 16.3. Parallel imaging using generalized autocalibrating partially parallel acquisitions (GRAPPA) algorithm was applied at an acceleration factor of $R = 3$, with 32 reference lines and auto-matrix coil mode.

Three anatomical levels were subsequently examined ascending from the lower leg, knee, and upper leg as well as hand and distal forearm, elbow and arm-shoulder in the upper limb. The examination was repeated at least twice following the first run. Usually, exploration entire limb exploration is performed at 15, 30, and 45 min after contrast injection (Figure 16.15).

#### 16.3.2.2 Contrast-Medium Application

A mixture of 10 ml gadolinium-based contrast media (gadodiamide 0.5 mmol/ml) and local anesthesia was then prepared. Previous local asepsia, a 1-ml sample of this mixture was injected strictly intracutaneously (Figure 16.16) in each interdigital spaces as well as injection at the lateral border or of the medial thigh, both in the foot or hand (Figure 16.17a through b), using 24-gauge cannulae inserted about 5 mm at an angle of about 10°–15° forming a wheal containing the contrast medium. The contrast medium was massaged in for 2 min. MRL was started thereafter.

**TABLE 16.3**
Suggested MRI Protocol

| | **3T MR** |
|---|---|
| **Range** | **Pelvis Toes** |
| **Acquisition Parameters** | **$T_1$-Weighted 3DGRE Sequence (FLASH)** |
| Acquisition time | 149 s |
| Voxel size | 2.4 × 1.9 × 1.5 mm |
| Orientation | Coronal |
| Phase encode direction | Anterior >> posterior |
| FoV* read | 380 mm |
| FoV* phase | 87.5 |
| Slice thickness | 0.8 mm |
| TR | 4.13 ms |
| TE | 1.47 ms |
| Clusters | 3 |
| Slices per cluster | 180 |
| Flip angle | 25° |
| Interslice gap | 0.16 mm |
| In-plane resolution | 0.8 × 0.8 mm$^2$ |
| Fat Sat | |
| Base resolution | 256 |
| Phase resolution | 80% |
| Slice resolution | 63% |
| Bandwidth | 340 Hz/Px |

*Notes:* FoV, field of view; * indicates that these are suggested minimum values of FoV.

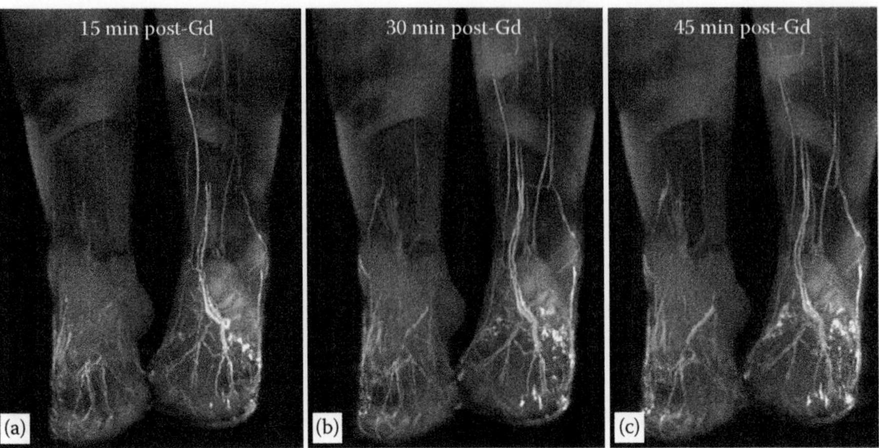


**FIGURE 16.15**
MRL anterior view coronal MIP images taken (a) 15, (b) 30, and (c) 45 min after contrast injection in a 25-year-old woman with history of trascient lymphedema in left foot where typical increased skin lymphatic and dermal back-flow in the media and lateral region of foot as well as delayed flow can be seen. Normal lymph vessels have a typically beaded appearance and a stronger contrast enhancement can be seen on rigth foot.

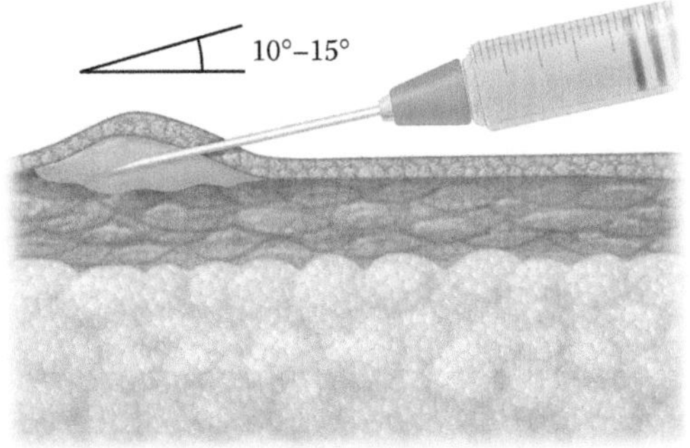


**FIGURE 16.16**
An intradermal injection is the injection of a small amount of fluid into the dermal layer of the skin. The needle, bevel up, has to be inserted into the skin at a 5 to 15-degree angle, and the apparition of a wheal reveal a correct technique.

The lymphatic vessels of the toes and the dorsum of the foot unite into a number of vessels that accompany the great saphenous vein and terminate in the distal group of superficial subinguinal lymph nodes. In order to detect these lymphatics, we consider as mandatory additional injection of contrast media at the lateral border of the foot or of the medial thigh.

Venous overlay may hamper the correct identification of lymph vessels since small veins may resemble lymph vessels (Figures 16.18 and 16.19). In lymphoscintigraphy, venous drainage is usually absent as the radioactive tracer is bound to colloid substances and does not enter blood vessels but is directly phagocytized by the deep lymphatic system. In our approach, injection is strictly intracutaneous, whereas radiotracers are

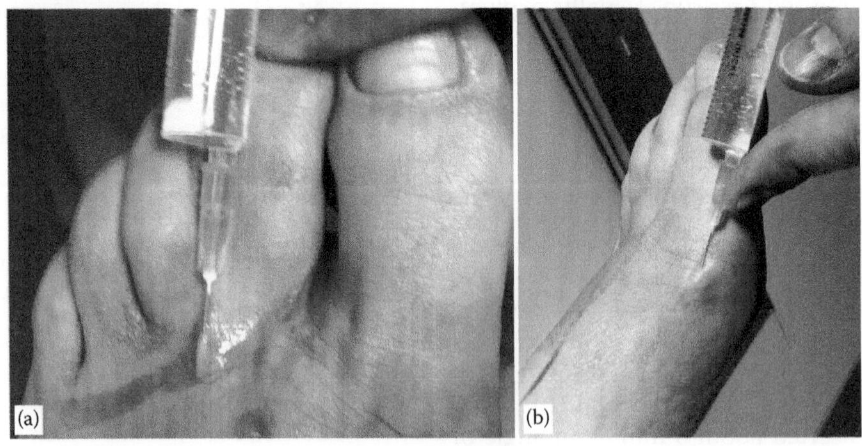


**FIGURE 16.17**
(a) Intracutenous injection of contrast media in each interdigital space. (b) Intracutenous injection of contrast media in both inner and lateral border of the feet.

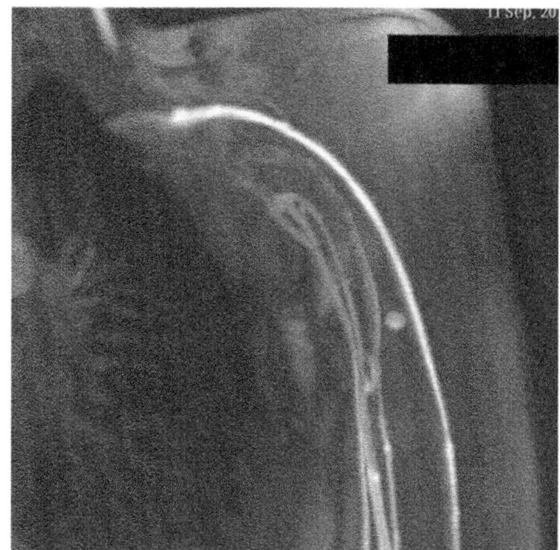

FIGURE 16.18
MRL images may mimmic venous system.

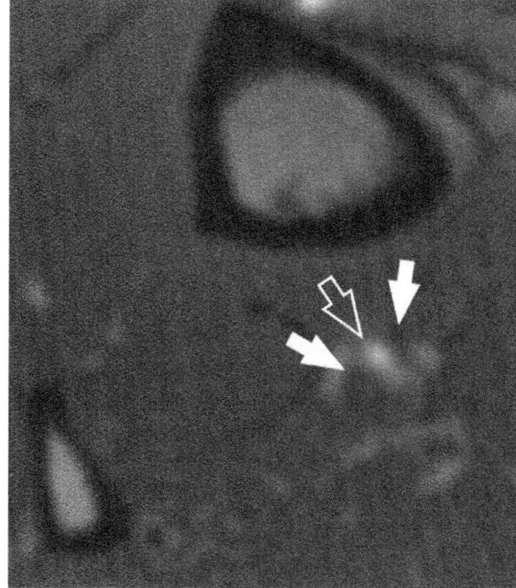

FIGURE 16.19
After intracutaneous contrast injection, posterior tibial vessel (arrows) maintain unopacified while lymphatics can be seen (open arrows).

typically injected subcutaneously. If intracutaneously injected, gadopentic-diethylene triamine pentaacetic acid (Gd-DTPA) is absorbed by the superficial lymphatic capillaries and transported into deeper dermal layers where it is transported centrally to the deep lymphatic system (Figure 16.3).

### 16.3.3 Visual Image Analysis

MIPs are reconstructed for a comprehensive visualization of the lymph vessels in a long-leg display of all four levels generated using composing software. 5- and 15-mm-thick MIPs considerably facilitated the assessment of the delicate lymph vessels. Normal lymph vessels have a typically beaded appearance and a stronger contrast enhancement, which distinguishe them from veins, which show smooth contours, a straight outline, and a lower contrast enhancement (Figure 16.15). The option to perform free 3D MPR in any virtual orientation and to obtain planes in angulations focusing on structures of interest improved the analysis and understanding of difficult anatomical conditions.

VR reformatted images in any MR study are much less *spectacular* than those obtained from CT studies. However, expending not much time in the workstation, good VR galleries can be reached (Figure 16.20). VR images are very useful for an entire comprehension of the limb and the display of the lymphatics, but also to achieve those where skin or bony landmarks can be applied for the location (Figure 16.21).

Surgeon and radiologist must work together and as a sole team in this. Misunderstanding, leak of communication or poor confidence will spoil any case as

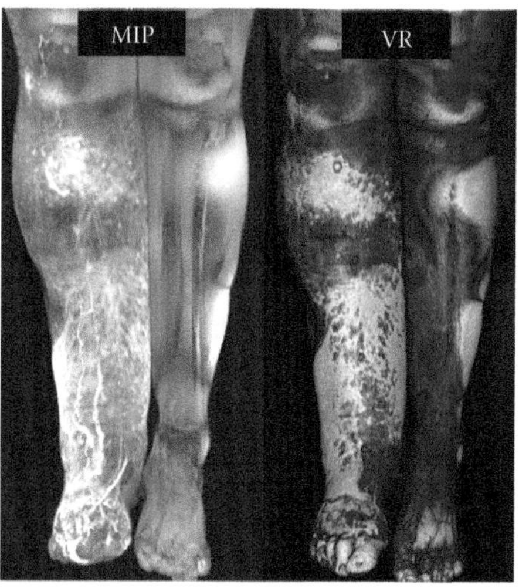


FIGURE 16.20
MIP and VR MRL reformatted images where tiny skin lymphatic vessels and severe dermal backflow can be seen in a patient with secondary right lower limb lymphedema.

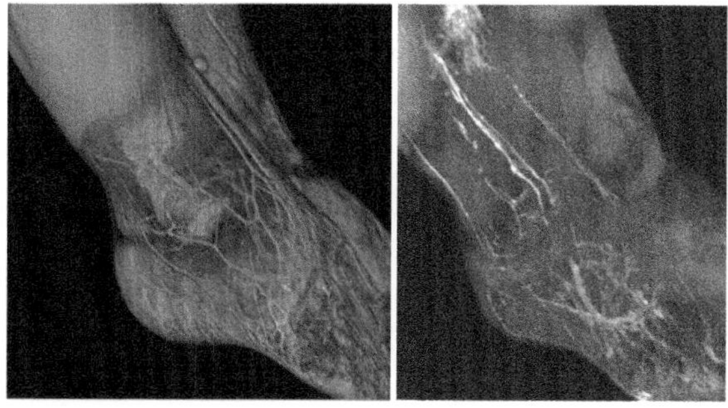

FIGURE 16.21
VR and MIP MRL reformatted images (left and right figures, respectively) where the lymphatics can be assessed. In this case, only bony landmarks are enough for the location.

a bad selection criteria or a bad imaging technique. Co-interpretation of the imaging data in a 3D format is mandatory. That virtual surgery is indeed performed. This also saves time.

## 16.4 Radiological Findings

The number of contrast-enhanced lymphatic vessels in the primary lymphedematous limbs varies and is non-specific. Lymphatic vessels that could be visualized had diameters of 1.2–8 mm. However, it has to be considered that as in any angiographic MR study, this technique oversizes the caliber of the vessel.

Lymphatic vessels in edematous limbs were also irregularly shaped, had an uneven diameter or were twisted, and could be easily distinguished from venous vessels (Figures 16.15 and 16.19). Patterns seen in the lymphatic pathways of patients with primary lymphedema included (both in upper and lower limbs)

- Numerous tiny skin lymphatic vessels and dermal backflow in the lower part of the leg (Figures 16.20 and 16.22) followed by one or two dilated lymphatic collectors in the upper part of the limb
- Bunches of extremely dilated and significantly enhanced lymphatic vessels located mainly in the medial and the lateral portion of the proximal limb (Figure 16.23)

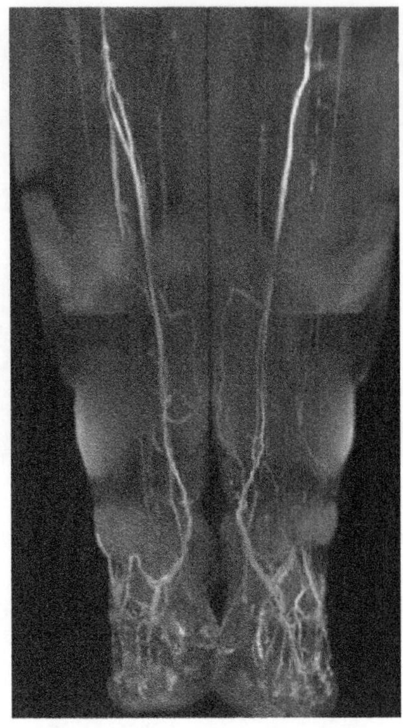

**FIGURE 16.22**
MRL coronal MIP image taken 45 min after contrast injection in a 35-year-old woman with history of primary lymphedema. One dilated deep lymphatic collectors in the upper part of each limb.

- Radiating enhanced vessels in lower part of the limb that run up to the medial portion and finally went up to (Figures 16.24 and 16.25)
- Discontinuous and lightly enhanced but dilated vessels in the medial portion of the limb

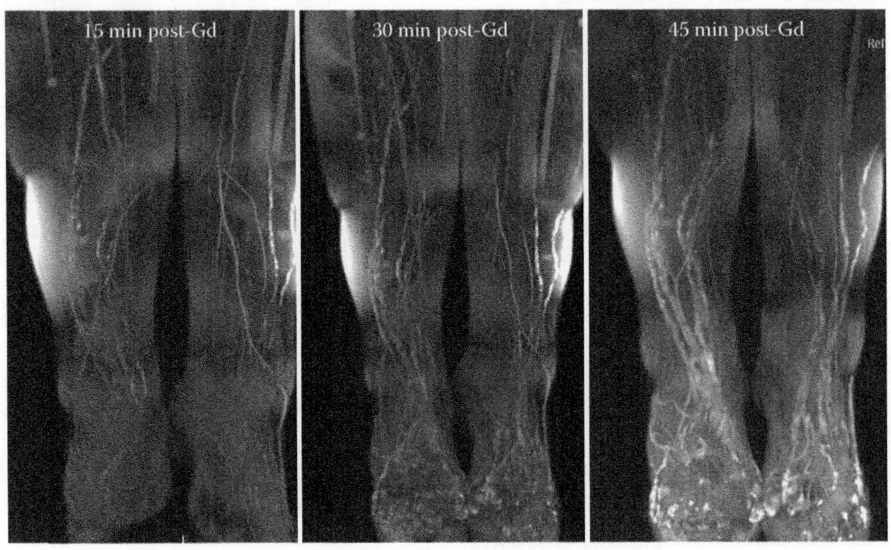


**FIGURE 16.23**
MRL coronal MIP images taken 15, 30, and 45 min after contrast injection in a 31-year-old woman with history of primary lymphedema. Dilated and significantly enhanced lymphatic vessels can be seen in both lower limbs.

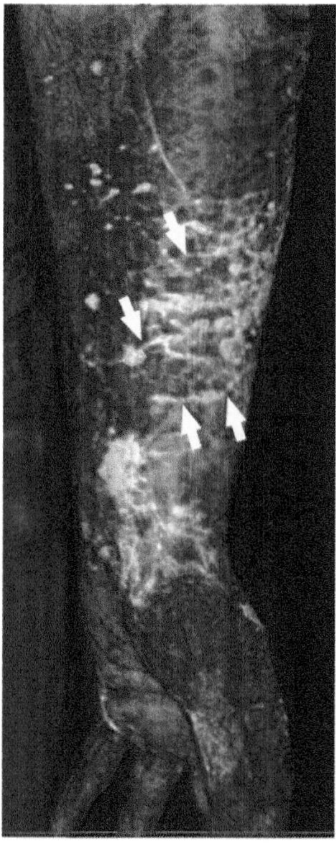

**FIGURE 16.24**
MRL VR image taken 15 min after contrast injection in a 48-year-old woman with history of left upper limb secondary lymphedema after breast cancer surgery. A subcutaneous lymphatic vessels can be observed with a transverse course (arrows).

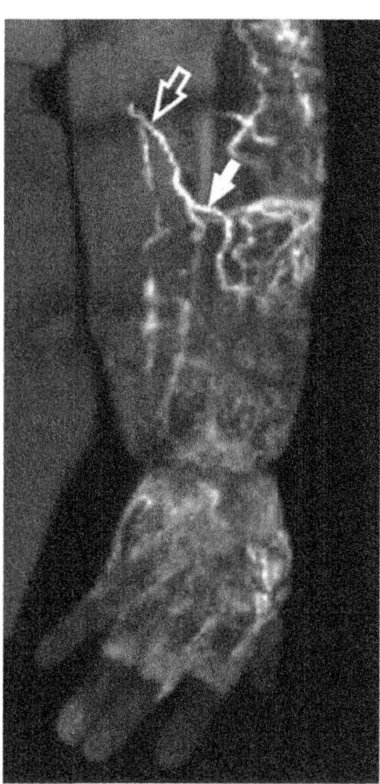

**FIGURE 16.25**
MRL MIP image taken 15 min after contrast injection in a 73-year-old woman with history of left upper limb secondary lymphedema after breast cancer surgery. A subcutaneous lymphatic vessels can be observed with a transverse course (arrow) followed by an ascending medial course (open arrow).

At last, contrast-enhanced lymph flow speed varies from 0.3 to 1.48 cm/min in normal limbs. This will be diminished in a limb with lymphedema where one of the most characteristic find will be a delayed flow pattern (Figures 16.15 and 16.23).

## 16.5 Limitations to MRL

Regarding to well-known limitations of any MR study, there are some limitations to MRL:

- Fat suppression has proven practical for MRL by allowing time-efficient acquisition and preserved lymph vessel signal without image subtraction. However, heterogeneous fat suppression may mimic diffuse subcutaneous enhancement by contrast medium. An increased difference in resonance frequency between water and fat at 3.0 T is then advantageous because of a better separation of the fat and water peak and better or faster fat suppression. Advanced techniques such as Dixon fat saturation may further improve the uniformity of fat suppression.
- Intracutaneous injection of gadolinium derivatives is still regarded as an off-label use. However, MRL with perivascular gadolinium injection has been found to be relatively safe in previous studies and experimental animal models, with less tissue damage compared to conventional radiographic contrast media, but is not without at least a temporary adverse effect.
- The detailed anatomic information provided by MRL suggests high impact on future pre-surgical planning of lymph vessel reconstruction and their postoperative follow-up.

## Bibliography


Baumeister RG, Siuda S. Treatment of lymphedemas by microsurgicallymphatic grafting: What is proved? *Plast Reconstr Surg*. 1990; 85: 64.

Calderon G, Roberts B, Johnson LL. Experimental approach to the surgical creation of lymphatic-venous communications. *Surgery*. 1967; 61: 122–128.

Campisi C. Lymphatic microsurgery: A potent weapon in the war on lymphedema. *Lymphology*. 1995; 28: 110–112.

Campisi C, Boccardo F, Lymphedema and microsurgery. *Microsurgery*. 2002; 22: 74–80.

Campisi C, Boccardo F, Microsurgical techniques for lymphedema treatment: Derivative lymphatic-venous microsurgery. *World J Surg*. 2004; 28: 609–613.

Campisi C, Davini D, Bellini C, Taddei G, Villa G, Fulcheri E, Zilli A, Da Rin E, Eretta C, Boccardo F. Lymphaticmicrosurgery for the treatment of lymphedema. *Microsurgery*. 2006; 26: 65–69.

Charles H. Elephantiasis of the leg. In: Latham A, English T, eds. *A System of Treatment*. London: Churchill; 1912.

Connell F, Brice G, Jeffery S, Keeley V, Mortimer P, Mansour S. A new classification system for primary lymphatic dysplasias based on phenotype. *Clin Genet*. 2010; 77: 438–452.

Grotte G. The discovery of the lymphatic circulation. *Acta Physiol Scand Suppl*. 1979; 463: 9–10.

Guermazi A(1), Brice P, Hennequin C, Sarfati E. Lymphography: An old technique retains its usefulness. *RadioGraphics*. 2003; 23: 1541–1558.

Hoerauf A, Pfarr K, Mand S, Debrah AY, Specht S. Filariasis in Africa-treatment challenges and prospects. *Clin Microbiol Infect*. 2011; 7: 977–985.

International Society of Lymphology. The diagnosis and treatment of peripheral lymphedema: Consensus document of the International Society of Lymphology. *Lymphology*. 2003; 36: 84.

Liu NF, Lu Q, Jiang ZH, Wang CG, Zhou JG. Anatomic and functional evaluation of the lymphatics and lymph nodes in diagnosis of lymphatic circulation disorders with contrast magnetic resonance lymphangiography. *J Vasc Surg*. 2009; 49: 980–987.

Lohrmann C, Foeldi E, Langer M. MR imaging of the lymphatic system in patients with lipedema and lipo-lymphedema. *Microvasc Res*. 2009; 77: 335–339.

Lohrmann C(1), Felmerer G, Foeldi E, Bartholomä JP, Langer M. MR lymphangiography for the assessment of the lymphatic system in patients undergoing microsurgical reconstructions of lymphatic vessels. *Microvasc Res*. 2008; 76: 42–45.

Lowry F. Study finds genetic link to lymphedema. April 22, 2013. *Medscape Med News*. http://www.medscape.com/viewarticle/802874. Accessed April 29, 2013.

Matsubara S, Sakuda H, Nakaema M, Kuniyoshi Y. Long-term results of microscopic lymphatic vessel-isolated vein anastomosis for secondary lymphedema of the lower extremities. *Surg Today*. 2006; 36: 859–864.

Michaely HJ, Attenberger U, Kramer H, Reiser MF, Schoenberg SO. Gadofosveset-enhanced steady state MRA of the peripheral vessels with Dixon fat-saturation. *Proc Intl Soc Mag Reson Med*. 2008; 16: 112.

Notohamiprodjo M, Baumeister RG, Jakobs TF, Bauner KU, Boehm HF, Horng A, Reiser MF, Glaser C, Herrmann KA. MR-lymphangiography at 3.0 T: A feasibility study. *Eur Radiol*. 2009; 19: 2771–2778.

Notohamiprodjo M, Weiss M, Baumeister RG, Sommer WH, Helck A, Crispin A, Reiser MF, Herrmann KA. MR lymphangiography at 3.0 T: Correlation with lymphoscintigraphy. *Radiology*. 2012; 264: 78–87.

O'brien BM, Mellow CG, Khazanchi RK et al. Long-term results after micro lymphaticovenous anastomoses for the treatment of obstructive lymphedema. *Plast Reconstr Surg*. 1990; 85: 562.

Olszewski WL. The treatment of lymphedema of the extremities withmicrosurgical lympho-venous anastomoses. *Int Angiol*. 1988; 7: 312.

Schulte-Merker S, Sabine A, Petrova TV. Lymphatic vascular morphogenesis in development, physiology, and disease. *J Cell Biol*. 2011; 193: 607–618.

Warren AG, BA, Brorson H, Borud LJ et al. Lymphedema: A comprehensive review. *Ann Plastic Surg*. 2007; 59: 464–472.

Zuther JE. *Lymphedema management: The comprehensive guide for practitioners*. 2nd ed. New York: Thieme; 2009.

# 17

# *Pediatric Applications*

**Marilyn J. Siegel and Ellen M. Chung**

**CONTENTS**

## 17.1 Introduction

Magnetic resonance imaging (MRI) is an important imaging technique in nearly every part of the pediatric body. It offers superb soft tissue contrast without exposure to ionizing radiation, which is important in a pediatric population. The introduction of 3.0 T magnets, faster sequences, multichannel coils, and parallel imaging techniques allows for increased resolution and shorter image acquisition times [1,2]. MRI is increasingly utilized as the primary study to evaluate soft tissue masses as well as joint abnormalities, and its use for evaluating thoracic, abdominal and pelvic masses is increasing [3–14]. It also plays an important role in the staging and monitoring of tumor response. This chapter will highlight the diagnostic applications of MRI in a wide variety of disease processes of the chest, abdomen, pelvis, and musculoskeletal system in children.

## 17.2 MRI: Technical Considerations

Imaging pediatric patients has several inherent problems that are not present in adults, in particular patient motion with prolonged imaging times, which can degrade image quality, and small body size. These factors require the need for faster scanning techniques. The following is a brief review of MR parameters and sequences that are useful to obtain consistent diagnostic quality images, while minimizing scanning time. More details can be found in the corresponding technical chapters in this textbook.

### 17.2.1 Imaging Parameters

Lesion detectability is dependent on the signal-to-noise (S/N) ratio, spatial resolution, and contrast resolution. These parameters vary with the size of the receiver coil, slice thickness, field of view (FOV), matrix size, and number of acquisitions. For optimal S/N ratio and spatial resolution, MRI examinations should be performed with the smallest coil that fits tightly around the body part being studied [1,2]. A head coil or phased-array coil can usually be used to study infants and small children, while a phased array or whole-body coil is required for imaging larger children and adolescents. The phased array coils, which contain multiple elements in contrast to the single-channel coil, are preferred over body coils because they provide better anatomic resolution Surface coils can be useful in the evaluation of superficial structures, such as the spine, but the drop-off in signal strength with increasing distance from the center of the coil limits the value of these coils in the evaluation of deeper abdominal structures.

Slice thickness varies with patient size and the area of interest. Thinner slices (3–4 mm) are used in the evaluation of small lesions and through areas of maximum interest, whereas thicker slices (6–8 mm) are preferred for a general survey of the chest and abdomen and for larger lesions. Most MR examinations are performed with a 128 or 192 matrix and one or two signal acquisitions to shorten imaging time. A 286 × 286 matrix may be needed in areas where more anatomic detail is desired.

The FOV can have a square or rectangular shape. A square shape is used when the body part being examined fills the FOV. An asymmetric rectangular FOV is ideal for body parts that are narrow in one direction, such as the abdomen in a thin patient. A large FOV is preferred over a smaller one to achieve a better S/N ratio. Decreasing the FOV improves spatial resolution, but it decreases the pixel size, which increases noise.

Parallel imaging has emerged as a method of increasing imaging speed. This technique relies on exploiting the inherent spatial sensitivities of phased-array coils to extract spatial information that traditionally was obtained through magnetic gradient encoding. The applications in pediatric imaging that are well-suited to parallel imaging are highlighted: contrast-enhanced dynamic imaging, breathhold $T_1$-weighted, and 3D volumetric imaging of the abdomen and pelvis.

### 17.2.2 Pulse Sequences

#### *17.2.2.1 Spin-Echo Techniques*

$T_1$-weighted ($T_1$W) and $T_2$-weighted ($T_2$W) sequences are obtained in virtually all patients. $T_1$-weighted sequences (short TR, short TE) provide excellent contrast between soft-tissue structures and fat and thus, help in tissue characterization (i.e., fluid, fat, or blood). Fat-suppressed $T_1$-weighted images are useful to improve conspicuity of diseased tissues.

$T_2$-weighted sequences (long TR, long TE) provide excellent contrast between lesions and adjacent soft tissues and thus, also help in tissue characterization. The $T_2$-weighted sequences can be acquired with either conventional or fast/turbo techniques. The latter are advantageous because they shorten imaging time. The fast spin-echo techniques, however, may result in some loss of contrast between fat and other high-signal-intensity tissues or fluid and thus, they should be used in combination with fat-suppression techniques. Two basic methods of fat suppression are widely available: short-tau inversion recovery (STIR) and radio-frequency presaturation of the lipid peak (fat saturation).

Most lesions have low signal intensity on $T_1$-weighted images and high signal intensity on $T_2$-weighted images. The $T_1$ signal can increase if the lesion contains fat, blood, proteinaceous fluid, or cartilage. Gadolinium chelate enhancement also results in high $T_1$ signal intensity. A decrease in $T_2$ signal intensity is seen with mineralization (calcification), hemosiderin and other blood products, iron oxide, and fibrosis. Signal from fat is nulled or very dark on STIR and fat saturated images, while most pathologic lesions, with increased free water and prolonged $T_1$ and $T_2$ values, are bright on the fat-suppressed sequences.

#### 17.2.2.2 Gradient-Echo Technique

In this technique, gradients, as opposed to RF pulses, are used to dephase (negative gradient) and rephase (positive gradients) transverse magnetization [15]. Gradient-echo (GE) or time-of-flight images result in high signal in flowing blood and are used to assess flow in vascular structures. By comparison, on spin-echo sequences flowing blood appears as a flow void or decreased signal within the vessel lumen. The flip angles determine the contrast of the images. Images acquired with large flip angles (>45°) are $T_1$-weighted, while those acquired with small flip angles (30°) are $T_2$-weighted. The relatively short acquisition time required to obtain GRE images also allows serial dynamic imaging immediately following intravenous (IV) administration of gadolinium chelate agents. Arterial and venous phases can be acquired and reformatted in a 3D display [15,16]. This technique provides thin 2.5 to 3.0 mm sections which can then be reconstructed as 3D or volumetric images.

#### 17.2.2.3 In-Phase and Out-of-Phase Imaging (Chemical Shift Imaging)

In- and out-of-phase imaging is routinely used to depict and characterize lesions suspected of containing fat. With this sequence, which is usually a GE or spoiled GRE sequence, fat and water are imaged when their $^1$H nuclei are spinning in phase with each other (TE = 4.2 ms at 1.5 T) and out of phase with each other (TE = 2.1 ms at 1.5 T). The shortest possible TE should be used for out-of-phase imaging to reduce $T_2$* effects. If microscopic fat is present, its signal is nulled (canceled out) on the out-of-phase images [15].

#### 17.2.2.4 Single-Shot Fast Spin-Echo Imaging

Single-shot fast spin-echo techniques, like half-Fourier acquisition single-shot turbo spin echo (HASTE) are used in MR cholangiopancreatography (MRCP) [17,18] and MR urography [19,20]. They are well adapted to imaging non-circulating fluid filled structures, producing a $T_2$-weighted image. This is a slice-by-slice technique. Each slice requires approximately 1.2 to 1.5 s before acquisition of the next slice. This sequence is relatively insensitive to breathing motion artifact compared to conventional spin-echo sequences.

Thick and thin slice thicknesses are acquired. The thick slabs are composed of 30 to 40 mm thick sections. The thicker slab approach has the advantage of displaying convoluted structures such as the ureter of bile ducts, in one image. The second approach uses thin, 3 to 4 mm sections. The thinner slab approach has the advantage of higher resolution and in addition, it enables images to be reconstructed as maximal intensity projections for volumetric imaging.

#### 17.2.2.5 Diffusion-Weighted MRI

Diffusion-weighted imaging (DWI) has been applied in the evaluation of thoracoabdominal tumors [21–24]. Primary applications include tissue characterization to help in differentiation of benign and malignant tumors and monitoring treatment response. DWI provides information at a microscopic level on diffusion of water protons, enabling separation between rapid diffusion of protons (unrestricted diffusion) and slow diffusion of protons (restricted diffusion). The technique uses either echo-planar or fast GRE sequences and diffusion gradients, termed *b*-values, to measure water motion. In general, two gradients are applied: one at 50–100 s/mm$^2$ and the second at 800–1000 s/mm$^2$. Low *b*-values null vascular signal. High *b*-values null signal in fluid and also enhance signal in cellular tissue due to restricted water motion. The quantitative measure is the apparent diffusion coefficient (ADC). Low ADC values are seen in every cellular tissues and tumors, whereas higher ADC values are noted in edematous tissue and tumors responding to therapy [21–24].

### 17.2.3 Gadolinium-Based Contrast-Enhanced MR Angiography

Gadolinium chelates are extracellular contrast agents that cause $T_1$ shortening of blood (i.e., hasten the recovery of longitudinal magnetization), which subsequently results in high signal intensity on $T_1$-weighted images [15]. The extracellular properties allow for imaging during the arterial and venous phases of contrast enhancement. The usual gadolinium dose is 0.1 mmol/kg. A 3D spoiled gradient-echo sequence is used and the data obtained in a volume, typically during suspended respiration. $T_1$-weighted fat-suppressed sequences after the administration of IV gadolinium chelates can help define areas of necrosis and cyst formation and may improve contrast between tumor and normal tissues. Although the incidence of renal failure, and hence the risk of nephrogenic systemic fibrosis, is lower in the pediatric population, it can occur and thus, it is imperative to follow guidelines similar to those used for adults [25].

### 17.2.4 1.5 and 3 T Magnets

MRI can be performed with 1.5 or 3 T magnets. Although imaging at 3 T can improve the S/N ratio and spatial resolution, it may be limited by dielectric effects, banding, and other pulse sequence-related artifacts. At 3 T, specific absorption rate levels increase relative to 1.5 T imaging. These rates can be lowered by reducing the number of sections, flip angle, or echo train length, having a delay between sequences, or using parallel imaging techniques. Overall, both are feasible in pediatric patients.

### 17.2.5 Optimizing Image Quality

As a result of the relatively long time required to perform abdominal MRI in children, gross voluntary motion or physiologic motion, such as respiration and blood flow, can produce artifacts that degrade the MR image. Voluntary motion can be minimized or eliminated by the use of sedation, whereas physiologic motion, and its resultant artifacts—ghosting and blurring—can be suppressed by the use of faster imaging sequences. These fast techniques (gradient echo, fast or turbo spin-echo (TSE), single-shot fast spin-echo such as HASTE) are discussed above. Another motion artifact reduction strategy is to use fat suppression of the abdominal wall fat and saturation of the signal intensity of the vascular structures. Other recent developments for reducing physiologic and voluntary motion artifacts are balanced steady-state techniques, navigator techniques, and PROPELLER imaging [2].

Electrocardiographic (ECG)-gating reduces motion unsharpness and is used in cardiac MRI examinations. This technique entails an increase in scan time, but it markedly improves the image quality.

### 17.2.6 Sedation

Voluntary motion can be minimized or eliminated by the use of sedation. Children over 6 years of age will usually cooperate for the MR examination after an explanation of the procedure and reassurance. It also helps to ensure that the patient is comfortable, free of pain, and has an empty bladder. Newborns and young infants may tolerate MRI without the need for sedation if they are recently fed and comfortably swaddled. For other children under 6 years of age, sedation usually is essential. Sedation for imaging examinations is nearly always conscious sedation. Conscious sedation is defined as a minimally depressed level of consciousness that retains the patient's abilities to maintain a patent airway, independently and continuously, and respond appropriately to physical stimulation and/or verbal command. Sedative agents that are used for healthy infants and children include IV infusion of pentobarbital, propofol, and dexmedetomidine [26–28]. Oral chloral hydrate has been a common sedative in children younger than 18 months. Midazolam can also be given for minimal sedation and anxiolysis. General anesthesia is used when a pediatric patient has comorbidities that are contraindications to conscious or moderate sedation. Prior to sedation, children should be nothing by mouth (NPO) for at least 4 to 6 h prior to the examination, depending on patient age.

Monitoring of vital signs during the MR procedure is essential for sedated patients. When the MRI is completed, the patient should be taken to a recovery area and monitored until fully awake before being discharged. Recovery time varies depending on the type and amount of drugs used. Patients should return to their baseline status with intact protective reflexes prior to discharge. The caregivers should be given verbal and written discharge instructions prior to leaving the recovery area with a contact phone number if questions arise.

## 17.3 Chest

The common clinical indications for MRI of the pediatric chest are (1) evaluation of mediastinal masses, (2) characterization of vascular anomalies, and (3) evaluation of congenital heart diseases. The role of MRI in imaging mediastinal masses is discussed below. MRI features of vascular and cardiac lesions are described in Chapters 1 through 14 in *Imaging of the Cardiovascular System, Thorax, and Abdomen*.

### 17.3.1 Normal Thymus

In the pediatric population, the normal thymus is seen in virtually every patient. In individuals under age 20, there are variations in size and shape of the normal thymus [29–31]. In patients under 5 years of age, the thymus usually has a quadrilateral shape with convex or straight lateral margins (Figure 17.1). Later in the first decade, the thymus is triangular or arrowhead-shaped with straight or concave margins, and by 15 years of age, it is triangular in nearly all individuals (Figure 17.2). In general, in the first two decades of life, the thymus abuts the sternum, separating the two lungs. A distinct anterior junction line between the lungs is usually not seen until the third decade of life.

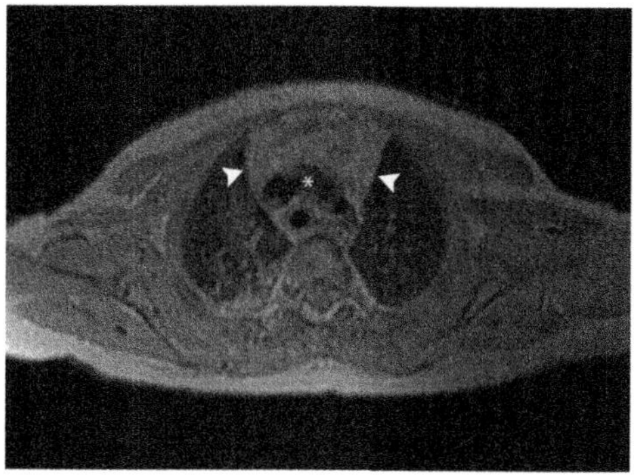

**FIGURE 17.1**
Normal MRI appearance of the thymus, 12-month-old girl. $T_1$-weighted axial MR image shows a quadrilateral-shaped thymus (arrowheads) anterior to the left innominate vein (*). The signal intensity is equal to that of chest wall musculature but less than that of subcutaneous fat.

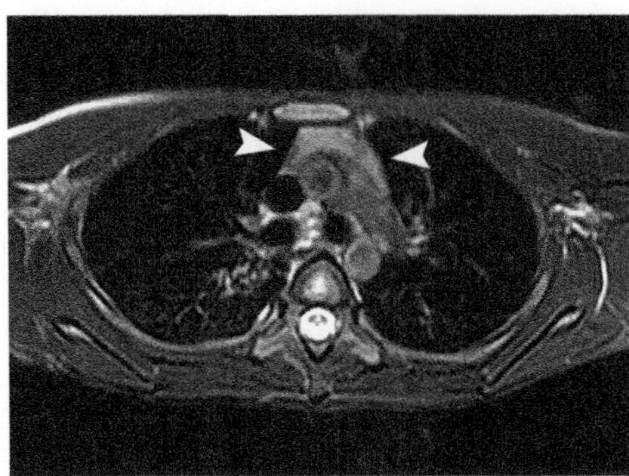

**FIGURE 17.2**
Normal MRI appearance of the thymus, 9-year-old boy. $T_2$-weighted axial MR image with fat suppression shows a triangular configuration of the thymus (arrowheads).

On MRI, the normal thymus in children and younger adolescents characteristically shows homogeneous signal intensity, which is slightly greater than that of muscle on $T_1$-weighted images, slightly less than or equal to that of fat on $T_2$-weighted images, and greater than that of fat on fat-suppressed $T_2$-weighted images. After puberty, the thymus becomes more heterogeneous and the $T_1$- and $T_2$-weighted signal intensities of the thymus increase with age since the thymus begins to involute and is replaced by fat.

### 17.3.2 Mediastinal Pathology

A widened mediastinum in infants and children often is due to a mass, usually a lymphoma, neurogenic tumor, teratoma, or cyst of foregut origin. Abundant mediastinal fat, aneurysms or tortuosity of the mediastinal vessels are rare in children. MRI has the capability of differentiating among lesions composed predominantly of fat, water, or soft tissues and therefore, can often provide a definitive diagnosis.

Lesions that can present with signal intensities near those of water include lymphangiomas, and duplication cysts of foregut origin. Bronchogenic cysts or duplication cysts may have signal intensity equal to that of soft tissue, because they contain thick viscid contents, rather than simple serous fluid. Fat-containing masses in children usually are teratomas. Rarely, they represent thymolipomas or herniation of omental fat through the foramen of Morgagni. Common soft tissue masses include lymphoma, thymic hyperplasia, and neuroblastoma.

#### *17.3.2.1 Anterior Mediastinal Masses*

##### *17.3.2.1.1 Lymphoma*

Lymphoma is the most common cause of a mediastinal mass in children, with Hodgkin disease occurring 3 to 4 times more frequently than non-Hodgkin lymphoma [32,33]. Approximately 65% of pediatric patients with Hodgkin disease have intrathoracic involvement at clinical presentation, and 90% of the chest involvement is mediastinal. In contradistinction, about 40% of pediatric patients with non-Hodgkin lymphoma have chest disease at diagnosis, and only 50% of this disease involves the mediastinum.

Lymphomatous involvement is most common in the anterior mediastinum and can appear as lymphadenopathy or infiltration and enlargement of the thymus. The enlarged thymus has a quadrilateral shape with convex, lobular lateral borders. The MR signal intensity is slightly greater than that of muscle on $T_1$-weighted pulse sequences and similar to or slightly greater than that of fat on $T_2$-weighted pulse sequences [29–31,34–37] (Figure 17.3). Calcifications or cystic areas, due to ischemic necrosis consequent to rapid tumor growth, can be seen within the tumor. Intrathoracic lymphadenopathy can range from mildly enlarged nodes in a single area to large conglomerate soft tissue masses in multiple regions. Hodgkin disease usually causes enlargement of the thymus or anterior mediastinal nodes, whereas non-Hodgkin lymphoma predominantly affects middle mediastinal lymph nodes. Additional findings include hilar lymph node enlargement, airway narrowing, and compression of vascular structures.

Successfully treated lymphomas usually decrease in size, but residual mediastinal masses may remain, especially in patients with a bulky initial mass. Differential diagnostic considerations include fibrosis versus persistent or recurrent lymphoma. In general, fibrosis has low-signal intensity (similar to that of muscle) on $T_1$- and $T_2$-weighted and fat-suppressed sequences, whereas active neoplasm has high signal intensity on $T_2$-weighted sequences [38]. In addition, gadolinium enhancement of residual mediastinal masses decreases after treatment in patients in complete remission but not in patients with relapse [39].

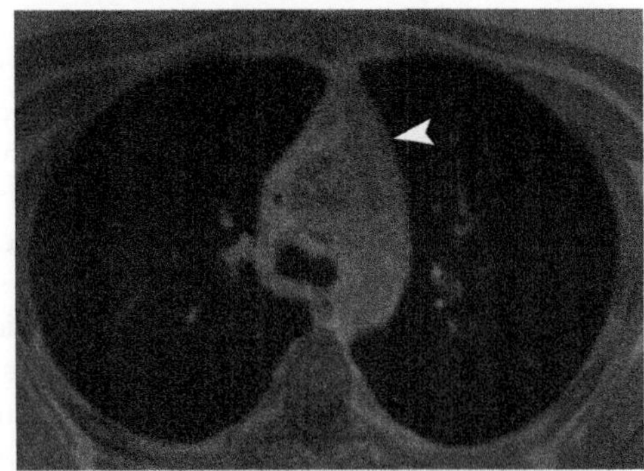

**FIGURE 17.3**
Thymic Hodgkin disease, nodular sclerosing type, 15-year-old girl. Axial $T_1$-weighted image shows thymic infiltration (arrowhead), which is isointense compared to the intensity of muscle.

*17.3.2.1.2 Thymic Hyperplasia*

In childhood, thymic hyperplasia is most often *rebound* hyperplasia associated with chemotherapy. Rebound hyperplasia may be observed during the course of chemotherapy or after the completion of therapy. Rare causes of hyperplasia include myasthenia gravis, red cell aplasia, and hyperthyroidism. On MRI, hyperplasia appears as diffuse enlargement of the thymus with preservation of the normal triangular shape (Figure 17.4). The signal intensity is similar to that of normal thymus [30,31]. Chemical shift MRI may help to diagnose thymic hyperplasia [40]. Patients with hyperplastic thymus show decrease in the signal intensity of the thymus at opposed-phase images in contrast to in-phase images while patients with thymic tumors will not show decrease in signal intensity at opposed-phase images [40]. The absence of other findings of active disease and a gradual decrease in size of the thymus on serial MRI studies also support the diagnosis of rebound hyperplasia as the cause of thymic enlargement.

*17.3.2.1.3 Thymic Epithelial Tumors*

Thymic epithelial tumors, including thymoma and thymic carcinoma, are rare in children. On MR images, thymomas typically appear as round, oval, or lobulated masses with low signal intensity on $T_1$-weighted images similar to or slightly higher than that of muscle and relatively high signal intensity on $T_2$-weighted images [41] (Figure 17.5). Diagnosis requires tissue sampling.

*17.3.2.1.4 Thyroid Masses*

Thyroid abnormalities are a rare cause of an anterior mediastinal mass in children. In childhood, intrathoracic thyroid gland is more likely to represent true ectopic thyroid tissue rather than substernal extension of a cervical thyroid gland. MR features of intrathoracic thyroid are a well-defined, intensely enhancing soft tissue mass anterior to the trachea.

*17.3.2.1.5 Germ Cell Tumors*

Germ cell tumors are the second most common cause of an anterior mediastinal mass in children and the

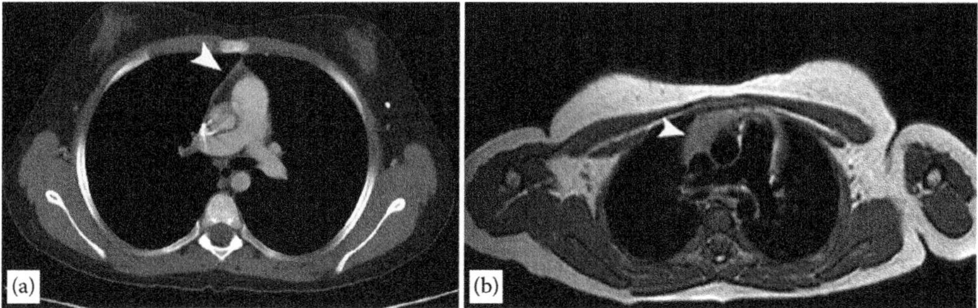


**FIGURE 17.4**
Rebound thymic hyperplasia, 16-year-old girl treated for Hodgkin disease. (a) Axial CT image of the chest soon after treatment shows a normal fatty replaced thymus (arrowhead). A central line is noted in the superior vena cava. (b) Postcontrast MR image obtained 3 months after therapy shows diffuse thymic enlargement with homogeneous intermediate signal intensity (arrowhead). (Courtesy of Rajesh Krishnamurthy, MD, Texas Children's Hospital, Houston, TX.)

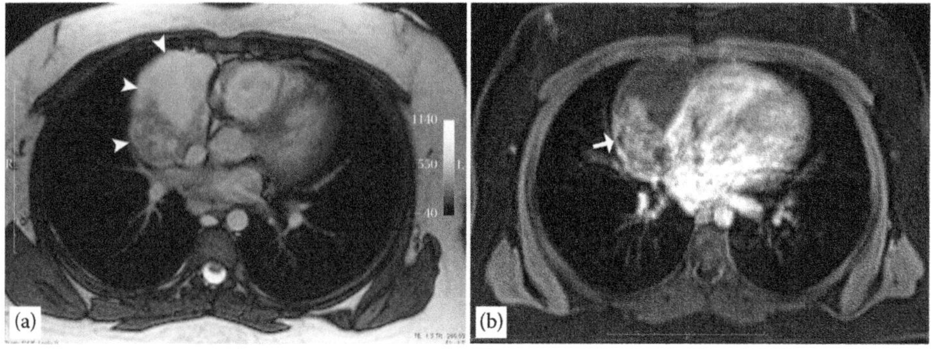


**FIGURE 17.5**
Thymoma, 10-year-old girl. (a) Axial $T_2$-weighted images demonstrates a heterogeneous mass (arrowheads) in the anterior mediastinum with fluid signal in the anterior portion. (b) Axial $T_1$-weighted image after gadolinium-DTPA injection with fat suppression shows enhancement of the posterior solid portion of the mass (arrow) but not the anterior cystic component.

most common cause of a fat-containing lesion. They are derived from one or more of the three embryonic germ cell layers and usually arise in the thymus. Germ cell tumors can be classified into three categories: (1) teratoma (mature teratoma, immature teratoma, and teratoma with malignant transformation); (2) nonseminomatuos malignant germ cell tumors (embryonal carcinoma, endodermal sinus tumor, choriocarcinoma, and mixed type) and (3) seminoma. More than 80% of germ cell tumors are benign, with the majority being mature teratomas.

On MRI, germ cell tumors are heterogeneous masses with variable signal intensities depending on the relative amounts of fluid and fat. Cystic components show low intensity on $T_1$-weighted images and high intensity on $T_2$-weighted images. Fatty tissue appears as high intensity areas on $T_1$-weighted images and shows signal loss on fat saturation image and out-of-phase chemical-shift images. Soft tissue elements are isointense with muscle. Calcification and bone have low signal intensity on all imaging sequences.

Benign cystic teratoma appears heterogeneous and contains cyst, fat, and calcification (Figure 17.6). A malignant teratoma generally appears as a predominantly soft tissue mass, sometimes containing calcification and fat. Local infiltration into the adjacent mediastinum with encasement or invasion of mediastinal vessels or airways also is frequent. The malignant nonseminomatous germ cell tumors are usually heterogeneous masses with areas of high signal intensity reflecting hemorrhage on $T_1$-weighted images and degenerative cystic changes on $T_2$-weighted images. The seminomatous tumors typically have homogeneous signal intensity. Both types of tumors can show minimal enhancement after administration of contrast material [37].

#### 17.3.2.1.6 Thymolipoma

Thymolipoma is a rare benign tumor that contains mature fat and strands of normal thymic tissue. Most cases in the pediatric population occur in the second decade. On MRI, the tumor appears as a heterogeneous mass with fatty areas having high signal intensity on $T_1$-weighted images and low signal intensity on fat-saturated images with soft tissue components showing intermediate signal intensity. Thymolipoma does not compress or invade adjacent structures [42]. The tumor may extend caudally to the diaphragm and mimic cardiomegaly or a cardiophrenic mass.

#### 17.3.2.1.7 Thymic Cysts

Thymic cysts are usually congenital lesions resulting from persistence of the thymopharyngeal duct. Typically, they are thin-walled, homogeneous masses with low signal intensity on $T_1$-weighted MR images and high signal intensity on $T_2$-weighted and fat-saturated sequences.

#### 17.3.2.1.8 Lymphangiomas

Lymphangiomas, also referred to as cystic hygromas are developmental tumors of the lymphatic system that almost always occur in the neck and occasionally extend into the anterosuperior mediastinum. On MRI, lymphangiomas have signal intensity equal to or slightly less than that of muscle on $T_1$-weighted images and greater than that of fat on $T_2$-weighted and fat-saturated images [43] (Figure 17.7). The surrounding fascial planes are obliterated if the tumor infiltrates the adjacent soft tissues. Hemorrhage can increase the signal intensity on $T_1$-weighted MR images.

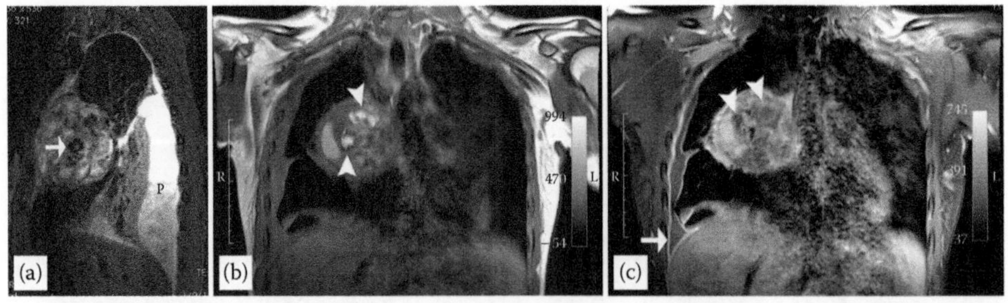


**FIGURE 17.6**
Mature teratoma, 16-year-old girl. (a) Sagittal $T_2$-weighted image with fat suppression shows a markedly heterogeneous anterior mediastinal mass containing dark signal foci corresponding to calcifications seen on CT and at pathologic examination (arrow). Posterior complex pleural fluid collection (P) and atelectasis of the right lower lobe are also noted. (b) Coronal $T_1$-weighted image demonstrates a heterogeneous mass with hyperintense foci (arrowheads) corresponding to fat on pathologic examination. (c) Contrast-enhanced axial $T_1$-weighted image with fat suppression demonstrates enhancement of the solid portions of the mass but not of the fatty components, which are hypointense with fat suppression (arrowheads). Enhancement of the pleura around the effusion is also noted (arrow).

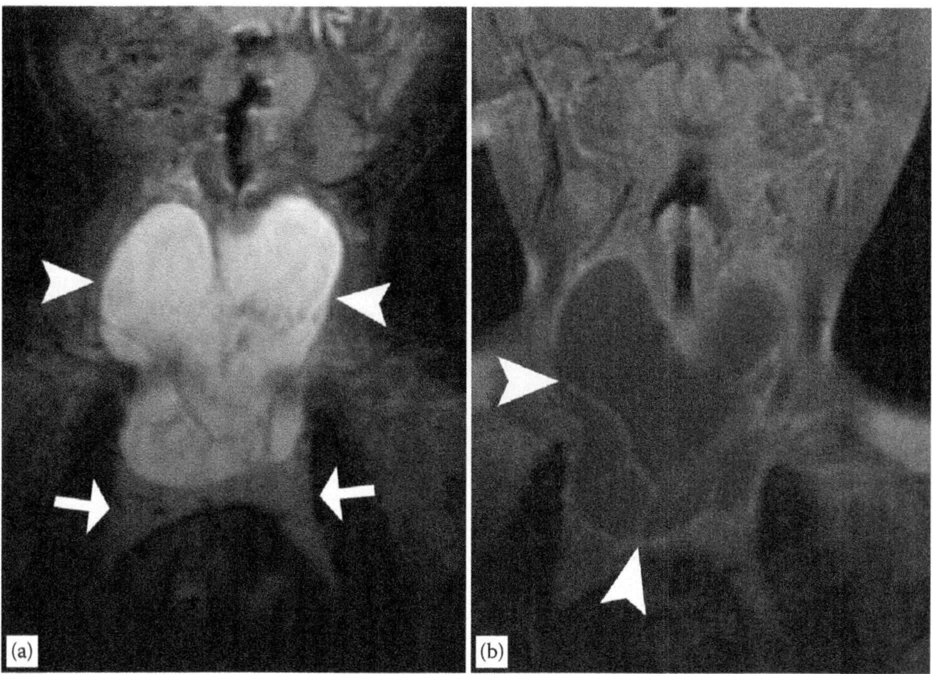


**FIGURE 17.7**
Cystic hygroma, 15-month-old boy. (a) $T_2$-weighted coronal MR image shows a homogeneous high-signal-intensity mass (arrowheads) in the neck extending into the superior mediastinum and pushing the thymus caudally (arrows). (b) Axial $T_1$-weighted image after gadolinium-DTPA injection with fat suppression demonstrates enhancement of only the walls and the septa (arrowheads).

#### *17.3.2.2 Middle Mediastinal Masses*

##### *17.3.2.2.1 Foregut Cysts*

Foregut cysts usually occur in the middle mediastinum and are classified as either bronchogenic or enteric, depending on their histology [44]. Bronchogenic cysts are lined by respiratory epithelium, and most are located in the subcarinal or right paratracheal area. Enteric cysts are lined by gastrointestinal mucosa and are located in a paraspinal position in the middle to posterior mediastinum. Foregut cysts are discovered because they produce symptoms of airway or esophageal compression or they may be detected incidentally on a chest radiograph.

On MRI, bronchopulmonary-foregut cysts typically have a low signal intensity on $T_1$-weighted images and high signal intensity on $T_2$-weighted and fat-saturated images (Figure 17.8). $T_1$-weighted signal intensity can increase if the fluid contains protein, calcium or blood [45].

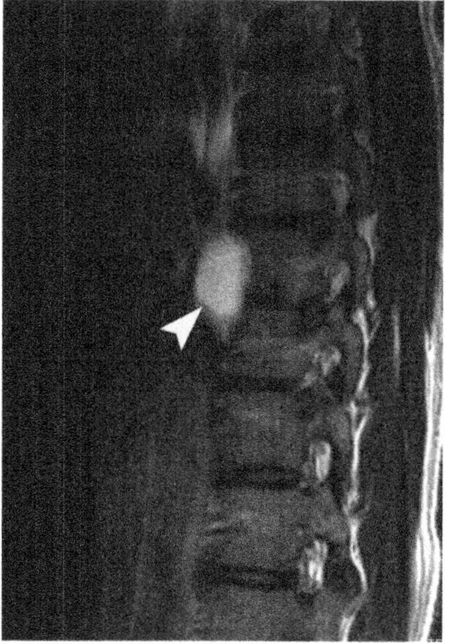

**FIGURE 17.8**
Bronchogenic cyst. $T_2$-weighted sagittal MR image shows a well-circumscribed, homogenous fluid-signal intensity mass in the posterior mediastinum (arrowhead).

#### *17.3.2.3 Posterior Mediastinal Masses*

Posterior mediastinal masses are of neural origin in approximately 95% of cases and may arise from sympathetic ganglion cells (neuroblastoma, ganglioneuroblastoma, or ganglioneuroma) or from nerve sheaths (neurofibroma or schwannoma). Rarer causes of posterior mediastinal masses in children include neurenteric cyst, lateral meningocele, hemangioma, and extramedullary hematopoiesis.

On MR, ganglion cell tumors appear as fusiform, paraspinal masses, extending over the length of several vertebral bodies. They are of intermediate, soft tissue signal intensity and contain calcifications in up to

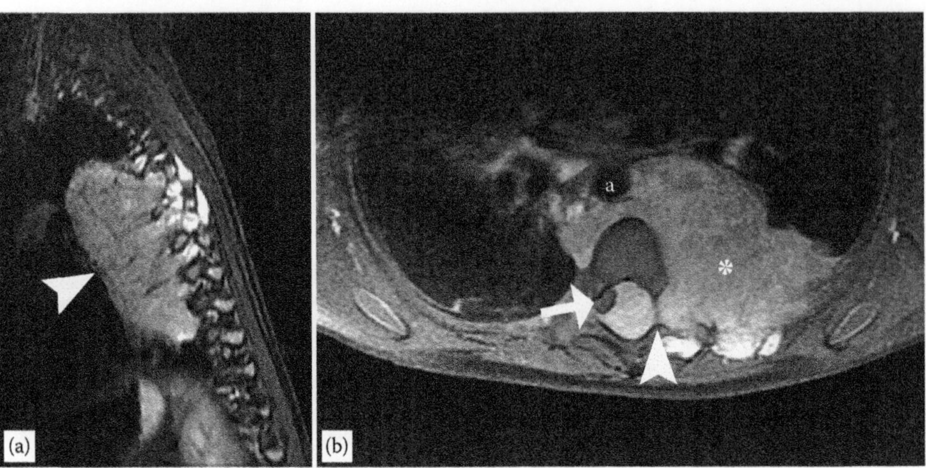


**FIGURE 17.9**
Neuroblastoma, 19-month-old girl. (a) Coronal $T_2$-weighted image shows a large posterior mediastinal mass (arrowhead) with high signal intensity entering the spinal canal through multiple neuroforamina. (b) Axial $T_1$-weighted image with gadolinium enhancement and fat saturation again demonstrates the tumor (*) extending into the chest wall and entering a left neuroforamen (arrowhead) and deflecting the spinal cord (arrow) to the right. Note that the mass crosses the midline behind the aorta (a).

50% of cases (Figure 17.9). Nerve root tumors tend to be smaller, spherical, and occur near the junction of a vertebral body with an adjacent rib. Both types of tumors may cause pressure erosion of a rib. Because of their origin from neural tissue, neurogenic tumors have a tendency to invade the spinal canal. Intraspinal extension is extradural in location, displacing and occasionally compressing the cord. Recognition of intraspinal invasion is critical because such involvement usually requires radiation therapy or a laminectomy prior to tumor debulking.

Neurenteric cysts result from incomplete separation of endoderm from notochord and usually have either a fibrous connection to the spine or an intraspinal component. The cysts are associated with midline vertebral anomalies, such as hemivertebra, butterfly vertebra, or spina bifida. Lateral meningoceles are herniations of meninges and cerebrospinal fluid. Both lesions demonstrate low signal intensity on $T_1$-weighted images, and very high signal intensity on $T_2$-weighted and fat-suppressed images. In neurenteric cysts, midline defect can be seen in one or more vertebral bodies.

Extramedullary hematopoiesis with a predominance of erythropoietic marrow shows intermediate signal intensity on both $T_1$- and $T_2$-weighted MR images, while those lesions with more fatty marrow show high signal intensity on $T_1$-weighted images and low signal on fat-suppressed images. Both types show minimal contrast enhancement. Hemangioma has low or intermediate signal on $T_1$-weighted images and high signal intensity on $T_2$-weighted images, and enhances after gadolinium administration.

### 17.3.3 Lungs

#### *17.3.3.1 Congenital Anomalies*

Congenital lung anomalies include a variety of conditions involving the pulmonary parenchyma, the pulmonary vasculature, or a combination of both. Chest CT is still the study of choice for evaluating most congenital lung anomalies, but those with associated vascular anomalies, such as sequestration and hypogenetic lung disease can be evaluated as well by MRI.

A pulmonary sequestration is characterized by lung that has no normal connection with the tracheobronchial tree and is supplied by an anomalous artery, usually arising from the aorta. When the sequestered lung is confined within the normal visceral pleura and has venous drainage to the pulmonary veins, it is termed *intralobar*. The sequestered lung is termed *extralobar* when it has its own pleura and venous drainage to systemic veins. Chronic or recurrent segmental or subsegmental pneumonitis in children, especially at a lung base, is a finding suggestive of sequestration. On MRI, the feeding vessel appears as an area of signal void on $T_1$-weighted spin-echo images and as a hyperintense area on gradient-echo and contrast-enhanced sequences. The parenchymal portion of the sequestration appears as an area of intermediate or high signal intensity (Figure 17.10).

MR findings of hypogenetic lung or scimitar syndrome include a small right lung, ipsilateral mediastinal displacement, a corresponding small pulmonary artery, and partial anomalous pulmonary venous return, usually from the right lung to the inferior vena cava (Figure 17.11). Other associated anomalies include

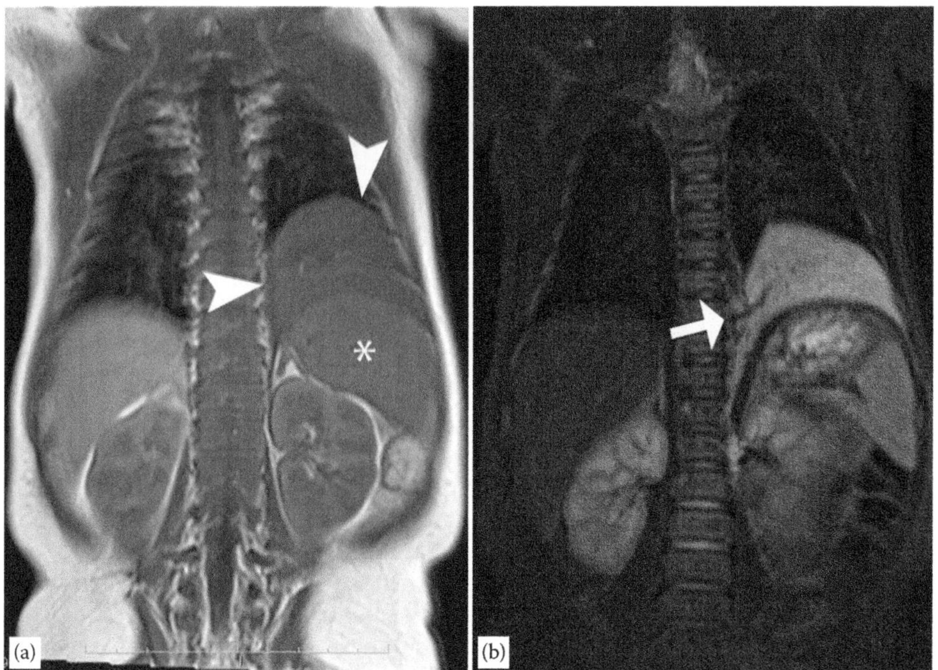


**FIGURE 17.10**
Pulmonary sequestration, 3-month-old boy. (a) Coronal $T_1$-weighted image reveals a well-circumscribed, homogeneous mass (arrowheads) at the base of the left hemithorax above the spleen (asterisk). (b) Coronal $T_2$-weighted image shows a bright signal mass with signal void representing systemic arterial supply (arrow).

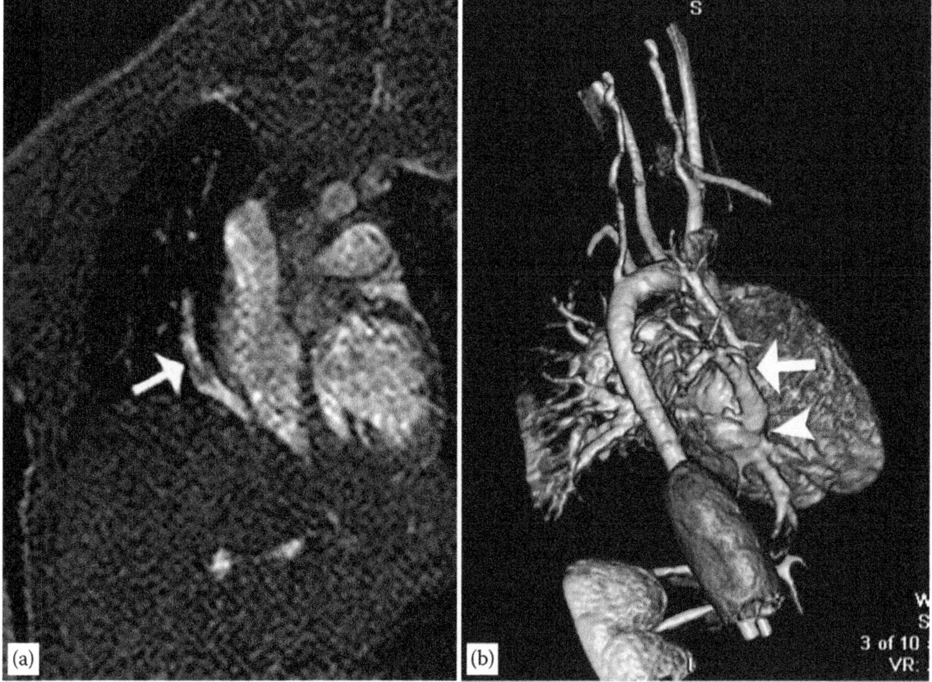


**FIGURE 17.11**
Scmitar syndrome, 4-year-old girl. (a) Coronal bright-blood image shows mesocardia and right scimitar vein (arrow) draining into the suprahepatic IVC at the level of the diaphragm. (b) 3D volume rendered oblique image from a contrast-enhanced MR angiogram shows the superior and inferior pulmonary veins joining to form the scimitar vein (arrow), which inserts into the IVC just above the confluence of the hepatic veins (arrowhead).

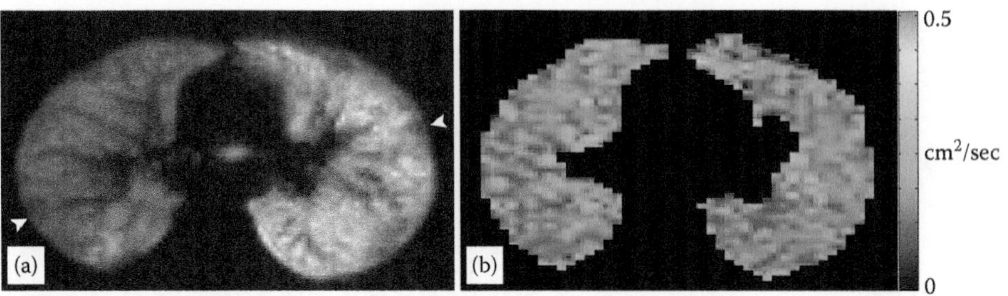


**FIGURE 17.12**
Hyperpolarized helium imaging, patient's status post lung transplant. (a) The ventilation image is minimally abnormal with a difference in signal between the two lungs and a few regions along the edge of slightly reduced signal (arrowheads). (b) The ADC map shows normal diffusivity of helium. Findings consistent with bronchiolitis obliterans.

systemic arterial supply to the hypogenetic lung, accessory diaphragm, and horseshoe lung. Horseshoe lung is a rare anomaly in which the posterobasal segments of both lungs are fused behind the pericardial sac.

#### *17.3.3.2 Pulmonary Metastases*

MRI can detect large parenchymal nodules, but it is not as sensitive as CT in detecting nodules less than 5 mm in diameter because of its poorer spatial resolution. Hence, CT remains the imaging method of choice for detecting and characterizing pulmonary nodules.

#### *17.3.3.3 Diffuse Parenchymal Disease*

Chest radiography and CT remain the imaging study of choice for evaluating diffuse parenchymal lung disease. However, there has been some success using MR to image bronchiectasis and air trapping in the older child associated with cystic fibrosis [46–49] and primary ciliary dyskinesia [50].

Gases such as $^{3}$He and $^{129}$Xe are being increasingly used to study lung function and pathology [2,51–54] (Figure 17.12). Hyperpolarized MRI relies on polarizing a molecular probe outside the patient, injecting or inhaling the hyperpolarized probe, then using MRI to receive signals from the probe as the probe either localizes to a target or becomes metabolized. Conventional $T_1$- and $T_2$-weighted images may then be fused with the functional information. Hyperpolarized MRI has been used to assess regional lung ventilation in diseases such as such as cystic fibrosis, asthma and small airways disease [51–54].

## 17.4 Abdomen

The appearance of the abdomen on MR examinations is similar in adults and children, except for the smaller size of the structures being examined and the relative paucity of perivisceral fat. The clinical questions usually prompting MR examination of the abdomen are: (1) pretreatment determination of the site of origin, extent and character of an abdominal mass as well as posttreatment assessment; (2) characterization of pancreaticobiliary anomalies; (3) evaluation of urinary tract abnormalities, particularly hydronephrosis; and (4) assessment of abnormalities of the major abdominal vessels (MR angiography [MRA]) [55,56].

### 17.4.1 Renal Masses

In evaluation of renal masses, the basic MR sequences include axial $T_1$-weighted, in and opposed phase, coronal and sagittal $T_2$-weighted fast/turbo spin echo, axial $T_2$-weighted fast/turbo spin echo (with and without fat suppression), and axial diffusion-weighted images. After contrast injection, axial $T_1$-weighted images with fat-saturation are obtained during arterial and venous phases.

#### *17.4.1.1 Solid Malignant Renal Tumors*

##### *17.4.1.1.1 Wilms Tumor*

Wilms tumor is the most common primary malignant renal tumor of childhood [13,57–59]. Affected patients generally are under 5 years of age, with a mean age between 3 and 4 years. They present most often with a palpable abdominal mass, and less often with abdominal pain, fever, and microscopic or gross hematuria. Approximately, 10% have metastatic disease at presentation. Metastases are typically to lungs and less frequently to liver. Several congenital abnormalities, including hemihypertrophy and sporadic aniridia, and syndromes, including Beckwith–Wiedemann (hemihypertrophy, omphalocele, macroglossia, and visceromegaly), WAGR (Wilms tumor, aniridia, genitourinary malformation, mental retardation), Drash syndrome (male pseudohermaphroditism and nephritis), and Perlman syndrome (visceromegaly, gigantism, cryptorchidism, polyhydramnios, classic facies) predispose to Wilms tumor [58–62].

Wilms tumor characteristically appears as a large (mean diameter 11 cm), spherical, at least partially intrarenal mass with an MR signal intensity equal to or lower than normal renal cortex on $T_1$-weighted images and higher than normal parenchyma on $T_2$- and diffusion-weighted images [3,13,63–65]. The tumor enhances after IV administration of contrast medium, but usually to a lesser extent than the adjacent parenchyma (Figure 17.13). Approximately 80% of tumors are heterogeneous, because they contain areas of necrosis or hemorrhage. Bilateral synchronous tumors occur in 5% to 10% of patients.

Wilms tumor spreads intra-abdominally either by direct extension into perinephric tissues, lymph nodes or adjacent organs, or by invasion of vessels with extension into the renal vein or inferior vena cava. Perinephric extension may be seen as a thickened renal capsule or as nodular or streaky densities in the perinephric fat. Lymph nodes have intermediate signal intensity on $T_1$-weighted sequences and high signal intensity on $T_2$-weighted and gadolinium enhanced sequences. Any identified retroperitoneal lymph node, regardless of size, should be regarded with suspicion. Although normal size nodes are commonly demonstrated on abdominal MRI in adults, such nodes are rarely, if ever, seen in infants and young children.

The presence of IVC invasion is an important determinant of the surgical approach. A thoracoabdominal approach is required for removal of tumor thrombus extending to or above the confluence of the hepatic veins, whereas an abdominal approach alone is satisfactory for intravascular thrombus below the hepatic veins. Tumor thrombus is hyperintense to flowing blood on spin-echo sequences and hypointense to flowing blood on gradient-echo and enhanced sequences (Figure 17.14).

After therapy, MRI can be used to detect local recurrence and hepatic metastases. Features that suggest localized recurrence are a soft tissue mass in the empty renal fossa and ipsilateral psoas muscle enlargement.

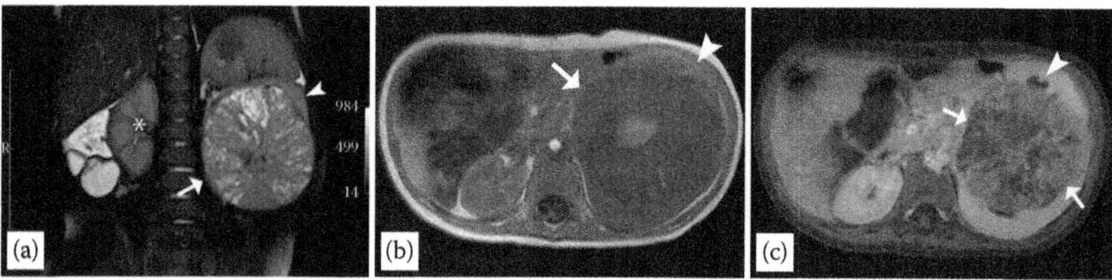


**FIGURE 17.13**
Wilms tumor, 1-year-old girl. (a) Coronal $T_2$-weighted image shows a heterogeneous high-signal-intensity mass (arrow) with surrounding left renal tissue (arrowhead). The right kidney appears normal (asterisk). (b) Axial $T_1$-weighted image shows the mass (arrow) is hypointense to renal cortex (arrowhead) with a central focal area of hyperintense hemorrhage. (c) Axial $T_1$-weighted image after gadolinium-DTPA injection shows heterogeneous enhancement of the mass (arrows). Arrowhead indicates dilated calyx.

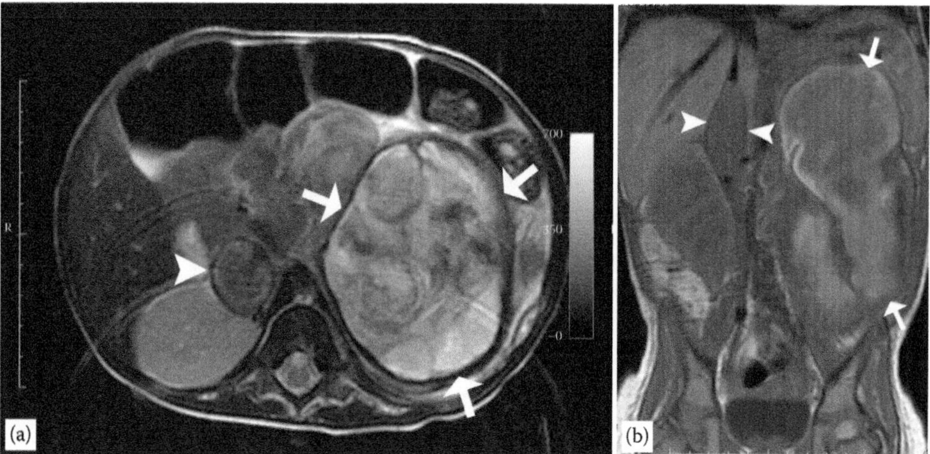


**FIGURE 17.14**
Wilms tumor with intracaval extension, 2-year-old girl. (a) Axial $T_2$-weighted image shows a markedly heterogeneous mass replacing the left kidney (arrows) and heterogeneous signal in an enlarged IVC (arrowhead) consistent with tumor thrombus. (b) Coronal $T_1$-weighted image demonstrates a large, heterogeneous mass in the left kidney with bright signal consistent with subacute hemorrhage (arrows). The caval tumor thrombus is of intermediate signal intensity (arrowheads).

*17.4.1.1.2 Nephroblastomatosis*

Nephroblastomatosis is an abnormality of nephrogenesis characterized by persistence of rests of fetal renal blastema beyond 36 weeks of intrauterine gestation [66]. Npehroblastomatosis or nephrogenic rests are not malignant tumors *per se*, but they have potential for malignant transformation into Wilms tumor. Nephrogenic rests are found in the cortex and may diffusely replace the parenchyma or present as focal or multifocal masses. On both $T_1$-weighted images and contrast-enhanced sequences nephroblastomatosis is hypointense relative to normal renal tissue [3,13,63,64,67]. On $T_2$-weighted images, nephrogenic rests usually are iso- or slightly hyperintense to renal cortex, but occasionally they can be hypointense. They enhance less than adjacent parenchyma (Figure 17.15).

*17.4.1.1.3 Lymphoma*

Lymphoma is more often associated with the non-Hodgkin than with the Hodgkin form of disease and usually affects patients over 5 years of age. The most common imaging appearance is that of multiple bilateral nodules [3,13,64,65,68]. Less common presentations include a solitary intraparenchymal mass (Figure 17.16), diffuse renal enlargement, and direct invasion from adjacent lymph nodes or a retroperitoneal mass. Secondary findings include renal fascial thickening and tumor encasement of the renal pelvis and proximal ureter with obstructive hydronephrosis. Lymphomatous masses are usually homogeneous, hypointense on $T_1$-weighted images, and hyperintense on $T_2$-weighted, contrast enhanced and high *b*-value diffusion-weighted images compared with normal parenchyma.

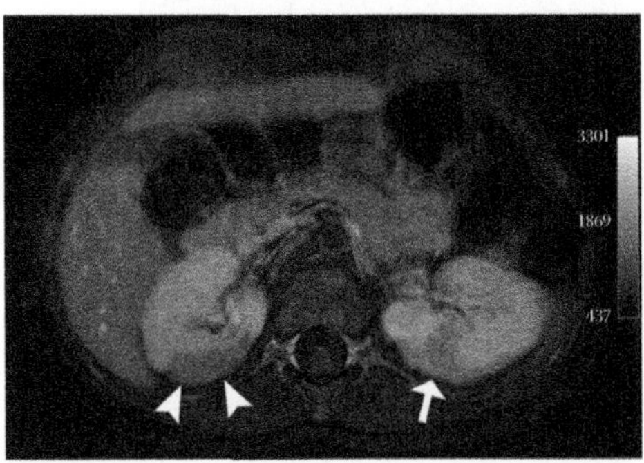

**FIGURE 17.15**
Nephroblastomatosis, 4-year-old girl. Axial fat-saturated $T_1$-weighted image after gadolinium administration shows a peripheral rind in the posterior kidneys (arrowheads on right) and a triangular-shaped mass in the left (arrow), which enhance less than the adjacent normal renal parenchyma.

*17.4.1.1.4 Rare Malignant Renal Tumors*

Clear-cell sarcoma and rhabdoid tumor are rare, highly aggressive renal malignancies in childhood [3,13,65,69,70]. The former tends to affect children between 1 and 4 years of age, whereas the latter is more frequent in infants under 2 years of age. Presenting signs are similar to those of Wilms tumor. Concomitant primary tumors of the posterior cranial fossa, soft tissues, and thymus occur in association with malignant rhabdoid tumor. Rhabdoid tumor metastasizes to lymph nodes, liver, lung, bone, and brain. Clear-cell sarcoma of the kidney can also metastasize to bone and is known as the *bone metastasizing tumor of childhood.*

At MRI, rhabdoid tumor of the kidney appears as a heterogeneous tumor with indistinct margins involving the renal hilum. Rhabdoid tumor has low or intermediate signal intensity on $T_1$-weighted sequences, high signal intensity on fat-suppressed $T_2$-weighted and diffusion-weighted sequences and enhances after gadolinium injection (Figure 17.17). A subcapsular fluid collection, representing necrosis and/or hemorrhage, is often seen and helps to differentiate this tumor from Wilms tumor.

Clear-cell sarcoma of the kidney (also known as bone-metastasizing renal tumor) also is a heterogeneous tumor with indistinct margins. Cystic areas

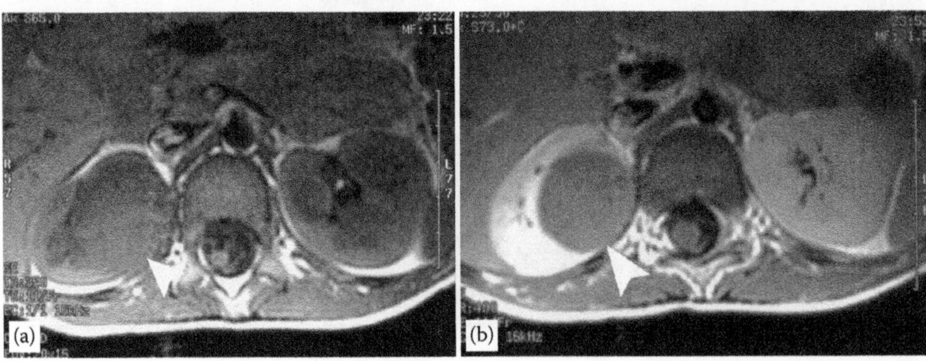

**FIGURE 17.16**
Renal Burkitt lymphoma in a 4-year-old boy. (a) Axial $T_1$-weighted image shows a well-circumscribed right renal mass which is homogeneously hypointense to renal cortex (arrowhead). (b) Contrast-enhanced axial $T_1$-weighted image shows that the mass (arrowhead) enhances less than normal renal parenchyma.

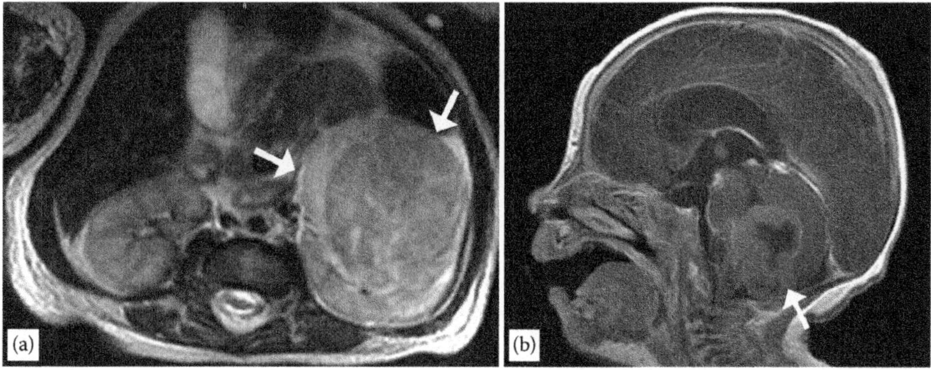

**FIGURE 17.17**
Rhabdoid tumor, 11-week-old boy. (a) $T_2$-weighted axial image demonstrates an intermediate signal intensity, heterogeneous tumor with ill-defined margins arising in the left kidney (arrows) the parenchyma of the left kidney. (b) Sagittal $T_1$-weighted MR following intravenous gadolinium administration shows a synchronous posterior fossa atypical teratoid rhabdoid tumor (arrow).

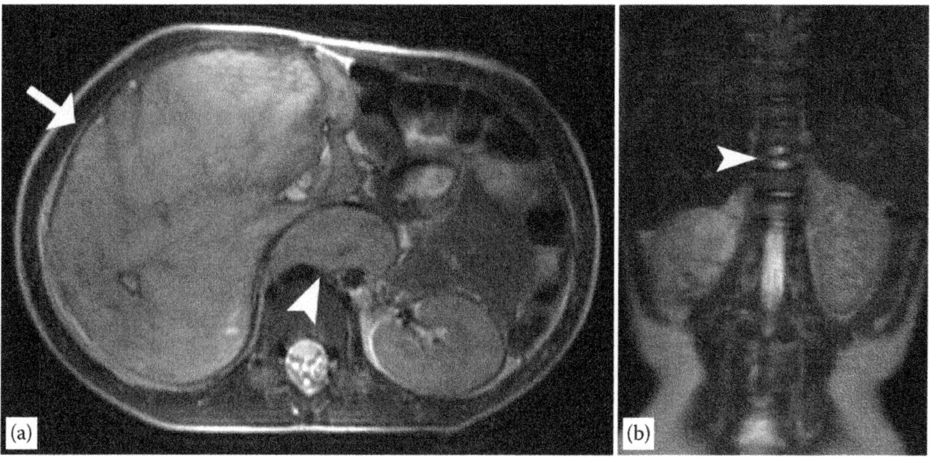

**FIGURE 17.18**
Clear-cell sarcoma in a 2-year-old girl. (a) Axial $T_2$-weighted image shows replacement of the right kidney by a mass of heterogeneous signal intensity (arrow). A lymph node metastasis (arrowhead) of similar signal intensity is noted near the aorta. (b) Coronal $T_2$-weighted image shows a compression fracture of T-10 due to metastatic disease (arrowhead).

representing necrosis and hemorrhage are common. It appears as a $T_1$-weighted hypointense mass and $T_2$ and diffusion-weighted hyperintense mass and enhances after gadolinium administration. The finding of bone metastasis suggests clear-cell sarcoma rather than Wilms tumor, which most commonly metastasizes to lung (Figure 17.18).

Renal cell carcinoma affects children between 9 and 15 years of age, with a mean patient age at presentation of 10 years [3,13,65]. Presenting signs and symptoms are nonspecific and include mass, pain, and hematuria. The MRI appearance of renal cell cancer is similar to that of Wilms tumor, except for a smaller size (mean diameter 4 cm). Renal cell carcinoma appears as a nonspecific solid intrarenal mass with well-circumscribed or ill-defined margins (Figure 17.19) [71]. Hemorrhage, necrosis, and calcification are common. The tumor is $T_1$-weighted hypointense, $T_2$-weighted and diffusion-weighted hyperintense and enhances but to a lesser degree than the normal renal parenchyma. Like Wilms tumor, renal cell carcinoma may spread to retroperitoneal lymph nodes or may invade the renal vein and metastasize to lung and liver.

Renal medullary carcinoma is an unusual tumor associated with sickle cell trait [70,72]. Most patients are diagnosed in the second or third decades of life and present with flank or abdominal pain or gross hematuria and less commonly with a palpable mass, weight loss, or fever. Distant metastases are to lung and liver. The tumor is typically heterogeneous, containing hemorrhage and extensive necrosis. It appears as a $T_1$-weighted hypointense and $T_2$-weighted hyperintense mass that is located centrally, deep in the parenchyma, invading and encasing the renal pelvis and causing caliectasis and

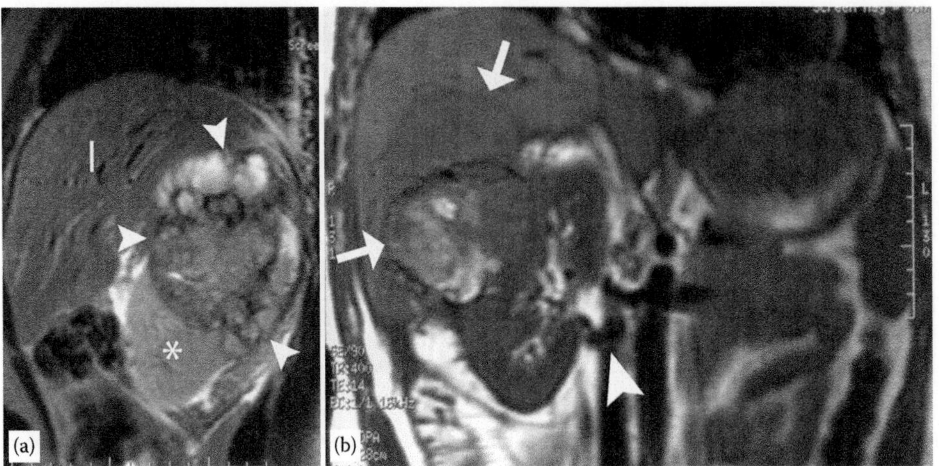

**FIGURE 17.19**
Renal cell carcinoma in 17-year-old boy. (a) Sagittal $T_2$-weighted image shows a very heterogeneous-signal mass arising in the upper lateral right kidney (arrowheads), which abuts the liver (l). Asterisk indicates adjacent normal kidney (b) Coronal $T_1$-weighted image demonstrates heterogeneous-signal mass (arrows) with hyperintense foci which correlated with hemorrhage at pathologic examination. The right renal vein is shown to be patent (arrowhead).

reniform enlargement. Contrast enhancement is heterogeneous, reflecting tumor necrosis.

#### 17.4.1.2 Solid Benign Renal Tumors

##### 17.4.1.2.1 Mesoblastic Nephroma

Mesoblastic nephroma, also termed fetal renal hamartoma, usually presents in neonates or young infants within the first 3 months of life as an abdominal mass [3,13,65,73]. It is nearly always benign although rare cases with cellular atypia and malignant potential have been described. The MR appearance is nonspecific and it most often appears as a fairly uniform mass, replacing a large part of the renal parenchyma (Figure 17.20). It tends to be iso- or hypointense to surrounding tissues on $T_1$-weighted sequences and hyperintense on $T_2$-weighted sequences and shows some enhancement, although not to the extent of normal renal parenchyma. Highly cellular tumors may show restricted diffusion [74]. Although this tumor cannot be reliably distinguished from Wilms tumor by imaging studies, the diagnosis should be suspected in an infant with a solid renal mass.

#### 17.4.1.3 Cystic Renal Masses

##### 17.4.1.3.1 Multilocular Cystic Renal Tumor

Multilocular cystic renal tumor (also termed cystic hamartoma, cystic lymphangioma, and partial polycystic kidney) is a unilateral, nonhereditary cystic mass with septations composed of fibrous tissue. The septations may contain mature elements or immature blastemal elements. The tumor has a male predominance, affecting boys under 4 years of age. Presenting signs are a nonpainful abdominal mass and hematuria. On MRI, the lesion appears as a well-defined intrarenal mass with multiple cystic areas and interspersed septa [3,13,65,75,76]. The cystic spaces have low $T_1$-weighted

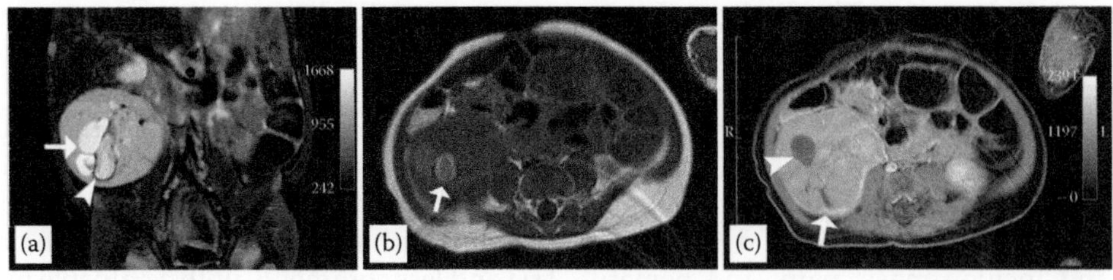

**FIGURE 17.20**
Mesoblastic nephroma, 2-month-old boy. (a) Coronal short-tau-inversion-recovery (STIR) MR image shows a heterogeneous mass replacing the right kidney with hyperintense foci (arrow) and a dark rim (arrowhead) consistent with hemosiderin. (b) Axial $T_1$-weighted image shows the mass is predominantly isointense compared to muscle with hyperintense foci (arrow) one with a dark rim consistent with hemorrhage found at gross pathologic exam. (c) Axial gadolinium-enhanced $T_1$-weighted image with fat suppression at a higher level than (b) shows the large heterogeneously enhancing mass. The cystic portions (arrowhead) do not enhance while the solid portions enhance less than the rim of adjacent renal parenchyma (arrow).

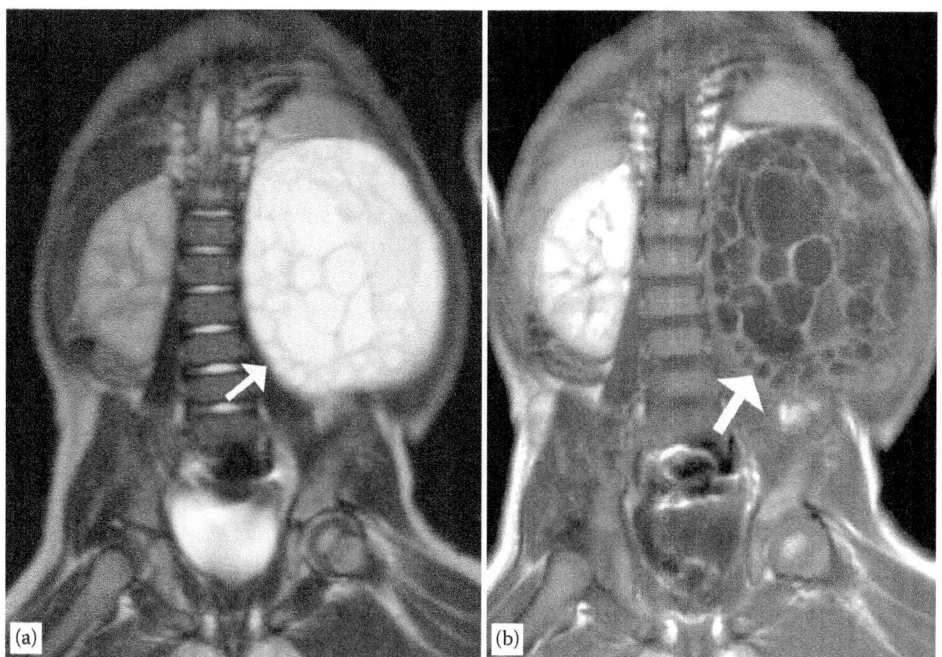

**FIGURE 17.21**
Cystic nephroma in 1-year-old girl. (a) Coronal $T_2$-weighted MR image shows a fluid-signal mass with multiple thin septa replacing the left kidney (arrow). (b) Coronal $T_1$-weighted MR image after intravenous gadolinium-DTPA enhancement of the wall and septa only (arrow).

and high $T_2$-weighted signal intensity and do not enhance after administration of gadolinium chelate agents (Figure 17.21). The septa are usually thin and may show minimal enhancement. Curvilinear calcifications may be seen within the wall or the septa.

#### 17.4.1.4 Fatty Renal Masses

##### 17.4.1.4.1 *Angiomyolipoma*

Angiomyolipoma is a benign renal tumor composed of angiomatous, myomatous, and lipomatous tissue. It is rare as an isolated lesion in the general pediatric population, but is present in as many as 80% of children with tuberous sclerosis. The lesions usually are detected as an incidental finding, but some patients present with abdominal pain or with renal failure because of extensive parenchymal replacement by tumor. Angiomyolipomas are typically small, multiple, and bilateral (Figure 17.22). Rarely, a large dominant mass is seen. Angiomyolipomas demonstrate high signal intensity on $T_1$- and $T_2$-weighted sequences and low signal intensity on fat-suppressed and opposed phase images. Coexistent cystic renal lesions are common.

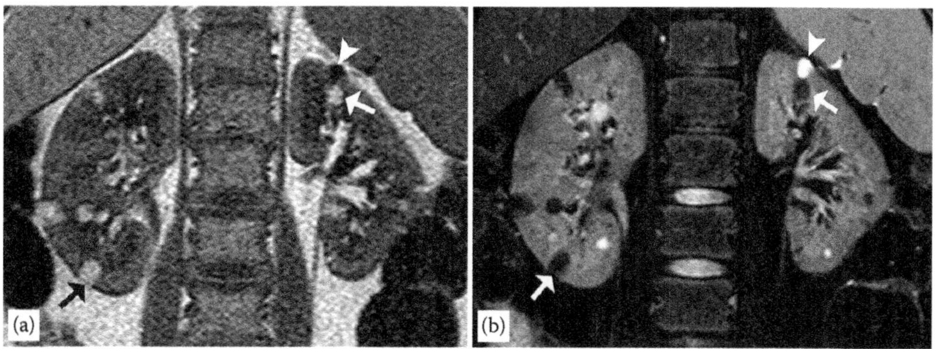

**FIGURE 17.22**
Angiomyolipomas in tuberous sclerosis, 7-year-old girl. (a) Coronal $T_1$-weighted image shows small hyperintense masses (arrows) in both kidneys. A hypointense cyst is also noted (arrowhead). (b) Coronal $T_2$-weighted image with fat demonstrates that the $T_1$-hyperintense masses are dark consistent with fat (arrows). The cyst is fluid-signal intensity (arrowhead). (Courtesy of Rajesh Krishnamurthy, MD, Texas Children's Hospital, Houston, TX.)

### 17.4.2 Adrenal Masses

The MR imaging protocol is the same as that use in imaging of renal tumors.

#### *17.4.2.1 Hemorrhage*

Hemorrhage is the most common cause of an adrenal mass in the neonate, occurring as a result of birth trauma, septicemia or hypoxia. Adrenal hemorrhage is less frequent in infants and children and is usually the result of trauma [77,78]. Affected newborns often present with a palpable abdominal mass, jaundice, and anemia. In contrast to tumors where the shape of the gland is distorted, the triangular shape of the adrenal is usually preserved in hemorrhage. The signal characteristics of the blood vary with the age of the hemorrhage. Acute blood has a low signal intensity on $T_1$-weighted images and high signal intensity on $T_2$-weighted images (Figure 17.23). Subacute hemorrhage has high signal intensity on $T_1$- and $T_2$-weighted sequences. As the blood clots and lyses, the intensity of the hemorrhage decreases over time. A chronic hematoma appears hypointense on both $T_1$- and $T_2$-weighted sequences.

The primary diagnostic problem in the neonate is congenital neuroblastoma. Differentiating between these two conditions is possible when there are hepatic metastases or when serum vanillylmandelic acid (VMA) levels are elevated. Serial imaging also can help in differentiation. A hematoma decreases in size over 1 to 2 weeks, whereas neuroblastoma either remains the same size or enlarges.

#### *17.4.2.2 Neuroblastoma*

Neuroblastoma is the most common malignant abdominal tumor in children, usually affecting children under the age of 4 years. More than half of all neuroblastomas originate in the abdomen, and two-thirds of these arise in the adrenal gland [12,78–80]. The extraadrenal tumors originate in the sympathetic ganglion cells or paraaortic bodies and may be found anywhere from the cervical region to the pelvis. Most patients present with a palpable abdominal mass. More than half of all patients have bone marrow, skeletal, liver, or skin metastases when initially diagnosed. Lung metastases are rare.

At MRI, neuroblastoma appears as an extrarenal or paraspinal mass [12,78,80–82]. Adrenal tumors displace the kidney inferiorly and laterally, while paraspinal tumors displace the kidney superiorly and laterally. Neuroblastoma is iso- or hypo-intense to surrounding soft tissues on $T_1$-weighted spin-echo images and hyperintense on $T_2$- and diffusion-weighted sequences and shows some enhancement after administration of IV gadolinium chelate agents (Figure 17.24) [83]. Heterogeneity is common due to hemorrhage, necrosis, and calcification, the latter occurring in about 85% of tumors. Hemorrhage results in variable signal intensity, depending on the age of the blood. Necrotic foci usually appear hypointense on $T_1$-weighted sequences and hyperintense on $T_2$-weighted sequences. Calcification has low signal intensity on both pulse sequences.

Local spread of tumor may take the form of midline extension, vessel encasement (Figure 17.25), regional lymph node involvement and/or intraspinal invasion (Figure 17.26). Hepatic metastases may also be identified [84].

Following surgery or chemotherapy, MRI can be used to monitor treatment response and detect recurrent disease. Demonstration of a residual mass with low signal intensity on $T_1$-weighted and fat-suppressed $T_2$-weighted imaging and DWI favors the diagnosis of fibrosis, whereas high signal on $T_2$- and DWI suggests residual tumor.

#### *17.4.2.3 Ganglioneuroblastoma and Ganglioneuroma*

Ganglioneuroblastoma tends to occur later in the first decade of life and ganglioneuroma tends to occur in the second decade of life. They have imaging findings

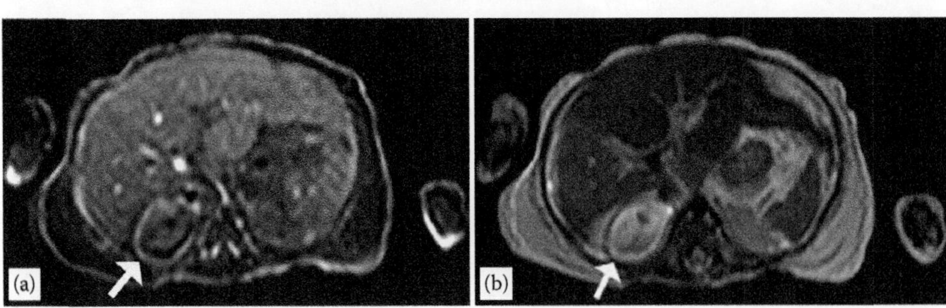


**FIGURE 17.23**
Subacute adrenal hemorrhage, neonate. (a) Axial $T_1$-weighted image through the upper abdomen shows a heterogeneous mass in the right adrenal gland (arrow). The central portion of the mass has low-to-intermediate signal intensity, while the periphery of the mass has high signal intensity. (b) On the axial $T_2$-weighted image, the mass is heterogeneous and predominantly bright with a hypointense rim (arrow). Ascites and anasarca are also noted. (Courtesy of Rajesh Krishnamurthy, MD, Texas Children's Hospital, Houston, TX.)

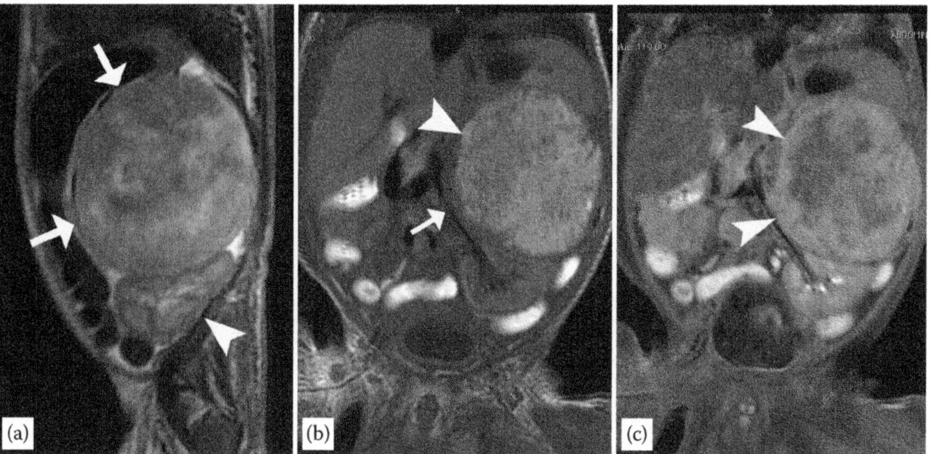


**FIGURE 17.24**
Adrenal neuroblastoma, 11-day-old boy. (a) $T_2$-weighted sagittal MR image demonstrates a large, heterogeneous left suprarenal mass (arrows) separate from the kidney (arrowhead). (b) Coronal $T_1$-weighted image demonstrates central high signal consistent with old hemorrhage (arrowhead). Arrow indicates left renal artery. (c) On a $T_1$-weighted image with gadolinium enhancement and fat suppression, the tumor shows some peripheral enhancement (arrowhead).

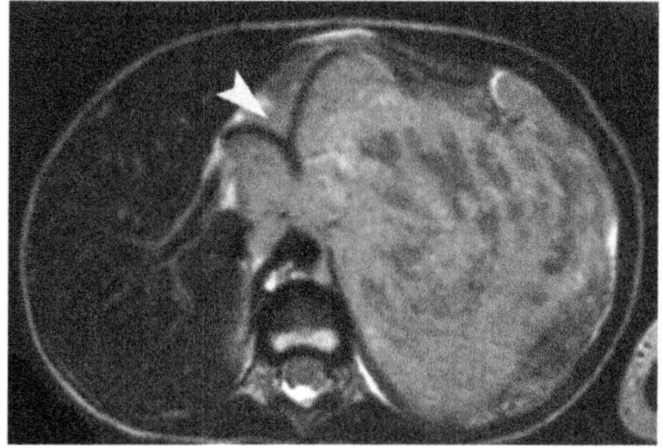

**FIGURE 17.25**
Neuroblastoma, 4-year-old girl. Axial $T_2$-weighted image demonstrates the mass crossing midline and encasing the celiac artery and its branches (arrowhead).

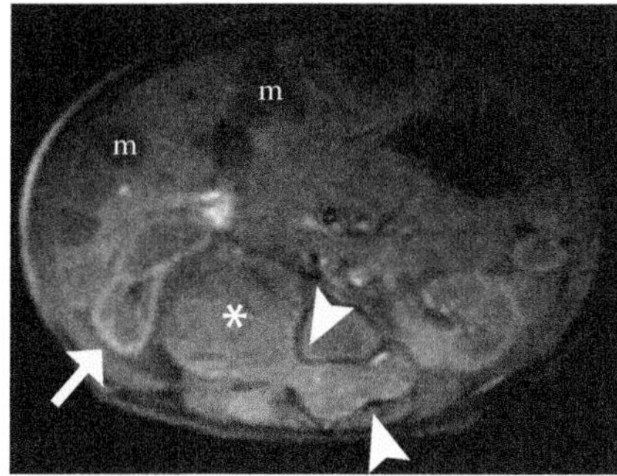


**FIGURE 17.26**
Neuroblastoma, intraspinal extension, 8-week-old girl. Axial $T_1$-weighted image shows a homogeneous intermediate signal mass (asterisk) in the right retroperitoneum deflecting the right kidney laterally (arrow). There is a smaller mass on the left and both extend through neuroforamina into the spinal canal (arrowheads). Liver metastases are indicated by (m).

similar to those of neuroblastoma. The diagnosis is made histologically by the degree of cellular maturation and differentiation.

#### *17.4.2.4 Adrenocortical Neoplasms*

Adrenal lesions, other than neuroblastomas, are rare in childhood, accounting for 5% or less of all adrenal tumors [85–87]. Of these, carcinoma is the most common, followed in frequency by adenoma. The mean ages at presentation of patients with carcinoma and adenoma are approximately 9 and 3 years, respectively. Adrenal carcinomas are usually hormonally active, producing virilization, feminization, or Cushing syndrome. Adenomas can cause Cushing syndrome or primary aldosteronism, but they also may be detected incidentally.

Adrenal carcinomas are typically large masses, greater than 4 cm in diameter (Figure 17.27). They exhibit low signal intensity on $T_1$-weighted and high signal intensity on fat-suppressed $T_2$-weighted and diffusion-weighted images. Heterogeneous contrast enhancement is typical [85,87,88]. There may be local invasion of adjacent structures, such as the inferior vena cava, liver, or lymph nodes. Distant metastases are to lung, liver, and lymph nodes. Adenomas can have high signal intensity on $T_1$- and $T_2$-weighted sequences and low signal intensity on out-of-phase images because of their high lipid content.

#### *17.4.2.5 Pheochromocytoma*

Pheochromocytomas are catecholamine-producing tumors that cause paroxysmal hypertension in

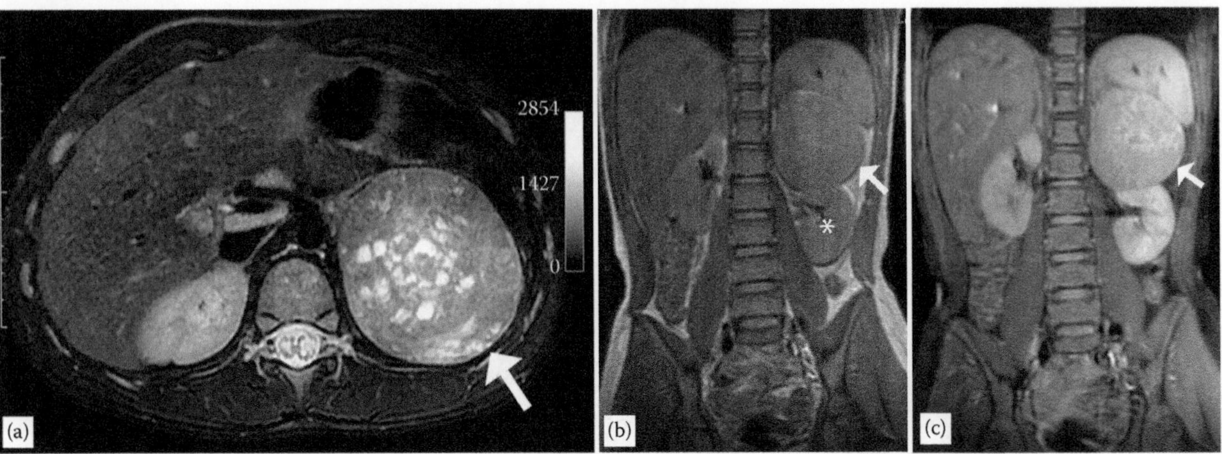


**FIGURE 17.27**
Adrenocortical neoplasm, 6-year-old girl with virulization. (a) $T_2$-weighted axial MR image with fat saturation shows a heterogeneous left adrenal tumor (arrow) with foci of fluid signal intensity. (b) Coronal $T_1$-weighted image shows a homogeneous mass (arrow), which is isointense to spleen and separate from the left kidney (asterisk). (c) $T_1$-weighted axial MR image with fat saturation following intravenous gadolinium demonstrates the mass is separate from the kidney and enhances homogeneously (arrow).

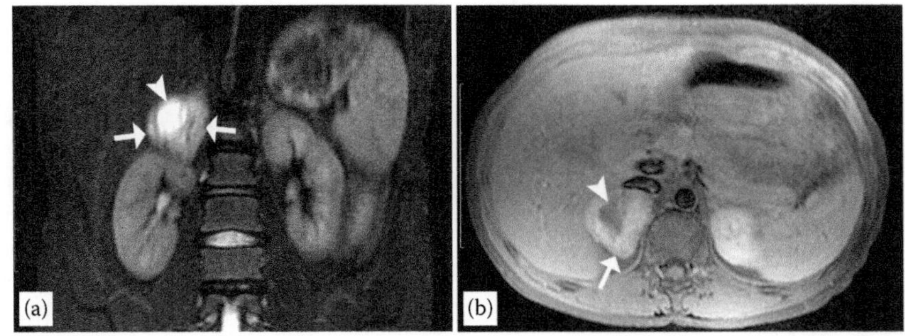


**FIGURE 17.28**
Adrenal pheochromocytoma, 11-year-old girl. (a) Coronal $T_2$-weighted image with fat suppression shows a small right suprarenal mass (arrows) that is isointense to spleen containing a focus of fluid-signal intensity (arrowhead) confirmed at pathology to represent at cystic component. (b) Axial $T_1$-weighted gradient-echo image with gadolinium enhancement and fat suppression demonstrates the enhancing mass (arrow) in the right adrenal gland with no enhancement of the cystic component (arrowhead).

children [89]. Most are sporadic; however, they may be associated with multiple endocrine neoplastic (MEN) syndromes and the phakomatoses, including neurofibromatosis, tuberous sclerosis, von Hippel–Lindau disease, and Sturge–Weber disease. Approximately 75% of childhood pheochromocytomas arise in the adrenal medulla; the remainder occurs in the sympathetic ganglia adjacent to the vena cava or aorta, near the organ of Zuckerkandl, or in the wall of the urinary bladder. Up to 70% are bilateral and about 5% to 10% are malignant [89].

Pheochromocytomas have low signal intensity on $T_1$-weighted images and high signal intensity on $T_2$-weighted images (Figure 17.28) [90]. Moderate to intense enhancement after administration of gadolinium chelate agents is common. Small tumors often are homogeneous, whereas larger tumors appear heterogeneous with cystic and solid components. Calcifications are rare. Signs of malignancy include local invasion, lymph node enlargement and distant metastases.

### 17.4.3 Retroperitoneal Masses

Although rare, both benign and malignant primary tumors occur in the retroperitoneal soft tissues. Benign tumors include teratoma, lymphangioma, neurofibroma and lipomatosis. Teratomas appear as well-defined, fluid-filled masses with a variable amount of fat or calcium (Figure 17.29). Lymphangiomas are well-circumscribed, multiloculated fluid-filled masses. Neurofibromas are usually well-defined, cylindrical, soft tissue lesions with a characteristic location in the neurovascular bundle. Lipomatosis appears as a diffuse, infiltrative mass with signal intensity equal to fat; it grows along fascial planes and may invade muscle.

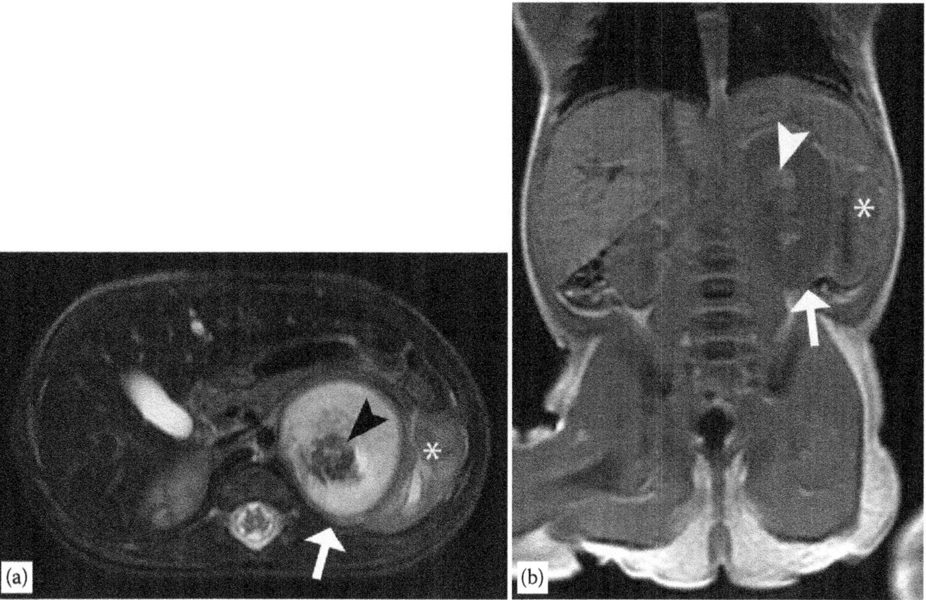

**FIGURE 17.29**
Retroperitoneal mature teratoma, 2-week-old boy. (a) $T_2$-weighted axial MR image with fat saturation shows a well-circumscribed fluid-signal-intensity mass (arrow) confirmed as a cyst at pathology. The cyst contains foci of low signal (arrowhead) confirmed pathologically to represent calcification/ossification. (b) Coronal $T_1$-weighted image shows most of the mass is of low signal consistent with fluid (arrow). The central high signal areas correspond to calcification (arrowhead). Asterisk indicates the displaced left kidney in both images.

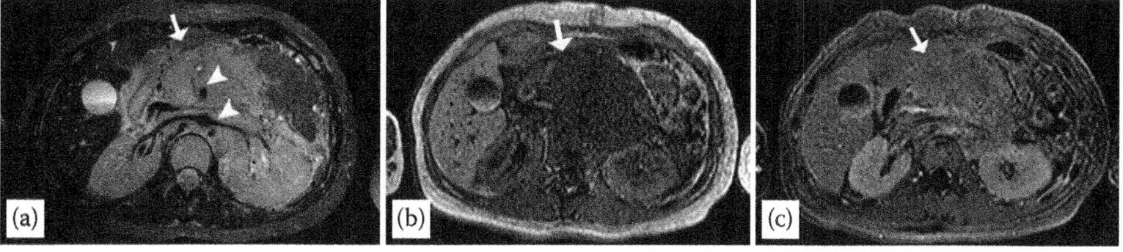

**FIGURE 17.30**
Retroperitoneal rhabdomyosarcoma, 14-year-old girl. (a) Axial $T_2$-weighted image shows a large predominantly bright signal mass in the retroperitoneum (arrow) encasing the superior mesenteric artery, aorta, and renal arteries (arrowheads). (b) Axial $T_1$-weighted image demonstrates hypointense signal in the mass (arrow). (c) Post-gadolinium $T_1$-weighted axial image shows diffuse enhancement of the mass (arrow). (Courtesy of Rajesh Krishnamurthy, MD, Texas Children's Hospital, Houston, TX.)

Malignant retroperitoneal tumors include rhabdomyosarcoma (Figure 17.30), neurofibrosarcoma, fibrosarcoma, and extragonadal germ cell tumors. These tumors appear as bulky soft tissue masses. Malignant tumors have low signal intensity on $T_1$-weighted images and high signal intensity on $T_2$-weighted, diffusion-weighted and contrast-enhanced images. Vessel displacement and encasement sometimes occur.

### 17.4.4 Hepatic Masses

Routine liver MRI includes coronal and axial $T_1$-weighted, axial fat-saturated $T_2$-weighted, and axial in-phase and out-of-phase imaging. Axial fat-saturated $T_1$-weighted sequences are obtained in arterial, portal venous and delayed phases after conventional gadolinium administration. Gadolinium disodium (Gd-EOB-DTPA, Eovist, Bayer HealthCare, Leverkusen, Germany), which is an organ-specific gadolinium-containing agent that is taken up by hepatocytes and excreted via the biliary system, can be used to help characterize hepatic tumors [91].

#### 17.4.4.1 Malignant Tumors

##### *17.4.4.1.1 Hepatoblastoma*

Hepatoblastoma is the third most common pediatric liver malignancy and the third most frequent solid abdominal mass in children, following Wilms tumor and neuroblastoma [4,11,92–95]. It typically occurs in children under the age of 5 years, with the peak age of presentation between 18 and 24 months. The tumor is discovered as asymptomatic upper abdominal mass, occasionally associated with anorexia and weight loss. Hepatoblastoma has been associated with Beckwith-Widemann syndrome, Gardner syndrome, glycogen storage disease, and trisomy 18 [91–95]. Invasion of the portal or hepatic veins is frequent.

Hepatoblastoma is usually confined to a single lobe, with the right lobe affected twice as often as the left, but it may involve both lobes and it may be multifocal. At MRI, it typically has a heterogeneous appearance and it predominantly hypointense relative to normal liver on $T_1$-weighted images and hyperintense on $T_2$-weighted and diffusion-weighted images (Figure 17.31) [4,11,91,96,97]. Contrast-enhanced sequences show early enhancement during the arterial phase with rapid wash out and hypointensity in the portal venous phase. Smaller lesions may enhance homogeneously, whereas larger lesions enhance heterogeneously. On gadolinium-enhanced, hepatocyte-specific imaging, the tumor is hypointense to normal liver on all phases of enhancement [91]. Other findings include a large feeding artery, mosaic pattern, and a tumor capsule. The mosaic pattern is characterized by a pattern of small nodules that are surrounded by thin hypo- or hyperintense septa.

Internal heterogeneity can be due to areas of hemorrhage, fat, necrosis, or calcification [4,11,92,96,97]. Tumor hemorrhage can appear hypo- or hyperintense on $T_1$-weighted sequences, depending on the age of the blood; it usually is hyperintense on $T_2$-weighted images. Focal steatosis produces signal hyperintensity on $T_1$-weighted sequences and low signal intensity on fat-suppressed and out-of-phase images. Calcifications are hypointense on all sequences.

Acute tumor thrombus will have high signal on $T_1$- and $T_2$-weighted images and signal void on gradient-echo images. Chronic thrombus may have low signal intensity. On gadolinium-enhanced images, tumor thrombus can show enhancement in the arterial phase.

#### *17.4.4.1.2 Hepatocellular Carcinoma*

Hepatocellular carcinoma is more frequent in older children with more than 65% of cases affecting children older than 10 years of age. Risk factors for hepatocellular carcinoma include biliary atresia, glycogen storage disease, alpha 1-antitrypsin deficiency, Alagille syndrome, and cirrhosis related to chronic hepatitis B or C infection [92–95].

MRI features of hepatocellular carcinoma are similar to those of hepatoblastoma. The tumor typically is hypointense to normal liver on $T_1$-weighted images and hyperintense on $T_2$-weighted images and it classically has arterial phase enhancement and portal venous phase washout (Figure 17.32). On hepatocyte-specific imaging, hepatocellular carcinoma is hypointense to adjacent liver [91]. Mosaic pattern, tumor capsule, and invasion of hepatic and portal veins are associated findings [92,96,97].

#### *17.4.4.1.3 Fibrolamellar Hepatocellular Carcinoma*

Fibrolamellar hepatocellular carcinoma is a variant of conventional hepatocellular carcinoma that occurs in adolescents and young adults. On MRI, it is hypointense to normal liver on $T_1$-weighted images and hyperintense on $T_2$-weighted and diffusion-weighted images [98]. The fibrous central scar is hypointense on both $T_1$- and $T_2$-weighted images. After contrast administration, the tumor shows heterogeneous arterial phase

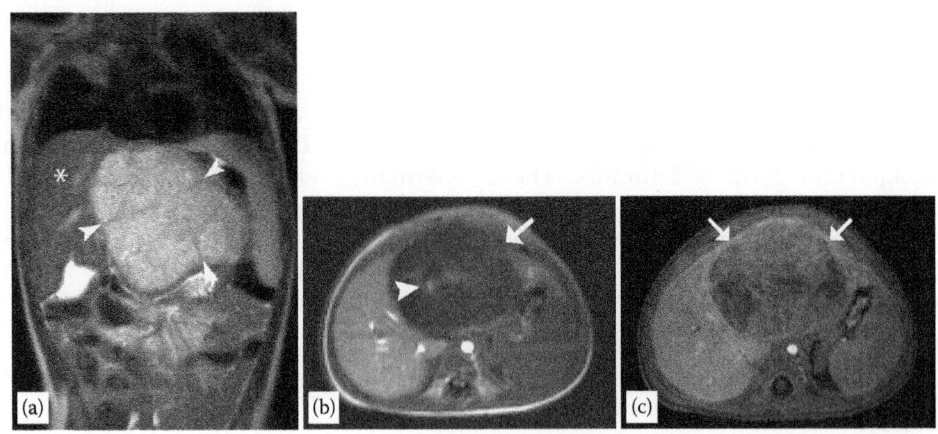


**FIGURE 17.31**
Hepatoblastoma, 11-month-old girl. (a) Coronal $T_2$-weighted MR image with fat suppression shows a well-circumscribed hyperintense mass involving the left lobe of the liver separated into lobules by hypointense septa (arrowheads). Asterisk indicates the adjacent liver. (b) On an axial $T_1$-weighted MR image with fat suppression, the tumor (arrow) is hypointense relative to adjacent parenchyma. Central foci of bright signal (arrowhead) corresponded to hemorrhagic components at pathology. (c) Delayed gadolinium-enhanced $T_1$-weighted image with fat saturation demonstrates that the mass (arrows) enhances somewhat heterogeneously and less than adjacent liver.

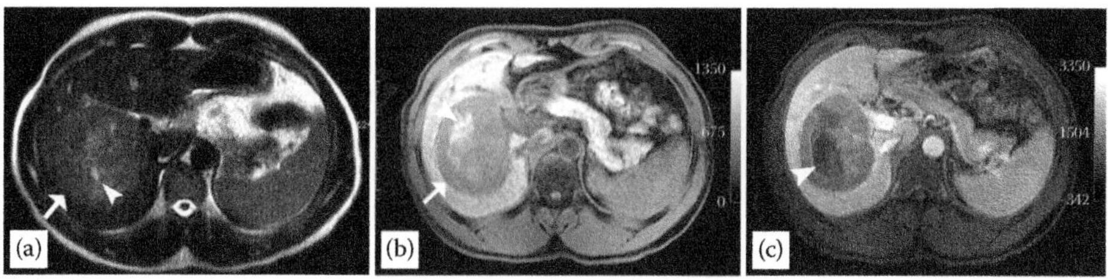

**FIGURE 17.32**
Hepatocellular carcinoma in a patient with hepatitis B. (a) Axial $T_2$-weighted fat-suppressed image shows a well circumscribed and heterogeneously hyperintense mass (arrow) with a focal area of high signal intensity (arrowhead) consistent with hemorrhage/necrosis. (b) Axial $T_1$-weighted image shows a large hypointense tumor (arrow) in the right hepatic lobe with a focal area of hyperintense signal (arrowhead) correlating with hemorrhage at gross examination. (c) Axial $T_1$-weighted image with fat suppression following administration of intravenous gadolinium-based contrast shows the tumor enhances less than adjacent liver. Central area of non-enhancement corresponds to central necrosis/old hemorrhage (arrowhead). (Courtesy of Rajesh Krishnamurthy, MD, Texas Children's Hospital, Houston, TX.)

enhancement, becoming hypo- or isointense on portal venous and delayed phases. The central scar does not enhance.

##### 17.4.4.1.4 *Undifferentiated Embryonal Sarcoma*

Undifferentiated embryonal sarcoma, also known as mesenchymal sarcoma, embryonal sarcoma, and malignant mesenchymoma, usually occurs in children between 6 and 10 years of age [99]. The usual presenting features are abdominal mass and pain. The MRI appearance is that of a heterogeneous, mixed solid, and cystic mass with predominantly hypointense contents on $T_1$-weighted images and hyperintense contents on $T_2$-weighted images (Figure 17.33) [92,100]. The fibrous rim has low signal intensity on $T_1$- and $T_2$-weighted sequences. Restricted diffusion and decreased ADC value can be seen on diffusion-weighted sequences. After gadolinium administration, the solid areas enhance, whereas the cystic spaces remain hypointense. Metastases are to lung and bone.

##### 17.4.4.1.5 *Hepatic Metastases*

The malignant tumors of childhood that most frequently metastasize to the liver are Wilms tumor, neuroblastoma, and lymphoma. Clinically, patients with hepatic metastases present with hepatomegaly, jaundice, abdominal pain or mass, or abnormal hepatic function tests.

Hepatic metastases are typically multiple, hypointense on $T_1$-weighted images, and hyperintense on $T_2$-weighted images (Figure 17.34). Other findings include central necrosis and mass effect with displacement of vessels. Hypervascular metastases may exhibit arterial enhancement with rapid wash out on contrast-enhanced images. Restricted lesion diffusion and decreased ADC value can be seen on diffusion-weighted sequences.

#### 17.4.4.2 Benign Hepatic Neoplasms

##### 17.4.4.2.1 *Infantile Hemangioma*

Infantile hemangiomas, previously known as *hemangioendothelioma*, is the most common benign hepatic tumor of childhood. It is derived from endothelial cells and exhibits initial rapid growth followed by a slow spontaneous involution over several months to years. Based on a unique pathologic marker, glucose transporter-1 (GLUT1) immunoreactivity, there are

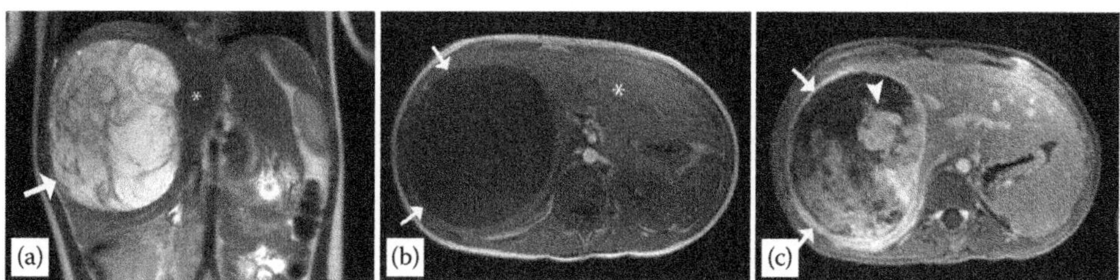

**FIGURE 17.33**
Hepatic embryonal sarcoma in a 6-year-old boy. (a) Coronal $T_2$-weighted image demonstrates a heterogeneous, predominantly fluid-signal mass (arrow) in the right lobe of the liver containing hypointense septa. Asterisk indicates the adjacent liver. (b) Axial $T_1$-weighted image demonstrates a well circumscribed, homogeneous, hypointense mass (arrows). Asterisk indicates the adjacent normal liver. (c) Delayed gadolinium-enhanced axial $T_1$-weighted image reveals enhancement of the rim (arrows) and solid portions (arrowheads) but not of the cystic components.

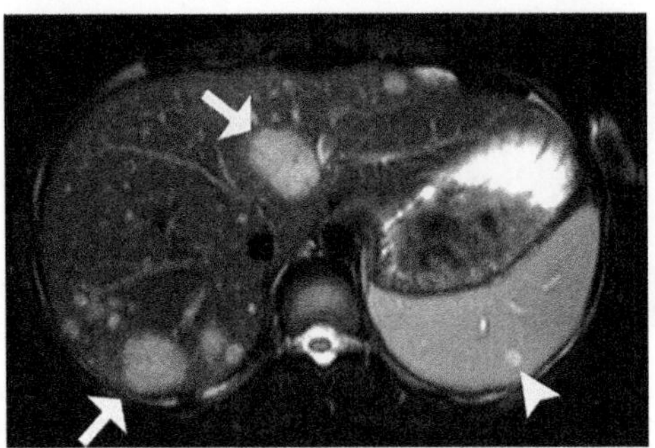

**FIGURE 17.34**
Hepatic metastases, 16-year-old girl with renal cell carcinoma. Axial $T_2$-weighted image with fat suppression shows multiple hyperintense masses in the liver (arrows). A small splenic lesion is also noted (arrowhead). (Courtesy of Rajesh Krishnamurthy, MD, Texas Children's Hospital, Houston, TX.)

two distinct clinical forms of hemangioendothelioma: GLUT-1–positive infantile hemangioendothelioma and GLUT-1–negative hepatic vascular malformation [101,102]. Infantile hemangioendotheliomas are usually diagnosed in the first few weeks or months of life. They are asymptomatic and are detected on screening sonography performed for evaluation of hepatomegaly or because of cutaneous hemangiomas, which may accompany visceral lesions. By comparison, congenital vascular malformations are commonly symptomatic and become apparent at or soon after birth. Presenting findings include cardiac failure, thrombocytopenia with consumptive coagulopathy (Kasabach–Merritt syndrome), and hemoperitoneum due to spontaneous tumor rupture [103,104]. At gross examination, both forms of hemangioendothelioma are composed of vascular channels lined by plump endothelial cells that are supported by reticular fibers.

GLUT-1–positive infantile hemangioendotheliomas usually appear as multiple small, nodules (Figure 17.35). GLUT-1–negative congenital vascular malformations commonly appear as a large complex mass (Figure 17.36). Both types are usually well circumscribed with round or lobular borders. They are characteristically hypointense to normal liver on $T_1$-weighted images and markedly hyperintense on fat-suppressed $T_2$-weighted images. Smaller lesions appear more homogeneous and larger ones have a more complex appearance, due to the presence of hemorrhage, calcification, necrosis, fibrosis, or thrombosis [4,11,92,105,106]. Images after administration of gadolinium chelates demonstrate centripetal enhancement with variable degrees of delayed central enhancement (see Figure 17.36). Large lesions (>4.0 or 5.0 cm) may not fill in on delayed enhanced images because of central fibrosis or hemorrhage. Small lesions (<1.5 cm) may fill rapidly and not show the typical centripetal pattern. A secondary finding is a small infrahepatic aorta distal to the level of the celiac artery, related to shunting of blood into the tumor via the celiac artery.

#### 17.4.4.2.2 *Cavernous Hemangioma*

Cavernous hepatic hemangioma is unusual in the pediatric population and when it occurs, it is common in older children and adolescents than in neonates and infants. They are often solitary lesions but can be multiple in 10%–20% of cases [107]. The hemangioma is composed of large blood-filled channels lined by a single layer of mature, flat endothelial cells, and separated by fibrous septa. Areas of fibrosis, calcification, hemorrhage, and cystic degeneration are frequent. Most are small and asymptomatic and are incidental findings on imaging studies. Hemangiomas are typically hypointense on $T_1$-weighted images and markedly hyperintense on $T_2$-weighted images. Areas of low signal intensity can be seen on $T_2$-weighted MR images if fibrosis is present. These lesions demonstrate early peripheral enhancement on contrast-enhanced arterial images, which is often nodular, with delayed centripetal filling

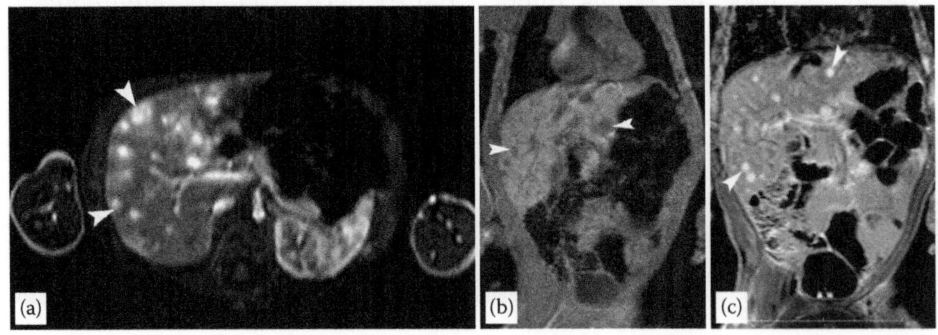


**FIGURE 17.35**
Glut-1 positive multifocal hemangioendotheliomas in an 8-month old girl. (a) Axial $T_2$-weighed image of the liver with fat suppression shows multiple round hyperintense lesions (arrowheads). (b) Coronal $T_1$-weighted image shows multiple small hypointense hepatic masses (arrowheads). (c) Coronal $T_1$-weighted image with fat saturation after intravenous administration of gadolinium-DTPA shows diffuse homogeneous intense enhancement of the masses (arrowheads).

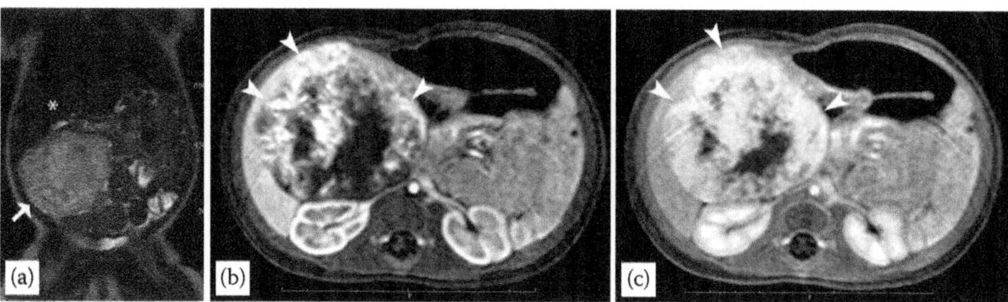

**FIGURE 17.36**
Glut-1 negative focal hemangioendothelioma, 3-month-old girl. (a) Coronal $T_2$-weighted image with fat suppression shows a heterogeneous, predominantly hyperintense pedunculated mass arising from the right lobe of the liver (arrow). Asterisk indicates adjacent liver. (b) Enhanced axial $T_1$-weighted image with fat suppression in the arterial phase demonstrates peripheral, nodular, intense enhancement of the mass (arrowheads). (c) Delayed phase axial image shows centripetal fill-in of enhancement (arrowheads).

in on venous phase imaging [107]. On hepatocyte phase imaging, there are typically hypointense.

#### *17.4.4.2.3 Mesenchymal Hamartoma*

After the vascular lesions, mesenchymal hamartoma is the next most common benign hepatic tumor of childhood. It is the benign counterpart to the undifferentiated embryonal sarcoma. This tumor usually is found as an asymptomatic mass in children less than 2 years of age. Rarely, the hamartoma has a large vascular component and produces arteriovenous shunting, leading to congestive heart failure. Malignant transformation of a mesenchymal hamartoma into an undifferentiated embryonal sarcoma has been reported, but this is extremely rare [108]. The cystic locules are hypointense on $T_1$-weighted sequences and hyperintense on $T_2$-weighted sequences (Figure 17.37) [4,11,92,109]. The signal intensity on the $T_1$-weighted images can increase if the tumor contains large amounts of protein or debris. On gadolinium-enhanced images, the solid components, but not the cystic spaces, enhance.

#### *17.4.4.2.4 Epithelial Tumors*

Focal nodular hyperplasia (FNH) and hepatic adenomas account for less than 5% of hepatic tumors in children. FNH is characterized by normal hepatocytes, bile ducts, Kupffer cells, and a central scar. FNH has an increased incidence in patients previously treated for malignancies, including neuroblastoma, Wilms tumor, rhabdomyosarcoma, lymphoma, and leukemia [110,111]. Hepatic adenoma contains normal hepatocytes and it often has hemorrhage, fat, or necrosis. Adenomas in childhood have been associated with type I glycogen storage disease (von Gierke disease), Fanconi anemia, and galactosemia.

In patients with no history of malignancy, FNH is typically solitary and relatively large (mean diameter 5.3 cm). It is iso- or hypointense on $T_1$-weighted images and iso- or hyperintense on $T_2$-weighted images and strongly enhances in the arterial phase, becoming hypointense during the portal venous phase. The central scar is hypointense on $T_1$-weighted images and hyperintense on $T_2$-weighted images, and does not enhance during the arterial phase, but may show delayed enhancement in the portal venous phase. In comparison, in patients with treated malignancy, FNH is generally smaller (<3 cm), multiple and may be hyperintense on $T_1$- and $T_2$-weighted sequences, although they may show signal intensity similar to FNH in the general population. The lesions enhance homogeneously and lack a central scar [110,111]. Recognizing these features of FNH is important, lest they be misdiagnosed as metastatic disease.

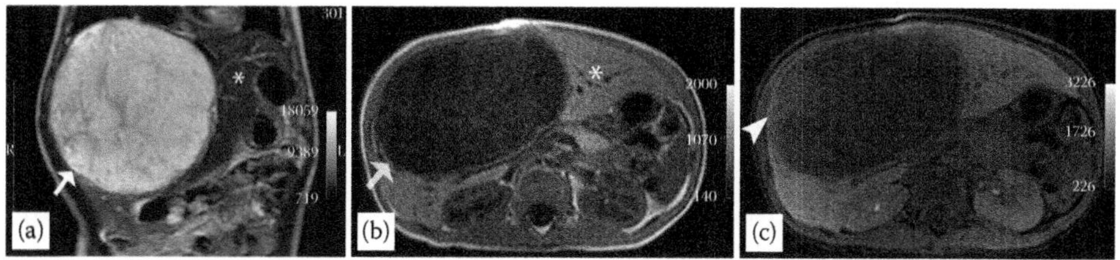

**FIGURE 17.37**
Mesenchymal hamartoma, 2-year-old boy. (a) Coronal $T_2$-weighted image shows a well-circumscribed hyperintense mass (arrow) containing dark septations in the right lobe of the liver. Asterisk indicates adjacent normal liver. (b) Axial unenhanced $T_1$-weighted image shows the tumor (arrow) is well circumscribed and hypointense compared to adjacent liver. Asterisk indicates normal liver. (c) Delayed (15 min) gadolinium-enhanced axial $T_1$-weighted image demonstrates minimal enhancement of the periphery and septations of the mass (arrowhead).

Hepatic adenomas are usually hyperintense on $T_1$-weighted images (related to fat or acute hemorrhage) and iso- or hyperintense on $T_2$-weighted, fat-suppressed and out-of-phase sequences. Most are heterogeneous because of the presence of hemorrhage, necrosis, glycogen, or fat. On contrast-enhanced MRI, adenomas enhances during the arterial phase, becoming iso- or hypointense in the venous phase.

### 17.4.5 Biliary Masses

#### 17.4.5.1 MR Cholangiopancreatography

MRCP is a fast, safe, accurate, noninvasive alternative to endoscopic retrograde cholangiogram in the evaluation of the biliary system in children [1,17,18,112]. The technique utilizes heavily $T_2$-weighted sequences because bile has a high water content and appears bright on these sequences with its long $T_2$ relaxation times, in contradistinction to the surrounding solid organs, which are relatively dark [1,17,18,112]. The protocols for MRCP include both thick-slab (30 to 80 mm thickness) $T_2$-weighted TSE and thin slab (3 to 4 mm thickness) HASTE in the coronal, oblique, and axial planes. The thinner slabs are reviewed as maximum and minimum intensity projections.

Four hours prior to the study, the patient should have NPO to decrease bowel motion. Negative oral contrast agents that have superparamagnetic effects can improve the visualization of the biliary ducts by nulling the inherent high signal of gastric juices in the stomach and duodenum. Commercially available agents or fruit juices, such as blueberry or pineapple juices, with high levels of manganese are also effective.

Conventional unenhanced $T_1$-and $T_2$-weighted imaging is also routinely performed in evaluation of the pancreaticobiliary system. $T_1$- and $T_2$-weighted images are used to assess abnormalities of the pancreas, extrahepatic causes of biliary obstruction, and the liver parenchyma. If indicated, gadolinium-enhanced MR images can be obtained to further evaluate the liver parenchyma.

#### 17.4.5.2 Choledochal Cyst

Choledochal cyst is the most common mass arising in the biliary ductal tree [1,17,18,112–115]. Classically, patients present with jaundice, pain, and a palpable abdominal mass, although the complete triad is present in only about one-third of patients. The diagnosis usually can be made by sonography. MR cholangiography has proven to be a useful, noninvasive alternative to endoscopic retrograde pancreatography to delineate the anatomy of the biliary system for surgical planning.

There are 5 types based on the Todani classification system and all have been well-described by MRCP [112–116]. The majority are type 1 which is diffuse involvement of the common bile duct (CBD) and common hepatic duct (Figure 17.38). Type 2 involves isolated cysts which protrude exophytically from the CBD. Type 3 is the choledochocele, which is a focal dilatation of the intraduodenal portion of the CBD. Type 4a involves dilatation of the intra and extrahepatic ducts, while type 4b involves only extrahepatic ducts and type 5 (Caroli disease) involves only intrahepatic ducts.

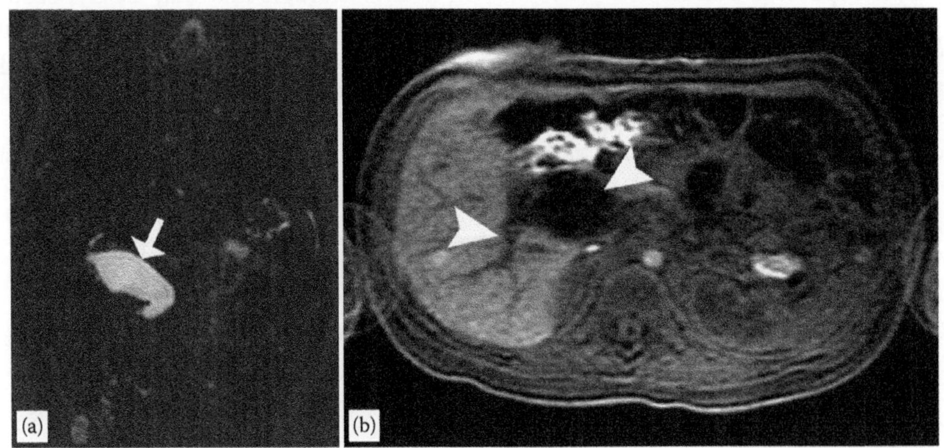


**FIGURE 17.38**
Choledochal cyst, 2-month-old girl with jaundice. (a) Coronal maximal intensity projection of a respiratory-triggered $T_2$-weighted 3D fast spin-echo image demonstrates fusiform extrahepatic dilatation of the common bile duct (arrow) which tapers distally, consistent with type I choledochal cyst. (b) Axial spoiled gradient-echo (SPGR) image following intravenous gadolinium administration shows the hypointense cyst in the porta hepatis communicates with the right bile duct (arrowheads).

#### 17.4.5.3 Embryonal Rhabdomyosarcoma

Rhabdomyosarcoma of the biliary tract, albeit rare, is the most common malignant neoplasm of the biliary tract in children, affecting children between 3 and 5 years of age [117]. It is a slowly growing tumor that arises from the CBD in the porta hepatis and grows along the bile ducts. Imaging findings are intra- and extrahepatic ductal dilatation and a low $T_1$-weighted and high $T_2$-weighted signal intensity mass in the porta hepatis (Figure 17.39) [117,118].

### 17.4.6 Pancreatic Masses

Pancreatic tumors are rare in children and most of these are exocrine tumors [119,120]. Pancreatoblastoma and solid pseudopapillary tumors are the most common exocrine pancreatic neoplasms. Pancreatoblastoma is an encapsulated, epithelial tumor composed of tissue resembling fetal pancreas and usually arises in the pancreatic head. It is a low-grade malignancy and usually has a favorable outcome. Mean patient age at diagnosis is 4.5 years, but the tumor can present in the fetus and neonate as well as older patients. Pancreatoblastoma is usually heterogeneous and has low to intermediate signal intensity on $T_1$-weighted images and high signal intensity on $T_2$-weighted images [119,121,122] (Figure 17.40). Heterogeneous enhancement is common after gadolinium administration. Secondary signs include hepatic and lymph node metastases and vascular encasement.

Solid pseudopapillary tumor of the pancreas (previously known as solid and papillary epithelial neoplasm, solid-cystic papillary tumor, papillary cystic tumor, and Frantz tumor) is a low-grade malignant tumor seen in adolescent girls and young women with a mean age of 22 years [119,120]. The tumor is typically large (mean diameter of 9.0 cm), well circumscribed and usually arises in the pancreatic tail. It is usually heterogeneous with solid and cystic components. $T_1$- and $T_2$-weighted images show a hypointense rim, representing the fibrous capsule. The tumors are either entirely hypointense or partially hypointense on $T_1$-weighted images and heterogeneously hyperintense on $T_2$-weighted images. Heterogeneous or peripheral enhancement is common (Figure 17.41). Rare exocrine

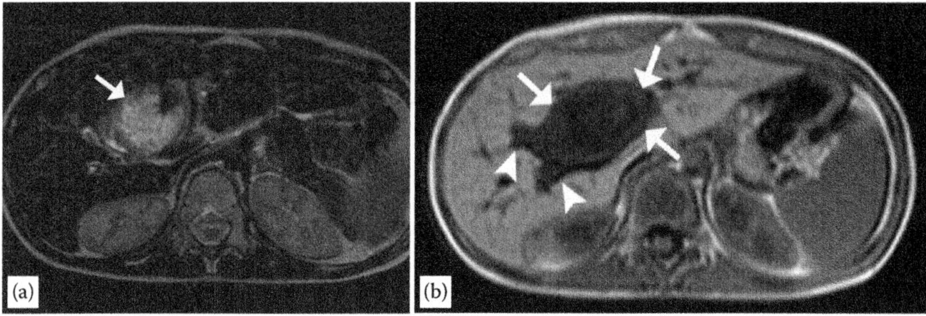


**FIGURE 17.39**
Biliary rhabdomyosarcoma, 3-year-old girl with jaundice. (a) Axial $T_2$-weighted image demonstrates a heterogeneous hyperintense mass in the porta hepatis (arrow). (b) Axial $T_1$-weighted image shows the hypointense mass in the porta hepatis (arrows) causing mild dilatation of the right hepatic ducts (arrowheads).

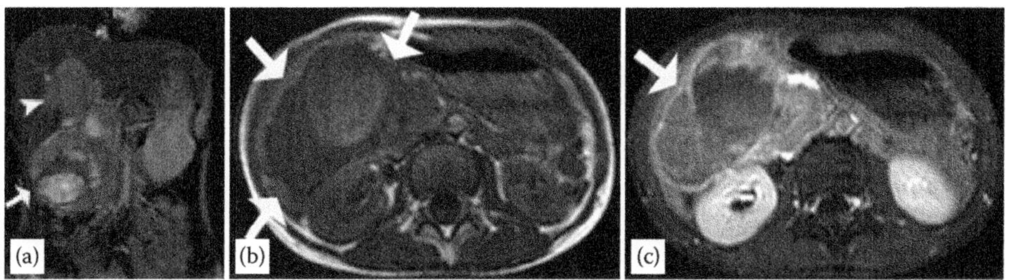


**FIGURE 17.40**
Pancreatoblastoma, 11-year-old malnourished girl. (a) Coronal SSFP image demonstrates a markedly heterogeneous mass arising in the head of the pancreas arrow. A large liver metastasis is also noted near the porta hepatis (arrowhead). (b) Axial $T_1$-weighted image shows that the mass (arrows) is predominantly hypointense with a hyperintense focus. (c) Enhanced axial $T_1$-weighted image demonstrates peripheral rim enhancement (arrow) suggesting central necrosis/old hemorrhage. (Courtesy of Rajesh Krishnamurthy, MD, Texas Children's Hospital, Houston, TX.)

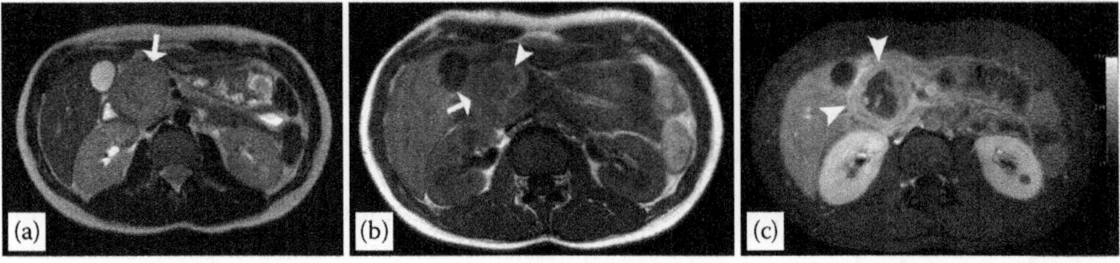


**FIGURE 17.41**
Pancreatic solid pseudopapillary neoplasm, 15-year-old girl. (a) Axial $T_2$-weighted image demonstrates a well-circumscribed, intermediate signal-intensity mass in the head of the pancreas (arrow). (b) Axial $T_1$-weighted image shows that the mass is of low signal (arrow) with irregular central bright signal corresponding to subacute hemorrhage at pathologic examination (arrowhead). (c) Enhanced axial $T_1$-weighted images shows enhancement of the periphery of the mass (arrowheads) but not of the necrotic center. Incidentally noted is a simple cyst in the left kidney.

tumors are lymphoma adenocarcinoma, and hemangioendothelioma. Imaging features are similar to pancreatoblastoma.

Endocrine tumors include insulinoma, gastrinoma, VIPoma, and glucagonoma. They are usually hypointense on $T_1$-weighted series and hyperintense on $T_2$-weighted images and enhance on contrast-enhanced studies.

### 17.4.7 Splenic Masses

Focal splenic lesions in children include abscess, cysts, neoplasms (most commonly lymphoma and rarely hamartoma), and vascular malformations (lymphangioma, hemangioma) [123–125]. Abscesses, vascular malformations, and cysts have low signal intensity on $T_1$-weighted MR images and high signal intensity greater than fat on $T_2$-weighted images. Bacterial abscesses, which are usually larger than fungal ones, typically show peripheral and perilesional enhancement on contrast-enhanced sequences. The splenic cyst is hypointense on $T_1$-weighted images and hyperintense on $T_2$-weighted images, with no enhancement after the injection of contrast medium.

Hemangioma is the most common benign tumor of the spleen [123–125]. Most hemangiomas are small (<2 cm), well-defined, homogeneous, hypo- to isointense on $T_1$-weighted images, and hyperintense on $T_2$-weighted images compared with splenic parenchyma [125] (Figure 17.42). They show early peripheral nodular enhancement with progressive fill-in and become more homogeneous on delayed images. Lymphangioma is another benign tumor. It appears as a thin-walled multilocular lesion with low signal intensity on $T_1$-weighted imaging and high signal intensity on $T_2$-weighted imaging. The septations may enhance, but the fluid contents do not enhance [125,126]. Hamartoma is a rare benign tumor that appears as a sharply defined, rounded, lesion [123,124]. It is hypointense on $T_1$-weighted images and heterogeneous and slightly hyperintense on fat-saturated $T_2$-weighted images and shows diffuse heterogeneous enhancement on contrast-enhanced images [124,125].

Lymphoma is the most common splenic malignancy and can present as multiple focal lesions or diffuse involvement or, occasionally, as a solitary mass. MRI findings are nonspecific and similar to those of metastases from primary tumors. Typically, lymphoma is iso- or hypointense to splenic parenchyma on $T_1$-weighted images and iso- to hyperintense on fat-saturated $T_2$-weighted images and enhances after contrast administration.

In some children, the spleen is highly mobile, due to failure of fusion of the gastric mesentery with the dorsal peritoneum, and presents as a mass in the anterior

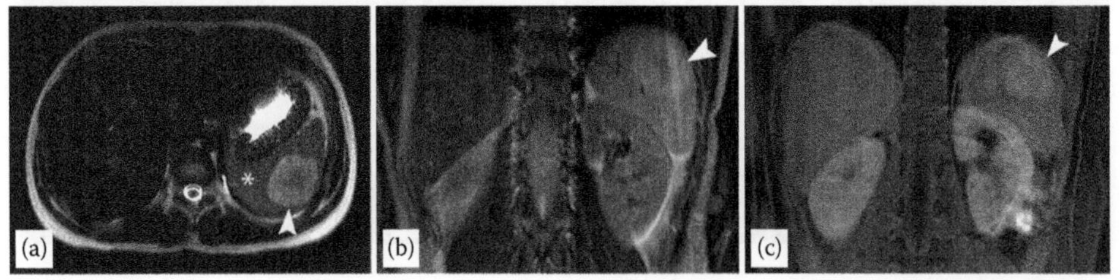


**FIGURE 17.42**
Splenic hemangioma. (a) $T_2$-weighted image shows a large predominantly hyperintense mass in the spleen (arrowhead) which is centrally hypointense. Asterisk indicates adjacent normal spleen. (b) The mass shows peripheral enhancement (arrowhead) on the coronal postcontrast $T_1$-weighted image with fat saturation. (c) There is complete fill-in of enhancement on the delayed postcontrast image (arrowhead). A small left renal cyst is noted. (Courtesy of David M. Biko, MD, David Grant Medical Center, Travis AFB, CA.)

abdomen. MRI demonstrates absence of the spleen in the left upper quadrant and a lower abdominal or pelvic soft tissue mass. The mobile spleen enhances after administration of IV contrast medium unless it has undergone torsion with resultant vascular compromise [127].

### 17.4.8 Gastrointestinal and Peritoneal Masses

Cystic gastrointestinal/mesenteric masses are predominantly lymphangiomatous malformations (also termed mesenteric cysts) and enteric duplications. A mesenteric cyst is a well-circumscribed cystic mass containing multiple septations (Figure 17.43). An enteric duplication cyst typically is unilocular and is usually associated with bowel wall. The ileum is the most commonly involved segment of bowel, although enteric duplication cysts can arise anywhere along the gastrointestinal tract. Both lesions are hypo- or isointense to muscle on $T_1$-weighted images, although the signal intensity may be higher if the lesions contain blood or proteinaceous material. The signal intensity increases and is greater than that of fat on fat-saturated $T_2$-weighted images. The walls of both lesions may enhance but the fluid does not enhance.

Lymphoma is the most common malignant neoplasm of the bowel and mesentery. The MR features of bowel lymphoma include bowel wall thickening greater than 1 cm in diameter, extraluminal soft tissue mass, and mesenteric invasion. Lymphomatous masses of bowel and mesentery are isointense to muscle on $T_1$-weighted images and show intermediate to high signal intensity on fat-saturated $T_2$-weighted images and some enhancement after gadolinium administration [128]. Rare solid tumors include inflammatory pseudotumor, which is probably an occult infection due to minor trauma, desmoid tumor (also known as intra-abdominal fibromatosis, and desmoblastic small round cell tumors. These tumors have an appearance similar to lymphoma.

Fatty masses include lipoma and lipoblastoma. Lipoblastoma is a benign tumor that occurs almost exclusively in infants and young children less than 5 years of age [129]. Pathologically, it contains fat, mesenchymal cells, myxoid matrix, and fibrous trabeculae. Lipoma and lipoblastoma are hyperintense on $T_1$ and $T_2$-weighted images and hypointense on fat-suppressed and out-of-phase images. Lipoblastoma will show some areas of hypointensity on $T_1$-weighted images, depending on the proportions of mesenchymal and fibrous tissue. Liposarcomas also contain soft tissue elements in addition to fatty tissue, but they are exceedingly rare in young children. Definitive differentiation between lipoblastoma and liposarcomas requires tissue sampling.

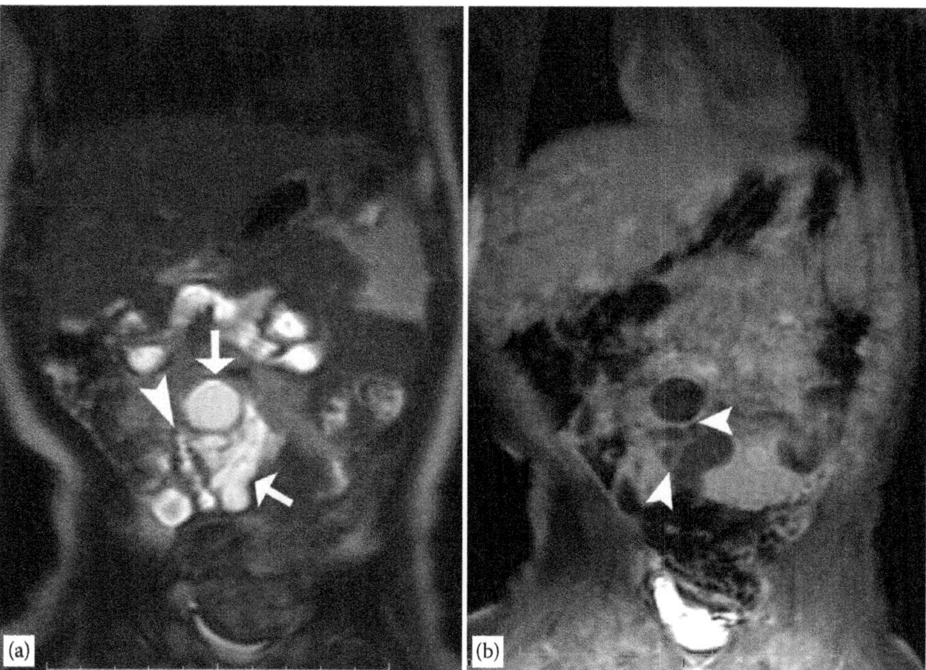


**FIGURE 17.43**
Mesenteric lymphangioma, 21-month-old girl. (a) Coronal $T_2$-weighted image reveals a multilocular fluid-signal intensity mass (arrows) adjacent to bowel (arrowhead). (b) Coronal $T_1$-weighted gradient-echo image obtained after administration of intravenous gadolinium-based contrast shows hypointense fluid signal intensity in the cysts and enhancement of only the walls and septa (arrowheads).

### 17.4.9 Urinary Tract: Hydronephrosis and Congenital Anomalies

Hydronephrosis is the most common indication for MR urography in infants and children [19,20,130–134]. MR urography can provide both anatomic and functional data in a single test without exposure to ionizing radiation.

MR urography requires patient hydration and administration of a diuretic agent. Oral or IV hydration (normal saline solution) and diuretic administration (furosemide) are used to improve ureteral distention and ensure uniform contrast distribution. Furosemide (Lasix) is given intravenously at a dose of 1 mg/kg (maximum 20 mg), 10 min before the examination starts. Hydration is started 30 to 40 min prior to the examination. A bladder catheter is useful in patients who are going to be sedated. With hydration and diuretic administration, the bladder fills quickly and a full bladder can cause discomfort and result in a suboptimal quality study. This discomfort can be obviated or minimized with catheter placement.

$T_2$-weighted TSE images with fat-saturation or HASTE sequences are obtained through the kidneys and pelvis, providing detailed morphologic information. 3D $T_2$ images acquired with fat saturation and thin slice section are used to generate maximum intensity projections and volume-rendered reconstructions [19,20,130–134]. These provide an overview of the urinary tract anatomy from multiple angles and improve depiction of complex urinary tract anomalies. Finally, a coronal, postcontrast $T_1$-weighted sequence with fat saturation is performed with images acquired in the corticomedullary, nephrographic, and excretory phases. In general, imaging is performed in the axial and coronal planes. Post-processing functional evaluation includes evaluation of renal transit time (RTT), which provides a measure of drainage, and calyceal transit time (CTT) and differential renal function with both volumetric (vDRF) and Patlak (pDRF) techniques, which provide measures of physiologic changes within the kidney [130–134].

Ureteropelvic junction (UPJ) obstruction is the most common cause of urinary tract obstruction in children. MR findings are dilated renal pelvis and calyces, cortical thinning, and non-visualization of the ureter (Figure 17.44). Other causes of obstruction include ureterovesical junction obstruction, renal duplication, and posterior urethral valves. Renal duplication can be partial, with the ureters joining above the bladder, or complete, with two ureters inserting separately. In complete duplication, the upper pole ureter typically inserts inferior and medial to the lower pole ureter and is more prone to obstruction. The upper pole ureter can insert either ectopically, with development of an ureterocele, or in an extravesicular location. The lower pole ureter of a duplicated kidney has a tendency to reflux. Thus, at MR urography both upper and lower pole moieties are often dilated. In posterior urethral valves, the upper urinary tract and ureters bilaterally are usually dilated and the bladder outlet is obstructed. Associated signs suggesting severe uropathy and permanent damage include architectural disorganization with loss of corticomedullary differentiation, small subcortical cysts and low cortical signal intensity on $T_2$-weighted images.

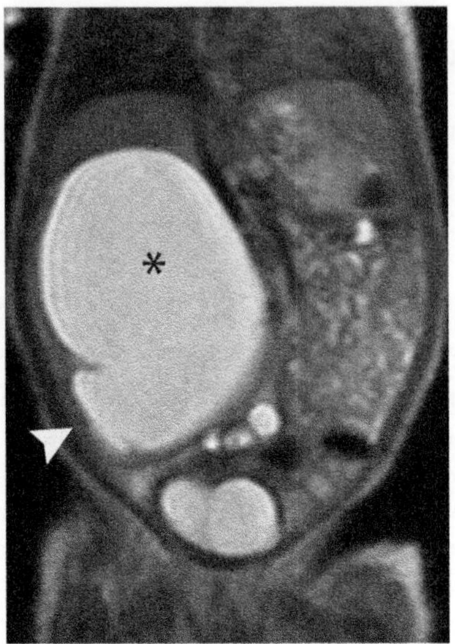

**FIGURE 17.44**
Obstructive uropathy in an infant boy with ureteropelvic junction obstruction. Coronal $T_2$-weighted images shows a markedly dilated renal pelvis (asterisk) with compressed, thinned renal parenchyma (arrowhead). (Courtesy of David M. Biko, MD, David Grant Medical Center, Travis AFB, CA.)

MR urography is able to identify the acquired segmental scars associated with vesicoureteral reflux and infection. These scars are characterized by volume loss and a contour defect of the kidney on $T_2$-weighted images, dilatation of the adjacent calyx indicating parenchymal loss, and no appreciable contrast enhancement. Anomalies of renal position and rotation are well demonstrated with MR urography. Horseshoe, ectopic kidneys (Figure 17.45) and pelvic kidneys are especially well demonstrated with MR urography.

### 17.4.10 Chronic Diffuse Liver Diseases

Chronic liver diseases encompass many different causes, including viral infections, nonalcoholic fatty liver disease, metabolic diseases, biliary cirrhosis, and hemochromatosis. Chronic liver diseases can lead to hepatic fibrosis, cirrhosis, end-stage liver disease, portal hypertension, and hepatocellular carcinoma.

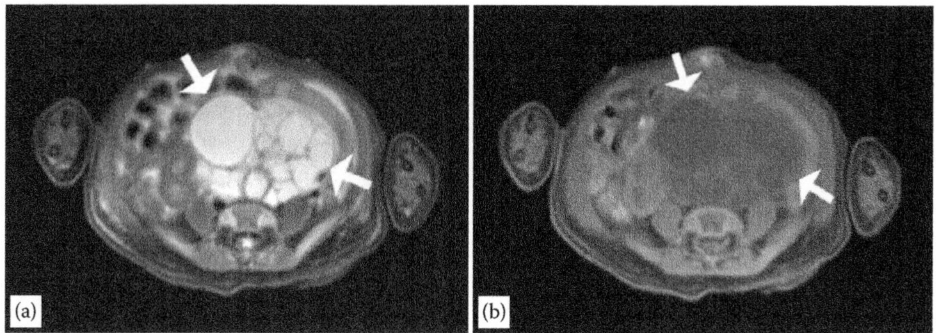

**FIGURE 17.45**
Pelvic multicystic dysplastic kidney in a neonate with abnormal *in utero* ultrasound. (a) Axial $T_2$-weighted image shows a pelvic mass consisting of fluid-signal intensity cysts of varying sizes (arrows). A left kidney was not visualized in the expected position. (b) Axial $T_1$-weighted image reveals hypointense signal in the cysts (arrows).

#### *17.4.10.1 MR Techniques*

Recent advances in MRI have led to a growing interest in applying functional MRI methods for assessment of chronic liver disease (fibrosis/cirrhosis). These methods include DWI and MR elastography (MRE) [135–139].

DWI is performed usually using a single-shot echoplanar imaging (EPI) sequence and ADC calculation. Several studies have shown that the ADC values of cirrhotic liver are lower than that of normal liver [135–139]. MRE is an emerging diagnostic imaging technique for quantitatively assessing the mechanical properties of tissue. MRE measures hepatic stiffness, which increases with fibrosis and cirrhosis [136,137,139]. With a shear stiffness cutoff value of 2.93 kPa, the predicted sensitivity and specificity for detecting liver fibrosis are 98% and 99% [136,137,139] (Figure 17.46).

#### *17.4.10.2 Cirrhosis*

Cirrhosis is the result of diffuse, irreversible hepatocyte damage and replacement by fibrosis. The causes in infants and children include chronic hepatitis, biliary cirrhosis secondary to biliary atresia, cystic fibrosis, metabolic diseases (Wilson disease, glycogen storage disease, tyrosinemia, galactosemia, alpha-1-antitrypsin deficiency), prolonged total parenteral nutrition, Budd–Chiari syndrome, and drugs. Characteristic findings include a small right hepatic lobe and medial segment of the left lobe, enlargement of the caudate lobe and lateral segment of the left lobe, and heterogeneous nodular parenchyma with nodular hepatic margins. Regenerative and dysplastic nodules are iso- or hypointense to liver on $T_1$- and $T_2$-weighted images. Regenerative nodules usually do not show arterial enhancement, whereas dysplastic nodules may show arterial enhancement.

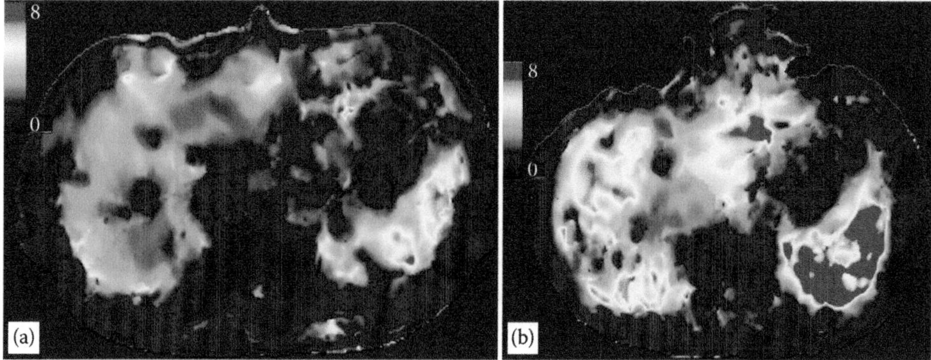

**FIGURE 17.46**
MR elastography. (a) Elastogram of the liver in a healthy 8-year-old female volunteer and (b) of a 15-year-old male with fibrosis. The elastograms show that the mean shear stiffness of the fibrotic liver is much higher than that of the normal liver (5.19 ± 0.95 kPa vs. 2.7 ± 0.25 kPa, respectively). The fibrotic liver is also markedly heterogeneous with increased stiffness manifested by red and yellow color-coded areas. The color-coded scale on the left is the kPa spectrum ranging from 0 (purple) to 8 (red).

Extrahepatic findings include ascites, splenomegaly and collateral vessels in the porta hepatis, umbilical, and splenic regions, indicative of portal hypertension.

#### *17.4.10.3 Hepatic Hemochromatosis*

Hemochromatosis is characterized by an increased hepatic iron deposition. Causes in children include a primary genetic disorder and transfusions for treatment of anemia. Iron overload causes magnetic susceptibility artifact, which leads to spin dephasing ($T_2$*-related signal loss), which in turn leads to low signal intensity on MR images. If the hepatic signal intensity is equal to or lower than that of skeletal muscle on $T_2$-weighted gradient-echo or $T_2$-weighted spin-echo images, increased iron accumulation can be diagnosed. Tissue iron can also be detected with in-phase imaging [140]. In hepatic iron overload, the signal intensity of the liver decreases on in-phase imaging because of longer echo times.

#### *17.4.10.4 Steatosis*

Fatty hepatic change in children is associated with obesity, fulminant liver diseases, severe malnutrition, cystic fibrosis and chemotherapy. Out-of-phase and proton MR spectroscopy are standard methods for assessing hepatic fat [141–143].

In the setting of hepatic steatosis, the measurement of abdominal wall fat is another important application of MR imaging. Multisection and single-section balanced GRE (true FISP) or $T_1$-weighted images are reliable methods for quantifying subcutaneous and visceral fat and its distribution in children and adolescent [144].

### 17.4.11 Biliary Diseases

Biliary atresia and the neonatal hepatitis syndrome are the common causes of conjugated hyperbilirubinemia and jaundice in the newborn. Biliary atresia is thought to be the sequel of a destructive *in utero* neonatal hepatitis inflammatory process leading to ductal fibrosis. Neonatal hepatitis is the term given to nonspecific hepatic inflammation which develops secondary to several different causes, including infection (cytomegalovirus, herpes simplex, toxoplasmosis, protozoa, syphilis) and metabolic defects (alpha 1-antitrypsin deficiency, galactosemia, glycogen storage disease, tyrosinosis). Both biliary atresia and the neonatal hepatitis syndrome usually present at 3 to 4 weeks of life with cholestasis and jaundice. Distinguishing between neonatal hepatitis and biliary atresia is important, because neonatal hepatitis is managed medically, whereas biliary atresia requires early surgical intervention to prevent biliary cirrhosis. MRI, including cholangiopancreatography, is used to help in differentiating these two conditions.

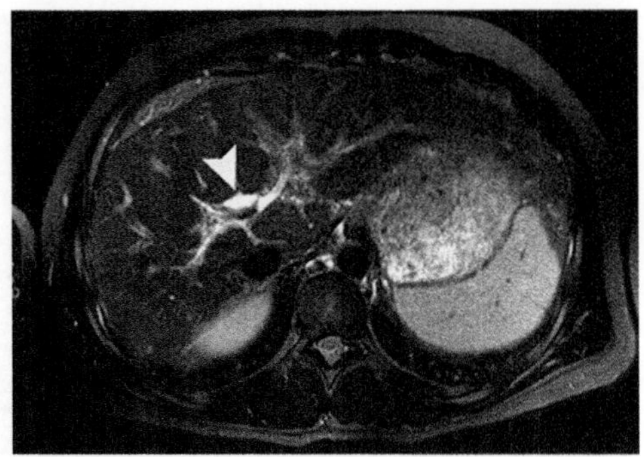

**FIGURE 17.47**
Biliary atresia, 7-week-old boy. Axial $T_2$-weighted image of the liver shows a triangular focus of periportal bright signal (arrowhead) representing atretic bile duct.

MR findings of biliary atresia include periportal thickening, which represents fibrosis, absence of intrahepatic ducts, and an absent or small gallbladder [11,145]. A bile duct remnant may be noted in the porta hepatis in extrahepatic biliary atresia. This finding has been termed the *triangular cord* and correlates with fibrous tissue in the porta hepatis at histologic examination. The ductal remnant is hyperintense on $T_2$-weighted MR cholangiography [146] (Figure 17.47). The MR diagnosis of neonatal hepatitis is based on demonstrating the presence of the extrahepatic ducts. These are usually well visualized in patients with neonatal hepatitis, thereby excluding biliary atresia as a diagnosis. In neonatal hepatitis, the gallbladder may be large, normal, or small.

Spontaneous perforation of the extrahepatic bile ducts affects is a rare cause of jaundice and ascites in young infants. The site of perforation is almost always at the junction of the cystic and common hepatic ducts. MRCP can be used to show the loculated fluid collection in the porta hepatis at the site of ductal perforation and biliary ductal anatomy [147].

### 17.4.12 Bowel Diseases

Currently, MRI is mainly used to evaluate inflammatory bowel disease in the pediatric age group. More recently, it use has increased for the diagnosis of appendicitis.

#### *17.4.12.1 MR Techniques*

MR enterography with oral contrast administration is the preferred technique for MR evaluation of bowel in children [148–151]. Although MR enteroclysis enables better distention of bowel compared with MR enterography, it is more difficult to implement in children. MR enteroclysis requires a fluoroscopic-guided tube insertion for oral contrast administration, which is invasive and results in radiation exposure. In addition it requires MR monitoring of small bowel filling using single-shot sequences, leading to longer examination times.

The bowel cleansing preparation includes a low-residue diet, ample fluids, a laxative on the day prior to the examination, and preprocedural fasting for at least 4 to 6 h. Bowel distension with oral contrast is required for a high quality diagnostic examination as collapsed bowel can obscure lesion detection or create the false appearance of wall thickening. Scanning is initiated within one hour of oral contrast administration. Rectal contrast material is not given routinely for MR enterography, but it is useful in specific scenarios, such as assessment of rectal stenosis or fistulas or evaluation of the distal pouch after colectomy. An antiperistaltic drug is given to transiently reduce peristaltic motion artifact and thus reduce image degradation.

Imaging is optimally performed with the patient in the prone position. This results in increased separation of the bowel loops and avoids susceptibility artifacts from bowel gas in the anterior abdomen [148–151]. However, the supine position affords greater patient comfort and may be better tolerated in patients with abdominal pain and stomas.

MR enterography is performed with both noncontrast and contrast-enhanced images. Precontrast sequences include coronal and axial fast $T_2$-weighted sequences and diffusion-weighted images. $T_2$-weighted images are acquired with both steady-state-free-precision or balanced GRE sequences and single-shot TSE sequences. The former sequence has high spatial and contrast resolution, while the latter is less affected by susceptibility artifacts due to intraluminal air. Diffusion-weighted MRI can provide useful information about disease activity and response to treatment [152–154]. Fat-saturated $T_1$-weighted images are routinely acquired after IV gadolinium administration.

#### 17.4.12.2 Congenital Anomalies

Anorectal malformations are characterized by varying degrees of atresia of the distal hindgut and the levator sling. Preoperative MRI can provide information about the level of atresia, the thickness of the puborectalis muscle and external anal sphincter, and associated anomalies, such as a tethered spinal cord, which is present in as many as 35% of patients [155–157]. Postoperatively, MRI can be useful to confirm the position of the neorectum in the levator ani sling. The neorectum needs to be positioned within both the puborectalis and external sphincter muscles if rectal continence is to be achieved [158].

#### 17.4.12.3 Inflammation

Crohn disease is the most frequent inflammatory condition affecting the small bowel in children. MRI can help to determine disease activity and response to therapy [159–163]. MR findings in the early stage of Crohn disease include circumferential bowel wall thickening (>3 mm), mucosal hyperenhancement, bowel wall stratification, prominent vasa recta (comb sign), and mesenteric fat stranding inflammation of the adjacent mesenteric fat, and enlarged regional lymph nodes [160,163] (Figure 17.48). Mural hyperenhancement is the most sensitive imaging finding of active disease. Findings of advanced fibrotic disease include decreased signal intensity of bowel wall on $T_2$-weighted images related to

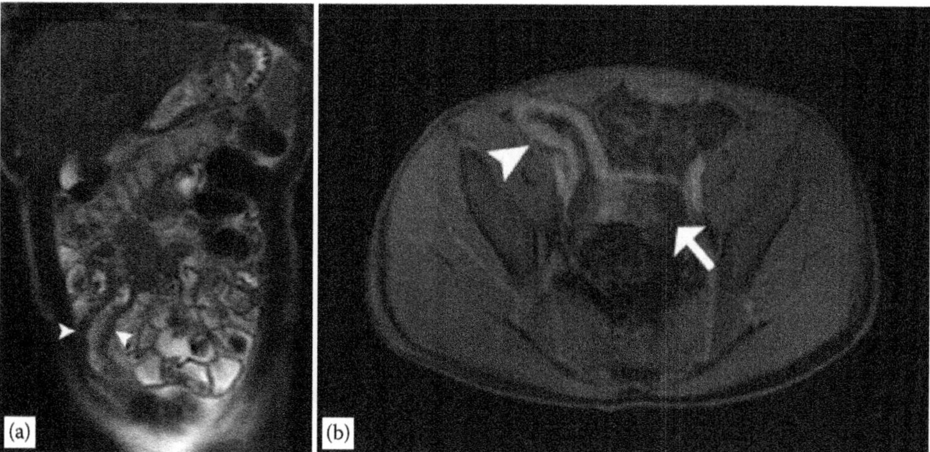


**FIGURE 17.48**
Crohn disease. MR enterography. (a) Coronal $T_2$-weighted image with fat suppression shows hypointense, thickened wall of the terminal ileum (arrowheads). (b) Axial $T_1$-weighted image with fat suppression after intravenous gadolinium-based contrast demonstrates enhancement of the thickened wall of the terminal ileum (arrowhead) consistent with active inflammation. Note the mild dilation of the adjacent proximal bowel (arrow).

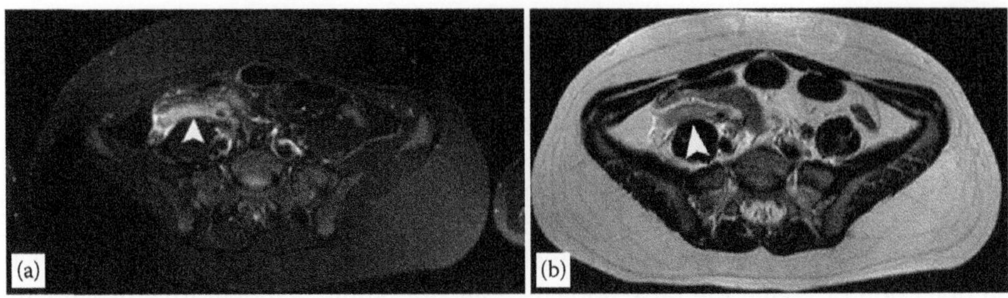


**FIGURE 17.49**
Acute appendicitis, 13-year-old boy. Axial $T_2$-weighted image with fat suppression (a) and without fat suppression (b) show dilated tubular fluid-filled appendix (arrowhead) with inflammatory changes in the surrounding mesentery.

absence of inflammation and edema, increased amounts of mesenteric fat, and segmental narrowed areas of bowel with absence of peristalsis. Fistulas or sinus tracts and abscesses may be associated findings [162].

Inflammatory disease of the appendix also can be diagnosed by MRI [164–169]. $T_1$- and $T_2$-weighted sequences and STIR sequences in axial and coronal planes are acquired for evaluation of appendicitis. The advantage of IV contrast agent is not yet clear. Oral contrast is not generally administered. MRI criteria for diagnosing appendicitis include: appendiceal diameter >6 mm), hyperintense wall on $T_2$-weighted sequences, lumen dilatation with hyperintense contents on $T_2$-weighted sequences, intraluminal rounded structures with low signal intensity on any sequence consistent with appendicolith or intraluminal air, periappendiceal hyperintensity on $T_2$-weighted sequences, and periappendiceal fluid collections (Figure 17.49). An abscess, which can be seen with perforation, appears as a walled-off fluid collection. The sensitivity of MRI for diagnosis of appendicitis is reported to be 97%–100% [164–169].

### 17.4.13 Vascular Lesions

MRA in children is performed with both noncontrast and contrast-enhanced techniques. Noncontrast sequences include black blood, time of flight, phase contrast and steady-state free precession imaging. Contrast-enhanced imaging is performed with fat-suppressed $T_1$-weighted datasets (also termed time-resolved MRA) [170–172].

MRA can be used to diagnose congenital anomalies or acquired lesions of the vascular structures [170–174]. Abdominal aortic narrowing, particularly suprarenal, typically manifests with hypertension, which can be diagnosed with MRA. Narrowing of the aorta can be caused by infectious, inflammatory or idiopathic causes and frequently involves branch vessels as well the aortic lumen. In Takayasu arteritis, which is the most common large-vessel arteritis, luminal narrowing is associated with vessel wall thickening (Figure 17.50). Aneurysms are rare in children and occur most frequently in association with Marfan syndrome, Ehlers–Danlos syndrome, Kawasaki disease or polyarteritis nodosa, sepsis, or trauma (Figure 17.51) [173,174]. Bland venous thrombus usually is a complication of an indwelling aortic catheter or the result of severe illness associated with intense dehydration, or trauma [175]. Tumor thrombus can occur in the setting of invasion of the renal veins by Wilms tumor or invasion of the hepatic veins by hepatoblastoma. In addition, developmental anomalies of the venous system, particularly interruption of the vena cava with azygos continuation, can occur, and need to be recognized lest they are misdiagnosed as pathology.

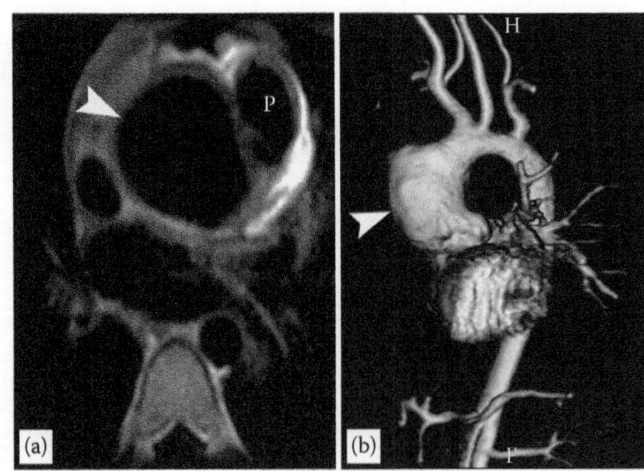


**FIGURE 17.50**
Takayasu arteritis, 12-year-old girl. (a) Axial black-blood image shows severe dilation of the aortic root (arrowhead). P indicates the pulmonary outflow tract. (b) 3D-volume rendered image anterior view shows marked dilation of the aortic root and ascending aorta (arrowhead). (Courtesy of Rajesh Krishnamurthy, MD, Texas Children's Hospital, Houston, TX.)

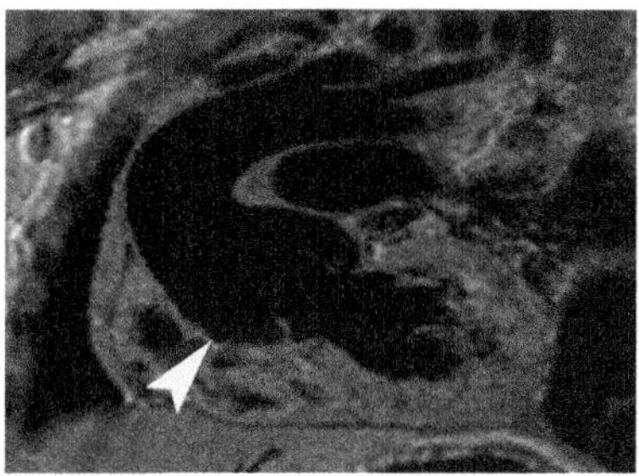

FIGURE 17.51
Marfan syndrome, neonate. Coronal black-blood image shows marked dilation of the aortic root (arrowhead). (Courtesy of Rajesh Krishnamurthy, MD, Texas Children's Hospital, Houston, TX.)

## 17.5 Pelvis

Ultrasonography is usually used initially in the evaluation of children with suspected pelvic masses, but MRI is a useful adjunct for further characterization and determination of the extent of the lesion and its relationship to adjacent structures, which can help in diagnosis, preoperative planning, and staging when malignancy is suspected [176–179]. MRI can also be used to characterize congenital uterine malformations, and localize nonpalpable testes [179,180].

A routine pelvic mass protocol includes coronal $T_1$-weighted fast spin-echo; coronal STIR; axial $T_2$-weighted fat-suppressed fast spin-echo; axial $T_1$-weighted dual-gradient echo in-phase and opposed-phase; axial diffusion-weighted; axial 3D $T_1$-weighted fat-suppressed images before IV contrast; and axial, sagittal, and coronal postcontrast fat-suppressed $T_1$-weighted acquisitions. If a detailed evaluation of the uterus is indicated, $T_2$-weighted fast spin-echo sequences in short and long axis relative to the uterus, and sagittal $T_2$-weighted fast spin-echo sequence are added.

### 17.5.1 Functional Cysts

Non-neoplastic functional cysts can result from exaggerated development of a Graafian follicle or corpus luteum. They are the most common ovarian mass in adolescent girls, but they can occasionally be encountered in the neonate, due to stimulation by maternal hormones [180]. The cysts may be incidental findings on sonography or if they achieve a large size, they may present as a pelvic or pelvic-abdominal mass or produce pain. A functional cyst appears as a large (>3 cm), unilocular, thin-walled mass that has very low signal intensity on $T_1$-weighted images and high signal intensity on $T_2$-weighted images (Figure 17.52). Intracystic bleeding can increase the signal intensity on $T_1$-weighted images [181]. Acute blood will have high signal on $T_1$-weighted images and low signal intensity on $T_2$-weighted images. Subacute blood will be bright on both $T_1$- and $T_2$-weighted images. In some cases, layering of fluid and high-signal-intensity blood can be observed. Peripheral enhancement of normal ovarian parenchyma can be noted after contrast administration. If the cyst is very large, it can extend into the upper abdomen. Most cysts resolve spontaneously in 3 to 4 months.

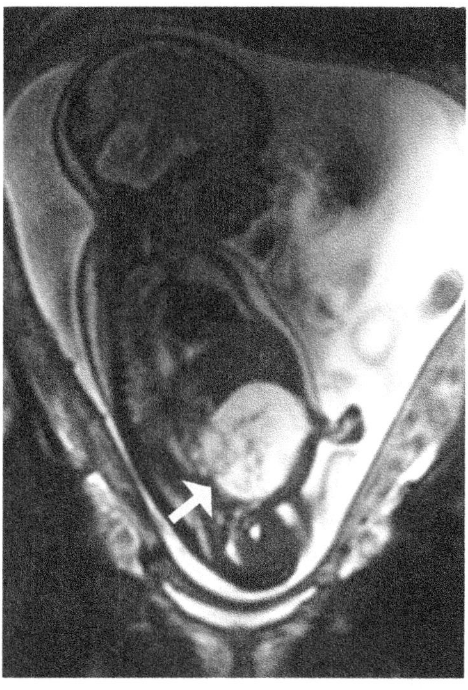

FIGURE 17.52
Hemorrhagic ovarian cyst, 33-week female gestation. Sagittal $T_2$-weighted prenatal MR image shows a large, well-circumscribed, predominantly fluid-signal intensity mass in the abdomen (arrow). Internal foci of dark signal correspond with old hemorrhage noted at pathologic examination.

### 17.5.2 Benign Ovarian Neoplasms

Benign ovarian neoplasms constitute about two-thirds of ovarian masses in children, with malignant neoplasms constituting the remainder of tumors [176,177,179,182–184]. Tumors of the ovary can arise from germ cells, stroma, or surface epithelium. Most ovarian tumors in the pediatric age group occur in the second decade of life.

#### 17.5.2.1 Cystic Teratoma

Cystic teratoma (also termed dermoid cyst) accounts for more than 90% of all benign ovarian neoplasms [182–184]. Approximately 90% of teratomas are benign and 10% are malignant. The cystic teratoma appears as a unilocular fluid-filled mass and typically has a peripheral nodule (Rokitansky nodule) that can contain

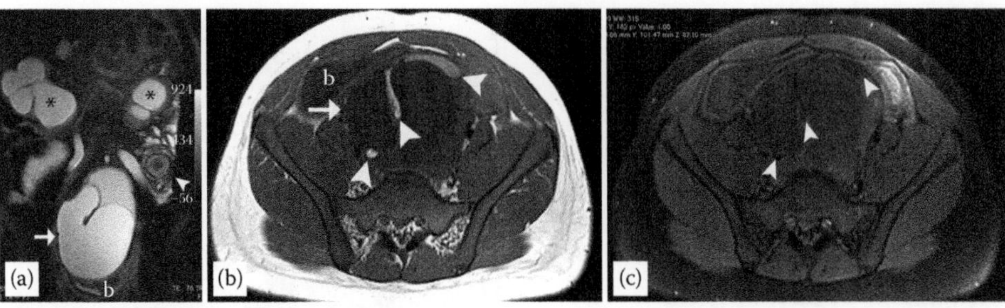

**FIGURE 17.53**
Benign mature teratoma of the ovary, 12-year-old girl. (a) Coronal $T_2$-weighted image shows a septated, fluid-signal intensity mass (arrow) superior to the bladder (b) causing dilation of the renal collecting systems (*). Arrowhead indicates the left ovary. No normal right ovary was identified. (b) Axial $T_1$-weighted image shows a predominantly fluid-signal intensity mass (arrows) containing foci of bright signal (arrowheads) consistent with fat. (c) Axial $T_1$-weighted image with fat suppression at the same level as (b) demonstrates loss of signal in the fat-containing components of the tumor (arrowheads).

hair, fat, soft tissue, and calcific components. MRI findings vary with the tissue composition. On $T_1$-weighted images, fat appears as an area of high signal intensity, whereas serous fluid and calcifications have low signal intensity. On $T_2$-weighted non-fat-suppressed images, fat and serous fluid show high signal intensity, whereas calcifications, bone, and hair demonstrate low-signal intensity (Figure 17.53). In general, benign cystic teratomas contain less than 50% soft tissue elements [185].

Ovarian cystadenomas comprise <5% of ovarian neoplasms in the pediatric population and are usually benign. Similar to cystadenomas in adults, serous cystadenomas are likely to be unilocular masses and mucinous ones are more often mulitloculated, septated masses. Both forms have low signal intensity on $T_1$-weighted images and high signal intensity on $T_2$-weighted images. When contents contain thick mucin or hemorrhage the signal intensity can increase on $T_1$-weighted images.

### 17.5.3 Malignant Ovarian Lesions

Ovarian tumors account for 1% to 2% of all malignant neoplasms in children less than 17 years of age [182–184]. Germ cell tumors (teratoma, dysgerminoma, endodermal sinus tumor, mixed malignant germ cell tumor, embryonal carcinoma) account for 60% to 90% of malignant neoplasms; sex cord stromal tumors have a 10% to 13% incidence; and epithelial carcinomas account for <5% of malignant ovarian lesions [182–184]. Rare tumors (<1%) include choriocarcinoma, polyembryoma, and leiomyosarcoma.

#### *17.5.3.1 Germ Cell Tumors*

Malignant germ cell tumors are usually larger than 10 cm in diameter and contain necrosis, calcifications, septations, and/or papillary projections [182–184]. Alpha-fetoprotein levels can be elevated in patients with immature teratomas and endodermal sinus tumors, whereas beta human choriogonadotropin levels increase in embryonal carcinomas and mixed germ cell tumors. At MRI, the malignant germ cell tumors show low or intermediate signal intensity on $T_1$-weighted images and intermediate or high signal intensity on $T_2$-weighted images (Figure 17.54) [176–179,186]. Ascites, peritoneal implants, lymphadenopathy, and hepatic metastases also may be noted.

#### *17.5.3.2 Sex Cord-Stromal Tumors*

Sex cord-stromal tumors arise from the sex cords (granulosa and Sertoli cells) of the embryonic gonad and are more common in premenarcheal than pubertal girls. They are considered to be low-grade malignancies [182–184]. The stromal tumors are commonly symptomatic. Granulosa-theca cell tumors cause isosexual precocity due to excessive production of estrogen, whereas Sertoli–Leydig cell tumors cause virilization due to androgen production. Sex cord-stromal tumors are usually complex masses containing multiple cystic areas due to hemorrhage or necrosis. They are predominantly hypointense on $T_1$-weighted images and hyperintense on $T_2$-weighed images (Figure 17.55) [187].

#### *17.5.3.3 Ovarian Cancer*

Epithelial tumors, such as ovarian cancer, are exceedingly rare in the pediatric population. The MRI appearance is that of a solid or complex mass. Areas of necrosis, thick irregular septations and papillary excrescences are common. Intraperitoneal seeding, that is, omental and mesenteric implants, is typical of epithelial neoplasms. The MR appearance is similar to that of other malignant ovarian tumors described above.

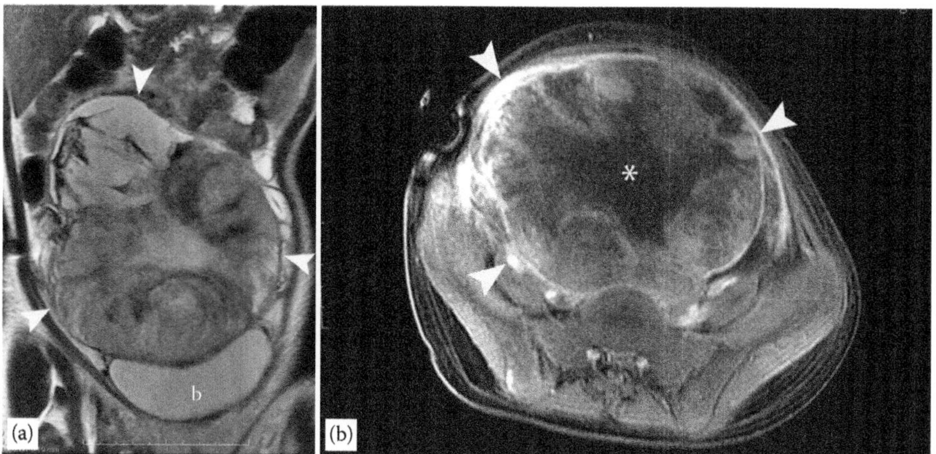


**FIGURE 17.54**
Endodermal sinus tumor of the ovary, 10-year-old girl. (a) Axial $T_2$-weighted image shows a large, heterogeneous predominantly hyperintense mass (arrowheads) superior to the bladder. (b) Axial $T_1$-weighted image with fat suppression following intravenous gadolinium-based contrast reveals peripheral enhancement (arrowheads). The central fluid-signal portion (*) corresponded to necrosis at pathologic examination.

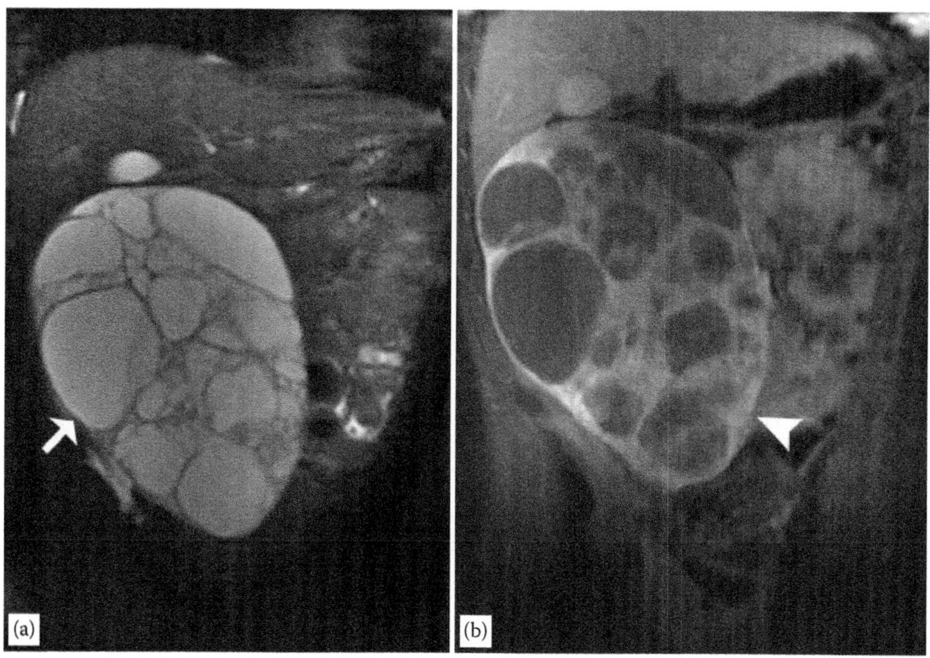


**FIGURE 17.55**
Juvenile granulosa cell tumor of the ovary, 9-year-old girl with premature menarche. (a) Coronal $T_2$-weighted image demonstrates a large abdominal mass containing numerous fluid-signal cysts (arrow). Only the left ovary was identified (not shown). (b) Coronal $T_1$-weighted image shows enhancement of the solid portions (arrowhead) but not the cysts.

### 17.5.4 Vaginal/Uterine Masses

Rhabdomyosarcoma is the most common vaginal tumor in childhood. It has a peak incidence between 2 and 4 years of life with a second peak in adolescence [188]. Patients come to clinical attention because of vaginal bleeding or a vaginal, perineal, or vulvar mass. Rhabdomyosarcoma demonstrates low to intermediate signal intensity on $T_1$-weighted sequences, intermediate to high signal intensity on $T_2$-weighted images, and restricted diffusion [177,179,189] (Figure 17.56). Central necrosis and calcification are common. The tumor enhances after IV administration of gadolinium compounds. Vaginal tumors may invade the bladder base and obstruct the ureters, causing hydronephrosis. Metastases to pelvic lymph nodes can be seen if the involved nodes are enlarged. Adenomyosis and leiomyomas are rare uterine tumors of childhood.

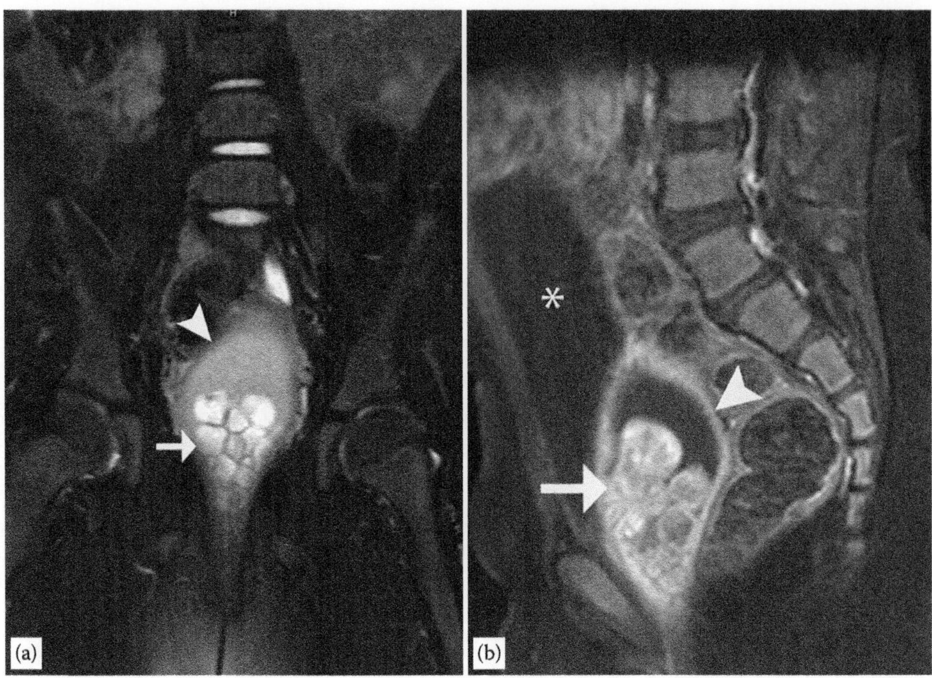


**FIGURE 17.56**
Vaginal rhabdomyosarcoma, 2-year-old girl. (a) Coronal $T_2$-weighted image shows a cluster of hyperintense masses (arrow), the botryoid appearance, obstructing the vagina (arrowhead). (b) Sagittal $T_1$-weighted image after intravenous gadolinium-based contrast reveals avid enhancement of the mass (arrow) in the base of the distended vagina (arrowhead). Asterisk indicates the bladder.

Hydrocolpos or vaginal enlargement is not uncommon in the pediatric population and can be caused by vaginal atresia or stenosis or an imperforate membrane. Hydrocolpos refers to vaginal distention by serious fluid. Hematocolpos refers to vaginal distention by blood products. There can be associated dilatation of the uterine cavity. Affected neonates present with a pelvic or lower abdominal mass or associated anomalies, including imperforate anus and urogenital sinus. Adolescent girls present with a history of absent menses, a pelvic mass, and/or cyclic pelvic pain. On MRI, the dilated vagina and uterus appear as midline, fluid-filled masses with low signal intensity on $T_1$-weighted images and high signal intensity on $T_2$-weighted images (Figure 17.57). The signal intensity increases on $T_1$-weighted images if the contents are hemorrhagic. Other findings include hydro- or hematosalpinx and hydronephrosis, resulting from extrinsic compression by the fluid-filled vagina.

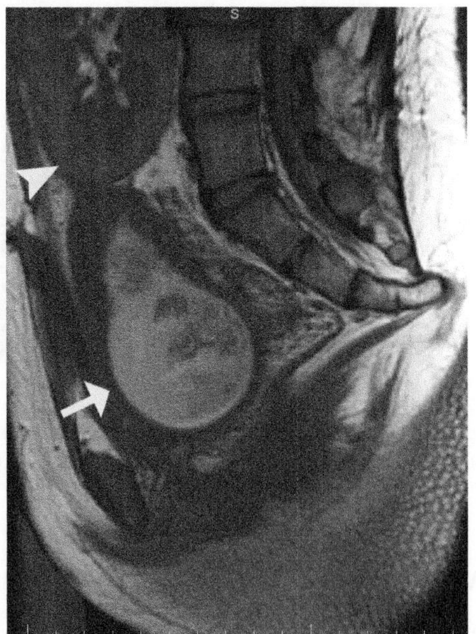


**FIGURE 17.57**
Vaginal atresia, hematometrocolpos, 9-year-old girl with uterine fusion anomaly. Sagittal $T_1$-weighted image shows a markedly dilated hemivagina filled with hyperintense blood products (arrow). Arrowhead indicates pelvic kidney. Incidentally noted is a sacral anomaly. (Courtesy of David M. Biko, MD, David Grant Medical Center, Travis AFB, CA.)

### 17.5.5 Bladder and Prostate Masses

Rhabdomyosarcoma accounts for most neoplasms of the bladder and prostate in children. It has a bimodal age distribution, typically children under 4 years of age and adolescents [190]. Patients come to clinical attention

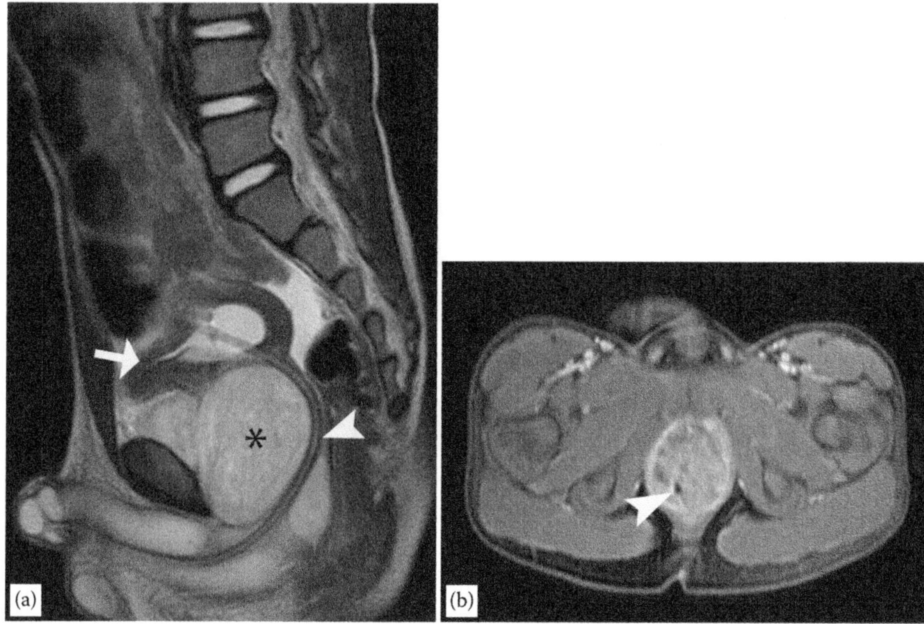


**FIGURE 17.58**
Prostatic rhabdomyosarcoma, 4-year-old boy with urinary retention. (a) Sagittal $T_2$-weighted image shows a hyperintense prostatic mass (*) elevating the bladder (arrow). Arrowhead indicates Foley catheter. (b) Gadolinium-enhanced, fat-saturated $T_1$-weighted image demonstrates somewhat heterogeneous enhancement of the mass surrounding the dark-signal catheter (arrowhead) in the urethra.

because of urinary retention, abdominal pain, dysuria, signs of urinary tract infection, or hematuria. The common histologic subtype is embryonal [191]. Bladder rhabdomyosarcoma presents with hematuria or urinary retention due to bladder outlet obstruction. MRI features are similar to those of vaginal rhabdomyosarcoma (Figure 17.58).

Less common bladder neoplasms include hemangioma, neurofibroma, pheochromocytoma, leiomyoma, and transitional cell carcinoma. On MRI, these appear as pedunculated or sessile soft tissue masses, projecting into the bladder lumen. Based on imaging findings alone, it is usually impossible to differentiate a benign lesion from a malignant one. However, when there is extension into the perivesical fat or adjacent structures, malignancy should be suspected.

### 17.5.6 Presacral Masses

Sacrococcygeal teratomas are the most common tumor of the presacral space in children [192]. Four types of teratomas have been described based on the relative amounts of internal and external tumor: type I, predominantly external (47%); type II, external and intrapelvic (34%); type III, external, pelvic, and abdominal (9%); and type IV, purely presacral (10%) [193]. They are usually benign in the first year of life, but if diagnosis or treatment is delayed, the frequency of malignant increases. Affected children usually present in the neonatal period with a large soft tissue mass in the sacrococcygeal or gluteal region and less often in childhood with constipation or pelvic pain.

In general, benign sacrococcygeal teratomas are predominantly cystic and contain some mature tissues, including fat, calcification, and minimal soft tissue (Figure 17.59). Cystic teratomas have low signal intensity

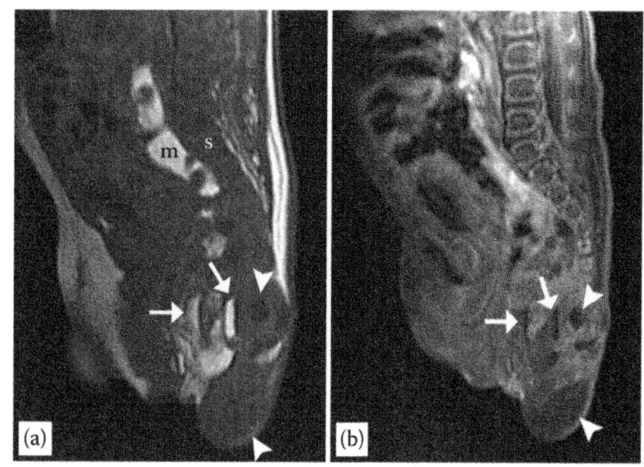


**FIGURE 17.59**
Benign sacrococcygeal teratoma, 2-day-old girl. (a) Sagittal $T_1$-weighted MR image shows a heterogeneous presacral mass, which is partially internal and partially external. Hyperintense foci represent fat (arrows). Foci of fluid-signal intensity are noted (arrowheads). The sacrum (s) is normal. Hyperintense meconium (m) is seen in the sigmoid colon. (b) Sagittal fat-suppressed $T_1$-weighted image after intravenous gadolinium shows signal dropout of the fat components (arrows) and lack of enhancement of the cystic components (arrowheads).

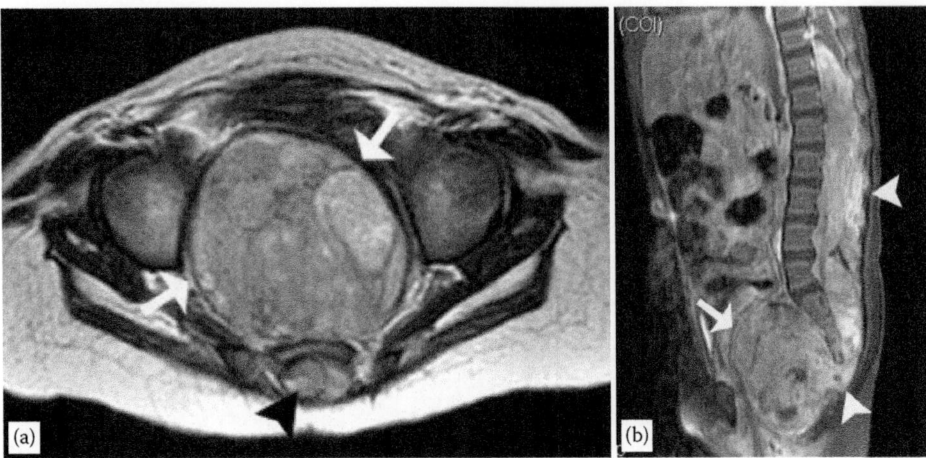


**FIGURE 17.60**
Malignant presacral immature teratoma, 11-month-old girl. (a) Axial $T_2$-weighted image shows a large presacral mass with heterogeneous, predominantly hyperintense signal (arrow). Note mass in spinal canal (arrowhead). (b) Sagittal $T_1$-weighted post-gadolinium image demonstrates large presacral enhancing mass (arrow) with extension into the spinal canal (arrowheads).

on $T_1$-weighted and high signal intensity on $T_2$-weighted MR images. Fat appears as high-signal-intensity foci on $T_1$-weighted images, and calcification, bone or hair as foci of low signal intensity on both $T_1$- and $T_2$-weighted images [177,178,192]. The soft tissue components as well as the cyst walls may enhance after gadolinium administration. Benign lesions usually do not have associated osseous anomalies Tumors containing predominantly solid components are usually malignant. Large solid components are highly vascular and demonstrate heterogeneous enhancement (Figure 17.60). Signs of local extension include intraspinal involvement, lymph node enlargement, and sacral destruction.

Other causes of presacral masses are anterior meningoceles and neuroblastoma. Anterior meningoceles are herniations of spinal contents through a congenital defect in the vertebral body (anterior dysraphism). The mass is termed a myelomeningocele when the contents of the herniated sac contain neural elements in addition to meninges and cerebrospinal fluid, and a lipomeningocele when fat and cerebrospinal fluid are present. The fluid-filled sac of the meningocele, the bony defect in the lumbar vertebra or sacrum, the presence of a tethered cord, the fatty elements of a lipomeningocele, and the relationship of the lesion to other structures of the pelvic cavity are well demonstrated with MRI.

Approximately, 5% of neuroblastomas arise in the pelvis. Pelvic neuroblastoma appears as a heterogeneous soft tissue mass with areas of necrosis and amorphous, coarse calcifications. It has isointense or slightly hyperintense to muscle on $T_1$-weighted images and hyperintense to muscle on $T_2$-weighted images and show variable degrees of enhancment Because of their neural origin, neuroblastomas can invade the spinal canal.

### 17.5.7 Impalpable Testes

Identification of an undescended testis or cryptorchidism is important because of the increased risk of infertility if the testis remains undescended and because of the increased incidence of malignancy, particularly with an intra-abdominal testis. Early surgery, either orchiopexy in younger patients or orchiectomy in patients past puberty, limit but do not eliminate these risks. MRI is used to localize the undescended testis, which can expedite surgical management and shorten the anesthesia time. Undescended testes can be found anywhere from the renal hilum to the inguinal canal (i.e., the course of normal testicular descent *in utero*). The majority (90%) are palpable in the inguinal canal. The remaining ones are in the abdomen or are nonexistent.

MRI is performed from the level of the kidneys to the level of the pelvic outlet. The pulse sequences are $T_1$- and $T_2$-weighted images and postgadolinium $T_1$-weighted images in the axial and coronal planes. The MRI diagnosis of an undescended testis is based on detection of an oval mass in the course of testicular descent with low signal on $T_1$-weighted images and high signal on $T_2$-weighted images [194,195]. Identification of the mediastinum testis is helpful in diagnosis. The more normal the testis is in size and shape, the lower is its signal intensity on $T_1$-weighted MR images. A very atrophic testis appears as a small focus of soft tissue with signal intensity similar to that of abdominal wall musculature on $T_1$- and $T_2$-weighted images. The diagnosis of an undescended testis is easier if the testis is in the inguinal canal or lower pelvis, where structures usually are symmetrical. Differentiation of an undescended testis from adjacent structures, such as bowel loops, vessels, and lymph nodes, is more of a problem in the upper pelvis and lower abdomen. Use of diffusion-weighted MRI with a high *b*-value can yield information that complements conventional MRI findings, and may improve identification and location of nonpalpable undescended testes [196].

## 17.6 Musculoskeletal System

Indications for MRI include (1) evaluation of the extent of skeletal and soft tissue neoplasms; (2) determination of the extent of infection; (3) evaluation of sequela of skeletal trauma; (4) assessment of intra-articular derangement; (5) definition of anatomy in selected congenital anomalies; (6) assessment of possible bone infarction and osteonecrosis; (7) evaluation of unexplained pain in patients with normal conventional imaging studies; and (8) assessment of the response of malignant lesions to treatment.

In general, imaging of musculoskeletal structures includes conventional spin-echo $T_1$-wieghted and proton density, fast/turbo proton density and $T_2$-weighted spin echo with fat suppression or STIR, and gradient recalled echo sequences (GRE). GRE images are useful for imaging cartilage or for delineation of hemorrhage. Contrast-enhanced sequences are used primarily for imaging infections, neoplastic processes, vascular lesions, synovial diseases, and in cases, where the vascular supply or drainage is of clinical interest [197,198].

### 17.6.1 Bone Marrow

The appearance of the normal bone marrow varies with the age of the patient [199–201]. At birth, hematopoietic or red marrow predominates. Shortly thereafter, an orderly conversion of hematopoietic to fatty marrow begins, proceeding from the appendicular skeleton to the axial skeleton. Within an individual long bone, conversion occurs first in the epiphysis, followed by the diaphysis, then in the distal metaphysis and finally in the proximal metaphysis. The adult pattern of marrow distribution is present by the mid part of the third decade. At this time, red marrow persists mainly in the skull, spine, flat bones, and proximal ends of the humeri and femora, while yellow marrow predominates in the remainder of the long bones and the epiphyses and apophyses [199–201]. One common exception to the normal pattern of complete conversion in the epiphysis is found in the proximal humeral epiphysis where red marrow may persist in adolescents and adults.

On both $T_1$- and $T_2$-weighted spin-echo images, hematopoietic marrow has signal intensity equal to or slightly higher than that of muscle and lower than that of fatty marrow. Yellow marrow demonstrates signal intensity identical to that of subcutaneous fat on $T_1$-weighted images and equal to or slightly less than that of subcutaneous fat on $T_2$-weighted images. On fat-saturated images, fatty marrow appears black; hematopoietic marrow exhibits intermediate signal intensity equal to that of muscle.

Disorders that affect marrow production can be divided into four categories: reconversion or hyperplasia, replacement disorders, depletion disorders, and myelofibrosis. Marrow reconversion refers to the repopulating of yellow marrow by hematopoietic cells. Causes include severe chronic anemia, treatment with granulocyte-macrophage colony stimulating factor (GM-CSF) during chemotherapy, and circumstances where there is an increased oxygen requirement (rigorous athletics such as marathon running and high altitudes). Hyperplastic marrow has signal intensity similar to that of normal red marrow.

Infiltrative marrow disorders include tumor, infection, and edema. In the oncologic evaluation of tumor extent, whole-body turbo-STIR and diffusion-weighted MRI can be used to detect tumor spread to skeleton and marrow, as well as extra skeletal tissues, including abdominal viscera, lung, and soft tissues [24,84,202–207]. On MRI, marrow infiltration produces low to intermediate signal intensity on $T_1$-weighted images and high signal intensity on $T_2$-weighted images, fat-suppressed $T_2$-weighted and diffusion-weighted images (Figure 17.61).

Myeloid depletion refers to replacement of red marrow by fat cells. Causes include viral infections, drugs, chemotherapy, radiation therapy, and idiopathic therapy. The marrow shows diffusely high signal intensity on $T_1$- and $T_2$-weighted images and low signal relative to muscle on fat-saturated images. Myelofibrosis is characterized by replacement of normal marrow cells by fibrotic tissue and in children usually is the result of radiation- or chemotherapy. Fibrotic marrow has low signal intensity on both $T_1$- and $T_2$-weighted images. The signal intensity may be slightly greater than that of muscle on fat-saturated images.

### 17.6.2 Malignant Osseous Neoplasms

#### *17.6.2.1 Osteosarcoma and Ewing Sarcoma*

Approximately 90% of malignant tumors in children are either osteosarcoma or Ewing sarcoma [208,209]. Both have peak incidences in the second decade of life and present with bone pain with or without soft tissue swelling. Over 90% of osteosarcomas are intramedullary in origin, arising in the metaphysis. The rest are juxtacortical forms (paraostial, periosteal, and high grade surface types). Ewing sarcoma most often arises in the

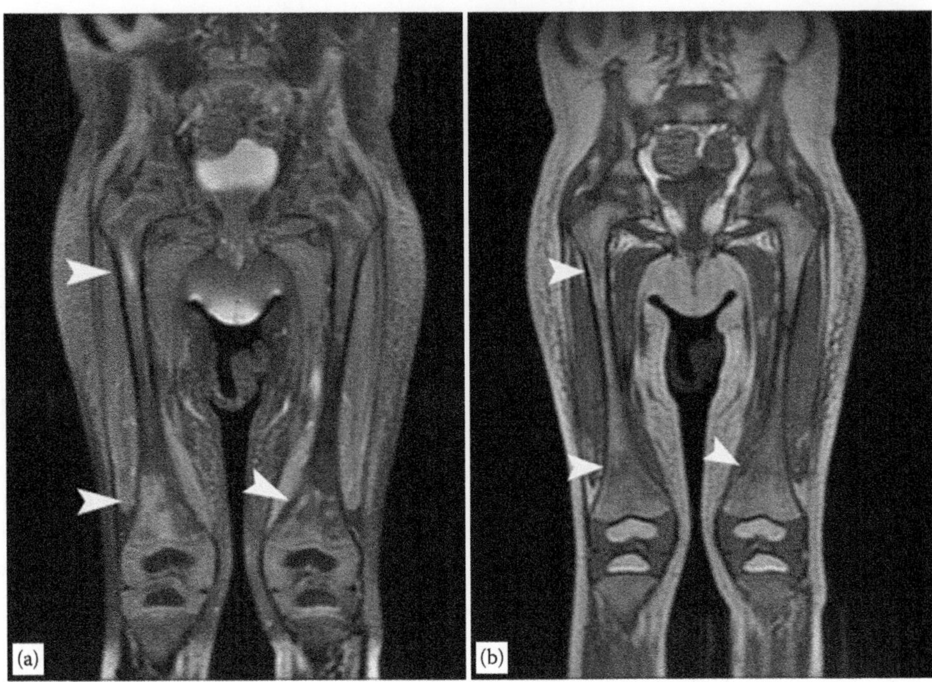


**FIGURE 17.61**
Leukemic infiltration of bone marrow, 2-year-old boy. (a) Coronal $T_2$-weighted image of the femora with fat suppression demonstrates ill-defined hyperintense foci (arrowheads). (b) Coronal $T_1$-weighted image reveal these foci are hypointense to fatty marrow (arrowheads).

metaphysis of long bones. The pelvis is the next most common site. The primary role of MRI in evaluating malignant tumors is establishing the extent of marrow involvement, soft tissue and intra-articular extension, and neurovascular encasement for staging [8,210,211].

Both osteosaracoma and Ewing saracoma have a nonspecific MR appearance with lesions displaying low signal intensity on $T_1$-weighted MR images, high signal intensity greater than that of fat on fat-saturated $T_2$-weighted images, restricted diffusion, and variable enhancement on gadolinium-enhanced $T_1$-weighted images depending on the extent of necrosis (Figure 17.62) [8,210–213]. Low signal intensity on $T_2$-weighted images suggests sclerosis, partially ossified matrix, tumor hypocellularity, or large amounts of collagen, whereas marked hyperintensity suggests highly cellular tumors with a high water content or hemorrhage. Fluid–fluid levels due to layering of new and old hemorrhage can be found in telangiectatic osteosarcomas, although they are not specific and also occur in aneurysmal bone cysts, fibrous dysplasia, and giant cell tumors.

#### 17.6.2.2 Langerhans Cell Histiocytosis

Langerhans cell histiocytosis (LCH) is a disorder of unknown cause, characterized by granuloma formation

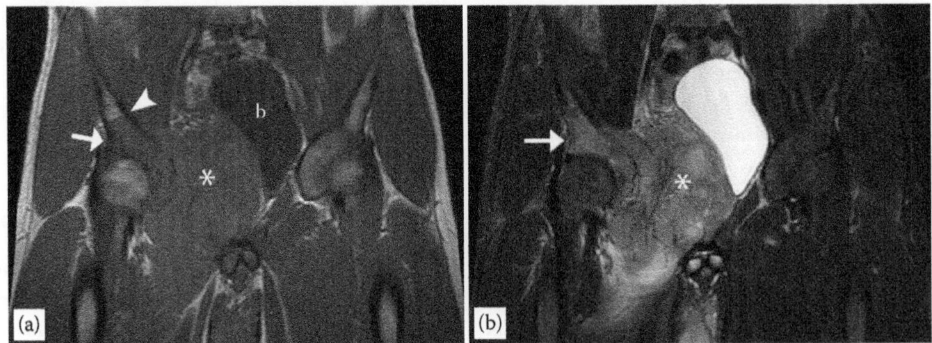


**FIGURE 17.62**
Ewing sarcoma, 16-year-old boy. (a) Coronal $T_1$-weighted MR image shows low-intensity tumor in the right ilium (arrow) with an associated soft tissue mass (*), which displaces the bladder (b) to the left. Arrowhead indicates the border between infiltrated bone marrow inferiorly and normal marrow superiorly. (b) Coronal fat-saturated $T_2$-weighted MR image shows increased signal intensity in the intramedullary portion of the tumor (arrow) and the extramedullary soft tissue component of the tumor (*).

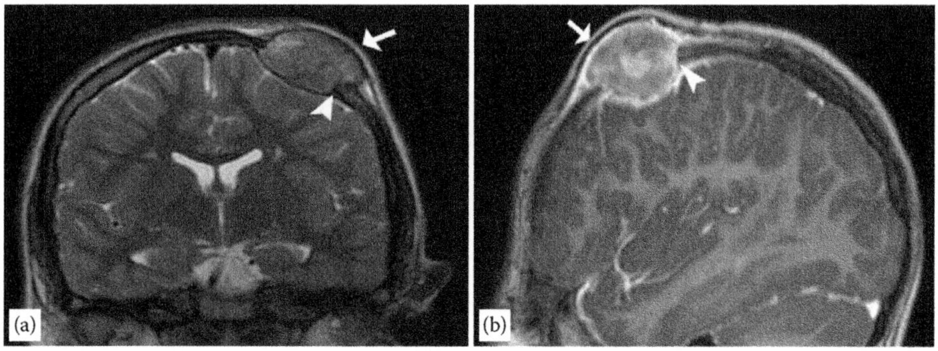


**FIGURE 17.63**
Langerhans cell histocytosis, 12-year-old boy. (a) Coronal $T_2$-weighted image of the head demonstrates a well-circumscribed, intermediate signal intensity mass (arrow) in the skull. Arrowhead indicates the asymmetric erosion of the inner and outer table of the skull (beveled edge). (b) Sagittal $T_1$-weighted image with fat suppression following intravenous gadolinium-based contrast shows enhancement of the mass (arrow) and the beveled edge of bone erosion (arrowhead).

due to proliferation of histiocytes of the Langerhans cell type [214]. The peak incidence is between 1 and 4 years, but the tumor can present in neonates. It has three forms: (1) a localized form (70%) with a single lesion that needs minimal or no therapy and undergoes spontaneous healing, (2) a chronic recurring form (20%) that is associated with diabetes insipidus and dermatitis and eventually burns out, and (3) a fulminant form (10%) that involves skeleton and multiple organs (liver, spleen, lungs, lymph nodes, soft tissues) and can be a life-threatening disease. In all forms, the skeleton is the organ most often involved and the skull is the common site of involvement followed by long bones, spine, ribs, and pelvis. In the acute phase, the lesions are usually hypointense on $T_1$-weighted images and hyperintense on $T_2$-weighted, fat-suppressed images and gadolinium-enhanced images (Figure 17.63) [215–217]. As lesions heal, they show a decrease in $T_2$-weighted fat-suppressed signal intensity.

### 17.6.3 Benign Tumors

Benign lesions include bone cysts (aneurysmal and unicameral), cartilaginous lesions (chondroblastoma, enchondroma, osteochondroma), bone-forming tumors (osteoid osteoma and osteoblastoma), and fibrous lesions (fibrous dysplasia, fibrous cortical defect, and ossifying fibroma). The signal intensity is generally nonspecific so that correlation with radiographs or tissue sampling is needed for final diagnosis.

Unicameral bone cysts are of fluid filled and of near-water signal intensity (Figure 17.64). The lesions usually involve the full diameter of the bone and expand the cortex without reactive sclerosis. Aneurysmal bone cysts are eccentric in location and show multiple compartments and fluid–fluid levels. Both tumors are hypointense on $T_1$-weighted images, hyperintense on $T_2$-weighted images and do not enhance. Increased $T_1$-weighted signal intensity, indicative of blood, and pathologic fracture or periosteal reaction may be present.

Chondroblastoma arises in the epiphyses or apophyses of long bones (Figure 17.65). Enchondroma is a cartilaginous tumor arising in the medulla of long bones, usually small tubular bones of the hand. Osteochondroma (also termed exostosis) is common in the femur, proximal tibia, and proximal humerus. It is a cartilage-capped bony spur that arises from the cortical surface of a bone. The cartilage cap is thick in young children (>2 cm) and becomes thinner in adolescents and adults (<1 cm in adults). Cartilaginous lesions show low $T_1$-signal intensity, heterogeneous, intermediate $T_2$-signal intensity, and minimal gadolinium enhancement.

Osteoid osteoma tends to arise in the cortex of long bone and is diaphyseal or metadiaphyseal in location. Less commonly it arises in cancellous bone (usually femoral neck) or intra-articular. The nidus has an intermediate signal intensity on $T_1$-weighted MR images and a high signal intensity on $T_2$-weighted images. The surrounding reactive sclerosis and calcifications within the nidus show low signal intensity on $T_1$- and $T_2$-weighted sequences. Marrow and soft tissue edema are common. Osteoblastoma usually arises in the spine and less often in the femur and tibia and appears similar to osteoid osteoma, except they are larger (>2 cm diameter) and there have less surrounding reactive bone formation.

Fibrous dysplasia can affect one bone (monostotic form) or several bones (polyostotic form). The polyostotic form can be associated with endocrinopathies, such as precocious puberty (McCune–Albright syndrome), Cushing syndrome, acromegaly, hyperthyroidism, or hyperparathyroidism. Fibrous cortical defect, also called nonossifying fibroma, are small, eccentric, expansile lesions with sclerotic borders located in the

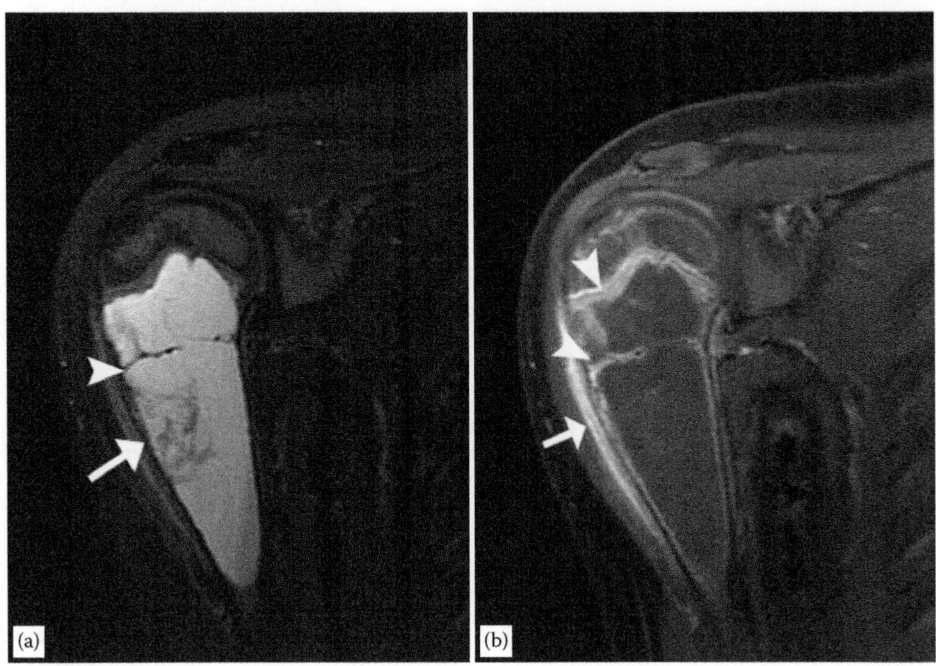


**FIGURE 17.64**
Unicameral bone cyst, 7-year old girl with acute shoulder pain. (a) Coronal $T_2$-weighted image with fat suppression demonstrates an expansile, fluid-signal mass in the metaphysis of the right humerus (arrow). (b) Coronal $T_1$-weighted image obtained after intravenous gadolinium shows only rim and septal enhancement (arrowhead) and enhancement of the periosteum related to pathologic fracture seen on plain radiographs.

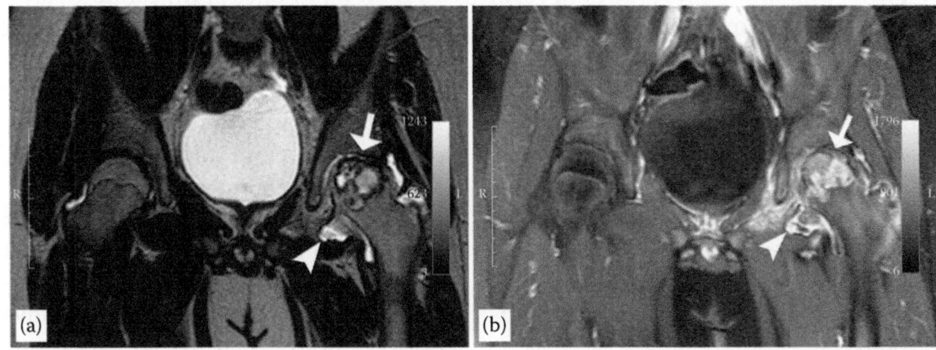


**FIGURE 17.65**
Chondroblastoma proximal femoral head epiphysis, 16-year-old boy. (a) Coronal $T_2$-weighted image demonstrates a heterogeneous, predominantly hyperintense mass in the epiphysis and metaphysis of the left femoral head (arrow). Arrowhead indicates associated joint effusion. (b) Coronal post-gadolinium $T_1$-weighted image with fat suppression shows enhancement of portions of the mass (arrow) and of the synovium (arrowhead).

metaphyseal cortex of long tubular bones. Ossifying fibroma, also known as osteofibrous dysplasia, is a multilocular lesion located in the anterior diaphyseal cortex of the tibia or fibula and causes anterior or anterolateral bowing. Fibrous lesions show hypointense or intermediate signal on $T_1$-weighted images and are hyperintense on $T_2$-weighted images.

### 17.6.4 Soft Tissue Neoplasms

The most common benign soft tissue masses in children are the congenital vascular tumors (e.g., hemangioma, venous malformations, and cystic hygroma), neurogenic tumors (neurofibroma, schwannoma), fatty tumors (lipoma, lipoblastoma), fibroblastic tumors, ganglion cyst, hematoma, and abscess [210,218–220]. The most frequent malignant mass is rhabdomyosarcoma. Rarer malignancies include synovial sarcoma, fibrosarcoma, neurofibrosarcoma, malignant fibrous histiocytoma, leiomyosarcoma, alveolar part sarcoma, and liposarcoma.

Sonography is the initial examination of choice for the study of most soft tissue masses to determine whether they are cystic or solid. MRI, however, has become the examination of choice for large lesions to characterize and define the extent of the mass and its local relationships [8,210,218–220]. Small size, well-defined margins with a capsule, homogenous matrix on $T_2$-weighted MR images and absence of edema suggest a benign lesion. Poorly-defined margins and a heterogeneous matrix on $T_2$-weighted images favor an aggressive malignant process. Bone erosion and infiltration of the neurovascular bundles are confirmatory evidence of malignancy. Unfortunately, some acute hematomas, abscesses, and benign neoplasms can have an aggressive appearance, whereas some malignant neoplasms can have a benign appearance. Tissue sampling is needed for a specific histologic diagnosis.

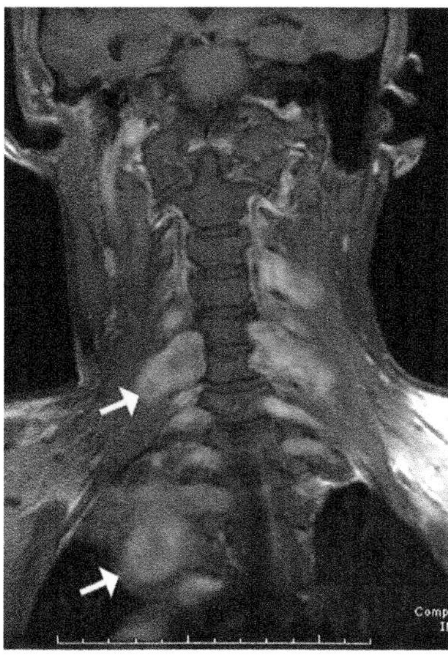

**FIGURE 17.66**
Plexiform neurofibromas, 18-year-old girl with neurofibromatosis type 1. Contrast-enhanced coronal $T_1$-weighted MR image with fat saturation demonstrates the typical target appearance of benign neural tumors with a central zone of low signal intensity and a peripheral zone of high signal intensity (arrows).

Certain non-vascular soft tissue lesions can have specific MR characteristics [8,210,218–221]. Lipomas appear as well-defined masses with signal intensity equal to that of subcutaneous fat on $T_1$- and $T_2$-weighted images. Ganglion cysts have a signal intensity lower than that of skeletal muscle on $T_1$-weighted images and greater than that of fat on $T_2$-weighted and fat-suppressed images. Neurofibromas are well-circumscribed, round or ovoid masses with low to intermediate signal intensity on $T_1$-weighted images and high signal intensity on $T_2$-weighted and gadolinium-enhanced images. Most benign neural tumors exhibit a target sign, characterized by a low intensity center and hyperintense rim on $T_2$-weighted and contrast-enhanced $T_1$-weighted images (Figure 17.66).

Rhabdomyosarcomas and other sarcomas have relatively nonspecific imaging characteristics. Like most soft-tissue tumors, they have intermediate $T_1$-weighted signal intensity and intermediate to high $T_2$-weighted signal intensity (Figure 17.67) [222,223]. Associated findings are bone erosion and neurovascular encasement.

#### 17.6.4.1 Congenital Vascular Lesions

Congenital vascular lesions are subdivided into hemangiomas and vascular malformations [219–221,224–229]. Hemangiomas are endothelial-lined neoplasms that have proliferating and involuting stages. Vascular malformations are composed of dysplastic vessels that show no cellular proliferation or regression. The latter are further categorized by their flow (high- or slow-flow malformations) and by the predominant vascular channel (arteriovenous, venous, capillary, or lymphatic). MR imaging sequences include fast spin-echo $T_1$-weighted images, $T_2$-weighted, fat-suppressed images, gradient-echo images, and dynamic post-contrast imaging to depict the pattern of arterial and venous enhancement.

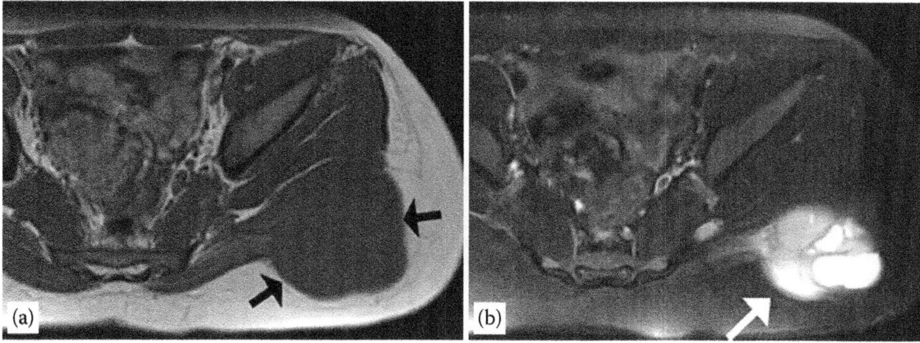


**FIGURE 17.67**
Rhabdomyosarcoma in a 5-year-old boy. (a) Axial $T_1$-weighted MR image shows a mass (arrows) which is isointense to muscle arising from the left gluteus maximus. (b) On a fat-saturated $T_2$-weighted image the mass has high signal intensity (arrow).

#### 17.6.4.2 Hemangioma

Infantile hemangioma is the most common vascular tumor of infancy. It is a benign, slow-growing lesion that may also contain nonvascular elements such as fat, fibrous tissue, and smooth muscle. It can arise within superficial or deep soft tissues. Hemangiomas usually appear in the first week or few months of life, presenting as blue or red colored masses and involute by the 3rd to 5th year. There are congenital forms of infantile hemangioma, which may involute rapidly or not at all. MRI shows a well-defined mass that is $T_1$ isointense or hypointense and $T_2$ hyperintense, often with flow voids. MRA may show large feeding vessels. Large lesions show peripheral enhancement with rapid centripetal fill-in. Small lesions may show flash fill-in or early complete enhancement. Focal heterogeneities are common, related to hemosiderin deposits, fibrosis, fat, calcification, thrombosis, or stagnant blood in involuting hemangiomas (Figure 17.68). Associated findings include the presence of feeding or draining vessels in the subcutaneous tissues and muscle atrophy.

#### 17.6.4.3 Vascular Malformations

##### *17.6.4.3.1 High-Flow Malformations*

Arteriovenous malformations and fistulas are high-flow vascular lesions [219,224–229]. Histologically, arteriovenous malformations are characterized by multiple abnormal vascular channels (termed a *nidus*) interposed between enlarged feeding arteries and draining veins. The arteriovenous fistula is a form of arteriovenous malformation, which has a single communication interposed between a feeding artery and a draining vein. The normal intervening capillary bed is absent in both lesions. Clinical findings include a pulsatile mass with a bruit or thrill. MRI/MRA shows abnormal vascular channel(s) connecting arteries and veins (Figure 17.69). They are $T_1$ isointense or hypointense and $T_2$ hyperintense and show early filling of the venous component.

##### *17.6.4.3.2 Slow Flow Malformations*

*Venous malformations* are slow-flow lesions characterized by dilated venous spaces and a normal arterial component. They may involve skin, muscle, or both tissues. Clinically, they present as soft, compressible masses usually of bluish color, in later childhood or adulthood. They grow with the child [224–229]. Venous malformations are $T_1$ isointense or hypointense and $T_2$ hyperintense fill and show contrast enhancement on delayed imaging. Thrombi and phleboliths are common and indicate slow flow (Figure 17.70).

*Capillary malformations* are characterized by a collection of small vascular channels in the dermis and are usually clinically evident at birth. The most common capillary malformation is the port-wine stain. There are usually no findings on imaging studies, although increased thickness of the subcutaneous fat and prominent venous channels may be seen in some patients.

*Lymphatic malformations,* also termed cystic hygroma and lymphangioma, are congenital lesions composed of dilated lymphatic channels that fail to communicate with peripheral draining channels [224–229]. There are two subsets: microcystic variety with numerous septations

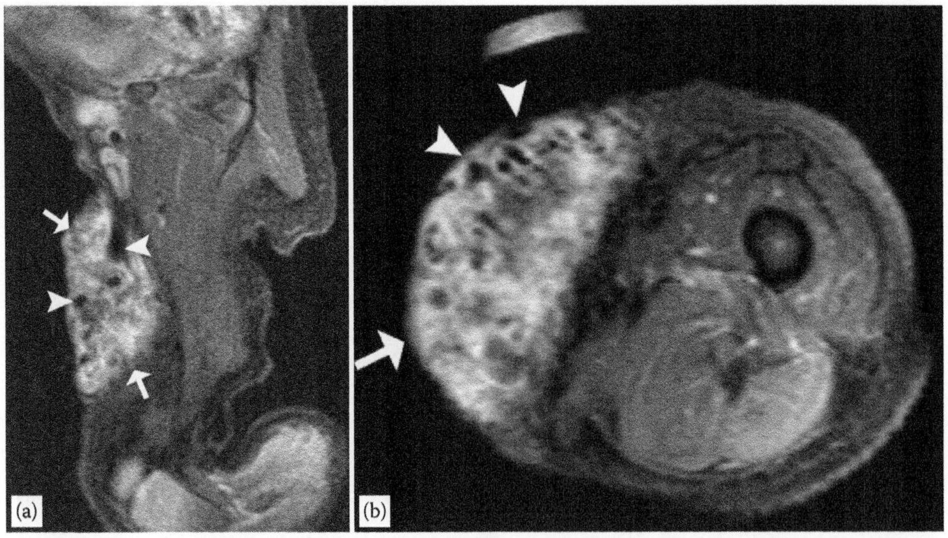


**FIGURE 17.68**
Rapidly involuting congenital hemangioma, 12-day-old girl. (a) Sagittal $T_2$-weighted image with fat saturation shows a predominantly hyperintense mass (arrows) in the anterior thigh with flow voids representing dilated feeding and draining vessels (arrowheads). Interspersed throughout the lesion is low signal intensity fibrous tissue. (b) Axial $T_1$-weighted image with fat saturation after intravenous gadolinium demonstrates a heterogeneously enhancing mass (arrow) with flow voids (arrowheads). Fibrotic areas don't enhance.

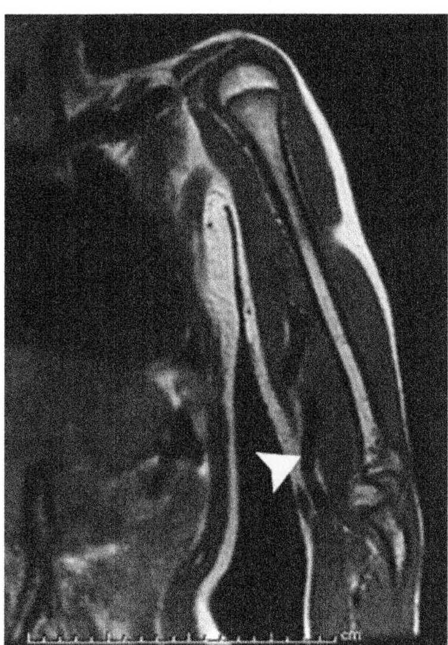

**FIGURE 17.69**
Arteriovenous malformation in a child with Parkes–Weber syndrome. Coronal $T_1$-weighted MR image demonstrates a large tortuous flow void in the medial upper left arm (arrowhead).

and macrocystic with relatively few septations [225]. Lymphatic malformations usually present at birth as soft masses and grow with the child. Most of which are found in the neck have an association with Turner syndrome. Next most common site is the axilla and rarely, they can affect soft tissues, bowel wall, and lung. Smaller lesions are well marginated, while larger lesions often are infiltrative and ill-defined. The lesions have $T_2$ hyperintensity and may show fluid–fluid levels [230]. $T_1$ signal intensity is typically hypointense, but it can increase if the cysts contain proteinaceous material or blood products. Peripheral and septal enhancement is common (Figure 17.71).

Limb overgrowth and vascular lesions may be observed in the following syndromes:

1. Klippel–Trenaunay syndrome (capillary port wine stain, varicosities, and lymphangiomas)
2. Parkes–Weber syndrome (capillary port wine stain, varicosities, and arteriovenous fistulas)
3. Proteus syndrome (port wine stains; lymphangiomas and lipohemangiomas; lipomas; epidermal nevi; partial gigantism with cerebroid gyriform hypertrophy of the hands and/or feet; asymmetric macrocephaly; and intra-abdominal lipomatosis)
4. Maffucci syndrome (venous malformations, lymphangiomas, enchondromas, and exostoses)
5. Blue rubber bleb nevus syndrome (venous malformations in the skin and gastrointestinal tract)

Hematomas can be distinguished from hemangioma on the basis of signal intensity characteristics. Very acute hematomas (one to 24 h of age) are generally hypo- or

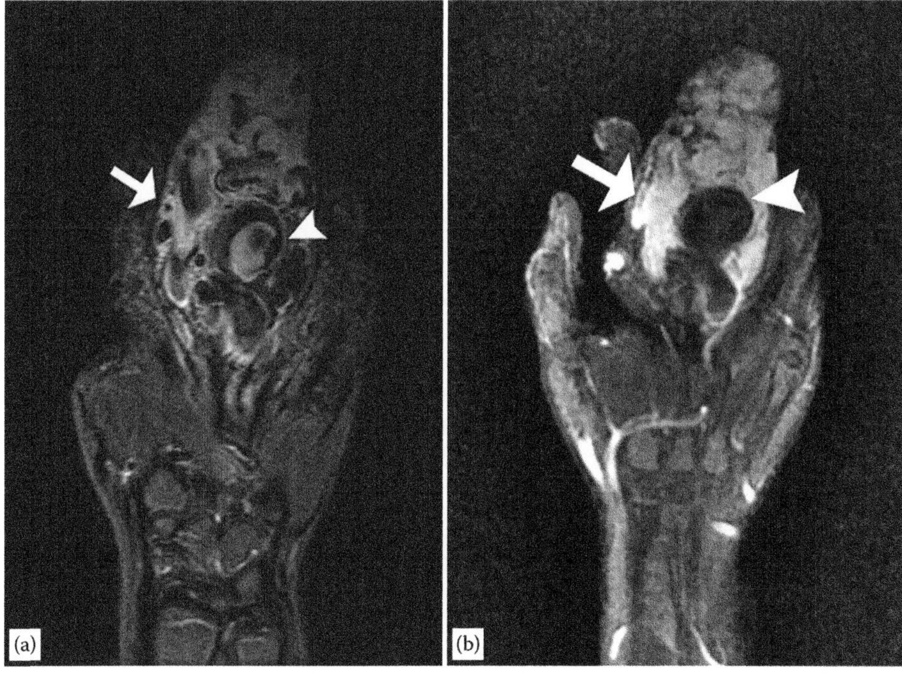


**FIGURE 17.70**
Venous malformation present in a 7-year-old boy since birth but now enlarging. (a) Coronal $T_2$-weighted image with fat suppression shows a heterogeneous mass of the middle finger with fluid signal tubular structures (arrow) and dark foci (arrowhead) corresponding to phleboliths and thrombi on pathologic examination. (b) Coronal $T_1$-weighted image with fat suppression following gadolinium-based contrast administration demonstrates enhancement of the tubular structures (arrow) but not the dark thrombi (arrowhead).

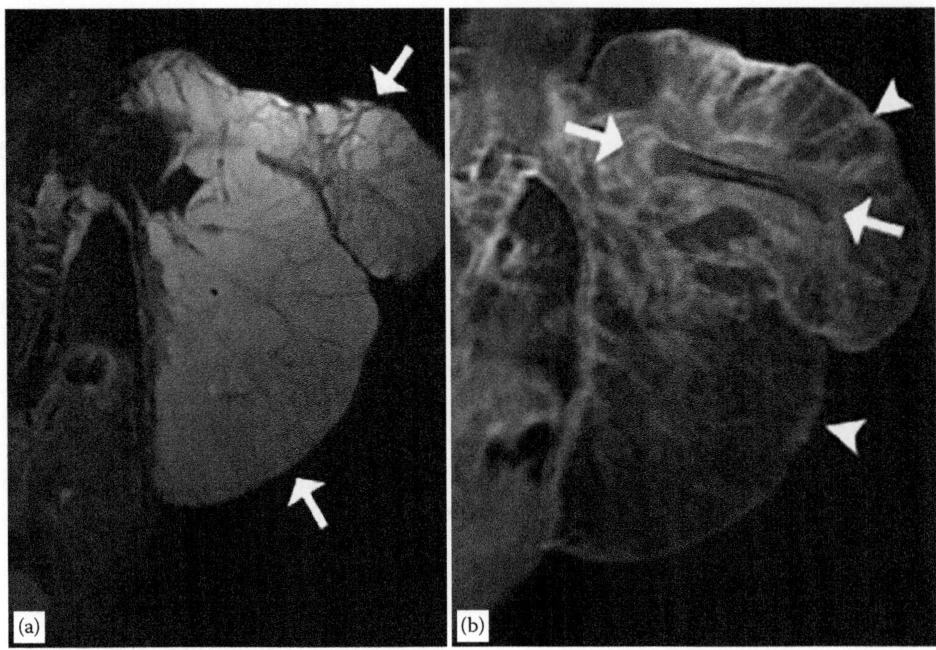


**FIGURE 17.71**
Lymphatic malformation, 25-day-old. (a) Coronal $T_2$-weighted image with fat saturation demonstrates a multiseptated predominantly fluid signal intensity mass of the left upper extremity and left trunk (arrows). (b) Coronal post-gadolinium $T_1$-weighted image with fat saturation shows enhancement of the walls and septa (arrowheads) but not the cyst contents. Arrows indicate the humerus.

isointense to muscle on $T_1$-weighted images and hyperintense to muscle on $T_2$-weighted images. Acute hematomas (one to 7 days of age) tend to be hypo- or isointense to muscle on $T_1$-weighted images and hypointense on $T_2$-weighted images. Chronic hematomas have MR characteristics similar to those of other fluid collections, which include low signal intensity on $T_1$-weighted images and high signal intensity on $T_2$-weighted and fat-suppressed images.

### 17.6.5 Infection

Acute osteomyelitis in the pediatric population is most often secondary to hematogenous spread and associated with a known focus of infection, such as a skin or upper respiratory infection. It initially affects the metaphyses of bones and then can extend into the subperiosteal space and the soft tissues. In patients under 1 year of age, nutrient vessels penetrate the physis and allow extension of infection into the epiphyses and joints.

Acute osteomyelitis has low $T_1$-weighted signal intensity and high $T_2$-weighed signal intensity (Figure 17.72). Marrow infection enhances after gadoliunium injection. MRI is especially well-suited for the detection of Brodie's abscess, sequestrum, and sinus tract and extension into the periosteum and soft tissues [231–232]. Brodie's abscess appears as a well-defined area of bone destruction surrounded by reactive bone formation. The periphery but not the center of the lesion, enhances after contrast administration. A sequestrum appears as an isolated segment of bone in a region of marrow infection or within a Brodie's abscess. A sinus tract or cloaca appears as a channel extending between an area of marrow infection and the cortex or the subcutaneous tissues.

#### 17.6.5.1 Chronic Recurrent Multifocal Osteomyelitis

Chronic recurrent multifocal osteomyelitis (CRMO) is an inflammatory disorder of bone in children with most patients presenting between 9 and 14 years of age. It is thought to be the result of an immune system disorder or occult viral or bacterial infection. Skeletal involvement is multifocal and typically involves the metaphyseal regions of long bones and medial clavicles [233–236]. CRMO does not respond to antibiotic therapy and treatment includes glucocorticoids and nonsteroidal anti-inflammatory agents. Recurrences are common.

During the active phase of the disease, MR imaging shows typical findings of marrow edema, which appears hypointense on $T_1$-weighted images and hyperintense on $T_2$-weighted images. Other findings include periostitis, soft tissue inflammation, transphyseal disease, and joint effusion. Presence of a large fluid collection or abscess, fistulous tract, or sequestrum makes the diagnosis of infectious osteomyelitis more likely than CRMO. During the healing phase, both $T_1$- and $T_2$-weighted images show markedly hypointense areas, reflecting the presence of sclerosis. Over time, marrow

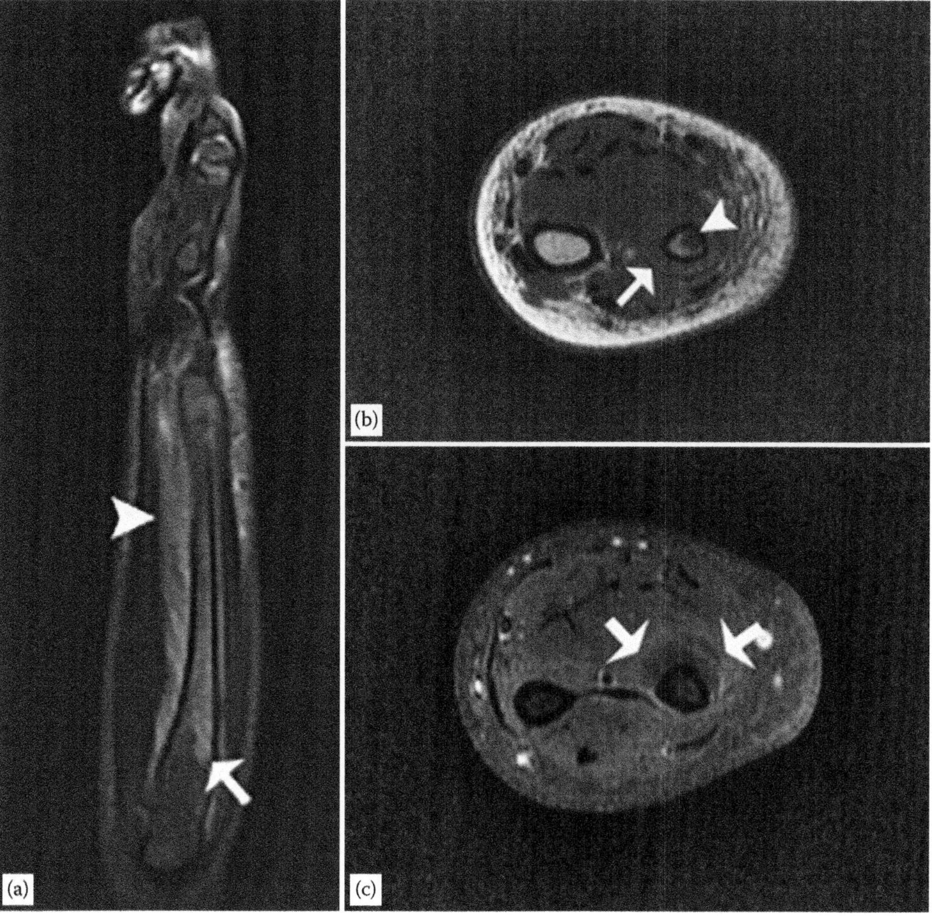


FIGURE 17.72
Osteomyelitis of the ulna, 2-year-old boy. (a) Sagittal $T_2$-weighted image reveals the abnormal hyperintense signal in the marrow (arrow) and in the adjacent muscles (arrowhead) as well as the subcutaneous fat. (b) Axial $T_1$-weighted image shows low signal in the bone marrow of the ulna (arrowhead) and elevation of the periosteum (arrow). Surrounding low signal in the subcutaneous fat due to edema is also noted. (c) Axial post-gadolinium $T_1$-weighted image with fat suppression shows enhancement of the elevated periosteum (arrows) but no enhancement of the small subperiosteal abscess.

signal can return to normal. In general, the features of CRMO and pyogenic osteomyelitis overlap and biopsy is needed for definitive diagnosis.

### 17.6.6 Chronic Joint and Synovial Disorders

Juvenile rheumatoid arthritis (JRA) is the most common chronic arthritis of children. MR, particularly with gadolinium enhancement, is used to show the extent of synovial inflammation and hypertrophy and demonstrate the status of articular cartilage and overall joint integrity. MRI findings of early JRA include thickened synovium, which has low signal on $T_1$-weighted images and mixed intensity on $T_2$-weighted images, reflecting the presence of inflammation and hemosiderin deposition, and joint effusion. Inflamed synovium shows rapid enhancement after the injection of gadolinium. Late changes include cartilage loss and bone erosions (Figure 17.73) [237,238]. Chronic fibrous synovial thickening enhances either heterogeneously or poorly.

Other causes of joint and synovial-based processes include hemophilic arthropathy, pigmented villonodular synovitis, and infection [239,240]. MRI findings of hemophilia and villonodular synovitis are synovial thickening, areas of hemosiderin deposition appearing as low $T_1$- and $T_2$-weighted signal intensity, and joint effusion. In infectious arthritis, the findings can be similar to JRA and clinical correlation is needed.

### 17.6.7 Skeletal Muscle Diseases

The pediatric skeletal muscle disorders include Duchenne-type muscular dystrophy, other upper and lower extremity muscular dystrophies, and Werdnig-Hoffman spinal muscular atrophy [241,242]. Duchenne muscular dystrophy is an inherited primary degeneration of skeletal muscle, predominantly affecting boys. It involves proximal pelvic girdle muscles in the early stage and the calf and proximal shoulder girdle muscles in later stage. Spinal muscular atrophy is an inherited disorder affecting

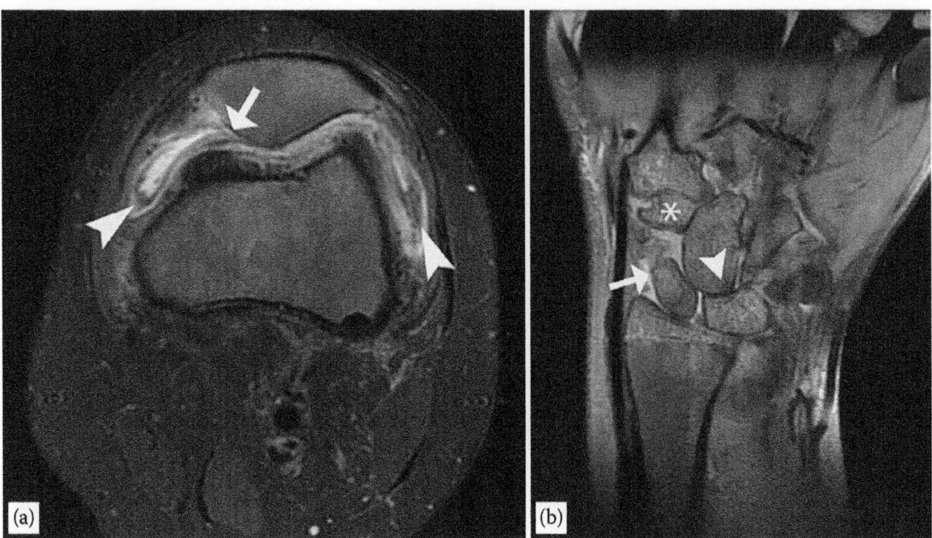


**FIGURE 17.73**
Juvenile idiopathic arthritis in a 19-year-old male. (a) Axial PD-weighted image with fat suppression shows thickening hypointense synovium (arrowheads) surrounding a small effusion. Marked thinning of the patellar articular cartilage is noted (arrow). (b) Coronal PD-weighted image with fat suppression of the wrist shows abnormal high signal around the carpal bones (arrow) and marked cartilage loss between the capitate and lunate (arrowhead) and capitate and scaphoid. Diffuse abnormal marrow signal and an erosion of radial side of the trapezium (*) are also seen.

anterior horn cells in the spinal cord and brain stem nuclei. It is characterized by asymmetric lower extremity muscle involvement. In these disorders, the muscle fascicles are replaced by fat and connective tissue. Muscle thickness is either normal or increased in the muscular dystrophies, whereas it is usually decreased in spinal muscular atrophy. $T_1$-weighted MR images reveal hyperintense fatty infiltration within and between diseased muscles. Fatty infiltration nulls on fat-suppressed images.

Other muscle disorders include Pompe disease, which is a glycogen storage disease characterized by abnormal metabolism of glycogen, glucose, or both, and dermatomyositis. In Pompe disease, involvement is greater in proximal than distal limb muscles. $T_1$-weighted images show diffuse hypertrophy of muscle groups without evidence of fatty infiltration. Childhood dermatomyositis is characterized by diffuse nonsuppurative inflammation of muscle fibers and skin. Initially, there is proximal lower extremity weakness, followed by proximal upper limb weakness. Involved muscles show high $T_2$-weighted signal intensity related to increased water content of the infarcted muscle because of vasculitis.

### 17.6.8 Bone and Joint Trauma

Traditional radiographs remain the initial examination of choice in the evaluation of acute trauma. When radiographs are not confirmatory, MRI can be used to establish the presence and extent of occult and subtle fractures, and in particular, stress fractures and epiphyseal and growth plate injuries [243–245]. The latter are of particular importance because if unrecognized they can lead to growth disturbance. After the injury has healed, MR imaging can be used to define size and location of a posttraumatic bony bridge and the severity of associated growth deformity (Figure 17.74).

MRI is the examination of choice for detecting meniscal and ligamentous tears [243–245]. Tears appear as alterations in morphology and signal intensity within the substance of the meniscus or ligament. MRI has also been shown to be an effective method for evaluating bone marrow edema associated with ligamentous and cartilaginous injuries. In addition, muscle and tendon injures can be evaluated by MRI.

### 17.6.9 Bone Infarction

Legg–Calve–Perthes (LCP) disease is the common cause of skeletal avascular necrosis in childhood. LCP refers to the idiopathic necrosis of the immature proximal femoral epiphysis. MRI has become the imaging examination for diagnosis of LCP and the assessment of the degree of epiphyseal involvement, femoral head containment, physeal bony bridging, and evidence of reperfusion and healing.

MR findings of acute or early LCP are low or intermediate $T_1$-weighted signal, variable $T_2$-weighted signal, including hypointensity related to necrosis and hyperintensity representing bone marrow edema, and nonenhancement of the proximal femoral epihpysis (Figure 17.75) [246]. A curvilinear subchondral defect (crescent sign) may be noted, suggesting subchondral fracture. In the

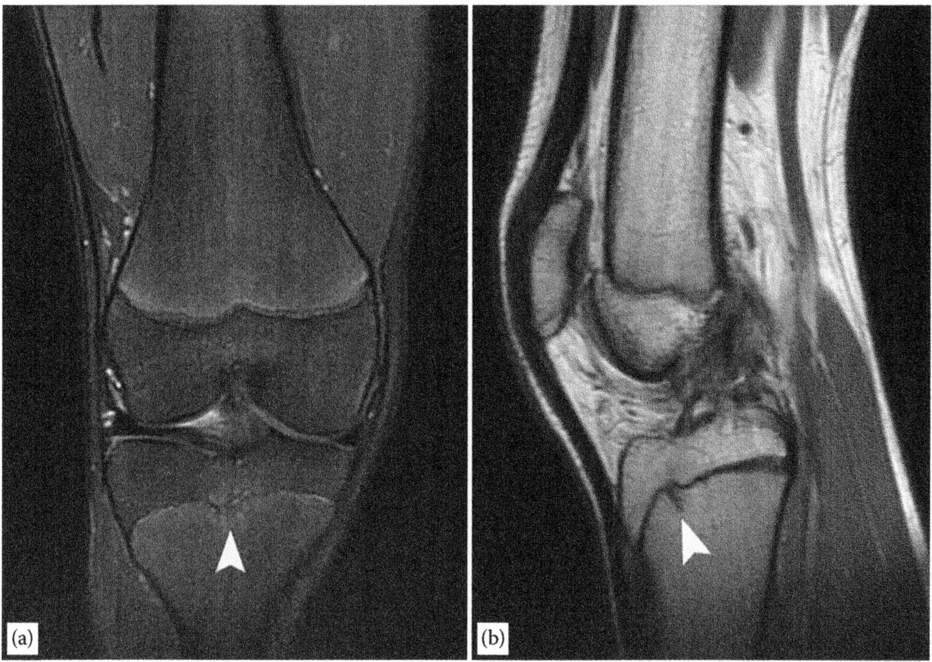

**FIGURE 17.74**
Physeal bar, 14-year-old girl with history of Salter–Harris V fracture of the proximal tibia. (a) Coronal PD-weighted image with fat suppression shows focal abnormal course and loss of the normal layered signal in the proximal tibial physis (arrowhead). (b) Sagittal PD-weighted image demonstrates focal hypoitense signal around the physis (arrowhead).

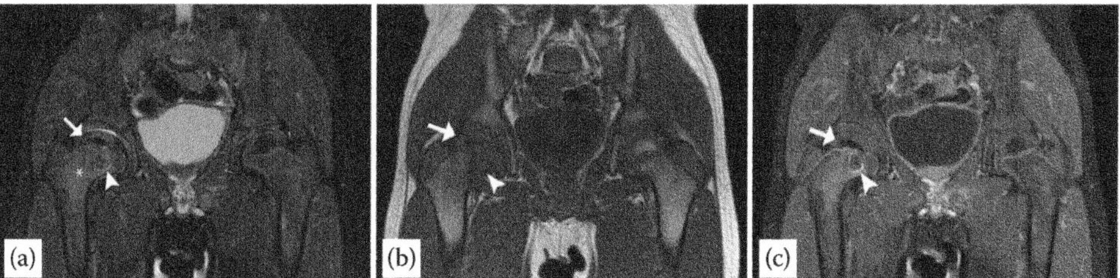

**FIGURE 17.75**
Legg–Calve–Perthes disease, 3-year-old boy. (a) Coronal $T_2$-weighted image with fat suppression shows flattening and widening of the right femoral head and dark signal in the femoral head epiphyseal ossification center (arrow). There is also hyperintense edema in the marrow of the femoral neck (*) and a characteristic metaphyseal cystic lesion (arrowhead). (b) Coronal $T_1$-weighted images demonstrates abnormal low signal in the femoral head epiphysis (arrow) compared to the normal bright fatty marrow signal on the left. The metaphyseal lesion is also hypointense with a dark rim (arrowhead). Hypointense marrow edema is also noted in the femoral neck. (c) Coronal $T_1$-weighted image with fat suppression following intravenous gadolinium-based contrast shows no enhancement of the right femoral head epiphysis (arrow) and predominantly rim enhancement of the metaphyseal cystic lesion (arrowhead). Enhancement of the marrow in the femoral neck is also noted

revascularization/reparative phases, the proximal femoral epiphysis typically shows $T_2$-weighted signal hyperintensity and contrast enhancement. Associated findings include a flattened articular surface, thickening of femoral head articular cartilage, joint effusion, and metaphyseal cysts. Other causes of avascular necrosis in children include sickle cell disease, steroid therapy, and trauma. Imaging findings are similar to those of LCP.

Osteochondritis dissecans is a subarticular osteonecrosis, likely secondary to chronic microtrauma. It is most often encountered in the femoral condyles, particularly the lateral aspect of the medial condyle, talar dome, and the capitellum [247]. MRI is used to determine the stability of the fragment. An unstable or loose fragment may need to be surgically removed, whereas a stable or attached fragment can be treated conservatively. The necrotic fragment of bone has low signal intensity on both $T_1$- and $T_2$-weighted MR sequences (Figure 17.76). The most reliable MRI sign of unstable osteochondritis dissecans is a high-signal-intensity line on $T_2$-weighted images between the bone fragment and native bone [243,247,248]. Other findings of an unstable lesion are a high signal defect ≥5 mm in the articular surface and >5 mm high-signal intensity, homogeneous cyst ≥5 mm beneath the lesion.

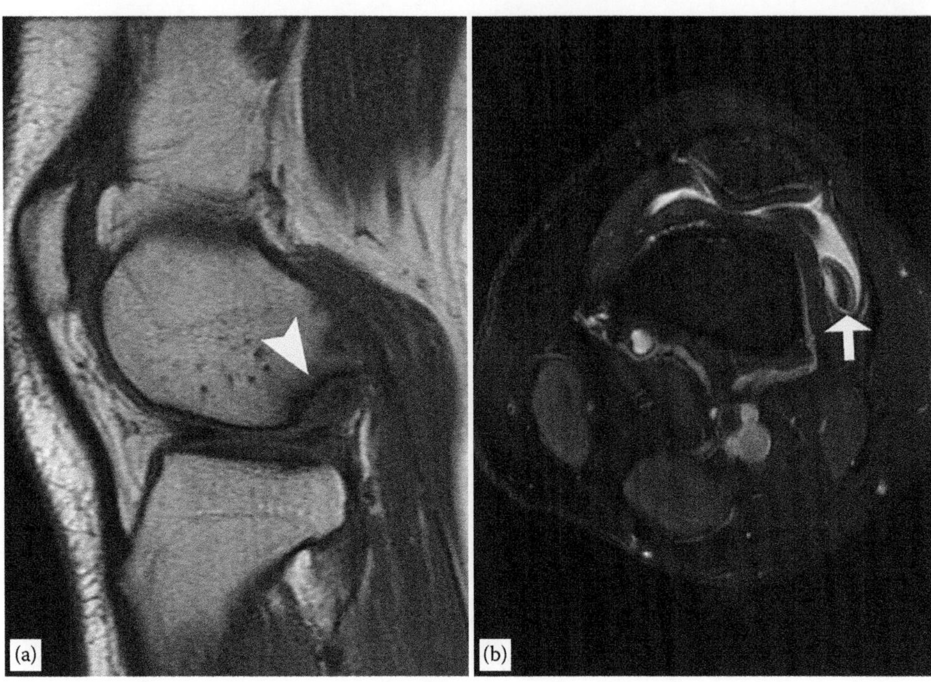


**FIGURE 17.76**
Osteochondral defect, 15-year-old girl. (a) PD-weighted oblique sagittal MR image shows a low signal intensity saucer-shaped defect (arrowhead) in the posterior medial femoral condyle of the left femur. (b) Axial fat-suppressed $T_2$-weighted image reveals the dark-signal osteochondral free fragment (arrow) with overlying slightly hyperintense layer of articular cartilage in the lateral joint space.

## References

1. Anupindi S, Jaramillo D. Pediatric magnetic resonance imaging techniques. *Magn Reson Imaging Clin N Am* 2012; 10; 189–207.
2. MacKenzie JD, Vasanawala SS. Advances in pediatric MR imaging. *Magn Reson Imaging Clin N Am* 2008; 16: 385–402.
3. Gee MS, Bittman M, Epelman M et al. Magnetic resonance imaging of the pediatric kidney: Benign and malignant masses. *Magn Reson Imaging Clin N Am* 2013; 21: 697–716.
4. Keup CP, Ratnaraj F, Pooja R et al. Magnetic resonance imaging of the pediatric liver: Benign and malignant masses. *Magn Reson Imaging Clin N Am* 2013; 21: 645–668.
5. Kim HK, Lindquist DM, Serai SD, Magnetic resonance imaging of pediatric muscular disorders: Recent advances and clinical applications. *Radiol Clin N Am* 2013; 51: 721–743.
6. Liszewski MC, Hersman FW, Altes TA et al. Magnetic resonance imaging of pediatric lung parenchyma, airways, vasculature ventilation and perfusion: State of the art. *Radiol Clin N Am* 2013; 51: 555–582.
7. Manson D. MR imaging of the chest in children. *Acta Radiol* online 25 July 2013; 1–11.
8. Meyer JS, Jaramillo D. Musculoskeletal MR imaging at 3-T. *Magn Reson Imaging Clin N Am* 2008; 16: 533–545.
9. Meyer JS, Harty MP, Khademian Z. Imaging of neuroblastoma and Wilms' tumor. *MRI Clin N Am* 2002; 10: 175–302.
10. Pai DR, Ladino-Torres MF. Magnetic resonance imaging of pediatric pelvic masses. *Mag Reson Imaging Clin N Am* 2013; 21: 751–773.
11. Siegel MJ, Chung EM, Conran RM. Pediatric liver: Focal masses. *Magn Reson Imaging Clin N Am* 2008; 12: 437–452.
12. Siegel MJ, Alok J. MRI of neuroblastic masses. *Magn Reson Imaging Clin N Am* 2008; 12: 499–514.
13. Siegel MJ, Chung EM. Wilms' tumor and other pediatric renal masses. *Magn Reson Imaging Clin N Am* 2008; 16: 479–497.
14. Siegel MJ. MR imaging of pediatric abdominal neoplasms. *Magn Reson Imaging Clin N Am* 2000; 8: 837–851.
15. Bital R, Leung G, Perng R et al. MR pulse sequences: What every radiologist wants to know but is afraid to ask. *RadioGraphics* 2006; 26: 513–537.
16. Haliloglu M, Hoffer FA, Gronemeyer SA, Furman WL, Shochat SJ. Three dimensional gadolinium-enhanced MR angiography: Evaluation of hepatic vasculature in children with hepatoblastoma. *J Magn Res Imaging* 2000; 11: 65–68.
17. Anupindi SA, Victoria T. Magnetic resonance cholangiopancreatography: Techniques and applications. *Magn Reson Imaging Clin N Am* 2008; 16; 453–466.
18. Egbert ND, Bloom DA, Dillman JR. Magnetic resonance of the pediatric pancreaticobiliary system. *Magn Reson Imaging Clin N Am* 2013; 21: 681–696.

19. Darge K, Higgins M, Hwang TJ et al. Magnetic resonance and computed tomography in pediatric urology. An imaging overview for current and future daily practice. *Radiol Clin N Am* 2013; 583–398.
20. Grattan-Smith JD, Little SB, Jones RA. MR urography in children: How we do it. *Pediatr Radiol* 2008; 38 (Suppl): S3–S17.
21. Gawande RS, Gonzalez SM, Khurana A, Daldrup-Link HE. Role of diffusion-weighted imaging in differentiation benign and malignant pediatric abdominal tumors. *Pediatr Radiol* 2013; 43: 836–845.
22. Humphries PD, Sebire NJ, Siegel MJ, Olsen OE. Tumors in pediatric patients at diffusion-weighted MR imaging: Apparent diffusion coefficient and tumor cellularity. *Radiology* 2007; 245: 848–854.
23. Kocaoglu M, Bulakbasi N, Sanal HT et al. Pediatric abdominal masses: Diagnostic accuracy of diffusion weighted MRI. *Magn Reson Imaging* 2010; 28: 629–636.
24. Siegel MJ, Jokerst CE, Rajderkar D et al. Diffusion-weighted MRI for staging and evaluating response in diffuse large B-cell lymphoma: A pilot study. *NMR Biomed* 2014; 27: 681–691.
25. Nardone B, Saddleton E, Laumann AE et al. Pediatric nephrogenic systemic fibrosis is rarely reported: A RADAR report. *Pediatr Radiol* 2014; 44: 173–180.
26. Coté CJ, Wilson S. American Academy of Pediatrics, Guidelines for monitoring and management of pediatric patients during and after sedation for diagnostic and therapeutic procedures: An update. *Pediatrics* 2006; 118: 2587–2602.
27. Dalal PG, Murray D, Cox T et al. Sedation and anesthesia protocols used for magnetic resonance imaging studies in infants: Provider and pharmacologic considerations. *Anesth Analg* 2006; 103: 863–868.
28. Krauss B, Green SM. Procedural sedation and analgesia in children. *Lancet* 2006; 367: 766.
29. Siegel MJ. Diseases of the thymus in children and adolescents. *Postgrad Radiol* 1993; 13: 106–132.
30. Siegel MJ, Glazer HS, Wiener JI, Molina PL. Normal and abnormal thymus in childhood: MR imaging. *Radiology* 1989; 172: 367–371.
31. Siegel MJ, Luker GD. Pediatric chest MR imaging. Noncardiac clinical uses. *Magn Reson Imaging Clin N Am* 1996; 4: 599–613.
32. Metzger M, Krasin MJ, Hudson MM, Onciu M. Hodgkin lymphoma. In: Pizzo PA, Poplack DG, eds. *Principles and Practice of Pediatric Oncology*, 6th ed. Philadelphia, PA: Lippincott Williams & Wilkins, 2011; 638–662.
33. Gross TG, Perkins SL. Malignant non-Hodgkin lymphoma in children. In: Pizzo PA, Poplack DG, eds. *Principles and Practice of Pediatric Oncology*, 6th ed. Philadelphia, PA: Lippincott Williams & Wilkins, 2011; 663–682.
34. Anthony E. Imaging of pediatric chest masses. *Paediatr Respir Rev* 2006; 75: 539–540.
35. Franco A, Mody NS, Meza M. Imaging evaluation of pediatric mediastinal masses. *Radiol Clin N Am* 2005; 43: 325–339.
36. Newman B. Thoracic neoplasms in children. *Radiol Clin N Am* 2011; 49: 6336–6364.
37. Takahashi K, Al-Janabi NJ. Computed tomography and magnetic resonance imaging of mediastinal tumors. *Magn Reson Imaging* 2010; 32: 1325–1339.
38. Rahmouni A, Tempany C, Jones R, Mann R, Yang A, Zerhouni E. Lymphoma: Monitoring tumor size and signal intensity with MR imaging. *Radiology* 1993; 188: 445–451.
39. Rahmouni A, Divine M, Lepage E et al. Mediastinal lymphoma: Quantitative changes in gadolinium enhancement at MR imaging after treatment. *Radiology* 2001; 219: 621–628.
40. Inaoka T, Takahashi K, Mineta M et al. Thymic hyperplasia and thymus gland tumors: Differentiation with chemical shift MR imaging. *Radiology* 2007; 243: 869–876.
41. Inoue A, Tomiyama N, Fujimoto K et al. MR imaging of thymic epithelial tumors: Correlation with World Health Organization classification. *Radiat Med* 2006; 24: 171–181.
42. Rosado-de-Christenson ML, Pugatch RD, Moran CA, Galobardes J. Thymolipoma: Analysis of 27 cases. *Radiology* 1994; 193: 121–126.
43. Siegel MJ, Glazer HS, St. Amour TE, Rosenthal DD. Lymphangiomas in children: MR imaging. *Radiology* 1989; 170: 467–470.
44. Haddon MJ, Bowen A. Bronchopulmonary and neurenteric forms of foregut anomalies. Imaging for diagnosis and management. *Radiol Clin N Am* 1991; 29: 241–254.
45. McAdams HP, Kirejczyk WM, Posado-de-Christenson ML et al. Bronchogenic cyst: Imaging features with clinical and histopathologic correlation. *Radiology* 2000; 217: 441–446.
46. Eichinger M, Heussel CP, Kauczor HU et al. Computed tomography and magnetic resonance imaging in cystic fibrosis lung disease. *J Magn Reson Imaging* 2010; 32: 1370–1378.
47. Puderbach M, Eichinger M, Haeselbarth J et al. Assessment of morphological MRI for pulmonary changes in cystic fibrosis (CF) patients: Comparison to thin-section CT and chest x-ray. *Invest Radiol* 2007; 42: 715–724.
48. Puderbach M, Kauczor HU. Can lung MR replace lung CT? *Pediatr Radiol* 2008; 38: 5439–5451.
49. Failo R, Wielopolski PA, Harm AWM et al. Lung morphology assessment using MRI: A robust ultra-short TR/TE 2D steady state free precession sequence used in cystic fibrosis patients. *Magn Reson Med* 2009; 61: 299–306.
50. Montella S, Santamria F, Salvatore M et al. Lung disease assessment in primary ciliary dyskinesia: A comparison between high magnetic resonance imaging and high resolution tomography findings. *Ital J Pediatr* 2009; 6: 24.
51. Fain S, Schiebler ML, McCormack DG, Parraga G. Imaging of lung function using hyperpolarized helium-3 magnetic resonance imaging: Review of current and emerging translational methods and applications. *J Magn Reson Imaging* 2010; 32: 1398–1408.
52. Koumellis P, van Beek EJR, Woodhouse N et al. Quantitative analysis of regional airways obstruction using dynamic hyperpolarized $^3$He MRI—Preliminary results in children with cystic fibrosis. *J Magn Reson Imaging* 2005; 22: 420–426.

53. Van Beek EJ and Hoffman EA. Functional imaging: CT and MRI. *Clin Chest Med* 2008; 29: 195–216.
54. Kauczor H, Ley-Zaporozhan J, Ley S. Imaging of pulmonary pathologies: Focus on magnetic resonance imaging. *Proc Am Thorac Soc* 2009; 6: 458–463.
55. Olsen OE. Imaging of abdominal tumours: CT or MRI? *Pediatr Radiol* 2008; 38 (Suppl 3): S452–458.
56. Riccabona M. Potential of MR-imaging in the pediatric abdomen. *Eur J Radiol* 2008; 68: 2235–2244.
57. Bernstein L, Linet M, Smith MA, Olshan AF. Renal tumors. In: Ries LAG, Smith MA, Gurney JG, eds. *Cancer Incidence and Survival among Children and Adolescents: United States SEER Program*, 1975–1995, National Cancer Institute, SEER Program. Bethesda, MD: NIH Publications No. 99—4649, 1999, 79–90.
58. Fernandez C, Geller JI, Ehrlich PF et al. Renal tumors. In: Pizzo PA, Poplack DG, eds. *Pediatric Oncology*, 6th ed. Philadelphia, PA: Lippincott Williams & Wilkins, 2011; 861–885.
59. Perlman EJ. Pediatric renal tumors: Practical updates for the pathologist. *Pediatr Dev Pathol* 2005; 8: 320–338.
60. Choyke PL, Siegel MJ, Craft AW, Green DM, DeBaun MR. Screening for Wilms tumor in children with Beckwith-Wiedemann syndrome and hemihypertrophy. *Med Ped Oncol* 1999; 32: 196–200.
61. DeBaun MR, Siegel MJ, Choyke PL. Nephromegaly in infancy and early childhood: A risk factor for Wilms tumor in Beckwith-Wiedemann syndrome. *J Pediatr* 1998; 132: 401–404.
62. Miller RW, Fraumeni JF Jr, Manning MD. Association of Wilms's tumor with aniridia, hemihypertrophy and other congenital malformations. *N Engl J Med* 1964; 270: 922.
63. Gylys-Morin V, Hoffer FA, Kozakewich H et al. Wilms tumor and nephroblastomatosis: Imaging characteristics at gadolinium-enhanced MR imaging. *Radiology* 1993; 188: 517–521.
64. Lonergan GJ, Martinez-Leon MI, Agrons GA et al. Nephrogenic rests, nephroblastomatosis, and associated lesions of the kidney. *RadioGraphics* 1998; 18: 947–968.
65. Lowe LH, Isuani BH, Heller RM et al. Pediatric renal masses: Wilms tumor and beyond. *RadioGraphics* 2000; 20: 1585–1603.
66. Beckwith JB, Kiviat NB, Bonadio JF. Nephrogenic rests, nephroblastomatosis, and the pathogenesis of Wilm tumor. *Pediatr Pathol* 1990; 10: 1–36.
67. Rohrschneider WK, Weirich A, Rieden K et al. US, CT and MR imaging characteristics of nephroblastomatosis. *Pediatr Radiol* 1998; 28: 435–443.
68. Sheth S, Ali S, Fishman E. Imaging of renal lymphoma: Patterns of disease with pathologic correlation. *RadioGraphics* 2006; 26: 1151–1168.
69. Argons GA, Kingsman KD, Wagner BJ, Sotelo-Avila C. Rhabdoid tumor of the kidney in children: A comparison of 21 cases. *AJR Am J Roentgenol* 1997; 168: 447–451.
70. Prasad SR, Humphrey PA, Menias CO et al. Neoplasms of the renal medulla: Radiologic-pathologic correlation. *RadioGraphics* 2005; 25: 369–380.
71. Downey RT, Dillman JR, Ladino-Torres MF et al. CT and MRI appearances and radiologic staging of pediatric renal cell carcinoma. *Pediatr Radiol* 2012; 42: 410–417.
72. Blitman NM, Berkenblit RG, Rozenblit AM, Levin TL. Renal medullary carcinoma: CT and MRI features. *AJR Am J Roentgenol* 2005; 185: 268–272.
73. Chaudry G, Perez-Atayde AR, Ngan BY, Gundogan M, Daneman A. Imaging of congenital mesoblastic nephroma with pathological correlation. *Pediatr Radiol* 2009; 39: 1080–1086.
74. Ko S-M, Kim M-J, Im Y-J et al. Cellular mesoblastic nephroma with liver metastasis in a neonate; prenatal and postnatal diffusion-weighted MR imaging. *Korean J Radiol* 2013; 14: 361–365.
75. Agrons GA, Wagner BJ, Davidson AJ, Suarez ES. Multilocular cystic renal tumor in children: Radiologic-pathologic correlation. *RadioGraphics* 1995; 15: 653–669.
76. Silver IM, Boag AH, Soboleski DA. Best cases from the AFIP: Multilocular cystic renal tumor: Cystic nephroma. *RadioGraphics* 2008; 28: 1221–1226.
77. Paterson A. Adrenal pathology in childhood: A spectrum of disease. *Eur Radiol* 2002; 12: 2491–2508.
78. Westra SJ, Zaninovic AC, Hall TR et al: Imaging of the adrenal gland in children. *RadioGraphics* 1994; 14: 1323–1340.
79. Brodeur GM, Hogarty MD, MosseYP, Maris JM. Neuroblastoma. In: Pizzo, Poplack, eds. *Principles and Practice of Pediatric Oncology*, 6th ed. Philadelphia, PA: Lippincott Williams & Wilkins, 2011; 886–922.
80. Lonnergan GJ, Schwab CM, Suarez ES, Carlson CL. Neuroblastoma, ganglioneuroblastoma and ganglioneuroma: Radiologic-pathologic correlation. *RadioGraphics* 2002; 22: 911–934.
81. Nour-Eldin A, Abdelmonem O, Tawfik AM et al. Pediatric primary and metastatic neuroblastoma: MRI findings: Pictorial review. *Magn Reson Imaging* 2012: 30: 893–906.
82. Papaioannou G, McHugh K. Neuroblastoma in childhood: Review and radiological findings. *Cancer Imaging* 2005; 5: 116–127.
83. Uhl M, Altehoefer C, Kontny U et al. MRI-diffusion imaging of neuroblastomas: First results and correlation to histology. *Eur Radiol.* 2002; 12: 2335–2338.
84. Siegel MJ, Ishwaran H, Fletcher B et al Staging of neuroblastoma by imaging: Report of the Radiology Diagnostic Oncology Group. *Radiology* 2002; 223: 168–175.
85. Argons GA. Lonergan GJ, Dickey GE, Perez-Monte JE. Adrenocortical neoplasms in children: Radiologic-pathologic correlation. *RadioGraphics* 1999; 19: 9898–1008.
86. Lack EE, ed. Adrenal cortical neoplasms in childhood. In: *Atlas of Tumor Pathology: Tumors of the Adrenal Gland and Extra Adrenal Paraganglia, Fasc 19, ser 3.* Washington, DC: Armed Forces Institute of Pathology, 1997; 153–168.
87. Ribeiro J, Ribeiro RC, Fletcher BD. Imaging findings in pediatric adrenocortical carcinoma. *Pediatr Radiol* 2000; 30: 45–51.
88. Bharwani N, Rockall AG, Sahdev A et al. Adrenocortical carcinoma: The range of appearances on CT and MRI. *AJR Am J Roentgenol* 2011; 196: W706–W714.

89. Reddy VS, O'Neill JA, Jr., Holcomb GW, 3rd et al. Twenty-five-year surgical experience with pheochromocytoma in children. *Am Surg* 2000; 66: 1085–1091; discussion 1092.
90. Elsayes KM, Narra VR, Leyendecker JR et al. MR of adrenal and extraadrenal pheochromocytoma. *AJR Am J Roentgenol* 2005; 184: 860–867.
91. Meyers AM, Towbin AJ, Serai S et al. Characterization of pediatric liver lesions with gadoxetate disodium. *Pediatr Radiol* 2011; 41: 1183–1187.
92. Chung E, Lattin GE, Jr., Cube R et al. From the archives of the ARIP: Pediatric liver masses: Radiologic-pathologic correlation. Part 2. Malignant tumors. *RadioGraphics* 2011; 31: 483–507.
93. Jha P, Chawla SC, Tavri S et al. Pediatric liver tumors-a pictorial review. *Eur Radiol* 2009; 19: 209–219.
94. Meyers RL, Aronson DC, Von Schweinitz D et al. Pediatric liver tumors. In: Pizzo PA, Poplack DG, eds. *Principles and Practice of Pediatric Oncology*, 6th Ed. Philadelphia, PA: Lippincott Williams & Wilkins, 2011; 838–860.
95. Meyers RL. Tumors of the liver in children. *Surg Oncol* 2007; 16: 195–203.
96. Hussain SM, Semelka RC. Liver masses. *Magn Reson Imaging Clin N Am* 2005; 13: 255–275.
97. Siegel MJ, Luker GD. MR imaging of the liver in children. *MRI Clin N Am* 1996; 4: 637–656.
98. Ganeshan D, Szklaruk J, Kundra V et al. Imaging features of fibrolamellar hepatocellular carcinoma. *AJR Am J Roentgenol* 2014; 203: 544–552.
99. Buetow PC, Buck JL, Pantongrag-Brown L et al. Undifferentiated (embryonal) sarcoma of the liver: Pathologic basis of imaging findings in 28 cases. *Radiology* 1997; 203: 779–783.
100. Psatha EA, Smeelka RC, Fordham L et al. Undifferentiated (embryonal) sarcoma of the liver (USL): MRI findings including dynamic gadolinium enhancement. *Magn Reson Imaging* 2004; 22: 897–900.
101. Chung E, Cube R, Lewis B, Conran RM. From the archives of the ARIP: Pediatric liver masses: Radiologic-pathologic correlation. Part 1. Benign tumors. *RadioGraphics* 2011; 30: 801–826.
102. Dimashkieh HH, Bove KE. GLUT1 endothelial reactivity distinguishes hepatic infantile hemangioma from congenital hepatic vascular malformation with associated capillary proliferation. *Hum Pathol* 2004; 35: 200–209.
103. Hernandez F, Navarro M, Encinas JL et al. The role of GLUT1 immunostaining in the diagnosis and classification of liver vascular tumors in children. *J Pediatr Surg* 2005; 40: 801–804.
104. Kassarjian A, Zurakowski D, Dubois J et al. Infantile hepatic hemangiomas: Clinical and imaging findings and their correlation with therapy. *AJR Am J Roentgenol* 2004; 182: 785–795.
105. Burrows PE, Dubois J, Kassarjian A. Pediatric hepatic vascular anomalies. *Pediatr Radiol* 2001; 31: 535–545.
106. Roos JE, Pfiffner R, Stallmach T et al. Infantile hemangioendothelioma. *RadioGraphics* 2002; 23: 1649–1655.
107. Ramji FG. Pediatric hepatic hemangioma. *RadioGraphics* 2004; 24: 1719–1724.
108. Ramanujam TM, Goh DW, Wong KT et al. Malignant transformation of mesenchymal hamartoma of the liver: Case report and review of the literature. *J Pediatr Surg* 1999; 334: 1684–1686.
109. Ros PR, Goodman AD, Ishak KG et al. Mesenchymal hamartoma of the liver: Radiologic-pathologic correlation. *Radiology* 1986; 158: 619–624.
110. Do RK, Shaylor DS, Shia J et al. Variable MR imaging appearances of focal nodular hyperplasia in pediatric cancer patients. *Pediatr Radiol* 2011; 41: 335–340.
111. Towbin SJ, Luo GG, Yin H, Mo JQ. Focal nodular hyperplasia in children, adolescents and young adults. *Pediatr Radiol* 2011; 41: 341–349.
112. Chavhan GB, Babyn PS, Manson D, Vidarsson L. Pediatric MR cholangiopancreatography: Principles, technique and clinical applications. *RadioGraphics* 2008; 28: 1951–1962.
113. Delaney L, Applegate K, Karmazyn B et al. MR cholangiopancreatography in children: Feasibility, safety, and initial experience. *Pediatr Radiol* 2008; 38: 64–75.
114. Mortele KJ, Rocha TC, Streeter JL et al. Multimodality imaging of pancreatic and biliary congenital anomalies. *RadioGraphics* 2006; 26: 715–731.
115. Prabhakar PD, Prabhadar AM, Pradhakar HB et al. Magnetic resonance choangiopancreatography of benign disorders of the biliary system. *Magn Reson Imaging Clin N Am* 2010; 18: 597–514.
116. Savader SJ, Benenati JF, Venbrux AC et al. Choledochal cysts: Classification and cholangiographic appearance. *AJR Am J Roentgenol* 1991; 156: 327–331.
117. Heller SL and Lee VS. MR imaging of the gallbladder and biliary system. *Magn Reson Imaging Clin N Am* 2005; 13: 295–311.
118. Donnelly LF, Bisset GS, 3rd, Frush DP. Diagnosis please. Case 2: Embryonal rhabdomyosarcoma of the biliary tree. *Radiology* 1998; 208: 621–623.
119. Ahmed TS, Chavhan GB, Navarro OM, Traubici J. Imaging features of pancreatic tumors in children: 13-year experience at a tertiary hospital. *Pediatr Radiol* 2013; 43: 1435–1443.
120. Chung EM, Travis MD, Conran RM. Pancreatic tumors in children: Radiologic-pathologic correlation. *RadioGraphics* 2006; 26: 1211–1238.
121. Montemarano H, Lonergan GJ, Bulas DI, Selby DM. Pancreatoblastoma: Imaging findings in 10 patients and review of the literature. *Radiology* 2000; 214: 476–482.
122. Xinghui Y, Xiqun W. Imaging features of pancreatoblastoma in 4 children including a case of ectopic pancreatoblastoma. *Pediatr Radiol* 2010; 40: 1609–1614.
123. Abbott RM, Levy AD, Aguilera NS et al. From the archives of the AFIP: Primary vascular neoplasms of the spleen: Radiologic-pathologic correlation. *RadioGraphics* 2004; 24: 1137–1163.
124. Kaza RK, Azar S, Al-Hawary MM, Francis IR. Primary and secondary neoplasms of the spleen. *Cancer Imaging* 2010; 10: 173–182.
125. Luna A, Ribes R, Caro P et al. MRI of focal splenic lesions without and with dynamic gadolinium enhancement. *AJR Am J Roentgenol* 2006; 186: 1533–1547.

126. Chang W-C, Lious C-H, Kao H-W et al. Solitary lymphangioma of the spleen: Dynamic MR findings with pathological correlation. *Br J Radiol* 2007; 80: e4–e6.
127. Arda K, Kizikanat K, Celik M, Turkalp E. Intermittent torsion of a wandering spleen in a child: The role of MRI in diagnosis. *JBR-BTR* 2004; 87: 70–72.
128. Cruso F, Pugliese F, Maselli A et al. Malignant small-bowel neoplasms: Spectrum of disease on MR imaging. *Abdominal Radiology* 2010; 115: 1279–1291.
129. Moholkar S, Sebire NH, Roebuck DJ. Radiological-pathological correlation in lipoblastoma and lipoblastomatosis. *Pediatr Radiol* 2006; 36: 851–856.
130. Grattan-Smith JD, Little SB, Jones RA. MR urography evaluation of obstructive uropathy. *Pediatr Radiol* 2008; 38 (Suppl 1): S49–S69.
131. Khrichenko D, Darge K. Functional analysis in MR urography: Made simple. *Pediatr Radiol* 2010; 40: 182–199.
132. Jones RA, Easley K, Little SB et al. Dynamic contrast-enhanced MR urography in the evaluation of pediatric hydronephrosis, Part 1, functional assessment. *AJR Am J Roentgenol* 2005; 185: 1598–1607.
133. McDaniel BB, Jones RA, Scherz H et al. Dynamice contrast-enhanced MR urography in the evaluation of pediatric hydronephrosis: Part 2, anatomic and functional assessment of ureteropelvic junction obstruction. *AJR Am J Roentgenol* 2005; 185: 1608–1614.
134. Leyendecker JR, Barnes CE, Zagoria RJ. MR urography: Techniques and clinical applications. *RadioGraphics* 2008; 28: 23–46; discussion 46–47.
135. Mortele KJ, Ros PR. Imaging of diffuse liver disease. *Semin Liver Dis* 2001; 21: 195–212.
136. Talwalkar JA, Yin M, Fidler JL et al. Magnetic resonance imaging of hepatic fibrosis: Emerging clinical applications. *Hepatology* 2008; 47: 332–342.
137. Taouli B, Ehman RL, Reeder SB. Advanced MRI methods for assessment of chronic liver disease. *AJR Am J Roentgenol* 2009; 193: 14–27.
138. Towbin AJ, Serai SD, Podberesky DJ., Magnetic resonance imaging of the pediatric liver: Imaging of steatosis, iron deposition, and fibrosis. *Magn Reson Imaging Clin N Am* 2013; 21: 669–680.
139. Huwart L, Sempoux C, Vicaut E et al. Magnetic resonance elastography for the noninvasive staging of liver fibrosis. *Gastroenterology* 2008; 135: 32–40.
140. Darge K, Anupindi SA, Jaramillo D. MR imaging of the abdomen and pelvis in infants, children and adolescents. *Radiology* 2011; 261: 12–29.
141. Ma X, Holalkere N-S, Kambadakone A et al. Imaging-based quantification of hepatic fat: Methods and clinical applications. *RadioGraphics* 2009; 29: 1253–1280.
142. Mehta SR, Thomas EL, Patel N et al. Proton magnetic resonance spectroscopy and ultrasound for hepatic fat quantification. *Hepatol Res* 2010; 40: 399–406.
143. Van Werven JR, Marsman HA, Nederveen AJ et al. Assessment of hepatic steatosis in patients undergoing liver resection: Comparison of US, CT, $T_1$-weighted dual-echo MR imaging and point-resolved 1H MR spectroscopy. *Radiology* 2010; 256: 159–168.
144. Siegel MJ, Hildebolt CF, Bae KT, Hong C, White NH. Measurement of total and intraabdominal fat distribution in adolescents by magnetic resonance imaging. *Radiology* 2007; 242: 846–856.
145. Guibaud L, Lachaud A, Touraine R et al. MR cholangiography in neonates and infants: Feasibility and preliminary applications. *AJR Am J Roentgenol* 1998; 170: 27–31.
146. Kim M-J, Park YN, Han SJ et al. Biliary atresia in neonates and infants: Triangular area of high signal intensity in the porta hepatis at $T_2$-weighed MR cholangiography with US and histopathologic correlation. *Radiology* 2000; 215: 395–401.
147. Lee M-J, Kim M-J, Youn C-S. MR cholangiopancreatography findings in children with spontaneous bile duct perforation. *Pediatr Radiol* 2010; 40: 687–692.
148. Anupindi SA, Darge K. Imaging choices in inflammatory bowel diease. *Pediatr Radiol* 2009; 39: S149–S152.
149. Anupindi SA, Terreblanche O, Courtier J. Magnetic resonance enterography: Inflammatory disease and beyond. *Magn Reson Imaging Clin N Am* 2013; 21: 731–750.
150. Cronin CG, Lohan DG, Browne AM et al. MR small-bowel follow-through for investigation of suspected pediatric small-bowel pathology. *AJR Am J Roentgenol* 2009; 192: 1239–1245.
151. Darge K, Anupinid S, Jaramilo D. MRI of the bowel: Pediatric applications. *Magn Reson Imaging Clin N Am* 2008; 16: 467–478.
152. Neubauer H, Pabst T, Dick A et al. Small-bowel MRI in children and young adults with Crohn disease: Retrospective head-to-head comparison of contrast-enhanced and diffusion-weighted MRI. *Pediatr Radiol* 2013; 43: 103–114.
153. Ream JM, Dillman JR, Adler J et al. MRI diffusion-weighted (DWI) in pediatric small bowel Crohn disease: Correlation with MRI findings of active bowel wall inflammation. *Pediatr Radiol* 2013; 43: 1077–1085.
154. Sinha R, Rajiah P, Ramachandran I et al. Diffusion-weighted MR imaging fo the gastrointestinal tract: Technique, indications, and imaging findings. *RadioGraphics* 2013; 33: 655–676.
155. Golonka NR, Haga LJ, Keating RP et al. Routine MRI evaluation of low imperforate anus reveals unexpected high incidence of tethered spinal cord. *J Pediatr Surg.* 2002; 37: 966–969.
156. Podberesky DJ, Towbin AJ, Eltonmey, Levitt MA. Magnetic resonance imaging of anorectal malformations. *Magn Reson Imaging Clin N Am* 2013; 21: 791–812.
157. Raschbaum GR, Bleacher JC, Grattan-Smith JD, Jones RA. Magnetic resonance imaging–guided laparoscopic-assisted anorectoplasty for imperforate anus. *J Pediatr Surg* 2010; 45: 220–233.
158. Eltomey MA, Donnelly LF, Emery KH et al. Postoperative pelvic MRI of anorectal malformations. *AJR Am J Roentgenol* 2008; 191: 1469–1476.
159. Duigenan S, Gee MS. Imaging of pediatric patients with inflammatory bowel disease. *AJR Am J Roentgenol* 2012; 199: 907–915.
160. Masselli G, Gualdi G. MR imaging of the small bowel. *Radiology* 2012; 264: 333–348.

161. Gee MS, Minkin K, Jsu M et al. Prospective evaluation of MR enterography as the primary imaging modality for pediatric Crohn disease assessment. *AJR Am J Roentgenol* 2011; 197: 224–231.
162. Sinha R, Verma R, Verma S, Rajesh A. MR enterography of Crohn disease: Part 2, Imaging and pathologic findings. *AJR Am J Roentgenol* 2011; 197: 8085.
163. Tolan DJM, Greenhalgh R, Zealley IA et al. MR enterographic manifestations of small bowel Crohn disease. *RadioGraphics* 2010; 30: 367–338.
164. Baldisserotto M, Valduga S, da Cunha CF. MR imaging evaluation of the normal appendix in children and adolescents. *Radiology* 2008; 249: 278–284.
165. Cobben L, Groot I, Kingma L et al. A simple MRI protocol in patients with clinically suspected appendicitis: Results in 138 patients and effect on outcome of appendectomy. *Eur Radiol* 2009; 19: 1175–1183.
166. Herliczek TW, Swenson DW, Mayo-Smith WW. Utility of MRI after inconclusive ultrasound in pediatric patients with suspected appendicitis: Retrospective review of 60 consecutive patients. *AJR Am J Roentgenol* 2013; 200: 969–973.
167. Johnson AK, Filippi CG, Andrews T et al. Ultrafast 3-T MRI in the evaluation of children with acute lower abdominal pain for the detection of appendicitis. *AJR Am J Roentgenol* 2012; 198: 1424–1430.
168. Moore MM, Brian JM, Methratta ST et al. MRI for clinically suspected pediatric appendicitis: An implemented program. *Pediatr Radiol* 2012; 42: 1056–1063.
169. Nitta N, Takahashi M, Furukawa A et al. MR imaging of the normal appendix and acute appendicitis. *J Magn Reson Imaging* 2005; 21: 156–65.
170. Krishnamurthy R, Muthupillai R, Chung T. Pediatric body MR angiography. *Magn Reson Imaging Clin N Am* 2009; 17: 133–144.
171. Strouse PJ. Magnetic resonance angiography of the pediatric abdomen and pelvis. *MR Clin N Am* 2002; 10; 345–361.
172. Vellody R, Liu PS, Sada DM. Magnetic resonance angiography of the pediatric abdomen and pelvis: Techniques and imaging findings. *Magn Reson Imaging Clin N Am* 2013; 21: 843–860.
173. Scuccimarri R. Kawasaki disease. *Pediatri Clin N Am* 2012; 59: 424–445.
174. Weiss P. Pediatric vasculitis. *Pediatr Clin N Am* 2012; 59: 407–423.
175. Macartney CA, Chan AK. Thrombosis in children. *Semin Thromb Hemost* 2011; 37: 763–771.
176. Epelman M, Chikwava KR, Chauvin N et al. Imaging of pediatric ovarian neoplasms. *Pediatr Radiol* 2011; 41: 1085–1099.
177. Paj DR, Ladno-Torres MF. Magnetic resonance imaging of pediatric pelvic masses. *Magn Reson Imaging Clin N Am* 2013; 21: 751–772.
178. Siegel MJ, Hoffer FA. Magnetic resonance imaging of nongynecologic pelvic masses in children. *Magn Reson Imaging Clin N Am* 2002; 10: 325–344.
179. Siegel MJ. Magnetic resonance imaging of the adolescent female pelvis. *Magn Reson Imaging Clin N Am* 2002; 10: 303–324.
180. Epelman M, Dinan D, Gee MS et al. Mullerian duct and related anomalies in children and adolescents. *Magn Reson Imaging Clin N Am* 2013; 21: 773–790.
181. Kanso HN, Hachern K, Aoun NJ et al. Variable MRI findings in ovarian functional hemorrhagic cysts. *J Magn Reson Imaging* 2006; 24: 356–361.
182. Lack EE, Youong RH, Scully RE. Pathology of ovarian neoplasms in childhood and adolescence. *Pathol Ann* 1992; 27: 281–356.
183. Olson TA, Schneider DT, Perlman EJ. Germ cell tumors. In: Pizzo PA, Poplack DG et al., eds. *Principles and Practice of Pediatric Oncology*. 6th ed. Philadelphia, PA: Lippincott Williams & Wilkins, 2011; 1045–1067.
184. Scully RE, Young RH, Clement PB. *Atlas of Tumor Pathology: Tumors of the Ovary, Maldeveloped Gonads, Fallopian Tube, and Broad Ligament*. Washington, DC: Armed Forces Institute of Pathology, 1998.
185. Quillin SP, Siegel MJ. CT features of benign and malignant teratomas in children. *J Comput Assist Tomogr* 1992; 16: 722–726.
186. Shah RU, Lawrence C, FIckenscher KA et al. Imaging of pediatric pelvic neoplasm. *Radiol Clin N Am* 2011; 49: 729–748.
187. Jung SE, Rha SE, Lee JM et al. CT and MRI findings of sex cord-stromal tumor of the ovary. *AJR Am J Roentgenol* 2005; 185: 207–215.
188. Wexler LH, Meyer WH, Helman LJ. Rhabdomyosarcoma. In: Pizzo PA, Poplack DG, eds. *Principles and Practice of Pediatric Oncology*. Philadelphia, PA: Lippincott Williams & Wilkins, 2011; 923–953.
189. Argons GA, Wagner BJ, Lonergan GJ et al. From the archives of the AFIP. Genitourinary rhabdomyosarcoma in children: Radiologitc-pathologic correlation. *RadioGraphics* 1997; 17: 919–937.
190. Wu JU, Snyder JM. Pediatric urologic oncology: Bladder, prostate, testis. *Urol Clin N Am* 2004; 31: 619–627.
191. Grimsby GJ, Ritchey ML. Pediatric urologic oncology. *Pediatr Clin N Am* 2012; 59: 947–959.
192. Kocaoglu M, Frush DP. Pediatric presacral masses. *RadioGraphics* 1006; 23: 295–303.
193. Altman RP, Randoph JG, Lilly JR. Sacrococcygeal teratoma: American Academy of Pediatrics Surgical Section Survey-1973. *J Pediatr Sur* 1974; 9: 369–398.
194. Kanemoto K, Hayashi Y, Kojima Y et al. Accuracy of ultrasonography and magnetic resonance imaging in the diagnosis of nonpalpable testis. *Int J Urol* 2005; 12: 668–672.
195. Krishnaswami S, Fonnesbeck C, Penson D, McPheeters ML. Magnetic resonance imaging for locating nonpalpable undescended testicles: A meta-analysis. *Pediatrics*. 2013; 131: e1908–e1916.
196. Kantarci M, Doganay S, Yalcin A et al. Diagnostic performance of diffusion-weighted MRI in the detection of nonpalpable undescended testes: Comparison with conventional MRI and surgical findings. *AJR Am J Roentgenol* 2010; 195: W268–W273.
197. Jaramillo D, Laor T. Pediatric musculoskeletal MRI. Basic principles to optimize success. *Pediatr Radiol* 2008; 38: 379–391.

198. Laor T, Chung T, Hoffer FA, Haramillo D. Musculoskeletal magnetic resonance imaging: How we do it. *Pediatr Radiol* 1996; 26: 695–700.
199. Vogler JB III, Murphy WA. Bone marrow imaging. *Radiology* 1998; 168: 679–693.
200. Siegel MJ, Luker GD. Bone marrow imaging in children. *MRI Clin N Am* 1996; 4: 771–796.
201. Siegel MJ. MR imaging of paediatric haematologic bone marrow disease. *J Hong Kong Coll Radiol* 2000; 3: 38–50.
202. Darge K, Jaramillo D, Siegel MJ. Whole body MRI in children: Current status and future applications. *Eur J Radiol* 2008; 68: 289–298.
203. Kellenberger CJ, Miller SF, Khan M et al. Initial experience with FSE STIR whole-body MR imaging for staging lymphoma in children. *Eur Radiol* 2004; 14: 1829–1841.
204. Mazumdar A, Siegel MJ, Narra V, Luchtman-Jones L. Whole-body fast inversion recovery NR imaging of small cell neoplasms in pediatric patients. *AJR Am J Roentgenol* 2002; 179: 1261–1266.
205. Padhani AR, Koh D-M, Collins DJ. Whole-body diffusion-weighted MR imaging in cancer: Current status and research directions. *Radiology* 2011; 261: 700–718.
206. Punwani S, Taylor SA, Bainbridge A et al. Pediatric and adolescent lymphoma: Comparison of whole-body STIR half-Fourier RARE MR imaging with an enhanced PET/CT reference for initial staging. *Radiology* 2010; 255: 182–190.
207. Siegel MJ, Acharyya S, Hoffer FA et al. Whole-body MRI in the staging of pediatric malignancies: Results of the American College of Radiology Imaging Network 6660 Trial. *Radiology* 2013; 266: 599–609.
208. Hawkins DS, Bolling T, Dubois S et al. Ewing Sarcoma. In: Pizzo PA, Poplack DG et al., eds. *Principles and Practice of Pediatric Oncology*, 6th ed. Philadelphia, PA: Lippincott Williams & Wilkins, 2011; 987–1014.
209. Gorlick R, Bielack S, Teot L et al. Osteosarcoma: Biology, diagnosis, treatment and remaining challenges. In: Pizzo PA, Poplack DG et al., eds. *Principles and Practice of Pediatric Oncology*, 6th ed. Philadelphia, PA: Lippincott Williams & Wilkins, 2011: 1015–1044.
210. Meyer JS, Dormans JP. Differential diagnosis of pediatric musculoskeletal masses. *Magn Reson Imaging Clin N Am* 1998; 6: 561–577.
211. Meyer JS, Nadel HR, Marina N et al. Imaging guidelines for children with Ewing sarcoma and osteosarcoma: A report from the Children's Oncology Group Bone Tumor Committee. *Pediatr Blood Cancer* 2008; 51: 163–170.
212. Mar WA, Taljanovic MS, Bagatell R et al. Update on imaging and treatment of Ewing sarcoma family tumors: What the radiologist needs to know. *J Comput Assist Tomogr*. 2008; 32: 108–118.
213. Suresh S, Saifuddin A. Radiological appearances of appendicular osteosarcoma: A comprehensive pictorial review. *Clin Radiol* 2007; 62: 314–323.
214. Azouz EM, Saigal G, Rodriguez MM, Podda A. Langerhans' cell histiocytosis: Pathology, imaging and treatment of skeletal involvement. *Pediatr Radiol*. 2005; 35: 103–115.
215. Goo HW, Yang DH, Ra YS et al. Whole-body MRI of Langerhans cell histiocytosis: Comparison with radiography and bone scintigraphy. *Pediatr Radiol* 2006; 36: 1019–1031.
216. Herman TE, Siegel MJ. Langerhans cell histiocytosis: Radiographic images in pediatrics. *Clin Pediatr* (Phila) 2009; 48: 228–231.
217. Hoover KB, Rosenthal DI, Mankin H. Langerhans cell histiocytosis. *Skeletal Radiol* 2007; 36: 95–104.
218. Laffan EE, Ngan B-Y, Navarro OM. Pediatric soft-tissue tumors and pseudotumors: MR imaging features with pathologic correlation. 2. Tumors of fibroblastic/myfibroblastic, so-called fibrohistiocytic, muscular, lymphomatous, neurogenic, hair matrix, and uncertain origin. *RadioGraphics* 2009; 29: 906–941.
219. Navarro OM, Laffan EE, Ngan BY. Pediatric soft-tissue tumors and pseudotumors: MR imaging features with pathologic correlation. I. Imaging approach, pseudotumors, vascular lesions, and adipocytic tumors. *RadioGraphics* 2009; 29: 887–906.
220. Siegel MJ. MRI of pediatric soft tissues masses. *MR Clin N Am* 2002; 39: 701–720.
221. Wu JS. Soft-tissue tumors and tumorlike lesions: A systematic imaging approach. *Radiology* 2009; 253: 297–316.
222. Van Rijn RR, Wilde JCH, Bras J et al. Imaging findings in noncraniofacial childhood rhabdomyosarcoma. *Pediatr Radiol* 2008; 38: 616–634.
223. Viry F, Orbach D, Klijanienko J et al. Alveolar soft part sarcoma-radiologic patterns in children and adolescents. *Pediatr Radiol* 2013; 43: 1174–1181.
224. Behr GG, Johnson C. Vascular anomalies: Hemangioma and beyond: Part 1, Fast-flow lesions. *AJR Am J Roentgenol* 2013; 200: 414–422.
225. Behr GG, Johnson C. Vascular anomalies: Hemangioma and beyond-Part 2, Slow-flow lesions. *AJR Am J Roentgenol* 2013; 200: 423–436.
226. Flors L, Leiva-Salina C, Maged IM et al. MR imaging of soft-tissue vascular malformations: Diagnosis, classification and therapy follow-up. *RadioGraphics* 2011; 31: 1321–1340.
227. Kollipara R, Dinneen L, Rentas KE et al. Current classification and terminology of pediatric vascular anomalies. *AJR Am J Roentgenol* 2013; 201: 1124–1135.
228. Lowe L, Marchant T, Rivard D et al. Vascular malformations: Classification and terminology the radiologist needs to know. *Semin Roentgenol* 2012; 47: 106–117.
229. Mulliken JB, Fishman SJ, Burrows PE. Vascular anomalies. *Curr Probl Surg* 2000; 37: 517–584.
230. Siegel MJ, Glazer HS, St. Amour TE et al. Lymphangiomas in children: MR imaging. *Radiology* 1989; 170: 467–470.
231. Pineda C, Vargas A, Rodriguez AV. Imaging of osteomyelitis: Current concepts. *Infect Dis Clin N Am* 2006; 20: 789–825.
232. Averill LW, Hernandez A, Gonzalez L, Pena AH, Jaramillo D. Diagnosis of osteomyelitis in children: Utility of fat-suppressed contrast-enhanced MRI. *AJR Am J Roentgenol* 2009; 192: 1232–1238.
233. Falip C, Alison M, Bourtyr N et al. Chronic recurrent multifocal osteomyelitis (CRMO): A longitudinal case series review. *Pediatr Radiol* 2013; 43: 355–375.

234. Fritz J, Tzaribatchev N, Claussen CD et al. Chronic recurrent multifocal osteomylelitis: Comparison of whole-body MR imaging with radiography and correlation with clinical and laboratory data. *Radiology* 2009; 252: 842–851.
235. Khanna G, Sato TS, Ferguson P. Imaging of chronic recurrent multifocal osteomyelitis. *RadioGraphics* 2009; 29: 1159–1177.
236. Iyer RS, Thapa MM, Chew FS. Chronic recurrent multifocal osteomyelitis: Review. *AJR Am J Roentgenol* 2011; 196: S87–S91.
237. Johnson K. Imaging of juvenile idiopathic arthritis. *Pediatr Radiol* 2006; 36: 743–758.
238. Malattia C, Damasio MB, Basso C. Dynamic contrast-enhanced magnetic resonance imaging in the assessment of disease activity in patients with juvenile idiopathic arthritis. *Rheumatology* 2010; 49: 178–185.
239. Kim HK, Zbojniewicz AM, Merrow AC et al. MR findings of synovial disease in children and young adults, Part 1. *Pediatr Radiol* 2011; 41: 495–511.
240. Kim HK, Zbojniewicz AM, Merrow AC et al. MR findings of synovial disease in children and young adults, Part 2. *Pediatr Radiol* 2011; 41: 512–524.
241. Chan WP, Liu G-C. MR imaging of primary skeletal muscle diseases in children. *AJR Am J Roentgenol* 2002; 179: 989–997.
242. Marden FA, Connolly AM, Siegel MJ, Rubin DA. Compositional analysis of muscle in boys with Duchenne muscular dystrophy using MR imaging. *Skeletal Radiol* 2005; 34: 140–148.
243. Jaimes C, Chauvin NA, Delgado J, Hramillo D. MR imaging of normal epiphyseal development and common epiphyseal disorders. *RadioGraphics* 2014; 34: 449–471.
244. Pai DR, Strouse PJ. MRI of the pediatric knee. *AJR Am J Roentgenol* 2011; 196: 1019–1027.
245. Zobel MS, Borrello JA, Siegel MJ et al. Pediatric knee. MR imaging pattern of injuries in the immature skeleton. *Radiology* 1994; 190: 397–401.
246. Dillman JR, Hernandez RJ. MRI of Legg-Calve-Perthes disease. *AJR Am J Roentgenol* 2009; 193: 1394–1407.
247. DeSmet AA, Risher DR, Graf BK et al. Osteochondritis disecans of the knee: Value of MR imaging in determining lesion stability and the presence of articular cartilage defects. *AJR Am J Roentgenol* 1990; 155: 549–553.
248. Flynn JM, Kocher MS, Ganley TJ. Osteochondiritis dissecans of the knee. *J Pediatr Orthop* 2004; 24: 434–432.

# 18

# *Fetal MRI*

**Francisco Sepulveda, Kelsey Budd, Peter Brugge, and Daniela Prayer**

**CONTENTS**

## 18.1 Introduction

The use of magnetic resonance imaging (MRI) for the prenatal diagnosis of fetal malformations was first reported by Smith in 1983 [1], although it was only in the 1990s that new technological advances including the development of fast sequences allowed this method to become incorporated into clinical practice. During the last decade, the use of this technique has increased exponentially due to the improved resolution and greater accessibility to the pregnant population. MRI has been recommended as an adjunct method to evaluate the fetal anatomy, particularly when suspected or established fetal anomalies have been detected by prenatal ultrasound [2]. Currently, fetal MRI requires highly expensive equipment and is carried out in a dedicated radiology facility only for specific medical indications, primarily the evaluation of an ultrasound-detected abnormality.

In this chapter, we will review the main indications and technique for fetal MRI, as well as the image characteristics of normal and abnormal fetal anatomy as determined by this method, with special emphasis on the central nervous system (CNS).

## 18.2 Indications

Indications for fetal MRI are triggered by the suspicion of a fetal abnormality and can be classified into general and specific. There are three general indications: maternal obesity (increase in body mass index) preventing the delineation of normal or abnormal fetal anatomy, oligohydramnios or anhydramnios which frequently are associated with renal abnormalities and suboptimal ultrasound examination of the fetus due to reduced amniotic fluid and a persistently unfavorable fetal position preventing the examination of a given fetal organ. MRI is specifically indicated when a severe single anomaly or a complex fetal anomaly is detected with ultrasound, and MRI will likely better characterize the finding or assess for other associated anomalies. The latter indications may include cerebral or non-cerebral malformations [3].

Cerebral indications for fetal MRI include congenital malformations or acquired lesions. A large number of congenital malformations can benefit from an MRI study. For instance, the prenatal detection of ventriculomegaly is one of the most frequent indications of fetal MRI with the purpose of detecting associated anomalies. Other examples are abnormal cortical folding (e.g., lissencephaly), spina bifida with Chiari II malformation, and Dandy–Walker malformation. On the other hand, acquired brain lesions that can be studied with MRI are mainly intraventricular hemorrhage and ischemic lesions.

Extra-CNS indications are analyzed according to anatomical location. In the face and neck, the main indications are orofacial defects, cervical tumors (e.g., lymphangioma, teratoma, and goiter), and tracheal and esophageal malformations. In the thorax, MRI is used to characterize solid, cystic, and mixed intrapulmonary lesions and to perform lung volumetry in the cases of fetal lung masses such as cystic adenomatoid malformation, pulmonary sequestration, and congenital diaphragmatic hernia. Gastrointestinal malformations include intestinal malrotation and bowel atresia/stenosis, whereas genitourinary malformations include horseshoe kidney, ectopic kidneys, renal hypoplasia agenesia, hydronephrosis, and cloacal dysgenesis. Regarding the musculoskeletal system, MRI may be indicated to characterize complex musculoskeletal syndromes such as skeletal dysplasia and skeletal effects of CNS alterations. An additional non-fetal indication is the antenatal study of abnormal placental implantation such as placenta accreta [3].

## 18.3 Technique

Although there are no formal contraindications for fetal MRI in early pregnancy, this technique is usually performed after 18 gestational weeks (GW), because during this period this technique can provide useful information for pregnancy management and fetal prognosis. The final diagnosis of the fetal pathology must be made as soon as possible, but it must be considered that it is better to study certain pathologies at a specific moment during pregnancy. For example, cortical folding alterations (e.g., lissencephaly), may best be characterized at a gestational age of at least 28 weeks which allows sufficient development of the main folds and the size of the fetus is adequate enough to obtain good image quality.

Regarding equipment, both 1.5T or 3T can be used. However, higher magnetic fields (3T) show greater susceptibility to movement, producing artifacts and deteriorating image quality. Also, in 3T MRI sequences generally require longer acquisition times. To obtain quality fetal MRI images, surface coils—either a body coil or a cardiac coil—must be used; the latter has had better results as it focuses the signal in a smaller zone.

The pregnant patient can enter either feet first or head first, depending on the technical staff's habit and the doctors' preference. Whenever possible, the patient must be in supine position because the surface coils must be as close to the region of interest as possible, and only when this position is not tolerated, can the patient be in lateral supine, as in this case the coils are located further away from the region of interest.

Given the capacity of current MR equipment to perform ultra-fast sequences, it is not necessary to administer sedatives routinely. Contraindications for fetal MRI are the same as for non-pregnant population.

The MRI study begins with a survey scan image, and then the fetal head and body are studied separately. Primary attention is given to the region of interest, but it is always necessary to assess all fetal anatomy to search for associated anomalies and syndromes. As the fetus could be in continuous movement, the MRI study must be done having the last sequence obtained as a reference. It is important to repeat the study as many times as necessary to obtain symmetrical images for their correct analysis. Regarding image quality, the duration of the sequence is the factor with greater influence as it is the objective to obtain sequences with a good signal-to-noise ratio as fast as possible. To reach a correct final diagnosis, different types of sequences should be used. The main sequences used are $T_2$W, $T_1$W, steady-state free precession (SSFP) sequence, echo-planar sequence, dynamic SSFP, single-shot magnetic resonance cholangiopancreatography and diffusion-weighted imaging (DWI). Additional sequences may be used when required depending on the pathology to be examined, as is the case for fluid-attenuated inversion recovery, diffusion tensor imaging, and spectroscopy. The radiologist must be present as the study is made, and must guide the study according to the clinical question and evaluate which sequences are to be used. Images have a slice thickness ranging from 3 to 6 mm, as in single-shot magnetic resonance cholangiopancreatography and a greater slice thickness as in dynamic SSFP sequences. The gap between slices is generally 0 or 10% of slice thickness, but overlap between slices is possible to achieve thinner slices.

In the images obtained, the face and profile should be systematically analyzed, as well as the cerebral parenchyma and cerebral development, the lungs and lung development in relation with the size and signal intensity, the abdomen with attention to the position of the stomach and meconium distribution, the kidneys regarding their number and position, the characteristics of the external genitalia, the position of the limbs and the number of segments in each one, the evaluation of the umbilical cord and its number of vessels, the assessment of the amount of amniotic fluid, and the position of the placenta [3].

## 18.4 Sequences

*$T_1$W sequence*: This sequence is particularly sensitive to movement, and the acquisition should be performed with breath hold. For this reason, a rapid sequence is required (approximately 20 s). In this sequence, meconium acts as a natural contrast, allowing good visualization of the bowel. This sequence is also useful for the visualization of the pituitary and thyroid glands; these show a hyperintense signal allowing excellent contrast with the surrounding structures (Figure 18.1). The liver can also be studied and shows a hyperintense signal on $T_1$W. In addition, blood-breakdown products such as methemoglobin can be detected and show a hyperintense signal in this sequence.

*$T_2$W sequence*: This allows detailed morphological characterization due to contrast in the liquid within fetal structures and analysis of the fetal surface (e.g., the fetal profile) and CNS. It is also useful to study lung maturation, and is used to calculate lung volume. The fluid-filled organs such as the gallbladder, stomach, trachea, and bladder can be studied in detail. It is used as a complementary sequence in the study of the bowel.

*SSFP sequence*: As with $T_2$W sequence, this sequence also allows for morphological analysis. The fluid is hyperintense and provides excellent contrast in the CNS. This sequence is less sensitive to motion and allows to obtain high resolution images (3 mm) with good signal-to-noise ratio. Vascular structures show hyperintense signal on this sequence and can be studied in detail. It is also the appropriate sequence to study the fetal heart [4].

*Echo-planar sequence*: This sequence allows the study of the fetal skeleton useful to evaluate long bones, spinal defects, and skeletal dysplasias [5]. It is also the most sensitive sequence for the detection of intracerebral hemorrhage due to its high susceptibility to blood-breakdown products.

*Dynamic sequence*: This allows evaluation of fetal movements in utero [6] and thus indirectly evaluates CNS function. Also, it allows the study of the swallowing process in cases of polyhydramnios, esophageal stenosis/atresia, and brainstem abnormalities.

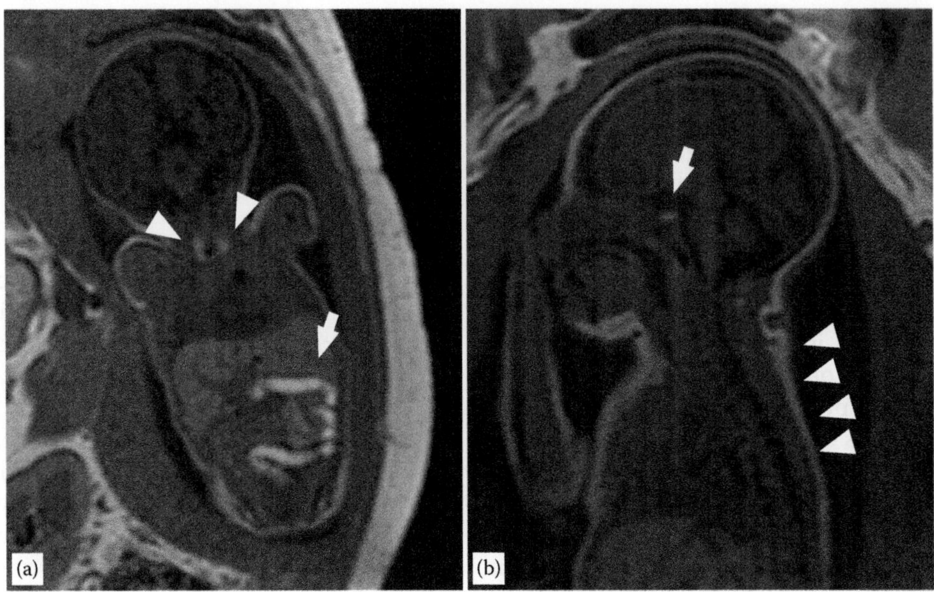


**FIGURE 18.1**
Two different normal fetuses on $T_1$W sequences. (a) Coronal sequence shows a thyroid gland localized on the anterior region of the neck, with hyperintense signal on $T_1$W sequence (arrowhead), also depicts in meconium (arrow). (b) Sagittal sequence shows hyperintense signal of a pituitary gland (arrow) and hyperintense signal of a subcutaneous fatty tissue (arrowheads).

*DWI*: DWI has been used to study the layering of the brain, lung development, and for the assessment of the temporal evolution of brain lesions. It is also useful for the evaluation of the kidneys.

*Diffusion tensor imaging*: This sequence allows the tracking of white matter fibers, which is particularly useful in the case of commissural agenesis, although it is difficult to obtain high-quality images due to its long acquisition time.

*MR spectroscopy*: This sequence allows the study of metabolites in the brain. By evaluating different metabolic peaks (choline, *N*-acetyl aspartate, lactate), it can provide information about the neuronal population and its biochemical production (Figure 18.2) which may help characterize a pathologic condition such as hypoxic-ischemic encephalopathy. At present, the use of this technique is limited due to its long acquisition time.

## 18.5 Fetal CNS Malformations

### 18.5.1 General Considerations

Fetal MRI is playing an increasingly important role in the evaluation of congenital CNS abnormalities and has become the method of choice to characterize congenital malformations as well as acquired conditions [7]. In addition, acute changes can be detected [8,9], and chronic lesions can be visualized, allowing a prognosis estimation [9]. Moreover, the use of MRI has proved to be superior to ultrasound in establishing abnormalities of the posterior fossa, brainstem, corpus callosum, and overall cortical development.

### 18.5.2 Fetal MRI of Brain Development

With fetal MRI it is possible to study white matter development, the brain surface, and cortical layering. In contrast to ultrasound technology, MRI allows a whole view of the brain without acoustic shadowings, making it ideal for this purpose.

In connection with cortical sulcation, gyration may be seen on MRI at 18 GW with the shallow bitemporal indentation of the Sylvian fissures. The first two fissures to appear are the parieto-occipital and calcarine fissures (Figure 18.3). The parieto-occipital fissure separates the parietal from the occipital lobe. This fissure is visible at 18–19 weeks and must be present at 23 weeks [10]. The calcarine fissure is seen on the medial surface of the occipital lobe. It starts from the medial part of the parieto-occipital fissure and extends posterior to the occipital lobe [10]. This fissure is first seen at 18–19 weeks and is present in 75% of the fetuses at 24–25 weeks. Subsequently, the cingulate sulcus, central sulcus, and convexity sulci develop are seen. The cingulate sulcus is seen at the medial surface of the brain. It begins below the anterior end of the corpus callosum and runs upward and forward, nearly

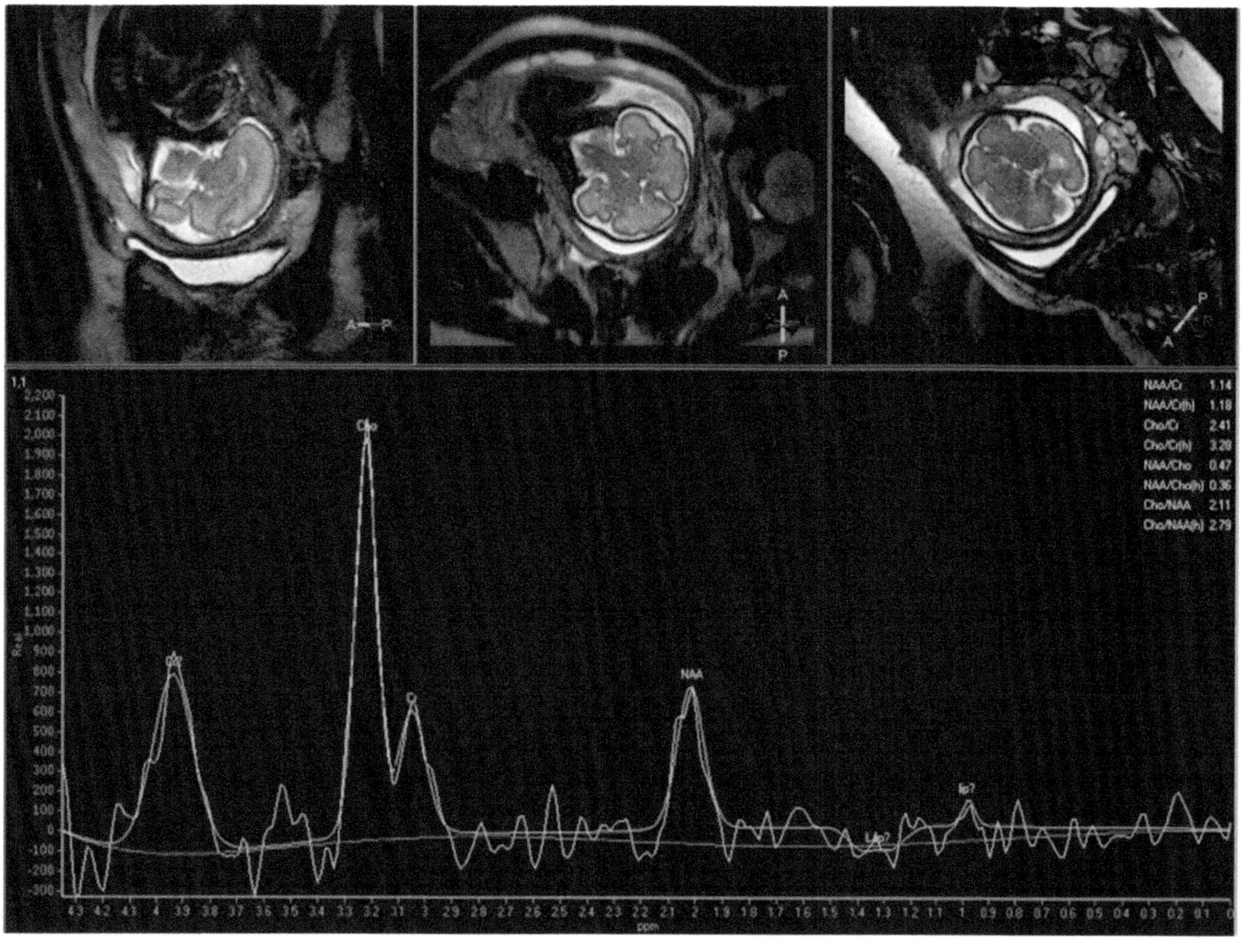

**FIGURE 18.2**
Single voxel spectroscopy whit long echo time of 144 ms in a normal fetus at 25 GW. Choline (Cho) dominates the spectrum. *N*-acetyl aspartate (NAA) also appears in the spectrum but with a smaller peak.

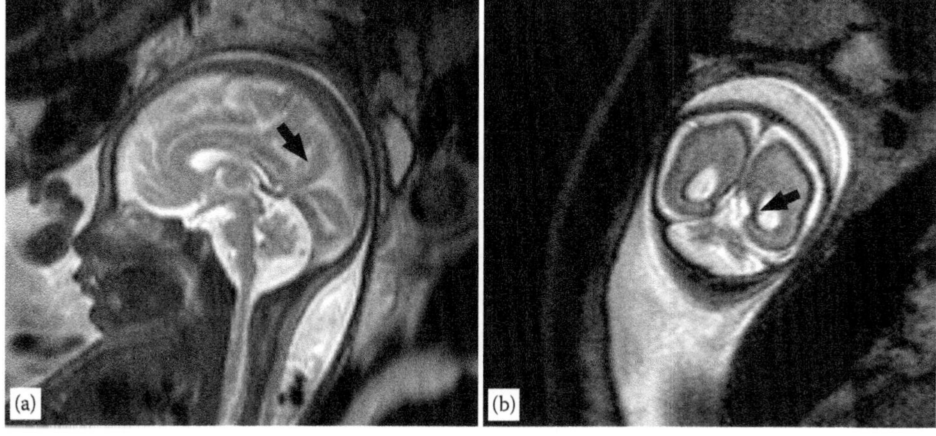


**FIGURE 18.3**
Two different fetuses at 30 GW on sagittal (a) and at 20 GW on coronal (b) $T_2$W sequences. (a) Normal parieto-occipital fissure (arrow). (b) Normal calcarine fissure (arrow).

paralleling the rostrum, and then ascends in parallel with the body of the corpus callosum to the superomedial border of the hemisphere [10]. It is first appreciated with MRI at 24–25 weeks. The last two sulci to appear are the central sulcus and the convexity sulci (Figure 18.4). These sulci can be seen on the lateral surface of the cerebral hemispheres. The convexity sulci include the central (Rolandic) sulcus, superior temporal sulcus, and any sulcus on the lateral hemispheric surface. In a normal fetus, the central sulcus is identified on MRI at 26–27 weeks [10]. The Sylvian fissure and insula demonstrate a characteristic pattern of development. In early pregnancy, the Sylvian fossa appears as a smooth-margined indentation on the lateral surface of the cerebral hemisphere. A suspected sulcal abnormality should be assessed ideally after 28 weeks, when most of the sulci should be visible.

Fetal MRI can also display information about both gross anatomical structures and histological microstructure [11]. In this way, it is possible to differentiate layers in the brain mantle (Figure 18.5). At this histological microstructure, it is possible to find ventricular, periventricular, intermediate, subplate, and cortical plate areas. The ventricular zone is seen as hypointense on $T_2$W and hyperintense on $T_1$W as it contains the germinal matrix. It also presents a hyperintense signal on DWI with low apparent diffusion coefficient values due to the cellularity and the microvascularity present in the area. The periventricular zone present in the frontal region is easily distinguishable from the adjacent areas. The intermediate zone is seen on $T_1$W and $T_2$W as hypointese relative to its external layer (subplate layer). The subplate zone is hyperintense on $T_2$W, due to its high water content, and hypointense on $T_1$W, while on DWI sequence, it is hypointense. The cortical plate is the most external layer, it is hypointense on $T_2$W and hyperintense on $T_1$W relative to the subplate. On DWI the cortical plate appears hyperintense (anisotropic behavior). The layered appearance of the brain persists until around 28 weeks, when $T_2$W intensity of the subplate zone decreases and becomes isointense to the intermediate zone, and thus cannot be differentiated (Figure 18.5b). When this loss of differentiation occurs earlier (20–25 weeks), it indicates that lamination pathology is in progress [11].

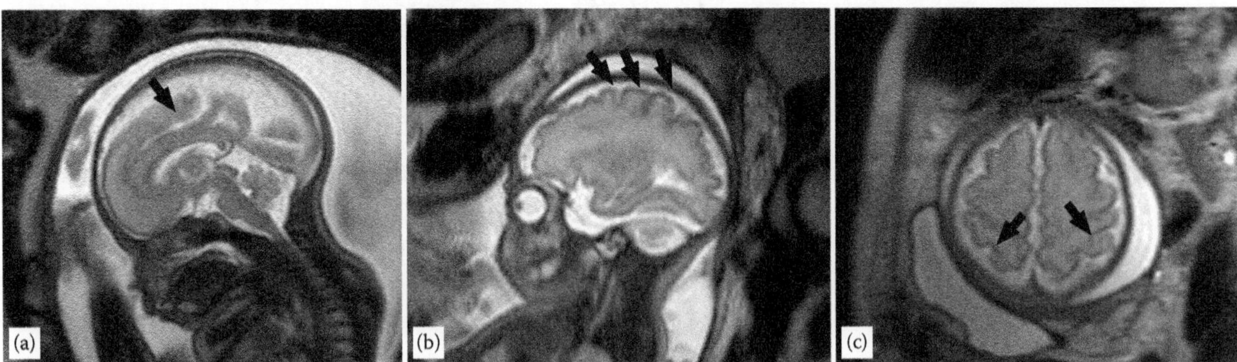


**FIGURE 18.4**
Normal fetus at 29 GW. (a) Middle line sagittal $T_2$W sequence shows the cingulate sulcus (arrow). (b) Parasagittal $T_2$W sequence shows convexity sulci (arrows). (c) Axial $T_2$W sequence shows a central sulcus (arrows).

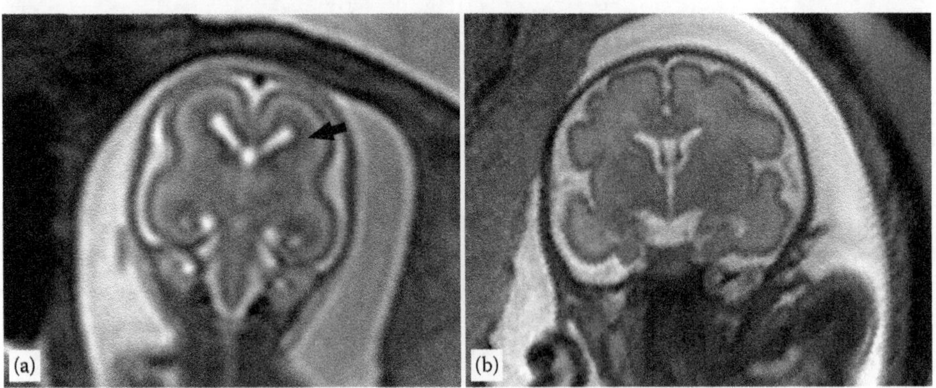


**FIGURE 18.5**
Fetuses at 20 GW (a) and 32 GW (b) on coronal $T_2$W sequences. (a) The subplate zone and intermediate zone have different signal intensity (arrow). (b) Subplate becomes isointense and cannot be differentiated with intermediate zone.

### 18.5.3 Open Neural Tube Defects

#### *18.5.3.1 Acrania and Anencephaly Sequence*

Acrania is a lethal condition characterized by the absence of the cranial vault. It can be diagnosed with ultrasound from week 11 and usually evolves to anencephaly, in which the cerebral tissue is fully destroyed as the exposed brain tissue is disrupted, leaving only variable amounts of angiomatous stroma at the base of the skull [12,13]. In a small proportion of cases, some disorganized cerebral tissue persists, a condition called *exencephaly* [12,13]. This is frequently seen in association with amniotic bands where the head is attached to the placenta, preventing fetal head movement and leading to brain destruction. The invariably lethal outcome of this condition and the confident diagnosis provided by ultrasound make the use of MRI unnecessary in the prenatal management of these conditions.

#### *18.5.3.2 Cephalocele*

Cephalocele is a condition characterized by protrusion of brain tissue (encephalocele) or meninges (cranial meningocele) (Figure 18.6) through a defect in the skull, frequently at the level of the occipital bone [12,13]. When the protruded tissue contains a portion of the ventricles, it is called *meningoencephalocystocele*. The diagnosis of small cranial defects might be difficult by ultrasound, in particular when the fetus position is not favorable. In these cases, MRI represents the ideal complementary method, not only for identifying the cephalocele, but also for its ability to characterize associated abnormalities, because of its high sensitivity, and provides important prognostic information for perinatal management. Fetal MRI is also useful in differentiating small cephaloceles from subcutaneous skull cysts and cranial hemangiomas [14]. The herniated neural tissue may show a heterogeneous signal due to necrosis presence, neoformation vessels, and calcifications.

Cephaloceles can be associated with different syndromes and other abnormalities, including Chiari type III, Von Voss–Cherstovoy syndrome, frontonasal dysplasia, iniencephaly, Meckel–Gruber syndrome, and Knobloch syndrome. Agenesis of the corpus callosum is an associated malformation in up to 80% of these cases, and although less frequently can be associated with cleft lip and palate, microphthalmia, hypotelorism, and hypertelorism [15]. The MRI report should be systematic in reporting the location of the lesion, the contents of the sac, and associated anomalies.

#### *18.5.3.3 Spina Bifida and Chiari II Malformation*

Spina bifida is a fusion defect of the posterior arches of the spine. It is the most common neural tube defect with a reported incidence of 1 in every 1000 deliveries. Chiari II syndrome is associated with open spinal defects, myelomeningocele being the main type and exceptionally found in closed defects. Chiari II malformation has a small posterior fossa with a descended tentorium associated with caudal displacement of the cerebellar tonsils, vermis, and brainstem. This syndrome is present clinically as a brainstem dysfunction (difficulty in swallowing, polyhydramnios, and apnea).

Myelomeningocele is strongly associated with neurodevelopmental sequelae, lower-limb paralysis, and urinary and rectal incontinence. Two-dimensional ultrasound is highly specific and sensitive as a screening method for this condition, although visualization of the posterior fossa is usually suboptimal due to limited evaluation

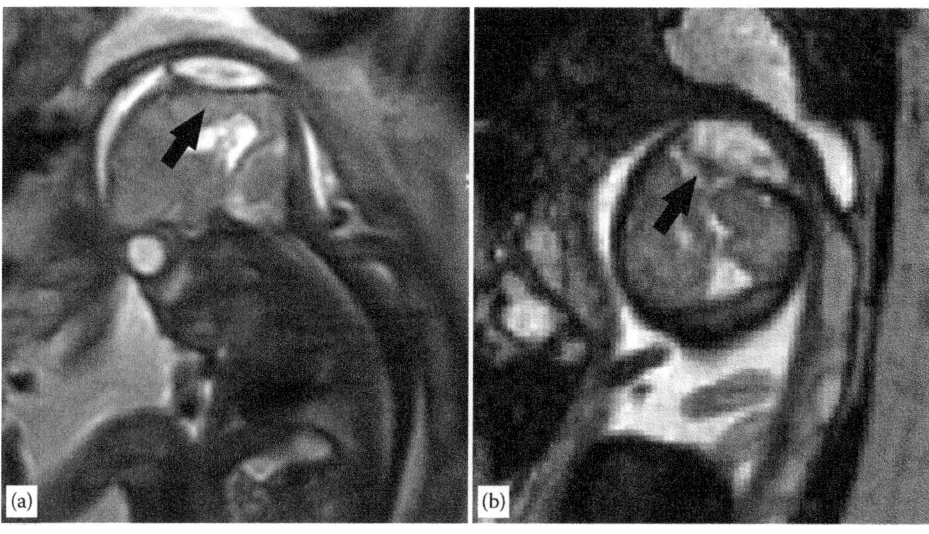


**FIGURE 18.6**
Fetus at 33 GW with cranial meningocele. Sagittal $T_2$W (a) and coronal SSFP (b) sequences show a meninges and cerebrospinal fluid protruding through a cranial defect (arrow) on the left parietal bone.

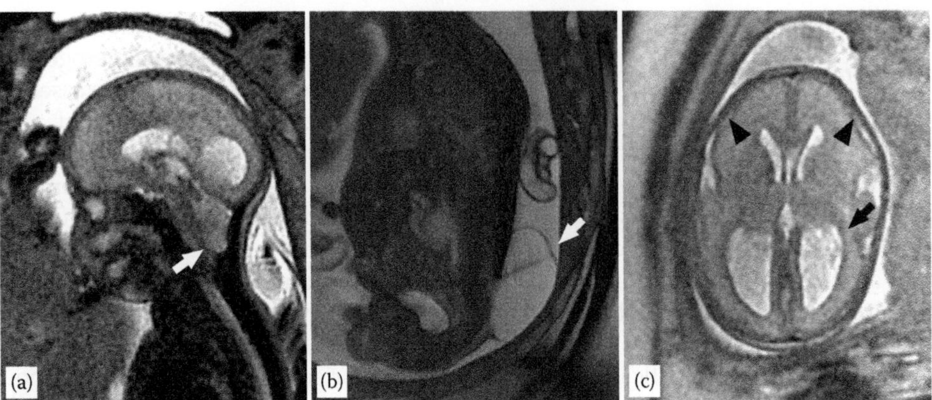

**FIGURE 18.7**
Fetus at 34 GW with a spina bifid—Chiari II malformation. (a) Sagittal SSFP sequence shows a small posterior fosa, lower implantation tentorium, and caudal displacement of the cerebellar tonsils (arrow). (b) Sagittal SSFP sequence shows a spine defect with myelomeningocele (arrow). (c) Axial $T_2$W sequence shows an associated ventriculomegaly (black arrow) and scalloping of the frontal bones or lemon sign (arrowhead).

of the cerebellar and brainstem size and shape, and the degree of cerebellar displacement into the spinal canal. Fetal MRI as a complementary method provides superior evaluation of the spinal defect together with the overall size of the posterior fossa, location of the tentorium, brainstem development, and degree of cerebellar herniation (Figure 18.7a). It can also depict subtle associated anomalies, such as beaking of the tectum, heterotopias, small subarachnoid space, and callosal dysgenesis [12]. In addition, both the spinal canal and cord can be examined thoroughly, targeting the exact level of the defect, determining the number of vertebrae affected, and detecting a tethered spinal cord if present [16–20].

The first step in the analysis of the congenital spine defect is the classification between closed defects (covered by skin) or open defects (not covered by skin). In the case of open defects, the differential diagnoses, depending on the tissue content of the sac which protrudes through the spinal canal, include myelomeningocele (meninges, neural tissue, and cerebrospinal fluid) (Figure 18.7b), myelocele (meninges and cerebrospinal fluid), hemimyelocele (affects one of the hemicords in diastematomyelia cases), and hemimyelomeningocele [15]. On the other hand, closed defects have a long list of anomalies and could be associated with a subcutaneous mass or cutaneous stigma. In the cases with a subcutaneous mass, the differential diagnoses mainly include lipomyelomeningocele, lipomyelocele, meningocele, and terminal myelocystocele [15].

### 18.5.4 Non-Open Neural Tube Defects

#### *18.5.4.1 Holoprosencephaly*

Holoprosencephaly is a condition characterized by abnormal cleavage of the prosencephalon, resulting in variable degrees of fusion of the lateral cerebral ventricles and thalami, absent midline structures (such as corpus callosum and septum pellucidum) and maldevelopment of the middle segment of the face [12,13]. It is classified into alobar, semilobar, and lobar types. The alobar type (Figure 18.8) is the most common and

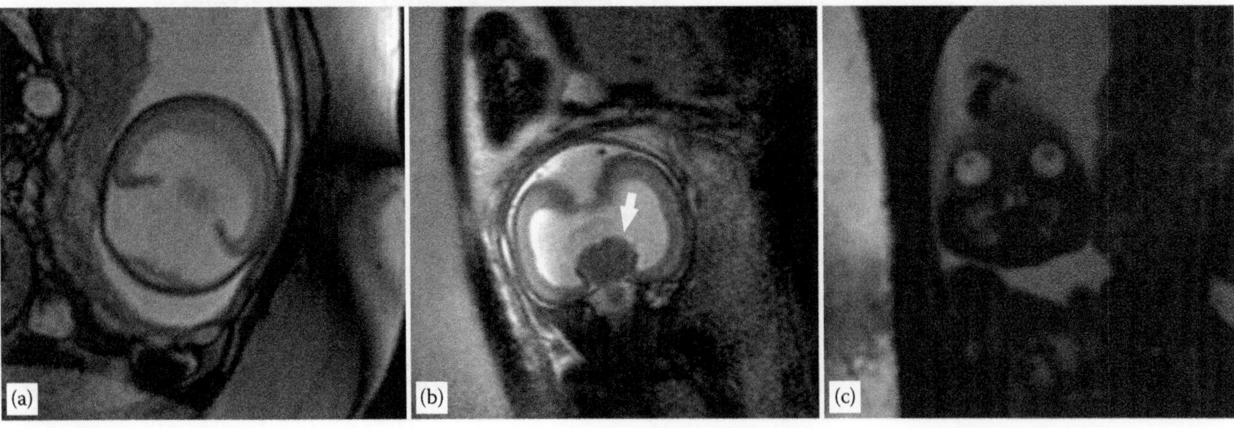

**FIGURE 18.8**
Fetus at 24 GW with alobar holoprosencephaly. (a) Axial SSFP sequence shows a monoventricular cerebral cavity and absent interhemispheric fissure. (b) Coronal $T_2$W shows fused thalami (arrow). (c) Coronal SSFP sequence shows a hypertelorism.

severe form and is associated with the most severe facial malformations, including cyclopia, hypotelorism, arrhinia, and midfacial cleft. The alobar and semilobar types have characteristic cerebral findings, including an anterior monoventricular cerebral cavity, fused thalami, absent partial, or complete interhemispheric fissure-falx cerebri, and dysgenesis of the corpus callosum. In cases of alobar and semilobar types with partial callosum agenesis, usually the posterior region of the callosum is present, in contrast to the isolated partial agenesis [12,13] where the anterior part of the corpus callosum body is present.

Distinctive findings in the alobar type are the complete absence of the interhemispheric fissure, the inability to distinguish the temporal lobes, and lack of well-formed ventricular temporal horns [15]. In the semilobar type, the interhemispheric fissure is present in the posterior portion, the ventricular temporal horns are rudimentary, and in most cases, the corpus callosum is present only in the posterior portion [15].

On the other hand, the lobar type is the most developed, and the least common type, and therefore, the prenatal diagnosis is often missed on ultrasound. The use of fetal MRI as a complementary method has proved valuable in these less severe forms of the anomaly. The demonstration of failure of cleavage in the anterior part of the frontal lobes, the basal ganglia, and the thalami, as well as the absent cavum septum pellucidum and the fused frontal horns are characteristic findings of lobar holoprosencephaly. Hippocampal malrotation is a frequent finding. Associated conditions that can be found in holoprosencephaly are neural tube defects and anterior cephaloceles, which are more frequently seen in the alobar type [15].

#### 18.5.4.2 Posterior Fossa Malformations

Fetal MRI is useful in evaluating more subtle anomalies of the cerebellar vermis and hemispheres and the brainstem, as well as to determine the exact location of the tentorium. It is also useful for evaluating associated anomalies such as callosal abnormalities and cerebral heterotopias [17,20–23].

By ultrasound, with increasing ossification of the skull, the brainstem is very difficult to examine. In these cases, and when pathology is suspected, MRI is particularly useful because it allows an easy morphological evaluation of the brainstem and its respective measurements.

Throughout pregnancy, the normal cisterna magna should not be greater than 10 mm. Abnormalities of the posterior fossa usually present as either a small or enlarged cisterna magna. A small cisterna magna is typically seen in association with the Arnold–Chiari malformation (Figure 18.7), pontocerebellar hypoplasia, and rhombencephalosynapsis (vermian agenesis with fused cerebellar hemispheres) [24]. Conditions associated with an enlarged cisterna magna include megacisterna magna, vermian hypoplasia (Dandy–Walker variant), and the classic Dandy–Walker malformation [12,13]. An enlarged cisterna magna can be a normal or an abnormal finding, as this can represent a partial vermian agenesis or hypoplastic vermis, which are pathological conditions, or a persisting Blake's pouch cyst, which is a benign finding. Megacisterna magna in isolation is considered a normal variant as the cerebellar vermis and hemispheres are both normal. Dandy–Walker malformation, on the other hand, is always pathological and is characterized by cystic dilatation of the fourth ventricle, vermian agenesis, splitting of the cerebellar hemispheres, severe enlargement of the posterior fossa, and elevation of the tentorium (Figure 18.9). It is frequently associated with hydrocephaly [12,13].

A condition of great importance is persisting Blake's pouch cyst (Figure 18.10), given its excellent prognosis in spite of its worrisome prenatal ultrasound appearance. This must be differentiated from a hypoplastic vermis. It is characterized by an upward rotation of the vermis and an enlarged posterior fossa, but the diameters of the vermis, cerebellum, and brainstem must be normal. It can be associated with ventriculomegaly due to the underdeveloped Luskca–Magendie foramen causing a non-communicating hydrocephalus [14].

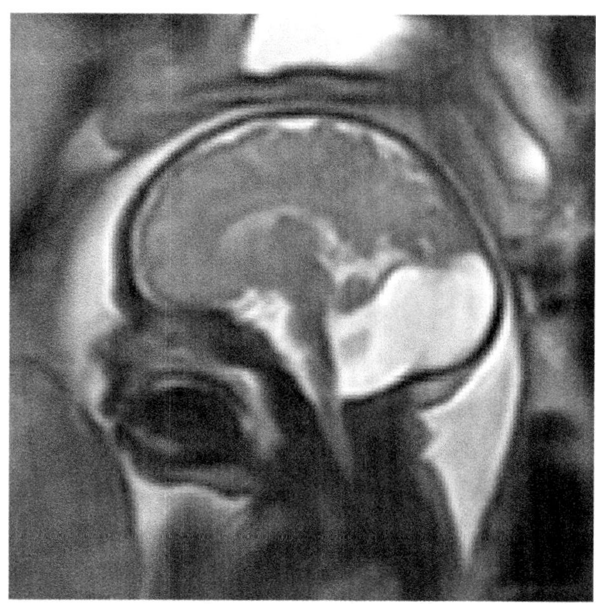

**FIGURE 18.9**
Sagittal $T_2$W sequence in a fetus at 32 GW with Dandy–Walker malformation.

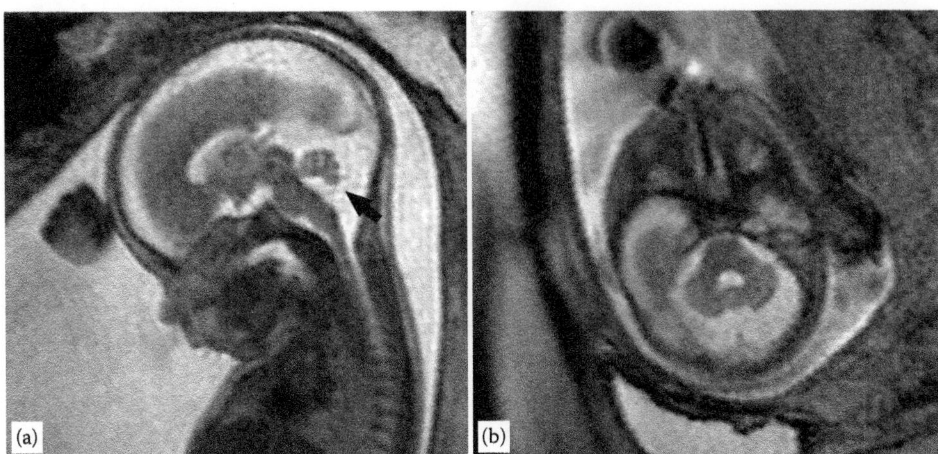

**FIGURE 18.10**
Fetus at 21 GW with persisting Blake's pouch cyst. (a) Sagittal $T_2$W shows an upward rotation of the vermis (arrow) and an enlarged posterior fossa. (b) Axial $T_2$W shows a normal fourth ventricle and cerebellar hemispheres.

#### 18.5.4.3 Commissural Agenesis and Agenesis of the Corpus Callosum

The corpus callosum is a flat, wide bundle of white matter fibers that connects the right and left cerebral hemispheres (Figure 18.11). It is concave in shape and is formed by four anatomical segments (rostrum, genu, body, and splenium) that are fully developed by 18–20 weeks. Agenesis of the corpus callosum (ACC) can be complete or partial (Figure 18.12) and is frequently associated with interhemispheric cysts and lipomas. Certain findings can suggest the diagnosis of ACC, including absent cavum septum pellucidum, colpocephaly (dilatation of the occipital horns with narrow, pointing frontal horns, or *teardrop* ventricles), and parallel orientation of the lateral ventricles. Another clue is the radial arrangement of the cerebral convolutions in both the coronal and sagittal planes, known as the *rising sun* sign.

A significant diagnostic challenge by ultrasound is partial ACC. It can be caused by the absence of one or more segments, particularly the splenium or the posterior part of the body. Cases of dysgenesis, in which the corpus callosum is thinner or smaller than normal, are also difficult to determinate with ultrasound. Detection of these lesions, however, is relatively easy with MRI. Fetal MRI also provides crucial information about associated anomalies such as gray matter heterotopia and cortical dysplastic patterns of the affected fetus with ACC [16,20–23].

Commissural agenesis is complete when there are no white matter fibers communicating between the cerebral hemispheres. Thus both are completely separate and not only the corpus callosum being absent, but also the anterior white and hippocampal commissures. It is very important to distinguish between complete and

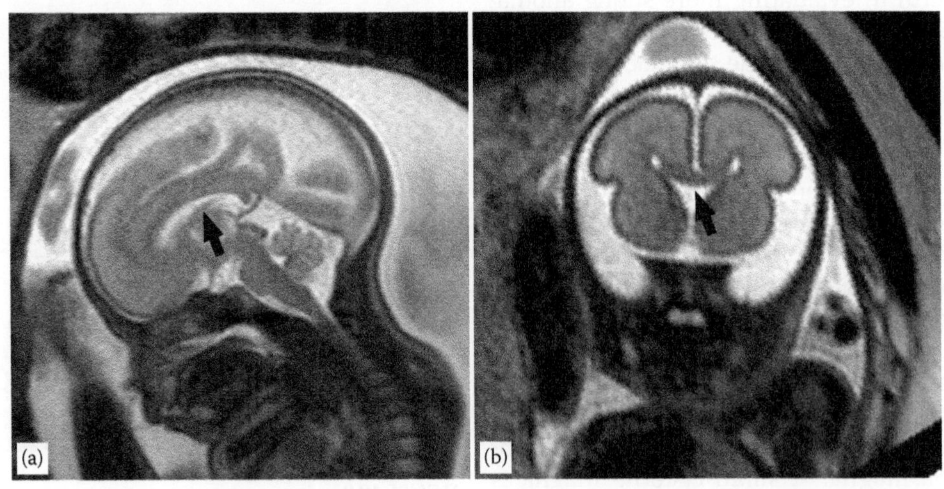

**FIGURE 18.11**
Normal fetus at 30 GW on sagittal (a) and coronal (b) $T_2$W sequences showing a normal corpus callosum.

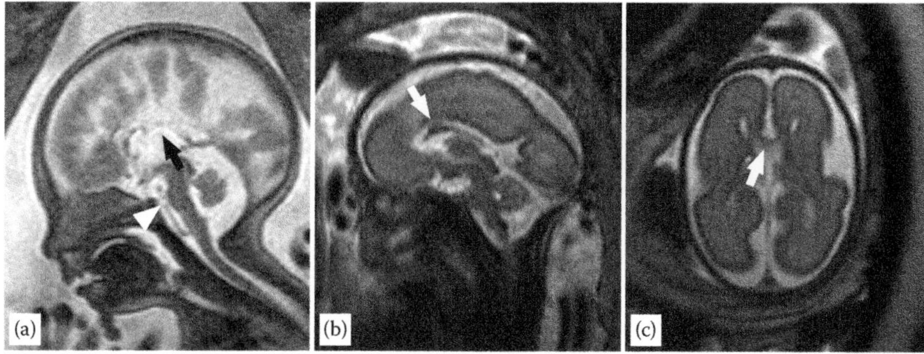


FIGURE 18.12
Different fetuses with complete (a) and partial (b and c) agenesia of corpus callosum. (a) Sagittal $T_2$W sequence shows a complete absence of corpus callosum (arrow). Note associated pontocerebellar hypoplasia (arrowhead). Sagittal (b) and axial (c) $T_2$W sequences show a partial agenesia of corpus callosum, only the anterior part of the body can be visualized.

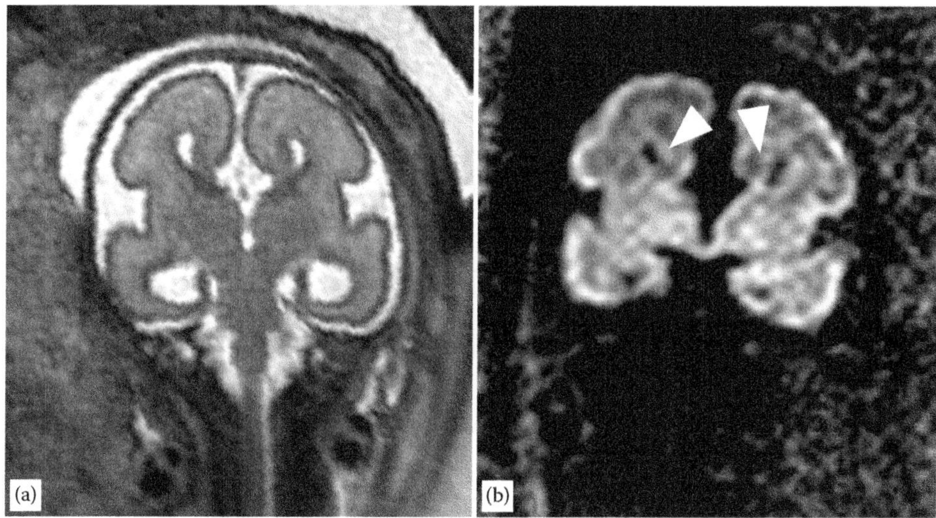


FIGURE 18.13
Fetus at 29 GW with agenesis of the corpus callosum with the presence of Probst bundle. (a) Coronal $T_2$W shows a typical shape of lateral ventricles (parallel arrangement) in the case of agenesis of corpus callosum. (b) Fractional anisotropy map shows an anisotropic behavior of Probst bundles (arrowhead) located medial to the lateral ventricle.

incomplete commissural agenesis to establish fetal cognitive prognosis, with incomplete commissural agenesis having a better prognosis.

A useful tool for the assessment of incomplete commissural agenesis is diffusion tensor imaging. This technique can provide information about the white matter tracts and has been used *in utero* for the assessment of callosal agenesis [25]. Particularly important is the localization of the Probst bundles, which contains white matter pathways passing in an anteroposterior orientation, located medial to the lateral ventricle and rostral to the fornix and represent an aberrant callosal axon (Figure 18.13) [25]. As the Probst bundle may be regarded as functional, with electrophysiological properties similar to those of the intact corpus callosum (Lelkowitz), its complete absence in the cases of complete or incomplete callosal agenesis may indicate a severe abnormality with high risk of neurodevelopment deficit [25].

#### 18.5.4.4 Ventricular Dilatation

*Ventriculomegaly*, defined as a posterior atrial width greater than 10 mm, is a relatively frequent prenatal finding detected in about 1% of prenatal scans [12,13]. It can affect one or both cerebral ventricles and be symmetric or asymmetric [26]. Fetuses with ventricular measurement greater than 15 mm are considered to have severe ventriculomegaly [11,12].

Ventricular dilatation can be caused by many underlying conditions, including chromosomal abnormalities, genetic syndromes, brain or spine anomalies, intracranial hemorrhage, aqueductal stenosis (Figure 18.14), and

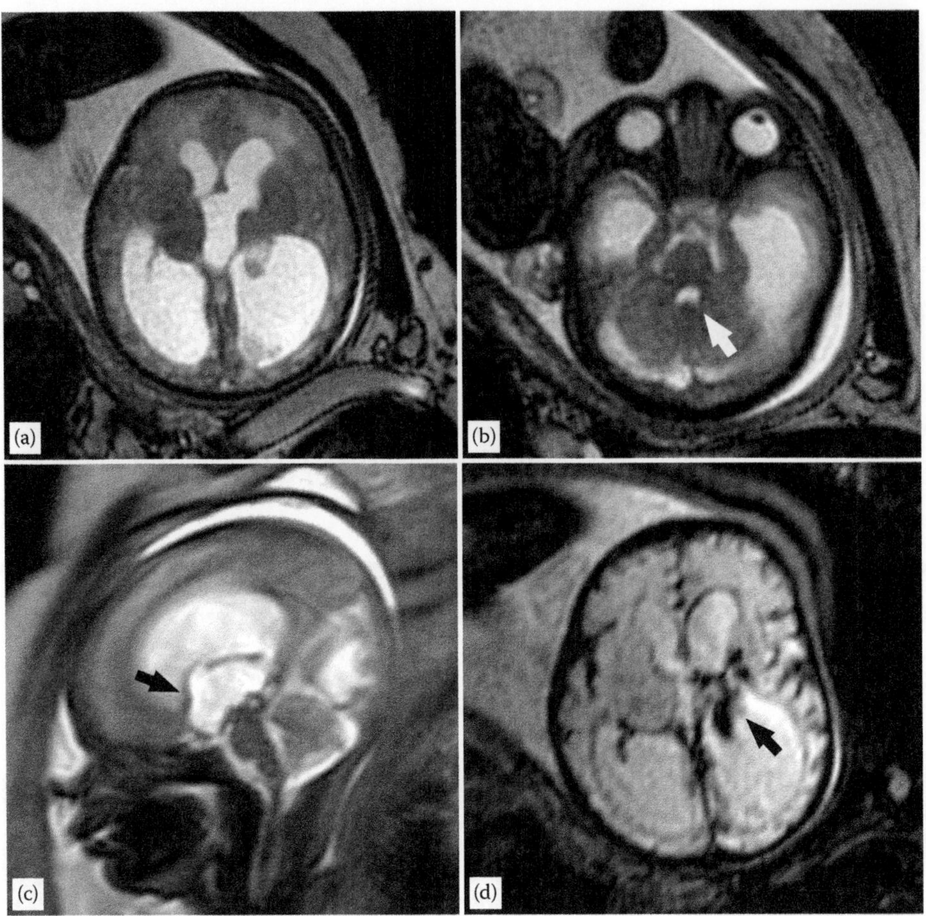


**FIGURE 18.14**
Fetus at 35 GW with ventriculomegaly and acquired aqueductal stenosis. (a) Axial SSFP shows a severe supratentorial ventriculomegaly. (b) Axial SSFP shows a normal fourth ventricle (arrow). (c) Sagittal $T_2$W shows dilatation of the lateral and third (arrow) ventricles, with a normal fourth ventricle. (d) Axial echo-planar sequence shows an intraventricular hemorrhage (arrow).

intrauterine infection [12,13]. Although the main prognostic factor is the presence or absence of associated anomalies, isolated ventriculomegaly is associated with an abnormal neurological outcome in up to 10%–15% of cases [25]. Fetal MRI has proven crucial in determining the extent of the ventriculomegaly and the presence of associated intracranial abnormalities, including subtle structural defects and abnormal cortical development [27,28].

#### 18.5.4.5 Schizencephaly

Schizencephaly is an abnormality characterized by a gray matter lined cleft that communicates with the cerebral ventricles and the subarachnoid space [11,12]. Two types of schizencephaly are recognized: fused also known as closed lip, and open lip [29]. They can be unilateral, bilateral, and multiple (associated with septo-optic dysplasia). The open lip type (Figure 18.15a, b) shows variable separation of the cleft wall with cerebrospinal fluid in its interior. The cleft walls are lined with gray matter, so this finding helps differentiate it from porencephalic cavities. The closed lip type (Figure 18.15a, c) has a better prognosis but its prenatal diagnosis is difficult. There is also presence of gray matter lining both sides of the cleft. A very useful finding for diagnosis is a dimple along the lateral ventricular wall that corresponds to the site where it enters the cleft. Fetal MRI provides delineation of the lesion: given its excellent tissue contrast, it can identify the signals of gray matter and detect associated anomalies such as those of cortical development, and more commonly polymicrogyria, which is habitually localized in the contralateral hemisphere [15].

#### 18.5.4.6 Arachnoid Cysts

Arachnoid cysts are caused by entrapment of cerebrospinal fluid between the arachnoid membranes, probably as the result of abnormal splitting or tearing of the intervening space between the arachnoid and piamater membranes [30]. They can be single or multiple, and their location and size vary greatly. In the fetus, arachnoid cysts are usually found incidentally in the third trimester,

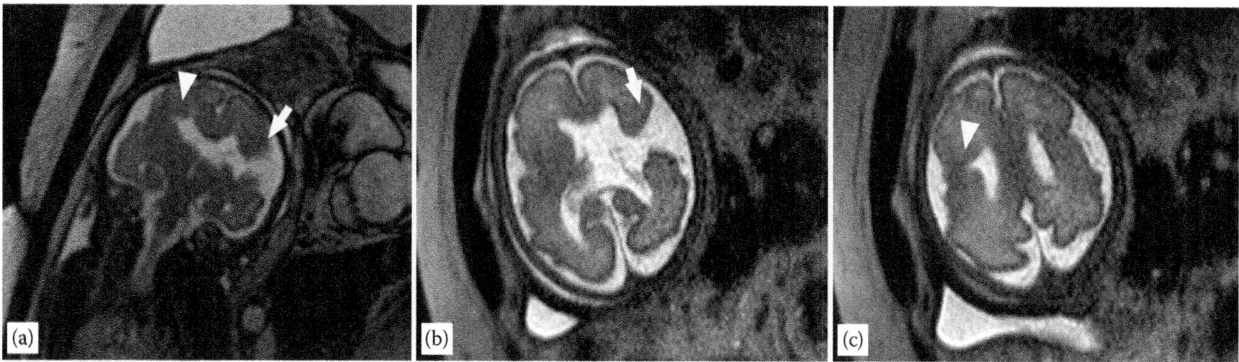


**FIGURE 18.15**
Fetus at 30 GW with a bilateral schizencephaly. Coronal (a) and axial (b) SSFP show an open lip schizencephaly with a cleft filled with cerebrospinal fluid (arrow). (c) Close lip schizencephaly with a ventricular wall dimple (arrowhead in a and c).

at which time a well-circumscribed cystic lesion with adjacent mass effect is identified. It is occasionally associated with hydrocephaly and, infrequently, intracystic hemorrhage can occur. Fetal MRI can clearly and precisely pinpoint the lesion as well as assess the cortical folds and overall maturation, which is important to rule out alteration of cortical development.

#### *18.5.4.7 Arteriovenous Malformations*

Aneurysm of the vein of Galen is the most common arteriovenous malformation seen in fetal life [27]. It is a complex vascular anomaly, thought to be the result of increased blood flow through arteriovenous shunting into the vein of Galen in association with venous ectasia or obstruction of a dural sinus distal to the aneurysm [31]. Owing to the high output flow shunted into the brain during fetal life, affected fetuses usually develop cardiac insufficiency and multiorgan failure, which may eventually lead to perinatal demise. Most cases are diagnosed in the third trimester, at which time the vascular lesion shows the typical *keyhole* or *comet tail* sign representing the aneurysmal dilatation and the dilated sagittal sinus (Figure 18.16).

Diagnosis by means of ultrasound is relatively simple. Fetal MRI with $T_2$W sequence demonstrates low signal intensity due to flow void and allows to be assurance that the image is vascular in origin. Fetal MRI also allows assessment of the complications caused by this type of malformation. The echo-planar sequence is particularly helpful for detection of blood-breakdown products. In addition, intracerebral lesions resulting from defective blood supply (the so-called melting brain) may be detected [32].

#### *18.5.4.8 Acquired Brain Lesions*

Acquired fetal brain damage is suspected in cases of destruction of previously normally formed tissue. The morphological presentations of *malformative* and *destructive* lesion patterns may overlap [33,34]. Because of its ability to discriminate slight differences in tissue composition and to depict small details, MRI has become the method of choice to characterize acquired brain lesions [35,36]. The primary reason for acquired fetal brain damage is hypoxia [9]. Hypoxic endangerment of the fetus is readily determined by means of ultrasound: the biophysical profile uses dynamic ultrasound variables and amniotic volumetry to assess fetal well-being and Doppler values of umbilical and fetal arteries, to show the condition of fetal blood supply. Biometric measurements may point toward intrauterine growth restriction [37].

There are numerous causes for acquired fetal brain damage; among them are maternal diseases, acute maternal problems (cardiovascular collapse and anaphylactic reaction), infections, toxic agents, metabolic diseases of the fetus, mechanical factors involving the placenta (abruption and placenta previa), iatrogenic factors (amniocentesis), feto-fetal transfusion syndrome, and space-occupying lesions (cysts and tumors). Note that study of extra-CNS organs can provide clues to make the correct diagnosis; for example, infections are known to cause lesions in the cornea, liver, bowels, and heart [7].

For the correct characterization of acquired brain lesions, the acquisition of DWI, $T_2$W, echo planar, and SSFP sequences is especially important. It is necessary to study the delineation of parenchymal defects, the form and width of the cerebrospinal fluid spaces, the lamination of the parenchyma, the cortical folding, and the presence of blood-breakdown products in the brain parenchyma.

MRI findings in acquired brain lesion include intra-axial or extra-axial hemorrhage, thrombosis, acute local ischemia (visible on DWI) (Figure 18.16), loss of lamination of the brain parenchyma, and general edema. In a chronic fetal brain injury, MRI may show parenchymal defects, unilateral or bilateral ventricular enlargement, distortion of the ventricular shape, irregular margins of the ventricle, irregular gyration, pathological signals in the parenchyma, irregularities in or premature loss of the germinal zone, calcifications or hemorrhagic residuals, laminar necrosis, malformations, pathologic

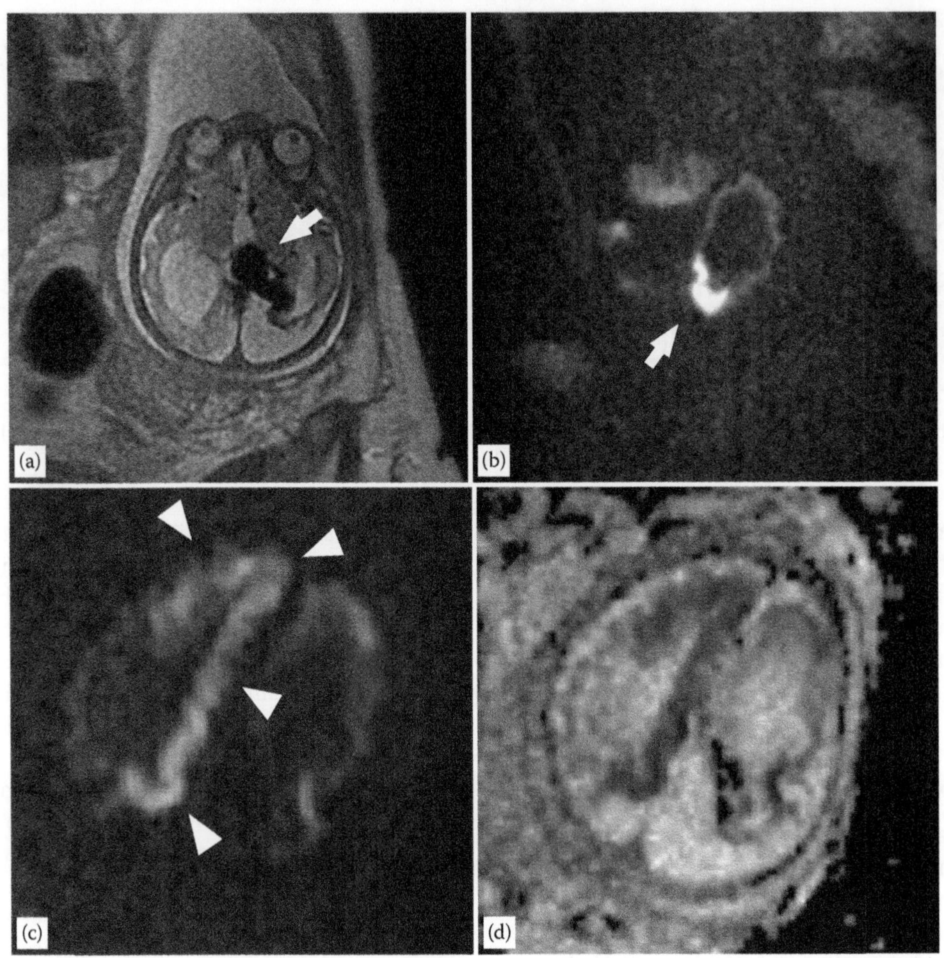


**FIGURE 18.16**
Fetus at 32 GW with an aneurysm of vein of Galen and acute stroke. (a) Axial $T_2$W sequence shows a hypointense signal of the vascular malformation (arrow). Coronal (b) and axial (c) DWI sequences show a hyperintense signal on occipital cortex (b, arrow) and fronto-parietal cortex (c, arrowheads). (d) Hypointense signal on ADC-MAP establish the presence of acute stroke injury. (Courtesy of Claudia Cejas, FLENI Foundation, Buenos Aires, Argentina.)

presentation of premyelinating structures, space-occupying lesions, and atrophy [7].

Destructive cerebral lesions include porencephaly and hydranencephaly. Porencephaly usually develops as a result of localized stroke or vascular hypoperfusion, and it is frequently seen in the setting of a single fetal demise in monochorionic twin pregnancies. As it is an evolving condition, the appearances vary depending on the stage of the disease. The prognosis depends on the cause as well as on the location and extent of the affected area. Fetal MRI can be useful in counseling parents of affected fetuses by determining the exact location of the lesion, as well as the mass effect on the surrounding cerebral tissues and the depiction of associated cortical lesions, which will help us to predict the outcome.

*Hydranencephaly*, on the other hand, is a lethal condition and the most severe form of an encephaloclastic lesion. It is caused by an acute vascular insult at the level of the internal carotid arteries, leading to massive necrosis of the cerebral hemispheres. The occipital lobes and posterior fossa structures are partially preserved because of their separate vascular supply from the vertebral-basilar system. The severe distortion of the intracranial anatomy allows straightforward prenatal diagnosis with ultrasound [38]. Fetal MRI can help in the evaluation of alternate diagnoses such as extreme forms of hydrocephaly and alobar holoproscencephaly.

#### 18.5.4.9 Intracranial Hemorrhage

*Intracranial hemorrhage* refers to hemorrhage anywhere within the fetal cranium. It is usually identified after birth and ascribed to the labor process, particularly in preterm infants. Prenatally, it is a rare condition that almost invariably presents as intraventricular hemorrhage with associated hydrocephaly. Hemorrhage can also occur in other sites, such as in subarachnoid, subdural, and intraparenchymal locations. The blood clots can produce variable degrees of mass effect and hydrocephaly, depending on size. Fetal MRI is crucial in determining the extent of the lesion by defining the hemorrhage in the germinal matrix with high definition, and in the assessment for delayed cortical sulcation or associated porencephalic lesions [21,23]. MRI is also of invaluable use in the detection of incidentally detected hemorrhages by allowing a better prediction of the condition's prognosis. Because fetal hemoglobin has a different composition than adult hemoglobin, conventional sequences such as $T_2$W and $T_1$W have a low sensitivity for its detection. Echo-planar sequence is the most sensible and allows finding previously unsuspected bleeding (Figure 18.17).

#### *18.5.4.10 Cortical Maldevelopment*

With increasing awareness of cerebral lesions due to abnormal neuronal cell migration, the ability to diagnose certain subtle, but neurologically significant conditions such as pachygyria, lissencephaly, microcephaly, and cerebral heterotopias during the prenatal period is becoming necessary. Ultrasound imaging shows the cerebral cortex poorly, and although the degree of cerebral sulcation may be appreciated, it is often difficult to evaluate with this technique. In contrast, fetal MRI can delineate the normal cerebral mantle and sulcation, making the prenatal diagnosis of abnormal cortical development easier (Figure 18.18).

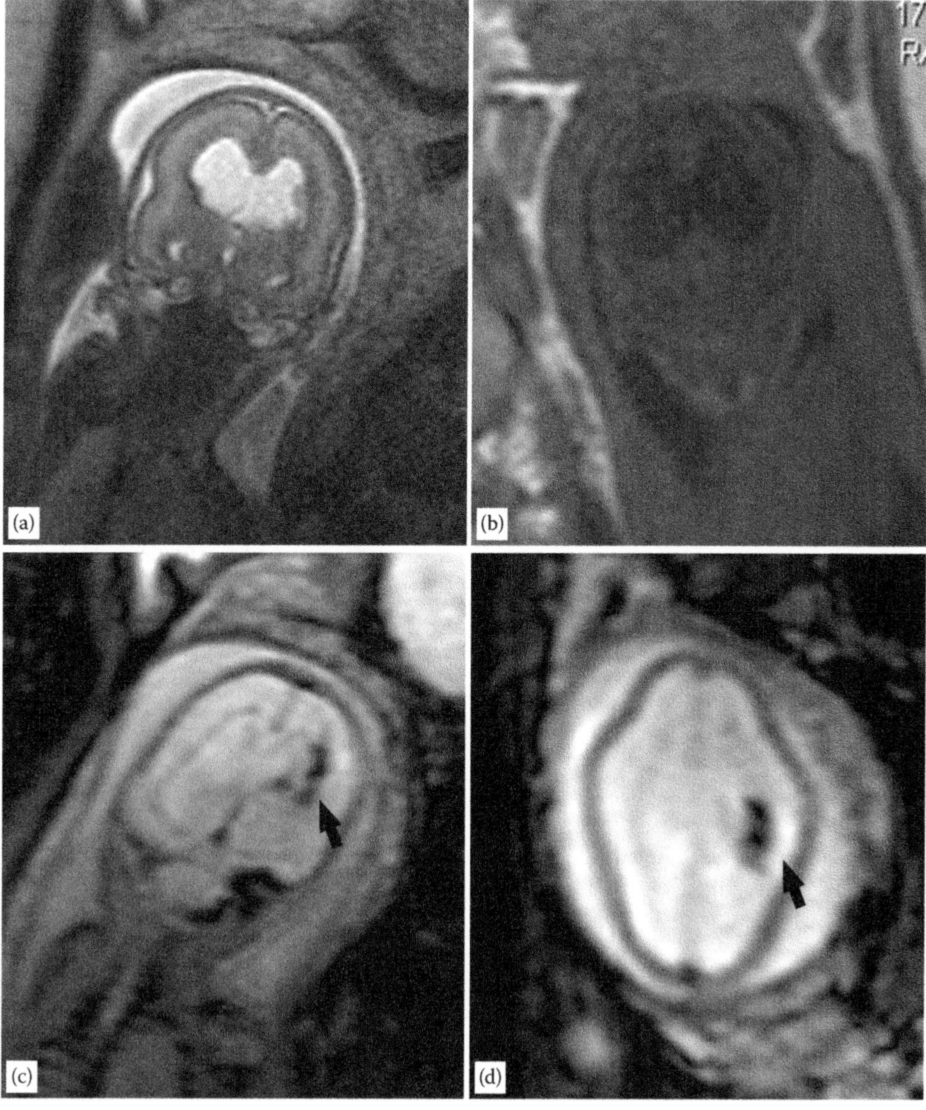


**FIGURE 18.17**
Fetus with a spina bifida—Chiari II malformation and unknown hemorrhage. Coronal $T_2$W (a) and coronal $T_2$W (b) sequences show ventriculomegaly. Coronal (c) and axial (d) echo-planar sequences show a hypointense signal of unsuspected intraventricular hemorrhage (arrow).

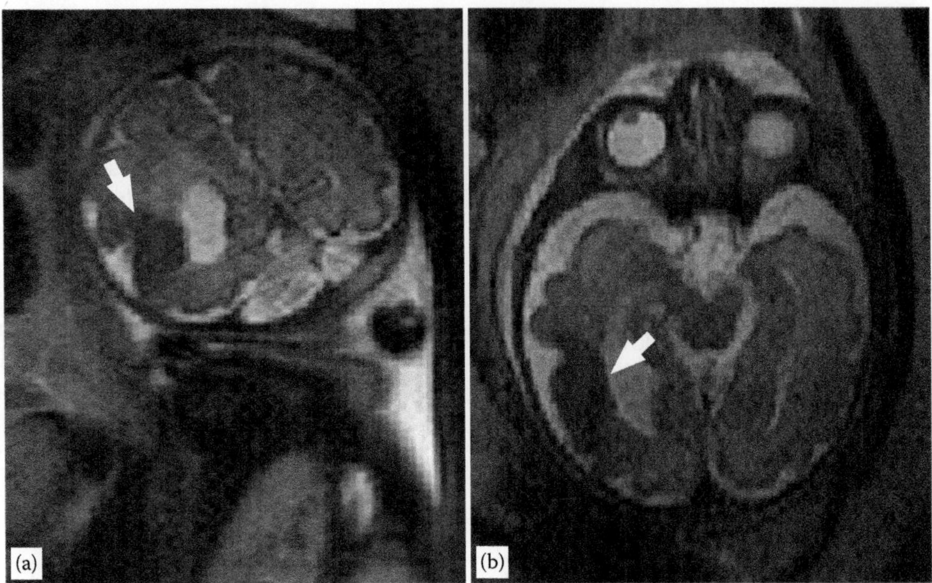


**FIGURE 18.18**
Fetus at 34 GW with a cortical malformation. Coronal (a) and axial (b) $T_2$W sequences show a hypointense cerebral cortex without differentiation between gray and white matter.

## 18.6 Fetal Non-CNS Malformations

### 18.6.1 Face and Palate

Recent experience with MRI has shown an increasing role in the evaluation of facial defects [39,40], as well as in the evaluation of associated anomalies such as brain defects [41]. MRI is a valuable complement to ultrasound [42], adding useful information about the maxillofacial anatomy and allowing precise evaluation of the primary (cleft lip/alveolus) and secondary palate (cleft lip and palate) (Figure 18.19) [43]. The overall prognosis of facial clefts depends on whether associated severe additional malformations exist, which occur in 21% of cases of cleft lip and palate and in 8% of cases of isolated cleft lip [12,13].

The sequences used for the study of the defects in face and palate are $T_2$W sequence, echo-planar sequence, and SSFP sequence with 2–6 mm slice thickness. They must be studied in coronal, axial, and sagittal planes. Axial planes are of great use to study cleft lips and alveoli, and coronal sequences to assess palate clefts. The sagittal view can be used complementarily in primary and secondary palates [12,13].

### 18.6.2 Neck

Most abnormalities of the fetal neck are caused by the presence of cystic, solid-cystic, or solid masses; all are easily detected with ultrasound when large and cause significant distortion of the fetal anatomy [12,13]. As many of these masses involve the anterior and lateral aspects of the neck, they often impair the swallowing process leading to polyhydramnios [12,13].

If the neck mass compresses the upper airway, perinatal asphyxia leading to early neonatal demise is anticipated. This complication can be prevented with the *ex utero* intrapartum treatment procedure which aims to secure the airway during delivery while the fetus is still connected to the placental circulation. This, however, requires an accurate prenatal diagnosis as well as a planned delivery, considerable technical expertise, and a multidisciplinary approach [44]. Fetal MRI in these cases is, therefore, critical in determining the size, location, and internal characteristics of the neck mass, as well as the degree of tracheal deviation, compression of the upper airway, and mass extension into the oropharynx [45]. Identification of the texto feeding vessels is also important in order to facilitate postnatal resection.

Fetal MRI has an excellent tissue contrast that allows for mass characterization, and when topographic location is added, an etiological diagnosis can often be reached. In the next paragraphs, the main pathologies found in the neck are described.

#### *18.6.2.1 Cystic Hygroma*

*Cystic hygroma* is the cervical mass most frequently seen. It originates from a defect in the communication between the lymphatic sacs and the jugular veins at around 7 weeks [12,13]. The most frequent location is the posterior triangle of the neck. In $T_2$W MRI, it can be seen as a hyperintense mass, acquiring a cystic aspect; it is

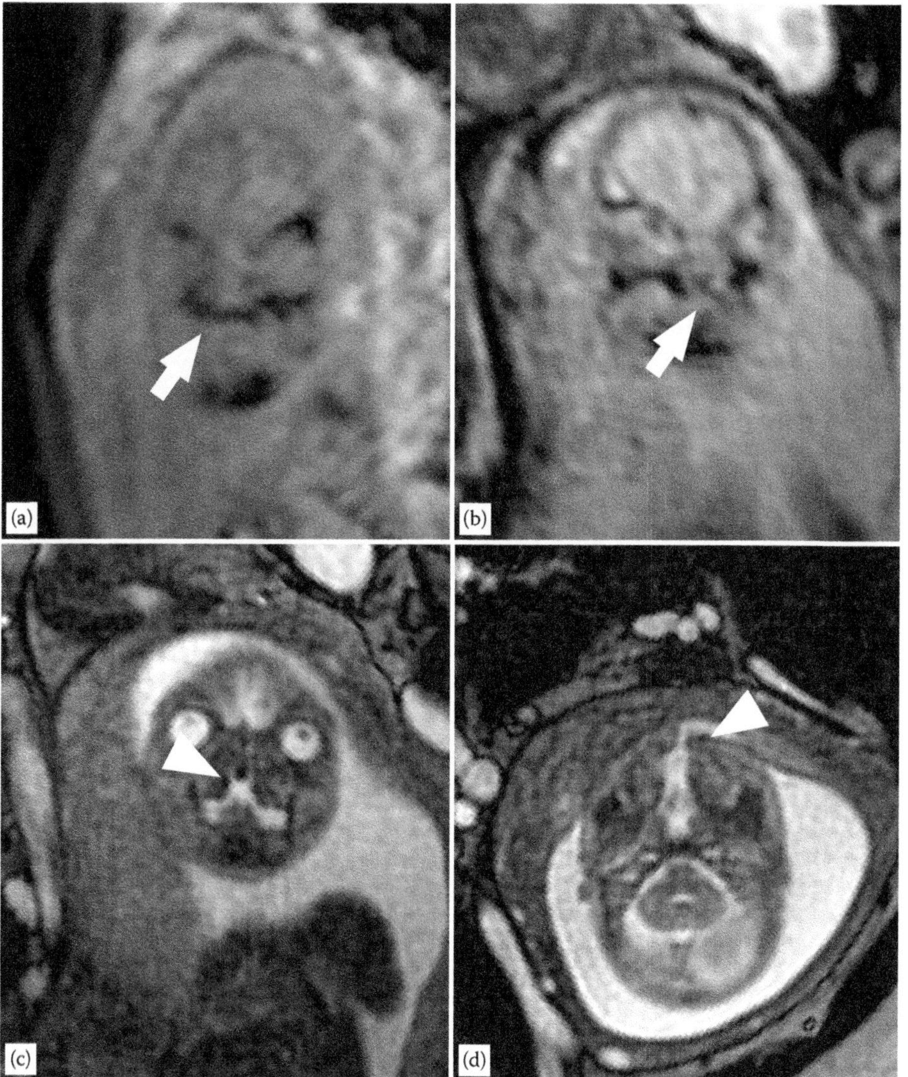


**FIGURE 18.19**
Fetus at 21 GW with a lip and palate cleft (b–d). (a) Echo-planar sequence shows a normal hypointense signal of palate (arrow). (b) Echo-planar sequence shows palate cleft with absence of normal hypointense signal (arrow). Coronal $T_2$W and axial SSFP sequences (c and d) show a lip cleft (arrowhead).

generally homogeneous even though it may present as heterogeneous signal when hemorrhagic complications occur. The SSFP sequence is useful to visualize the thin septae that characterize this condition. Cystic hygroma can be associated with several syndromes, most commonly Turner and Noonan syndromes.

#### *18.6.2.2 Cervical Teratoma*

A cervical teratoma is a tumor constituted by the three germinal layers (endoderm, mesoderm, and ectoderm). According to the degree of differentiation, they can be divided into mature and immature tumors, with the majority of them appearing as benign and mature [12,13]. It is composed mostly of tissue originanting from mesoderm. On MRI, these masses appear heterogeneous, because they are comprised of different kinds of tissue (e.g., fatty, thyroid, and neural). $T_1$W and $T_1$W fat sat sequences can show fatty tissue and also thyroid tissue, with both appearing hyperintense on $T_1$W, and fatty tissue appearing hypointense on T1W fat sat. The $T_2$W sequence is useful to show liquid and solid lesion contents as well.

#### *18.6.2.3 Lymphangioma*

A cervical lymphangioma is a benign mass of lymphatic vessels. It originates due to the obstruction of the lymphatic vessels that drain into the jugular vein [12,13]. On $T_2$W MRI, they present as intense signal, with a thin-walled multicystic aspect; if associated with pleural spillage and ascitis, they have a poor prognosis (Figure 18.20).

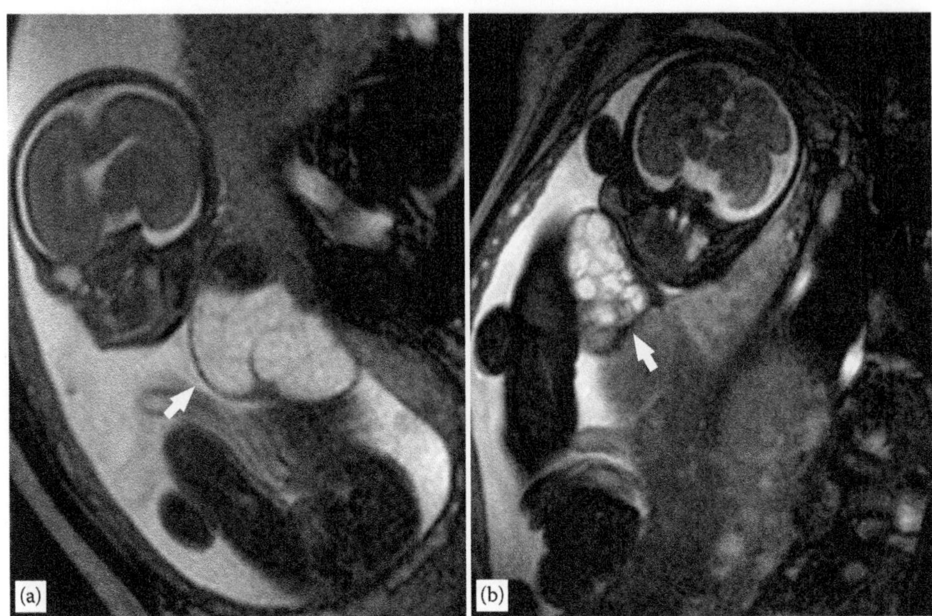

**FIGURE 18.20**
Fetus at 31 GW with cervical lymphangioma. Sagittal $T_2$W (a) and coronal SSFP (b) show a thin-walled multicystic lesion in lateral aspect of the neck (arrow).

#### 18.6.2.4 Fetal Goiter

The thyroid gland is easily recognized given its placement in the anterior region of the fetal neck and its hyperintense signal on $T_1$W, which provides an excellent contrast with the neighboring tissue. Fetal goiter can be defined as an enlargement of the thyroid gland, which can be caused by either lack of or excess of iodine exposure to drugs, including lithium and amiodarone, or as a result of a congenital alteration of thyroid hormone synthesis [12,13]. On MRI, it can be seen as a solid hyperintense mass on $T_1$W in the anterior region of the neck (Figure 18.21). It can cause neck hyperextension and polyhydramnios due to esophageal compression.

### 18.6.3 Thorax

Fetal MRI provides detailed visualization of the thoracic organs irrespective of conditions known to interfere with ultrasound image quality, such as maternal obesity, advanced ribs ossification, or oligo and anhydramnios. MRI can study structural and biochemical changes of developmental processes, thus, the pulmonary tissue is subject to substantial structural changes.

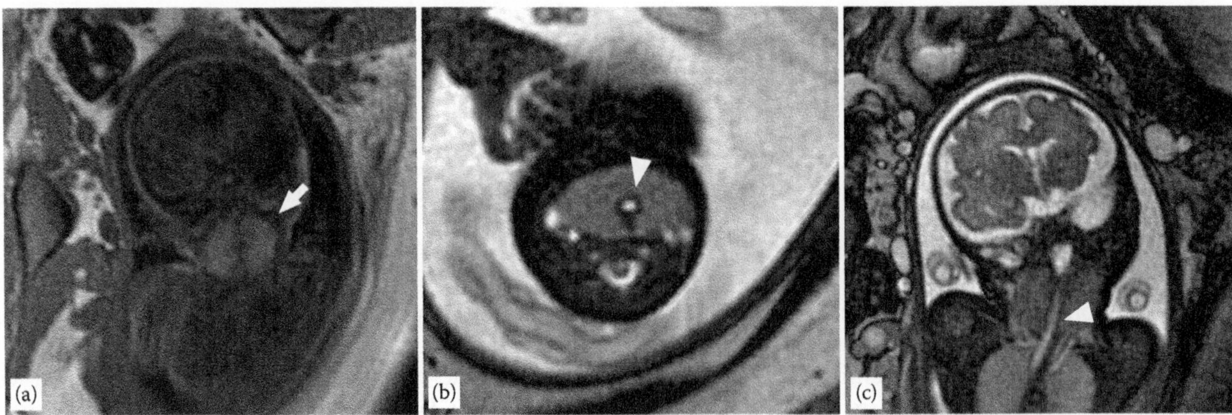

**FIGURE 18.21**
Fetus at 35 GW with fetal goiter. (a) Coronal $T_1$W shows a hyperintense enlarged thyroid in the anterior region of the neck (arrow). Axial SSFP (b) and coronal SSFP (c) sequences show a hyperintense signal fluid of an airway (arrowhead).

MRI allows evaluation of the pulmonary maturation process because of the signal behavior in different sequences, with $T_2$W being the most commonly used sequence for this purpose. $T_1$W provides valuable information too and can depict the lipid and protein content. Signal intensity of the fetal lung parenchyma on $T_2$W increases with gestational age, while $T_1$W signal of the fetal lung decreases between 20 and 30 GW [11]. DWI has also been used to monitor lung maturation; but, its high susceptibility to motion makes it a second-line technique for lung maturation study. Thus, an abnormally low signal on $T_2$W and an abnormally high signal in $T_1$W are a reliable indicator of abnormal maturation process.

Surfactant production is a biochemical marker of fetal lung maturity. Classically, one invasive test, amniocentesis, can provide an amniotic fluid sample to measure the lecithin/sphingomyelin ratio. Surfactant is composed of proteins and lipids (90%); this composition makes MR spectroscopy a potential method to detect and quantify surfactant in the amniotic fluid and fetal lungs. Phosphatidylcholine, the main component of surfactant, can be characterized with spectroscopy in the amniotic fluid, with a characteristic peak of choline at 3.2 ppm, and thus, the choline/creatine ratio can be calculated in the second and third trimester of pregnancy [11]. However, this method has great technical challenges (fetal movement and time of acquisition), and also the low signal/noise ratio and contamination of maternal tissue and fetal fatty make spectroscopy a technique hard to interpret and to use in clinical practice.

#### 18.6.3.1 Lung Volumetry

Pulmonary volumetry constitutes the most important prognostic parameter for postnatal respiratory outcome and is the test of choice to use when pulmonary hypoplasia is suspected. The final pulmonar volumen has more impact on the outcome than the etiological factor that produce the pulmonar hypoplasia. Quantification of fetal lung volume is performed between 17 and 30 GW, after that period fetal lung volume shows a wide range of individual variability making it a non-useful technique [11].

Fetal lung volumetry is a complicated and time-consuming procedure, and manual tracing is the only reliable tool to assess fetal lung volumes on MRI. The most important requisite for valid MR data volumetry is a good quality image without evident fetal motion artifacts. In this sense, it is essential that the chosen sequence does not present motion artifacts, because the pulmonary volume could be over or underestimated. To ensure that no fetal movements occur during sequence acquisition, a reformatting in an orthogonal plane can be done. To perform a pulmonary volumetry, it is preferable to perform acquisition in the axial plane, even though the coronal and sagittal planes can be used as well. We prefer to use a $T_2$W sequence because it provides better differentiation of the mediastinal structures and the liver from the pulmonary parenchyma. The fetal lung volumetry reference data available in the literature have been obtained in the axial slice orientation using $T_2$W sequences [11]. The post-processing of fetal image is performed by specific software, which allows the volumetric quantification of lungs and other fetal organs.

#### 18.6.3.2 Congenital Diaphragmatic Hernia

One of the most important fetal defects is congenital diaphragmatic hernia (CDH); it is caused by a defect in the diaphragm, through which abdominal contents enter the thoracic cavity. The left side is more frequently affected, resulting in the stomach, loops of bowel, and occasionally the spleen to enter the left hemithorax. In right-sided CDH, the liver is invariably located in the thorax. The presence of ectopic abdominal organs produces mediastinal shift, deviation and compression of the heart, and extrinsic compression of the ipsilateral as well as the contralateral lungs. The main pregnancy complication is polyhydramnios secondary to fetal esophageal compression [12,13]. Although the prenatal diagnosis of CDH is relatively easy with ultrasound [12,13], it is very difficult to establish the degree of pulmonary hypoplasia with this technique [46]. Fetal MRI, however, can provide better resolution to assess the size of the diaphragmatic defect, determine the organs involved (Figure 18.22), and allow for volumetric calculation of the lungs [46–52].

To diagnose CDH, $T_1$W sequences are particularly important because they easily recognize bowel loops by the meconium content and, thus have a hyperintense signal [46–51]. With MRI, it is also possible to assess pulmonary maturity of both lungs on $T_2$W and $T_1$W sequences.

In CDH, the development of both lungs is impaired, but to a greater degree in the ipsilateral than in the contralateral lung [52]. In infants who die after repair of CDH, quantitative analysis of the bronchi, arteries, and alveoli has also confirmed that while these structures are reduced in both lungs, the ipsilateral lung is more severely affected [53].

In fetuses with CDH, lung volume measurements obtained with ultrasound and those obtained with MRI have shown good correlation for the contralateral lung, but poorer correlation for the ipsilateral lung. For the contralateral lung, ultrasound yielded an estimate of lung volume that was 25% lower than that estimated with MRI; this difference is independent of the lung size, the side of CDH, and the presence of intrathoracic herniation of the liver [53].

The reproducibility of fetal lung volume measurements with MR imaging is high for both lungs, but with ultrasound, good reproducibility is obtained for the contralateral lung only [53].

#### 18.6.3.3 Lung Lesions

Lung lesions include cystic, solid cystic, or solid lung mass. Cystic lesions are easy to identify on MRI, whereas solid lesions sometimes are not. As most solid lung lesions are unilateral, their detection is facilitated by comparing the signal intensity between the two lungs. The most common types of intrinsic lung masses are congenital cystic adenomatoid malformation (Figure 18.23) and pulmonary sequestration (Figure 18.24) [54,55]. Not infrequently, histological analysis of the resected specimen shows a mixture of these two entities, the so-called hybrid lesions. An additional diagnosis to consider when bilateral $T_2$W hyperintensity and enlargement lungs are visualized is tracheal or bronchial atresia.

Fetal MRI can provide an excellent delineation of lung masses, presence of feeding vessels, and degree of lung compression [56,57]. Volume of the ipsilateral and contralateral lungs, as well total lung volume, can be calculated. Most of the information on lung volume and the correlation with perinatal outcome has been obtained with fetal MRI [57,58]. It is also important in the evaluation of rare mediastinal masses, including lymphangiomas, gastrointestinal duplication cysts involving the foregut and esophagus, and mediastinal teratomas [58].

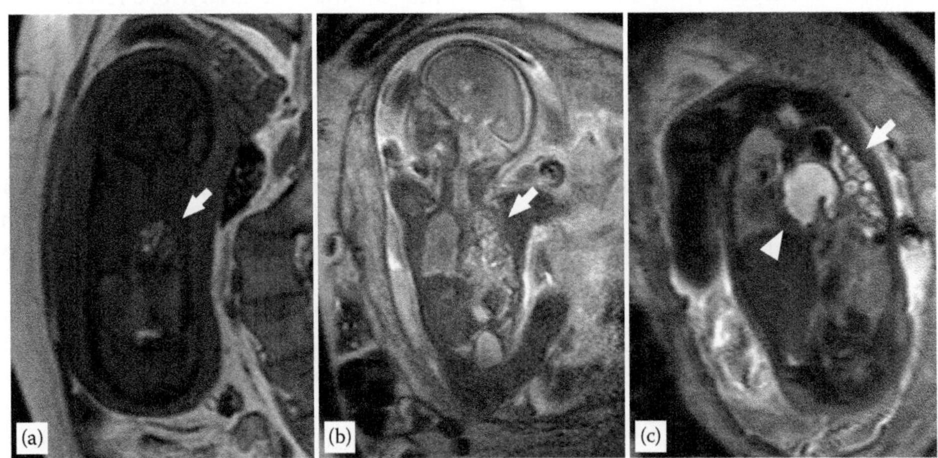


**FIGURE 18.22**
Two different fetuses at 22 GW (a, b) and 32 GW (c) with congenital diaphragmatic hernia. Coronal $T_1$W (a) and $T_2$W (b) show a hyperintense signal of bowel loops in the fetal thorax (arrows). (c) Fetus with bowel loops (arrow) and stomach (arrowhead) in the fetal thorax.

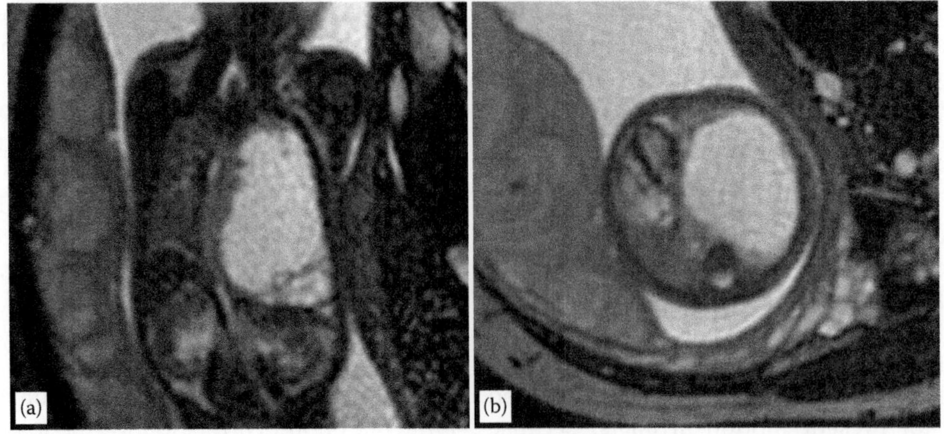


**FIGURE 18.23**
Fetus at 25 GW with congenital cystic adenomatoid malformation. (a) Coronal SSFP sequence shows an extensive malformation that expands the thorax. (b) Axial SSFP sequence shows mediastinal shift and the heart in abnormal right position.

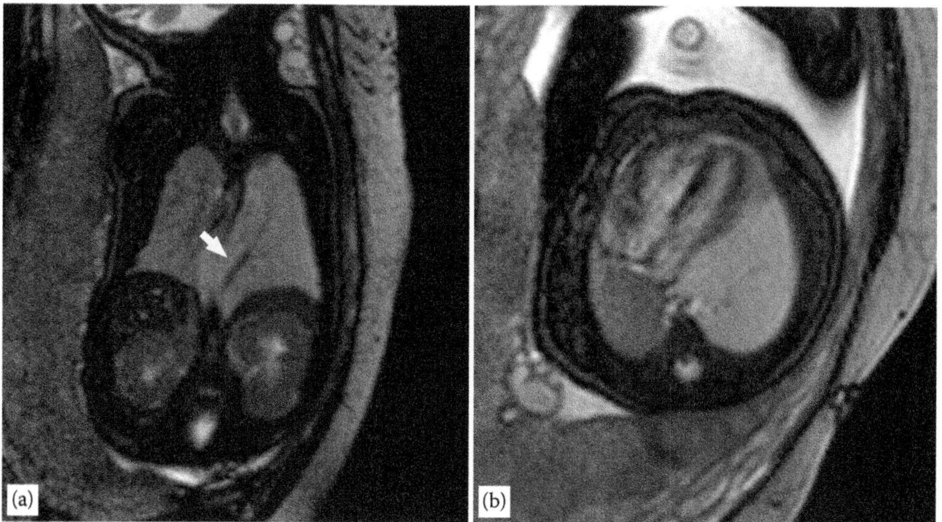


**FIGURE 18.24**
Fetus with pulmonary sequestration. Coronal (a) and axial (b) SSFP sequence show a hyperintense lesion with an abnormal vessel (arrow) and mediastinal shift.

### 18.6.4 Gastrointestinal Tract

Fetal gastrointestinal tract (GIT) abnormalities involve a wide range of conditions, most commonly anterior abdominal wall defects (Figure 18.25), such as omphalocele and gastroschisis. Less common anomalies include GIT obstruction, such as esophageal atresia, duodenal obstruction, and small and large bowel atresia. Intra-abdominal cystic masses and hepatic tumors are rare. The use of MRI to show maturation of the GIT by assessing the intraluminal meconium, which presents as hyperintense signal in $T_1$W sequences, has been suggested as important for determining the bowel function and patency in some GIT disorders, such gastroschisis and bowel atresia [59,60].

Atresia or stenosis of the small intestines leads to dilation of the bowel proximal to the side of obstruction. Duodenal stenosis or atresia is easily recognized by the presence of the characteristic *double bubble* signs: proximal to the obstruction, the distended duodenum may have similar dimensions as the stomach, which is also dilated. The content of the distended duodenal loop shows the same $T_2$W and $T_1$W signal intensities as the stomach [11].

Atresia or stenosis of the duodenum or proximal jejunum is usually associated with polyhydramnios; this may not be the case in distal atresias. In distal

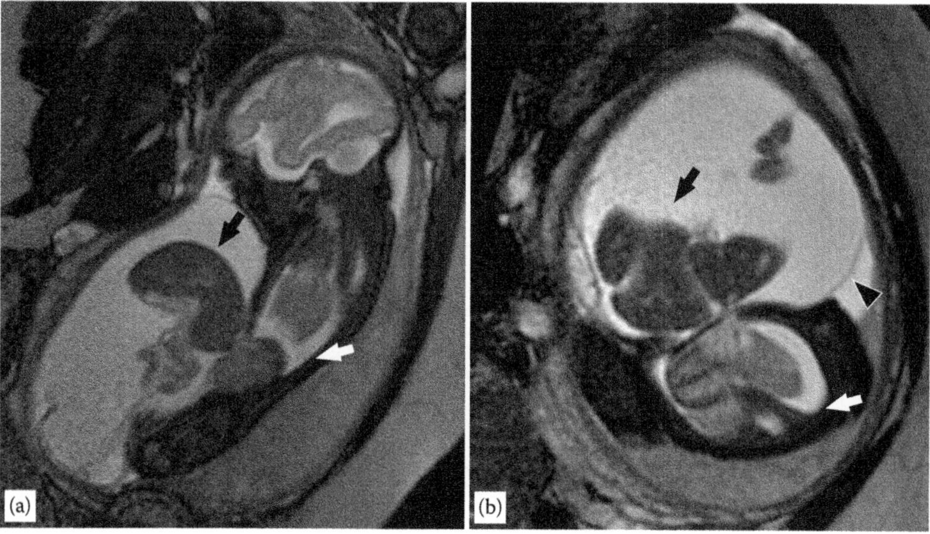


**FIGURE 18.25**
Fetus at 32 GW with a large abdominal wall defect. Sagittal (a) and axial (b) SSFP sequences show the liver outside abdominal cavity (black arrow). Notice the presence of hydrothorax (a and b, white arrows) and membranous covering (b, arrowhead).

ilial atresia, the content of the distal prestenotic bowel reportedly presents with a meconium-like signal intensity [11]. Less frequently causes of intestinal stenosis include gastroschisis and volvulus. Intestinal atresias may be complicated by perforation of the small bowel, which results in meconium peritonitis [11].

### *18.6.4.1 Gastroschisis*

Gastroschisis is a well-recognized congenital defect of the ventral abdominal wall which is usually on the right side of the umbilical cord insertion, although a few cases with left-sided gastroschisis have been recorded [12]. This entity can be easily recognized by its salient feature of bowel loops floating in the amniotic cavity lacking a membranous covering (Figure 18.26). The extra-abdominal bowel usually shows $T_1$-weighted hyperintensity and can easily be recognized. In fetuses with gastroschisis assessed in the second trimester, the colon entering the abdomen usually doesn't show pronounced changes in caliber and the urinary bladder is in the median position. In

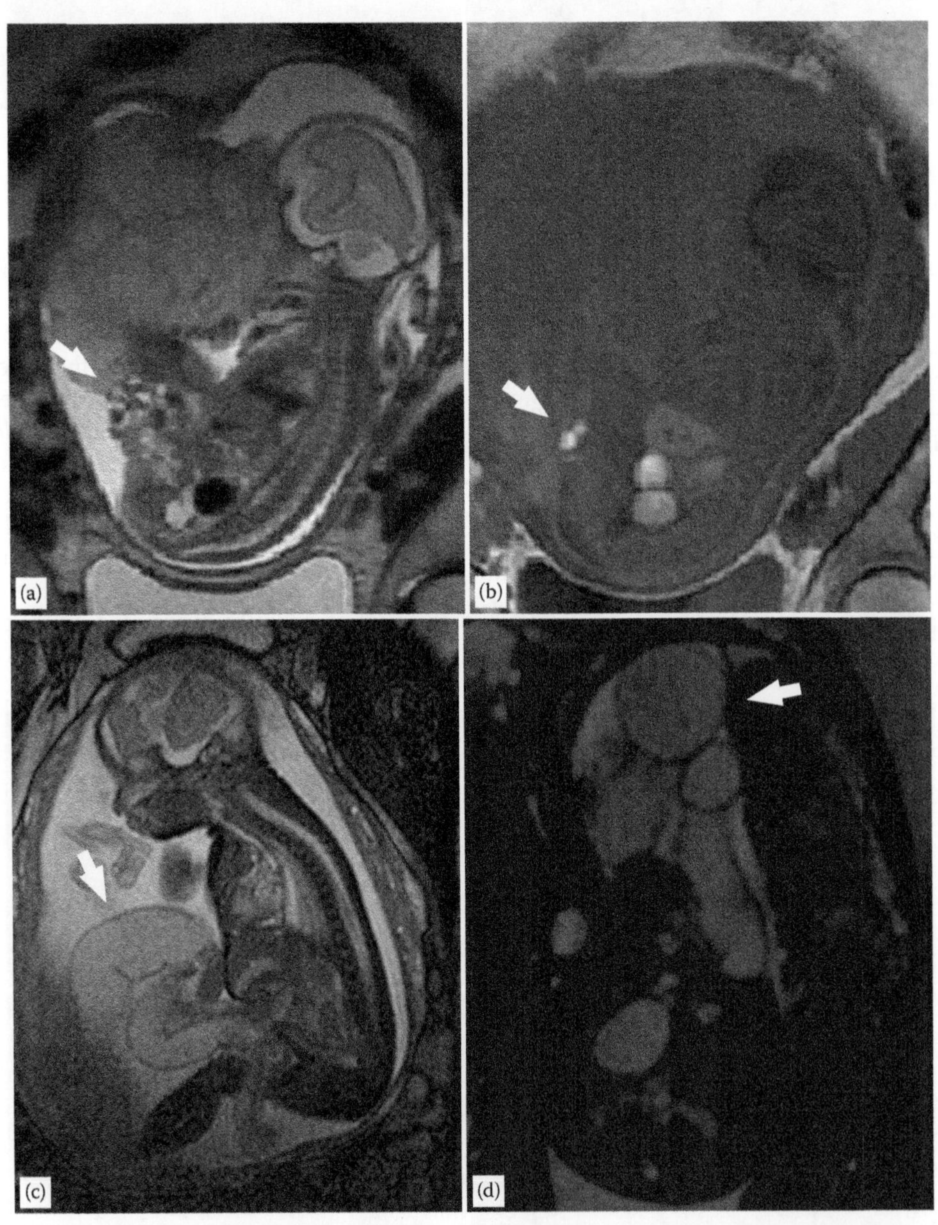


**FIGURE 18.26**
Fetus at 24 GW (a, b) and 30 GW (c, d) with gastroschisis. (a) Sagittal $T_2$W show bowel loops floating in the amniotic cavity (arrow). (b) Sagittal $T_1$W shows a hyperintense signal of meconium in the amniotic cavity (arrow). Sagittal (c) and axial (d) SSFP sequences show a distended small bowel in the amniotic cavity (arrows) because of an abdominal narrow defect.

some cases, we can see that there exists a partial contact between the stomach and urinary bladder, but it is more frequent to find small bowel loops otherwise positioned between them [60].

In some cases, a very narrow defect in the abdominal wall can result in typical signs of small-bowel obstruction. The duodenum and variable lengths of the jejunum are found to be distended intra-abdominally. The content of these dilated bowel loops shows moderate hyperintense signal intensity on $T_2$W images, shows intermediate signal intensity on SSFP, and has a variable appearance on $T_1$W images. In these cases, there is no contact between the stomach and urinary bladder, as the markedly dilated intra-abdominal bowel loops fill the abdominal cavity [60].

#### *18.6.4.2 Omphalocele*

Omphalocele is the most common type of defect in the abdominal wall and is located in the region of the umbilical cord. It is possible to find bowel loops, liver, and less frequently other abdominal organs, in a sac outside the abdominal cavity. It is associated with other anomalies in 70% of the cases [61]. The most widely accepted etiologic theory is a failure of the bowel to return back into the abdominal cavity which typically occurs at 10–12 GW.

Fetal MRI should be performed after the 18 GW; at this point, the presence of bowel loops and liver out of the abdominal cavity, including the umbilical cord, should suggest the diagnosis of omphalocele (Figure 18.27). By MRI, it is possible to visualize the intestinal loops because of the hyperintense signal on $T_1$W of meconium. The liver also shows a hyperintense signal on this sequence, and vascular structures can be identified by a hyperintense signal on SSFP sequence. MRI can be useful in finding anomalies associated with omphalocele, such as gastrointestinal malformation as well as genitourinary, and neural tube defects.

### 18.6.5 Genitourinary Tract

Genitourinary tract malformations are among the most frequent defects that are diagnosed prenatally [12,13]. They vary from minor defects, such as renal pelvis dilatation, to lethal conditions, such as multicystic dysplastic kidney, renal agenesis, (Figure 18.28) and lower urinary tract obstruction. Frequently, the detection of decreased or absent amniotic fluid and dilatation of the urinary tract are important clues for diagnosis. MRI has an ability to clearly depict fluid-filled cavities and detect associated malformations in the setting of severely decreased amniotic fluid volume [62–66], a setting in which ultrasound is particularly limited. In cases of renal agenesis, the confirmation of diagnosis with MRI may be necessary for medico-legal purposes.

By means of DWI, particularly with high $b$-values (e.g., $b$ = 700), the kidneys can easily be recognized

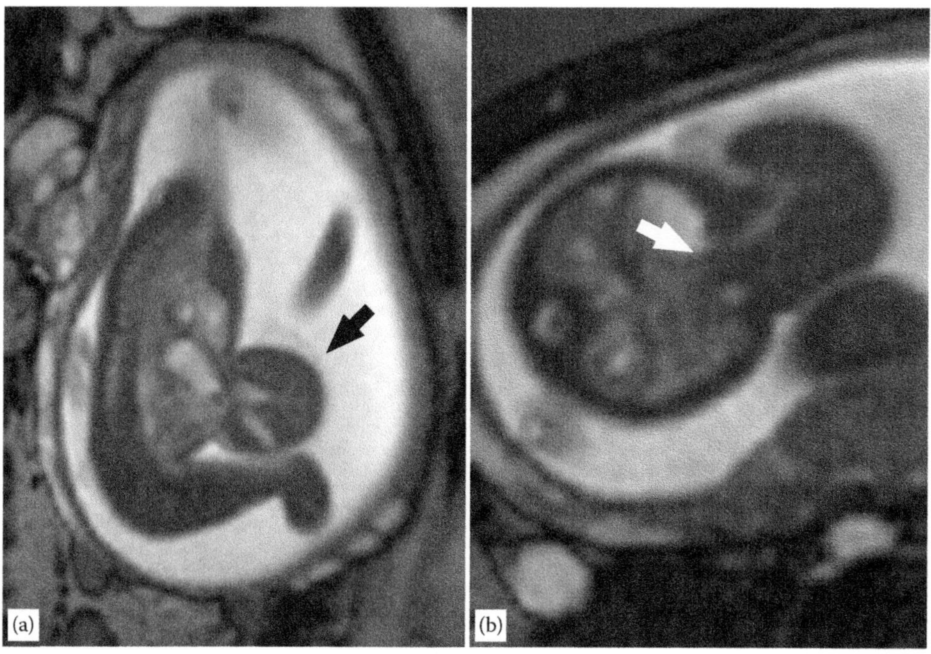


**FIGURE 18.27**
Fetus at 29 GW with omphalocele. (a) Sagittal SSFP sequence shows a liver outside of the abdominal cavity (arrow). (b) Axial SSFP sequence shows an abdominal defect in the anterior abdominal wall (arrow) for which protrudes the liver.

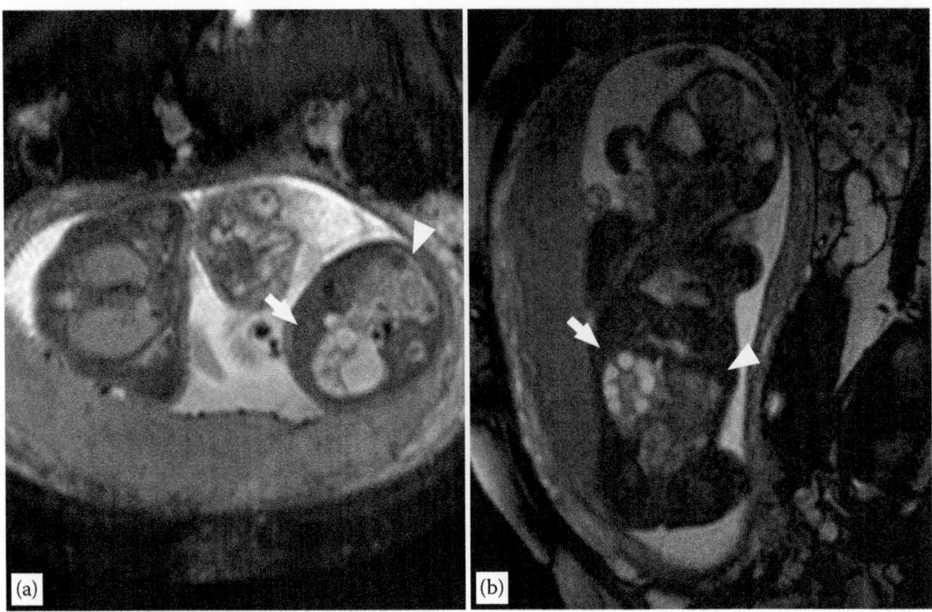

**FIGURE 18.28**
Twins fetuses at 25 GW, one of them with multicystic dysplastic kidney at the right side and agenesia of left kidney. T2 fat sat (a) and coronal SSFP (b) sequences show a multicystic and enlarged right kidney (arrow) and absence of the kidney in the left side (arrowhead).

(Figure 18.29). This can be very useful in cases of renal hypoplasia, agenesia, and ectopic and malformed kidneys (e.g., horseshoe kidney), cases which are often difficult to diagnose with ultrasound. Another feature of the DWI sequence is its ability to assess renal tissue functionality. With normally functioning kidneys, the renal medulla and the renal cortex present a hyperintense signal on DWI, while with inadequately functioning kidneys it presents a hypointense signal on DWI.

MRI can be of value in the prenatal evaluation of complex malformations, such as cloacal dysgenesis (Figure 18.30) [67]. In these cases, $T_1$W sequence allows

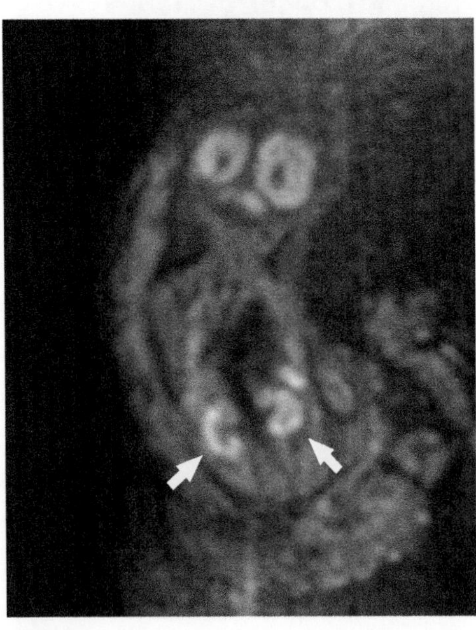

**FIGURE 18.29**
Coronal DWI sequence with high *b*-value (b700) shows normal hyperintense kidneys (arrows).

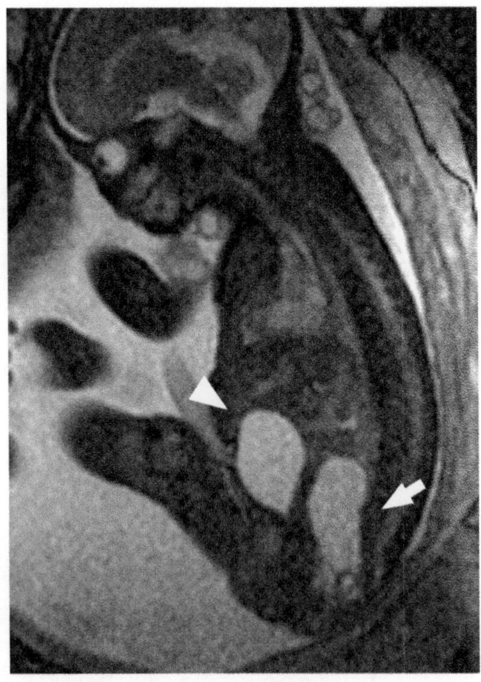

**FIGURE 18.30**
Fetus at 29 GW with cloacal dysgenesis. (a) Sagittal SSFP sequence cannot differentiate between the rectum and bladder (arrow). Note a cystic structure in the abdominal cavity (arrowhead).

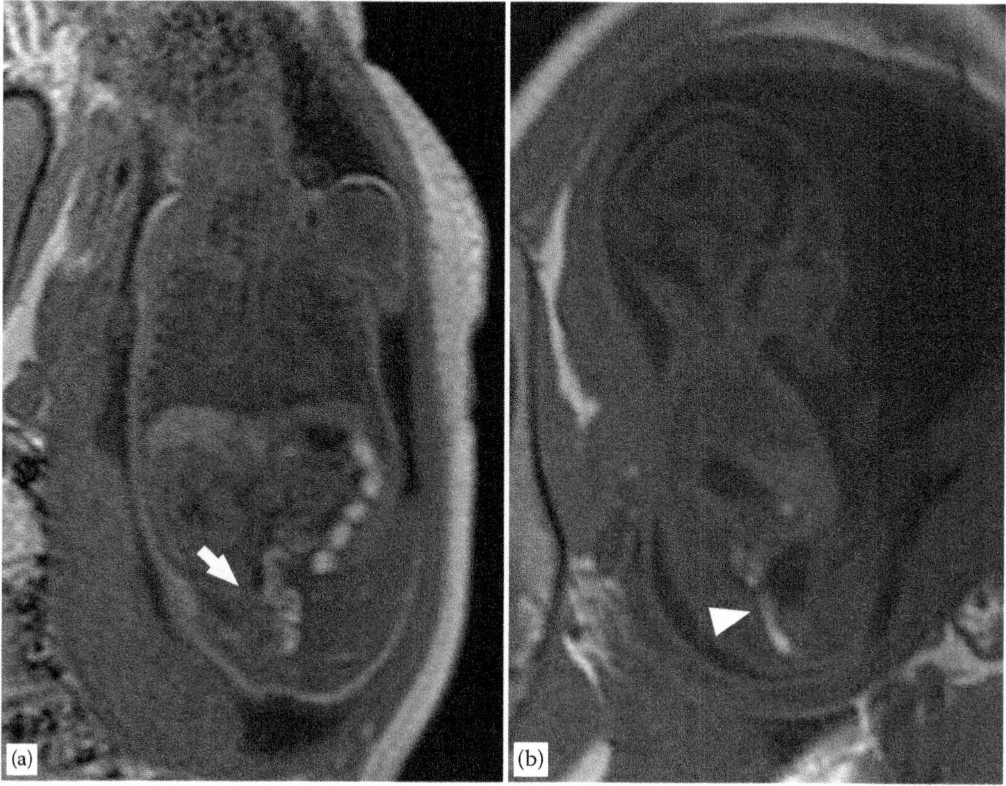


**FIGURE 18.31**
Different fetuses at 29 GW (a) and 27 GW (b). (a) Coronal $T_1$W sequence shows a normal meconium with a hyperintense signal. Rectum is in an anatomic position (arrow). (b) The normal rectum (arrowhead) can be visualized behind bladder.

the rectum to be studied, which is key in this pathology. In fetuses which present cloacal anomalies, the rectum cannot be distinguished in its anatomical position on $T_1$W and sequence. It is always useful to assess the rectum in the coronal and sagittal planes to perfectly observe its position in the pelvis and its relationship with the bladder which presents a hypointense signal on $T_1$W (Figure 18.31). A hyperintense signal on $T_1$W of the bladder indicates a cloacal anomaly associated with a rectal-bladder fistula.

## 18.7 The Skeleton and Musculature

In the last years, fetal MRI has showed its ability to study musculoskeletal abnormalities [68], thus positioning itself as a first-line method and free of the use of X-radiation.

To study the bones, the elected sequence is echo-planar image, with which an excellent delineation of the bone structures that show a hypointense signal is obtained. This sequence also provides valuable information about long bone epiphyses, being the proximal femoral the one mostly studied (Figure 18.32) [68], thus obtaining information about the development of bone structures not only from the femoral length, but also about the morphological changes suffered by the distal femoral epiphysis in established gestational ages.

Thick-slab $T_2$W image can show a three-dimensional impression of the fetus [10]; thus, a global vision of fetal proportion is obtained and thus easily evidence asymmetries and alterations, such as foot deformity.

Dynamic sequences, which are useful to establish fetal activity and vitality, can be very helpful to interpret the complex position of the fetal limbs, to discard polydactyly, and to assess swallowing movements in the presence of polyhydramnios.

The analysis of muscular structures is based on their morphology and signals; in this way, we can find atrophy which may indicate the presence of a neuromuscular disease, muscular hypertrophy, or signal alterations. The normal signal of muscular tissue is hypointense on $T_2$W, and a hyperintense signal on $T_2$W can be seen due to the presence of edema in neuromuscular diseases.

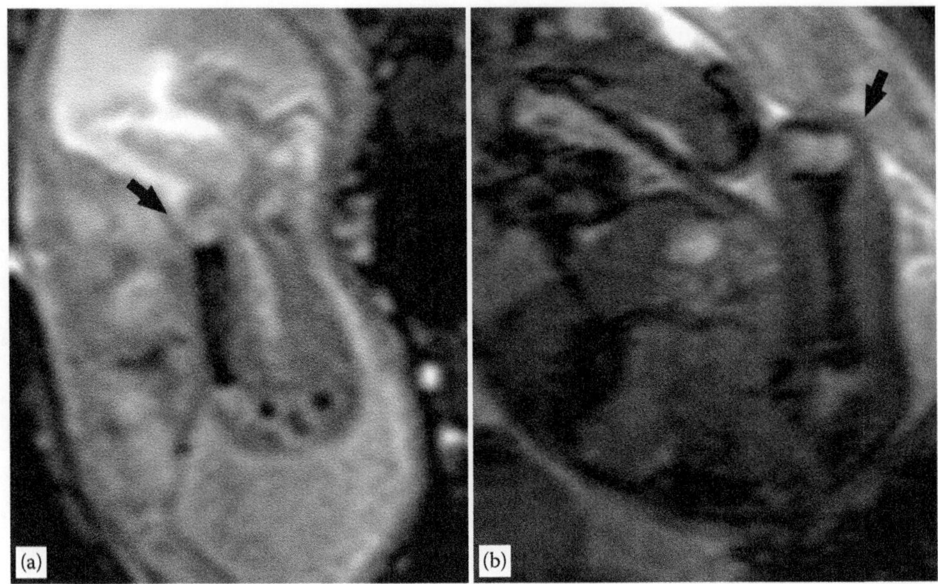


**FIGURE 18.32**
Echo-planar sequence of fetuses at 18 GW (a) and 26 GW (b) shows the changes in epiphyseal shape. (a) Coronal sequence of a fetus at 18 GW shows spherical epiphyseal shape (arrow). Coronal sequence of a fetus at 26 GW shows hemispheric epiphyseal shape (arrow).

## 18.8 Conclusion

Due to the increasing use of high-resolution ultrasound as a screening method in pregnancy, it is currently common and necessary to use other complementary methods such as MRI in the evaluation of selected cases. This method can be used when certain conditions require deeper anatomical characterization or when there are technical limitations of ultrasound.

Particularly important is the detection of anomalies associated with CNS anomalies which have been previously detected by ultrasound and the quantification of lung volumetry in fetal malformations which cause hypoplasia, as these factors have a great impact on fetal prognosis. Finally, fetal MRI provides excellent tissue characterization and additional assessment of function and metabolism. These tools make MRI a powerful tool as a complementary method to ultrasound in prenatal diagnosis and evaluation of fetal anomalies.

## References

1. Smith F, Adam A, and Phillips W. NMR imaging in pregnancy. *Lancet*. 1983; 1(8314–5): 61–62.
2. Reddy U, Filly R, and Copel J. Prenatal imaging: Ultrasonography and magnetic resonance imaging. *Obstet Gynecol* 2008; 112: 145–157.
3. Asenbaum U, Brugger P, Prayer D et al. Indikationen und Technik der fetalen Magnetresonanztomographie. *Radiologe* 2013; 53: 109–115.
4. Brugger P, Prayer D. Actual imaging time in fetal MRI. *Eur J Radiol* 2012; 81: 194–196.
5. Nemec U, Nemec S, Krakow D et al. The skeleton and musculature on foetal MRI. *Insights Imaging* 2011; 2: 309–318.
6. Nemec S, Nemec U, Brugger P et al. MR imaging of the fetal musculoskeletal system. *Prenat Diagn* 2012; 32: 205–213.
7. Prayer D, Brugger PC, Langer M et al. MRI of fetal acquired brain lesions. *Eur J Radiol* 2006; 57: 233–249.
8. Baldoli C, Righini A, Parazzini C et al. Demonstration of acute ischemic lesions in the fetal brain by diffusion magnetic resonance imaging. *Ann Neurol* 2002; 52: 243–246.
9. Girard N, Gire C, Sigaudy S et al. MR imaging of acquired fetal brain disorders. *Childs Nerv Syst* 2003; 19: 490–500.
10. Ghai S, Fong KW, Blaser S et al. Prenatal US and MR imaging findings of lissencephaly: Review of fetal cerebral sulcal development. *RadioGraphics* 2006; 26: 2, 389–405.
11. Prayer D. *Fetal MRI*. Berlin, Germany: Springer, 2011.
12. Nyberg D, McGahan J, Pretorius D et al. (eds.). *Diagnostic Imaging of Fetal Anomalies*. Philadelphia, PA: Lippincott Williams & Wilkins, 2003.
13. Bianchi D, Crombleholme T, D'Alton M et al. *Fetology: Diagnosis and Management of the Fetal Patient*. 2nd ed. New York: McGraw-Hill, 2010.
14. Sepulveda W, Wong AE, Sepulveda S et al. Fetal scalp cyst or small meningocele: Differential diagnosis with three-dimensional ultrasound. *Fetal Diagn Ther* 2011; 30: 77–80.
15. Tortori-Donati P. *Pediatric Neuroradiology*. Berlin, Germany: Springer, 2005.

16. Bulas D. Fetal magnetic resonance imaging as a complement to fetal ultrasonography. *Ultrasound Q* 2007; 23: 3–22.
17. Bulas D. Fetal evaluation of spine dysraphism. *Pediatr Radiol* 2010; 40: 1029–1037.
18. Von Koch CS, Glenn OA, Goldstein RB et al. Fetal magnetic resonance imaging enhances detection of spinal cord anomalies in patients with sonographically detected bony anomalies of the spine. *J Ultrasound Med* 2005; 24: 781–789.
19. Appasamy M, Roberts D, Pilling D et al. Antenatal ultrasound and magnetic resonance imaging in localizing the level of lesion in spina bifida and correlation with postnatal outcome. *Ultrasound Obstet Gynecol* 2006; 27: 530–536.
20. Pistorius LR, Hellmann PM, Visser GH et al. Fetal neuroimaging: Ultrasound, MRI, or both? *Obstet Gynecol Surv* 2008; 63: 733–745.
21. Guibaud L. Contribution of fetal cerebral MRI for diagnosis of structural anomalies. *Prenat Diagn* 2009; 29: 420–433.
22. Glenn OA. MR imaging of the fetal brain. *Pediatr Radiol* 2010; 40: 68–81.
23. Kline-Fath BM and Calvo-Garcia MA. Prenatal imaging of congenital malformations of the brain. *Semin Ultrasound CT MR* 2011; 32: 167–188.
24. Malinger G, Lev D, and Lerman-Sagie T. The fetal cerebellum. Pitfalls in diagnosis and management. *Prenat Diagn* 2009; 29: 372–380.
25. Kasprian G, Mitter C, Prayer D et al. Assessing prenatal white matter connectivity in commissural agenesis. *Brain* 2013; 136: 168–179.
26. Melchiorre K, Bhide A, Gika AD et al. Counseling in isolated ventriculomegaly. *Ultrasound Obstet Gynecol* 2009; 34: 212–224.
27. Benacerraf BR, Shipp TD, Bromley B et al. What does magnetic resonance imaging add to the prenatal sonographic diagnosis of ventriculomegaly? *J Ultrasound Med* 2007; 26: 1513–1522.
28. Yin S, Na Q, Chen J et al. Contribution of MRI to detect further anomalies in fetal ventriculomegaly. *Fetal Diagn Ther* 2010; 27: 20–24.
29. Oh KY, Kennedy AM, Frias AE et al. Fetal schizencephaly: Pre and postnatal imaging with a review of the clinical manifestations. *RadioGraphics* 2005; 25: 647–657.
30. Malinger G, Corral-Sereno E, and Lerman-Sagie T. The differential diagnosis of fetal intracranial cystic lesions. *Ultrasound Clin* 2008; 3: 553–558.
31. Sepulveda W, Vanderheyden T, Pather J et al. Vein of Galen malformation: Prenatal evaluation with three-dimensional power Doppler angiography. *J Ultrasound Med* 2003; 22: 1395–1398.
32. Rutherford M. Chapter 12, Vascular malformations of the neonatal brain. In: *MRI of the Neonatal Brain*. New York: W.B. Saunders. eBook EG Systems, 2002.
33. Guibaud L, Attia-Sobol J, Buenerd A et al. Focal sonographic periventricular pattern associated with mild ventriculomegaly in foetal cytomegalic infection revealing cytomegalic encephalitis in the third trimester of pregnancy. *Prenat Diagn* 2004; 24: 727–732.
34. Sanchis A, Cervero L, Bataller A et al. Genetic syndromes mimic congenital infections. *J Pediatr* 2005; 146: 701–705.
35. Levine D, Barnes PD, Robertson RR et al. Fast MR imaging of fetal central nervous system abnormalities. *Radiology* 2003; 229: 51–61.
36. Elchalal U, Yagel S, Gomori JM et al. Fetal intracranial hemorrhage (fetal stroke): Does grade matter? *Ultrasound Obstet Gynecol* 2005; 26: 233–243.
37. Manning FA. Fetal biophysical profile: A critical appraisal. *Clin Obstet Gynecol* 2002; 45: 975–985.
38. Sepulveda W, Cortes-Yepes H, Wong AE et al. Prenatal sonography in hydranencephaly: Findings during the early phase of the disease. *J Ultrasound Med* 2012; 31: 799–804.
39. Mailath-Pokorny M, Worda C, Krampl-Bettelheim E et al. What does magnetic resonance imaging add to the prenatal ultrasound diagnosis of facial clefts? *Ultrasound Obstet Gynecol* 2010; 36: 445–451.
40. Wang G, Shan R, Zhao L et al. Fetal cleft lip with and without cleft palate: Comparison between MR imaging and US for prenatal diagnosis. *Eur J Radiol* 2011; 79: 437–442.
41. Rosen H, Chiou GJ, Stoler JM et al. Magnetic resonance imaging for detection of brain abnormalities infetuses with cleft lip and/or cleft palate. *Cleft Palate Craniofac J* 2011; 48: 619–622.
42. Ghi T, Tani G, Bovicelli L et al. Prenatal imaging of facial clefts by magnetic resonance imaging with emphasis on the posterior palate. *Prenat Diagn* 2003; 23: 970–975.
43. Pugash D, Brugger PC, Prayer D et al. Prenatal ultrasound and fetal MRI: The comparative value of each modality in prenatal diagnosis. *Europ J Radiol* 2008; 68: 214–226.
44. Lazar DA, Olutoye OO, Moise KJ et al. Ex-utero intrapartum treatment procedure for giant neck masses—Fetal and maternal outcomes. *J Pediatr Surg* 2011; 46: 817–822.
45. Dighe M, Dubinsky T, Cheng et al. EXIT Procedure: Technique and indications with prenatal imaging parameters for assessment of airway patency. *RadioGraphics* 2011; 31: 511–526.
46. Knox E, Lissauer D, Khan K et al. Prenatal detection of pulmonary hypoplasia in fetuses with congenital diaphragmatic hernia: A systematic review and meta-analysis of diagnostic studies. *J Matern Fetal Neonatal Med* 2010; 23: 579–588.
47. Ba'ath ME, Jesudason EC, and Losty PD. How useful is the lung-to-head ratio in predicting outcome in the fetus with congenital diaphragmatic hernia? A systematic review and meta-analysis. *Ultrasound Obstet Gynecol* 2007; 30: 897–906.
48. Kilian AK, Schaible T, Hofmann V et al. Congenital diaphragmatic hernia: Predictive value of MRI relative lung-to-head ratio compared with MRI fetal lung volume and sonographic lung-to-head ratio. *AJR Am J Roentgenol* 2009; 192: 153–158.
49. Alfaraj MA, Shah PS, Bohn D et al. Congenital diaphragmatic hernia: Lung-to-head ratio and lung volume for prediction of outcome. *Am J Obstet Gynecol* 2011; 205: 43.e1–43.e8.

50. Mayer S, Klaritsch P, Petersen S et al. The correlation between lung volume and liver herniation measurements by fetal MRI in isolated congenital diaphragmatic hernia: A systematic review and meta-analysis of observational studies. *Prenat Diagn* 2011; 31: 1086–1096.
51. Sandaite I, Claus F, De Keyzer F et al. Examining the relationship between the lung-to-head ratio measured on ultrasound and lung volumetry by magnetic resonance in fetuses with isolated congenital diaphragmatic hernia. *Fetal Diagn Ther* 2011; 29: 80–87.
52. Peralta CF, Jani J, Cos T, Deprest J et al. Left and right lung volumes in fetuses with diaphragmatic hernia. *Ultrasound Obstet Gynecol* 2006; 27: 551–554.
53. Jani J, Nicolaides KH, Dymarkowski S et al. Lung volumes in fetuses with congenital diaphragmatic hernia: Comparison of 3D US and MR imaging assessments. *Radiology* 2007; 244: 575–582.
54. Cavoretto P, Molina F, Poggi S et al. Prenatal diagnosis and outcome of echogenic fetal lung lesions. *Ultrasound Obstet Gynecol* 2008; 32: 769–783.
55. Sepulveda W. Perinatal imaging in bronchopulmonary sequestration. *J Ultrasound Med* 2009; 28: 89–94.
56. Epelman M, Kreiger PA, Servaes S et al. Current imaging of prenatally diagnosed congenital lung lesions. *Semin Ultrasound CT MR* 2010; 31: 141–157.
57. Bulas D and Egloff AM. Fetal chest ultrasound and magnetic resonance imaging: Recent advances and current clinical applications. *Radiol Clin N Am* 2011; 49: 805–823.
58. Deshmukh S, Rubesova E, and Barth R. MR assessment of normal fetal lung volumes: A literature review. *AJR Am J Roentgenol* 2010; 194: W212–W217.
59. Zizka J, Elias P, Hodik K et al. Liver, meconium, hemorrhage: The value of $T_1$-weighed images in fetal MRI. *Pediatr Radiol* 2006; 36: 792–801.
60. Brugger PC and Prayer D. Development of gastroschisis as seen by magnetic resonance imaging. *Ultrasound Obstet Gynecol* 2011; 37: 463–470.
61. Salihu H, Boos R, Schmidt W. Omphalocele and gastroschisis. *J Obstet Gynecol* 2002; 22: 489–492.
62. Cassart M, Massez A, Metens T et al. Complementary role of MRI after sonography in assessing bilateral urinary tract anomalies in the fetus. *AJR Am J Roentgenol* 2004; 182: 689–695.
63. Farhataziz N, Engels JE, Ramus RM et al. Fetal MRI of urine and meconium by gestational age for the diagnosis of genitourinary and gastrointestinal abnormalities. *AJR Am J Roentgenol* 2005; 184: 1891–1897.
64. Barseghyan K, Jackson H, Chmait R et al. Complementary roles of sonography and magnetic resonance imaging in the assessment of fetal urinary tract anomalies. *J Ultrasound Med* 2008; 27: 1563–1569.
65. Hawkins JS, Dashe JS, and Twickler DM. Magnetic resonance imaging diagnosis of severe fetal renal anomalies. *Am J Obstet Gynecol* 2008; 198: 328.e1–328.e5.
66. Alamo L, Laswad T, Schnyder P et al. Fetal MRI as complement to US in the diagnosis and characterization of anomalies of the genito-urinary tract. *Eur J Radiol* 2010; 76: 258–264.
67. Calvo-Garcia M, Kline-Fath BM, Levitt MA et al. Fetal MRI clues to diagnose cloacal malformations. *Pediatr Radiol* 2011; 41: 1117–1128.
68. Nemec U, Brugger P, Prayer D et al. Human long bone development in vivo: Analysis of the distal femoral epimetaphysis on MR images of fetuses. *Radiology* 2013; 267: 570–580.

# 19

# *Postmortem and Forensic Magnetic Resonance Imaging*

**Patricia Mildred Flach, Dominic Gascho, Thomas Daniel Ruder, Sabine Franckenberg, Steffen Günter Ross, Lukas Ebner, Michael Josef Thali, and Garyfalia Ampanozi**

CONTENTS



## 19.1 Postmortem and Forensic Magnetic Resonance Imaging

Imaging modalities such as plain radiography have been used successfully for more than a century to shed light on forensic investigations. In 1895, the first case involving an X-ray of a lower extremity was admitted to court in North America. This particular case involved a shooting in the leg of an individual named Mr. Tolson Cunning, and the bullet could be located between the fibula and tibia by X-ray [1,2]. The first attempted murder that was admitted to court with radiography as evidence was a case in the United Kingdom in which a man attempted to kill his wife with four shots to the head before he committed suicide [3–5].

In the mid-1940s, the use of X-ray for identification purposes was reported in a famous case involving an inglorious man, Adolf Hitler, of whom three X-rays of the skull were obtained while he was still alive that were used later by the Russians for the dental identification of his charred remains [2,6,7].

Publications dating back to the 1960s already valued the use of postmortem or forensic X-rays as an established method of identifying skeletal status (e.g., in child abuse or homicides) for the improved investigation of subsequent legal cases [3,8–15]. During that decade, probably the most famous case was the 1963 assassination of the President of the United States, John F. Kennedy, in which X-rays revealed that he was struck by two bullets fired from above and behind him [16–19]. Photographs and radiography that were obtained appeared to support the findings of the Warren Commission Report.

With the invention of cross-sectional imaging, these modalities also found their way into forensics, and newer techniques were occasionally adopted for use in postmortem examination. The performance of the first forensic or postmortem computed tomography (PMCT) was reported in the late 1970s in a case of a fatal gunshot wound to the head [20]. In 1994, Donchin et al. advocated the use of PMCT for trauma victims when conventional autopsy was unattainable [21]. Initial experience with the use of forensic or postmortem magnetic resonance (PMMR) imaging surprisingly arose almost parallel to the emergence of PMCT. The first publication by Ros et al. (1990) stated that pre-autopsy magnetic resonance imaging (MRI) may provide an alternate method in restricted or denied autopsies and may provide an additional MR research and educational tool [22]. Eight years later, Bisset et al. went so far as to state that radiological investigations—specifically PMMR—may be as valuable in death as they are in life for determining the underlying pathology, if autopsy is unavailable [23]. In the same year, the late Dr. Gil Brogdon published the first textbook on forensic radiology, dedicating an entire chapter to PMMR in child abuse cases [2].

In 2000, Prof. Dirnhofer, Prof. Thali, and Prof. Vock founded the Virtopsy project at the University of Berne, Switzerland. This was the first attempt to devise a multidisciplinary approach between forensic pathology and diagnostic radiology, leading to routine virtual autopsies. Nowadays, PMCT and PMMR, as well as angiographic imaging procedures, are an essential part of the forensic workup and serve as an adjunct to—or even a replacement for—autopsy [2,24–28].

Almost contemporarily, at the turn of the millennium, a U.S. research group from Boston deemed PMMR a useful adjunct to autopsy, particularly in decedents in whom autopsy is limited due to decedent/family consent, inoculation risks, and ethnic doctrines. PMMR was scrutinized because autopsy was considered the gold standard [29]. Since these earlier scarce publications, there has been a strong increase in their number during the last decade, leading to numerous publications in the field of forensic imaging, with contributions from more than 40 countries worldwide [30].

Currently, forensic institutes are increasingly embracing the new technical developments and implementing a growing number of dedicated scanners (predominantly PMCT) in the morgue worldwide or, at least, affiliating with clinical radiology to perform cross-sectional imaging [31–48]. This modality—especially if PMCT and PMMR are combined—enhances the quality of forensic death investigations. PMCT is an established tool that is used to detect foreign bodies, gas (e.g., gas embolism, pneumothorax, or perforation), and fractures and is used for human radiological identification in unknown cases [2,26,28,49–64]. PMMR is a modality that is complementary to PMCT, which is far superior to PMMR in depicting soft tissue lesions (e.g., hematoma) and parenchymal pathologies [2,28,65–81] because PMMR is more vulnerable to artifacts from gas and foreign bodies. In addition, PMMR should only be performed and read by radiologically trained personnel, as handling and reporting is more complex than for PMCT. The clear drawbacks of PMMR are the cost of equipment acquisition and maintenance and the time-consuming operation of this modality, which leads to less frequent usage than PMCT. Based on a survey conducted among experts and frequent users of forensic imaging in 2013, only 5% (55% for PMCT) are confident in their ability to read PMMR images and only 12% (42% for PMCT) routinely use PMMR for postmortem cases [78].

Both cross-sectional methods are usually used without contrast media in postmortem settings. Notably, several forensic institutes already use specifically designed contrast media mixtures (adapted from clinically used contrast media protocols) to display vascular and parenchymal pathologies by performing PMCT-angiography (PMCTA) and/or PMMR-angiography (PMMRA) [82–95]. However, it is crucial to read all imaging modalities, non-contrast-enhanced and contrast-enhanced PMCT and PMMR, to complete accurate forensic radiological reports, as each modality complements the other.

### 19.1.1 General Findings on PMMR

Postmortem imaging must address a wide variety of findings that are completely normal on postmortem scans but may mimic antemortem pathologies or lead to misinterpretation [2,96]. The investigator must be aware of these specific findings and consider the postmortem interval, body temperature, and case circumstances.

#### *19.1.1.1 Decomposition*

The decay of a body leads to gas accumulation, initially within the hepatobiliary tract and digestive system and later within the soft tissue, vasculature, and organs (Figure 19.1). Furthermore, intrahepatic gas may develop based on an underlying pathology such as organ injury, the result of unsuccessful cardiopulmonary resuscitation, a gas embolism, or iatrogenic manipulation [2,28]. However, decompositional gas will also be influenced by internal factors, such as underlying disease (e.g., sepsis), the corpse's position, and external factors such as ambient temperature, coverage of the body, and position-dependency of the gas, which tends to accumulate in opposition to gravity. The intestines begin to bloat, the abdomen (and scrotum) appears gaseous, and maggots or other animal infestations may be present.

Later, decomposition leads to fluid accumulation under the influence of gravity (e.g., when the corpse lies in supine position, fluid will accumulate in the posterior parts of the body), as well as in the soft tissue and body cavities. In the course of decay, organs will liquefy until there is little parenchymal material left within the body cavities.

#### *19.1.1.2 Gas*

Gas is a frequent finding in cases of decomposition, but it may also be produced in cases of fatal and non-fatal gas embolism, or simply as pathology (e.g., in extensive soft tissue emphysema, pneumothorax, hollow organ perforation, or intrathecal gas) (Figure 19.2). Gas within the deceased body may have a multifactorial etiology and may not necessarily be linked to putrefaction. Cases of sepsis show gas accumulation within the entire body that may not fit the time course of the postmortem interval, effectively indicating accelerated putrefaction on imaging. However, this phenomenon has not yet been

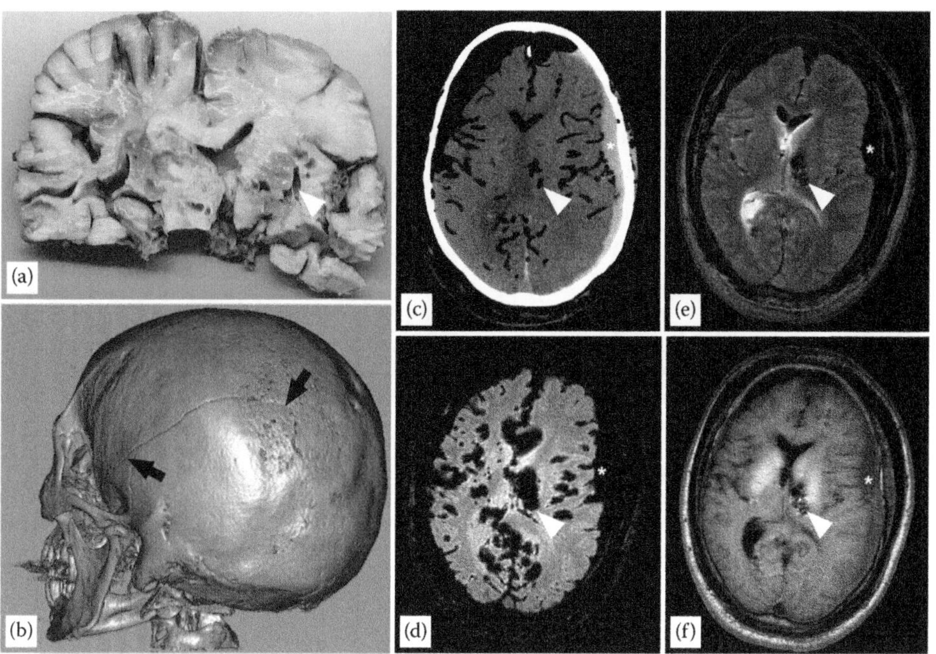

**FIGURE 19.1**
(a) Macroscopic specimen from a coronal cut of the formalin fixed brain. The decomposed brain was fixed to achieve a better tissue consistency for later dissection. Note the gas within the soft tissue (arrowhead). The extent of the pneumoencephalon and intravascular gas is hardly detectable. (b) Three-dimensional reconstruction of the skull displaying the left-sided fracture (black arrows) in this case of homicide with blunt force. (c) PMCT image in axial orientation. The pneumoencephalon results from a combination of the skull fracture and decompositional collapse of the brain. There is gas within the entire intracranial vasculature, in the soft tissue such as the galea and in the brain tissue, especially within the basal ganglia on the left (arrowhead). Concomitantly, there is a left-sided subdural hematoma (asterisk) and slight subarachnoid hemorrhage. (d) Venous BOLD sequence (susceptibility-weighted sequence [SWI]) alone does not allow for differentiation between gas, calcification, or blood products. This sequence is impaired by decompositional gas (arrowhead) and must be read in combination with morphological sequences and, optimally, with PMCT. The subdural hematoma (asterisk) presents with signal intensities similar to gas. (e) The $T_2$-weighted sequence of the brain allows for precise differentiation between the gray and white matter, even in a decomposed state. Gas appears to be hypointense (arrowhead), as does the intracranial hemorrhage (asterisk). (f) Note the hyperintense signal within the basal ganglia (postmortem temperature dependency) in the $T_1$-weighted sequence of the brain.

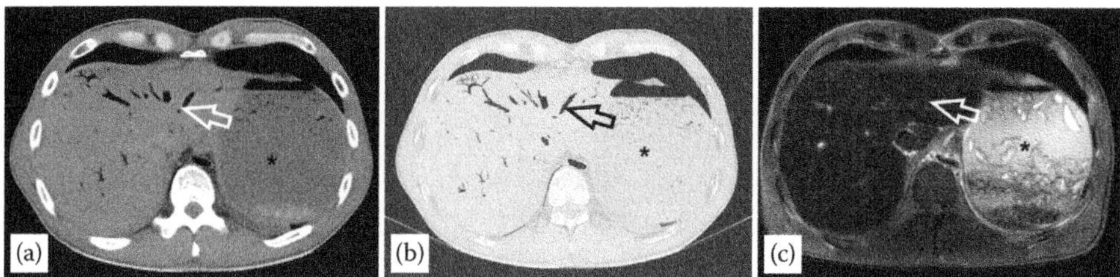

**FIGURE 19.2**
(a) PMCT at the level of the liver in a soft tissue window. (b) PMCT at the same level in lung window. (c) PMMR ($T_2$-weighted SPAIR sequence, fat-saturated) at the same level. Note the left-sided pneumothorax and gas within the abdominal aorta. This case exhibits intrahepatic vascular gas (arrow). Initial decomposition may present similarly, but the combination of the other traumatic findings with only intravascular gas leads to the diagnosis of a fatal gas embolism. The stomach is distended (asterisk) due to chyme with position-dependent sedimentation of rice.

evaluated systematically. Gas accumulation, regardless of the etiology, may very well impair image quality on PMMR to a much greater extent than on PMCT.

#### 19.1.1.3 Motion Artifacts

Postmortem imaging is not impeded by motion artifacts, such as arterial pulsation, heartbeat, or breathing movements, as is clinical radiology. However, there are some artifacts induced by positional changes of a corpse as the body gradually settles under the influence of gravity, for example, when the body is turned from the supine to the prone position (to overcome position-dependent artifacts) on the gurney. This settling may create some artifacts during the first few minutes of scanning with PMMR (Figure 19.3). Therefore,

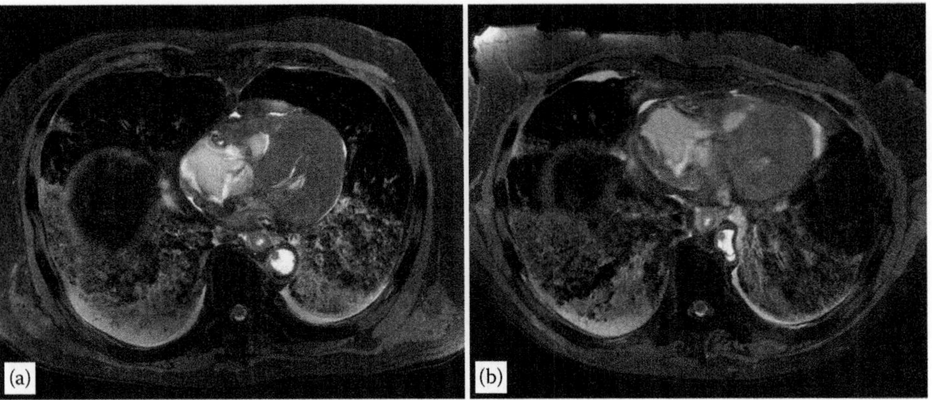

**FIGURE 19.3**
(a) Fat-saturated $T_2$-weighted sequence in supine position. (b) The same case in prone position with the equivalent imaging parameters. In the prone image there are motion artifacts of the mediastinum, predominantly visible at the heart, leading to blurring and loss of image quality caused by gravity and subsidence of the body. Concomitantly there is a slight bilateral pleural effusion and internal livores.

it is advisable to begin initially with short sequences or to wait several minutes to allow the body to settle in its new position. On PMCT, these artifacts are usually irrelevant, as the scan is too fast to allow such artifacts to occur.

#### *19.1.1.4 Sedimentation*

Given that there is no circulation in a corpse, intravascular position-dependent sedimentation occurs. Corpuscular blood particles sediment in the posterior areas, whereas the plasma settles on top of the descended blood particles, creating a fluid–fluid level (Figure 19.4). This phenomenon is very well observed on $T_2$-weighted or short-tau inversion recovery (STIR) (steady-state inversion recovery) images: the position-dependent particles appear posteriorly as a dark, homogeneous hypointensity, whereas the fluid blood components are layered on top as a homogeneous, bright, hyperintense accumulation. This often also applies for wound cavities in the soft tissue (similar to a clinical case with a Morel–Lavallée lesion, which is a closed degloving injury that typically occurs when skin and subcutaneous tissue are abruptly severed from the fascia underneath; this injury is often observed at the hip) and may also be observed in large intracranial hemorrhages or hemothorax, hemoperitoneum, and hemopericardium [72]. The sedimentation effects of corpuscular blood components also occur

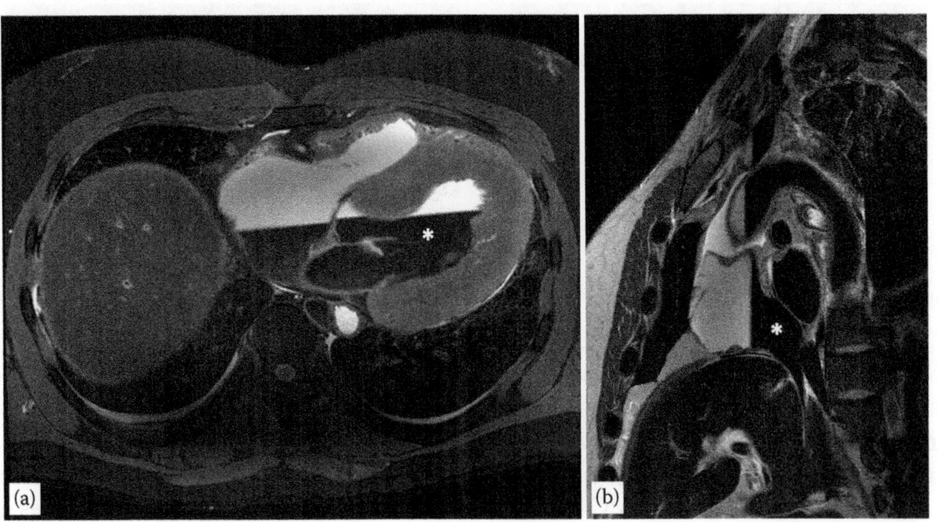

**FIGURE 19.4**
(a) Axial $T_2$-weighted imaging (fat-saturated) displaying typical sedimentation of corpuscular blood particles (asterisk) in the posterior parts and the plasma settling on top of the descended blood particles, creating a fluid–fluid level. (b) Sagittal $T_2$-weighted image with a dark, homogeneous, hypointense level (asterisk) posteriorly with fluid blood components layering on top as a homogeneous, bright, hyperintense accumulation.

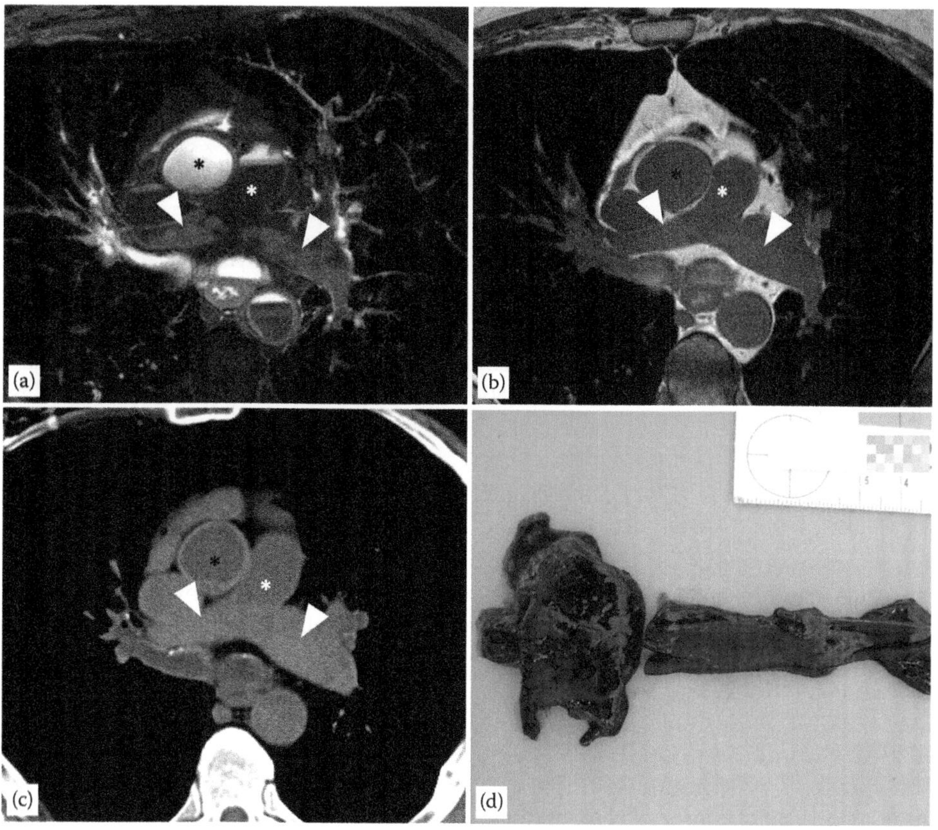


**FIGURE 19.5**
(a) Detailed image of the pulmonary trunk and both pulmonary arteries in $T_2$-weighting (fat-saturated). (b) Same level in $T_1$-weighting. (c) PMCT at the same level. (d) Autopsy specimen. The dark reddish parts correspond to currant jelly clot components, and the brighter parts to chicken fat clot. This case exhibits postmortem clots (arrowheads) with homogeneous isointense–hypointense (currant jelly clot) with little heterogeneous isointense–hyperintense signal changes consistent with components of a chicken fat clot. On PMCT, there is almost no precise discrimination feasible; only with slight alteration of the window may Hounsfield unit (HU) changes become visible. There is concomitant sedimentation of corpuscular cells (white asterisk) and, on top, a fluid level (black asterisk) consistent with plasma.

in cases of internal hemorrhage and may be useful as an indirect sign for detection of parenchymal or vascular lesions [2,96].

In the heart and blood vessels, depending on the case circumstances, this settling of blood may be impaired by postmortem clotting (cruor) that should not be confused with antemortem clotting (Figure 19.5). These postmortem clots may (rarely) occur to a vast extent, presenting with homogeneous isointense–hypointense (known as a *currant jelly clot*, which is a non-adherent coagulum composed of red cells and scattered white cells) to heterogeneous isointense–hyperintense signal changes (known as a *chicken fat clot*, formed from leukocytes and plasma of sedimented blood, which usually forms more slowly than a currant jelly clot), depending on the components of the postmortem clot. This postmortem clotting may even mimic antemortem clotting and is an issue that must be taken into account when ruling out thrombosis or pulmonary thromboembolism, which may look similar.

Hypostasis of the lungs is another sedimentation effect similar to external visible livor mortis. Livor mortis (or postmortem lividity) occurs in the course of the settling of blood in the lower position-dependent body parts, when red blood cells start to sink under the influence of gravity (Figure 19.6). Inner livores in the lungs seem to occur similar to external postmortem lividity and are able to be repositioned within a certain time frame. This hypostasis of the lungs should not be mistaken for pathology and may aggravate detection of other pulmonary pathologies such as infiltrates, aspiration, or pulmonary edema. The same phenomenon has been described within the liver parenchyma and is best observed on PMMR [72].

#### 19.1.1.5 Metal Objects and Artifacts

Forensic cases that involve retained bullets, pellets, or shrapnel—leading to potential heating, dislocation, or artifacts in a corpse—represent a vast portion of the forensic material being radiologically investigated in several countries. Ballistic projectiles are ferromagnetic if they contain steel (e.g., iron), but most bullets are made

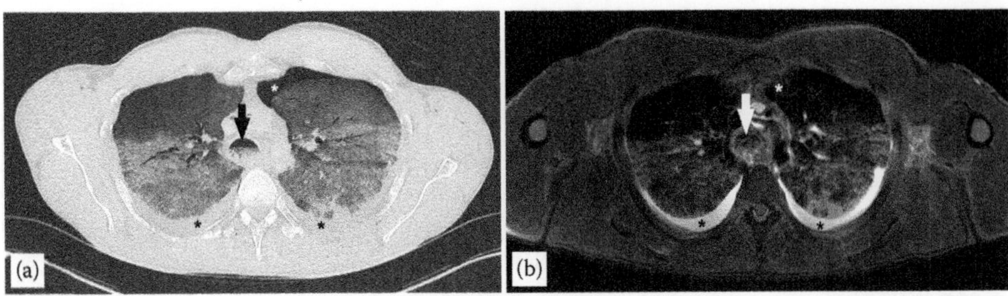


**FIGURE 19.6**
(a) PMCT in pulmonary window. (b) $T_2$-weighted sequence (fat-saturated). Both images display typical inner livores (postmortem lividity) represented by opacity of the lower position-dependent parts of the lung due to settling of blood. Note additional pneumothorax on the left (white asterisk) in a case of central regulatory failure due to a bicycle accident. Additional froth or foamy fluid within the airways is a frequent postmortem finding (arrow).

of lead (which is inexpensive), or are jacketed or plated with gilding metal, cupronickel, copper alloys, or brass and are therefore not ferromagnetic.

General metal artifacts may present with signal extinction and distortion of the surrounding tissue. Therefore, awareness of the limitations in imaging metal objects, how to eventually overcome these, and the advantages of PMMR relative to PMCT is desirable for postmortem cases, as well as in clinical cases. On PMMR, there might not only be gunshots with retained bullets or bullet fragment but also typical day-to-day problems, such as knee or hip replacements, spondylodesis, and dental work, that impair image quality but do not pose a liability risk. It is advisable to always image corpses by PMCT before placing the body in the PMMR suite to gain information about any metallic objects or debris in or on the body.

PMCT is an established imaging modality in the depiction and localization of metallic foreign bodies. In addition, the osseous findings, especially in the skull, allow for proper differentiation between the entry and exit defects or identification of ricochets [97]. Scattered metal abrasions or bony fragments along the wound channel may also aid in the detection of the bullet path. However, PMMR is certainly superior for the depiction of soft tissue lesions and defects in the brain parenchyma (Figure 19.7) [52]. Still, the imaging of potentially dislocated metallic objects such as bullets is infrequently performed, as there are potential liability issues regarding the integrity of the PMMR suite [78,98]. Furthermore, there are concerning factors such as susceptibility artifacts and heating of the surrounding tissue that lead to reluctance to perform PMMR in cases of penetrating and non-perforating bullets [99–104].

Prior to placing the corpse into the magnet, there are several possibilities for differentiating ferromagnetic from non-ferromagnetic projectiles to improve safety issues. Ferromagnetic objects in superficial locations may be tested by placing a magnet over the suspected area (known from PMCT) and ruling out a potential magnetic torque or by pretesting magnetism with a similar type of object, if the intra-corporal one is known [99]. Ferromagnetic detection systems may be able to detect ferromagnetic bullets [102]. However, a simple metal detector may be placed before the entrance door of the PMMR suite, giving an audio and visual alarm, although it does not differentiate between ferromagnetic and non-ferromagnetic objects. Pretesting may also be triaged by the usage of dual-energy PMCT to detect ferromagnetic projectiles [105].

There are three major considerations that have to be made regarding lodged projectiles or ferromagnetic objects: first, dislocation or migration of the object; second, susceptibility and how to reduce this; and finally, thermal effects on the surrounding tissue.

##### *19.1.1.5.1 Dislocation*

Ferromagnetic, diamagnetic (e.g., copper and lead), and paramagnetic (e.g., aluminum) materials interact slightly with the static magnetic field due to the Lenz effect. This clinical effect may impact the function of a non-ferromagnetic cardiac valve in high magnetic fields due to continuous deceleration [106]. The current literature does not contain a single report of a case of magnetic torque in a clinical MR suite leading to ferromagnetic extra-corporal dislocation of a bullet or fragment. Nevertheless, there are publications describing the alignment of a ferromagnetic object along its long axis parallel to the $B_0$ field and a 90° rotation under maximum magnetic torque in a phantom, but not with migration of the object [103]. Other experimental research setups claim (still potentially patient-safe) excess movement of steel objects in 1.5–7 Tesla scanners [98]. Magnet-induced bullet dislocation must be differentiated from other bullet migration etiologies. It is important to note that bullet migration (ferromagnetic or not) may also occur

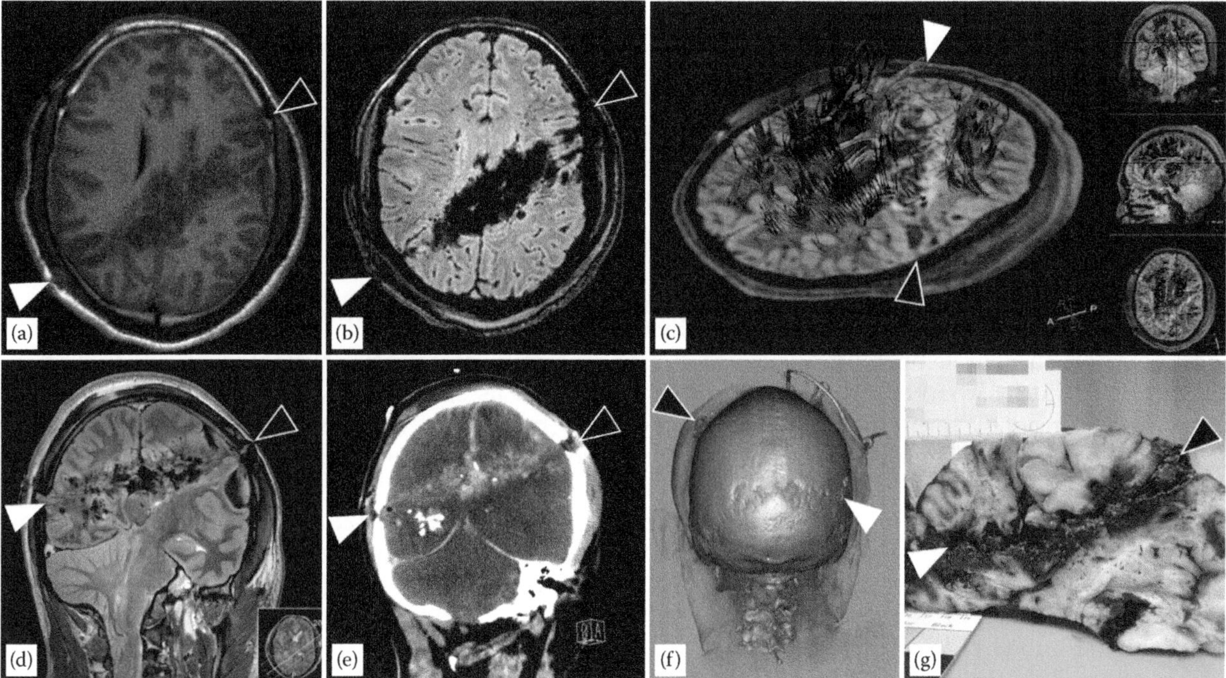


**FIGURE 19.7**
(a) $T_1$-weighted sequence. (b) Venous BOLD (SWI). (c) Fiber tracking in diffusion tensor imaging. (d) $T_2$-weighted imaging aligned to the perforating gunshot to the head. (e) PMCT aligned to the shot canal, as in image d. (f) Three-dimensional reconstruction of the gunshot wound of the PMCT, viewed from the posterior. (g) Fixed autopsy specimen. The white arrowhead represents the entry wound, and the black one denotes the exit wound. PMCT (e) precisely displays the scattered metal abrasions and bony fragments along the wound channel. The PMMR superiorly sequences the soft tissue lesions and the defects in the brain parenchyma. There is a vast signal void in the SWI (b) along the intra-axial hemorrhage that had time to develop due to a prolonged agonal interval. Image (c) displays diffusion tensor imaging with the traumatic lesions and potential fiber tract rupture.

independent of magnetic torque in cases of ricochet injury or large wound cavities.

#### *19.1.1.5.2 Susceptibility*

Most projectiles lack ferromagnetic properties and, therefore, do not exhibit the potential for heating effects or migration. Those types of bullets usually present with few susceptibility artifacts, producing only a focal *small, black hole*, therefore allowing virtual autopsy diagnosis on PMMR that is superior to PMCT, while they would introduce extensive streak artifacts on PMCT that could be minimized with the use of PMMR [105,107]. The severity of susceptible artifact impact on MR obviously depends on the material compounds in the projectiles. The literature on lodged projectiles in living patients describes highly susceptible ferromagnetic projectiles without complaints about any discomfort during or after the 1.5 T examination [103]. There are influencing factors additional to the metallic ferromagnetic compounds, including the shape and length of the projectile, the position of the bullet along the magnetic field, and the conducting characteristics of the surrounding tissue [103,108].

The employment of a specific metal artifact sequence reduces susceptibility artifacts resulting from magnetic field distortion. Spin-echo sequences are less vulnerable to artifacts; STIR sequences with an increased inversion pulse and new warp sequences (STIR-warp and $T_1$-warp), as well as parameter adjustments such as shorter echo spacing, smaller water-fat shift, thinner slices, and high SNR (signal-to-noise ratio), all lead to reduced metal artifacts [108–111].

#### *19.1.1.5.3 Thermal Effects*

Excessive thermal effects resulting from surgical implants have not been described [112]. Another study showed that there was no detectable magnetic field interaction at all at field strengths up to 3 T in both steel-containing and non-steel-containing bullets [98]. In addition, the literature explains that lodged metallic objects deep inside the human body may rarely be heat excessive due to the surrounding soft tissue, which does not serve as a conducting medium [98,113].

However, caution is advised if metallic objects are located superficially, as heating has been described in clinical cases. The relevance of these effects is unknown [113].

#### 19.1.1.6 Temperature Dependency

Temperature is not as influencing a factor on PMCT (unless the bodies are frozen or charred) as it is on PMMR. Image contrast on PMMR is greatly dependent on temperature, as $T_1$ and $T_2$ relaxation times are parameters that are strongly influenced by the temperature of the imaging object [114,115]. Corpses seldom present with physiological antemortem body temperatures (such as 36°C) but, rather, with much lower body temperatures due to algor mortis (cooling of the corpse after death). Therefore, image contrast is altered significantly in corpses with lower temperatures. This fact compels examiners to take rectal temperature measurements prior to and after PMMR imaging to enable retrospective evaluation of the correlation between core temperature and PMMR in both routine imaging and research settings.

Ruder et al. described a lower image contrast between fat tissue and muscle tissue in bodies at low temperature, whereas the contrast between fat tissue and fluid increased on $T_2$-weighted images in a 1.5 T scanner [116]. The authors stated that the influence of the body temperature became visible below 20°C on $T_2$-weighted images and that imaging on PMMR below 10°C on $T_1$-weighted images should be avoided, as the technical quality became too low. Below this temperature of 10°C, PMMR image contrast is insufficient to allow for diagnostic radiological interpretation. Similar specific cerebral anatomical findings have been described by Kobayashi et al. in corpses that have been stored at 4°C with a mean postmortem interval of 26 h. According to this research group, the basal ganglia and thalamus showed a higher signal intensity on $T_1$-weighted images and a decrease in signal on diffusion-weighted images (Figure 19.1) [117]. In addition, if clinical settings for sequences are used for a fluid-attenuated inversion recovery (FLAIR) sequence, the suppression of the cerebrospinal fluid (CSF) will be unsuccessful. Image parameters of a FLAIR must be adapted to an inversion time of 1500 to 1700 ms (on 1.5 T), which is lower than that in a living person [114,117,118]. Tofts et al. described the temperature influence on PMMR as *cold brain* effects, and they determined that diffusion coefficient values could serve as a noninvasive brain core temperature measurement to within 1°C [118]. In addition, they advise an inversion time in FLAIR sequences of 1500 ms (on 1.5 T), to return the CSF suppression [118]. This observation was also later confirmed by Kobayashi et al. [117,119]. A Japanese research group also describes how the $T_1$ value of subcutaneous fat in PMMR correlates linearly with the body temperature and notes that this correlation may be used to determine the inversion time needed for sufficient fat suppression on STIR sequences on PMMR [119]. PMMR offers a great number of unexplored possibilities for postmortem imaging determining body temperature, pathology, and advanced techniques such as spectroscopy, diffusion-weighted and tensor imaging, and, eventually, functional MRI in the future.

#### 19.1.1.7 Advanced Techniques

The field of diffusion-weighted and tensor imaging, as well as spectroscopy and functional MRI, has not yet been fully explored. There are several publications regarding *in situ* neuroimaging (not a harvested human brain fixed in formalin) for PMMR; Yen et al. describes the potential of diffusion tensor imaging for traumatic fiber tract rupture in a case with a brain stem lesion and two controls (Figure 19.7) [120]. Scheurer et al. investigated 20 cases with diffusion-weighted and tensor imaging to determine the postmortem interval. According to this group, the apparent diffusion coefficient (ADC) could serve as an indicator for the assessment of the postmortem interval [121].

Attempts to determine the postmortem interval have also been made by means of PMMR spectroscopy. One research group measured five different metabolites (acetate, alanine, butyrate, free trimethyl ammonium, and propionate) as parameterized functions of the time course, mainly for intervals after time of death (up to 250 h) in sheep and human subjects, to determine the specific postmortem interval [122–124]. There have also been attempts to determine the forensically crucial time of death by phosphorus metabolite measurements in the skeletal muscles on PMMR spectroscopy [125]. The authors found a significant correlation between the time of death and the ratio of ATP and phosphate in the adductor magnus muscle that aided in the determination of the time of death [125].

There is ongoing research in this technically advanced field of metabolic, diffusion, and functional PMMR. However, research teams should always include multidisciplinary expertise, such as physicists, physicians, and radiological technical personnel who are well trained in MRI, to avoid inaccurate results in this complex sub-specialty. Bennett et al. won in 2012 the Ig Nobel Prize in neuroscience with the *dead salmon* study. The authors examined a mature postmortem Atlantic salmon by means of visual stimuli on functional PMMR, discovering active voxels within the salmon's brain cavity [126]. The authors interpreted this activity as random noise and showed that it was necessary to employ multiple comparison corrections as standard statistical practice to avoid analyzing error measurements [126]. This study clearly shows that flawed results may be produced if not interpreted, amended, or analyzed correctly.

#### *19.1.1.8 Forensic Sentinel Sign*

As previously described by Ruder et al., the predominant sequence used in unenhanced PMMR is a heavily $T_2$-weighted sequence, such as the STIR sequence [78]. This sequence uses a fat-suppression technique in which the signal of fat is zero. This approach allows for a distinct depiction of fluid accumulations in the body. Forensic indications usually include detection of soft tissue hematomas (e.g., in traffic accidents, homicide, or suicide), parenchymal lesions, bone marrow contusions, and edema in ischemia of the heart or any other pathology with fluid accumulation (Figure 19.8) [49,65,69,71,72,76,77,79,80,127–134]. Therefore, the fluid accumulations—best detected with STIR and other $T_2$-weighted fat-suppressed sequences—are to be considered a forensic sentinel sign and comprise the key aspect of forensic PMMR in screening for major pathology.

#### *19.1.1.9 Radiological Interpretation of PMMR*

##### *19.1.1.9.1 Head*

The postmortem brain generally presents on PMCT with a blurring of the sulci effacement and loss of the cortico-medullary differentiation. This fact strongly aggravates the detection of agonal or antemortem brain edema. On PMMR, this effect is visible to a lesser degree. Despite the above-mentioned postmortem signal changes, there is still a clear depiction of the gray and white matter and a distinct superior soft tissue contrast, allowing for better and more precise radiological interpretation of pathology (Figure 19.9) [114,117,119]. However, research

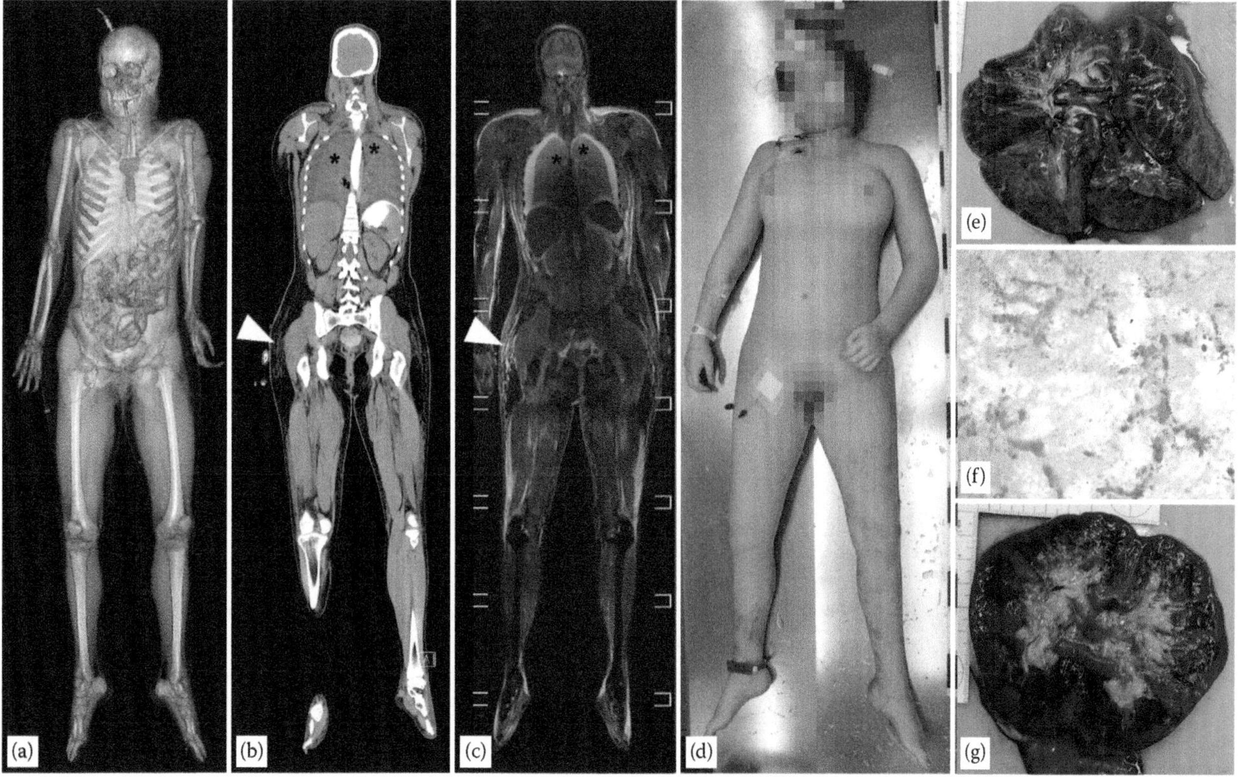


**FIGURE 19.8**
(a) Three-dimensional reconstruction of the PMCT. Note the left-sided humeral fracture and residual oral contrast media in this case of an accident. (b) Coronal PMCT. (c) $T_2$-weighted STIR sequence (fat saturated) of the whole body on PMMR. (d) Photograph of the external inspection. (e) Autopsy specimen of the lung. (f) Histological specimen in Sudan staining. (g) Autopsy specimen of the kidney. This case sustained severe head trauma with central regulatory failure and subsequent accident-related pathologies. Images (b) and (c) show vast subcutaneous fluid accumulations: the forensic sentinel sign (specifically indicated by the arrowhead) due to intensive care resulting in vast anasarca. On external inspection, there were signs of general soft tissue swelling, according to the imaging findings. On the coronal PMCT and PMMR, both lungs appear with consolidation (black asterisk) and adjacent pleural effusions in a case of shock lung with extensive fat embolism (as shown in image f). Both kidneys showed signs of acute renal failure (g).

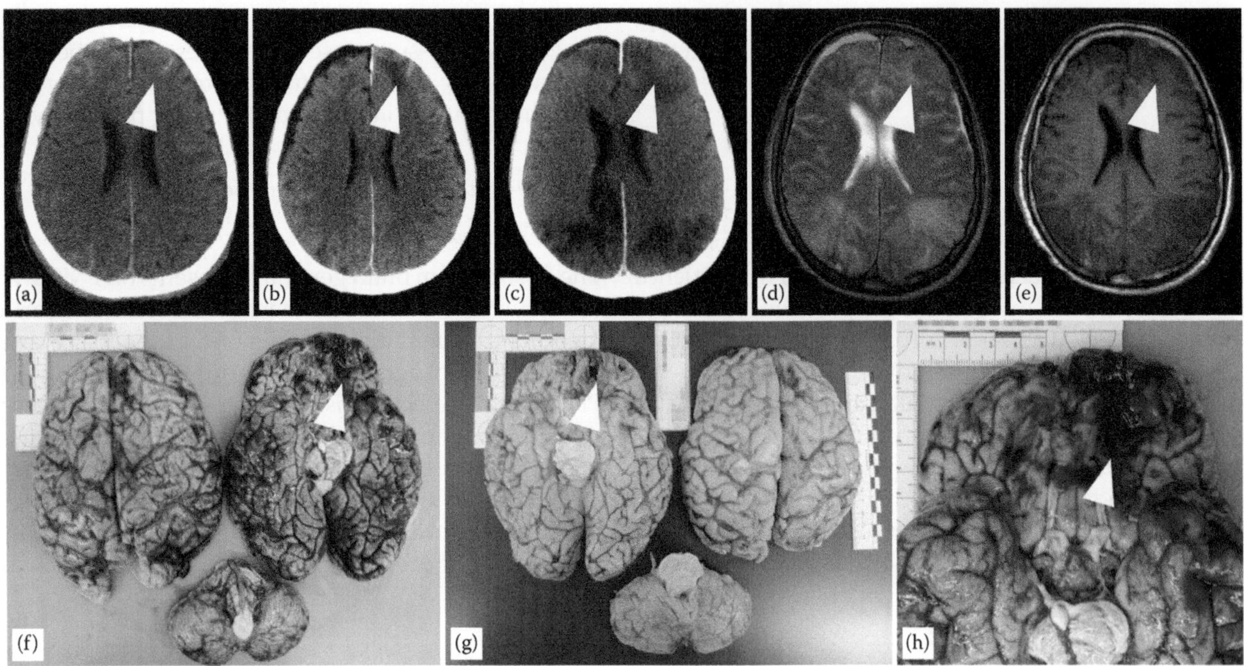

**FIGURE 19.9**
(a) Clinical head scan on CT after trauma (car-to-pedestrian collision). (b) Clinical head scan on CT at one week follow-up. (c) PMCT 14 days after the follow-up clinical head scan. (d) PMMR ($T_2$-weighted) directly after PMCT. (e) PMMR ($T_1$-weighted) directly after PMCT. (f) Autopsy photograph of the brain. (g) Fixed brain specimen. (h) Detailed view on the contusion of the frontal lobe from below. Initial head scan (a) revealed subarachnoid (arrow) and slight subdural hemorrhage with concomitant occipital skull fractures extending into both petrosal bones. On initial head scan the contusions were not yet demarcated. The follow-up scan (b) one week later revealed resorption of the left-sided fronto-polar contusion (arrow), residual subarachnoid hemorrhage and subacute hematoma on the right. A postmortem head scan (c) and PMMR $T_2$-weighted (d) and $T_1$-weighted (e) displayed extensive resorption of the primary impact with hypodense appearance of the bifronto-polar (arrow) and bioccipital contusions. Note the improved depiction of the gray and white matter of the brain and the subacute subdural hematoma compared to PMCT. During autopsy an extensive parenchymal defect with visible hemorrhage is discovered (arrow) (f–h). Fixation of the horizontal cut brain (Flechsig cut) allows for better dissection due to the soft consistency of such a traumatic brain. This individual died due to an accident that caused central regulatory failure.

on diagnostic issues for PMMR is still lacking, and any diagnosis that is made is merely based on experience rather than on knowledge gained from the literature.

Agonal and antemortem edema may be differentiated from typical postmortem swelling, as stated by Aghayev et al., who described PMMR as being sensitive in postmortem diagnostics of tonsillar herniation and subsequent brain edema [68]. PMMR surpasses PMCT in detecting lesions of the brain stem and the posterior fossa. PMCT usually displays increased noise at this level, and streak artifacts due to dental work often tremendously impair image quality.

Position-dependent stasis within the intracranial vessels (especially the veins and venous sinus) occurs frequently, as does the accumulation of corpuscular blood components, dependent on the position of the head. This phenomenon may particularly affect susceptibility-weighted or hemo-sequences (susceptibility-weighted imaging [SWI] and venous blood oxygenation level dependent imaging [venBOLD]) and does not necessarily imply pathology.

Mild subarachnoid hemorrhage is difficult to detect on PMCT due to several artifacts that cause sulcal hyperdensities (e.g., putrefaction, proteinaceous inspissated fluid, prominent dammed vessels, and position-dependency). On PMMR, the signal alterations in $T_1$W, $T_2$W images, and hemo-sensitive sequences may assist differentiation between pathology and postmortem artifacts.

There are no postmortem publications dedicated solely to the topic of the aging of intracranial hemorrhages. Based on current knowledge, the signal changes caused by an intracranial bleed in a corpse are evaluated as if in living subjects, and further research must be performed [135,136]. Recent publications investigated PMCT and PMMR correlated with autopsy regarding postmortem forensic neuroimaging and yielded an overall specificity of 94% for intra- and extra-axial findings in PMCT and PMMR [137]. Furthermore, the lowest specificity was 70% in subarachnoid and ventricular hemorrhages [137]. Añon et al. stated that discrete extra-axial hemorrhages may escape on both PMCT and PMMR, particularly in the basilar or posterior fossa of the skull. [69].

Brain ischemia due to antemortem or agonal infarction is well depicted on PMMR and presents with perifocal edema and swelling in the affected territory [138]. Gradual changes in signal intensities, especially in acute stroke, are more precisely detected on PMMR, and radiological interpretation is better than on PMCT.

In the depiction of a bullet path, PMMR is especially valuable combined with the osseous findings on PMCT. The intracranial intra-axial path is usually better depicted than on PMCT alone; however, the entry and exit wounds (if present) are much easier to assess on PMCT. Advanced techniques such as postmortem diffusion tensor imaging with fiber tracking allow even laypersons to easily assess the bullet path (Figure 19.7).

Diffusion-weighted imaging (DWI) is clearly restricted in diagnostics as the brain tissue presents as if a generalized stroke occurred after cessation of circulation and oxygen supply. On DWI, the brain displays a high signal intensity—most prominent at the cerebral cortex and the ventricular wall and without a significant effect on the white matter—whereas the corresponding ADC map shows a hypointense and very low signal within the entire brain parenchyma, with the greatest decrease being within the vermis [117,121]. It is advisable to acquire DWI single-shot sequence with a short TE time to obtain the highest signal-to-noise ratio.

From the forensic perspective, galeal and/or facial hematomas, as well as orbital and paranasal sinus pathologies, are of interest for the investigation. The same applies for the area at the base of the skull where autopsy is limited. PMMR combined with PMCT plays a major role in detecting lesions (Figure 19.10). PMMR may depict retinal hemorrhage in non-accidental injury (hypointense signal on $T_2$W, non-accidental injury), forensically important galeal hematomas, or even vascular occlusion [60,139,140]. In vascular occlusion, signal alterations may be observed that must be differentiated from postmortem clotting and sedimentation effects for proper readout. Table 19.1 summarizes a potential head PMMR protocol.

*19.1.1.9.2 Neck*

In a postmortem setting, imaging of the neck region is mainly applicable to cases of strangulation. Kempter et al. evaluated the accuracy of the laryngeal fracture detection rate by PMCT compared to autopsy within a

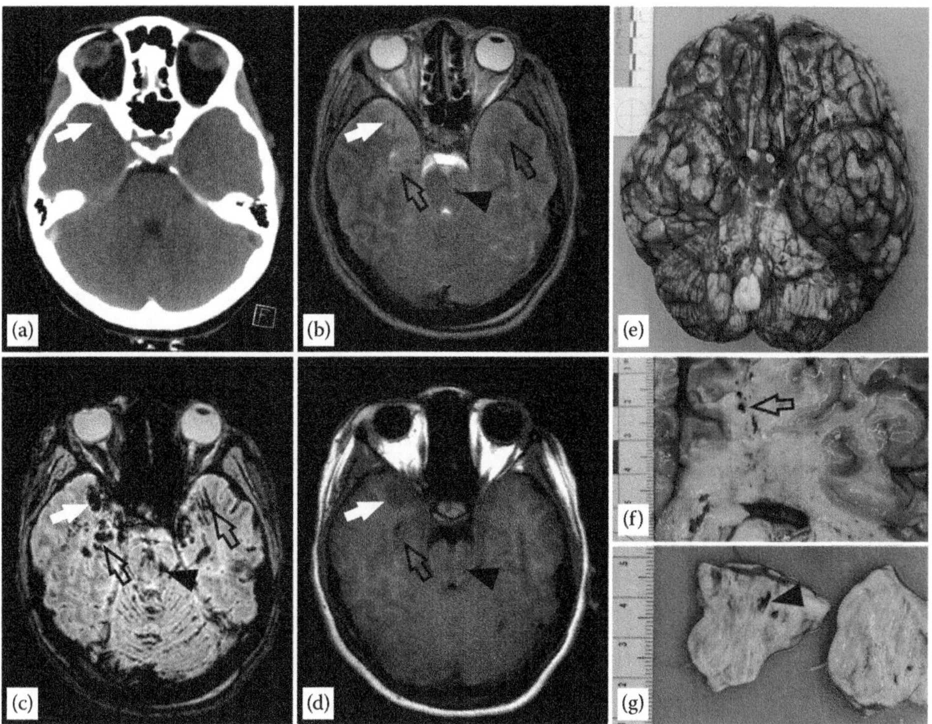


**FIGURE 19.10**
(a) PMCT after an accident, displaying dot-like hyperdense lesions in the right temporal lobe (white arrow). (b) PMMR $T_2$W already showing more detailed lesions than PMCT. Note the slight edema in the right temporal lobe with multiple shearing injuries (white arrow), also visible on the contralateral side (black arrows). PMMR additionally exhibits hemorrhage due to shearing in the pons (arrowhead). (c) PMMR SWI (SWI) clearly showing the extent of the shearing injuries and pontine hemorrhage (arrows). (d) PMMR $T_1$-weighted corresponding to the $T_2$W image (b). (e) Autopsy specimen of the brain, viewed from below. (f) Detailed photograph of the cerebrum showing shearing injuries (black arrow) according to the PMMR images. (g) Section cut of the pons with the parenchymal hemorrhage (arrowhead) correlating well to the PMMR images.

**TABLE 19.1**

Comprehensive Summary of Potential Standard Sequences

| Head: PMMR Imaging Protocol | | | |
|---|---|---|---|
| **Sequence** | **Slice Thickness** | **Plane** | **Diagnostics** |
| $T_2$W | 4 mm<br>(neonates: 2 mm) | Axial | Forensic sentinel<br>Edema<br>Hemorrhage |
| $T_1$W | 4 mm<br>(neonates: 2 mm) | Axial | Morphology<br>Hemorrhage |
| Hemosequence (e.g., SWI, SWIp, venBOLD, $T_2$*) | 0.5 mm (+ MIP) | Axial | Hemorrhage<br>Shearing injury |
| FLAIR (3D) | 0.5 mm<br>(neonates: 2 mm) | Sagittal | Adapted IR for CSF suppression<br>Multiplanar view |
| $T_2$ (3D) | 0.5 mm | Sagittal | Multiplanar view |

*Notes:* SWI, susceptibility-weighted imaging; SWIp, phase; venBOLD, venous blood-oxygen-level dependent; 3D, 3 dimensional; MIP, maximum intensity projection; IR, inversion recovery; CSF, cerebrospinal fluid.

small study population and found that PMCT detected all fractures observed during autopsy and two more that the autopsy did not detect. Based on this result, PMCT is clearly a valuable modality to detect laryngeal lesions in strangulation. However, PMCT failed to detect lesions of the soft tissue or vasculature in 5 of 6 cases. This limitation may be overcome by performing PMMR of the neck (Figure 19.11). Particularly with knowledge of external findings (e.g., the level of the strangulation mark or bruises) and the specific fracture location on PMCT,

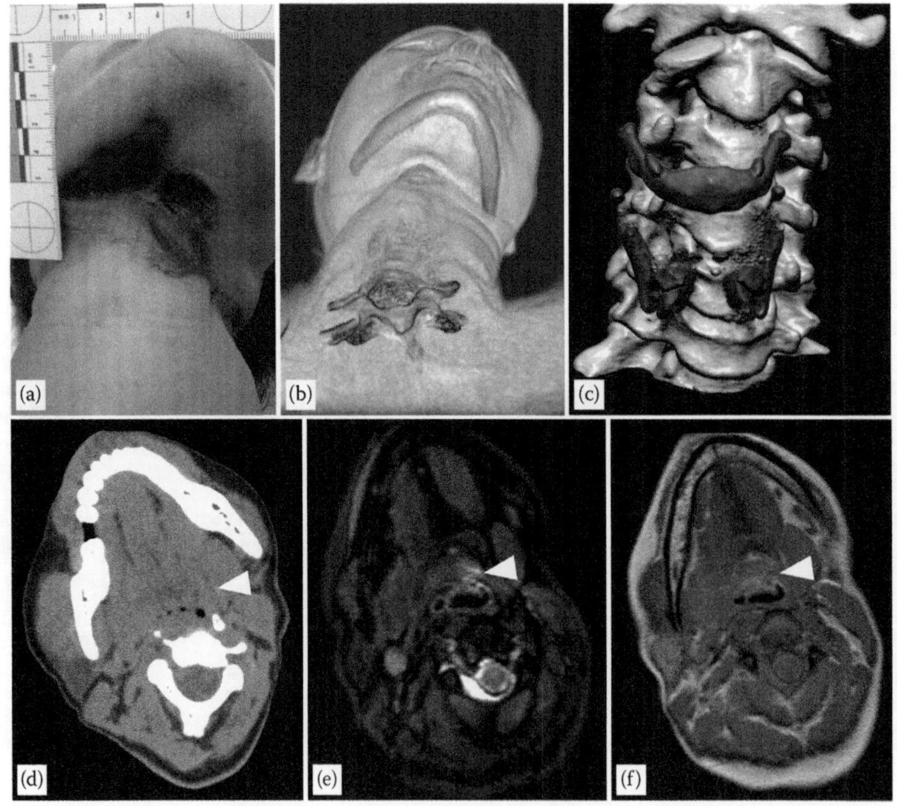


**FIGURE 19.11**

(a) Photograph of the neck of the deceased showing strangulation marks. (b) PMCT 3-dimensional reconstruction displaying the indentation of the strangulation proximal to the laryngeal structures. (c) PMCT 3-dimensional reconstruction of the ossified parts of the larynx (colored in blue). (d) PMCT of the neck, corresponding to images (e) and (f). (e) PMMR $T_2$-weighted (fat saturated). (f) PMMR $T_1$-weighted. Suffocation due to incomplete hanging in suicide. Note the $T_2$W hyperintense, $T_1$-weighted hypointense edema or hemorrhage (arrowhead) in the laryngeal musculature that was not detectable on PMCT. There were no fractures in the laryngeal skeleton. Autopsy confirmed the radiological findings.

PMMR imaging will allow for more *in situ* detail and adjacent hemorrhage perifocal to the fracture in the musculature and the surrounding soft tissue [65]. Yen et al. investigated nine deceased cases of hanging by PMCT, PMMR and autopsy and came to the conclusion that PMCT and PMMR together revealed strangulation signs concordant with forensic autopsy, apart from a vocal cord hemorrhage that was only identified during autopsy [141]. Still, PMMR allowed for statistically significant detection of traumatic lymph node hemorrhage [141].

On PMMR, the forensically critical sites for the detection of strangulation-related lesions are clearly primary findings at the site of the strangulation mark and concomitant subcutaneous desiccation or soft tissue alteration. In addition, the investigator wants to search for subcutaneous, intramuscular, platysmal and lymph node hemorrhage. Congestion or even hemorrhage of the salivary glands may occur, as may lesions of the (especially deep) vasculature. Fractures of the larynx itself are better observed on PMCT. It is advisable to acquire a head scan concomitant with neck PMMR sequences, as some findings are linked to one another. Therefore, at best, the neck protocol should also include a standard protocol of the head on PMMR.

Additional forensic inquiries by neck PMMR may be conducted in cases with sharp or blunt force against the neck, for example, in traffic accidents, homicide or suicide [58]. A depiction of vascular transection at the neck is hardly feasible on PMCT, and angiographic procedures frequently result in substantial extravasation of contrast media, impairing the detection of the precise location of the vascular lesion. It is advisable to perform an unenhanced PMMR of the neck initially. Soft tissue hemorrhage may even allow matching with an external patterned injury and help in reconstructing subsequent events.

Another application of neck PMMR is the depiction of laryngeal obstructing objects. In adults, especially the elderly, bolus death is often termed *café coronary syndrome* (lodging of chyme in the airways, causing complete airway obstruction and resembling the symptoms of a coronary syndrome) [142,143]. Bolus death may lead to asphyxia in a complete obstruction of the laryngeal inlet or complete airway obstruction, mainly in the elderly or in very young subjects, or due to a *reflex-related* process of bolus death. Other foreign material that has been swallowed must be differentiated from chyme or primary pathology (e.g., tumor). PMCT has a high diagnostic value among noninvasive procedures and for the fast *in situ* depiction of foreign material (especially radio-dense material), but PMMR allows for much better discrimination between soft tissue structures and radio-isodense or radio-hypodense foreign bodies lodged in the pharynx [144]. The combination of PMCT and PMMR will clearly present greater diagnostic value [144].

Fluid accumulation or froth in the pharynx (as well as in the paranasal sinus and mastoid cells) is a frequent finding on imaging. The etiology may vary, but this finding is often observed in deceased individuals who underwent intensive care prior to death, in drowning and in pulmonary edema with subsequent fluid in large parts of the airways. PMMR helps in discriminating the components of this accumulated fluid. Hemorrhageous fluid will present with $T_2$W hypointense components settling in the posterior position-dependent areas within a fluid accumulation and aid in determining potential etiology. Cases dependent on such a fluid level must be differentiated from other settling materials, such as sand, in cases of drowning that may look alike. Therefore, knowledge of the case circumstances is always paramount for a correct interpretation of radiological findings. Table 19.2 summarizes a potential comprehensive neck protocol.

#### *19.1.1.9.3 Whole Body*

There are several possibilities for obtaining a whole-body PMMR, depending on the scanner used. Some companies offer whole-body surface coils (e.g., total imaging matrix [TIM]) or allow an upgrade to a whole-body imaging package, with the drawback of losing other functionality, for example, applications such as spectroscopy. The TIM comes with a surface-coil design that combines seamless integrated coil elements [133]. This type of system allows for a fast whole body scan in a single examination, which is favorable in a postmortem setting.

If the PMMR scanner does not offer such a package, the exam must be divided into several single stacks that must be acquired separately and are later merged into a whole-body image. There are 3 axial stacks that cover the thorax and abdomen and 8 stacks that cover the whole body in the coronal plane (Table 19.3). The usage of a body coil (stationary coil) is advisable. Additional scans, for example, the head or spine or extremities, may be included in the whole-body protocol, based on the case. The axial sequences should be acquired after the coronal images, as these images already allow for a quick overview to screen for pathology (forensic sentinel sign), which allows the investigator to tailor subsequent axial cross sections (e.g., in head trauma, the investigator might prefer a head scan rather than axial slices on the torso). In routine examination, the investigator usually has limited scan time and therefore must adjust the protocol based on the pathology from case to case.

Despite an entire overview of pathologies with various etiologies (homicide, natural cause of death, or suicide), the major application for postmortem whole-body imaging is blunt trauma, usually homicide or traffic accidents, especially in children who sustained patterned injuries after being run over by a car. The soft tissue hemorrhages, combined with additional surface scanning, may help in reconstructing the incident and vastly enhance

**TABLE 19.2**

Comprehensive Summary of Potential Standard Sequences

| Neck: PMMR Imaging Protocol | | | |
|---|---|---|---|
| **Sequence** | **Slice Thickness** | **Plane** | **Diagnostics** |
| $T_2W$ | 3 mm | Axial | Forensic sentinel<br>Edema<br>Hemorrhage |
| $T_2W$_STIR (fat saturated) | 3 mm | Axial | Forensic sentinel<br>Edema<br>Hemorrhage |
| $T_2W$ | 3 mm | Coronal | Second plane for confirmation or detection of a lesion |
| $T_2W$_STIR (fat saturated) | 3 mm | Coronal | Forensic sentinel<br>Second plane for confirmation or detection of a lesion |
| $T_1W$ | 3 mm | Axial | Morphology<br>Hemorrhage |
| $T_1W$_SPIR (fat saturated) | 3 mm | Axial | Morphology<br>Hemorrhage |
| $T_1W$ | 3 mm | Coronal | Second plane for confirmation or detection of a lesion |
| $T_2W$_FFE (fat saturated) | 3 mm | Axial | Hemorrhage<br>Susceptibility |

*Notes:* STIR, steady-state inversion recovery; SPIR/SPAIR, spectral attenuated inversion recovery; FFE, fast field echo; syn. FLASH, fast low angle shot; $T_1W$, $T_1$-weighted; $T_2W$, $T_2$-weighted.

**TABLE 19.3**

Comprehensive Summary of Potential Standard Sequences

| Whole-Body: PMMR Imaging Protocol | | | |
|---|---|---|---|
| **Sequence** | **Slice Thickness** | **Plane** | **Diagnostics** |
| $T_2W$ | 5 mm (infants: 3 mm) (neonates: 2 mm) | Axial | Cross sections for the torso<br>3 stacks to cover the entire thorax and abdomen |
| $T_2W$_STIR (fat saturated) | 5 mm (infants: 3 mm) (neonates: -) | Axial | Forensic sentinel<br>Cross sections for the torso<br>3 stacks to cover the entire thorax and abdomen |
| $T_2W$ | 5 mm (infants: 3 mm) (neonates: 2 mm) | Coronal | 8 stacks for a true whole-body |
| $T_2W$_STIR (fat saturated) | 5 mm (infants: 3 mm) (neonates: -) | Coronal | Forensic sentinel<br>8 stacks for a true whole-body |
| $T_1W$ (neonates: $T_1W$_IR) | 5 mm (infants: 3 mm) (neonates: 2 mm) | Axial | Morphology<br>Hemorrhage |
| $T_1W$ (neonates: $T_1W$_IR) | 5 mm (infants: 3 mm) (neonates: 2 mm) | Coronal | 8 stacks for a true whole-body |
| $T_2W$ | neonates: 2 mm | Sagittal | 3rd plane for interpretation |
| $T_2W$_FFE (fat saturated) | 3 mm | Axial | |

*Note:* The single stacks need to be combined after acquisition STIR, steady-state inversion recovery.

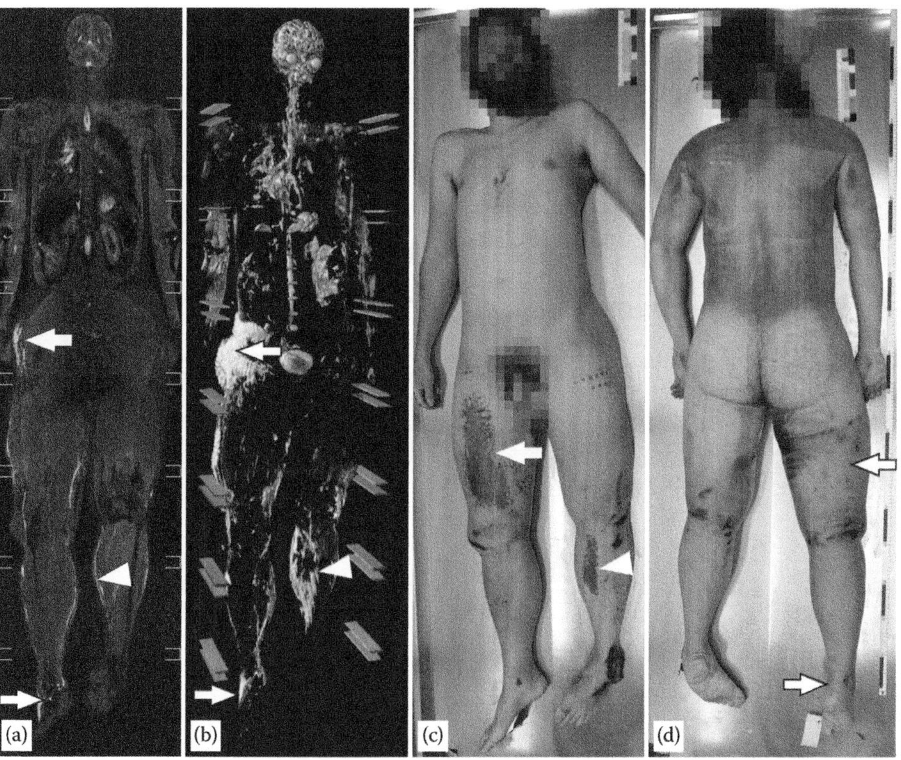


**FIGURE 19.12**
(a) Coronal whole-body PMMR, STIR sequence. (b) Volume rendering of the coronal PMMR. (c) Anterior external inspection of the decedent. (d) Posterior external inspection of the decedent. All arrows mark a hyperintense fluid accumulation within the soft tissue in a case of bicycle-to-truck collision. The direct impact is visible on the right side with extensive hemorrhage and skin abrasions. Note the open comminuted fracture of the left ankle. The coronal whole-body PMMR allows for a quick overview and detection of the forensic sentinel sign, in this case consistent with multiple soft tissue hematomas due to direct impact.

a postmortem exam (Figure 19.12) [24,71,145–147]. Axial slices of the chest and/or abdomen, accompanying the coronal whole-body scan, provide much more detail than would PMCT alone (Figure 19.13). Organ laceration and tissue properties are better assessed by PMMR and allow for more precise diagnosis, which is especially valuable if autopsy is not performed or when the investigators are dealing with a high-profile case (Figure 19.14 through 19.16). This utility also applies to bony structures and the evaluation of bone bruises, which may allow for answering a prolonged agonal or antemortem interval [70,129].

The protocol for neonates must be adjusted with thinner slices (2 mm). In addition, the $T_2$-weighted STIR sequences (coronal and axial) are replaced by $T_2$-weighted sequences, and $T_1$-weighted IR (inversion recovery) sequences are replaced by $T_1$-weighted sequences, as these give a better contrast. Moreover, a sagittal $T_2$-weighted whole-body sequence is preferable to the standard protocol in neonates (Table 19.3).

#### 19.1.1.9.4 Heart

Natural causes of death are part of a forensic investigation and must be differentiated from non-natural causes of death. Unenhanced PMCT is often insufficient for diagnosing natural causes such as sudden cardiac death. PMMR offers a more sensitive detection method to determine cardiovascular disease (e.g., myocardial infarction) and, therefore, plays a pivotal role in postmortem imaging (Figure 19.17). If myocardial infarction is observed on PMMR, an additional image- or robotic-guided biopsy may be taken to prove the cause of death histologically, and autopsy may be avoided because the forensic questions have been sufficiently answered.

From a forensic perspective, sudden cardiac death (myocardium) is difficult, as macroscopic evidence is often not visible during the first 12 hours of survival before death occurs [78,148]. In contrast, microscopic detection of ischemia-induced alterations is possible after 4 h, which means that cases of sudden cardiac death lack precise macro- and microscopic detection for the first 4 h [78,148].

A PMMR study by Ruder et al. investigated cardiac PMMR and reported that edema from ischemia/reperfusion lesions can be detected within 3 h after the onset of vascular occlusion [80]. Jackowski et al. recently stated that PMMR (3 T) was able to detect hyperacute

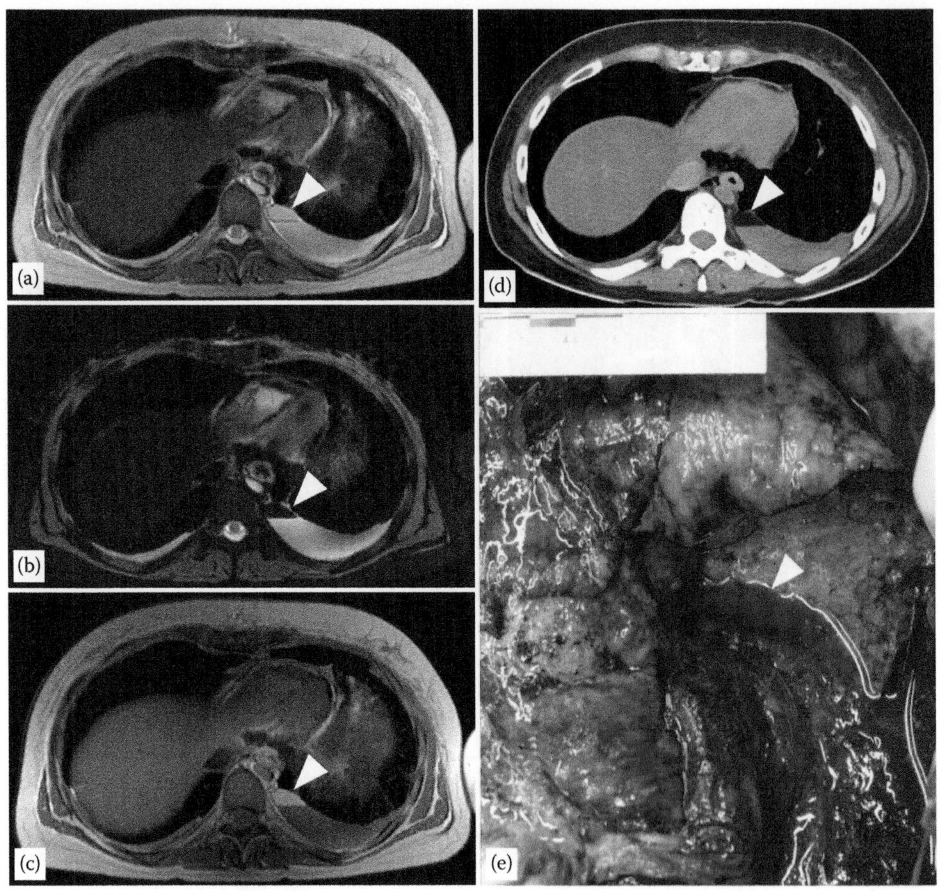


**FIGURE 19.13**
(a) $T_2$-weighted imaging at the level of the diaphragm. (b) $T_2$-weighted sequence (fat-saturated), same level. (c) $T_1$-weighted sequence, same level. (d) PMCT, axial orientation according to PMMR imaging. (e) Autopsy photograph with caudal-anterior view into the opened chest and abdominal cavity, with the diaphragm destructed. Note the fluid–fluid level within the left thoracic cavity (white arrow). The upper layer in the pleural effusion represents liquefied fat (best observed in PMCT and the STIR, fat-suppressed sequence, as well as in the $T_1$-weighted image). This case is a suicide by formic acid with vast necrosis and frank dissolution of the diaphragm, the distal esophagus, parts of the stomach and digestion of the thoracic wall, left lung, and perigastric organs. These findings could all be detected by radiological imaging, but only with concomitant PMMR and not solely PMCT.

infarction (minutes up to 1 h) on imaging that was not visible during macroscopic assessment [76]. In this study, chronic, subacute, and acute infarction on PMMR correlated well with autopsy [76].

Still, this topic has not yet been fully investigated and the radiological interpretation of cardiac PMMR is often complex. Nevertheless, PMMR of the heart has been reported to allow for heart weight measurements by using circumferential area measurements of the left ventricle with excellent intra- and inter-reader reliability [149].

If there is clear edema with a high signal on $T_2$-weighted images, the etiology is likely to be an ischemic lesion. If there is a patchy pattern, focal or several signal loss areas within the myocardium on $T_2$-weighted images, early myocardial infarction or even myocarditis may be among the probable etiologies [73,75,81,127,131,150]. In SWIs, differentiation between dot-like or small myocardial hemorrhage and vascular congestion is feasible when the investigator observes the appearance, be it focal or tubular (Figure 19.18). Abrupt small signal loss is more likely to represent hemorrhage, and vascular congestion is more ubiquitous throughout the myocardial muscle. Gas inclusion (e.g., in decomposition) should be verified on PMCT (Figure 19.17). Influencing factors such as postmortem interval, cardiovascular pathology and duration, collateral vascular supply, reperfusion despite the corpse's temperature, and autolysis or decomposition may play a diagnostic role through cardiac PMMR. A potential standard protocol for cardiac PMMR that includes anatomically aligned planes within a small field of view (e.g., acquired with a torso coil) is listed in Table 19.4. It is advisable to add axial $T_1$- and $T_2$-weighted images covering the chest for a full diagnostic overview, as described in the thorax protocol.

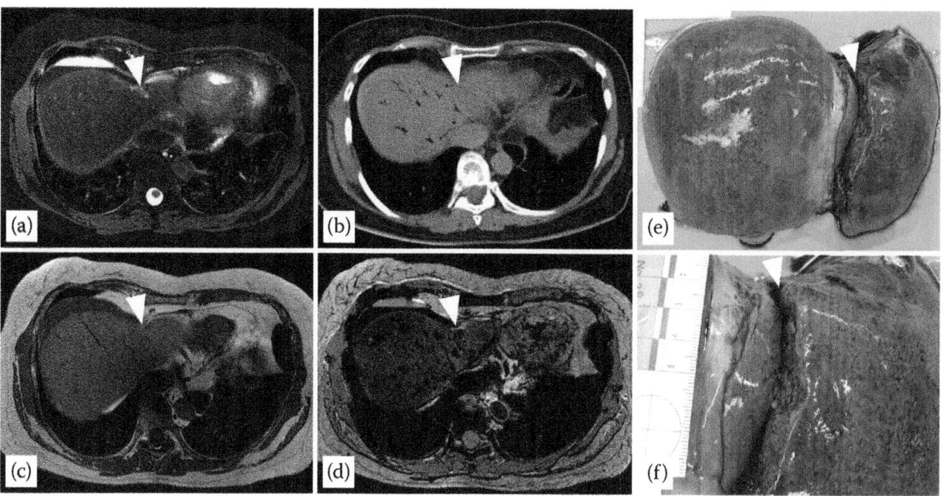


**FIGURE 19.14**
(a) $T_2$-weighted STIR sequence with fat suppression at the diaphragmatic level of the liver. (b) PMCT at the same level. (c) $T_1$-weighted imaging, same level. (d) SWI (venBOLD) of the liver, same level. (e) Autopsy photograph with anterior view of the liver. Note the subcapsular hematoma despite the vast laceration of the left lobe. (f) Detailed photograph of the laceration in the left lobe. PMCT (b) shows hyperdense fluid surrounding the liver and gas within the parenchyma. However, the liver laceration (arrowhead) is not clearly visible on PMCT. PMMR, in contrast, clearly depicts the laceration of the left lobe (arrow). In addition, the sedimentation of the perihepatic blood is easily assessed on PMMR, as is the slight subcapsular hematoma on the diaphragmatic side of the left liver lobe (left to the arrowhead). Note the patchy appearance of the liver in (d) that is consistent with the sustained liver injury.

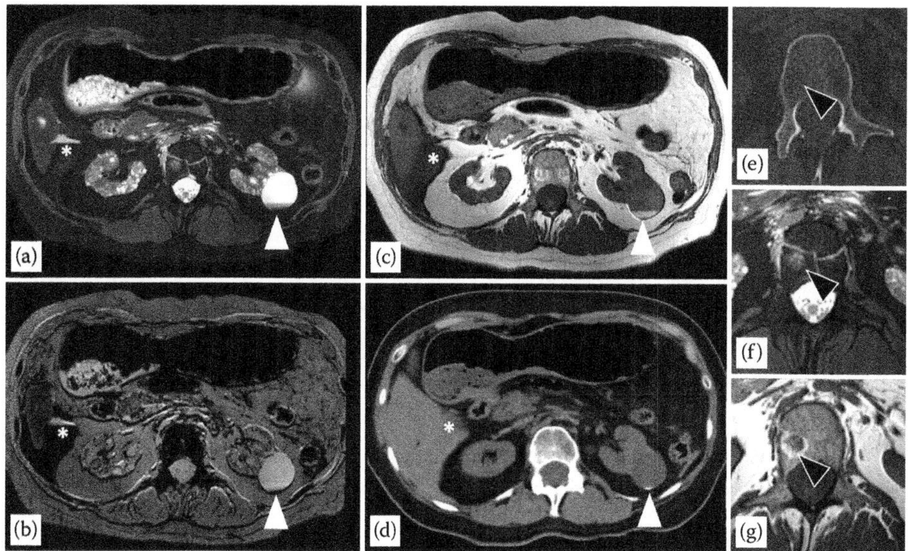


**FIGURE 19.15**
(a) $T_2$-weighted STIR sequence. (b) $T_1$-weighted sequence. (c) VenBOLD sequence. (d) PMCT. Soft kernel, abdominal window. (e) PMCT reconstruction in axial orientation of the spine. Hard kernel, osseous window. (f) Detailed view of image (a). (g) Detailed view of image (b). This case exhibits perihepatic fluid due to liver injury during resuscitation attempts. These findings are scarcely assessable on PMCT (d), but are clearly depicted on PMMR (white asterisk). Both kidneys display multiple cystic lesions partially containing sediment within the cyst according to proteinaceous/bloody sediment (not very assessable on PMCT). On PMCT the tiny calcification is distinct (arrowhead), while on PMMR this can only be presumed [images (a) and (c)]. PMCT depicts a slight salt and pepper appearance within the vertebra. PMMR distinctly displays the vertebral hemangioma (black arrowhead). This case of natural death again indicates the supplemental character of both modalities, PMCT and PMMR.

#### *19.1.1.9.5 Chest*

Imaging of the chest combined with cardiac PMMR may serve as a triple rule-out protocol (to simultaneously rule out myocardial infarction, pulmonary thromboembolism, and vascular disease). Cardiac PMMR may aid in diagnosing myocardial pathology or even eventually coronary occlusion, whereas chest PMMR may aid in diagnosing pulmonary thromboembolism (Figure 19.19). The problem with diagnosing pathology within the pulmonary arteries is the

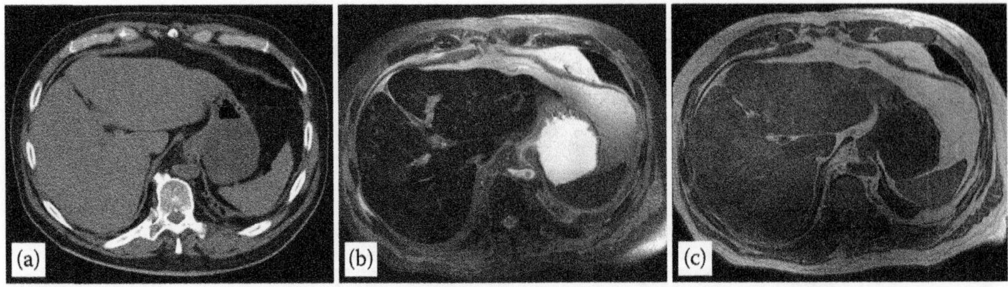


**FIGURE 19.16**
(a) PMCT. (b) $T_2$-weighted imaging. (c) $T_1$-weighted imaging. All images display the liver at the same level. PMCT does not allow for precise assessment of the liver parenchyma. PMMR reveals perfusion deficits with heterogeneous liver parenchyma that was already described in the antemortem clinical data.

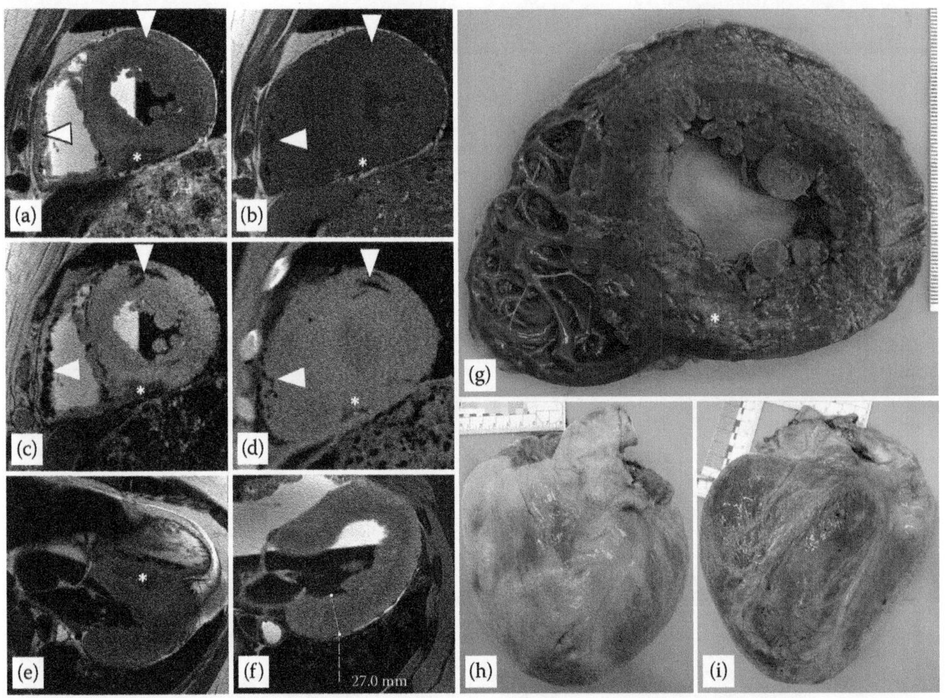


**FIGURE 19.17**
(a) $T_2$-weighted PMMR, short axis. (b) $T_2$* sequence PMMR, short axis. (c) $T_1$-weighted PMMR, short axis. (d) PMCT, angulated according to image (a–c). (e) $T_2$-weighted PMMR, four-chamber view. (f) $T_2$-weighted PMMR in axial orientation. (g) Autopsy specimen of the heart, dissected according to the short axis view on PMMR. (h) Anterior view of the entire heart. (i) Posterior view of the entire heart. The $T_2$W hypointense signal alteration in close proximity to the septum at the posterior wall (indicated by an asterisk, a–e and g) corresponds well to the pale or clay colored area of the autopsy specimen and represents a fresh myocardial infarction. Note the gas accumulation within the right cardiac chamber (position-dependent anteriorly) and within the tissue of the left myocardium (white arrowheads) that may easily be mistaken for hemorrhage (especially on hemo-sequences, b) if no correlation to PMCT (d) is feasible. The myocardium is hypertrophic (diameter 2.7 cm, f) which shows a relatively increased heart weight of 127% (according to Zeek) with a weight of 770 grams (h, i).

presence of postmortem clotting. Sedimentation usually appears as a homogeneous level with a $T_2$-weighted hypointense signal (Figure 19.4). Within the sediment or even protruding into the cardiac cavity or inferior caval vein, heterogeneous to homogeneous material may be present. The signal intensity of postmortem clots is greatly influenced by the type of cruor that is present (currant jelly or chicken fat clot). Similarly, the appearance of pulmonary thromboembolisms differs depending on the age (hyperacute, acute, subacute, or chronic) of the vitally embolized clot. Consequentially, there is high diagnostic insecurity in determining precisely if the clot is a true pulmonary thromboembolism or only postmortem clot. Jackowski et al. investigated a relatively small study population of 8 corpses and claimed to have evaluated no false-positive findings by chest PMMR [74]. The authors describe pulmonary thromboembolism as presenting with homogeneous intermediate signal intensity on $T_2$-weighted images and allowing for the differentiation of postmortem

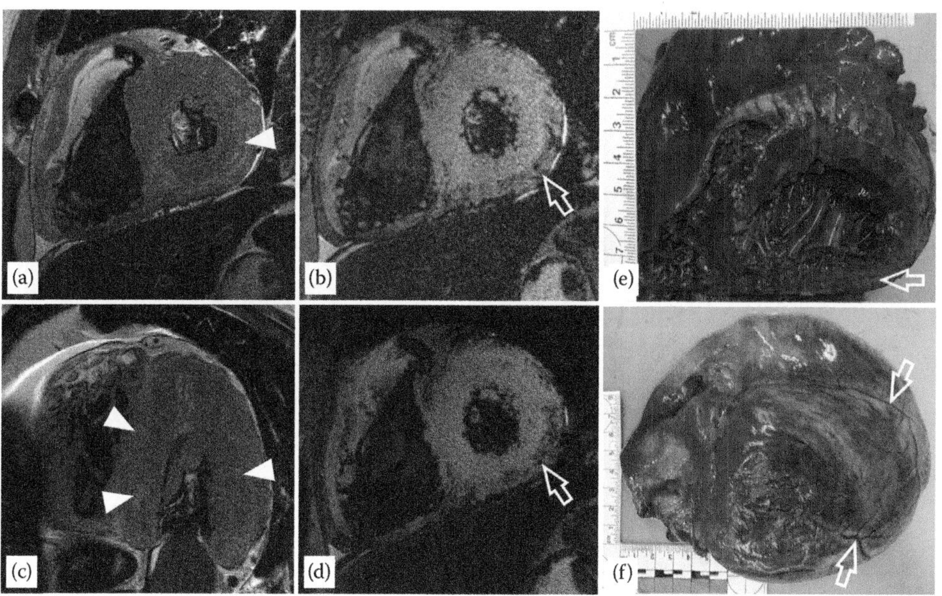


**FIGURE 19.18**
(a) Short axis view in $T_2$-weighted imaging. (b) SWI with short axis view. (c) Four-chamber view in $T_2$-weighted imaging. (d) MIP (maximum intensity projection) of the heart of the hemo-sequence, short axis view. (e) Photograph of the autopsy specimen cut in short axis corresponding to PMMR. (f) Photograph of the entire heart with anterior view. The white arrowhead displays a patchy hyperintense signal within the myocardium consistent with edema in a case of myocarditis. Vascular congestion was present concomitant with the inflammatory process and is best observed in image (b) and (d), indicated by signal decrease due to congested vasculature (white arrow). The autopsy photographs (e, f) also display the vascular congestion in this case of histology-proven myocarditis (white arrows).

**TABLE 19.4**
Comprehensive Summary of Potential Standard Sequences

| Cardiac: PMMR Imaging Protocol | | | |
|---|---|---|---|
| **Sequence** | **Slice Thickness** | **Plane** | **Diagnostics** |
| $T_2$W | 3 mm | 4-chamber view | Coronaries<br>Myocardium |
| $T_2$W | 3 mm | Short axis | Forensic sentinel<br>Cross sections compared to 4-chamber view |
| $T_2$W_SPIR (fat saturated) | 3 mm | Short axis | Forensic sentinel<br>Cross sections compared to 4-chamber view |
| PD | 3 mm | Short axis | Cross sections compared to 4-chamber view |
| $T_1$W | 3 mm | Short axis | Coronaries<br>Morphology |
| $T_2$W_FFE ($T_2$* weighted) | 3 mm | Short axis | Hemorrhage<br>Vascular congestion |

*Notes:* STIR, steady-state inversion recovery; PD proton density; FFE, fast field echo; $T_1$W, $T_1$-weighted; $T_2$W, $T_2$-weighted.

clots [74]. In addition, this study also investigated the deep veins of the lower extremities for thrombosis in three cases, which is a logical approach to doing so—either by whole-body PMCT or by means of dedicated scans of the lower extremities [74].

Another approach includes enhancing PMCT with PMCTA, starting with a venous contrast media mixture application to visualize the pulmonary trunk prior to viewing the arterial system. However, biopsied specimens should still be used to validate a definite diagnosis, as PMCTA combined with PMCT does not yet allow for accurate differentiation based on densities. Table 19.5 summarizes a potential thorax protocol for a triple rule-out, if combined with cardiac PMMR.

#### 19.1.1.10 PMMR-Angiography

Postmortem cross-sectional imaging (i.e., PMCT and PMMR), is being enhanced by angiographic visualization

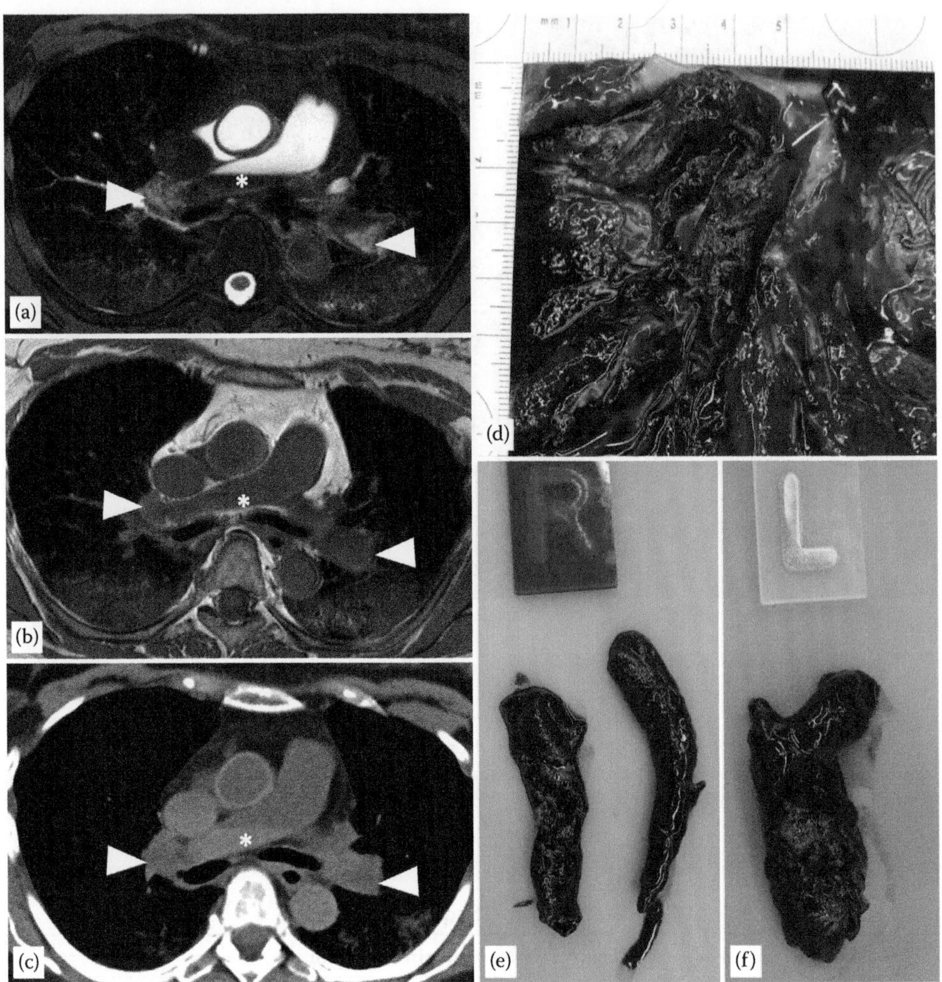


**FIGURE 19.19**
(a) $T_2$-weighted imaging of the chest, axial orientation. (b) $T_1$-weighted imaging of the chest, axial orientation. (c) PMCT with the same level as in (a) and (b). (d) Autopsy photograph with the depiction of pulmonary clotting. (e) Specimen photograph of the pulmonary thromboembolism in the right pulmonary artery. (f) Specimen photograph of the pulmonary thromboembolism in the left pulmonary artery. All postmortem images (a–c) exhibit distinct sedimentation (white asterisk) due to settling of the corpuscular particles. Note the signal alterations of the heterogeneous hyperintense material (arrowheads) within the pulmonary artery branching on both sides on $T_2$-weighted imaging. According to this, the material appears hypodense and inhomogeneous on PMCT. The findings correlated to fatal pulmonary embolism with cardiac arrest as confirmed during autopsy (d–f) and by histology. Note the acute on chronic appearance with different consistency and coloring of the antemortem clotting.

**TABLE 19.5**
Comprehensive Summary of Potential Standard Sequences

| Thorax: PMMR Imaging Protocol | | | |
|---|---|---|---|
| **Sequence** | **Slice Thickness** | **Plane** | **Diagnostics** |
| $T_2$W_SPAIR (fat saturated) | 3 mm | Axial | Morphology<br>Signal of potential PE |
| $T_1$W | 3 mm | Axial | Sediment<br>Signal of potential PE |
| $T_2$W (fat saturated) | 1.5 mm | Aligned to the pulmonary arteries | High-resolution scan<br>Small stack covering the pulmonary trunk and the left and right pulmonary artery |

*Notes:* SPAIR/SPIR, spectral attenuated inversion recovery; PE, pulmonary thromboembolism; $T_1$W, $T_1$-weighted; $T_2$W, $T_2$-weighted.

in several forensic institutes [82–88,91,92,95,151–154]. The approach is usually performed by PMCT-angiography (PMCTA), either in a minimally invasive manner by cannulation of a larger artery and vein (e.g., femoral vessels) or by selective insertion of a catheter (e.g., in the supra-aortal vessels for cardiac access), with the approaches differing between institutes [83,84,87,92,95,155,156]. Every angiographic procedure, be it PMCTA or PMMRA, must be accompanied by an unenhanced dataset of the specific modality to allow for an accurate radiological interpretation. PMMRA remains a partially neglected forensic imaging field, mainly due to time and financial issues. Therefore, the literature is lacking regarding the utilization of PMMRA. Ruder et al. published in 2012 a study dedicated to PMMRA in comparison with PMCTA, investigating 4 human cadavers [94]. The intravascular contrast was provided by the polyethylene glycol of the contrast media mixture and the iodine contrast media, both administered during PMCTA [84,92,95]. The polyethylene glycol allows for high signal on $T_1$-weighted images, and no gadolinium (as performed in clinical practice) was used for PMMRA [94]. This study showed that the results of PMMRA were technically equal in quality compared to PMCTA in the majority of the evaluated vessels; best results were achieved in the head and chest region [94]. When performing PMMRA, the authors advise the use of as little ferro- or paramagnetic material as possible, especially when clamps are used. Clamps and even cannulas may produce artifacts and therefore impair the image quality of PMMRA. In addition, if no roller pump is used during or directly prior to PMMRA, it is advisable to minimize the time elapsed between a prior PMCTA scan and a subsequent PMMR to avoid settling of the intravascular fluid due to gravity, luminal volume loss and passive diffusion over time. Clearly, PMCTA may be sufficient in most cases, but if an autopsy is denied and imaging is the sole investigation method, PMMRA may add valuable information to that provided by PMCTA.

##### *19.1.1.10.1 Forensic MR in Living Victims*

Forensic applications for MR comprise of two major topics that are eligible for submission as court evidence in the majority of countries:

1. MR in strangulation victims
2. MR in non-accidental injury

*19.1.1.10.1.1 MR in Strangulation Victims* Strangulation includes hanging, ligature, and manual strangulation. Manual strangulation, in particular, plays a major role in domestic violence, followed by ligature with a variety of tools and, rarely, hanging. On external inspection, there are often few or no signs of the alleged violence against the neck, and a forensic determination of life-threatening or non-life-threatening strangulation is difficult based only on these externally visible findings (Figure 19.20). Pathophysiologically, the mechanisms of strangulation may lead to death either by venous or even arterial obstruction, arterial spasm in the course of carotid pressure, vagal collapse due to pressure on the carotid sinus, or even compression of the airways. The forensic pathologist relies on objective signs such as petechial hemorrhages, strangulation marks, hematomas (and their coloring), erythema, abrasions, and swelling of the craniocervical region (Figure 19.20). In addition, a forensic investigation includes subjective symptoms such as loss of consciousness, incontinence of urine and feces, dysphagia, pain, sore throat or hoarseness, or sustained dyspnea or hallucinations. Usually, injuries that are life threatening are concluded when congestive ophthalmic petechial hemorrhages are present or substantial objective and subjective signs of strangulation are evaluated [141,157–159]. The ascertainment of danger to an individual's life is crucial, as in many countries the degree of the penalty for the perpetrator depends on whether the event was life threatening. Neck MR of living individuals allows for objective evidence of injuries and the subsequent assessment of a danger to life, not merely based on the patient's history and objective external findings but also on morphologic findings in the affected area (Figure 19.20).

It is advisable to perform a neck MR as soon as possible after the incident, as hematomas may vanish in a short time span in living victims. Therefore, the elapsed time should be, at best, shorter than 72 h for proper MR imaging to detect incident-related lesions.

Christe et al. investigated strangulation victims for life-threatening or non-life-threatening events and found that cases with danger to life showed distinct MR findings in the superficial soft tissue (subcutaneous hemorrhage/edema in 80%, intracutaneous hemorrhage/edema in 53%) and deep soft tissue (intramuscular hemorrhage/edema and swelling of the platysma, both 53%), as well as hemorrhage and edema in the lymph nodes in 40% [157]. The authors divided the neck into three distinct danger zones using a scoring system: the superficial zone A includes the skin and subcutaneous fatty tissue; the middle zone B includes muscles, vessels, salivary glands, and lymph nodes; and the deep zone C comprises the larynx and perilaryngeal tissue [157]. Based on the outcome of this study, the significant danger zone was zone B. The positive predictive value (meaning the risk for a life-threatening event) was 50% for zone B, followed by 44% for the deep zone C and, finally, only 20% for zone A [157].

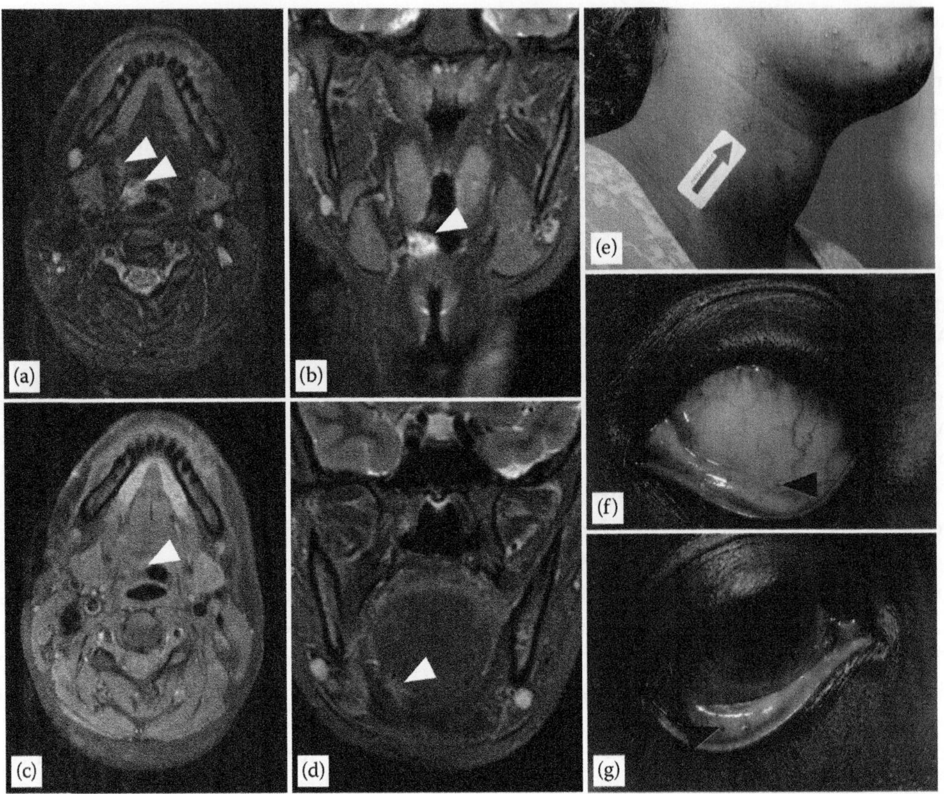


**FIGURE 19.20**
The victim has a darker skin color, aggravating the detection of bruises (labeled) or slight hematomas on the skin. External inspection showed tiny petechiae (black arrowhead). The clinical MR scan of the neck reveals a hyperintense signal (corresponding with edema/hemorrhage) within the area of the lateral and median glossoepiglottic plica on the right and the extrinsic musculature of the tongue on image (a) (axial $T_2$-weighted, fat-suppressed sequence), (d) (coronal STIR sequence) as well on (b) ($T_1$-weighted sequence). This attack was deemed a threat to life, and imaging was conducted based on the combination of external inspection and clinical MR in strangulation.

Currently, MR is a routine procedure supplementing forensic investigation of surviving strangulation victims in Switzerland and is acknowledged by the court.

*19.1.1.10.1.2 MR in Non-Accidental Injury* Clinical examinations of suspected child abuse include external findings (skin lesions such as abrasions and hematomas) and, usually, an entire skeletal survey by X-ray and in an emergency setting, ultrasound (if feasible) or CT of the head, usually followed by head MR, which may even be used as the first-line examination to search for intracranial findings. Bone scintigraphy may be used in several countries due to its higher diagnostic yield in anatomically complex locations (e.g., pelvis and feet); however, this modality has been found to be less sensitive for metaphyseal lesions [160].

Forensic investigation of the suspected abused child includes documentation of the entire external findings and a multidisciplinary approach between the treating physicians, radiologists, and authorities. However, the complex topic of non-accidental injury is beyond the scope of this book chapter [15,139,140,160–164].

## 19.2 Conclusion

PMCT and PMMR (and both PMCTA and PMMRA) are modalities that supplement each other. Therefore, a full imaging workup should include both modalities to provide a complete overview of a forensic case and to be truly comparable to a traditional autopsy. Screening for forensic sentinel signs is paramount. Histology and toxicology are supplementary examinations that may be undertaken either under image-guidance or during dissection for both virtopsy and autopsy. Virtopsy, however, has the drawback of providing limited bioptic access to samples within the skull. Recent research with robotic-guided access, for example, has already begun, and the future will surely show procedures that cannot yet be imagined for full postmortem examinations during the next decades [145–147,165–167]. Moreover, there is a clear trend toward increased forensic-indicated clinical examinations to serve as evidence in court, not only in strangulation or child abuse cases. Forensic radiology is clearly a forward-looking and visionary subspecialty that will be an integral part of any forensic investigation in the future, with as yet unknown potential for medico-legal cases, and it may even obviate autopsy.

## References


1. Cox J, Kirkpatrick RC. The new photography [microform] : With report of a case in which a bullet was photographed in the leg. *Montreal Med. J.*; 1896. http://archive.org/details/cihm_01784.
2. Thali MJ, Brogdon BG, Viner MD. *Brogdon's Forensic Radiology*, 2nd ed. Boca Raton, FL: CRC Press; 2010.
3. Evans KT, Knight B. Forensic radiology. *Br. J. Hosp. Med.* 1986;36(1):14–20.
4. Eckert WG, Garland N. The history of the forensic applications in radiology. *Am. J. Forensic Med. Pathol.* 1984;5(1):53–6.
5. Withers S. The story of the first roentgen evidence. *Radiology*. 1931;17(1):99–103.
6. Sognnaes RF. Dental evidence in the postmortem identification of Adolf Hitler, Eva Braun, and Martin Bormann. *Leg. Med. Annu.* 1977;1976:173–235.
7. Sognnaes RF, Ström F. The odontological identification of Adolf Hitler. Definitive documentation by x-rays, interrogations and autopsy findings. *Acta Odontol. Scand.* 1973;31(1):43–69.
8. Brogdon BG. The scope of forensic radiology. *Clin. Lab. Med.* 1998;18(2):203–40.
9. Deadman WJ. The identification of human remains. *Can. Med. Assoc. J.* 1964;91(15):808.
10. Jablonski NG, Shum BS. Identification of unknown human remains by comparison of antemortem and postmortem radiographs. *Forensic Sci. Int.* 1989;42(3):221–30.
11. Ruddiman RA, Kerr NW, Gillanders LA. Forensic odontology. Identification by comparison. *Br. Dent. J.* 1969;127(11):505–7.
12. Mann GT, Fatteh AB. The role of radiology in the identification of human remains: Report of a case. *J. Forensic Sci. Soc.* 1968;8(2):67–8.
13. Messmer JM, Fierro MF. Radiologic forensic investigation of fatal gunshot wounds. *RadioGraphics*. 1986;6(3):457–73.
14. Singleton AC. The roentgenological identification of victims of the "Noronic" disaster. *AJR Am. J. Roentgenol. Radium Ther.* 1951;66(3):375–84.
15. Norman MG, Smialek JE, Newman DE, Horembala EJ. The postmortem examination on the abused child. Pathological, radiographic, and legal aspects. *Perspect. Pediatr. Pathol.* 1984;8(4):313–43.
16. Lattimer JK. Observations based on a review of the autopsy photographs, x-rays, and related materials of the late President John F. Kennedy. *Med. Times.* 1972; 100(6):33–64.
17. www.jfklancer.com. 1968 Panel Review of Photographs, X-Ray Films, Documents and Other Evidence Pertaining to the Fatal Wounding of President John F. Kennedy on November 22, 1963, in Dallas, TX. 2014.
18. Rohrich RJ, Nagarkar P, Stokes M, Weinstein A. The Assassination of John F. Kennedy: Revisiting the medical data. *Plast. Reconstr. Surg.* 2013;132(5):1340–50.
19. Jones RC. The President's been shot and they are bringing him to the emergency room. *J. Am. Coll. Surg.* 2014;218(4):856–68.
20. Wüllenweber R, Schneider V, Grumme T. A computer-tomographical examination of cranial bullet wounds (author's transl). *Z. Für Rechtsmed. J. Leg. Med.* 1977;80(3):227–46.
21. Donchin Y, Rivkind AI, Bar-Ziv J, Hiss J, Almog J, Drescher M. Utility of postmortem computed tomography in trauma victims. *J. Trauma.* 1994;37(4):552–5; discussion 555–6.
22. Ros PR, Li KC, Vo P, Baer H, Staab EV. Preautopsy magnetic resonance imaging: Initial experience. *Magn. Reson. Imaging.* 1990;8(3):303–8.
23. Bisset R. Magnetic resonance imaging may be alternative to necropsy. *BMJ.* 1998;317(7170):1450.
24. Thali MJ, Braun M, Buck U, Aghayev E, Jackowski C, Vock P et al. VIRTOPSY—Scientific documentation, reconstruction and animation in forensic: Individual and real 3D data based geo-metric approach including optical body/object surface and radiological CT/MRI scanning. *J. Forensic Sci.* 2005;50(2):428–42.
25. Thali MJ, Jackowski C, Oesterhelweg L, Ross SG, Dirnhofer R. VIRTOPSY—The Swiss virtual autopsy approach. *Leg. Med. Tokyo Jpn.* 2007;9(2):100–4.
26. Thali MJ, Schweitzer W, Yen K, Vock P, Ozdoba C, Spielvogel E et al. New horizons in forensic radiology: The 60-second digital autopsy—full-body examination of a gunshot victim by multislice computed tomography. *Am. J. Forensic Med. Pathol.* 2003;24(1):22–7.
27. Thali MJ, Yen K, Schweitzer W, Vock P, Boesch C, Ozdoba C et al. Virtopsy, a new imaging horizon in forensic pathology: Virtual autopsy by postmortem multislice computed tomography (MSCT) and magnetic resonance imaging (MRI)—A feasibility study. *J. Forensic Sci.* 2003;48(2):386–403.
28. Thali MJ, Dirnhofer R, Vock P. *The Virtopsy Approach.* Boca Raton, FL: CRC Press; 2009. http://www.crcpress.com/product/isbn/9780849381782.
29. Patriquin L, Kassarjian A, O'Brien M, Andry C, Eustace S. Postmortem whole-body magnetic resonance imaging as an adjunct to autopsy: Preliminary clinical experience. *J. Magn. Reson. Imaging.* 2001;13(2):277–87.
30. Baglivo M, Winklhofer S, Hatch GM, Ampanozi G, Thali MJ, Ruder TD. The rise of forensic and postmortem radiology—Analysis of the literature between the year 2000 and 2011. *J. Forensic Radiol. Imaging.* 2013;1(1):3–9.

31. Pollanen MS, Woodford N. Virtual autopsy: Time for a clinical trial. *Forensic Sci. Med. Pathol.* 2013;9(3):427–8.
32. Wichmann D, Obbelode F, Vogel H, Hoepker WW, Nierhaus A, Braune S et al. Virtual autopsy as an alternative to traditional medical autopsy in the intensive care unit: A prospective cohort study. *Ann. Intern. Med.* 2012;156(2):123–30.
33. Westphal SE, Apitzsch J, Penzkofer T, Mahnken AH, Knüchel R. Virtual CT autopsy in clinical pathology: Feasibility in clinical autopsies. *Virchows Arch. Int. J. Pathol.* 2012;461(2):211–9.
34. Scandurra I, Forsell C, Ynnerman A, Ljung P, Lundström C, Persson A. Advancing the state-of-the-art for virtual autopsies–Initial forensic workflow study. *Stud. Health Technol. Inform.* 2010;160(Pt 1):639–43.
35. Blaauwgeers JLGH, van Rijn RR. Virtual autopsy—Why not?. *Ned. Tijdschr. Geneeskd.* 2012;156(19):A4786.
36. Levy AD, Harcke HT, Mallak CT. Postmortem Imaging: MDCT features of postmortem change and decomposition. *Am. J. Forensic Med. Pathol.* 2010;31(1):12–7.
37. Daly B, Abboud S, Ali Z, Sliker C, Fowler D. Comparison of whole-body post mortem 3D CT and autopsy evaluation in accidental blunt force traumatic death using the abbreviated injury scale classification. *Forensic Sci. Int.* 2013;225(1–3):20–6.
38. Takahashi N, Higuchi T, Shiotani M, Hirose Y, Shibuya H, Yamanouchi H et al. The effectiveness of postmortem multidetector computed tomography in the detection of fatal findings related to cause of non-traumatic death in the emergency department. *Eur. Radiol.* 2012;22(1):152–60.
39. Watts G. Imaging the dead. *BMJ.* November 22, 2010;341(2): c6600.
40. Roberts ISD, Benamore RE, Benbow EW, Lee SH, Harris JN, Jackson A et al. Post-mortem imaging as an alternative to autopsy in the diagnosis of adult deaths: A validation study. *Lancet.* 2012;379(9811):136–42.
41. Leth PM. Computerized tomography used as a routine procedure at postmortem investigations. *Am. J. Forensic Med. Pathol.* 2009;30(3):219–22.
42. Capuani C, Guilbeau-Frugier C, Dedouit F, Rougé D, Delisle M-B, Telmon N. Forensic scientist's implication regarding medical autopsies: Experience in a French university hospital (CHU Toulouse). *Ann. Pathol.* 2013;33(2):87–92.
43. Woźniak K, Moskała A, Urbanik A, Kłys M. Usefulness of preliminary evaluation of postmortem CT as an extension of diagnostic capabilities of conventional forensic autopsy. *Arch. Med. Sądowej Kryminol.* 2010;60(1):27–37.
44. Bedford PJ, Oesterhelweg L. Different conditions and strategies to utilize forensic radiology in the cities of Melbourne, Australia and Berlin, Germany. *Forensic Sci. Med. Pathol.* 2013;9(3):321–6.
45. Rutty GN, Morgan B, O'Donnell C, Leth PM, Thali M. Forensic institutes across the world place CT or MRI scanners or both into their mortuaries. *J. Trauma.* 2008;65(2):493–4.
46. Burton EC, Mossa-Basha M. To image or to autopsy? *Ann. Intern. Med.* 2012;156(2):158–9.
47. Thomsen AH, Jurik AG, Uhrenholt L, Vesterby A. An alternative approach to computerized tomography (CT) in forensic pathology. *Forensic Sci. Int.* 2009; 183(1–3):87–90.
48. Gorincour G, Tassy S, Bartoli C. Why the United States should have virtopsied Osama Bin Laden. *AJR Am. J. Roentgenol.* 2012;198(3):W323.
49. Aghayev E, Christe A, Sonnenschein M, Yen K, Jackowski C, Thali MJ et al. Postmortem imaging of blunt chest trauma using CT and MRI: Comparison with autopsy. *J. Thorac. Imaging.* 2008;23(1):20–7.
50. Aghayev E, Yen K, Sonnenschein M, Jackowski C, Thali M, Vock P et al. Pneumomediastinum and soft tissue emphysema of the neck in postmortem CT and MRI; a new vital sign in hanging? *Forensic Sci. Int.* 2005;153(2–3):181–8.
51. Ampanozi G, Schwendener N, Krauskopf A, Thali MJ, Bartsch C. Incidental occult gunshot wound detected by postmortem computed tomography. *Forensic Sci. Med. Pathol.* 2013;9(1):68–72.
52. Andenmatten MA, Thali MJ, Kneubuehl BP, Oesterhelweg L, Ross S, Spendlove D et al. Gunshot injuries detected by post-mortem multislice computed tomography (MSCT): A feasibility study. *Leg. Med.* 2008;10(6):287–92.
53. Berger N, Ross SG, Ampanozi G, Majcen R, Schweitzer W, Gascho D et al. Puzzling over intracranial gas: Disclosing a pitfall on postmortem computed tomography in a case of fatal blunt trauma. *J. Forensic Radiol. Imaging.* 2013;1(3):137–41.
54. Bolliger SA, Thali MJ, Ross S, Buck U, Naether S, Vock P. Virtual autopsy using imaging: Bridging radiologic and forensic sciences. A review of the Virtopsy and similar projects. *Eur. Radiol.* 2008;18(2):273–82.
55. Dedouit F, Savall F, Mokrane F-Z, Rousseau H, Crubézy E, Rougé D et al. Virtual anthropology and forensic identification using multidetector CT. *Br. J. Radiol.* 2014;87(1036):20130468.
56. Dirnhofer R, Jackowski C, Vock P, Potter K, Thali MJ. VIRTOPSY: minimally invasive, imaging-guided virtual autopsy. *Radiogr. Rev. Publ. Radiol. Soc. N. Am. Inc.* 2006;26(5):1305–33.
57. Egger C, Bize P, Vaucher P, Mosimann P, Schneider B, Dominguez A et al. Distribution of artifactual gas on post-mortem multidetector computed tomography (MDCT). *Int. J. Legal Med.* 2012;126(1):3–12.
58. Flach PM, Ampanozi G, Germerott T, Ross SG, Krauskopf A, Thali MJ et al. Shot sequence detection aided by postmortem computed tomography in a case of homicide. *J. Forensic Radiol. Imaging.* 2013;1(2):68–72.
59. Flach PM, Ross SG, Bolliger SA, Ampanozi G, Hatch GM, Schön C et al. Massive systemic fat embolism detected by postmortem imaging and biopsy*: Systemic fat embolism on postmortem CT. *J. Forensic Sci.* 2012;57(5):1376–80.
60. Flach PM, Egli TC, Bolliger SA, Berger N, Ampanozi G, Thali MJ et al. "Blind spots" in forensic autopsy: Improved detection of retrobulbar hemorrhage and orbital lesions by postmortem computed tomography

(PMCT). *Leg. Med.* 2014. Available from: http://linkinghub.elsevier.com/retrieve/pii/S1344622314000893.

61. Hatch GM, Dedouit F, Christensen AM, Thali MJ, Ruder TD. RADid: A pictorial review of radiologic identification using postmortem CT. *J. Forensic Radiol. Imaging.* 2014;2(2):52–9.
62. Ruder TD, Thali Y, Bolliger SA, Somaini-Mathier S, Thali MJ, Hatch GM et al. Material differentiation in forensic radiology with single-source dual-energy computed tomography. *Forensic Sci. Med. Pathol.* 2013;9(2):163–9.
63. Thali MJ, Markwalder T, Jackowski C, Sonnenschein M, Dirnhofer R. Dental CT imaging as a screening tool for dental profiling: Advantages and limitations. *J. Forensic Sci.* 2006;51(1):113–9.
64. Levy AD, Harcke HT. Essentials of Forensic Imaging, A Text-Atlas, First edition. Boca Raton, FL: CRC Press; 2011. Available from: http://www.crcpress.com/product/isbn/9781420091113.
65. Aghayev E, Jackowski C, Sonnenschein M, Thali M, Yen K, Dirnhofer R. Virtopsy hemorrhage of the posterior cricoarytenoid muscle by blunt force to the neck in postmortem multislice computed tomography and magnetic resonance imaging. *Am. J. Forensic Med. Pathol.* 2006;27(1):25–9.
66. Aghayev E, Jackowski C, Thali MJ, Yen K, Dirnhofer R, Sonnenschein M. Heart luxation and myocardium rupture in postmortem multislice computed tomography and magnetic resonance imaging. *Am. J. Forensic Med. Pathol.* 2008;29(1):86–8.
67. Aghayev E, Thali MJ, Sonnenschein M, Hurlimann J, Jackowski C, Kilchoer T et al. Fatal steamer accident; blunt force injuries and drowning in post-mortem MSCT and MRI. *Forensic Sci. Int.* 2005;152(1):65–71.
68. Aghayev E, Yen K, Sonnenschein M, Ozdoba C, Thali M, Jackowski C et al. Virtopsy post-mortem multi-slice computed tomography (MSCT) and magnetic resonance imaging (MRI) demonstrating descending tonsillar herniation: Comparison to clinical studies. *Neuroradiology.* 2004;46(7):559–64.
69. Añon J, Remonda L, Spreng A, Scheurer E, Schroth G, Boesch C et al. Traumatic extra-axial hemorrhage: Correlation of postmortem MSCT, MRI, and forensic-pathological findings. *J. Magn. Reson. Imaging JMRI.* 2008;28(4):823–36.
70. Buck U, Christe A, Naether S, Ross S, Thali MJ. Virtopsy—noninvasive detection of occult bone lesions in postmortem MRI: Additional information for traffic accident reconstruction. *Int. J. Legal Med.* 2009;123(3):221–6.
71. Cha JG, Kim DH, Kim DH, Paik SH, Park JS, Park SJ et al. Utility of postmortem autopsy via whole-body imaging: Initial observations comparing MDCT and 3.0 T MRI findings with autopsy findings. *Korean J. Radiol. Off. J. Korean Radiol. Soc.* 2010;11(4):395–406.
72. Jackowski C, Thali M, Aghayev E, Yen K, Sonnenschein M, Zwygart K et al. Postmortem imaging of blood and its characteristics using MSCT and MRI. *Int. J. Legal Med.* 2006;120(4):233–40.
73. Jackowski C, Christe A, Sonnenschein M, Aghayev E, Thali MJ. Postmortem unenhanced magnetic resonance imaging of myocardial infarction in correlation to histological infarction age characterization. *Eur. Heart J.* 2006;27(20):2459–67.
74. Jackowski C, Grabherr S, Schwendener N. Pulmonary thrombembolism as cause of death on unenhanced postmortem 3T MRI. *Eur. Radiol.* 2013;23(5):1266–70.
75. Jackowski C, Hofmann K, Schwendener N, Schweitzer W, Keller-Sutter M. Coronary thrombus and peracute myocardial infarction visualized by unenhanced postmortem MRI prior to autopsy. *Forensic Sci. Int.* 2012;214(1–3):e16–9.
76. Jackowski C, Schwendener N, Grabherr S, Persson A. Post-mortem cardiac 3-T magnetic resonance imaging: Visualization of sudden cardiac death? *J. Am. Coll. Cardiol.* 2013;62(7):617–29.
77. Ruder TD, Germerott T, Thali MJ, Hatch GM. Differentiation of ante-mortem and post-mortem fractures with MRI: A case report. *Br. J. Radiol.* 2011;84(1000):e75–8.
78. Ruder TD, Thali MJ, Hatch GM. Essentials of forensic post-mortem MR imaging in adults. *Br. J. Radiol.* 2014;87(1036):20130567.
79. Ruder TD, Bauer-Kreutz R, Ampanozi G, Rosskopf AB, Pilgrim TM, Weber OM et al. Assessment of coronary artery disease by post-mortem cardiac MR. *Eur. J. Radiol.* 2012;81(9):2208–14.
80. Ruder TD, Ebert LC, Khattab AA, Rieben R, Thali MJ, Kamat P. Edema is a sign of early acute myocardial infarction on post-mortem magnetic resonance imaging. *Forensic Sci. Med. Pathol.* 2013;9(4):501–5.
81. Shiotani S, Yamazaki K, Kikuchi K, Nagata C, Morimoto T, Noguchi Y et al. Postmortem magnetic resonance imaging (PMMRI) demonstration of reversible injury phase myocardium in a case of sudden death from acute coronary plaque change. *Radiat. Med.* 2005;23(8):563–5.
82. Chevallier C, Christine C, Doenz F, Francesco D, Vaucher P, Paul V et al. Postmortem computed tomography angiography vs. conventional autopsy: advantages and inconveniences of each method. *Int. J. Legal Med.* 2013;127(5):981–9.
83. Jackowski C, Persson A, Thali MJ. Whole body postmortem angiography with a high viscosity contrast agent solution using poly ethylene glycol as contrast agent dissolver. *J. Forensic Sci.* 2008;53(2):465–8.
84. Ross S, Spendlove D, Bolliger S, Christe A, Oesterhelweg L, Grabherr S et al. Postmortem whole-body CT angiography: Evaluation of two contrast media solutions. *AJR Am. J. Roentgenol.* 2008;190(5):1380–9.
85. Flach PM, Ross SG, Bolliger SA, Preiss US, Thali MJ, Spendlove D. Postmortem whole-body computed tomography angiography visualizing vascular rupture in a case of fatal car crash. *Arch. Pathol. Lab. Med.* 2010;134(1):115–9.
86. Grabherr S, Djonov V, Friess A, Thali MJ, Ranner G, Vock P et al. Postmortem angiography after vascular perfusion with diesel oil and a lipophilic contrast agent. *AJR Am. J. Roentgenol.* 2006;187(5):W515–23.
87. Grabherr S, Djonov V, Yen K, Thali MJ, Dirnhofer R. Postmortem angiography: Review of former and current methods. *AJR Am. J. Roentgenol.* 2007;188(3):832–8.

88. Grabherr S, Doenz F, Steger B, Dirnhofer R, Dominguez A, Sollberger B et al. Multi-phase post-mortem CT angiography: Development of a standardized protocol. *Int. J. Legal Med.* 2011;125(6):791–802.
89. Grabherr S, Gygax E, Sollberger B, Ross S, Oesterhelweg L, Bolliger S et al. Two-step postmortem angiography with a modified heart–lung machine: Preliminary results. *AJR Am. J. Roentgenol.* 2008;190(2):345–51.
90. Grabherr S, Hess A, Karolczak M, Thali MJ, Friess SD, Kalender WA et al. Angiofil®-mediated visualization of the vascular system by microcomputed tomography: A feasibility study. *Microsc. Res. Tech.* 2008;71(7):551–6.
91. Jackowski C, Bolliger S, Aghayev E, Christe A, Kilchoer T, Aebi B et al. Reduction of postmortem angiography-induced tissue edema by using polyethylene glycol as a contrast agent dissolver. *J. Forensic Sci.* 2006;51(5):1134–7.
92. Ross SG, Thali MJ, Bolliger S, Germerott T, Ruder TD, Flach PM. Sudden death after chest pain: Feasibility of virtual autopsy with postmortem CT angiography and biopsy. *Radiology.* 2012;264(1):250–9.
93. Ruder TD, Ross S, Preiss U, Thali MJ. Minimally invasive post-mortem CT-angiography in a case involving a gunshot wound. *Leg. Med.* 2010;12(3):154–6.
94. Ruder TD, Hatch GM, Ebert LC, Flach PM, Ross S, Ampanozi G et al. Whole body postmortem magnetic resonance angiography: Postmortem MR Angiography. *J. Forensic Sci.* 2012;57(3):778–82.
95. Ross SG, Bolliger SA, Ampanozi G, Oesterhelweg L, Thali MJ, Flach PM. Postmortem CT angiography: Capabilities and limitations in traumatic and natural causes of death. *Radiogr. Rev. Publ. Radiol. Soc. N. Am. Inc.* 2014;34(3):830–46.
96. Christe A, Flach P, Ross S, Spendlove D, Bolliger S, Vock P et al. Clinical radiology and postmortem imaging (Virtopsy) are not the same: Specific and unspecific postmortem signs. *Leg. Med. Tokyo Jpn.* 2010;12(5): 215–22.
97. Thali MJ, Yen K, Vock P, Ozdoba C, Kneubuehl BP, Sonnenschein M et al. Image-guided virtual autopsy findings of gunshot victims performed with multi-slice computed tomography (MSCT) and magnetic resonance imaging (MRI) and subsequent correlation between radiology and autopsy findings. *Forensic Sci. Int.* 2003;138(1–3):8–16.
98. Dedini RD, Karacozoff AM, Shellock FG, Xu D, McClellan RT, Pekmezci M. MRI issues for ballistic objects: information obtained at 1.5-, 3- and 7-Tesla. Spine J. *Off. J. North Am. Spine Soc.* 2013;13(7):815–22.
99. Hess U, Harms J, Schneider A, Schleef M, Ganter C, Hannig C. Assessment of gunshot bullet injuries with the use of magnetic resonance imaging. *J. Trauma.* 2000;49(4):704–9.
100. Eshed I, Kushnir T, Shabshin N, Konen E. Is magnetic resonance imaging safe for patients with retained metal fragments from combat and terrorist attacks? *Acta Radiol. Stockh. Swed.* 1987. 2010;51(2):170–4.
101. Jourdan P, Cosnard G. MRI: Projectiles, bullets and counter-indications. *J. Radiol.* 1989;70(12):685–9.
102. Karacozoff AM, Pekmezci M, Shellock FG. Armor-piercing bullet: 3-T MRI findings and identification by a ferromagnetic detection system. *Mil. Med.* 2013;178(3):e380–5.
103. Teitelbaum GP, Yee CA, Van Horn DD, Kim HS, Colletti PM. Metallic ballistic fragments: MR imaging safety and artifacts. *Radiology.* 1990;175(3):855–9.
104. Smith AS, Hurst GC, Duerk JL, Diaz PJ. MR of ballistic materials: Imaging artifacts and potential hazards. *AJNR Am. J. Neuroradiol.* 1991;12(3):567–72.
105. Winklhofer S, Stolzmann P, Meier A, Schweitzer W, Morsbach F, Flach P et al. Added value of dual-energy computed tomography versus single-energy computed tomography in assessing ferromagnetic properties of ballistic projectiles: Implications for magnetic resonance imaging of gunshot victims. *Invest. Radiol.* 2014;49(6):431–7.
106. Condon B, Hadley DM. Potential MR hazard to patients with metallic heart valves: The Lenz effect. *J. Magn. Reson. Imaging JMRI.* 2000;12(1):171–6.
107. Winklhofer S, Benninger E, Spross C, Morsbach F, Rahm S, Ross S et al. CT metal artefact reduction for internal fixation of the proximal humerus: Value of mono-energetic extrapolation from dual-energy and iterative reconstructions. *Clin. Radiol.* 2014;69(5):e199–206.
108. Stradiotti P, Curti A, Castellazzi G, Zerbi A. Metal-related artifacts in instrumented spine. Techniques for reducing artifacts in CT and MRI: State of the art. *Eur. Spine J. Off. Publ. Eur. Spine Soc. Eur. Spinal Deform. Soc. Eur. Sect. Cerv. Spine Res. Soc.* 2009;18(Suppl 1):102–8.
109. Sutter R, Ulbrich EJ, Jellus V, Nittka M, Pfirrmann CWA. Reduction of metal artifacts in patients with total hip arthroplasty with slice-encoding metal artifact correction and view-angle tilting MR imaging. *Radiology.* 2012;265(1):204–14.
110. Ulbrich EJ, Sutter R, Aguiar RF, Nittka M, Pfirrmann CW. STIR sequence with increased receiver bandwidth of the inversion pulse for reduction of metallic artifacts. *AJR Am. J. Roentgenol.* 2012;199(6):W735–42.
111. Lee MJ, Janzen DL, Munk PL, MacKay A, Xiang QS, McGowen A. Quantitative assessment of an MR technique for reducing metal artifact: Application to spin-echo imaging in a phantom. *Skeletal Radiol.* 2001;30(7):398–401.
112. Davis PL, Crooks L, Arakawa M, McRee R, Kaufman L, Margulis AR. Potential hazards in NMR imaging: Heating effects of changing magnetic fields and RF fields on small metallic implants. *AJR Am. J. Roentgenol.* 1981;137(4):857–60.
113. Dempsey MF, Condon B. Thermal injuries associated with MRI. *Clin. Radiol.* 2001;56(6):457–65.
114. Kobayashi T, Isobe T, Shiotani S, Saito H, Saotome K, Kaga K et al. Postmortem magnetic resonance imaging dealing with low temperature objects. *Magn. Reson. Med. Sci. MRMS Off. J. Jpn. Soc. Magn. Reson. Med.* 2010;9(3):101–8.
115. Nelson TR, Tung SM. Temperature dependence of proton relaxation times in vitro. *Magn. Reson. Imaging.* 1987;5(3):189–99.

116. Ruder TD, Hatch GM, Siegenthaler L, Ampanozi G, Mathier S, Thali MJ et al. The influence of body temperature on image contrast in post mortem MRI. *Eur. J. Radiol.* 2012;81(6):1366–70.
117. Kobayashi T, Shiotani S, Kaga K, Saito H, Saotome K, Miyamoto K et al. Characteristic signal intensity changes on postmortem magnetic resonance imaging of the brain. *Jpn. J. Radiol.* 2010;28(1):8–14.
118. Tofts PS, Jackson JS, Tozer DJ, Cercignani M, Keir G, MacManus DG et al. Imaging cadavers: Cold FLAIR and noninvasive brain thermometry using CSF diffusion. *Magn. Reson. Med. Off. J. Soc. Magn. Reson. Med. Soc. Magn. Reson. Med.* 2008;59(1):190–5.
119. Kobayashi T, Monma M, Baba T, Ishimori Y, Shiotani S, Saitou H et al. Optimization of inversion time for postmortem Short-tau Inversion Recovery (STIR) MR imaging. *Magn. Reson. Med. Sci. MRMS Off. J. Jpn. Soc. Magn. Reson. Med.* 2014;13(2):67–72.
120. Yen K, Weis J, Kreis R, Aghayev E, Jackowski C, Thali M et al. Line-scan diffusion tensor imaging of the posttraumatic brain stem: Changes with neuropathologic correlation. *AJNR Am. J. Neuroradiol.* 2006;27(1):70–3.
121. Scheurer E, Lovblad K-O, Kreis R, Maier SE, Boesch C, Dirnhofer R et al. Forensic application of postmortem diffusion-weighted and diffusion tensor MR imaging of the human brain in situ. *AJNR Am. J. Neuroradiol.* 2011;32(8):1518–24.
122. Ith M, Bigler P, Scheurer E, Kreis R, Hofmann L, Dirnhofer R et al. Observation and identification of metabolites emerging during postmortem decomposition of brain tissue by means of in situ 1H-magnetic resonance spectroscopy. *Magn. Reson. Med. Off. J. Soc. Magn. Reson. Med. Soc. Magn. Reson. Med.* 2002;48(5):915–20.
123. Scheurer E, Ith M, Dietrich D, Kreis R, Hüsler J, Dirnhofer R et al. Statistical evaluation of time-dependent metabolite concentrations: Estimation of postmortem intervals based on in situ 1H-MRS of the brain. *NMR Biomed.* 2005;18(3):163–72.
124. Ith M, Scheurer E, Kreis R, Thali M, Dirnhofer R, Boesch C. Estimation of the postmortem interval by means of $^{1}$H MRS of decomposing brain tissue: influence of ambient temperature. *NMR Biomed.* 2011;24(7):791–8.
125. Schmidt TM, Wang ZJ, Keller S, Heinemann A, Acar S, Graessner J et al. Postmortem 31P magnetic resonance spectroscopy of the skeletal muscle: α-ATP/Pi ratio as a forensic tool? *Forensic Sci. Int.* 2014;242:172–6.
126. Bennett CM, Miller MB, Wolford GL. Neural correlates of interspecies perspective taking in the post-mortem Atlantic Salmon: An argument for multiple comparisons correction. *NeuroImage.* 2009;47(Suppl 1):S125.
127. Abdel-Aty H, Cocker M, Meek C, Tyberg JV, Friedrich MG. Edema as a very early marker for acute myocardial ischemia: a cardiovascular magnetic resonance study. *J. Am. Coll. Cardiol.* 2009;53(14):1194–201.
128. Aghayev E, Thali MJ, Jackowski C, Sonnenschein M, Dirnhofer R, Yen K. MRI detects hemorrhages in the muscles of the back in hypothermia. *Forensic Sci. Int.* 2008;176(2–3):183–6.
129. Berger N, Paula P, Gascho D, Flach PM, Thali MJ, Ross SG et al. Bone marrow edema induced by a bullet after a self-inflicted accidental firing. *Leg. Med. Tokyo Jpn.* 2013;15(6):329–31.
130. Christe A, Ross S, Oesterhelweg L, Spendlove D, Bolliger S, Vock P et al. Abdominal trauma—Sensitivity and specificity of postmortem noncontrast imaging findings compared with autopsy findings. *J. Trauma.* 2009;66(5):1302–7.
131. Jackowski C, Schweitzer W, Thali M, Yen K, Aghayev E, Sonnenschein M et al. Virtopsy: Postmortem imaging of the human heart in situ using MSCT and MRI. *Forensic Sci. Int.* 2005;149(1):11–23.
132. Yen K, Vock P, Tiefenthaler B, Ranner G, Scheurer E, Thali MJ et al. Virtopsy: Forensic traumatology of the subcutaneous fatty tissue; multislice computed tomography (MSCT) and magnetic resonance imaging (MRI) as diagnostic tools. *J. Forensic Sci.* 2004;49(4):799–806.
133. Ross S, Ebner L, Flach P, Brodhage R, Bolliger SA, Christe A et al. Postmortem whole-body MRI in traumatic causes of death. *AJR Am. J. Roentgenol.* 2012;199(6):1186–92.
134. Ampanozi G, Preiss U, Hatch GM, Zech WD, Ketterer T, Bolliger S et al. Fatal lower extremity varicose vein rupture. *Leg. Med.* 2011;13(2):87–90.
135. Bradley Jr WG. MR appearance of hemorrhage in the brain. *Radiology.* 1993;189(1):15–26.
136. Ruder TD, Zech W-D, Hatch GM, Ross S, Ampanozi G, Thali MJ et al. Still frame from the hour of death: Acute intracerebral hemorrhage on post-mortem computed tomography in a decomposed corpse. *J. Forensic Radiol. Imaging.* 2013;1(2):73–6.
137. Yen K, Lövblad K-O, Scheurer E, Ozdoba C, Thali MJ, Aghayev E et al. Post-mortem forensic neuroimaging: Correlation of MSCT and MRI findings with autopsy results. *Forensic Sci. Int.* 2007;173(1):21–35.
138. Jones NR, Blumbergs PC, Brown CJ, McLean AJ, Manavis J, Perrett LV et al. Correlation of postmortem MRI and CT appearances with neuropathology in brain trauma: A comparison of two methods. *J. Clin. Neurosci. Off. J. Neurosurg. Soc. Australas.* 1998;5(1):73–9.
139. Altinok D, Saleem S, Zhang Z, Markman L, Smith W. MR imaging findings of retinal hemorrhage in a case of nonaccidental trauma. *Pediatr. Radiol.* 2009;39(3):290–2.
140. Demaerel P, Casteels I, Wilms G. Cranial imaging in child abuse. *Eur. Radiol.* 2002;12(4):849–57.
141. Yen K, Thali MJ, Aghayev E, Jackowski C, Schweitzer W, Boesch C et al. Strangulation signs: Initial correlation of MRI, MSCT, and forensic neck findings. *J. Magn. Reson. Imaging JMRI.* 2005;22(4):501–10.
142. Jacob B, Wiedbrauck C, Lamprecht J, Bonte W. Laryngologic aspects of bolus asphyxiation-bolus death. *Dysphagia.* 1992;7(1):31–5.
143. Wick R, Gilbert JD, Byard RW. Café coronary syndrome-fatal choking on food: an autopsy approach. *J. Clin. Forensic Med.* 2006;13(3):135–8.
144. Oesterhelweg L, Bolliger SA, Thali MJ, Ross S. Virtopsy: Postmortem imaging of laryngeal foreign bodies. *Arch. Pathol. Lab. Med.* 2009;133(5):806–10.

145. Ebert LC, Ptacek W, Naether S, Fürst M, Ross S, Buck U et al. Virtobot—A multi-functional robotic system for 3D surface scanning and automatic post mortem biopsy. *Int. J. Med. Robot. Comput. Assist. Surg. MRCAS.* 2010;6(1):18–27.
146. Ebert LC, Ptacek W, Breitbeck R, Fürst M, Kronreif G, Martinez RM et al. Virtobot 2.0: The future of automated surface documentation and CT-guided needle placement in forensic medicine. *Forensic Sci. Med. Pathol.* 2014;10(2):179–86.
147. Buck U, Naether S, Braun M, Bolliger S, Friederich H, Jackowski C et al. Application of 3D documentation and geometric reconstruction methods in traffic accident analysis: With high resolution surface scanning, radiological MSCT/MRI scanning and real data based animation. *Forensic Sci. Int.* 2007;170(1):20–8.
148. Saukko P, Knight B. *Knight's Forensic Pathology*, 3rd ed. New York: CRC Press; 2004.
149. Ruder TD, Stolzmann P, Thali YA, Hatch GM, Somaini S, Bucher M et al. Estimation of heart weight by post-mortem cardiac magnetic resonance imaging. *J. Forensic Radiol. Imaging.* 2013;1(1):15–8.
150. Jackowski C, Warntjes MJB, Berge J, Bär W, Persson A. Magnetic resonance imaging goes postmortem: Noninvasive detection and assessment of myocardial infarction by postmortem MRI. *Eur. Radiol.* 2011;21(1):70–8.
151. Saunders SL, Morgan B, Raj V, Rutty GN. Postmortem computed tomography angiography: Past, present and future. *Forensic Sci. Med. Pathol.* 2011;7(3):271–7.
152. Vogel B, Heinemann A, Tzikas A, Poodendaen C, Gulbins H, Reichenspurner H et al. Post-mortem computed tomography (PMCT) and PMCT-angiography after cardiac surgery. Possibilities and limits. *Arch. Med. Sądowej Kryminol.* 2013;63(3):155–71.
153. Nakazono T, Suzuki M, White CS. Computed tomography angiography of coronary artery bypass graft grafts. *Semin. Roentgenol.* 2012;47(3):240–52.
154. Morgan B, Adlam D, Robinson C, Pakkal M, Rutty GN. Adult post-mortem imaging in traumatic and cardiorespiratory death and its relation to clinical radiological imaging. *Br. J. Radiol.* 2014;87(1036):20130662.
155. Jolibert M, Cohen F, Bartoli C, Boval C, Vidal V, Gaubert J-Y et al. Postmortem CT-angiography: Feasibility of US-guided vascular access. *J. Radiol.* 2011;92(5):446–9.
156. Rutty G, Saunders S, Morgan B, Raj V. Targeted cardiac post-mortem computed tomography angiography: A pictorial review. *Forensic Sci. Med. Pathol.* 2012;8(1):40–7.
157. Christe A, Thoeny H, Ross S, Spendlove D, Tshering D, Bolliger S et al. Life-threatening versus non-life-threatening manual strangulation: Are there appropriate criteria for MR imaging of the neck? *Eur. Radiol.* 2009;19(8):1882–9.
158. Christe A, Oesterhelweg L, Ross S, Spendlove D, Bolliger S, Vock P et al. Can MRI of the neck compete with clinical findings in assessing danger to life for survivors of manual strangulation? A statistical analysis. *Leg. Med.* 2010;12(5):228–32.
159. Yen K, Vock P, Christe A, Scheurer E, Plattner T, Schön C et al. Clinical forensic radiology in strangulation victims: Forensic expertise based on magnetic resonance imaging (MRI) findings. *Int. J. Legal Med.* 2007;121(2):115–23.
160. Van Rijn RR, Sieswerda-Hoogendoorn T. Educational paper: Imaging child abuse: The bare bones. *Eur. J. Pediatr.* 2012;171(2):215–24.
161. Dedouit F, Guilbeau-Frugier C, Capuani C, Sévely A, Joffre F, Rougé D et al. Child abuse: Practical application of autopsy, radiological, and microscopic studies. *J. Forensic Sci.* 2008;53(6):1424–9.
162. Dwek JR. The radiographic approach to child abuse. *Clin. Orthop.* 2011;469(3):776–89.
163. Kleinman PK. *Diagnostic Imaging of Child Abuse.* 2nd ed. St. Louis, MO: Mosby; 1998.
164. Radiology S on. Diagnostic imaging of child abuse. *Pediatrics.* 2000;105(6):1345–8.
165. Ebert LC, Thali MJ, Ross S. Getting in touch—3D printing in forensic imaging. *Forensic Sci. Int.* 2011;211(1–3):e1–6.
166. Ebert LC, Ptacek W, Fürst M, Ross S, Thali MJ, Hatch G. Minimally invasive postmortem telebiopsy*: Minimally invasive postmortem telebiopsy. *J. Forensic Sci.* 2012;57(2):528–30.
167. Ebert LC, Hatch G, Ampanozi G, Thali MJ, Ross S. You can't touch this: Touch-free navigation through radiological images. *Surg. Innov.* 2012;19(3):301–7.

# 20

# *Magnetic Resonance–Guided Focused Ultrasound*

**Alessandro Napoli and Gaia Cartocci**

**CONTENTS**



In the past decade, three new minimally invasive procedures surfaced: thermal ablations, high-intensity focused ultrasound (HIFU), and magnetic resonance imaging (MRI)-guided interventions. Each has the potential to transform or replace invasive tumor surgeries. Thermal ablations using surgically or percutaneously inserted probes became an alternative to surgical resections. In the early 1990s, MRI guidance for interventions and surgeries was introduced and became well accepted intraoperatively in neurosurgery, primarily for maximizing the effectiveness of resections of low-grade glioma. MRI-based thermometry was also introduced for monitoring thermal ablations and controlling energy deposition. Finally, after half of a century of research and technology development efforts, HIFU or focused ultrasound (FUS) surgery became recognized as a noninvasive extracorporeal thermal ablation method and has been tested for the ablative treatment of benign lesions and malignancies.

For guiding and monitoring HIFU ablation, MRI offers clear advantages over other imaging modalities. First, with its unparalleled soft tissue contrast, MRI provides high-resolution imaging in any orientation for planning treatment and evaluating treatment effects. In contrast, ultrasound (US) imaging, without the ability to clearly define tumor margins, may limit application of HIFU technology. Second, MRI is the only currently available technique with proven capabilities to create quantitative temperature maps. MRI thermal imaging is the only method that provides a means to ensure that, without effects to surrounding tissues, the proper US exposures are being applied for a safe and effective ablation of the target volume. Combining thermal ablation using advanced acoustic transducer technology with the anatomic, functional, and thermal guidance possible with MRI methods allows accurate targeting, real-time temperature monitoring, and closed-loop control of energy deposition. The result of this integration is arguably the most complex image-guided and controlled therapy delivery system available today: MRI-guided FUS (MRgFUS). So far, only MRI offers the best combination of tumor margin detection, depictions of details of surrounding anatomy, and the monitoring of temperature changes during therapy.

Although the cost of integrated therapy delivery systems is high, it can be offset by the elimination of the need for an operating room, the absence of expensive hospitalization and anesthesia, and a reduced complication rate. An MRgFUS system provides safe, real-time monitored, controlled, and repeatable treatment for benign tumors that does not necessitate an aggressive approach. For malignancies, it can be used either for complete *in situ* tissue destruction or for debulking prior to chemo- or radiation therapy. As such, MRgFUS is safe and effective for noninvasive surgery that can replace invasive surgery and ionizing radiation-based therapies, such as radiosurgery or brachytherapy, and can provide treatment for patients for whom invasive surgery or radiation may not be options. The feasibility and effectiveness of MRgFUS are being tested in several clinical applications, which include the ablation of benign and malignant tumors and palliative therapy of bone pain due to metastasis. Nonthermal effects of FUS are also being utilized for various novel clinical applications from targeted drug delivery to gene therapy. Because of these multiple uses, future advances of this method will likely have a major impact on several medical fields, such as surgery, oncology, and radiation therapy.

## 20.1 Generation of US

US is a pressure wave with a frequency above the audible range of a human ear (18–20 kHz); it is generated by a mechanical motion that induces the molecules in a medium to oscillate around their rest positions. Due to the bonding between the molecules, the disturbance is transmitted to neighboring molecules. The motion causes compressions and rarefactions of the medium and thus a pressure wave travels with the mechanical disturbance. As a result, an US wave requires a medium for propagation. In most cases, the molecules vibrate along the direction of the propagation (longitudinal wave), but in some instances, the molecular motion is across the direction of the wave propagation (shear wave). Shear waves propagate in solids such as bone but are quickly attenuated in soft tissues. Therefore, most current medical US methods utilize longitudinal waves.

US is generated by applying radio-frequency (RF) voltage across a material that is piezoelectric; it expands and contracts in proportion to the applied voltage. This phenomenon is the inverse of the piezoelectric effect, which was discovered by Jacques and Pierre Curie in natural quartz crystals in 1880. Since then, many piezoelectric materials have been discovered and developed. From these materials, a group of artificial piezoelectric materials known as *polarized polycrystalline ferroelectrics* (e.g., lead zirconate titanate or PZT) is used for medical US applications. The piezoelectric property is lost above a material-specific temperature—the Curie point (e.g., 328°C for PZT-4). Also, piezoelectric material rods or grains can be placed into a polymer matrix to have more control over the acoustic and electrical properties of the material. These so-called piezo composite materials are used especially in phased array transducers.

For many applications of US therapy, transducers capable of producing high-power, single-frequency, continuous waves are needed.

The US wave is generated by a piezoelectric plate of uniform thickness that has electrodes on its front and back surfaces. The electrodes are connected to the driving RF-line. Maximum power from a transducer can be delivered when it is operated close to its resonant frequency, which is achieved when the thickness of the plate is equal to the wavelength/2. However, a range of frequencies can be used with piezo composite materials. The frequency, which corresponds to the half-wavelength thickness, is called the *fundamental resonant frequency* of the transducer and it gives the maximum displacement amplitude at the transducer faces. The transducer can be driven at a frequency which is three, five, or so on times its fundamental frequency. The conversion efficiency is, however, reduced when compared with the fundamental frequency operation. US transducers can be manufactured in practically any desired shape and size. Spherically curved focused transducers of various sizes up to 30 cm diameter hemispherical transducer arrays have been manufactured. Both nonfocused and focused, single and multielement transducers and arrays have been manufactured for endocavity use. Interstitial applicators inserted directly into the tissue via catheter have been constructed down to the size of 1 mm in diameter (Figure 20.1).

## 20.2 US Propagation through Tissue

In order to be able to use US for therapy, it is essential to know the ultrasonic properties of tissues. For instance, the ultrasonic velocity that determines the field shape and the amount of reflected energy at tissue interfaces is dependent on the acoustical impedance (1⁄4 speed of sound × tissue density) differences between two neighboring tissues. The temperature elevation induced at the focus is partially dependent on the US attenuation, while the beam propagates through the overlying tissues, and the tissue absorption coefficient at the target site.

### 20.2.1 Speed of Sound

The speed of US is not frequency-dependent and has a similar average magnitude of 1550 m/s in all soft tissues (excluding lung). The velocity in fatty tissues is less than that in other soft tissues, being about 1480 m/s,

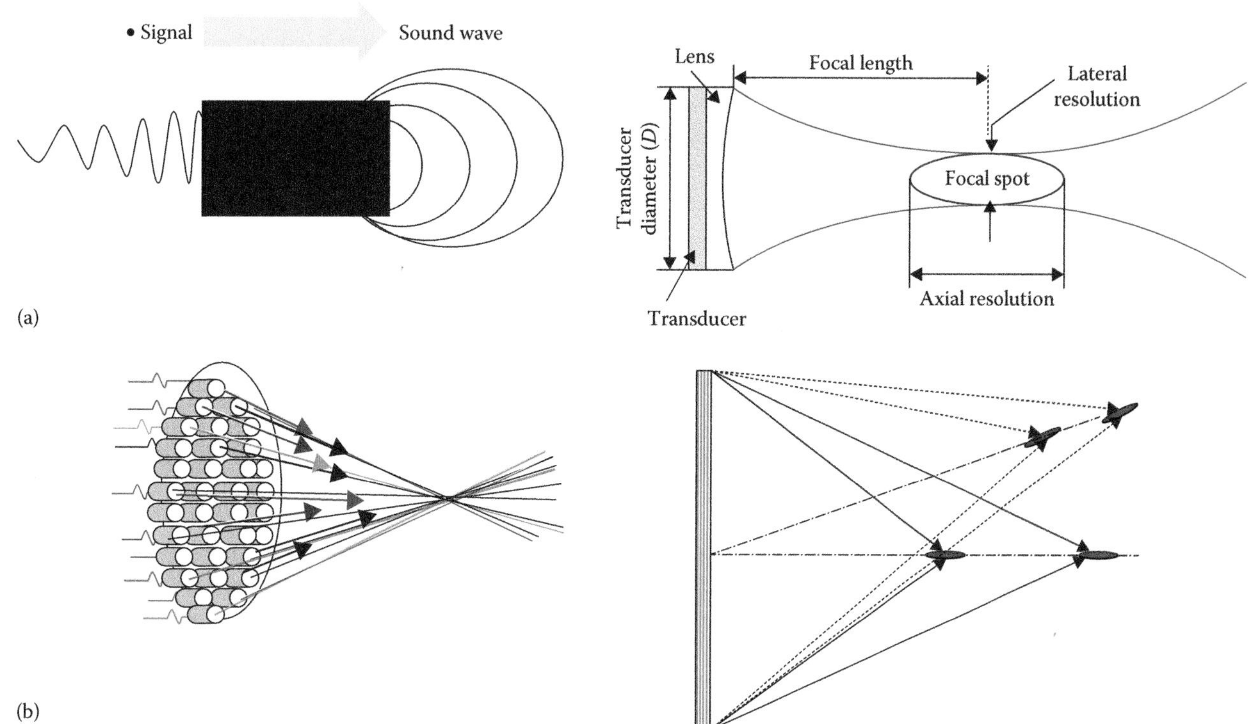


**FIGURE 20.1**
Principal types of transducers: (a) curved or (b) phased array. In (b), each element has its specific phase and amplitude. The phases of different elements can be adjusted in order to create focus at various distances.

while in the lungs the air spaces reduce the velocity to about 600 m/s. The highest values have been measured in bones, between 1800 and 3700 m/s depending on the density, structure, and frequency of the wave. In various soft tissues, the speed of US increases gradually as a function of temperature, with the slope between 0.04 and 0.08% $K^{-1}$. In fatty tissues, the speed of US decreases as the temperature increases. The effect of the temperature-dependent sound speed is small on the field shape and can be ignored when sharply focused fields are used.

### 20.2.2 Absorption and Attenuation

Ultrasonic attenuation in tissues is a sum of the losses due to absorption and scattering, and it determines the penetration of the beam into the tissue. In experimental studies, attenuation has been found to be dominated by absorption and thus follows a frequency dependence similar to that of absorption. Therefore, the amount of scattered energy is small and it will also be absorbed by the tissue, although it may broaden the energy distribution beyond what is expected from free field measurements. US absorption in a viscous medium is well understood and is a result of viscous forces between the moving particles that cause a lag between the particle pressure and velocity (or change in density). Therefore, an energy loss during each cycle will result. However, the tissue viscosity can explain only part of the energy loss experienced by US while propagating through soft tissues. In tissues, there is energy absorption due to a relaxation mechanism that can be briefly described as follows. During the compressive part of the cycle, energy is stored in the medium in a number of forms, such as lattice vibrational energy, molecular vibrational energy, and translational energy. During the expansion part of the US wave cycle, this stored energy is returned to the wave and the temperature of the medium returns to the original level. In tissue, the increased kinetic energy of the molecules is not in balance with the environment and the system tries to redistribute the energy. This transfer of energy takes time, and thus, during the decompression cycle, kinetic energy will return out of phase to the wave and absorption results.

### 20.2.3 Characteristic Acoustic Impedance

The acoustic impedance of a tissue is the product of the speed of sound and the density of the medium. Generally, most soft tissues have an impedance roughly equal to that of water, having a density around 1000 kg/m$^3$ and an acoustical impedance $1.6 \times 10^6$ kgm$^{-2}$ s$^{-1}$. Fat has a slightly lower impedance value of $1.35 \times 10^6$ kgm$^{-2}$ s$^{-1}$ due to its lower density and lower speed of sound. Bone and lung have impedances significantly higher and lower, respectively. In practice, these impedance differences mean that an US beam suffers little reflection loss while penetrating from one soft tissue to another, unless the angle of incidence is large. This may become an issue in strongly curved tissues such as breast. Soft tissue bone is an exception with 30%–40% reflection at the normal incidence of the wave and total reflection of the longitudinal wave at angles larger than 25°–30°. At a tissue–gas interface, all the energy is reflected back into the tissue.

## 20.3 Biological Effects of US

US interacts with tissue through the particle motion and pressure variation associated with wave propagation. First, all US waves are continuously losing energy through absorption resulting in an increase in temperature within the tissue. If the temperature elevation is large enough and is maintained for an adequate period, the exposure causes tissue damage. This thermal effect that can be used for tissue coagulation or ablation is similar to that obtained using other heating methods with equal thermal exposure. Second, at high-pressure amplitudes, the pressure wave can cause formation of small gas bubbles that concentrate acoustic energy. Similar focusing of energy can be induced by the oscillation of small bubbles already present. This type of interaction between a sound wave and a gas body is called *cavitation,* and it can cause a multitude of bioeffects from cell membrane permeability changes to complete destruction of tissue. Finally, the mechanical stress and strain associated with wave propagation may sometimes cause direct changes in a biological system.

## 20.4 Thermal Ablation

The thermal effects produced by US have been utilized in hyperthermia as a cancer therapy as well as in many US surgery applications. If live tissue is heated beyond the threshold for protein denaturation (57°C–60°C) for a few seconds, coagulation necrosis occurs. Lower temperatures can also kill tissues if the exposure time is lengthened. Freezing below −20°C, especially if it is repeated multiple times, results in irreversible cell damage. Because heating or cooling above or below these critical levels is not selective and kills both normal and neoplastic cells, thermal ablations are comparable to surgery rather than to more selective hyperthermia. The safe and effective use of heat or cold has required improvements in energy-delivery technologies and in treatment monitoring. Technical difficulties, particularly the lack of small, percutaneously introducible probes that ablate relatively large tissue volumes, slowed down clinical introduction. The development of energy-delivery devices significantly advanced thermal ablation applications by allowing percutaneous treatment, especially with MRI guidance. Heat-conducting probes and the way they deposit and distribute energy, however, are suboptimal. Using a single source of heat or cold limits the size and shape of the resulting thermal tissue injury. This then necessitates long exposures to increase the ablation volume, which result in a relatively shallow temperature gradient within the treated volume that causes uncertainty in tissue-killing effects and ambiguity about the demarcation zone between irreversible tissue death and reversible tissue damage. The resulting geometry may not exactly correspond to the size and shape of the treated tumor. The lesion size cannot be increased indefinitely because of the developing steady state between the rate of energy deposition and heat sink effects influenced mainly by perfusion and blood flow. These limitations can cause either undertreatment with insufficient results or overtreatment with higher complication rates. In thermal ablations, therefore to achieve the threshold of cell destruction inside the targeted tumor volume, spatial and temporal monitoring of temperature distribution within the targeted tumor volume is important. A temperature-sensitive imaging method is required. As part of MRgFUS, this ability allows the complete ablation of the gross target volume, which includes the tumor with additional margins as deemed appropriate by the treating physician, while sparing the nontargeted tissue. As in radiosurgery, to optimize energy delivery, preoperative treatment planning is essential, but unlike radiosurgery, MRgFUS can be monitored and controlled in real time by MRI thermometry (Figure 20.2). Thermal ablation also enables immediate evaluation of the treatment response and allows repeated treatments. In FUS therapy delivery systems, acoustic energy is generated by piezoelectric transducers that, depending on the application, are typically operated between 200 kHz and 4 Mhz. Such transducers are a flexible platform for which one can produce a wide range of devices that are tailored for particular applications. The simplest focused devices use a single-element spherically curved transducer or a flat transducer with a spherically curved acoustic lens. Such transducers are moved mechanically (sometimes with robots) to ablate multiple locations, and they can either be fairly large, to allow high powers that permit focusing deep within the body, or relatively small to fit on an intracavitary (e.g., transurethral or transrectal) device. Greater control over the acoustic field can be achieved by transducers

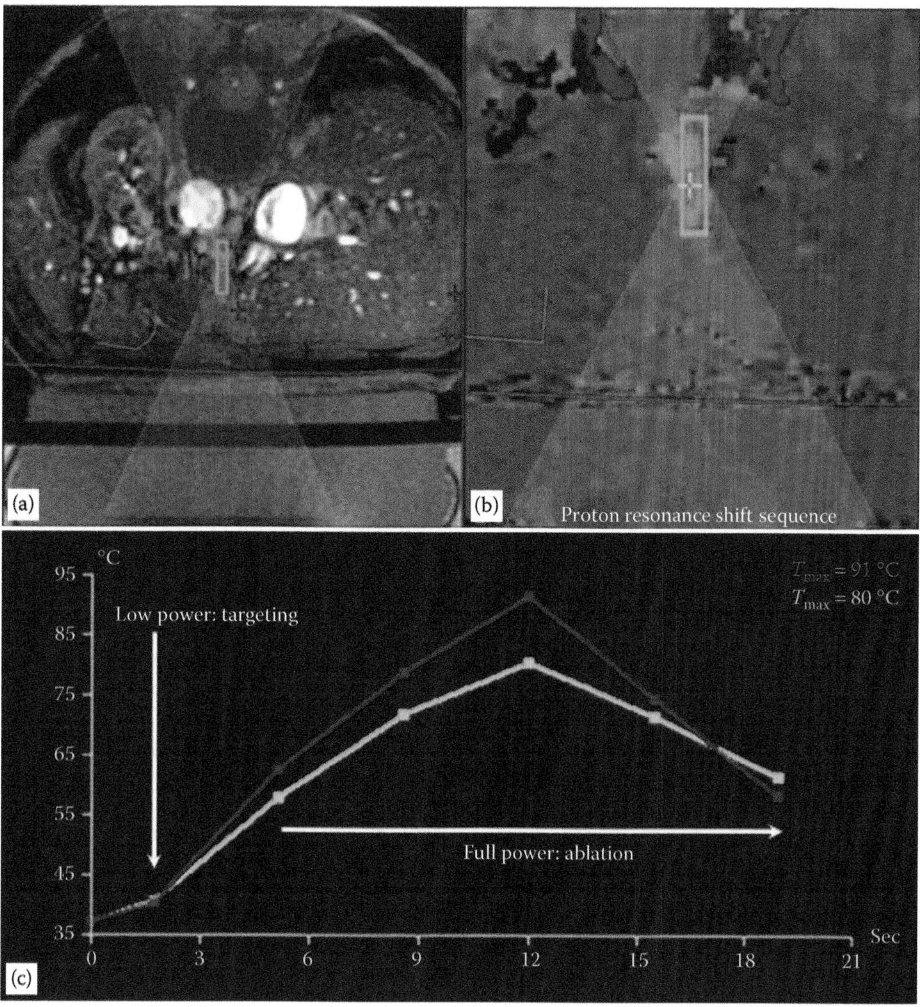


**FIGURE 20.2**
Therapy control graphic user interface during treatment with volumetric MRI-guided high-intensity focused ultrasound ablation therapy in a 39-year-old woman with a uterine fibroid. (a) Coronal and (b) sagittal images show the MR temperature maps during high-intensity focused ultrasound ablation with a MRgFUS. (c) The second table shows in real time the change in the temperature, at the site of treatment. This makes the method perfectly safe.

that consist of multiple elements with individual driving signals. Such phased arrays can steer the focal point electronically by modulating the phase of the individual driving signals. They are useful to target different locations and to increase the focal volume per sonication and could eliminate the need for mechanical motion. Another strategy to increase the focal volume with such arrays is to simultaneously create multiple focal points. Commercially available FUS systems have phased-array transducers with several hundreds to thousands of piezoelectric elements. The higher number of elements provides greater flexibility in both targeting and shaping the focus, a useful aspect for the treatment of larger or complex tumor volumes. The ability to steer the focal point in real time may also permit the treatment of moving organs without inducing imaging artifacts.

## 20.5 Image Guidance and Control: MRI versus US

As a noninvasive procedure, FUS requires image guidance for each step of the procedure: anatomic imaging for accurate target localization, tumor margin, and surrounding anatomy definition (to plan the safe trajectory of FUS beam through acoustic windows); thermal imaging to verify the focal coordinate before ablation and to monitor the temperature elevation to ensure a sufficient thermal dose is delivered only to the focal zone; and posttreatment imaging to verify the ablation. Only if all of these conditions are met can a complete and safe treatment be performed. US imaging, which has been used for guidance, does not currently fulfill these requirements. In most cases, especially in malignancies, diagnostic US is not sensitive enough to detect exact tumor margins, and it may fail to identify essential anatomical details, such as the exact location of nerves. Bone and air cause serious artifacts that can impede visualization. The ability to localize the thermal spot at low energy levels requires thermal sensitivity that is currently possible only with MRI. Efforts to develop other noninvasive techniques to visualize the focus have not materialized to a useful level. Although tissue temperature is accepted as a surrogate measure of tissue viability, direct measurement of tissue viability by functional or metabolic nuclear tracers is not yet performed. MRI is able to measure tissue temperature with sensitivities of better than ±2°C, whereas US imaging cannot measure temperature with needed accuracies and sensitivities, despite a significant research effort. At present, without sufficient thermal sensitivity of US imaging, the focal spot cannot be localized and the temperature elevations cannot be measured. Because of the limitations of US guidance, neither the thoroughness of treatment nor the safety of normal tissue can be guaranteed. US image-guided FUS has been applied for the treatment of benign and malignant tumors with some success in China but without widespread acceptance elsewhere. Many advocates of US guidance believe that MRI is needed only for temperature monitoring and that a temperature-sensitive US method will eventually be developed. However, the localization of tumor margins and the three-dimensional definition of the targeted tumor volume are the most essential requirements for noninvasive image-guided thermal therapies. Localization defines the exact spatial extent of the targeted tissue and demonstrates the related critical anatomic structures around it. For both purposes, MRI is superior to US as a diagnostic imaging method. For example, in prostate cancer diagnosis, compared to transrectal US, MRI has improved the evaluation of cancer location, size, and extent and has allowed concurrent evaluation of the details of prostate, periprostatic, and pelvic anatomy. In determining the extent of breast cancer, contrast-enhanced diagnostic MRI was proven to be more sensitive than all other imaging techniques. In addition, MRI can detect intraductal spread more accurately than other imaging modalities. MRI appears to be indispensable for accurate tumor localization in radiation therapy or for minimizing local recurrence in breast-conserving surgery. MRI can also be used to reduce high reexcision rates for lumpectomies and margin-positive partial mastectomies. High-resolution MRI of the prostate and breast at the higher field strength of 3T can further improve the quality and usefulness of imaging not only for diagnosis but also for surgical and ablation therapy. Increased signal-to-noise allows imaging at higher resolution, which may improve the definition of tumor margins over 1.5 T. Additional advantages of higher field strength, such as improved sensitivity of temperature measurements (which can enable multi-slice or three-dimensional thermometry) well justify the integration of FUS with 3 T MRIs. Despite the improving ability of MRI to detect cancerous tissue, the full extent of most malignant tumors is not revealed by MRI because of ill-defined tumor margins and the method's limited sensitivity to relatively small amounts of any disseminated tumor cells infiltrating normal tissue. The direct consequence of this inaccurate target definition is the frequent failure of surgical and thermal therapy methods. Molecular imaging with tumor seeking contrast agents or biomarkers may improve detection of tumors and better define their margins, resulting in better targeting and, in some cases, complete removal. Moreover, developments in MRI hardware and software (3 T vs. 1.5 T imaging) continue to improve spatial and temporal resolution and the signal-to-noise ratio of MRI examinations. MRgFUS has been used extensively in the successful treatment of uterine fibroids and has been shown to be an effective treatment option for the palliative ablation of the painful bone metastases (Figure 20.3).

### 20.5.1 Treatment Monitoring

The chemical environment and relaxation properties of nuclei are the source of the signal in MR. MR is also sensitive to Brownian motion and associated molecular tumbling rates; thus, MRI techniques are intrinsically sensitive to temperature. By choosing correct parameters, MR thermometry can provide all the requirements for thermal imaging during ablation, including temperature sensitivity, linearity of temperature effect, and adequate spatial and temporal resolution, with minimal artifact. The proton resonance frequency (PRF) method is the most appropriate parameter for quantitative temperature monitoring. In this method, hydrogen bonds between local water molecules bend, stretch, and break as the temperature increases and the decreased net hydrogen bond strength results as the strength of the covalent bond between the water proton and its oxygen increases. That effect better shields the proton from the external magnetic field and causes a PRF shift. The PRF has an approximately linear relationship with the change in temperature and very little tissue dependency. Magnetic resonance PRF thermometry allows noninvasive temperature monitoring during ultrasound thermal ablation. The method uses the temperature dependence of the PRF, which can be determined from the phase in gradient-echo images. In order to provide volumetric and rapid thermometry, the acquisition

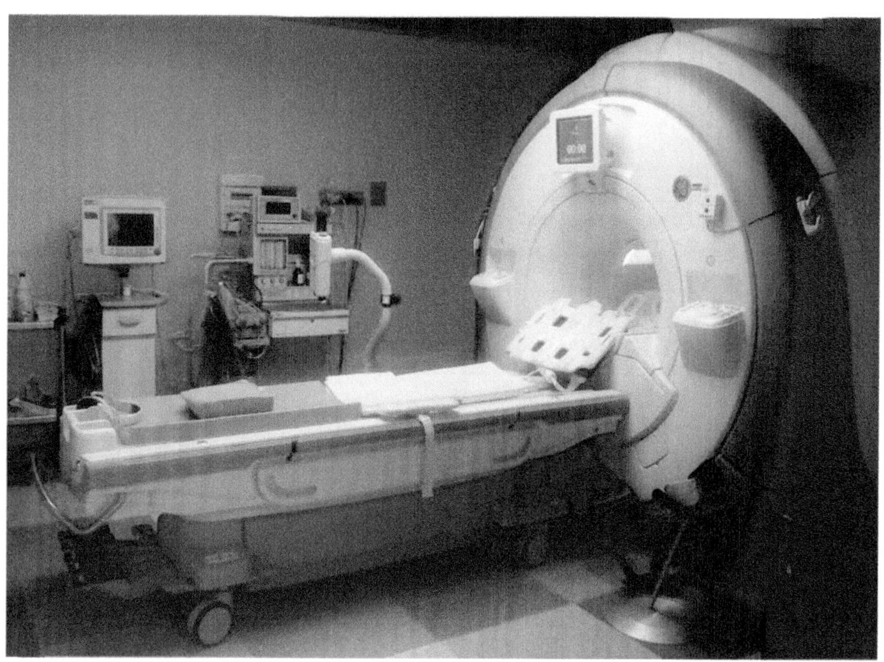

**FIGURE 20.3**
Examples of currently available commercial MR-guided focused ultrasound systems. Insightec OR MR-guided system. (Courtesy of Radiological, Oncological and Anatomopathological Sciences, Sapienza University of Rome, Policlinico Umberto I, Rome, Italy.)

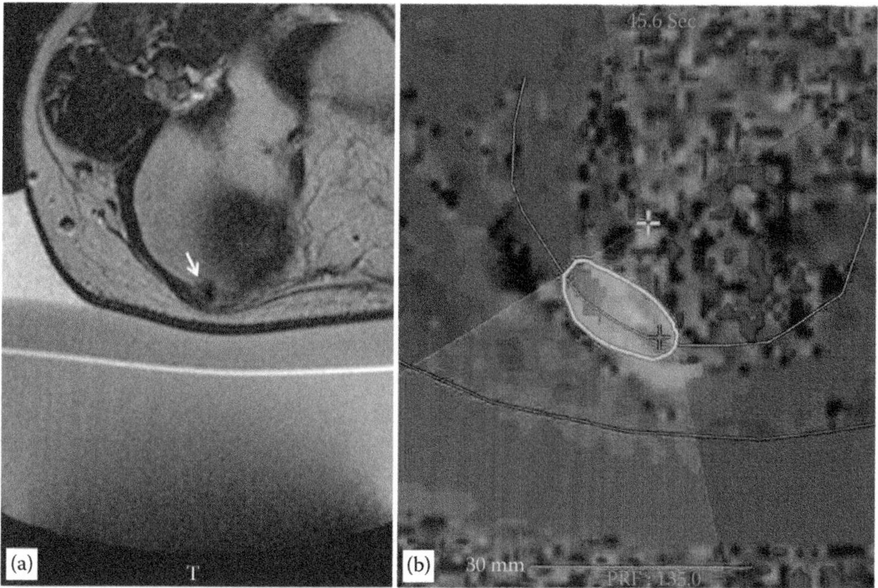


**FIGURE 20.4**
MR-guided focused ultrasound for osteoid osteomas. Images documenting the treatment of a patient with an osteoid osteoma treated with MRgFUS about 1 month ago. (a) $T_1$ spin-echo coronal image of a patient with small osteoid osteoma at the level of the olecranon. (b) The picture above shows the procedure in real time, with the control of temperature *in vivo* after treatment, the patient has shown no more pain at that level.

sequence is multislice, gradient-recalled, single-shot, echo-planar imaging (EPI) (Figure 20.4). MRgFUS clinical applications to date involve targeted tissue volumes being sonicated with the intensity of US at between 500 and 20,000 W/cm$^2$ and a duration of 1–60 s. Shorter times are preferred to circumvent perfusion cooling effects that increase over time. After each sonication, a cooling period lasts until the temperature returns to baseline. This prevents the buildup of heat in the nontargeted tissue volume (beam path zone).

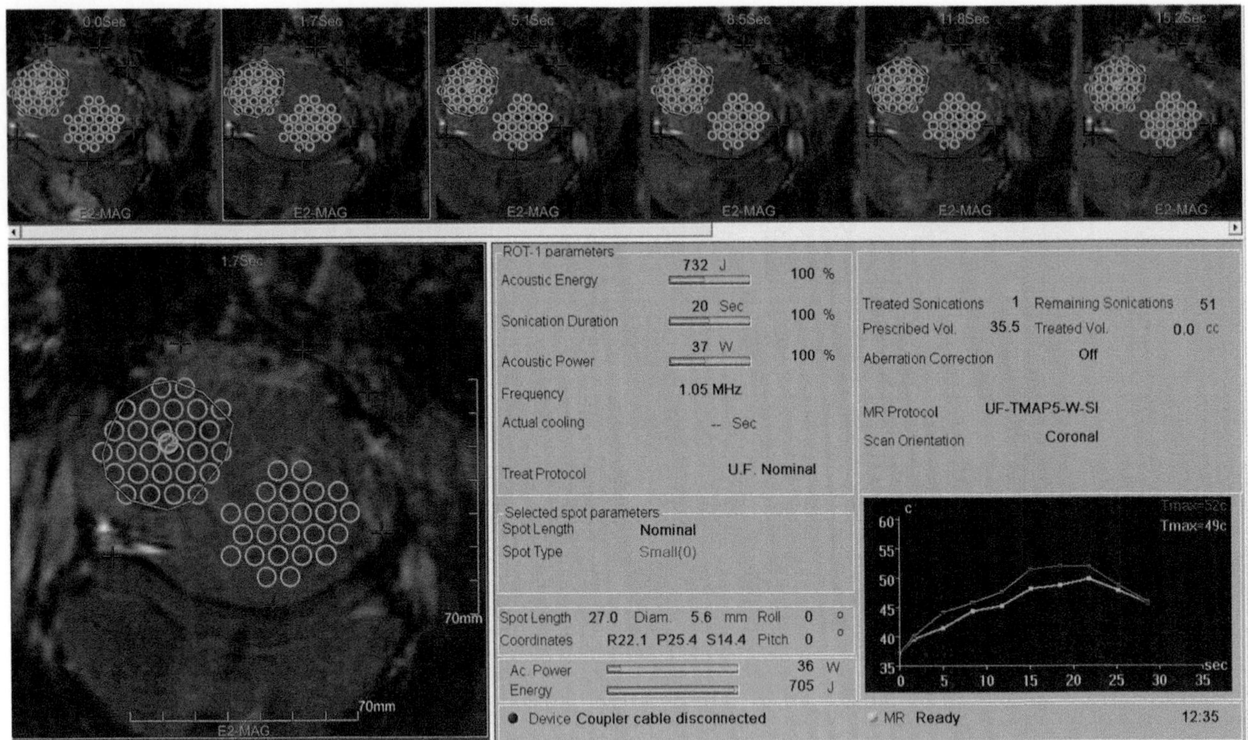


**FIGURE 20.5**
Examples of treatment planning systems for MRgFUS. Treatment planning for the InSightec Exablate. Planning screens allow the operator to set treatment parameters, monitor beam paths per transducer, thermal lesion location, time/temperature graphs, and ultrasound frequency spectrum. This system allows the operator to monitor real-time temperature rise at the target, as well as in near-field and far-field regions.

Using MRI thermometry, the focal spot is visible on the images, and the temperature rise and cooling are measurable. Therefore, the spatiotemporal control of the sonication by MRI guidance is precise, predictable, and reproducible. Before the actual treatment, MRI defines the target by detecting the focal spot with a noncoagulative low-power test pulse. When targeting is accomplished, the power level can be increased to achieve irreversible tissue damage by protein denaturation and capillary bed destruction. Sequential imaging steps follow the progressive tissue coagulation by overlapping focal spots. The coagulative effect is instant and the related tissue phase transitions are detectable by MRI with contrast administration immediately after the sonication ceases (Figure 20.5). In addition to image guidance, it is important to be in constant contact with the patient. In most situations, the patient is awake, though sedated, during the procedure and can notify the doctor about unanticipated heating-related pain before irreversible damage is done. He or she holds a panic button that can instantly stop the treatment if discomfort or pain occurs.

## 20.6 Clinical Applications

### 20.6.1 Uterine Fibroids

Uterine fibroids are the most common benign neoplasm of the uterus. Pathologically, they are smooth muscle neoplasms with some fibrous connective tissue. Histologically, they are nonencapsulated but demarcated from surrounding myometrium by a pseudocapsule of light areolar tissue or compressed myometrial tissue. They are categorized by their location in the uterus as intramural (entirely or mostly contained within the myometrium), submucosal (projecting into the endometrial cavity and may also be pedunculated), or subserosal (projecting outward from the serosal surface of uterus and may be pedunculated). Uterine fibroids are estrogen dependent and contain estrogen receptors; thus, they can enlarge during pregnancy or with the use of oral contraceptive pills and shrink after menopause. Fibroids can be subject to a wide variety of degenerative phenomena, especially during rapid growth including myxoid, hyaline, cystic, red (hemorrhagic), and fatty degeneration as well as calcification and necrosis. These all contribute to the complexity and variability of fibroid imaging appearance. Approximately 25% of women with uterine fibroid have pelvic pain; menorrhagia; dysmenorrhagia; dyspareunia; pressure-related symptoms, such as pelvic fullness; and urinary frequency. They are thought to contribute to infertility and can be associated with early pregnancy losses and difficulty in maintaining a viable pregnancy. In general, the larger the fibroids, the more likely they are to be symptomatic; therefore, it is believed that by reducing fibroid size symptoms may improve. Interestingly, this does not always hold; even small fibroids such as small submucosal ones that bleed a lot and can be very symptomatic. Current medical treatment of uterine fibroids mainly involves hormonal manipulation. Gonadotrophin-releasing hormone (GnRH) analogues have been the most widely used for temporarily reducing fibroid size; their use may influence the type of surgical approach (abdominal vs. vaginal hysterectomy) and decrease blood loss during operation. Hysterectomy has always been the ultimate way of resolving all fibroid-associated symptoms. Laparoscopic myomectomy is used frequently for intramural and subserosal fibroids, and hysteroscopic myomectomy is used if the fibroid is submucosal. Advancements in medical technology have made less invasive treatment options available like uterine artery embolization, laser ablation, cryotherapy, RF ablation, and MRgFUS. The ExAblate 2000 (InSightec Inc., Haifa, Israel) is the first MRgFUS device approved by the FDA in 2004 for fibroid treatments, and some studies have proposed the use of this technique for treatment of adenomyosis. To date, no precise appropriateness criteria have been defined to establish treatment indications and to assess response to treatment. In the majority of cases, treatment indications are performed after the assessment of complex clinical and imaging variables, including fibroid size, localization, number, signal intensity on MR images, symptoms, and desire for future pregnancy. The initial trials proving the efficacy and safety of this new approach were restricted to premenopausal women with no desire of future pregnancy; however, since safety and efficacy of treatment have been successfully assessed, the preservation of fertility in patients with uterine fibroids recently became one of the indications to MRgFUS. In particular, a study performed by Rabinovici et al. showed 54 pregnancies in 51 women after MRgFUS with live births in 41% of them, 28% spontaneous abortion rate, 11% rate of elective pregnancy termination, and 11% ongoing pregnancies. Other authors reported also the feasibility to treat adenomyosis using MRgFUS: in 2006, nine patients with adenomyosis were treated by the team of Rabinovici with one of them having spontaneous menstrual cycles after MRgFUS and a successfully and uneventful pregnancy course. In 2008, 20 premenopausal women affected by adenomyosis were treated with a significant improvement in symptom severity during a 6-month follow-up. In light of these data, MRgFUS may be considered a safe and effective method to treat adenomyosis, enabling to obtain large ablative volumes and significant pain relief using low energy protocols (420 W). Treatment is performed in prone position under light sedation with active monitoring of vital signs. Rectal and bladder fill (US gel and saline, respectively) is considered after evaluating the position and eventual mobility of the uterus with low-resolution fast-acquired localizer images. If patient positioning and alignment (transducer-fibroid) is considered to be adequate for treatment, full resolution $T_2$-weighted images are obtained for ablation planning (Figure 20.6). Prior to treatment start, low-energy sonications are delivered in order to verify the correct

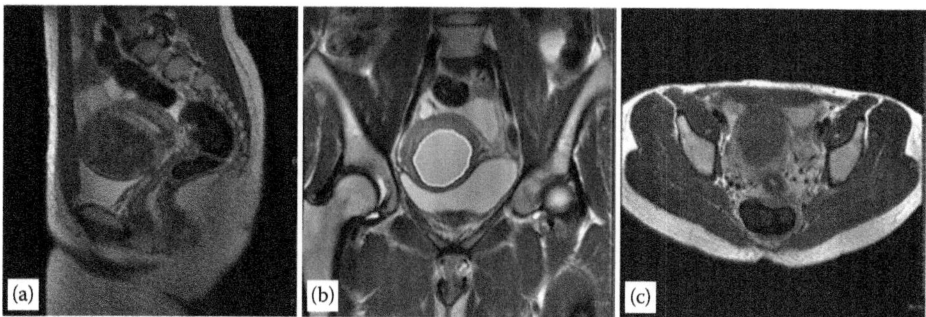


**FIGURE 20.6**
$T_2$-weighted images used to plan fibroid treatment. Sagittal plane shows a hypointense fibroid of the anterior wall of the uterus (a); this plan is used to identify bowel loops with limited energy density line; coronal plane is used to determine the region of treatment (b) (green circle); axial plane (c) is used to identify skin line and possible air bubbles at the skin surface.

position of the focus and the absorption rate of the fibroid. When these elements are verified, the energy can be increased and the real treatment begins. At the end of treatment, $T_1$-weighted postcontrast images are acquired to assess the resulting necrosis within the fibroid, calculated as non-perfused volume (NPV). Despite a mild to moderate reduction in the size, all patients felt a significant reduction in fibroids-related symptoms and a better quality of life.

### 20.6.2 Bone Tumors

Even if the rate of success of combined treatment for bone tumors, ranging from surgery to chemoradiation and various form of percutaneous ablation, is more than acceptable, a significant percentage of patients doesn't benefit from symptoms relief or face symptoms recurrence in the short term. In these cases, MRgFUS can represent a safe and effective approach for both pain palliation and tumor control.

To understand the theoretical basis of these two different clinical approaches, it should be taken in consideration that cortical bone has high acoustic absorption and low thermal conduction rates, and hence, the focused US energy is absorbed by the cortical surface with none or little penetration into the medullary bone. When pain palliation is the planned objective, the interaction between focused USs and intact cortical bone can be used to produce a temperature increase over the periosteal surface of the target area, finally causing thermal damage to the periosteal nerves, responsible for nociception (Figure 20.7). Gianfelice et al. treated patients with localized painful bone metastases, evaluating symptoms relief with a Visual Analog Scale (VAS) score and reporting progressive pain reduction in treated regions and decrease in pain medication usage during a 3-month follow-up period. Furthermore, MRgFUS can be a feasible option to obtain pain palliation in patients with benign bone tumors (such as osteoid osteoma) or non-tumoral conditions. On the other hand, when tumor control or tumor debulking is the primary clinical intent, focused US should be applied over a damaged (severely thinned or eroded) bony cortex, thus allowing thermal damage to lesions located deep into the medullary bone. MRgFUS is clinically approved in the EU for palliative treatment of bone metastases in patients who have exhausted or refused all other pain palliation methods including EBRT. If compared to other non-surgical treatment options for both pain palliation and tumor control, and most of all with radiotherapy, a relevant advantage of MRgFUS ablation is represented by the fact that treatment can be repeated indefinitely until the clinical aim is achieved, without issue related to radiation absorption or other toxicity effects. Similar considerations

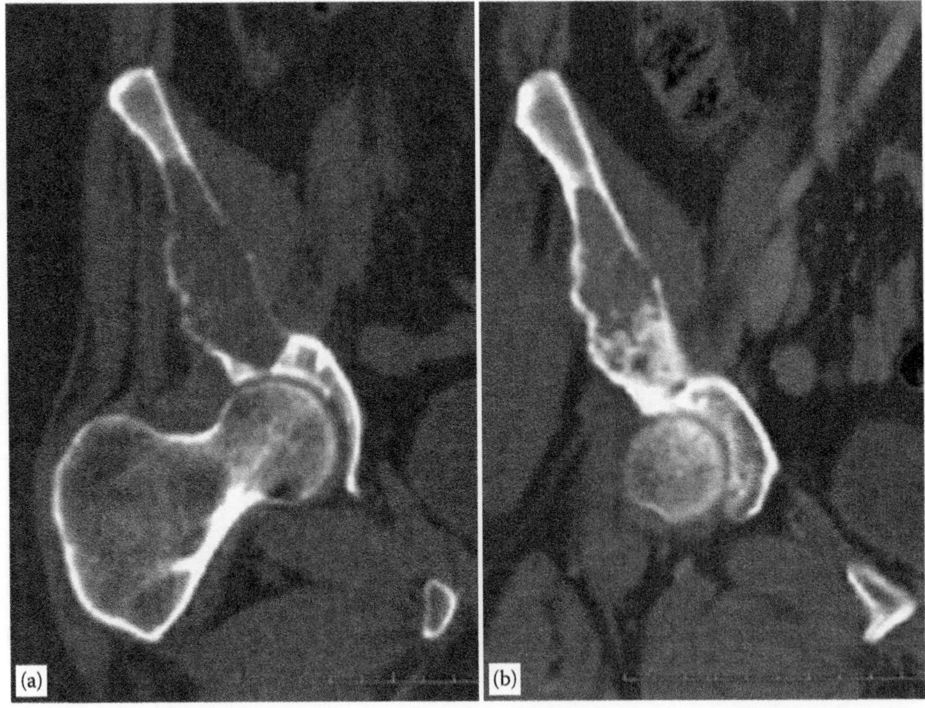


**FIGURE 20.7**
Bone rearrangement followed at MRgFUS pain palliation treatment; coronal CT images performed (a) before and (b) at 3 months' follow-up revealed an increase of bone density with restoration of the cortical border as a sign of *de novo* mineralization.

could be applied to the treatment of benign bone lesions that often represent a clinical challenge for surgeons. During the treatment, patients are variably positioned according to tumor location, and a general, epidural, or peripheral anesthesia was chosen based on different parameters (i.e., lesion location, patient's age, and clinical conditions). In all cases, pre-treatment planning is performed with $T_1$-weighted, $T_2$-weighted contrast-enhanced sequences acquired on multiple planes with and without fat saturation pulses. Treatment is started after delivering low-energy sonications to verify the correct position of the focus; the ablation is therefore performed over the periosteal surface for osteoid osteoma and bone metastases with intact cortex or through the eroded cortex in cases of lytic metastases. Although MR thermal maps cannot be measured directly from the bone itself (due to the absence of moving protons in the cortical zone), heating due to conductive processes from the bone surface within the adjacent soft tissue is measurable and considered adequate for treatment monitoring. To evaluate treatment efficacy in terms of pain palliation, the Brief Pain Inventory-Quality of Life (BPI-QOL) criteria were used to calculate pain severity score. To evaluate treatment efficacy in terms of local tumor control, lesion changes were evaluated according to MD Anderson (MDA) criteria.

### 20.6.3 Breast Cancer

MRgFUS is considered a safe and feasible noninvasive alternative to surgical or radiotherapy treatment of breast benign and malignant tumors. This treatment is thought to be more psychologically and cosmetically acceptable to patients and more suitable for treating patients who are at high risk for operation. However, because follow-up studies evaluating the rate of disease progression and recurrence after treatment are still far from being completed, the long-term efficacy of MRgFUS is still under investigation. Treatment is performed with the patient in prone position. The breast is placed under moderate compression within a dedicated breast coil. Pretreatment planning is performed on $T_1$-weighted contrast-enhanced sequences, acquired in multiple planes with and without fat saturation pulses. Similar to what happens for uterine fibroids, low-energy sonications are delivered in order to verify the correct position of the focus and the absorption rate of the lesion before treatment start at full energy. Tumor necrosis is calculated as NPV on contrast-enhanced $T_1$-weighted sequences.

### 20.6.4 Prostate

Wide experience obtained with US-guided HIFU suggests that this kind of focal therapy is emerging as an alternative to active surveillance for management of low-risk prostate cancer, in particular for patients with localized disease (stage T1–T2 Nx–N0 M0), those with disease not suitable for radical prostatectomy or those who refuse surgery. It could also be used as a salvage therapy for locally proven recurrence of prostate cancer after radiotherapy or brachytherapy failures. There are, however, some relevant limits in treatment efficacy assessment, substantially related to the intrinsic technical nature of conventional US. Transrectal conventional US is able to identify index lesions as hyperechoic areas, but it is not able to provide real-time controls on ablative procedure or to define treatment effects with high spatial resolution. On the other hand, MRI provides a thermometric monitoring and a superior anatomic imaging with feasible multiparametric evaluation of prostatic neoplasm (including dynamic contrast-enhanced imaging, proton spectroscopy, and diffusion-weighted imaging). Patients underwent peripheral block by spinal anesthesia, and a urinary catheter was positioned to ensure urine flow during the procedure. Treatment is performed with a transrectal MRgFUS transducer device covered with a plastic balloon filled with cooled, degassed water to reduce thermal dispersion at the rectal wall interface and to preserve mucosal and submucosal layers from collateral damage during prostate treatment. As in standard MRgFUS ablative protocol, low acoustic energy administration precisely defined the targeted lesion, and high-energy ablation was performed once focus targeting was confirmed. After treatment, tumor necrosis is calculated as NPV on contrast-enhanced $T_1$-weighted sequences (Figure 20.8). MRgFUS feasibility, safety, and efficacy remain under investigation.

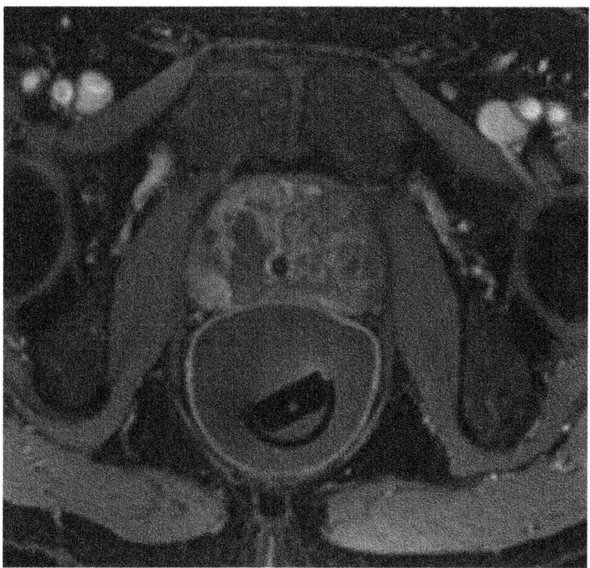

**FIGURE 20.8**
Contrast-enhanced $T_1$-weighted sequences acquired after MRgFUS shows tissue necrosis in the treatment area.

### 20.6.5 New Clinical Applications

MRgFUS has the potential to be a therapeutic option in the field of brain oncology, and first studies demonstrate tumor ablation through craniotomy windows. However, effective penetration of focalized HIFU beam through the intact skull still represents a challenge for scientist. At present, the development of a dedicated brain ablation device is based on the combination of three following technologies in a single unit: (1) thermal ablation with HIFU, (2) intraoperative guidance by MRI and real-time thermal monitoring, and (3) full hemispheric US-phased arrays to correct cranial bone distortion and focus the beam deeply into the brain. In addition to neoplastic indications, today there are now several modern applications of HIFU in the field of neurological disease. The capability of occluding vessels could make focused US a therapeutic tool for the treatment of vascular malformation. Furthermore, lesions can be induced using MRI targeting to treat movement disorders (Parkinson's disease) or epilepsy. Moreover, another new clinical application concerns abdominal moving-organ ablation. Real-time liver motion compensation has been developed and tested in healthy volunteers, potentially providing a chance for more accurate MRI guidance for abdominal tumor ablation. In conclusion, HIFU has been proven to be an effective, noninvasive ablation technique for the treatment of both benign and malignant tumors, with a well-established clinical experience under conventional US guidance. Recent introduction of MRI guidance systems featuring real-time thermal mapping technology, as well as the development of advanced focused US transducers, can significantly improve the efficacy of this modality, mostly in consideration of new clinical applications, such as transcranial brain ablation or moving-organ ablation. Although the MRgFUS procedure has high initial costs, it provides rapid gains in quality of life and shortens the rehabilitation time after treatment compared to surgery. Moreover, MRgFUS has been demonstrated to reduce the length of hospitalization for treated subjects; it is feasible on an outpatient basis and requires no specific care on an inpatient basis and severe complications are virtually absent.

## Bibliography

1. DeVita VT Jr, Lawrence TS, Rosenberg SA (2001) *Cancer: Principles and Practice of Oncology*. Lippincott Williams & Wilkins, Philadelphia, PA.
2. Brown JM, Giaccia AJ (1998) The unique physiology of solid tumors: Opportunities (and problems) for cancer therapy. *Cancer Res* 58:1408–1416.
3. Lee SH, Lee JM, Kim KW et al. (2011) Dual-energy computed tomography to assess tumor response to hepatic radiofrequency ablation: Potential diagnostic value of virtual noncontrast images and iodine maps. *Invest Radiol* 46:77–84.
4. Goldberg SN, Grassi CJ, Cardella JF et al. (2009) Image-guided tumor ablation: Standardization of terminology and reporting criteria. *J Vasc Interv Radiol* 20(7 suppl):S377–S390.
5. Goldberg SN, Gazelle GS, Mueller PR (2000) Thermal ablation therapy for focal malignancy: A unified approach to underlying principles, techniques, and diagnostic imaging guidance. *AJR Am J Roentgenol* 174: 323–331.
6. Lynn JG, Zwemer RL, Chick AJ (1942) The biological application of focused ultrasonic waves. *Science* 96(2483):119–120.
7. Fry WJ, Fry FJ (1960) Fundamental neurological research and human neurosurgery using intense ultrasound. *IRE Trans Med Electron* ME-7:166–181.
8. Wang ZB, Wu F, Wang Z et al. (1997) Targeted damage effects of high intensity focused ultrasound (HIFU) on liver tissues of Guizhou Province miniswine. *Ultrason Sonochem* 4:181–182.
9. Yu T, Luo J (2011) Adverse events of extracorporeal ultrasound-guided high intensity focused ultrasound therapy. *PLoS One* 6:e26110.
10. Jolesz FA (2009) MRI-guided focused ultrasound surgery. *Annu Rev Med* 60:417–430.
11. Jolesz FA, Hynynen K (2002) Magnetic resonance image-guided focused ultrasound surgery. *Cancer J* 8(suppl 1): S100–S112.
12. Simon CJ, Dupuy DE, Mayo-Smith WW (2005) Microwave ablation: Principles and applications. *RadioGraphics* 25 (suppl 1):S69–S83.
13. Ter Haar G (2010) Ultrasound bioeffects and safety. *Proc Inst Mech Eng H* 224:363–373.
14. Jolesz FA, McDannold N (2008) Current status and future potential of MRI-guided focused ultrasound surgery. *J Magn Reson Imaging* 27:391–399.
15. Rabinovici J, David M, Fukunishi H et al. (2010) Pregnancy outcome after magnetic resonance–guided focused ultrasound surgery (MRgFUS) for conservative treatment of uterine fibroids. *Fertil Steril* 93:199–209.
16. Schmitt F, Grosu D, Mohr C et al. (2004) 3 Tesla MRI: Successful results with higher field strengths (review). *Radiologe* 44:31–47.
17. Stewart EA, Gedroyc WM, Tempany CM et al. (2003) Focused ultrasound treatment of uterine fibroid tumors: Safety and feasibility of a noninvasive thermoablative technique. *Am J Obstet Gynecol* 189:48–54.
18. Harding G, Coyne KS, Thompson CL, Spies JB (2008) The responsiveness of the Uterine Fibroid Symptom and Health-Related Quality of Life questionnaire (UFS-QOL). *Health Qual Life Outcomes* 6:99.
19. Hesley GK, Felmlee JP, Gebhart JB et al. (2006) Noninvasive treatment of uterine fibroids: Early Mayo Clinic experience with magnetic resonance imaging–guided focused ultrasound. *Mayo Clin Proc* 81:936–942.

20. Funaki K, Fukunishi H, Sawada K (2009) Clinical outcomes of magnetic resonance-guided focused ultrasound surgery for uterine myomas: 24-month follow-up. *Ultrasound Obstet Gynecol* 34:584–589.
21. Behera MA, Leong M, Johnson L, Brown H (2010) Eligibility and accessibility of magnetic resonance-guided focused ultrasound (MRgFUS) for the treatment of uterine leiomyomas. *Fertil Steril* 94:1864–1868.
22. Kim YS, Kim JH, Rhim H et al. (2012) Volumetric MR-guided high-intensity focused ultrasound ablation with a one-layer strategy to treat large uterine fibroids: Initial clinical outcomes. *Radiology* 263:600–609.
23. Rabinovici J, Stewart EA (2006) New interventional techniques for adenomyosis. *Best Pract Res Clin Obstet Gynaecol* 20:617–636.
24. Fukunishi H, Funaki K, Sawada K et al. (2008) Early results of magnetic resonance-guided focused ultrasound surgery of adenomyosis: Analysis of 20 cases. *J Minim Invasive Gynecol* 15:571–579.
25. Dong X, Yang Z (2010) High-intensity focused ultrasound ablation of uterine localized adenomyosis. *Curr Opin Obstet Gynecol* 22:326–330.
26. Hynynen K, Pomeroy O, Smith DN et al. (2001) MR imaging-guided focused ultrasound surgery of fibroadenomas in the breast: A feasibility study. *Radiology* 219:176–185.
27. Gianfelice D, Khiat A, Amara et al. (2003) MR imaging-guided focused ultrasound surgery of breast cancer: Correlation of dynamic contrast-enhanced MRI with histopathologic findings. *Breast Cancer Res Treat* 82:93–101.
28. Furusawa H, Namba K, Nakahara H et al. (2007) The evolving non-surgical ablation of breast cancer: MR guided focused ultrasound (MRgFUS). *Breast Cancer* 14:55–58.
29. Wu F, Wang ZB, Zhu H et al. (2005) Extracorporeal high intensity focused ultrasound treatment for patients with breast cancer. *Breast Cancer Res Treat* 92:51–60.
30. Mundy GR (2002) Metastasis to bone: Causes, consequences and therapeutic opportunities. *Nat Rev Cancer* 2:584–593.
31. Saarto T, Janes R, Tenhunen M, Kouri M (2002) Palliative radiotherapy in the treatment of skeletal metastases. *Eur J Pain* 6:323–330.
32. Catane R, Beck A, Inbar Y et al. (2007) MR-guided focused ultrasound surgery (MRgFUS) for the palliation of pain in patients with bone metastases—Preliminary clinical experience. *Ann Oncol* 18:163–167.
33. Gianfelice D, Gupta C, Kucharczyk W et al. (2008) Palliative treatment of painful bone metastases with MR imaging-guided focused ultrasound. *Radiology* 249:355–363.
34. Liberman B, Gianfelice D, Inbar Y et al. (2009) Pain palliation in patients with bone metastases using MR-guided focused ultrasound surgery: A multicenter study. *Ann Surg Oncol* 16:140–146.
35. Weeks EM, Platt MW, Gedroyc W (2012) MRI-guided focused ultrasound (MRgFUS) to treat facet joint osteoarthritis low back pain-case series of an innovative new technique. *Eur Radiol* 22:2822–2835.
36. Warmuth M, Johansson T, Mad P (2010) Systematic review of the efficacy and safety of high-intensity focused ultrasound for the primary and salvage treatment of prostate cancer. *Eur Urol* 58:803–815.
37. Crouzet S, Poissonnier L, Murat FJ et al. (2011) Outcomes of HIFU for localised prostate cancer using the Ablatherm Integrate Imaging(R) device. *Prog Urol* 21:191–197.
38. Blana A, Rogenhofer S, Ganzer R et al. (2008) Eight years' experience with high-intensity focused ultrasonography for treatment of localized prostate cancer. *Urology* 72:1329–1333.
39. Ahmed HU, Hindley RG, Dickinson L et al. (2012) Focal therapy for localised unifocal and multifocal prostate cancer: A prospective development study. *Lancet Oncol* 13:622–632.
40. Nau WH, Diederich CJ, Ross AB et al. (2005) MRI-guided interstitial ultrasound thermal therapy of the prostate: A feasibility study in the canine model. *Med Phys* 32:733–743.
41. Pauly KB, Diederich CJ, Rieke V et al. (2006) Magnetic resonance-guided high-intensity ultrasound ablation of the prostate. *Top Magn Reson Imaging* 17:195–207.
42. Ram Z, Cohen ZR, Harnof S et al. (2006) Magnetic resonance imaging-guided, high-intensity focused ultrasound for brain tumor therapy. *Neurosurgery* 59:949–955.
43. McDannold N, Clement GT, Black P et al. (2010) Transcranial magnetic resonance imaging-guided focused ultrasound surgery of brain tumors: Initial findings in 3 patients. *Neurosurgery* 66:323–332.
44. Hynynen K, Colucci V, Chung A, Jolesz F (1996) Noninvasive arterial occlusion using MRI-guided focused ultrasound. *Ultrasound Med Biol* 22:1071–1077.
45. Hynynen K, McDannold N, Vykhodtseva N, Jolesz FA (2001) Noninvasive MR imaging-guided focal opening of the blood-brain barrier in rabbits. *Radiology* 220:640–646.
46. McDannold N, Vykhodtseva N, Hynynen K (2006) Targeted disruption of the blood-brain barrier with focused ultrasound: Association with cavitation activity. *Phys Med Biol* 51:793–807.
47. Kinoshita M, McDannold N, Jolesz FA, Hynynen K (2006) Noninvasive localized delivery of Herceptin to the mouse brain by MRI-guided focused ultrasound-induced blood-brain barrier disruption. *Proc Natl Acad Sci USA* 103:11719–11723.
48. Kinoshita M, McDannold N, Jolesz FA, Hynynen K (2006) Targeted delivery of antibodies through the blood-brain barrier by MRI-guided focused ultrasound. *Biochem Biophys Res Commun* 340:1085–1090.
49. Broggi G, Dones I, Ferroli P, Franzini A, Pluderi M (2000) Contribution of thalamotomy, cordotomy and dorsal root entry zone Caudalis trigeminalis lesions in the treatment of chronic pain. *Neurochirurgie* 46: 447–453.
50. Jeanmonod D, Werner B, Morel A et al. (2012) Transcranial magnetic resonance imaging-guided focused ultrasound: Noninvasive central lateral thalamotomy for chronic neuropathic pain. *Neurosurg Focus* 32:E1.

51. Orsi F, Zhang L, Arnone P et al. (2010) High-intensity focused ultrasound ablation: Effective and safe therapy for solid tumors in difficult locations. *AJR Am J Roentgenol* 195:W245–W252.
52. Wu F, Wang Z, Chen W (2001) Pathological study of extracorporeally ablated hepatocellular carcinoma with high-intensity focused ultrasound. *Zhonghua Zhong Liu Za Zhi* 23:237–239.
53. Sung HY, Jung SE, Cho SH et al. (2011) Long-term outcome of high-intensity focused ultrasound in advanced pancreatic cancer. *Pancreas* 40:1080–1086.
54. Ritchie RW, Leslie T, Phillips R et al. (2010) Extracorporeal high intensity focused ultrasound for renal tumours: A 3-year follow-up. *BJU Int* 106:1004–1009.
55. Napoli A et al. (2013) MR-guided High Intensity Focused Ultrasound: Current status of an emerging technology. *Cardiovasc Intervent Radiol* 36(5):1190–1203.
56. Napoli A et al. (2013) Focused ultrasound therapy of the prostate with MR guidance. *Curr Radiol Rep* 1:154–160.
57. Napoli A et al. (2014) High-intensity focused ultrasound in breast pathology: Non-invasive treatment of benign and malignant lesions. *Expert Rev Med Devices* 24:1–9.
58. Napoli A et al. (2014) Magnetic resonance-guided high-intensity focused ultrasound treatment of locally advanced pancreatic adenocarcinoma: Preliminary experience for pain palliation and local tumor control. *Invest Radiol* 49(12):759–765.
59. Geiger D et al. (2014) MR-guided focused ultrasound (MRgFUS) ablation for the treatment of nonspinal osteoid osteoma: A prospective multicenter evaluation. *J Bone Joint Surg Am* 96(9):743–751.
60. Anzidei M et al. (2014) Magnetic resonance-guided focused ultrasound ablation in abdominal moving organs: A feasibility study in selected cases of pancreatic and liver cancer. *Cardiovasc Intervent Radiol* 37(6):1611–1617.
61. Napoli A et al. (2013) MR imaging-guided focused ultrasound for treatment of bone metastasis. *RadioGraphics* 33(6):1555–1568.
62. Pediconi F et al. (2012) MRgFUS: From diagnosis to therapy. *Eur J Radiol* 81(Suppl 1):S118–S120.
63. Napoli A et al. (2014) Primary pain palliation and local tumor control in bone metastases treated with magnetic resonance-guided focused ultrasound. *Invest Radiol.* 48(6):351–358.
64. Napoli A et al. (2013) Osteoid osteoma: MR-guided focused ultrasound for entirely noninvasive treatment. *Radiology* 267(2):514–521.
65. Napoli A et al. (2013) Real-time magnetic resonance-guided high-intensity focused ultrasound focal therapy for localised prostate cancer: Preliminary experience. *Eur Urol* 63(2):395–398.

# 21

## PET/MRI: Concepts and Clinical Applications

**Mateen C. Moghbel, Abass Alavi, and Drew A. Torigian**

**CONTENTS**



### 21.1 Introduction

The concept of merging structural and molecular imaging originated in the early 1990s, nearly a decade before the advent of single photon emission computed tomography/computed tomography (SPECT/CT) in 1998 and positron emission tomography/CT (PET/CT) in 2000 [1,2]. These developments in hybrid imaging paved the way for later innovations, including PET/magnetic resonance imaging (PET/MRI). Although CT and MRI play comparable roles as structural imaging modalities within these hybrid systems, there are distinct advantages and disadvantages to each of them. MRI is favored in terms of safety, particularly in pediatric and pregnant patients, as it utilizes radiowaves and magnetism rather than ionizing radiation, thereby avoiding the potential risk of mutagenic and carcinogenic side effects. Furthermore, MRI demonstrates superior contrast for visualizing soft tissues and benefits from the ability to acquire functional information through specialized techniques including diffusion-tensor imaging, diffusion-weighted imaging, MR elastography, MR spectroscopy, and perfusion-weighted imaging [3]. When combined with PET, MRI has the ability to greatly enhance image reconstruction, partial volume correction, and motion compensation. On the other hand, it also presents challenges related to electromagnetic interference between the PET and MRI systems. This chapter will briefly explore the design of modern hybrid PET/MRI systems and review their potential clinical applications.

### 21.2 Instrumentation and Design

There are two general approaches to performing PET/MRI—the data can be acquired either sequentially or concurrently. With sequential PET/MRI, the PET and MRI machines are mounted back to back, with the patient being shuttled between the two using a mobile bed. This method sidesteps the issue of interference between the two systems, but is particularly susceptible to motion artifacts as a result of transporting the patient between machines and acquiring the images non-simultaneously. With concurrent PET/MRI, there is integration of the two systems into a single gantry, which is a complex task raising a host of challenges. Foremost among the potential problems is the interference between the modalities. The photodetectors of the PET system can be affected by the strong magnetic field created by the MRI machine. Conversely, the MRI system can also experience electronic interference from some of the components contained within the PET machine. Moreover, the size of the MRI magnet limits the space available for the PET apparatus, necessitating an efficiently designed hybrid machine.

Although solutions to many of these limitations hindering concurrent PET/MRI acquisition were soon engineered, the obstacle of electromagnetic interference demanded novel technological innovations to overcome this particular challenge. Christiansen et al. first attempted to minimize interference purely through design by placing the crystals of the PET system inside the

MRI apparatus and positioning conventional photomultiplier tubes 5 m away from the crystals [4]. A more recent approach has been the utilization of solid-state detectors such as avalanche photodiodes, which are insensitive to magnetic fields [5]. The recent development of multicell Geiger-mode avalanche photodiodes, or silicon photomultiplier tubes, may soon allow PET/MRI machines to be designed with even better gain, signal-to-noise ratio, and timing resolutions. These characteristics make these new hybrid systems that utilize solid state photodiodes ideal for time-of-flight imaging [6].

There are currently three designs of concurrent PET/MRI systems. The first is a PET insert, where the PET imager is placed between the radio-frequency coil and gradient set of the MRI system. The second is fully integrated PET/MRI, where both systems are on the same gantry. The third involves a split-field magnet and low-field-strength MRI system, allowing a large number of PET detectors to be placed within the MRI apparatus [7]. The capability of these hybrid systems to acquire PET and MR images simultaneously optimizes the temporal and spatial alignment of structural, functional, and molecular data, while also reducing acquisition time to 45–70 min. These improvements, coupled with the inherent advantages of the MRI modality, may make PET/MRI a valuable clinical complement—or perhaps even a replacement—for PET/CT in certain clinical applications.

## 21.3 Clinical Applications in Oncology

The high spatial resolution of the MRI system enables provision of anatomical information that is a useful complement of PET imaging in oncology, as it allows for detailed delineation and staging of tumors. MRI proves to be especially valuable for assessment of tissues of the brain, head and neck, spinal cord, liver, pelvic organs, breasts, and musculoskeletal system, which CT is not able to visualize as effectively [3]. (Please see Figures 21.1 through 21.4 for examples of PET/MRI assessment of various malignancies.) For instance, MRI is useful to assess for chest wall invasion by primary lung tumors

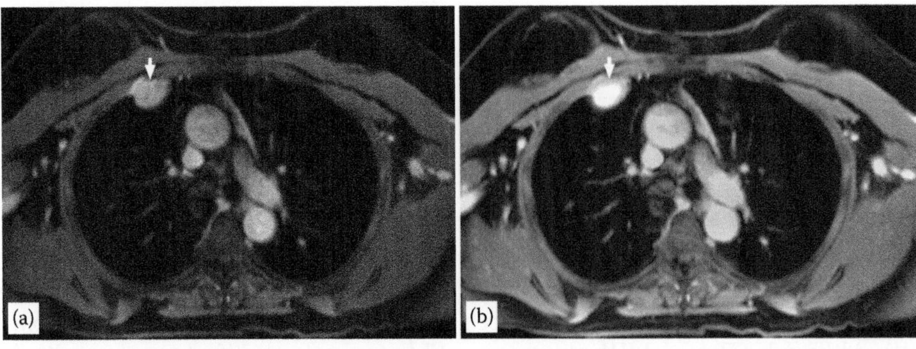


**FIGURE 21.1**
A 70-year-old woman with history of abnormal chest radiograph undergoing further evaluation. Axial fat-suppressed $T_1$-weighted post-contrast MR image (a) and axial software-fused FDG-PET/MR image (b) through chest show 2.7 cm FDG avid enhancing right pleural-based mass (arrows) due to focal malignant pleural mesothelioma.

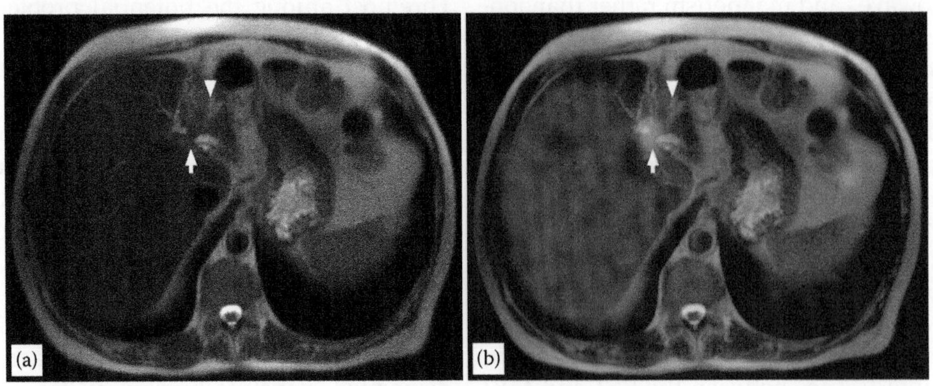


**FIGURE 21.2**
A 71-year-old man with history of liver mass undergoing further evaluation. Axial heavily $T_2$-weighted MR image (a) and axial software-fused FDG-PET/MR image (b) through abdomen reveal 3.0 cm FDG avid mass (arrows) in left hepatic lobe with associated peripheral intrahepatic biliary ductal dilation (arrowheads) due to intrahepatic cholangiocarcinoma.

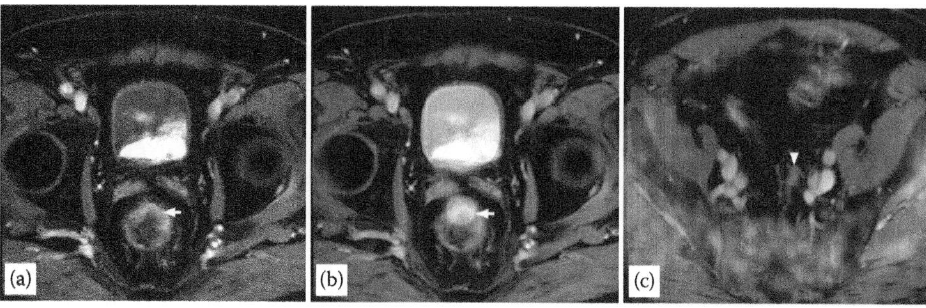

**FIGURE 21.3**
A 77-year-old man with history of rectal carcinoma undergoing staging assessment. Axial fat-suppressed $T_1$-weighted postcontrast MR image (a) and axial software-fused FDG-PET/MR image (b) through pelvis demonstrate FDG avid enhancing nodular thickening of anterior wall of distal rectum (arrows) due to primary malignancy. Axial software-fused FDG-PET/MR image (c) through more superior aspect of pelvis shows minimally FDG avid subcentimeter round presacral lymph nodes (arrowhead) due to regional lymph node metastatic disease.

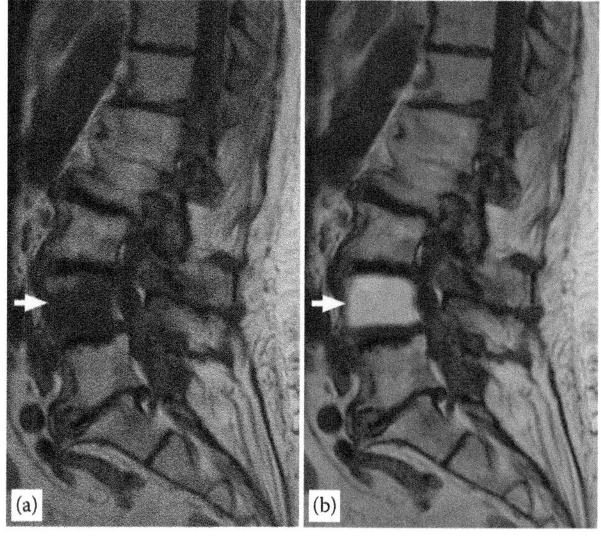

**FIGURE 21.4**
A 88-year-old woman with history of non-small cell lung carcinoma undergoing restaging assessment. Sagittal $T_1$-weighted MR image (a) and sagittal software-fused FDG-PET/MR image (b) through lumbar spine reveal diffusely low signal intensity of bone marrow of L4 vertebral body with avid FDG uptake (arrows) due to bone marrow metastasis.

and is superior to CT for detecting and characterizing metastases in the brain and adrenal glands [8]. These findings were corroborated by a study performed by Kim et al. on a cohort of 49 patients with non-small cell lung cancer [9]. The authors compared the diagnostic performance of [18F]-2-fluoro-2-deoxy-D-glucose (FDG)-PET/CT with and without diffusion-weighted and $T_2$-weighted MR imaging for the preoperative detection and characterization of regional lymphadenopathy. The sensitivity, specificity, and accuracy of FDG-PET/CT alone were found to be 46%, 96%, and 87%. By comparison, these figures rose to 69%, 93%, and 89% for FDG-PET/CT combined with MRI, indicating that addition of the MRI modality in non-small cell lung cancer improves the detection of nodal masses. Another study on non-small cell lung cancer in 15 patients using FDG-PET/MRI and FDG-PET/CT found a large degree of agreement between the modalities on primary tumor staging [10]. This led the authors, Heusch et al., to conclude that simultaneous PET and non-Gaussian diffusion-weighted MRI acquisition is a feasible approach for the assessment of non-small cell lung cancer. Finally, a study by Kohan et al. that evaluated lymph node involvement in 11 lung cancer patients using PET/CT and PET/MRI found that the modalities performed similarly in terms of measurement of the maximum standardized uptake value (SUVmax), interobserver agreement, and diagnostic performance for regional nodal staging [11].

MRI may confer an advantage in the imaging of tumors of the liver, another organ whose soft tissue is suboptimally visualized by CT. A retrospective study carried out by Donati et al. compared the detection of lesions using FDG-PET/CT, gadolinium-ethoxybenzyl-diethylenetriamine pentaacetic acid (Gd-EOB-DTPA)-enhanced MRI, and FDG-PET/MRI [12]. The results showed a significant difference between the first two modalities in the detection rate of liver lesions: 64% in FDG-PET/CT and 85% in Gd-EOB-DTPA-enhanced MRI. FDG-PET/MRI was found to have a significantly higher sensitivity than FDG-PET/CT (93% vs. 76%), as well as a somewhat higher specificity (92% vs. 90%) and accuracy (94% vs. 85%). Overall, the diagnostic confidence for liver lesions larger than 1 cm was shown to be significantly higher with FDG-PET/MRI. Another retrospective study that was performed by Yong et al. evaluated the sensitivity of CT, MRI, FDG-PET, PET/CT, and PET/MRI in 24 patients with liver metastases from colorectal cancer [13]. The sensitivities of these various stand-alone and hybrid imaging modalities were 64.5%, 80.2%, 54.5%, 84.2%, and 98.3%, respectively, suggesting that PET/MRI is highly useful for detecting metastatic lesions under 1 cm in size. Yet another study by Beiderwellen et al. compared PET/MRI and PET/CT in the depiction and characterization of liver metastases in

70 patients and found that PET/MRI offered significantly higher lesion conspicuity and diagnostic confidence in both malignant and benign lesions [14].

Tumors of the head and neck are also more appropriate for MRI. Platzek et al. evaluated the feasibility of PET/MRI for this indication and found that PET/MRI demonstrated high image quality without artifacts due to intermodality interference [15]. Furthermore, PET/MRI was more sensitive to tumors than either of the modalities individually. Similarly, Boss et al. compared FDG-PET/CT and FDG-PET/MRI in 8 patients with head and neck tumors and reported that the PET/MRI system produced better spatial resolution and image contrast than PET/CT without demonstrating any of the artifacts caused by interference between the PET and MRI modalities [16]. Another study by Kanda et al. using retrospective image fusion of FDG-PET and MRI for preoperative staging of head and neck cancer revealed that FDG-PET/MRI was significantly more accurate than FDG-PET/CT (87% vs. 67%) in staging primary tumors [17]. However, a significant difference was not found between these modalities for assessment of the extent of metastasis to regional lymph nodes.

A study of 33 patients with high-risk differentiated thyroid carcinoma who had undergone thyroidectomy and received radioiodine therapy was carried out by Nagarajah et al., who used [124I]-sodium iodine PET/CT and software-registered [124I]-sodium iodine PET/MRI for the diagnosis and dosimetry of lymph node metastases and thyroid remnant tissues [18]. Out of the 106 lesions that were detected on PET, there were 23 lesions that were morphologically correlated on MRI but not on CT, 15 of which were smaller than 1 cm. The authors concluded that PET/MRI was superior to PET/CT for tracking a PET focus to a morphological correlate. An early study by Seiboth et al. found that fused PET/MRI can offer vital information in surgical planning, radioactive iodine therapy decision making, and follow-up for patients with recurrent or persistent thyroid cancer [19]. The authors reported that coregistered PET/MRI data provided additional information that altered the treatment plan in 46% of cases. A case study by Ceriani et al. found that PET/MRI is also able to detect unusual muscular metastases secondary to papillary thyroid carcinoma that PET/CT was unable to visualize [20].

The same conclusion was reached in a study of 17 patients with advanced buccal squamous cell carcinoma, where Huang et al. compared the performance of FDG-PET/MRI and FDG-PET/CT, along with standalone MRI and CT [21]. After finding that the likelihood ratio was significantly higher for FDG-PET/MRI (42.56) than for FDG-PET/CT (25.02), MRI (22.94), and CT (8.6), the authors concluded that FDG-PET/MRI can better detect tumor involvement in both bone and muscle and is especially effective at identifying invasion of the masseter muscle. Moreover, the sensitivity, specificity, and diagnostic confidence of FDG-PET/MRI (90%, 91%, 85.9%) were found to be higher than those of FDG-PET/CT (80%, 84%, 70.3%), as well as those of MRI and CT individually. These results overwhelmingly support FDG-PET/MRI as a more appropriate imaging modality for the delineation of tumor size and assessment of tumor invasion in advanced buccal squamous cell carcinoma.

The literature also includes numerous reports on the clinical application of PET/MRI in breast cancer. A study by Park et al. compared diffusion-weighted MRI and FDG-PET/CT to predict pathological complete response to neoadjuvant chemotherapy in 34 patients with invasive breast cancer [22]. The areas under the receiver operating characteristic curves were comparable for diffusion-weighted MRI, FDG-PET/CT, and a combination of the two: 0.910, 0.873, and 0.944, respectively. The authors concluded that diffusion-weighted MRI and PET/CT demonstrate similar diagnostic accuracy in predicting the response to neoadjuvant chemotherapy in breast cancer patients. Another study by Treglia et al. reported that FDG-PET/MRI is able to detect metastatic brachial plexopathies secondary to breast cancer [23].

The feasibility of applying PET/MRI to prostate cancer has also been investigated by a number of studies. The current modality of choice for detection is multiparametric MRI, although PET/CT is also used clinically, particularly for identifying and staging distant metastases and recurrent disease. However, recent additions to the literature have suggested that PET/MRI may ultimately supplant these modalities for the detection of prostate cancer [24]. Wetter et al. evaluated a group of 15 biopsy-positive, biopsy-negative, and suspected cases of prostate cancer using [18F]-fluorocholine PET/MRI and found excellent image quality that allowed exact morphological correlation of elevated focal or diffuse radiotracer uptake on PET to $T_2$-weighted lesions in the prostate [25]. In another study, Park et al. used parametric fusion PET/MRI based on [11C]-choline PET/CT and apparent diffusion coefficient (ADC) maps derived from diffusion-weighted MRI to detect primary prostate cancer in 17 patients [26]. Tumor-to-background ratios were calculated for [11C]-choline uptake on PET/CT and for the ratio of [11C]-choline uptake to ADC, revealing significantly higher tumor-to-background ratios in patients with prostate cancer with a Gleason score of 3 + 4 or above than those with a score of 3 + 3 or below, leading the authors to conclude that parametric MRI fused with [11C]-choline PET would improve tumor-to-background contrast. The findings of these two studies suggest that PET/MRI is a viable diagnostic tool for the assessment of patients with prostate cancer. A case study by Afshar-Oromieh et al. provided additional evidence in support of this by demonstrating that PET/$T_2$-weighted MRI with a [68Ga]-labeled prostate-specific membrane antigen

(PSMA) ligand can reveal important morphological and molecular information about soft tissue in the detection of prostate cancer that PET/CT is not able to replicate [27].

The ability of PET/MRI to detect and characterize gynecological tumors has already been explored by multiple studies. Kim et al. retrospectively compared the diagnostic ability of FDG-PET/CT and FDG-PET/MRI for identifying metastatic lymph nodes in 79 patients with stage IB-IVA cervical cancer [28]. A significant difference was found between the areas under the receiver operating characteristic curves for PET/MRI (0.735) and PET/CT (0.690), indicating that the former had improved performance for diagnosing lymph node metastases secondary to cervical cancer. A comparable study by Nakajo et al. compared FDG-PET/CT and fused PET/MRI in 31 patients with gynecological tumors [29]. The authors reported significantly higher rates of detection of uterine and ovarian lesions using $T_2$-weighted MRI as compared to $T_1$-weighted MRI or CT. A study by Kitajima et al. also compared the value of FDG-PET/contrast-enhanced CT and FDG-PET/dynamic contrast-enhanced MRI in 30 patients with biopsy-confirmed cases of endometrial cancer [30]. Both hybrid modalities detected the vast majority of primary tumors (96.7% for PET/MRI and 93.3% for PET/CT). However, PET/MRI demonstrated a sizeable advantage over PET/CT (80% vs. 60%) in terms of staging of primary tumors.

Reports in the literature on the application of PET/MRI to pancreatic cancer include a retrospective study performed by Tatsumi et al., who compared FDG-PET/CT with fused PET/MRI in 47 patients with known or suspected pancreatic cancer [31]. The authors reported a significantly higher confidence score for $T_1$-weighted MR image quality as compared to CT, as well as higher diagnostic accuracy of PET/$T_1$-weighted MRI (93%) and PET/$T_2$-weighted MRI (90.7%) compared to PET/CT (88.4%). A larger study carried out by Nagamachi et al. involved evaluation of 119 patients, 96 with pancreatic cancer, and 23 with benign lesions, using FDG-PET/MRI and FDG-PET/CT [32]. For differentiation of malignant from benign lesions, FDG-PET/MRI was shown to have significantly superior accuracy as compared to FDG-PET/CT (96.6% vs. 86.6%). The authors also noted that PET/$T_2$-weighted MRI was able to detect a number of benign cystic lesions that were not visible on PET/CT images. These findings indicate that PET/MRI possesses a distinct advantage in the detection and diagnosis of pancreatic tumors.

Finally, studies investigating the performance of PET/MRI in skin cancer include one performed by Laurent et al., who used whole-body MRI and FDG-PET/CT to stage melanoma in 35 patients [33]. Using histology, imaging follow-up, or clinical follow-up as reference standards, the authors determined the sensitivity and specificity of whole-body MRI (82% and 97%, respectively) to be superior to those of FDG-PET/CT (72.8% and 92.7%, respectively). Similarly, a case study of a patient with primary meningeal melanomatosis compared the diagnostic value of FDG-PET/MRI and FDG-PET/CT. The authors, Lee et al., found that PET/MRI demonstrated presence of hypermetabolic lesions more accurately than PET/CT, aiding in the determination of the course of treatment [34].

## 21.4 Clinical Applications in Neurological Disorders

Both individually and in tandem, PET and MRI have been applied for the evaluation a host of neurological conditions, including neurodegenerative, vascular, oncological, traumatic, psychiatric, behavioral, epileptic, congenital, and age-related disorders [3]. (Please see Figure 21.5 for an example of PET/MRI assessment of a cerebrovascular disorder.) However, in addition to its diagnostic indications, PET/MRI also has the potential to validate the efficacy of treatments targeting these disorders, including cellular and gene therapies [35]. The accurate alignment of structural, functional, and molecular data is particularly vital for informing biopsies and treatment planning in neurological conditions.

The viability of PET/MRI with diffusion-tensor MRI in neuro-oncology was demonstrated by Boss et al. in a study involving 4 patients with brain tumors and 7 controls [36]. Though stronger rim artifacts were found on fractional anisotropic images computed from diffusion-tensor MRI, the computation of the direction of the principal eigenvector and fractional anisotropic values at the region of interest were not affected. The authors concluded that diffusion-tensor MRI combined with PET offers valuable morphological and functional information for treatment planning in patients with brain tumors.

In another study of 10 patients with intracranial masses, Boss et al. compared the efficacy of PET/MRI and PET/CT imaging, using [11C]-methionine for glial tumors and [68Ga]-D-phenylalanine(1)-tyrosine(3) octreotide for meningiomas [37]. In addition to calculating ratios of tumor-to-gray matter and tumor-to-white matter, the authors compared radiotracer uptake in meningiomas to the reference tissue of nasal mucosa. These tumor-to-reference count density tissue ratios revealed a high degree of correlation between the two hybrid imaging modalities ($R = 0.98$). PET/MRI and PET/CT also showed comparable image quality, where artifacts produced by the latter were eliminated by prefiltering images with a 4 mm Gaussian filter at a resolution comparable to PET/CT. Another study performed by Schwenzer et al. imaged 50 patients with intracranial masses, head and neck tumors, and neurodegenerative diseases with

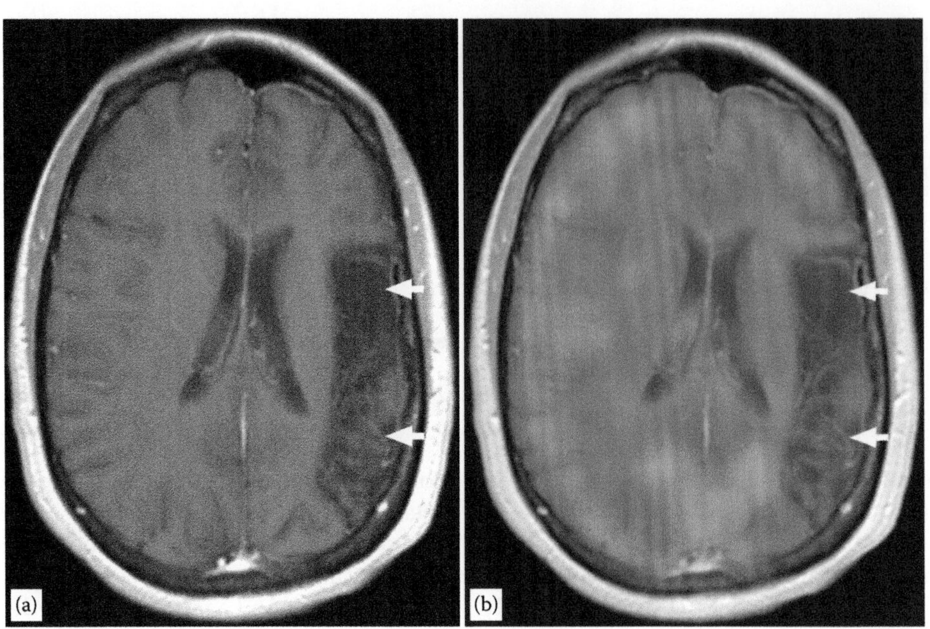


**FIGURE 21.5**
A 51-year-old woman with history of prior cerebrovascular accident and endometrial carcinoma undergoing restaging evaluation. Axial $T_1$-weighted postcontrast MR image (a), and axial software-fused FDG-PET/MR image (b) through brain demonstrate region of left cerebral parenchymal atrophy in left middle cerebral artery vascular distribution with associated decreased FDG uptake in gray matter due to encephalomalacia from prior cerebrovascular accident. No metastatic disease was seen.

simultaneous PET/3 T (T) MRI and PET/CT using FDG, [11C]-methionine, and [68Ga]-D-phenylalanine(1)-tyrosine(3) octreotide [38]. PET/MRI showed high concordance with PET/CT in terms of image quality, tumor delineation, frontal and parietal-occipital ratios, and left–right asymmetry indices.

There are also reports in the literature of the application of PET/MRI for the evaluation of various neurodegenerative disorders. Vercher-Conejero et al. published two case studies of patients with dementia who were imaged with [18F]-florbetapir PET/MRI to assess for amyloid deposition [39]. Similarly, Moodley et al. reported a case study of a patient with frontotemporal dementia who was evaluated simultaneously with 3 T MRI and FDG-PET on an integrated scanner [40]. The precise temporal and spatial alignment of the structural, functional, and molecular data afforded by MRI and PET offer invaluable insight into the neuropathology of patients with neurodegenerative disorders, which are often difficult to distinguish through any individual imaging modality. These early case studies suggest that hybrid PET/MRI may offer greater diagnostic accuracy in this field.

Other indications of PET/MRI in neurology include the localization of seizure foci in the brain. This application was investigated retrospectively by Salamon et al. in patients with cortical dysplasia [41]. The authors concluded that FDG-PET/MRI would enhance the noninvasive identification and surgical treatment of seizure foci in patients with this condition.

Cho et al. have conducted a number of neuroimaging studies looking into the sequential combination of FDG-PET with the ultra-high resolution structural information provided by 7 T MRI. The first, which measured glucose metabolism in the hippocampal structures of five subjects, demonstrated excellent spatial resolution and was able to discern that the dentate gyrus and cornu ammonis were the anatomical regions with the highest uptake [42]. Another study using a similar methodology produced detailed anatomical, functional, and molecular data about the thalamus and subthalamic structures, indicating that the medial dorsal thalamic nucleus had the highest glucose metabolism [43]. A third study on individual raphe nucleus groups produced equally high-resolution hybrid images and hinted at future approaches to investigating a host of neurological disorders [44].

## 21.5 Clinical Applications in Cardiovascular Disorders

The potential clinical applications of PET/MRI in the field of cardiology are myriad; it can provide a more accurate method for assessing myocardial viability and infarction, ventricular function, cardiomyopathy, myocarditis, atherosclerotic disease, vasculitis, and cardiac and pericardiac tumors, among other cardiopathologies.

(Please see Figure 21.6 for an example of PET/MRI assessment of a cardiac neoplasm.) The utility of PET/MRI in patients with dilated cardiomyopathy and recipients of cardiac transplants was demonstrated by Bengel et al. in a study involving [11C]-hydroxyephedrine PET and MRI [45]. The authors reported that they were able to visualize cardiac regional innervations and control mechanisms using this hybrid imaging approach. In a more recent study, Higuchi et al. found that PET/MRI can assess cell and gene therapies for cardiovascular disorders, including cardiac stem cell replacement [46]. A case study performed by Schneider et al. in a patient with cardiac sarcoidosis illustrated the diagnostic value of PET/MRI for identifying myocardial infiltration and inflammation of the cardiac walls, as well as for determining and following changes that occur with the course of treatment [47].

Multiple studies on cardiac and pericardiac tumors have also shown that MRI is especially useful for detecting, localizing, and characterizing metastases, whereas FDG-PET is able to evaluate tumor responses to treatment [48,49]. Another set of studies have demonstrated that PET/MRI of the vascular system is useful for the regional and global detection, characterization, and quantification of atherosclerosis [50–52].

## 21.6 Clinical Applications in Musculoskeletal Disorders

The potential uses of PET/MRI in musculoskeletal disorders are numerous; the hybrid imaging modality may prove diagnostically beneficial in various conditions including musculoskeletal infection, the diabetic foot, painful arthroplasty, metabolic bone marrow disorders, and arthritis [53]. (Please see Figure 21.7 for an example

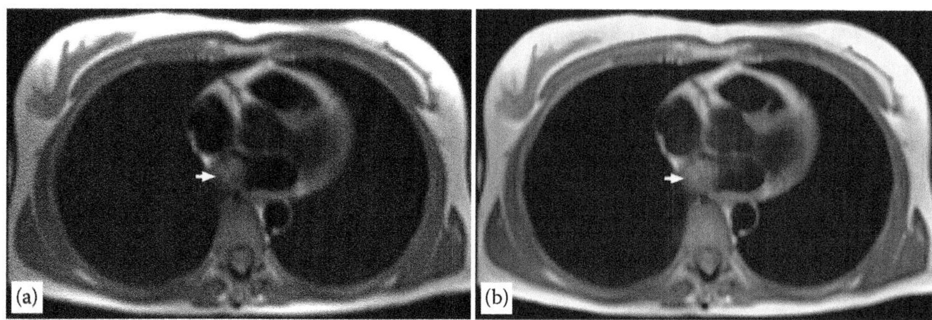


**FIGURE 21.6**
A 41-year-old woman with history of melanoma and cardiac mass undergoing further evaluation. Axial black-blood MR image (a) and axial software-fused FDG-PET/MR image (b) through heart show 2.9 cm mildly FDG avid left atrial mass (arrows) due to atrial myxoma. No metastatic disease was seen.

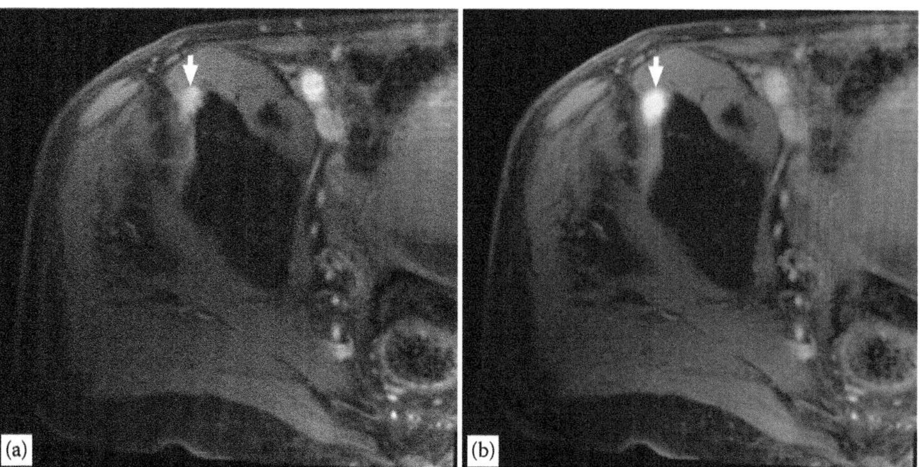


**FIGURE 21.7**
A 67-year-old man with history of lymphoma and right groin pain undergoing further assessment. Axial fat-suppressed $T_1$-weighted post-contrast MR image (a) and axial software-fused FDG-PET/MR image (b) through right pelvis reveal thickening, enhancement, and FDG uptake in proximal right rectus femoris tendon (arrows) due to chronic partial tear.

of PET/MRI assessment of a musculoskeletal disorder.) Nawaz et al. performed an FDG-PET/MRI study of complicated diabetic foot and osteomyelitis in a cohort of 110 patients [54]. The reported sensitivity and specificity were 81% and 93%, respectively, for FDG-PET and 91% and 78%, respectively, for MRI. The complementary nature of the data from these two modalities suggests that hybrid PET/MRI would improve diagnostic accuracy for evaluation of the complicated diabetic foot and osteomyelitis. A study by Sauter et al. used simultaneous PET/MRI to monitor treatment responses in six patients with chronic sclerodermatous graft-versus-host disease [55]. The authors reported that the complementary morphological and functional information of the hybrid imaging technique provided useful in the assessment of inflammation. A series of other studies have explored applications of PET/MRI in a variety of specific cases. El-Haddad et al. used hybrid imaging to localize tears of the meniscus related to synovitis [56]. Miese et al. reported that PET/MRI is valuable as an imaging tool for rheumatoid arthritis of the hand [57]. Finally, Blebea et al. used PET/MRI to quantitatively assess bone marrow activity in both normal and abnormal states [58].

## 21.7 Future Challenges

The fairly limited scope of studies that have thus far investigated the diagnostic abilities of PET/MRI suggest that additional research is needed to establish what this new hybrid imaging modality can offer, especially in relation to existing alternative diagnostic tools that have already been validated. From a technical perspective, one of the main challenges to be overcome is the improvement of MRI-based attenuation correction for PET data, which currently limits the accuracy of reconstructed images and SUVs. The susceptibility of MRI to artifacts due to motion and magnetic field inhomogeneities also poses a challenge to the visualization and assessment of a number of organs. Nevertheless, MRI remains more accurate than CT for the evaluation of many of the anatomical regions explored in this chapter: the head and neck, brain, breast, musculoskeletal system, abdominal organs, and pelvic organs, among others.

Yet, further comparative effectiveness studies are necessary to justify the use of simultaneous PET/MRI over software-based PET/MRI coregistration, PET/CT, and stand-alone MRI. Retrospectively fused PET/MRI is a far cheaper alternative to simultaneous image acquisition, but lacks the temporal and spatial alignment of structural, functional, and molecular information. PET/CT, which is currently the most widely used hybrid imaging modality of its kind, is also less expensive to perform than PET/MRI and has high spatial resolution due to the CT component. However, CT falls short of MRI in terms of soft tissue contrast and exposes the patient to ionizing radiation.

In addition to its comparative cost effectiveness, the safety of PET/MRI must be demonstrated prior to its clinical use. Recent evidence that static low frequency magnetic fields may enhance the genotoxic potential of ionizing radiation indicates that more biological studies are required to assess the long-term safety of this hybrid imaging approach [59–64]. Research on the adverse effects and cost effectiveness of the use of higher field strength MRI as part of PET/MRI will also be necessary in the years to come [44,65]. Once these requirements are met, the standardized training of physicians, technologists, and physicists in the operation and interpretation of PET/MRI, along with the Centers for Medicare & Medicaid Services' establishment of reimbursement guidelines for various clinical applications of this hybrid imaging modality, will be the final hurdles to its full adoption in the clinical setting.

## 21.8 Conclusion

Though still in its nascency, PET/MRI has demonstrated the potential for clinical utility across a diverse range of medical conditions. This hybrid imaging modality holds promise for improving the detection and quantification of lesions, as well as for improving assessment of the response of lesions to therapy. Moreover, the excellent alignment of anatomical, functional, and molecular information will further improve diagnostic accuracy by allowing for partial volume correction, which is of great importance for the accurate quantification of neurological, cardiovascular, and oncological disorders [66]. As compared to PET/CT, PET/MRI will limit exposure to ionizing radiation, which is a particular concern for children and pregnant women, as well as in patients who will undergo repeated imaging assessments. If further studies validate the diagnostic performance, technical feasibility, and cost-effectiveness of PET/MRI relative to existing alternatives, then PET/MRI may become a mainstay of diagnostic evaluation in clinical medicine.

## References

1. Bolus, N.E. et al., PET/MRI: The blended-modality choice of the future? *J Nucl Med Technol*, 2009. **37**(2): pp. 63–71; quiz 72–3.
2. Townsend, D.W., Combined positron emission tomography-computed tomography: The historical perspective. *Semin Ultrasound CT MR*, 2008. **29**(4): pp. 232–5.
3. Torigian, D.A. et al., PET/MR imaging: Technical aspects and potential clinical applications. *Radiology*, 2013. **267**(1): pp. 26–44.
4. Christensen, N.L. et al., Positron emission tomography within a magnetic field using photomultiplier tubes and lightguides. *Phys Med Biol*, 1995. **40**(4): pp. 691–7.
5. Catana, C. et al., Simultaneous acquisition of multislice PET and MR images: Initial results with a MR-compatible PET scanner. *J Nucl Med*, 2006. **47**(12): pp. 1968–76.
6. Zaidi, H. and A. Del Guerra, An outlook on future design of hybrid PET/MRI systems. *Med Phys*, 2011. **38**(10): pp. 5667–89.
7. Poole, M. et al., Split gradient coils for simultaneous PET-MRI. *Magn Reson Med*, 2009. **62**(5): pp. 1106–11.
8. Akata, S. et al., Evaluation of chest wall invasion by lung cancer using respiratory dynamic MRI. *J Med Imaging Radiat Oncol*, 2008. **52**(1): pp. 36–9.
9. Kim, Y.N. et al., A proposal for combined MRI and PET/CT interpretation criteria for preoperative nodal staging in non-small-cell lung cancer. *Eur Radiol*, 2012. **22**: pp. 1537–46.
10. Heusch, P. et al., Hybrid [F]-FDG PET/MRI including non-Gaussian diffusion-weighted imaging (DWI): Preliminary results in non-small cell lung cancer (NSCLC). *Eur J Radiol*, 2013. **82**: pp. 2055–60.
11. Kohan, A.A. et al., N staging of lung cancer patients with PET/MRI using a three-segment model attenuation correction algorithm: Initial experience. *Eur Radiol*, 2013. **23**: pp. 3161–9.
12. Donati, O.F. et al., Value of retrospective fusion of PET and MR images in detection of hepatic metastases: Comparison with 18F-FDG PET/CT and Gd-EOB-DTPA-enhanced MRI. *J Nucl Med*, 2010. **51**(5): pp. 692–9.
13. Yong, T.W. et al., Sensitivity of PET/MR images in liver metastases from colorectal carcinoma. *Hell J Nucl Med*, 2011. **14**(3): pp. 264–8.
14. Beiderwellen, K. et al., Depiction and characterization of liver lesions in whole body [F]-FDG PET/MRI. *Eur J Radiol*, 2013. **82**: pp. e669–e675.
15. Platzek, I. et al., PET/MRI in head and neck cancer: Initial experience. *Eur J Nucl Med Mol Imaging*, 2013. **40**(1): pp. 6–11.
16. Boss, A. et al., Feasibility of simultaneous PET/MR imaging in the head and upper neck area. *Eur Radiol*, 2011. **21**(7): pp. 1439–46.
17. Kanda, T. et al., Value of retrospective image fusion of F-FDG PET and MRI for preoperative staging of head and neck cancer: Comparison with PET/CT and contrast-enhanced neck MRI. *Eur J Radiol*, 2013. **82**: pp. 2005–10.
18. Nagarajah, J. et al., Diagnosis and dosimetry in differentiated thyroid carcinoma using 124I PET: Comparison of PET/MRI vs PET/CT of the neck. *Eur J Nucl Med Mol Imaging*, 2011. **38**(10): pp. 1862–8.
19. Seiboth, L. et al., Utility of PET/neck MRI digital fusion images in the management of recurrent or persistent thyroid cancer. *Thyroid*, 2008. **18**(2): pp. 103–11.
20. Ceriani, L. et al., Unusual muscular metastases from papillary thyroid carcinoma detected by fluorine-18-fluorodeoxyglucose PET/MRI. *J Clin Endocrinol Metab*, 2013. **98**(6): pp. 2208–9.
21. Huang, S.H. et al., A comparative study of fused FDG PET/MRI, PET/CT, MRI, and CT imaging for assessing surrounding tissue invasion of advanced buccal squamous cell carcinoma. *Clin Nucl Med*, 2011. **36**(7): pp. 518–25.
22. Park, S.H. et al., Comparison of diffusion-weighted MR imaging and FDG PET/CT to predict pathological complete response to neoadjuvant chemotherapy in patients with breast cancer. *Eur Radiol*, 2011. **22**: pp. 18–25.
23. Treglia, G. et al., Metastatic brachial plexopathy from breast cancer detected by F-FDG PET/MRI. *Rev Esp Med Nucl Imagen Mol*, 2014. **33**(1): pp. 54–5.
24. Rothke, M.C., A. Afshar-Oromieh, and H.P. Schlemmer, Potential of PET/MRI for diagnosis of prostate cancer. *Radiologe*, 2013. **53**(8): pp. 676–81.
25. Wetter, A. et al., Simultaneous 18F choline positron emission tomography/magnetic resonance imaging of the prostate: Initial results. *Invest Radiol*, 2013. **48**(5): pp. 256–62.
26. Park, H. et al., Introducing parametric fusion PET/MRI of primary prostate cancer. *J Nucl Med*, 2012. **53**(4): pp. 546–51.
27. Afshar-Oromieh, A. et al., PET/MRI with a Ga-PSMA ligand for the detection of prostate cancer. *Eur J Nucl Med Mol Imaging*, 2013. **40**: pp. 1629–30.
28. Kim, S.K. et al., Additional value of MR/PET fusion compared with PET/CT in the detection of lymph node metastases in cervical cancer patients. *Eur J Cancer*, 2009. **45**(12): pp. 2103–9.
29. Nakajo, K. et al., Diagnostic performance of fluorodeoxyglucose positron emission tomography/magnetic resonance imaging fusion images of gynecological malignant tumors: comparison with positron emission tomography/computed tomography. *Jpn J Radiol*, 2010. **28**(2): pp. 95–100.
30. Kitajima, K. et al., Value of fusion of PET and MRI for staging of endometrial cancer: Comparison with F-FDG contrast-enhanced PET/CT and dynamic contrast-enhanced pelvic MRI. *Eur J Radiol*, 2013. **82**(10): pp. 1672–6.
31. Tatsumi, M. et al., 18F-FDG PET/MRI fusion in characterizing pancreatic tumors: comparison to PET/CT. *Int J Clin Oncol*, 2011. **16**(4): pp. 408–15.

32. Nagamachi, S. et al., The usefulness of F-FDG PET/MRI fusion image in diagnosing pancreatic tumor: Comparison with F-FDG PET/CT. *Ann Nucl Med*, 2013. **27**(6): pp. 554–63.
33. Laurent, V. et al., Comparative study of two whole-body imaging techniques in the case of melanoma metastases: Advantages of multi-contrast MRI examination including a diffusion-weighted sequence in comparison with PET-CT. *Eur J Radiol*, 2010. **75**(3): pp. 376–83.
34. Lee, H.J. et al., F-18 fluorodeoxyglucose PET/CT and post hoc PET/MRI in a case of primary meningeal melanomatosis. *Korean J Radiol*, 2013. **14**(2): pp. 343–9.
35. Heiss, W.D., The potential of PET/MR for brain imaging. *Eur J Nucl Med Mol Imaging*, 2009. **36**(Suppl 1): pp. S105–12.
36. Boss, A. et al., Diffusion tensor imaging in a human PET/MR hybrid system. *Invest Radiol*, 2010. **45**(5): pp. 270–4.
37. Boss, A. et al., Hybrid PET/MRI of intracranial masses: Initial experiences and comparison to PET/CT. *J Nucl Med*, 2010. **51**(8): pp. 1198–205.
38. Schwenzer, N.F. et al., Simultaneous PET/MR imaging in a human brain PET/MR system in 50 patients-Current state of image quality. *Eur J Radiol*, 2012. **81**(11): pp. 3472–8.
39. Vercher-Conejero, J.L. et al., Amyloid PET/MRI in the differential diagnosis of dementia. *Clin Nucl Med*, 2014. **39**(6): pp. e336–9.
40. Moodley, K.K. et al., Simultaneous PET/MRI in frontotemporal dementia. *Eur J Nucl Med Mol Imaging*, 2013. **40**(3): pp. 468–9.
41. Salamon, N. et al., FDG-PET/MRI coregistration improves detection of cortical dysplasia in patients with epilepsy. *Neurology*, 2008. **71**(20): pp. 1594–601.
42. Cho, Z.H. et al., Substructural hippocampal glucose metabolism observed on PET/MRI. *J Nucl Med*, 2010. **51**(10): pp. 1545–8.
43. Cho, Z.H. et al., Observation of glucose metabolism in the thalamic nuclei by fusion PET/MRI. *J Nucl Med*, 2011. **52**(3): pp. 401–4.
44. Cho, Z.H. et al., In-vivo human brain molecular imaging with a brain-dedicated PET/MRI system. *Magma*, 2013. **26**(1): pp. 71–9.
45. Bengel, F.M. et al., Myocardial efficiency and sympathetic reinnervation after orthotopic heart transplantation: A noninvasive study with positron emission tomography. *Circulation*, 2001. **103**(14): pp. 1881–6.
46. Higuchi, T. et al., Combined reporter gene PET and iron oxide MRI for monitoring survival and localization of transplanted cells in the rat heart. *J Nucl Med*, 2009. **50**(7): pp. 1088–94.
47. Schneider, S. et al., Utility of multimodal cardiac imaging with PET/MRI in cardiac sarcoidosis: Implications for diagnosis, monitoring and treatment. *Eur Heart J*, 2014. **35**(5): p. 312.
48. Syed, I.S. et al. MR imaging of cardiac masses. *Magn Reson Imaging Clin N Am*, 2008. **16**(2): pp. 137–64, vii.
49. Probst, S. et al., The appearance of epidural extranodal marginal zone lymphoma (MALToma) on F-18 FDG PET/CT and post hoc PET/MRI fusion. *Clin Nucl Med*, 2011. **36**(4): pp. 303–4.
50. Bural, G.G. et al., Quantitative assessment of the atherosclerotic burden of the aorta by combined FDG-PET and CT image analysis: A new concept. *Nucl Med Biol*, 2006. **33**(8): pp. 1037–43.
51. Tahara, N. et al., Simvastatin attenuates plaque inflammation: Evaluation by fluorodeoxyglucose positron emission tomography. *J Am Coll Cardiol*, 2006. **48**(9): pp. 1825–31.
52. Fayad, Z.A. et al., Safety and efficacy of dalcetrapib on atherosclerotic disease using novel non-invasive multimodality imaging (dal-PLAQUE): A randomised clinical trial. *Lancet*, 2011. **378**(9802): pp. 1547–59.
53. Chen, K. et al., Evaluation of musculoskeleta disorders with PET, PET/CT, and PET/MRI. *PET Clin*, 2008. **3**(3): pp. 451–465.
54. Nawaz, A. et al., Diagnostic performance of FDG-PET, MRI, and plain film radiography (PFR) for the diagnosis of osteomyelitis in the diabetic foot. *Mol Imaging Biol*, 2010. **12**(3): pp. 335–42.
55. Sauter, A.W. et al., Imaging findings and therapy response monitoring in chronic sclerodermatous graft-versus-host disease: Preliminary data of a simultaneous PET/MRI approach. *Clin Nucl Med*, 2013. **38**(8): pp. e309–17.
56. El-Haddad, G. et al., PET/MRI depicts the exact location of meniscal tear associated with synovitis. *Eur J Nucl Med Mol Imaging*, 2006. **33**(4): pp. 507–8.
57. Miese, F. et al., Hybrid 18F-FDG PET-MRI of the hand in rheumatoid arthritis: Initial results. *Clin Rheumatol*, 2011. **30**(9): pp. 1247–50.
58. Blebea, J.S. et al., Structural and functional imaging of normal bone marrow and evaluation of its age-related changes. *Semin Nucl Med*, 2007. **37**(3): pp. 185–94.
59. Brix, G. et al., Risks and safety aspects related to PET/MR examinations. *Eur J Nuc Med Mol Imaging*, 2009. **36**(Suppl 1): pp. 131–138.
60. Koyama, S. et al., Combined exposure of ELF magnetic fields and x-rays increased mutant yields compared with x-rays alone in pTN89 plasmids. *J Radiat Res*, 2005. **46**(2): pp. 257–64.
61. Miyakoshi, J., Effects of static magnetic fields at the cellular level. *Prog Biophys Mol Biol*, 2005. **87**(2–3): pp. 213–23.
62. Miyakoshi, J. et al., Exposure to strong magnetic fields at power frequency potentiates X-ray-induced DNA strand breaks. *J Radiat Res (Tokyo)*, 2000. **41**(3): pp. 293–302.
63. Walleczek, J., E.C. Shiu, and G.M. Hahn, Increase in radiation-induced HPRT gene mutation frequency after nonthermal exposure to nonionizing 60 Hz electromagnetic fields. *Radiat Res*, 1999. **151**(4): pp. 489–97.
64. Hintenlang, D.E., Synergistic effects of ionizing radiation and 60 Hz magnetic fields. *Bioelectromagnetics*, 1993. **14**(6): pp. 545–51.
65. Theysohn, J.M. et al., Subjective acceptance of 7 Tesla MRI for human imaging. *Magma*, 2008. **21**(1–2): pp. 63–72.
66. Erlandsson, K. et al., A review of partial volume correction techniques for emission tomography and their applications in neurology, cardiology and oncology. *Phys Med Biol*, 2012. **57**(21): pp. R119–59.

# 22

## *Computer-Aided Diagnosis with MR Images of the Brain*

**Mark A. Haidekker and Geoff Dougherty**

**CONTENTS**



### 22.1 Introduction

Magnetic resonance imaging (MRI), with its superior soft tissue contrast, is the imaging modality of choice for the brain. Imaging sequences exist that make perfusion and diffusion visible, and MRI not only provides a static, structural image, but with limitations provides functional images. Over recent years, both the resolution and the signal-to-noise ratio (SNR) have improved. Contrast, resolution, and SNR have experienced another boost with the wider availability of stronger magnets (often 3 T or higher) and with specialized head coils. Moreover, fast acquisition sequences have been introduced, which can acquire a three-dimensional image of the entire head within minutes.

With these features, MRI provides an excellent platform for computer-aided design (CAD), that is, computerized image analysis where the computer provides quantitative metrics that support a diagnosis. CAD is a natural extension of the image formation process, because MRI *requires* computerized data processing for image formation, and images are immediately available in digital form. Additional image processing and analysis steps can be directly integrated into the MRI workstation, although standalone processing is currently more common.

CAD needs to be broadly defined in the sense that the computer does not necessarily have diagnostic capabilities. Even with accelerating recent development and exponentially increasing computing power, the final decision will for the foreseeable future remain in the purview of the radiologist. However, image processing is usually a sequence of several steps (i.e., operations) in which the image is manipulated or information extracted. Those steps have by themselves the potential to aid in the diagnosis of diseases. It is therefore worth introducing key individual steps in automated image analysis and examining their potential to facilitate the diagnostic process.

A typical image analysis chain, sketched in Figure 22.1, can be subdivided into several stages:

- *Image preprocessing and enhancement*: In this step, desired image features are enhanced and undesired ones suppressed. Examples include selective contrast enhancement or noise reduction.

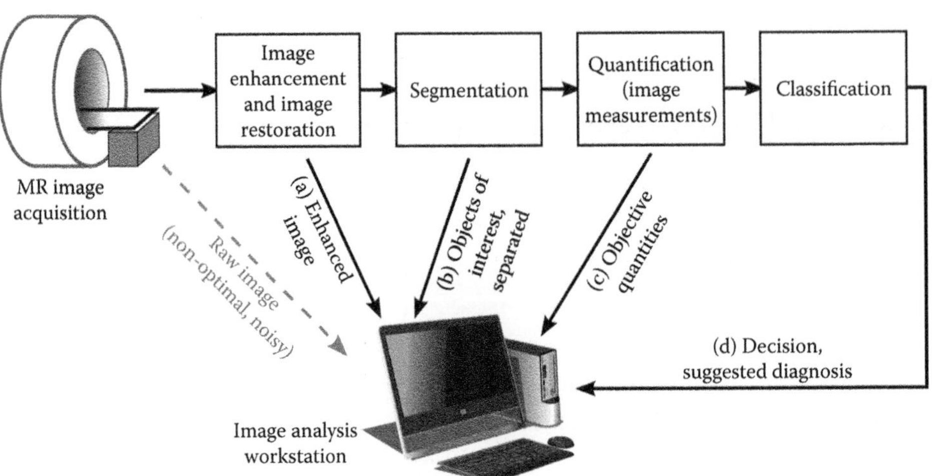


**FIGURE 22.1**
Schematic of image analysis workflow. In conventional radiology, the raw MR image is examined (gray dashed line), but it is not optimal due to acquisition artifacts and noise. Each step of the image analysis chain provides improved value for the radiologist: (a) The enhanced image can provide better detail, contrast, and higher SNR. (b) After segmentation, isolated regions of interest (e.g., ventricles and subcortical regions) can be examined without the surrounding tissue. (c) Quantification methods provide objective metrics (e.g., ventricle volume) without time-consuming manual measurements. (d) Classification can potentially provide a decision (e.g., diseased vs. healthy). (Adapted from Haidekker, M.A., *Medical Imaging Technology*, Springer Briefs in Physics, Springer, New York, 2013. With permission.)

After this step, an image can be easier to read for the radiologist, and imaging workstations usually feature several image-enhancing filter operations.

- *Image segmentation*: This step refers to the isolation of regions of interest from other regions that are usually referred to as background. In MR images of the brain, the segmentation step can differentiate white matter, gray matter, and cerebrospinal fluid (CSF), and even separate individual subcortical regions. A segmented image can be helpful, for example, by highlighting the position and size of a tumor.
- *Image measurements*: In this step, the computer takes measurements and provides quantitative metrics that serve as a basis for classification. Examples include the ventricular volume, morphological symmetry, or the dynamics of uptake of a contrast agent. Measurements can often be taken by hand, but the computer automates this process and thus accelerates it and reduces observer-dependent variability.
- *Classification*: In the classification steps, different quantitative metrics can be used to identify a region as normal or diseased.

This chapter has two major sections. The first of these, Section 22.2, covers the steps described above in more detail and presents some technical information regarding how those image operations work. The second section, Section 22.3, presents those techniques from the application side, that is, introduces several examples that highlight the necessary image processing steps to reach a specific analysis goal.

## 22.2 Technical Aspects of CAD in the Brain

### 22.2.1 MR Image Acquisition for CAD in the Brain

The central image quality criteria that are important to the radiologist, high spatial resolution, high contrast, and low noise, are even more important for CAD. Generally, a high image quality reduces the demand on first-level image enhancement filters, such as noise reduction. Moreover, CAD relies on three-dimensional images more often than conventional diagnosis, and a balance needs to be found between low-noise and high-resolution 3D image sequences and image acquisition time.

The physical processes involved in the excitation and reception of MR signals are inherently slow. The time taken to invoke a specific sequence of pulses, repeated for each line of a reconstructed 2D image, and to measure the signals that are generated can be significant. To acquire sufficient measurements to reconstruct a 3D image of the brain will take many minutes, and with some sequences it may take up to an hour. This time cannot be reduced by improving the hardware (i.e., electronics) of the scanner: it needs to be addressed by shortening the acquisition or invoking so-called *sparse* or *compressive* sampling [2,3].

The readout MR signal is stored in Fourier space (in MR terminology known as *k*-space). Frequency-encoding and phase-encoding ensure that the data is spatially encoded by differences in frequency and phase, amenable to analysis by Fourier transform. The location of the data in *k*-space depends on the net strength and duration of the phase-encoding gradient and frequency-encoding gradient. A low-amplitude or short-duration gradient event encodes low spatial frequency information, while a high-amplitude or long-duration gradient event encodes high spatial frequency information. Low spatial frequency information is mapped near the center of *k*-space and the high spatial frequency information is mapped to the periphery of *k*-space. Most MR image information (contrast and general shape) is contained in the center of *k*-space, that is, at low spatial frequencies. High spatial frequencies encode the edges of an object. The farther from the center of *k*-space the data are collected, the higher is the spatial-frequency information and the better the spatial resolution will be. The classic spin-echo (SE) sequence fills *k*-space line by line. One line of *k*-space is fully acquired at each excitation, containing low and high (horizontal) spatial frequency information (contrast and resolution in the horizontal direction). Between each repetition, there is a change in phase-encoding gradient strength. This allows filling of all the lines of *k*-space from top to bottom.

The *SE* sequence was designed to minimize the signal-reducing effect of $T_2^*$ decay (which is caused by magnetic field inhomogeneities) by introducing a 180° radiofrequency (RF) pulse to re-phase, or re-focus, spins in the transverse plane. The spins are flipped 180° so that the phase-position of each spin is inverted, that is, spins that were precessing faster are now *behind* spins that were precessing at a slower rate. A finite time later the spins will have caught up with each other, and a SE is formed: this is a signal peak which forms at the echo time, TE. Of course, $T_2$ decay continues unabated throughout the time taken for the sequence. The total acquisition time for the sequence is given by the product of the repetition time, TR, the number of phase-encoding steps, *NPE* (the number of pixels or matrix size in the phase direction) and the number of times the sequence is repeated, *NEX*, in order to acquire an average with reduced noise. *Multislice imaging* is achieved by making use of the time between the end of echo collection and the next 90° excitation pulse (TR–TE), referred to as *dead time*. In this period the next slice can be excited. The scanner will determine how many more slices will fit into the sequence. Another consideration is the *cross-talk* (or more correctly *cross-excitation*) which occurs between adjacent slices due to imperfect slice profiles. This is accounted for by leaving gaps, or by interleaving slices, so that even slices are excited first, followed by the odd slices.

The SE sequence has drawbacks for clinical imaging, its main limitation being its slow acquisition time. Faster sequences would be more comfortable for the patient, would increase throughput, and would open up possibilities for 3D imaging and for imaging of moving organs in the body. The *gradient-echo* (GE) sequence (also known as gradient recalled echo [GRE] sequence) was developed to be a faster alternative to the SE sequence. A much smaller flip angle, $\alpha$, is used instead of the 90° flip of the SE sequence, so that it is not necessary to wait a long time (usually five times $T_1$ of the tissue of interest when a 90° radio-frequency (RF) pulse is used) for the longitudinal component of the magnetization, $M_z$, to recover sufficiently for another repetition. This means that the total acquisition time can be less, leading to the possibility of rapid imaging. However, the signal measured will be smaller because less magnetization is tipped into the transverse plane, and SNR decreases. A further reduction in scan time is obtained by avoiding the use of the 180° refocusing RF pulse. Instead, the GRE sequence uses a bipolar frequency-encoding gradient to dephase and then re-phase the spins. This does not eliminate non-random magnetic field inhomogeneities, which contribute to a lower SNR and its vulnerability to chemical shift and metal-induced susceptibility artifacts. Faster acquisition is the payoff for the loss of SNR.

The signal amplitude in GRE imaging depends on TR, TE, the flip angle and the $T_1$ value of the tissue. For any $T_1$ value, at a given $\text{TR}(\ll T_1)$, there is a flip angle that results in the maximum signal for that tissue. This angle is known as the Ernst angle, $\alpha_{\text{Ernst}}$. It can be shown that

$$\cos(\alpha_{\text{Ernst}}) = \exp\left(-\frac{\text{TR}}{T_1}\right) \tag{22.1}$$

The majority of the many other pulse sequences are variations of the SE and GRE sequences. Each sequence is a subtle combination of RF pulses and gradients. The aims are to favor the signal of a particular tissue (contrast), and to acquire the images as quickly as possible (speed), while limiting artifacts and without altering the SNR. SE sequences are typically black-blood techniques, while GE sequences are typically bright-blood techniques.

In fast spin-echo sequences, the interval of time after the first echo is used to receive the echo train, and thus to fill the other *k*-space lines in the same slice. Because of the reduced number of repetitions required, the *k*-space is filled faster and slice acquisition time is reduced. This is done by applying new 180° pulses to obtain a SE train. After each echo, the phase-encoding is cancelled and a different phase-encoding is applied to the following echo. The number of echoes received in the same repetition (during TR time) is called the turbo factor or echo train length (ETL). A multi-echo SE uses more than one refocusing pulse to create separate echo images at increasingly longer echo times.

In GE, TR reduction may cause permanent (steady-state) residual transverse magnetization in TR below $T_2$: the transverse magnetization will not have completely disappeared at the onset of the following repetition and will also be submitted to the flip caused by the excitation pulse. The steady state can be detrimental, particularly for obtaining $T_1$-weighted sequences. To resolve this problem, the spoiled GE pulse sequence uses gradients or RF pulses (spoilers) to eliminate residual transverse magnetization. Ultrafast GE sequences use a small flip angle, a very short TR and optimized *k*-space filling to reduce acquisition time to roughly one second per slice. The drawback of a small flip angle and very short TR is poor $T_1$-weighting. To preserve $T_1$ contrast, a 180° inversion pulse prepares magnetization before repetitions of the ultrafast GE imagery sequence. Effective inversion time will correspond to the delay between the inversion pulse and acquisition of the central *k*-space lines. To obtain $T_2$-weighting, the preparatory pattern is of the SE type (90°–180°) so that the ultrafast GRE imaging sequence can start with longitudinal magnetization whose amplitude depends on $T_2$.

*Echo-planar imaging* (EPI) is the fastest acquisition method in MRI (100 ms/slice), but with limited spatial resolution. It can be adapted to SE-EPI or GRE-EPI sequences. EPI sequences are the basis for advanced MRI applications such as diffusion, perfusion, and functional imagery.

The *inversion recovery* (IR) sequence starts with a 180° RF inversion pulse which flips longitudinal magnetization, $M_z$, into the opposite (negative) direction. Due to longitudinal relaxation, longitudinal magnetization will increase to return to its initial value, passing through a null value. A 90° RF pulse is applied to obtain the transverse magnetization for measurement. The delay between the 180° RF inversion pulse and the 90° RF excitation pulse is referred to as the inversion time $T_I$. The IR technique allows the signal of a given tissue to be suppressed by selecting a $T_I$ value for which that tissue recovers to $M_z = 0$, that is, when it equals $T_1$ ln2, and, hence, there is no signal when flipped into the transverse plane. IR can be combined with either SE or GE sequences. In particular, it can be used with fast SE sequences, for faster acquisition times since IR requires relatively long TR to allow magnetization the time to regrow. Fluid-attenuated IR (FLAIR) is used to suppress liquid signals by IR at an adapted $T_I$. Nulling of the water signal is seen at $T_I$ of approximately 2000 ms. As in the case of the other IR sequences, an imaging sequence of the fast SE type is preferable to compensate the long acquisition time. These sequences are routinely used in brain imaging to suppress CSF, and to bring out hyper-intense lesions, such as multiple sclerosis (MS) plaques. FLAIR can be used with both three-dimensional imaging (3D FLAIR) and two-dimensional imaging (2D FLAIR).

Good spatial resolution will allow fine details to be seen and minimize the partial volume effect. The spatial resolution of a digital image is synonymous with the size of the pixel (or *voxel* in 3D), which is equal to the length *L* of the field of view (FoV) in that direction divided by the number of pixels (*N*) spanning it. The scan parameters that are responsible for spatial resolution are related to the bandwidth of the RF pulse, the bandwidth of the RF receiver, and the gradients, but not to the timing parameters of the sequence itself. In the frequency-encoding direction, the pixel size $\Delta_{FE}$ is given by

$$\Delta_{FE} = \frac{L_{FE}}{N_{FE}} = \frac{RBW}{\gamma G_{FE} N_{FE}} \tag{22.2}$$

where:

- RBW is the RF receiver bandwidth
- $\gamma$ is the gyromagnetic ratio (42.6 MHz $T^{-1}$ for the hydrogen nucleus)
- $G_{FE}$ is the frequency-encoding gradient strength
- $N_{FE}$ is the number of pixels in this direction

In the phase-encoding direction, the pixel size $\Delta_{PE}$ is given by

$$\Delta_{PE} = \frac{L_{PE}}{N_{PE}} = \frac{1}{\gamma \Delta G_{PE} N_{PE} t_{PE}} \tag{22.3}$$

where:

- $\Delta G_{PE}$ is the increment in field strength between successive phase-encoding gradients
- $t_{PE}$ is the time for which each gradient is applied
- $N_{PE}$ is the number of pixels in this direction (which is equal to the number of phase-encoding steps)

In the slice selection direction, the pixel size $\Delta_{SS}$ is given by

$$\Delta_{SS} = \frac{TBW}{\gamma G_{SS}} \tag{22.4}$$

where:

- TBW is the bandwidth of the RF pulse
- $G_{SS}$ is the slice selection gradient strength

For a fixed number of pixels in an image (say, 512 × 512) the best spatial resolution is obtained with long RF pulses (i.e., small TBW)—although this requires a compromise with scan time—and large field gradients.

A small FoV will result in increased spatial resolution (i.e., smaller pixel size), but when the FoV is smaller than the object or region being imaged wraparound or aliasing will occur.

The main source of noise in MR imaging is due to thermal fluctuations of electrolytes in the region of interest. Electronic noise is also present but is usually of lesser order. The noise in an image is related to the number of contributing spins within each pixel (for 2D images), or voxel (for 3D images). Generally, the SNR increases linearly with the voxel size and as the square root of the acquisition time. Thus, any moves to either improve the imaging resolution or reduce the acquisition time inevitably lead to a reduction of the SNR. Because of the many trade-offs involved in a single acquisition, usually two or more images are acquired during the pulse sequence and added together. Uncorrelated noise, such as thermal noise from the patient's body and the RF coils, will be averaged and therefore reduced in comparison with the signal. The SNR improves by a factor $\sqrt{N_{EX}}$, where $N_{EX}$ is the number of identical images averaged. However, the increased time needed to acquire the multiple images increases the probability of patient motion and consequent blurring.

One of the great advantages of MRI is its excellent soft-tissue contrast which can be widely manipulated. In a typical image acquisition the basic unit of each sequence (i.e., the 90°–180°-signal detection sequence) is repeated hundreds of times over. By altering the echo time (TE) or repetition time (TR), that is, the time between successive 90° pulses, the signal contrast can be altered or weighted. For example, the influence of $T_1$ can be minimized by using a long TR. If a long TE is used as well, inherent differences in $T_2$ times of tissues will become apparent. Tissues with a long $T_2$ (e.g., water) will take longer to decay and their signal will be greater (and appear brighter in the image) than the signal from tissue with a short $T_2$ (fat). Such images are referred to as *$T_2$-weighted images*, and they have a high fluid-to-tissue contrast. Many tumors have higher water content and show good $T_2$ contrast with surrounding tissue. A disadvantage is that the long TR results in a longer scan time. Conversely, the effect of $T_2$ can be minimized, resulting in $T_1$ contrast or *$T_1$-weighted images*: TR should be kept short (in order to see differences in $T_1$) and $T_2$ should be short (to minimize the effect of $T_2$). Tissues with a long $T_1$ will take a long time to recover back to the equilibrium magnetization value, and a short TR interval will make this tissue appear dark compared to tissue with a short $T_1$. $T_1$-weighted images have good contrast between gray and white matter, and the CSF can be easily segmented due to its low signal. The short TR and TE make it suitable for many fast 3D sequences, such as GE and low-angle sequences. When TE and TR are chosen to minimize both these weightings, that is, long TR and short TE, the signal contrast is primarily derived from the number or density of spins in a given tissue. This image is said to be *proton-density weighted*. Soft tissue contrast is generally lower than in $T_1$- and $T_2$-weighted images, and consequently proton density weighting is less suitable for CAD.

Although MRI delivers excellent soft-tissue contrast sometimes there is a need to further differentiate different regions of tissue by using contrast agents to make the signal higher or lower than it would otherwise have been. MRI contrast agents are usually administered intravenously, and they affect the relaxation times of the tissues in which they are taken up. Contrast agents based on gadolinium (Gd) are the most commonly used in $T_1$-weighted imaging. They shorten the $T_1$ of tissues where they are present, so that these tissues appear brighter than they would have otherwise. Tumors, for example, are usually highly vascularized and therefore take up more of the agent than surrounding tissue. Gadolinium agents can cross the blood–brain barrier if it is compromised, for example, in the presence of intracranial lesions. Figure 22.2 shows brain images both

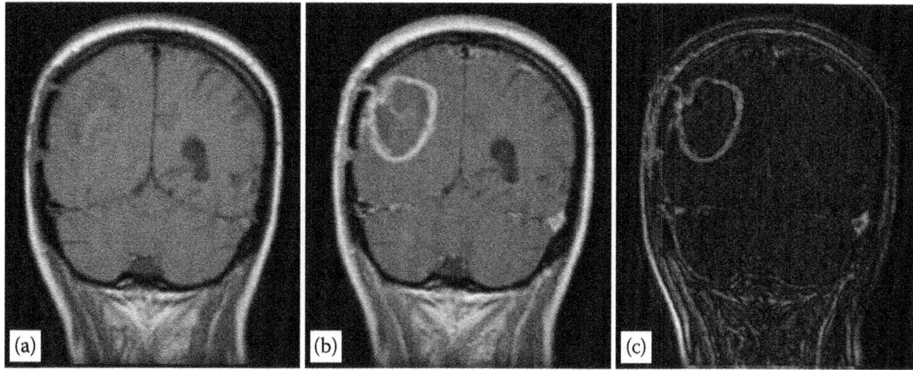


**FIGURE 22.2**
$T_1$-weighted images of the brain before (a) and after (b) the uptake of a Gd-based contrast agent. (c) shows the difference between (b) and (a) to further enhance the contrast (see Section 22.2.2).

before and after Gd contrast agent uptake. The increased vascularity of tumors produces a preferential uptake of contrast agent and allows them to be more easily distinguished from surrounding normal tissue. Furthermore if MR scans are repeatedly acquired following the contrast injection, the dynamic nature of contrast uptake can be examined (dynamic contrast-enhancement), which may improve the differentiation of benign and malignant disease.

Other contrast agents, such as superparamagnetic iron oxide (SPIO), shorten the $T_2$ time so that the signal on a $T_2$-weighted image is reduced. In the liver, for example, SPIO accumulates in normal tissue so that the signal from healthy tissue is suppressed and a lesion appears brighter. However, use of SPIO particles in brain imaging is less widely used, because it is not known whether SPIO is transported into the brain primarily by macrophages, diffuses passively through a disrupted blood-brain barrier, or circulates mostly within the vascular system [4].

*Functional MRI* (fMRI) is a technique for examining brain activity, by detecting associated changes in blood flow. This technique relies on the fact that cerebral blood flow and neuronal activation are coupled. When an area of the brain is in use, blood flow to that region also increases. The most common method utilizes a technique called BOLD (blood oxygen level-dependent) contrast. This is an example of endogenous contrast, making use of the change in magnetization between oxygen-rich and oxygen-poor blood. In the normal resting state, a high concentration of deoxyhemoglobin attenuates the MR signal due to its paramagnetic nature. However, neuronal activity, in response to some task or stimulus, creates a local demand for oxygen supply which increases the fraction of oxyhemoglobin causing a signal increase in $T_2$ or $T_2^*$-weighted images. In a typical experiment the patient is subjected to a series of rest and task intervals, during which MR images are repeatedly acquired. The signal changes during this time course are then examined on a pixel-by-pixel basis to test how well they correlate with the known stimulus pattern. Pixels that demonstrate a statistically significant correlation are highlighted in color and overlaid on a grayscale MRI to create an activation map of the brain. The location and extent of activation is linked to the type of task or stimulus performed, for example a simple thumb-finger movement task will produce activation in the primary motor cortex. The technique can localize activity to within millimeters but, using standard techniques, no better than within a window of a few seconds.

A major disadvantage of BOLD imaging is that it is only an indirect measure of neural activity which is also biased toward one type of neural processing, namely input and intraregional processing. In addition, it is susceptible to several imaging artifacts and has limited temporal resolution. Although this can be improved to a certain degree, increasing temporal resolution comes at the expense of decreasing the spatial resolution. Due to the comparatively low SNR of fMRI, analysis of the acquired data is rather extensive, complex, and occasionally based on rather simplified assumptions, for example, those related to the properties of hemodynamic response. For these reasons it is less suitable for CAD.

*Diffusion imaging* extracts information on the random or Brownian motion of the water molecules within voxels. This motion is restricted by different obstacles in the body (e.g., cell membranes, proteins, macromolecules, fibers), which vary according to the tissues and certain pathological modifications (e.g., intracellular edema, abscesses, tumors). Diffusion data provides indirect information about the structure surrounding these water molecules. Diffusion-weighted MR imaging (DWI) aims at highlighting the differences in water molecule mobility, irrespective of their direction of displacement, whereas diffusion tensor imaging (DTI) studies the directions of water molecule motion to determine, for example, whether or not they diffuse in all directions (fractional anisotropy), or attempts to render the direction of a particular diffusion.

The diffusion-weighted sequences used in DWI are based on a $T_2$-weighted SE sequence, usually of the SE-EPI type (SE ultrafast echo-planar imaging preparation). Additional unipolar gradients are added to cause phase loss in diffusing molecules. The degree of diffusion weighting, expressed as the $b$-factor (in s mm$^{-2}$) depends on the diffusion gradients—their amplitude, application time, and the time between the two gradients. The loss in signal due to diffusion weighting using a non-diffusion-weighted image with $b = 0$ and a diffusion-weighted image with $b \approx 1000$ enables the calculation of the apparent diffusion constant, ADC, in the direction of the unipolar gradients. If TE is kept constant, ADC is independent of proton density and $T_2$. The sequence must be repeated by applying gradients in at least three spatial directions. It is common to acquire several diffusion-weighted images for each direction in order to improve the random error in ADC and therefore improve its SNR. Once the diffusion constants are determined for each pixel a diffusion constant image can be obtained, in which brightness is proportional to the diffusion constants.

The microarchitecture specific to nerve tissues causes diffusion anisotropy in the white matter of the brain: water molecule diffusion preferably follows the direction of the fibers and is restricted perpendicularly to the fibers. By performing diffusion-weighted acquisitions in at least six directions (and far more in angular high resolution imaging), it is possible to extract the diffusion tensor which synthesizes all the data. Different images can be obtained from this DTI depending on the complexity of the post-processing of this data: the main diffusion direction, fractional anisotropy (null when diffusion is isotropic, and of increasing value when diffusion becomes anisotropic), and fiber tracking. The limitations of the tensor model in precisely determining the directions of water molecule displacements in the case of fiber crossing, for example, have led to the development of new models, which require a greater number of measurements. *Q*-space (diffusion spectrum) imaging is able to describe fiber crossings, but requires a high number of acquisitions (129 to 515!) in different directions.

Diffusion-weighted imaging (DWI) is used in applications to diagnose stroke. Stroke manifests itself in the acute phase as a drop in ADC translating an ischemic cytotoxic edema. It is also used to date the stroke event and to distinguish between acute and sub-acute strokes. Diffusion imaging also participates in diagnosis in different categories of brain pathology: tumors (cerebral lymphoma [reduced ADC], epidermoid and cholesteatoma cysts [hypersignal in diffusion]), infectious conditions such as pyogenic brain abscess (reduced ADC, providing differential diagnosis from a necrotic tumor in which the ADC is increased) and herpes encephalitis, degenerative conditions such as Creutzfeldt–Jakob's disease, and inflammatory conditions such as MS.

DTI enables the in vivo study of tissue microstructure. It can indicate possible nerve fiber anomalies in white matter or the spinal cord that are not visible in conventional imaging. Fiber tractography is the only method giving an indirect, in vivo view of the nerve fiber trajectory [5]. It can be associated with fMRI to study the interconnections between nerve centers, used to analyze brain maturation and development, and assist in the preoperative check-up for brain tumors or for medullary compression. DTI can also be used in exploring Alzheimer's disease, certain psychiatric conditions, inflammatory, tumoral, vascular, traumatic (irreversible comas) pathologies, or drug-resistant epilepsies.

### 22.2.2 Image Enhancement Strategies

Raw MR images are usually unsuited for any form of automated image analysis, including CAD. As will be discussed in this section, the image acquisition process introduces three sources of degradation, and image enhancement strategies are used to reverse or partly reverse the degradation:

- *Blur*: Small detail features, for example, a very small area with large RF signal, are spread over multiple pixels, with the image intensity decaying with the distance from the center. Details appear blurred. Since a idealized point is spread over multiple pixels, the amount of blur is mathematically described with the *point-spread function*.
- *Noise*: Quantum processes in the tissue and the amplifier electronics introduce random deviations from the idealized image value. Often (but not always), the actual pixel value $g(x,y)$ can be described as the idealized pixel value $f(x,y)$ to which a random offset $n(x,y)$ was added. A large cohort of these values $n(x,y)$ is often assumed to have zero mean value and a Gaussian distribution.
- *Contrast*: The FoV of the MR image usually includes regions that are not of diagnostic relevance. The skull is a prominent example. The raw MR image captures the entire value range of the RF echo, and the relative difference between tissues is reduced as a consequence.

Noise results in a degradation of image quality; since it is a stochastic process it is not possible to predict its values precisely. It is common practice to assume that the noise in MRIs can be described by a Gaussian distribution, and it has been shown that the theoretical noise distribution reduces to the Gaussian distribution for even small SNR [6].

One of the challenges in the post-acquisition removal of noise from an image is to preserve the edges in the image. Additive (or *non-coherent*) noise can be removed by convolving the image with a linear mask, for example, a Gaussian mask, which produces weighted averaging with minimal ringing because of its gradually tapering profile. The process can be performed more efficiently in the Fourier (frequency) domain, because the convolution process is replaced by multiplication (of the Fourier transform of the image by the Fourier transform of the mask, now known as a filter).

Multiplicative noise, also known as *coherent noise* or *speckle noise* [7], is signal-dependent and more difficult to remove. When an image, $g(x,y)$, originates from an idealized, noise-free image $f(x,y)$ and is contaminated by multiplicative noise $n(x,y)$,

$$g(x,y) = f(x,y) \cdot n(x,y) \tag{22.5}$$

*homomorphic filtering* can be used. This comprises taking the logarithm of the image values to yield an additive linear result,

$$\log\left[g(x,y)\right] = \log\left[f(x,y)\right] + \log\left[n(x,y)\right] \tag{22.6}$$

followed by conventional linear filtering to reduce the log noise component, and then taking the exponential after filtering. Although Equation 22.6 appears straightforward, it is numerically problematic, because negative values are not allowed, and values close to zero need to be treated separately due to the highly nonlinear mapping of the dynamic range by the log function.

Unfortunately, linear masks generally do not give a satisfactory performance because both noise and edges contain high frequencies, and linear noise reduction is associated with a loss of detail (blurring). Nonlinear masks perform much better at preserving edges but cannot be used in the Fourier domain. The classical median mask is nonlinear but performs better at removing shot noise (also known as salt-and-pepper noise) than Gaussian noise (Figure 22.3). Wavelet-based masks [8] and nonlinear diffusion masks that are based on solving partial differential equations [9], including the total variation model (TVM) [10–13], have proved useful. A number of other denoising methods have been used specifically with MR images [14–16].

An improvement over linear filters is those that adapt their behavior to the image properties, for example, smoothing filters that automatically weaken in the presence of a large discontinuity (edge). Those filters are referred to as *adaptive filters* or adaptive masks. More specifically, adaptive masks change their behavior based on the statistics of the pixels within a defined local neighborhood for each pixel. They are inherently nonlinear masks. Their performance in reducing noise is superior to that of global masks, but there is an increase in filter complexity.

Consider the case of additive, Gaussian noise that corrupts an image $f(x,y)$ to form the noisy image $g(x,y)$: The variance of the noise component is $\sigma_N^2$, and we assume that it can be estimated from the noisy image. Once a neighborhood size is chosen, the local mean, $m_L$ and local variance, $\sigma_L^2$, of the pixels within their respective neighborhood can be calculated. An adaptive mask would then produce an estimate, $\hat{f}(x,y)$, of the original image pixels using

$$\hat{f}(x,y) = g(x,y) - \frac{\sigma_N^2}{\sigma_L^2}\left[g(x,y) - m_L\right] \quad (22.7)$$

This produces an output close to $g(x,y)$ if the local variance is high; this is appropriate because high variance implies the presence of image detail (such as edges), which should be preserved. Conversely, if the local variance is low (i.e., approaching $\sigma_N^2$), such as in background

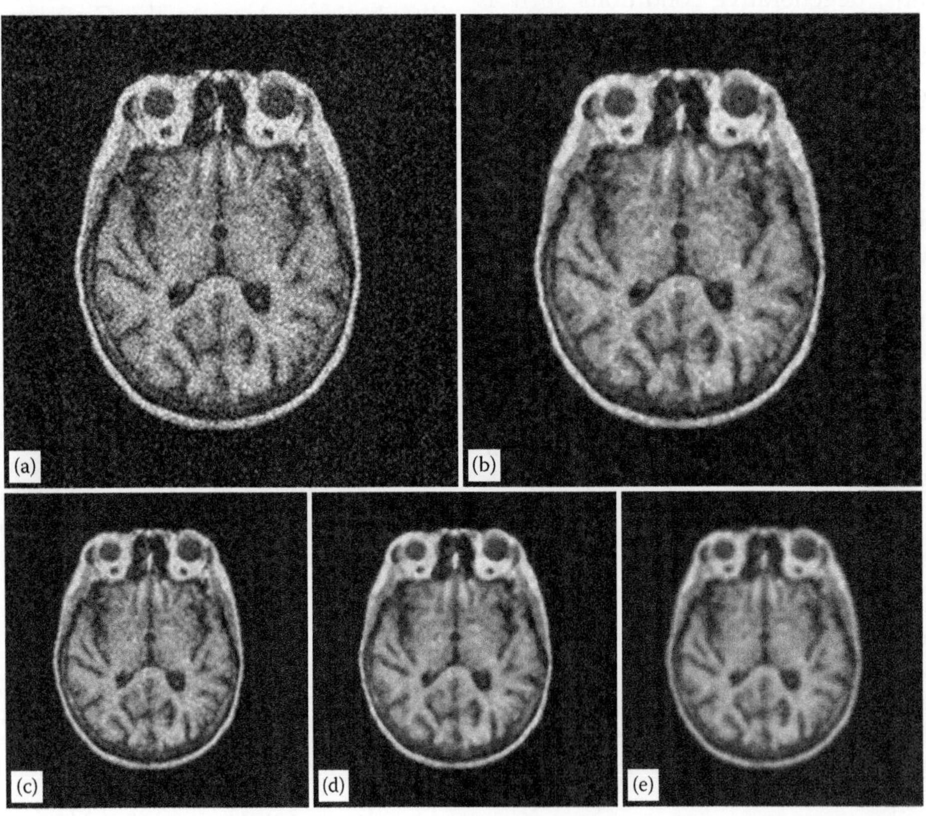


**FIGURE 22.3**
(a) An image degraded by Gaussian noise ($\sigma = 40$) and the effect of convolution with (b) a 3 × 3 median mask, (c) a 5 × 5 median mask, (d) a 3 × 3 averaging mask, and (e) a 5 × 5 averaging mask. (From Dougherty, G., *Digital Image Processing for Medical Applications*, Cambridge University Press, Cambridge, 2009. With permission.)

areas of the image, the output will be close to the local mean value, $m_L$. This reduces noise while preserving edges. This mask is known as the *minimum mean-square error* (MMSE) mask. A disadvantage of this mask is its poor noise reduction near edges, although it can be modified [18] to improve this behavior.

There is a very powerful adaptive mask that operates by numerically simulating anisotropic diffusion [19]. Image intensity values are interpreted as local concentrations caught inside the pixel. If the pixel boundaries are considered semi-permeable, the pixel intensity would diffuse over time into neighboring pixels of lower intensity (i.e., follow the negative gradient). The diffusion process can be modeled by a partial differential equation,

$$\frac{\partial I(x,y)}{\partial t} = c \cdot \Delta I(x,y) \tag{22.8}$$

where:
- $I(x,y)$ is the image intensity
- $\Delta$ indicates the Laplacian operator
- $c$ is the diffusion constant

For anisotropic diffusion, $c$ is no longer constant, but becomes a function that monotonically falls with the image gradient, $c = g\left[\nabla I(x,y)\right]$. The function $g$ ensures that diffusion across large gradients is reduced, and the smoothing operation is therefore stronger in regions with low contrast. Despite the comparatively complex theory behind the anisotropic diffusion mask, a numerical approximation of the diffusion equation is surprisingly simple [20], and the outcome of the mask depends on an adjustable steepness of the function $g$ and the number of iterations, optimum values of which are normally determined experimentally. The steepness of $g$ can be automatically tied to the noise component in the image (by, e.g., adjusting it to be the 90% quantile of the image gradient magnitude at each iteration). Further modifications to the algorithm [21,22] have enabled it to work well even in the presence of a weak multiplicative noise component.

*Restoration techniques* attempt to model the degradation that is caused by the image formation process (in this case, the MR image acquisition) and apply the inverse process to recover the original image. The degradation model is the application of a blurring process that is represented by its point-spread function $h(x,y)$ or modulation transfer function $H(u,v)$, followed by the addition of independent Gaussian, zero-mean noise (Figure 22.4).

Restoration techniques are most effective when the point spread function or modulation transfer function is known and the nature of noise component is well understood. In many cases however, we will only have limited statistical knowledge of the degradation process, and the inverse transform will be correspondingly ill-conditioned. If the noise is additive, as indicated in Figure 22.4, then the degraded image from a linear, space-invariant (LSI) imaging system is given by

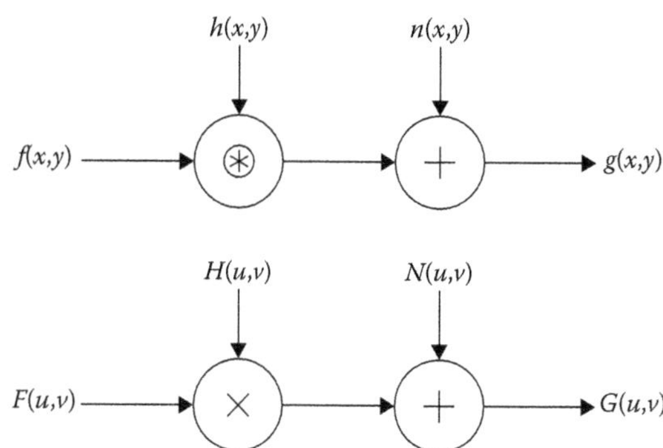


**FIGURE 22.4**
Imaging by an imperfect, but linear, imaging system that adds noise as well as blurring, as seen in both the spatial (top) and frequency (bottom) domains. The symbol ⊛ indicates the convolution operation. (From Dougherty, G., *Digital Image Processing for Medical Applications*, Cambridge University Press, Cambridge, 2009. With permission.)

$$g(x,y) = f(x,y) \circledast h(x,y) + n(x,y) \tag{22.9}$$

where the symbol ⊛ represents the convolution operation. Taking the Fourier transform of Equation 22.9 and re-arranging to solve for $F(u,v)$ gives a theoretical solution to restore the original image from the degraded image

$$F(u,v) = \frac{G(u,v)}{H(u,v)} - \frac{N(u,v)}{H(u,v)} \tag{22.10}$$

However, the term involving the noise is problematic; noise is random and generally broadband, while the modulation transfer function of the imaging system, $H(u,v)$, drops off rapidly with increasing $u$ and $v$, and falls to zero beyond its cut-off frequency. The outcome is that the noise is amplified in the restoration for spatial frequencies below the system cut-off, even when the noise power is small, and numerical overflow results from divisions by zero at spatial frequencies above the cutoff. The classic remedy is to employ *Wiener filtering* in the Fourier domain to remove those frequencies that would be dominated by noise. The Wiener filter is an optimal filter in the sense that it delivers the best estimate of the original, undegraded image in a least squares sense for additive Gaussian noise, that is, it finds an estimate, $\hat{f}(x,y)$, of the uncorrupted image, $f(x,y)$, such that the mean square error between them is minimized. The Wiener filter is, in fact, MMSE mask (Equation 22.7), implemented in the Fourier domain

$$\hat{F}(u,v) = \frac{|H(u,v)|^2}{H(u,v)\left(|H(u,v)|^2 + \left(|N(u,v)|^2 / |F(u,v)|^2\right)\right)} G(u,v) \quad (22.11)$$

where $N(u,v)^2$ and $F(u,v)^2$ are the power spectra of the noise and the undegraded image respectively. This realization requires the SNR to be known at every frequency, but an approximation exists,

$$\hat{F}(u,v) \approx \frac{|H(u,v)|^2}{H(u,v)\left(|H(u,v)|^2 + k\right)} G(u,v) \quad (22.12)$$

where $k$ is the inverse of the SNR of the image averaged over all frequencies: more conveniently, $k$ can be considered as an adjustable empirical parameter chosen to balance sharpness against noise. However, this basic form of the Wiener filter is not spatially adaptive: its characteristics remain the same over the entire image.

Adaptive variations of the filter are possible: a frequency-adaptive form with $k$ as a function of frequency, that is, $k(u,v)$, and a spatially adaptive form where $k$ is locally dependent, that is, $k(x,y)$. However, since spatial information is lost in the frequency domain, adaptive frequency-domain filters require a separate inverse transform for each region of the image where statistical properties differ: they are, consequently, computationally inefficient. Alternatively, a (nonadaptive) parametric form of the Wiener filter [23] can be used. The Wiener filter is the optimum filter for known image and noise variance, but if either is unknown it is more efficient to adopt an empirical approach with the MMSE mask in the spatial domain. The adaptive nature can result from image characteristics in a fixed window [24], an adaptive-sized window [25], or even an adaptive-neighborhood window which varies in size and shape for each pixel [26]. A side-by-side comparison of conventional nonadaptive Gaussian smoothing, Wiener filtering, and anisotropic diffusion is shown in Figure 22.5.

Software for viewing images makes it possible to easily manipulate contrast. The simplest option is to use a linear *histogram stretch* (or contrast stretch), which maintains the shape of the raw pixel value histogram. The pixel values, $p_i$, are remapped by a linear lookup table to gray values, $g_i$, spanning a larger dynamic range, $D$, which can be as large as the full dynamic range available ($2^n$, for an $n$-bit image):

$$g_i = \frac{D}{p_{\max} - p_{\min}} p_i \quad (22.13)$$

The stretching can be limited to a sub-range of pixel values, $(p_2 - p_1)$, in which case the range is known as the *window width* (WW), and the midpoint of the range is known as the *window level* (WL) or *center*. The number of gray levels is the same as the number of pixel values, but they span a larger range, enhancing the contrast (but providing no additional information). If the sub-range is expanded to the full dynamic range available, pixel values outside the window will become saturated, that is, accumulate as white or black levels, and convey no information.

When it is necessary to compare several images, which may have been acquired under differing conditions, on a specific basis, such as for quantitative texture measurement, it is common practice to try and standardize their histograms. The most common standardization technique is histogram equalization, where one attempts to change the histogram into a flat, uniform histogram, in which every pixel value occurs equally frequently. The expectation is that this maximizes the information conveyed in the image (expressed in quantitative terms as its *entropy*), and that the transformed image has an enhanced appearance. In this case the lookup table is nonlinear and is obtained from the cumulative density function (CDF), suitably scaled, of the raw pixel values for that specific image. A range of raw pixel values which occur frequently will result in a steep slope in the CDF: when this range is looked up using the CDF it will be spread over a larger range, resulting in a smaller density of transformed values. Hence, peaks in the original histogram will be reduced in the transformed histogram. Similarly, a range of less frequent raw pixels result in a shallow slope on the CDF, a narrower range of transformed values and a consequently larger density. These changes tend to equalize the transformed histogram. However, there are limitations in the discrete implementation. The lookup table (from the CDF) may yield values that are unavailable and which will have to be quantized to the nearest available levels, resulting in pixel values which were different in the original image being quantized to the same level in the transformed image, with a consequent loss of local contrast and reduction in overall entropy. This can be troublesome for images with limited quantization levels, but the effect decreases for images with greater pixel depth, for example, 12-bit or 16-bit images. An advantage of histogram equalization is that it is fully automatic, without the need for specifying any parameters. The computation is simple, since it only involves repeated additions of the raw histogram values to produce the lookup table.

Attempts have been made to address the issue of merging levels by applying histogram equalization on a local basis (e.g., 8 × 8 for a 512 × 512 image), so-called local-area histogram equalization. The image can be divided

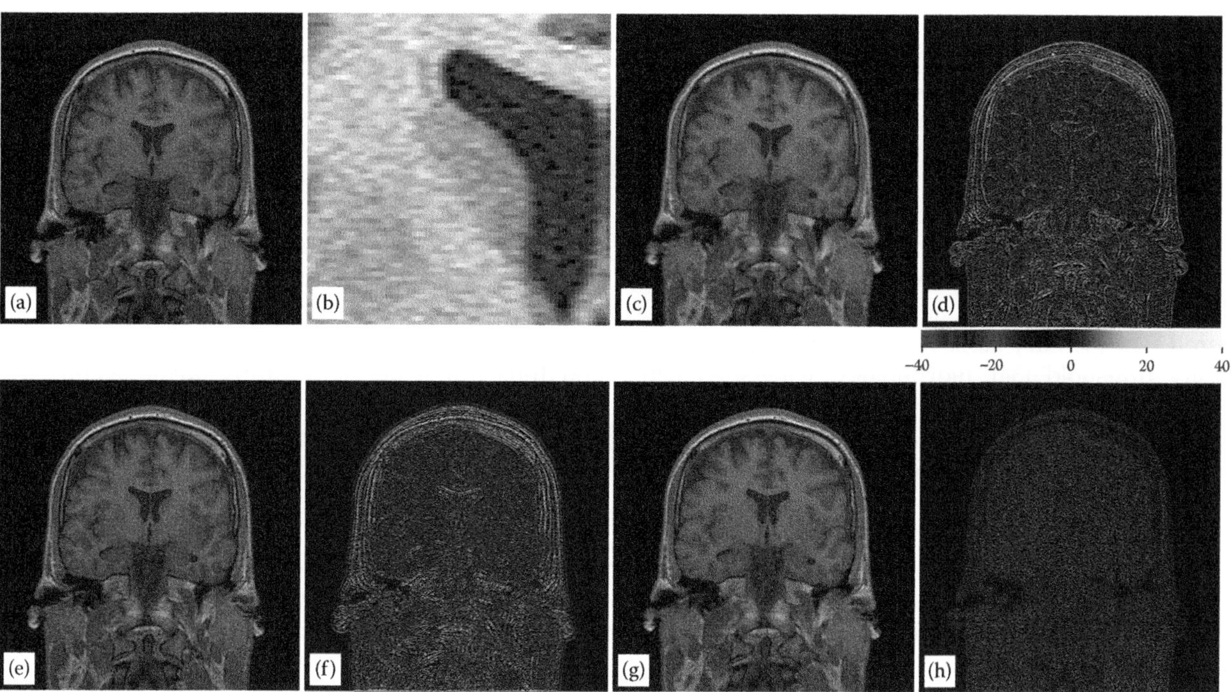


**FIGURE 22.5**
Comparison of filtering methods. (a) One slice of a $T_1$-weighted image of the head, acquired with a Turbo-FLASH sequence. (b) A magnified and contrast-enhanced section reveals strong anisotropic noise. (c) Linear smoothing with a Gaussian filter kernel with $\sigma = 1$. (d) The difference between smoothed and original image shows the removed noise, but also significant structural elements that show edge degradation. (e) Application of a Wiener filter with $k = 4$ (Equation 22.12). (f) The difference image contains structural elements at a lower level, but some ringing is visible. (g) Application of anisotropic diffusion, which shows the edge-preserving properties of the adaptive filter. (h) The difference image is almost entirely dominated by noise, and structural elements are much less prominent. The histogram is almost perfectly Gaussian with $\sigma = 3.9$, which coincides with the estimated noise level. The color bar underneath (d) is valid for (d), (f), and (h).

into a grid of square regions and histogram equalization performed on each pixel within each region: bilinear interpolation can be used to reduce the visibility of the region boundaries. A sliding region/window is preferred since this will avoid edge artifacts. The histogram of the pixels within the sliding window, centered on the current pixel being processed, is applied only to that pixel and the process repeated for every pixel in the image. The method is computationally more expensive, and can introduce artifacts if the sliding window is restricted to being rectangular in shape. Adaptive neighborhood histogram equalization (AHE) has also been used, where the window used for a particular pixel is not constrained to a particular shape or size but can adapt to its environment. The neighborhoods can be produced using a region-growing technique based on a simple graphical seed-fill algorithm, also known as pixel aggregation [27]. Histogram equalization can result in increased noise in homogeneous regions of the image. This can be controlled by clipping the original histogram of raw pixel values in a region to reduce the slope of the CDF; this is known as contrast limited adaptive histogram equalization, or CLAHE [28,29]. There is a tradeoff between reducing noise and increasing contrast. Reducing noise followed by enhancement often makes the edges in an image weaker. Contrast enhancement increases noise, so that the subsequent denoising effect can be poor. One approach to bridging these conflicting outcomes is to combine anisotropic diffusion with CLAHE using a differential model [30,31].

Subtraction is particularly useful in contrast-enhanced MRI examinations of the brain. The precontrast images are subtracted from the images after an injection of contrast agent as shown in Figure 22.2. Although subtracting similar images has the unfortunate consequence of producing a noisier subtracted image, the resulting increase in conspicuity results in small vessels becoming more easily noticeable and hence, for example, better tumor detection (think: spot-the-difference pictures in magazines/newspapers). The technique is sensitive to patient motion, which would cause misalignment, and would need to be corrected before subtraction [32,33]. In Figure 22.2, this misalignment is particularly prominent in the lower left neck region.

### 22.2.3 Segmentation of Regions in the Brain

Among the steps that have been widely automated is the segmentation of parts of the brain. The segmentation usually begins with the removal of the skull from the image (a process often referred to as *skull stripping*), followed by the separation of the brain into regions of gray matter, white matter, and CSF. Often, intensity- and region-based methods are applicable, and those are introduced below. Skull stripping is explained separately.

Manual skull-stripping is very time-consuming and can be inconsistent even after rigorous training. A fully automatic method is highly desirable, but at the least an accurate method with minimal operator dependence is required. The primary bases for skull-stripping include intensity threshold [34], morphology, watershed [35], surface-modeling [36], and hybrid methods [37–39]. Threshold methods define minimum and maximum values along one or more axes representing voxel intensities for univariate or multivariate histograms [40]. Morphology or region-based methods rely on connectivity between regions, such as similar intensity values, and often are used with intensity thresholding methods (e.g., AFNIs 3dIntracranial [41]). Other approaches combine morphological methods with edge detection and anisotropic filtering (e.g., brain surface extractor [BSE], [36]). Although watershed algorithms use image intensities, they operate under the assumption of white matter connectivity [35]. Watershed algorithms try to find a local optimum of the intensity gradient for pre-flooding of the defined basins to segment the image into brain and non-brain components. The volume is separated into regions connected in 3D space, and basins are filled to a preset height. Surface-model-based methods, in contrast, incorporate shape information through modeling the brain surface with a smoothed deformed template (e.g., brain extraction tool [BET], [39]). The hybrid watershed algorithm, HWA [38], incorporates both watershed techniques and surface-based methods. The resulting method relies on white matter connectivity to build an initial estimate of the brain volume and applies a parametric deformable surface model, integrating geometric constraints and statistical atlas information, to locate the brain boundary.

A comparison of the performance of automated skull-stripping algorithms on $T_1$-weighted images [42] suggested that HWA may remove substantial non-brain tissue from the difficult face and neck regions, carefully preserving the brain, although the outcome often would benefit from further stripping of other non-brain regions. BSE, in contrast, more clearly reaches the surface of the brain, although, in some cases, some brain tissue may be removed. 3dIntracranial and BET often left large non-brain regions or removed some brain regions, particularly in images of older populations. Both aging and common neurodegenerative diseases, such as Alzheimer's disease, reduce image contrast and adversely homogenize histograms, create partial volume effects, and obscure edges. Images from patients with Alzheimer's disease are often the most difficult to skull strip accurately. A skull-stripping approach that combines several methods, either sequentially or in parallel, may be beneficial. For example, HWA simplifies the problem of stripping away non-brain image regions while proving to be quite sensitive, and following the application of HWA with BSE may improve the specificity of the final result. Another approach [43] pursued the possibility of combining methods within a single meta-algorithm to optimize results. BSE, BET, and 3dIntracranial are also applicable to $T_2$-weighted images, and thus might be significantly advantageous under such circumstances. A region growing method, starting from two seed regions in the brain and non-brain automatically identified by a mask employing morphological operations has been shown to be robust in 2D [44]. A more recent study [45], based on deformable models and histogram analysis, showed a better performance than BET, BSE, and HWA on images from commonly used databases (BrainWeb, Internet Brain Segmentation Repository [IBSR], and Segmentation Validation Engine [SVE]—see also Section 22.2.7). It uses a rough initial segmentation step to find the optimal starting point for the deformation based on thresholds and morphological operators. Thresholds are computed using comparisons with an atlas, and modeling by Gaussian functions. The deformable model is based on a simplex mesh and its deformation is controlled by the image local gray levels and the information obtained on the gray level modeling of the rough-segmentation.

Once the skull has been stripped away, segmentation of the brain tissue from surrounding CSF can proceed. Because of issues such as spatial resolution (and partial volume effects), limited contrast, diffuse organ boundaries, noise and acquisition artifacts, segmentation—especially automated segmentation—remains a challenging task. In medical imaging, semi-automated methods (i.e., human guided) may be sufficient: a reasonable approximation is obtained quickly, and then adjusted by an operator, who has knowledge of the anatomy and may be assisted by atlas-based methods. Simple thresholding, where pixels are selected based on whether their values are greater or smaller than a threshold value, is very crude but can often help. Multiple thresholding using several values, including histogram-based metrics, rather than a single value may improve the result. Optimal thresholding (such as the isodata method [46] or Otsu thresholding [47]) renders the method automatic. An adaptive variable window size can be incorporated iteratively [48]. Since it is known that human vision relies heavily on contours, contour tracking based on edges (obtained from, say, Canny edge detection [49]) has also been used. A problem with the edge-based approach is that there is no optimality condition in 2D, no topology or connectivity constraints and it is difficult to impose higher-level knowledge on the results. Furthermore, the approach does not extend well to 3D. The opposite approach is to work on the objects or regions themselves, that is, region-based segmentation. This includes region growing (Figure 22.6) and splitting and merging techniques. Statistical methods, such as the Markov random field (MRF) model, can be a good choice in some cases but they cannot be recommended for general work, since simpler and faster methods often are sufficient.

*Active contours* are a potential solution to combine edge-based approaches with topological and connectivity constraints: Starting from an initial user-supplied closed contour, active contours can be made to evolve in such a way as to converge on the true contours of an image [50]. Additional examples for of deformable models can be found in [51] and Chapter 6 of [20]. The active contour model is very flexible: it contains internal terms, image data terms and constraint terms. In practice, however, there are some difficulties. It is sensitive to its parameterization and to initialization. During contour evolution, points tend to spread out or bunch up, requiring regular and frequent resampling. While it can be extended to 3D via triangulation, such extensions are complicated, and topological problems become more difficult. One way to avoid some of the problems brought about by the way active contours are discretized is to embed the contour into a higher dimensional manifold. This idea gave rise to *level sets* [52]. A contour is represented on the surface of an evolving function by its zero level-set, which is simply the threshold of the function at zero. The advantages of the level set method are that contour resampling is no longer necessary, level sets can change topology easily, and the formulation is independent of dimensions so 3D work is easily implemented. Although iterative solving of partial differential equations is expensive in terms of computer time, level sets remain a popular solution and have been used frequently in brain segmentation [53,54]. The geodesic active contour, GAC [55], using level sets, is widely used and implemented in many software packages such as the Insight Toolkit (ITK).

The *Random Walker* (RW) algorithm [56] is similar in some ways to classical segmentation procedures such as seeded region growing, but has some interesting differentiating properties and characteristics. It is quite robust to noise and can cope well with weak boundaries. Watershed segmentation works by analogy to the topography feature in geography [57]. It is simple and intuitive, can be parallelized, and always produces a complete division of the image. However, when applied to medical image analysis, it has important drawbacks such as over-segmentation and sensitivity to noise. An interesting and efficient implementation uses the maximum spanning tree [58]. Improvements using probability calculations and atlas registration have rendered it successful for the challenging problem of gray matter/white matter segmentation in MR images [59]. The *Power Watershed* (PW) algorithm [60] incorporates an energy interpretation and can find a globally optimal result in the presence of multiple labels. While it is slightly slower than standard Watershed segmentation, it is much faster than the RW, especially in 3D.

Although there are numerous segmentation methods they are not all equally applicable to medical image segmentation. Seeded segmentation is more robust than model-based segmentation [61], and flexibility and the ability to cope with noise and weak edges is paramount. The RW and (Power) Watershed Transform are among the most suitable for brain segmentation.

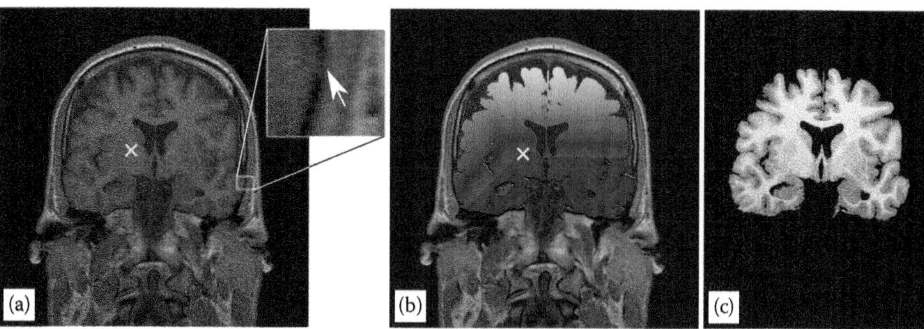


**FIGURE 22.6**
Basic skull stripping performed with an user-assisted region growing method. (a) shows the unsegmented head: a thin line separates the cortex from the skull, which invites intensity-based segmentation. However, many skull regions have similar intensity values as the brain tissue, and a connectivity criterion needs to be used. For this purpose, a seed point (marked ×) is provided by the user. (b) From the seed point, neighbors are tested whether they fall inside the intensity window. If so, these neighbors are added to the region and treated as seed points in turn. The false colors show the progression of the growth process in one direction (blue hues), and later in the other direction (red-orange hues). (c) Segmented (skull-stripped) brain region. The inset in (a) shows a critical section where a connecting *bridge* (arrow) exists that allows the region growing algorithm to *leak*. Leakage can be prevented by including additional threshold criteria, such as the local mean value.

### 22.2.4 Registration and Template Matching

Many recent studies make use of templates to facilitate segmentation of brain regions beyond gray and white matter. A template could be, for example, a randomly selected brain image from a study cohort, a preselected reference image, or a so-called *brain atlas*. A brain atlas is a model of a brain, in which individual structures have been labeled. The challenge associated with templates is the huge inter- and intra-subject variability of the brain, and the possibility that the head orientation of individual images does not match the template. Two steps are necessary to match a brain image to a template (i.e., to bring the two images into congruence, which also means that key landmarks can be found at the same places in a coordinate system): First, linear transformations, such as scaling, rotating, and translating the brain image are applied to match the template's size, orientation, and position. Second, nonlinear warping is used to bring the landmarks into congruence. For a detailed overview of the mathematical formulation of these operations, see Chapter 11 in [20].

For the first step, we need to define a coordinate system. One of the first comprehensive atlases was the Talairach atlas [62], and standard orientation is often defined with respect to this atlas. The origin of the coordinate system defined by Talairach is the anterior commissure, which can easily be found by visual inspection of the images. The $x$-, $y$-, and $z$-axes point right, forward, and up, respectively, from the patient's perspective (Figure 22.7), whereby the $y$-axis is defined in a more precise fashion as connecting the anterior and posterior commissures. A rigorous definition of the reference coordinate system is of fundamental importance, because all subsequent alignment and mapping steps depend on the initial alignment.

Initial alignment is performed with *rigid-body transformations*, that is, translation, rotation, and scaling. With respect to Figure 22.7, we could interpret the alignment step as (a) using translation and rotation operations to have all images conform to the same coordinate system, irrespective of the patient's position during imaging, and (b) using scaling operations to have the same voxel distance AC–PC in all images. The process of aligning two images can widely be automated by the computer, provided that either the intensity values are similar between two images, or specific landmark points can be extracted. Both the distribution of landmark points and the similarity of intensity values distributed in 3D space lead to possible formulations of a *quality metric* that quantitatively indicates how well an image—referred to as the *floating image*—matches the reference image. Once it is possible to quantify how well two images are matched, the process can be automated following the schematic in Figure 22.8. Iteratively, the quality metric is optimized, for example, with gradient descent methods or simplex optimization. At the end of the process, an optimal transformation is found, which minimizes the mismatch under the given set of transformations.

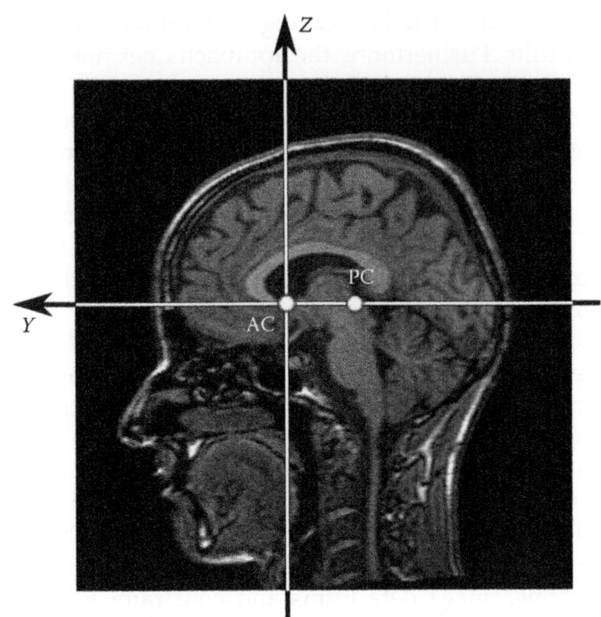


**FIGURE 22.7**
Talairach coordinate system. The origin is the anterior commissure (white circle AC). The $z$-axis points up, and the $y$-axis points forward (from the patient's perspective) and passes through the posterior commissure (white circle PC). The $x$-axis points right (i.e., into the plane of the page).

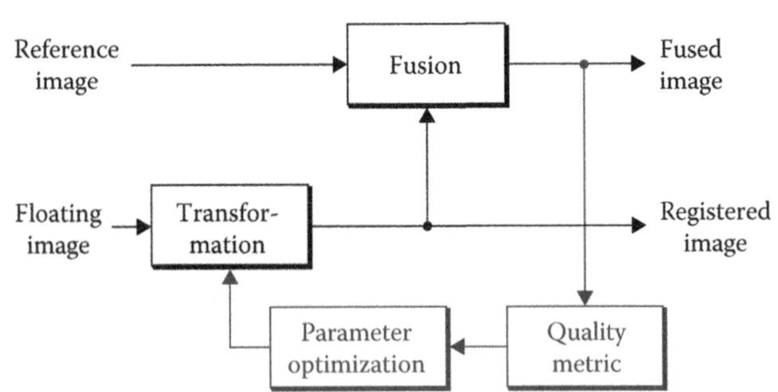


**FIGURE 22.8**
Schematic diagram of the registration process. The goal is to bring a floating image into congruence with a reference image by means of linear or nonlinear transformations. A registration quality metric (such as the spatial distribution of landmarks) drives the optimization of the transformation parameters, and the transformation is improved iteratively through the blue-shaded feedback loop until the quality metric is optimal. (Adapted from Haidekker, M., *Advanced Biomedical Image Analysis*, John Wiley & Sons, Hoboken, NJ, 2011.)

The initial alignment, in which only rigid-body transformations are allowed, leads to matching orientation and overall brain volume. It is now possible, for example, to separate the right and left hemispheres with the same plane. However, the *relative volumes*, for example, the ventricle volumes, will generally not match between the floating image and the reference image. At this point, nonlinear transformations (*warping*) are employed. The general optimization principle shown in Figure 22.8 still applies, but the transformation operation differs fundamentally. We can intuitively access the steps as follows. First, a grid of orthogonal, equidistant lines is superimposed over the floating image and the reference image. Landmark points (e.g., points along prominent sulci, boundary points between brain matter and ventricles, apex points) are identified within this grid. The grid of the reference image is now deformed by displacing the grid vertices. The optimum direction of the displacement can be determined by computing the gradient of the quality metric with respect to the displacement vector at each vertex. At the end of the process, the landmark points of the floating image and the reference image have the same positions inside the coordinate system, but the initially rectilinear grid of the floating image is now bent or warped. An example registration process is shown in Figure 22.9.

At this point, an initial segmentation can be applied to both the floating and the reference image. For example, if the ventricles of the reference image have been previously segmented, the same boundary now applies to the floating image. If the reference image is an atlas, and the volumes of individual entities (e.g., hippocampus, corpus callosum, and substantia nigra) are known, the boundaries can be applied to the floating image as well. As a consequence, the floating image is now fully segmented. The warping process can now be reversed, and the boundaries *unwarp* with the image.

Reversal of the warping process is not always necessary, because the amount of warping contains information about pathologies. We can interpret the amount of warping as a form of local energy when the grid is interpreted as a mesh of springs, and the displacement of a vertex is associated with work. For example, a tumor can be suspected when a local energy maximum is found (i.e., unusual local warping indicates an anomalous region). Specific examples are given in Sections 22.3.1 and 22.3.3.

The main advantage of image registration methods for template matching is that automated and unsupervised algorithms are available. For this reason, registration methods have become very popular tools for both segmentation and volumetry. Registration methods can be considered to be a key element of emerging CAD toolboxes.

### 22.2.5 Image Quantification and Feature Extraction

The idealized goal of any CAD system is to make a decision. The decision is not necessarily limited to *healthy* versus *diseased*. Rather, a decision can be part of the processing chain, for example, during subcortical segmentation, where a decision could be *voxel is part of ventricle* or *voxel is part of substantia nigra*. A simple example based on pure image intensities is shown in Figure 22.10. Such a branching chain of decisions is referred to as *decision tree*.

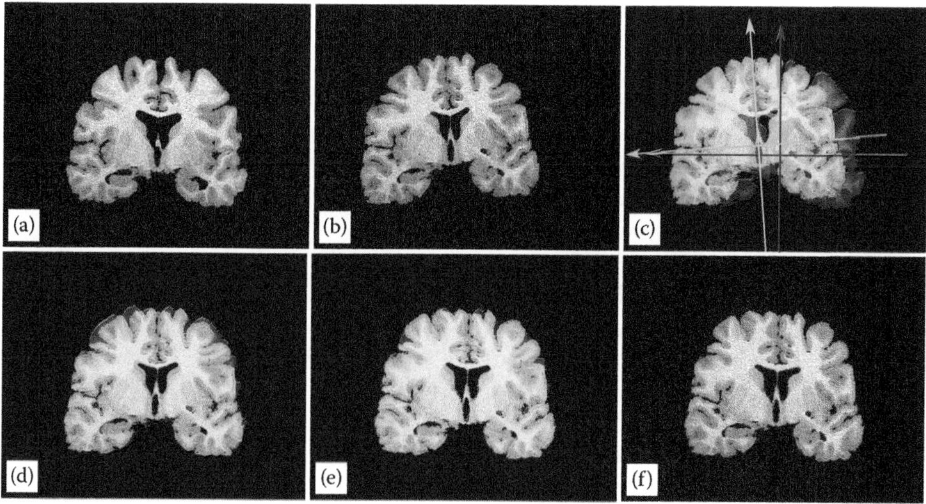


**FIGURE 22.9**
Example of a registration process. (a) One coronal slice of a $T_1$-weighted image of the brain after skull-stripping. This image is defined as the reference image. (b) Coronal slice of a different brain. This image is the floating image that needs to be registered to the reference. (c) In false colors, the floating image (green) overlaid over the reference image (red) reveals the position mismatch. The mismatch is further highlighted by the superimposed Talairach axes. (d) Rigid-body transforms (translation to match centroids, followed by rotation) remove the position mismatch, but significant differences between the brain images remain. (e) Nonlinear registration (warping) of the floating image brings both brain slices in congruence. (f) Floating image after nonlinear registration, shown for comparison with reference image (a) and original image (b). The process, shown here in 2D, is normally performed in three dimensions.

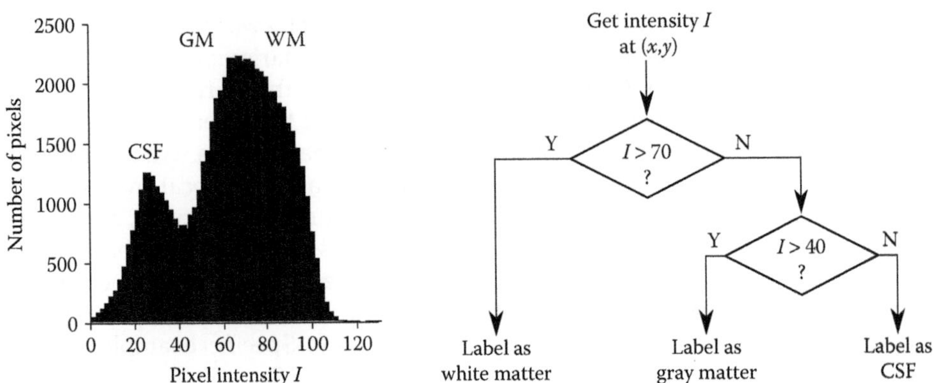


**FIGURE 22.10**
Simplified example of a decision tree. In this case, a pixel is assigned to any of the three classes (white matter, gray matter, CSF) based on its intensity value. The decision tree in this example is one formal way to represent thresholding with multiple thresholds, but decision trees of higher complexity, or those that use multiple metrics, are a straightforward extension.

Decision trees are attractive, because the outcome can be directly and unambiguously related to some property of the image. However, the challenge becomes immediately visible, too: in this example, partial-volume effects, noise, or a bias field create random or systematic deviations from the idealized values. Deviating voxels would be misclassified. The same type of challenge holds for any other form of decision, for example, a diagnosis based on ventricle volume or shape, where interindividual variability plays a major role.

Any decision mechanism relies on quantifiable properties of the image. A single metric, however, as explained in the example above, may overlap and lead to misclassification. One strategy to overcome this shortcoming is to extract multiple metrics that are—ideally—independent from each other (they are said to be *orthogonal*). Those metrics form a *feature vector*, which can be seen as a point in $n$-dimensional space, where $n$ is the number of elements in the feature vector. A hypothetical example is shown in Figure 22.11 with two metrics. Each of the two metrics shows a statistically significant ($p < .0001$) difference between diseased and control groups, but any single threshold would yield a very low accuracy due to the large overlap in values. When the two metrics are combined to form a 2-dimensional feature vector, a diagonal line can be used to separate the two groups with much higher accuracy than each individual metric could provide. The idea of finding a $n-1$-dimensional hyperplane (in 2D, this is a line) gives rise to support vector machines (SVMs), which are explained in Section 22.2.6.

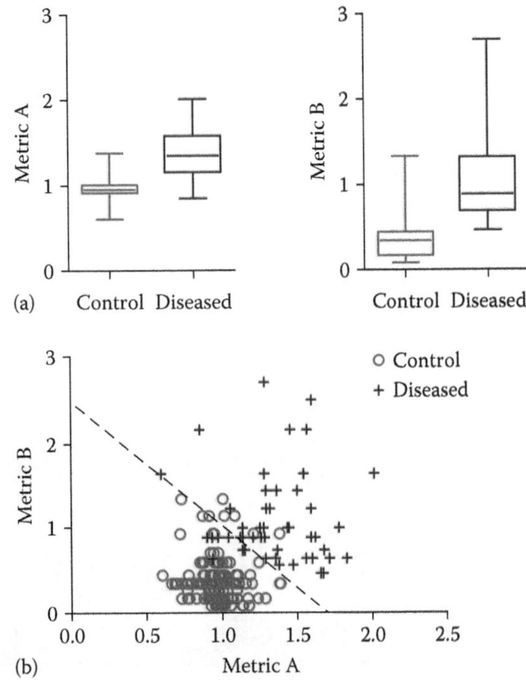


**FIGURE 22.11**
Combining two metrics to improve decision accuracy. (a) Two hypothetical metrics, A and B, show statistically significant differences, yet the overlap of individual values is large. (b) When both metrics are plotted in two dimensions, a diagonal line can be found that separates the groups with much fewer false positives and false negatives than any single metric would be capable of.

One simple example how a feature vector can be constructed is to combine image values from matching voxels in registered $T_1$-, $T_2$-, and FLAIR images from the same patient. It is also possible to obtain voxel-by-voxel feature metrics by examining the local texture of the image. *Texture* can be defined as any systematic intensity variation of the voxels in an image. To illustrate this idea, let us assume that we compute the gray-value gradient of a brain image slice. The gradient is large near transitions from one tissue class to another. For example, we see high gradient values at the boundary between white matter and CSF, that is, in all those voxels that could be affected by partial volume artifacts and are at risk of being misclassified by pure intensity thresholding. With this information, we could now limit the initial segmentation to voxels in homogeneous regions and assign voxels with high gradient values in a second step based on a different criterion, such as proximity to already-classified voxels.

Several established texture quantification metrics are available, and most of them can be applied to a moving window around the voxel, leading to local texture metrics. The simplest of these are the *moments of the histogram*. The first moment is the local mean value and corresponds to the value seen in a smoothed (blurred) version of the image. The second moment is the variance and indicates the roughness of the pixels. The third and fourth moments are known as skew and kurtosis and relate to the degree that the pixel values follow a Gaussian distribution.

The gradient operator was presented as an example above, and computation of the gradient requires the convolution with edge-enhancing kernels. Laws [63] proposed four one-dimensional convolution kernels that enhance certain texture properties: edge, point, ripple, and wave. A fifth kernel would have Gaussian smoothing properties. Each kernel can be combined with any of the five kernels in the orthogonal direction to yield a total of 25 convolution kernels. Laws defined the texture *energy* per pixel as the absolute value of the convolution result, averaged over a small neighborhood. The energy metrics obtained from the 25 kernels would provide a 25-dimensional feature vector.

Three more series of metrics require a transformation: First, the *co-occurrence matrix* [64] is a joint histogram of the image values with the image values of a shifted version of the image. Therefore, the co-occurrence matrix contains information about how similar adjacent regions are. The co-occurrence matrix itself is a 2D image, and single-value metrics need to be extracted. Numerous such metrics are defined, and they include values such as energy, entropy, correlation, and statistical moments. The displacement vector for the shifted image is variable (both direction and length), and a very high number of feature values can be extracted. Second, the run-length histogram [65] is the joint probability of $n$ voxels having the same gray value $g$ in one direction (i.e., the *run*). In the presence of noise, gray value binning is performed, and gray value bands are used. Once again, the joint histogram $P(n, g)$ is a 2D image in the more general definition, and single-value metrics need to be applied. Those metrics give an indication whether long or short runs dominate, and whether they occur at high or low intensities. Third, texture metrics can be obtained in the Fourier domain, where once again single-valued metrics need to be extracted from the 2D Fourier transform (e.g., isotropy, decay, dominance of high frequencies).

All three methods (co-occurrence matrix, run-lengths, Fourier-domain metrics) have in common that they are difficult to apply locally. When the transformations are limited to small regions around a pixel of interest, computational expense increases steeply. Complexity increases further when those methods—originally designed for 2D images—are extended into three-dimensional space. The feature vectors become huge (e.g., Law's energy metrics would extend to 125 possible combinations, and 26 single-voxel displacements for the co-occurrence matrix are possible), and a reduction of dimensionality is necessary. None the less, interest in texture analysis has recently attracted interest (see, for example, [66–68]).

Another set of possible metrics emerges from a feature's *shape*. Shape refers more to segmented regions, such as a tumor or ventricle. When an object is segmented, a quantitative description of its shape can be used as a metric that is orthogonal to size, much like texture is orthogonal to intensity. Therefore, automated shape description can improve an automated diagnosis (see [69–71] for some examples). Intuitively, we expect shape-descriptive metrics to be invariant under translations, rotations and scaling. Some of the simplest shape descriptors show this property: Aspect ratio (ratio of the radii of the circumscribed sphere and the inscribed sphere) and circularity (ratio of the volume, multiplied by $4\pi$, to the squared surface area). The most relevant information lies in the boundary pixels. If those are expressed relative to the shape's centroid, the boundary voxels (and metrics derived from them) become translation-invariant. Statistical moments once again provide meaningful metrics with implicit rotation invariance, but a normalization is necessary to ensure scaling invariance: for example, the second moment needs to be expressed relative to the first moment, and the result, that is, the coefficient of variation, emerges as a descriptor of shape irregularity.

There are a number of shape metrics that are well-defined in 2D images, but would be difficult to extend into three dimensions, such as chain codes. Fourier descriptors (i.e., the discrete Fourier transform of the unrolled radius) are straightforward in 2D images, but would become a 2D spectrum for volumes, with the associated challenge to extract meaningful single-value metrics. On the other hand, elastic contours provide a good opportunity for shape quantification, because the deformation of a reference shape to match the shape under examination is associated with deformation energy. Deformation energies are particularly suitable for the quantification of the brain morphology, notably the pattern of gyri and sulci.

A popular family of shape metrics is related to the similarity of structures on different scales, which is expressed as the *fractal dimension*. Intuitively, the fractal dimension can be seen as a metric of complexity of a shape. A well-known example for the use of the fractal dimension is the coastline of England. How long is it? It certainly depends on the scale of the map that is used to measure it. As a larger-scale map reveals more and more details (inlets, channels, bays, etc.), the length appears to grow. In fact, it can be demonstrated that the estimated length $l$ increases with the map scale $s$ in a power-law fashion, $l \propto s^D$ where $D$ is the fractal dimension. Estimators of the fractal dimension attempt to identify such power-law scaling behavior by examining structures in the image on different scales. For many images, apparent fractal properties are limited to a range of scales. Although fractal dimension estimation can be used as a texture metric (see, for example, Chapter 10 of [20]), estimation of the fractal dimension is more commonly found in shape quantification [72–75]. For texture analysis, the frequency domain (Fourier transform) presents a fractal signature. Fractal dimension does not provide a complete characterization of spatial arrangement. *Lacunarity* is a scale-dependent measure of the deviation of a geometric structure from translational invariance, and can be thought of as a measure of *gappiness* [76]. It is not predicated on self-similarity and can discriminate between images with the same fractal dimension [75,77]. Both fractal dimension and lacunarity can serve as metrics in a feature vector.

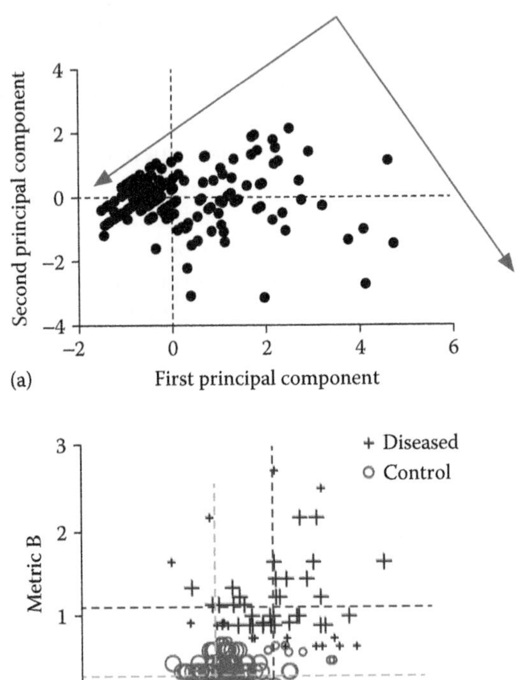


**FIGURE 22.12**
(a) Principal component analysis of the data in Figure 22.11b. The relative orientation of the data points remains the same, but the principal component axis points in the direction of highest variance. The original coordinate system is indicated in blue. (b) Classification through clustering. An automated decision process has found two cluster centroids at (0.9, 0.3) and (1.3, 1.14) for control and diseased, respectively (dashed lines). Individual data points are assigned based on their Euclidean distance, and large symbols indicate a higher certainty (75% or greater membership) than small symbols (50% to 75% membership).

### 22.2.6 Classifiers and Decision Mechanisms

One simple decision mechanism was presented at the outset of Section 22.2.5. The example also highlighted the need to obtain multiple descriptive metrics that are independent from each other (and thus add information). Multiple metrics form a feature vector, which is the basis for classification. Feature vectors are multidimensional, and depending on the application, high-dimensional feature vectors are obtained. Typically, not all metrics are orthogonal. Returning once again to the example of a feature vector made of $T_1$-, $T_2$-, and FLAIR intensity values, $T_1$ and $T_2$ relaxation times are to some extent anticorrelated, and $T_1$ and FLAIR correlated. *Principal component analysis* (PCA) can be used to reduce the correlation between feature elements. We can think of PCA as a method to place a new coordinate system in the cloud of data points in such a fashion that the variance along one axis is maximized. The other axes are orthogonal to the first axis, and a $n$-dimensional feature vector becomes again a $n$-dimensional feature vector. However, after the transformation, its first element (i.e., the principal component) has the highest variance (or in image terms, the largest contrast). When the contrast along non-principal axes is very low, the elements can be dropped, because they carry little information. An example, based on the hypothetical dataset in Figure 22.11 is shown in Figure 22.12.

*Discriminant analysis* is a supervised method that explicitly attempts to optimize class separability, whereas PCA finds directions which are efficient for representing the total, pooled dataset. Supervised in this context means that discriminant analysis requires a labeled dataset for training. The basis vectors of this transformation are known as canonicals, and the resulting canonical directions produce the optimal discrimination (i.e., separation) of the projected classes (see, e.g., [78]). For $k$ classes, the $n$-dimensional data is projected into $k-1$-dimensional space, which is useful for dimensionality reduction. Linear discriminant analysis (LDA) uses a pooled covariance matrix for all the classes, and results in a decision boundary that is a hyperplane (viz., linear for two classes). If a separate covariance matrix is used for each class, which is preferable but requires larger datasets, the analysis is known as quadratic discriminative analysis (QDA) and the decision boundaries are quardics (viz., quadratics for two classes). Discriminant analysis depends on prior probabilities; these can either be taken as proportional to the occurrence of the classes in the sample set, or, preferably, if these are known, to the occurrences in the whole population.

One of the most popular classification methods is *clustering*, notably, $k$-means clustering and fuzzy $c$-means clustering. In both cases, $k$ classes are given (e.g., two classes, healthy and diseased). Cluster centroids are seeded, either by user intervention, prior knowledge, random seeding, or some property of the data, such as distribution probabilities. Each data point is now assigned to one centroid based on its Euclidean distance, and the centroids recomputed from the current class membership. The process is repeated until class membership remains unchanged. The difference between $k$-means clustering and fuzzy $c$-means clustering is that each data point can be assigned to only one cluster in $k$-means clustering, but belongs to some distance-dependent extent to all clusters in fuzzy $c$-means clustering. When the process converges, the strength of class membership in fuzzy $c$-means clustering allows to quantify the degree of confidence as demonstrated in Figure 22.12b.

The remaining classifiers described in this section are based on artificial intelligence methods, and they have in common that they need training. The *Bayesian classifier* uses a training dataset of known mappings of a feature vector $\vec{f}_T$ to a class C to build a probabilistic model of the training data. The probabilistic model consists of conditional probabilities, such as the joint probability that a segmented shape is a tumor when its shape irregularity is greater than some value $v$. Here, the first part is the outcome (i.e., the class)—known when we are using the training set—and the second part is a feature. Based on the probabilistic model, the probability of an unknown feature vector $\vec{f}$ to belong to class $C_i$ can be computed, and the feature is assigned to that class where the probability is highest.

Conceptually similar, but mathematically very different is the $k$-nearest neighbor classifier. Let us assume that a scatter plot, such as the one in Figure 22.11b, has been created by using training feature vectors $\vec{f}_T$ with known class membership (i.e., the training set). A new and unknown feature vector $\vec{f}$ is now placed in this diagram, and $k$ neighbors with the smallest Euclidean distance are found. $k$ is usually a low, odd number. A majority decision assigns the unknown feature vector to that class to which the majority of its neighbors belong.

The idea of SVMs was briefly mentioned at the outset of Section 22.2.5. SVMs use a training set to find a hyperplane in $n$-dimensional space that optimally separates the classes. For two-component feature vectors, the hyperplane is a line, and for three-component feature vectors, it is a plane. Once again, a training set of feature vectors $\vec{f}_T$, known to belong to class C, is used to find that hyperplane that optimally separates the classes. Placement of an unknown feature vector in $n$-dimensional space separated by those hyperplanes immediately yields its class membership. The advantage of SVMs over $k$-nearest neighbor methods is that the separating boundary is regular and parametric, and that the entire training set is used for the determination of the boundaries, rather than a few (random) neighbors.

Possibly the most abstract and opaque of the artificial intelligence-based methods are *artificial neural networks* (ANN). ANN consists of computational neurons, each of them with two computational stages: weighted summation of the inputs and computation of the output with the activation function $h$ (Figure 22.13a). The actual neural network is composed of at least two or three layers (input layer, hidden layer, output layer). The computational neurons in the output layer can have a yes/no activation function (thresholding) and are in that case referred to as *perceptrons*. A complete neural network with one hidden layer is shown in Figure 22.13b.

Training of the neural network consists of setting appropriate weight coefficients $w_{ij}$ and $w_{jk}$. One commonly used method is *backpropagation*, whereby initial weight coefficients are randomly set and then modified in the learning process: A feature vector is applied to the inputs and the network allowed to predict the outcome. The outcome is compared to (subtracted from) the known outcome, providing a prediction error. From the input signal and the prediction error, the *gradient* of the weight coefficients with respect to the input can be computed, and the weight coefficients updated with the negative gradient. In this fashion, the prediction error

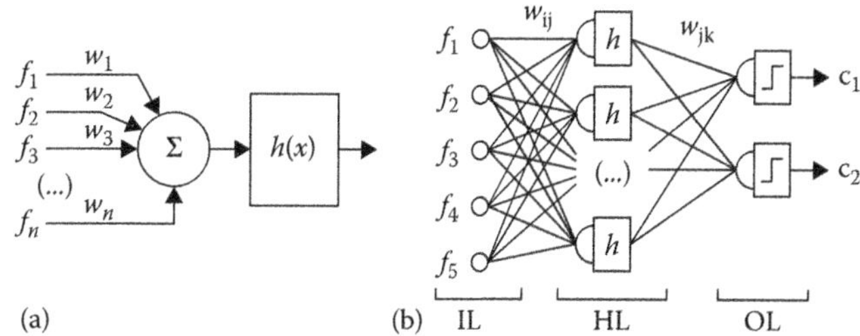


**FIGURE 22.13**
Artificial neural networks: (a) Schematic of an individual computational neuron. Its input signals (such as the feature vector components $f_i$) are summed up with individual weight factors $w_i$ and subjected to the activation function $h(x)$ to provide the neuron's output level. (b) A complete neural network with an input layer (IL), hidden layer (HL), and output layer (OL). Each of the connections carries an individual weight function $w_{ij}$ for the hidden-layer inputs and $w_{jk}$ for the output-layer inputs. The output layer perceptrons provide a yes/no answer of, for example, class membership.

decreases with the amount of training (although it is possible to over-train a network). Neural networks are a topic of appreciable complexity, and the underlying theory needs to be well understood for its successful application.

### 22.2.7 Software for CAD of the Brain

Clearly, specialized software is needed to perform some or all of the methods listed above. Many functions, such as noise filtering, contrast enhancement, and simple intensity-based segmentation are usually included in the control software that comes with the MRI scanner. CAD of diseases in the brain, however, can be thought of as an ongoing development project, and no established and clinically accepted CAD method, to the authors' knowledge, has emerged. Frequently, research groups make their software publicly available. In fact, a review of the literature shows preference for a few emerging MRI software suites that are continuously improved by the community of researchers and programmers around the world. Its source code is openly available, which is a prerequisite for the distributed improvement by the community. A large movement has evolved around such *open source* software, which can be seen as the community-driven alternative to commercial, black-box-type software in the sense that distribution, usually free of charge, occurs through the Internet; both customer service and software quality control (i.e., bug reporting) is available through Internet-based forums of users and developers. In this respect, *free software* refers not necessarily to the absence of license fees, but rather to certain *rights* of the user, first and foremost the right to access, study, share, and modify the source code. Due to the open nature of this software, verification of the functionality is possible for all interested parties, which is an enormous advantage over black-box software in a novel and rapidly evolving field. In this section, some popular software suites are briefly introduced.

A very comprehensive resource, and an excellent starting point for anything related to computers in MRI analysis, is the Web site of the Neuroimaging Informatics Tools and Resources Clearinghouse (NITRC) at http://www.nitrc.org. The Web site not only contains links to the most relevant software suites, but also allows access to MR study data and images as well as atlas data. A second comprehensive resource that warrants mentioning is the Neuro-Debian project [79] at http://neuro.debian.net/. Debian is a well-known and popular distribution of the Linux/GNU operating system, and it is interesting in itself as it meets the criteria of free software—in its entirety—as set forth above. In other words, a *distribution* in this context is a complete operating system combined with a highly comprehensive set of software packages, including, for example, developer tools, office and productivity suites, and Internet applications. Neuro-Debian is an add-on collection for the Debian distribution and its derivatives (e.g., Ubuntu, Linux Mint). It can be downloaded as a stand-alone installation, or it can be added to an existing installation of Debian, Ubuntu, Mint, and so on in the form of *repositories*. Downloading and installing additional software and datasets is as easy as a single mouse click. The list of software packages and datasets included in Neuro-Debian is too comprehensive to be included here, and the reader is referred to the Web site. Most of the software packages described below are included in Neuro-Debian.

FSL—abbreviation for the FMRIB Analysis Group's Software Library [80]—is a collection of software tools for processing and analysis of structural MRI, fMRI, and DTI data. It is maintained by the University of Oxford, UK, and its Web site is http://fsl.fmrib.ox.ac.uk/fsl/fslwiki/. FSL has been used in several of the studies cited in Section 22.3. FSL is a collection of several disjunct, but interacting, software programs that automate numerous crucial image processing tasks. Among these is BET to perform skull-stripping, FLIRT and FNIRT (FSL linear/nonlinear image registration tool), FAST (a segmentation tool that separates gray matter, white matter, and CSV in an automated fashion), and FIRST (to segment subcortical structures with the help of an atlas). Furthermore, there are multiple tools for voxel- and model-based analysis of image data. Although the individual modules are generally command-line driven, an overarching graphical user interface exists.

FreeSurfer is another comprehensive software suite that is referenced in several publications in Section 22.3. FreeSurfer is maintained by Harvard University's Martinos Center for Biomedical Imaging, and its web page can be found at http://surfer.nmr.mgh.harvard.edu/. FreeSurfer has capabilities somewhat similar to FSL; these include skull stripping, registration, segmentation (including subcortical regions), region- and voxel-based analysis, and tractography. FreeSurfer has its own image format for data storage, but it can read DICOM data. A registration key needs to be requested to run the software, and access to the source code is granted after registration.

In this context, Slicer (http://www.slicer.org) is worth mentioning. Although more a general-purpose medical image analysis software, Slicer has a number of MRI-relevant functions, including a fairly advanced DWI/tractography package. Compared to other packages, Slicer is fairly easy to use. Moreover, Slicer features many of the basic image processing functions that were mentioned in the sections above. Slicer has the potential to solve many basic tasks, and its comparatively easy extensibility makes it suitable for the development of new algorithms.

The Brain Imaging Center at McGill University offers the Brain Imaging Software Toolbox (http://www.bic.mni.mcgill.ca/software/). The Brain Imaging Software Toolbox is similar to FSL in that it is a collection of individual software utilities that interact with each other. The scope of implemented functions is similar to FSL as well. The individual modules are purely command-line driven, which may at first appear uncomfortable to users of GUI-based software. However, one strength is the pipelining tool, which allows to combine the modules to perform a sequence of processing steps on a large set of MR images. Since some steps, such as cortex extraction, can take hours to complete, pipelining tools that automate the process are desirable.

BrainVisa is another set of brain MR image processing tools. The BrainVisa contributors are predominantly French research organizations, and BrainVisa can be accessed at http://brainvisa.info. At the core of BrainVisa are toolboxes for MRI segmentation, cortical sulcus recognition and morphometry, and cortical surface parameterization and analysis. BrainVisa can interface with FSL and FreeSurfer, and some add-on toolboxes for fMRI analysis, inclusion of EEG and MEG, and for cortical thickness maps, gyrification index, sulcal length and depth are available. BrainVisa is a fairly young project, which may be a reason why there is not as much functionality implemented as in FSL or FreeSurfer. However, Brain Visa is completely GUI-based and advertises itself for users who feel uncomfortable with command-line interfaces.

Brainsuite is the brain MRI toolbox from the University of Southern California. It can be accessed at http://brainsuite.org. Its functionality includes bias field correction, skull-stripping, brain surface and cortical surface extraction, and the labeling of subcortical structures, including registration functions. In addition, specialized functions for fMRI and DTI exist, such as registration with structural MRI, orientation distribution function and tensor fitting, tractography, and connectivity. Brainsuite is a mix of GUI- and command-line programs. The Brainsuite collection is incomplete for Windows and Linux, and only executable files are available. This is not a free software.

Lastly, it is worth including the image processing toolbox from the University of Utah Center for Integrative Biomedical Computing, available at http://www. sci.utah.edu/cibc-software.html. Its scope is different from the examples above: The CIBC software tools are not specialized MRI tools. Rather, they include frequently used image processing steps, manual and semi-assisted segmentation, and surface parametrization. In this respect, CIBC software complements the functionality of, for example, FSL or FreeSurfer, and allows more specialized operations. One strength is its surface mesh generation module, in which the surface of a segmented volume is represented by a series of vertices. This representation allows interaction with modeling and simulation software and with 3D printers.

Without exception, the software suites listed in this section are very complex, and accordingly complex to operate. Although the software is free of license fees, it has in common with commercial software that a considerable time investment is needed to master even the most basic functions, such as skull stripping. Also without exception, the software listed here comes with extensive documentation and tutorials. It is worth the time to closely study the functions included in the software and decide on one software suite. At this point, a good approach might be to use the sample datasets that are used in the tutorials and process an independent dataset in parallel. The initial time investment will then rapidly translate into the productive use of the software suite.

## 22.3 Practical Aspects and Case Examples

### 22.3.1 Detection and Classification of Tumors

Magnetic resonance is the modality of choice for tumor diagnosis both by a radiologist and with computer-aided diagnostic tools. Important targets are gliomas and meningiomas, which comprise the majority of aggressive brain tumors. In recent years, the automated detection of brain tumors in MR images has gained some momentum with the availability of more advanced segmentation and classification methods. For the purpose of computer-aided tumor detection, MR offers several unique advantages: $T_1$-weighted sequences are frequently used, often with the application of contrast agents. Computer tools can be used to quantify the dynamics of contrast agent uptake. Brain tumors show higher contrast because they tend to spread across the blood–brain barrier and more readily accumulate contrast agent than healthy brain tissue. Fluid-attenuation sequences (FLAIR) are frequently employed to distinguish fluid-rich regions of the lesion, such as the surrounding edema, from CSF. Computer-aided diagnostic tools are often built on obtaining large amounts of information from different sequences, including no-contrast and contrast-enhanced $T_1$, complementary $T_2$, and sometimes FLAIR.

As described in the previous sections, image processing typically begins with noise suppression and is followed by the isolation of the brain region, often referred to as *skull-stripping* [42,81]. Moreover, multiple images taken with different sequences need to be carefully aligned as the patient could have moved between, for example, the initial $T_1$ scan of the head and the subsequent FLAIR scan or time-resolved contrast scans.

Segmentation of the brain tissue is widely automated (see, e.g., references [42,82]), and registration of images taken in a relatively short time interval can usually be performed with conventional rigid-body transformations guided either by anatomical landmarks, mutual information, or fiducial markers. A very important step in the preparation of MR images for automated image processing is intensity normalization [83], because absolute image intensity values are not standardized, nor do they represent accurate values of the tissue parameters (such as $T_1$ relaxation time). The primary purpose of the normalization step is to give similar image values similar tissue meaning. The step primarily involves piecewise linear mapping of the intensity values [83]: in its simplest form, histogram modes of an individual image are mapped to the modes of the average histogram of a training set, but more advanced normalization steps also allow for bias field correction.

After these preparation steps, the tumor tissue needs to be segmented. The methods for this step diverge, however, as they strongly depend on the imaging sequence. In its simplest form, seed points [84] or a bounding box [85] provide input to a supervised method. With suitable transforms, the tumor region can be grown from its seed points [84] or narrowed down from the bounding box [85] based on pixel intensity values and similarity metrics. Such user-assisted methods can provide objective criteria to detect the actual tumor boundary and provide a speed improvement over fully manual tumor delineation, especially when large datasets need to be processed.

One representative example how unsupervised tumor detection can be realized was presented by Hsieh et al. [86]. The method is based on a combination of $T_1$- and $T_2$-weighted images without contrast application. Application of a clustering algorithm on the two-dimensional $T_1$–$T_2$ intensity histogram provides initial clusters that serve as seeds for similarity-based region growing. In the next step, *a priori* knowledge of tumor size and shape is used to exclude false positives. The attractiveness of this algorithm is based on the need to only obtain two images, $T_1$- and $T_2$-weighted, and the use of widely known image processing steps.

The symmetry of the brain hemispheres can be used for robust and unsupervised tumor detection. Pedoia et al. [87] proposed a FLAIR-based method to detect gliomas whereby the symmetry of the brain hemispheres is quantified: One hemisphere is mirrored and brought into congruence with the other hemisphere with standard registration methods. The intensity difference between the two hemispheres highlights the anomalies. A combination of clustering and graph-cut segmentation then isolates the suspected tumor. Following a similar idea, Bach Cudara et al. [88] made use of the deformation imposed on healthy tissue by large-scale meningiomas. In this approach, simulated variational flow is used to quantify tissue deformation between healthy subjects or other atlas-type reference images with patients with a tumor, thereby revealing presence and size of the tumor.

Automated shape recognition can give important clues as to the presence of tumors or lesions. One method was presented by Descombes et al. to detect Virchow–Robin spaces [89]. Although Virchow–Robin spaces are not malignant lesions, the method serves as a typical example for shape-based recognition of brain anomalies. *A priori* knowledge of the shape that can be expected is referred to as *prior*. Descombes et al. defined their prior as an elongated, connected group of voxels that are darker then the surrounding tissue and form a tubular shape whose diameter is typically close to the voxel resolution. The challenge is the presence of hundreds of Virchow–Robin spaces that can be partly intersecting or overlapping. Descombes et al. solved the problem with a modified genetic algorithm where a large number of candidates for the detected Virchow–Robin spaces can be subjected to any of the moves: generation (*birth*), removal (*death*), segment displacement, end point displacement, splitting, and merging. Intensity metrics along the segments drive the decision to accept or reject any randomly selected move.

A different shape recognition method was presented by Ambrosini et al. [90], who described a metastasis prior as spherical or spheroid structures with well-circumscribed appearance in contrast-enhanced $T_1$-weighted images. Ambrosini et al. used a normalized cross-correlation coefficient of voxel-by-voxel intensity values to provide a metric of *trust* that a region of the brain MR image matches the template. In this example, the absolute value of the normalized cross-correlation coefficient was used as threshold value for the decision whether a region in the brain volume is identified as metastatic lesion or not. With a training set of images, the threshold was optimized with respect to the algorithm sensitivity. In contrast to Ambrosini et al., Sanjuán et al. [91] suggested the use of intensity- and similarity-based priors such as those used to segment the image of a healthy brain into three classes, that is, gray matter, white matter, and CSF. The presented method is an extension of the unified segmentation method by Ashburner and Frison [92], in which Bayesian classification and clustering are combined to perform the segmentation of gray and white matter, and CSF. Sanjuán et al. then add a fourth, unknown, class as a *catch-all* for diseased tissue that can be considered to be an outlier with respect to the three known classes. The algorithm gains in complexity as data from healthy training sets are used to more accurately identify outlier tissue and thus minimize false-negatives.

One final example serves as representative example for the use of dynamic contrast-enhanced MRI [93].

The underlying idea is that vascularization of tumor tissue changes the arterial input function, that is, the kinetics of contrast agent uptake. Initially, voxels are selected based on the rise of contrast-enhanced intensity over time. In a subsequent step, connected regions are identified and selected based on shape criteria. The resulting regions are then used to calculate the arterial input function of the entire region, which allows conclusions on the permeability of the tissue.

It becomes evident from these examples and other literature references that all proposed methods are, to some extent, work in progress. Notably, no single method is universally applicable. Some methods require adjustable parameters that are determined with training sets (e.g., [86,91,93]). Different instrument calibration and different sequences may lead to highly divergent results. Prior-based methods are highly dependent on the choice of prior. In the context of tumor detection and classification, therefore, CAD needs to be seen predominantly as supporting the observation by a trained radiologist. Although attempts at fully automated detection exist in the literature, most computer methods require some form of user intervention. However, computer-aided diagnostic methods hold the promise of providing an objective and comparatively fast way to detect, measure, and classify tumors, and therefore are not only a diagnostic aid, but also allow the radiologist to monitor tumor progression or recession during treatment.

### 22.3.2 Computer-Assisted Ventricle Volumetry

Volume changes of the brain ventricles are known to be associated with aging [94], Alzheimer's disease [95], schizophrenia [96], and MS [97]. In addition, studies have shown ventricle enlargement in vitamin D deficiency [98] and in dehydration [99], to name a few examples. In a 3D MR image, segmentation of the ventricles should be straightforward in both $T_1$- and $T_2$-weighted images, because the relaxation times of CSF are much longer than in the surrounding tissue, thus ensuring ample contrast. Moreover, the lateral ventricles are major structures that can be easily found in a 3D image of the head. Histogram-based thresholding [100] and knowledge-based region growing [101] were among the earlier methods proposed. Both methods have limitations due to leakage, that is, inclusion of non-ventricle CSF. Moreover, morphometry of the ventricles places a high demand on the accurate determination of the ventricle boundary. Slice orientation, voxel size, voxel anisotropy, and the presence of noise influence the result. For example, a phantom study has shown that the apparent ventricle volume increases by roughly 3% per mm increase of MR slice thickness [102]. In addition, a study has shown that a left–right bias exists in user-guided segmentations [103], although the ventricles themselves were reported to be relatively robust against this bias.

For reasons of objectivity and reproducibility, some degree of automation of the segmentation process is desirable. Two approaches continue to dominate ventricle segmentation. The first is a continued refinement of intensity-based segmentation, and the second is a shape-based approach to match the ventricles of an individual image to a reference image or an atlas.

Intensity-based methods are often user-assisted, for example, with the user providing seed points for region growing or by delineating a region of interest. In one particularly straightforward example [104], the user needs to provide three circular regions that lie fully within the lateral ventricles. A threshold is computed from the mean value and standard deviation of the three regions, which is then used to separate CSF from gray matter. This step is followed by a contour tracing step, where the contour is limited by a user-defined region of interest. A relatively simple, yet clever extension of the pure intensity-based thresholding method involves the exclusion of boundary voxels [105]: in a $T_1$-weighted image of the brain, the intensity histogram is trimodal (CSF, gray matter, white matter), but the histogram peaks are poorly separated because of partial-volume effects near the boundary. In a two-pass segmentation, likely boundary pixels were excluded and the histogram recomputed. The resulting histogram showed better separation of the peaks and thus allowed to more reliably find the threshold values in the valleys between the histogram peaks.

Intensity-based segmentation can be improved by the inclusion of fuzzy connectedness criteria [102,106,107]: Both region growing and clustering can be steered with distance metrics, either spatial distance or intensity distance. Even with these improvements, however, intensity variations introduced by inhomogeneous magnetic fields or shift artifacts can significantly influence the segmentation result. Early attempts to use local histogram correction [108] or homomorphic high-pass filtering [109] have widely been removed from the image processing chain as they can influence useful intensity information and actually *cause* misclassified voxels.

The second approach is to use a known reference shape for the ventricles, either from a model, from a segmented reference image, or from an atlas. For this purpose, several brain atlas datasets are available, among them the Talairach atlas [62], the MNI-Colin27 brain phantom [110], which was created from 27 repeated and registered scans of the same individual, an improved, higher-resolution version of the MNI-Colin27 phantom [111], and the MNI-305 dataset [112], which was created from 305 individuals. Any model- or atlas-based approach is a multi-pass approach, because it requires an initial segmentation, followed by atlas-based correction. In many cases, the segmentation process is further subdivided into a sequence of segmentation steps, where segmentation results from more reliable structures are used to segment more difficult structures. One example of such a segmentation pipeline is given by Liu et al. [113]: In the first step, the image is aligned with the standard coordinate system as defined by Talairach. From the atlas, the locations of the lateral ventricles are estimated, and the ventricles are segmented by hysteresis thresholding and clustering. The lateral ventricles are then removed from the image to avoid biasing the next step, that is, the segmentation of the third ventricle. This step is again based on hysteresis thresholding, but this time followed by a comparison to landmarks in the atlas, which leads to the trimming of regions that have a low probability of belonging to the third ventricle. Finally, the fourth ventricle is identified and segmented with the help of hysteresis thresholding. Its shape is then corrected by shape-based removal of a narrow channel that connects it with the basal cistern.

Nonlinear registration methods can also be used to segment the ventricles and thus calculate their volume. In one example [114], images were first resampled to $1 \times 1 \times 1$ mm$^3$ and preprocessed. One image was randomly selected, and the ventricles segmented manually. All other images were aligned with the first through nonlinear registration techniques in such a way that the manual delineation of the segmented ventricle in the reference image, subjected to the same transformation, would delineate the new ventricle. Interestingly, this technique can be used for any other structure in the brain, such as the hippocampus [115], although the segmentation quality strongly depends on the manually segmented reference image.

At the other end of the complexity scale, a recent study [116] suggested that measurements in 2D correlate highly with 3D volumetry. Notably, in a population of 14 patients, correlation coefficients of 0.9 to 0.95 were found between 3D ventricle volume and the maximum width of the body of the lateral ventricles. These examples show that even an apparently simple task—based on the high fluid-tissue contrast—can lead to image analysis algorithms of appreciable complexity; sometimes, however, it may be simpler to use proxy measurements if the accuracy requirements can be met.

### 22.3.3 Degenerative Diseases

Degenerative diseases, most prominently Alzheimer's disease, Parkinson's disease, MS, dementia, and amyotrophic lateral sclerosis, have distinct signatures in MR images of the brain. The diagnostic challenge, however, is based on the fact that anomalies in the image of the brain are secondary effects, because the primary mechanism of nerve degeneration is a molecular process. Two examples are the amyloid-β buildup in Alzheimer's disease and the accumulation of α-synuclein combined with dopamine dysfunction in Parkinson's disease. Correspondingly, the emphasis of diagnostic methods lies on biomarkers rather than structural imaging, and radionuclide imaging (e.g., positron emission tomography) is frequently employed. For these reasons, MRI is often used as an initial diagnostic tool with ambiguous outcomes (e.g., are white matter abnormalities seen in MRI related to traumatic injuries or dementia? [117]), and additional tests beyond MRI are needed to narrow down the diagnosis. In this section, some of the abnormalities are presented that are visible in the MR image, and representative studies that show the potential methods that could be used in CAD are highlighted.

Alzheimer's disease has been linked both to abnormalities of white matter [118], atrophy of the hippocampus [119], reduced size of the caudate nucleus [120], and enlarged ventricle volume [119,121]. In addition, volumetry of the hippocampus was identified as a predictor for Alzheimer's disease [122]. Moreover, Querbes et al. proposed a metric of cortical thickness to predict early stage Alzheimer's disease, although the authors carefully account for similar, age-related reduction in cortical thickness.

Automated analysis of MR scans for the early detection of Alzheimer's disease was proposed by Klöppel et al. [123]. Image processing steps included the segmentation of the brain (i.e., separation of white matter, gray matter, and CSF volumes from each other). Image registration methods were then used to match individual brain images with a template without changing the overall volume of each tissue class [124]. Klöppel et al. then identified those gray matter regions where atrophy (i.e., volume reduction) was more indicative of the disease, as opposed to those regions where enlargement was indicative of the disease [123].

The above study shares with more recent studies that computer-based diagnostic tools for Alzheimer's disease use large areas of the brain image. Since a large number of voxels (with their spatial position and intensity) are involved, computer learning methods are popular. In the study by Klöppel et al. [123], a SVM was used. Yan et al. used a special form of feature transform, referred to as *kernel density estimation*, which has at its core multiscale Gaussian blurring and the difference-of-Gaussians (DoGs) operator to identify regions of interest in the brain volume. A probabilistic classification of the features follows, for which prior training is necessary. Liu et al. [125] propose a hierarchical subdivision of the entire brain volume down to the individual voxels. Only gray matter voxels were used for feature extraction, but the result is a high-dimensional feature vector. To reduce dimensionality, a penalty function was included in the feature extraction that encouraged the grouping of similar voxels and thus the use of a higher level in the hierarchy. Once again, a SVM was trained to perform the final classification.

Similar to Alzheimer's disease, patients with Parkinson's disease also show localized atrophy of the gray matter in MR images [126,127], and atrophy of the hippocampus region has been reported [128]. More frequently than in the diagnosis of Alzheimer's disease, however, functional imaging and diffusion imaging are employed. For example, Long et al. acquired resting-state fMRI with a fast echo-planar sequence to obtain time-resolved changes in the blood oxygenation levels. Noise reduction and drift correction was performed with a bandpass filter on a voxel-by-voxel basis. As features, the amplitude of low-frequency fluctuations, the regional homogeneity, and the regional functional connectivity strength (which is explained as the correlation of any region with all other regions) were extracted. An SVM was used as classifier. A different approach was presented by Haller et al. [129], who based their study on DTI. The premise is that the *directionality* of diffusion can be measured and quantified with a metric referred to as *fractional anisotropy* [130]. Fractional anisotropy (i.e., a directional preference) is highest—close to unity—in white matter, lower in gray matter, and almost zero in CSF, and that degenerative diseases are accompanied by reduced fractional anisotropy in specific regions. Haller et al. used an SVM with suitable training to distinguish Parkinson patients from healthy controls and found that frontal white matter showed reduced fractional anisotropy [129]. Interestingly, anomalous diffusion patterns have also been reported in specific brain regions (fornix and splenium) in Alzheimer's disease [131].

A trend emerges in which even more information is collected and merged, for example, $T_1$-weighted images, FLAIR images, and diffusion images. When such images are brought into congruence, a large number of quantitative metrics can be extracted to create a voxel-by-voxel feature vector. Computer learning methods can then be used to find even subtle differences between images from patients with a degenerative disease and healthy controls. Some examples include signal intensities in $T_1$- and $T_2$-weighted images, GRE images, and ADC [132], and combined structural imaging and DTI [133,134].

Even though a substantial role of MRI in the diagnosis of degenerative diseases is widely acknowledged (see, e.g., [135–137] for reviews), SPECT and PET imaging, EEG, and cognitive test play a significant role in the diagnosis of such diseases. Because of the various forms of degenerative diseases and the mostly subtle changes in MR images of the brain, automated diagnosis is out of reach for the foreseeable future. The increased recent interest in advanced MR sequences, most notably diffusion images, however, indicates that there is considerable promise in computerized analysis when structural and functional image information is combined.

### 22.3.4 Quantification of the Brain Morphology to Detect Psychiatric Disorders

There is strong interest in the relationship between brain morphology and psychiatric diseases, especially schizophrenia. This interest has been intensified by the introduction of tomographic modalities, including CT and MRI. One optimistic review predicted that pathology and pathophysiology of schizophrenia would be understood by the early 2000s [138]. Although some morphological changes, most importantly an enlargement of the ventricles, seems to be widely agreed upon [138], morphological abnormalities and their relationship to psychiatric diseases is the subject of intense research and debate.

Review papers [139,140] summarize brain morphological changes as reduction of whole brain volume, predominantly seen as reduction of gray matter volume, but with a possible increase in white matter volume. In addition, abnormalities of the frontal and temporal lobes, amygdala, hippocampus, and other brain structures were listed. One specific study [141] listed localized volume reductions of the frontal, parietal, and occipital gyri, and the limbic system. Moreover, a reduction of the normal asymmetry between brain hemispheres, or even a reversal (right hemisphere larger than the left) was found to accompany schizophrenia [142–145]. Interestingly, structural changes in the opposite direction—increased total brain volume, increased amygdala volume, increased cortical thickness in the parietal lobes—were found in autism-spectrum disorders [146]. A rough sketch of the location of these structures is provided in Figure 22.14.

One intriguing feature is the complex pattern of the brain surface, that is, the complex folding structure of the gyri and sulci. Zilles [147] defined a gyrification index as the ratio between total and superficially exposed brain surface. In a 2D slice, the gyrification index could be interpreted as the ratio of the inner and outer brain contours as illustrated in Figure 22.15a. In one study, the gyrification index was found to be reduced in the left hemisphere of schizophrenic patients although no differences in brain volume could be seen [148]. The gyrification index is particularly attractive for unsupervised computer analysis, because it can be derived from the surface area of active contours at different stages of their evolution as demonstrated in Figure 22.15. One author of this chapter has developed a method to compare the three-dimensional pattern of the sulci between individuals, and between

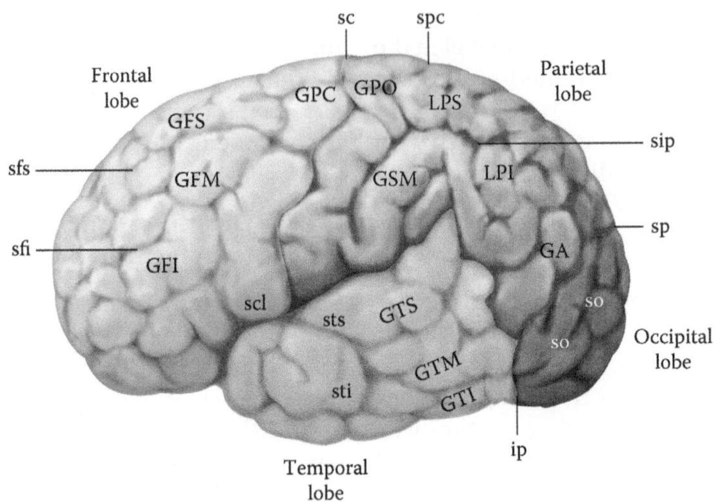


**FIGURE 22.14**
Sketched side view of the brain with some major structures indicated. The major gyri in the frontal lobe are the gyrus frontalis superior (GFS), gyrus frontalis medius (GFM), gyrus frontalis inferior (GFI), and the gyrus praecentralis (GPC) with the sulci frontales superior (sfs) and inferior (sfi) separating them. The sulcus centralis (sc) separates frontal and parietal lobes. In the parietal lobe, prominent gyri are the gyrus postcentralis (GPO), lobulus parietalis superior (LPS) and lobulus parietalis inferior (LPI), the gyrus supramarginalis (GSM), and the gyrus angularis (GA). Prominent parietal sulci are the sulcus postcentralis (spc) and the sulcus intraparietalis (sip). The temporal lobe can be broadly divided into the gyri temporales superior (GTS), medius (GTM), and inferior (GTI), separated by the sulci temporales superior (sts) and inferior (sti). Both amygdala and hippocampus are found in the temporal lobe. The temporal lobe is separated from the frontal lobe by the sulcus cerebri lateralis (scl) and from the occipital lobe by the incisura praeoccipitalis (ip). The occipital lobe contains the sulci occipitales (so) and is separated from the parietal lobe by the sulcus parietooccipitalis (sp).

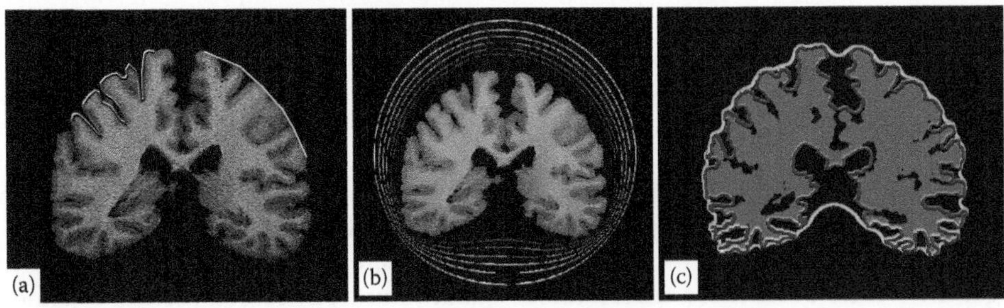


**FIGURE 22.15**
Gyrification index computed with elastic contours. (a) coronal slice of a $T_1$-weighted image after skull-stripping. For illustration purposes, part of the inner contour (red) was traced manually on the left side, and part of the outer contour (green) was traced on the right side. (b) Brain region after anisotropic diffusion. Also shown are iterations of an initially circular elastic contour that contracts around the brain area. (c) Two elastic contours after convergence. The red contour is softer and therefore pulled further into the sulci. The green contour is more rigid and does not enter the sulci. Both contours approximate the outer and inner contours that are used to compute the gyrification index.

hemispheres of a single individual [149]. The underlying principle was to use a simulated gradient field to project the hidden interior volume of the sulci onto a simulated sphere that enclosed the brain volume. The sulcal pattern was therefore recorded on a normalized geometric shape, that is, the sphere, and a straightforward geometric transformation was used to reduce the sphere surface to a 2D plane. At this point, a conventional 2D cross-correlation metric was be used to quantify the similarity between the patterns.

Brain morphology, due to its complex and irregular nature, advertises itself for analysis with fractal methods. These were briefly introduced in Section 22.2.5 as a metric of *complexity* in an abstract sense. King et al. found a reduced fractal dimension in the cortical ribbon and the interface between gray and white matter in Alzheimer's patients and noted a similar trend as seen with the gyrification index [150]. Ha et al. found reduced asymmetry of the fractal dimension between hemispheres in schizophrenic patients and (to a lesser extent) in those with obsessive-compulsive disorder when compared to healthy controls [151]. Similar observations, but with a stronger focus on more localized structures were also made by Yotter et al. [152]. However, Nenadic et al. later made the observation that trends, especially on a more localized scale, are not as uniform, and that a large variability between subgroups of schizophrenia exists [153]. The estimation of the fractal dimension is fairly sensitive to the scales examined and sometimes even to the relative position of the structure. It is very difficult to account for all factors that bias the fractal dimension estimator, and the fractal dimension is best used as one metric in a higher-dimensional classifier.

Although relevant in a research setting, the diagnostic value of structural imaging is limited. Differences can be seen at the group level, but inter-individual variability precludes pure morphology from being useful as a diagnostic tool [154]. The field is still evolving, however, because additional information can be gained from functional and diffusion imaging. For example, fMRI revealed reduced cognitive activation in the anterior cingulate gyrus [155]. Although one group described a predictive mechanism based on DTI [156], a related study did not reveal differences in diffusivity between patients and controls [157]. It is, of course, possible that some of the differences seen are caused by, or correlated with, uncontrolled-for variables, such as smoking [158] or drug abuse [159].

These examples highlight the challenges that need to be overcome before MRI can be used to assist the diagnosis of psychiatric disorders. A single voxel represents tens of thousands of neurons, and even with functional imaging, huge clusters of neurons are combined and averaged in each voxel. Morphological changes may be correlated to psychiatric disorders, but neither functional nor structural imaging helps understanding of the cause at the neuronal or even molecular level. In this respect, MRI can assist a diagnosis, and—even more likely—can reveal some macro-scale correlates of such diseases, but not present a reliable diagnosis.

On the other hand, color-coded orientation maps from DTI are powerful tools for visualizing white matter anatomy/morphometry. The resulting vector information on diffusion is assumed to represent local fiber orientation, and where there is a group of voxels that share similar fiber orientations these voxels are taken to be part of a specific white matter tract. Computer-aided 3D tract tracking techniques (called *tractography*) can be very useful to understand tract trajectories and their spatial relationships with other white matter tracts or gray matter structures. These tracts are not individual axons (those have a diameter of approximately 1–10 microns), but bundles of axons, since DTI voxel resolution is on the order of 1–5 mm. Nevertheless, DTI gives powerful information on white matter organization in roughly 10 min of scanning time which cannot be obtained by other scanning modalities or by invasive chemical tracer technologies. DTI tractography can, for example, be used to delineate the deformation of white matter tracts in brain tumor patients. DTI has also shown its value as a diagnostic tool in schizophrenia and related psychiatric disorders [160] to the extent that the term *connectomics* has been introduced [161] to highlight the ability of DTI to image the interconnectivity of the brain network.

### 22.3.5 Advantages and Limitations of the Computer as Diagnostic Tool

Despite recent progress, the holy grail of a fully unsupervised computerized diagnosis remains elusive. The missing link it probably best captured in the term *image understanding*. A trained observer is very adept at recognizing meaningful features, even those with low contrast in the presence of noise. The human eye easily recognizes the elements of a scene (or, in this case, an image), because the human eye immediately recognizes the broader spatial relationships and works its way down to the details. Moreover, a trained radiologist puts the features of the image into the context of anatomy, pathology, and physiology, and the patient's symptoms and history. Thus, we see *understanding* in the broadest sense.

Conversely, the computer examines an image on a pixel-by-pixel basis, working its way up to the larger features. This bottom-up approach to image analysis becomes particularly apparent in the older image processing operations that were developed 20–30 years ago, such as convolutions with small kernels, or pixel-by-pixel segmentation. Since then, newer algorithms have been developed that process images on a higher abstraction level. Good examples are active contours (those detect and parametrize edges even in the presence of noise and if some parts of the edge are absent) and non-rigid registration methods (those lead to a global, but locally adaptive, deformation). Both examples build on simulated physical systems, for which solving partial differential equations is necessary. This evolution has been helped by the incredible increase in computing power over the last decades. Integer convolution of a $256 \times 256$ pixel image with a $3 \times 3$ kernel (approximately 600,000 multiplications and additions, 128 kB memory requirement) was a task that brought 1980s computers to their performance limits. Today, such an operation would not even show on a memory-usage monitor. It is today possible to solve equations such as Equation 22.8 in a few seconds, despite the fact that its numerical implementation requires two convolutions (the Laplacian operation and the gradient operation) to be repeated iteratively many times to advance the model in time. Even more importantly, computing speed and memory are essential to the manipulation of tensors in DTI, and the availability of increased computing power has helped enable this new technology.

One genuine advantage of computerized image processing is the ability to perform most operations in three dimensions. A human observer would still have to display a volumetric image slice-by-slice, or view parts of the anatomy in a 3D rendering process. However, 3D rendering is a non-standardized process, which is additionally associated with a loss of detail, because the large number of volume elements cannot be displayed in a reasonable time even with today's powerful graphics processing units. For the computer, on the other hand, computing the ventricle volume is not fundamentally different from computing its area in an image slice.

A clever trick has been used to help the computer with the single most difficult task—segmentation: subcortical regions are extremely difficult to segment, and registration of a given image with a pre-segmented atlas allows to transfer the segmentation boundaries. Clearly, the final segmentation result depends both on the quality of the atlas and on the quality of the registration, but a certain level of objectivity and repeatability has been achieved. Building on a reliable segmentation, volumetry and morphology are goals within reach. However, many studies have focused on basic MRI sequences, with the $T_1$-weighted image being the most popular. Many segmentation and quantification techniques are limited to the one type of sequence, because they rely on a specific form of contrast.

Arguably the most critical limitation for the computer is brought by the limited information in an image. Quantifiable observations (such as enlarged ventricles) can be associated with several disease forms. Many studies show significant differences between cohorts, but individual variability poses a major obstacle towards an individualized diagnosis. In other words, the present capabilities of computerized image analysis are limited to findings, such as perhaps "On average, ventricles are enlarged in patients with Alzheimer's disease compared to healthy controls," as opposed to "enlarged ventricles and other morphological findings indicate an 88% probability of Alzheimer's disease for this patient." At present, the computer acts mostly in a supporting role, assisting the radiologist with tedious tasks, such as segmentation, measurements, or aligning images.

## 22.4 Conclusions and Future Directions

Computer-assisted diagnosis using MRI brain images is a formidable pursuit, especially with generally huge 3D datasets. Low SNR and the potential presence of artifacts make the analysis particularly challenging. The task has mobilized cutting-edge ideas in segmentation, registration and template matching, and (statistical) classification algorithms. In the near future it is likely that we will see methods unifying all these different aspects and aligning them more specifically to imaging protocols. In order to assist radiologists and clinicians more user-friendly software tools will need to be developed, enabling them to extract the relevant information easily.

Statistical brain atlases, which aim at gathering and modeling more completely and efficiently the characteristics of a whole population of patients, will need to be developed further. They can incorporate known shape information, and fuse data from other modalities (e.g., CT and PET) ideally acquired simultaneously. Combining non-rigid registration with information on the medial axes and orientation, can lead to an atlas with vectorial attributes. Moreover, DTI holds out the promise of correlating anatomy with function without the need for radionuclide use (e.g., PET).

The significant time taken to acquire volume MR images has led to innovations in pulse sequences. Faster methods reduce scanning time, and therefore improve patient comfort and reduce the adverse effect of patient motion during imaging, improve the efficiency in the use of expensive scanning facilities, and improve the temporal resolution of dynamic studies. Efforts are also being directed toward developing smarter algorithms to reconstruct images from fewer measurements (i.e., sparse or compressive sampling [162]) and therefore in less time, without compromising image quality. Compressive transforms such as the discrete cosine transform (DCT) and discrete wavelet transform (DWT) have been used, and it seems likely that others will be developed allowing further gains to be made. Indeed it is hoped that manufacturers of MRI scanning systems will soon incorporate algorithms based on sparse sampling.

To increase the impact of the computer on the diagnostic process, progress must be made beyond image analysis and MR sequences. The unique capabilities of the radiologist involve combining findings in the image with knowledge of the disease and the patient. It can be envisioned that a diagnosis on the level of the individual patient requires combining advanced image analysis with expert systems such that abnormalities found in the image can be combined with knowledge of image-independent symptoms. Viable expert systems of this nature are not visible on the horizon at present.

## References

1. Haidekker MA. *Medical Imaging Technology*. Springer Briefs in Physics, Springer, New York; 2013.
2. Donoho D. Compressed sensing. *IEEE Transactions on Information Theory* 2006;52(4):1289–1306.

3. Bones PJ, Wu B. Sparse sampling in MRI. In: *Medical Image Processing Techniques and Applications*. Springer, New York; 2011. pp. 319–339.
4. Stoll G, Bendszus M. Imaging of inflammation in the peripheral and central nervous system by magnetic resonance imaging. *Neuroscience*. 2009; 158(3): 1151–1160.
5. Mori S. Three-dimensional tract reconstruction. In: Mori S, editor. *An Introduction to Diffusion Tensor Imaging*. Elsevier, Boston, MA; 2007.
6. Gudbjartsson H, Patz S. The Rician distribution of noisy MRI data. *Magnetic Resonance in Medicine*. 1995;34(6):910–914.
7. Goodman JW. Some fundamental properties of speckle. *JOSA*. 1976;66(11):1145–1150.
8. Shen L, Papadakis M, Kakadiaris IA, Konstantinidis I, Kouri D, Hoffman D. Image denoising using a tight frame. *IEEE Transactions on Image Processing*. 2006;15(5):1254–1263.
9. Kim S. PDE-based image restoration: A hybrid model and color image denoising. *IEEE Transactions on Image Processing*. 2006;15(5):1163–1170.
10. Rudin LI, Osher S, Fatemi E. Nonlinear total variation based noise removal algorithms. *Physica D: Nonlinear Phenomena*. 1992;60(1):259–268.
11. Babacan SD, Molina R, Katsaggelos AK. Variational Bayesian blind deconvolution using a total variation prior. *IEEE Transactions on Image Processing*. 2009;18(1):12–26.
12. Beck A, Teboulle M. Fast gradient-based algorithms for constrained total variation image denoising and deblurring problems. *IEEE Transactions on Image Processing*. 2009;18(11):2419–2434.
13. Drapaca CS. A nonlinear total variation-based denoising method with two regularization parameters. *IEEE Transactions on Biomedical Engineering*. 2009;56(3):582–586.
14. Vemuri BC, Liu M, Amari SI, Nielsen F. Total Bregman divergence and its applications to DTI analysis. *IEEE Transactions on Medical Imaging*. 2011;30(2):475–483.
15. Ramani S, Thévenaz P, Unser M. Regularized interpolation for noisy images. *IEEE Transactions on Medical Imaging*. 2010;29(2):543–558.
16. Zibetti MVW, De Pierro AR. A new distortion model for strong inhomogeneity problems in echo-planar MRI. *IEEE Transactions on Medical Imaging*. 2009;28(11): 1736–1753.
17. Dougherty G. *Digital Image Processing for Medical Applications*. Cambridge University Press, Cambridge, 2009.
18. Rangayyan RM, Ciuc M, Faghih F. Adaptive-neighborhood filtering of images corrupted by signal-dependent noise. *Applied Optics*. 1998;37(20):4477–4487.
19. Perona P, Malik J. Scale-space and edge detection using anisotropic diffusion. *IEEE Transactions on Pattern Analysis and Machine Intelligence*. 1990;12(7):629–639.
20. Haidekker M. *Advanced Biomedical Image Analysis*. John Wiley & Sons, Hoboken, NJ; 2011.
21. Yu Y, Acton ST. Speckle reducing anisotropic diffusion. *IEEE Transactions on Image Processing*. 2002;11(11):1260–1270.
22. Krissian K, Westin CF, Kikinis R, Vosburgh KG. Oriented speckle reducing anisotropic diffusion. *IEEE Transactions on Image Processing*. 2007;16(5):1412–1424.
23. Abramatic JF, Silverman LM. Nonlinear restoration of noisy images. *IEEE Transactions on Pattern Analysis and Machine Intelligence*. 1982;4(2):141–149.
24. Hadhoud MM, Thomas DW. The two-dimensional adaptive LMS (TDLMS) algorithm. *IEEE Transactions on Circuits and Systems*. 1988;35(5):485–494.
25. Pearlman WA, Mahesh B, Song WJ. Adaptive estimators for filtering noisy images. *Optical Engineering*. 1990;29(5):488–494.
26. Rangayyan RM, Das A. Filtering multiplicative noise in images using adaptive region-based statistics. *Journal of Electronic Imaging*. 1998;7(1):222–230.
27. Morrow WM, Paranjape RB, Rangayyan RM, Desautels JEL. Region-based contrast enhancement of mammograms. *IEEE Transactions on Medical Imaging*. 1992;11(3):392–406.
28. Cromartie R, Pizer SM. Edge-affected context for adaptive contrast enhancement. In: *Information Processing in Medical Imaging*. Springer, New York; 1991. pp. 474–485.
29. Caselles V, Lisani JL, Morel JM, Sapiro G. Shape preserving local histogram modification. *IEEE Transactions on Image Processing*. 1999;8(2):220–230.
30. Jia D, Han F, Yang J, Zhang Y, Zhao D, Yu G. A synchronization algorithm of MRI denoising and contrast enhancement based on PM-CLAHE model. *JDCTA*. 2010;4(6):144–149.
31. Alemán-Flores M, Álvarez-León L, Alemán-Flores P, Fuentes-Pavón R, Santana-Montesdeoca JM. Medical image noise reduction and region contrast enhancement using partial differential equations. In: *Proceedings of the IADIS International Conference on Applied Computing, Rome, Italy, November 19–21*, 2009. pp. 222–226.
32. Meijering EH, Zuiderveld KJ, Viergever MA. Image registration for digital subtraction angiography. *International Journal of Computer Vision*. 1999;31(2–3):227–246.
33. Meijering EH, Niessen WJ, Bakker J, van der Molen AJ, de Kort GA, Lo RT et al. Reduction of patient motion artifacts in digital subtraction angiography: Evaluation of a fast and fully automatic technique. *Radiology*. 2001;219(1):288–293.
34. Dale AM, Fischl B, Sereno MI. Cortical surface-based analysis: I. Segmentation and surface reconstruction. *NeuroImage*. 1999;9(2):179–194.
35. Hahn HK, Peitgen HO. The skull stripping problem in MRI solved by a single 3D watershed transform. In: *Medical Image Computing and Computer-Assisted Intervention –MICCAI 2000*. Springer, New York; 2000. pp. 134–143.
36. Sandor S, Leahy R. Surface-based labeling of cortical anatomy using a deformable atlas. *IEEE Transactions on Medical Imaging*. 1997;16(1):41–54.
37. Shattuck DW, Sandor-Leahy SR, Schaper KA, Rottenberg DA, Leahy RM. Magnetic resonance image tissue classification using a partial volume model. *NeuroImage*. 2001;13(5):856–876.
38. Ségonne F, Dale A, Busa E, Glessner M, Salat D, Hahn H et al. A hybrid approach to the skull stripping problem in MRI. *NeuroImage*. 2004;22(3):1060–1075.
39. Smith SM. Fast robust automated brain extraction. *Human Brain Mapping*. 2002;17(3):143–155.

40. DeCarli C, Maisog J, Murphy DG, Teichberg D, Rapoport SI, Horwitz B. Method for quantification of brain, ventricular, and subarachnoid CSF volumes from MR images. *Journal of Computer Assisted Tomography*. 1992;16(2):274–284.
41. Cox RW. AFNI: Software for analysis and visualization of functional magnetic resonance neuroimages. *Computers and Biomedical Research*. 1996;29(3):162–173.
42. Fennema-Notestine C, Ozyurt IB, Clark CP, Morris S, Bischoff-Grethe A, Bondi MW et al. Quantitative evaluation of automated skull-stripping methods applied to contemporary and legacy images: Effects of diagnosis, bias correction, and slice location. *Human Brain Mapping*. 2006;27(2):99–113.
43. Rex DE, Shattuck DW, Woods RP, Narr KL, Luders E, Rehm K et al. A meta-algorithm for brain extraction in MRI. *NeuroImage*. 2004;23(2):625–637.
44. Park JG, Lee C. Skull stripping based on region growing for magnetic resonance brain images. *NeuroImage*. 2009;47(4):1394–1407.
45. Galdames FJ, Jaillet F, Perez CA. An accurate skull stripping method based on simplex meshes and histogram analysis for magnetic resonance images. *Journal of Neuroscience Methods*. 2012;206(2):103–119.
46. Ridler TW, Calvard S. Picture thresholding using an iterative selection method. *IEEE Transactions on Systems, Man and Cybernetics*. 1978;8(8):630–632.
47. Otsu N. A threshold selection method from gray-level histograms. *Automatica*. 1975;11(285–296):23–27.
48. Bieniecki W, Grabowski S. Multi-pass approach to adaptive thresholding based image segmentation. In: *Proceedings of the 8th International IEEE Conference CADSM*, Lviv-Slavske, Ukraine, February 23-26, 2005. pp. 418–423.
49. Canny J. A computational approach to edge detection. *IEEE Transactions on Pattern Analysis and Machine Intelligence*. 1986;6:679–698.
50. Kass M, Witkin A, Terzopoulos D. Snakes: Active contour models. *International Journal of Computer Vision*. 1988;1(4):321–331.
51. Alfiansayah A. Deformable models and level sets in image segmentation. In: Dougherty G, editor. *Medical Image Processing Techniques and Applications*. Springer, New York; 2011. pp. 59–88.
52. Osher S, Sethian JA. Fronts propagating with curvature-dependent speed: Algorithms based on Hamilton-Jacobi formulations. *Journal of Computational Physics*. 1988;79(1):12–49.
53. Suri JS, Liu K, Reden L, Laxminarayan S. A review on MR vascular image processing algorithms: Acquisition and prefiltering: Part I. *IEEE Transactions on Information Technology in Biomedicine: A Publication of the IEEE Engineering in Medicine and Biology Society*. 2002;6(4):324.
54. Suri JS, Liu K, Singh S, Laxminarayan SN, Zeng X, Reden L. Shape recovery algorithms using level sets in 2-D/3-D medical imagery: A state-of-the-art review. *IEEE Transactions on Information Technology in Biomedicine*. 2002;6(1):8–28.
55. Caselles V, Kimmel R, Sapiro G. Geodesic active contours. *International Journal of Computer Vision*. 1997;22(1):61–79.
56. Grady L. Multilabel random walker image segmentation using prior models. In: *IEEE Computer Society Conference on Computer Vision and Pattern Recognition*, San Diego, CA, 2005. pp. 763–770.
57. Beucher S, Lantuejoul C. Use of watersheds in contour detection. In: *International Workshop on Image Processing*. CCETT/IRISA, Rennes, France; 1979.
58. Cousty J, Bertrand G, Najman L, Couprie M. Watershed cuts: Minimum spanning forests and the drop of water principle. *IEEE Transactions on Pattern Analysis and Machine Intelligence*. 2009;31(8):1362–1374.
59. Grau V, Mewes AUJ, Alcaniz M, Kikinis R, Warfield SK. Improved watershed transform for medical image segmentation using prior information. *IEEE Transactions on Medical Imaging*. 2004;23(4):447–458.
60. Couprie C, Grady L, Najman L, Talbot H. Power watershed: A unifying graph-based optimization framework. *IEEE Transactions on Pattern Analysis and Machine Intelligence*. 2011;33(7):1384–1399.
61. Couprie C, Najman L, Talbot H. Seeded segmentation methods for medical image analysis. In: Dougherty G, editor. *Medical Image Processing Techniques and Applications*. Springer, New York; 2011. pp. 27–58.
62. Talairach J, Tournoux P. *Co-Planar Stereotactic Atlas of the Human Brain, 1988*. Thieme, New York; 1988.
63. Laws KI. Texture energy measures. *Proceedings of the DARPA Image Understanding Workshop*. 1979; pp. 47–51.
64. Haralick RM, Shanmugam K, Dinstein IH. Textural features for image classification. *IEEE Transactions on Systems, Man and Cybernetics*. 1973;(6):610–621.
65. Galloway MM. Texture analysis using gray level run lengths. *Computer Graphics and Image Processing*. 1975;4(2):172–179.
66. Nachimuthu DS, Baladhandapani A. Multidimensional texture characterization: On analysis for brain tumor tissues using MRS and MRI. *Journal of Digital Imaging*. 2014;27:496–506.
67. Iftekharuddin KM, Ahmed S, Hossen J. Multiresolution texture models for brain tumor segmentation in MRI. In: *Annual International Conference of the IEEE Engineering in Medicine and Biology Society*, Boston, MA, August 30–September 2, 2011. pp. 6985–6988.
68. Gutierrez DR, Awwad A, Meijer L, Manita M, Jaspan T, Dineen RA et al. Metrics and textural features of MRI diffusion to improve classification of pediatric posterior fossa tumors. *American Journal of Neuroradiology*. 2014;35:1009–1015.
69. Buckley PF, Dean D, Bookstein FL, Friedman L, Kwon D, Lewin JS et al. Three-dimensional magnetic resonance-based morphometrics and ventricular dysmorphology in schizophrenia. *Biological Psychiatry*. 1999;45(1):62–67.
70. Narr KL, Thompson PM, Sharma T, Moussai J, Cannestra AF, Toga AW. Mapping morphology of the corpus callosum in schizophrenia. *Cerebral Cortex*. 2000;10(1):40–49.
71. Uchiyama Y, Kunieda T, Asano T, Kato H, Hara T, Kanematsu M et al. Computer-aided diagnosis scheme for classification of lacunar infarcts and enlarged Virchow-Robin spaces in brain MR images. In:

*Proceedings of the 30th Annual International Conference of the IEEE Engineering in Medicine and Biology Society*. IEEE; 2008. pp. 3908–3911.

72. Sandu AL, Rasmussen Jr IA, Lundervold A, Kreuder F, Neckelmann G, Hugdahl K et al. Fractal dimension analysis of MR images reveals grey matter structure irregularities in schizophrenia. *Computerized Medical Imaging and Graphics*. 2008;32(2):150–158.
73. Shyu KK, Wu YT, Chen TR, Chen HY, Hu HH, Guo WY. Analysis of fetal cortical complexity from MR images using 3D entropy based information fractal dimension. *Nonlinear Dynamics*. 2010;61(3):363–372.
74. Farahibozorg S, Hashemi-Golpayegani SM, Ashburner J. Age and sex-related variations in the brain white matter fractal dimension throughout adulthood: An MRI study. *Clinical Neuroradiology*. 2014;pp. 1–14.
75. Jayasuriya SA, Liew AWC, Law NF. Brain symmetry plane detection based on fractal analysis. *Computerized Medical Imaging and Graphics*. 2013;37(7):568–580.
76. Gefen Y, Meir Y, Mandelbrot BB, Aharony A. Geometric implementation of hypercubic lattices with noninteger dimensionality by use of low lacunarity fractal lattices. *Physical Review Letters*. 1983;50(3):145.
77. Allain C, Cloitre M. Characterizing the lacunarity of random and deterministic fractal sets. *Physical Review A*. 1991;44(6):3552.
78. Dougherty G. *Pattern Recognition and Classification*. Springer, New York; 2013.
79. Möller S, Krabbenhöft HN, Tille A, Paleino D, Williams A, Wolstencroft K et al. Community-driven computational biology with Debian Linux. *BMC Bioinformatics*. 2010;11:S5.
80. Jenkinson M, Beckmann CF, Behrens TE, Woolrich MW, Smith SM. FSL. *NeuroImage*. 2012;62(2):782–790.
81. Bauer S, Fejes T, Reyes M. A skull-stripping filter for ITK. *Insight Journal*. 2012; http://hdl.handle.net/10380/3353:(accessed 06/2014).
82. Pham DL, Xu C, Prince JL. Current methods in medical image segmentation 1. *Annual Review of Biomedical Engineering*. 2000;2(1):315–337.
83. Nyúl LG, Udupa JK. On standardizing the MR image intensity scale. *Magnetic Resonance in Medicine*. 1999;42:1042–1081.
84. Xu T, Mandal M. Automatic brain tumor extraction from $T_1$-weighted coronal MRI using fast bounding box and dynamic snake. In: *Proceedings of the Annual International Conference of the IEEE Engineering in Medicine and Biology Society*. IEEE; 2012. pp. 444–447.
85. Saha BN, Ray N, Greiner R, Murtha A, Zhang H. Quick detection of brain tumors and edemas: A bounding box method using symmetry. *Computerized Medical Imaging and Graphics*. 2012;36(2):95–107.
86. Hsieh TM, Liu YM, Liao CC, Xiao F, Chiang IJ, Wong JM. Automatic segmentation of meningioma from noncontrasted brain MRI integrating fuzzy clustering and region growing. *BMC Medical Informatics and Decision Making*. 2011;11(1):54.
87. Pedoia V, Binaghi E, Balbi S, De Benedictis A, Monti E, Minotto R. Glial brain tumor detection by using symmetry analysis. In: *Proceedings of SPIE*, Vol. 8314, Medical Imaging 2012: Image Processing, 2012. pp. 831445–831445.
88. Bach Cuadra M, De Craene M, Duay V, Macq B, Pollo C, Thiran JP. Dense deformation field estimation for atlas-based segmentation of pathological MR brain images. *Computer Methods and Programs in Biomedicine*. 2006;84:66–75.
89. Descombes X, Kruggel F, Wollny G, Gertz HJ. An object-based approach for detecting small brain lesions: Application to Virchow-Robin spaces. *IEEE Transactions on Medical Imaging*. 2004;23(2):246–255.
90. Ambrosini RD, Wang P, O'Dell WG. Computer-aided detection of metastatic brain tumors using automated threeâdimensional template matching. *Journal of Magnetic Resonance Imaging*. 2010;31(1):85–93.
91. Sanjuán A, Price CJ, Mancini L, Josse G, Grogan A, Yamamoto AK et al. Automated identification of brain tumors from single MR images based on segmentation with refined patient-specific priors. *Frontiers in Neuroscience*. 2013;7:241.
92. Ashburner J, Friston KJ. Unified segmentation. *NeuroImage*. 2005;26(3):839–851.
93. Chen J, Yao J, Thomasson D. Automatic determination of arterial input function for dynamic contrast enhanced MRI in tumor assessment. In: *Medical Image Computing and Computer-Assisted Intervention*. Springer, New York; 2008. pp. 594–601.
94. Anderton BH. Changes in the ageing brain in health and disease. *Philosophical Transactions of the Royal Society of London B: Biological Sciences*. 1997;352:1781–1792.
95. Yücel F, Yaman SO, Özbabalık D, Özkan S, Ortuğ G, Özdemr G. Morphometric measurements of MRI findings in patients with Alzheimer's disease. *Advances in Clinical and Experimental Medicine*. 2014;23:91–96.
96. Puri BK. Progressive structural brain changes in schizophrenia. *Expert Review of Neurotherapeutics*. 2010;10(1):33–42.
97. Benedict RH, Bobholz JH. Multiple sclerosis. *Semin Neurol*. 2007;27:78–85.
98. Annweiler C, Montero-Odasso M, Hachinski V, Seshadri S, Bartha R, Beauchet O. Vitamin D concentration and lateral cerebral ventricle volume in older adults. *Molecular Nutrition & Food Research*. 2013;57(2):267–276.
99. Streitbürger DP, Möller HE, Tittgemeyer M, Hund-Georgiadis M, Schroeter ML, Mueller K. Investigating structural brain changes of dehydration using voxel-based morphometry. *PLoS One*. 2012;7(8):e44195.
100. Worth AJ, Makris N, Patti MR, Goodman JM, Hoge EA, Caviness VS et al. Precise segmentation of the lateral ventricles and caudate nucleus in MR brain images using anatomically driven histograms. *IEEE Transactions on Medical Imaging*. 1998;17(2):303–310.
101. Schnack HG, Hulshoff Pol HE, Baaré WFC, Viergever MA, Kahn RS. Automatic segmentation of the ventricular system from MR images of the human brain. *NeuroImage*. 2001;14(1):95–104.
102. Khan AF, Drozd JJ, Moreland RK, Ta RM, Borrie MJ, Bartha R. A novel MRI-compatible brain ventricle phantom for validation of segmentation and volumetry methods. *Journal of Magnetic Resonance Imaging*. 2012;36(2):476–482.

103. Maltbie E, Bhatt K, Paniagua B, Smith RG, Graves MM, Mosconi MW et al. Asymmetric bias in user guided segmentations of brain structures. *NeuroImage*. 2012;59(2):1315–1323.
104. Saeed N, Puri BK, Oatridge A, Hajnal JV, Young IR. Two methods for semi-automated quantification of changes in ventricular volume and their use in schizophrenia. *Magnetic Resonance Imaging*. 1998;16(10):1237–1247.
105. Wang D, Doddrell DM. A segmentation-based and partial-volume-compensated method for an accurate measurement of lateral ventricular volumes on $T_1$-weighted magnetic resonance images. *Magnetic Resonance Imaging*. 2001;19(2):267–273.
106. Saha PK, Udupa JK. Fuzzy connected object delineation: axiomatic path strength definition and the case of multiple seeds. *Computer Vision and Image Understanding*. 2001;83(3):275–295.
107. Zhang DQ, Chen SC. A novel kernelized fuzzy c-means algorithm with application in medical image segmentation. *Artificial Intelligence in Medicine*. 2004;32(1):37–50.
108. DeCarli C, Murphy DGM, Teichberg D, Campbell G, Sobering GS. Local histogram correction of MRI spatially dependent image pixel intensity nonuniformity. *Journal of Magnetic Resonance Imaging*. 1996;3:519–528.
109. Johnston B, Atkins MS, Mackiewich B, Anderson M. Segmentation of multiple sclerosis lesions in intensity corrected multispectral MRI. *IEEE Transactions on Medical Imaging*. 1996;15(2):154–169.
110. Holmes CJ, Hoge R, Collins L, Woods R, Toga AW, Evans AC. Enhancement of MR images using registration for signal averaging. *Journal of Computer Assisted Tomography*. 1998;22(2):324–333.
111. Aubert-Broche B, Evans AC, Collins L. A new improved version of the realistic digital brain phantom. *NeuroImage*. 2006;32(1):138–145.
112. Evans AC, Collins DL, Mills SR, Brown ED, Kelly RL, Peters TM. 3D statistical neuroanatomical models from 305 MRI volumes. In: *Nuclear Science Symposium and Medical Imaging Conference*, IEEE Service Center, Piscataway, NJ, 1993. pp. 1813–1817.
113. Liu J, Huang S, Nowinski WL. Automatic segmentation of the human brain ventricles from MR images by knowledge-based region growing and trimming. *Neuroinformatics*. 2009;7(2):131–146.
114. Carmichael OT, Kuller LH, Lopez OL, Thompson PM, Dutton RA, Lu A et al. Cerebral ventricular changes associated with transitions between normal cognitive function, mild cognitive impairment, and dementia. *Alzheimer Disease and Associated Disorders*. 2007;21(1):14.
115. Carmichael OT, Aizenstein HA, Davis SW, Becker JT, Thompson PM, Meltzer CC et al. Atlas-based hippocampus segmentation in Alzheimer's disease and mild cognitive impairment. *NeuroImage*. 2005;27(4):979–990.
116. Bourne SK, Conrad A, Neimat JS, Davis TL. Linear measurements of the cerebral ventricles are correlated with adult ventricular volume. *Journal of Clinical Neuroscience*. 2013;20(5):763–764.
117. Fakhran S, Yaeger K, Alhilali L. Symptomatic white matter changes in mild traumatic brain injury resemble pathologic features of early Alzheimer dementia. *Radiology*. 2013;269(1):249–257.
118. Radanovic M, Pereira FRS, Stella F, Aprahamian I, Ferreira LK, Forlenza OV et al. White matter abnormalities associated with Alzheimer's disease and mild cognitive impairment: A critical review of MRI studies. *Expert Review of Neurotherapeutics*. 2013;13(5):483–493.
119. De Leon MJ, DeSanti S, Zinkowski R, Mehta PD, Pratico D, Segal S et al. MRI and CSF studies in the early diagnosis of Alzheimer's disease. *Journal of Internal Medicine*. 2004;256(3):205–223.
120. Jiji S, Smitha KA, Gupta AK, Pillai VPM, Jayasree RS. Segmentation and volumetric analysis of the caudate nucleus in Alzheimer's disease. *European Journal of Radiology*. 2013;82(9):1525–1530.
121. Giorgio A, De Stefano N. Clinical use of brain volumetry. *Journal of Magnetic Resonance Imaging*. 2013;37(1):1–14.
122. Gosche KM, Mortimer JA, Smith CD, Markesbery WR, Snowdon DA. Hippocampal volume as an index of Alzheimer neuropathology Findings from the Nun Study. *Neurology*. 2002;58(10):1476–1482.
123. Klöppel S, Stonnington CM, Chu C, Draganski B, Scahill RI, Rohrer JD et al. Automatic classification of MR scans in Alzheimer's disease. *Brain*. 2008;131(3):681–689.
124. Ashburner J. A fast diffeomorphic image registration algorithm. *NeuroImage*. 2007;38(1):95–113.
125. Liu M, Zhang D, Yap PT, Shen D. Tree-guided sparse coding for brain disease classification. *Medical Image Computing and Computer-Assisted Intervention*, Nice, France, October 1-5, 2012. pp. 239–247.
126. Tessitore A, Amboni M, Cirillo G, Corbo D, Picillo M, Russo A et al. Regional gray matter atrophy in patients with Parkinson disease and freezing of gait. *American Journal of Neuroradiology*. 2012;33(9):1804–1809.
127. Rosenberg-Katz K, Herman T, Jacob Y, Giladi N, Hendler T, Hausdorff JM. Gray matter atrophy distinguishes between Parkinson disease motor subtypes. *Neurology*. 2013;80(16):1476–1484.
128. Weintraub D, Doshi J, Koka D, Davatzikos C, Siderowf AD, Duda JE et al. Neurodegeneration across stages of cognitive decline in Parkinson disease. *Archives of Neurology*. 2011;68(12):1562–1568.
129. Haller S, Badoud S, Nguyen D, Garibotto V, Lovblad KO, Burkhard PR. Individual detection of patients with Parkinson disease using support vector machine analysis of diffusion tensor imaging data: Initial results. *American Journal of Neuroradiology*. 2012;33(11):2123–2128.
130. Smith SM, Jenkinson M, Johansen-Berg H, Rueckert D, Nichols TE, Mackay CE et al. Tract-based spatial statistics: Voxelwise analysis of multi-subject diffusion data. *NeuroImage*. 2006;31(4):1487–1505.
131. Nowrangi MA, Lyketsos CG, Leoutsakos JMS, Oishi K, Albert M, Mori S et al. Longitudinal, region-specific course of diffusion tensor imaging measures in mild cognitive impairment and Alzheimer's disease. *Alzheimer's & Dementia*. 2013;9(5):519–528.
132. Wadia PM, Howard P, Ribeirro MQ, Robblee J, Asante A, Mikulis DJ et al. The value of GRE, ADC and routine MRI in distinguishing Parkinsonian

disorders. *The Canadian Journal of Neurological Sciences.* 2013;40(3):389–402.
133. Cherubini A, Morelli M, Nistic R, Salsone M, Arabia G, Vasta R et al. Magnetic resonance support vector machine discriminates between Parkinson disease and progressive supranuclear palsy. *Movement Disorders.* 2013;29(2):266–269.
134. Ota M, Nakata Y, Ito K, Kamiya K, Ogawa M, Murata M et al. Differential diagnosis tool for Parkinsonian syndrome using multiple structural brain measures. *Computational and Mathematical Methods in Medicine.* 2013;2013:571289.
135. Mascalchi M, Vella A, Ceravolo R. Movement disorders: Role of imaging in diagnosis. *Journal of Magnetic Resonance Imaging.* 2012;35(2):239–256.
136. Hotter A, Esterhammer R, Schocke MF, Seppi K. Potential of advanced MR imaging techniques in the differential diagnosis of parkinsonism. *Movement Disorders.* 2009;24(S2):S711–S720.
137. Weiller C, May A, Sach M, Buhmann C, Rijntjes M. Role of functional imaging in neurological disorders. *Journal of Magnetic Resonance Imaging.* 2006;23(6):840–850.
138. Hyde TM, Weinberger DR. The brain in schizophrenia. *Seminars in Neurology.* 1990;10(3): 276–286.
139. Lawrie SM, Abukmeil SS. Brain abnormality in schizophrenia. A systematic and quantitative review of volumetric magnetic resonance imaging studies. *The British Journal of Psychiatry.* 1998;172(2):110–120.
140. Shenton ME, Dickey CC, Frumin M, McCarley RW. A review of MRI findings in schizophrenia. Schizophrenia research. 2001;49(1):1–52.
141. Davatzikos C, Shen D, Gur RC, Wu X, Liu D, Fan Y et al. Whole-brain morphometric study of schizophrenia revealing a spatially complex set of focal abnormalities. *Archives of General Psychiatry.* 2005;62(11):1218–1227.
142. Petty RG, Barta PE, Pearlson GD, McGilchrist IK, Lewis RW, Tien AY et al. Reversal of asymmetry of the planum temporale in schizophrenia. *American Journal of Psychiatry.* 1995;152(5):715–721.
143. Falkai P, Bogerts B, Greve B, Pfeiffer U, Machus B, Fölsch-Reetz B et al. Loss of sylvian fissure asymmetry in schizophrenia: A quantitative post mortem study. *Schizophrenia Research.* 1992;7(1):23–32.
144. Kleinschmidt A, Falkai P, Huang Y, Schneider T, Fu¨rst G, Steinmetz H. In vivo morphometry of planum temporale asymmetry in first-episode schizophrenia. *Schizophrenia Research.* 1994;12(1):9–18.
145. Honer WG, Bassett AS, Squires-Wheeler E, Falkai P, Smith GN, Lapointe JS et al. The temporal lobes, reversed asymmetry and the genetics of schizophrenia. *Neuroreport.* 1995;7(1):221.
146. Chen R, Jiao Y, Herskovits EH. Structural MRI in autism spectrum disorder. *Pediatric Research.* 2011;69:63R–68R.
147. Zilles K, Armstrong E, Schleicher A, Kretschmann HJ. The human pattern of gyrification in the cerebral cortex. *Anatomy and Embryology.* 1988;179(2):173–179.
148. Kulynych JJ, Luevano LF, Jones DW, Weinberger DR. Cortical abnormality in schizophrenia: An in vivo application of the gyrification index. *Biological Psychiatry.* 1997;41(10):995–999.
149. Haidekker MA, Evertsz CJ, Fitzek C, Boor S, Andresen R, Falkai P et al. Projecting the sulcal pattern of human brains onto a 2D plane—A new approach using potential theory and MRI. *Psychiatry Research: Neuroimaging.* 1998;83(2):75–84.
150. King RD, Brown B, Hwang M, Jeon T, George AT. Fractal dimension analysis of the cortical ribbon in mild Alzheimer's disease. *NeuroImage.* 2010;53(2):471–479.
151. Ha TH, Yoon U, Lee KJ, Shin YW, Lee JM, Kim IY et al. Fractal dimension of cerebral cortical surface in schizophrenia and obsessive-compulsive disorder. *Neuroscience Letters.* 2005;384(1):172–176.
152. Yotter RA, Nenadic I, Ziegler G, Thompson PM, Gaser C. Local cortical surface complexity maps from spherical harmonic reconstructions. *NeuroImage.* 2011;56(3): 961–973.
153. Nenadic I, Yotter RA, Sauer H, Gaser C. Cortical surface complexity in frontal and temporal areas varies across subgroups of schizophrenia. *Human Brain Mapping.* 2014;35:1691–1699.
154. Haukvik UK, Hartberg CB, Agartz I. Schizophrenia-what does structural MRI show? *Tidsskrift for den Norske laegeforening.* 2013;133(8):850.
155. Schultz CC, Koch K, Wagner G, Nenadic I, Schachtzabel C, Güllmar D et al. Reduced anterior cingulate cognitive activation is associated with prefrontal–temporal cortical thinning in schizophrenia. *Biological Psychiatry.* 2012;71(2):146–153.
156. Ingalhalikar M, Kanterakis S, Gur R, Roberts TP, Verma R. DTI based diagnostic prediction of a disease via pattern classification. In: *Medical Image Computing and Computer-Assisted Intervention.* Springer, New York; 2010. pp. 558–565.
157. Murakami M, Takao H, Abe O, Yamasue H, Sasaki H, Gonoi W et al. Cortical thickness, gray matter volume, and white matter anisotropy and diffusivity in schizophrenia. *Neuroradiology.* 2011;53(11):859–866.
158. Schneider CE, White T, Hass J, Geisler D, Wallace SR, Roessner V et al. Smoking status as a potential confounder in the study of brain structure in schizophrenia. *Journal of Psychiatric Research.* 2014;50:84–91.
159. Welch KA, Moorhead TW, McIntosh AM, Owens DGC, Johnstone EC, Lawrie SM. Tensorbased morphometry of cannabis use on brain structure in individuals at elevated genetic risk of schizophrenia. *Psychological Medicine.* 2013;10;43:2087–2096.
160. Shizukuishi T, Abe O, Aoki S. Diffusion tensor imaging analysis for psychiatric disorders. *Magnetic Resonance in Medical Sciences: MRMS: An Official Journal of Japan Society of Magnetic Resonance in Medicine.* 2013;12:153–159.
161. Fornito A, Bullmore ET. Connectomics: A new paradigm for understanding brain disease. *European Neuropsychopharmacology.* 2015;25(5):733–748.
162. Candès EJ, Romberg J, Tao T. Robust uncertainty principles: Exact signal reconstruction from highly incomplete frequency information. *IEEE Transactions on Information Theory.* 2006;52(2):489–509.

# Index

**Note**: Locators followed by *'f'* and *'t'* denote figure and table in the text

**M**

For Product Safety Concerns and Information please contact our EU representative GPSR@taylorandfrancis.com Taylor & Francis Verlag GmbH, Kaufingerstraße 24, 80331 München, Germany

Printed and bound by CPI Group (UK) Ltd, Croydon, CR0 4YY
01/05/2025
01858597-0001